Klaus Mollenhauer · Helmut Tschoeke

Handbook of Diesel Engines

Klaus Mollenhauer · Helmut Tschoeke

Handbook of Diesel Engines

With 584 Figures and 86 Tables

Editors
Prof. Dr.-Ing. Klaus Mollenhauer
Orber Str. 25
14193 Berlin
Germany
Klamoll@aol.com

Prof. Dr.-Ing. Helmut Tschoeke
Otto von Guericke University Magdeburg
Institute of Mobile Systems
Universitätsplatz 2
39106 Magdeburg
Germany
helmut.tschoeke@ovgu.de

Translator
Krister G. E. Johnson
Otto-von-Guericke-Strass 56 b
39104 Magdeburg
Germany

ISBN 978-3-540-89082-9   e-ISBN 978-3-540-89083-6
DOI 10.1007/978-3-540-89083-6
Springer Heidelberg Dordrecht London New York

Library of Congress Control Number: 2010924045

© Springer-Verlag Berlin Heidelberg 2010
This work is subject to copyright. All rights are reserved, whether the whole or part of the material is concerned, specifically the rights of translation, reprinting, reuse of illustrations, recitation, broadcasting, reproduction on microfilm or in any other way, and storage in data banks. Duplication of this publication or parts thereof is permitted only under the provisions of the German Copyright Law of September 9, 1965, in its current version, and permission for use must always be obtained from Springer. Violations are liable to prosecution under the German Copyright Law.
The use of general descriptive names, registered names, trademarks, etc. in this publication does not imply, even in the absence of a specific statement, that such names are exempt from the relevant protective laws and regulations and therefore free for general use.

Cover design: WMXDesign GmbH, Heidelberg

Printed on acid-free paper

Springer is part of Springer Science+Business Media (www.springer.com)

Preface

This machine is destined to completely revolutionize engine engineering and replace everything that exists. (From Rudolf Diesel's letter of October 2, 1892 to the publisher Julius Springer.)

Although Diesel's stated goal has never been fully achievable of course, the diesel engine indeed revolutionized drive systems. This handbook documents the current state of diesel engine engineering and technology. The impetus to publish a Handbook of Diesel Engines grew out of ruminations on Rudolf Diesel's transformation of his idea for a rational heat engine into reality more than 100 years ago. Once the patent was filed in 1892 and work on his engine commenced the following year, Rudolf Diesel waited another 4 years until the Association of German Engineers provided him a platform to present his engine to the public at its convention in Kassel on June 16, 1897. The engine came to bear the name of its ingenious inventor soon thereafter.

The editors and publisher intend this English edition of the handbook to furnish readers outside German-speaking regions a scholarly and practical presentation of the current state of the diesel engine and its large range of applications. The handbook has not only been conceived for diesel experts but also "diesel laypersons" with prior knowledge of engineering or at least an interest in technology. Furthermore, it is intended to benefit students desiring a firsthand comprehensive and sound overview of diesel engine engineering and technology and its state of development.

These aims are reflected in the book's five-part structure. Part I provides a brief history of the diesel engine followed by sections on the fundamentals, including supercharging systems, diesel engine combustion, fuels and modern injection systems. Parts II–IV treat the loading and design of selected components, diesel engine operation, the pollution this causes and the increasingly important measures to reduce it. Part V presents the entire range of engines from small single cylinder diesel engine up through large low speed two-stroke diesel engines. An appendix lists the most important standards and regulations for diesel engines.

Further development of diesel engines as economizing, clean, powerful and convenient drives for road and nonroad use has proceeded quite dynamically in the last twenty years in particular. In light of limited oil reserves and the discussion of predicted climate change, development work continues to concentrate on reducing fuel consumption and utilizing alternative fuels while keeping exhaust as clean as possible as well as further increasing diesel engine power density and enhancing operating performance. Development is oriented toward the basic legal conditions, customer demands and, not least, competition with gasoline engines, which are still considered the benchmark car engine in many sectors.

The topics to be treated were weighed with all this in mind: In addition to engine internal measures that reduce exhaust emissions with the aid of new combustion systems and new fuels, the section on *Exhaust Gas Aftertreatment* deserves particular mention. The oxidation catalytic converters introduced in the car sector as standard in the 1990s will soon no longer meet the mounting requirements for air hygiene; particulate filters and nitrogen oxide reduction systems, e.g. SCR and storage catalysts, have become standard.

New combustion systems with a larger share of premixed, homogeneous combustion than normal diffusion combustion are just as much the subject of this handbook as the refinement of supercharging to enhance the power output, increase the peak cylinder pressure and thus limit load as the brake mean effective pressure increases. Quickly emerging as the optimal injection system when the car sector switched from indirect to direct injection at the end of the 1990s, the common rail system also came to be used – initially only experimentally – for larger diesel engines at the start of the new millennium. The common rail system is now standard in diesel engines of virtually every size. Hence, reflecting current but by far not yet finalized development, this handbook treats the different designs, e.g. with solenoid valve-controlled or piezo-actuated injectors, in detail. Ample space has accordingly also been given to electronics with its diverse options to control processes in the engine.

To be able meet the expectations and demands connected with a Handbook of Diesel Engines, we relied as much on the collaboration of outstanding engineers from the engine industry as on the research findings of professors at universities of applied sciences and universities. After all, a particularly close connection has existed between theory and practice, between academia and industry, in engine research since Diesel's day, his invention itself being based on the engineering of his day.

Thanks to the work of many generations of engineers, scientists, researchers and professors, the diesel engine continues to be the most cost effective internal combustion engine and has evolved into an advanced high-tech product.

We would like to thank all the authors – whether experts working in industry where the utmost dedication is demanded or our colleagues in academia where the days of creative leisure have long since become a thing of the past – for their collaboration, their ready acceptance of our ideas and the many fruitful discussions. We would also like to extend our gratitude to the companies that allowed their employees to work on the side, supported the compilation of texts and master illustrations and provided material. Acknowledgement is also due the many helpers at companies and institutes for their contributions without which such an extensive book manuscript could never have been produced.

Particularly special thanks go to the Diesel Systems Division at Robert Bosch GmbH for the technical and financial support, which made it possible to complete this extensive work in the first place.

Despite the sometimes hectic pace and considerable additional work, the editors tremendously enjoyed their collaboration with the authors, the publisher and all the other collaborators.

Berlin, Germany,
Magdeburg, Germany
September 2009

Klaus Mollenhauer
Helmut Tschoeke

My engine continues to make great advances. . . . (From Rudolf Diesel's letter of July 3, 1895 to his wife.)

Contents

Contributors IX

Part I   The Diesel Engine Cycle 1

1 History and Fundamental Principles of the Diesel Engine *(Klaus Mollenhauer and Klaus Schreiner)* 3
1.1 The History of the Diesel Engine 3
1.2 Fundamentals of Engine Engineering 7
1.3 Combustion Cycle Simulation 18
Literature 29

2 Gas Exchange and Supercharging *(Helmut Pucher)* . . 31
2.1 Gas Exchange 31
2.2 Diesel Engine Supercharging 38
2.3 Programmed Gas Exchange Simulation 56
Literature 59

3 Diesel Engine Combustion *(Klaus B. Binder)* 61
3.1 Mixture Formation and Combustion 61
3.2 Design Features 69
3.3 Alternative Combustion Processes 73
3.4 Process Simulation of Injection Characteristic and Rate of Heat Release 74
Literature 75

4 Fuels *(Gerd Hagenow, Klaus Reders, Hanns-Erhard Heinze, Wolfgang Steiger, Detlef Zigan, and Dirk Mooser)* 77
4.1 Automotive Diesel Fuels 77
4.2 Alternative Fuels 94
4.3 Operation of Marine and Stationary Engines with Heavy Fuel Oil 103
4.4 Fuel Gases and Gas Engines 114
Literature 124

5 Fuel Injection Systems *(Walter Egler, Rolf Jürgen Giersch, Friedrich Boecking, Jürgen Hammer, Jaroslav Hlousek, Patrick Mattes, Ulrich Projahn, Winfried Urner, and Björn Janetzky)* 127
5.1 Injection Hydraulics 127
5.2 Injection Nozzles and Nozzle Holders . . . 129
5.3 Injection Systems 137
5.4 Injection System Metrology 170
Literature 173
Further Literature 173
Further Literature on Section 5.2 174

6 Fuel Injection System Control Systems *(Ulrich Projahn, Helmut Randoll, Erich Biermann, Jörg Brückner, Karsten Funk, Thomas Küttner, Walter Lehle, and Joachim Zuern)* 175
6.1 Mechanical Control 175
6.2 Electronic Control 176
6.3 Sensors 184
6.4 Diagnostics 186
6.5 Application Engineering 189
Literature 191
Further Literature 191

Part II   Diesel Engine Engineering . . 193

7 Engine Component Loading *(Dietmar Pinkernell and Michael Bargende)* 195
7.1 Mechanical and Thermal Loading of Components 195
7.2 Heat Transfer and Thermal Loads in Engines 202
Literature 217
Further Literature 219

8 Crankshaft Assembly Design, Mechanics and Loading *(Eduard Köhler, Eckhart Schopf, and Uwe Mohr)* 221
8.1 Designs and Mechanical Properties of Crankshaft Assemblies 221
8.2 Crankshaft Assembly Loading 228
8.3 Balancing of Crankshaft Assembly Masses . 236
8.4 Torsional Crankshaft Assembly Vibrations 250
8.5 Bearings and Bearing Materials 259
8.6 Piston, Piston Rings and Piston Pins 270
Literature 287
Further Literature 290

9 Engine Cooling *(Klaus Mollenhauer and Jochen Eitel)* 291
   9.1 Internal Engine Cooling 291
   9.2 External Engine Cooling Systems 309
   Literature 336

10 Materials and Their Selection *(Johannes Betz)* 339
   10.1 The Importance of Materials for Diesel Engines 339
   10.2 Technical Materials for Engine Components . 339
   10.3 Factors for Material Selection 348
   10.4 Service Life Concepts and Material Data . . . 348
   10.5 Service Life Enhancing Processes 349
   10.6 Trends in Development 352
   Literature 354
   Further Literature 355

Part III Diesel Engine Operation . . . 357

11 Lubricants and the Lubrication System *(Hubert Schwarze)* 359
   11.1 Lubricants 359
   11.2 Lubrication Systems 370
   Literature 376

12 Start and Ignition Assist Systems *(Wolfgang Dressler and Stephan Ernst)* 377
   12.1 Conditions for the Auto-Ignition of Fuel . 377
   12.2 Fuel Ignition Aids 378
   12.3 Start and Ignition Assist Systems 379
   12.4 Cold Start, Cold Running Performance and Cold Running Emissions for Cars . . 383
   12.5 Conclusion 386
   Literature 386
   Further Literature 386

13 Intake and Exhaust Systems *(Oswald Parr, Jan Krüger, and Leonhard Vilser)* 387
   13.1 Air Cleaners 387
   13.2 Exhaust Systems 393
   Literature 398
   Further Literature 399

14 Exhaust Heat Recovery *(Franz Hirschbichler)* 401
   14.1 Basics of Waste Heat Recovery 401
   14.2 Options of Waste Heat Recovery 404
   Literature 413

Part IV Environmental Pollution by Diesel Engines 415

15 Diesel Engine Exhaust Emissions *(Helmut Tschoeke, Andreas Graf, Jürgen Stein, Michael Krüger, Johannes Schaller, Norbert Breuer, Kurt Engeljehringer, and Wolfgang Schindler)* 417
   15.1 General Background 417
   15.2 Emission Control Legislation 426
   15.3 Pollutants and Their Production 443
   15.4 In-Engine Measures for Pollutant Reduction 449
   15.5 Exhaust Gas Aftertreatment 455
   15.6 Emissions Testing 469
   Literature 483
   Further Literature 485

16 Diesel Engine Noise Emission *(Bruno M. Spessert and Hans A. Kochanowski)* 487
   16.1 Fundamentals of Acoustics 487
   16.2 Development of Engine Noise Emission 487
   16.3 Engine Surface Noise 489
   16.4 Aerodynamic Engine Noises 498
   16.5 Noise Reduction by Encapsulation 499
   16.6 Engine Soundproofing 502
   Literature 502

Part V Implemented Diesel Engines 505

17 Vehicle Diesel Engines *(Fritz Steinparzer, Klaus Blumensaat, Georg Paehr, Wolfgang Held, and Christoph Teetz)* 507
   17.1 Diesel Engines for Passenger Cars 507
   17.2 Diesel Engines for Light Duty Commercial Vehicles 521
   17.3 Diesel Engines for Heavy Duty Commercial Vehicles and Buses 528
   17.4 High Speed High Performance Diesel Engines 544
   Literature 556
   Further Literature 557

18 Industrial and Marine Engines *(Günter Kampichler, Heiner Bülte, Franz Koch, and Klaus Heim)* . . . 559
   18.1 Small Single Cylinder Diesel Engines 559
   18.2 Stationary and Industrial Engines 568
   18.3 Medium Speed Four-Stroke Diesel Engines 576
   18.4 Two-Stroke Low Speed Diesel Engines 592
   Literature 607

Standards and Guidelines for Internal Combustion Engines . 609

Index 621

As light as a feather

This "featherlight" unit weighs 40% less than conventional balance shafts, while retaining optimum function without increased noise.

In addition, every INA balance shaft supported by rolling contact enables friction losses to be reduced by half in comparison with balance shafts supported by plain bearings. A further advantage is that both the shaft and the bearings are supplied from a single source. Reason enough for leading automotive manufacturers to use this INA solution in their vehicles.

Make use of our engineering know-how on bearing supports in engines, work with us to push forward developments in fuel-saving solutions.

Schaeffler Technologies GmbH & Co. KG · www.ina.com

SCHAEFFLER GROUP
AUTOMOTIVE

www.hjs.com Diesel Exhaust Systems

Environment Protection Technologies

As a medium-sized company based in Menden in central Germany, HJS Fahrzeugtechnik GmbH & Co KG has many years of experience and expertise in the field of exhaust-gas aftertreatment. Some 400 employees develop, produce and market modular systems for reducing pollutant emissions. The innovative environment protection technologies can be used either as original equipment or for retrofitting in passenger cars, commercial vehicles and various of non-road applications.

With a wide range of patents for DPF® (Diesel Particulate Filter) and SCRT® (Selective Catalytic Reduction Technology), HJS sets benchmarks both nationally and globally.

SCRT®-System with Self-Regeneration Unit

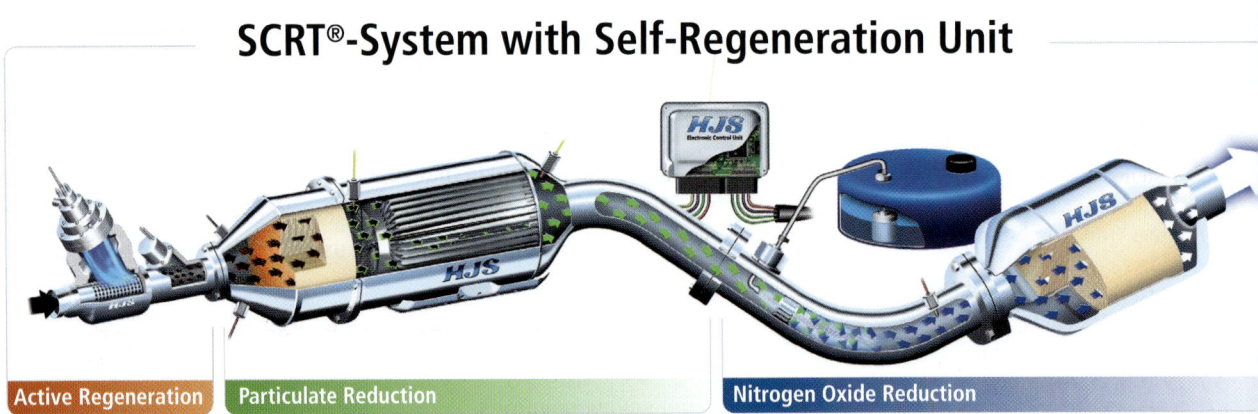

HJS Fahrzeugtechnik GmbH & Co KG
Dieselweg 12 • D-58706 Menden/Sauerland

Telefon +49 2373 987-0
Telefax +49 2373 987-199

hjs@hjs.com
www.hjs.com

HJS
Abgas-Systeme • Exhaust System

Contributors

Michael Bargende, Prof. Dr.-Ing., Universität Stuttgart, Stuttgart, Germany, michael.bargende@ivk.uni-stuttgart.de

Johannes Betz, MTU Friedrichshafen GmbH, Friedrichshafen, Germany, johannes.betz@mtu-online.com

Erich Biermann, Dr.-Ing., Robert Bosch GmbH, Diesel Systems, Stuttgart, Germany, erich.biermann@de.bosch.com

Klaus B. Binder, Prof. Dr.-Ing., Deizisau, Germany, klaus.b.binder@t-online.de

Klaus Blumensaat, Volkswagen AG, Wolfsburg, Germany, klaus.blumensaat@volkswagen.de

Friedrich Boecking, Robert Bosch GmbH, Diesel Systems, Stuttgart, Germany, friedrich.boecking@de.bosch.com

Norbert Breuer, Dr.-Ing., Robert Bosch GmbH, Diesel Systems, Stuttgart, Germany, norbert.breuer@de.bosch.com

Jörg Brückner, Dr., Robert Bosch GmbH, Diesel Systems, Stuttgart, Germany, joerg.brueckner@de.bosch.com

Heiner Bülte, Dr.-Ing., Deutz AG, Köln, Germany, buelte.h@deutz.com

Wolfgang Dressler, Dr., Robert Bosch GmbH, Diesel Systems, Stuttgart, Germany, wolfgang.dressler@de.bosch.com

Walter Egler, Dr.-Ing., Robert Bosch GmbH, Diesel Systems, Stuttgart, Germany, walter.egler@de.bosch.com

Jochen Eitel, Behr GmbH & Co. KG, Stuttgart, Germany, jochen.eitel@behrgroup.com

Kurt Engeljehringer, AVL List GmbH, Graz, Austria, kurt.engeljehringer@avl.com

Stephan Ernst, Dr.-Ing., Robert Bosch GmbH, Diesel Systems, Stuttgart, Germany, stephan.ernst@de.bosch.com

Karsten Funk, Dr.-Ing., Robert Bosch GmbH, Diesel Systems, Stuttgart, Germany, karsten.funk@de.bosch.com

Rolf Jürgen Giersch, Dipl.-Ing., Robert Bosch GmbH, Diesel Systems, Stuttgart, Germany, juergen.giersch@de.bosch.com

Andreas Graf, Dipl.-Ing., Daimler AG, Stuttgart, Germany, andreas.g.graf@daimler.com

Gerd Hagenow, Dr., Shell Global Solutions (Deutschland) GmbH, Hamburg, Germany

Jürgen Hammer, Dr.-Ing., Robert Bosch GmbH, Diesel Systems, Stuttgart, Germany, juergen.hammer@de.bosch.com

Klaus Heim, Wärtsilä Switzerland Ltd, Winterthur, Switzerland, klaus.heim@wartsila.com

Hanns-Erhard Heinze, Dr.-Ing., Magdeburg, Germany, cheheinze@gmx.de

Wolfgang Held, Dr.-Ing., MAN Nutzfahrzeuge AG, Nürnberg, Germany, wolfgang.held@man.eu

Franz Hirschbichler, Dr., München, Germany, franz.hirschbichler@gmx.de

Jaroslav Hlousek, Dipl.-Ing., KEFICO Co, Gunpo, Korea (RoK), jaroslav.hlousek@kr.bosch.com

Björn Janetzky, Dr.-Ing., Robert Bosch GmbH, Diesel Systems, Stuttgart, Germany, bjoern.janetzky@de.bosch.com

Günter Kampichler, Dipl.-Ing., Ruhstorf, Germany

Franz Koch, Dr.-Ing., MAN Diesel & Turbo SE, Augsburg, Germany, franz.koch@man.eu

Hans A. Kochanowski, Dr.-Ing., Ruhstorf, Germany

Eduard Köhler, Prof. Dr.-Ing. habil., KS Aluminium Technologie GmbH, Neckarsulm, Germany, eduard.koehler@de.kspg.com

Jan Krüger, Dr.-Ing., J. Eberspächer GmbH & Co. KG, Esslingen, Germany, jan.krueger@eberspaecher.com

Michael Krüger, Dr.-Ing., Robert Bosch GmbH, Diesel Systems, Stuttgart, Germany, michael.krueger2@de.bosch.com

Thomas Küttner, Dipl.-Ing., Robert Bosch GmbH, Diesel Systems, Stuttgart, Germany, thomas.kuettner@de.bosch.com

Walter Lehle, Dr. rer. nat., Robert Bosch GmbH, Diesel Systems, Stuttgart, Germany, walter.lehle@de.bosch.com

Patrick Mattes, Dr., Robert Bosch GmbH, Diesel Systems, Stuttgart, Germany, patrick.mattes@de.bosch.com

Uwe Mohr, Dr., Mahle GmbH, Stuttgart, Germany, uwe.mohr@mahle.com

Klaus Mollenhauer, Prof. Dr.-Ing., Berlin, Germany, klamoll@aol.com

Dirk Mooser, Dr.-Ing., Caterpillar Motoren GmbH & Co. KG, Kiel, Germany, mooser_dirk@CAT.com

Georg Paehr, Dr., Volkswagen AG, Wolfsburg, Germany, georg.paehr@volkswagen.de

Oswald Parr, Dr.-Ing., Ludwigsburg, Germany

Dietmar Pinkernell, MAN Diesel & Turbo SE, Augsburg, Germany

Ulrich Projahn, Dr.-Ing., Robert Bosch GmbH, Diesel Systems, Stuttgart, Germany, ulrich.projahn@de.bosch.com

Helmut Pucher, Prof. Dr.-Ing., Technische Universität Berlin, Berlin, Germany, hegre.pucher@t-online.de

Helmut Randoll, Dr. rer. nat., Robert Bosch GmbH, Diesel Systems, Stuttgart, Germany, helmut.randoll@de.bosch.com

Klaus Reders, Dipl.-Ing., Shell Global Solutions (Deutschland) GmbH, Hamburg, Germany, klausredershh@t-online.de

Johannes Schaller, Dr., Robert Bosch GmbH, Diesel Systems, Stuttgart, Germany, johannes.schaller@de.bosch.com

Wolfgang Schindler, Dr., AVL List GmbH, Graz, Austria, wolfgang.schindler@avl.com

Eckhart Schopf, Dr.-Ing., Wiesbaden, Germany

Klaus Schreiner, Prof. Dr.-Ing., HTGW Konstanz (University of Applied Sciences), Konstanz, Germany, schreiner@htgw-konstanz.de

Hubert Schwarze, Prof. Dr.-Ing., TU Clausthal, Clausthal-Zellerfeld, Germany, schwarze@itr.tu-clausthal.de

Bruno M. Spessert, Prof. Dr.-Ing., FH Jena (University of Applied Sciences), Jena, Germany, bruno.spessert@fh-jena.de

Wolfgang Steiger, Dr.-Ing., Volkswagen AG, Wolfsburg, Germany, wolfgang.steiger@volkswagen.de

Jürgen Stein, Daimler AG, Stuttgart, Germany, hj.stein@daimler.com

Fritz Steinparzer, BMW Group, München, Germany, fritz.steinparzer@bmw.com

Christoph Teetz, Dr.-Ing., MTU Friedrichshafen GmbH, Friedrichshafen, Germany, christoph.teetz@mtu-online.de

Helmut Tschoeke, Prof. Dr.-Ing., Otto von Guericke Universität Magdeburg, Magdeburg, Germany, helmut.tschoeke@ovgu.de

Winfried Urner, Robert Bosch GmbH, Diesel Systems, Stuttgart, Germany, winfried.urner@de.bosch.com

Leonhard Vilser, Dr.-Ing., J. Eberspächer GmbH & Co. KG, Esslingen, Germany, leonhard.vilser@eberspaecher.com

Detlef Zigan, Dr.-Ing., Kiel, Germany, detlefzigan@hotmail.com

Joachim Zuern, Dipl.-Ing., Robert Bosch GmbH, Diesel Systems, Stuttgart, Germany, joachim.zuern@de.bosch.com

Units and Conversion Factors

Quantity	Symbol	Unit		Conversion factors	
		US Customary	Metric (SI)	Metric → US	US → Metric
Force	F	Pound-force (lbf)	Newton (N)	1 N = 0.2248 lbf	1 lbf = 4.4482 N
Weight		Pound-weight (lbw)			
Length	s	inch (in)	mm	1 mm = 0.03937 in	1 in = 25.4 mm
		foot (ft)	Meter (m)	1 m = 3.2808 ft	1 ft = 0.3048 m
		mile (mi)	km	1 km = 0.62137 mi	1 mi = 1.6093 km
Mass	m	pound (lbm)	Kilogramm (kg)	1 kg = 2.2046 lb	1 lb = 0.4536 kg
Power	P	horsepower (hp)	Kilowatt (kW)	1 kW = 1.341 hp	1 hp = 0.7457 kW
Pressure	p	lb/sq in (psi)	bar, Pascal (Pa)	1 bar = 14.504 psi	1 psi = 6895 Pa = 0.06895 bar
Specific fuel consumption	sfc	lbm/hp h	g/kWh	1 g/kWh = 0.001644 lb/hp h	1 lb/hp h = 608.277 g/kWh
Temperature	T	°Fahrenheit (°F)	°Celsius (°C)	$T_{°C} = 5/9\,(T_{°F} - 32)$	$T_{°F} = 9/5\,T_{°C} + 32$
		Rankine (R)	Kelvin (K)		
Torque	T	ft lbf	Nm	1 Nm = 0.7376 ft lbf	1 ft lbf = 1.3548 Nm
Velocity	v	ft/s	m/s	1 m/s = 3.28 ft/s	1 ft/s = 0.305 m/s
		mi/h	km/h	1 km/h = 0.6214 mi/h	1 mi/h = 1.6093 km/h
Volume	V	gallon (gal)	liter (l)	1 l = 0.2642 gal	1 gal = 3.785 l
			cm³		
			m³		
Work, Energy	W	ft lbf	Joule (J)	1 J = 0.7376 ft lbf	1 ft lbf = 1.3558 J
		British thermal unit (Btu)	kWh	1 kWh = 3412.1 Btu	1 Btu = 2.9307 × 10⁻⁴ kWh

Part I The Diesel Engine Cycle

1 History and Fundamental Principles of the Diesel Engine 3

2 Gas Exchange and Supercharging 31

3 Diesel Engine Combustion 61

4 Fuels 77

5 Fuel Injection Systems 127

6 Fuel Injection System Control Systems 175

1 History and Fundamental Principles of the Diesel Engine

Klaus Mollenhauer and Klaus Schreiner

1.1 The History of the Diesel Engine

On February 27, 1892, the engineer Rudolf Diesel filed a patent with the Imperial Patent Office in Berlin for a "new rational heat engine". On February 23, 1893, he was granted the patent DRP 67207 for the "Working Method and Design for Combustion Engines" dated February 28, 1892. This was an important first step toward the goal Diesel had set himself, which, as can be gathered from his biography, had preoccupied him since his days as a university student.

Rudolf Diesel was born to German parents in Paris on March 18, 1858. Still a schoolboy when the Franco-Prussian War of 1870–1871 broke out, he departed by way of London for Augsburg where he grew up with foster parents. Without familial and financial backing, young Rudolf Diesel was compelled to take his life into his own hands and contribute to his upkeep by, among other things, giving private lessons. Scholarships ultimately enabled him to study at the Polytechnikum München, later the Technische Hochschule, from which he graduated in 1880 as the best examinee ever up to that time.

There, in Professor Linde's lectures on the theory of caloric machines, the student Diesel realized that the steam engine, the dominant heat engine of the day, wastes a tremendous amount of energy when measured against the ideal energy conversion cycle formulated by Carnot in 1824 (see Sect. 1.2). What is more, with efficiencies of approximately 3%, the boiler furnaces of the day emitted annoying smoke that seriously polluted the air.

Surviving lecture notes document that Diesel already contemplated implementing the Carnot cycle as a student, if possible by directly utilizing the energy contained in coal without steam as an intermediate medium. While working at Lindes Eismaschinen, which brought him from Paris to Berlin, he also ambitiously pursued the idea of a rational engine, hoping his invention would bring him financial independence together with social advancement. He ultimately filed and was granted the aforementioned patent [1-1] with the following claim 1:

Working method for combustion engines characterized by pure air or another indifferent gas (or steam) with a working piston compressing pure air so intensely in a cylinder that the temperature generated as a result is far above the ignition temperature of the fuel being used (curve 1-2 of the diagram in Fig. 2), whereupon, due to the expelling piston and the expansion of the compressed air (or gas) triggered as a result (curve 2-3 of the diagram in Fig. 2), the fuel is supplied so gradually from dead center onward that combustion occurs without significantly increasing pressure and temperature, whereupon, after the supply of fuel is terminated, the mass of gas in the working cylinder expands further (curve 3-4 of the diagram in Fig. 2).

Once the gas has been decompressed to the discharge pressure, heat dissipates along the isobars 4-1 (Fig. 1-1), thus ending the cycle.

A second claim asserts patent protection of multistage compression and expansion. Diesel proposed a three cylinder compound engine (Fig. 1-2). Adiabatic compression occurs in two high pressure cylinders 2, 3 operating offset at 180° and the fuel (Diesel initially spoke of coal dust) supplied by the hopper B in top dead center auto-ignites so that isothermal combustion and expansion occur, which turns adiabatic after combustion ends. The combustion gas is transferred into the double-acting center cylinder 1 where it completely expands to ambient pressure and is expelled after the reversal of motion at the same time as the isothermal precompression by water injection or the preceding intake of the fresh charge for the second engine cycle that runs parallel. Thus, one cycle occurs per revolution.

To implement the Carnot cycle, Diesel reverted to the four-stroke cycle considered "state-of-the-art" since Nikolaus Otto's day. He believed isothermal combustion at a maximum of 800°C would enable him to keep the thermal load in the engine low enough that it would run without cooling. This limiting temperature requires compressions of approximately 250 at with which Diesel far surpassed the "state-of-the-art": On the one hand, this gave the "outsider" Diesel the naïveté

K. Mollenhauer (✉)
Berlin, Germany
e-mail: Klamoll@aol.com

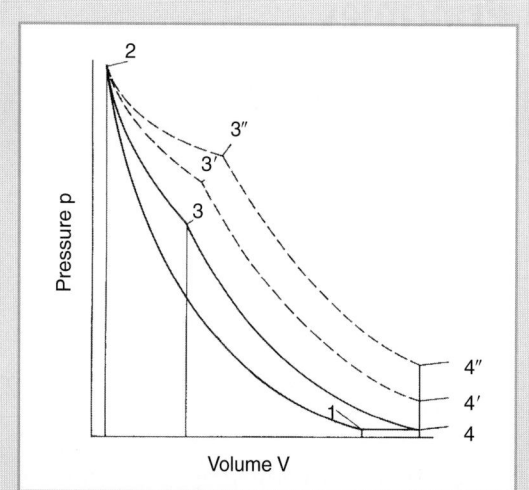

Fig. 1-1 Ideal diesel engine process (1-2-3-4) based on Fig. 2 in [1-1], supplemented by modified "admission periods" (1-2-3'-4' and 1-2-3''-4'') according to Diesel's letter to Krupp of October 16, 1893 [1-4, p. 404]

necessary to implement his idea. On the other hand, firms experienced in engine manufacturing such as the Deutz gas engine factory shied away from Diesel's project.

Conscious that "an invention consists of two parts: The idea and its implementation" [1-2], Diesel wrote a treatise on the "Theory and Design of a Rational Heat Engine" [1-3] and sent it to professors and industrialists as well as Deutz at the turn of 1892–1893 to propagate his ideas and win over industry: With a Carnot efficiency of approximately 73% at 800°C, he expected maximum losses of 30 to 40% in real operation, which would correspond to a net efficiency of 50% [1-3, p. 51].

After nearly a year of efforts and strategizing, Diesel finally concluded a contract in early 1893 with the renowned Maschinenfabrik Augsburg AG headed by Heinrich Buz, a leading manufacturer of steam engines. The contract contained Diesel's concessions to an ideal engine: The maximum pressure was lowered from 250 at to 90 at and later 30 at, the compound engine's three cylinders were reduced to one high pressure cylinder and coal dust was abandoned as the fuel. Two other heavy machinery manufacturers, Krupp and, soon thereafter, Sulzer entered into the contract, which was lucrative for Diesel.

Construction of the first uncooled test engine with a stroke of 400 mm and a bore of 150 mm was begun in Augsburg early in the summer of 1893. Although petroleum was the intended fuel, gasoline was first injected in a powered engine on August 10, 1893 under the misguided assumption it would ignite more easily: The principle of auto-ignition was indeed confirmed even though the indicator burst at pressures over 80 bar!

Selected indicator diagrams (Fig. 1-3) make it possible to follow the further developments: Once the first engine, which was later provided water cooling, had been modified, the fuel could no longer be injected directly. Rather, it could only be injected, atomized and combusted with the aid of compressed air. The first time the hitherto powered engine idled on February 17, 1894, it became autonomous. Finally, a first braking test was performed on June 26, 1895: Using petroleum as fuel and externally compressed injection air, an indicated efficiency of η_i = 30.8% and a net efficiency of η_e = 16.6% were measured at a consumption of 382 g/HPh.

Only a revised design, the third test engine [1-4] furnished with a single stage air pump, delivered the breakthrough though: Professor Moritz Schröter from the Technische Hochschule München conducted acceptance tests on February 17, 1897. Together with Diesel and Buz, he presented the results at a general meeting of the Association of German Engineers in Kassel on June 16, 1897, thus introducing the first heat engine with an efficiency of 26.2%, which was sensational in those days [1-5]!

It necessitated abandoning the isothermal heat input claimed in the original patent: In light of the narrow region of the diagram proportional to indicated work and the frictional losses to be expected as a result of the high pressures, even Diesel must have realized no later than when he plotted the theoretical indicator diagrams (Fig. 1-4) that the engine would not perform any effective work. Taking great pains not to jeopardize the basic patent, he gave thought early on to prolonging the "admission period", i.e. raising the line of isothermal heat input in the p, V diagram (Fig. 1-1). A second application for a patent (DRP 82168) on November 29, 1893 also cited the constant pressure cycle, which was considered consistent with the basic patent because of its "insubstantial pressure increase". The patent granted overlooked the fact that, contrary to the basic patent, both the mass of the fuel and the maximum temperature increased!

Unsurprisingly, Diesel and the Diesel consortium were soon embroiled in patent disputes in Kassel. According to the charge, Diesel's engine did not fulfill any of his patent claims: The engine was unable to run without cooling and expansion did not occur without substantially increasing pressure and temperature as a function of compression. Only the auto-ignition mentioned in claim 1 took place. Yet, just as Diesel never admitted that his engine did not complete any phase of the Carnot cycle, he vehemently denied to the end that auto-ignition was a basic characteristic of his invention [1-4, p. 406].

The additional charge that coal dust was not employed was less weighty [1-5, 1-6]: Especially since his engine was intended to replace the steam engine, Diesel, a nineteenth century engineer, was at first unable to circumvent coal, the primary source of energy in his day. However, he did not rule out other fuels as later tests, even with vegetable oils among other things, prove [1-2]. Measured against the "state-of-the-art" of the day, nobody, not even Diesel, could have known which fuel was best suited for the Diesel engine. Documented by many draft designs, his ingeniously intuitive grasp of the

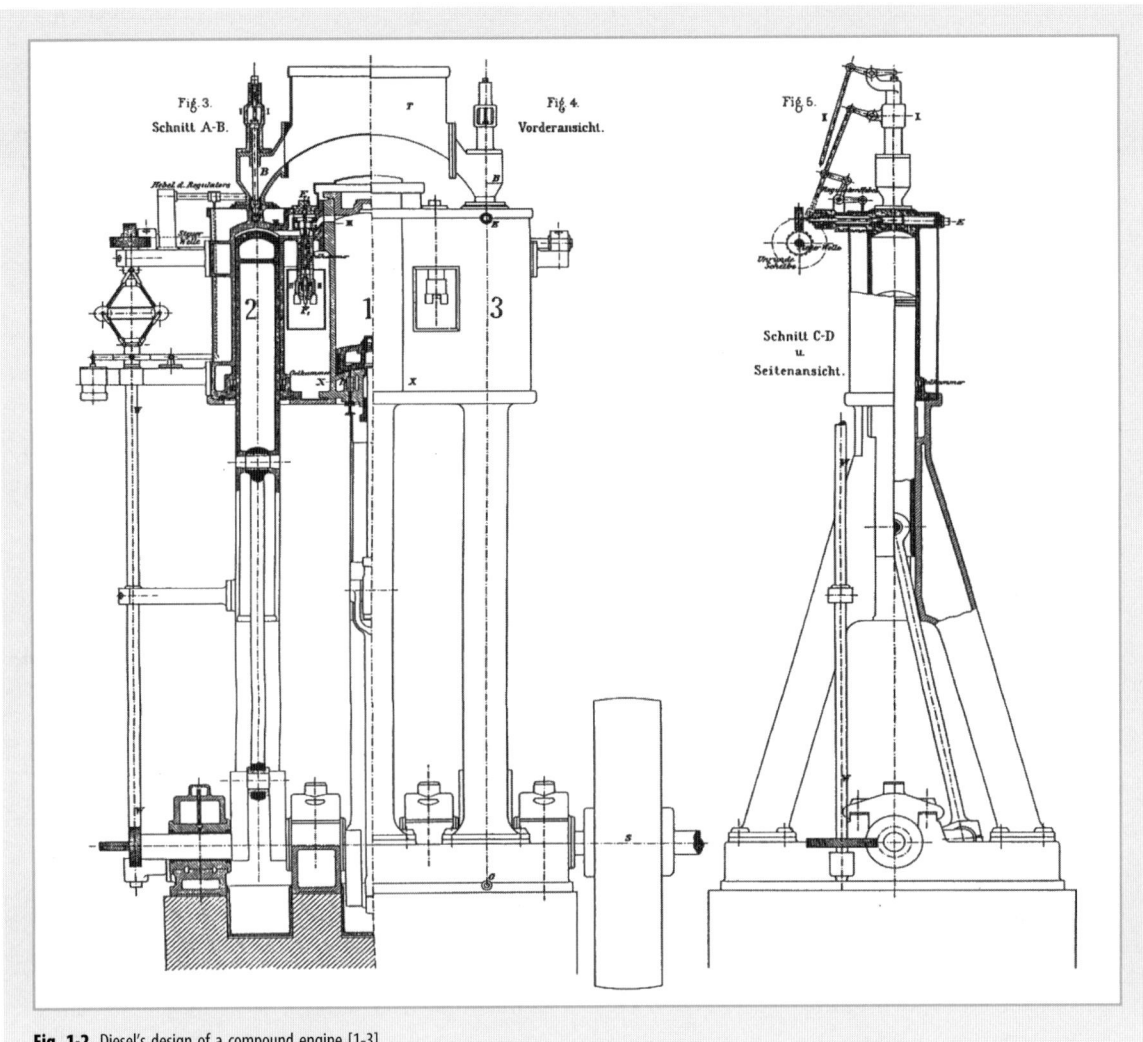

Fig. 1-2 Diesel's design of a compound engine [1-3]

diesel engine's combustion cycles, which were largely unfamiliar to him then and are often only detectable with advanced measurement and computer technology today (see Sect. 3), is all the more admirable (Fig. 1-5).

In addition to successfully weathered patent disputes, the Diesel engine's path continued to be overshadowed by conflicts between the inventor and the Diesel consortium: The latter was interested in profitably "marketing" the engine intended to replace stationary and ship steam engines as soon as possible [1-7]. First, the marketability prematurely asserted in Kassel had to be established. This was done, above all, thanks to the skill and dogged commitment of Immanuel Lauster in Augsburg. It also presaged the line of development of "high performance diesel engines" (Table 1-1).

On the other hand, principally interested in distributed energy generation [1-3, pp. 89ff] and thus anticipating cogeneration unit technology and modern developments in railroad engineering [1-8] that quite realistically envision satellite remote controlled, automatically guided boxcars [1-3], Rudolf Diesel considered the heavy test engine with its A-frame borrowed together with the crosshead engine from steam engine engineering to only be a preliminary stage on the way to a lightweight "compressorless" diesel engine.

The end of Diesel's development work at Maschinenfabrik Augsburg was marked by the reluctantly conceded construction of a compound engine unable to fulfill the hopes placed in it and a few tentative tests of coal dust and other alternative fuels.

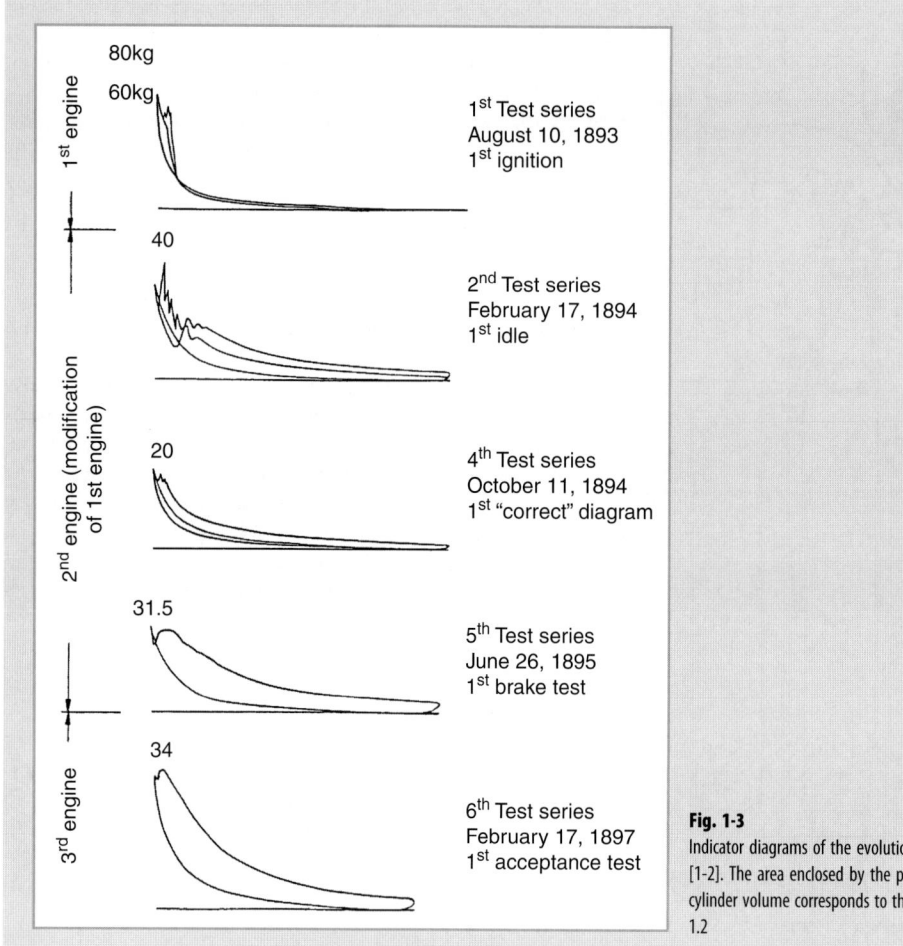

Fig. 1-3
Indicator diagrams of the evolution of the diesel engine based on [1-2]. The area enclosed by the pressure curve as a function of the cylinder volume corresponds to the engine's internal work, see Sect. 1.2

One of Diesel's later tests together with the small firm Safir intended to bring about general acceptance of the line of "vehicle diesel engine" development failed, among other things, because of the poor fuel metering. This problem was first solved by Bosch's diesel injection system [1-9].

Rudolf Diesel met his fate during a crossing from Antwerp to Harwich between September 29 and 30, 1913, just a few weeks after the appearance of his book: "The Origin of the Diesel Engine". After years of struggle and exertion had strained his mental and physical powers to their limit, financial collapse was threatening despite his vast multimillion earnings from his invention: Too proud to admit he had speculated badly and made mistakes or to accept help, Diesel, as his son and biographer relates, saw suicide as the only way out [1-10].

Left behind is his life's work, the high pressure engine that evolved from the theory of heat engines, which bears his name and, 100 years later, is still what its ingenious creator Rudolf Diesel intended: The most rational heat engine of its and even our day (Fig. 1-6). Compared to 1897, its efficiency has approximately doubled and corresponds to the approximation of Carnot efficiency estimated by Diesel. Maximum cylinder pressure p_{Zmax} has more than quintupled and, at 230 bar in present day high performance engines (MTU 8000, see Sect. 17.4), nearly achieves the maximum value Diesel proposed for the Carnot cycle at more than ten times the power density P_A.

Measured by the *ecological imperative*, the diesel engine's high efficiency and multifuel compatibility conserves limited resources and reduces environmental pollution by the greenhouse gas carbon dioxide. However, only consistent development that continues to further reduce exhaust and noise emissions will ensure the diesel engine is accepted in the future too. At the same time, it might also be possible to fulfill Diesel's vision [1-10]:

"That my engine's exhaust gases are smokeless and odorless".

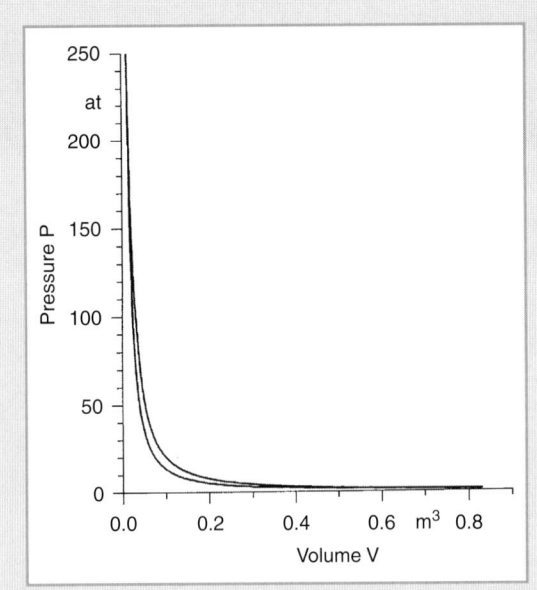

Fig. 1-4 Theoretical indicator diagrams of the Carnot cycle based on [1-3]

1.2 Fundamentals of Engine Engineering

1.2.1 Introduction

Just like gasoline engines, diesel engines are, in principle, energy converters that convert chemically bound fuel energy into mechanical energy (effective work) by supplying the heat released by combustion in an engine to a thermodynamic cycle.

As a function of the system boundaries of the converter represented as a "black box", the energy balance (Fig. 1-7) is:

$$E_B + E_L + W_e + \Sigma E_V = 0.$$

If the energy of the combustion air relative to the ambient state is $E_L = 0$, then the energy supplied with the fuel m_B is equal to the effective work W_e and the total of all energy losses ΣE_V.

The technical system of a "diesel engine" is also part of a widely networked global system defined by the concepts of "resources" and "environmental pollution". A view based purely on energy and economics aimed at minimizing the losses ΣE_V fails to satisfy present day demands specified by the ecological imperative according to which energy and material must always be converted with *maximum efficiency while minimally polluting the environment*. The outcome of the complex research and development work made necessary by these demands is the diesel engine of our day, which has evolved from a simple engine into a complex engine system consisting of a number of subsystems (Fig. 1-8). The increased integration of electrical and electronic components and the transition from open control systems to closed control loops are characteristic of this development. Moreover, international competition is making minimum manufacturing costs and material consumption imperative. Among other things, this requires fit-for-purpose designs that optimally utilize components.

1.2.2 Basic Engineering Data

Every reciprocating engine's geometry and kinematics are clearly specified by the geometric parameters of the:
- stroke/bore ratio $\zeta = s/D$,
- connecting rod ratio $\lambda_{Pl} = r/l$ and
- compression ratio $\varepsilon = V_{max}/V_{min} = (V_c + V_h)/V_c$.

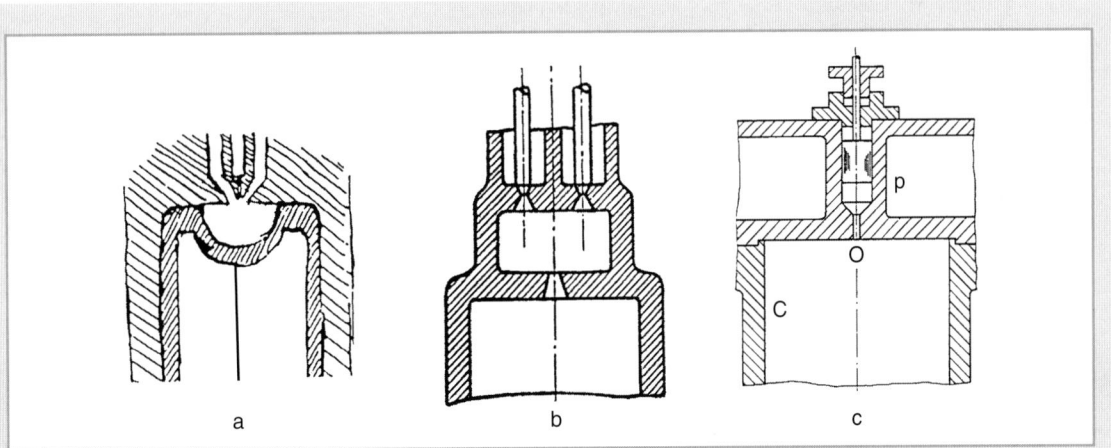

Fig. 1-5 Diesel's proposals for a combustion system. (**a**) Piston with piston crown bowl (1892); (**b**) secondary combustion chamber (1893); (**c**) Pump-nozzle unit (1905), see Sect. 5.3

Table 1-1 Milestones in the development of the diesel engine

Line of "high performance large diesel engine" development

1897	First run of a diesel engine with an efficiency of $\eta_c = 26.2\%$ at Maschinenfabrik Augsburg
1898	Delivery of the first two-cylinder diesel engine with 2×30 HP at 180 rpm to the Vereinigte Zündholzfabriken AG in Kempten
1899	First two-stroke diesel engine from MAN by Hugo Güldner (unmarketable)
1899	First diesel engine without a crosshead, model W, from Gasmotorenfabrik Deutz
1901	First MAN trunk-piston diesel engine by Imanuel Lauster (model DM 70)
1903	First installation of a two cylinder four-stroke opposed piston diesel engine with 25 HP in a ship (the barge Petit Pierre) by Dyckhoff, Bar Le Duc
1904	First MAN diesel power station with 4×400 HP starts operation in Kiev
1905	Alfred Büchi proposes utilizing exhaust gas energy for supercharging
1906	Introduction of the first reversible two-stroke engine by the Sulzer and Winterthur brothers for a marine engine 100 HP/cyl. ($s/D = 250/155$)
1912	Commissioning of the first seagoing ship MS Selandia with two reversible four-stroke diesel engines from Burmeister & Wain each with 1,088 HP
1914	First test run of a double acting six-cylinder two-stroke engine with 2,000 HP/cyl. from MAN Nürnberg ($s/D = 1050/850$)
1951	First MAN four-stroke diesel engine (model 6KV30/45) with high-pressure supercharging: $\eta_e = 44.5\%$ at $w_{emax} = 2.05$ kJ/l, $p_{Zmax} = 142$ bar and $P_A = 3.1$ W/mm^2
1972	Hitherto largest two-stroke diesel engine ($s/D = 1,800/1,050$, 40,000 HP) commences operation
1982	Market launch of super long stroke, two-stroke engines with $s/D \approx 3$ (Sulzer, B & W)
1984	MAN B & W achieves consumption of 167.3 g/kWh ($\eta_e = 50.4\%$)
1987	Commissioning of the largest diesel-electric propulsion system with MAN-B & W four-stroke diesel engines and a total output of 95,600 kW to drive the Queen Elizabeth 2
1991/92	Two-stroke and four-stroke experimental engines from Sulzer (RTX54 with $p_{Zmax} = 180$ bar, $P_A = 8.5$ W/mm^2) and MAN B & W (4T50MX with $p_{Zmax} = 180$ bar, $P_A = 9.45$ W/mm^2)
1997	Sulzer12RTA96C ($s/D = 2,500/960$): two-stroke diesel engine, $P_e = 65,880$ kW at $n = 100$ rpm commences operation
1998	Sulzer RTX-3 research engine to test common rail technology on large two-stroke diesel engines
2000/01	MAN B & W 12K98MC-C ($s/D = 2,400/980$): The currently most powerful two-stroke diesel engine with $P_e = 68,520$ kW at $n = 104$ rpm
2004	First four-stroke medium speed diesel engine MAN B & W 32/40, $P_e = 3,080$ kW, common rail (CR) injection in real use on a container ship
2006	With a consumption of $b_e = 177$ g/kWh, the MaK M43C is the leading four-stroke medium speed marine engine with a cylinder output of 1,000 kW ($s/D = 610/430$, $w_e = 2.71$ kJ/dm^3, $c_m = 10.2$ m/s)
2006	Wärtsilä commissions the world's first 14 cylinder two-stroke engine and thus the most powerful diesel engine: Wärtsilä RTA-flex96C, CR injection, $P_e = 80,080$ kW, $s/D = 2,500/900$, $c_m = 8.5$ m/s, $w_e = 1.86$ kJ/dm^3 ($p_e = 18.6$ bar)

Line of "high-speed vehicle diesel engine" development

1898	First run of a two cylinder four-stroke opposed piston engine ("5 HP horseless carriage engine") by Lucian Vogel at MAN Nürnberg (test engine, unmarketable)
1905	Test engine by Rudolf Diesel based on a four cylinder Saurer gasoline engine with air compressor and direct injection (unmarketable)
1906	Patent DRP 196514 by Deutz for indirect injection
1909	Basic patent DRP 230517 by L'Orange for a prechamber
1910	British patent 1059 by McKenchie on direct high pressure injection
1912	First compressorless Deutz diesel engine, model MKV, goes into mass production
1913	First diesel locomotive with four cylinder two-stroke V-engine presented by the Sulzer brothers (power 1,000 HP)
1914	First diesel-electric motor coach with Sulzer engines for the Prussian and Saxon State Railways
1924	First commercial vehicle diesel engines presented by MAN Nürnberg (direct injection) and Daimler Benz AG (indirect injection in prechamber)
1927	Start of mass production of diesel injection systems at Bosch
1931	Prototype test of the six-cylinder two-stroke opposed piston aircraft diesel engine JUMO 204 of Junkers-Motorenbau GmbH: power 530 kW (750 HP), power mass 1.0 kg/HP
1934	V8 four-stroke diesel engines with prechambers from Daimler-Benz AG for LZ 129 Hindenburg with 1,200 HP at 1,650 rpm (power mass: 1.6 kg/HP including transmission)
1936	First production car diesel engines with prechambers from Daimler-Benz AG (car model 260 D) and Hanomag
1953	First car diesel engine with swirl chamber from Borgward and Fiat
1978	First production car diesel engine with exhaust gas turbocharging (Daimler-Benz AG)
1983	First production high-speed high-performance diesel engine from MTU with twin-stage turbocharging: $w_{emax} = 2.94$ kJ/l bei $p_{Zmax} = 180$ bar, power per unit piston area $P_A = 8.3$ W/mm^2
1986/87	First ever electronic engine management (ECD) for vehicle diesel engines implemented (BMW: car, Daimler-Benz: commercial vehicle)
1988	First production car diesel engine with direct injection (Fiat)
1989	First production car diesel engine with exhaust gas turbocharging and direct injection at Audi (car Audi 100 DI)
1996	First car diesel engine with direct injection and a four-valve combustion chamber (Opel Ecotec diesel engine)
1997	First supercharged car diesel engine with direct common rail high pressure injection and variable turbine geometry (Fiat, Mercedes-Benz)

Table 1-1 (Continued)

1998	First V8 car diesel engine: BMW 3.9 l DI turbodiesel, $P_e = 180$ kW at 4,000 rpm, $M_{max} = 560$ Nm (1,750...2,500 rpm)
1999	Smart cdi, 0.8 dm³ displacement, currently the smallest turbodiesel engine with intercooler and common rail high pressure injection: $P_e = 30$ kW at 4,200 rpm with 3.4 l/100 km first "3 liter car" from DaimlerChrysler
2000	First production car diesel engines with particulate filters (Peugeot)
2004	OPEL introduces a Vectra OPC study suitable for everyday with a 1.9 liter CDTI twin turbo unit with a specific power output of $P_V = 82$ kW/dm³
2006	At the 74th 24 h Le Mans race, an AUDI R10 TDI with a V12 diesel engine ($P_e > 476$ kW at $n = 5,000$ rpm, $V_H = 5.5$ dm³, $w_e = 2.1$ kJ/dm³ with a biturbo boost pressure of $p_L = 2.94$ bar) wins the race

V_{min} corresponds to the compression volume V_c and the maximum cylinder volume V_{max} corresponds to the total from V_c and the cylinder displacement V_h, to which the following with the cylinder bore D and piston stroke s applies:

$$V_h = s \cdot \pi \cdot D^2/4.$$

Accordingly, $V_H = z \cdot V_h$ is the displacement of an engine with z cylinders.

The trunk-piston engine (Fig. 1-9) has established itself. Only large two-stroke engines (see Sect. 18.4) have a crosshead drive to relieve the piston from cornering forces (see Sect. 8.1). Both types are still only used with a unilaterally loaded piston. A standardized time value, the crank angle φ and the rotational speed ω have the following relationship:

$$\omega = d\varphi/dt = 2 \cdot \pi \cdot n.$$

When the speed n is not denoted as engine speed (s⁻¹) but rather, as is customary in engine manufacturing, in revolutions per minute (rpm), then ω is $= \pi \cdot n/30$.

An internal combustion engine's combustion cycle proceeds in the hermetic cylinder volume V_z, which changes periodically with the piston motion z_K within the boundaries V_{max} and V_{min}:

$$V_z(\varphi) = V_c + z_K(\varphi) \cdot \pi \cdot D^2/4.$$

With the crank radius r as a function of the instantaneous crank position φ in crank angle degree (°CA) and top dead center TDC ($\varphi = 0$) as the starting point, the following applies to the piston stroke:

$$z_K = r \cdot f(\varphi),$$

The following approximation function is usually applied:

$$f(\varphi) = 1 - \cos \varphi + (\lambda_{Pl}/4) \cdot \sin^2 \varphi.$$

The following ensues for instantaneous piston velocity c_K and acceleration a_K:

$$c_K = dz_K/dt = r \cdot \omega \cdot [\sin \varphi + (\lambda_{Pl}/2) \cdot \sin 2\varphi]$$
$$a_K = d^2z_K/dt^2 = -r \cdot \omega \cdot [\cos \varphi + \lambda_{PL} \cdot \cos 2\varphi].$$

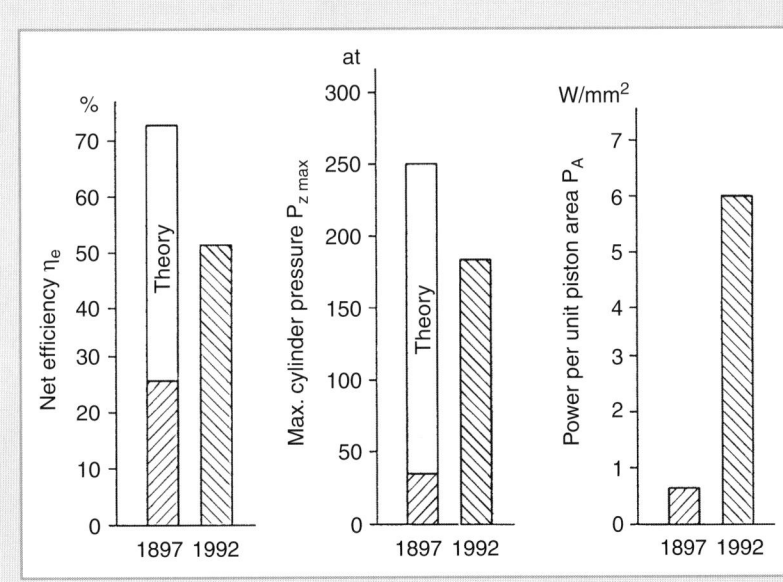

Fig. 1-6

Optimum net efficiency η_e, maximum cylinder pressure p_{Zmax} and power per unit piston area P_A for production engines approximately 100 years after the introduction of the first diesel engine (see also Fig. 1-13 and Table 1-3)

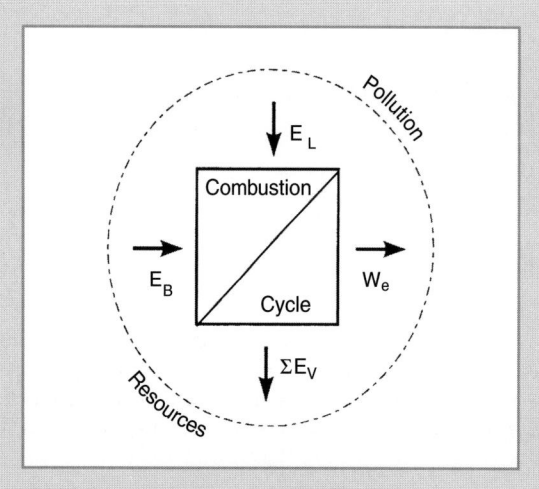

Fig. 1-7 The diesel engine as an energy converter

Following from the piston stroke s in m and engine speed n in s^{-1}, the *mean piston velocity*

$$c_m = 2 \cdot s \cdot n \quad [\text{m/s}] \tag{1-1}$$

is an important *parameter for kinematic and dynamic engine performance*. As it increases, inertial forces ($\sim c^2_m$), friction and wear also increase. Thus, c_m may only be increased to a limited extent. Consequently, a large engine runs at low speeds or a high-speed engine has small dimensions. The following correlation to engine size is approximated for diesel engines with a bore diameter of $0.1\ m < D < 1$ m:

$$c_m \approx 8 \cdot D^{-1/4}. \tag{1-2}$$

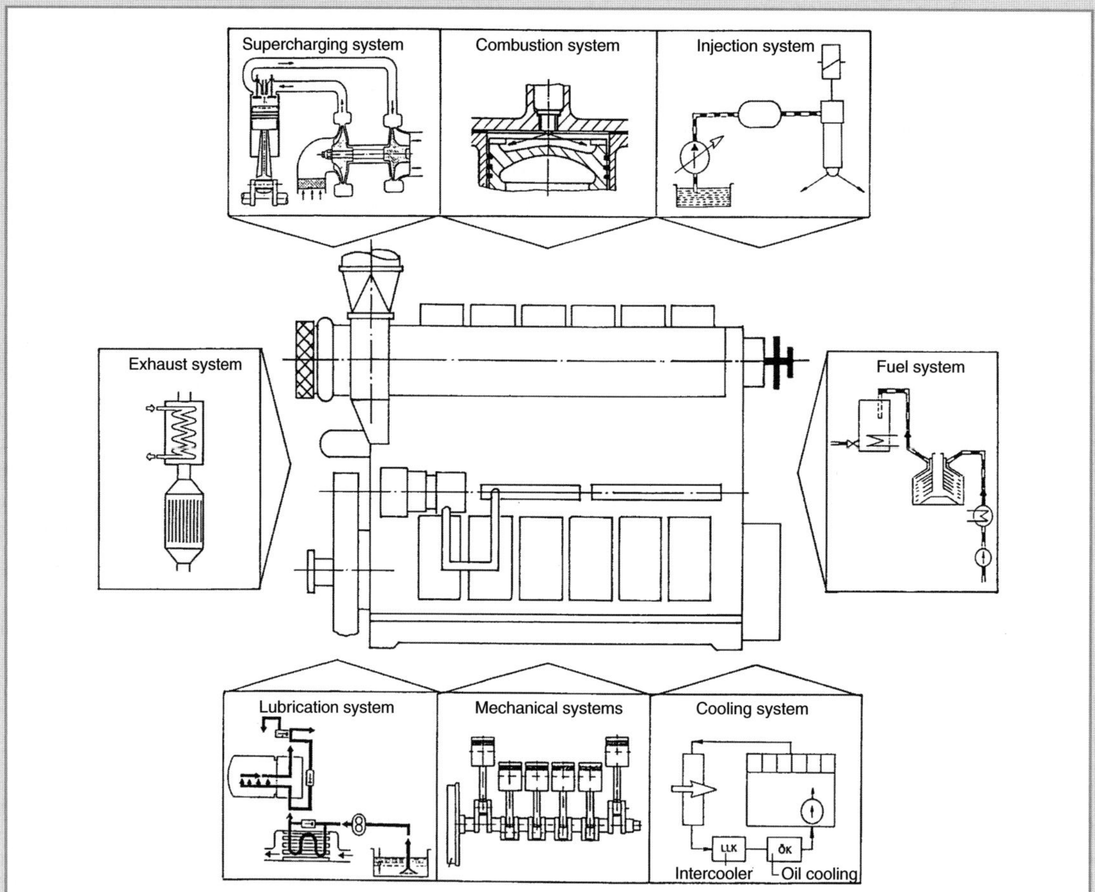

Fig. 1-8 The modern diesel engine as a complex of subsystems

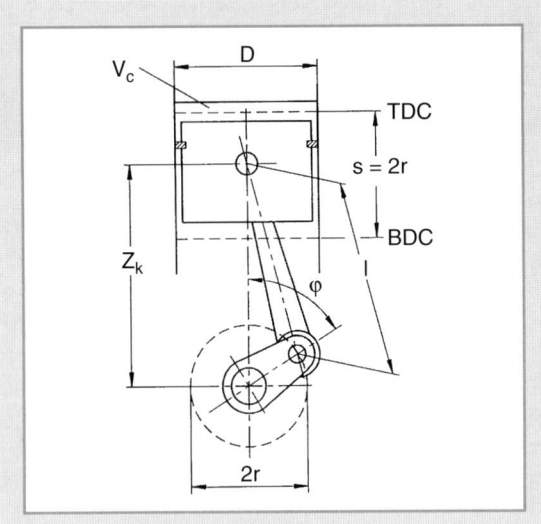

Fig. 1-9 Basic engineering data of a trunk-piston engine

1.2.3 Engine Combustion

1.2.3.1 Fundamentals of Combustion Simulation

Chemically, combustion is the oxidation of fuel molecules with atmospheric oxygen as the oxidant. Thus, the maximum convertible fuel mass m_B is limited by the air mass present in the engine cylinder. Using the fuel-specific stoichiometric air requirement L_{min} (kg air/kg fuel) for complete combustion, the air/fuel ratio λ_V specifies the ratio of "supply to demand" in combustion:

$$\lambda_v = m_{LZ}/(m_B \cdot L_{min}). \tag{1-3}$$

The following applies to the "supply" of the air mass m_{LZ} of all cylinders ($V_Z = z \cdot V_z$) contained in the entire engine:

$$m_{LZ} = V_Z \cdot \rho_Z = \lambda_l \cdot \rho_L \cdot V_H. \tag{1-4}$$

Since the density ρ_Z of the cylinder charge is usually unknown, the definitions of the volumetric efficiency λ_l (see Sect. 2.1) and the density ρ_L of the fresh charge directly at the inlet to the cylinder head are generally reverted to:

$$\rho_L = p_L/(R_L \cdot T_L). \tag{1-5}$$

The air requirement ensues from the elemental analysis of the fuel: A petroleum derivative, diesel fuel (DF) is a conglomerate of hydrocarbons and primarily consists of carbon C, hydrogen H and sulfur S with usually insignificant fractions of oxygen O and nitrogen N. Thus, the balance equation for the complete oxidation of a generic fuel molecule $C_xH_yS_z$ into carbon dioxide CO_2, water H_2O and sulfur dioxide SO_2

$$C_xH_yS_Z + [x + (y/2) + z] \cdot O_2 \rightarrow$$
$$x \cdot CO_2 + (y/2) \cdot H_2O + z \cdot SO_2 + Q_{ex},$$

yields the stoichiometric air/fuel ratio L_{min} corresponding to the oxygen content of the air and the particular number of moles:

$$L_{min} = 11.48 \cdot (c + 2.98 \cdot h) + 4.3 \cdot s - 4.31 \cdot o \quad [kg/kg]$$

(c, h, s, o: mass fraction of 1 kg fuel according to the elemental analysis. Reference value for DK: $L_{min} = 14.5$ kg/kg).

The heat Q_{ex} released during combustion corresponds to the fuel-specific *calorific value* H_u, which may also be calculated [1-11] from the elemental analysis as:

$$H_u = 35.2 \cdot c + 94.2 \cdot h + 10.5 \cdot (s - o) \quad [MJ/kg]$$

The following approximate relation is based on the fuel density ρ_B at 15°C:

$$H_u = 46.22 - 9.13 \cdot \rho_B^2 + 3.68 \rho_B \quad [MJ/kg].$$

Thus, the following applies to the heat supplied to the combustion cycle by "internal combustion":

$$E_B = Q_{zu} \leq m_B \cdot H_u.$$

1.2.3.2 Comparison of Engine Combustion Systems

Combustion is preceded by the preparation of usually liquid fuel to obtain a combustible mixture of fuel vapor and air. This process proceeds differently in diesel and gasoline engines (Table 1-2).

Internal mixture formation in diesel engines (see Chap. 3) begins with the injection of the fuel into the highly compressed and thus heated air shortly before TDC, whereas *external mixture* in classic gasoline engines is formed outside the working chamber by a carburetor or by injection into the intake manifold and often extends through the induction and compression stroke.

While gasoline engines have a *homogeneous fuel/air mixture*, diesel engines have a *heterogeneous mixture* before ignition, which consists of fuel droplets with diameters of a few micrometers distributed throughout the combustion chamber. They are partly liquid and partly surrounded by a fuel vapor/air mixture.

Provided the *air/fuel ratio* of the homogeneous mixture lies within the ignition limits, combustion in gasoline engines is triggered by controlled spark ignition by activating an electrical discharge in a spark plug. In diesel engines, already prepared droplets, i.e. droplets surrounded by a combustible mixture, *auto-ignite*. Ignition limits in the stoichiometric mixture range ($\lambda_V = 1$) only exist for the micromixture in the region of the fuel droplets (see Chap. 3).

Table 1-2 Comparison of features of engine combustion

Feature	Diesel engine	Gasoline engine
Mixture formation	Inside V_z	Outside V_z
Type of mixture	Heterogeneous	Homogenous
Ignition	Auto-ignition with excess air	Spark ignition within ignition limits
Air/fuel ratio	$\lambda_V \geq \lambda_{min} > 1$	$0.6 < \lambda_V < 1.3$
Combustion	Diffusion flame	Premix flame
Torque change through fuel	Variable l_V (quality control)	Variable mixture quantity (quantity control)
	Highly ignitable	Ignition resistant

Diesel engines require excess air ($\lambda_V \geq \lambda_{min} > 1$) for normal combustion. Consequently, the *supply of energy is adapted to the engine load* in diesel engines by the air/fuel ratio, i.e. the *mixture quality* (quality control) and, in light of the ignition limits, by the *mixture quantity* (quantity control) in gasoline engines by throttling that entails heavy losses when a fresh charge is aspirated.

The type of ignition and mixture formation determines the *fuel requirements*: Diesel fuel must be *highly ignitable*. This is expressed by its *cetane number*. Gasoline must be *ignition resistant*, i.e. have a high *octane number*, so that uncontrolled auto-ignition does not trigger uncontrolled combustion (detonation). The latter is ensured by low boiling, short chain and thus thermally stable hydrocarbons (C_5 through C_{10}). Diesel fuel, on the other hand, consists of high boiling, long chain hydrocarbons (C_9 through C_{30}) that disintegrate earlier and form free radicals that facilitate auto-ignition (see Chap. 3).

1.2.4 Fundamentals of Thermodynamics

1.2.4.1 Ideal Changes of States of Gases

The state of a gas mass m is determinable by two thermal state variables by using the general equation of state for ideal gases:

$$p \cdot V = m \cdot R \cdot T$$

(p absolute pressure in Pa, T temperature in K, V volume in m³, R specific gas constant, e.g. for air $R_L = 287.04$ J/kg · K). Ideal gases are characterized by a constant isentropic exponent κ (air: $\kappa = 1.4$; exhaust gas: $\kappa \approx 1.36$) as a function of pressure, temperature and gas composition.

Consequently, the state of a gas can be represented in a p, V diagram with the variables p and V and tracked. Changes of state are easy to calculate by setting constants of a state variable for which simple closed equations exist for isobars ($p = $ const.), isotherms ($T = $ const.) and isochors ($V = $ const.) [1-12]. The adiabatic change of state is a special case:

$$p \cdot V^\kappa = \text{const.,}$$

heat not being transferred between the gas and the environment. When this cycle is reversible, it is called an isentropic change of state. However, just as the real isentropic exponent depends on the state and composition of a gas, this is never actually the case in reality [1-13].

1.2.4.2 Ideal Cycle and Standard Cycle

In an ideal cycle, the gas undergoes a self-contained change of state, returning to its initial state once it has completed the cycle. Thus, the following applies to the internal energy $U = U(T)$:

$$\oint dU = 0.$$

Hence, from the first law of thermodynamics that describes the conservation of energy in closed systems:

$$\partial Q = dU + p \cdot dV,$$

it follows that the heat Q converted during the cycle accumulates as mechanical work:

$$\oint \partial Q = \oint p \cdot dV = W_{th},$$

i.e. the change in pressure and volume corresponds to the theoretically useful work W_{th} of the ideal cycle.

An ideal cycle becomes the standard cycle for a thermal machine once it has been adjusted for reality. For a reciprocating piston engine, this means that the ideal cycle and real combustion cycle proceed similarly between two volume and pressure limits specified by V_{max} and V_{min} and p_{max} and p_{min}. The upper pressure limit p_{max} corresponds to the maximum cylinder pressure p_{Zmax} allowable for reasons of stability and p_{min} the air pressure p_L before intake into the engine (Fig. 1-10a). Other specifications that must agree are the compression ratio ε and input heat Q_{zu} or Q_B:

$$Q_{zu} = Q_B = m_B \cdot H_u.$$

When the mass of the fresh charge m_{LZ} is given (Eq. (1-4)), the fuel mass m_B is limited by the air/fuel ratio λ_V and the *calorific value of the mixture* h_u:

$$h_u = Q_{zu}/(m_B + m_{LZ}) = H_u/(1 + \lambda_v \cdot L_{min}).$$

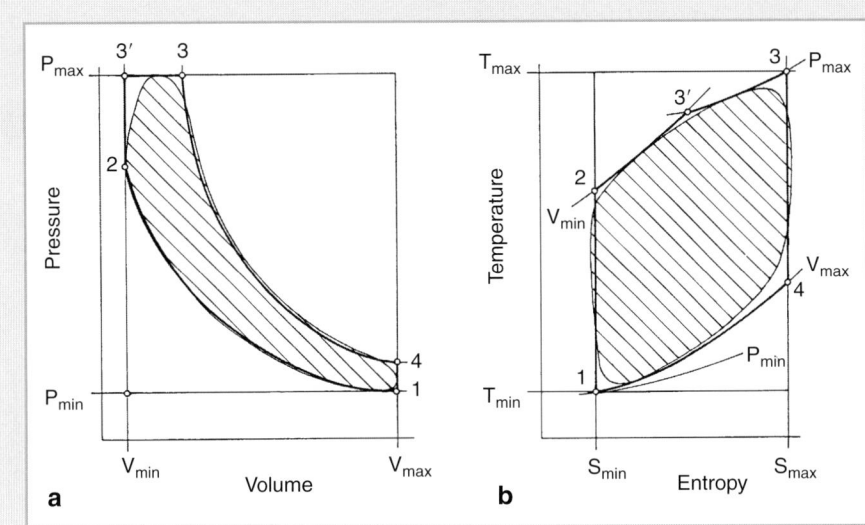

Fig. 1-10
The Seiliger cycle as standard cycle for internal combustion engines in a p, V diagram (**a**) and T, s diagram (**b**)

Based on the diesel engine's combustion phase and assuming the gas exchange phase is free of losses along the isobars p_{min} (see Sect. 2.1), the standard cycle in 1 begins with adiabatic compression to $p_2 = p_c = p_1 \cdot \varepsilon^\kappa$ (Fig. 1-10a). Afterward, heat is transferred: first, isochorically until it reaches the limit pressure p_{max} in 3' and, then, isobarically up to 3. The adiabatic expansion that follows ends in 4. The cycle concludes when heat begins to dissipate along the isochors V_{max} afterward. The area of the region 1-2-3'-3-4-1 corresponds to the theoretical work:

$$W_{th} = h_{th} \cdot Q_{zu},$$

Applying the charging ratio $\delta = V_3/V_2$ and the pressure ratio $\psi = p_3/p_2$ makes it possible to specify a closed expression (provided that $\kappa =$ const.) for the thermal efficiency η_{th} of the Seiliger cycle described here

$$\eta_{th} = (1 - \varepsilon^{1-\kappa}) \cdot (\delta^\kappa \cdot \psi - 1)/[\psi = 1 + \kappa \cdot (\delta - 1)],$$

The Seiliger cycle's conversion of energy can be followed in the temperature entropy (T, s-) diagram (Fig. 1-10b): Since the areas of the regions s_{min}-1-2-3'-3-4-s_{max} and s_{min}-1-4-s_{max} correspond to the heat supplied Q_{zu} and extracted Q_{ab} respectively, the difference corresponds to the theoretical effective work. Thus, the following applies to thermal efficiency:

$$\eta_{th} = (Q_{zu} - Q_{ab})/Q_{zu} = W_{th}/Q_{zu}. \qquad (1\text{-}6)$$

The rectangles formed by the limit values in both diagrams correspond to the maximum useful work in each case, yet with different efficiencies: The full load diagram of an ideal reciprocating piston steam engine with moderate efficiency in the p, V diagram is presented alongside the Carnot efficiency with real non-useable work (see Sect. 1.1). The temperature difference $T_{max} - T_{min}$ is crucial to the Carnot cycle's efficiency η_c:

$$\eta_c = (T_{max} - T_{min})/T_{max}.$$

The T, s diagram reveals that high temperatures (up to 2,500 K) occur even during real combustion (see Sect. 1.3). Since combustion is intermittent, the engine components fall below the temperatures critical for them when they have been designed appropriately (see Sect. 9.1). The lowest possible temperature $T_{min} \leq T_L$ is also conducive to this.

Since it can be adapted to the real engine process, the Seiliger cycle corresponds to the most general case of a standard cycle. It also encompasses the limit cases of the constant volume cycle ($\delta \to 1$) and the constant pressure cycle ($\psi \to 1$), which are often referred to as the ideal gasoline engine or diesel engine process even though combustion in gasoline engines does not occur with an infinitely high combustion rate and combustion in diesel engines is not isobar (Fig. 1-11).

Allowing for the real gas behavior, i.e. $\kappa \neq$ const., the compression ratio's influence on the thermal efficiency η^*_{th} for a pressure ratio $p_{max}/p_{min} = 60$ is evident in Fig. 1-12: The allowable maximum pressure in the constant volume cycle is already exceeded for $\varepsilon \approx 9$ at an air/fuel ratio of $\lambda_V = 2$. A Seiliger cycle allows higher compression ratios but transforms into a constant pressure cycle for $\varepsilon \approx 19.7$.

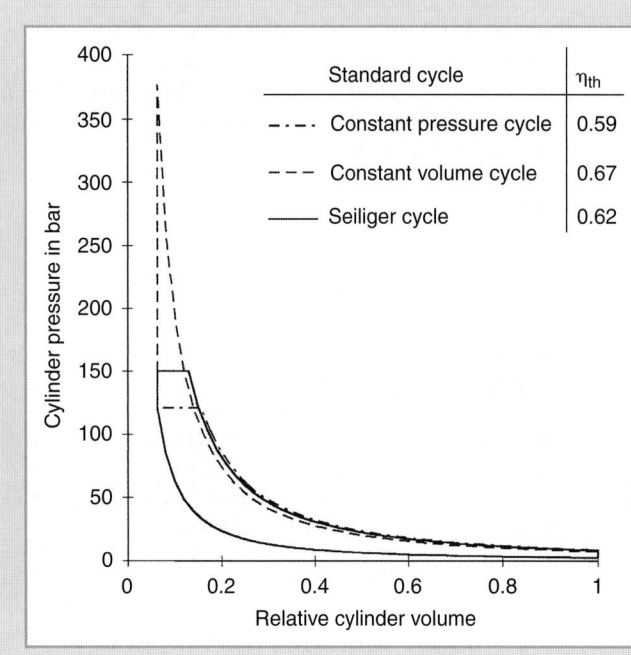

Fig. 1-11 Ideal cycle as standard cycle: Seiliger cycle (p_{Zmax} = 150 bar), constant pressure and constant volume cycle for p_1 = 2.5 bar, T_1 = 40 °C, ε = 16, λ_v = 2 and H_u = 43 MJ/kg

Engine process simulation (Sect. 1.3) has eliminated the idealized standard cycle in the field, yet has retained its worth for quick "upward" estimates, e.g. when engine process control is varied.

1.2.5 The Diesel Engine Process

1.2.5.1 Two-stroke and Four-stroke Cycle

Unlike the idealized cycle with external heat input, internal combustion requires the exchange of the charge after every combustion phase by a gas exchange phase (see Sect. 2.1). To do so, a four-stroke engine requires two additional strokes or cycles as the motion from one dead center to the other is termed. Hence, by expelling the exhaust gas and aspirating the fresh charge after the expansion stroke (compression as well as combustion and expansion), the entire working cycle is comprised of two revolutions or 720°CA. Consequently, a frequency ratio exists between the speed and the working cycle frequency n_a:

$$a = n/n_a, \tag{1-7}$$

which is often denoted as the "cycle rate" without specifying the actual number of cycles with $a = 2$ (four-stroke cycle) or $a = 1$ (two-stroke cycle).

1.2.5.2 Real Engine Efficiencies

Along with the thermal efficiency, which enables an upward estimate, the net efficiency is of prime interest:

$$\eta_e = W_e/(m_B \cdot H_u) = \eta_{th} \cdot \eta_u \cdot \eta_g \cdot \eta_m = \eta_i \cdot \eta_m, \tag{1-8}$$

It can also be specified as the product of the thermal efficiency and the reference variable that describes the percentage loss.

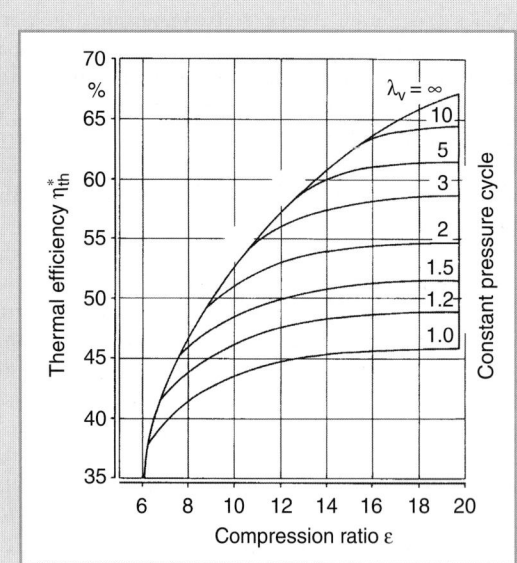

Fig. 1-12 Thermal efficiency η^*_{th} incorporating real gas behavior (based on [1-13])

Losses by incomplete combustion are included by the conversion factor:

$$\eta_u = Q_{zu}/(m_B \cdot H_u).$$

The efficiency factor:

$$\eta_g = W_i/W_{th}$$

describes the real cycle's deviations from the ideal cycle by employing
- a real instead of an ideal working gas,
- wall heat losses instead of the adiabatic change of state,
- real combustion instead of idealized heat input and
- gas exchange (throttling, heating and scavenging losses).

In accordance with DIN 1940, mechanical efficiency

$$\eta_m = W_e/W_i$$

encompasses the frictional losses at the piston and in the bearings, the heat loss of all the assemblies necessary for engine operation and the aerodynamic and hydraulic losses in the crankshaft assembly.

Referred to as indication, measurement of the cylinder head pressure curve can be used to determine the indicated work W_i (the hatched area in Fig. 1-10a) present in a piston and thus the internal (indicated) efficiency

$$\eta_i = W_i/(m_B \cdot H_u) = \oint p_z \cdot dV/(m_B \cdot H_u).$$

1.2.5.3 Engine Operation and Engine Parameters

Effective Brake Work and Torque

The effective brake work W_e ensues from the torque M and the "cycle rate" a measurable at the engine's output shaft:

$$W_e = 2 \cdot \pi \cdot a \cdot M = w_e \cdot V_H. \quad (1\text{-}9)$$

When W_e is related to displacement V_H, then the *specific work w_e in kJ/dm³* denotes the effective work obtained from one liter of displacement. Thus, along with the *mean piston velocity c_m* Eq. (1-1), it is the most important engine parameter that characterizes the "state-of-the-art". Engine companies with a sense of tradition often still apply the parameter p_e, "brake mean effective pressure", which, despite being specified in "bar", does not correspond to any *measurable pressure*. Rather, it is rooted in the history of mechanical engineering.[1] The following applies to conversions:

1 bar "brake mean effective pressure" = 0.1 kJ/dm³.

---
[1] The inability to reach a consensus among the many authors from industry and academia must be borne in mind in the individual sections, especially when numerical data is provided.

Based on Eq. (1-9) with $M/V_H = w_e/(2 \cdot \pi \cdot a)$, the term volume-specific torque M/V_H in Nm/dm³ sometimes employed for vehicle engines likewise equals the specific effective work, $w_e \approx 0.0125 \cdot (M/V_H)$ applying to four-stroke engines.

Fundamental Diesel Engine Equation

With the net efficiency η_e and the air/fuel ratio λ_V, Eq. (1-8), (1-3), the following ensues for the effective work:

$$W_e = \eta_e \cdot m_B \cdot H_u = \eta_e \cdot m_{LZ} \cdot H_u/(\lambda_V \cdot L_{min}). \quad (1\text{-}10)$$

The fresh air mass m_{LZ} in an engine, Eq. (1-4), is defined by the volumetric efficiency λ_l and charge density ρ_L, Eq. (1-5), so that the following ensues for the specific effective work:

$$w_e = \eta_e \cdot \lambda_l \cdot (p_L/(R_L \cdot T_L)) \cdot (H_u/(\lambda_V \cdot L_{min})). \quad (1\text{-}11)$$

If the specific fuel parameters are regarded as given just like the indirectly influenceable efficiency, then only increasing the pressure p_L by compression, e.g. by exhaust gas turbocharging with intercooling (see Sect. 2.2), remains a freely selectable option to increase effective work since limits exist for both the volumetric efficiency with $(\lambda_l)_{max} \to \varepsilon/(\varepsilon-1)$ and the air/fuel ratio $\lambda_V \to \lambda_{min} > 1$.

Engine Power

The combustion cycle frequency n_a and specific effective work w_e, Eq. (1-7), (1-9), yield the following for the net power:

$$P_e = W_e \cdot n_a = w_e \cdot z \cdot V_h \cdot n/a, \quad (1\text{-}12)$$

and the following with the mean piston velocity c_m, Eq. (1-1):

$$P_e = C_o \cdot w_e \cdot C_m \cdot z \cdot D^2 \quad (1\text{-}13)$$

($C_0 = \pi/(8 \cdot a) \approx 0.2$ or 0.4 for four and two-stroke respectively).
With its quadratic dependence on bore diameter D, the second form of the power equation suggests a large engine is another option to boost power. At the same time, engine torque (Eq. (1-9)) increases:

$$M \sim W_e \sim w_e \cdot z \cdot D^3.$$

Accordingly, when the cylinder dimensions are retained, comparable engine power through specific work w_e can only be achieved by maximum supercharging (see Sect. 17.4).

When speed is specified in rpm, displacement in dm³ and specific work in kJ/dm³, the following is obtained for practical calculations

$$P_e = w_e \cdot z \cdot V_h \cdot n/(60.a)$$

or, using the brake mean effective pressure p_e in bar, the following

$$P_e = p_e \cdot z \cdot V_h \cdot n/(600.a)$$

in kW in each case.

The *fundamental diesel engine equation*, Eq. (1-11), reveals that engine power is a function of the ambient condition: A diesel engine run at an altitude of 1,000 m cannot produce the same power as at sea level. Hence, set reference conditions (x) for performance comparisons and acceptance tests for users' specific concerns have been defined to convert the power P measured into the power P_x applicable to the reference condition.[2] Generally, the following applies:

$$P_x \cong \alpha^\beta \cdot P.$$

In addition to air pressure and temperature, influencing variables for α and β are relative humidity, coolant inlet temperature in the intercooler and engine mechanical efficiency ($\eta_m = 0.8$ if unknown). Since a danger of overcompensation has been proven to often exist, some vehicle engine manufacturers have switched to measuring power in air conditioned test benches with ambient conditions that conform to standards. Since diesel engines have low overload capacity, the *blocked ISO net power* that may not be exceeded or the *ISO standard power* that may be exceeded depending on the engines' use is specified with the defined magnitude and duration of their extra power [1-14]. At 10% overload, it corresponds to the CIMAC recommendation for "continuous brake power" for marine engines.

Power-Related Engine Parameters

Frequently applied to vehicle engines, the displacement specific power output

$$P_V = P_e/V_H = w_e \cdot n/a. \quad (1\text{-}14)$$

is a function of the speed and thus also engine size. On the other hand, the specific power per unit piston area:

$$P_A = P_e/(z \cdot A_k) = w_e \cdot c_m/(2 \cdot a), \quad (1\text{-}15)$$

(with w_e in kJ/dm^3, c_m in m/s, $2 \cdot a = 4$ results for a four-stroke engine and $2 \cdot a = 2$ for a two-stroke engine, where P_A is in W/mm^2) is independent of engine size if the correlation from Eq. (1-2) is disregarded this once. The product of *mechanical and thermal* (w_e) as well as *dynamic load* (c_m) characterizes the "state-of-the-art" for two-stroke or four-stroke engines and large or vehicle engines in equal measure as the following example makes clear:

In a comparison of two production engines, the low speed two-stroke Wärtsilä RT96C diesel engine [1-15] with an MCR cylinder output of 5,720 kW, specific effective work of $w_e = 1.86$ kJ/dm^3 and a mean piston velocity of $c_m = 8.5$ m/s and the currently most powerful BMW diesel engine for cars (BMW 306 D4: $w_e = 1.91$ kJ/dm^3, $c_m = 13.2$ m/s [1-16]), the following ensues for the power per unit piston area and displacement specific power output:
- Wärtsilä $P_A = 7.91$ W/mm^2 and $P_V = 3.16$ kW/dm^3,
- BMW $P_A = 6.31$ W/mm^2 and $P_V = 70.2$ kW/dm^3.

The comparison of the powers per unit piston area clearly demonstrates that even the low speed two-stroke engine, sometimes derogatorily called a "dinosaur", is a "high-tech" product that even leaves the BMW 306 D4, a "powerhouse" with 210 kW rated power, far behind.[3]

A car diesel engine has to deliver its full load power on the road only very rarely, whereas a marine diesel engine – barring a few maneuvers – always runs under full load, not infrequently up to 8,000 h a year.

The development of the variable P_A presented in Fig. 1-13 reveals that the potential for diesel engine development has apparently not been exhausted yet! However, current emphases of development are geared less toward enhancing performance than reducing fuel consumption and improving exhaust emission in light of rising fuel prices.

Specific Fuel Consumption

The fuel mass flow $\dot{m}_B$ yields the performance-related specific fuel delivery rate or fuel consumption:

$$b_e = \dot{m}_B/P_e = 1/(\eta_e \cdot H_u).$$

Accordingly, comparative analyses require identical calorific values or fuels. When alternative fuels are used (see Sect. 4.2), the quality of energy conversion cannot be inferred from consumption data. Thus the specification of net efficiency is fundamentally preferable. Standard ISO fuel consumptions relate to a fuel (DF) with $H_u = 42$ MJ/kg. This allows the following conversion for specifications of fuel consumption in g/kWh:

$$\eta_e = 85.7/b_e \quad \text{and} \quad b_e = 85.7/\eta_e.$$

Specific Air Flow Rate or Air Consumption

Analogous to specific consumption, the total air flow rate $\dot{m}_L$ (see Sect. 2.1.1) yields an engine's specific air flow rate or consumption (see Table 1-3):

$$l_e = \dot{m}_L/P_e,$$

Thus, the following applies to the total air/fuel ratio:

$$\lambda = l_e/(b_e \cdot L_{min}).$$

---

[2] Common standards include Part 1 of DIN ISO 3046, DIN 70020 (11/76) specifically for motor vehicle engines and ECE Regulation 120 for "internal combustion engines to be installed in agricultural and forestry tractors and in non-road mobile machinery".

[3] Occasionally applied, the term "$p_e \cdot c_m$" [1-17] yields 158 (bar · m/s) for the low speed engine and 252 (bar · m/s) for the BMW engine. Since the different working processes are disregarded, the product "$p_e \cdot c_m$" is not a real variable. Moreover, the specification in (bar · m/s) defies any sound analysis.

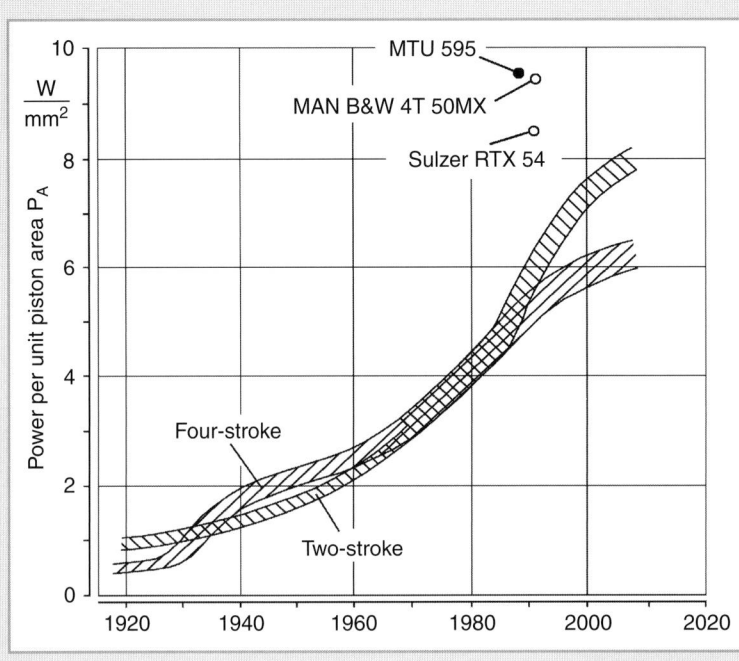

Fig. 1-13
Development of the power per unit piston area P_A of large diesel engines; maximum values of two-stroke (Sulzer RTX54) and four-stroke (MAN B&W 4T 50MX) experimental engines and a production engine (MTU 595)

Engine Characteristic Map

As a rule, the use of an engine to drive stationary systems or vehicles requires adjusting the engine characteristic map of the curve of torque M as a function of speed: As full load torque is approached, the air/fuel ratio λ_V drops so that the smoke limit is reached at $\lambda_V \rightarrow \lambda_{min}$. This corresponds to a smoke number still considered acceptable. As the speed limits n_A and n_N (starting and rated speed) increasingly spread, vehicle engines exibit a spike in the average speed range, which gives them far more flexible responsiveness (Fig. 1-14).

Table 1-3 Operating values of diesel engines at nominal load

Engine type	Specific fuel consumption b_e [g/kWh]	Spec. flow rate l_e [kg/kWh]	Air/fuel ratio λ_V	Specific oil consumption $b_ö$ [g/kWh]	Exhaust gas temperature T_A after turbine [°C]
Car diesel engines:					
- without supercharging	265	4.8	1.2	<0.6	710
- with exhaust gas turbocharging	260	5.4	1.4	<0.6	650
Commercial vehicle diesel engines* with exhaust gas turbocharging and intercooling	205	5.0	1.6	<0.2	550
High performance diesel engines	195	5.9	1.8	<0.5	450
Medium speed four-stroke diesel engines	180	7.2	2.2	0.6	320
Low speed two-stroke diesel engines	170	8.0	2.1	1.1	275

* for heavy commercial vehicles and buses.
Note: While the specific air flow rate l_e not only includes the combustion air but also the scavenging air, the combustion air ratio λ_V, only incorporates the mass of the combustion air. The specified mean values cover a range of approximately ± 5%.

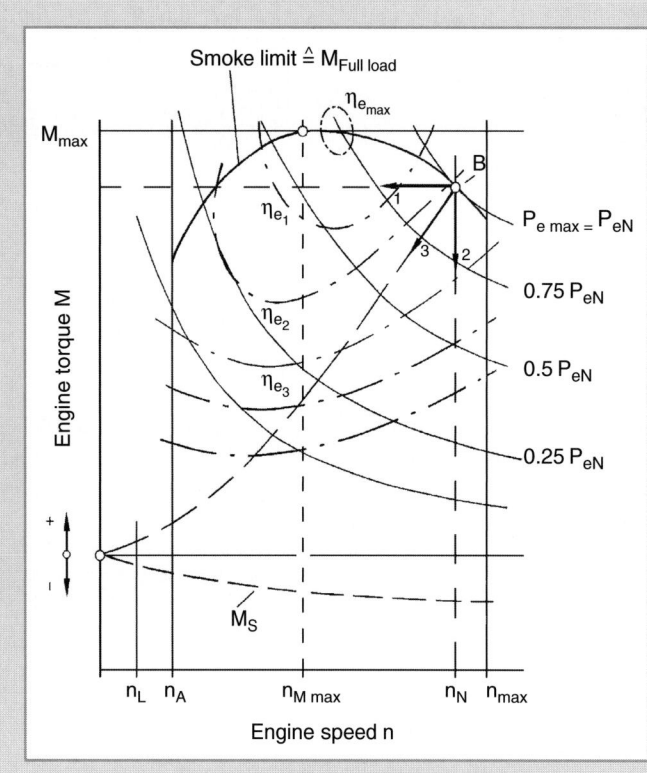

Fig. 1-14
Map representation of engine torque M with lines P_e = const. and η_e = const. and specification of selected engine characteristic maps. *1* speed decrease at rated engine torque, *2* generator operation and *3* propeller curve

Apart from the smoke limit and the power hyperbolae (curves of constant power), such engine maps also often include curves of constant efficiency and specific fuel consumption or other engine parameters. Some specific engine characteristic maps are:

1. speed decrease at rated engine torque: M = const. and n = variable,
2. generator operation: M = variable and n = const.,
3. propeller operation: $M \sim n^2$.

Depending on the rolling resistance, the entire map range can be covered for the vehicle drive including motored operation with drag torque M_s. "Speed decrease at rated engine torque" ought to be avoided in supercharged engines, since the decreasing air/fuel ratio can cause thermal overloading in limit loaded engines (see Sect. 2.2).

Corresponding to the engine characteristic map specified by the rolling resistance curves, the vehicle drive requires an adjustment of the characteristic map by converting the torque with a transmission system (Fig. 1-15). The map is limited by the maximum transferrable torque M_{Rmax} at the gear wheel ("slip limit"), the maximum engine or gear wheel speed n_R and the torque characteristic for P_{max} = const. denoted as the "ideal traction hyperbola". The following applies to the torque M_R acting on the gear wheel incorporating the reduction i_{ges} (gear stage, axel drive and differential) and all mechanical losses with η_{ges}:

$$M_R = i_{ges} \cdot \eta_{ges} \cdot M.$$

At a running speed of $c_F = 2 \cdot \pi \cdot R \cdot n_R$, the following applies to driving performance that overcomes all rolling resistance ΣF_W:

$$P_R = 2 \cdot \pi \cdot n_R \cdot M_R = C_F \cdot \Sigma F_W.$$

Transmission design matched to the consumption map (Fig. 1-14) can achieve favorable fuel consumption with good driving comfort (see Sect. 17.1).

1.3 Combustion Cycle Simulation

1.3.1 Introduction

The processes in a diesel engine cylinder run intensely transiently since the working cycles of compression, combustion, expansion and charge follow one another in fractions of a second. Hence, it is impossible to use the simple means of the ideal standard cycle to simulate a diesel engine accurately

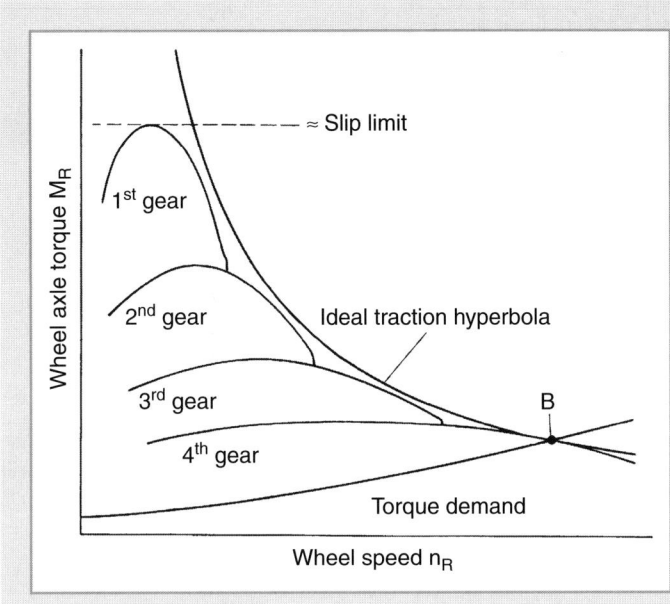

Fig. 1-15
Engine characteristic maps for a vehicle engine with four-speed transmission

enough for engine development. Rather, the differential equations of mass and energy conservation must be solved mathematically, incorporating thermal and calorific state equations.

The rapid development of data processing made it possible to solve these differential equations mathematically for the first time in the 1960s [1-18]. Since the mathematical work helped reduce high test bench costs, the first tests were performed in the large engine industry.

In the meantime, combustion cycle simulation has become a standard tool in engine development and will continue to gain importance in the future [1-19]. Applications range from simple descriptions of the processes in a cylinder up through the complex, transient processes for transient additional loading of diesel engines with two-stage sequential turbocharging allowing for dynamic user behavior [1-20–1-22].

Thermodynamic analysis of the cylinder pressure curve constitutes state-of-the-art testing today, Thanks to advanced computers, it is not only able to ascertain the instantaneous combustion characteristic but also other operating parameters such as the residual exhaust gas in a cylinder in real time [1-23]. This can be built upon for control based on cylinder pressure for new combustion systems, e.g. the HCCI system, for use in mass production, provided accurate and stable pressure sensors are available.

Naturally, an introduction to engine process simulation cannot treat every thermodynamically interesting engine assembly such as the cylinder, exhaust gas turbocharger or air and exhaust manifold systems. Hence, taking modeled thermodynamic processes in a cylinder without a divided combustion chamber as an example, the following sections only explain the fundamentals of engine process simulation. The individual sections provide references to more detailed literature.

1.3.2 Thermodynamic Foundations of Engine Process Simulation

1.3.2.1 General Assumptions

Thermodynamic Cylinder Model

The assumptions put forth in Sect. 1.2 to analyze the ideal combustion cycle can no longer be retained by engine process simulation that simulates the change of state of the cylinder charge (pressure, temperature, mass, composition, etc.) during a combustion cycle. Suitable thermodynamic models must be defined for both the individual cylinder and the process boundary conditions such as energy release by combustion, wall heat losses or the conditions before and after the cylinder (Table 1-4).

System boundaries are set for the cylinder's working chamber (Fig. 1-16). To this end, the pressure, temperature and composition of the gases in the cylinder are generally assumed to be alterable as a function of time and thus crank angle, yet independent of their location in the cylinder. Consequently, the cylinder charge is considered homogeneous. This is referred to as a single zone model. Naturally, this premise does not correspond to the actual processes in a diesel engine cylinder; however, it yields computerized results that are accurate enough for most development

Table 1-4 Differences of different submodels in the ideal and real cycle

Submodel	Ideal cycle	Real cycle
Physical properties	Ideal gas	Real gas; composition changes during the cycle
	$c_p, c_v, \kappa =$ constant	Physical properties are a function of pressure, temperature and composition
Gas exchange	Gas exchange as heat dissipation	Mass exchange through the valves, residual exhaust gas remains in the cylinder
Combustion	Complete combustion based on specified, idealized regularity	Different combustion characteristics possible depending on mixture formation and combustion cycle; fuel burns partially or incompletely
Wall heat losses	Wall heat losses are ignored	Wall heat losses are factored in
Leaks	Leaks are ignored	Leaks are partially factored in, however ignored in this introduction

work as long as there is no intention to simulate concentrations of pollutants. The formation mechanisms of pollutants, especially nitrogen oxides, are highly dependent on temperature and require the temperature in the burned mixture (post-flame zone) as an input value. It is significantly higher than the energy-averaged temperature of the single zone model. In this case, the cylinder charge is divided into two zones (two zone model [1-24–1-27]). One zone contains the unburned components of fresh air, fuel and residual exhaust gas (relatively low temperature), the other zone the reaction products of exhaust gas and unutilized air (high temperature). Both zones are separated by an infinitesimally small flame front in which the (primary) fuel oxidizes. Figure 1-17 presents both the average temperature of the single zone model and the temperatures of both zones of the two zone model for a diesel engine's high pressure cycle. Clearly, the burned zone has a significantly higher temperature level than in the single zone model. Nonetheless, only the single zone model will be examined to introduce the methods of real cycle simulation.

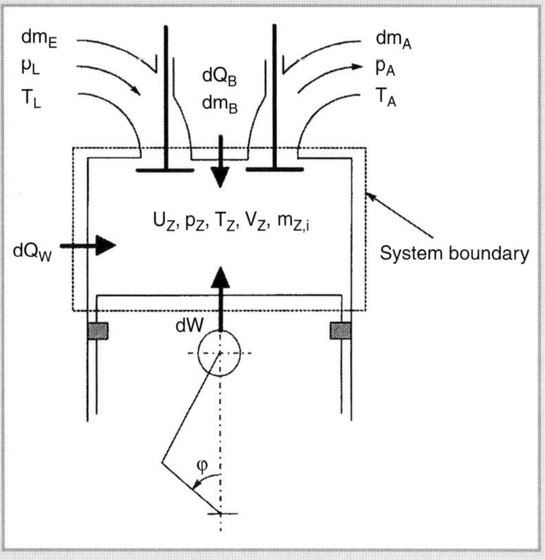

Thermal and Calorific Equations of State

The state of the charge in the cylinder is described by pressure p, temperature T, volume V and composition (masses m_i of the components i). A physical relation exists between these values, the thermal equation of state. In the case of the ideal gas, it is:

$$p \cdot V = m \cdot R \cdot T \tag{1-16}$$

where R is the gas constants and the total mass m is the total of the partial masses of the individual components:

$$m = m_1 + m_2 + \ldots + m_i \tag{1-17}$$

If the cylinder charge is regarded as a real gas, then Eq. (1-16) must be replaced by one of the many equations of state for real gases familiar from the literature (e.g. Justi [1-28], Zacharias [1-29]).

With the aid of the calorific equation of state, the state variables p and T yield the specific internal energy u_i for every component i. In the case of the ideal gas, this is

$$du_i(T) = c_{v,i}(T) \cdot dT \tag{1-18}$$

where c_v is the specific isochoric thermal capacity. As a rule, c_v is a function of temperature. In the case of real gases, c_v and thus u is a function of temperature and pressure and can, for example, be taken from appropriate collections of tables or calculated [1-30–1-32, 1-27, 1-29].

Laws of Mass and Energy Conservation

The cylinder contains a charge mass m with a certain composition. The mass can change by being fed in or drawn off through the intake valve dm_E, the exhaust valve dm_A or the injection valve dm_B. (Gas losses by leaks are disregarded

Fig. 1-16 Thermodynamic model of the cylinder

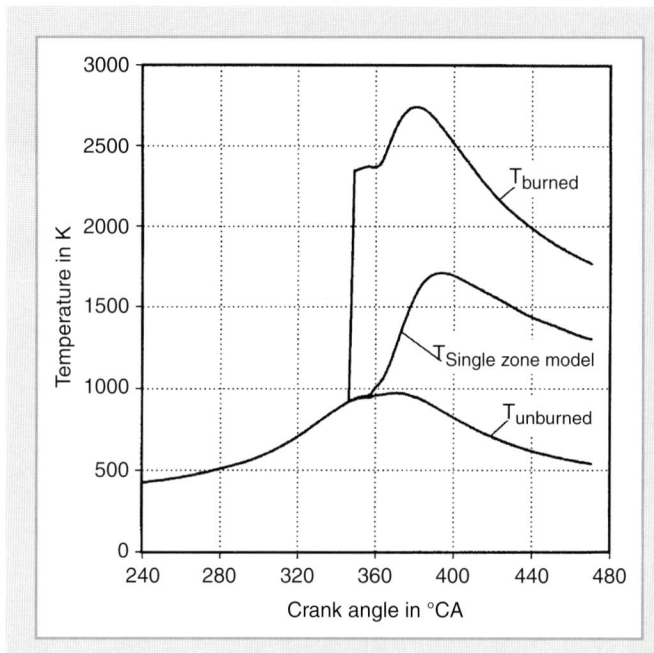

Fig. 1-17 Temperature curves for single and two zone models

here). The law of mass conservation yields the following equation:

$$dm = dm_E + dm_A + dm_B \quad (1\text{-}19)$$

The influent masses are entered positively in Eq. (1-19) and the effluent masses negatively.

The first law of thermodynamics describes the conservation of energy. It states that the energy inside the cylinder U can only change when enthalpy dH is supplied or removed through the system boundary in conjunction with mass dm, heat dQ_W or work ($dW = -p \cdot dV$). The energy released by the combustion of the injected fuel is regarded as internal heat input dQ_B. The law of energy conservation from the first law of thermodynamics describes the correlation between the individual forms of energy:

$$dW + dQ_W + dQ_B + dH_E + dH_A + dH_B = dU \quad (1\text{-}20)$$

When the individual terms of this differential equation are known, then it can be solved with appropriate mathematical methods. The Runge-Kutta method (cf. [1-33]) or algorithms derived from it are usually applied. First, the initial state in the cylinder when "intake closes" is estimated and then Eq. (1-20) is integrated with the algorithm selected for one combustion cycle in small crank angle steps. A check at the end of the combustion cycle to determines whether the estimated initial state is produced when the "intake closes". If this is not the case, improved estimated values are employed to calculate combustion cycles until the estimated values are reproduced with sufficient accuracy.

Wall Heat Losses

Wall heat losses are calculated according to the relationship

$$\frac{dQ_W}{d\varphi} = \frac{1}{\omega} \cdot h \cdot A \cdot (T - T_W) \quad (1\text{-}21)$$

where the angular velocity is $\omega = 2 \cdot \pi \cdot n$, the mean heat transfer coefficient α, the heat transferring area A and the wall temperature T_W. The heat transferring area consists of the areas of the cylinder head, piston crown and cylinder liner enabled for the particular crank angle. Either known from measurements or estimated, an average wall temperature is needed for every area. Since exhaust valves have a significantly higher temperature than the cylinder head, the area of the cylinder head is normally divided into the exhaust valve area and remaining area.

Many authors already deal with calculating the heat transfer coefficient before beginning to simulate the engine process. The equation used most today stems from Woschni [1-34] (see Sect. 7.2 and [1-26, 1-35, 1-27]).

Gas Exchange

The enthalpies of the masses flowing in and out through the intake and exhaust valves are produced from the product of specific enthalpy h and mass change dm:

$$\begin{aligned} dH_E &= h_E \cdot dm_E \\ dH_A &= h_A \cdot dm_A \end{aligned} \quad (1\text{-}22)$$

The specific enthalpies are simulated with the respective temperature before, in or after the cylinder. The following equation yields the mass elements that cross the system boundary [1-26, 1-27]:

$$\frac{dm_{id}}{d\varphi} = \frac{A(\varphi)}{2 \cdot \pi \cdot n} \cdot \frac{p_v}{(R \cdot T_v)^{0.5}}$$
$$\cdot \left\{ \frac{2 \cdot \kappa}{\kappa - 1} \left[\left(\frac{p_n}{p_v}\right)^{2/\kappa} - \left(\frac{p_n}{p_v}\right)^{(\kappa+1)/\kappa} \right] \right\}^{0.5}$$

(1-23)

The states "v" and "n" each relate to the conditions before or after the valve analyzed (e.g. to the state of the charge air and the state in the cylinder for an intake valve with normal inflow, i.e. without backflow).

Equation (1.23) is based on the flow equation for the isentropic (frictionless and adiabatic) port flow of ideal gases [1-26, 1-27]. Area A denotes the geometric flow cross-section enabled by the valve at a given instant (see Sect. 2.1).

In real engine operation, frictional losses and spray contraction generate a mass flow that is reduced compared to the ideal value. Defining a standard cross section, this is factored into the flow factors μ determined in experiments on a swirl flow test bench [1-36]. Therefore, the related reference areas must always be known whenever flow factors (also denoted with μ, σ or α) are compared (see Sect. 2.1).

The pressures before the intake valve and after the exhaust valve are needed to calculate the charge flow according to Eq. (1-23). In the case of an exhaust gas turbocharged engine, boost pressure und exhaust gas pressure before the turbine are produced by balancing the exhaust gas turbocharger with the aid of measured compressor and turbine maps (see Sect. 1.3.2).

Combustion Characteristic

In addition to the physical models treated thus far, a description of combustion is required to model a cylinder. The energy released by combustion is produced by the specific calorific value H_u and the unburned fuel mass dm_B:

$$dQ_B = H_u \cdot dm_B \quad (1-24)$$

while disregarding the enthalpy of the injected fuel dH_B.

A function of time or the crank angle, the curve of the energy released by combustion (combustion characteristic $dQ_B/d\varphi$) is one of the most important set parameters for engine process simulation. In contrast to ideal cycles (Sect. 1.2) in which the combustion characteristic directly results from the desired pressure curve (e.g. constant pressure or constant volume cycle), a real engine's combustion characteristic depends on many parameters.

An approach that follows Fig. 1-18 would be optimal: The injection pump's delivery curve is the sole set parameter. The injection system (injection pump, rail and nozzle) is then simulated with suitable models to calculate the

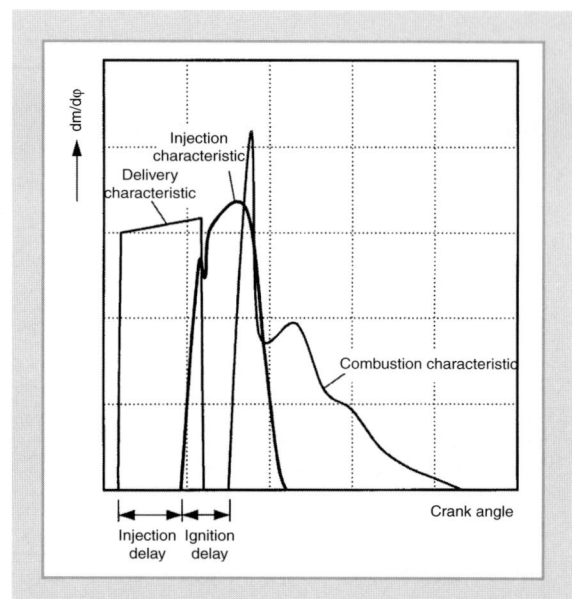

Fig. 1-18 Delivery curve, injection characteristic and combustion characteristic

injection characteristic from the delivery curve. If there were sufficient knowledge of the physical processes during spray disintegration, evaporation and mixture formation, it would be possible to simulate the ignition delay and combustion (combustion characteristic) with mathematical models [1-37, 1-38].

However, the models and methods developed so far are not yet able to predetermine diesel engine combustion with the accuracy desired to simulate the engine process. Knowledge about the combustion processes in internal combustion engines mostly comes from cylinder pressure indication. These measurements supply information on the processes inside a cylinder, piezoelectric pressure gauges being applied when the resolution is high, e.g. every 0.5°CA or less [1-39–1-41]. When the cylinder pressure curve is known, the combustion characteristic can be determined by inverting Eq. (1-18) (pressure curve analysis). This delivers insight into the conversion of energy in the engine.

Instead of the combustion characteristic calculated from the pressure curve analysis, a simple mathematical function, the rate of heat release, is usually employed to simulate the engine process. Optimization methods can be applied to select this function's parameters so that the combustion characteristic known from the pressure curve analysis is optimally reproduced. If cylinder pressure indication is unavailable, measured engines are simulated on a test bench with the aid of engine process simulation and a selected rate of heat release is estimated so that the measured and calculated parameters concur.

The most commonly applied rate of heat release goes back to the work of Vibe [1-42] who uses an exponential function to specify the integral of the rate of heat release (total combustion characteristic or combustion function $Q_B(\varphi)$):

$$\frac{Q_B(\varphi)}{Q_{B,O}} = 1 - \exp\left[-6.908 \cdot \left(\frac{\varphi - \varphi_{VB}}{\varphi_{VE} - \varphi_{VB}}\right)^{m+1}\right] \quad (1\text{-}25)$$

φ_{VB}: start of combustion
φ_{VE}: end of combustion
$Q_{B,0}$: total energy $Q_B(\varphi_{VE})$ released at the end of combustion
m: form factor

The factor -6.908 is produced by calibrating the asymptotic exponential function moving toward zero to a numerical value of 0.001 at the end of combustion.

The rate of heat release $dQ_B/d\varphi$ is obtained as a derivative of Eq. (1-25):

$$\frac{dQ_B}{d\varphi} = \frac{Q_{B,0}}{\varphi_{VE} - \varphi_{VB}} \cdot 6.908 \cdot (m+1) \cdot \left(\frac{\varphi - \varphi_{VB}}{\varphi_{VE} - \varphi_{VB}}\right)^m \cdot$$
$$\exp\left[-6.908 \cdot \left(\frac{\varphi - \varphi_{VB}}{\varphi_{VE} - \varphi_{VB}}\right)^{m+1}\right]$$

$$(1\text{-}26)$$

An equation called the Vibe function describes combustion with three parameters: the start of combustion φ_{VB}, the duration of combustion $\Delta\varphi_{BD} = \varphi_{VE} - \varphi_{VB}$ and the form factor m. As can be gathered from Fig. 1-19, the form factor defines the relative position of the maximum of the Vibe function.

One important task of engine process simulation is to ascertain the influence of changed boundary conditions, e.g. ambient conditions, on the combustion cycle in parameter studies (Sects. 1.3.3.3 and 1.3.3.4). The prerequisite for such simulation is knowledge of the influence of significant engine parameters on the rate of heat release as boundary conditions.

Woschni/Anisits [1-43] calculated the following dependencies of the air/fuel ratio λ, conversion factor η_u, speed n and state during "intake closed" (index IC) for the Vibe function:

$$\frac{\Delta\varphi_{BD}}{\Delta\varphi_{BD,0}} = \left(\frac{\lambda_V}{\lambda_{V,0}}\right)^{-0.6} \cdot \left(\frac{n}{n_0}\right)^{0.5} \cdot \eta_u^{0.6} \quad (1\text{-}27)$$

$$\frac{m}{m_0} = \left(\frac{\Delta\varphi_{BD}}{\Delta\varphi_{BD,0}}\right)^{-0.5} \cdot \frac{p_{Es}}{p_{Es,0}} \cdot \frac{T_{Es}}{T_{Es,0}} \left(\frac{n}{n_0}\right)^{-0.3} \quad (1\text{-}28)$$

Index 0 refers to the known initial operating point.

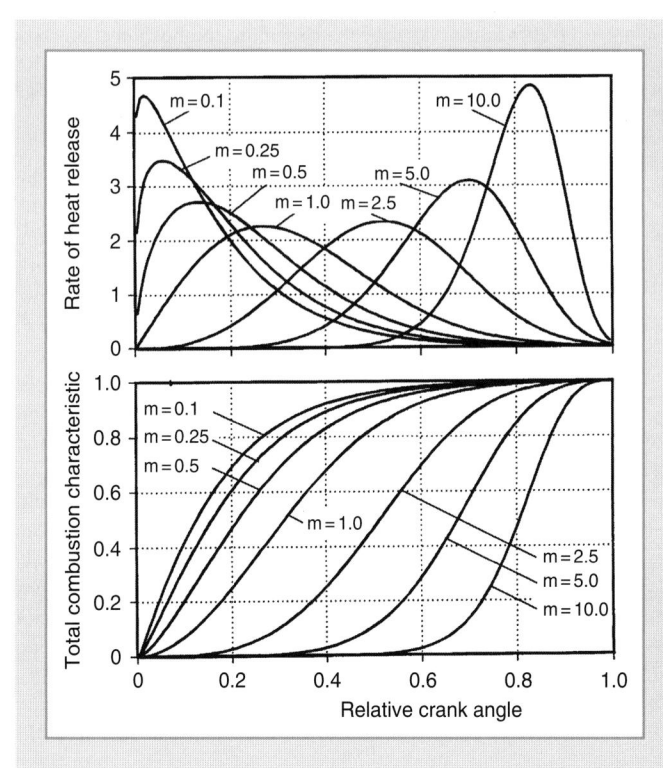

Fig. 1-19
Vibe function for different form factors (m)

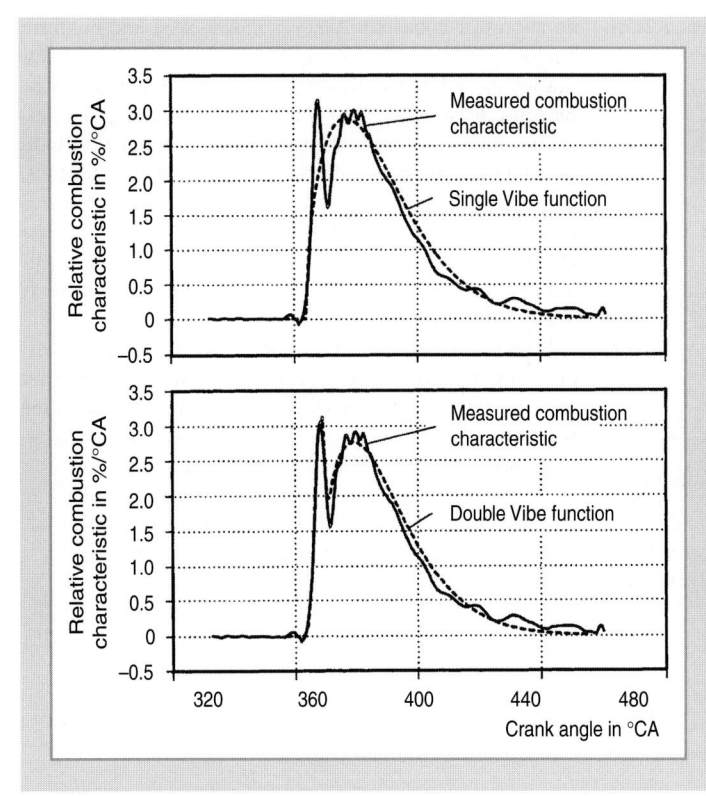

Fig. 1-20 Approximation of a combustion characteristic by the double Vibe function

The start of combustion ensues from the start of delivery φ_{FB}, the injection delay $\Delta\varphi_{EV}$ and the ignition delay $\Delta\varphi_{ZV}$:

$$\frac{\Delta\varphi_{EV}}{\Delta\varphi_{EV,0}} = \frac{n}{n_0} \qquad (1\text{-}29)$$

$$\frac{\Delta\varphi_{ZV}}{\Delta\varphi_{ZV,0}} = \frac{n}{n_0} \cdot \frac{\exp\left(\dfrac{b}{T_{ZV}}\right)}{\exp\left(\dfrac{b}{T_{ZV,0}}\right)} \cdot \left(\frac{p_{ZV}}{p_{ZV,0}}\right)^{-c} \qquad (1\text{-}30)$$

p_{ZV}: pressure in the ignition delay phase
T_{ZV}: temperature in the ignition delay phase
b, c: from parameters of the equation that have to be determined from measurements.

Other approaches to ignition delay can be found in [1-44–1-46].

Since its simple mathematical form prevents the Vibe function from reproducing combustion characteristics with sufficient accuracy, especially for high speed direct injection diesel engines, two Vibe functions are sometimes combined as a "double Vibe function" [1-47]. Figure 1-20 presents an operating point for a high performance high speed diesel engine reproduced by the Vibe function and the double Vibe function. Clearly, the simple Vibe function (Eq. (1-29)) cannot describe the rise of the combustion characteristic at the start of combustion ("premixed peak").

1.3.2.2 Indicated and Effective Work

The solution to the differential equation Eq. (1-20) for the first law of thermodynamics delivers the pressure curve in the cylinder and thus the indicated work W_i from which the mean indicated pressure p_i and the specific indicated work w_i can be derived as engine parameters (see Sect. 1.2). However, in general, the so-called brake mean effective pressure p_e and the specific effective work w_e are usually of interest. Hence when simulating the real cycle, model statements are used to simulate frictional losses expressed by the mean friction pressure p_r as the difference of p_i and p_e ($p_r = p_i - p_e$) if unknown from measurements.

The literature contains various suggestions for calculating frictional work [1-38, 1-48, 1-49], which, depending on the author, may be a function of speed, load, engine geometry, boost pressure and water and oil temperatures. Starting from the friction pressure (index 0 in Eq. (1-31)) determined in a design point according to [1-38] for example, the following applies

$$\frac{p_r - p_{r,0}}{p_{r,0}} = 0.7 \cdot \frac{n - n_0}{n_0} + 0.3 \cdot \frac{p_i - p_{i,0}}{p_{i,0}} \qquad (1\text{-}31)$$

according to which only speed and mean indicated pressure have to be specified.

1.3.2.3 Modeling the Complete Engine

A cylinder was modeled as an example in Sects. 1.3.2.1 and 1.3.2.2. Naturally, every significant engine component must be modeled to simulate a complete engine. Likewise, the basic physical equations of mass conservation (continuity equation), impulse (conservation of impulse) and energy (first law of thermodynamics) as well as the second law of thermodynamics must be solved for every flow process outside the cylinder in such engine components as the intake and exhaust manifolds, intercooler, catalytic converter or exhaust gas turbocharger. The first simulations of this type were performed with the program system PROMO [1-50, 1-51]. Commercial programs such as GT-Power (a component of the GT-Suite [1-52]) or Boost (from AVL [1-53]) are generally used today.

Various methods, which also describe real conditions more accurately as complexity increases, lend themselves to the simulation of air and exhaust manifolds. In the simplest case, pressures in the charge air and exhaust manifolds (i.e. infinitely large reservoirs) are assumed to be constant (so-called zero dimensional models). The so-called filling and emptying method models the manifolds as reservoirs of finite volume, which are transiently filled and emptied by the cylinders and continuously filled and emptied by the supercharger or the turbine. In this method, the pressure in the manifolds varies temporally but not locally (i.e. the sound of speed is infinitely great).

When the equations are employed for nonstationary, one-dimensional and compressible pipe flow, the changes of state in the intake and exhaust system are captured with the methods of transient gas dynamics (characteristic method or simplifying acoustic method [1-54]). This one-dimensional method can also simulate local pressure differences and pipe branches, the mathematical work required being far greater than the filling and emptying method.

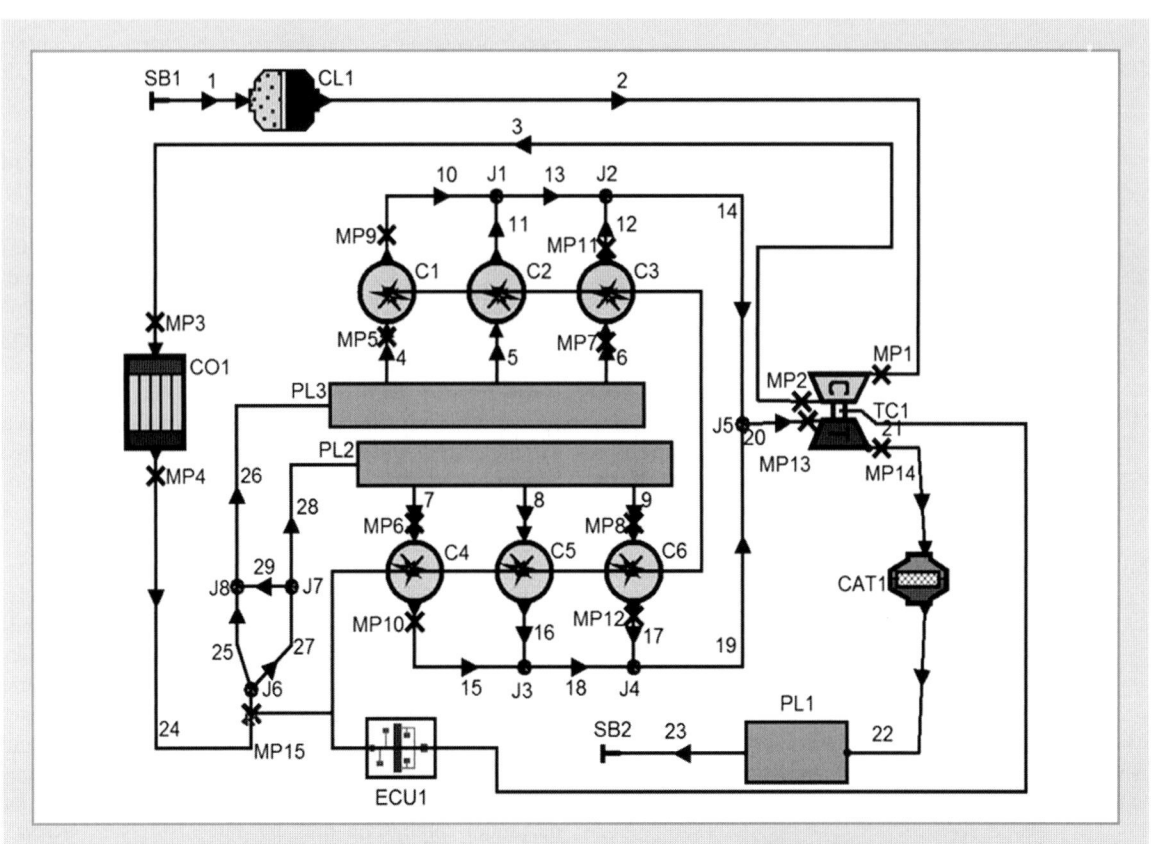

Fig. 1-21 Model for simulating a supercharged V6 diesel engine: (CL1) air filter, (TC1) exhaust gas turbocharger, (CO1) intercooler, (CAT1) catalytic converter, (PL1) muffler, (PL2, PL3) V engine intake manifold, (C1 to C6) engine cylinders, (J) connections and branches, (ECU1) engine electronics to control injection and the turbocharger's wastegate [1-39]

More recently, quasidimensional models have been developed in which variables that are a function of position and time factor in local phenomena. Examples include flow cycle, combustion and heat transfer models [1-27].

Figure 1-21 presents a model of an exhaust gas turbocharged six cylinder diesel engine simulated with the Boost program as an example [1-53]. It incorporates all the significant components attached to the engine, beginning with the air filter to the exhaust gas turbocharger up through the catalytic converter and exhaust muffler. The modeled engine electronics (ECU1), which controls injection and the turbocharger's wastegate, is added to this.

So as not to prolong this introduction to engine process simulation, the authors refer readers to further literature (e.g. [1-26, 1-27] or [1-55, 1-56]) and Sect. 2.2.

1.3.3 Typical Examples of the Application of Engine Process Simulation

1.3.3.1 Introduction

Basically, two types of application are distinguished:
1. Measured data is already available for the engine operating point being simulated.

   In this case, engine process simulation determines parameters that are already known from measurement. Comparing the computed and measured results allows checking the plausibility of the measured results for instance or computing the physical submodels of the process simulation (e.g. rate of heat release, wall heat transfer and mean effective friction pressure).

2. Measured data is not yet available for the engine operating point being simulated.

   In this case, the engine process simulation is a projection of a yet unknown operating point. First, the parameters of the physical submodels must be estimated, i.e. they are adopted from a similar operating point and, if necessary, corrected with the conversion equations specified in Sects. 1.3.2.2 and 1.3.2.3. Naturally, the results of this process simulation are only as precise as the relationships used for the conversion. Therefore, in practice, they are checked and calibrated for the particular engines by resimulating as many measured operating points as possible with the engine process simulation and comparing the results with the measured values.

Thus, the difference between a process simulation to design new engines and a resimulation of already measured engines is insubstantial.

1.3.3.2 Results of Engine Process Simulation

Based on Sect. 1.3.2, typical input variables for engine process simulation are:

Engine geometry, valve lift curve, valve flow coefficients, speed, engine power, mechanical efficiency, rate of heat release, coefficients for the heat transfer equation, component wall temperatures, charge air pressure and temperature before the cylinder and pressure after the cylinder.

Typical Results Are

Pressure curve, temperature curve, wall heat losses, effective fuel consumption, net efficiency, internal efficiency, maximum combustion pressure, maximum final compression pressure, maximum rate of pressure rise, maximum cycle temperature (average energy value), temperature before the exhaust gas turbine, gas exchange losses, charge flow and the air/fuel ratio.

When, in addition to the cylinder, the exhaust gas turbocharger can also be described thermodynamically as well, e.g. by employing appropriate compressor and turbine maps, then the engine process simulation ascertains the pressure before and after the cylinder. Then, the ambient condition counts as an input variable. If the potentially present intercooler is also modeled in the further course of process simulation, then, when the ambient temperature is given, the engine process simulation also yields the charge air temperature in addition to the water temperature when the charge air is water cooled.

As an example, Fig. 1-22 presents the results of a process simulation of an engine with a cylinder displacement of 4 dm^3 and a specific effective work w_e of 2 kJ/dm^3 for the operating point $n = 1,500$/min. It represents pressure, temperature, combustion characteristic, mass flow in the cylinder and valves as well as the wall heat losses as a function of the crank angle. The correspondence between the engine process simulation and reality is verified by comparing global values, e.g. exhaust gas temperature, charge flow or boost pressure, with measured values. When they correspond, it is assumed that even the unverifiable values such as temperature curve or mass curve have been simulated correctly.

1.3.3.3 Parameter Studies

A significant field of application for engine process simulation is parameter studies that analyze the influence of boundary conditions on the combustion cycle in depth. The results are needed in the design phase of new engines or to optimize or enhance the performance of existing engines. Parameter studies may have optimum fuel consumption, power and torque values as possible target parameters. Optimizations may be implemented so that engineering or legal limits for maximum combustion pressure, rate of pressure increase, exhaust gas temperature or pollutant emissions are not exceeded.

The outcome of a typical parameter study is presented below. Since the maximum combustion pressure limited

Fig. 1-22
Results of the engine process simulation

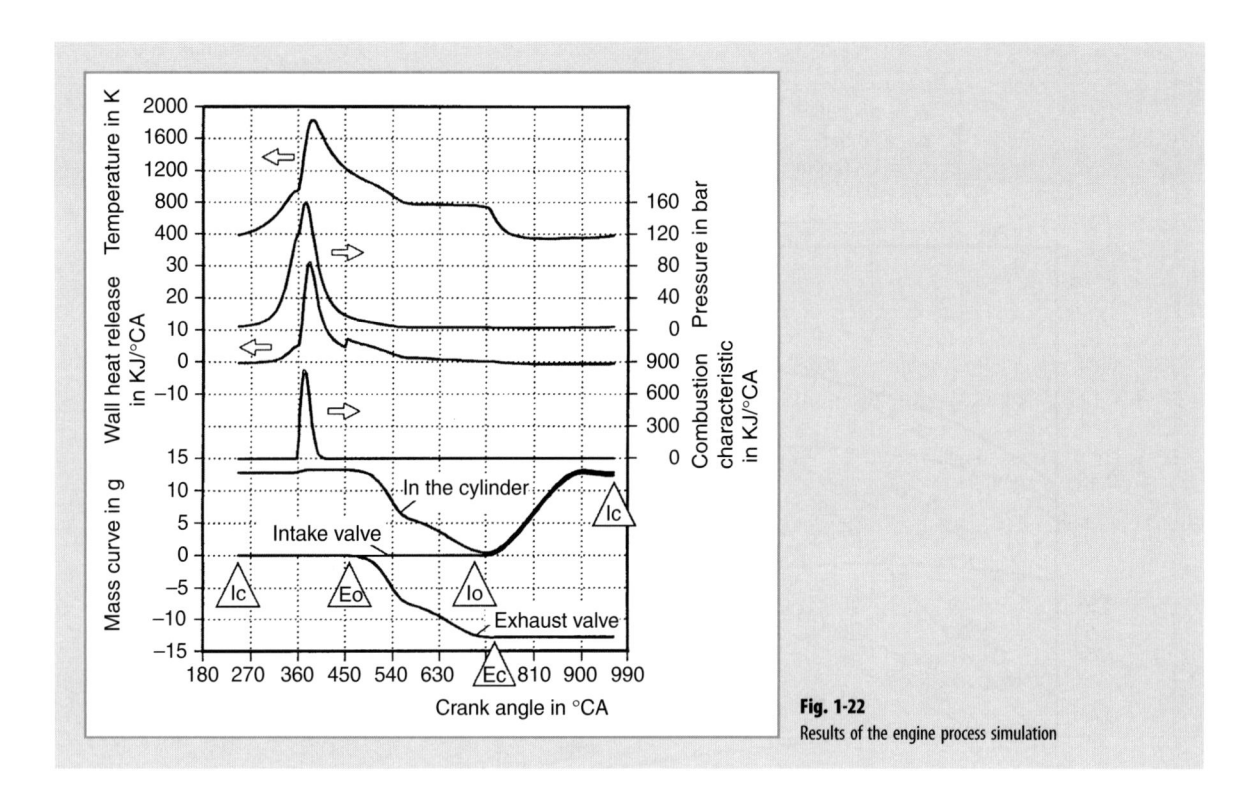

by engineering greatly influences a cycle's net efficiency and thus its fuel consumption, one of the most important parameter studies serves to determine the dependence of net efficiency η_e on maximum combustion pressure p_{Zmax}. When the combustion characteristic is given, the maximum combustion pressure is defined by the parameters of start of delivery (and thus start of combustion) and final compression pressure. The latter, in turn, primarily depends on the compression ratio and the boost pressure. The boost pressure is itself essentially defined by the exhaust gas turbocharging efficiency and the desired air/fuel ratio. Figure 1-23 presents the net efficiency η_e as a function of the compression ratio ε for various maximum combustion pressures at constant numerical values for the exhaust gas turbocharging efficiency η_{TL} and the air/fuel ratio λ. The dotted lines indicate the position of the start of combustion. An optimum compression ratio with a maximum efficiency η_e exists for every maximum combustion pressure. (By contrast, theoretical standard cycles state that maximum net efficiency is attained at the maximum compression ratio.)

1.3.3.4 Other Examples of Application

Apart from parameter studies, engine process simulation also serves many other purposes.

- heat balance, loss analysis: simulation of heat balances and analyses of losses to assess engines (development potential, optimization, cooling system design)
- design of exhaust gas turbocharging groups: simulation of the energy supply available for supercharging (exhaust gas mass flow and temperature) and the boost pressure requirement and air flow rate [1-57]
- optimization of valve lift and valve gear timing: simulation of the gas exchange with the goal of low gas exchange losses and large volumetric efficiencies
- temperature field simulation: simulation of engines' heat balance and thermal load (input variables for simulating the temperature fields in the cylinder, cylinder liner, piston and valves) (see Sect. 7.1)
- gas pressure curves for further studies: simulation of the gas pressure curves as input variables for further studies such as strength simulation, torsional vibration analysis, piston ring movement simulation
- wet corrosion: analysis of the danger of wet corrosion (undershooting of the exhaust gas dew-point temperature)
- nitrogen oxide emissions: application of a combustion model (e.g. two zone model) to analyze nitrogen oxide emissions [1-24–1-27]
- ambient conditions: determination of engines' operating values when ambient conditions (pressure and temperature) change

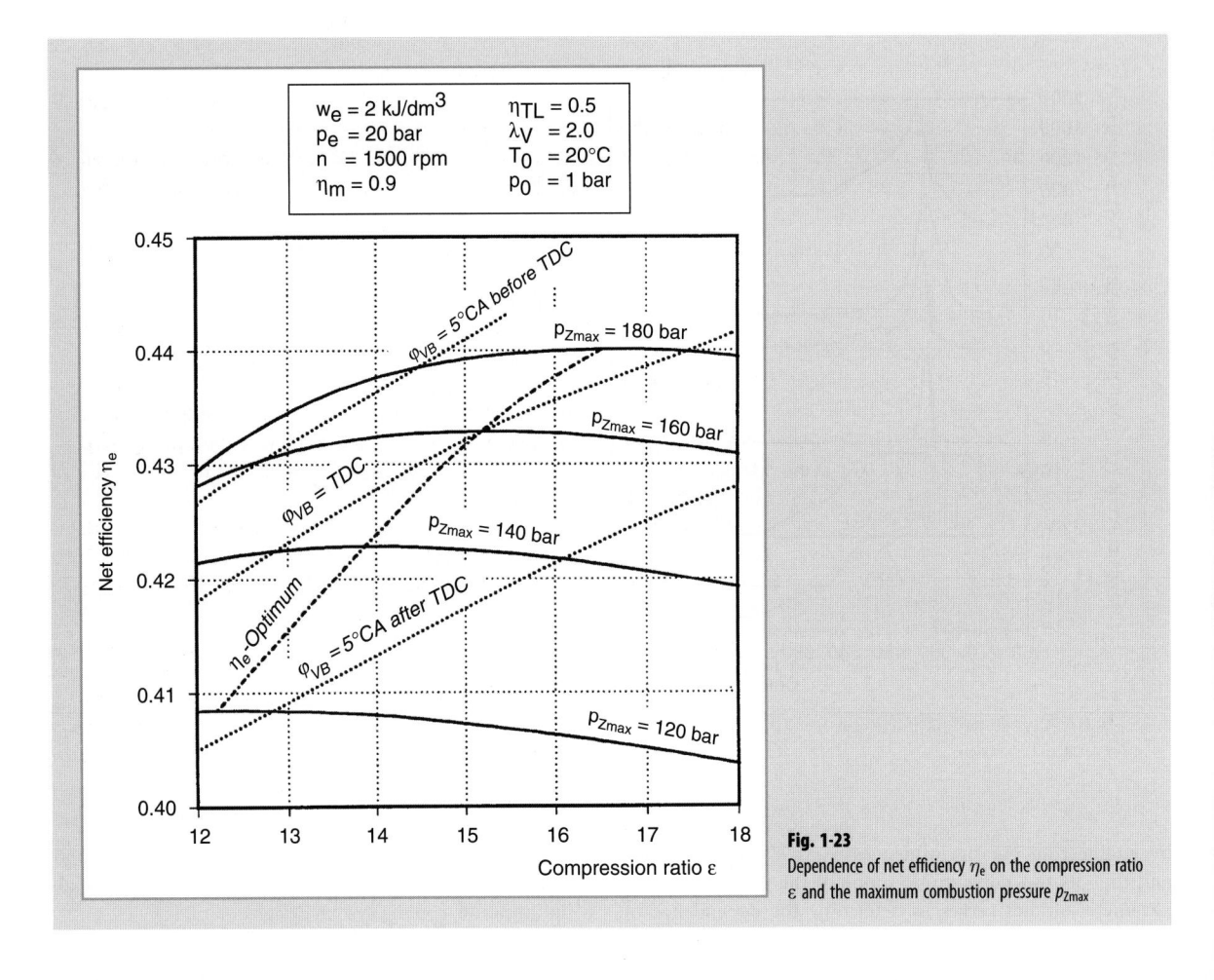

Fig. 1-23 Dependence of net efficiency η_e on the compression ratio ε and the maximum combustion pressure p_{Zmax}

- plausibility check: plausibility check of measured values or hypotheses for a damage analysis
- transfer of experimental single cylinder results to multi-cylinder engines: conversion of the operating values measured in an experimental single cylinder engine to conditions for multi-cylinder engines.

1.3.4 Future Studies/Work in the Field of Engine Process Simulation

Engine process simulation is an instrument suitable for relative statements (e.g. parameter studies). The requirements for submodel accuracy are not as great. Absolute statements (e.g. supercharger or cooling systems design, comparisons of different engines) require far more accurate submodels. Therefore, many and diverse efforts are being made to further improve the models.

When maximum combustion pressures are elevated above 200 bar, the cylinder charge may no longer be considered an ideal gas and the real gas properties of the components involved have to be factored in [1-29].

Present heat transfer models calculate wall heat losses in the part load region too small. Furthermore, they only allow for heat losses radiating from soot particulates during combustion imprecisely or not at all. Consequently, the calculated losses are too small, especially when combustion is poor. Pressure curve analysis yields apparent energy losses in the compression phase, which may be due to inaccuracies in the dependence of the crank angle in the wall heat models employed. New heat transfer models may be found in [1-58–1-63].

Simulation of the charge flow can be improved by applying computationally very intensive simulations of three-dimensional flow fields. Such simulations enable optimizing flow conditions in cylinder heads for example. Flow field simulations may also be applied to analyze mixture formation and in part to already simulate the combustion cycle [1-26, 1-27].

Work is being done on simulating the combustion characteristic directly from injection data [1-64, 1-65, 1-37, 1-38, 1-24, 1-66, 1-67]. On the other hand, the models of the rate of heat release and its conversion in the map have to be improved [1-68] in order to be able to describe combustion with better precision when simulating nitrogen oxide emissions [1-24].

Other studies also being performed on converting the mean effective friction pressure in the map are aimed at determining the individual assemblies' contribution to the total frictional losses [1-48].

Literature

1-1 DRP Nr. 67207: Arbeitsverfahren und Ausführungsart für Verbrennungskraftmaschinen. To: R. Diesel as of February 28, 1892
1-2 Diesel, R.: Die Entstehung des Dieselmotors. Berlin/Heidelberg/New York: Springer (1913)
1-3 Diesel, R.: Theorie und Konstruktion eines rationellen Wärmemotors zum Ersatz der Dampfmaschinen und der heute bekannten Verbrennungsmotoren. Berlin/Heidelberg/New York: Springer (1893), Reprint: Düsseldorf: VDI-Verlag (1986)
1-4 Sass, F.: Geschichte des deutschen Verbrennungsmotorenbaus von 1860–1918. Berlin/Göttingen/Heidelberg: Springer (1962)
1-5 Reuss, H.-J.: Hundert Jahre Dieselmotor. Stuttgart: Franckh-Kosmos (1993)
1-6 Adolf, P.: Die Entwicklung des Kohlenstaubmotors in Deutschland. Diss. TU Berlin (D83) (1992)
1-7 Knie, A.: Diesel – Karriere einer Technik: Genese und Formierungsprozesse im Motorenbau. Berlin: Bohn (1991)
1-8 Heinisch, R.: Leichter, komfortabler, produktiver – die technologische Renaissance der Bahn. Mobil 1 (1994) 3. Also: Heinrich, J.: Flinker CargoSprinter hilft der Deutschen Bahn. VDInachr. (1996) 41, pp. 89 ff.
1-9 Diesel, E.: Die Geschichte des Diesel-Personenwagens. Stuttgart: Reclam (1955)
1-10 Diesel, E.: Diesel. Der Mensch, das Werk, das Schicksal. Stuttgart: Reclam (1953)
1-11 Boie, W.: Vom Brennstoff zum Rauchgas. Leipzig: Teubner (1957)
1-12 Schmidt, E.: Einführung in die technische Thermodynamik. 10th Ed. Berlin/Heidelberg/New York: Springer (1963)
1-13 Pflaum, W.: I, S-Diagramme für Verbrennungsgase, 2nd Ed. Part I and II. Düsseldorf: VDI-Verlag (1960, 1974)
1-14 DIN ISO 3046/1: Reciprocating internal combustion engines – Performance – Part 1: Declarations of power, fuel and lubricating oil consumptions, and test methods
1-15 Technical Review Wärtsilä RT-flex 96C / Wärtsilä RTA 96C. Wärtsilä Corporation publication (2006)
1-16 Steinparzer, F.; Kratochwill, H.; Mattes, T.; Steinmayr, T.: Der neue Sechszylinder-Dieselmotor von BMW mit zweistufiger Abgasturboaufladung – Spitzenstellung bezüglich effizienter Dynamik im Dieselsegment. Proceedings of the 15th Aachen Colloquium Automobile and Engine Technology 2006, pp. 1281–1301
1-17 Groth, K.; Syassen, O.: Dieselmotoren der letzten 50 Jahre im Spiegel der MTZ – Höhepunkte und Besonderheiten der Entwicklung. MTZ 50 (1989) pp. 301–312
1-18 Woschni, G.: Elektronische Berechnung von Verbrennungsmotor-Kreisprozessen. MTZ 26 (1965) 11, pp. 439–446
1-19 Woschni, G.: CIMAC Working Group Supercharging: Programmiertes Berechnungsverfahren zur Bestimmung der Prozessdaten aufgeladener Vier- und Zweitaktdieselmotoren bei geänderten Betriebsbedingungen. TU Braunschweig. Forschungsvereinigung Verbrennungskraftmaschinen e.V. (FVV) Frankfurt (1974)
1-20 Albers, W.: Beitrag zur Optimierung eines direkteinspritzenden Dieselmotors durch Variation von Verdichtungsverhältnis und Ladedruck. Diss. Universität Hannover (1983)
1-21 Schorn, N.: Beitrag zur rechnerischen Untersuchung des Instationärverhaltens abgasturboaufgeladener Fahrzeugdieselmotoren. Diss. RWTH Aachen (1986)
1-22 Zellbeck, H.: Rechnerische Untersuchung des dynamischen Betriebsverhaltens aufgeladener Dieselmotoren. Diss. TU München (1981)
1-23 Friedrich, I.; Pucher, H.; Roesler, C.: Echtzeit-DVA – Grundlage der Regelung künftiger Verbrennungsmotoren. MTZ-Konferenz-Motor Der Antrieb von Morgen. Wiesbaden: GWV Fachverlage 2006, pp. 215–223
1-24 Hohlbaum, B.: Beitrag zur rechnerischen Untersuchung der Stickstoffoxid-Bildung schnellaufender Hochleistungsdieselmotoren. Diss. Universität Karlsruhe (TH) (1992)
1-25 Krassnig, G.: Die Berechnung der Stickoxidbildung im Dieselmotor. Habilitation TU Graz (1976)
1-26 Merker, G.; Schwarz, C.; Stiesch, G.; Otto, F.: Verbrennungsmotoren: Simulation der Verbrennung und Schadstoffbildung. Wiesbaden: Teubner-Verlag (2004)
1-27 Pischinger, R.; Klell, M.; Sams, T.: Thermodynamik der Verbrennungskraftmaschine – Der Fahrzeugantrieb. Wien: Springer (2002)
1-28 Justi, E.: Spezifische Wärme, Enthalpie, Entropie und Dissoziation technischer Gase. Berlin: Springer (1938)
1-29 Zacharias, F.: Analytische Darstellung der thermodynamischen Eigenschaften von Verbrennungsgasen. Diss. Berlin (1966)
1-30 Heywood, J.B.: Internal Combustion Engine Fundamentals. New York: McGraw-Hill Book Company (1988)
1-31 NIST/JANAF: Thermochemical Tables Database. Version 1.0 (1993)
1-32 Pflaum, W.: Mollier-(I, S-)Diagramme für Verbrennungsgase, Teil II. Düsseldorf: VDI-Verlag (1974)

1-33 Zurmühl, R.: Praktische Mathematik für Ingenieure und Physiker. Berlin/Heidelberg/New York: Springer (1984)
1-34 Woschni, G.: Die Berechnung der Wandverluste und der thermischen Belastung der Bauteile von Dieselmotoren. MTZ 31 (1970) 12, pp. 491–499
1-35 Pflaum, W.; Mollenhauer, K.: Wärmeübergang in der Verbrennungskraftmaschine. Vienna: Springer (1977)
1-36 Frank, W.: Beschreibung von Einlasskanaldrallströmungen für 4-Takt-Hubkolbenmotoren auf Grundlage stationärer Durchströmversuche. Diss. RWTH Aachen (1985)
1-37 Constien, M.; Woschni, G.: Vorausberechnung des Brennverlaufs aus dem Einspritzverlauf für einen direkteinspritzenden Dieselmotor. MTZ 53 (1992) 7/8, pp. 340–346
1-38 Flenker, H.; Woschni, G.: Vergleich berechneter und gemessener Betriebsergebnisse aufgeladener Viertakt-Dieselmotoren. MTZ 40 (1979) 1, pp. 37–40
1-39 Hohenberg, G.; Möllers, M.: Zylinderdruckindizierung I. Abschlussbericht Vorhaben No. 362. Forschungsvereinigung Verbrennungskraftmaschinen (1986)
1-40 Nitzschke, E.; Köhler, D.; Schmidt, C.: Zylinderdruckindizierung II. Abschlussbericht Vorhaben Nr. 392. Forschungsvereinigung Verbrennungskraftmaschinen (1989)
1-41 Thiemann, W.: Verfahren zur genauen Zylinderdruckmessung an Verbrennungsmotoren. Part 1: MTZ 50 (1989), Vol. 2, pp. 81–88; Part 2: MTZ 50 (1989) 3, pp. 129–134
1-42 Vibe, I.I.: Brennverlauf und Kreisprozess von Verbrennungsmotoren. Berlin: VEB Verlag Technik (1970)
1-43 Woschni, G.; Anisits, F.: Eine Methode zur Vorausberechnung der Änderung des Brennverlaufs mittelschnelllaufender Dieselmotoren bei geänderten Randbedingungen. MTZ 34 (1973) 4, pp. 106–115
1-44 Hardenberg, H.; Wagner, W.: Der Zündverzug in direkteinspritzenden Dieselmotoren. MTZ 32 (1971) 7, pp. 240–248
1-45 Sitkei, G.: Kraftstoffaufbereitung und Verbrennung bei Dieselmotoren. Berlin/Göttingen/Heidelberg: Springer (1964)
1-46 Wolfer, H.: Der Zündverzug im Dieselmotor. VDI Forschungsarbeiten 392. Berlin: VDI Verlag GmbH (1938)
1-47 Oberg, H.J.: Die Darstellung des Brennverlaufs eines schnellaufenden Dieselmotors durch zwei überlagerte Vibe-Funktionen. Diss. TU Braunschweig (1976)
1-48 Schwarzmeier, M.: Der Einfluss des Arbeitsprozessverlaufs auf den Reibmitteldruck. Diss. TU München (1992)
1-49 Thiele, E.: Ermittlung der Reibungsverluste in Verbrennungsmotoren. MTZ 43 (1982) 6, pp. 253–258
1-50 Seifert, H.: Erfahrungen mit einem mathematischen Modell zur Simulation von Arbeitsverfahren in Verbrennungsmotoren. Part 1: MTZ 39 (1978) 7/8, pp. 321–325; Part 2: MTZ 39 (1978) 12, pp. 567–572
1-51 Seifert, H.: 20 Jahre erfolgreiche Entwicklung des Programmsystems PROMO. MTZ 51 (1990) 11, pp. 478–488
1-52 Simulation program GT-Power: www.gtisoft.com
1-53 Simulation program Boost: www.avl.com
1-54 Seifert, H.: Instationäre Strömungsvorgänge in Rohrleitungen an Verbrennungskraftmaschinen. Berlin/Heidelberg/New York: Springer (1962)
1-55 Hiereth, H.; Prenninger, P.: Aufladung der Verbrennungskraftmaschine – Der Fahrzeugantrieb. Vienna: Springer (2003)
1-56 Zinner, K.: Aufladung von Verbrennungsmotoren. Berlin/Heidelberg/New York: Springer (1985)
1-57 Golloch, R.: Downsizing bei Verbrennungsmotoren. Berlin/Heidelberg/New York: Springer 2005
1-58 Bargende, M.: Ein Gleichungsansatz zur Berechnung der instationären Wandwärmeverluste im Hochdruckteil von Ottomotoren. Diss. TH Darmstadt (1991)
1-59 Boulouchos, K.; Eberle, M.; Ineichen, B.; Klukowski, C.: New Insights into the Mechanism of In Cylinder Heat Transfer in Diesel Engines. SAE Congress February 27–March 3, (1989)
1-60 Huber, K.: Der Wärmeübergang schnellaufender, direkteinspritzender Dieselmotoren. Diss. TU München (1990)
1-61 Kleinschmidt, W.: Entwicklung einer Wärmeübergangsformel für schnellaufende Dieselmotoren mit direkter Einspritzung. Zwischenbericht zum DFG Vorhaben K1 600/1 1 (1991)
1-62 Kolesa, K.: Einfluss hoher Wandtemperaturen auf das Betriebsverhalten und insbesondere auf den Wärmeübergang direkteinspritzender Dieselmotoren. Diss. TU München (1987)
1-63 Vogel, C.; Woschni, G.; Zeilinger, K.: Einfluss von Wandablagerungen auf den Wärmeübergang im Verbrennungsmotor. MTZ 55 (1994) 4, pp. 244–247
1-64 Barba, C.; Burckhardt, C.; Boulouchos, K.; Bargende, M.: Empirisches Modell zur Vorausberechnung des Brennverlaufs bei Common-Rail-Dieselmotoren. MTZ 60 (1999) 4, pp. 262–270
1-65 Chemla, F.; Orthaber, G.; Schuster, W.: Die Vorausberechnung des Brennverlaufs von direkteinspritzenden Dieselmotoren auf der Basis des Einspritzverlaufs. MTZ 59 (1998) 7/8
1-66 Witt, A.: Analyse der thermodynamischen Verluste eines Ottomotors unter den Randbedingungen variabler Steuerzeiten. Diss. TU Graz (1999)
1-67 De Neef, A.T.: Untersuchung der Voreinspritzung am schnellaufenden direkteinspritzenden Dieselmotor. Diss. ETH Zürich (1987)
1-68 Schreiner, K.: Untersuchungen zum Ersatzbrennverlauf und Wärmeübergang bei schnellaufenden Hochleistungsdieselmotoren. MTZ 54 (1993) 11, pp. 554–563

2 Gas Exchange and Supercharging

Helmut Pucher

2.1 Gas Exchange

2.1.1 General

Beginning at the conclusion of the expansion stroke, the *gas exchange phase* basically performs two functions, namely:
- replacing the utilized cylinder charge (exhaust) with fresh gas (air in a diesel engine), a basic prerequisite for an internal combustion engine, and
- dissipating the heat as required to conclude the thermodynamic cycle.

Gas exchange can proceed based on the four-stroke or the two-stroke cycle. Regardless, the outcome of gas exchange can be characterized and evaluated by a series of dimensionless parameters, the following being stipulated first:

m_Z total mass of working gas in the cylinder at the end of gas exchange,
m_L total air mass that has flowed into the cylinder through the intake element,
m_{LZ} mass of fresh air in the cylinder at the end of gas exchange,
m_{RG} mass of residual exhaust gas in the cylinder at the end of gas exchange,
ρ_L density of the air before the intake elements
m_{Ltheor} theoretical air mass.

The theoretical air mass

$$m_{Ltheor} = \rho_L \cdot V_h \tag{2-1}$$

corresponds to the mass of air with the density ρ_L brought into the cylinder to exactly fill the cylinder displacement V_h.

The air efficiency

$$\lambda_a = \frac{m_L}{m_{Ltheor}} \tag{2-2}$$

is a measure of the total quantity of air relative to the theoretical air mass that flows into the cylinder during gas exchange. For a steady-state engine operating point, it corresponds to the measured air flow rate.

The volumetric efficiency

$$\lambda_l = \frac{m_{LZ}}{m_{Ltheor}} \tag{2-3}$$

specifies the quantity of inflow air mass remaining in the cylinder relative to the theoretical air mass.

Accordingly, the *retention rate* is defined as:

$$\lambda_z = \frac{m_{LZ}}{m_L} = \frac{\lambda_l}{\lambda_a}. \tag{2-4}$$

Boost efficiency and scavenging efficiency play a role as well, especially for two-stroke engines.

Boost efficiency

$$\lambda_t = \frac{m_Z}{m_{Ltheor}} \tag{2-5}$$

specifies the amount of working gas mass located in the cylinder at the end of gas exchange relative to the theoretical air mass. The proportion of the working gas mass m_Z that consists of fresh air m_{LZ} is expressed by the *scavenging efficiency*:

$$\lambda_s = \frac{m_{LZ}}{m_Z} \tag{2-6}$$

The following applies to the residual gas m_{RG}, i.e. residual working gas from the preceding combustion cycle remaining in the cylinder:

$$m_{RG} = m_Z - m_{LZ}. \tag{2-7}$$

2.1.2 Four-Stroke Cycle

2.1.2.1 Control Elements

Gas exchange in four-stroke reciprocating piston engines is now almost exclusively controlled by valves. Although formerly used in vehicle gasoline engines, slide valve control [2-1] failed to even establish itself in gasoline engines because

---

H. Pucher (✉)
Technische Universität Berlin, Berlin, Germany
e-mail: hegre.pucher@t-online.de

of sealing problems resulting from the variable cold and hot clearance as well as thermally induced warping. It is thoroughly unfeasible for diesel engines, particularly since they have higher in-cylinder pressures. Universally common today, a valve shape with a conical seat (Fig. 18-31) perfectly seals a cylinder chamber against maximum in-cylinder pressures because the increased internal pressure also directly generates increased contact pressure and thus a sealing effect in the valve seat. The lift produced by the cam in cam-actuated valve drives, which are still primarily in use, is transmitted to the valve
- by a tappet, push rod and rocker arm in *underhead camshafts,*
- by a rocker arm or cam follower or by a bucket tappet in *overhead camshafts* and the valve is lifted against the valve spring's force.

Since they have single-cylinder heads, large diesel engines down to and including commercial vehicle engines are equipped with underhead camshafts. Car diesel engines (block cylinder heads) predominantly have overhead camshafts because this can lessen the valve gear masses being moved.

2.1.2.2 Valve Lift Curves and Timing

Theoretically, a four-stroke engine requires one entire crankshaft revolution for its gas exchange. According to the premise of the theoretical engine process, the gas exchange elements (valves) must open and close exactly in their dead centers and have a rectangular lift characteristic (Fig. 2-1).

However, the valves in a real engine are only able to open and close gradually because of the accelerations in the valve gear. Since not only the valve gear masses but also each of the gas columns flowing in and out at the cylinder first have to be accelerated and the flow initially stops when a piston reverses, the opening time is set before and the closing time is set after the respective dead center. In detail, this so-called *timing* is defined as follows:

Exhaust Opens (EO)

When the exhaust valve opens, a supercritical pressure ratio normally exists between the pressure in the cylinder and the pressure in the exhaust line. Therefore, the exhaust initially flows out through the narrowest cross section (at the valve seat) at sonic speed. As a result, the pressure in the cylinder decreases relatively rapidly. Thus, the work of expulsion the piston has to apply during its subsequent upward stroke is none too great. However, timing EO very advanced to minimize this work of expulsion would accordingly reduce the work of expansion transferrable to the piston by the working gas during the expansion stroke. Therefore, EO is optimally set when the total power losses (loss of work of expansion and work of expulsion) reach a minimum or the indicated work reaches a maximum. This optimum is relatively "flat" and set in the range of 40–60° CA before BDC.

Valve Overlap

Since the exhaust valve only opens after the intake valve but already before top dead center (TDC in Fig. 2-1), so-called valve overlap ensues.

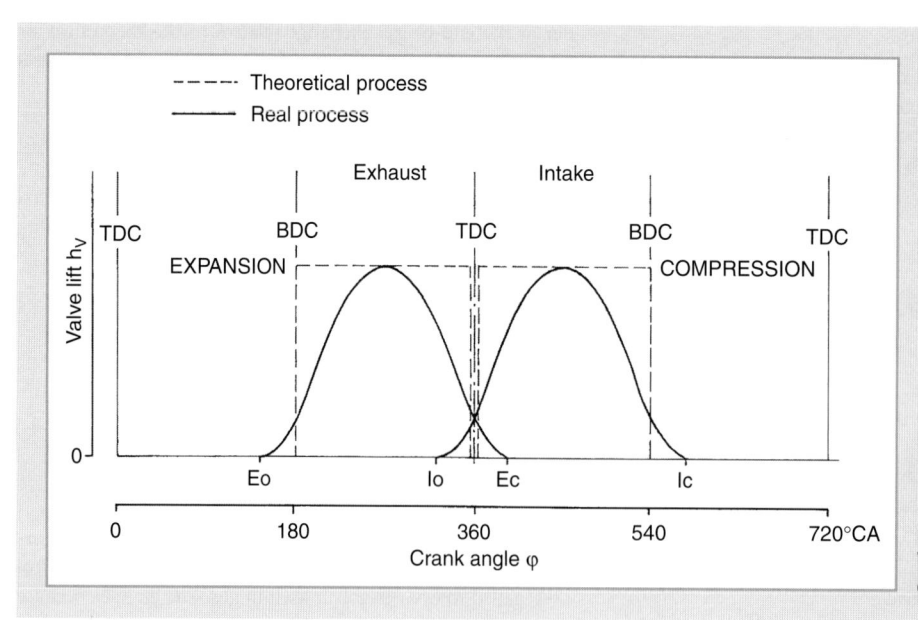

Fig. 2-1 Valve lift curves of a four-stroke engine

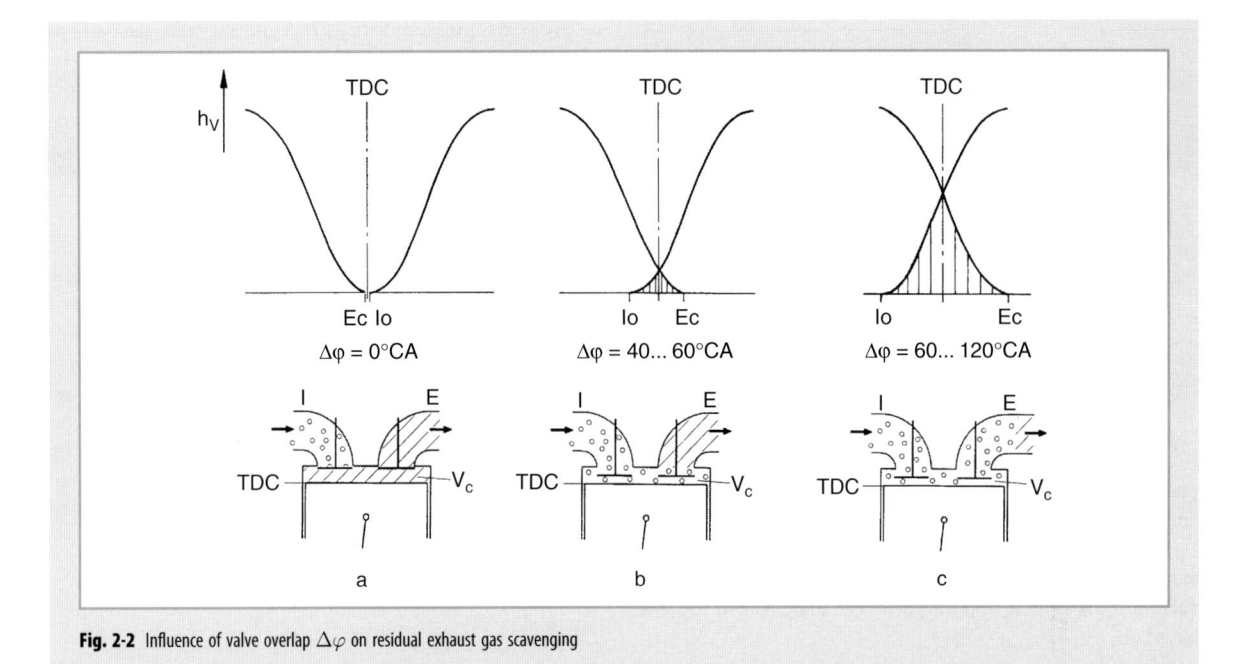

Fig. 2-2 Influence of valve overlap $\Delta\varphi$ on residual exhaust gas scavenging

Without any valve overlap, the case presented schematically in Fig. 2-2a would occur. If the exhaust valve were closed in TDC, the compression volume V_c would still be filled with exhaust, which would appear in the subsequent combustion cycle as residual exhaust gas.

Close to top dead center, the piston moves with very low velocity. Thus, it is practically unable to exert any expulsion or intake effect on the working fluid during this phase. If, however, the intake valve already opens before top dead center and the exhaust valve also remains open beyond top dead center, the dispersing exhaust column exerts a suction effect on the cylinder and the connected intake port, causing fresh gas (air) to flow into the cylinder and residual exhaust gas to be scavenged out (Fig. 2-2b). Only valve overlap of up to approximately 40–60°CA may be implemented in a naturally aspirated engine. However, this suffices to largely scavenge the residual exhaust gas. If it were selected substantially larger, essentially symmetrical to TDC, exhaust would also be forced into the intake manifold during the expulsion stroke and exhaust aspirated out of the exhaust manifold during the subsequent intake stroke. The valve overlap in highly supercharged diesel engines (see Sect. 2.2) in which the mean pressure before intake is higher than the mean pressure in the exhaust manifold may be selected to be significantly longer (up to 120°CA). This is employed to scavenge fresh air through the cylinder by purging residual exhaust gas (Fig. 2-2c). On the one hand, this relieves the thermal load of the components adjacent to the combustion chamber. On the other hand, it allows keeping the exhaust temperature before the turbocharger turbine below a certain limit. This is especially important for the operation of large diesel engines that run on heavy fuel oil (see Sect. 3.3).

Intake Closes (IC)

Intake closes has to be set so that the charge with fresh air, i.e. the volumetric efficiency λ_l, becomes a maximum since it determines the drivable engine load for a specific air/fuel ratio. Thus the specification of IC has very special significance. IC is normally set after BDC (see Fig. 2-1) because the inertia of the influent air causes the inflow into the cylinder to continue when the piston practically no longer exerts any suction effect upon reaching the proximity of BDC. If IC is set too far after BDC, an undesired back thrust of inflow air into the intake line occurs. Typical values are IC = 20–60°CA after BDC. Just like volumetric efficiency, the optimum value for IC is primarily a function of engine speed. At a fixed value for IC, the volumetric efficiency as a function of speed assumes a curve as in Fig. 2-3 (solid line). This curve's drop to the left or right of the peak mainly depends on IC being set too late or too early in the respective speed range. If, for example, an engine is intended to be operated chiefly in the upper speed range in the future, then it would make sense to set IC suitably later (dashed curve in Fig. 2-3).

2.1.2.3 Valve Cross Section and Flow Coefficient

When the valve and valve port are designed to be stationary, the gas mass that flows through during the opening duration

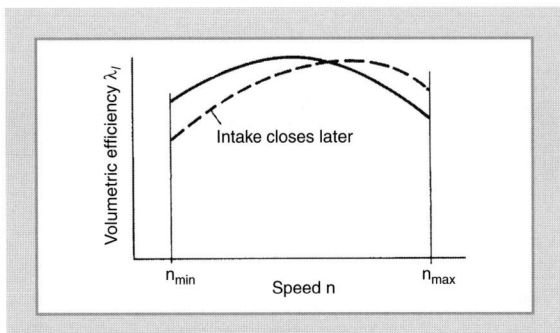

Fig. 2-3 Volumetric efficiency as a function of speed

is not only a function of the adjacent pressure ratio but primarily also the valve lift characteristic. This directly influences the free flow cross section.

The geometric valve cross section $A_V(\varphi)$ enabled by the valve during a particular lift $h_V(\varphi)$ is calculated with the inner valve seat diameter d_i and the valve seat angle β as in Fig. 2-4 based on:

$$A_v(\varphi) = \pi \cdot h_v(\varphi) \cdot \cos\beta \cdot [d_i + 0,5 \cdot h_v(\varphi) \cdot \sin 2\beta]. \tag{2-8}$$

The effective valve cross section $A_{Veff}(\varphi)$ actually available for the flow during a specific valve lift $h_V(\varphi)$ is normally smaller than $A_V(\varphi)$. It corresponds to the cross section that has to be inserted in the Saint-Venant equation (Eq. (2-9)) to obtain the actual mass flow $\dot{m}$ for given values for the total state of the gas on the inflow side (p_{01}, T_{01}) and the static pressure p_2 on the outflow side

$$A_{Veff} = \frac{\dot{m}\sqrt{R \cdot T_{01}}}{p_{01} \cdot \sqrt{\frac{2\kappa}{\kappa - 1}\left[\left(\frac{p_2}{p_{01}}\right)^{\frac{2}{\kappa}} - \left(\frac{p_2}{p_{01}}\right)^{\frac{\kappa+1}{\kappa}}\right]}}. \tag{2-9}$$

To experimentally determine $A_{Veff}(\varphi)$ on a stationary flow test bench as in Fig. 2-5, the cylinder head under test is placed on a tube with an inner diameter equal to the cylinder's bore D and a tube with an inner cross section equivalent to a continuation of the valve port cross section A_K is connected to the outer port end of the valve under test.

The valve port's effective cross section A_{Veff} can then be determined for both possible directions of flow as a function of valve lift. To this end, each of the following must be measured for discrete valve lifts over the entire valve lift range:
- total state at position 1 (p_{01}, T_{01}),
- static pressure at position 2 (p_2) and
- mass flow $\dot{m}$.

By definition, all flow losses that occur between control points 1 and 2 are incorporated in A_{Veff}. Hence, as a rule, A_{Veff} will always be smaller than the related geometric port cross section A_K (Fig. 2-4a). Such stationary flow measurements deliver the boundary conditions relevant for engine process simulation (see Sect. 1.3) as well as direct information on the aerodynamic quality of a given valve and valve port design. When the A_{Veff} characteristic, a dashed line in Fig. 2-4a, progresses

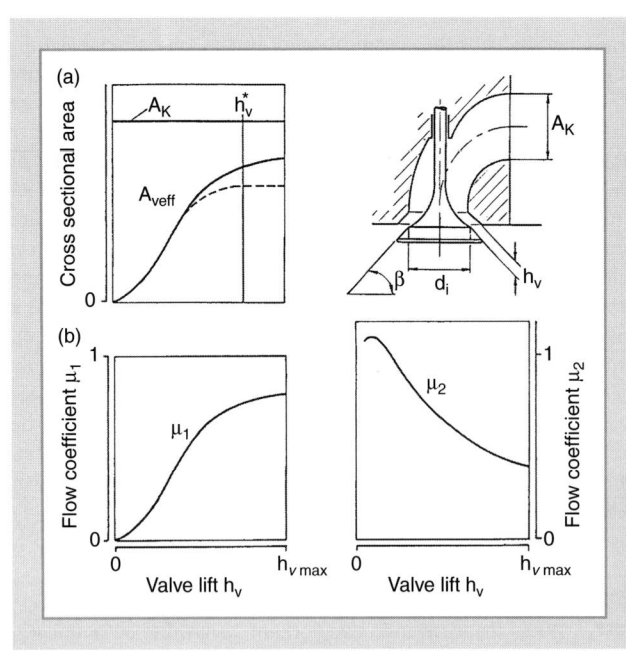

Fig. 2-4
Effective valve cross section $A_{Veff}(\varphi)$ and valve flow coefficients

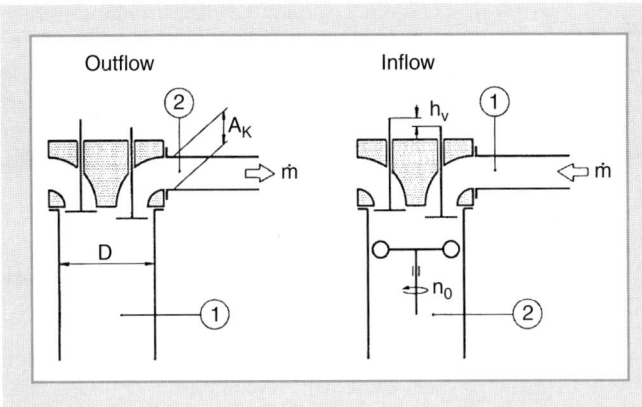

Fig. 2-5 Measurement of the valve flow coefficients and the swirl number in a steady-state flow test

horizontally before the maximum valve lift is reached, an excessively large lift has been selected, provided the volumetric efficiency is satisfactory. If the volumetric efficiency is not high enough, the valve port is too narrow for the maximum valve lift. Valve and valve port geometry and the maximum valve lift have to be matched so that the narrowest flow cross section is always in the valve seat for the entire range of the valve lift.

In practice, valve flow properties are usually not represented by the effective valve cross section $A_{Veff}(h_V)$ but rather by a valve flow coefficient $\mu(h_V)$, which according to

$$\mu(h_V) = \frac{A_{Veff}(h_V)}{A_{bez}} \quad (2\text{-}10)$$

is equal to the effective valve cross section relative to a reference cross section A_{bez}. In practice, two versions are primarily used for the reference cross section:

Version 1: $A_{bez} = A_K$
The constant port cross section A_K (see Fig. 2-4) is used as the reference cross section to produce a flow coefficient μ_1 that is in the range of $0 \leq \mu_1 \leq 1$ for the entire range of valve lift (Fig. 2-4b).

Version 2: $A_{bez} = A_V(h_V)$
The thusly defined flow coefficient $\mu_2(h_V)$ is undefined for $h_V = 0$ and may also assume values larger than 1 for low h_V values (Fig. 2-4c).

2.1.2.4 Intake Swirl

As a rule, high-speed direct injection diesel engines depend on intake swirl for satisfactory mixture formation and combustion, i.e. the rotation of the influent air in the cylinder around the cylinder's axis normally generated by a *swirl port*, which is usually further intensified by compression (see also Sect. 3.1).

Since both an unduly weak swirl and an unduly intense swirl can be detrimental to a given combustion system, swirl intensity needs to be specified objectively (swirl number). To do so, a stationary flow test bench (Fig. 2-5) is equipped with a vane anemometer that has defined dimensions and is installed in a defined position. To characterize the swirl, the vane speed n_D is placed in relation to an intended engine speed n obtained when the measured mean axial flow velocity c_a is equated with the mean piston velocity c_m, i.e.

$$c_a = c_m = 2s \cdot n \quad (2\text{-}11)$$

and thus

$$n = \frac{c_a}{2 \cdot s}. \quad (2\text{-}12)$$

The ratio n_D/n changes with the valve lift and must be averaged as a function of it when the mean swirl (swirl number D) representative for a particular valve port is sought [2-2]:

$$D = \left(\frac{n_D}{n}\right)_m = \frac{1}{\pi} \int_{BDC}^{TDC} \frac{n_D}{n} \cdot \left(\frac{c_k}{c_m}\right)^2 d\varphi. \quad (2\text{-}13)$$

In Eq. (2-13), c_K signifies instantaneous piston velocity assigned to the particular crank position at speed n calculated according to Eq. (2-12).

2.1.2.5 Influence of the Intake Manifold

Along with the geometry and aerodynamic engineering of the intake tract inside the cylinder head (valve and intake port), the volumetric efficiency λ_l defined in Eq. (2-3) also particularly depends on the geometry of the connected intake manifold.

Imagine a four-stroke single-cylinder engine with a smooth pipe connected on the intake side as schematically represented in Fig. 2-6. The pipe has the (elongated) length L and an "open pipe end" at the end opposite the cylinder. The pressure p inside the pipe directly before IO is equal to the external pressure p_0 ($p/p_0 = 1$). When the intake valve is opened, the vacuum that ensues from the intake stroke in

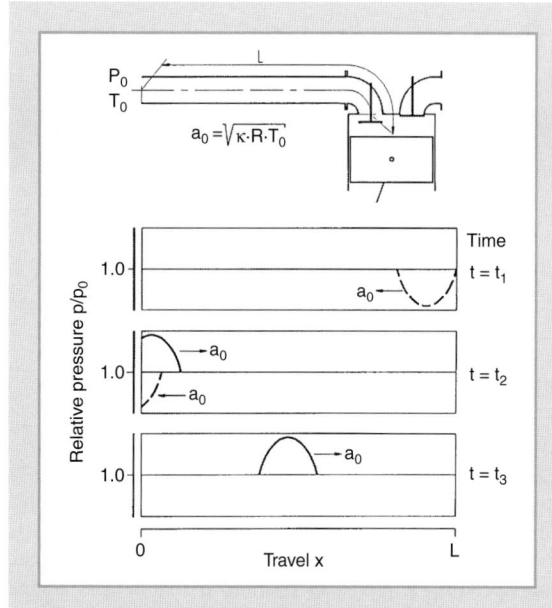

Fig. 2-6 Wave travel in the intake pipe based on acoustic theory

the cylinder causes a suction wave ($p/p_0 < 1$) to spread from the intake valve toward the open end of the pipe (t_1). According to acoustic theory on which this idea is based, this suction wave moves with sonic speed a_0, which in accordance with

$$a_0 = \sqrt{\kappa \cdot R \cdot T_0} \qquad (2\text{-}14)$$

directly depends on the gas temperature T_0 in the pipe, assumed to be constant here. After the time $t = L/a_0$, the suction wave reaches the open end of the pipe where it is reflected (t_2) as a pressure wave ($p > p_0$), which then returns (t_3) to the intake valve at the speed a_0. If the intake valve is still open when the pressure wave arrives, then this can boost volumetric efficiency. To this end, the entire wave travel time $\Delta t = 2L/a_0$ must be shorter than the valve opening duration $\Delta t_{\text{IO-IC}}$. This allows formulating the following condition

$$L \leq \frac{a_0}{720 \cdot n} \cdot \Delta\varphi_{\text{IO-IC}} \qquad (2\text{-}15)$$

Clearly, at the given intake valve timing and thus also the given intake valve opening duration $\Delta\varphi_{\text{IO-IC}}$ at a particular speed n, the intake pipe must have a particular length L to attain maximum volumetric efficiency. The converse is more important for practice however, i.e. at a given timing, a given intake pipe length L only produces maximum volumetric efficiency at a particular speed and usually only produces relatively high volumetric efficiency values within a relatively narrow span of speed. Logically, the intake manifold's fundamental influence on volumetric efficiency elucidated with the example of a single-cylinder engine can be transferred to multi-cylinder engines and systematically employed to set the maximum cylinder charge for a very definite speed range (intake manifold tuning and variable length intake manifold).

2.1.3 Two-Stroke Cycle

2.1.3.1 Distinctive Features of Two-stroke and Four-stroke Gas Exchange in Comparison

The two-stroke working cycle corresponds to two piston strokes or one crankshaft revolution (= 360°CA). Gas exchange has to occur in the temporal environment of bottom dead center (BDC). Two consequences can be inferred directly from this:
– Since gas exchange already commences before BDC and only terminates after BDC, a portion of the expansion and compression stroke is unusable.
– Instantaneous piston velocity is so low during the entire gas exchange phase that the piston is virtually unable to exert any intake or expulsion effect on the cylinder charge. Hence, gas exchange can only occur when a positive *scavenging gradient* exists, i.e. overpressure from the intake to the exhaust side, to which end, two-stroke engines must be equipped with a *scavenging blower* (or scavenging pump) as a matter of principle.

Both valves and ports are applied in both pure and mixed forms as gas exchange elements. In the case of port timing (ports in the surface of the cylinder liner), the piston also assumes the function of a control slide.

2.1.3.2 Scavenging Methods

All two-stroke scavenging processes implemented thus far can be classified in two basic categories
– loop scavenging and
– uniflow scavenging.

MAN reverse scavenging, loop scavenging, was important for large two-stroke diesel engines (with D = 250–900 mm) until the start of the 1980s. The stroke to bore ratio was just above s/D = 2. Fig. 2-7a presents a schematic of the arrangement of exhaust and inlet ports and the related timing diagram.

The pre-release phase (from EO to IO) begins when the downwardly moving piston enables the exhaust ports. During this phase, the pressure in the cylinder, which is still relatively high at the time of EO, ought to dissipate enough by the discharge of a portion of the cylinder charge that it is already lower than the scavenging pressure when the inlet ports subsequently open. Only then can fresh air already flow into the cylinder as of IO. This should develop a loop flow in the

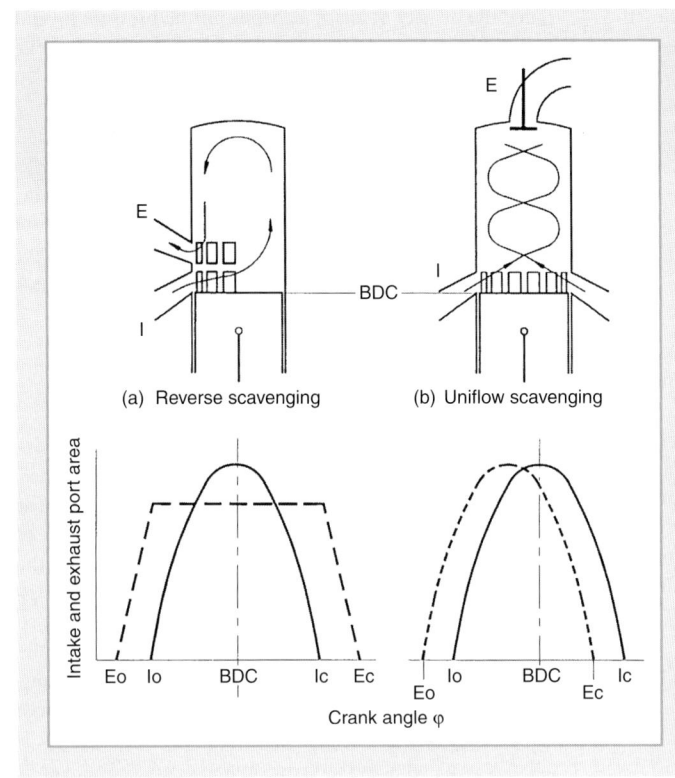

Fig. 2-7
Schematic scavenging flow and timing diagram for two basic scavenging processes of large two-stroke diesel engines

cylinder interacting with the still discharging cylinder gas as indicated in Fig. 2-7a. After reversing in BDC, the piston first closes the inlet ports during its upward stroke and again expels part of the cylinder charge through the exhaust ports as of IC. This loss of fresh gas connected with the so-called *post-exhaust* is a significant drawback of every loop scavenging system and has its origins in the symmetrical timing diagram (at BDC).

Uniflow scavenging has asserted itself over reverse scavenging for two-stroke large diesel engines since the early 1980s, primarily because of the necessity for larger stroke to bore ratios (s/D to > 4) (see Sect. 18.4). Uniflow scavenging is generally implemented with inlet ports and a single exhaust valve centered in the cylinder head (see Figs. 18-36 and 18-43). The free selection of exhaust valve timing this allows makes an asymmetrical timing diagram (see Fig. 2-7b) possible and eliminates post-exhaust. By appropriately tapering the inlet port's edges, a swirl can be superimposed on the longitudinal direction of scavenging running from bottom to top. This additionally stabilizes scavenging and also makes it possible to exert influence on mixture formation and combustion. The ability to control the exhaust valve's high thermal load is essential (see Sect. 6.1).

The parameters defined in Sect. 2.1.1 may be used to derive the correlation represented in Fig. 2-8 to assess two-stroke scavenging. The optimal scavenging characteristic, i.e. pure *displacement scavenging*, exists when the air efficiency λ_a only has to be maintained commensurately with the volume $V_h + V_c$ or $\lambda_a = \varepsilon/(\varepsilon - 1)$ to obtain the scavenging efficiency $\lambda_s = 1.0$, i.e. complete scavenging. The straight line $\lambda_s = 0$ corresponds to the *short-circuit flow*, i.e. no scavenging is obtained when the air efficiency λ_a is still so large. The curve denoted as a *total mixture* corresponds to scavenging during which every mass element of the fresh charge entering the cylinder completely mixes with the total charge mass instantaneously

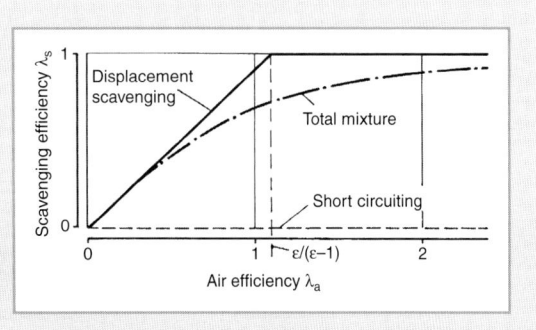

Fig. 2-8 Scavenging efficiency as a function of the air efficiency for idealized scavenging sequences

located in the cylinder and only mass elements from the instantaneous mixture charge flow through the exhaust.

Understandably, of all the scavenging systems, uniflow scavenging comes closest to pure displacement scavenging.

2.2 Diesel Engine Supercharging

2.2.1 General

2.2.1.1 Definition and Goals of Supercharging

As the following remarks are intended to illustrate, combustion engine supercharging is primarily a method to enhance power density.

In accordance with the definition of effective efficiency (Eq. (1-8)), net engine power increases with the fuel mass $\dot{m}_B$ converted in the unit of time. Depending on the combustion system, a particular air mass flow $\dot{m}_{LZ}$ is required for combustion. Applying the air/fuel ratio λ_V and the stoichiometric air/fuel ratio L_{min} as well as the volumetric efficiency delivers the conditional equation for P_e in the form:

$$P_e = \frac{H_u}{L_{min}} \cdot \frac{V_H}{a} \cdot \frac{1}{\lambda_V} \cdot \lambda_l \cdot n_M \cdot \rho_L \cdot \eta_e \qquad (2\text{-}16)$$

where $a = 2$ for a four-stroke engine and $a = 1$ for a two-stroke engine (see Sect. 1.2). Applying a particular fuel (H_u, L_{min}) and a particular combustion system (λ_V), this indicates that, at a particular speed n_M ($\rightarrow \lambda_l = $ const.) and irrespective of the effective efficiency η_e, a particular engine's net power (V_H, a) is still only a function of the density ρ_L of the air before engine intake.

When the air before intake is supplied to the engine with a higher density than the ambient air, this is *supercharging*.

Since the density of air ρ_L depends on the pressure p_L and the temperature T_L calculated by the thermal equation of the state of a gas:

$$\rho_L = \frac{1}{R} \cdot \frac{p_L}{T_L} \qquad (2\text{-}17)$$

and T_L may normally not be lowered below the ambient temperature, supercharging is primarily an elevation of pressure before intake to a value above ambient pressure, the so-called *boost pressure* p_L. The unit employed to do this is called a *supercharger*. The options for supercharging are stipulated by DIN 6262.

2.2.1.2 Exhaust Gas Turbocharging and Mechanical Supercharging in Comparison

Since exhaust gas turbocharging and mechanical supercharging have attained the greatest practical significance, their different interactions with a basic engine shall be illustrated with the aid of the idealized cycle (Fig. 2-9). In the case of mechanical supercharging, the supercharger powered by the engine supplies the cylinders air with a pressure $p_2 = p_L$ identical to the cylinder pressure during the intake stroke so that compression starts at a higher pressure than in a naturally aspirated engine (1Z). Once the expansion stroke has concluded (5Z), the exhaust valve opens and the cylinder charge is expelled against the ambient pressure (p_1). This produces positive gas exchange work W_{LDW} (areas 1Z, 6Z, 7Z, 8Z, 1Z) in the sense of work delivered by the engine. The (isentropic) supercharger work W_L the engine has to produce is greater than its gas exchange work though. The perpendicularly hatched area corresponds to the loss of work that arises from the cylinder charge being throttled from the state 5Z to the pressure p_1 (after the exhaust valve) rather than being expanded isentropically (loss through incomplete expansion).

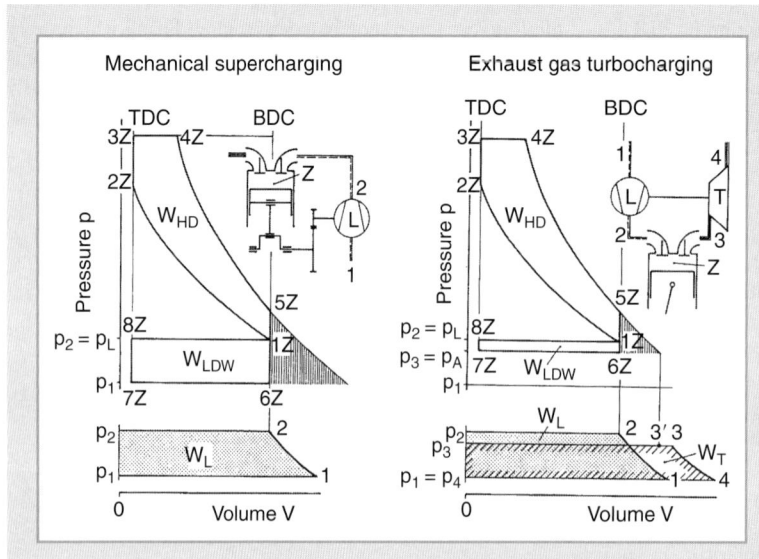

Fig. 2-9 Supercharging for the idealized engine process

Exhaust gas turbocharging also has to produce equally great supercharger work W_L at the same boost pressure p_2 and in the same high pressure phase of the engine process as mechanical supercharging. Further, the cylinder has the same state at the end of expansion (5Z). Rather than being diverted from the crankshaft work, supercharger work W_L is covered by the (equiareal) turbine work W_T, drawn from the exhaust energy. Since the exhaust temperature T_3 before the turbine is higher than T_2, under the condition $W_T = W_L$, the exhaust pressure p_3 before the turbine is lower than p_2. Thus, gas exchange work W_{LDW} is positive here too. In addition, the higher back pressure at the outlet makes the loss of work as a result of incomplete expansion of the cylinder charge (vertically hatched area) lower than in the case of mechanical supercharging. The higher temperature in point 3 than 3′, which would adjust to the exhaust pressure p_3 if the cylinder charge expanded isentropically, indicates that the exhaust turbine even recovers a portion of this loss of work on the cylinder side. Even though Fig. 2-9 only corresponds to idealized changes of state, it still indicates that exhaust gas turbocharging certainly provides better conditions for overall engine efficiency than mechanical supercharging.

2.2.2 Engine and Supercharger Interaction

2.2.2.1 Supercharger Types and Maps

Every familiar supercharger type can be classified in one of two groups based on the operating principle, namely:
– positive-displacement superchargers and
– turbo compressors.

Accordingly, two basic forms of supercharger map also exist. A supercharger map is a representation of the supercharger pressure ratio $\pi_L = p_2/p_1$ as a function of the volume flow rate $\dot{V}_1$ relative to a defined reference condition (p_1, T_1) with families of lines of constant supercharger speed n_L (supercharger characteristic curve) and constant isentropic supercharger efficiencies η_{sL} (Figs. 2-10 and 2-11).

Positive-Displacement Superchargers

Not only the reciprocating piston compressors formerly only used to supercharge large two-stroke engines but also Roots blowers, various rotary piston compressors (*rotary piston superchargers* and sliding-vane superchargers), spiral-type superchargers (*G superchargers*) and screw-type superchargers are positive-displacement superchargers (Fig. 2-10).

Supercharger characteristic curves (n_L = const.) develop relatively steeply according to the principle of positive displacement. They would even slope exactly vertically if a

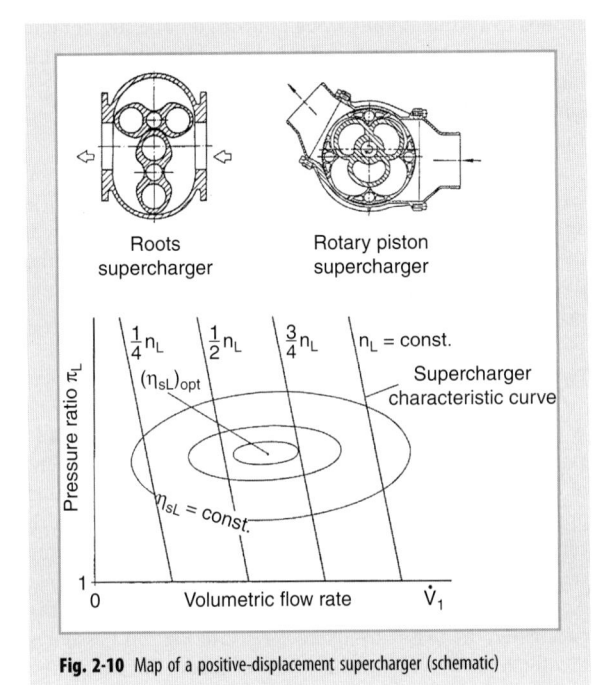

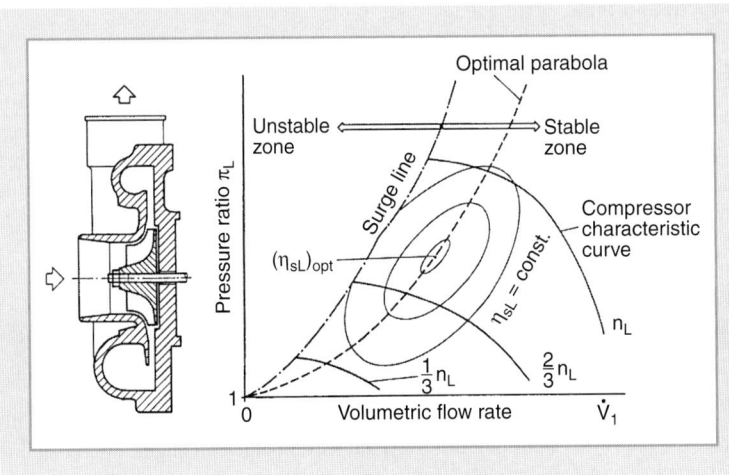

Fig. 2-10 Map of a positive-displacement supercharger (schematic)

Fig. 2-11 Map of a radial turbo compressor (schematic)

supercharger were loss-free, the flow rate only being a function of the speed and not the pressure ratio.

Depending on its design, a real positive-displacement supercharger's leakage losses or losses caused by re-expansion of the compressed medium in the dead space increase as the pressure ratio increases and (at n_L = const.) the volume flow rate correspondingly decreases. From this, it follows that:
– The attainable pressure ratio is not a function of speed. Therefore, high pressure ratios are possible even at low speeds and thus small volumetric flows.
– The volume flow rate $\dot{V}_1$ is practically only a function of speed.
– The map is stable in the entire region and consequently usable for supercharging.

Turbo Compressors

Turbo compressors include axial and radial compressors. Since, unlike axial compressors, radial compressors can deliver a high pressure ratio even in a one-stage design, they are used virtually exclusively for purposes of supercharging.

Figure 2-11 presents a schematic map of a radial compressor. The volume flow rate $\dot{V}_1$ increases roughly proportionally and the pressure ratio π_L roughly quadratically with the compressor speed n_L. The compressor characteristic curves reach the *surge line* as the volumetric flow rate decreases in the proximity of their respective peaks. This divides the compressor map into a stable (right) and an unstable (left) map range. A compressor must always be adjusted to an engine so that the anticipated operating range comes to be to the right of the surge line [2-3]. This prevents surging, which is manifested in pulsating pressure and a volume flow rate that – apart from transient delivery – excites blade vibrations in a compressor, which can damage it.

Three key points have been established:
– The attainable pressure ratio is a function of speed; high pressure ratios cannot be attained at low speeds and small volumetric flow rates.
– The volume flow rate is a function of the speed and the pressure ratio.
– The map region to the left of the surge line is an unstable region.

2.2.2.2 Engine Mass Flow Characteristics

Representing the engine as the "consumer" in the supercharger map is an expedient method to illustrate engine and supercharger interaction. The consumer characteristic is denoted as the *engine mass flow characteristics* and has a fundamentally different curve for four-stroke and two-stroke engines.

Four-Stroke Engine

Assuming scavenging losses are negligibly small, the following applies to the volumetric flow rate $\dot{V}_1$ "consumed" by an engine relative to the intake condition (p_1, T_1)

$$\dot{V}_1 = \frac{\dot{m}_L}{\rho_1} = V_H \cdot \lambda_l \cdot \frac{n_M}{2} \cdot \frac{\rho_L}{\rho_1} \quad (2\text{-}18)$$

By logically applying Eq. (2-18), Eq. (2-19) can be rendered in the following form for a particular engine (V_H = const.)

$$\pi_L = \frac{p_L}{p_1} \sim \frac{T_L}{\lambda_l \cdot n_M} \cdot \dot{V}_1 \quad (2\text{-}19)$$

This states that, at a particular engine speed n_M and thus also at a given volumetric efficiency λ_l, the boost pressure ratio π_L and volumetric flow rate $\dot{V}_1$ are directly proportional at a particular charge air temperature T_L. In Fig. 2-12, this corresponds to a straight line through the origin. The straight line exhibits a smaller slope for a higher speed ($n_{M2} > n_{M1}$). This family of straight lines only has practical meaning for pressure ratio values $\pi_L \geq 1$ through, $\pi_L = 1.0$, which correspond to a naturally aspirated engine.

Incorporating the increase of T_L with π_L produces the family of thick lines for $\pi_L \geq 1$. If valve overlap (VO) is additionally implemented in the engine being analyzed, then the mass flow characteristics take on the form of the dashed curves. This implies that the pressure p_1 is on the exhaust side.

Two-Stroke Engine

A pressure gradient from the intake to the exhaust side is essential for two-stroke cylinder gas exchange. Independent of engine speed, a two-stroke cylinder's intake and exhaust openings act like two throttles connected in a series, which can be replaced by a single throttle providing the same resistance to the flow in which the pressure ratio is p_L/p_A. Accordingly, using the pressure p_A after exhaust as the parameter, the mass flow characteristic in Fig. 2-13 assumes the form of a throttle characteristic curve in which the pressure ratio π_L increases approximately quadratically with the flow rate $\dot{V}_1$.

2.2.2.3 Engine Operating Lines

Every one of a supercharged engine's operating points appears in the supercharger map as the intersection between the related engine mass flow characteristic and the related supercharger characteristic curve. Connecting every possible intersection for a particular engine operating mode produces the *engine operating line*.

The engine operating line for some practically important supercharging processes is presented schematically below.

Mechanical Supercharging

When a four-stroke engine is mechanically supercharged with a positive-displacement supercharger, an engine operating line that only slightly slopes with the engine speed appears in the

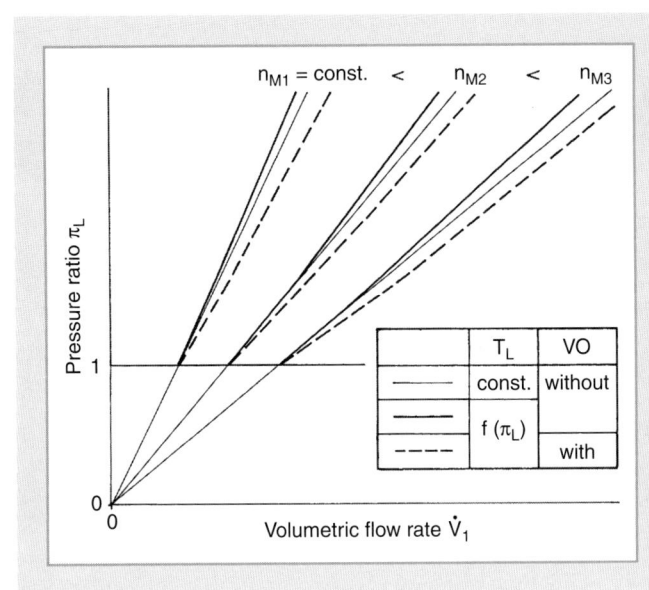

Fig. 2-12 Engine mass flow characteristics of a four-stroke engine (schematic). T_L charge air temperature, VO valve overlap

supercharger map for a constant gear ratio $ü$ between engine speed n_M and supercharger speed n_L. Thus, relatively high boost pressure is still on hand even in the lower speed range. This enables a car engine to have good accelerating performance for instance. Enlarging the gear ratio $ü$ can elevate the overall boost pressure level or, conversely, lower it (Fig. 2-14, top).

A turbo compressor on the other hand performs entirely differently when used for mechanical supercharging. At the same boost pressure for rated speed as a positive-displacement supercharger, boost pressure decreases more strongly as engine speed drops (Fig. 2-14, bottom). A variable gear ratio is always a fundamental feature of the concept anytime efforts are made in development to also implement turbo compressors for mechanical supercharging because of their high efficiency.

2.2.3 Exhaust Gas Turbocharging

Rather than being directly determined by engine speed as in mechanical supercharging, compressor speed in exhaust gas turbocharging is determined by the engine's instantaneous supply of exhaust gas energy (exhaust power) to the exhaust gas turbocharger.

The top part of Fig. 2-15 presents a design that attains the desired boost pressure in the rated power point for a four-stroke diesel engine. Taking this as the starting point, the operating point in the compressor map shifts depending on the load characteristics (see Sect. 1.2).

In *generator operation* ($n_M = n_{Nenn}$ = const.), the operating point moves downward along the engine mass flow characteristics as engine power decreases because the exhaust power also decreases with it. At least in steady-state operation, this drop in boost pressure is unproblematic because a lower engine load also only requires lower boost pressure.

If the engine power is reversed under the condition $M=$ const. *(speed decrease at rated engine torque)*, then a drop in boost pressure due to diminishing exhaust power must also be accepted, even when constant boost pressure would be desirable for M = const. In addition, the operating point reaches the surge line relatively quickly.

In the case of *propeller operation*, the correlation $M \sim n_M^2$ applies to a fixed propeller. The particular boost pressure at reduced engine speed is normally sufficiently high, at least in steady-state mode.

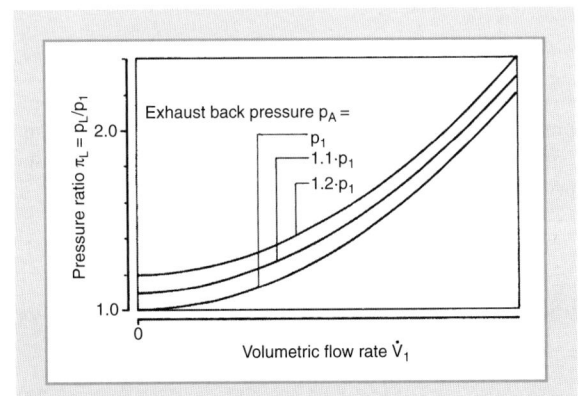

Fig. 2-13 Engine mass flow characteristics of a two-stroke engine at differing exhaust back pressure

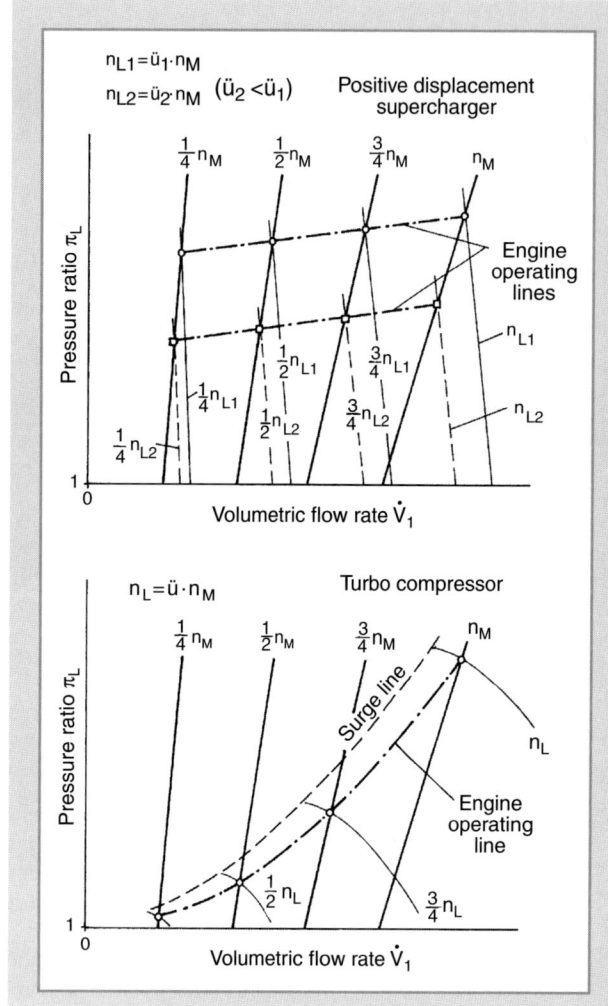

Fig. 2-14 Engine operating line of a mechanically supercharged four-stroke engine in the supercharger map

In a turbocharged two-stroke diesel engine, only one single engine operating line is produced for all three of the modes of operation considered. The exhaust back pressure p_A rises after the cylinders and before the turbocharger turbine as the flow rate ($\dot{V}_1$) increases. While every possible engine operating point lies on this engine operating line, the operating points of 50% power in propeller operation and 50% generator power, for example, are not located in the same point.

2.2.3.1 Turbocharger Fundamental Equations, Turbocharger Efficiency

The principle of exhaust gas turbocharging in steady-state operation presupposes that an equality of power always exists between the compressor and turbine:

$$P_L = P_T \qquad (2\text{-}20)$$

Under the generally accepted assumption that the compressor and turbine are adiabatic machines (*adiabatic* = without loss or gain of heat), the following applies to compressor power (see Fig. 2-16)

$$P_L = \dot{m}_L \cdot \Delta h_{sL} \cdot \frac{1}{\eta_{sL} \cdot \eta_{mL}} \qquad (2\text{-}21)$$

with the isentropic enthalpy difference Δh_{sL} of the compressor

$$\Delta h_{sL} = C_{pL} \cdot T_1 \left[\left(\frac{p_2}{p_1}\right)^{\frac{\kappa_L - 1}{\kappa_L}} - 1 \right] \qquad (2\text{-}22)$$

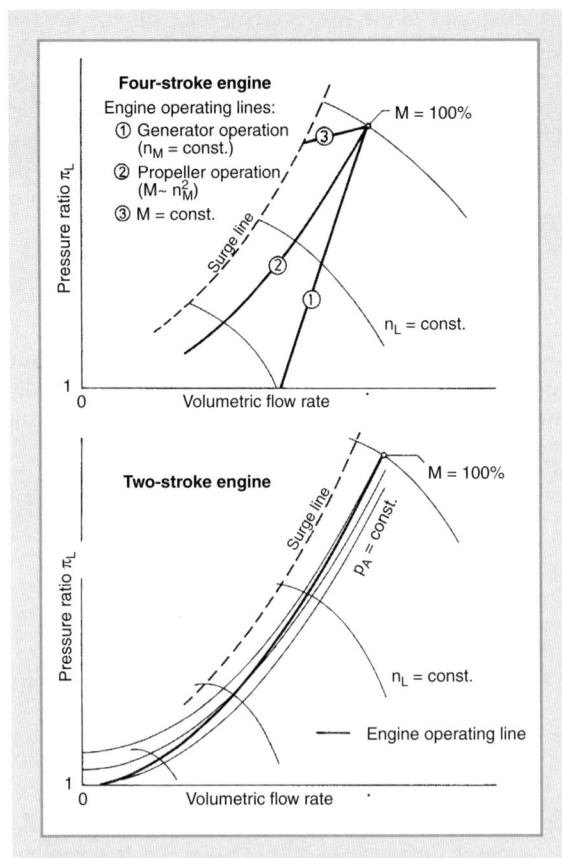

Fig. 2-15 Engine operating lines for turbocharging in the compressor map

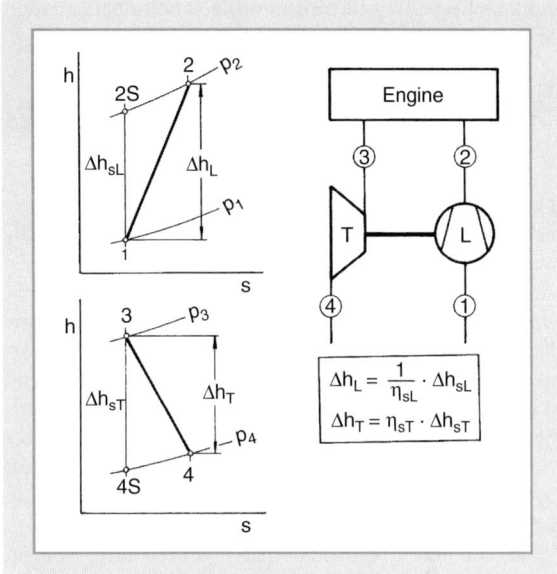

Fig. 2-16 Specific work of a supercharger and turbine

and, analogously, for turbine power

$$P_T = \dot{m}_T \cdot \Delta h_{sT} \cdot \eta_{sT} \cdot \eta_{mT} \quad (2\text{-}23)$$

$$\Delta h_{sT} = c_{pT} \cdot T_3 \left[1 - \left(\frac{p_4}{p_3}\right)^{\frac{\kappa_T - 1}{\kappa_T}}\right] \quad (2\text{-}24)$$

Thus, Eq. (2-20) can be rendered as

$$\dot{m}_L \cdot c_{pL} \cdot T_1 \left[\left(\frac{p_2}{p_1}\right)^{\frac{\kappa_L - 1}{\kappa_L}} - 1\right] = \eta_{sL} \cdot \eta_{mL} \cdot \eta_{mT} \cdot \eta_{sT} \cdot \dot{m}_T$$

$$\cdot c_{pT} \cdot T_3 \left[1 - \left(\frac{p_4}{p_3}\right)^{\frac{\kappa_T - 1}{\kappa_T}}\right] \quad (2\text{-}25)$$

Since the turbine and compressor share a shaft, the related mechanical losses can be summarized in the turbocharger's mechanical efficiency η_{mTL}:

$$\eta_{mTL} = \eta_{mL} \cdot \eta_{mT}. \quad (2\text{-}26)$$

In practice, the (complete) turbocharger's mechanical losses are superimposed on the isentropic turbine efficiency η_{sT}, thus transforming it into the turbine efficiency η_T:

$$\eta_T = \eta_{sT} \cdot \eta_{mTL}. \quad (2\text{-}27)$$

The product chain of overall efficiency in Eq. (2-25) is denoted as turbocharger efficiency η_{TL}:

$$\eta_{TL} = \eta_{sL} \cdot \eta_T. \quad (2\text{-}28)$$

Applying Eq. (2-28), Eq. (2-25) can be converted into the *1st fundamental turbocharger equation*:

$$\pi_L = \frac{p_2}{p_1}$$

$$= \left\{1 + \frac{\dot{m}_T}{\dot{m}_L} \cdot \frac{c_{pT}}{c_{pL}} \cdot \frac{T_3}{T_1} \cdot \eta_{TL} \cdot \left[1 - \left(\frac{p_4}{p_3}\right)^{\frac{\kappa_T - 1}{\kappa_T}}\right]\right\}^{\frac{\kappa_L}{\kappa_L - 1}}. \quad (2\text{-}29)$$

Apart from the subsequent increase of the boost pressure ratio as the exhaust pressure, temperature and turbocharger efficiency η_{TL} increase, it further follows that the transition to a turbocharger with higher efficiency η_{TL} for a desired boost pressure p_2 only requires a slight buildup of exhaust gases before the turbine (smaller values for p_3, T_3), whereupon the engine runs with lower residual exhaust gas and reacts with lower fuel consumption because of the lower work during expulsion. Equation (2-29) can additionally be transformed into the conditional equation for turbocharger efficiency:

$$\eta_{TL} = \frac{\dot{m}_L}{\dot{m}_T} \cdot \frac{c_{pL}}{c_{pT}} \cdot \frac{T_1}{T_3} \cdot \frac{\left(\frac{p_2}{p_1}\right)^{\frac{\kappa_L-1}{\kappa_L}} - 1}{1 - \left(\frac{p_4}{p_3}\right)^{\frac{\kappa_T-1}{\kappa_T}}}. \quad (2\text{-}30)$$

It is simultaneously the conditional equation for so-called *supercharging efficiency* too. This term may be applied quite generally to *supercharging systems with exhaust gas energy recovery*.

The supercharging system is viewed as a "black box" into which exhaust gas with the state p_3, T_3 is fed. It expands in the "black box" to the back pressure p_4 and, in the return, compresses air from its intake state p_1, T_1 to the boost pressure p_2. In addition to an (one-stage) exhaust gas turbocharger, this "black box" may also stand for a pressure-wave supercharger (Comprex, see Sect. 2.2.5) or a two-stage turbocharger unit together with an intercooler between the low pressure and high pressure compressor.

The conditional equations that produced them and, in particular, incorporation of the state variables (p, T) for the conditions 1–4 as static values or as total state variables constitute important information when concrete numerical values are adopted for η_{TL}.

In steady-state operation, an engine's exhaust mass flow corresponds to that supplied to the turbine. This produces the *2nd fundamental turbocharger equation*:

$$\dot{m}_T = A_{\text{Teff}} \cdot \frac{p_{03}}{\sqrt{R \cdot T_{03}}} \cdot \sqrt{\frac{2\kappa_T}{\kappa_T-1} \left[\left(\frac{p_4}{p_{03}}\right)^{\frac{2}{\kappa_T}} - \left(\frac{p_4}{p_{03}}\right)^{\frac{\kappa_T+1}{\kappa_T}} \right]}. \quad (2\text{-}31)$$

It corresponds to the flow equation for a throttle with an effective cross section equal to the effective turbine cross section A_{Teff}. In addition to the geometry of the turbine guide vane and rotor, it particularly depends on the turbine pressure ratio p_{03}/p_4. The larger it becomes, the larger A_{Teff} becomes.

Thus, for axial turbocharger turbines, the following empirical correlation applies to A_{Teff}:

$$A_{\text{Teff}} \sim \left(\frac{p_{03}}{p_4}\right)^{0.204} \quad \text{for} \quad \frac{p_{03}}{p_4} \geq 1. \quad (2\text{-}32)$$

Not only the turbine efficiency η_T but also the turbine flow rate of vehicle turbocharger turbines, which are virtually always radial turbines, is usually plotted in the turbine map as a reduced mass flow $\dot{m}_{\text{Tred}}$

$$\dot{m}_{\text{Tred}} = \frac{\dot{m}_T \cdot \sqrt{T_{03}}}{p_{03}}$$

$$= \frac{A_{\text{Teff}}}{\sqrt{R}} \cdot \sqrt{\frac{2\kappa_T}{\kappa_T-1} \left[\left(\frac{p_4}{p_{03}}\right)^{\frac{2}{\kappa_T}} - \left(\frac{p_4}{p_{03}}\right)^{\frac{\kappa_T+1}{\kappa_T}} \right]}.$$

$$(2\text{-}33)$$

as a function of the turbine pressure ratio p_3/p_{04} (see Fig. 2-17).

If, assuming an operating point of the engine and thus also the turbocharger was hitherto steady-state ((P_T, P_L, n_{TL})=const.), the supply of exhaust gas energy to the turbocharger is changed, then the turbine power P_T increases or decreases compared to the instantaneous compressor power P_L. As a result, the turbocharger speed n_{TL} changes

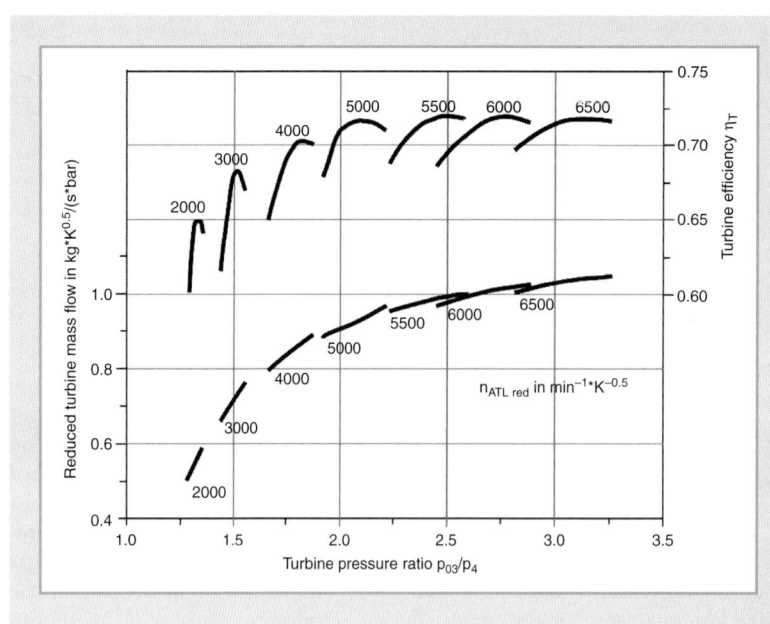

Fig. 2-17
Turbine map of a vehicle turbocharger

according to the principle of conservation of angular momentum (Eq. (2-34)).

$$\frac{dn_{TL}}{dt} = \frac{1}{4\pi^2 \cdot \Theta_{TL} \cdot n_{TL}} (P_T - P_L). \quad (2\text{-}34)$$

The smaller the turbocharger rotor's mass moment of inertia Θ_{TL}, the larger this change of speed turns out to be. Above all, this is important for good turbocharger acceleration performance (*response*).

2.2.3.2 Pulse Turbocharging and Constant Pressure Turbocharging

Influence of the Exhaust Manifold

The exhaust manifold is particularly important for exhaust gas turbocharging. It should be engineered so that
- the connected cylinders do not interfere with each other during the exhaust process,
- the technically utilizable exhaust energy is transported from the cylinder to the turbine with minimal losses and
- the exhaust energy is supplied to the turbine over a period that ensures it is converted into turbine work with maximum efficiency.

Two important basic forms are distinguished, pulse turbocharging and constant pressure turbocharging.

Pulse Turbocharging

Pulse turbocharging goes back to Büchi's patent of 1925 for a so-called *pressure wave system* [2-4]. According to it, in piping with a flow cross section equal to the outlet cross section of the cylinder head, only the exhaust gases from each of only those cylinders of a cylinder bank that are far enough apart in the ignition interval that they do not interfere with each other during their outlet phase should be combined into a common subbranch of the exhaust manifold and conducted to a separate turbine inlet. In turn, the ignition interval of two of a subbranch's successively expelling cylinders also may not be so large that the exhaust pressure in the related subbranch of the exhaust manifold drops to the turbine back pressure between two exhaust pulses.

These requirements are best met by *triple-pulse turbocharging* with an ignition interval of 3 × 240°CA in four-stroke engines and 3 × 120°CA in two-stroke engines. In pulse turbocharging, the exhaust pressure p_A drops below the boost pressure p_L during the valve overlap phase as is evident in Fig. 2-18.

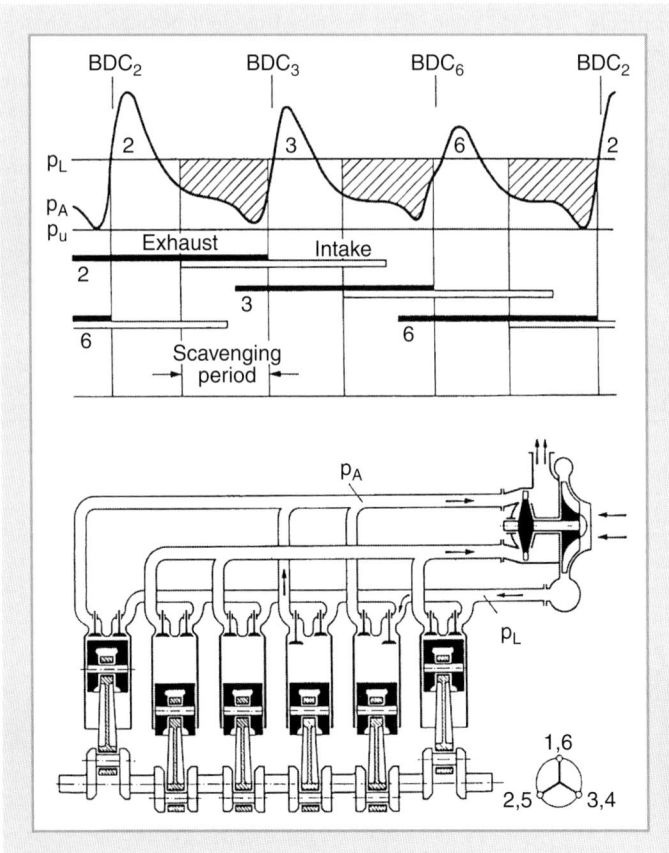

Fig. 2-18
Exhaust gas pressure curves and exhaust gas line guidance for a six-cylinder engine with pulse turbocharging based on [2-5]

As a result, the cylinder is still scavenged even when the mean exhaust pressure is as high or even higher than the boost pressure: However, the triple-pulse turbocharging considered ideal can only be implemented when the numbers of cylinders is $z = 3n$ ($n = 1, 2, \ldots N$) per cylinder bank.

Symmetrical double-pulse turbocharging is an option for other numbers of cylinders, provided they are whole numbers divisible by two. Since the ignition interval (for a four-stroke engine) is then $2 \times 360°$CA, the exhaust pressure drops to the turbine back pressure after every exhaust pulse.

On the other hand, the single "pressure elevation" decays more slowly than with the triple-pulse because the partial cross section of the turbine inlet assigned to each subbranch of the exhaust manifold only has to accommodate the exhaust from two cylinders and is accordingly smaller than in the case of a triple-pulse (see Fig. 2-19). The latter's effect of impeding scavenging can be largely eliminated by setting the valve overlap later.

Pulse turbocharging may only be implemented in engines with five or seven cylinders per cylinder bank by combining each double or triple *asymmetrical double-pulse* with a *single-pulse*. The aforementioned disadvantages of the symmetrical double-pulse over the triple-pulse become increasingly apparent then. In a five-cylinder engine, the cylinders of each of the two double combinations have ignition intervals of 288 and 432°CA. The ignition interval for the single-pulse is 720°CA (Fig. 2-20).

In addition to the symmetrical double-pulse, a quadruple pulse is also a possible variant of pulse turbocharging for numbers of cylinders per cylinder bank that are whole numbers divisible by four. The ignition interval is then $4 \times 180°$CA.

Since the exhaust pulse of the subsequently expelling cylinder consequently already reaches the exhaust valve

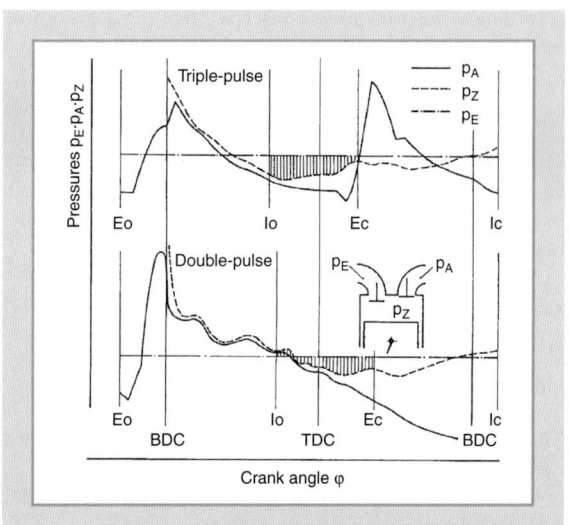

Fig. 2-19 A medium-speed diesel engine's gas exchange pressure curves for triple-pulse and symmetrical double-pulse turbocharging

of the cylinder during its scavenging phase, a quadruple-pulse would be particularly disadvantageous for the preceding reasons alone. This was also true until perhaps twenty-five years ago. Given today's tremendously increased turbocharger efficiencies, all that is needed for a desired boost pressure is an exhaust gas pressure low enough to prevent exhaust gas pulsations that develop during the valve overlap phase in quadruple-pulse turbocharging from reaching boost pressure

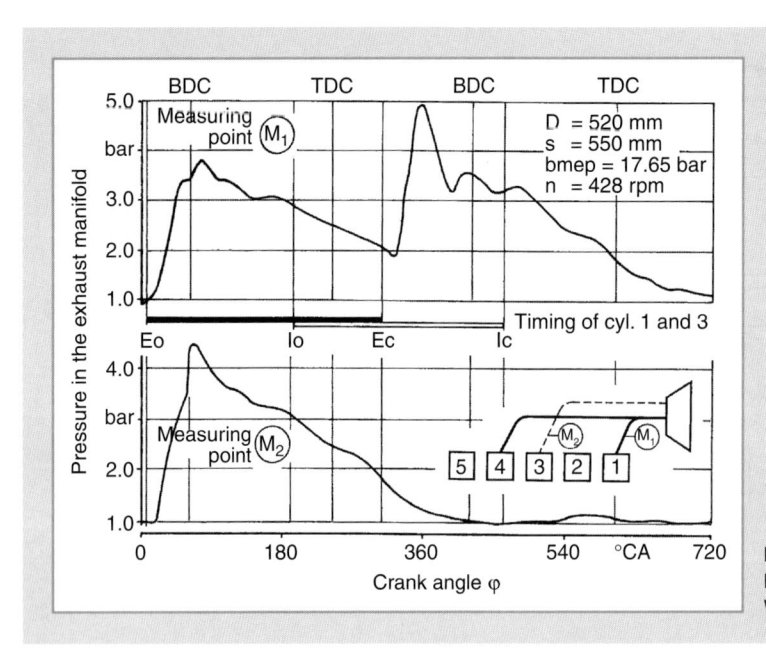

Fig. 2-20 Pulse turbocharging of a 10 cylinder medium speed V engine

and, thus, a positive scavenging gradient during the entire scavenging phase. The exhaust pressure curve for quadruple-pulse turbocharging still has only relatively low amplitudes because of the short ignition interval and the relatively large inlet cross section of the turbine per subbranch of the exhaust manifold and delivers a largely steady supply of exhaust gas energy to the turbine over time.

Constant Pressure Turbocharging

In constant pressure turbocharging, relatively short connecting pipes connect every cylinder of a cylinder bank on the exhaust side to a manifold pipe that is guided along the cylinder bank and connected with the turbocharger turbine on one end. The inner cross section of the manifold is usually selected somewhat smaller than the cylinder bore (see Fig. 19-35). The exhaust manifold's relatively large volume ensures that the exhaust energy flow at the turbine is largely uniform despite intermittent pressurization by the cylinders. This turbocharging system is also called a *constant pressure turbocharger* because the exhaust pressure before the turbine only fluctuates just slightly. Consequently, the number of cylinders no longer plays any appreciable role in a constant pressure turbocharged engine, i.e. an inline five-cylinder engine no longer differs from a six-cylinder in terms of its supercharging. There is also a structural advantage. The exhaust manifold for differing numbers of cylinders can be assembled of equally many subsections. Among other things, this simplifies spare parts stocking and, all in all, can be considered a significant cost argument for constant pressure turbocharging.

From the perspective of thermodynamics, the advantage of largely continuous turbine pressurization in constant pressure turbocharging is offset by the disadvantage of larger throttling losses than in pulse turbocharging during the discharge from the cylinder because the exhaust pressure in the manifold remains at a nearly constant level over time, while the exhaust back pressure "felt" by the cylinder in pulse turbocharging rapidly climbs close to the level of the instantaneous cylinder pressure because the exhaust pipes of the manifold are narrower.

Generally, the view on the question of whether the use of constant pressure turbocharging or pulse turbocharging is more advantageous in steady-state medium-speed diesel engines as a function of the supercharging rate is now that constant pressure turbocharging is advantageous as of brake mean effective pressures of $p_e \approx 18$ bar or specific work $w_e \approx 1.8$ kJ/dm^3, which corresponds to boost pressures of approximately 3.4 bar [2-6].

Pulse turbocharging basically always outperforms constant pressure turbocharging in terms of part load performance and acceleration performance. In both cases, the turbine operates with very low efficiency when engine power is low and the engine cylinders' supply of exhaust gas energy is correspondingly low. However, since its throttling losses at the cylinder are lower, pulse turbocharging delivers a bit more usable exhaust energy to the turbine. In addition, since the principle of pulse turbocharging causes the exhaust pressure to "undershoot" the boost pressure during valve overlap, residual exhaust gas scavenging is improved and thus more oxygen content is in the cylinder at the end of the gas exchange phase. However, its level also determines the amount of the fuel rate increased during the initiation of the acceleration process that is really convertible into acceleration power.

Since good acceleration performance is especially important in vehicle engines, their exhaust manifolds should be largely engineered based on the requirements of pulse turbocharging. Disregarding any company traditions this once, not only a type engine's supercharging rate but also its primary use in particular should determine any decision for pulse or constant pressure turbocharging for all other categories of engines.

It has generally been possible to improve turbocharger efficiencies tremendously in recent years. This defuses the

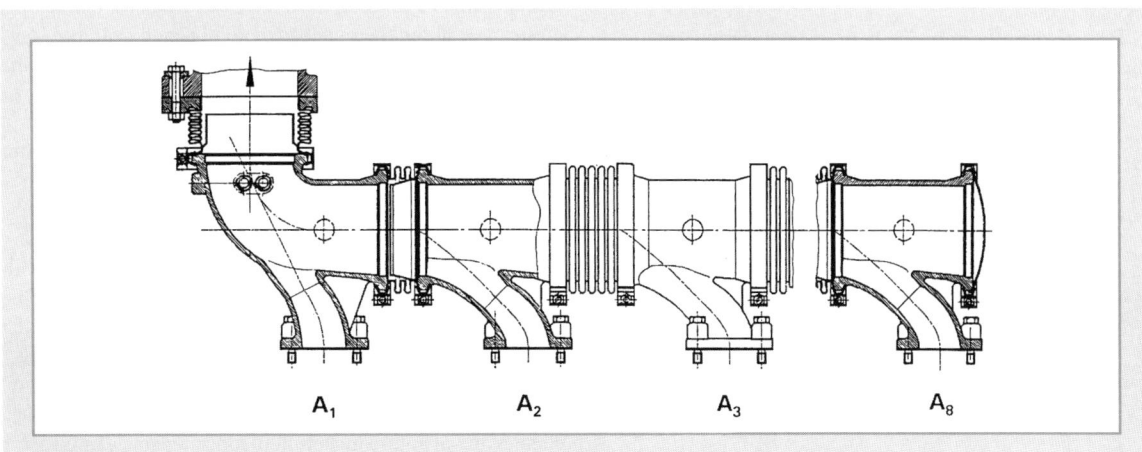

Fig. 2-21 Exhaust pipe of an MWM TBD 604 BV 16 engine ($D = 170$ mm, $s = 195$ mm)

conflict of objectives and permits exhaust manifold designs that may be considered mixed forms and optimal compromises between pulse and constant pressure turbocharging in every case (see Fig. 2-21).

2.2.3.3 Intercooling

When the air in a supercharger compresses from state 1 to pressure p_2, the temperature at the supercharger outlet (T_2), which is normally higher than the corresponding temperature T_{2S} for isentropic compression, also rises with the pressure.

When the isentropic supercharger efficiency η_{sL} determined in testing according to

$$\eta_{sL} = \frac{T_{2s} - T_1}{T_2 - T_1} \quad \text{with} \quad T_{2s} = T_1 \cdot \left(\frac{p_2}{p_1}\right)^{\frac{\kappa - 1}{\kappa}} \quad (2\text{-}35)$$

or taken from the supercharger map is known, then the temperature T_2 can be calculated (see Fig. 14-3):

$$T_2 = T_1 \cdot \left\{ 1 + \frac{1}{\eta_{sL}} \cdot \left[\left(\frac{p_2}{p_1}\right)^{\frac{\kappa - 1}{\kappa}} - 1 \right] \right\} \quad (2\text{-}36)$$

Undesirable in the engine, the temperature increase in the supercharger can partly be canceled by isobaric recooling (at p_2 = const.), i.e. *intercooling* in an *intercooler*.

Apart from the temperature level of the available cooling medium (coolant inlet temperature T_{Ke}), the intercooler's effectiveness determines the potential reduction of temperature in the intercooler (see Fig. 2-22). This is expressed in the heat recuperation rate η_{LLK}, also called *intercooler efficiency*:

$$\eta_{LLk} = \frac{T_2 - T_{2*}}{T_2 - T_{Ke}} \quad (2\text{-}37)$$

Large engine intercoolers that operate according to the countercurrent principle attain the highest values for η_{LLK} (> 0.90) (see Sect. 14.3).

Disregarding the complexity of its engineering, intercooling only brings benefits, namely

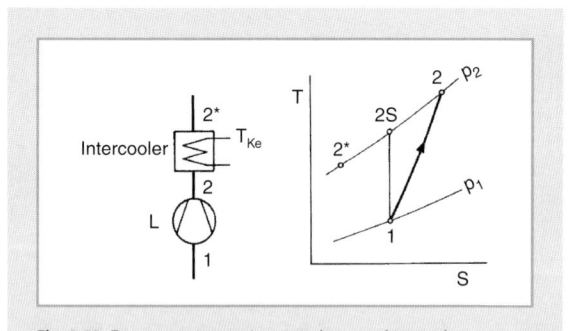

Fig. 2-22 Temperature increase in a supercharger and intercooler

- lower thermal loading of the engine,
- lower mechanical loading of the engine because intercooling attains a desired value of cylinder charge density at a lower boost pressure and
- lower NO_x emission.

2.2.3.4 Steady-State and Dynamic Engine Drivability during Exhaust Gas Turbocharging

Boost Pressure Control

Like every turbomachine, an exhaust gas turbocharger's compressor and turbine are each designed for a particular operating point (design point) in which each operates under optimal conditions.

A turbocharger has been tuned to a particular engine operating point when the exhaust gas energy flow from the turbocharger delivered by the engine in this operating point provides the desired boost pressure (see point A in Fig. 2-23).

Every operating point in the engine map under the *tractive force hyperbola* (line of constant rated power) corresponds to a lower engine power than the rated power. Since it also only delivers lower exhaust power to the turbocharger, the related boost pressure is also lower than in point A.

The engine operating lines for the three operating modes singled out:
- generator operation (n_M = const.)
- propeller operation ($M \sim n_M^2$)
- speed decrease at rated engine torque (M = const.)

have been plotted in the engine map (Fig. 2-23, left) and (for a four-stroke engine) in the compressor map (Fig. 2-23, right) (see also Fig. 2-15).

Provided the engine operating points change quasistatically (steady-state drivability), then, as explained in Sect. 2.2.2.3, the drop in boost pressure during generator and propeller operation is unproblematic. However, a speed decrease at rated engine torque can only be implemented by decreasing the air/fuel ratio. In addition to an increase of exhaust blackening, an increase of the thermal load of combustion chamber components can be expected. A speed decrease at rated engine torque additionally holds the risk of dangerously nearing or even exceeding the surge line (Fig. 2-23, right).

Since, taking the rated power point as the starting point, not only a horizontal but also even an ascending full load line is required as speed drops, the turbocharger must, however, be adapted differently for vehicle engines than for large engines. A distinction is made between turbochargers with fixed and with variable turbine geometry. A turbocharger with fixed turbine geometry (Fig. 2-24) is selected so that it already produces the boost pressure required for the maximum engine torque M_{max} at the related partial engine speed n_2. This is set at approximately 60% of the rated engine speed n_3 for commercial vehicle engines and at approximately 40% for car engines. In turn, for the aforementioned reasons, the boost pressure and the related full load torque for lower engine speeds ($n < n_2$) based on it

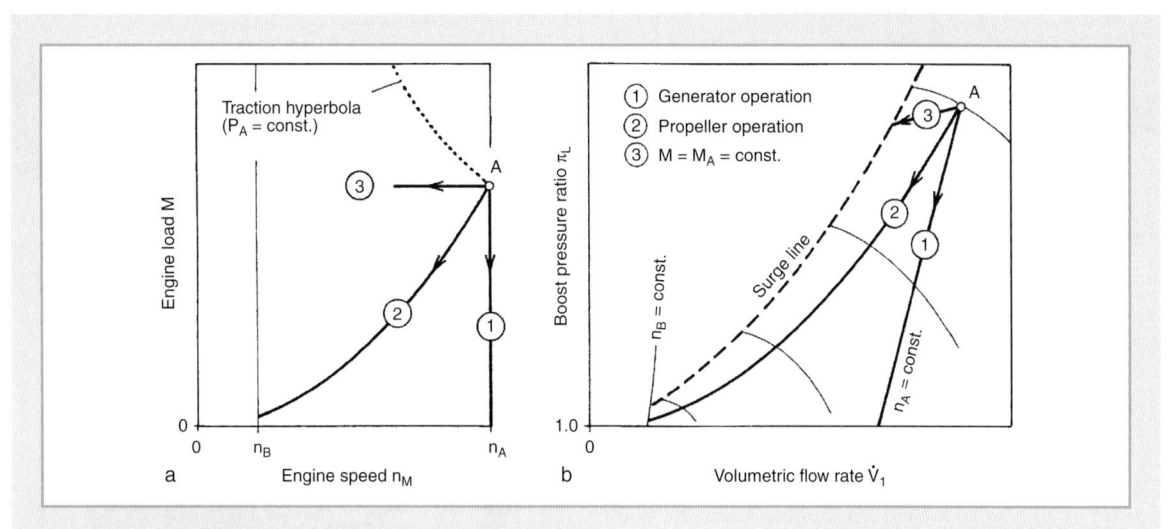

Fig. 2-23 Engine operating lines in an engine map *(left)* and a compressor map *(right)* with engine derating; four-stroke diesel engine with unregulated exhaust gas turbocharging (EGT)

decrease more intensely relative to M_{max} than in a naturally aspirated engine, which naturally has the maximum "boost pressure" at every speed.

Without any control intervention in the turbocharger, the full load boost pressure steadily increases at engine speeds that are increasingly higher than those at maximum engine torque ($n > n_2$). This full load boost pressure increase can usually still be accepted in commercial vehicle engines since they are basically constructed more ruggedly and is used to produce a correspondingly large air/fuel ratio since the full load torque curve slopes toward rated speed.

The full load boost pressure in a car engine would rise particularly steeply for speeds $n > n_2$ because of the altogether greater range of speed to be covered and $n_2 \approx 0.40 \cdot n_3$. The high mechanical load of the crankshaft assembly connected with this makes this unacceptable simply because of the high compression ratio required in car diesel engines.

Hence, boost pressure control or at least boost pressure limiting by an *open loop wastegate* is imperative in turbocharged car engines (Fig. 2-25). A boost pressure-pressurized diaphragm opens the wastegate valve against a spring and

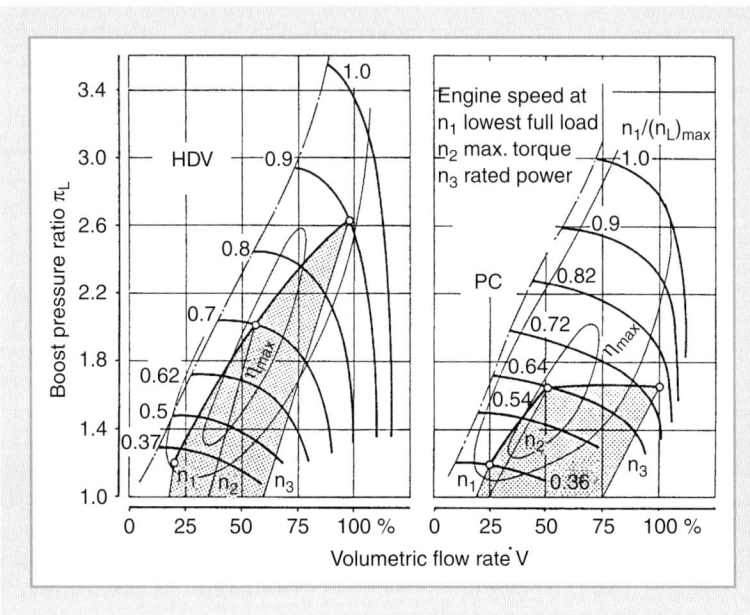

Fig. 2-24
Full load engine operating line in a supercharger map for a commercial vehicle (HDV) and a car engine (PC)

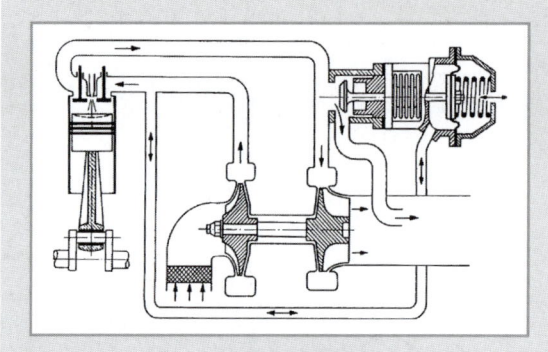

Fig. 2-25 Boost pressure limiting with a wastegate [2-5]

exhaust is conducted around the turbine when the boost pressure has reached the allowable upper limit. Prompted by the special requirements of gasoline engines in this respect (quantity control), gasoline and diesel engines now predominantly have electronic boost pressure control systems. A controller compares the actual value of the boost pressure with the operating point-dependent nominal value stored in the engine control unit and adjusts it, in this case by electro-pneumatic control of the wastegate (*closed loop wastegate*). The lower the boost pressure adjusted at part load is than the boost pressure at full load, the lower the exhaust pressure before the turbine is too. Not least, this is reflected in correspondingly lower engine fuel consumption.

Variable turbine geometry (VTG) can also obtain a convergence of the turbocharger's supply of air with the engine's operating point-dependent air requirement, which continues to diverge when an uncontrolled turbocharger is employed, especially as engine speed decreases. The turbocharger turbine is selected with a cross section so (large) that the desired boost pressure is attained in the engine's rated power point. As engine speed decreases, a drop in boost pressure is counteracted by constricting the inflow cross section to the turbine rotor, preferably by adjustable turbine guide blades (Fig. 2-26). This generates an increased buildup of exhaust before the turbine – higher values for p_3 and T_3 in Eq. (2-29) – and thus higher boost pressure. However, the constriction of the cross section of the turbine's guide vane system may not be allowed to excessively degrade the flow conditions in the rotor. Otherwise, an excessive decrease of η_T and, further, of η_{TL} (see Eqs. (2-28) and (20-29)) makes too little of the supply of exhaust gas energy useable to increase boost pressure.

The mechanically more dependable solution of partially covering the turbine rotor's inflow cross section with a sleeve able to slide in the axial direction of the rotor (see Fig. 2-27) must be classified as poorer in terms of efficiency because a reduction of the turbine cross section is connected with partial pressurization of the turbine rotor.

Fig. 2-26 Vehicle turbocharger with variable turbine geometry (adjustable guide blades) (Source: BorgWarner Turbo Systems)

Turbochargers with adjustable turbine guide blades, which allow adjusting the turbine cross section in a range between 100% and approximately 70% of the full cross section, were already developed for large diesel engines in the 1970s [2-7]. However, they are no longer used in production since the generally high level of turbocharger efficiencies now attained makes other solutions preferable for large diesel engines [2-8].

Vehicle diesel engines on the other hand are predominantly equipped with VTG turbochargers. In addition to the pneumatic actuators that were solely common initially, electric actuators are now increasingly being used because their adjustment speed is higher by a factor of 10 [2-9] and the control characteristic is improved significantly.

Sequential turbocharging provides another solution, i.e. adapting the turbocharger turbine cross section to the engine's demand [2-10]. To this end, an engine is equipped with several turbochargers connected in parallel, each of which can be engaged or disengaged by flaps on the air and exhaust side so that each "correct" turbine cross section is available to the engine as a function of the operating point. Since this enables operating every single turbocharger relatively closely to its best point, relatively good supercharging efficiency of the engine is produced even in operating ranges (low load and low speed) in which a single large turbocharger operates far from its best point.

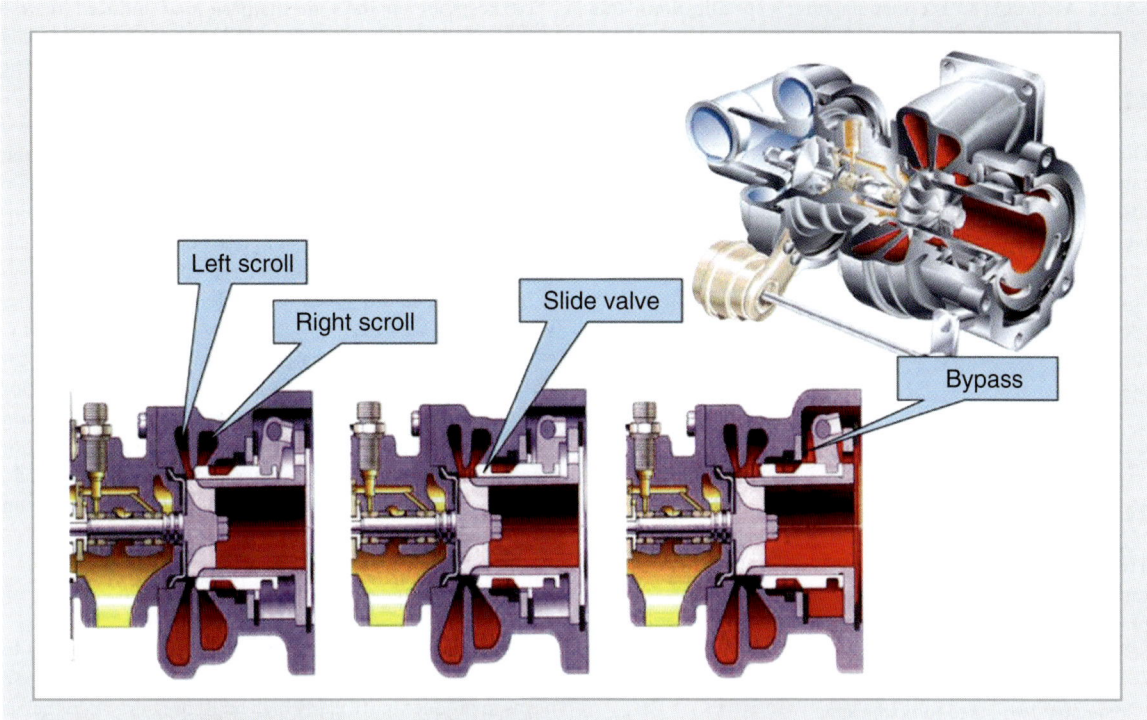

Fig. 2-27 VST turbocharger with a variable slider ring turbine (Source: BorgWarner Turbo Systems)

The advantages of sequential turbocharging become particularly apparent during engine acceleration. At the start of the acceleration stage, the entire available exhaust energy is supplied to only one single turbocharger of a total of four for example. Consequently, its speed revs up very high, not least because its rotor assembly also has a lower mass moment of inertia than a single large turbocharger. The faster buildup of boost pressure connected with this allows increasing the amount of fuel injected more rapidly.

Since the engine revs up to high speed faster and delivers an increased exhaust energy rate, the other turbochargers can be engaged sequentially and the operating point desired at a higher level is obtained far faster than when a single (large) turbocharger is employed. A significant disadvantage of sequential turbocharging is the supercharging system's more complex configuration (several turbochargers, choke and exhaust flaps and their control) and the increased capital expenditure connected with this.

Sequential turbocharging has been effectively implemented in high-speed high performance diesel engines with two-stage supercharging for many years (see Sect. 18.4). It is now also being applied to vehicle engines, albeit usually only with two turbochargers [2-11]. When two differently sized turbochargers are used, engaging and disengaging them may give an engine a total of three differently sized overall turbine cross sections.

The VMP (variable multi-pulse) system produced by MaK (today Caterpillar Engines) provides another interesting method to produce a variable turbine cross section for medium speed engines for marine applications. However, it is no longer used in the current engine program [2-12].

In the power range below 75% of the rated power, a slide valve called a *variator* closes part of the cross section of the axial turbine's nozzle ring (Fig. 2-28). Thus, the turbine builds up higher exhaust pressure and temperature and, consequently, higher charge air pressure is available for the engine

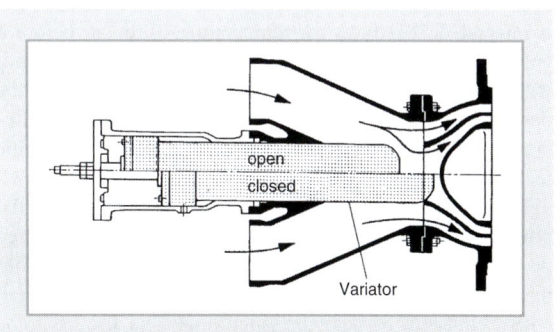

Fig. 2-28 Adjustment of an exhaust gas turbocharger turbine cross section based on MaK's variable multi-pulse (VMP) method

[2-12]. According to the manufacturer's specifications, this can cut fuel consumption by up to 10 g/kWh.

2.2.3.5 Downsizing

Downsizing refers to concepts intended to produce a desired rated engine power through a smaller engine (smaller total displacement, possibly also a smaller number of cylinders), which is correspondingly highly supercharged. The smaller engine's lower friction increases its mechanical and thus also effective efficiency. In addition, the engine's weight is lowered. Therefore, downsizing has become an important feature of design, above all in the development of vehicle engines.

2.2.4 Special Forms of Exhaust Gas Turbocharging

2.2.4.1 Two-Stage Turbocharging

Two-stage turbocharging connects two freewheeling exhaust gas turbochargers in a series, the one designated as a low pressure turbocharger and the other as a high pressure turbocharger. There are two variants;
- unregulated two-stage turbocharging and
- regulated two-stage turbocharging.

Unregulated Two-Stage Turbocharging

Boost pressure ratios of up to six and above are a basic requirement for unregulated two-stage turbocharging when brake mean effective pressures in the range of p_e = 30 bar and greater have to be generated [2-10].

If two superchargers (compressors) are connected in a series and, for example, each of them builds up a pressure ratio of π_L = 2.5 at an isentropic supercharger efficiency of $\eta_s\text{L}$ = 80%, then both supercharging stages attain a total boost pressure ratio of $\pi_{L\text{ges}}$ = 6.25 with an isentropic efficiency of at least still 77.5%. In fact, a single radial compressor stage could possibly still produce a boost pressure ratio of 6.25, yet only with significantly lower efficiency. This advantage of two-stage compression is further heightened when an intercooler is integrated. It reduces the temperature of the air before it enters the high pressure supercharger and thus, in accordance with Eq. (2-22), the compressor power to be applied for the desired pressure ratio. Each of the two effects positively influences supercharging efficiency and thus an engine's specific fuel consumption.

The more the instantaneous boost pressure (at partial power) deviates from the maximum boost pressure (at rated power), the smaller the positive effect of intercooling becomes on supercharging efficiency and thus directly on overall engine efficiency because correspondingly less also has to be cooled then.

To keep an engine's mechanical load induced by the maximal cylinder pressure (firing pressure) under control at such high supercharging rates as two-stage turbocharging permits – values already run up to $p_{Z\text{max}} \approx 200$ bar [2-13] – the compression ratio ε is lowered significantly compared with values common for one-stage turbocharging. However, a drop of ε as a single factor always represents a deterioration of engine efficiency. This is a basic reason for the fundamentally higher specific fuel consumptions of high-performance, high-speed two-stage turbocharged engines (with $p_e \approx$ 30 bar) over medium-speed one-stage turbocharged engines (with $p_e \approx 21...24$ bar).

Thus, two-stage turbocharging – applications were prototyped in the 1970s – has not been able to establish itself for medium-speed diesel engines. On the other hand, maximum values for power density as produced by two-stage turbocharged high-speed diesel engines, e.g. speedboat engines, are given top priority and low specific fuel consumption only secondary priority.

Regulated Two-Stage Turbocharging

Regulated two-stage turbocharging has come to be applied to commercial vehicle and car diesel engines not necessarily as a means to produce particularly high boost pressure but rather as an alternative to twin turbocharger sequential turbocharging.

The basic difference to unregulated two-stage turbocharging is one controllable bypass apiece around the high pressure turbine and the high pressure compressor (see Fig. 2-29) as well as a wastegate in the LP turbine in applications for cars. Vehicle engines also already need maximum boost pressure in the lower engine speed range for high acceleration power. Simple exhaust gas turbocharging (without boost pressure control) cannot produce this because the engine's exhaust mass flow is low. Both bypass valves are kept closed in this phase of operation. This causes the entire exhaust mass flow and the entire exhaust energy flow to be conducted to the (smaller) HP turbine, which, similar to a narrowly set VTG turbine, rotates very quickly so that the HP compressor generates the desired high boost pressure. The downstream LP turbine obtains merely a small remainder of exhaust energy in this phase of engine operation. Thus, it rotates correspondingly slowly and, although it forces the entire air mass flow through the LP compressor, only builds up a very small pressure ratio.

As the working gas mass flow through the engine and the exhaust energy flow from the engine increase as engine speed increases and load rises, the two bypass valves open ever further until the HP and LP turbochargers operate as a type of mixed form of series and parallel connections. Engines equipped with such a supercharging system, exhibit tremendously good response [2-14].

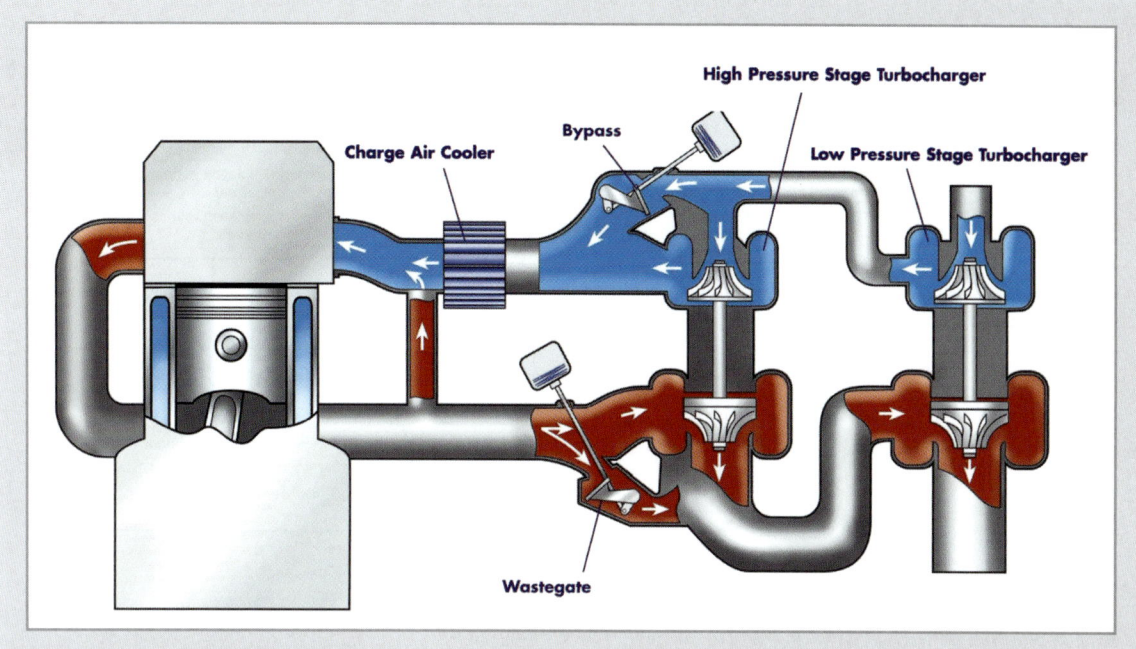

Fig. 2-29 Controlled two-stage supercharging, block diagram (Source: BorgWarner Turbo Systems)

2.2.4.2 Miller System

The prerequisite for the Miller system [2-15] is a four-stroke engine with exhaust gas turbocharging and intercooling in which the timing of "intake closes" (IC) can be adjusted in operation. The goal is to reach a lower cylinder temperature at a desired cylinder pressure at the start of compression than the given intercooler usually can in the normal case. To this end, the turbocharger has to be tuned so that, corresponding to the early closing of the intake valve (yet before BDC) and the continued expansion of the cylinder charge afterward, it delivers such high boost pressure that the initial compression pressure stipulated for the normal case is attained nevertheless. The attendant expansion cooling of the cylinder gas causes its temperature in BDC to drop to values below what the temperature would be in the normal case, i.e. when "intake closes" (IC) later.

This method allows shifting the detonation limit in supercharged gasoline engines to higher engine loads. The methods of *early intake closing (EIC)* or *late intake closing (LIC)* already being applied to gasoline engines with and without supercharging can be used to largely put the throttle "out of work". This is referred to as *dethrottling* a gasoline engine. In a diesel engine, the larger cylinder charge (than in the normal process) at the start of compression obtainable by the Miller system at equal cylinder pressure can either enhance performance or result in operation with a larger air/fuel ratio. Since the cylinder pressure level is lower, it can also be used to cut NO_x emission.

However, any assessment of the Miller system in its pure form must consider that the higher boost pressure generated by the compressor than in the normal process is obtained by the engine's increased work of expulsion. This alone affects engine efficiency adversely.

The Miller system could also open the possibility of controlling an engine's boost pressure with a fixed geometry turbocharger. To this end, the turbocharger and, in particular, its turbine would have to be adapted to the engine (*turbocharger matching*) so that it already produces the requisite boost pressure for the desired full load characteristic in the low engine speed range, the engine being operated here with normal IC (after BDC). This turbocharger design's boost pressure and, consequently, cylinder pressure level is kept from growing too large in the upper engine speed range and thus at high flow rates by closing the intake valve correspondingly early (before BDC) with a positive secondary effect of in-cylinder expansion cooling.

2.2.4.3 Electrically Assisted Supercharging

Electrical power temporarily taken from a vehicle's electrical system can also be used to obtain a faster buildup of boost pressure, which turbocharged vehicle engines need for better acceleration performance, than what the turbocharger is capable of generating on its own. So far, this has been prototyped in two forms, namely the eBooster and the electrically assisted turbocharger (EAT).

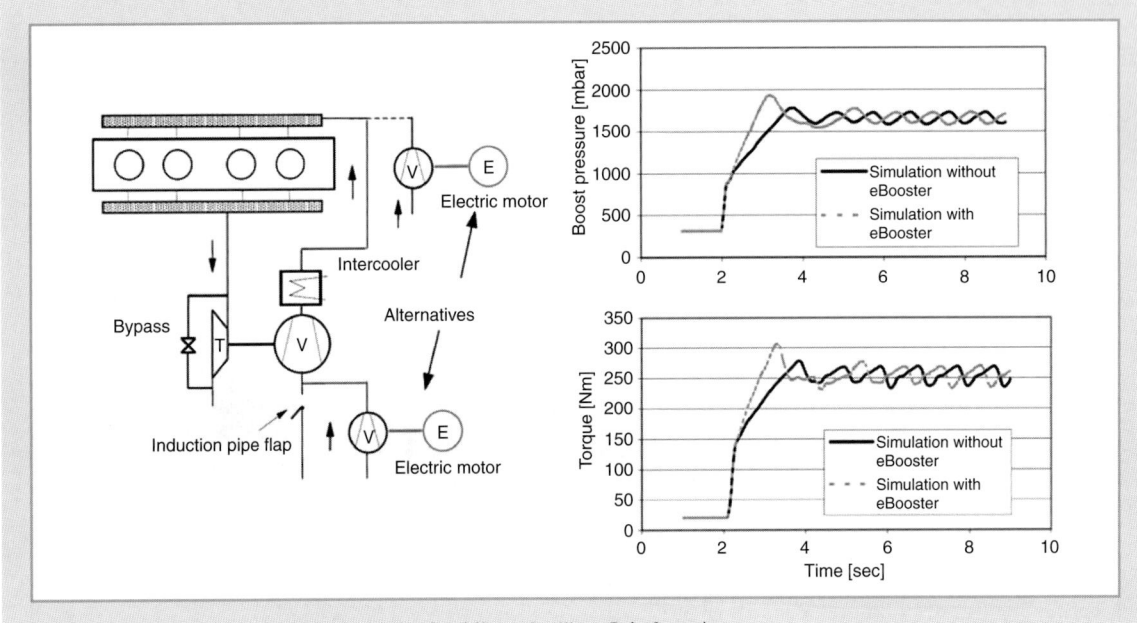

Fig. 2-30 Exhaust gas turbocharger with eBooster, based on [2-16] (Source: BorgWarner Turbo Systems)

eBooster

Usually placed before the turbocharger compressor, an electrically powered turbo compressor (radial compressor) is serially connected to the exhaust gas turbocharger's compressor. This additional electrically powered compressor is only activated at the start of the engine acceleration stage though. The rest of the time, it remains switched off and a bypass around the additional compressor feeds the intake air directly to the turbocharger compressor (see Fig. 2-30). The limits on this process for a 12 V electrical system are the maximum potential propulsion power and the maximum electrical power extractable per acceleration sequence for the given battery capacity.

EAT

A boost pressure deficit in the lower engine speed range can also be reduced by using a separate electric drive to bring the turbocharger rotor to a higher speed than can be done by the instantaneous exhaust energy flow from the engine alone. To this end, the electric motor's rotor is integrated in the turbocharger rotor in addition to the respective compressor and turbine rotors (see Fig. 2-31). However, this system has only been possible since the existence of electric motors with speed ranges that extend into those of vehicle engine turbochargers, i.e. at least up to 100,000 rpm.

The same upper power limits provided by the electrical system that apply to the eBooster apply to EAT. However, EAT has a crucial disadvantage over the eBooster, namely it increases the turbocharger rotor's mass moment of inertia. While the exhaust energy flow alone might indeed accelerate the basic turbocharger (without an electric motor on the rotor) just enough in a particular situation of engine operation, this would no longer be possible (without electrical assistance) if the electric motor's rotors were additionally present on the rotor (see the influence of Θ in Eq. (2-34)).

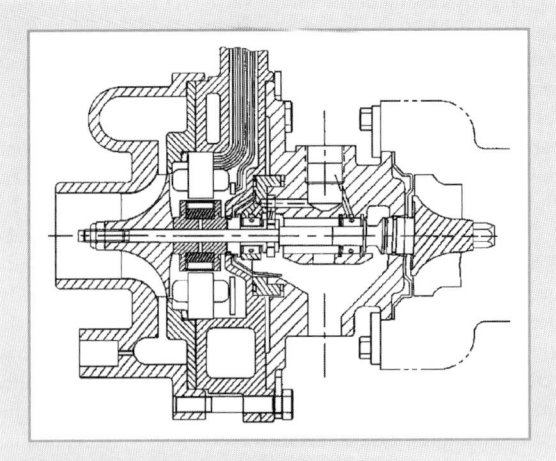

Fig. 2-31 Electrically assisted turbocharger (EAT) (Source: BorgWarner Turbo Systems)

The electric motor integrated in the turbocharger rotor could however also be utilized as a generator if the intention were to convert any excess exhaust energy flow not needed for supercharging into electrical power and resupply it to the electrical system (see Sect. 2.2.4.4 and [2-17]).

2.2.4.4 Turbocompounding

Turbocompounding refers to the operation of an internal combustion engine together with one or more gas turbines, the brake horsepower being drawn from not only the engine but also at least one of the turbines. Of the types of circuits [2-18] presented in Fig. 2-32, variant 4 has attained practical significance for large diesel engines and variant 1 for commercial vehicle diesel engines.

A turbocharged engine constitutes the initial version in both cases. Once again, although around for decades, this method [2-5, 2-19] has only been being implemented for the few years since turbocharger efficiencies increased tremendously, an absolute prerequisite for turbocompounding.

According to [2-18], a commercial vehicle engine can cut fuel consumption by up to 5% by utilizing a downstream power turbine (variant 1 in Fig. 2-32) with a fixed gear ratio. Moreover, the boost pressure and thus the air/fuel ratio are higher toward lower load and speed than in the basic engine. Since an engine normally operates with a relatively small air/fuel ratio at lower speed and full load, this is a positive side effect of turbocompounding. However, the power turbine causes an increase in specific fuel consumption below brake mean effective pressures of 5 bar because – intensified by the fixed gear ratio – it runs far afield from its design point and its efficiency is commensurately low. However, simulations of consumption per distance in [2-18] have demonstrated that the improvement obtainable by switching off the power turbine in the lower load range is not worth the effort.

Better acceleration performance than that of the basic engine because the turbocharger turbine is narrower is another positive effect of a downstream power turbine.

Turbocompounding with large four-stroke and two-stroke diesel engines (Fig. 2-32, variant 4) processes up to 12.5% of the exhaust gas flow in correspondingly smaller power turbines connected in parallel. (Since turbocharger efficiency is a function of size, variant 4 is out of the question for commercial vehicle engines.)

Thus, medium speed constant pressure turbocharged diesel engines can achieve an additional brake horsepower of approximately 4% while lowering specific fuel consumption by 4.5 g/kWh, which is still approximately 2.5 g/kWh for 40% propeller power [2-20]. Even more significant cuts in fuel consumption can be achieved when the power turbine is switched off in the (propeller) power range below 75% because the entire exhaust energy flow supplied by the engine is then supplied to the turbocharger turbine and a higher

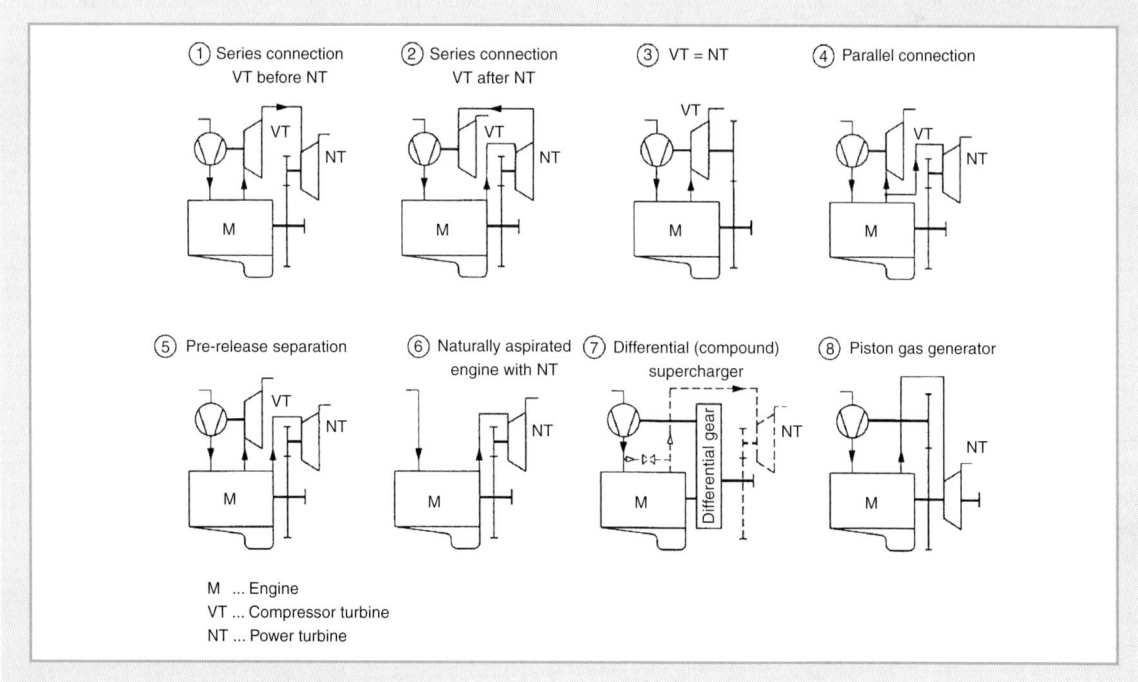

Fig. 2-32 Types of circuits for compound operation [2-18]

boost pressure is generated, especially since a turbocompound engine's turbocharger turbine must generally be designed more narrowly than that of the basic engine (without turbocompounding). In principle, a power turbine ought to remain switched off below 40% engine power.

Low speed two-stroke diesel engines achieve comparable results [2-21]. Higher turbocharger efficiency is the basic prerequisite for an appreciable increase of overall engine efficiency by turbocompounding. Today, large diesel engines' turbochargers can achieve values of over 70% [2-22, 2-23].

2.2.4.5 Turbobrake

A *turbobrake* is not a supercharging system in the narrower sense. Rather, it additionally utilizes a commercial vehicle engine's VTG turbocharger to increase engine braking power. *Mercedes Benz* and *Iveco* were the first commercial vehicle manufacturers to mass produce this system for heavy duty commercial vehicles [2-24].

This system is preferably implemented on the basis of turbochargers with turbines with a twin-scroll inflow housing. A slide valve that can be shifted toward the rotor shaft can close a maximum of one of the two scrolls at the inlet to the turbine rotor (see Fig. 2-27). Should the braking effect be activated – the engine operating in overrun condition and without combustion – the slide valve closes one of the two turbine scrolls. With the relatively high mass flow from the engine, the then narrower turbine produces high turbocharger speed and correspondingly high boost pressure. This demands correspondingly high compression output from the engine during the compression stroke, which acts as braking power. The exhaust valve already opens at the end of the compression stroke so that the compressed cylinder gas is not fully discharged to the piston as (positive) work of expansion, which would counteract the braking effect. This produces a negative high pressure loop in the engine's indicator diagram. A blow-off mechanism before the turbine, i.e. a wastegate, ensures that excessively high boost pressure does not overload the engine during either braking or operation.

A turbobrake can be used to produce significantly higher engine braking power than the particular rated engine power.

2.2.5 Pressure-Wave Supercharging (Comprex)

Just like an exhaust gas turbocharger, the pressure-wave supercharger known by its brand name Comprex also uses the exhaust energy supplied by the engine to generate boost pressure but, unlike an exhaust gas turbocharger, by directly transmitting energy from the exhaust gas to the air being compressed. Building upon Burghard's (1912) and Seippel's patents (1940) and Berchtold's seminal work [2-25], BBC (today ABB) developed this supercharging unit, primarily designed for vehicle diesel engines, for mass production in the 1960s and 1970s. Its functional principle is fascinating.

The pressure-wave supercharger's particular strength is its capability derived from its principle to promptly convert an exhaust energy rate abruptly increased by the engine into increased boost pressure. In such a case, a turbocharger must first overcome the mass inertia of the turbocharger's rotor assembly (turbo lag). This feature made the pressure-wave supercharger appear particularly predestined for vehicle engines and engines to which high load steps are applied [2-26]. Turbochargers have now caught up to Comprex though, at least in terms of acceleration performance. Moreover, they cost less to manufacture, weigh less and furnish greater freedom of installation in an engine. Hence, Comprex is no longer installed as standard.

2.2.6 Mechanical Supercharging

As explained in Sect. 2.2.2, common mechanical supercharging already has a relatively high boost pressure in the lower speed range, as a result of the full load torque shifts roughly parallel to higher values compared with a basic naturally aspirated engine (Fig. 2-33). The coupling of boost pressure and engine speed causes this to happen not only in steady-state but also in dynamic operation (acceleration). As much as this speaks for applying mechanical supercharging in car diesel engines too, its higher specific fuel consumption than exhaust gas turbocharging's argues against it.

The incorporation of a boost pressure control system is required to minimize this disadvantage, i.e. the supercharger may only be permitted to generate as much boost pressure and consume the corresponding propulsion power at any time as the engine actually requires as a function of the operating point. In addition to a variable ratio drive supercharger, a controllable bypass in the supercharger can also do this. A magnetic clutch would only allow switching in the supercharger when the "naturally aspirated engine's supply of air" to the engine is depleted.

An expedient application of mechanical supercharging for diesel engines might be a mechanically driven positive-displacement supercharger placed in a commercial vehicle engine's air path before the turbocharger compressor, which a (magnetic) clutch only switches in in each acceleration phase. This corresponds to an eBooster's function (see Sect. 2.2.4.3).

2.3 Programmed Gas Exchange Simulation

In addition to the simulation of in-cylinder changes of state, the simulation of changes of state in the intake and exhaust lines, so-called gas exchange simulation, constitutes the heart of engine process simulation (see Sect. 1.3).

Disregarding methods of simulation that take an extremely simplified premise of a spatiotemporal constant gaseous state in the intake and exhaust line as their point of departure, the computer programs in use can be divided into two groups.

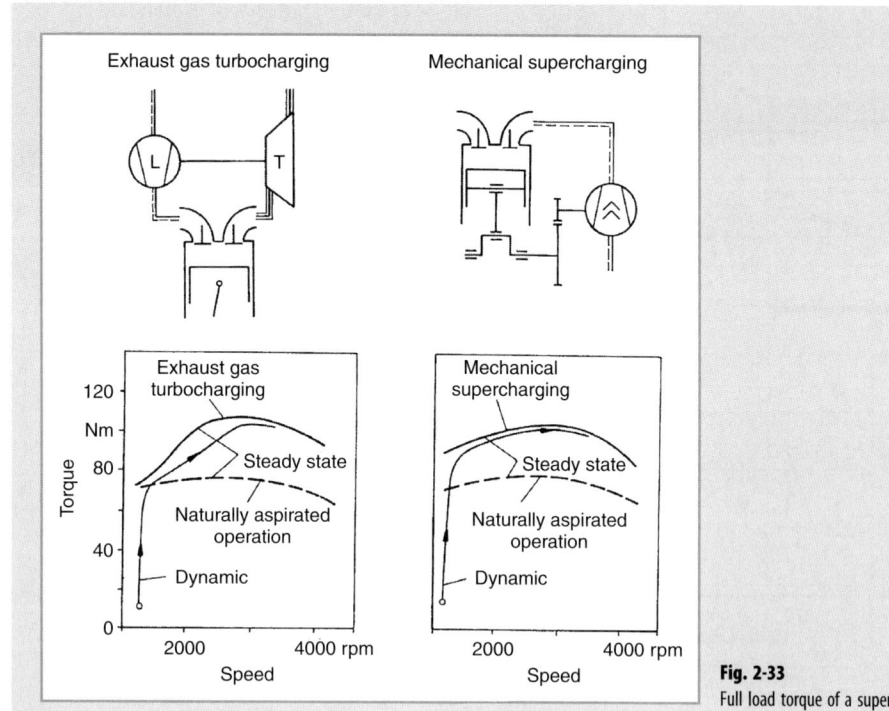

Fig. 2-33
Full load torque of a supercharged vehicle diesel engine

Programs based on the quasi steady-state procedure of the *filling and emptying method* only simulate a temporal characteristic of the changes of state in the gas exchange system. Hence, they are also regarded as *zero-dimensional* methods. To this end, a turbocharged multi-cylinder engine's exhaust pipe, for example, is regarded as a reservoir of constant volume intermittently charged with exhaust gas according to the connected cylinder's firing sequence and continuously discharged by the exhaust gas turbine. Basically, applying the equation of mass and energy balance and the equation of state of a gas to this control volume allows simulating the characteristics of the pressure and temperature and thus, among other things, the supply of exhaust gas energy to the turbocharger turbine as well. Their maps are used to enter a turbocharger's turbine and compressor into the simulation as boundary conditions. The comparison of simulation and measurements in Fig. 2-34 verifies that thoroughly realistic results are obtainable.

The higher the level of speed and the longer and narrower the lines of the gas exchange system of the engine being simulated, the less the filling and emptying method's basic conditions are fulfilled. The local dependence of the state variables in the lines of the gas exchange system must also be allowed for by incorporating appropriate methods of unsteady simulation, e.g. *the method of characteristics*. The flow in the lines of the gas exchange system is treated as a *one-dimensional* unsteady pipe flow [2-28, 2-29].

Pucher [2-27] compares the application of the quasi steady-state filling and emptying method and the method of characteristics to different four-stroke diesel engines (Fig. 2-35): The method of characteristics can reproduce the intensely gas-dynamic exhaust gas pressure curve for a high-speed high performance diesel engine with symmetrical double-pulse turbocharging relatively well; the filling and emptying method can only reproduce its general tendency.

Since its beginnings in the 1960s and as computer technology rapidly evolved, engine process simulation has become an indispensible tool of engine development. In addition to simulations of the changes of state in the cylinders and in the lines of gas exchange, it can also incorporate modeling of a driven vehicle (vehicle longitudinal dynamics) and the driver. Thus, for instance, operating strategies can be developed to use a vehicle engine with controlled two-stage turbocharging in a very particular type of vehicle [2-30]. Figure 2-36 presents the engine speed and boost pressure characteristics of a two-stage turbocharged car diesel engine at full load acceleration from 0 to 100 km/h. In particular, the speed characteristic reveals the time of a particular gear shift accompanied by a steep drop in speed. The real-time capability of engine process simulation that has been attained [2-31] allows applying it as HIL (*hardware in the loop*) among other things.

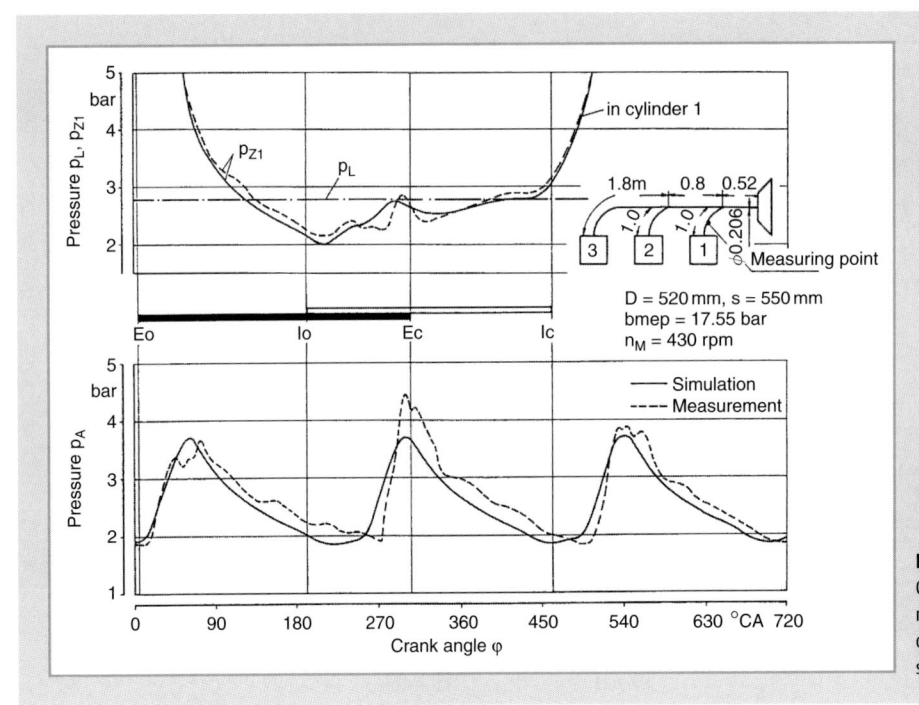

Fig. 2-34 Comparison of simulation and measurement of the charging and discharging method for a medium-speed diesel engine [2-27]

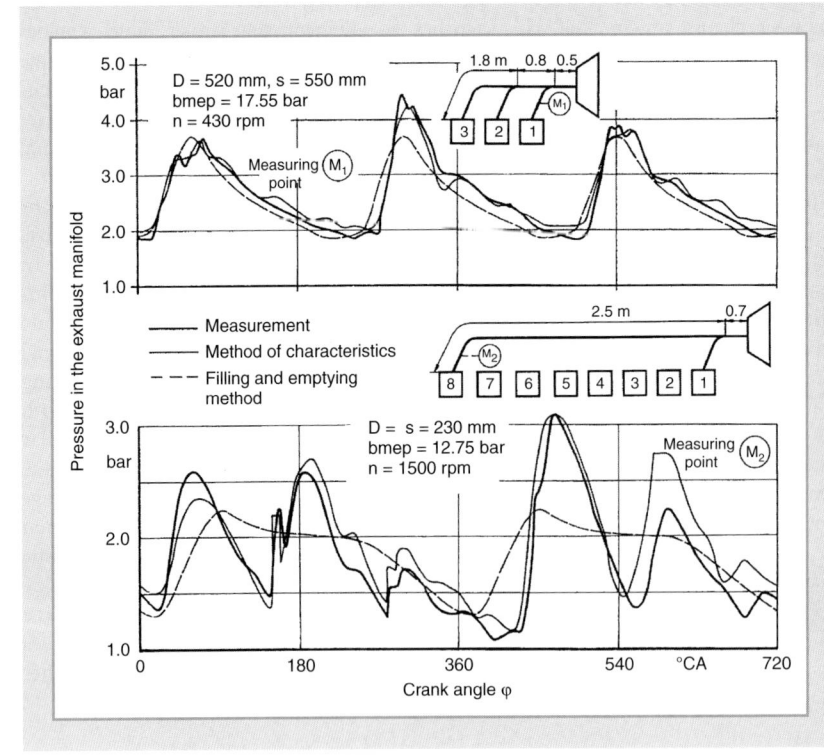

Fig. 2-35 Simulation of quasi steady-state and transient gas exchange compared to a measurement

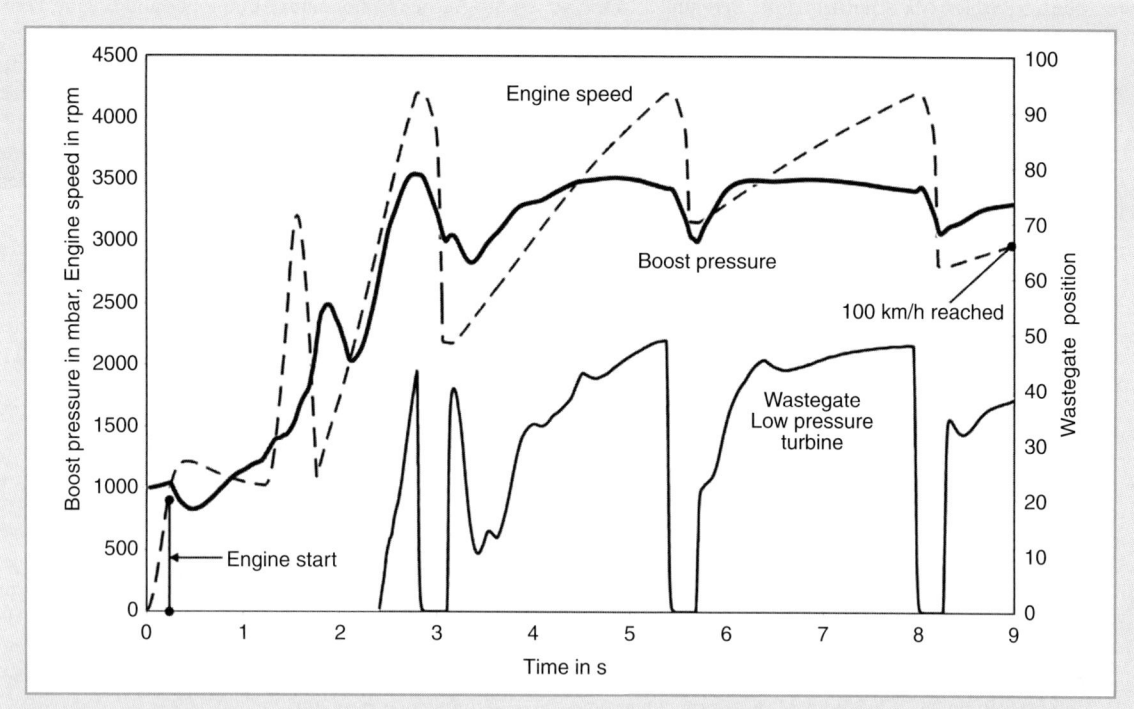

Fig. 2-36 Car diesel engine with controlled two-stage supercharging, boost pressure and speed increase at full load acceleration from 0 to 100 km/h, simulated with THEMOS®

Literature

2-1 Bensinger, W.-D.: Die Steuerung des Gaswechsels in schnellaufenden Verbrennungsmotoren. Berlin/Göttingen/Heidelberg: Springer (1955)

2-2 Pischinger, F.: Entwicklungsarbeiten an einem Verbrennungssystem für Fahrzeugdieselmotoren. ATZ 65 (1963) 1, pp. 11–16

2-3 Petermann, H.: Einführung in die Strömungsmaschinen. Berlin/Heidelberg/New York: Springer (1974)

2-4 DRP Nr. 568855

2-5 Zinner, K.: Aufladung von Verbrennungsmotoren. 3rd Ed. Berlin/Heidelberg/New York: Springer (1985)

2-6 Zapf, H.; Pucher, H.: Abgasenergie-Transport und Nutzung für Stoss- und Stau-Aufladung. HANSA Schiffahrt – Schiffbau – Hafen 114 (1977) 14, pp. 1321–1326

2-7 Bozung, H.-G.: Die M.A.N.-Turboladerbaureihe NA und NA-VP für ein- und zweistufige Aufladung. MTZ 41 (1980) 4, pp. 125–133

2-8 Holland, P.; Wachtmeister, G.; Eilts, P.: Untersuchungen zum Einfluss des Aufladesystems auf das dynamische Verhalten mittelschnelllaufender Viertakt-Dieselmotoren. Proceedings of the 8th Aufladetechnische Konferenz Dresden 2002, pp. 31–40

2-9 Anisits, F. et al.: Der erste Achtzylinder-Dieselmotor mit Direkteinspritzung von BMW. MTZ 60 (1999) 6, pp. 362–371

2-10 Deutschmann, H.: Neue Verfahren für Dieselmotoren zur Mitteldrucksteigerung auf 30 bar und zur optimalen Nutzung alternativer Kraftstoffe. In: Pucher, H. et al.: Aufladung von Verbrennungsmotoren. Sindelfingen: Expert (1985)

2-11 Borila, Y.G.: Sequential Turbocharging. Automotive Engineering, Vol. 34 (1986) 11, pp. 39–44

2-12 Zigan, D.; Heintze, W.: Das VMP-Verfahren, eine neue Aufladetechnik für hohe Drehmomentanforderung. 5th Aufladetechnische Konferenz Augsburg October 11–12, 1993

2-13 Rudert, W.; Wolters, G.-M.: Baureihe 595 – Die neue Motorengeneration von MTU; Part 2. MTZ 52 (1991) 11, pp. 538–544

2-14 Stütz, W.; Staub, P.; Mayr, K.; Neuhauser, W.: Neues 2-stufiges Aufladekonzept für PKW-Dieselmotoren. Proceedings of the 9th Aufladetechnische Konferenz Dresden 2004, pp. 211–228

2-15 Miller, R.; Liebherr, H.U.: The Miller Supercharging System for Diesel and Gas Engines Operating Conditions. CIMAC Congress 1957 Zurich, pp. 787–803

2-16 Münz, S.; Schier, M.; Schmalzl, H.-P.; Bertolini, T.: Der eBooster – Konzeption und Leistungsvermögen eines fortgeschrittenen elektrischen Aufladesystems. Firmenschrift der 3 K-Warner Turbosystems GmbH (2002) 9

2-17 Hopmann, U.: Ein elektrisches Turbocompound Konzept für NFZ Dieselmotoren. Proceedings of the 9th Aufladetechnische Konferenz Dresden 2004, pp. 77–87

2-18 Woschni, G.; Bergbauer, F.: Verbesserung von Kraftstoffverbrauch und Betriebsverhalten von Verbrennungsmotoren durch Turbocompounding. MTZ 51 (1990) 3, pp. 108–116

2-19 Khanna, Y.K.: Untersuchung der Verbund- und Treibgasanlagen mit hochaufgeladenen Viertaktdieselmotoren. MTZ 21 (1960) 1, pp. 8–16, 3, and 73–80

2-20 Pucher, H.: Analyse und Grenzen der Kraftstoffverbrauchsverbesserung bei Schiffsdieselmotoren im Turbocompoundbetrieb. Jahrbuch der Schiffbautechnischen Gesellschaft Vol. 82. Berlin/Heidelberg/New York/London: Springer (1988)

2-21 Meier, E.: Turbocharging Large Diesel Engines – State of the Art and Future Trends. Brochure from ABB Turbo Systems Ltd. Baden (Switzerland) (1994)

2-22 Appel, M.: MAN B&W Abgasturbolader und Nutzturbinen mit hohen Wirkungsgraden. MTZ 50 (1989) 11, pp. 510–517

2-23 Nissen, M.; Rupp, M.; Widenhorn, M.: Energienutzung von Dieselabgasen zur Erzeugung elektrischer Bordnetzenergie mit einer Nutzturbinen-Generator-Einheit. HANSA Schiffahrt – Schiffbau – Hafen 129 (1992) 11, pp. 1282–1287

2-24 Flotho, A.; Zima, R.; Schmidt, E.: Moderne Motorbremssysteme für Nutzfahrzeuge. Proceedings of the 8th Aachen Colloquium Automobile and Engine Technology 4.-06.10.1999, pp. 321–336

2-25 Berchtold, M.: Druckwellenaufladung für kleine Fahrzeug-Dieselmotoren. Schweizerische Bauzeitung 79 (1961) 46, pp. 801–809

2-26 BBC Brown Boveri, Baden (Switzerland): Erdbewegungsmaschinen mit Comprex-Druckwellenlader. Comprex Bulletin 7 (1980) 1

2-27 Pucher, H.: Ein Rechenprogramm zum instationären Ladungswechsel von Dieselmotoren. MTZ 38 (1977) 7/8, pp. 333–335

2-28 Seifert, H.: 20 Jahre erfolgreiche Entwicklung des Programmsystems PROMO. MTZ 51 (1990) 11, pp. 478–488

2-29 N.N.: GT-Power – User's Manual and Tutorial, GT-Suite TM Version 6.1, Gamma Technologies Inc. Westmont IL (2004)

2-30 Birkner, C.; Jung, C.; Nickel, J.; Offer, T.; Rüden, K.v.: Durchgängiger Einsatz der Simulation beim modellbasierten Entwicklungsprozess am Beispiel des Ladungswechselsystems: Von der Bauteilauslegung bis zur Kalibrierung der Regelalgorithmen. In: Pucher, H.; Kahrstedt, J. (Eds.): Motorprozesssimulation und Aufladung. Haus der Technik Fachbuch Bd. 54, Renningen: expert 2005, pp. 202–220

2-31 Friedrich, I.; Pucher, H.: Echtzeit-DVA – Grundlage der Regelung künftiger Verbrennungsmotoren. Tagungsband der MTZ-Konferenz – Motor 2006. (June 1-2, 2006, Stuttgart) Der Antrieb von morgen, Wiesbaden: Vieweg Verlag 2006, pp. 215–224

3 Diesel Engine Combustion

Klaus B. Binder

3.1 Mixture Formation and Combustion

3.1.1 Process Characteristics

The preferred drive engines for motor vehicles are based on combustion engines. They utilize the oxygen in the combustion air to convert the fuel-based chemical energy that predominantly consists of hydrocarbons into heat, which in turn is transferred to the engine's working medium. The pressure in the working medium rises and, by exploiting the expansion, can be converted into piston motion and thus into mechanical work.

The replacement of the working medium, also designated as a working gas, after expansion and combustion takes place inside the engine's combustion chamber is referred to as "open process control with internal combustion" [3-1]. It applies to both gasoline engines and diesel engines. By contrast, a Stirling engine, for example, is described as an engine with closed process control and external combustion.

In conventional gasoline engines, the air/fuel mixture forms in the intake manifold. A predominantly homogeneous mixture forms during the intake and compression cycle, which is ignited by a spark plug. This combustion system is also characterized by "external mixture formation", a homogeneous mixture and spark ignition. Starting at the spark plug, energy is released as the flame propagates and is therefore proportional to the surface area of the flame front. The flame speed depends on the fuel, the mixture temperature and the air/fuel ratio. The rate of combustion is additionally influenced by the surface area of the flame front. "Flame folding" induced by turbulences in the mixture causes it to increase with the engine speed. Flows of the mixture caused by the intake process and compression as well as combustion itself are a significant factor influencing flame folding. The fuel must be ignition resistant (detonation resistant) to prevent auto-ignition or premature ignition. The compression ratio is limited by "knocking" combustion or premature ignition. In knocking combustion, conditions for ignition are obtained in the entire mixture not yet reached by the flame, the so-called "end gas". The highly compressed and therefore energy rich mixture burns nearly isochronously without controlled flame propagation. This produces steep pressure gradients with characteristic pressure oscillations and causes very high local thermal and mechanical stress of components. Longer operation during knocking combustion results in complete engine failure and must therefore be strictly avoided. Limited compression, necessary load control systems (quantity or throttle control) and limited supercharging capability diminish the efficiency of the process with external mixture formation and spark ignition. However, this process does not have any fuel induced particulate emission since no regions with a rich mixture appear in the combustion chamber because operation is homogeneous when $\lambda = 1$. Modern gasoline engines also operate with direct fuel injection and, depending on the injection timing, may form a homogeneous or inhomogeneous mixture. In diesel engines, this is referred to as "internal mixture formation".

Air rather than a mixture is compressed in a diesel engine. The fuel is highly compressed shortly before top dead center and hot combustion air is injected in with it. Thus, mixture formation elapses in the engine's combustion chamber extremely quickly and ignition occurs without any external ignition source solely by transferring the heat from the compressed air to the fuel. Therefore, a diesel engine is an engine with "internal mixture formation" and "auto-ignition". Highly ignitable fuels must be used and the necessary temperatures guaranteed to ensure ignition is initiated. The latter is effected by high compression (compression ratio $12 < \varepsilon < 21$) and, if necessary, by additionally heating the air (e.g. with a glow plug). Ignition problems can occur especially when starting an engine. The low starting speed eliminates any boosting effects in the intake system. Therefore, in its compression phase, the piston forces already aspirated air back into the intake manifold. This process only concludes when the intake valves close and compression may only begin at this late time. As a result, the effective compression ratio and thus the compression temperature drop starkly. The increased loss of heat from the working gas air to the cold walls of the combustion chamber during a cold start intensifies this problem with starting.

K.B. Binder (✉)
Deizisau, Germany
e-mail: klaus.b.binder@t-online.de

The injection rate and the speed of mixture formation influence energy conversion in diesel engines. Since mixture formation is heterogeneous, the flame propagation typical in gasoline engines is absent and any danger of "knocking combustion" is eliminated. Therefore, high compression ratios and boost pressures can be produced in diesel engines. Both benefit efficiency as well as an engine's torque characteristic. The limit of compression and boost pressure is not predetermined by "knocking combustion" – as in gasoline engines – but rather by the maximum allowable cylinder pressure, which is why modern diesel car engines operate in ranges of approximately 160–180 bar and commercial vehicle engines in ranges of approximately 210–230 bar. The low compression ratio range specified here applies to highly supercharged large diesel engines.

Since the mixture formation is internal, the time required for fuel evaporation and mixture formation limits a diesel engine's speed. Therefore, even high speed diesel engines seldom operate at speeds above 4,800 rpm. Resultant disadvantages in power density are compensated by their particular suitability for supercharging.

The injection of the fuel into a secondary chamber of the main chamber, a "swirl chamber" or "prechamber", is referred to as "indirect fuel injection". It was formerly used to better form the mixture and utilize air in the main chamber as well as to control combustion noise. Advanced diesel combustion systems, i.e. direct injection engines, inject the fuel directly into the main combustion chamber.

Internal mixture formation and the attendant retarded injection of fuel into the combustion chamber produce distinct air/fuel gradients (λ gradients) in the combustion chamber. While virtually no oxygen is present in the core of the fuel spray ($\lambda \approx 0$), there are zones in the combustion chamber with pure air ($\lambda = \infty$) too. Every range between $\infty > \lambda > 0$ exists more or less pronouncedly in a diesel engine's combustion chamber during injection. Complete air utilization is virtually impossible in heterogeneous mixture formation. The time is far too short to produce and completely burn a homogeneous mixture. Therefore, diesel engines also operate at full load with excess air of 5–15%. Large low speed diesel engines must be operated with even far greater excess air because of the thermal loading of components.

This affects any potentially required exhaust gas aftertreatment systems. Three way catalysts (TWC), operated homogeneously in gasoline engines at $\lambda = 1.0$, cannot be employed since an "oxidizing" atmosphere is always present in the exhaust.

The air/fuel gradient is not only responsible for differences in mixture quality but also local differences in temperature in a combustion chamber. The highest temperatures appear outside the fuel spray in ranges of $\infty > \lambda$, the lowest in the spray core in ranges of $\lambda \approx 0$. As Fig. 3-1 illustrates, nitrogen oxides form in the zones with excess air and high temperatures. Combustion temperatures in the lean outer flame zone are so low that the fuel cannot completely oxidize. This is the source of unburned hydrocarbons. Soot particulates and their precursor carbon monoxide form in air deficient zones in the spray core. Since the rich mixture region makes it impossible to prevent soot formation in a heterogeneous mixture, modern diesel systems aim to oxidize particulates in the engine. This can be improved substantially by maintaining or generating greater turbulence during the expansion stroke. Consequently, modern diesel systems burn up 95% of the particulates formed in the engine.

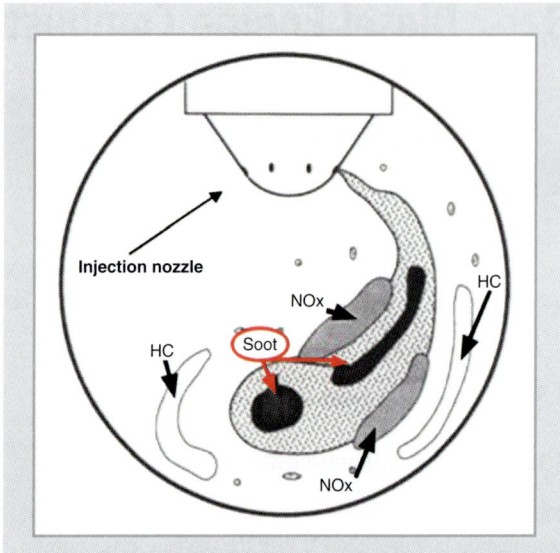

Fig. 3-1 Regions of pollutant production in a combustion chamber with a heterogeneous mixture

Internal mixture formation involving high compression and a method of load control (quality control) is the basis of excellent overall diesel engine efficiency.

3.1.2 Mixture Formation

3.1.2.1 Main Influencing Variables

Apart from the air movement in the combustion chamber (squish or squish flow and air swirl), which can be shaped by the design of the combustion chamber and the intake port, internal mixture formation is essentially dominated by the injection. An injection system must perform the following tasks: Generate the required injection pressure, meter the fuel [3-2], ensure spray propagates, guarantee rapid spray breakup, form droplets and mix the fuel with the combustion air (see also Chap. 5).

3.1.2.2 Air Swirl

Air swirl is essentially a "rotary flow of solids" around the axis of the cylinder, the rotational speed of which can be shaped by the design of the intake port and increases with the engine speed because the piston velocity increases. A basic function of the air swirl is to break up the compact fuel spray and to mix the air sectors located between the fuel sprays with the fuel. Obviously, the swirl requirement decreases as the number of nozzle holes increases. This is advantageous because an increasing level of swirl causes increased wall heat losses and swirl generation must be obtained with charge losses. While supercharging can compensate for charge losses, the adverse effects of swirl generation on gas exchange efficiency and thus fuel consumption persist.

Swirl can be generated quite easily when the air flows into the cylinder tangentially. However, this arrangement produces extremely high charge losses, disproportionately increases swirl as a function of engine speed and is extremely critical for manufacturing tolerances.

Helical ports, in which the air already spirals in the port, are better suited in this regard (Fig. 3-2). This facilitates a nearly linear increase of the level of swirl as a function of engine speed and thus a constant ratio of swirl speed to engine speed (n_{Drall}/n_{Mot}) and can produce a good compromise between the necessary level of swirl and acceptable loss of volumetric efficiency.

Beveling one side of the annular valve seat can facilitate the discharge of the air in the direction of swirl and increase swirl in the lower valve lift range. This measure can also be cleverly used in combination with valve timing to decrease swirl as a function of engine speed.

Shrouded valves – another method to generate swirl – must be installed in a fixed position and are consequently unsuited as a standard method to generate swirl for reasons of wear but are optimally suited for basic tests.

Since air swirl generated by the port geometry swirls faster as engine speed increases, the air segments blown over per degree of crank angle also grow larger. However, this self-regulating effect, which accelerates mixture formation with the speed, can only be utilized when the injection time also acts similarly. When the injection time defined in °CA (e.g. at full load) as a function of speed is constant, swirl and injection time can be adjusted optimally in the engine's entire speed range. However, when the injection time (in °CA) as a function of speed increases, the swirl in the lower speed range is too low and the air utilization unsatisfactory or it is too high in the upper speed range and the individual spray regions are blown over. Both eventualities reduce the attainable brake mean effective pressure and cause increased emissions. This problem crops up in any injection system that operates with constant nozzle hole diameters, i.e. every standard system today. A high speed ratio (n_{max}/n_{min}) and/or a large mass ratio of full load fuel delivery to idle fuel delivery complicate engine design. Variable injection pressure in the map and the use of so-called register nozzles constitute attempts to solve this problem.

Although the swirl speed as a function of engine speed does exactly what it should and the swirl/engine speed ratio is constant in a helical port at least, in the absence of suitable starting points in the injection system, attempts are made to mistune the swirl and adjust it to the injection system's "misbehavior". In engines with two or more intake valves, the swirl can be boosted in the lower speed range by disengaging an intake valve (intake port shut-off IPSO) and thus adjusted to the short injection time common in this range. Infinitely variable throttle adjustment even allows adjusting the level of swirl for every map point as a function of the opening angle (Fig. 3-3). However, this is suboptimal because it accelerates mixture formation in an engine's lower speed range and retards it in the upper speed range.

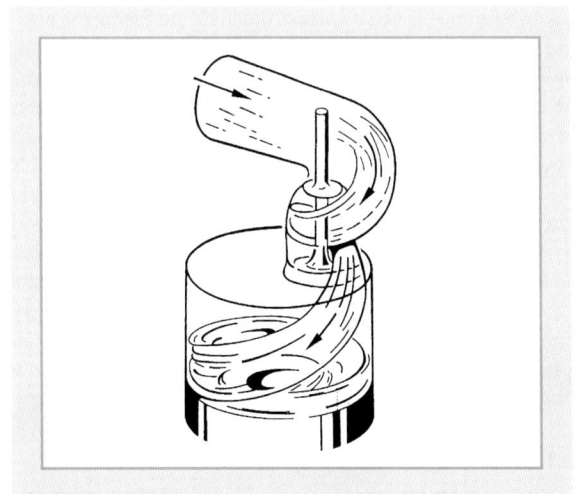

Fig. 3-2 Swirl design with a helical port designed as an intake port

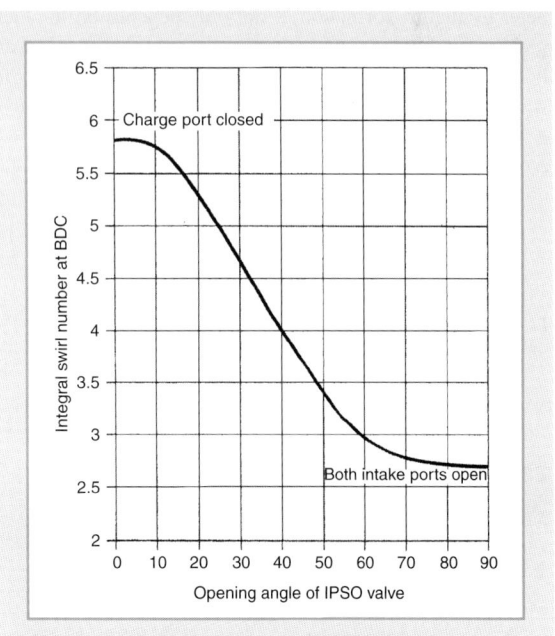

Fig. 3-3 Level of swirl as a function of the position of the IPSO valve [3-3]

Flexible injection systems are also only able to counter the conflict of objectives between swirl and injection time to a limited extent by adjusting the injection pressure. This measure would cause the pressure to drop in the lower speed range. The optimal solution would be a nozzle with a variable flow cross section that increases with the speed.

3.1.2.3 Squish

Air is increasingly squished into the piston bowl during the compression stroke. This increases the air swirl. The smaller a piston bowl is, the greater the swirl becomes.

A squish flow increasingly interferes with the aforementioned air swirl generated by the intake flow into the cylinder or combustion chamber bowl as the piston approaches top dead center. The squish flow is generated by the displacement of the air located between the piston crown and the cylinder head into the piston bowl (Fig. 3-4).

The squish flow counteracts the propagation of the fuel spray and consequently supports the exchange of momentum between the combustion chamber air and injection spray, which is important for mixture formation.

The direction of flow reverses when the expansion stroke begins. Appropriately designing the bowl geometry, especially the bowl rim, enables generating a highly turbulent flow in the piston gap, which facilitates mixture formation and accelerates combustion.

3.1.2.4 Kinetic Energy of the Fuel Spray

The kinetic energy of the fuel spray is the dominant parameter in mixture formation. It not only depends on the fuel mass in the injection spray but also the pressure gradient at the injection nozzle. Together with the spray cone angle, it determines the exchange of momentum between the combustion chamber air and fuel spray as well as the size range of the droplet diameters. Above all, the spray cone angle depends on the internal nozzle flow and thus the nozzle design and the adjacent pressure as well as the air density. As the cavitation in a nozzle hole increases, the spray cone angle grows larger and the exchange of momentum with the air intensifies. In cam driven injection systems, the injection pump's delivery rate and the flow cross sections in the injection nozzle can influence the spray energy. Rail pressure is the crucial parameter in accumulator injection systems.

The injection spray transports the fuel to the outer regions of a combustion chamber. Since the air is highly compressed, hot and thus highly viscous, this function ought not to be underestimated. The pressure curve at the nozzle hole is critically important. Rising or at least constant pressure as a function of injection time is advantageous [3-4]. Pressure that drops during injection fails to facilitate any interaction of the individual fuel zones in the spray and therefore ought to be avoided as far as possible. Coverage of the outer regions of a combustion chamber is the prerequisite for good utilization of the combustion chamber air and consequently high engine power density. Only a limited number of nozzle holes and thus a high fuel mass in the spray can accomplish this when injection pressure is limited. The number of injection sprays may be increased as injection pressure rises and thus fuel distribution in the combustion chamber improved without impairing spray propagation. This is manifested by the correlation between spray penetration depth, the injection pressure or the pressure gradient at the nozzle outlet and the hole diameter. Accordingly, spray penetration depth is a function of the pressure adjacent to the nozzle hole, the diameter of one nozzle hole, the fuel density, the reciprocal value of the air density and the time after the start of injection [3-5]. Provided the injection time is approximately identical, the diameter of the individual hole and thus the fuel mass in the individual spray decrease as the number of nozzle holes increases. Thus, the spray momentum decisive for spray propagation must again be compensated by proportionately increasing the pressure. The entrance of air into the spray (air entrainment) intensifies as the injection pressure rises and thus the local λ in the spray increases. According to Wakuri [3-5], the greater spray penetration depth alone already increases the local air/fuel ratio as injection pressure increases.

The importance of swirl and squish decreases as injection pressure increases. Modern commercial vehicle combustion systems operate in conjunction with nozzles with eight to ten holes at injection pressures above 2,000 bar and are virtually free of swirl and squish. Faster running car engines use the very stable swirl flow to oxidize particulates during the expansion phase. Further, the larger speed range requires a relatively deep and thus narrow bowl, which inevitably produces a squish flow.

3.1.2.5 Spray Breakup

Direct injection requires the completion of evaporation and mixture formation in a few milliseconds. This necessitates

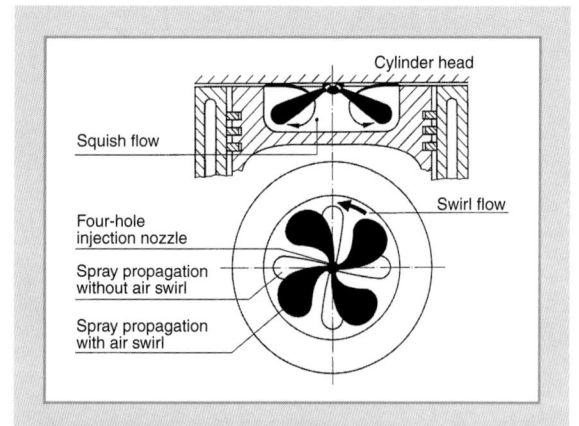

Fig. 3-4 Overlaid flow processes in an engine's combustion chamber influence fuel spray propagation and mixture formation

extremely rapid breakup of the compact spray and the formation of many droplets with large surface areas.

Two mechanisms ensure that the fuel spray breaks up rapidly and the surface area of the fuel created is large: "primary breakup" in the region adjacent to the nozzle induced by the turbulent flows and cavitation in the nozzle and "secondary breakup" induced by aerodynamic forces in the nozzle's far field.

3.1.2.6 Primary Breakup

The primary breakup of a fuel spray injected into highly compressed, highly viscous combustion chamber air is influenced by the redistribution of the velocity profile inside the spray (interaction of different segments in the spray), the surface tension, the aerodynamic forces (exchange of momentum between moving spray and "resting" air), the turbulence (largely induced by the spray momentum) and the cavitation [3-6]. Cavitation is produced by the movement of the fuel flowing turbulently in the nozzle. Strong deflections influenceable by the ratio of the radii of the nozzle curvatures to the radius of the hole, hydrodynamic flow effects and hole shape and conicity play an important role. Knowledge of both the velocity and turbulence parameters and the vapor and gas content by volume can be applied to determine the size and number of the cavitation bubbles. Cavitation bubbles in a nozzle hole influence spray breakup, spray propagation and droplet formation as well as the accumulation of deposits in a hole and a nozzle's durability. Figure 3-5 illustrates the breakup phenomena of an injection spray adjacent to a nozzle.

Fuel temperature and composition determine the property of volatility and play a key role in spray breakup since the formation of cavitation nuclei is influenced by the degassing of gases dissolved in the fuel as a result of the saturated vapor pressure fallen below locally [3-7].

Initially, a compact liquid spray core is observed at close range to the nozzle. However, air and fuel vapor bubbles already subject it to strong breakup at a distance from the nozzle outlet, which is five to ten times the nozzle hole diameter. The relation of the aerodynamic forces to the surface forces, i.e. the Weber number, describes droplet size and droplet distribution:

$$We = \rho_k \cdot v_{inj}^2 \cdot d \cdot \sigma^{-1}$$

where ρ_K signifies the fuel density, v_{inj} the spray velocity at the nozzle hole, d the nozzle hole diameter and σ the surface tension. The Weber number denotes the relation of the kinetic energy of the continuous spray column emerging from the nozzle hole per unit of time and the energy of the free surface area generated per unit of time.

3.1.2.7 Secondary Breakup

The secondary breakup actually "atomizes" the injection spray from coarse ligaments into medium sized droplets by wavy disintegration and into microdroplets by atomization. The formation of the latter is essential for fast heating and evaporation – and thus to shorten the physical ignition delay. Aerodynamic forces play the crucial role in secondary atomization. The injection pressure, injection pressure curve, spray cone angle and air density are significant influencing parameters.

Two effects that proceed simultaneously during secondary breakup deserve attention:

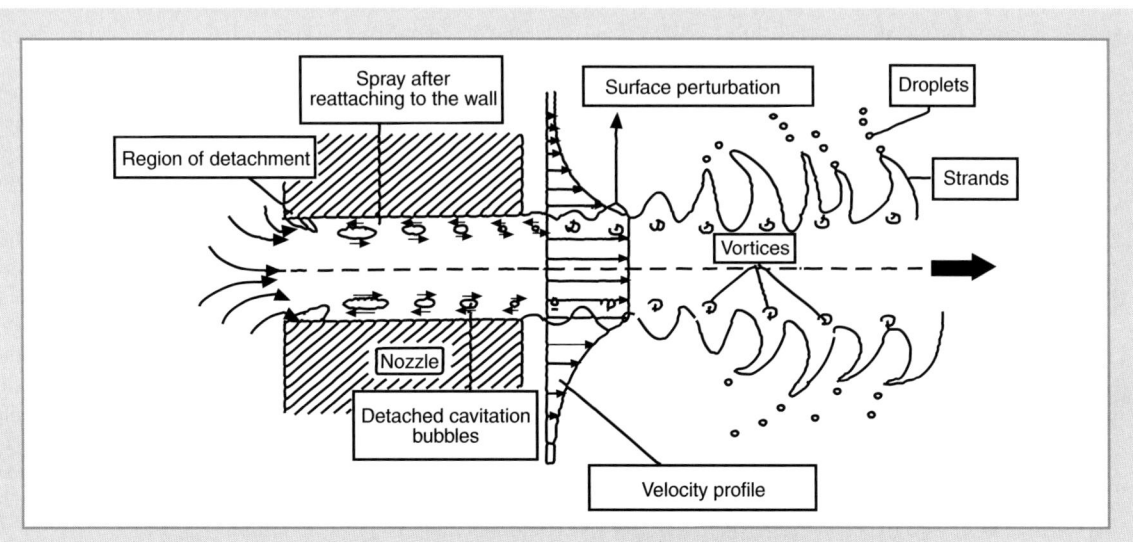

Fig. 3-5 Spray disintegration and breakup adjacent to the nozzle [3-7]

(a) the deformation of the primary droplets decelerated by the frictional forces as a result of the greater inertia of the spray core than the edge of the spray and
(b) the shearing of droplets in the μm range as a result of the wavy disintegration of the flanks of the edge of the spray.

Here too, the parameter defined above, i.e. the Weber number, is a characteristic value used to calculate the density of the spray's ambient air.

Pressure at the nozzle hole that rises as a function of injection time facilitates the exchange of momentum between the combustion chamber air and fuel spray and is therefore conducive to rapid spray breakup.

Not only does more air reach the spray but the droplet diameters also grow smaller as the injection pressure rises. According to Sauter, the statistically mean droplet diameter d_{32} (the Sauter mean diameter) is a function of the Weber number described above, the Reynolds number and the nozzle hole diameter or pressure gradient at the nozzle outlet Δp, the fuel density ρ_K, the air density ρ_L and the fuel viscosity v_K.

$$d_{32} = f\left(\frac{1}{\Delta p, \rho_k, \rho_L, v_K}\right)$$

3.1.2.8 Fuel Evaporation

The fuel must be vaporous so that chemical reactions can proceed in the thusly formed heterogeneous mixture of air and liquid fuel droplets of varying size and distribution.

Crucial importance is attached to the heat transport of the air heated by compression to the liquid fuel. This process is fundamentally influenced by the fuel spray's kinetic energy and thus, in turn, by the injection pressure (Fig. 3-6). A high relative velocity between droplets and their environment

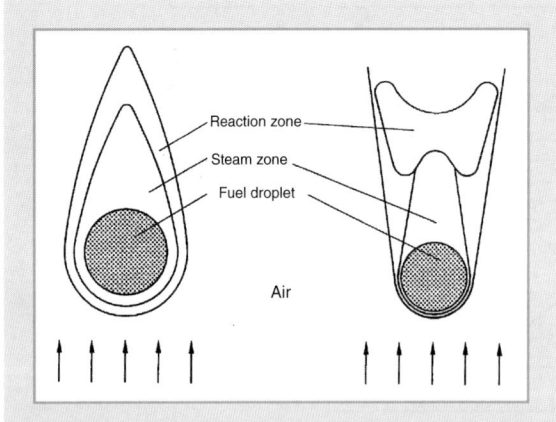

Fig. 3-6 Preparation of a fuel droplet at low (*left*) and high (*right*) inflow velocity

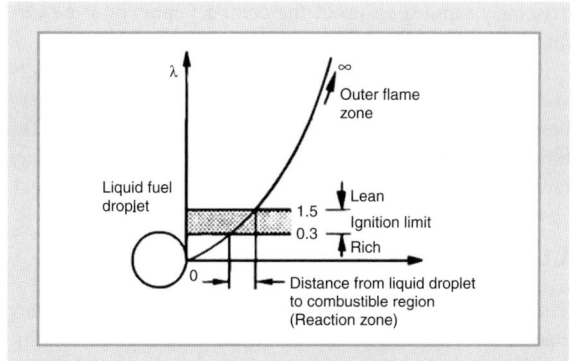

Fig. 3-7 Schematic of the air/fuel ratio as a function of the distance to the fuel droplet

facilitates both the creation of free droplet surfaces and mass transport and heat transfer. The finer the atomization of the droplets and the higher the relative speed of the dispersed fuel phase and the continuous charging phase of the combustion chamber, the sooner temperatures that lead to perceptible evaporation are reached in the outer shell of the droplet surface, the surface film. The air/fuel mixture in the thusly forming diffusion and reaction zones is combustible once the air/fuel ratio λ is in a range between $0.3 < \lambda < 1.5$ (Fig. 3-7).

3.1.3 Ignition and Ignition Delay

The ignition performance of the fuel injected into the compressed and therefore hot combustion chamber air depends on the rate of the reaction that forms ignition radicals as a result of thermal excitation of the molecules. Dependent on the previously described heating and diffusion processes following the secondary breakup, both the thermodynamic conditions in the combustion chamber, i.e. pressure and local temperatures, and the local concentration of vapor, determine the conditions for auto-ignition. Naturally, the fuel itself plays an important role. The cetane number CN describes its ignition quality. Extremely highly ignitable n-hexadecane (cetane) is assigned an index of 100 and slow-to-ignite methyl naphthalene an index of 0. The higher its cetane number, the more highly ignitable a fuel is. Cetane numbers $CN > 50$ are desirable for compliance with extremely stringent exhaust and noise regulations (see Chap. 4).

The time interval between the start of injection and the start of ignition is crucially important for efficiency, pollutant emission, combustion noise and component load. Usually calculated from the nozzle needle lift and the indication of the combustion chamber pressure, the time between these two events is referred to as ignition delay and is a significant feature of diesel engine combustion (Fig. 3-8).

A distinction is made between a physically and a chemically induced component in ignition delay. Physical ignition

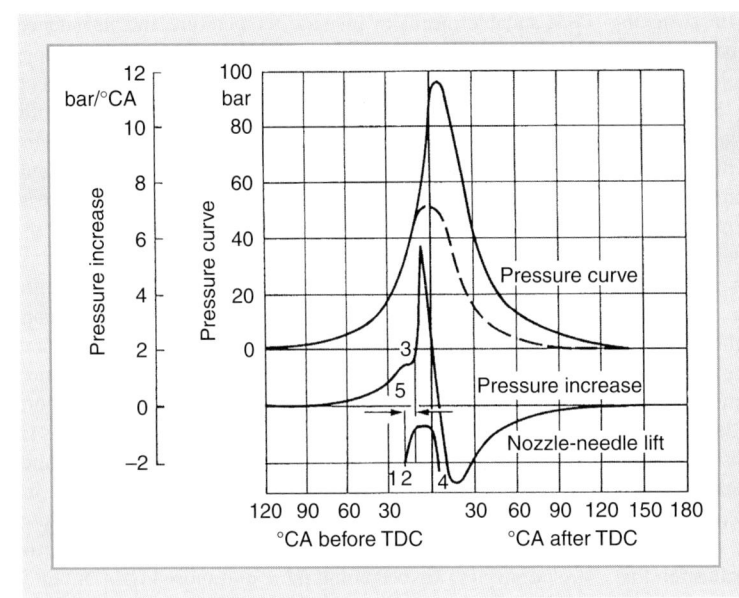

Fig. 3-8 Ignition delay for a direct injection diesel engine. *1* Start of delivery, *2* Start of injection, *3* Start of ignition, *4* End of injection, *5* Ignition delay

delay encompasses the processes of primary and secondary spray breakup described above, the evaporation of the fuel and the processes that generate a reactive air/fuel mixture. Chemical ignition delay specifies the timeframe during which the ignition radicals (e.g. OH) form in a pre-reaction.

Modern highly supercharged diesel engines that operate with injection pressures of up to 2,000 bar have ignition delays of between 0.3 and 0.8 ms. Naturally aspirated engines, have ignition delays of between 1 and 1.5 ms with correspondingly low injection pressures.

Apart from the cetane number, values that describe the temperature at the start of injection (compression ratio, intake air temperature, injection time) and the state of the air in the cylinder (boost pressure, air swirl, squish flow, piston velocity) also enter into the complex calculation of ignition delay.

Numerous empirical formulas have been developed over the course of time to describe ignition delay [3-8 – 3-10].

Usually, the prepared fuel initially ignites at the edge of the spray in the lee side (low air penetration) of the injection spray. The λ gradients that appear are significantly lower in this region of the diffusion zone, i.e. there are fewer inhomogeneities in the composition of the mixture, than on the luff side or at the spray tip. As a result, higher temperatures than in the zones with high λ gradients are also possible here. Therefore, ignition is impossible at close range to the injection nozzle because, as described above, a compact fuel spray dominates.

3.1.4 Combustion and Rate of Heat Release

A fundamental feature of diesel engines is their ability to use the time and the manner fuel is introduced into the combustion chamber (rate shaping) to control combustion and thus energy conversion. While this facilitates efficiency, it is also responsible for the conflicts of objectives between particulate emission (particulate matter PM) and nitrogen oxide (NO_X) on the one hand and fuel consumption and NO_X on the other hand. The less fuel reaches the combustion chamber walls in liquid form and the better liquid fuel is retained in the combustion air and evaporates rapidly, the more the injection rate shapes the rate of heat release. The penetration velocity and evaporation rate of the spray play an important role. Optionally even combined with accumulator injection systems (common rail systems with injection pressure as a function of load and speed), electronically activated control elements can be used to implement injection rate shaping, multiple pilot injections and/or multiple post-injections (Fig. 3-9). Along with increased switching frequency, the advance from solenoid valves to piezo injectors basically

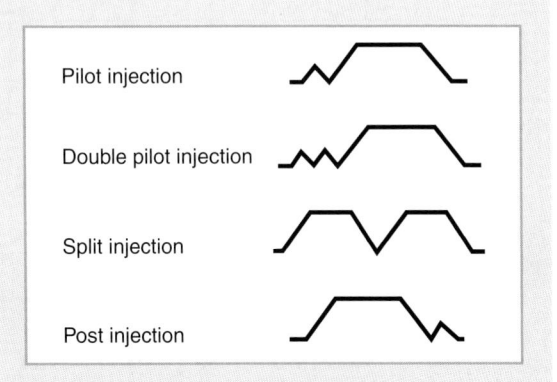

Fig. 3-9 Schematic of potential injection characteristics for multiple injection [3-11]

provides an option to specify defined paths for the switching element (variable throttle). However, since mixture formation profits from high pressure in the nozzle hole, throttling must be prevented in the needle seat (see Chap. 5).

Modern direct injection diesel engines have very short ignition delays in the range of 0.3–0.5 ms at full load and 0.6–0.8 ms at lower part load. Since the injection time is thus longer than the ignition delay in a broad load range, only a small fraction of the fuel is injected before the start of ignition. This fuel fraction is mixed with the combustion air very well and has high λ values and a low λ gradient. While this prevents the formation of soot particulates in these mixture regions, a significant portion of the nitrogen oxides originates inside the engine in this "premixed flame" phase of combustion. What is more, the proportion of the premixed flame influences combustion noise and fuel consumption. A larger proportion of a premixed flame (constant volume combustion) increases noise but benefits fuel consumption (Fig. 3-10).

The majority of the fuel is injected during combustion that is already underway. Given the locally low λ values, little NO_X but much soot forms in this "diffusion flame" (constant pressure combustion). Thus, soot oxidation inside engines is one emphasis of modern diesel engine engineering. Excellent boundary conditions for oxidation prevail since small still uncoagulated particulates with large surface areas are present when temperatures are sufficiently high. The necessary turbulence can be provided, for instance, by high spray kinetics based on the high injection pressures. Further, post-injection is a suitable means to increase temperature and turbulence and thus to reduce particulates by oxidation inside the engine.

Energy conversion should be completed early on to prevent undesired heat losses. Increasingly necessary to comply with future exhaust standards, exhaust gas aftertreatment systems impose particular requirements on exhaust temperature and exhaust composition. Consequently, ever more importance is being attached to the analysis of the overall engine/aftertreatment system.

Engine components are increasingly being furnished with ever more variable parts. In principle, fully flexible injection systems, variable valve timing, variable swirl, variable turbine geometry, variable compression and/or coolant temperature control provide many parameters to optimize a combustion cycle for different requirements. Increasing importance is being attached to the sensor and actuator systems and engine management. When aftertreatment systems are included in the analysis, then even regulated systems are usually too slow in transient engine operation. "Model-based closed loop control strategies" are becoming more and more important and achieve the desired objectives.

3.1.5 Pollutant Production

In principle, the air/fuel mixture can be ignited in a relatively broad air/fuel ratio range of $1.5 > \lambda > 0.5$. However, the best ignition conditions exist at the edge of the spray. While the temperature in the spray core is low (fuel temperature), the temperatures at the edge of the spray are virtually those of the

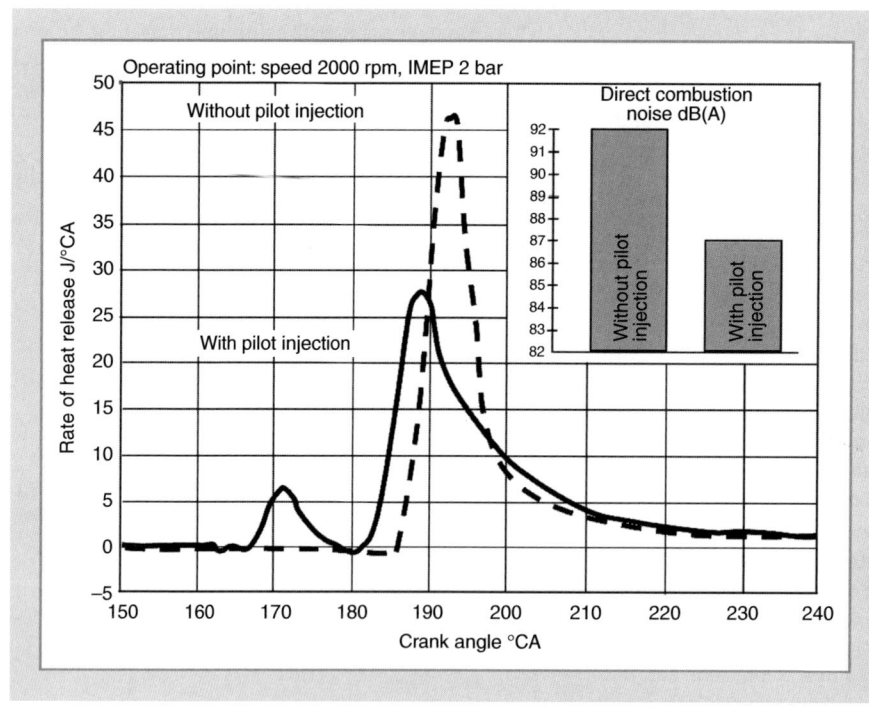

Fig. 3-10 Rate of heat release and combustion noise with and without pilot injection [3-12]

compressed air. Consequently, the mixture begins to ignite in the rather lean region at the edge of the spray. The highest combustion temperatures develop in zones around $\lambda = 1.1$ (see Fig. 3-1). These are the favored regions of nitrogen oxide formation since not only oxygen but also nitrogen is present.

Figure 3-11 indicates that nitrogen oxides form in the lean mixture region at temperatures above 2,000 K. Not only the local air/fuel ratio but also the formation of thermal nitrogen monoxide described by Zeldovich is a function of the residence time and increases exponentially with the local temperature. Some regions along the outwardly diluting mixture are so lean that combustion fails to ignite the mixture despite the rising temperature. Unburned hydrocarbons form in this "lean outer flame zone".

Soot particulates form in rich mixture zones at temperatures above 1,600 K. Such high temperatures only occur after ignition has been triggered. Rich regions are primarily found in the spray core or in the region of fuel spray blocked up by the bowl wall. The fuel must be vaporous for soot to form in the first place. Thus, the size of the fuel droplets has no direct influence on the size distribution of the particulates emitted.

In the phase of ignition delay, the spray has time to propagate at temperatures below the temperature of soot formation, to dilute and to deplete the rich regions. Thus, in principle, a longer ignition delay is better for a low rate of soot formation. However, the dilution of the mixture also produces large λ regions in which nitrogen monoxide forms (premixed flame). The lean outer flame zones in which unburned hydrocarbons are produced also expand. Since fuel burns very quickly in a premixed flame, a long ignition delay also affects combustion noise adversely. Therefore, a shorter ignition delay is striven for in modern combustion systems.

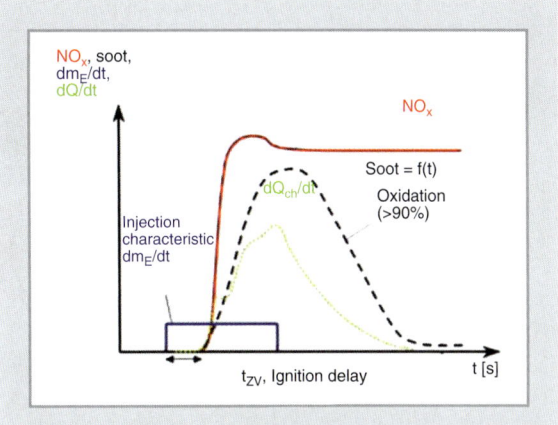

Fig. 3-12 NO_X and soot concentrations in a combustion chamber as a function of piston position in °CA

Figure 3-12 presents a schematic of the injection process as a function of time as a rectangle. With a more or less steep gradient determined by the premixed flame, the heat release begins after ignition delay. Controlled by diffusion combustion, the second phase of energy conversion elapses substantially slower. Clearly, the nitrogen oxides form in the first phase of combustion and, over time, are reduced negligibly little by the hydrogen or carbon monoxide that is briefly present.

Since the rich mixture must be subjected to appropriately high temperatures, soot particulates also only begin to form as energy is converted. However, the soot concentration in the combustion chamber perceptibly decreases as combustion proceeds. Up to approximately 95% of the soot formed in the combustion chamber is reoxidized in the expansion phase. High temperatures and turbulences facilitate soot oxidation. Since they only coagulate to larger particles over time, the continued smallness of the particles has a beneficial effect. The conditions for soot oxidation grow poorer as expansion proceeds because the pressure, temperature, turbulence and surface area of the particulates decrease (as a result of coagulation). Therefore, the time required to oxidize particulates increases with expansion. The oxidation conditions in the exhaust become so unfavorable that soot oxidation is not worth mentioning. The particulates must be trapped in a filter, a diesel particulate filter or DPF, to produce the required residence time. Since it is still impossible to guarantee an adequate temperature in diesel engines' broad operating ranges, additional measures are required in the DPF to reach the combustion temperature of the soot particulates.

3.2 Design Features

3.2.1 Combustion Chamber Design

A distinction in the structural design of combustion chambers is made between engines with undivided combustion

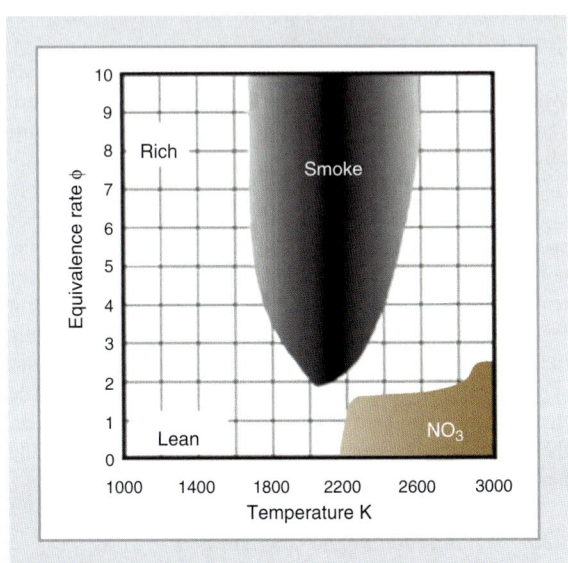

Fig. 3-11 Air/fuel ratio and temperature ranges of NO_X and soot formation ($\varphi = 1/\lambda$)

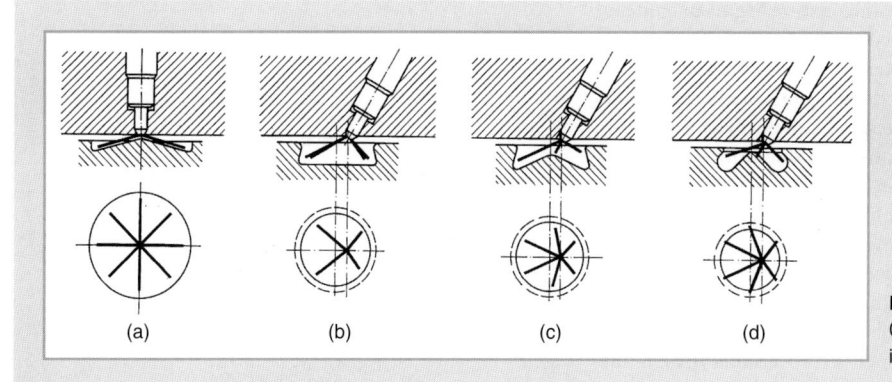

Fig. 3-13 Combustion chamber bowls of direct injection diesel engines

chambers in which the fuel is injected directly into the main combustion chamber (direct injection) and engines with divided combustion chambers. In the latter, the fuel is injected into a prechamber (or swirl chamber). Therefore, this is referred to as "indirect" injection. Part of the fuel is burned in the prechamber. The rise in pressure blows vaporized fuel and/or elements of partially oxidized fuel into the main chamber where it continues burning with the air there. Thus, partial combustion of the fuel generates the energy necessary to cover the main chamber. This two-stage combustion generates advantages in terms of combustion noise but disadvantages in terms of fuel consumption because of the long duration of combustion and the increase in wall heat losses. Especially in the lower speed range, the combustion in the main chamber and the attendant rise in pressure prolong the escape of the gases from the prechamber. The transfer cross sections from the prechamber to the main chamber determine the duration of combustion in the upper speed range. A compromise has to be found between the pressure needed in the chamber to utilize the main chamber air and the duration of combustion.

As injection technology for higher injection pressures and options for pilot injection and injection rate shaping have been refined, more fuel efficient direct injection engines have increasingly edged prechamber engines from the market. Therefore, the features of prechamber engines are not treated any further here.

Only a combustion chamber that is compact and thus simultaneously good in terms of wall heat losses can produce the high compression ratio diesel engines need to vaporize and ignite fuel ($15 < \varepsilon < 20$; large engines $12 < \varepsilon < 16$). Therefore, diesel engines usually have a flat cylinder head with parallel overhead and recessed valves. Since the gap between the cylinder head and the piston is kept as small as possible (< 1 mm), the combustion chambers of direct injection diesel engines are nearly exclusively formed by a bowl located in the piston. Thus, the air in the region of top dead center available for fuel combustion is overwhelmingly concentrated (80–85%) in the piston bowl. A squish flow directed into the bowl is generated during compression as well as after the initiation of ignition and a turbulent flow back into the piston gap during expansion. The design of the piston bowl, especially the edge of the bowl, can influence both flows. They facilitate the mixing of air and fuel substantially. The effect can be intensified with "retracted" bowls. However, the edges of such piston bowls are highly loaded mechanically and thermally.

Air swirl furnishes another option to facilitate mixture formation. Describable as solid body rotation of the combustion air chiefly around the cylinder axis, this flow is influenced by the design of the intake port, the diameter of the bowl and the engine's stroke and serves to cover the air sectors between the injection sprays. The level of swirl required can be lowered as the number of nozzle holes increases. Long-stroke engines operate at higher piston velocity and thus at higher inlet velocity and therefore run with a lower level of intake port swirl. Bowls with a small diameter also increase the swirl generated by the ports.

The intake ports of direct injection engines that generate swirl are designed as helical ports or tangential ports. Both variants may be used in four-valve technology or also combined with a charge port.

The design of the piston bowl should always be viewed in conjunction with the injection nozzle's design and the engine's speed range. Liquid fuel has to be prevented from reaching the bottom of the bowl. Therefore, engines with a large speed range tend to operate with narrow and deep piston bowls (*b* to *d* in Fig. 3-13). A very starkly retracted bowl generates a highly turbulent flow in the piston gap. This accelerates diffusion combustion and shortens the duration of combustion. The thermomechanical load of the bowl rim places constraints on these measures. Low-swirl commercial vehicle engines usually have broad and flat piston bowls (*a* in Fig. 3-13).

3.2.2 Injection Nozzle Configuration

The injection nozzle significantly influences spray breakup, droplet formation and the coverage of the combustion air by the injection spray. Two-valve technology's configuration of the gas exchange valves entails positioning the injection nozzle eccentrically relative to the cylinder and the piston bowl. For optimal air coverage, eccentrically positioned nozzles should be designed with differing hole diameters and an asymmetrical distribution of holes on the circumference. This is usually foregone though for reasons of cost and manufacturing. Further, since the fuel sprays ought to strike the bowl wall at the same height, the holes have to be configured with a different angle relative to the nozzle axis, i.e. the angle of the nozzle hole cone deviates from that of the nozzle axis. This adversely affects flow conditions in the nozzle tremendously and, despite great efforts in the field of nozzle design and nozzle manufacturing, the properties of individual sprays vary widely.

Four-valve technology allows centering the nozzle relative to the cylinder and thus facilitates symmetrical conditions for fuel sprays. This benefits mixture formation and thus the characteristic engine parameters of consumption, combustion noise and emissions and enables optimizing the partly countervailing influences.

Together with the injection timing and the spray velocity, nozzle projection and nozzle hole cone angle determine the fuel sprays' point of impact on the bowl rim (Fig. 3-14). The point of impact ought to be as high as possible. Design work must allow for a potentially present squish flow's influence on the point of impact as a function of speed and the increasing air density's interference with spray propagation. Figure 3-15 presents an option for central nozzle configuration for four-valve technology and the configuration of a glow plug necessary for cold start assist.

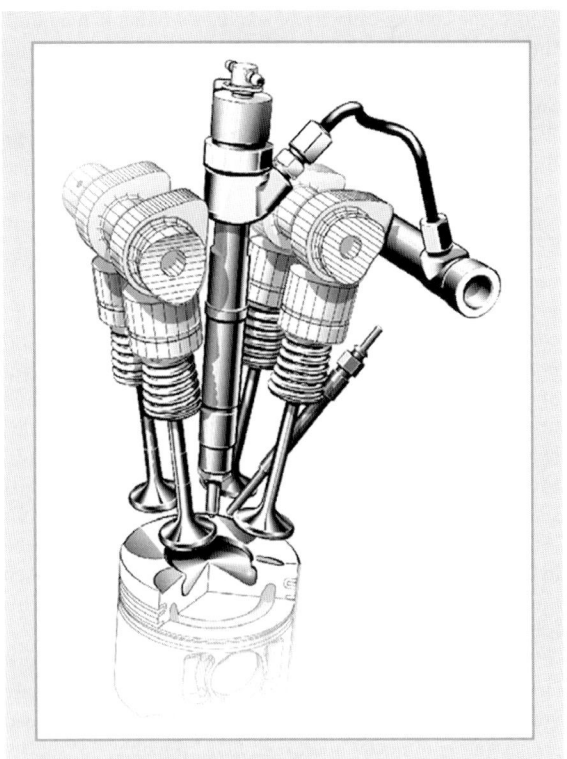

Fig. 3-15 Configuration of a central injection nozzle for four-valve technology

3.2.3 Exhaust Gas Recirculation, Combustion Temperature Decrease

Very inert once formed, nitrogen monoxide hardly reforms in the expansion phase (see Sect. 3.1.1.5). Even the additional introduction of hydrogen, carbon monoxide or hydrocarbons has little effect. If nitrogen oxides are not prevented from forming, only exhaust gas aftertreatment can effectively reduce them. Familiar from gasoline engines, established three-way catalyst technology cannot be implemented because excess air is always present.

Since the nitrogen monoxide produced according to the Zeldovich mechanism – also called the thermal NO mechanism – forms very quickly (prompt NO) and the local λ zones prevalent during heterogeneous mixture formation and conducive to NO formation cannot be prevented, lowering the combustion temperature furnishes a technically effective approach to reducing NO formation.

The best known method to lower temperature is exhaust gas recirculation (EGR), which has been in use in car diesel engines for a long time. Essentially, the increased heat capacity of the inert combustion products of vapor and carbon dioxide affect the local temperature. Cooled EGR is particularly effective and diminishes the adverse effects on fuel consumption but stresses the heat balance of a vehicle's radiator.

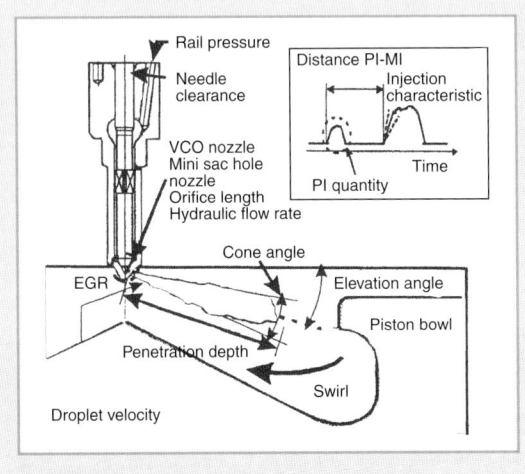

Fig. 3-14 Spray propagation and its influencing factors [3-13]

Hence, the cooling capacity of a vehicle's radiator may limit the potential EGR rate in some load ranges. Since exhaust gases are aggressive, radiators must be made of stainless steel. Supercharged diesel engines especially lend themselves to EGR since they facilitate the transport of exhaust gas from the exhaust gas system into the intake tract. When exhaust gas is extracted before the turbine and fed to the air after intercooling, this is referred to as "high pressure EGR" (short path EGR). "Low pressure EGR" extracts the exhaust gas after the turbine or after the diesel particulate filter (DPF) and feeds it to the intake air before the compressor (long path EGR). This configuration stresses the compressor and intercooler and is not as good in terms of efficiency but has advantages in terms of the mixing of the exhaust gas with the combustion air and uniform distribution in the cylinder. A pneumatically, hydraulically or electromagnetically activated EGR valve controls the recirculation rate in every case. Since the exhaust gas supplied usually replaces portions of the combustion air, this is referred to as "replaced EGR". The air/fuel ratio is reduced in this type of EGR. Increased boost pressure is required when the air/fuel ratio ought to be kept constant despite EGR. This is referred to as "additional EGR".

Turbochargers with variable turbine geometry are particularly suited for EGR since they enable regulating the pressure gradient necessary for exhaust gas transport in broad load ranges and even make "additional EGR" possible.

EGR transport necessitates considerable pressure buildup. Hence, the turbine can no longer be fully depressurized. A second, downstream turbine can utilize the residual pressure gradient remaining. It is either connected with a second compressor (two-stage turbocharging) or releases its energy to the crankshaft (turbocompounding TC).

Another option for exhaust gas transport involves taking advantage of the exhaust pressure peaks. The pressure peaks' overpressurization of a reed valve briefly enables a connection between the exhaust manifold and intake manifold.

Selecting a retarded start of injection is a simple method to lower combustion temperature since the rise in pressure and thus temperature caused by combustion conflicts with the drop in pressure caused by expansion. However, this effect adversely affects soot oxidation and engine efficiency.

The so-called Miller cycle can also be used to lower the combustion temperature. Early "intake closes" utilizes part of the exhaust cycle to expand the aspirated air. As a result, its temperature is lowered and combustion proceeds at a lower temperature level. This technique is primarily being researched for use in large diesel engines and is quite conducive to the objectives desired in the part load range, particularly since λ is also lowered. A disadvantage of this concept, early "intake closes" adversely affects volumetric efficiency and thus power and must be compensated by measures in the supercharger.

Water injection is also a means to lower the combustion temperature. The introduction of water in the intake manifold displays the least effect and additionally has the disadvantage of diluting the oil. While injecting water into a combustion chamber through a separate nozzle is better suited, cooling it is problematic. A "dual fuel nozzle" is the most effective but most elaborate method. During the pause in injection, a metering pump stores water in the injection nozzle. The water is positioned so that first diesel fuel, then water and once again diesel fuel are introduced into the combustion chamber. This has the big advantage of not prolonging ignition delay and the end of combustion and making the water available at the right place at the right time to lower the temperature.

In this respect, the introduction of water as a diesel/water emulsion is less suitable and supplying the water is problematic in every case.

3.2.4 Effect of Supercharging

As Fig. 3-16 indicates, exhaust gas recirculation can significantly reduce nitrogen oxide emission. However, the air/fuel ratio λ drops significantly with replaced EGR. The poorer conditions for soot oxidation cause an increase in particulates. The original particulate values can nearly be reached at an identical air/fuel ratio λ, i.e. with so-called additional EGR. At the same start of injection, NO_X emission only increases marginally. Additional EGR makes great demands on supercharger engineering, the heat balance and an engine's allowable peak pressure.

Since a reciprocating piston engine and a turbomachine or turbocharger's air flow rate as a function of speed exhibits different mass flow characteristics, turbine cross-sections either have to be sized excessively large in the lower speed range or excessively small in the upper. Rapid supercharger response requires a turbine designed for small mass flows. As a result, it clogs in the upper speed range. This impairs the efficiency of gas exchange and increases the internal exhaust gas recirculation. A pressure-controlled discharge valve can solve this problem but at the expense of optimal exhaust gas energy recovery. A turbine with "variable turbine geometry"

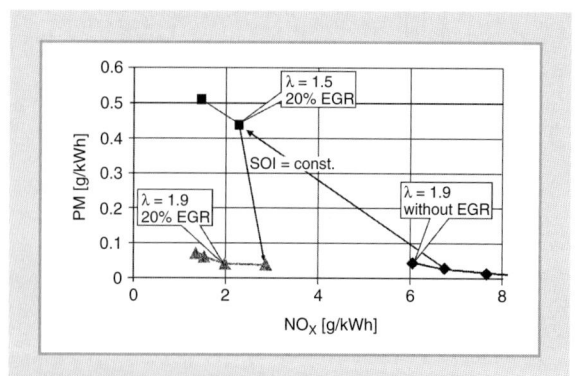

Fig. 3-16 Influence of the EGR rate and the air/fuel ratio on a direct injection diesel engine's NO_X and particulate emission

or sequential turbocharging are better alternatives where this is concerned [3-14].

Supercharging not only helps defuse the conflict of objectives between NO_X and PM. It is also tremendously important for increasing power density and adapting an engine's torque performance. Hence, modern European diesel engine concepts already include two-stage turbocharging with charge air intercooling.

3.3 Alternative Combustion Processes

The diesel engine's heterogeneous mixture formation causes conflicts of objective between PM and NO_X and between NO_X and consumption. A conventional diesel engine's heterogeneous mixture always contains temperature and λ ranges in which both nitrogen oxides and particulates can form. Since, unlike the particulates formed in the combustion chamber, nitrogen oxides can no longer be reduced in the engine once they have formed, modern combustion systems aim to prevent nitrogen oxides from forming in the first place by lowering the temperature (later start of injection, EGR, Miller cycle, water injection). Methods to oxidize soot (higher injection pressure, post-injection, supercharging) must be increasingly applied when a particular measure comes at the expense of soot formation.

A good approach is also provided by the fuel itself. Since aromatics exhibit the basic annular structure of soot particulates and thus deserve to be regarded as their precursors, aromatic-free fuels help ease the conflict of objectives between NO_X and PM. GTL (gas to liquid) fuels produced from methane (natural gas) by means of the Fischer-Tropsch process solely consist of paraffins and are thus ideal diesel fuels (see Chap. 4).

The oxygen atoms present in their molecules, keep oxygen-containing fuels such as methanol or dimethyl ether (DME) from forming any soot. However, their low ignition propensity (methanol) or their vapor (DME) makes them less suitable for conventional diesel injection.

Rape oil methyl ester (RME) is only approved by engine manufacturers to a limited extent. Oil change intervals are significantly shorter. The widely varying quality of commercially available RME also produces differences in viscosity, which influences mixture formation. Therefore, most engine manufacturers tend to advocate an unproblematic blend of up to 5% RME in conventional diesel fuel. Engine manufacturers view pressed rape oil (without conversion into methyl ester) very critically since it can lead to problems in an injection system and cause engine damage as a result.

Since only the fruit of the plant is used in RME, the latest approaches to biomass utilization are aimed at gasifying entire plants. The gas can especially be used in stationary plants or liquefied in a further step for mobile applications (as the example of GTL demonstrates).

Alternative combustion systems attempt to lower the combustion temperature and fully prevent the critical λ ranges around $1.3 > \lambda > 1.1$ (NO_X formation) or $0 < \lambda < 0.5$ (soot formation). The goal is to operate an engine substantially leaner, homogeneously and at low temperatures. Most approaches reach the time that is absolutely necessary for adequate homogenization by prolonging the phase of ignition delay.

The homogeneous charge late injection process (HCLI) comes closest to conventional diesel mixture formation. The process functions with somewhat more advanced injection timing than conventional diesel engines and thus a longer ignition delay. This is intended to prolong the time to reduce rich regions and increase the share of lean mixture regions. The process requires EGR rates in the magnitude of 50–80% to prevent premature ignition and therefore may only be applied in the part load range.

The highly premixed late injection process (HPLI) also functions with a long ignition delay but a moderate exhaust gas recirculation rate. As the name indicates, the long ignition delay is obtained by extremely retarding injection significantly after TDC. The process has drawbacks in terms of fuel consumption and the exhaust gas temperature limits the drivable map range.

In the dilution controlled combustion system (DCCS), EGR rates $> 80\%$ are intended to lower the temperature below the temperature of NO_X and soot formation at conventional injection timing.

Homogenization to lower NO_X and soot is crucially important to the classic homogeneous charge compression ignition process (HCCI). The mixture is given a great deal of time to homogenize and therefore injected in the compression cycle very early (90–140°CA before TDC) or even external mixture formation is worked with. Dilution of lubricating oil caused by poorly vaporized diesel fuel may cause problems. Combustion is initiated when the requisite ignition temperature has been reached by compressing the mixture. Control of the thermodynamically correct ignition point and the combustion cycle under the different boundary conditions is critically important to this concept, which is closely related to the principle of the conventional gasoline engine. The process requires lowering the compression ratio to $12:1 < \varepsilon < 14:1$ and employing higher EGR rates (40–80%) to prevent premature ignition. Higher exhaust gas recirculation rates are partly produced by employing a valve gear assembly with variable valve timing. Such valve gear assemblies also permit applying the Miller cycle with which the mixture temperature can be lowered. Nevertheless, the line between combustion, premature ignition and misfires is very fine. At extreme part load, the latter additionally requires throttling of the intake air for EGR. In light of these constraints, this process is also only feasible in the lower and medium load range.

Figure 3-17 presents the air/fuel ratio and temperature ranges favored and striven for in the processes described above. DCCS achieves the largest drivable λ range and the lowest temperatures. HCCI furnishes the smallest drivable λ

Fig. 3-17 Operating regions of concrete alternative combustion systems

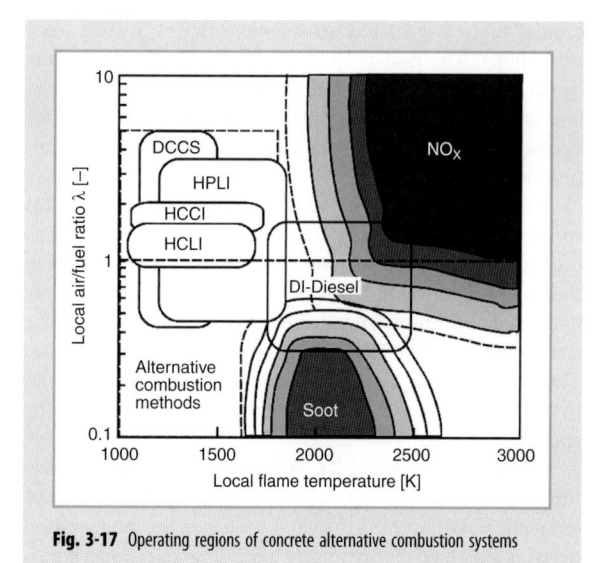

range. Intake air throttling is additionally required in the lowest load range for EGR. The ability of one of these processes to establish itself will not only depend on the drivable load range and the air management required but also on the effectiveness of the compromise of engine parameters at full load and part load. If these processes predominantly designed for the part load range bring excessive disadvantages in the full load range, the likelihood of implementation will diminish.

The hope of achieving better homogenization and more precise control of the ignition point by adapting the fuel to the changed boundary conditions might well be difficult to fulfill.

Multifuel engines do not place any requirements on fuel in terms of detonation limits (octane number ON) or ignition quality (cetane number CN), and therefore ought to be equally compatible with gasoline and diesel fuel. As explained in Sect. 3.1.1, ignition resistant gasolines require an external ignition source in the form of a spark plug. While highly ignitable fuels do not need any external ignition source, they may only be introduced into a combustion chamber very late in order to prevent premature ignition. Thus, they require late internal mixture formation. An engine run solely with gasoline could in turn dispense with this. A multifuel engine intended to run with both types of fuel necessitates internal mixture formation with retarded injection and additional spark ignition. The application of such processes is mostly limited to niche products (e.g. military vehicles).

3.4 Process Simulation of Injection Characteristic and Rate of Heat Release

In addition to experiments, programs that simulate process control in combustion engines have become indispensible tools to optimize frequently conflicting engine parameters [3-15]. Process simulation can significantly shorten development times and reduce test runs to finely optimize processes. Not only can the tools available today be used to perform sensitivity studies that forecast trends, they also provide detailed quantitative information in subareas. Such phenomenological approaches as 3D simulation are equally helpful. The entire process of engine combustion must be described from intake flow and injection to evaporation, mixture formation und combustion up through the formation and emission of pollutants. The extremely complex thermodynamic system of heterogeneous (diesel engine) combustion can only be controlled by breaking it down into subprocesses. Various simulation platforms for modeling are commercially available as computational fluid dynamics (CFD) software. Established programs include FLUENT, STAR CD and FIRE. The PROMO program system is also used to simulate gas exchange. So-called "zero-dimensional phenomenological models" are frequently applied to parameter studies.

Interactive simulation systems oriented toward the concrete geometries of the ports and the combustion chamber serve as the basis for 3D modeling to optimize combustion systems. Fundamental equations for flow processes are solved with the aid of mathematical techniques and facilitate accurately detailed representations of the intake and compression stage incorporating piston and valve movement. The basis for describing combustion processes is CFD modeling of fuel spray propagation, spray breakup and droplet formation, droplet evaporation, vapor mixture with combustion air, initiation of ignition and combustion and the production of exhaust as well as its reduction in the engine.

The aforementioned software also integrates the "discrete droplet model", the method of describing injection sprays based on statistical mechanics generally common today. It describes the dynamics of a multiparticulate system's probability distribution in which every particulate is subjected to a continuous process of change (e.g. evaporation and retardation for droplets as well as collisions or breakup and coagulation processes).

A simple mechanism of chemical reaction in which chemical species that exhibit identical or similar behavior are merged into "generic species" is applied to describe auto-ignition reactions in the gas phase [3-16]. This can significantly reduce the number of computation steps required in a multidimensional computer program.

Simulation of the combustion cycle builds upon model calculations of mixture formation since they significantly influence combustion and pollutant production. However, mathematical models harbor the risk that the model parameters may interact with the numerics and consequently produce false results. This was the experience with spray models and their use for injection systems, which permitted an ever higher injection pressure. CFD software abstracts the real geometry as the system configuration. 3D simulation is often unable to describe microgeometries, e.g. as they appear in nozzle holes, with sufficient resolution of the system configuration or to appropriately represent the relevant numerical values with system points

[3-17]. However, it is impossible to refine system configuration at the nozzle at will for various reasons. Hence, the need to verify the results from model calculations and check them against reliable measured results continues to persist.

Proceeding from the assumption that the oxidation process elapses significantly faster than the process of mixture formation between air and vaporized fuel, the "single-stage global reaction" model is used for the high temperature phase of combustion in which the fuel oxidizes with the oxygen of the combustion air to form carbon dioxide and water.

Models incorporating the subprocesses of nucleation and coagulation as well as oxidation also exist for soot formation. Nitrogen oxide formation is described with the aid of the Zeldovich mechanism. Since the formation of thermal NO takes significantly longer than the oxidation reactions in the flame, it may be treated uncoupled from the actual combustion reactions.

Literature

3-1 Schmidt, E.: Thermodynamische und versuchsmässige Grundlagen der Verbrennungsmotoren, Gasturbinen, Strahlantriebe und Raketen. Berlin/Heidelberg/New York: Springer (1967)

3-2 Krieger, K.: Dieseleinspritztechnik für PKW-Motoren. MTZ 60 (1999) 5, 308

3-3 Naber, D. et al.: Die neuen Common-Rail-Dieselmotoren mit Direkteinspritzung in der modellgepflegten E-Klasse. Part 2: MTZ 60 (1999) 9

3-4 Binder, K.: Einfluss des Einspritzdruckes auf Strahlausbreitung, Gemischbildung und Motorkennwerte eines direkteinspritzenden Dieselmotors. Diss. TU München (1992)

3-5 Wakuri et al.: Studies on the Fuel Spray Combustion Characteristics in a Diesel Engine by Aid of Photographic Visualisation. ASME ICE Vol. 10 Fuel Injection and Combustion. Book-No. G00505 (1990)

3-6 Leipertz, A.: Primärzerfall FVV. 730 (2002)

3-7 Ruiz, E.: The Mechanics of High Speed Atomisation. 3rd International Conference on Liquid Atomisation and Spray Systems London (1985)

3-8 Hardenberg, H. et al.: An empirical Formula for Computing the Pressure Rise Delay of a Fuel from its Cetane Number and from the Relevant Parameter of Direct Injection Diesel Engines. SAE 7900493 (1979)

3-9 Heywood, J.B.: Internal Engines Fundamentals. New York: McGraw Hill Book Company (1988)

3-10 Hiroyasu, H. et al.: Spontaneous Ignition Delay of Fuel Sprays in High Pressure Gaseous Environments. Trans. Japan Soc. Mech. Engrs. Vol. 41, 40345, pp. 24–31

3-11 Chmela, F. et al.: Emissionsverbesserung an Dieselmotoren mit Direkteinspritzung mittels Einspritzverlaufsformung. MTZ 60 (1999) 9

3-12 Kollmann, K.: DI-Diesel or DI-Gasoline Engines – What is the future of Combustion Engines. 4th Conference on Present and Future Engines for Automobiles Orvieto (1999)

3-13 Wagner, E. et al.: Optimierungspotential der Common Rail Einspritzung für emissions- und verbrauchsarme Dieselmotoren. Tagungsbericht AVL-Tagung Motor und Umwelt Graz (1999)

3-14 Zellbeck, H. et al.: Neue Aufladekonzepte zur Verbesserung des Beschleunigungsverhaltens von Verbrennungsmotoren. Vol. 1, 7th Aachen Colloquium Automobile and Engine Technology (1998)

3-15 Chmela, F. et al.: Die Vorausberechnung des Brennverlaufes von Dieselmotoren mit direkter Einspritzung auf Basis des Einspritzverlaufes. MTZ 59 (1998) 7/8

3-16 Winklhofer, E. et al.: Motorische Verbrennung – Modellierung und Modelverifizierung. Vol. 1, 7th Aachen Colloquium Automobile and Engine Technology (1998)

3-17 Krüger, C. et al.: Probleme und Lösungsansätze bei der Simulation der dieselmotorischen Einspritzung. Mess- und Versuchstechnik für die Entwicklung von Verbrennungsmotoren. Essen: Haus der Technik (2000) 9

4 Fuels

Gerd Hagenow, Klaus Reders, Hanns - Erhard Heinze, Wolfgang Steiger, Detlef Zigan, and Dirk Mooser

4.1 Automotive Diesel Fuels

4.1.1 Introduction

When Rudolf Diesel developed the first auto-ignition combustion engine at the close of the nineteenth century, he realized that gasoline's resistance to auto-ignition made it unsuitable as fuel. Comprehensive tests with various fuels revealed that so-called middle distillates were clearly more suitable. These are components that evaporate when crude oil is distilled at higher temperatures than gasoline. Until then, their potential uses had only been limited. In those days, they were typically used as lamp oil and as an additive to city gas, whence the still common designation of middle distillates as "gas oil" originates, which is still a standard customs designation.

Despite many technical problems at first, its better efficiency and the initially lower costs of diesel fuel production led to the diesel engine's commercial success. For a long time, diesel fuel was a byproduct of gasoline production.

In principle, auto-igniting engine combustion may employ widely differing fuels, provided they are highly ignitable enough and the engines and fuels are matched to each other (diesel fuels for road engines and residual oils for marine engines). Among other things, increasing demands on operational safety and exhaust and noise emissions have generated additional quality factors for diesel fuel for road vehicles, e.g.

– cleanliness,
– oxidation stability,
– flowability at low temperatures,
– lubrication reliability and
– low sulfur content.

Present day diesel fuels are specified just as precisely and strictly as gasoline. Partially contradictory requirements, e.g. ignition quality and winter capability or lubricity and low sulfur content, increasingly necessitate the use of additives.

Distinctly different fuels are basically used for the many applications for auto-ignition engines in road vehicles, locomotives or ships for economic reasons. Engineering measures can match particular engines to the different fuels.

A trend toward maximally standardized fuel is increasingly discernible on the global diesel vehicle market (cars and commercial vehicles). In fact, differences throughout the world are sometimes considerable. Gasoline engines and gasoline continue to have broad influence in the USA. Nonetheless, a trend toward advanced diesel cars and accordingly adapted diesel fuels can even be detected in the USA of late.

Unlike gasoline, usually only one grade of diesel fuel was formerly available for road traffic. Diesel fuels of differing grades, e.g. so-called truck diesel or premium diesel, have only recently also started being marketed in some countries with large populations of diesel vehicles.

Diesel fuels are still primarily produced from petroleum. Qualitatively high-grade components (with high ignition quality) have also been being produced from natural gas of late. Renewable raw material components (biofuel components) are blended with diesel fuel in low concentrations in some countries, e.g. in Germany since 2004 (DIN EN 590 [4-1]) and as a mandatory blend required by law since January 1, 2007.

4.1.2 Availability

As presently used throughout the world, diesel fuel is obtained almost exclusively from crude oil/petroleum. Reserves of some 150 billion tons of petroleum are exploitable at this time. However, this amount is dependent on the technology available, which has advanced dramatically, and on the capital available for exploration and extraction. The more crude oil costs on international markets, the more capital can be raised for exploration, development, production and transport. At the same time, this also increases the opportunities for alternative and hitherto uneconomical raw materials and manufacturing processes. The ratio between annual extraction and definitely exploitable reserves is a parameter that measures availability. At present, it is assumed that the reserves will last approximately 40 years.

H.-E. Heinze (✉)
Magdeburg, Germany
e-mail: cheheinze@gmx.de

Crude oil deposits and consumption are distributed very unevenly throughout the world. Most reserves lie in the Middle East. Current scenarios assume the production of petroleum will peak in the twenty-first century. Only the timeframe and the maximum producible quantity are in dispute, depending on the particular scenario [4-2]. The synthesis of alternative fuels from natural gas and renewable raw materials is of particularly great interest (see Sects. 4.2 and 4.4).

While worldwide consumption of petroleum products including diesel fuel will initially continue to rise, more intelligent utilization will cause consumption to decline in classic industrial nations. Approximately 200 million tons of diesel fuel were consumed in Europe in 2008.

4.1.3 Production

The yield from classic production of diesel fuel by distilling crude oil (Fig. 4-1) varies depending on the crude oil used (light, low viscosity or heavy, high viscosity) (Table 4-1).

Diesel fuel is produced from middle distillates. Along with simple atmospheric and vacuum distillation, a whole raft of other methods (thermal or catalytic cracking, hydrocracking) also exist, which increase both the yield from the crude oil and/or the quality of the components of the diesel fuel produced Fig. 4-2.

Every cracking process breaks up high boiling crude oil fractions and converts them into lower boiling hydrocarbons. Thermal cracking (visbreaker, coker) solely uses high pressure and high temperature. A catalyst is additionally present in catalytic cracking. As a result, the finished product's composition (molecular structure) can be controlled better and fewer unstable hydrocarbons form. Hydrocracking allows maximum flexibility in terms of the yield structure (gasoline or middle distillate). In this process, hydrogen (obtained from the catalytic reformer in gasoline production) is fed to the feedstock obtained from distillation at high pressure and high temperature. This process significantly reduces hydrocarbons with double bonds, e.g. olefins and aromatics, which are less suitable for diesel fuels.

Desulfurization is another important process. Depending on its provenience, crude oil contains varying quantities of chemically bonded sulfur. Concentrations between 0.1 and 3% are typical (Table 4-2).

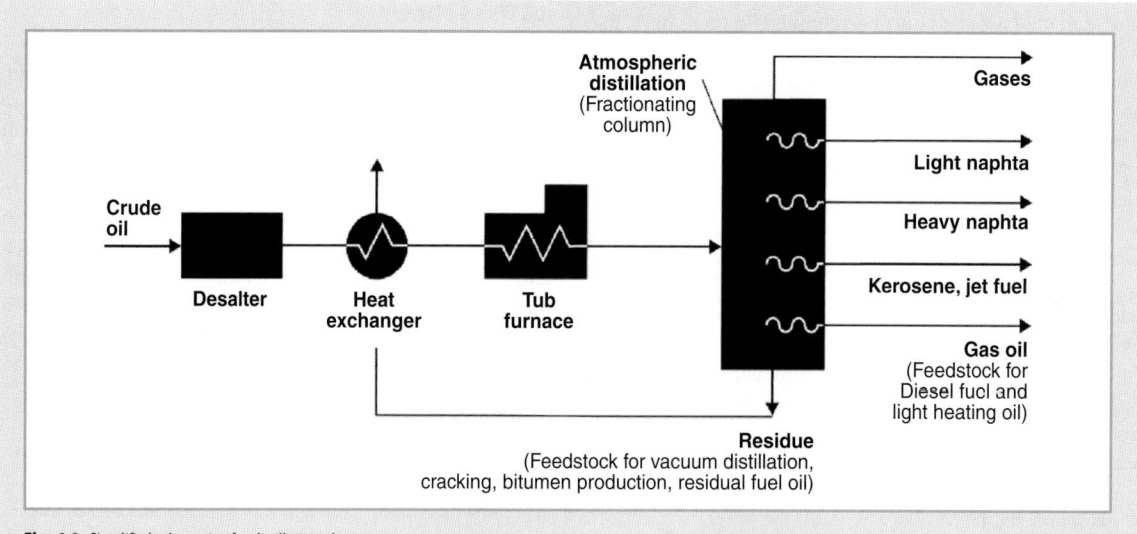

Fig. 4-1 Simplified schematic of a distillation plant

Table 4-1 Percentage distilled yield of various crude oils

Crude oil type	Middle East Arabia Light	Africa Nigeria	North Sea Brent	South America Maya
Liquefied gas	<1	<1	2	1
Naphta (gasoline fraction)	18	13	18	12
Middle distillate	33	47	35	23
Residual oils	48	39	45	64

Fig. 4-2
Simplified schematic of an oil refinery

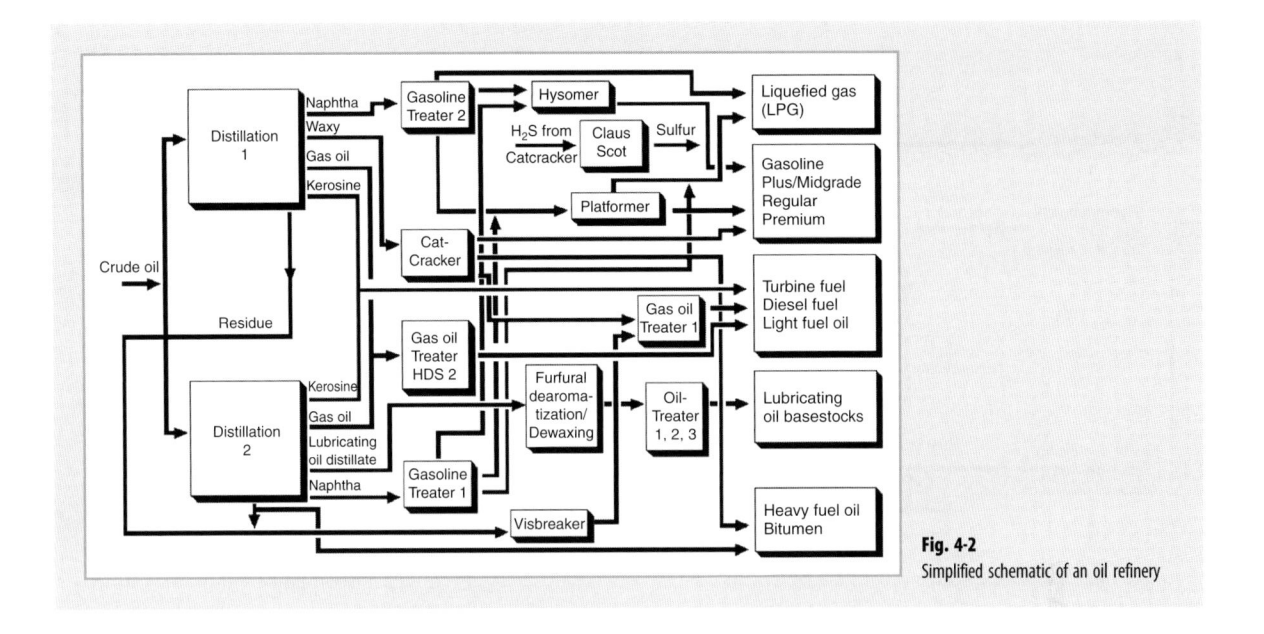

Extremely effective desulfurization is required to obtain the low sulfur content stipulated for diesel fuel (e.g. the EU only still allows sulfur-free fuels, their actual limit being a maximum of 10 ppm of sulfur). To this end, hydrogen-rich gas from a catalytic reformer is fed to the feedstock, e.g. from crude oil distillation, and conducted to a catalyst after being heated (Fig. 4-3). In addition to the desulfurized liquid phase (diesel fuel components), hydrogen sulfide is produced as an intermediate product from which a downstream system (Claus process) produces elementary sulfur.

Depending on the crude oil used and the refinery systems available, diesel fuel is produced from various components so that the product is a high grade fuel that is stable for engines in conformity with the quality requirements for summer and winter fuels stipulated in the standards. Common components from atmospheric and vacuum distillation are:
– kerosene,
– light gas oil,
– heavy gas oil and
– vacuum gas oil (basically as feedstock for downstream cracking processes).

Furthermore, components with names corresponding to their production process are also employed. While components from distillation have differing compositions that depend on the crude oil utilized, the composition of components from cracking largely depends on the process (Table 4-3).

Good filterability at low temperatures is usually the distinctive feature of low boiling components and components with higher aromatic contents (e.g. kerosene). Therefore, their concentration is increased in winter. However, their ignition quality is lower. High boiling components with low aromatic content (e.g. heavier gas oils) have higher ignition quality but poorer filterability in winter. Thus, their use must be reduced in winter. Where necessary, ignition improvers can compensate for winter fuels' potential loss of ignition quality (see Sect. 4.1.5).

Once fuels are blended from the basic components, special metering units blend in additives to obtain the standardized level of quality (low temperature performance, ignition quality and wear protection) or a specific brand's properties (injection nozzle cleanliness and foam inhibition).

Fuels with quality levels that significantly exceed the standard are produced by specially selecting the components and, of late, also by using synthetic components such as gas-to-liquid (GTL) from Shell Middle Distillate Synthesis (SMDS).

Table 4-2 Typical sulfur contents of some crude oils

Provenience	Designation	Percentage sulfur content by weight
North Sea	Brent	0.4
Middle East	Iran heavy	1.7
	Arabia light	1.9
	Arabia heavy	2.9
Africa	Libya light	0.4
	Nigeria	0.1 – 0.3
South America	Venezuela	2.9
Russia		1.5
Northern Germany		0.6 – 2.2

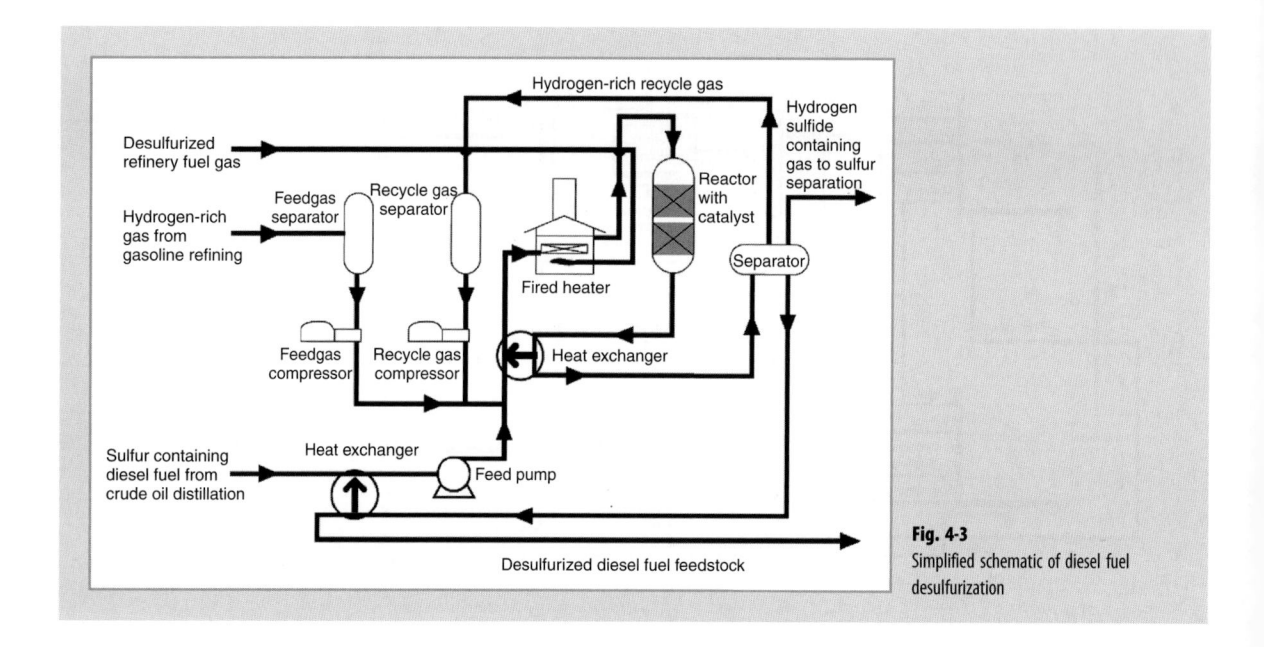

Fig. 4-3 Simplified schematic of diesel fuel desulfurization

Table 4-3 Composition of diesel fuel components produced in various processes

Diesel Fuel Components	Paraffins	Olefins	Aromatics
'Straight run' distilled gas oil	Variable: middle to high	Variable: low to very low	Variable: middle to low
Thermally cracked gas oil with hydrogenation	High	Very low	Low
Catalytically cracked gas oil	Low	None	High
Hydrocracker gas oil	Very high	None	Very low
Synthetic gas oil (SMDS)	Very high	None	Very low to none

This process synthesizes a diesel fuel component with very high ignition quality (a cetane number of roughly 80) from natural gas.

Synthetic diesel fuels are nothing new. The Fischer-Tropsch process was utilized during the Second World War to produce diesel fuel from coal. However, the low price of crude oil for many years made this process uneconomical. Developments in the synthesis of fuels from renewable raw materials have better prospects for commercial success at present. Unlike the "biodiesel" from plant seeds (fatty acid methyl ester or FAME) currently in use, future fuels will much more likely be produced from entire plants or even organic residues.

Low concentrations of FAME are presently employed as a blending component in commercial diesel fuel. The EU is aiming for a concentration equivalent to 6.25% of the fuel's energy content to reduce CO_2 emissions. However, this concentration is not yet achievable because of restrictions on availability. FAME from certain regions is unacceptable because its production endangers indigenous rain forests.

4.1.4 Composition

Crude oil and the fuels obtained from it are a mixture of different hydrocarbon compounds that may roughly be divided into paraffins, naphthenes, aromatics and olefins. Other elements, e.g. sulfur, are present in very low concentrations and must be removed as far as possible during fuel production.

The potential range of variations of individual hydrocarbons increases as the boiling range increases. While the lightest energy source, natural gas, only consists of very few and precisely defined hydrocarbons, basically methane, gasoline contains more than 200 various hydrocarbons and diesel fuels even significantly more because of the frequently present combination of types of hydrocarbons described below. Aromatics have the greatest variety of paraffinic or olefinic side chains. Hydrocarbons with approximately 10–20 carbon atoms are present in diesel fuels' boiling range since it is standardized.

Unlike spark ignited gasoline engines, diesel engines need hydrocarbons that facilitate good auto-ignition at high pressures and high temperatures. These are primarily normal paraffins.

Normal paraffins are chain-like hydrocarbons with a simple carbon bond. They represent a subgroup of saturated hydrocarbons (alkanes), the term saturated referring to the chemical bonding of hydrogen. The simple carbon bond in a paraffin molecule produces the maximum potential (saturated) hydrogen content. Normal paraffins are usually present in crude oil in high concentrations. However their poor flow characteristics when cold (paraffin separation) is disadvantageous for vehicle operation. Flow improvers (see Sect. 4.1.5) in modern fuels can compensate for this. Cetane $C_{16}H_{34}$ is one example (Fig. 4-4). Unlike normal paraffins, branched isoparaffins with identical empirical formulas are unsuitable for diesel engines because of their high resistance to auto-ignition (e.g. isocetane with a cetane number of 15, see Fig. 4-4). The molecules become more unstable as their chain length (molecule size) increases and thus their ignition quality increases (Fig. 4-5).

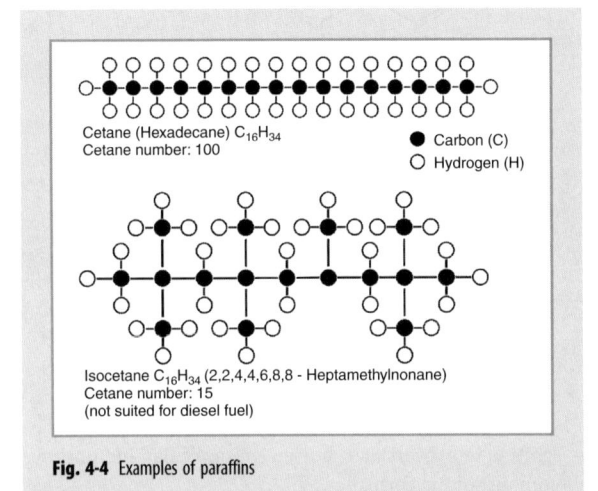

Naphthenes (Fig. 4-6) are cyclic saturated hydrocarbons (a simple bond between the carbon atoms in the molecule). Present in crude oil in varying quantities, naphthenes are produced by hydrogenating aromatic middle distillates. Naphthenes produce good low temperature performance in diesel fuel but only a medium cetane number (yet higher than aromatic hydrocarbons).

Olefins (alkenes) are simple or polyunsaturated, chain-like or branched chain hydrocarbons. While their cetane numbers are lower than those of n-paraffins, they are still quite high. For instance, the n-paraffin cetane $C_{16}H_{34}$ has a cetane number of 100, yet 1-cetene $C_{16}H_{32}$, a simple unsaturated olefin with the same number of carbon atoms, has a cetane number of 84.2. Unlike the short chain olefins used in gasoline, the simple double bond in long chain diesel fuel olefins only slightly influences physical properties and combustion performance.

Aromatics are cyclic hydrocarbon compounds in which the carbon atoms are alternately double bonded with one another, e.g. benzene, a compound not found in diesel fuel though because of its boiling point of 80°C. A distinction is generally made between mono, di, tri, and tri+aromatics based on the number of aromatic ring systems. Figure 4-7 presents a few examples relevant for diesel fuel. Since their boiling range is approximately 180–370°C, a large number of the most widely varying aromatic compounds may be present in diesel fuel. Predominantly monoaromatics with variously configured side chains (approximately 15–25% by weight) are present in conventional diesel fuels. Diaromatics account for approximately 5% by weight and triaromatics and higher polyaromatics usually make up less than 1% by weight.

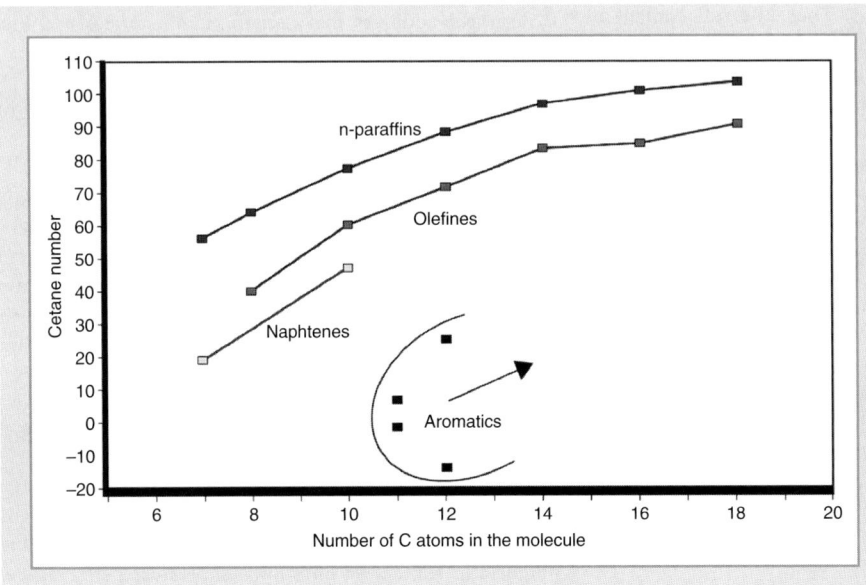

Fig. 4-4 Examples of paraffins

Fig. 4-5 Cetane numbers rise as molecule size increases

Fig. 4-6 Examples of naphtenes in diesel fuel

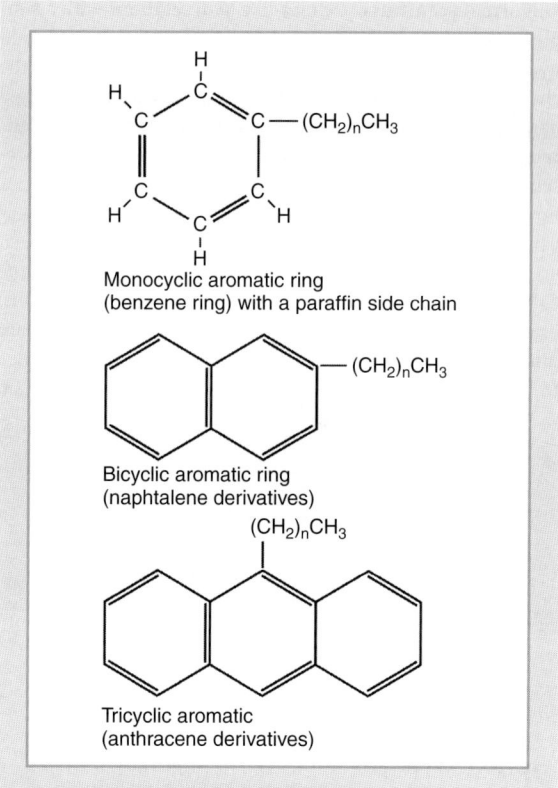

Fig. 4-7 Examples of aromatics in diesel fuel

The properties of monoaromatics with paraffinic side chains vary. While aromatic properties (e.g. high solubility of other products and low cetane number) are dominant in molecules with relatively short side chains, molecules with long side chains have properties that make them behave more like paraffins. Polynuclear aromatics and their derivatives (see Fig. 4-7) are undesired in diesel fuels because of their poor ignition quality. They also cause increased particulate emission and are therefore limited to a maximum of 11% in the European standard EN 590 in force since 2004. The actual concentrations are usually significantly smaller. One reason is that the standardized method of analysis (high pressure liquid chromatography or HPLC) detects entire molecules including their paraffinic side chains. Thus, the real content of aromatic rings is significantly lower. Fatty acid methyl esters (FAME) appear in low concentration. (For the properties of FAME, see Sect. 4.2 Alternative Fuels.)

4.1.5 Diesel Fuel Additives

The production of modern standard diesel fuels is all but impossible without additives. The partly conflicting properties of the individual components (molecular groups) must frequently be equalized by additives to satisfy the high requirements for operational safety, rate of heat release and exhaust emission throughout an engine's entire service life. This has undergone a definite change in recent years. As a rule, more varieties and quantities of diesel fuels than gasolines are additized. With the exception of antifoaming agents, all additives consist of purely organic compounds. Described in more detail below, the most important groups of additives, are:
– flow improvers and wax anti-settling additives that improve winter capability,
– ignition improvers that shorten ignition delay and improve combustion performance,
– anti-wear additives that protect injection nozzles and pumps;
– antifoaming agents that prevent foaming and spilling when pumping fuel,
– detergent additives that keep injection nozzles and fuel systems clean,
– anti-corrosion additives that protect fuel systems and
– antioxidants, dehazers and metal deactivators that improve fuels' storage stability.

Odor masking agents are also used occasionally. Antistatic additives are used during manufacturing and subsequent redistribution (logistics) to the extent they are necessary to prevent electrostatic charging at high pumping rates. The use of biocides to prevent fungus infestation on tank bottoms in the water/fuel phase may be foregone when fuel distribution systems are serviced regularly.

4.1.5.1 Flow Improvers and Wax Anti-settling Additives (WASA)

Flow improvers and wax anti-settling additives make it possible to utilize paraffinic components with high cetane numbers in winter, yet limit low temperature performance.

While flow improvers cannot inhibit the formation of paraffin crystals, they can however reduce their size and prevent them from coalescing.

Typical products are ethyl vinyl acetates (EVA). On their own, flow improvers can drop the CFPP (cold filter plugging point) below the cloud point by more than 10°C. The CFPP can be lowered even further by additionally using WASA. This simultaneously reduces the sedimentation of paraffin crystals when fuel is stored below its cloud point for a long time (Figs. 4-8, 4-9, 4-10 a: formation of paraffin crystals in fuel, b: fuel sample in a glass flask).

These (usually prediluted) cold flow additives must be meticulously worked into still warm fuel at the refinery. Subsequent blending into cold fuel in a vehicle tank is usually ineffective.

4.1.5.2 Ignition Improvers

Ignition improvers facilitate an economical increase of the cetane number with a correspondingly positive influence on combustion and exhaust emission. Organic nitrates are the active ingredients. Ethyl hexyl nitrate has especially proven itself as a commercial ignition improver.

4.1.5.3 Anti-wear Additives

Also known as "lubricity additives", these products became necessary when low-sulfur and sulfur-free diesel fuels were introduced. Diesel fuels lose their natural lubrication properties when hydrogenated to remove sulfur compounds. Fatty acid derivatives are typical products and were also already used earlier in aircraft turbine fuels.

Fig. 4-9 Diesel fuel solely with flow improvers at −22°C: Paraffin crystal sedimentation

Fig. 4-10 Diesel fuel with flow improvers and wax anti-settling additives at −22°C: No paraffin crystal sedimentation

4.1.5.4 Antifoaming Agents

Antifoaming agents assure easy refueling and safe automatic shutoff of the fuel nozzle at a gas station without the spilling frequently encountered earlier and thus lower environmental pollution at gas stations. Extremely low concentrations of silicon fluids are employed. They are also used in engine oils to prevent foaming in the crankcase (Fig. 4-11).

4.1.5.5 Detergent Additives

Diesel fuels can form carbonaceous deposits in injection nozzles and thus alter the injection characteristic while adversely influencing the combustion cycle and exhaust

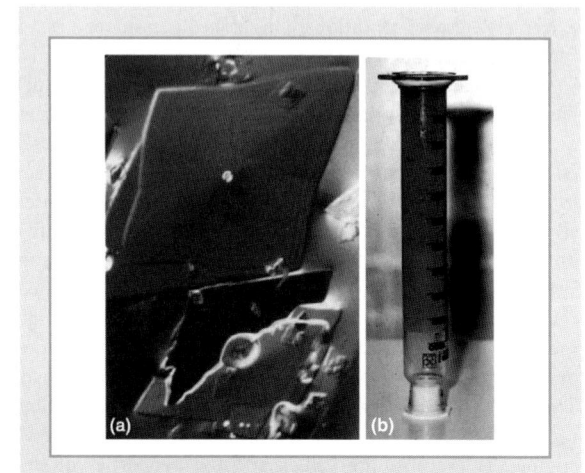

Fig. 4-8 Diesel fuel with flow improvers at −10°C: Solid, unpumpable (the glass flask is upside down)

Fig. 4-11 Fuel with antifoaming agents foams less when pumped

emission. A whole series of complex organic compounds is capable of preventing nozzle coking. The active ingredients continuously have to be adapted to injection technology. Developments of new additives necessitate elaborate tests in complete engines, preferably under realistic operating conditions. Suitable detergent additives are derived from groups of amines, amides, succinimides and polyetheramines for example. Requirements for "cleanliness" and a diesel fuel's detergent effect keep increasing as advances in diesel engine development continually increase injection pressures while reducing the size of nozzle holes

4.1.5.6 Corrosion Protection

Anti-corrosion additives are particularly necessary when small quantities of water have infiltrated the fuel, e.g. condensation during longer downtimes (Fig. 4-12). Polar molecular groups of esters or alkenoic succinimide acids build up a monomolecular protective coating on metallic surfaces and prevent direct contact with water and acids (Fig. 4-13).

4.1.5.7 Antioxidants, Dehazers and Metal Deactivators

These additives are primarily used to improve fuel storage stability. Initially, they do not directly influence a fuel's engine performance at all. However, they prevent fuel deterioration over long periods and thus ensure a fuel's filterability is good and deposits are minimal after extended vehicle downtimes. They were especially necessary in fuel components from thermal and catalytic cracking plants because fuel components from these plants had higher contents of unstable hydrocarbons. Antioxidants prevent oxidation and polymerization. In part, dehazers are required to facilitate rapid settling of finely dissolved water particles. Metal deactivators prevent catalytic effects of metals on fuel ageing. Now more advanced, the hydrogenation of diesel fuel (hydrocracking, desulfurization, limitation of polynuclear aromatics) has caused these kinds of additives to lose their importance.

4.1.6 Quality Requirements

In and of themselves, diesel engines seemingly make comparatively small demands on fuel quality and, unlike spark ignited engines, even allow operation with fuels with extremely widely varying properties, e.g. residual fuels in low speed marine diesel engines. However, other than in cars and trucks, the fuel is cleaned and heated on board before being fed to the engine. Qualitatively better fuel for starting and operating an engine is also frequently available in more environmentally sensitive regions.

Barring filtering for sporadically potential impurities, fuel preparation is impracticable in road vehicles. What is more, road vehicle fuel systems and engines must operate under extremely variable conditions. This alone necessitates stricter fuel quality requirements.

Unlike gasoline engines that can be destroyed when fuel has inadequate detonation limits, diesel engines also run on fuels with lower ignition quality, albeit not well. Diesel engines are expected to deliver optimal results with respect to power, consumption, noise, exhaust emission, etc. under all conditions. However, an engine can only be successfully optimized when the interaction of the fuel and combustion system is painstakingly matched and the fuel is specified relatively narrowly. Hence, any deviation from one of the fuel properties that influences optimization generates more or less substantial disadvantages in engine performance. Therefore, the stability and narrowest possible toleration of the fuel properties are just as important for diesel engines as the absolute values and the properties of the fuel.

While great importance has always been attached to the quality of gasoline, this was not the case for diesel fuels for a

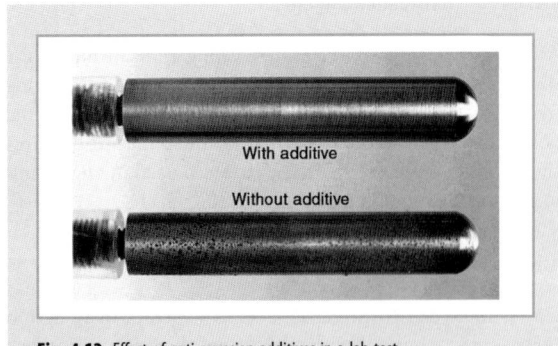

Fig. 4-12 Effect of anti-corrosion additives in a lab test

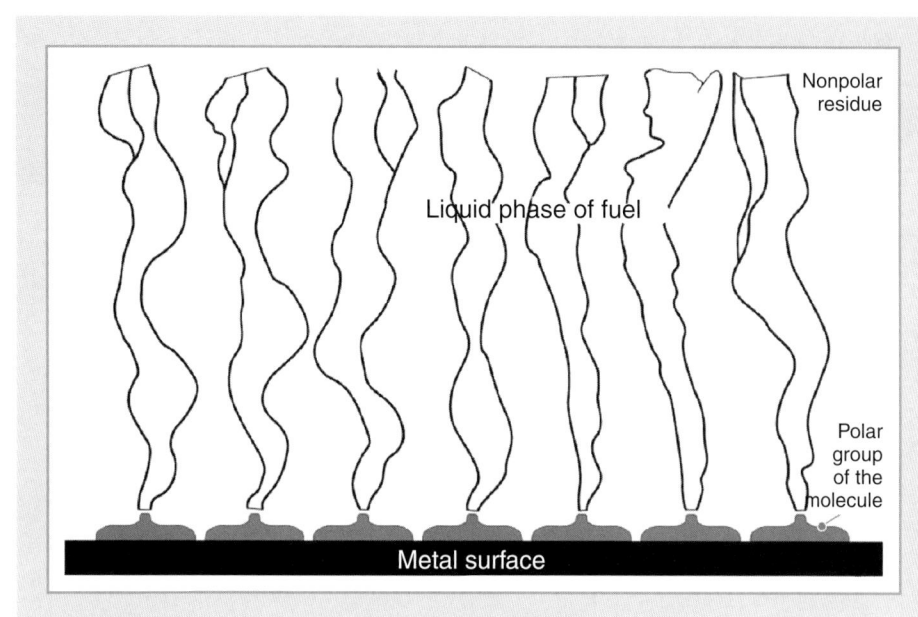

Fig. 4-13 Function of anti-corrosion additives (monomolecular protective coating)

long time. The cetane number was only considered important for cold starts. Inadequate cold flow properties made diesel operation a problem on cold days. Boiling characteristics were virtually unrestricted. The sulfur content was high. Additives were unknown. Purity was not particularly high. Since the quality of diesel fuel is now known to influence an engine's exhaust, noise, drivability and service life, diesel fuel quality has gained the attention it is due in large parts of the world.

In the past, quality differed considerably throughout the world. Even in such a highly industrialized country as the USA, diesel fuel and the network of filling stations was largely unsuited for cars and consumers did not accept diesel cars at first.

Historically, the quality of diesel fuel was initially largely dependent on the crude oil processed since practically only distillates were available. Quality also varied accordingly. As cracking plants that produce gasoline increased around 1970, the components they produced, which were unsuitable for gasoline because of their high boiling range, also began steadily accumulating. Since intensive hydrogenation was not yet available, the mineral oil industry increasingly considered blending these components into diesel fuel at that time. These components' aromatic and olefinic constituents would have degraded diesel fuel quality though. However, these ideas were not pursued any further because of resistance from the engine industry as well as mounting quality requirements ensuing from limits on exhaust emissions. On the contrary, diesel fuels were additized against deposits and foaming and even to improve odor. Ignition improvers were employed and, above all, cold drivability based on climatic conditions was guaranteed in winter, e.g. at $-22\,°C$ in Germany.

In the meantime, many countries have additional types of diesel fuel with significantly improved properties, which, among other things, even lower untreated emissions. One example and trendsetter is Sweden's "city fuel".

While lawmakers formerly only regulated maximum sulfur content, regulations corresponding to those for gasoline are now in force in the European Union. In addition, it has an act on flammable liquids (flash point), a customs tariff and fuel oil designations.

4.1.7 Fuel Standards

Standards that regulate minimum requirements are indispensible for communication between engine manufacturers and fuel producers, retailers and consumers. Here too, much has changed positively in terms of suitability and environmental compatibility. While every country formerly had its own specifications, the standard EN 590 has been in effect for all of Europe since 1993 [4-1] and applies to all the countries represented in the European Committee for Standardization (CEN). However, every country may enact (additional) individual regulations in a national appendix. Special classes specifying flow characteristics during cold weather with correspondingly low CFPP (cold filter plugging point) and CP (cloud point) were implemented for arctic countries. EN 590 has largely defined every relevant fuel parameter at a relatively high level of quality. The most recent version from 2008 is currently undergoing the process of adoption (Table 4-4). Section 4.1.8 outlines the individual parameters.

Regular revision of the standards and the test methods applied guarantees the inclusion of new or modified criteria

Table 4-4 Diesel fuel quality: Minimum requirements according to DIN EN 590: 2008–2009

	Unit	Minimum requirements		Test procedure
Density at 15°C	kg/m^3	Moderate climate 820–845	Arctic climate 800–840 or 845	EN ISO 3675 EN ISO 12185
Cetane number		Min. 51	Min. 47, 48 or 49	EN ISO 5165 EN 15195
Cetane index		Min. 46	Min. 43 or 46	EN ISO 4264
Distillation: percentage recovered				EN ISO 3405
to 180°C	Percent by volume		Max. 10	
to 250°C	Percent by volume	Max. 65		
to 340°C	Percent by volume		Max. 95	
to 350°C	Percent by volume	Min. 85		
95% recovered	°C	Max. 360		
Flash point	°C	Min. 55		EN ISO 2719
Viscosity at 40°C	mm^2/s	2.00–4.50	1.2; 1.4 or 1.5–4.0	EN ISO 3104
Filterability / CFPP				EN 116
Class A	°C	Max. +5		
Class B	°C	Max. 0		
Class C	°C	Max. −5		
Class D	°C	Max. −10		
Class E	°C	Max. −15		
Class F	°C	Max. −20		
Class 0	°C		Max. 20	
Class 1	°C		Max. 26	
Class 2	°C		Max. 32	
Class 3	°C		Max. 38	
Class 4	°C		Max. 44	
Cloud point				EN 23105
Class 0	°C		Max. 10	
Class 1	°C		Max. 16	
Class 2	°C		Max. 22	
Class 3	°C		Max. 28	
Class 4	°C		Max. 34	
Sulfur content	mg/kg	Max. 10		EN ISO 20846 EN ISO 20884
Carbon residue	Percent by weight	Max. 0–30		EN ISO 10370
Ash content	Percent by weight	Max. 0.01		EN ISO 6245
Cu corrosion	Corrosion rating	Class 1		EN ISO 2160
Oxidation stability	g/m^3 h	Max. 25 Min. 20		EN ISO 12205 pr EN 15751
Total impurities	mg/kg	Max. 24		EN 12662
Water content	mg/kg	Max. 200		EN ISO 12937
Lubricity (HFRR)	μm	Max. 460		EN ISO 12156-1
Polyaromatics	Percent by weight	Max. 11		EN 12916
Fatty acid methyl ester content (FAME)	Percent by volume	Max. 5		EN 14078

for requirements. For instance, wear protection (lubricity) and limits on polyaromatics have been newly incorporated in recent years.

The test methods applied in the standard and to ongoing quality control are also based on standardized laboratory methods. By performing extensive tests in complete engines for example, working groups from the engine and petroleum industry and independent institutes ensure that the standardized methods are relevant to the application of fuel in vehicles.

EN 590 contains legal regulations from EU directives and compromises negotiated between the automotive and petroleum industries in the CEN working group TC19/WG24.

Furthermore, the German government has enacted the Fuel Quality Act [4-3] and related provisions for its implementation [4-4]. State authorities are able and required to review the conformity of fuels offered for public purchase to the standards. These efforts are aimed at ensuring that customers who purchase fuel with an EN 590 designation can be confident that the fuel is suitable for their engines and is environmentally compatible.

4.1.7.1 Worldwide Fuel Charter

In the early 1990s, diesel engine manufacturers were universally unhappy with the minimum requirements specified for diesel fuel, the unduly slow progress of international standardization and the fragmentation stemming from national differences in quality. Peugeot and Renault compiled a *Cahier des Charges* that not only contained higher diesel fuel quality specified by lab values but also a method to test the cleanliness of injection nozzles (additivation).

The automotive associations in Europe, the USA and Japan have compiled performance specifications for "good" qualities of diesel fuel. A logical consequence of globalization was the drafting of a Worldwide Fuel Charter by the world's automotive associations, the fourth edition of which was published in 2006 [4-5]. It applies to both gasoline and diesel fuel and contains four categories for differently developed markets, which are defined by the requirements of emission control legislation.

The Fuel Charter outlines laboratory and engine test methods with related limits. Category 4 requires sulfur-free fuel, i.e. with a maximum sulfur content of 5–10 ppm. Commercially available diesel fuels now fulfill a large number of stipulated quality criteria.

4.1.7.2 Reference Fuels

Reference measurements, legal approval, engine devlopments, engine oil tests, etc. require defined reference fuels. Naturally, these can only be defined when everyone involved collaborates. The Deutscher Koordinierungsausschuss (DKA) is responsible for the development of test methods for the evaluation and development of fluids for combustion engines in Germany. The Coordinating European Council (CEC) is responsible for the same in Europe. A CEC reference fuel for diesel engine emission tests is legally stipulated. By definition, it is a weighted mean value of the European commodity.

Engine oil tests for diesel engines primarily check piston cleanliness, sludge formation and wear, which are also influenced by the fuel. Therefore, a CEC reference fuel is used in most diesel engines even for oil tests.

4.1.8 Basic Parameters and Test Methods

4.1.8.1 Ignition Quality (Cetane Number, Cetane Index)

Naturally, a fuel's ignition quality plays a prominent role in auto-ignition engines. Prior experience has demonstrated that laboratory equipment is unable to determine it accurately enough. Rather, it can only be reliably determined in an engine, preferably a modern multi-cylinder engine. However, ignition quality is determined in standardized single cylinder engines for reasons of standardization. The CFR engine compliant with EN ISO 5165 is used worldwide. The so-called BASF engine compliant with EN 15195 is also used, mainly in Germany. The cetane number specifies ignition quality. Where the CFR engine is used, it is calculated for a fuel under test by varying the engine's compression ratio for constant ignition delay (high fuel-ignition quality requires a drop in the compression ratio and vice versa). The air/fuel ratio in the BASF engine is varied to obtain a constant and defined ignition delay. Cetane and α-methylnaphthalene with defined cetane numbers of 100 and 0 respectively are used as reference fuels (Fig. 4-14). By definition, a fuel that produces the same engine settings in a test engine as a mixture of 52% cetane and 48% α-methylnaphthalene (percent by volume) has a cetane number of 52.

Fuels with insufficient ignition qualities produce greater ignition delays. This leads to poor cold starting, high pressure peaks and thus higher exhaust and noise emission.

Paraffins have high cetane numbers, while hydrocarbons with double bonds and aromatics in particular have low ignition quality. The cetane numbers of paraffins increase as the length of their chains increases, i.e. as their molecular weight and boiling point increase (Figs. 4-5 and 4-15).

EN 590 stipulates a minimum cetane number of 51. German diesel fuels have a cetane number of approximately 52 with a tendency toward higher values in summer fuels and lower values in winter fuels (because high boiling paraffins are eliminated partly to ensure safe low temperature filterability).

The ignition quality of conventional diesel fuels is adequately characterized by their cetane numbers. This is also true for fuels with ignition qualities enhanced by ignition improvers. By contrast, alcohol fuels with large quantities of

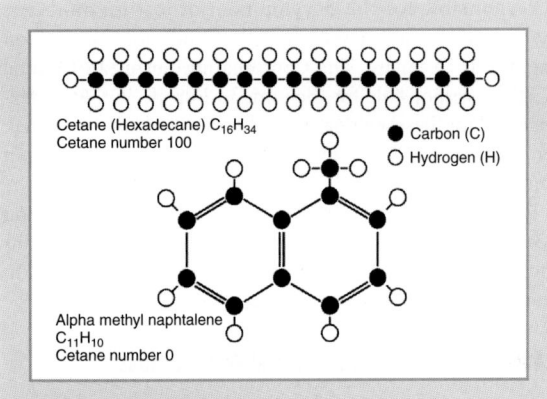

Fig. 4-14 Reference fuels for cetane number measurement

ignition improvers perform differently in a complete engine than their cetane number might lead one to expect.

Starting and noise properties and, in particular, gaseous exhaust emissions of HC and CO as well as NO_X improve in diesel engines as the cetane number increases.

High pressure and high temperature hydrogenation elevates the cetane number of refinery components. The refinery processes of hydrocracking and desulfurization are headed in this direction to some extent. Plants that produce synthetic diesel fuel components (e.g. from natural gas) for premium fuels have now become cost effective. The available quantities of such components are still limited however.

Alternatively, the cetane index may be calculated from the fuel density and boiling characteristics according to EN ISO 4264 to evaluate the ignition quality of conventional diesel fuels without measuring the cetane number. The empirical formula developed for typical commercial fuels is based on an analysis of approximately 1,200 diesel fuels. With a number of corrections, the fundamental statement asserts that the cetane number drops as density increases (i.e. the fraction of cracking products with double bonds or aromatics increases) and the cetane number increases as the high boiling components increase (the chain lengths of molecules grow).

The formula has been altered repeatedly to reflect longer-term changes in refinery structures and diesel fuel components. However, it is less suitable for individual fuel components. By the same token, it cannot specify the cetane number of fuels with ignition improvers.

The imprecision of the empirical formula (in particular the non-inclusion of fuels containing ignition improvers) and the relatively broad spread in the measurement of cetane numbers result in variations between the measured cetane number and the calculated cetane index of up to 3 units.

4.1.8.2 Boiling Characteristics

Diesel fuels consist of hydrocarbon mixtures that boil in a range of approximately 170–380°C. EN ISO 3405 defines distillation equipment and distillation conditions (among others, a variable supply of energy for a constant distillation

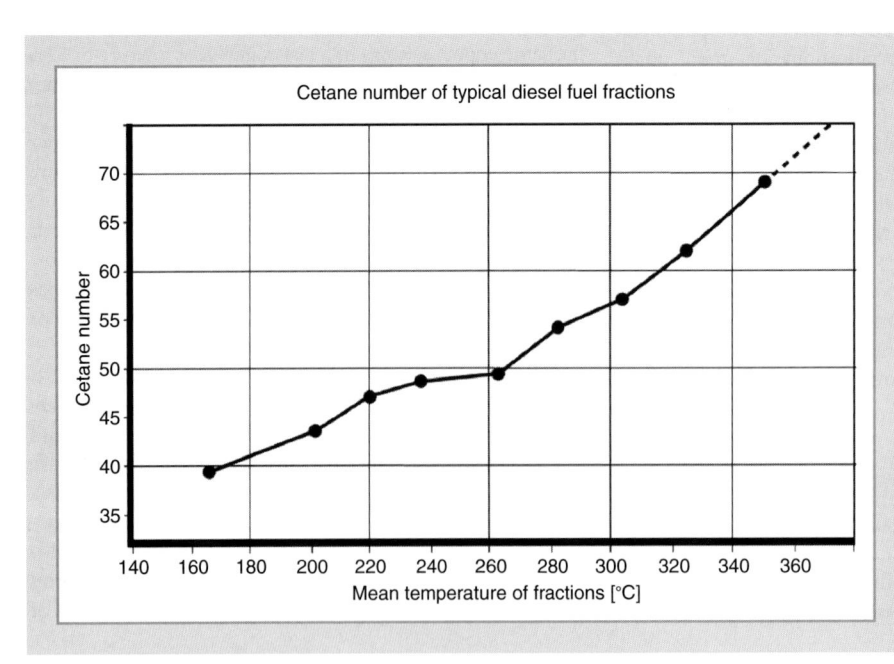

Fig. 4-15 The cetane number of typical diesel fuel fractions rises with the boiling temperature

rate of 4–5 ml/min). The method does not specify any physically exact boiling range but rather practical conditions of boiling characteristics approximating rapid vaporization. Low boiling components are partially retained when a product mixture's temperature rises, while high boiling components are already entrained. Therefore, the physically exact initial or final boiling point is lower or higher than specified. Nevertheless this standardized method is well suited for evaluating diesel fuels.

In principle, hydrocarbons outside the usual boiling range are also suitable for diesel engine combustion, e.g. marine engines run with fuels that evaporate at considerably higher temperatures. A number of other constraints (e.g. viscosity, low temperature flowability, density, ignition quality and flash point) limit the boiling range of fuels allowable for vehicle operation and fuel system design.

Therefore, EN 590 stipulates only three points that define fuel in the mid to final boiling point range (Fig. 4-16). However, the final boiling point of diesel fuels is not precisely determinable, in part because cracking processes may already commence in the distillation equipment at temperatures above 350°C when the final fuel fractions are being vaporized. This uncertainty is also the reason why EN 590 does not explicitly define the final boiling point.

Diesel engines react less critically to a fuel's boiling rate. Lowering the final boiling point has proven to enhance combustion performance and reduce exhaust emissions in high speed diesel engines. Therefore, the latest revision of EN 590 lowered the temperature for a 95% vaporized fraction from 370 to 360°C.

4.1.8.3 Sulfur Content

Crude oil already contains varying quantities of sulfur. SO_2 formation during combustion, engine oil acidification, sulfate emission, increased particulate emission and damage to exhaust gas aftertreatment systems (catalysts) make the desulfurization of diesel fuel essential. Even though road traffic contributes little directly to sulfur dioxide pollution, exhaust gas aftertreatment systems require sulfur-free fuel (S <10 ppm). The sulfur contents can be ascertained with different methods, e.g. UV fluorescence or X-ray fluorescence. EN ISO 20846 and EN ISO 20884 describe the appropriate methods for diesel fuels. The limits for the maximum allowable sulfur content have been lowered steadily. Diesel fuel's formerly higher sulfur content seriously affected engine service life. The acidic products produced during combustion caused corrosive wear, above all on the cylinder barrel and in the piston ring zone, which, in turn, caused increased oil consumption and loss of power. Cylinder wear particularly increases during stop-and-go operation. However, highly alkaline engine oils can neutralize the acidic combustion products that accrue. Reducing oil change intervals is another option when the sulfur content is higher. The EU has limited the maximum sulfur content to 50 ppm since 2005 and to 10 mg/kg (sulfur-free) since January 2009.

4.1.8.4 Low Temperature-Flowability

The hydrocarbons generally viewed as beneficial for operation in diesel engines, paraffins unfortunately frequently

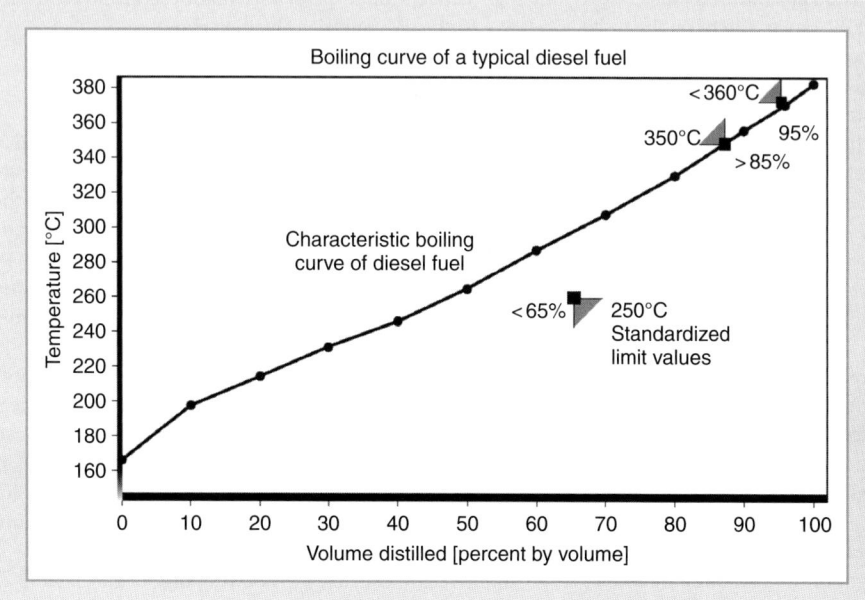

Fig. 4-16 Boiling characteristic of a typical diesel fuel

precipitate as paraffin crystals at low subzero temperatures, which coagulate and clog the fuel filter, lines and injection system and thus impede an engine's operation. Although a start is frequently still possible, the lack of fuel causes the engine to stop without suffering damage.

Diesel fuels' low temperature flowability is measured with the cold filter plugging point (CFPP), which ascertains the lowest temperature at which a diesel fuel still flows unimpeded and is filterable. EN 116 stipulates measuring this with a sieve with a mesh width of 45 µm and gauge of 32 µm at a cooling rate of approximately 1°C/min.

Fuels without cold flow additives have a CFPP just slightly below the cloud point, i.e. the start of paraffin separation. Depending on the type and quantity of additives, cold flow additives reduce paraffin crystals so that the CFPP can be lowered more than 20°C below the cloud point. The cloud point (CP) is the measured temperature at which the first crystallizing paraffins visibly cloud fuel. It is irrelevant for diesel fuel filterability in modern vehicles and therefore also not specified in DIN EN 590. CFPP values of up to approximately 15°C below the cloud point can solely be obtained by using flow improvers. Further reductions are possible by combining flow improvers and wax anti-settling additives.

EN 590 calls for low temperature resistances that correspond to ambient temperatures and are expressed by the CFPP. The values presented in Table 4-5 were selected for Germany, i.e. a Central European country. Arctic fuels require adequate filterability at lower temperatures. These fuels also have a cloud point specification with a difference of 10°C between the cloud point CP and cold filter plugging point CFPP.

The CFPP approximately describes a vehicle's operational reliability. Since the fuel contains small paraffin crystals below the cloud point and in the CFPP range, filters ought to be installed in a fuel system whenever the engine or other measures heat the filter once an engine starts. Vehicles with actively heated filters facilitate reliable operation even at temperatures lower than specified by the CFPP. The reliable operational temperature in vehicles with filters heated by engine heat approximately corresponds to the fuel's CFPP.

4.1.8.5 Density

Density is the mass of a certain volume of fuel specified in kg/m^3 at 15°C (EN ISO 3675, EN ISO 12185). Fuel density has traditionally been measured with aerometers or, more recently, on the basis of the principle of flexural vibration. A pipe bent into a U-shape is filled with a small amount of the fuel under test. The pipe is excited to vibrate and the resultant resonant frequency is measured and converted into a density value.

Diesel fuel's density increases as its carbon content increases, i.e. as the chain length of the paraffinic molecules increases and the proportion of double bonds increases

Table 4-5 Seasonal cold filter plugging point (CFPP) specification of diesel fuels according to EN 590 (in Germany)

Winter	November 16–February 28	Class F	Max. −20°C
Spring	March 1–April 14	Class D	Max. −10°C
Summer	April 15–September 30	Class B	Max. +/−0°C
Fall	October 1–November 15	Class D	Max. −10°C

(aromatics and olefins). Accordingly, an increasing hydrogen fraction in diesel fuel lowers its density. A fuel's volumetric calorific value also rises as its carbon content increases, i.e. increasing density is an indication of higher calorific value per volume. Hence, the energy supplied to an engine increases when the fuel density increases while the volume injected remains constant. Since the quantity of fuel injected in standard vehicle diesel engines is not yet controlled as a function of fuel, high fuel density in the full load range can simultaneously boost engine power and increase particulate and smoke emission. This is especially true of volume metering injection systems. Accordingly, when the same power is delivered, the volumetric fuel consumption drops as density increases. Decreasing density has the converse effect, i.e. higher volumetric consumption and less particulate emission with potential loss of power at full load. A recent revision of EN 590 lowered the maximum density from 860 to 845 kg/m^3 with the goal of reducing particulate emissions in the existing car fleet.

These correlations are only valid when fuels' combustion performance remains roughly constant. The higher hydrogen and thus energy content in synthetic (gas-to-liquid) fuel than in conventional diesel fuel manifests itself positively relative to the gravimetric calorific value.

4.1.8.6 Viscosity

Viscosity (resistance) is the capability of a flowable material to absorb stress while deforming. It is only dependent on the rate of deformation (see DIN 1342). Fuel viscosity influences the fuel delivery characteristic in delivery and injection pumps and the fuel's atomization by the injection nozzle. A distinction is made between dynamic viscosity η in Pa·s and kinematic viscosity v, the latter being the quotient of dynamic viscosity and density specified in m^2/s or in mm^2/s.

The Ubbelohde capillary viscosimeter (EN ISO 3104) measures diesel fuels' kinematic viscosity by measuring the time it takes a sample quantity of 15 ml to flow through a defined capillary tube at a temperature of 40°C.

EN 590 stipulates a viscosity range of 2.0–4.5 mm^2/s for diesel fuels. Lower minimum values down to minimum values of 1.2 mm^2/s apply to arctic fuels.

Commercially available fuels have a range of 2.0–3.6 mm^2/s at the stipulated temperature of 40°C. Viscosity is usually not

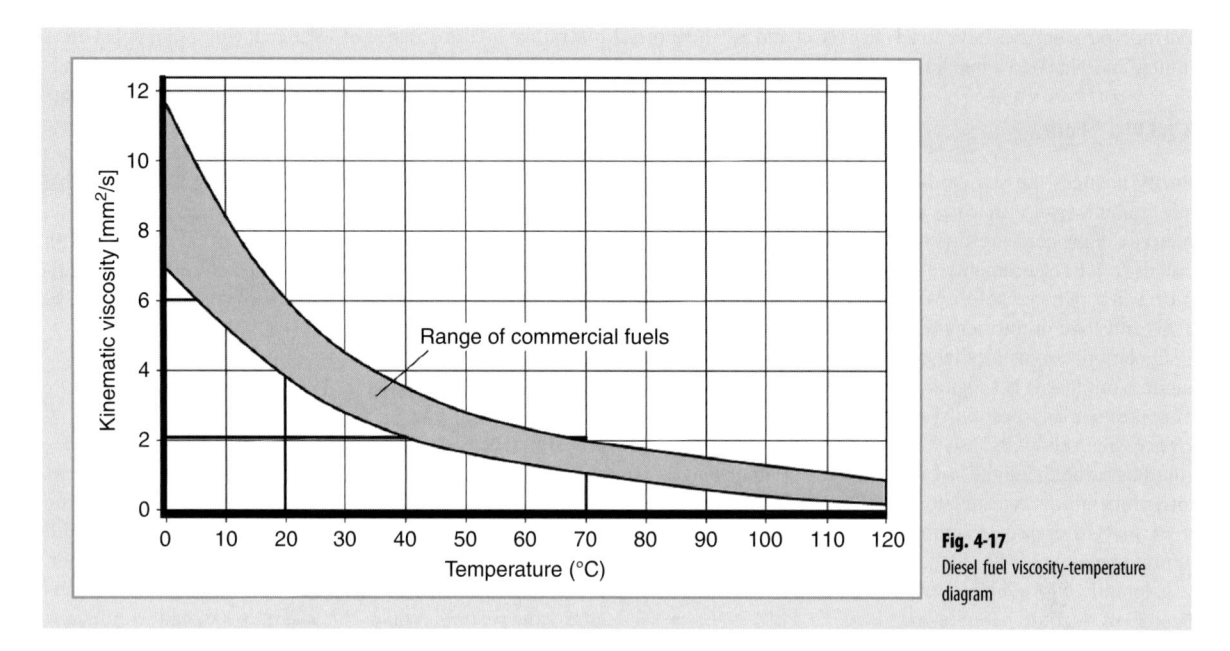

Fig. 4-17 Diesel fuel viscosity-temperature diagram

a primary criterion in fuel production. Rather, it results from other fuel parameters.

Viscosity increases as the temperature drops and the pressure rises. For instance, the viscosity of diesel fuel doubles when the temperature drops from 40 to 20 °C (Fig. 4-17) or the pressure increases to approximately 600 bar. Viscosity influences a fuel's flow and pump characteristics in a fuel system and the development of the injection spray in the combustion chamber, which is shaped by the injection nozzle. Excessively high viscosity impedes pumpability at low temperatures and causes cold start problems, while excessively low viscosity makes hot starting difficult and causes power losses at high temperatures and pump wear.

4.1.8.7 Flash Point

A liquid's flash point is the lowest temperature at normal pressure at which it vaporizes in a closed vessel in such quantity that a vapor-air mixture is produced, which external ignition can cause to ignite (as defined by EN ISO 2719). The flash point is a safety parameter. It is not important for combustion in engines.

The risk of ignition decreases as the flash points increase (in °C). A maximum of 55°C is specified for diesel fuel. A number of ordinances and technical regulations regulate the storage and transportation of inflammable liquids, the handling of hazardous substances and industrial safety. The flash point is one of the major criteria in most of these ordinances when they are applied to diesel fuel.

Gasoline has a much lower flash point (<2°C). The ingress of even small quantities of gasoline can reduce the flash point of diesel fuel below the limit of 55°C and represents a safety hazard, e.g. when alternately transported in the same tank chambers used for diesel fuel.

The flash point limits the use of low boiling components in the production of diesel fuel. The flash point's limit is one reason why the initial boiling point of diesel fuel does not have to be specified.

4.1.8.8 Aromatics

With their cyclic and double bonded molecular structures, aromatics are basically unsuitable for diesel engine combustion because of their poor ignition quality (see Sect. 4.1.4). However, the mononuclear aromatics with longer side chains present in diesel fuels have properties similar to long chain paraffins. Benzene (without side chains), toluene and xylene (ring benzene with short side chains) have boiling points that are too low for diesel fuel and are therefore not contained in it. Given their higher boiling point, polynuclear aromatics, e.g. naphthalene (2-ring) or anthracene (3-ring), may also be present without "thinning" side chains. 4-ring aromatics boil above 380°C, i.e. above diesel fuel's boiling range, and are therefore only allowed to be present in traces.

Finding a suitable method for continuous quality control that identifies the aromatics in diesel fuel was problematic. For a number of years, the total polyaromatics have been measured with high pressure liquid chromatography and refractive index detectors in accordance with EN 12916. This determines the mass of polyaromatic ring structures including their side chains, i.e. the content of aromatic ring structures actually present is lower than detected. The content of polynuclear aromatics in commercial diesel fuels drops as the number of rings increases. Mononuclear aromatics with long side chains are now known to behave similarly to paraffins, particularly with respect to exhaust emission and particulate formation.

Polynuclear aromatics have an adverse affect and are therefore limited in EN 590 to a maximum of 11% by weight.

4.1.8.9 Purity

Purity includes the standardized criteria of carbon residue, ash content, total impurities and water content. Since diesel fuel is a high grade resource and its use in diesel engines entails strict requirements, it is supposed to be clean and pure when pumped into a vehicle's tank. It must be free of acids and solid impurities and clear at room temperature.

In conjunction with other criteria, carbon residue (according to Conradsen) is a significant indicator of the formation of deposits in an injection system and combustion chamber. Hence, this residue is also limited. Carbon residue is measured by carbonizing the last 10% from the boiling analysis at low temperature according to EN ISO 10370. A maximum of 0.3% may be present. Commercially available fuels contain approximately 0.03%. Since some diesel fuel additives (e.g. ignition improvers) elevate carbon residue, only unadditized fuels ought to be investigated.

Ash content describes the content of inorganic contaminants in the fuel. It is ascertained by incinerating/ashing a fuel sample according to EN ISO 6245 and may not exceed 0.01% by weight. It is usually below the detection limit in typical commercial fuels.

Total impurities specify the total undissolved contaminants (sand, rust, etc.) in a fuel according to EN 12662. This is measured by filtering and weighing after washing with n-heptane. Twenty-four milligrams per kilogram are the permissible maximum. Values below 10 mg/kg are usually common. Higher values may cause problems, especially in winter when paraffin crystals additionally load filters that are partially clogged by contaminants.

Water is already present in crude oil and also winds up in fuel during some refinery processes. Hence, fuel is dried afterward. The amount of dissolved water decreases as the temperature and aromatics content decrease. Roughly between 50 and 100 mg/kg at 20°C, it is significantly lower than in gasoline. Two hundred milligrams per kilogram are the permissible maximum. Water content is measured by titration based on the Karl Fischer method in EN ISO 12937.

Any ingress of water in diesel fuel ought to be prevented, especially in winter, because ice crystals can quickly clog filters together with paraffin crystals.

The careful handling of diesel fuel required also entails preventing the formation of any algae, bacteria and fungi. This can usually be accomplished merely by cleaning and regularly dewatering the storage tanks in a distribution network.

4.1.8.10 Lubricity

Diesel fuels with very low sulfur content have proven to potentially cause high wear in fuel-lubricated injection pumps. This is caused not by the absence of sulfur as such but rather by the absence of wear reducing polar substances removed during desulfurization. Additives assure the requisite lubrication reliability. This can be ascertained by a fast mechanical test using a high frequency reciprocating rig (HFRR) based on EN ISO 12156-1. The HFRR test simulates sliding wear in an injection pump by rubbing a sphere (with a 6 mm diameter) on a polished steel plate under liquid at a constant pressure at a test temperature of 60°C (Fig. 4-18). The measured flattening of the sphere produced after 75 minutes is the test result (mean wear diameter in μm). EN 590 allows a maximum wear diameter of 460 μm.

4.1.8.11 Calorific Value

A distinction is made between the gross calorific value H_o and the net calorific value H_u (now only designated as calorific value H). The gross calorific value H_o, or calorific power, is ascertained by complete combustion in a bomb calorimeter with an oxygen atmosphere of 30 bar. Carbon dioxide and potentially present sulfur dioxide are gaseous after combustion, while the steam produced condenses. Since the steam does not condense during engine combustion, the calorific power is unrealistically high for a practical evaluation of fuels. Therefore, the hydrogen content is determined in an elemental analysis and the heat of condensation from the steam produced is calculated and subtracted from the calorific power to ultimately obtain the calorific value (DIN 51900).

Calorific value is a value that results from density, boiling characteristics and fuel composition. It is not measured for the purpose of quality control during fuel production. Only fuels used for particular research and development work must be measured. Table 4-6 presents values of some typical diesel fuels. Given their higher densities (higher carbon fraction), diesel fuels have volumetric calorific values that are approximately 15% higher than those of gasolines.

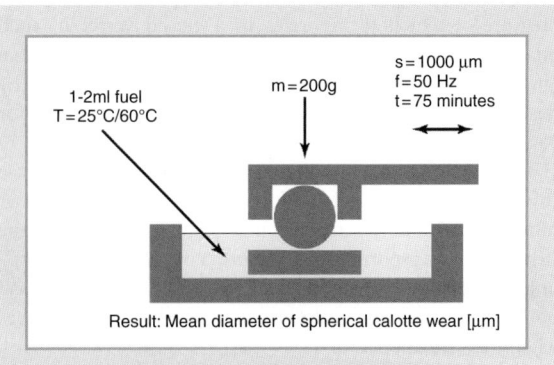

Fig. 4-18 Schematic lab apparatus for the determination of fuel lubricity (HFRR lubricity test)

Table 4-6 Calorific values und elemental analyses of commercially available diesel fuels (Source: DGMK, Hamburg)

Diesel fuel	Density at 15°C [kg/m^3]	Elemental analysis			H_o MJ/kg	H_u MJ/kg	H_u MJ/l
		C	H	O			
		Percent by weight					
A	829.8	86.32	13.18	–	45.74	42.87	35.57
B	837.1	85.59	12.7	–	45.64	42.9	35.91
C	828.3	86.05	13.7	–	46.11	43.12	35.72
Mean	831.7	85.99	13.19	–	45.84	42.96	35.73

4.1.8.12 Corrosiveness on Metals

Diesel fuel inevitably comes into contact with moisture and oxygen during its transport, storage and use in vehicles. Resultant temperature changes and condensation can corrode lines and reservoirs. Corrosion products can cause damage in a vehicle's distribution chain and fuel system including its sensitive injection nozzles.

Steel's corrosion behavior is tested according to DIN 51585. A steel rod is inserted in a mixture of fuel and distilled water (variant A) or artificial saltwater (variant B) with a 10:1 ratio for twenty-four hours at 60°C. Once the test has concluded, the formation of rust is evaluated visually. This method is used, for example, to test the effectiveness of anti-corrosion additives (see Fig. 4-12).

The corrosion of cupreous materials that come into contact with fuel (e.g. fuel pump components) is problematic for two reasons. On the one hand, components are damaged. On the other hand, dissolved copper is catalytically active and causes the formation of substantial molecular impurities in the fuel. A fuel's corrosiveness basically depends on its water content, oxygenic compounds, the type and quantity of sulfur compounds and, naturally, the anti-corrosion additive employed. A ground copper strip is brought into contact with diesel fuel at 50°C for three hours (EN ISO 2160) to test the corrosion limit stipulated in EN 590. Even under adverse conditions, additives largely protect all the metals that come into contact with fuel.

4.1.8.13 Oxidation Stability

Fuels may partially oxidize and polymerize when stored over long periods (> 1 year as strategic stocks/petroleum reserves). This can cause insoluble constituents to form and thus filters to clog later in vehicles. The chemical mechanism is the cleavage of hydrogen and the attachment of oxygen, especially to unsaturated olefinic fuel molecules. Antioxidants (additives) can prevent or effectively interrupt the process of oxidation and polymerization caused by the formation of "free radicals" during the process.

Oxidation stability is measured in a lab by artificially ageing fuel for sixteen hours in an open vessel aerated by pure oxygen (3 l/h) at a temperature of 95°C (EN ISO 12205). No more than 25 g/m^3 of soluble and insoluble resinous material may form. The quantity of resinous material this method measures in commercially available diesel fuels is substantially lower, generally below 1 g/m^3. A new test method has been developed for fuels containing FAME and was introduced in the European standard EN 590 and pr EN 15751. In this test, filtered air is fed through the fuel sample and the air-fuel vapor is passed through distilled water. The water's electrical conductivity is measured. The time measured until a steep rise in conductivity is observed is employed as a quality criterion. A minimum of twenty hours is required. At present, this test method is unsuitable for pure hydrocarbon fuels.

4.1.9 Future Trends

Their high efficiency, which enables fuel economizing, and the still competitive production costs of diesel fuels compared to gasoline will ensure diesel engines retain a large share of the engine market for road vehicles for a long time. Growing acceptance engendered by diesel engines' effective exhaust gas aftertreatment systems as well as good drivability, high performance and reduced noise emissions will encourage this trend and wrest more market share from gasoline engines. Recent successes in racing clearly document this development. Moreover, the consumption of diesel fuel is steadily increasing. Alternative resources for the production of diesel fuel from biomass are well established.

However, vegetable oils are unsuitable for current and future engines without chemical processing. Conversion enables producing products, e.g. vegetable oil methyl ester or fatty acid methyl ester (FAME), that may be used as fuels when they meet the requirements formulated by the working group TC19/WG24 (see Sect. 4.2). These products may be blended into petroleum-based diesel fuel.

A number of countries stipulate small concentrations of FAME as a blending compound to reduce CO2 emissions. Accordingly, the European standard EN 590 allows blending a maximum of 7% FAME (which in turn must comply with the requirements of EN 14214 [4-6]). The basis for this is the EU Biofuels Directive 2003/30/EG [4-7]. However, the quantities are small at present because availability is limited, in part because products from certain region are inacceptable since they potentially damage rain forests. In the past, only tax breaks made biodiesel competitive. However, the costs of biodiesel production were recently in ranges similar to conventional diesel fuels because of the significant rise in crude oil prices.

Quality requirements and monitoring are even more important for biodiesel since the quality of natural products varies more greatly. While the use of biomass to produce fuel holds promise in the medium and long-term, its short-term development is unclear. As tax privileges for and compulsory blending of conventional diesel fuel expire, biodiesel as a separate, pure variety may only be expected to retain a small market share, especially since its use in conjunction with state-of-the-art exhaust gas aftertreatment systems is not without problems.

Engines and fuels, especially their synthetic components and additives, are both being developed further. Developments in the production of fuel components from renewable raw materials are also promising. Unlike rape oil methyl ester, these, e.g. CHOREN's BTL process (see Sect. 4.2.2.4), are able to utilize entire plants or residues as raw material. Such developments will also further postpone the longer range depletion of conventional diesel fuel.

4.2 Alternative Fuels

4.2.1 Introduction

Globally increasing energy consumption, related emission of gases that affect the climate, the projected depletion of fossil fuels (see Sect. 4.1.2), unstable political situations in important petroleum and natural gas supplier countries and rising energy prices are intensifying the search for possible alternatives to conventional diesel fuels and gasolines produced from petroleum and for solutions for a secure and sustainable supply of fuel. The goal of sustainability has been outlined in the demand in [4-8] to "Limit transport related greenhouse gas emissions emissions to sustainable levels," i.e. "to eliminate transportation as a major source of greenhouse gas emissions.... This likely will require both the development of hydrogen as a major transport energy carrier and the development of advanced biofuels." However, sustainability also requires an ecological and economic balance that incorporates sociological aspects.

Figure 4-19 presents an overview of potential fuel paths.

Basically, two groups of fuel alternatives have been identified [4-9]:
– conventional fuels based on fossil fuels as well as synthetic fuels and hydrogen and

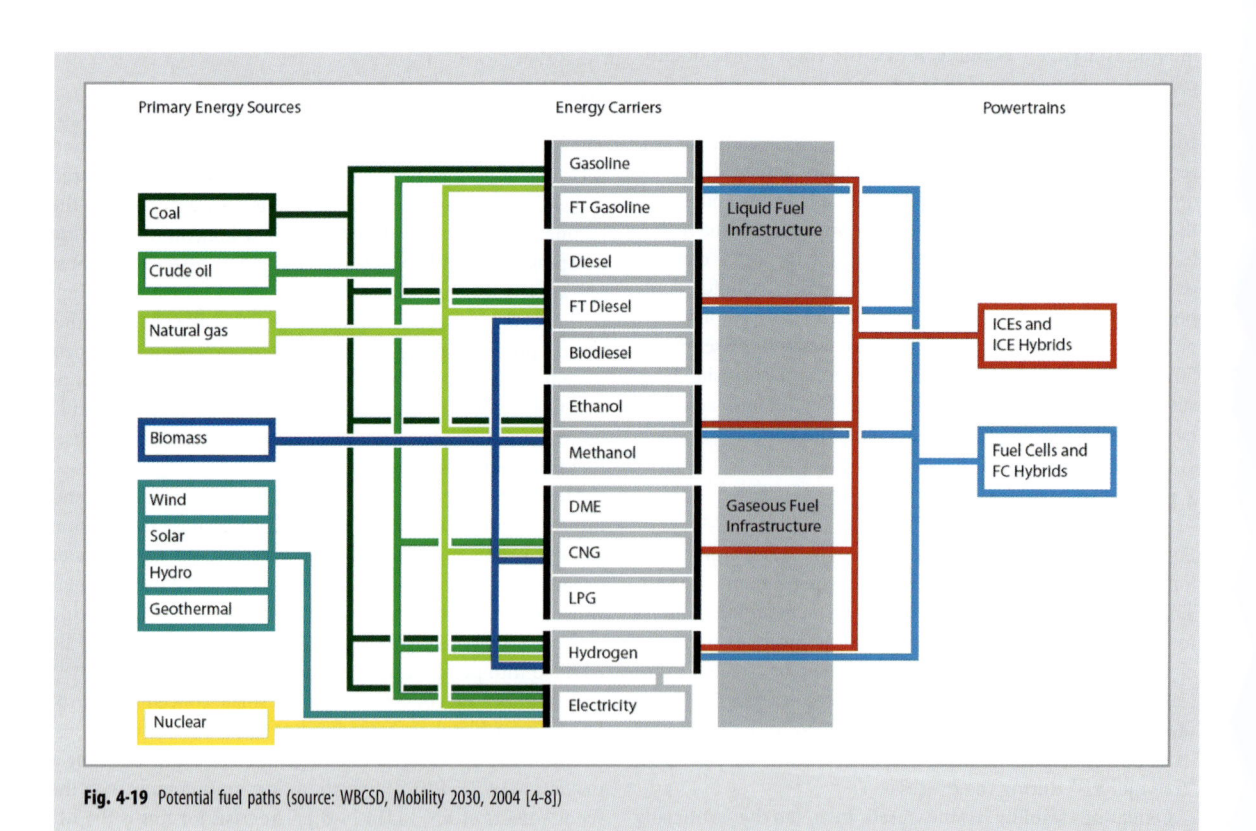

Fig. 4-19 Potential fuel paths (source: WBCSD, Mobility 2030, 2004 [4-8])

– renewable alternatives from, among others, biogenic energy, water power or photovoltaic.

Fuel contributes to meeting the stated goal in two ways:
1. Directly (in/at the engine):
   – improved energy conversion in conventional combustion systems,
   – prevention of the formation of pollutants and
   – new, alternative combustion systems and exhaust gas aftertreatment systems.

2. Indirectly:
   – closed CO_2 cycle for biofuels and
   – reduction of fuel's C/H ratio up through H_2.

Short to medium-term, this yields three groups of alternative fuels:
– synthetic fuels from fossil fuels (GTL),
– first and second generation biofuels and
– low to carbon-free fuels (CNG to pure hydrogen).

Hydrogen powered fuel cells presently have the potential for maximum efficiency as a single energy supply for vehicles. However, this is conditional on the availability of renewable hydrogen since hydrogen can only help reduce CO_2 emissions when it is also produced from primary energy free of CO_2. Three critical technological barriers are hampering this:
– the lack of a storage system for mobile use that is acceptable to customers,
– the lack of a distribution infrastructure and
– the lack of an economically feasible technology to produce renewable hydrogen.

Since a technological solution to these three barriers is not yet available, hydrogen and thus fuel cell drives as well only constitute a long-term solution.

If hydrogen will only be available in the long-term, then a short to medium-term solution that also has real prospects in the long term also becomes a challenge.

The four basic requirements for a future fuel for road vehicles (and in a broader sense also for ships, rail vehicles and aircraft) are:
– high power density,
– certain supply,
– overall economic feasibility and
– the incorporation of environmental and climate protection requirements.

No currently available energy source, not even hydrogen, fulfills these requirements.

A multitude of variants have been discussed and also researched in part [4-10]. A trend toward increased diversification of fuels is emerging. Figure 4-20 presents an overview of the most frequently discussed variants.

Alternative fuels may be classified according to their primary energy, type of production and properties, thus making it possible to define groups of liquid fuels made from vegetable oils and fats, alcohol fuels produced by fermentation and gaseous fuels. Alternative liquid fuels obtained from agricultural products are called first generation biofuels. Their use is controversial because their production competes with food production. Liquid synthetic fuels with properties that can be influenced by the production process are becoming increasingly important. Synthetic fuels made from renewable biomass (utilizing the entire plant) are second generation biofuels.

4.2.2 Liquid Fuels

4.2.2.1 Fuels from Vegetable Oils and Fats

Vegetable oils are pressed from oleaginous fruits. Rudolf Diesel already recognized their utilizability in diesel engines. Their high energy content approaching diesel fuel's is

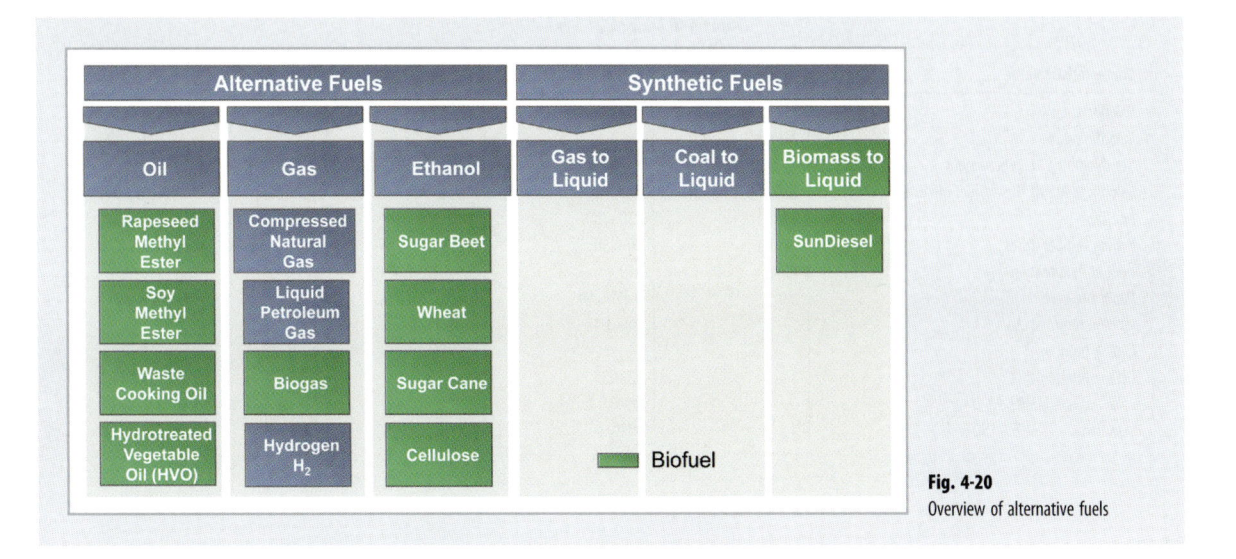

Alternative Fuels			Synthetic Fuels		
Oil	Gas	Ethanol	Gas to Liquid	Coal to Liquid	Biomass to Liquid
Rapeseed Methyl Ester	Compressed Natural Gas	Sugar Beet			SunDiesel
Soy Methyl Ester	Liquid Petroleum Gas	Wheat			
Waste Cooking Oil	Biogas	Sugar Cane			
Hydrotreated Vegetable Oil (HVO)	Hydrogen H_2	Cellulose			

Biofuel

Fig. 4-20
Overview of alternative fuels

advantageous. However, their inadequate physical properties such as viscosity and high boiling temperature that result in poor combustion are highly disadvantageous.

In addition, their cold start performance is unsatisfactory and they are only stable in storage for a limited time. Some characteristic values of vegetable oils are compared with diesel fuel in Table 4-7.

Pure Vegetable Oils

The fuel properties of vegetable oils necessitate modified engines. Rape oil has acquired limited importance for powering agricultural machinery and tractors [4-11]. Compliance with the fuel standard is important [4-12].

The automotive industry rejects the use of pure vegetable oils as well as their blending in diesel fuel because of their poor fuel properties. Another potential use of vegetable oils for fuel production might be their addition to the refinery process, the outcome being a higher quality diesel fuel partially produced from biomass [4-13].

Biodiesel

The term biodiesel (also called first generation BTL) originally denoted esterified rape oil (rape oil methyl ester or RME): Esterifying this vegetable oil fundamentally improved its properties. The term biodiesel has been expanded to include esterified fatty acids (vegetable oils, animal fats and used cooking oils).

A blend of up to 7% biodiesel (B7) by volume in diesel fuel is now accepted in accordance with EN 590 and DIN 51628. This corresponds to the European standard EN 14214 for fatty acid methyl ester (FAME) [4-1, 4-6]. Once larger areas of the market are supplied with a 7% blend (e.g. EU 25), the biodiesel content could conceivably be increased by up to 10% by volume (B10) at first and even by up to 20% by volume (B20) later, provided the appropriate compatibility tests have concluded successfully. Problems may well arise, specifically with the latest diesel engine engineering and particulate filter systems. Should acceptance nevertheless still be possible, EN 590 shall have to be revised in due time.

Pure biodiesel (B100) is unlikely to receive any further approval for car engines in the future for reasons of emissions.

4.2.2.2 Alcohols

While alcohols possess excellent combustion properties, their energy density (vehicle range), cold start performance and corrosion behavior toward metals and elastomers are significant disadvantages (Table 4-8). Hence, larger percentages of alcohol (>15% by volume) necessitate the development of special engines.

Table 4-7 Comparison of vegetable oil and diesel fuel specifications (DF)

Parameter	Unit	Diesel	Rape Oil	Sunflower Oil	Linseed Oil	Soy Oil	Olive Oil	Palm Oil
Density	g/cm^3	0.83	0.915	0.925	0.933	0.93	0.92	0.92
Viscosity (20°C)	mm^2/s	≈2	74	65.8	51	63.5	83.8	39.6
Calorific value	MJ/kg	43	35.2	36.2	37.0	39.4	(40.0)	35
Cetane number	–	50	40	35.5	52.5	38.5	39	42
Flash point	°C	55	317	316	320	330	325	267

Table 4-8 Fuel specifications of alcohols

Characteristic value		Unit	Gasoline	Diesel	Methanol	Ethanol
Calorific value		MJ/kg	≈42	42–43	19.7	26.8
Calorific value		MJ/dm^3	≈32	36	15.5	21.2
Stoichiometric air requirement		kg/kg	14–14.7	14.5	6.46	9.0
Calorific value of the air/fuel mixture		kJ/kg	≈2,740	2,750	2,660	2,680
Density		kg/m^3	730–780	810–855	795	789
Boiling temperatures		°C	30–190	170–360	65	78
Heat of evaporation		kJ/kg	419	544–785	1,119	904
Vapor pressure		bar	0.45–0.90	–	0.37	0.21
Ignition limits at λ_V			0.4–1.4	0.48–1.40	0.34–2.00	
Cetane number			–	45–55	–	–
Octane number	ROZ		98–92		114.4	114.4
	MOZ		88–82		94.6	94.0
Sensitivity	MOZ – ROZ		≈10	–	≈20	≈17

Methanol

Methanol is primarily produced from primary fossil fuels (natural gas, coal). However, renewable energy sources (biomass) may also be used. Methanol in the form of methyl tertiary butyl ether (MTBE) is primarily added to gasoline as an anti-detonation agent. EN 228 allows up to 15% MTBE. Up to 3% methanol may also be directly added secondarily. Larger percentages of methanol ought to be avoided for reasons of toxicity.

Ethanol

Ethanol is fermented directly from sacchariferous or amylaceous raw materials (grains, sugar beets, sugar cane, etc). Wood and culmiferous feedstocks may be used when the lignocellulose (e.g. straw) is pulped enzymatically. Such a process was developed by the Canadian company IOGEN [4-14] and has been implemented on an industrial scale (Fig. 4-21).

Ethanol is suitable for gasoline engines. First and foremost, EN 228 calls for up to 15% ethyl tertiary butyl ether (ETBE) by volume (47% ethanol content) [4-15]. Secondly, EN 228 allows directly using up to 5% ethanol (E5) by volume [4-16]. It is impossible to supply all of Europe at present because too little is available. Should the rapidly growing capacities make this possible in the future, it may initially increase to 10% by volume (E10) and later even up to 15% by volume (E15). Many new gasoline engines entering the market are already designed for a 10% blend. However, EN 228 shall have to be duly revised.

E85 and FFV (flexible fuel vehicles) will only be justified once large quantities of ethanol become available on the entire fuel market (e.g. in Brazil).

Blending ethanol with diesel fuel for cars is largely rejected because of the poor mixture stability and other problems, yet is locally important for commercial vehicles deployed in fleets (e.g. in the USA and Brazil) [4-17].

4.2.3 Gaseous Fuels

Gaseous under ambient conditions, these fuels (see Sect. 4.4) have extremely low energy density (relative to volume) and require considerable technology, in part to be stored on board a vehicle.

4.2.3.1 Natural Gas

Natural gas (predominantly methane) is a natural fossil fuel, the processing of which merely entails cleaning and the removal of sulfur and other disturbing components. In the future, it will be used increasingly in nonroad domains (power plants and CHP). Biogas processed into fuel gas with the quality of natural gas is a sustainable alternative with a potential that cannot be ignored [4-18].

Storing natural gas on board a car to guarantee acceptable ranges is very complex:

– Either it is stored liquefied at −167°C (liquefied natural gas or LNG). This is connected with high evaporation losses when a vehicle is stopped for longer periods and thus has not established itself for use in cars.
– Or it is stored gaseous at approximately 200–250 bar (compressed natural gas or CNG) in large, weight-optimized pressure tanks.

Natural gas has high detonation limits and is therefore particularly suited for gasoline engines.

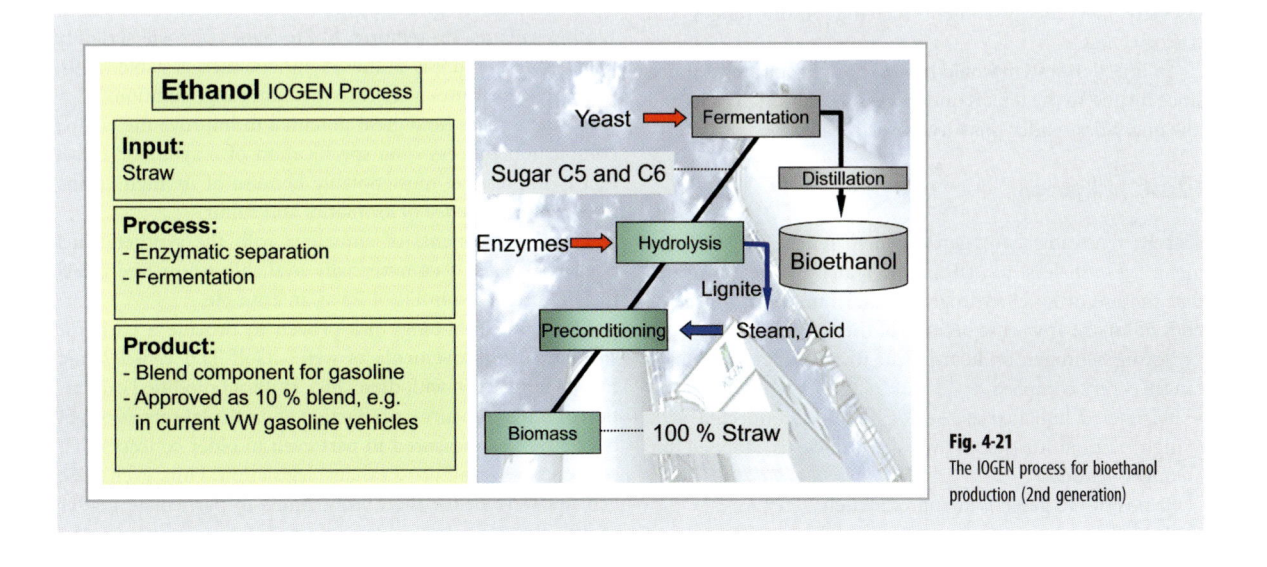

Ethanol IOGEN Process

Input:
Straw

Process:
- Enzymatic separation
- Fermentation

Product:
- Blend component for gasoline
- Approved as 10 % blend, e.g. in current VW gasoline vehicles

Yeast → Fermentation
Sugar C5 and C6 → Distillation
Enzymes → Hydrolysis → Bioethanol
Lignite
Preconditioning ← Steam, Acid
Biomass ⋯ 100 % Straw

Fig. 4-21 The IOGEN process for bioethanol production (2nd generation)

In Germany for instance, substantial tax cuts have ensured that natural gas will enjoy increasing use in the coming years. Natural gas can be and is used to power vehicles directly. However, in light of the known disadvantages of all gaseous fuels in terms of range, the space requirements for tanks and the increased work exhaust gas aftertreatment requires to comply with strict emission limits, natural gas may only be expected to supplement fuels to a limited degree, not replace them.

Natural gas has achieved a certain level of importance in road traffic in some markets such as Italy, Argentina and Russia (and is used in approximately four million vehicles worldwide) and is anticipated to have a market share of approximately 5% in Europe by 2020.

4.2.3.2 Liquefied Gas

A butane-propane mixture, liquefied gas or liquefied petroleum gas (LPG) is a companion product of petroleum production and processing and thus of fossil origin. Its limited sales volume has only enabled it to garner a niche market. It may be stored on board vehicles as a liquid at a pressure of approximately 5–10 bar. LPG is used practically only in conjunction with spark ignition.

Liquefied gas has achieved particular importance in road traffic especially in Italy, the Netherlands and Eastern Europe (and is used in approximately nine million vehicles worldwide). LPG vehicle lines are basically only retrofits.

4.2.3.3 Dimethyl Ether (DME)

DME can be produced from natural gas or biomass. Like LPG, it may be stored as a liquid on board a vehicle at a pressure of approximately 5–10 bar.

DME is a suitable fuel for diesel engines. Its excellent combustion properties (soot-free and low NO_x emissions) are undercut by the disadvantages of its low lubricity, low viscosity, low energy content (halving of the range) and corrosiveness.

An injection system would have to be appropriately refined since the fuel in the tank is under pressure. The high costs and low quantities render this idea unsustainable.

4.2.3.4 Hydrogen

Clearly, renewable hydrogen can help relieve the environment. Nevertheless, a well-to-wheel analysis demonstrates that the use of fossil hydrogen in the transportation sector does not make any sense because of the CO_2 emissions.

Storing hydrogen on board a car to guarantee acceptable ranges is very complex:
- It is stored liquefied at $-253°C$. This is connected with high evaporation losses when a vehicle is stopped for longer periods.
- Or it is stored gaseous at approximately 700 bar.
- Or it is stored in large, heavy metal hydride tanks.

A tank shape that is cost effective from the perspective of mass production is not in sight. The absence of cost effective renewable hydrogen production and the extremely capital-intensive infrastructure required to produce and market it make it improbable that both hydrogen and fuel cell technology will become marketable and competitive enough for mass production within the next two decades.

4.2.4 Synthetic Fuels

Parallel offerings of every fuel on the market, e.g. diesel, gasoline, methanol, ethanol, natural gas and other fuels, cannot be a cost effective solution since every one of these fuels not only necessitates the development of an independent engine but also an independent distribution infrastructure. Hence, blending alternative fuels into standard fuels within acceptable upper limits makes considerable sense. The sale and use of these blends are assured everywhere.

The introduction of other alternative energies on the market will require seeking an opportunity to diversify the primary energies and simultaneously restrict the energy sources being utilized for vehicles to a minimum of variants. Synthetic fuels such as gas-to-liquid (GTL) and biomass-to-liquid (BTL) provide such an opportunity.

4.2.4.1 Fossil Fuels (GTL)

Well known and industrially tested processes such as Shell Middle Distillate Synthesis (SMDS) can be used to produce other secondary energy carriers from natural gas.

At present crude oil price levels, such gas-to-liquid technologies are highly economical in many regions of the world with supplies of inexpensive natural gas or oil-associated gas. Companies such as Shell, Sasol, ConocoPhilips and Chevron have begun expanding their production capacities considerably. Nonetheless, given the investments and construction of synthesis plants this requires, five to eight years will certainly pass before the supply of these synthetic fuels is stable. Thus, this only constitutes a short to medium range solution.

Synthetic fuels hold great potential to improve the engine combustion process. The specification of a synthetic diesel fuel is impressive, most notably because of its high cetane number and absence of aromatics and sulfur.

The improvements of emissions with the synthetic fuel Shell GTL over a commercially available sulfur-free diesel fuel are presented in Fig. 4-22 as an example.

The latest technology has proven to provide a basis for substantial improvements as well – Golf TDI vehicles met Euro 4 limits even with diesel fuel. Without modified calibration or other measures, a reduction of particulate emissions of over 50% was measured in part even in older concepts that only comply with Euro 3 exhaust legislation. Thus, even these vehicles were below the Euro 4 limits of particulate matter [4-19].

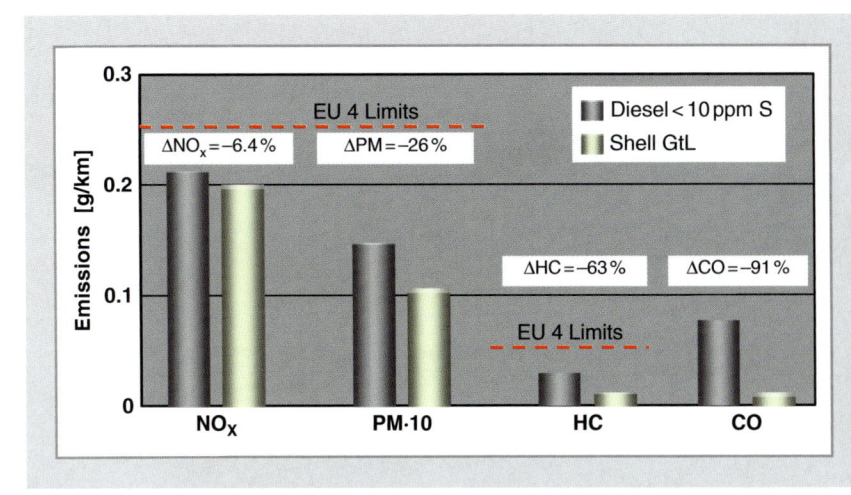

Fig. 4-22 Comparison of synthetic fuel (GTL) and diesel fuel emissions (fleet test with 25 VW Golf 1.9 l TDI in Berlin in 2003)

4.2.4.2 Renewable Sources (BTL)

Once the input stages have been modified accordingly, the production process for synthetic fuels will be able use the widest variety of primary energy sources, e.g. even renewable energy sources such as residual timber, residual straw, energy plants or biowaste. Even wastes largely ignored before will become highly interesting residual materials that may be supplied as more material and subsequently as energy. A crucial advantage is the independence of the final product's quality from the nature of the primary energy used. The energy stored in the world's annual crop growth corresponds to approximately fifty times humankind's energy consumption, i.e. tremendous potential for substitution exists. From a policy perspective as well, the use of biomass relieves the supply sector since, unlike fossil energy sources, biomass is distributed relatively uniformly all over the world. While, this does not reduce local CO_2 emissions to zero, it does create a nearly neutral CO_2 cycle for which the sun delivers the operating power (see Fig. 4-23). Thus, the fuel cycle is integrated in the natural CO_2 cycle into which approximately 98% of all CO_2 emissions are routed.

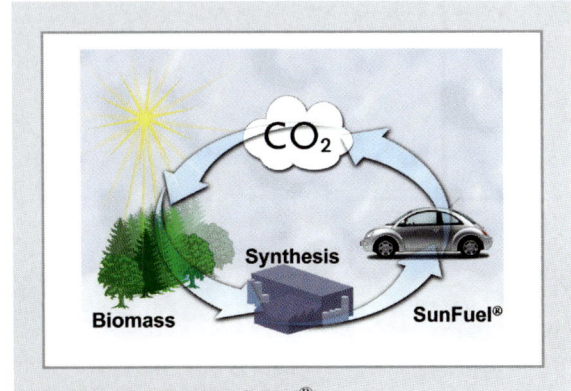

Fig. 4-23 CO_2 cycle with BTL (SunFuel®)

The biomass available is basically dispersed among residual materials and cultivated energy plants. Widely varying views exist about the potential to substitute existing fuels. Literature that postulates a potential of 10% to 15% in Europe takes the actual present situation as its point of departure. Agriculture will be able to massively increase yields as soon as the target of breeding and production is optimized quantity rather than quality food plants. Bearing these factors in mind, a potential for approximately 25% substitution could presumably be reached by 2030 [4-20].

Along with analyses of the cultivation of different fast growing timbers and special energy plants (see Fig. 4-24), residual materials and, in particular, the potential of industrial and municipal biowastes and other wastes deserve in depth study. In the future, all collections of biomass could be fed to the production of high grade fuels instead of composting.

Figure 4-25 schematically depicts a plant that produces SunFuels® based on the BTL process. Depending on the process, pyrolysis converts the biomass into gaseous, liquid and solid constituents in an initial stage. The CHOREN pyrolysis process [4-21] produces a pyrolysis gas and biocoke. Other processes of Canadian and American companies chiefly produce solid or liquid substances as the pyrolysis product, which are called biocrude. This pumpable primary product is similar to crude oil. When it is precompressed in smaller distributed units, biomass is particularly suited for transport to a large central plant. This can increase the efficiency of the actual central production plant considerably, without the biomass transport negating the efficiency of the entire product chain. Such pyrolysis plants remain within a financially manageable scope and may even be operated by communities or agricultural machinery cooperatives in place of composting plants. Another advantage is the usability of both moist and dry feedstocks as well as the potential to remove the mineral constituents and return them to the soil as fertilizer.

A second stage, the actual gasification, produces a synthesis gas. After appropriate cleaning, Fischer-Tropsch synthesis

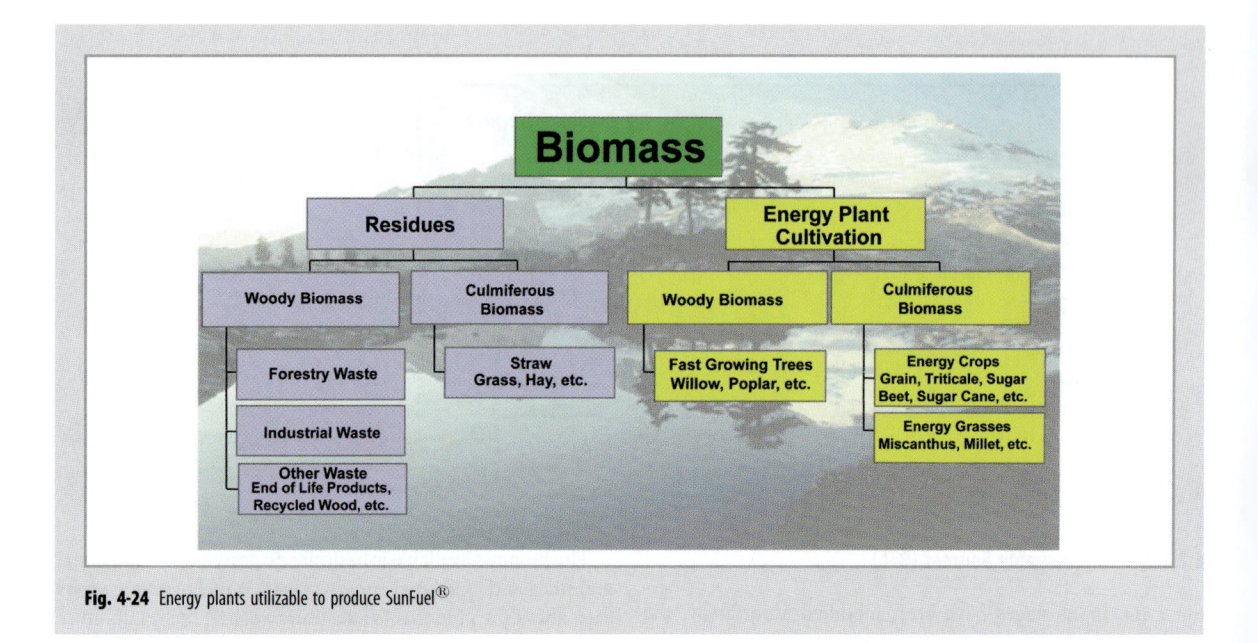

Fig. 4-24 Energy plants utilizable to produce SunFuel®

(FT synthesis) followed by hydrogenation and subsequent distillation converts the synthesis gas into high grade fuel and waxes. The waxes serve as a basis for the production of synthetic oils that are predominantly produced from petroleum and natural gas at present.

Since it is not yet economically feasible, this biomass-to-fuel solution (also called second generation BTL) also remains a medium range prospect. The pure production costs without taxes (based on a production plant sized for 200 MW_{th}) tally up to a cost disadvantage of approximately 20–30 cents per liter compared with fuel from fossil sources (with production costs of approximately 35 cents per liter at a crude oil price of $50 per barrel). However, the production costs are significantly lower than current service station prices. Thus, until its economic feasibility can be demonstrated, it lies in government's hand to promote the development of the process through appropriate tax laws and the initial introduction of the fuel. In and of itself, Germany's tax exemption of BTL fuels until 2015 will not be enough to ensure this.

The nearly unlimited possibilities to sell this fuel additionally furnish a tremendous opportunity to safeguard jobs in agriculture. Particularly in light of the restructuring of EU subsidy guidelines, the provision of energy plants could also stabilize income in the long-term. The development and construction of production plants will furnish industry a new source of revenue too.

As soon as it is available, less expensive renewable hydrogen could also be added to the SunFuel® production process.

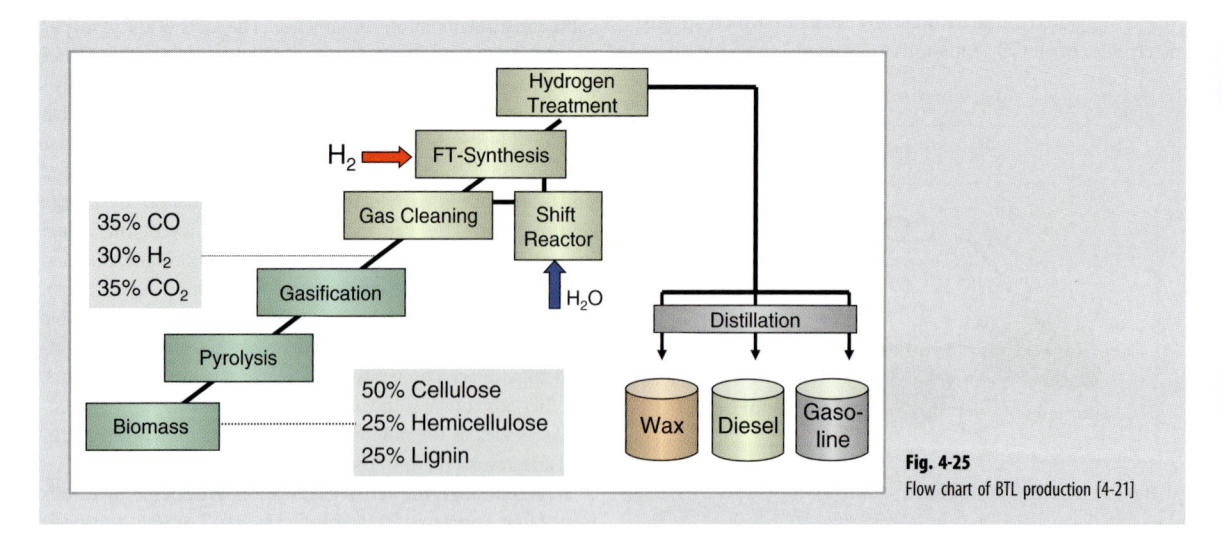

Fig. 4-25 Flow chart of BTL production [4-21]

This would allow nearly doubling the yield of fuel. It also means that the establishment of a hydrogen industry would not inevitably result in the utilization of hydrogen in the mobile economy. Viewed holistically, biomass-based synthetic fuels could prove to be expedient, especially in terms of sustainability.

Alternative energy sources will rapidly become more important in the future as the technical limits of petroleum production are approached and the world's demand for energy increases. The demands on conventional fuel's quality and purity are rising too. Related cost increases are encouraging the introduction of generally more expensive alternatives. Volkswagen AG anticipates an evolution of fuel in the coming years, which will proceed from conventional petroleum-based fuels to synthetic SynFuels produced from natural gas up through biomass-based SunFuel®. Hydrogen will only be able to play a role as an energy source in mobile applications in the distant future once every technological barrier has been eliminated. Figure 4-26 describes such a scenario.

The intermediate stage between first and second generation biofuels has been dubbed NExBTL [4-13]. This is a BTL fuel obtained from vegetable oils and animal fats by means of FT synthesis and, like GTL, usable as a blend component in conventional diesel. Table 4-9 compares the most important properties.

Given the constraints on present day knowledge about the utilization of agricultural lands and the efficiency of the processes, substituting 15 to 20% of the EU's fuel requirement with SunFuels® appears possible. Higher values also appear realizable by improving biomass production, processes and logistics. However, stable boundary conditions such as a sustainable commitment to biofuels on the part of government are also a prerequisite.

The EU's biofuel directive to promote the use of biofuels or other renewable fuels in the transportation sector [4-7] defines biofuel fractions as reference values based on an energy content of 2% in 2005 and 5.75% in 2010.

The EU intends to revise its biofuel directive by 2008. Among other things, issues of cost effectiveness and the environmental impact of biofuels shall be incorporated and objectives shall be formulated for the period after 2010. Improving the boundary conditions for higher blending ratios in conventional fuels will also be essential.

4.2.5 Life Cycle Assessment

An analysis of all the phases of the life cycle is particularly important for the development of future environmentally friendly propulsion and fuel concepts. Merely optimizing just one segment of the life, e.g. a car's use phase and its attendant emissions, does not always deliver the ecological optimum when the entire life cycle is analyzed.

Therefore, life cycle assessment is used as an instrument to analyze a product's environmental profile throughout its entire life cycle. Life cycle assessments identify the time and way one ecological problem is solved at the cost of another. Shifting problems become recognizable and environmental strategies can thus be formulated more reliably.

In a joint project, DaimlerChrysler and Volkswagen developed a comparative life cycle assessment of SunDiesel and conventional diesel to evaluate the environmental profile of BTL fuels throughout their entire life cycle. The cultivation of biomass, the synthesis of fuel from biomass and the use of the fuels in vehicles were analyzed. The BTL fuel analyzed had been produced from wood with the CHOREN process (Fig. 4-27).

The study came to the conclusion that SunDiesel can cut between 61 and 91% more greenhouse gases than conventional

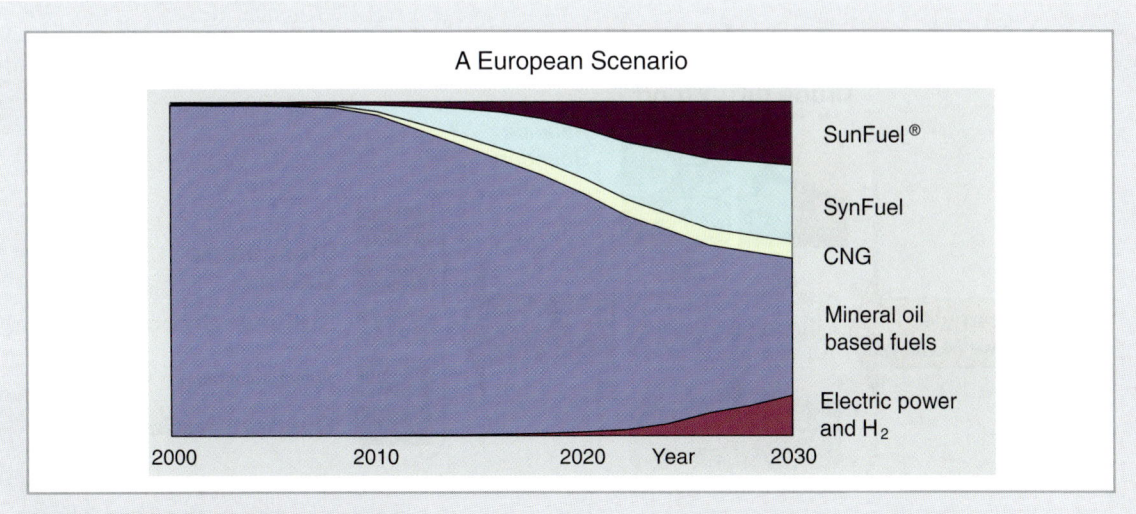

Fig. 4-26 European scenario for energy carriers in mobility

Table 4-9 Properties of alternative fuels [4-13]

	NExBTL	GTL	FAME (RME)	Swedish class 1 diesel	Summer DF (EN 590)
Density at +15°C (kg/m^3)	775 ... 785	770 ... 785	≈885	≈815	≈835
Viscosity at +40°C (mm^2/s)	2.9 ... 3.5	3.2 ... 4.5	≈4.5	≈1.8	≈3.5
Cetane number	≈80 ... 99	≈73 ... 81	≈51	≈53	≈53
Distillation 90% by volume (°C)	295 ... 300	325 ... 330	≈355	≈280	≈350
Cloud point (°C)	≈−5 ... −25	≈0 ... −25	≈−5	≈−30	≈−5
Calorific value (MJ/kg)	≈44.0	≈43	≈37.5	≈43	≈42.7
Calorific value (MJ/l)	≈34.4	≈34	≈33.2	≈35	≈35.7
Total aromatics (percent by weight)	0	0	0	≈4	≈30
Polyaromatics (percent by weight)	0	0	0	0	≈4
Oxygen content (percent by weight)	0	0	≈11	0	0
Sulfur content (mg/kg)	<10	<10	<10	<10	<10
Lubricity HFRR at +60°C (μm)	<460	<460	<460	<460	<460

diesel throughout the entire life cycle. These cuts are primarily based on the CO$_2$ emissions from normal driving being reabsorbed by growing plants and CO$_2$ consequently being routed into the cycle.

In addition, the use of SunDiesel also lowered the hydrocarbon emissions that contribute to summer smog approximately 90% over conventional diesel. The lower HC emissions during normal driving and the HC emissions prevented during drilling and refining crude oil for conventional diesels are responsible for this.

An analysis of the entire life cycle leads to holistic ecological product development. Precise analyses replace blanket assumptions. Thinking in life cycles identifies the concrete boundary conditions under which certain propulsion or fuel strategies may be formulated more environmentally compatibly.

The diesel fuel produced from wood with the BTL process is nearly CO$_2$-neutral and, based on the analysis, deserves to be called SunFuel® for good reason.

4.2.6 New Combustion Systems

All these considerations demonstrate that liquid hydrocarbons will presumably be available and dominant as fuel even in the coming 30 years. At the same time, synthetic fuels furnish potential to optimally adapt fuel properties to combustion. Further reducing exhaust emissions or reducing the considerable complexity of exhaust gas aftertreatment will only be feasible when, first and foremost, the raw NO$_X$ emissions of stratified combustion systems can be lowered. This means the production of NO$_X$ must be suppressed during combustion without diminishing engine efficiency, making it essential to retain direct injection quality control. This will

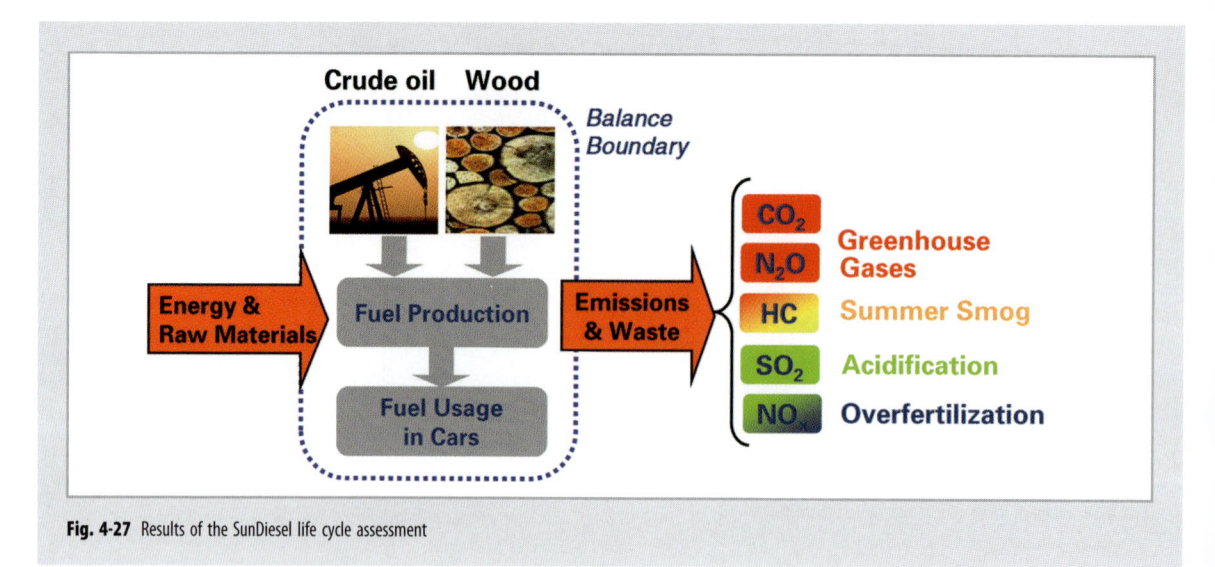

Fig. 4-27 Results of the SunDiesel life cycle assessment

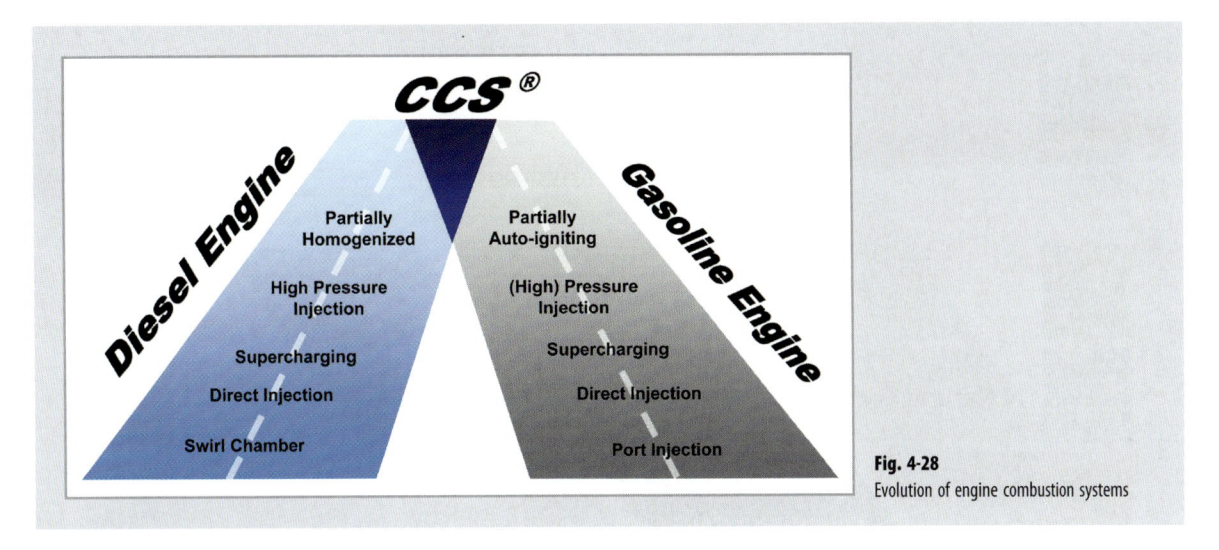

Fig. 4-28 Evolution of engine combustion systems

necessitate uniting the respective advantages of gasoline and diesel engines in one new process (see Sect. 3.3).

The combustion systems of both engine concepts had already grown significantly more alike when direct injection was introduced for gasoline engines too. The next stages in combustion system development will entail intensifying this trend (see Fig. 4-28).

The development of 'partially homogenized diesel combustion' and the current stage of development of "auto-igniting gasoline engines" in research and development labs are likewise already based on hardware with a comparable core. Thus, developing a new combined combustion system that subsumes the fundamental features of both systems is only logical. Volkswagen calls this a combined combustion system (CCS). The basis of the system is a new synthetic fuel with modified specific properties [4-22, 4-23].

Numerous obstacles still have to be overcome to implement the CCS. Hence, a market launch may no longer be expected in this decade.

4.2.7 Conclusion

Figure 4-29 presents a complete potential scenario for the development of future drives and related fuels [4-24].

Present day petroleum-based fuels and conventional engines will facilitate further decreases of specific CO_2 emissions that correspond with the automotive industry's voluntary commitments and technical advances. Engines with direct fuel injection will play a key role.

In addition to petroleum-based fuels, synthetic conventional fuels primarily based on natural gas will be launched on the market this decade. Since all the commercial features and distribution structures shall be retained, the introduction of synthetic fuels will not change anything for vehicle users. Synthetic fuels are free of sulfur and aromatics and their properties can be toleranced more narrowly than current fuels. These advantageous properties will enable carmakers to further develop their products to reduce fuel consumption and improve emissions, especially in diesel engines.

If, instead of being produced from fossil fuels, synthesis gas is produced on the basis of CO_2-free or CO_2-neutral energy, then the specific CO_2 emissions of vehicle operation will drop even when consumption remains unchanged. This holds true regardless of the type of fuel, i.e. even synthetic fuel made of renewable raw materials (SunFuel®). The great advantage of this route is its retention of the current fuel infrastructure even in this phase.

In the medium term, novel engine combustion systems will be implemented, which will combine the consumption advantages of present day diesel engines with the emission potential of gasoline engines and their exhaust gas aftertreatment. The appropriate fuels will have to be customized for these hybrid combustion systems. Synthetic fuels (SynFuel and later SunFuel®) provide the best conditions for this.

Above all, the electric motor is regarded as the optimal drive for sustainable mobility in the long term. Whether such vehicles will have advanced battery systems or hydrogen as the energy source with fuel cells as energy conversion systems is still hard to judge. This is unlikely in the next 20 years at the least. Whatever system establishes itself later, one undisputed advantage will be its ability to generate electrical power and hydrogen from renewable energies such as wind, water or sunlight. The fact that electric motors do not cause the local emissions combustion engines do is advantageous too.

4.3 Operation of Marine and Stationary Engines with Heavy Fuel Oil

4.3.1 Heavy Fuel Oil

Heavy fuel oil is a mixture of residual oils that accumulate from fractioned distillation during petroleum (crude oil) processing. Their main constituents are hydrocarbon

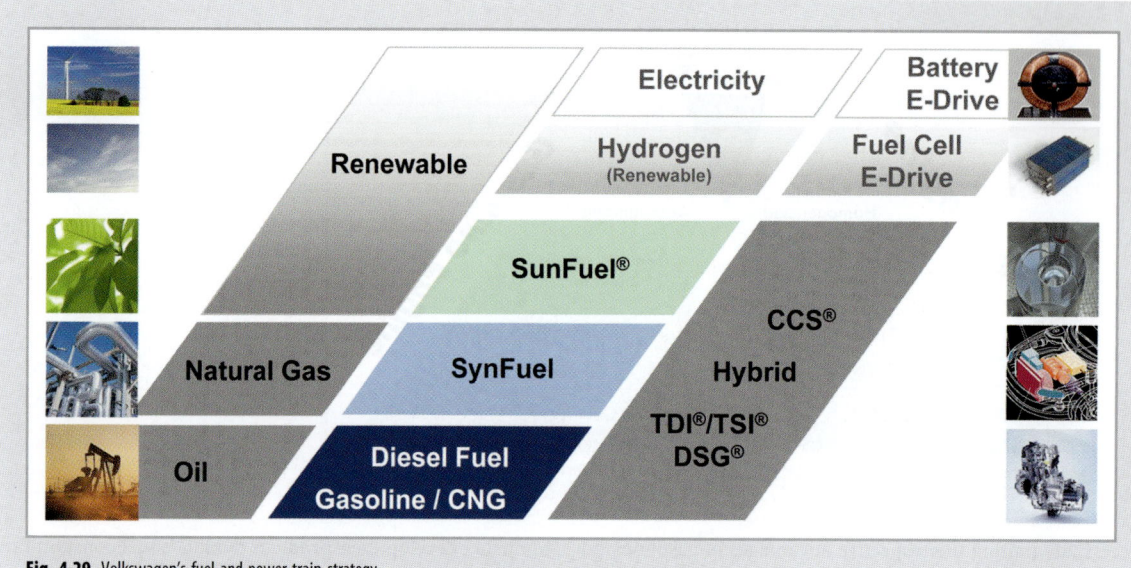

Fig. 4-29 Volkswagen's fuel and power train strategy

compounds that remain as high boiling fractions after crude oil has been distilled. Since residual oils are considerably less expensive than distillates, e.g. gasoline or light fuel oil, there is an economic incentive to use these components as fuel.

In most cases, distillates are blended into the residual oils to secure specific product properties and, in particular, to adhere to a predefined upper viscosity limit.

Heavy fuel oils have viscosities between 4.5 and 55 mm^2/s (cSt) relative to 100°C and are specified in the standard ISO 8217. On the basis of this ISO standard, the Conseil International des Machines a Combustion (CIMAC) classified heavy fuel oils based on their physical-chemical data with even more requirements. These "requirements for residual fuels for diesel engines" are part of the operating specifications issued by every manufacturer of diesel engines compatible with heavy fuel oil (see Fig. 4-30).

The properties or composition of heavy fuel oils vary within broad limits based on the crude oils' provenience and dependent on the different processing processes in the refineries (see also [4-25]). Apart from their significantly higher viscosities, heavy fuel oils are characterized by higher densities, higher sulfur contents and greater propensity to coking than distillates. The incombustible content (ash) is two orders of magnitude higher, the ignition and burning properties are poorer because of the higher aromatic and asphalt contents and appreciable quantities of water and solid contaminants that induce wear may also be present.

Heavy fuel oil's greater density than gas oil's (corresponding to DF) indicates a gain of the weight ratio of carbon to hydrogen. Regardless of the sulfur content, this reduces the net calorific value H_u (see Fig. 4-31).

The constant tendency of heavy fuel oils to diminish in quality is the result of the increasing spread of such conversion processes as catalytic and thermal cracking in modern refineries to better utilize crude oil. (The residue fraction is 32–57% in classic atmospheric distillation, 12–25% in a conversion refinery.) The high aromatic content connected with this lessens ignition quality (see Sect. 4.3.4.1), the increased asphaltene fraction the stability. Intensified sludge and resin formation can disrupt fuel processing.

Other adverse effects stem from an intensified trend toward disposing of used lubricating oils, organic solvents or chemical wastes in the residual oils.

Adapting diesel engines to the given conditions, assuring the quality of heavy fuel oils through standards (ISO, CIMAC, Fig. 4-30) and, in particular, optimally processing heavy fuel oil are instrumental in making cost effective, trouble-free use of heavy fuel oil possible in diesel engines under these conditions.

4.3.2 Heavy Fuel Oil Processing

Heavy fuel oil must be processed to be used as fuel in diesel engines. This eliminates or largely reduces undesired impurities such as water with any corrosive substances possibly dissolved in it as well as solid impurities such as coke, sand, rust, catalyst residues from the refinery and sludge-like constituents such as agglomerated asphaltenes.

Requirements (1990) for residual fuels for diesel engines (as delivered)

Related to ISO 8217 (87):	Designation:			CIMAC A 10	CIMAC B 10	CIMAC C 10	CIMAC D 15	CIMAC E 25	CIMAC F 25	CIMAC G 35	CIMAC H 35	CIMAC K 35	CIMAC H 45	CIMAC K 45	CIMAC H 55	CIMAC K 55
		F –		RMA 10	RMB 10	RMC 10	RMD 15	RME 25	RMF 25	RMG 35	RMH 35	RMK 35	RMH 45	RMK 45	RMH 55	–
Characteristic		Dim.	Limit													
Density at 15 °C		kg/m³	max	950	975	975	980	991	991	991	991	1010	991	1010	991	1010
Kinematic viscosity at 100 °C [1]		cSt²	max	10	10	10	15	25	25	35	35	35	45	45	55	55
			min [4]	6				15								
Flash point		°C	min	60	60	60	60	60	60	60	60	60	60	60	60	60
Pour point		°C	max	0 [3] / 6		24	30	30	30		30	30	30	30	30	30
Carbon Residue		% (m/m)	max	12	14	14	14	15	20	18	22	22	22	22	22	22
Ash		% (m/m)	max	0.10	0.10	0.10	0.10	0.10	0.15		0.15	0.15	0.15	0.15	0.15	0.15
Total sediment after ageing		% (m/m)	max		0.10	0.10	0.10	0.10	0.10		0.10	0.10	0.10	0.10	0.10	0.10
Water		% (V/V)	max	0.50	0.50	0.50	0.80	1.0	1.0		1.0	1.0	1.0	1.0	1.0	1.0
Sulphur		% (m/m)	max	3.5	3.5	3.5	4.0	5.0	5.0		5.0	5.0	5.0	5.0	5.0	5.0
Vanadium		mg/kg	max	150	300	300	350	200	500	300	600	600	600	600	600	600
Aluminium + Silicon		mg/kg	max	80	80	80	80	80	80		80	80	80	80	80	80
Ignition properties									see appendix, section 3								

[1] Approximate equivalent viscosities (for information only):

Kinematic viscosity (cSt) at	100 °C	6	10	15	25	35	45	55
Kinematic viscosity (cSt) at	50 °C	22	40	80	180	380	500	700
Sec. Redwood I at	100 °F	165	300	600	1500	3500	5000	7000

[2] $1 \text{ cSt} = 1 \text{ mm}^2/\text{sec}$
[3] Applies to region and season in which fuel is to be stored and used. (upper value winter quality, bottom value summer quality)
[4] Recommended value only. May be lower if density is also lower. See appendix, part 3

Fig. 4-30 Classification of heavy fuel oils according to CIMAC/ISO (except from the original)

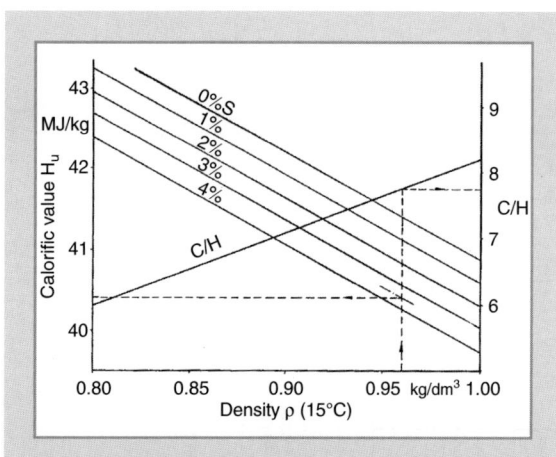

Fig. 4-31 Influence of the carbon/hydrogen ratio C/H on the density ρ and calorific value H_u of heavy fuel oils as a function of the sulfur content

and 160 (170)°C, depending on the initial viscosity (see Fig. 4-32).

Figure 4-33 presents the components used in an optimal heavy fuel oil processing system based on the current state-of-the-art.

The heavy fuel oil travels from the storage tanks to the settling tanks. A twenty-four hour residence time at temperatures around 70°C facilitates precleaning.

The next station is the centrifugal separators, which play a key role in the processing system. Connected in a series or in parallel, they act – depending on the setting – as a separating stage (also called a purificator) to remove water and foreign substances or as a clarifying stage (clarificator) to remove only foreign substances. To this end, the heavy fuel oil is preheated to approximately 95°C to obtain low viscosities and high density differences for high purification efficiency.

Modern separators independently adjust to changed heavy fuel oil parameters, e.g. density, viscosity and water content, and are self-emptying as well.

Designed as a buffer for at least eight hours of full load operation, the day tank supplies the heavy fuel oil to the booster system where, controlled by a viscosity controller, it is preheated to the requisite injection viscosity. The system pressure is higher than the evaporation pressure of water to prevent the formation of steam and gas.

An automatic backflush filter with a very fine mesh width (10 μm) provides for the final purification of the fuel and,

If these damaging impurities are not removed, corrosion and/or wear damage must be expected in the injection system (e.g. pumps and nozzles) and in the engine itself (e.g. on cylinder liners, pistons and piston rings) in the short-term with every aggravating consequence this entails.

Another function of processing is to provide the injection viscosity the heavy fuel oil needs for optimal engine operation. This requires preheating temperatures between 90

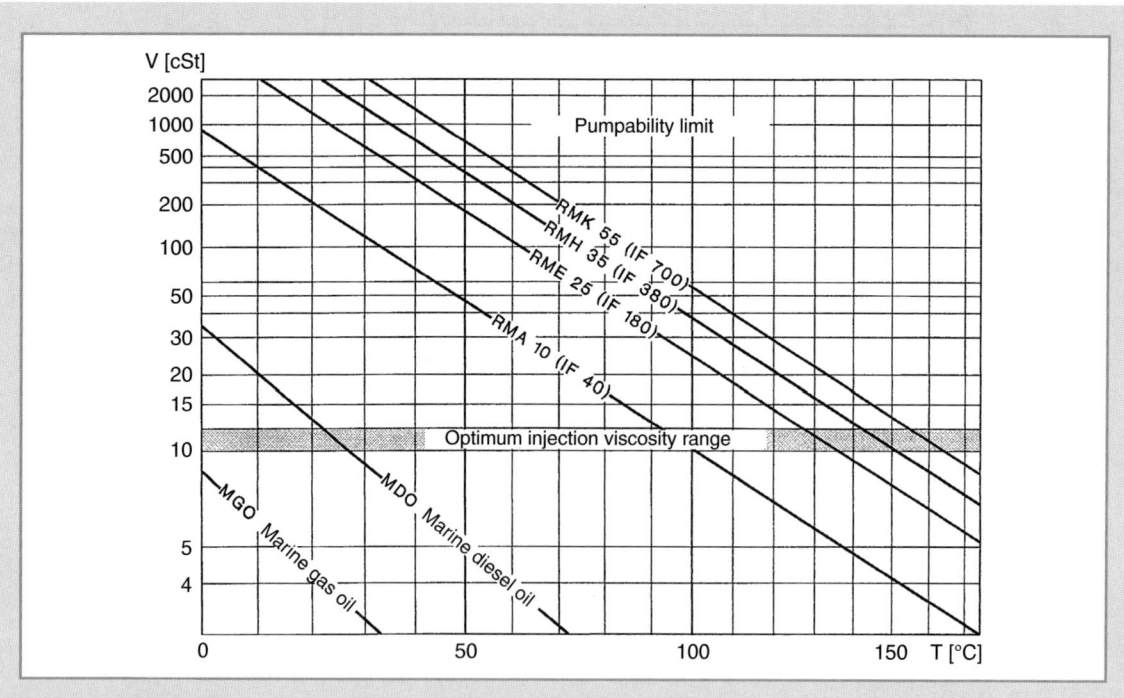

Fig. 4-32 Marine fuels' viscosity/temperature correlation. RMA: Residual marine fuel, class A, based on Fig. 4-30

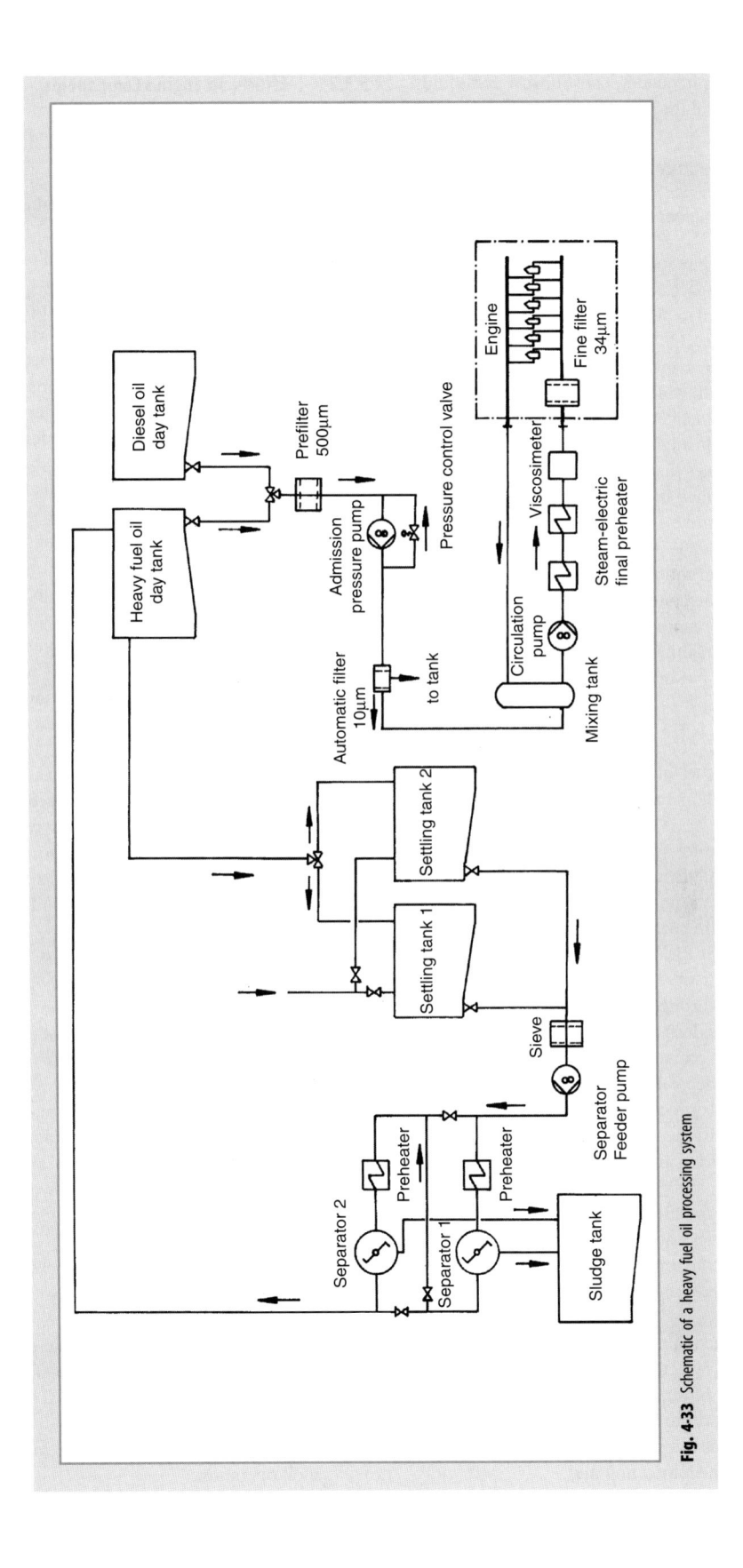

Fig. 4-33 Schematic of a heavy fuel oil processing system

based on its backflushing frequency, constitutes a useful indicator of the effectiveness of the upstream processing elements.

4.3.3 Distinctive Features of Heavy Fuel Engines

4.3.3.1 Heavy Fuel Engines and Their Problems

Heavy fuel oil processing requires both considerable investment and appropriate space for installation. Hence, heavy fuel oil operation is chiefly encountered on board ships and in large stationary engines. Only long operating periods with correspondingly high fuel consumption in conjunction with the price of heavy fuel oil, which is roughly 35% lower than DF, and the specific distinctive features of heavy fuel engines gone into below justify the high plant costs.

The two types of engines that may operate with heavy fuel oil have already been named indirectly:
- medium speed four-stroke engines with a piston diameter of approximately 200–600 mm and a speed of 1,000 to approximately 400 rpm and a power of 500 kW to approximately 18,000 kW per engine and
- low speed two-stroke engines with a piston diameter of approximately 250–900 mm and a speed of approximately 250–80 rpm and a power of 1,500 to approximately 70,000 kW per engine.

Apart from the processing (see Sect. 4.3.2) absolutely necessary for reliable heavy fuel oil operation, other distinctive features set heavy fuel oil operation apart from gas oil operation, namely the:
- risk of high temperature corrosion of the components surrounding the combustion chamber because of the fuel's vanadium and sodium content,
- risk of low temperature corrosion when the dew point is exceeded because of the fuel's sulfur content and the combustion gas' water content,
- risk of increased wear through abrasion because of the solid coke, sand and rust residues and catalyst residues as well as asphaltenes and water remaining in the fuel,
- risk of unacceptable deposits of combustion residues on combustion chamber components and in the exhaust lines and the exhaust gas turbine,
- risk of sticking, lacquering and congealing of fuel injection system components,
- risk of lubricating oil contamination through fuel leaks in injection pumps, nozzles and nozzle holders,
- risk of lubricating oil contamination because of insufficient combustion quality resulting in unreasonably short filter service life,
- risk of corrosion of the main and connecting rod bearings resulting from the fuel's reaction with the lubricating oil and
- risk of the formation of deposits in cooling ducts cooled with lubricating oil and the resultant diminished dissipation of heat, likewise as a result of the fuel's (asphalt content's) reaction with the lubricating oil.

4.3.3.2 Effects on Engine Components

Injection Equipment

Heavy fuel engines are equipped with single injection pumps. Common rail heavy fuel systems are in the implementation phase (see Sect. 18.3).

Injection equipment comes directly into contact with the preheated heavy fuel oil. Depending on the viscosity grade of the heavy fuel oil used, the preheating temperature is as high as 160°C to ensure the injection viscosity is the 12 cSt desired (see Fig. 4-32). This relatively high temperature level alone makes it necessary to solve several problems. Sealing elements such as O-rings have to be suitable. Plunger and pump cylinder clearance have to be appropriately designed to operate without seizing under higher temperature on the one hand and to ensure that leak rates are still permissible when switching to cold gas oil on the other hand.

Furthermore, specially heat treated materials have to be used for the injection elements to isolate unacceptable structural transformation resulting from the increased fuel temperature and to prevent pump seizing.

Other distinctive features of injection pumps are measures that prevent sticking and lacquering such as:
- Fuel rack lubrication: Separate lubrication protects the fuel rack and control sleeve from the ingress of fuel and keeps them clean and running smoothly.
- Leak fuel removal in the suction chamber: A circular groove in the piston guide and a connection to the suction chamber remove leak fuel in the fuel cycle.
- Leak fuel removal from the pump: Another groove in the piston guide collects the remaining leak fuel, which an additional connection to the pump conducts into the leak fuel tank.

In addition, leak oil and leak fuel must be prevented from reaching the engine's crankcase through the pump drive. Therefore, these leak quantities are trapped separately and also conducted back to the leak fuel tank through a leak oil line.

Another distinctive feature of heavy fuel oil operation is injection nozzle cooling to prevent carbonaceous deposits on the nozzle holes. This is done with a separate cooling circuit with supply and discharge through the nozzle holder. Lubricating oil, gas oil or water is employed as the coolant.

For trouble free operation, the heavy fuel oil in the injection system must be unable to cool down when the engine stops and block the injection pump for instance. The heated heavy fuel oil continues circulating through the engine when it is stopped for shorter periods. In order to start a cold engine without any problems, it has to be switched over to gas oil and then run "clean" before being shut down.

Figure 4-34 presents an example of an injection pump for heavy fuel oil operation.

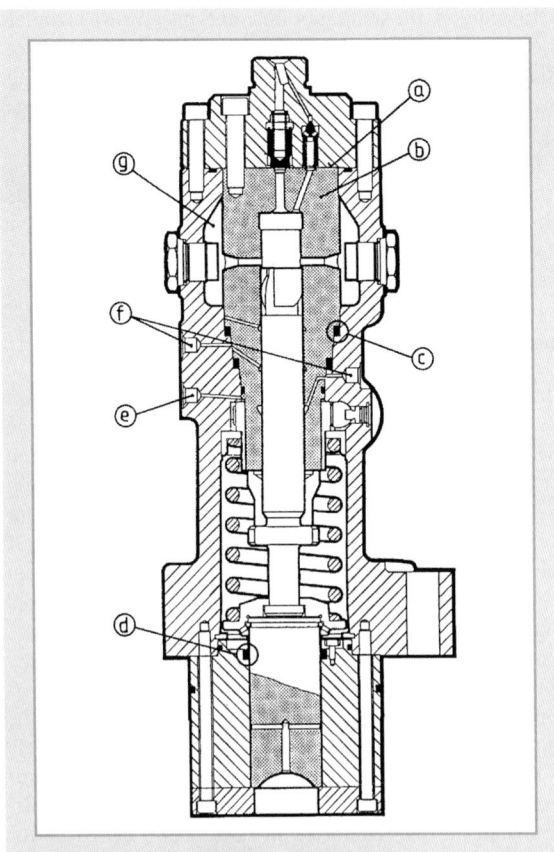

Fig. 4-34 Heavy fuel oil injection pump with fuel rack lubrication and leak fuel removal; blind hole element for high injection pressure (L'Orange). (**a**) Only a high pressure sealing face; (**b**) rigid monoelement for high pressures in heavy fuel oil operation; (**c**) special sealing elements for all types of heavy fuel oil; (**d**) central tappet with a hydraulic seal to protect the engine lubricating oil against fuel contamination; (**e**) scavenging connection for the governed range; (**f**) dual drainage for good sealing against heavy fuel oil; (**g**) large suction chamber – low pressure amplitudes to protect heavy fuel oil equipment on the low pressure side

For the heavy fuel oil to be distributed optimally in the combustion chamber and ignite well, it must be injected at very high pressure and finely atomized by many nozzle holes of small diameter. A medium speed engine with the following data serves as an example:
- brake mean effective pressure $p_e = 27$ bar (or $w_e = 2.7$ kJ/dm^3)
- injection pressure $p_E = 1,800$ bar
- number of holes $z = 10\text{--}14$

Turbocharging

Several specific factors must be considered when turbocharging heavy fuel engines.

The engines generally run with an excess of air so that the air/fuel ratio is $\lambda_V > 2$ in every operating state to prevent high temperature corrosion. Principally induced by the correspondingly high charge air pressure, the high air flow rate lowers the temperature during combustion and thus also at the components surrounding the combustion chamber. This prevents sodium-vanadium compounds from depositing as liquid ash, which causes high temperature corrosion. The critical temperature in valve seats, above which ash deposits, is approximately 450 °C. Thus, maintaining a sufficient safety margin is essential. Along with corrosion damage in the valve seat, which causes the valve disk to break in the final stage, corrosion damage on the underside of the valve must be prevented by keeping component temperatures sufficiently low.

A larger air flow rate is also beneficial for the exhaust temperature at the turbine inlet. Temperatures significantly below 550 °C prevent deposits of combustion residues in the turbine and thus a drop in turbine efficiency as well as boost pressure and air flow rate. Otherwise, the combustion chamber temperature and thus the risk of high temperature corrosion and, in turn, deposits in the turbine would increase.

This self-reinforcing effect and its adverse effects must be prevented at all costs. Therefore, both the compressor and the turbine must be cleaned at regular intervals. At reduced power, a special device sprays water into the turbine inlet so that deposits flake off the nozzle ring and rotor.

Special fuel consumption standards apply to heavy fuel diesel engines. Hence, the requisite boost pressure and air flow rate may not be obtained by increasing gas exchange losses (piston work), i.e. maximum efficiency is required from the exhaust gas turbocharger. This is the only way to generate high boost pressure at low exhaust back pressure (before the turbine). At the same time, a good scavenging gradient is produced during valve overlap. This is a prerequisite for low component temperatures and contaminant free operation of the gas exchange elements.

Among other things, supercharging also determines the capability to operate at part load with heavy fuel oil. This is addressed in detail in [4-26] (see Sect. 2.2.3.4).

Thermal Efficiency and Fuel Consumption

A combustion chamber's design codetermines mixture formation and combustion and ultimately determines an engine's capability to reliably burn heavy fuel oil in continuous operation. High efficiency requires rapid combustion of the prepared mixture and consequently a compact combustion chamber with a high compression ratio (see Fig. 4-35).

Uniform distribution of the fuel throughout the entire combustion chamber is essential. Four-stroke engines accomplish this with a maximum number of fuel sprays targeted far outwardly. Two-stroke engines have several injection nozzles distributed on the periphery. Fuel droplets should be as fine as possible. Injection pressure and small nozzle bores achieve this.

High Temperature Corrosion

Low combustion temperatures and intensive, steady cooling of components prevent high temperature corrosion on the components surrounding the combustion chamber. Among other things, the shape of the piston crown, the direction of spray and number of spray jets and the type of air movement (intake port) additionally influence the temperature of the exhaust valves.

Low Temperature Corrosion

Basically, low temperature corrosion can attack the cylinder liner, piston rings and ring grooves. A gentle combustion pressure curve and sufficiently high component temperatures at the pertinent points can help prevent such corrosion. Hence, an appropriate temperature profile for the cylinder liner is particularly important.

Cylinder Liner Wear and Oil Consumption

To keep cylinder liner wear low in the area of the gusset and thus assure lubricating oil consumption is low over a long period, hard oil coke that induces wear must be prevented from depositing on the piston crown. This is done by cooling the piston crown well and limiting its clearance to the cylinder liner. Furthermore, the piston edge is elevated to protect the cylinder wall so that the fuel sprays do not strike the cylinder bore surface. This may limit the direction of spray.

Flame or calibration rings arranged in the upper region of the cylinder liner are another means to reduce wear. The smaller diameter of the calibration ring and piston in this region is intended to prevent the coke deposited on the piston top land from abrading the cylinder liner's contact surface during the piston's upward stroke (see Fig. 4-35).

In addition, the mating of liner and piston ring materials decisively influences a cylinder liner's wear performance. Hardened liner surfaces (nitriding, induction hardening, laser hardening) are combined with piston ring coatings, e.g. chrome, chrome-ceramic and plasma coatings.

Constant oil consumption requires that pistons' ring grooves retain their original geometry over long periods. The distinctive feature of heavy fuel pistons is the placement of the hardened compression ring grooves in the composite piston's steel crown.

Compression Ratio

The ignition performance of occasionally ignition resistant heavy fuel oils can be positively influenced by fuel droplets on the injection side as well as by a maximum final compression temperature with a correspondingly high final compression pressure. This entails selecting a high compression ratio for the combustion chamber. As the stroke/bore ratio increases, ε values of approximately 13–16 are common for medium speed engines and values of 11–14 for low two-stroke speed engines.

Since, as the compression ratio increases, the air density in the combustion chamber increases when ignition is applied, the temperature peaks of the fuel gas drop at the sources of ignition. Furthermore, the pressure curve and rate of heat release can be shaped more smoothly as the compression ratio increases. This reduces nitrogen oxide emission. Thus, a high compression ratio is also beneficial for emission performance.

The aspects of combustion chamber design chiefly illustrated here with the example of a four-stroke engine, necessitate various compromises, which are easier to strike, the larger the engine's stroke/bore ratio is. Hence, modern medium speed four-stroke engines have relatively large stroke/bore ratios of up to $s/D = 1.5$ (see Fig. 18.35).

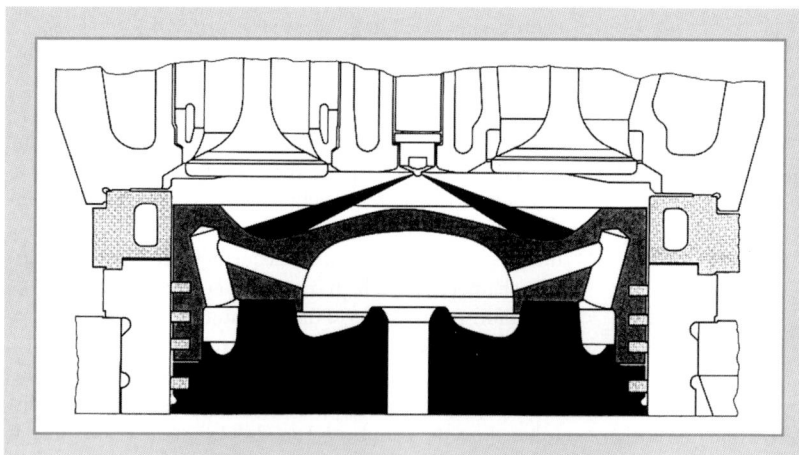

Fig. 4-35
A heavy fuel engine's combustion chamber

The rationale behind present day low speed two-stroke engines' extremely long strokes ($s/D \approx 4$) has less to do with their suitability for heavy fuel oil than efforts to run engines or propellers at minimum speed at the same mean piston velocity. This allows a maximum propeller diameter and thus maximum efficiency.

4.3.4 Running Properties of Heavy Fuel Engines

4.3.4.1 Ignition and Combustion Performance

The ignition and combustion performance of diesel engines that run on heavy fuel oil continues to be the subject of numerous research studies, the majority of which are cited in [4-25]. Fuels with a high percentage of aromatics have particularly proven to cause ignition difficulties. Given their molecular structure, aromatics resist thermal splitting in diesel engines far more than paraffins, olefins and naphthenes.

If the engine is not modified, then large ignition delays occur when fuels containing aromatics are burned. High slopes in the cylinder pressure curve are the consequence. In the extreme case, "detonating" combustion with mechanical and thermal overloading is observed [4-27, 4-28].

Thus, problem fuel must be identified to prevent engine damage. Aromatics' property of high density with low viscosity can be drawn on to do so. Empirical values served as the basis for Fig. 4-36, which provides information on the reliability of a fuel as a function of density and viscosity.

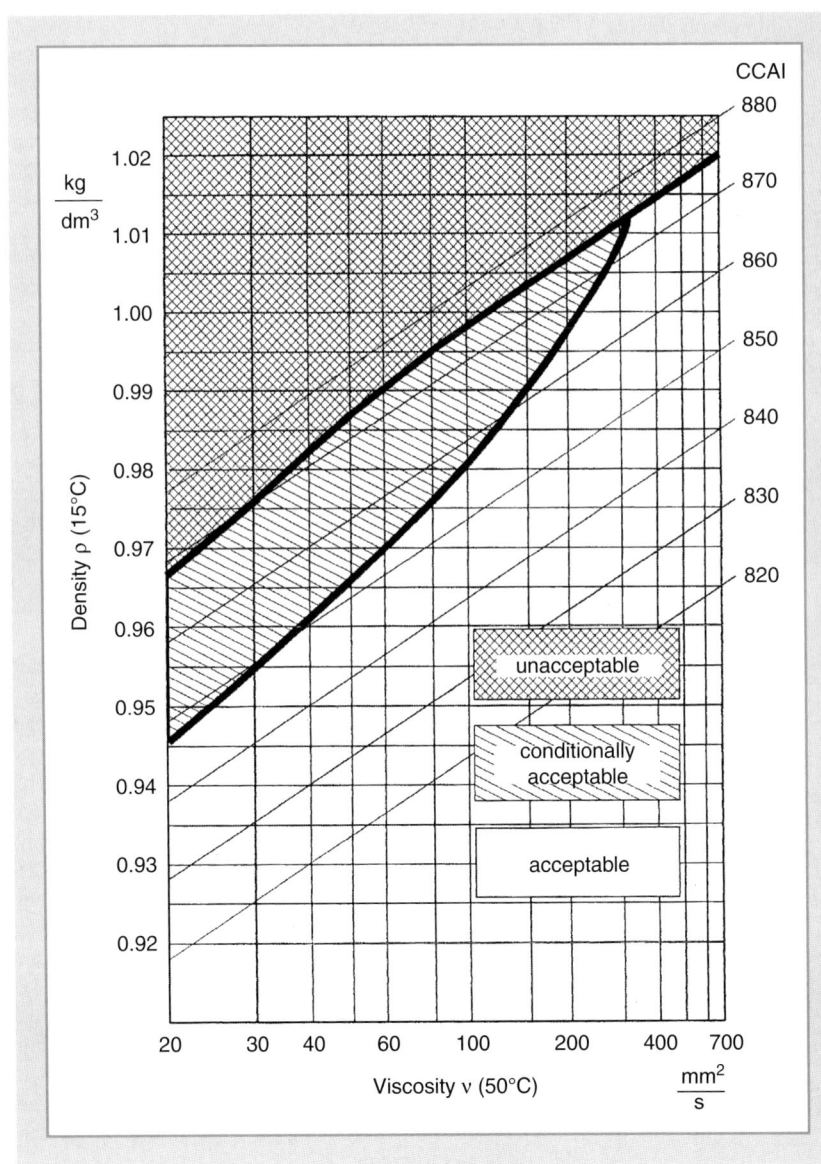

Fig. 4-36
Acceptability of fuels as a function of density and viscosity (according to the MaK operator manual)

The calculated carbon aromaticity index (CCAI) is a useful indicator for estimating ignition performance from density ρ (in kg/m^3 to 15°C) and viscosity v (in mm^2/s at 50°C) and is calculated with the following empirical relationship:

$$\text{CCAI} = \rho - 141 \log\log(v + 0.85) - 81.$$

Generally, a higher CCAI translates into poorer expected ignition performance.

In the vast majority of cases, heavy fuel oils do not present any problems for commercial combustion systems. The cylinder pressure curve for a modern medium speed engine only varies slightly when gas oil and heavy fuel oil are burned. The ignition delay for heavy fuel oil operation tends to be slightly larger and the maximum pressure somewhat lower.

The rate of heat release (Fig. 4-37) also differs only slightly. The start of combustion is delayed somewhat and the end of combustion is the same within the accuracy of evaluation. Even the maximum rates of combustion hardly vary.

4.3.4.2 Emission Performance

Heavy fuel engines are basically used as main and auxiliary marine engines and as stationary engines to generate electricity in countries with underdeveloped infrastructures and must be compliant with the emission regulations of the International Marine Organization (IMO) or the World Bank, which frequently finances stationary diesel power stations.

The IMO limit for nitrogen oxide emission is:

$$NO_x = 45 \cdot n^{-0.2} \quad \text{in g/kWh},$$

n being the engine speed in rpm.

The limit is constant in the speed range below 130 rpm (in two-stroke engines). The emission value ensues from weighting measurements in four power points according to ISO 8178-4. Preparations are underway to tighten the regulation and presumably lower the limit by 30%.

The IMO does not specify a limit for particulate emission. A requirement is imposed on ship operators based on invisible smoke emission, which corresponds to a Bosch smoke number of SN <0.5.

The World Bank's limit for nitrogen oxide emission is:

$$NO_x = 940 \, \text{ppm}$$

with 15% oxygen in the exhaust. A cut (NO_x = 710 ppm) is planned here too.

The World Bank's limit for particulate emission is PM = 50 mg/m^3.

In-engine measures that prevent nitrogen oxide emission in order to comply with the aforementioned limits are predominantly aimed at lowering the temperature level of the combustion gas during the formation of NO_x. Examples include high boost pressure, retarded start of delivery, high compression ratio, shaped and, where necessary, split injection and valve timing to apply the Miller cycle.

On the other hand, a reduction of visible smoke and particulates necessitates an increase of the fuel gas temperature in the critical operating range.

The more stringent emission requirements become, the more difficult it is to resolve the conflict between NO_x emission and particulate emission. Hence, engine manufacturers are also implementing variable injection systems that are suitable for heavy fuel oil operation, e.g. the common rail system and variable valve timing.

Other very effective measures that reduce particulate emission include lowering lubricating oil consumption and employing heavy fuel oil with low sulfur content.

The exhaust gas opacity and particulate emission (soot) of heavy fuel oil combustion is distinctive:

Using the filter paper method BOSCH designed for small vehicle engines (see Sect. 15.6.2.4) to determine exhaust gas opacity usually produces a very low value far below 1 for large diesel engines, inviting the assumption that the soot production is correspondingly low. Measurements taken on behalf of the FVV (Research Association for Combustion Engines) demonstrated that the low soot emission inferred only applied to operation with normal gas oil. The BOSCH smoke number SN is meaningless for heavy fuel oil operation (see Fig. 4-38).

The dust emission according to VDI 2066 is many times greater than the "soot emission" calculated from the smoke number SN by means of a correlation. The particulate emission measured according to ISO 8178 and resembling pollutant input indicates correspondingly high air pollution. The fuel's sulfur content is a fundamental influence because particulate emission increases approximately linearly with the sulfur flow in the engine.

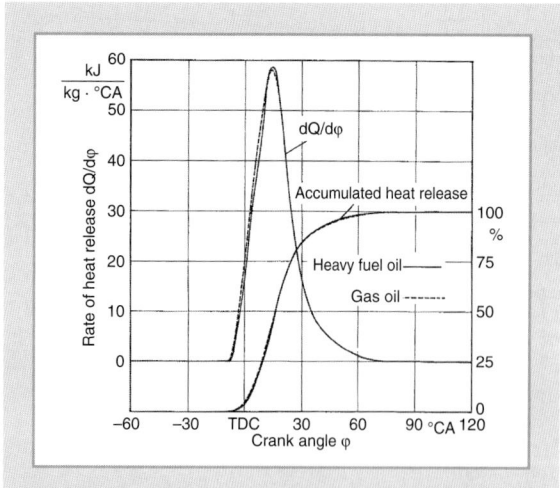

Fig. 4-37 Rate of heat release and accumulated heat release in gas oil and heavy fuel oil operation

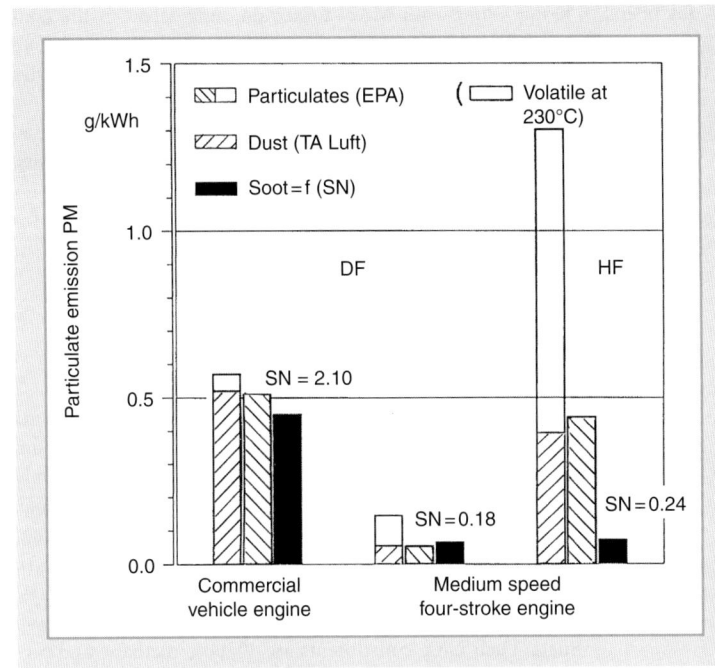

Fig. 4-38 Smoke number SN and the soot emission calculated with the SN, dust emission based on *TA Luft* and particulate emission measurement taken in conformance with the EPA. Medium speed engine (s/D = 320/240 mm/mm) at full load with gas oil (DF) and heavy fuel oil IF 380 (HF) compared to the full load values of an older model truck engine

Consequently, in addition to improving combustion, limiting heavy fuel oil's sulfur content like diesel fuel's (see Sect. 4.1) would seem the obvious course to pursue. In an initial step, the IMO has limited the sulfur fraction to 4.5% worldwide and to 1.5% for the Baltic Sea, the North Sea and the English Channel. As of 2010, auxiliary diesel engines for onboard power supply shall only be allowed to be run in harbors with fuels with a maximum sulfur fraction of 0.1%.

4.3.5 Lubricating Oil for Heavy Fuel Engines

The use of heavy fuel oil in diesel engines necessitates the use of engine lubricating oils specially developed for them.

Two fundamentally different engine designs are distinguished when defining the requirements for these oils:

Two-stroke crosshead engines have separate lubrication systems for the cylinder area and the crankcase because of the seal created by the piston rod stuffing box; the cylinder and crankcase in trunk piston engines are interconnected.

A crosshead engine's separate cylinder lubrication is pure loss lubrication, i.e. the cylinder lubrication system constantly replenishes the quantity of oil consumed. The cylinder oil must have the following qualities to perform or support the following important functions:
– good wetting and distribution capacity to ensure uniform distribution on the cylinder surface,
– high neutralization capacity to prevent corrosion resulting from the high sulfur content in the fuel,
– high cleaning capacity (detergent effect) to prevent deposits of combustion residues,
– high oxidation and thermal stability to prevent products from lubricating oil decomposition with attendant deposits and
– high load bearing capacity to prevent high wear and risk of seizing.

Measured in milligrams of potassium hydroxide per gram of lubricating oil (mgKOH/g), the Total Base Number (TBN) is applied as a measure of a lubricating oil's neutralizing capability.

Typical cylinder oils for crosshead engines have viscosities of SAE 50 and TBN values of 70–90.

Oils with comparatively low doping, a viscosity grade of SAE 30 and a TBN of approximately 6 mgKOH/g serve as engine oils for crosshead engines.

Unlike crosshead engines, *four-stroke trunk piston engines* have no partition between their cylinders and crankcases. Hence, aggressive combustion gases constantly load the lubricating oil in trunk piston engines. Acids, coke and asphalt-like residues load the lubricating oil together with abrasive solids from heavy fuel oil combustion, especially when specific quantities of circulating oil are low (i.e. relative to power output) and the consumption of lubricating oil is low (low refill quantities). Apart from the quality of the fuel, the loading of lubricating oil is significantly influenced by the operating conditions to which engine maintenance, the piston rings' sealing effect and the lubricating oils' processing contribute. In addition, the oil fill is normally not changed, i.e. it must perform its functions over many thousands of operating hours.

This generates the following requirements for trunk piston engine oil compatible with heavy fuel oil:
- high oxidation and thermal stability to prevent the oil from causing lacquering and coke-like deposits,
- high neutralization capacity of acidic combustion residues to prevent corrosion,
- particularly good cleaning and dirt suspending capacity (detergent/dispersant effect) to neutralize the coke and asphalt-like combustion residues, which increasingly form,
- particularly careful matching of the detergent/dispersant additives in order to effectively clean the lubricating oil in separators and filters as required and
- low propensity to emulsify and low sensitivity of the active ingredients to water to maintain effective oil care and prevent premature losses of additives.

In addition, high requirements are imposed on its load bearing capacity, good foaming tendency and low propensity to evaporate.

Typical lubricating oils for trunk piston engines powered by heavy fuel oil have viscosities of SAE 40 and TBN values of 30–40.

4.4 Fuel Gases and Gas Engines

4.4.1 Historical Review

Precursors of present day gasoline engines, the first combustion engines were spark ignited gas engines that utilized generator gas or city gas obtained from coal or wood as fuel gas. The brake mean effective pressures and efficiencies attained in those days were modest. An easier to store liquid fuel, gasoline was used only later as vehicles began being motorized. With that, the gas engine, only still in use on a small scale as a stationary engine, e.g. in iron and steel works to recover blast furnace gases, disappeared from collective memory. Rudolf Diesel [1-3] initially also envisioned gas operation for his new rational heat engine. This prompted Krupp along with MAN to acquire an interest in the construction of a test engine. More recent tests on the use of hydrogen in large diesel engines have made Diesel's idea timely again.

The gas engine has been able to catch up with the diesel engine, specifically to generate power in cogeneration units, only by fully taking advantage of lean burn combustion's potential to reduce exhaust emission (see Sect. 14.2 for more on CHP).

4.4.2 Fuel Gases, Parameters

4.4.2.1 Fuel Gases for Gas Engines

Table 4-10 contains the most important fuel gases used in gas engines with their characteristic combustion values. The pure gases solely appear in gas mixtures such as natural gases. Only hydrogen H_2 can be used as pure gas, provided it is available. Landfill and digester gas are some of the renewable fuel gases produced when biomass, e.g. feces, garden cuttings or straw, decomposes. This relieves the environment in two ways: Their release is prevented and their use conserves fossil fuels. Their basic constituents are 40–60% methane and carbon dioxide. In addition, they may also contain harmful substances, e.g. chlorosulfide, fluorosulfide und hydrogen sulfides, which can be removed from the environment when these fuels are burned. In principle, the same combustion characteristics used to calculate combustion with liquid fuels apply to gases (see Sect. 1.2.3.1). However, the gas volume in m^3_n under standard conditions is often selected as the reference value.

The hydrocarbons contained in natural gas are mostly paraffins (aliphates) with chain-like configurations of C and H atoms and the structural formula C_nH_{2n+2}. The simplest hydrocarbon, methane CH_4 is followed by ethane C_2H_6, propane C_3H_8 and butane C_4H_{10}. Like methane, they are gaseous under standard conditions. Called isomerism, the alteration of the molecules' chain-like structure already appears in butane. Thus, the physical properties of isobutane

Table 4-10 Parameters of the most important fuel gases relative to the volume of gas under standard conditions (Table 3-5, 2nd Ed.)

Type of gas	Calorific value H_u kWh/m^3_n	Minimum air requirement L_{min} m^3/m^3	Calorific value of the air/fuel mixture h_u kWh/m^3_n	Methane number MN –
Methane	9.97	9.54	0.95	100
Natural gas L	9.03	8.62	0.93	88
Natural gas H	11.04	9.89	0.96	70
Propane	25.89	23.8	1.03	34
N-butane	34.39	30.94	1.03	10
Landfill gas	4.98	4.73	0.86	>130
Digester gas	6.07	5.80	0.89	130
Coke ovengas	4.648	4.08	0.91	36
Carbon monoxide	3.51	2.381	1.038	75
Hydrogen	2.996	2.38	0.89	0

Table 4-11 Influence of isomerism on the characteristic combustion data of butane (Table 3-6, 2nd Ed.)

C_4H_{10}	λ_u Percentage of gas in air by volume	λ_o Percentage of gas in air by volume	Auto-ignition temperature K	Molar mass kg/kmol	Density kg/m^3_n	Boiling point at 101.325 kPa K
Isobutane	1.8	8.4	733	58.123	2.689	261.43
N-butane	1.9	8.5	678	58.123	2.701	272.65

change from the standard form (n-butane) while the molar mass remains identical (see Table 4-11). These factors must be allowed for, especially when chemical waste gases are utilized in gas engines.

4.4.2.2 Combustion Parameters

Calorific Value and the Calorific Value of the Air/Fuel Mixture

Butane's calorific value of $H_u = 34.4$ kWh/m^3_n, is the highest among the fuel gases used in gas engines (see Table 4-10). A single gas, hydrogen has the lowest calorific value of 2.99 kWh/m^3_n. The difference in these single gases' calorific value corresponds to a factor of more than 11. Mixing them with inert gases, e.g. carbon dioxide CO_2 or nitrogen N_2, further increases this difference. The "weakest" low energy content gas mixtures that gas engines are still able to utilize have calorific values of around 0.5 kWh/m^3_n with a sufficient H_2 fraction, i.e. the ratio of the calorific values of "poor gas to rich gas" can be as high as 1:60. These differences in calorific value result from the differing C and H^2 fractions, which also affect the differences when stoichiometric combustion has a minimum air requirement L_{min}. By contrast, pairing H_u and L_{min} causes the calorific values of the air/fuel mixture $h_u = H_u/(1 + \lambda_v L_{min})$, which are determinative for the energy yield, to display a rather uniform pattern of values with only slight differences.

Methane Number MN

The methane number MN is defined by the volumetric mixture ratio of methane (MN = 100) and hydrogen (MN = 0) and thus directly furnishes information on a gas mixture's detonation limits. A methane number near 100 signifies high detonation limits, a methane number near zero low detonation limits. Accordingly, a mixture of 20% hydrogen H_2 and 80% methane CH_4 has a methane number of 80. Cartellieri and Pfeifer, who simultaneously also determined the methane number of other gases and gas mixtures in tests on a one cylinder CFR test engine, formulated their definition at an air/fuel ratio of $\lambda = 1$ [4-29].

The methane number in three-component gas mixtures can be determined with the aid of ternary diagrams, which also contain parameter lines of constant methane numbers for the lines of the constant gas fraction. Figure 4-39 illustrates the method of reading such a diagram: Point P in the diagram to the left represents the mixture for the percentage fraction a, b and c of the three components A, B and C. The diagram to the right additionally contains the line of constant detonation limits or the methane number passing through P.

Figure 4-40 is a ternary diagram of the composition of digester gases' three main components methane, carbon dioxide and nitrogen with the resultant methane numbers.

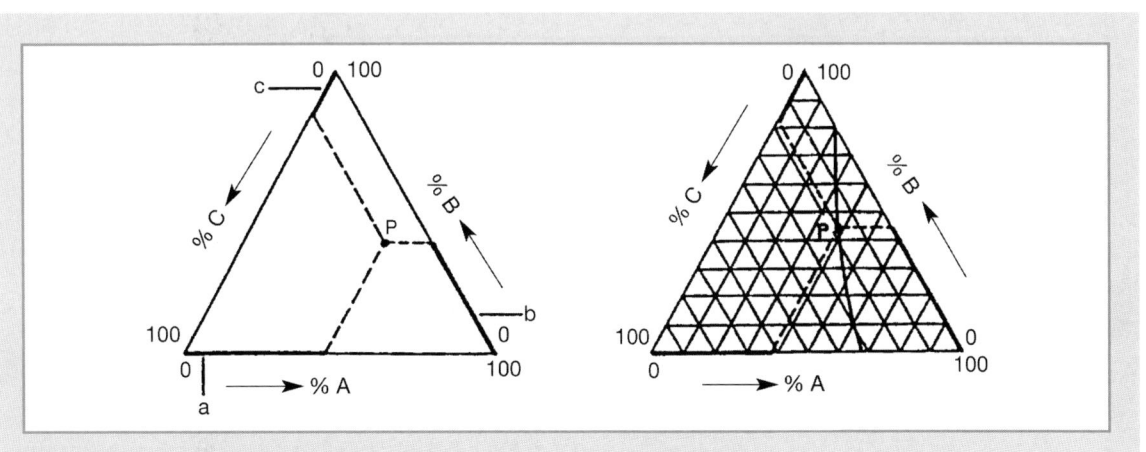

Fig. 4-39 Plotting of a ternary diagram to determine the methane number for a mixture of three gases [4-30]

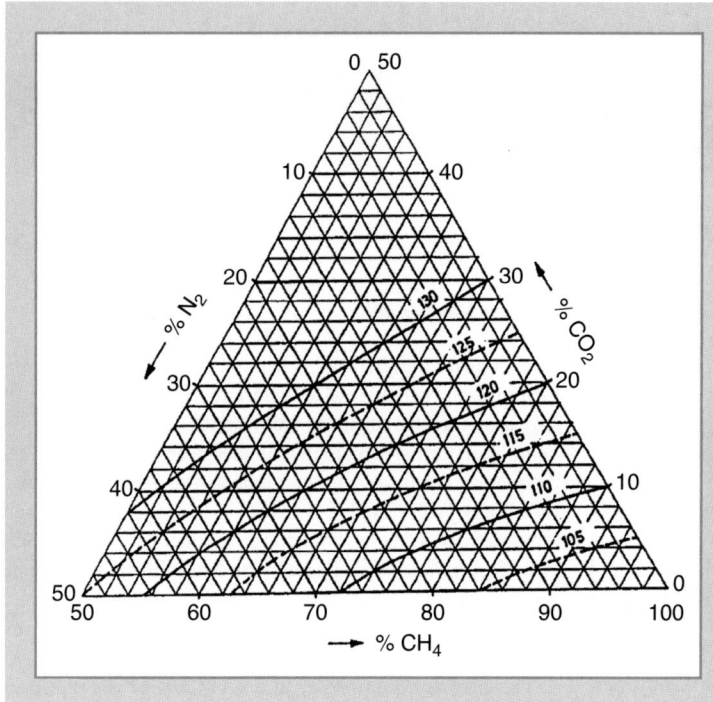

Fig. 4-40
Ternary diagram for digester gas composed of methane, carbon dioxide and nitrogen [4-30]

Various manufacturers sell programs that compute the methane number. The program developed by AVL is primarily used in Europe [4-30]. Contingent on the operating point's H_2 content, it can shift toward the "lean" side when applied to hydrogen gas mixtures. However, since the computation is based on an air/fuel ratio of $\lambda = 1$ it fails to state the detonation characteristics correctly.

Laminar Flame Velocity

Laminar flame velocity specifies the velocity with which the flame propagates in laminar flow conditions. It is largest in the range of stoichiometric conditions and decreases as the mixture is increasingly leaned as well as enriched ($\lambda < 1$). The individual fuel gases behave differently (see Fig. 4-41).

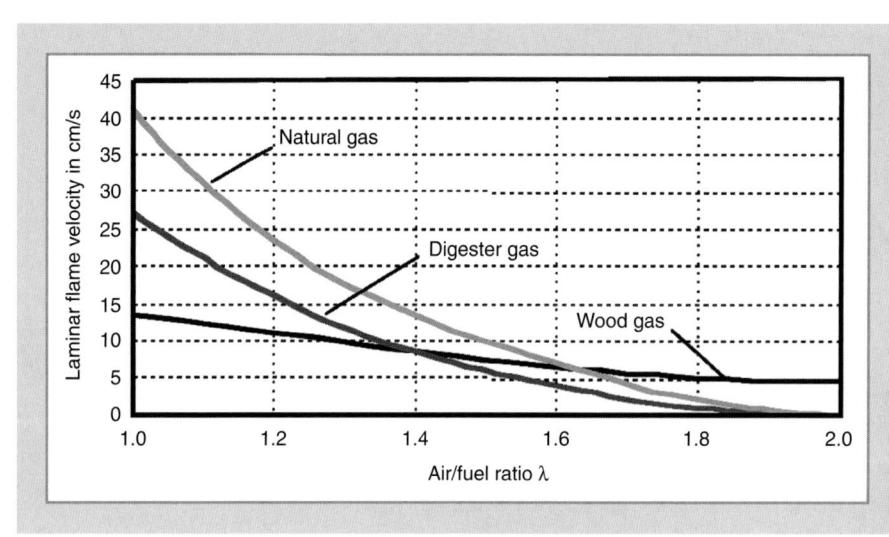

Fig. 4-41
Influence of the air/fuel ratio λ on laminar flame speed in cm/s

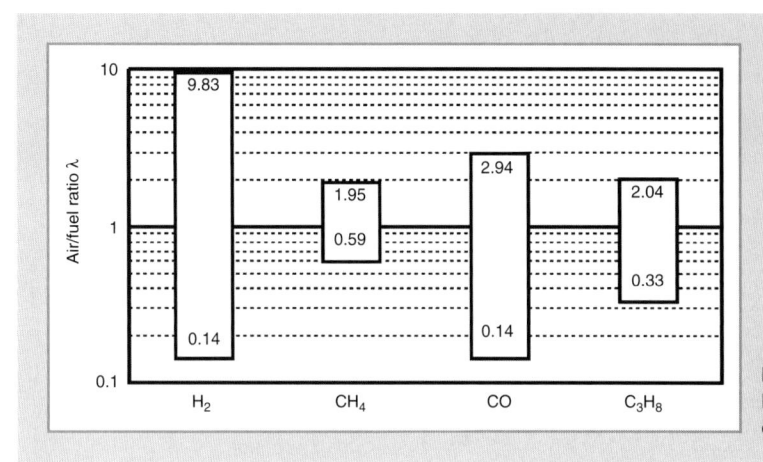

Fig. 4-42
Ignition limits of the pure gases hydrogen H_2, methane CH_4, carbon monoxide CO and propane C_3H_8

Ignition Limits

Single gases' ignition limits strongly influence fuel gases' mixture formation (Fig. 4-42). They specify the hypostoichiometric or stoichiometric air/fuel ratios at which ignition can still occur. Hydrogen has the broadest ignition limits, methane a relatively small ignition range. Hence, it is important to understand the processes during mixture formation in order to be able to implement appropriate measures for ignition in the combustion chamber. The requirements for mixture formation or the fuel admission system are rather moderate when hydrogen is used as fuel. Particularly high demands are made on the homogeneity of a natural gas mixture, above all for extremely lean operation.

4.4.2.3 Evaluation of Gas Quality

The quality of a gas intended for gas engines is evaluated primarily on the basis of the air/fuel mixture's calorific value and the methane number. While the calorific value of the air/fuel mixture defines a gas system's design (gas pressure and gas valve opening duration), the methane number determines the detonation limit and thus the maximum engine power.

Gases from different sources are blended together to ensure reliable gas delivery. Natural gas is blended into the base gas according to the allowable limits of the gas family. As a rule, gases are blended (usually a propane/butane gas mixture) in such a way that either the calorific value of the air/fuel mixture or the Wobbe index (characteristic value of the thermal load of gas burners) is constant. Even when the air/fuel mixture's calorific value is constant, this distinctly changes the methane number. Smooth engine operation without detonating combustion and resultant engine damage requires keeping the distance to the detonation limit sufficiently large. The losses of efficiency and power this causes are prevented by active detonation control, thus making it possible to operate an engine at the detonation limit under optimal conditions and at maximum engine efficiency.

4.4.3 Definition and Description of Gas Engines

4.4.3.1 Classification of the Combustion Systems

The combustion systems employed in gas engines are derived from the air/fuel mixture's type of gas blending and type of ignition. The generally accepted definition of combustion systems allows a division into spark ignited (SI) engines, dual fuel (DF) engines and gas diesel (DG) engines (Fig. 4-43).

Both SI and DF engines operate based on the gasoline engine system. Only their type of ignition differs. However, the heterogeneous mixture in gas diesel engines is ignited by auto-igniting the fuel gas with the aid of a small quantity of diesel fuel (pilot fuel) and thus corresponds to the diesel engine system.

Spark Ignited (SI) Engines

Based on the system of the gasoline engine, the homogeneous air/fuel mixture of spark ignited engines is produced outside the combustion chamber [4-31]. The mixture may form centrally in the engine air intake as well as after the compressor or before every intake valve in supercharged engines. Consequently, the requisite gas pressure must be higher than the intake air pressure or the charge air pressure.

A spark plug initiates ignition electronically. Glow plug ignition for prechamber engines might also be conceivable but would require a controlled supply of gas into the compressed mixture in the prechamber to trigger ignition [4-32]. Engine power is determined by the mass of the air/fuel mixture supplied and thus quantity-controlled.

The compression ratio in production diesel engines routinely used as SI engines must be adjusted to the detonation

Fig. 4-43 Definition of the combustion process in gas engines

Spark Ignited	Dual Fuel	Gas Diesel
External mixture formation Homogeneous mixture Quantity control Spark ignition		Internal mixture formation Heterogeneous mixture Quality control Autoignition
Electric - $p_{Gas} > p_1$ or p_3 $P \approx 0.8 \, P_{Diesel}$	Pilot fuel $x_{eB} = 0.5\% \ldots 100\%$ $p_{Gas} > p_1$ or p_3 $P \approx 0.9 \, P_{Diesel}$	Pilot fuel/gas $x_{eB} = 5\% \ldots 100\%$ $p_{Gas} > p_x$ $P = P_{Diesel}$

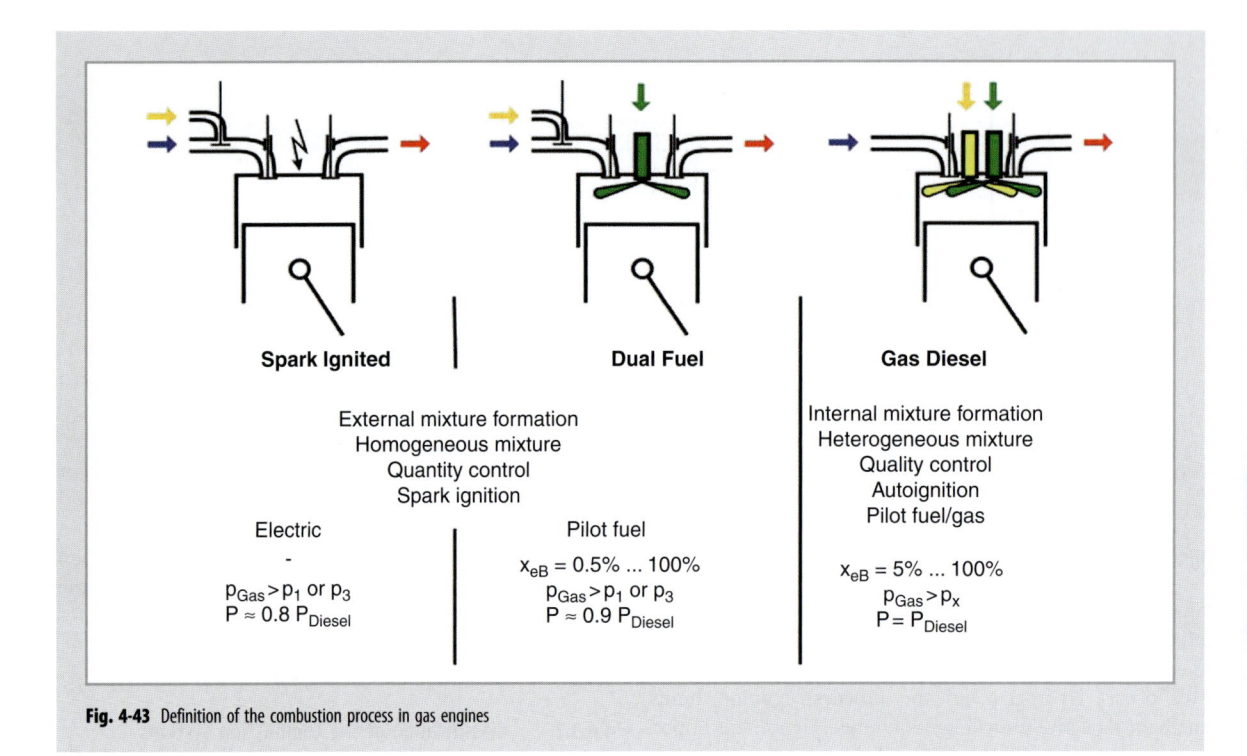

limits (methane number) of the fuel gas available so that the air/fuel mixture is unable to auto-ignite. This reduces the power by approximately 20% over the diesel system.

Dual Fuel (DF) Engine

Diesel and spark ignited engines resemble one another in terms of their type of ignition and mixture formation and thus operate on the basis of the system of the gasoline engine, the fundamental difference being the type of ignition source. Unlike electric ignition in SI engines, ignition in DF engines is initiated by injecting diesel fuel, so-called pilot fuel [4-33, 4-34]. In principle, it is possible to boost the quantity of pilot fuel to 100 in order to also run the engine in pure diesel operation (dual-fuel engines). A micropilot engine limits the pilot fuel to only the percentage that starts the engine in diesel operation and the quantity of pilot fuel required, approximately 10 to 15% of the quantity injected at full load. Since its mixture preparation and delivery resemble that of SI engines, this combustion system's engine power is also determined by the mass of the air/fuel mixture supplied and thus quantity-controlled.

Since, on the one hand, the pilot fuel must auto-ignite and, on the other hand, the homogeneous air/fuel mixture that delivers the main energy for engine operation may not tend to detonate, i.e. it must have a sufficiently high methane number, DF engines require a lower reduction in the compression ratio than SI engines. The application engineering of production diesel engines does not require replacing the injection equipment in dual fuel engines with an ignition system (injection pumps) and spark plugs (injection nozzles) as in spark ignited engines.

Gas Diesel (DG) Engines

Based on the system of the diesel engine, gas diesel engines are fundamentally auto-igniting, the mixture being produced by injecting the gases into the compressed combustion air. Thus, the mixture in the combustion chamber is inhomogeneous at the time of ignition [4-35]. Ignition is initiated by injecting an additional amount of diesel fuel. Engine power is only determined by the gas mass and is thus quality-controlled.

The compression ratio needs no significant adjustment to ensure the pilot fuel ignites. Thus, the same engine power can be produced as in the diesel system.

4.4.3.2 Spark Ignited (SI) Engines

Two basically different ignition concepts are distinguished, direct ignition and prechamber ignition. With a few exceptions, the direct ignition concept illustrated in Fig. 4-44a applies to high-speed engines (n $\geq$ 1,500 rpm) with bore diameters of up to 170 mm. This essentially corresponds to the ignition concepts employed in car gasoline engines. At lower speed (n = 800–1,200 rpm) and richer combustion (λ = 1), the flame speed in the combustion chamber is high

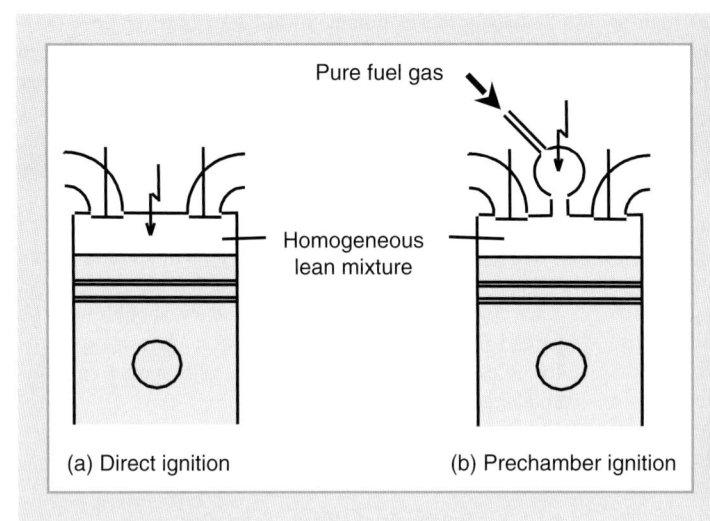

Fig. 4-44 Directed ignition (**a**) and precombustion chamber ignition (**b**) in spark ignited engines

(a) Direct ignition (b) Prechamber ignition

enough to even implement this concept for bore diameters of up to 250 mm. Lean burn engines ($\lambda \geq 1.7$) with bore diameters of 200 mm and larger require precombustion chamber ignition (Fig. 4-44b), i.e. the combustion chamber is divided into a main and secondary combustion chamber called the precombustion chamber, which has a volume of between 1 and 5% of the compression volume. The precombustion chamber has its own gas supply, which enriches the mixture in the chamber up to an air/fuel ratio in the range of $\lambda \approx 1$ to guarantee ignition.

Thus, the rich mixture from the precombustion chamber allows reliable ignition of extremely lean mixtures ($\lambda \geq 2.2$). High speed SI engines can achieve NO_X emissions that are compliant with the Technical Instructions on Air Quality Control (*TA Luft*) in the range of 250–500 mg/m^3_n with net efficiencies of over 45%.

Unscavenged precombustion chambers are also implemented in lean burn engines with bore diameters below 200 mm to enhance the ignition conditions at the spark plug (see Sect. 4.4.4.3).

4.4.3.3 Dual Fuel (DF) Engines

Auto-ignition of the injected diesel fuel, the pilot fuel, initiates the spark ignition of the gas mixture. Rudolf Diesel already proposed this system to burn difficult to ignite fuels such as coal.

The quantity of pilot fuel fundamentally influences the level of NO_x emission from dual fuel engines. Since the combustion cycle of the mixture ignited by the injection sprays nearly doubles the NO_X values (1,500–2,000 mg NO_x/m^3_n), NO_x limits can presently only be met with SCR catalysts when the quantities of pilot fuel are above 5% (dual-fuel engines). Reducing the quantity of pilot fuel (energetically) to approximately 0.5% of the full load quantity (as in micropilot engines) makes it possible to comply with NO_x values of 500 mg/m^3_n (*TA Luft* limit) even without exhaust gas aftertreatment.

The advantage of the dual-fuel engine concept is its capability to continue running as a diesel engine should the gas run out (mainly important for emergency power applications). The engine is started as a diesel engine so that gas, as the primary energy source, can then be successively mixed into the air flow. This type of engine is chiefly used for large quantities of gas with low calorific values and low flammability, e.g. gases with high percentages of inert gas.

The relatively high methane number required is a drawback since a compression ratio must be selected, which is sufficient for the diesel fuel to auto-ignite during a start. When gases or gas mixtures have low methane numbers, countermeasures, e.g. lead reduction or "leaning", must guarantee operation without detonation by replacing the fraction of fuel gas energy with larger quantities for pilot ignition. This is particularly problematic in hydrogen-rich synthesis gases since it means accepting either only moderate mean pressures (at the level of naturally aspirated engines) or high levels of diesel fuel consumption.

4.4.3.4 Gas Diesel Engines

The gas diesel engine concept is characterized by mixture formation in the combustion chamber. The high pressure fuel gas is upstream from the pilot ignition and then subsequently ignited by the diesel spray [4-36]. The ignition quality of the gas is of such secondary importance that this method can even use gases with low methane numbers without special measures. This development is based on the presence of such high pressure gas (approximately 200 bar) as accumulates as a "byproduct" on CNG tankers or is available in the petroleum industry on drilling platforms or in pumping stations. Since the requisite gas pressure must be higher than the final compression pressure, increased demands are placed on the gas system (compression resistance and safety).

This concept's efficiencies are practically just as high as those of diesel engines. However, the NO_x emissions of any engines operating on the basis of the pilot ignition concept require exhaust gas aftertreatment with a deNOx catalyst.

4.4.3.5 Hydrogen Operation of Gas Diesel Engines

In the future, efforts to utilize alternative fuels and reduce environmental pollution from engine exhaust gases will attach great importance to hydrogen [4-37], provided it is not obtained by electrolysis using electrical power generated with fossil fuels. With an eye toward even being able to use hydrogen in large diesel engines at some point, pertinent tests were conducted on a single cylinder diesel engine with a piston diameter of 240 mm [4-38]. In the current stage of development, hydrogen compressed to 300 bar is supplied to the engine shortly before top dead center by an injection valve with several spray holes, which replaces the centered injection nozzle. Then, the air/hydrogen mixture auto-ignites. In the course of the tests, the compression ratio was increased from its original $\varepsilon = 13.7$ in the diesel version to 16.8–17.6. The air/hydrogen mixture's sizeable ignition range largely allows quality control comparable to load adjustment common in diesel engines. The brake mean effective pressures of approximately 19 bar corresponding to specific work $w_e = 1.9$ kJ/dm^3 attained in test operation of a supercharged engine suggest advances may approach 25 bar in the future. This is comparable to the power level of present day marine diesel engines.

Combustion proceeds soot-free. The greenhouse gas carbon dioxide CO_2, carbon monoxide CO and hydrocarbon compounds HC only appear in insignificantly small concentrations. Nitrogen oxides are the sole quantitatively relevant components. Their concentration exceeds the emission level of emission-optimized diesel engines. The strong inhomogeneities of the air/fuel mixture and the extremely short duration of combustion with high conversion rates, which cause local high temperatures in the combustion chamber and thus increased NO formation, can be regarded as the reasons for this. Further work on development is needed here. Aftertreatment systems that reduce NO_x, which may become necessary, function extremely effectively because of the purity of the exhaust.

Even though a hydrogen powered diesel engine remains a future option from the current vantage point, it already fulfills two of Rudolf Diesel's visions, namely smoke-free operation of his rational heat engine and its suitability for fuel gases too!

4.4.3.6 Gas Engines as Vehicle Engines

Spark ignited engines derived from diesel engines on the basis of the $\lambda = 1$ concept are used in municipal vehicles, commuter vehicles and public transportation. Such vehicles use natural gas (CNG) compressed to 200 bar carried in roof tanks. They are easily modified and handle like a vehicle designed with a diesel engine.

Systematic development toward lean burn engines ($\lambda \geq 1.7$) is opening significantly expanded options with enhanced power, efficiencies and ranges [4-39]. The demand for higher power at lower fuel consumption and minimal emissions has generated numerous changes in the basic diesel engine, e.g. combustion chamber shape, piston design, intensive cooling of cylinder head and liner, lubricating oil consumption, fuel system and engine control. Thus, today's vehicle gas engines represent independent developments [4-40].

Gas engines in industrial locomotives should now be regarded more as an exception. However, they are sporadically necessary in areas of operation with special environmental regulations.

4.4.4 Mixture Formation and Ignition

4.4.4.1 Potentials of Mixture Formation

Similar to car gasoline engines, a distinction can be made between gas engines with:
- *central mixture formation* by means of mixture-forming equipment or a gas mixer (corresponding to single-point injection) and
- *individual mixture formation* before the intake into the combustion chamber or in the engine's *combustion chamber* (corresponding to the options of multi-point injection).

The central gas mixer in supercharged gas engines may be placed before or after the supercharger group's compressor. Placement on the suction side is advantageous since it only requires low gas pre-pressure, e.g. for digester or landfill gas (30–100 mbar). The homogenization of the mixture in the compressor generates another advantage. This provides every bank in V engines an exactly equal air/fuel ratio. Placing the gas mixer on the pressure side (p_2) requires a correspondingly higher gas pre-pressure, which must be generated by an internal gas compressor case by case. In addition, there is always a risk of disruptively rich stream filaments, making it expedient to connect a homogenizing unit downstream from the gas mixer.

Individual mixture formation is found in larger engines (D > 140 mm). Provided the engine is not naturally aspirated, the gas pressure level must be elevated commensurately with the charge air pressure. Additional gas intake valves are necessary in both cases. This requires design changes in the cylinder head. When a throttle is used to adjust the air supply to control the torque, the valve cross section in the gas valve must be adjusted so that the mixture continues to be ignitable. The requisite air mass flow in supercharged engines is adjusted by controlling the boost pressure with an exhaust bypass (wastegate), compressor bypass or variable turbine geometry (VTG), the torque being changed either by means of load-dependent gas pressure control or gas intake valves with variable opening durations.

Fig. 4-45 Venturi mixer

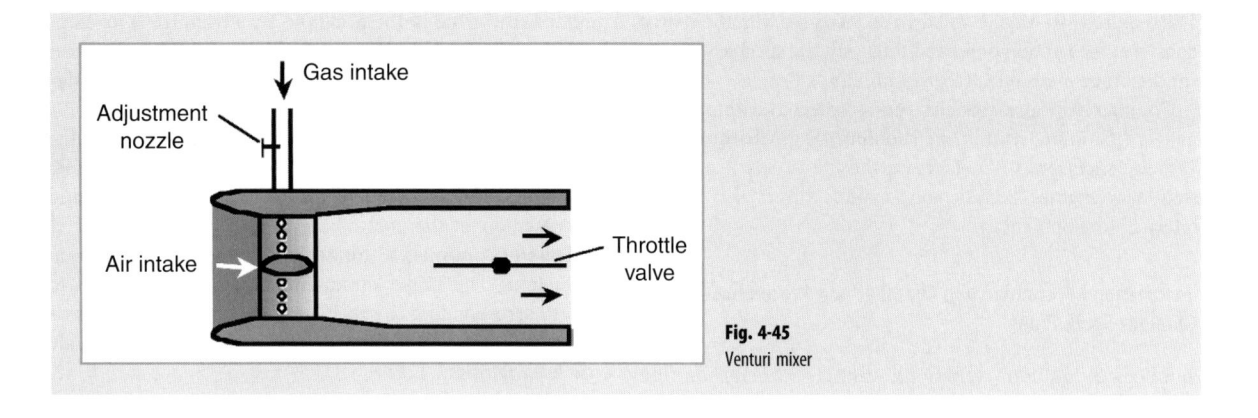

4.4.4.2 Gas Mixers

Most gas mixers function on the basis of the venturi principle: A venturi nozzle positioned in the intake air flow lowers the stationary pressure in the narrowest cross section where two bores interconnected by a annular channel that admits gas are located (Fig. 4-45). The differential pressure in the inlet and thus the influent gas mass flow also increase as the cylinder charge increases, e.g. when speed increases. Since the densities of gas and air change simultaneously, additional control functions are necessary to maintain the desired air/fuel ratio.

A variable restriction gas mixer (IMPCO) is frequently used in engines with low emission requirements. Figure 4-46 presents its functional principle. The distinctive feature of this type of gas metering valve is its external shape, which allows obtaining a mass flow-dependent air/fuel ratio curve relatively easily. When another characteristic is desired, then the shape of the control valve is adapted to the desired mixture ratio. The spring located in the vacuum chamber provides another potential control function. This type of gas mixer may be implemented on the suction or the pressure side. Since it has no electronic actuator, the simple mechanics are a significant advantage. However, its fixed setting

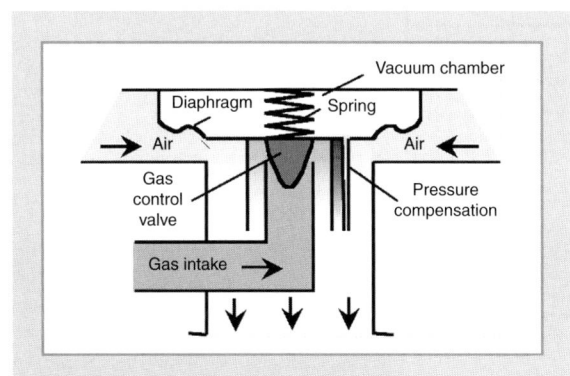

Fig. 4-46 Gas mixer: IMPCO variable restriction carburetor

corresponding to the geometry and thus a particular air/fuel ratio value when it is adjusted are a disadvantage. This concept precludes any control function to compensate for variable calorific values or changing ambient conditions (intake temperature and air pressure).

Its high flow losses and the attendant charge losses make this type of mixer unsuitable for supercharged gas engines with high specific power.

Another widespread basic concept utilizes the area ratio of the air and gas cross section to adjust the mixture ratio. Already relatively old, the underlying idea – corresponding to a variable venturi carburetor – was taken up by Ruhrgas AG in the mid 1980s and refined to its current state (with the brand name HOMIX).

The advantage of this concept for lean burn engines is the ability to regulate the air/fuel area ratio relatively easily during operation. This type of gas mixer always requires a pressure regulator (zero pressure regulator) that adjusts the pressure conditions in the gas feed to the pressure in the suction pipe before the compressor.

A so-called gas train is needed to adjust the pressure level in the gas systems to the pressure level of the mixture formation intended in the gas engine. In addition to adjusting the pressure, the gas train or "gas path" usually integrates the required safety equipment [4-41].

4.4.4.3 Electric Ignition Systems and Spark Plugs

As in car gasoline engines, the air/fuel mixture in spark ignited engines ignites with the aid of a spark plug. In use until the early 1990s, mechanically driven generator ignition units (e.g. from Altronic) have been abandoned in favor of electronic ignition in which a thyristor controls a capacitor's discharge to easily vary the ignition point, e.g. when there is a risk of detonating combustion. It is important to differentiate between the maximum potential gate-trigger voltage (gate-trigger voltage supplied) and the voltage required for arcing (gate-trigger voltage required). The amount is primarily determined by the gap size between the center and ground electrode.

The required gate-trigger voltage also increases with the charge density at the ignition point and thus with the specific work w_e (or the brake mean effective pressure Pm_i).

Commercially available electrodes protected with precious metals (platinum, iridium or rhodium), e.g. Champion RB 77WPC and Denso 3-1, increase the life of spark plugs in stationary engines. Some engine manufacturers also rely on internal developments.

Unscavenged Precombustion Chamber and Precombustion Chamber Spark Plugs

Since their ignition conditions strongly depend on the conditions in the combustion chamber, e.g. charge movement and mixture inhomogeneities, lean burn engines impose special requirements on ignition systems. One measure is the use of precombustion chamber spark plugs (unscavenged precombustion chambers without their own gas feed). The spark plug is surrounded by a chamber that relatively small passages (transfer bores) connect with the main combustion chamber. This establishes constant ignition conditions for the spark plug, which improve ignition conditions and thus the spark plug's life [4-42]. The mixture can be leaned out, thus producing lower NO_x emissions than in an undivided combustion chamber.

Laser Ignition

Concepts that employ a pulse laser to ignite the mixture in the combustion chamber are being developed as alternatives to electronic ignition systems with spark plugs [4-43]. Through an optical access to the combustion chamber, a fiber optic cable focuses light from a laser source anywhere in the combustion chamber in such a way that ignition of the mixture is initiated. By virtue of the principle, its advantage is its elimination of spark plug wear. Further, several ignition sources may be implemented in a combustion chamber, thus producing extremely rapid mixture conversion rates.

Goals of development are stable optical accessibility to the combustion chamber, the minimization of power losses between the laser source and the combustion chamber and the reduction of the cost of the overall system. The latter particularly continues to hinder the use of such systems in production. The brake mean effective pressures of up to 30 bar corresponding to specific work $w_e = 3.0$ kJ/dm^3 envisioned for future spark ignited engines are only conceivable with laser ignition.

4.4.5 Gas Engine Emissions

4.4.5.1 Combustion of Homogeneous Gas Mixtures

Presented as a function of the air/fuel ratio λ in Fig. 4-47, the emission characteristic of the regulated exhaust emissions of carbon monoxide CO, total hydrocarbons HC and nitrogen oxides NO_x is typical for the combustion of a homogeneous fuel-air mixture. While CO and HC strongly decrease as the excess air ($\lambda > 1$) increases after stoichiometric combustion, NO_x emission only reaches its maximum at $\lambda \approx 1.1$ because the temperatures in the combustion chamber connected with the oxygen supply are still very high at $\lambda > 1$. The emission level decreases when air/fuel ratios are increased further, reaching a level in the extreme lean range at $\lambda \approx 1.6$, which is acceptable for the emission control legislation in *TA Luft*.

The boundaries in the figure indicate the two types of operation for an emission-optimized gas engine: combustion of a stoichiometric ($\lambda = 1$) or a lean air/fuel mixture with $\lambda \geq 1.6$.

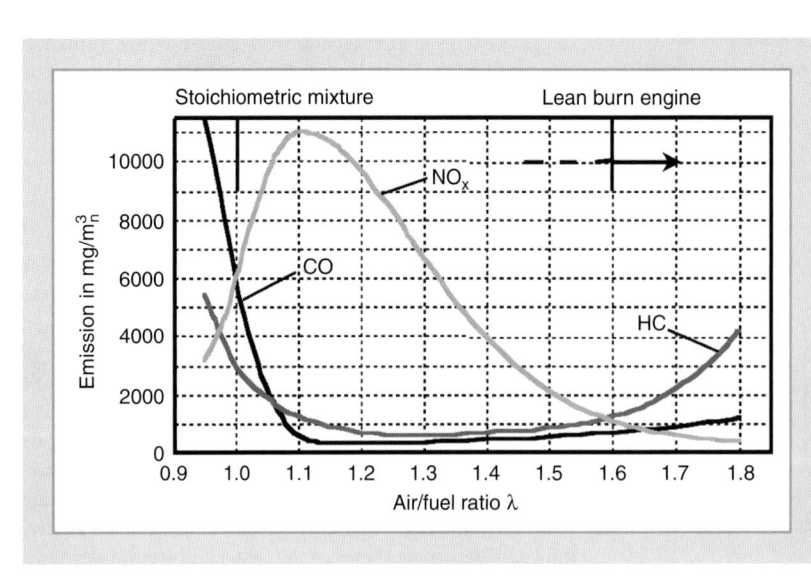

Fig. 4-47 Air/fuel ratio and exhaust emission, limits of the concept for NO_x reduction

European emission control legislation is essentially oriented toward Germany's *TA Luft* (see Sect. 15.2) with the noteworthy distinction that, instead of referring to power as is often common, the limits refer to the charge flow rate in m^3_n with an O_2 content of 5% in exhaust. Thus, the amount of power generated from the same quantity of fuel gas is inconsequential since the efficiency of utilization does not enter into the calculated emission values. Therefore, individual countries such as Denmark factor in net efficiency when establishing their limits for unburned hydrocarbons. Emission limits not only exist for NO_x, CO and HC but also non-methane hydrocarbons (NMHC), dust or particulate emission (based on stipulated measuring procedures, see Sect. 15.6) and C compounds classified as C_1, C_2, C_3, etc. as well as dioxins and furans.

4.4.5.2 Gas Engines with Stoichiometric Mixtures

As in car gasoline engines, a stoichiometric air/fuel ratio is the prerequisite for use of a three-way catalyst that simultaneously lowers CO, HC and NO_x emission by post-oxidation or reduction. The air/fuel ratio must be controlled with a lambda oxygen sensor that ensures the catalyst functions within the very narrow limits of the "lambda window" of 0.980–0.991. Only hypostoichiometric combustion ($\lambda \approx 0.997$) can effectively reduce NO_x.

In conjunction with the influences generated by oil ash and the problem gases contained in the fuel gas, the high thermal load diminishes catalyst efficiency. Hence, servicing and maintenance costs are relatively high. This limits the use of three-way catalysts to engines with lower power, predominantly naturally aspirated engines ($p_{me} \leq 8$ bar or $w_e \leq 0.8$ kJ/dm^3). However, a combination of exhaust gas turbocharging and exhaust gas recirculation can simultaneously lower the thermal load and raw NO_x emission and thus costs.

4.4.5.3 Lean Burn Engines

Mixture Control with a Lean Oxygen Sensor

Lean burn operation in compliance with *TA Luft* limits requires lean mixture supercharging to compensate for the losses of effective brake work connected with lean combustion by supercharging. A premixed air/fuel mixture has proven to be easily precompressible in the exhaust gas turbocharger's compressor with advantages for mixture homogenization (see Sect. 4.4.4.1). Only lean burn engines can be used for biogases loaded with such pollutants as chlorine, fluorine and silicon compounds and hydrogen sulfide H_2S since these "catalyst poisons" would render a three-way catalyst inoperable in no time.

A closed loop control maintains the requisite stability of the lean air/fuel ratio (comparable to air/fuel ratio 1 operation). The "lean oxygen sensors" employed to do this deliver an electrical signal usable by the control loop starting only at an air/fuel ratio of $\lambda > 1.6$. The sensors' limited lives, particularly when used in biogases with extremely harmful constituents (Cl, F, S, etc.), make this concept so extremely cost intensive for an engine operating time of 8,000 h/a that alternative concepts have been developed to control mixture composition.

Alternative Concepts for Lean Burn Engines

Combustion Chamber Temperature (TEM) Measurement

Deutz's [4-44] concept envisions measuring representative temperature in a volumetric element of the combustion chamber to maintain the air/fuel ratio. The relatively slow sensor (thermocouple) does not measure the true temperature in the volumetric element, a recess in the bottom of the cylinder head. Rather, it measures a mean temperature reached during the combustion cycle, to which the average NO_x emission measured at an operating point and stored in the controller (TEM) is assigned.

The sensor's decreasing sensitivity due to isolating deposits has an adverse effect since the mixture is enriched at a pseudo excessively low average temperature. Operation at the detonation range on the other hand produces better heat transfer in the sensor so that the controller readjusts toward "lean" as desired and thus prevents detonation.

LEANOX

Developed and patented by Jenbacher [4-45], the LEANOX concept employs the pressure and temperature values measured after the throttle, which correspond to the energy supply at a given engine setting. In conjunction with the related NO_X values, a correlation exists between the air/fuel ratio and NO_X emission. The advantage of this concept is its independence from an engine's operating life, which has no influence on the measurement. Moreover, the controller's interpretation of this as a deviation toward "lean" equalizes any decrease of calorific value. This produces the desired enrichment of the mixture.

Ionization Sensor

Caterpillar uses an ionization sensor for its large 3600 and G-CM34 series. The basic principle is based on capturing the speed of the flame front from the spark plug to the ionization sensor mounted close to the cylinder liner. The flame speed is assigned to an NO_X emission during "calibration". The signal detected by the sensor is relatively unclear and the control system no longer functions precisely enough when combustion is very lean ($\lambda > 2.5$).

Light Emission Measurement

Research findings have demonstrated that flame radiation can be attributed to the range of the OH band for 310 nm of

NO$_x$ emission. Thus, a distinct correlation exists. The contamination of the combustion chamber window that occurs as the running period lengthens is problematic for real engine operation. The frequent readjustment required is cost intensive and limits the system's precision and thus its application.

Cylinder Pressure Measurement

Cylinder-selective pressure measurement and the online thermodynamic analysis it enables provide an extremely elegant option. The advantage of this method is the ability to use other relevant parameters for control and monitoring, e.g. mean indicated pressure Pm$_i$, maximum cylinder pressure, ignition point, duration of combustion and even detonation phenomena. This (still very expensive) concept is implemented in large marine diesel engines for the purpose of engine management [4-46]. In conjunction with inexpensive powerful computers, new sensors and methods of pressure measurement [4-47] could facilitate future use of this concept for gas engines too [4-48].

Exhaust Gas Aftertreatment

Oxidation Catalyst

TA Luft limits gas engines' CO emission to ≤ 650 mg/m^3_n. State-of-the-art, efficiency-optimized gas engines have raw emission of approximately 800–1,100 mg CO/m^3_n. The CO primarily stems from incomplete reactions during combustion. Formaldehyde is an intermediate product of methane oxidation. Both emission components and higher hydrocarbons can be reduced greatly when the oxidation catalyst is adequately sized (space velocity and precious metal content).

Thermal Post-oxidation

Thermal post-oxidation is employed to reduce partially or unburned components in the exhaust of efficiency-optimized gas engines, which contains biogases with the aforementioned content of problem gases, which are "catalyst poisons" that render oxidation catalysts inoperable in no time. Sufficient oxygen O$_2$ is present in the exhaust depending on the combustion concept. However, an increase of the required oxidation temperature of $> 760°$C is necessary. Recuperative or renewable heat exchangers are used to reduce the energy required. GE Jenbacher's CL.AIR system has established itself for landfill gas plants.

Literature

4-1 DIN EN 590 Automotive fuels - Diesel - Requirements and test methods. German version of EN 590 (2006) 3
4-2 Gerling, P. et al.: Reserven, Ressourcen und Verfügbarkeit von Energierohstoffen 2005 – Kurzstudie. Hannover: Bundesanstalt für Geowissenschaften und Rohstoffe (2007) 2 (www.bgr.bund.de)
4-3 Bundes-Immissionsschutzgesetz in der Fassung der Bekanntmachung vom 26. September 2002 (BGBl. I S. 3830), zuletzt geändert durch Artikel 3 des Gesetzes vom 18. Dezember 2006 (BGBl. I S. 3180), Sect. 34, 37
4-4 Zehnte Verordnung zur Durchführung des Bundes-Immissionsschutzgesetzes: Verordnung über die Beschaffenheit und die Auszeichnung der Qualitäten von Kraftstoffen. BGBl. I (2004) 6, p. 1342
4-5 European Automobile Manufacturers Association (ACEA), Japan Automobile Manufacturers Association (JAMA), Alliance of Automobile Manufacturers (Alliance), Engine Manufacturers Association (EMA), Organisation Internationale des Constructeurs d'Automobiles (OICA) (Ed.): Worldwide Fuel Charter. 4th Ed. (2006) 9
4-6 DIN EN 14214 Automotive fuels - Fatty acid methyl esters (FAME) for diesel engines - Requirements and test methods. German version of EN 14214 (2004) 11
4-7 Richtlinie 2003/30/EG des Europäischen Parlaments und des Rates vom 8. Mai 2003 zur Förderung der Verwendung von Biokraftstoffen oder anderen erneuerbaren Kraftstoffen im Verkehrssektor. EU-ABl. L 123 (2003) 5, S. 42–46
4-8 World Business Council for Sustainable Development (WBCSD) (Ed.): Mobility 2030 – Meeting the challenges to sustainability. The Sustainable Mobility Project – Full Report 2004. Geneva (2004) 7 (www.wbcsd.org)
4-9 Geringer, B.: Kurz- und mittelfristiger Einsatz von alternativen Kraftstoffen zur Senkung von Schadstoff- und Treibhausgas-Emissionen. MTZ extra: Antriebe mit Zukunft. Wiesbaden: Vieweg (2006)
4-10 Schindler, V.: Kraftstoffe für morgen: eine Analyse von Zusammenhängen und Handlungsoptionen. Berlin/Heidelberg: Springer (1997)
4-11 Hassel, E.; Wichmann, V.: Ergebnisse des Demonstrationsvorhabens Praxiseinsatz von serienmässigen neuen rapsöltauglichen Traktoren. Abschlussveranstaltung des 100-Traktoren-Demonstrationsprojekts am 9. 11 2005 in Hannover. Universität Rostock 2005 (www.bio-kraftstoffe.info)
4-12 DIN V 51605 Fuels for vegetable oil compatible combustion engines - Fuel from rapeseed oil - Requirements and test methods. (2006) 7
4-13 Rantanen, L. et al.: NExBTL – Biodiesel fuel of the second generation. SAE Technical Papers 2005-01-3771 (www.nesteoil.com)
4-14 Bourillon, C. (Iogen Corporation): Meeting biofuels targets and creating a European Biofuels Industry. 4. Intern. Fachkongress "Kraftstoffe der Zukunft 2006". Berlin, November 2006
4-15 DIN EN 228 Automotive fuels - Unleaded petrol - Requirements and test methods. German version of EN 228, (2006) 3

4-16 DIN EN 15376 Automotive fuels - Ethanol as a blending component for petrol - Requirements and test methods. German version of EN 15376 (2006) 7
4-17 Fachagentur für Nachwachsende Rohstoffe (Ed.): Biokraftstoffe – eine vergleichende Analyse. Gülzow 2006 (www.bio-kraftstoffe.info)
4-18 Fachverband Biogas e.V. (Ed.): Biogas – das Multitalent für die Energiewende. Fakten im Kontext der Energiepolitik-Debatte. Freising 3/2006 (www.biogas.org)
4-19 Steiger, W.; Warnecke, W.; Louis, J.: Potentiale des Zusammenwirkens von modernen Kraftstoffen und künftigen Antriebskonzepten. ATZ 105 (2003) 2
4-20 N.N.: Biofuels in the European Union – A vision for 2030 and beyond (EUR 22066). European Commission, Office for Official Publications of the European Communities Luxembourg (2006)
4-21 CHOREN Industries GmbH: The Carbo-V Process (http://www.choren.com/en/biomass_to_energy/carbo-v_technology/)
4-22 Steiger, W.: Die Volkswagen Strategie zum hocheffizienten Antrieb. 22th Wiener Motorensymposium 2001, Vienna (4/2001)
4-23 Steiger, W.: Potentiale synthetischer Kraftstoffe im CCS Brennverfahren. 25th Wiener Motorensymposium 2004, Vienna (4/2001)
4-24 Steiger, W.: Evolution statt Revolution. Die Kraftstoff- und Antriebsstrategie von Volkswagen. Volkswagen AG Forschung Antriebe (2007)
4-25 Groth, K. et al.: Brennstoffe für Dieselmotoren heute und morgen: Rückstandsöle, Mischkomponenten, Alternativen. Ehningen: expert (1989)
4-26 Häfner, R.: Mittelschnelllaufende Viertaktmotoren im Schwerölbetrieb (im Teillastdauerbetrieb). Jahrbuch der Schiffbautechnischen Gesellschaft Bd. 77. Berlin: Springer (1983)
4-27 Zigan, D. (Ed.): Erarbeitung von Motorkennwerten bei der Verbrennung extrem zündunwilliger Brennstoffe in Dieselmotoren im Vollast- und Teillastbereich. Schlussbericht zum Teilvorhaben MTK 03367. Krupp MaK Maschinenbau GmbH, Kiel 1988. TIB/UB Hannover: Signatur: FR 3453
4-28 Wachtmeister. G.; Woschni, G.; Zeilinger, K.: Einfluss hoher Druckanstiegsgeschwindigkeiten auf die Verformung der Triebwerksbauteile und die Beanspruchung des Pleuellagers. MTZ 50 (1989) 4, pp. 183–189
4-29 Christoph, K.; Cartellieri, W.; Pfeifer, U.: Die Bewertung der Klopffestigkeit von Kraftgasen mittels Methanzahl und deren praktische Anwendung bei Gasmotoren. MTZ 33 (1971) 10
4-30 Zacharias, F.: Gasmotoren. Würzburg: Vogel (2001)
4-31 Mooser, D.: Caterpillar High Efficiency Engine Development – G-CM34. The Institution of Diesel and Gas Turbine Engineers (IDGTE) Paper 530. The Power Engineer 6 (2002) 5
4-32 Hanenkamp, A.; Terbeck, S.; Köbler, S.: 32/40 PGI – Neuer Otto-Gasmotor ohne Zündkerzen: MTZ 67 (2006) 12
4-33 Hanenkamp, A.: Moderne Gasmotorenkonzepte – Strategien der MAN B&W Diesel AG für wachsende Gasmärkte. 4th Dessauer Gasmotoren-Konferenz (2005)
4-34 Wideskog, M.: The Fuel Flexible Engine. 2nd Dessauer Gasmotoren-Konferenz (2001)
4-35 Wärtsilä VASA Gas Diesel Motoren. Brochure Wärtsilä NSD
4-36 Mohr, H.: Technischer Stand und Potentiale von Diesel-Gasmotoren. Part 1: BWK 49 (1997) 3; Part 2: BWK 49 (1997) 5
4-37 Wagner, U.; Geiger, B.; Reiner, K.: Untersuchung von Prozessketten einer Wasserstoffenergiewirtschaft. IfE Schriftenreihe. TU München: (1994) 34
4-38 Vogel, C. et al.: Wasserstoff-Dieselmotor mit Direkteinspritzung, hoher Leistungsdichte und geringer Abgasemission. Part 1: MTZ 60 (1999) 10; Part 2: MTZ 60 (1999) 12; Part 3: MTZ 61 (2000) 2
4-39 Knorr, H.: Erdgasmotoren für Nutzfahrzeuge. Gas-Erdgas GWF 140 (1999) 7, pp. 454–459
4-40 Geiger, J.; Umierski, M.: Ein neues Motorkonzept für Erdgasfahrzeuge. FEV Spectrum (2002) 20
4-41 Pucher, H. et al.: Gasmotorentechnik. Sindelfingen: expert (1986)
4-42 Latsch, R.: The Swirl-Chamber-Spark-Plug: A Means of Faster, More Uniform Energy Conversion in the Spark-Ignition Engine. SAE-Paper 840455, 1994
4-43 Herdin, G.; Klausner, J.; Winter, E.; Weinrotter, M.; Graf, J.: Laserzündung für Gasmotoren – 6 Jahre Erfahrungen. 4. Dessauer Gasmotoren-Konferenz (2005)
4-44 TEM-Konzept. Brochure Deutz Power Systems
4-45 LEANOX-Verfahren. Brochure GE Jenbacher
4-46 Pucher, H. et al.: In-Betrieb-Prozessoptimierung für Dieselmotoren – Arbeitsweise und Chancen. VDI Fortschrittsberichte. Düsseldorf: VDI-Verlag 12 (1995) 239, pp. 140-158
4-47 Raubold, W.: Online-Ermittlung von Zünddruck und Last aus der Zylinderkopfschraubenkraft. Diss. TU Berlin (1997)
4-48 Eggers, J.; Greve, M.: Motormanagement mit integrierter Zylinderdruckauswertung. 5. Dessauer Gasmotoren-Konferenz (2007)

5 Fuel Injection Systems

Walter Egler, Rolf Jürgen Giersch, Friedrich Boecking, Jürgen Hammer,
Jaroslav Hlousek, Patrick Mattes, Ulrich Projahn, Winfried Urner,
and Björn Janetzky

5.1 Injection Hydraulics

A description of the processes in injection systems entails the interdisciplinary application of methods of fluid mechanics, technical mechanics, thermodynamics, electrical engineering and control engineering. Pressures are very high and pilot injection to the main injection requires a minimum delivery of 1.5 mm³ of fuel per injection with a metering accuracy of ± 0.5 mm³ in flexibly selectable intervals. This imposes substantial demands on the quality of models and numerical methods. Moreover, the processes in the compressible fluid are profoundly transient. Thus, components are endangered by cavitation erosion and can be excited to oscillations with high mechanical loads. Substantially influencing fuel properties and bearing or plunger/liner clearances in pumps, the considerable heating of the fuel by throttling and frictional losses must be quantifiable as well.

5.1.1 Equation of State for Fuels

Knowledge of the fluid properties is the prerequisite for understanding and modeling hydraulic behavior. Density ρ and specific volume $v = 1/\rho$ describe the compressibility of fuel and test oil. The literature contains a number of approaches based on systematic measurements in high pressure labs that approximate the densities or specific volumes of fluids as a function of pressure p and temperature T [5-1]. One, the modified Tait equation

$$v(p,T) = v_0(T)\left(1 + C(T)\ln\frac{p+B(T)}{p_0+B(T)}\right)$$
$$v_0(T) = a_1 + a_2 T + a_3 T^2 + a_4 T^3$$
$$B(T) = b_1 + b_2/T + b_3/T^2 \quad (5\text{-}1)$$
$$C(T) = c_1 + c_2 T$$

has consistently proven itself for diesel fuels and test oils where p_0 is the ambient pressure.

U. Projahn (✉)
Robert Bosch GmbH, Diesel Systems, Stuttgart, Germany
e-mail: ulrich.projahn@de.bosch.com

Changes of state proceed so rapidly during an injection that they may be considered adiabatic. Compression and expansion processes with insignificant frictional and momentum losses may additionally be regarded as reversible and thus as isentropic. Since

$$a^2 = (\delta p/\delta \rho)_s = -1/\rho^2 (\delta p/\delta v)_s \quad (5\text{-}2)$$

applies to the speed of sound $a(p, T)$ at constant entropy s, it characterizes both the rate of linear pressure wave propagation and local isentropic pressure changes. According to [5-2], isentropic changes of state additionally correlate with the change of temperature

$$(\delta T/\delta p)_s = -T/(c_p \rho^2)(\delta \rho/\delta T)_p = T/c_p (\delta v/\delta T)_p \quad (5\text{-}3)$$

The following correlation relates the specific heat capacity c_p to the speed of sound at a known density $\rho(p, T)$

$$c_p = (\delta \rho/\delta T)_p^2 / [(\delta \rho/\delta p)_T - 1/a^2] T/\rho^2$$
$$= -(\delta v/\delta T)_p^2 / [(\delta v/\delta p)_T + (v/a)^2] \cdot T \quad (5\text{-}4)$$

Thus, the description of state behavior is theoretically completed when the correlations $\rho(p, T)$ or $v(p, T)$ and $a(p, T)$ or $c_p(p, T)$ are established. However, the partial derivatives make the equation extremely susceptible to error. Thus, an inherently consistent, accurate approximation is difficult to obtain by means of empirical equations of state with limited measurements. According to Davis and Gordon [5-3], if the speed of sound is known in the entire region of state of interest, $v(p, T)$ can be calculated for discrete points by integrating Maxwell's equation

$$(\delta c_p/\delta p)_T = -T(\delta^2 v/\delta T^2)_p \quad (5\text{-}5a)$$

and the equation derived from Eq. (5-4)

$$(\delta v/\delta p)_T = -[T/c_p (\delta v/\delta T)_p^2 + (v/a)^2] \quad (5\text{-}5b)$$

A further development proposed by Jungemann [5-4] calculates the selected equation of state's coefficients in a minimization problem so that the thusly ascertained fluid values optimally fulfill the system of partial differential equations (5-5) and the speed of sound and specific heat are expressed optimally. This method was used, for example, to determine the empirical equation of state based on

Eq. (5-1) for a common test oil compliant with ISO 4113. It was based on values for the speed of sound measured in the entire region of state and values for the specific volume and specific heat capacity measured along the isobars at ambient pressure. Figure 5-1 presents the calculated fluid properties.

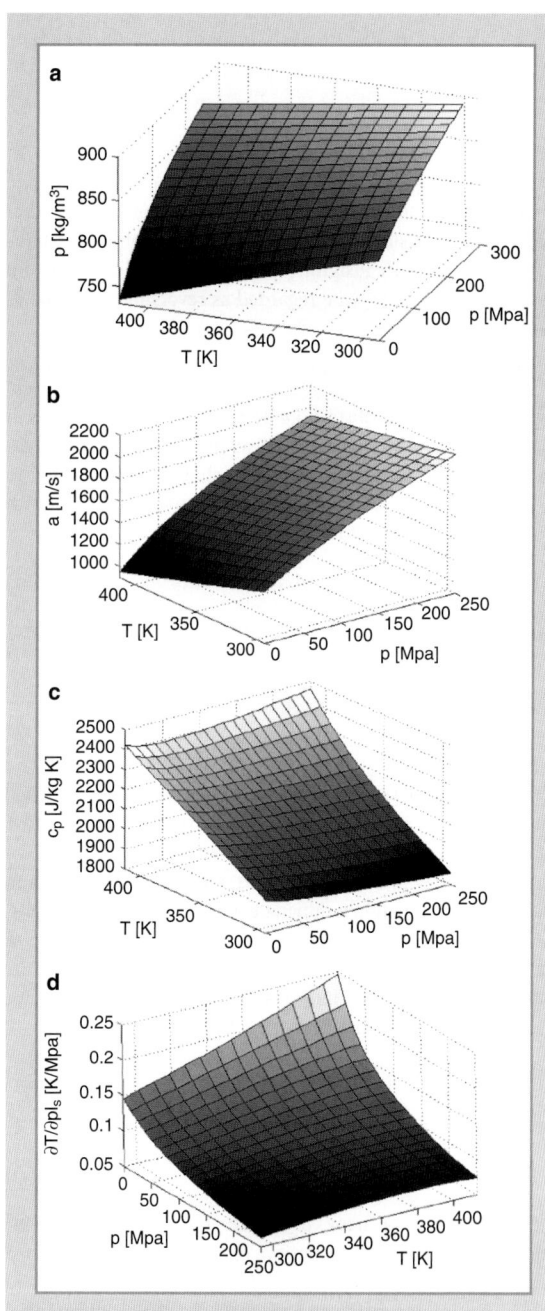

Fig. 5-1 Fluid properties of test oil according to ISO 4113. (a) Density ρ from Eq. (5-1); (b) speed of sound a from Eq. (5-4); (c) specific heat capacity c_p from Eq. (5-5a); (d) Isentropic temperature change from Eq. (5-3)

5.1.2 Modeling, Simulation and Design

Diesel injection systems are now substantially developed and designed with the aid of numerical simulation based on mathematical modeling, i.e. a description of reality with mathematical equations. It is expected to correctly describe mass flows, pressure oscillations and pressure losses over time. Target parameters, e.g. the characteristic of the rate of injection, depend on the complex interplay of various system components and necessitate both an analysis of the overall system that appropriately allows for local influences and individual analyses with very high local resolution. Computational fluid dynamics (CFD) may be employed to analyze local three-dimensional flows in detail. This is addressed in Sect. 5.2.

Current tools for system simulations (cf. [5-5–5-7]) allow selecting parameterizable model elements from a kit and merging them into a complete model that is usually supported by graphic interfaces. Incorporating every relevant 3D effect, the streamline theory is employed to model hydraulic zones. This produces a "hydraulic network model". The most important model elements are highlighted below.

Chamber: This model element has elementary significance as a node in the network. Chambers may be fuel-filled parts, injectors and high pressure pump barrels. By definition, local changes of pressure and temperature are disregarded. The contact surfaces of the allocated reference space can transport mass flows with the velocity vector $\vec{v}$ resulting from moving walls (so-called "body bound surfaces") and from the flow of free surfaces A. Using Eq. (5-2) to incorporate compressibility, a mass balance yields the following for pressure changes:

$$(V/a^2)(dp/dt) = -\rho(dV/dt) - \int_A \rho\vec{v}dA \qquad (5\text{-}6)$$

Lines: Linear wave propagation, e.g. between the pump and the high pressure accumulator or between the high pressure accumulator and the injectors, plays a significant role in highly transient processes. Corresponding to Eq. (5-6), the differential balance along a flow tube with the coordinate x and the velocity w is

$$\begin{aligned}(\delta p/\delta t + w\delta p/\delta x)/(a^2\rho) + \delta w/\delta x \\ +(\delta A/\delta t + w\delta A/\delta x)/A = 0.\end{aligned} \qquad (5\text{-}7)$$

Expanded by friction, Euler's equation of motion

$$\delta w/\delta t + w\delta w/\delta x + (1/\rho)(\delta p/\delta x) + r = 0 \qquad (5\text{-}8)$$

incorporates the fuel's inertia as a momentum balance. The partial differential equation system consisting of Eq. (5-7) and (5-8) is based on a very precise calculation using the model element of the line.

Short pipe: The fuel's inertia plays a significant role in relatively short segments, e.g. fuel bores in an injector's nozzle region, but the pressure wave propagation does not. Hence, a model of a short pipe dispenses with Eq. (5-7). Equation (5-8) makes allowance for flow pressure losses and the fluid's inertia. A short pipe requires substantially less computing time than a line. Even flows in longer bores or lines can be approximated well by connecting short pipes and chambers in series.

Throttles and valves: The volumetric flow in throttles and valve seats depends on the adjacent differential pressure and an effective cross section area determined by the geometric conditions. Bernoulli's equation expanded by friction is a suitable mathematical model:

$$\int_1^2 (\delta w/\delta t) dx + \int_1^2 (1/\rho) dp + (w_2^2 - w_1^2)/2 + \int_1^2 r dx = 0. \tag{5-9}$$

It is produced by integrating equation (5-8) between the cross sections "1" and "2". Phenomenological statements for pressure losses caused by wall friction and discontinuities enter into r. Since the lengths of the segments analyzed are extremely short, the inertia and thus the first term may usually be disregarded.

Gap flow: Leaks, e.g. in high pressure pumps or injector plunger guides, may be calculated as a planar flow with the Reynolds lubrication equation [5-8]. Additional influences such as gap expansion at high pressures and eccentric plunger position must be additionally incorporated in many cases. Since there is no guarantee that the flow remains laminar, further corrective interventions are necessary.

Pressure Forces: Pressure forces play a crucial role in moving components, e.g. pressure valves or nozzle needles. The modeling of this coupling of hydraulics and mechanics is based on the mathematical statement

$$\vec{F} = \int_O p d\vec{O} \tag{5-10}$$

Were the pressure distribution on the surface O known from a parallel coupled 3D flow simulation, Eq. (5-10) would deliver the exact force. This effort can be eliminated in most cases.

$$\vec{F} = \sum_i (p_i \cdot A_{F,i}) \tag{5-11}$$

reduces the calculation of the force to the adjacent chamber pressures p_i and allocated areas of effective pressure $A_{F,i}$. Local deviations of chamber pressure are accommodated with pressure and stroke-dependent maps of the areas of effective pressure.

Cavitation: Vapor pockets form, i.e. cavitation occurs, when the pressure reaches the vapor pressure locally. For instance, a throttle in an injector before the solenoid valve is intentionally designed so that cavitation occurs during injection, thus rendering the flow independent of back pressure and solenoid valve lift. This has substantial advantages for volumetric stability. In 1966, Schmitt [5-9] already identified a method to relatively simply link hypotheses with the energy Eq. (5-8) in order to also be able to compute the flow in throttles. What is more, the development of cavitation models and their integration in three-dimensional CFD requires great effort to assure the quality of projections of component damage [5-10].

Parameterization: The known dimensions, e.g. the nozzle needle's mass, the lines' lengths and diameters and the injector and the rail's trapped volumes, are used to parameterize injection system models. A lift and pressure-dependent map specifies the effective cross sectional areas of the valves and the effective areas of valve pressure. These maps are computed with 3D CFD or detailed streamline models [5-4] and inserted into the models in an appropriate form.

Given the high system pressures, component deformations may cause the hydraulic function to change. Influences of elasticities are determined with the aid of the finite element method (FEM) and integrated in the models.

Actuators: Precise reproduction of the actuating elements, e.g. the solenoid valve or piezoelectric actuator, also plays a very large role in precise metering of the injected fuel quantity. Electromagnetic models, which can be directly coupled with the hydraulic and mechanical models, are used to calculate the generation and abatement of the actuating force over time. Taking a piezoelectrically controlled system as an example, Fig. 5-2 presents typical simulation results for quintuple injection.

5.2 Injection Nozzles and Nozzle Holders

Injection nozzles are the interface between the injection system and combustion chamber. They significantly influence an engine's power, exhaust emissions and noise and seal the injection system from the combustion chamber between injections. They are incorporated in nozzle holder assemblies (NHA), unit injectors (UI) and common rail injectors (CRI) and mounted together with them in the cylinder head as a functional unit positioned very exactly to the combustion chamber (Fig. 5-3).

Pertinent subsections of Sect. 5.3 explain unit and common rail injectors.

The following functional principles apply to nozzle needle control:

Needle closes/seals the injection system from the combustion chamber: A mechanically or hydraulically generated closing force that acts on the end of the needle presses the nozzle needle into the nozzle seat.

Needle opens: The nozzle needle opens at the start of the injection phase as soon as the "hydraulic" force F_D on the seat side (the injection pressure acting on the annulus area between the needle guide and nozzle seat) becomes greater than the closing force F_S.

NHA and UI are cam-driven injection systems with pressure controlled nozzle needles, CRI are pressure accumulator

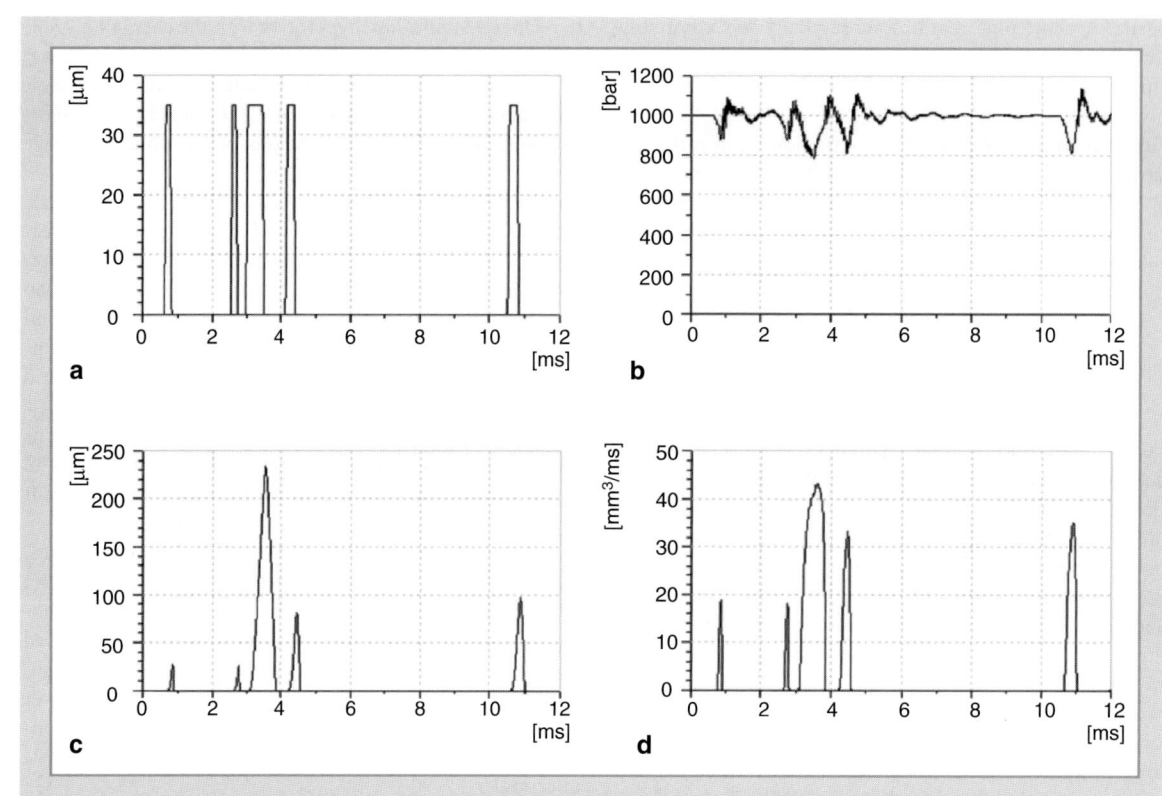

Fig. 5-2 Simulation results for a piezo controlled common rail injector for cars. (**a**) Valve lift; (**b**) Pressure in the injection nozzle; (**c**) Nozzle needle lift; (**d**) Injection characteristic

injection systems with lift controlled nozzle needles, i.e. a "hydraulic" closing force can be modulated as a function of a fuel map-dependent system pressure and, thus, the needle can be lift controlled (see also Sect. 5.3.1.1, Fig. 5-14).

5.2.1 Injection Nozzles

By systematically distributing and optimally atomizing the fuel in the combustion chamber, nozzles strongly influence mixture formation and they influence the injection characteristic.

5.2.1.1 Design, Types

A standard nozzle consists of a nozzle body with a high pressure inlet, needle guide, seat and spray hole zone and an inwardly opening needle. Three basic nozzle designs based on the combustion system and needle control are common (Fig. 5-4):
– Pintle nozzles for NHA applications in IDI engines are no longer significant in engine development.
– Hole-type nozzles are utilized in NHA, UI and CRI applications in DI engines.
– Nozzle modules, i.e. hole-type nozzles with an integrated hydraulic control chamber, are utilized for CRI applications in DI engines. Inlet and controlled outlet throttles in the injector modulate the pressure in the control chamber. The control chamber's volume is hydraulically rigid, i.e. designed with a small internal volume to obtain a low injected fuel quantity.

Nozzle size depends on cylinder displacement and the injected fuel quantity. Hole-type nozzles and nozzle modules are further divided into valve covered orifice or sac hole designs (Fig. 5-5).

Design: Every current DI engine concept employs hole-type nozzles and nozzle modules. The goal of nozzle design is to convert pressure energy into kinetic energy with optimized efficiency, i.e. into injection sprays with penetration, breakup and atomization characteristics optimally adjusted to the combustion system, combustion chamber geometry, injected fuel quantity, engine air management and the load and speed-dependent injection pattern.

Seat geometry: The seat's design incorporates the sealing function and its diameter determines the opening pressure.

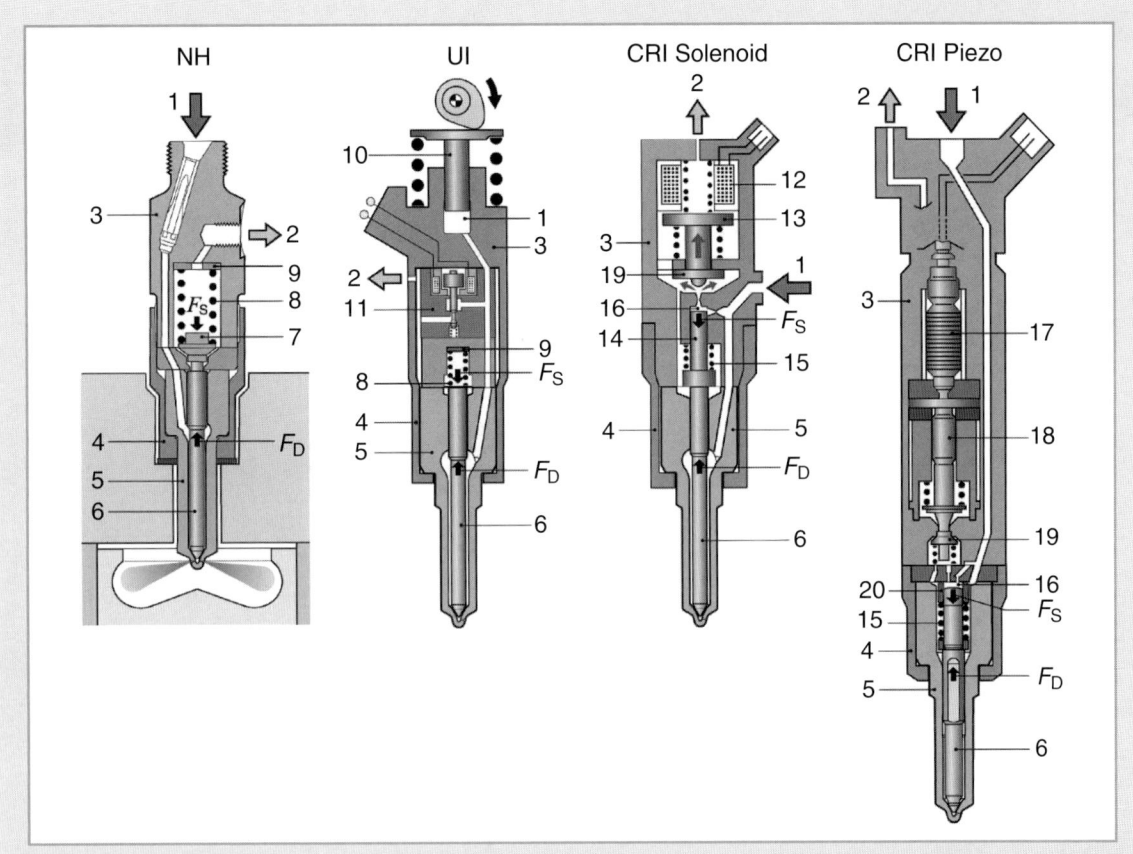

Fig. 5-3 Nozzle holders, injectors. *1* High pressure inlet; *2* Fuel return; *3* Nozzle holder; *4* Nozzle retaining nut; *5* Nozzle body; *6* Nozzle needle; *7* Thrust bolt; *8* Compression spring; *9* Adjusting shim; *10* Pump plunger; *11* Pressure control solenoid valve; *12* Solenoid coil; *13* Valve armature; *14* Push rod; *15* Return spring; *16* Control chamber; *17* Piezoelectric actuator; *18* Hydraulic coupler, *19* Control valve, *20* Control chamber sleeve

At smaller lifts, the seat gap acts as a flow throttle and influences the spray holes' inflow and thus the spray preparation as well as the needle's dynamics through the pressure fields in the gap induced by the flow. In terms of length and angle differences to the body, the needle seat and needle tip cone are designed based on the system and are a compromise of needle dynamics (injected fuel quantity and injection characteristic) and long-time stability (wear and resultant injected fuel quantity drift).

Needle guide: The needle guide in the nozzle body centers the needle on the body seat during injection and separates high and low pressure regions. (This does not apply to nozzle modules). Guide clearances are in the range of 1–5 μm. The higher the injection or system pressure, the smaller the guide clearance is in order to minimize leakage losses. Valve covered orifice nozzles often have a second guide in the nozzle shaft to improve needle centering on the seat and thus fuel distribution to spray holes and needle dynamics. Sac hole nozzles are more rugged in this respect since the flow in the seat does not directly influence the conditions in spray holes.

Needle lift: The hydraulic design keeps the throttling losses in the seat insignificant at full lift. The needle is either lifted ballistically or is limited by a solid stop. The advantage of ballistic lift is a nearly linear (smooth) fuel supply characteristic as a function of the duration of injection. However, it is only expedient for common rail injectors, which control the opening and closing time far more precisely than other systems.

Valve covered orifice, sac hole design and trapped volume (Fig. 5-5): The size of the so-called trapped volume remaining under the seat edge after the needle closes is the feature in valve covered orifice and sac hole nozzles relevant for emissions. The fuel content evaporates suboptimally burned and increases HC emissions. A valve covered orifice nozzle has the smallest possible trapped volume, followed by conical and cylindrical sac hole nozzles. The spray holes in valve covered

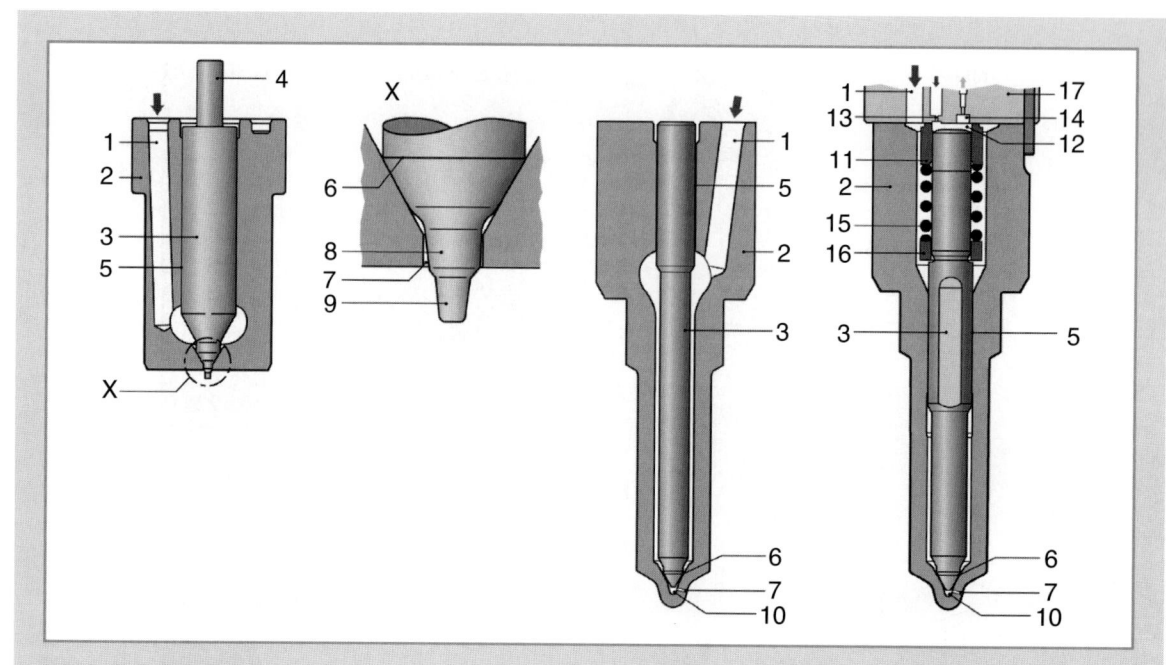

Fig. 5-4 Nozzle designs. *1* High pressure inlet; *2* Nozzle body; *3* Nozzle needle; *4* Pressure pin; *5* Needle guide; *6* Needle seat (−Ø); *7* Spray hole; *8* Throttling pintle; *9* Pintle; *10* Sac hole; *11* Control chamber sleeve; *12* Control chamber; *13* Inlet throttle; *14* Controlled outlet orifice; *15* Return spring; *16* Spring shim; *17* Restrictor plate (injector)

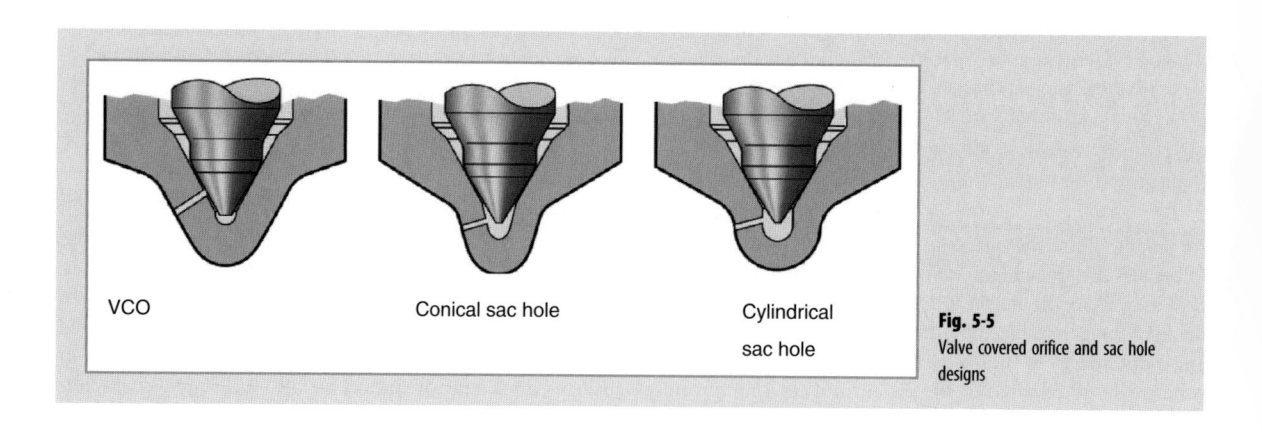

Fig. 5-5 Valve covered orifice and sac hole designs

orifice nozzles are arranged in single or multiple rows in the seat cone below the needle seat. A minimum residual wall thickness has to be maintained because of the spray holes' inflow and for reasons of strength. Sac hole nozzles require significantly smaller minimum distances between spray holes.

Spray hole length (Fig. 5-6): Current spray hole lengths are between 0.7 and 1 mm. Since the spray holes' are close to the point of attack of forces in the seat, they influence the spray as well as the tip strength, especially in valve covered orifice nozzles.

Hole geometry and spray: The goal is to produce optimal fuel distribution, atomization and mixture preparation in the combustion chamber. The number of spray holes and the direction of the sprays, each of which has a spatial correlation to the cylinder head, the glow plug and the combustion chamber bowl, are designed first.

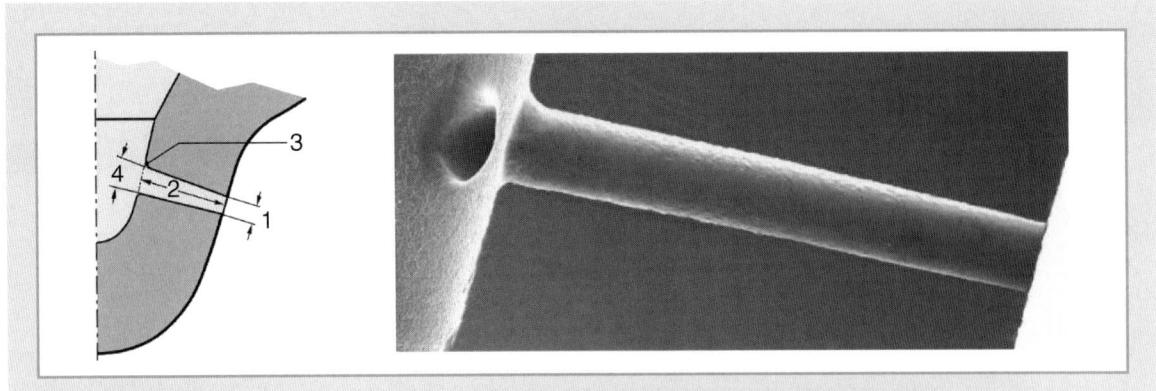

Fig. 5-6 Spray hole geometry. *1* Spray hole diameter; *2* Spray hole length; *3* Hydroerosive rounded spray hole inlet; *4* Tapered spray hole

The spray hole cross section is defined by the maximum injected fuel quantity, the related injection pressure and the acceptable duration of injection.

The number of spray holes depends on the combustion system and air management (including swirl). The sprays may not intermix. At present, seven to nine spray holes with diameters of 105–135 μm are employed in cars and six to eight spray holes with diameters of 150–190 μm in commercial vehicles.

Spray configuration: Certain parameters are available to optimally design a nozzle spray hole. Suitable parameters must be selected to allow for the high susceptibility of nozzles with efficiency-optimized, virtually cavitation-free spray hole flows to coking. Research is being done on the potential of hole-selective spray hole designs (i.e. every spray hole is designed individually), double hole configurations and even clusters of holes or combinations of valve covered orifice and sac hole designs with parallel, diverging or intersecting sprays (Table 5-1). This requires smaller spray hole diameters and a spark erosion or laser drilling system specifically developed for them.

5.2.1.2 Spray Analysis, Simulation

Optical spray pattern analysis: By comparing their spray contours, spray patterns recorded with a high speed camera quickly deliver information on spray shape, spray pattern symmetry, spray development at the start and end of injection and shot-to-shot variations (Fig. 5-8).

Spray force analysis: This analysis delivers precise information on efficiency, symmetry, spray breakup and structure. A pressure sensor scans the injection spray at varying distances to the nozzle and captures the unconditioned signal and the spray structure (Fig. 5-9).

Existing procedures are continuously being refined and new techniques tested (Fig. 5-10) in order to obtain more

Table 5-1 Overview of spray hole design

Parameter	Design targets
Number of nozzle holes	The number ought to be as high as possible, however intermingling of sprays is critical
Nozzle hole cross section	The smallest possible cross section is optimal for atomization and mixture formation
Hydroerosive rounding (Inflow edge)	Hydroerosive rounding anticipates wear and, depending on the extent of rounding, influences the internal spray hole flow (with/without cavitation) Figs. 5-6 and 5-7
Conicity	In conjunction with hydroerosive rounding and spray hole length, conicity influences the efficiency of the pressure conversion and the spray breakup
Spray hole length	The shorter the length, the smaller the depth of spray penetration (at equivalent efficiency)

Fig. 5-7 Flow simulation: Comparison of spray holes (rounded and unrounded inlet)

Fig. 5-8 Optical spray pattern analysis

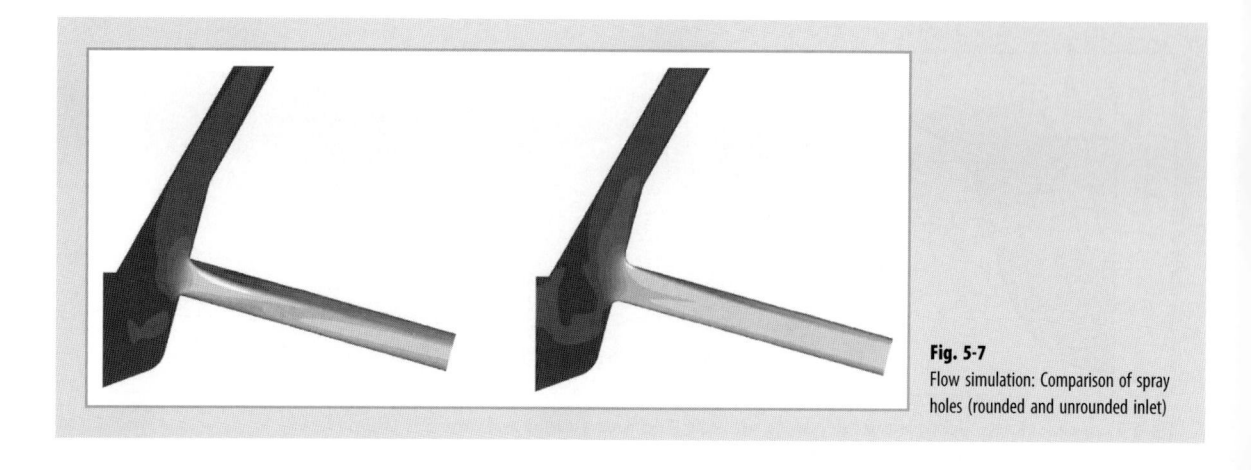

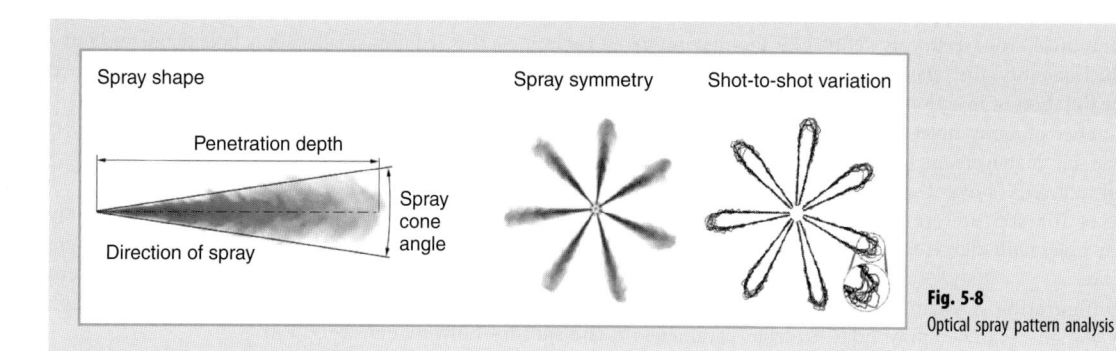

information on the spray, e.g. spray breakup, droplet size, vaporization, air entrainment, mixture formation and combustion.

Knowledge of these quantities is the prerequisite for simulation of the process chain from internal nozzle flow through combustion and for numerical simulation of emission (Fig. 5-11). Intensive research and development are being pursued in every domain of this process chain. Excellent models of injection systems already exist, which even reproduce the hydraulic effect of changes to components throughout the life cycle.

Materials, tip temperature: The pressure oscillations in common rail systems generate relative motions in the seat between the needle and body. Surface coatings are applied to reduce wear.

Nozzle tips are thermally highly loaded (up to approximately 300°C in cars and > 300°C in commercial vehicles). Heat resistant steels and thermally conductive bushings are employed for the higher temperature range.

Nozzles operated at low temperatures (< approximately 120°C) for a prolonged time may suffer corrosion in the entire seat/spray hole area (exhaust gas and water react and form sulfuric acid).

5.2.2 Nozzle Holders and Nozzle Holder Assemblies

Nozzles are combined with single spring and dual spring nozzle holders (Fig. 5-12) to form nozzle holder assemblies, which are used in cam-driven systems. Single spring holders (1SH) are used in commercial vehicles and dual spring holders (2SH) in cars since the added pilot injection may be used to reduce combustion noise.

Function of a single spring holder: See the introduction to Sect. 5.2 (needle closes and needle opens). The compression spring's load generates the needle's closing force.

Function of a dual spring holder: The nozzle needle initially opens against the force of the compression spring

Fig. 5-9 Spray force analysis

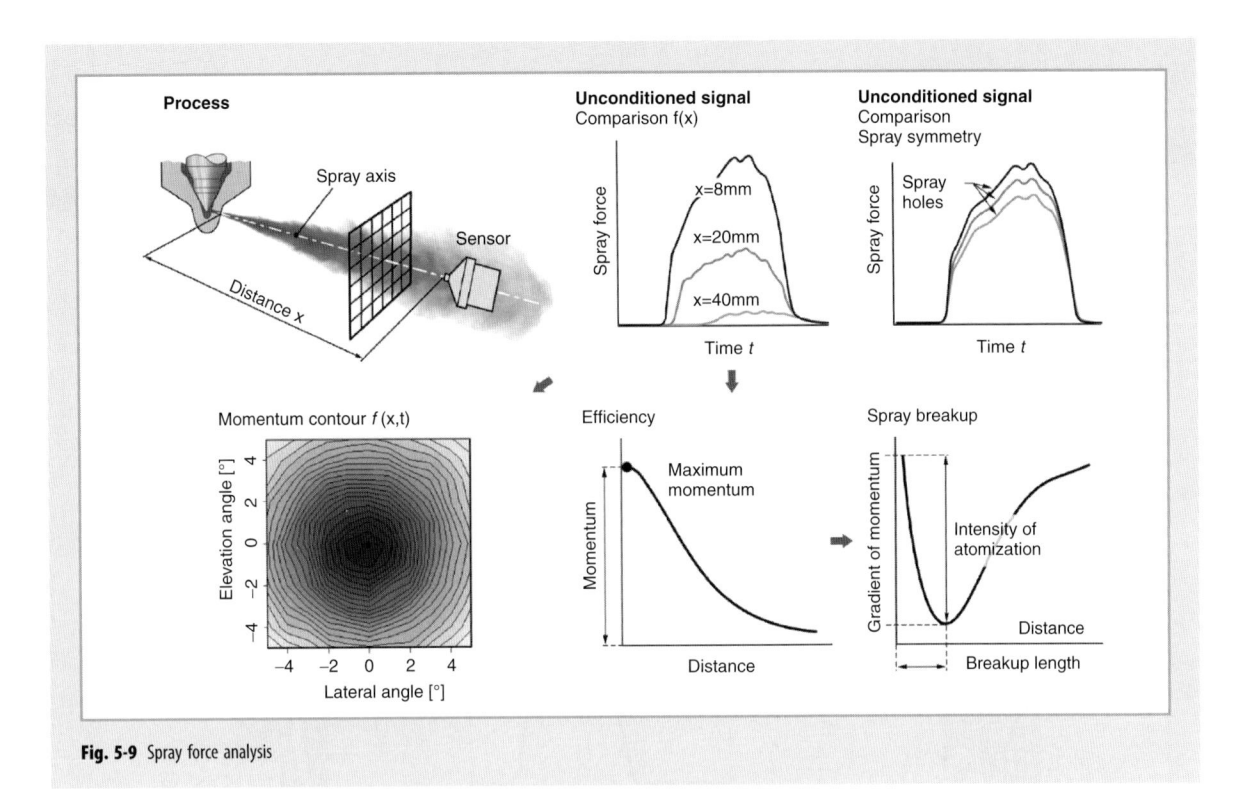

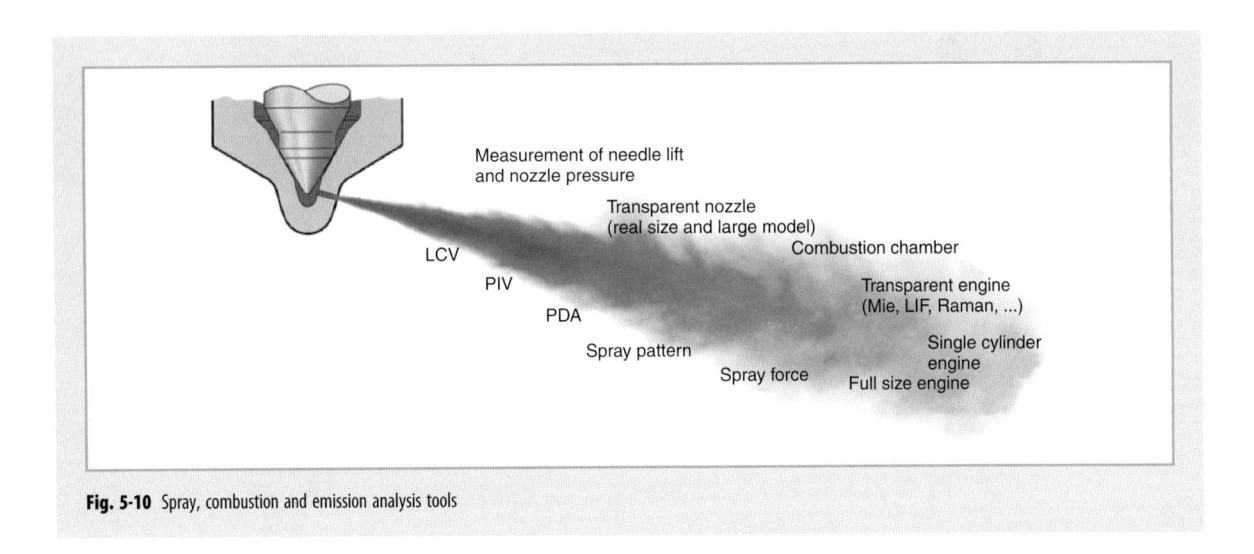

Fig. 5-10 Spray, combustion and emission analysis tools

1 (15, lift h1, pilot injection phase), which acts on the needle through a push rod. For the main injection (lift h2), the hydraulic force must use a lift adjustment sleeve to overcome the force of the compression spring 2 (18) additionally acting on the needle. The first stage is passed very quickly at high speeds.

The closing forces and the opening pressures are set by prestressing the compression springs with adjusting shims.

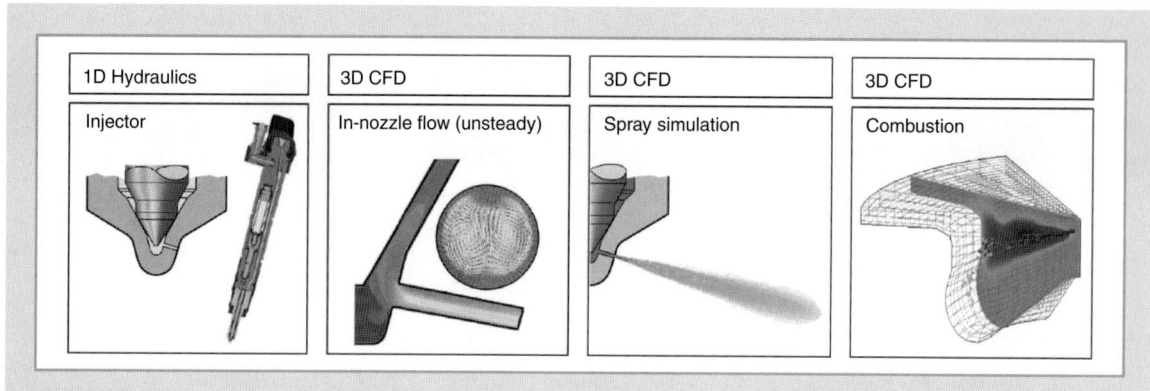

Fig. 5-11 Simulation process chain

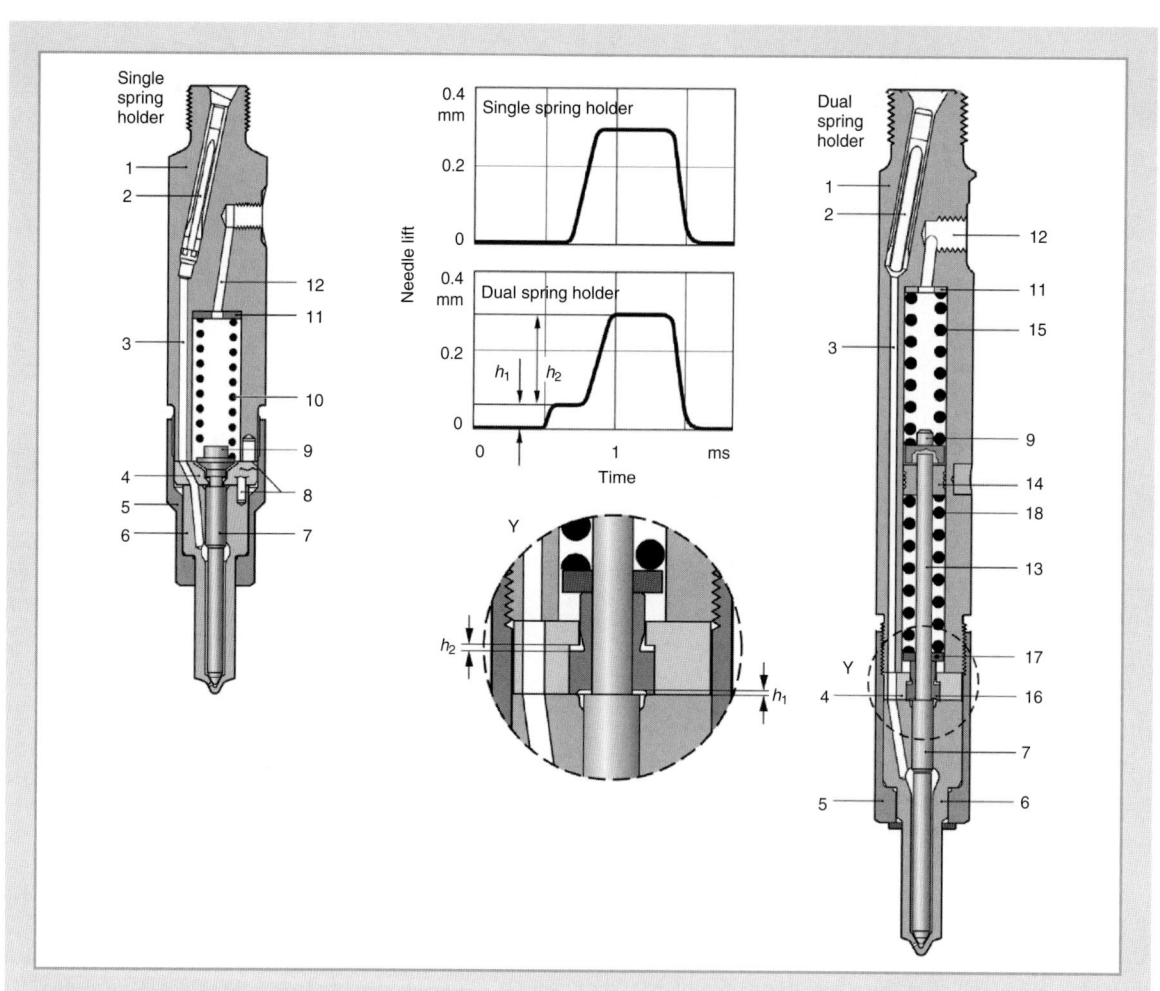

Fig. 5-12 Nozzle holder assembly. *1* Injector body; *2* Edge filter; *3* High pressure inlet; *4* Intermediate disk; *5* Nozzle retaining nut; *6* Nozzle body; *7* Nozzle needle; *8* Locating pin; *9* Push rod; *10* Compression spring; *11* Shim; *12* Fuel return; *13* Pressure pin; *14* Guide disk, *15* Compression spring 1; *16* Stroke adjustment sleeve; *17* Spring shim; *18* Compression spring 2

Electronically controlled injection systems employ nozzle holders with needle motion sensors. A pin connected with the push rod plunges into an induction coil in the nozzle holder and delivers signals for the start, end and frequency of injection.

5.3 Injection Systems

5.3.1 Basic Functions

The basic functions of a diesel injection system can be broken down into four subfunctions:
- **Fuel delivery** (low pressure side) from the tank through the fuel filter to high pressure generation. This function is assumed by the "low pressure circuit" subsystem, which is generally equipped with the components of pre-filter, main filter (heated if necessary), feed pump and control valves. The low pressure circuit connects the vehicle tank to the high pressure system feed and return by lines through the low pressure components. The functionally determinative pressure and flow specifications of the connected high and low pressure components must be observed.
- **High pressure generation and fuel delivery** (high pressure side) to the metering point or in an accumulator with high efficiency during compression. Optimal steady state and dynamic injection pressure both have to be provided as a function of the engine operating point. The required injected fuel quantity and system-dependent control and leak quantities have to be delivered. This function is assumed by the high pressure pump and, depending on the system, an accumulator. Valves are installed in the high pressure circuit to control the mass flows and pressures. In advanced injection systems, they are electronically actuated.
- **Fuel metering** that precisely meters the fuel mass into the combustion chamber as a function of speed and engine load and is supported by exhaust gas aftertreatment systems. Advanced injection systems meter fuel with the aid of electrically actuated solenoid or piezo valves mounted on the high pressure pumps or directly on the injectors.
- **Fuel preparation** by optimally utilizing the pressure energy for primary mixture formation for the purpose of a fluid spray that is optimally distributed in the combustion chamber in terms of time and location. The fuel is prepared in the injection nozzle. The metering valve's interaction with the nozzle needle control and the routing of the flow from the nozzle inlet until its discharge at the nozzle holes are of key importance.

5.3.1.1 Types

Overview

The basic functions described above are implemented differently depending on the type of injection system. Figure 5-13 presents an overview of commercially available injection systems and typical fields of application.

An initial distinction must be made between conventionally designed systems and systems with high pressure accumulators. Injection systems without accumulators always have high pressure pump plungers driven directly by a cam and thus generate a pressure wave in the high pressure system, which is directly utilized to open the injection nozzle and inject the fuel cylinder-selectively according to the firing sequence.

The next level of classification includes systems with a "central injection pump" that serves every cylinder and delivers and meters the fuel. Typical representatives are inline pumps and distributor pumps with axial and radial pump elements.

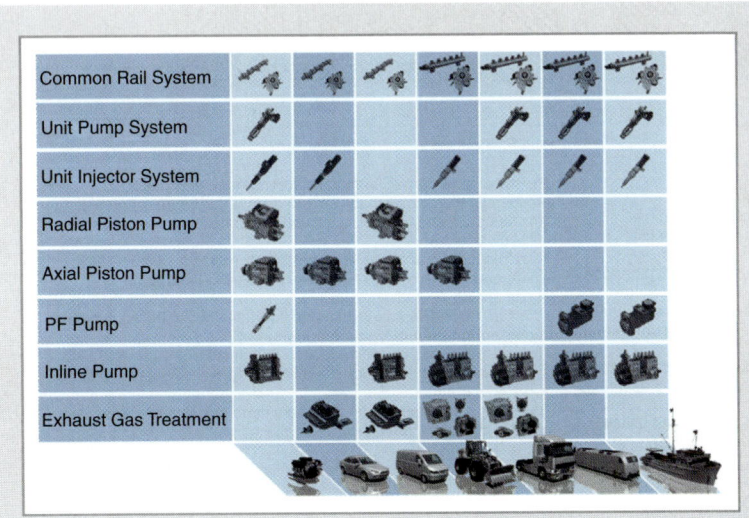

Fig. 5-13
Present injection system designs and their applications

The other design is characterized by "detached injection pumps per engine cylinder". One discrete pressure generation unit driven by the engine camshaft is attached for every one of the combustion engine's cylinders. The fuel is metered by rapidly switching solenoid valves integrated in the pump unit. The unit injector is one familiar example of this type of injection system.

Accumulator systems on the other hand have a central high pressure pump that compresses the fuel and delivers it to an accumulator at high pressure. Low pressure and high pressure valves control the pressure in the accumulator. The fuel is metered from the accumulator by injectors, which are in turn controlled by solenoid or piezoelectric valves. The name common rail system stems from the "common accumulator/distributor". Based on the type of actuator in the injectors, a distinction is made between "solenoid common rail" systems and "piezoelectric common rail" systems as well as special designs.

Nozzle needle lift/pressure control: All injection systems prepare the fuel independently of the design of the injection nozzle, which is either connected with the pump unit by a high pressure line or directly integrated in the pump unit housing or in the injector. The type of nozzle needle control is a main feature that sets conventional and common rail injection systems apart. While the nozzle needle in cam-driven injection systems is "pressure controlled", the injector in common rail systems is "lift controlled". Figure 5-14 compares the types of nozzle needle control and summarizes their main features.

The implementation of common rail systems in virtually every engine can be expected in the future. Since they are more flexible than conventional designs, these injection systems are already the main application for cars. The capability to freely select the pressure and the number of injections per working cycle as a function of engine speed and load and other parameters is indispensible to fulfilling the target engine parameters. Moreover, an accumulator makes it possible to situate injection very late relative to the engine crank angle to control exhaust gas aftertreatment. This will be essential for compliance with future emission standards. Although a pressure controlled nozzle needle also has advantages for emissions, it is foregone in favor of flexible multiple injections and the lift control of the nozzle needle in common rail injectors is relied on. In an analysis of the overall system, the advantages of lift controlled fuel metering in terms of precision, minimum injected fuel quantity and minimum spray intervals outweigh those of conventional systems with pressure controlled needles.

Features of lift controlled injection

- Nozzle may be actuated directly or with a servo valve
- Nozzle is power operated, cross section is variable
- Put simply, high pressure is constant up to the variable cross section; pressure is modulated by the drop in pressure in the nozzle seat
- Nozzle needle is operated ballistically (not to maximum lift) or non-ballistically (reaching maximum lift)
- Nozzle may close quickly independent of delivery pressure
- High pressure may be delivered by pump and line or accumulator and line
- Cross section control usually produces high seat throttling at part load at the nozzle seat; therefore, the usable differential pressure over the spray hole is small
- Rate shape causes poorer for emission; fewer noises and better multiple injection than pressure controlled injection

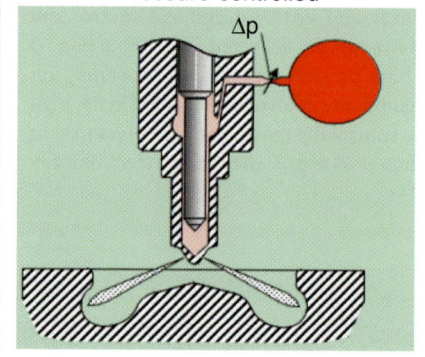

Features of pressure controlled injection

- Pressure wave coming from the metering valve elevates the pressure in the nozzle chamber until opening pressure is reached
- Nozzle opens and remains open until the pressure drops below the initial closing pressure
- Nozzle needle is only operated ballistically (not to maximum lift) when quantities are very small or non-ballistically (reaching maximum lift)
- Pressure is modulated by the pump through a line or by the metering valve through the line from the accumulator
- High needle speed usually produces high seat throttling at part load at the nozzle seat; therefore, the usable differential pressure over the spray hole is large
- Rate shape facilitates better emission; more noises and poorer multiple injection than lift-controlled injection

Fig. 5-14 Comparison of pressure and lift controlled injection

Inline Pumps (Fig. 5-15)

Main Features

- There is one pump element per engine cylinder and the elements are arranged inline.
- The plunger is driven by the pump camshaft and reset by the plunger spring and the plunger stroke is constant.
- Delivery starts when the plunger closes the spill ports.
- The plunger compresses fuel when it moves upward and delivers it to the nozzle.
- The nozzle operates pressure controlled.
- The inclined helix re-clears the connection to the spill port and thus reduces the load on the plunger chamber. The nozzle closes as a result.
- The effective stroke is the plunger stroke after the plunger chamber closes until shutoff. The effective stroke and thus also the injected fuel quantity may be varied by using the control to rotating the plunger.

Axial Distributor Pumps (Fig. 5-16)

Main Features

- There is one axial pump elements for all engine cylinders.
- The cam plate is driven by the engine camshaft and the number of cams equals the number of engine cylinders (≤ 6).
- Cam lobes roll on the roller ring and this generates rotary and longitudinal motion of the distributor plunger.
- A central distributor plunger opens and closes ports and bores.
- The fuel flow is distributed to outlets to the engine cylinders,
- The plunger compresses axially and delivers fuel to the pressure controlled nozzle.
- A control collar varies the effective stroke and thus the injected fuel quantity.
- The start of delivery is varied by an injection timing mechanism, which rotates the roller ring relative to the cam plate.

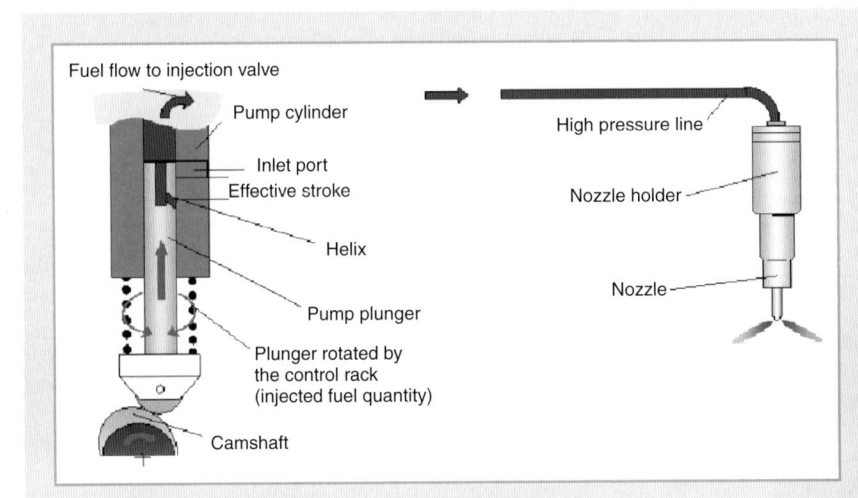

Fig. 5-15
Design and functional principle of an inline pump

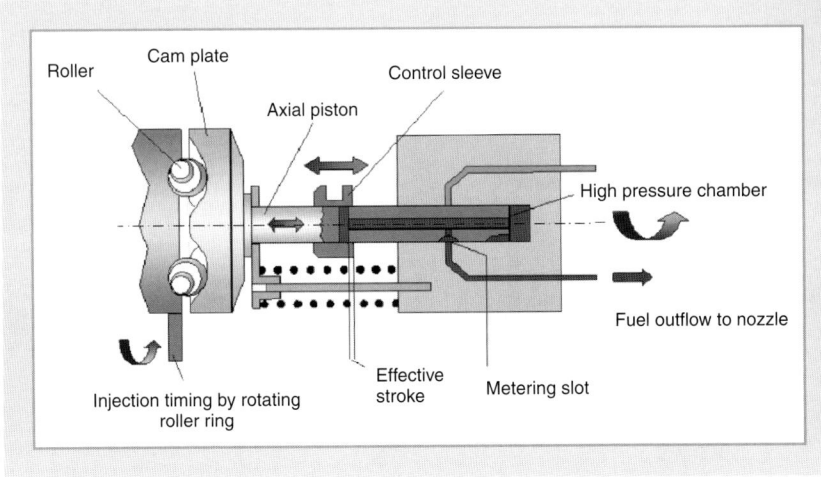

Fig. 5-16
Design and functional principle of an axial distributor pump

Radial Distributor Pumps (Fig. 5-17)

Main Features

- High pressure is generated by a radial plunger or one or two pairs of plungers or three independent plungers.
- The number of cam lobes on the cam ring equals the number of engine cylinders (≤ 6).
- A distributor shaft driven by the engine supports roller tappets.
- Roller tappets roll on the cam ring and generate pump motion.
- Plunger pairs compress fuel toward the center and deliver it to the pressure controlled nozzle.
- A central distributor shaft opens and closes ports and bores.
- The fuel flow is distributed to outlets to the engine cylinders.
- A solenoid valve controls the injected fuel quantity (and start of delivery).
- High pressure builds when the solenoid valve is closed.
- The start of delivery is varied by a solenoid controlled injection timing mechanism, which rotates the roller ring relative to the distributor shaft.

Unit Injectors (Unit Injectors) (Fig. 5-18)

Main Features

- One unit injector per engine cylinder is integrated in the engine's cylinder head.
- It is driven by the engine camshaft by means of an injection cam and tappet or roller rocker.
- High pressure is generated by a pump plunger with spring return.
- High pressure is locally generated directly before the nozzle. Hence, there is no high pressure line.
- The nozzle operates pressure controlled.
- A solenoid valve controls the injected fuel quantity and start of injection.
- High pressure builds when the solenoid valve is closed.
- A control unit computes and controls injection.

Unit Pump Systems (Unit Pumps) (Fig. 5-19)

Main Features

- Its principle is comparable to the unit injector system.
- However, a short high pressure line connects the nozzle in the nozzle holder with the pump.
- There is one injection unit (pump, line and nozzle holder assembly) per engine cylinder.
- It is driven by an underhead engine camshaft (commercial vehicles).
- The nozzle operates pressure controlled.
- A high pressure solenoid valve controls the injected fuel quantity and start of injection.

Common Rail System (Fig. 5-20)

Main Features

- It is an accumulator injection system.
- High pressure generation and injection are decoupled.
- A central high pressure pump generates pressure in the accumulator, which may be adjusted in the entire map independent of engine speed and load.
- Repeated extraction of fuel from the rail per working cycle of the engine allows high flexibility of the position, number and size of injections.
- One injector (body with nozzle and control valve [solenoid or piezo actuator]) is mounted per engine cylinder.
- The nozzle operates lift controlled.

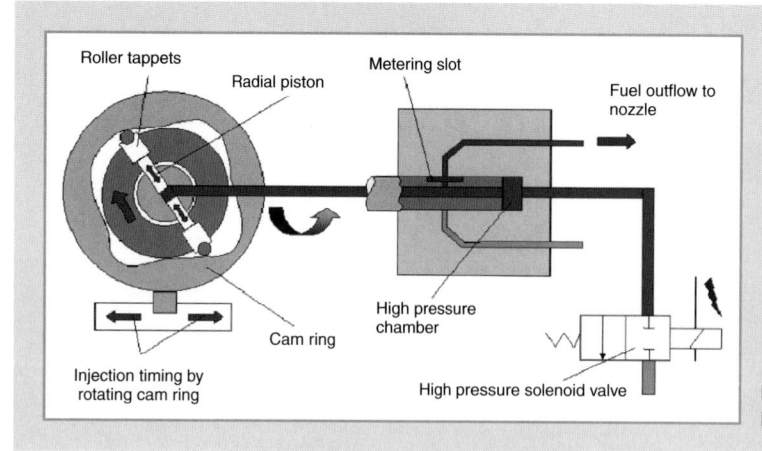

Fig. 5-17 Design and functional principle of a radial distributor pump

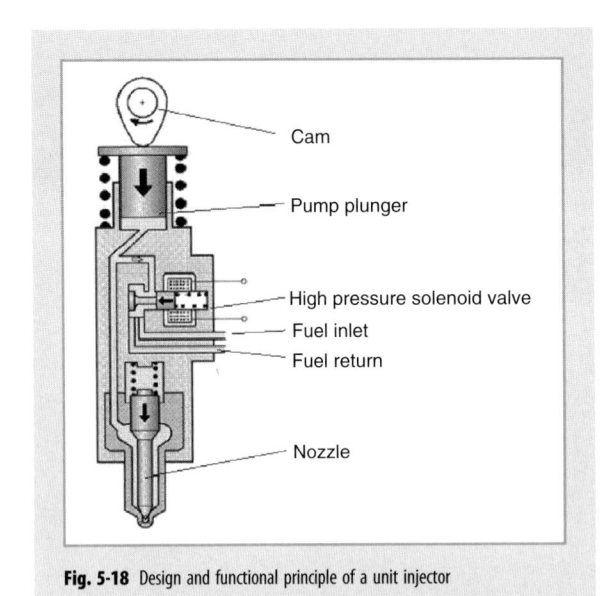

Fig. 5-18 Design and functional principle of a unit injector

– The injector operates time controlled and the injected fuel quantity is a function of the rail pressure and duration of control.
– A control unit controls the number and position of injections and the injected fuel quantity.

5.3.2 Inline Pumps

5.3.2.1 Configuration and Operation

Consisting of pump barrels and pump plungers corresponding to the number of engine cylinders, an inline pump's pump elements (Fig. 5-21) are typically united in their own housing. The pump's own camshaft, which is driven by the engine's timing gear drive, moves the pump plungers. The quantity of fuel is metered solely by helix control by rotating the pump plunger. Every pump plunger has an oblique helix so that, in conjunction with the spill port fixed on the cylinder side, a different delivery stroke and thus a differing injected fuel quantity can be delivered or adjusted as a function of the pump plunger's angular position. The complete plunger stroke is constant in each case and corresponds to the cam lift. A pressure valve in the inline pump's high pressure outlet separates the high pressure region in the pump from the injection line and the nozzle holder so that the fuel located in the line-nozzle system after injection remains pressurized, i.e. a certain static pressure exists there. A return flow restrictor that prevents any secondary injection merely with low injection pressure is often integrated in the pressure valve. A control sleeve that interlocks with a longitudinally moving control rack rotates the plunger for every barrel simultaneously. Thus, the delivery rate may be regulated between zero delivery and maximum quantity. The control rack itself is moved by a controller connected with the injection pump.

The controller may either be a mechanical centrifugal governor that slides the control rack as a function of speed and thus regulates full load speed in particular or an electronic controller that uses an electromagnetic actuator mechanism to act on the control rack. Mechanically controlled pumps require auxiliary devices, e.g. a boost pressure-dependent full load stop, to adjust the injected fuel quantity to the different operating conditions.

A low pressure feed pump actuated by a special cam on the pump's own camshaft is mounted on the inline pump to reliably supply the pump elements with fuel. This feed pump supplies the inline pump's fuel gallery with fuel under a pressure of up to approximately 3 bar.

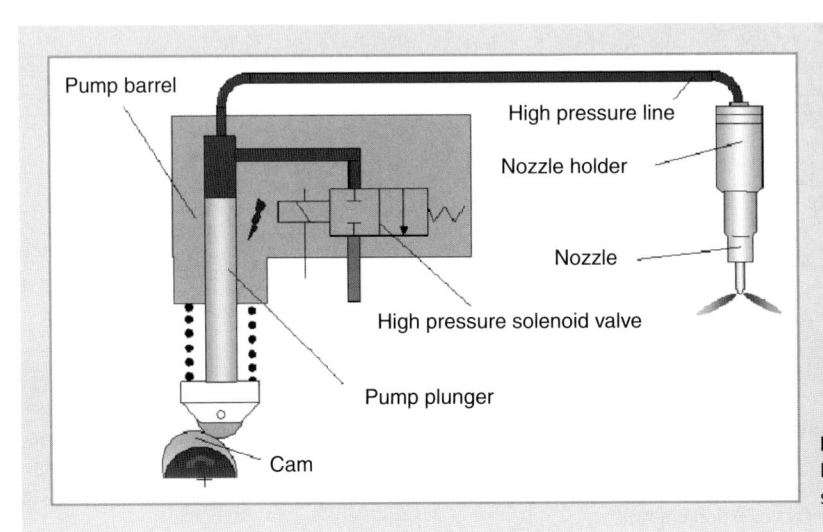

Fig. 5-19 Design and functional principle of a unit pump system

Fig. 5-20
Design and functional principle of a common rail system

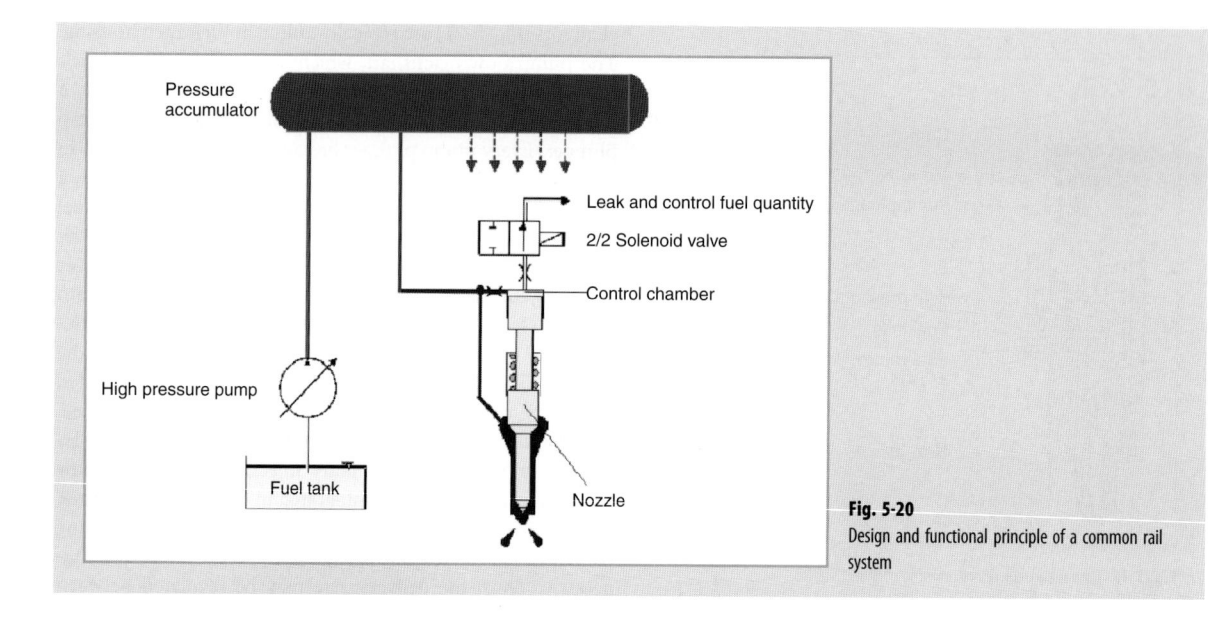

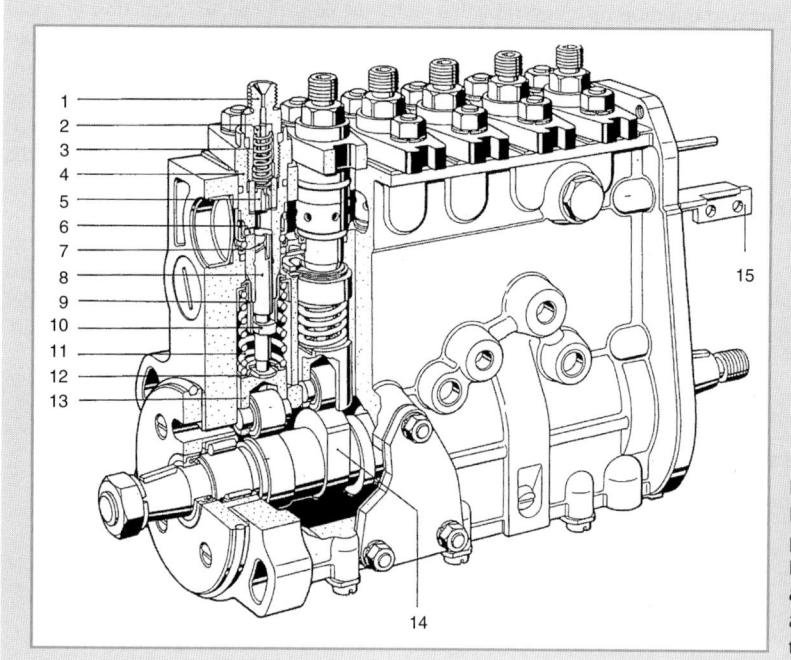

Fig. 5-21
P-type inline pump. *1* Pressure valve holder; *2* Filler piece; *3* Pressure valve spring; *4* Pump barrel; *5* Delivery valve cone; *6* Inlet and control port; *7* Helix; *8* Pump plunger; *9* Control sleeve; *10* Plunger control arm; *11* Plunger spring; *12* Spring seal; *13* Roller tappet; *14* Camshaft; *15* Control rack

Models, Variants

Various sizes of inline pumps have been adapted to appropriate engine powers. Inline pumps now have injection pressures of between 400 and 1,150 bar on the pump side, depending on whether they are used for chamber or direct injection engines.

The A and P pump are typical Bosch types. Various variants of the P type are implemented depending on the injection pressure, injected fuel quantity and duration of injection. The so-called control sleeve injection pump with variable settings for the start of delivery is used for commercial vehicle engines.

5.3.3 Distributor Injection Pumps

Distributor injection pumps [5-11–5-15] are compact, low cost pumps. Their chief field of application is direct injection car engines (formerly, IDI engines too) as well as commercial vehicle engines of up to approximately 45 kW/cylinder. A distributor pump usually consists of the following assemblies:
– high pressure pump with distributor,
– speed/fuel controller,
– injection timing mechanism,
– low pressure feed pump,
– electric cutoff device (pumps with control collars) and
– add-on functional units (mechanically controlled pumps).

Types include the axial plunger pump that generates up to 1,550 bar pressure in the injection nozzle and the radial plunger pump that generates up to 2,000 bar pressure (see Figs. 5-22 and 5-23). The utilizable number of engine cylinders is limited to six in both types. Their basic function is described in Sect. 5.3.1.

5.3.3.1 Adjusting the Start of Injection

Dependent on load and speed, the adjustment of the start of injection in a distributor pump is taken over by an injection timing mechanism that can rotate the roller ring in an axial piston pump or the cam ring in a radial piston pump to a cam angle of approximately 20° (from extremely retarded to extremely advanced). The rings are rotated by pressurizing or depressurizing the injection timing piston with a pressure that is proportional to the speed and generated by a vane presupply pump integrated in the high pressure pump. When necessary, the pressure in this piston can be set precisely with the aid of a pulse width modulation controlled solenoid valve in conjunction with an injection start sensor.

Along with this simple variant of an injection timing mechanism, there is also a follow-up movement injection timing device with a control piston integrated in the timing piston. This improves the control dynamics since the control piston reacts independently of the influences of friction on the roller ring and the injection timing piston.

5.3.3.2 Variants

Mechanically controlled distributor injection pumps: Purely mechanically controlled distributor pumps are characterized by various, individually configurable functional add-ons, e.g. to adjust start of delivery, to control idling and full load or to improve cold start performance. A centrifugal governor controls pump speed.

Electronically controlled distributor injection pumps: Since it controls the quantity of fuel and start of delivery and injection by sensors and the control unit, the design pictured in Fig. 5-22 does not require separate functional add-ons. The fuel quantity actuator is a rotary magnet actuator with an inductive sensor that delivers the engine control unit very precise information on the control collar's position. This facilitates precise and fully flexible fuel metering.

Solenoid controlled distributor injection pumps: Figure 5-23 pictures a solenoid controlled radial piston distributor pump.

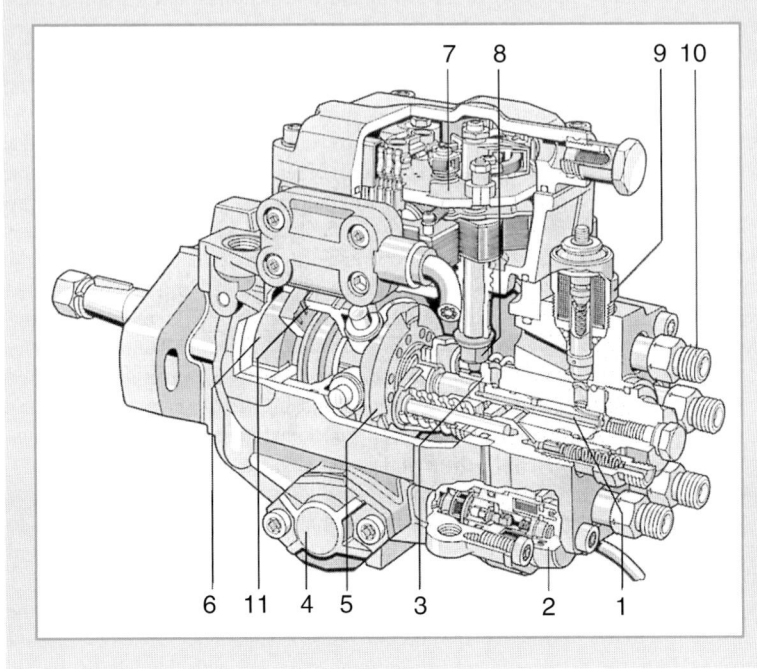

Fig. 5-22
Electronically controlled axial piston distributor injection pump, VE, Robert Bosch GmbH. *1* Distributor plunger; *2* Solenoid valve for injection timing; *3* Control collar; *4* Injection timing mechanism; *5* Cam plate; *6* Feed pump; *7* Electric fuel quantity actuator with feedback sensor; *8* Setting shaft; *9* Electric shutoff device; *10* Pressure valve holder; *11* Roller ring

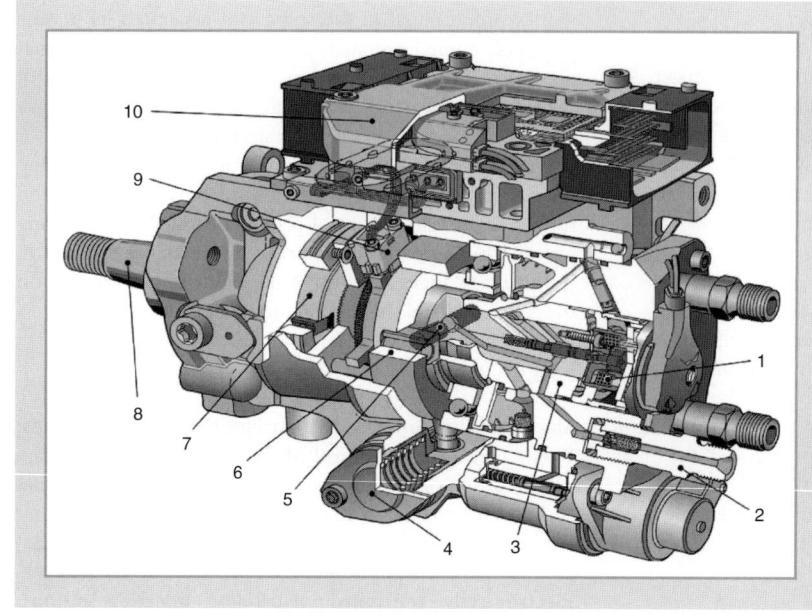

Fig. 5-23
Electronically controlled radial piston distributor injection pump with solenoid valve control, VP44, Robert Bosch GmbH. *1* Solenoid valve; *2* High pressure port; *3* Distributor shaft; *4* Injection timing mechanism; *5* Pump plunger; *6* Cam ring; *7* Vane pump; *8* Drive shaft; *9* Angle of rotation sensor; *10* Electronic pump control unit

The particular advantages of this type of control are its high precision (adjustable by the physical unit of the control unit and pump), high fuel dynamics (cylinder-individual fuel metering) and influence on the delivery rate by variable start of delivery.

The time of the 2/2 high pressure solenoid valve's closing determines the start of delivery. Its opening time defines the delivery rate through the piston stroke. The start and end of delivery (duration of delivery) are controlled in the pump control unit by processing angle and speed signals from internal pump and engine sensors. For one pilot injection, the pump control unit actuates the solenoid valve of the pump twice - first, for the pilot fuel injection (typically 1.5–2 mm^3/injection) and, then, for the main fuel injection.

5.3.4 Single Plunger Pump Systems

Basically, cam and time controlled single plunger fuel injection systems in which a pump is fitted to each engine cylinder may be classified as unit injector or unit pump systems. These injection systems are mainly used in heavy duty engine applications [5-16–5-21].

5.3.4.1 Design and Function

Unit injector systems: The pressure generating pump and the injection valve form a physical unit, which enables minimizing the trapped volume in the injection system and reaching very high injection pressures (over 2,000 bar at rated power). Each engine cylinder has its own unit injector installed directly in the cylinder head. The nozzle assembly is integrated in the unit injector and projects into the combustion chamber. The engine camshaft has an individual cam for each unit injector. A roller rocker transmits its stroke to the pump plunger. This moves it up and down with the aid of the return spring.

Modern single plunger pump systems are controlled with a high pressure solenoid valve or piezo actuator, opening and closing the connection between the low pressure circuit and the pump's plunger chamber. The fuel under constant excess pressure in the low pressure stage flows into the pump's plunger chamber during the suction stroke (when the pump plunger moves upward). The high pressure solenoid valve closes at a certain time determined by the control unit. The exact closing time ("electric start of injection" or start of delivery) is ascertained by analyzing the coil current in the solenoid valve and is applied to control the start of injection and correct the duration of the drive input pulse (reducing fuel tolerances). The pump plunger compresses the fuel in the plunger chamber until it reaches the nozzle opening pressure ("actual start of injection"). As a result, the nozzle needle is elevated and the fuel injected into the combustion chamber. The pump plunger's high delivery rate causes the pressure to continue to rise during the entire injection process. The maximum peak pressure is reached shortly after the solenoid is shut off. The pressure drops very rapidly afterward. The duration of solenoid valve control determines the injected fuel quantity.

Mechanically and hydraulically controlled by an accumulator plunger, pilot injection is integrated in unit injectors for cars to reduce noise and pollution (Fig. 5-24). Very small quantities of fuel (approximately 1.5 mm^3) may be pilot injected.

Unlike systems with mechanical metering by helices, the position of the injection piston in systems controlled by solenoid valves is not imperatively connected with the delivery stroke.

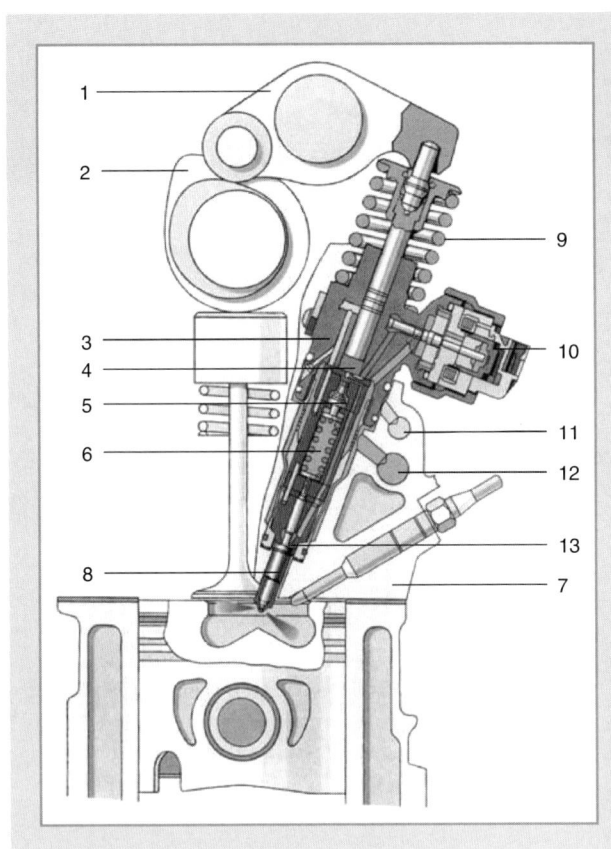

Fig. 5-24
Unit injector for cars mounted in the cylinder head. *1* Roller rocker; *2* Drive camshaft; *3* Pump body assembly; *4* High pressure chamber; *5* Accumulator plunger; *6* Needle-valve spring; *7* Cylinder head; *8* Nozzle; *9* Follower spring; *10* Solenoid valve; *11* Fuel return; *12* Fuel inlet; *13* Damping unit

Sensors capture the camshaft's angular position and the control unit computes the delivery stroke from the specified quantity. In order to minimize the tolerances of the start of injection and the quantity injected, the camshaft's angular position must exactly reproduce the geometric position of the pump plunger's stroke unrestricted by any cylinders. This requires a very rigid design of the engine and the unit injector's drive.

Unit pump systems: This system consists of a high pressure pump with an integrated solenoid valve similar to those in unit injectors, a short injection line, high pressure connectors and conventional nozzle holder assemblies (Fig. 5-25). The start of injection and quantity of fuel are controlled as in the unit injector system.

The modular design (pump, high pressure line and nozzle holder assembly) enables integrating the unit pump laterally in the engine. This eliminates reengineering of the cylinder head and simplifies customer service. The pumps are attached to the engine block above the camshaft by a flange integrated in the pump body. The camshaft and pump plunger are directly connected by a roller tappet.

Unit injector and unit pump systems have a triangular injection characteristic. Car unit injector systems additionally furnish the option of separated pilot injection.

In addition to solenoid controlled unit pump systems, mechanically controlled single plunger pumps (helix controlled) are also widely used in small and very large engines. Their basic function corresponds to that of mechanically controlled inline pumps with control racks.

5.3.5 Common Rail Systems

5.3.5.1 Design

Unlike cam-driven injection systems, the common rail system decouples pressure generation and injection. The pressure is generated independently from the injection cycle by a high pressure pump that delivers the fuel under injection pressure to an accumulator volume or rail. Short high pressure lines connect the rail with the engine cylinders' injectors. The injectors are actuated by electrically controlled valves and inject the fuel into the engine's combustion chamber at the desired time. The injection timing and injected fuel quantity are not coupled with the high pressure pump's delivery phase. Separating the functions of pressure generation and fuel injection renders the injection pressure independent of speed and load. This produces the following advantages over cam driven systems:

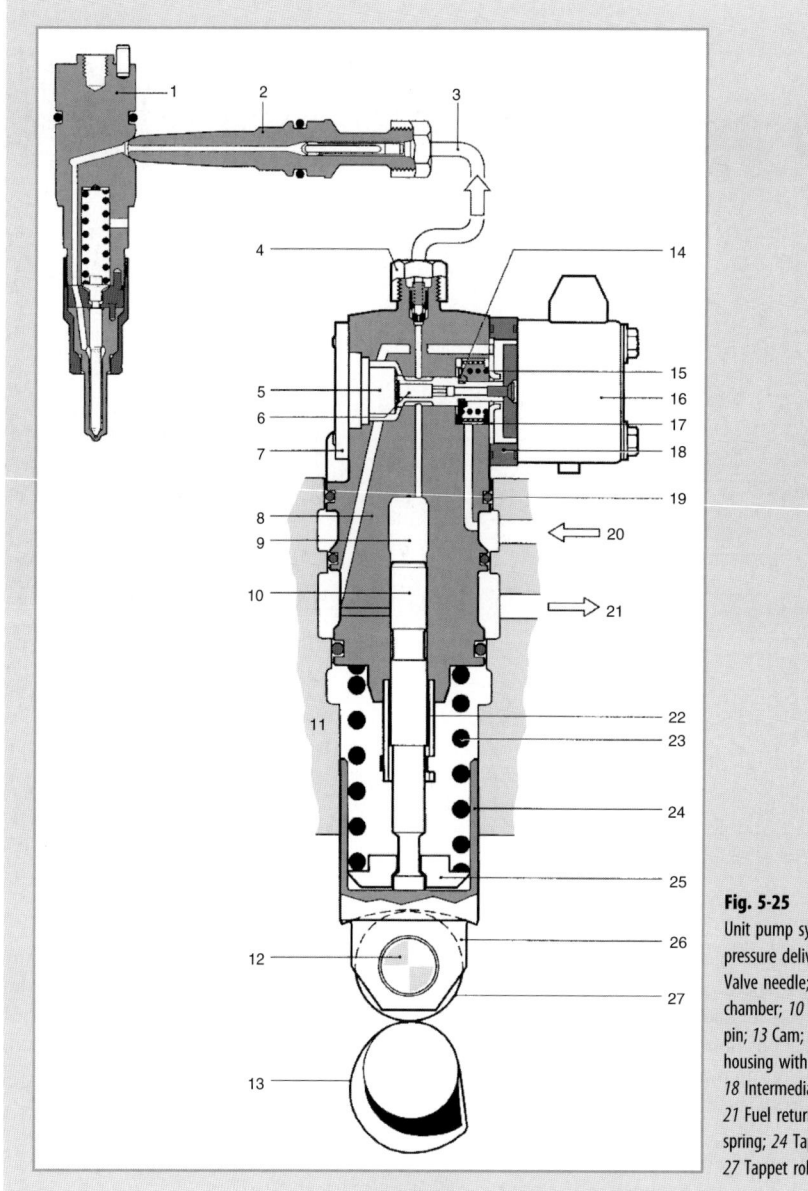

Fig. 5-25
Unit pump system. *1* Nozzle holder; *2* Pressure fitting; *3* High pressure delivery line; *4* Connection; *5* Stroke stop; *6* Solenoid Valve needle; *7* Plate; *8* Pump housing; *9* High pressure chamber; *10* Pump plunger; *11* Engine block; *12* Roller tappet pin; *13* Cam; *14* Spring seat; *15* Solenoid valve spring; *16* Valve housing with coil and magnet core; *17* Armature plate; *18* Intermediate plate; *19* Seal; *20* Fuel inlet (low pressure); *21* Fuel return; *22* Pump plunger retention device; *23* Tappet spring; *24* Tappet body; *25* Spring seal; *26* Roller tappet; *27* Tappet roller

- continually available speed and load-independent injection pressure allows flexibly selecting the start of injection, the injected fuel quantity and the duration of injection,
- high injection pressures and thus good mixture formation are possible even at lower speeds and loads,
- it provides high flexibility for multiple injections,
- it is easily mounted on the engine and
- drive torque peaks are significantly lower.

Common rail systems are employed in all DI engine applications for cars and commercial vehicles (on and off-highway).

Maximum system pressures are 1,800 bar. Systems for pressures > 2,000 bar are in development.

- A common rail system can be divided into the following subsystems (Fig. 5-26):
- low pressure system with the fuel supply components (fuel tank, fuel filter, presupply pump and fuel lines),
- high pressure system with the components of high pressure pump, rail, injectors, rail pressure sensor, pressure control valve or pressure limiting valve and high pressure lines and
- electronic diesel control with control unit, sensors and actuators.

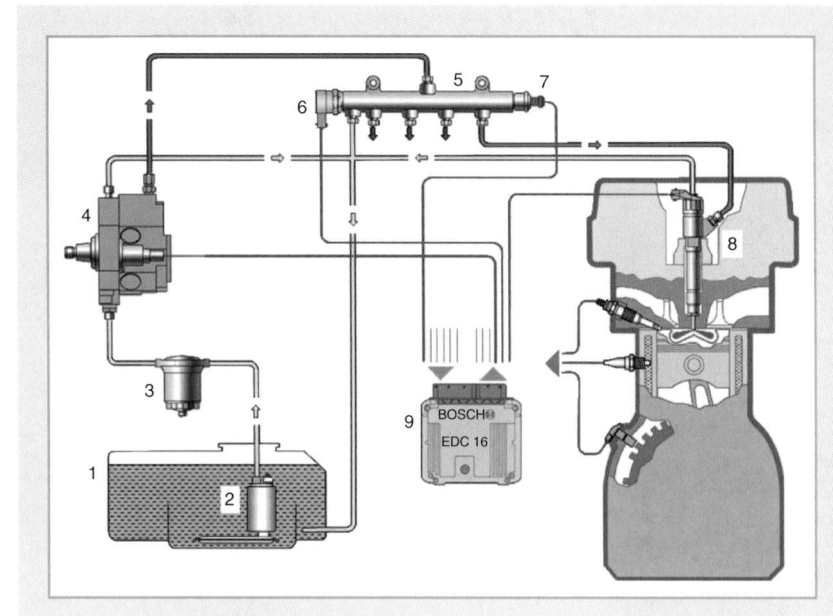

Fig. 5-26
Common rail system: *1* Fuel tank; *2* Presupply pump with sieve filter; *3* Fuel filter; *4* High pressure pump with metering unit; *5* Rail; *6* Pressure control valve; *7* Rail pressure sensor; *8* Injector; *9* Electronic control unit with inlets for sensors and outlets for actuators

Driven by the engine, the continuously operating high pressure pump generates the desired system pressure and maintains it largely independent of engine speed and the injected fuel quantity. Given its nearly uniform delivery, the pump is smaller in size and generates a smaller peak driving torque than pumps in other injection systems.

The high pressure pump is designed as a radial piston pump and, for commercial vehicles, partly as an inline or single plunger pump (driven by the engine camshaft). Various modes are employed to control rail pressure. The pressure may be controlled on the high pressure side by a pressure control valve or on the low pressure side by a metering unit integrated in the pump (housed in a separate component for single plunger pumps). Dual actuator systems combine the advantages of both systems. Short high pressure lines connect the injectors with the rail. The engine control unit controls the solenoid valve integrated in the injector to open and reclose the injection nozzle. The opening time and system pressure determine the injected fuel quantity. At constant pressure, it is proportional to the time the solenoid valve is switched on and thus independent of engine and pump speed.

A basic distinction is made between systems with and without pressure amplification. In systems with pressure amplification, a stepped piston in the injector amplifies the pressure generated by the high pressure pump. The injection characteristic can be shaped flexibly when the pressure intensifier is separately controllable by its own solenoid valve. The systems predominantly in use today operate without pressure amplification.

5.3.5.2 Low Pressure System

The high pressure pump's fuel supply from the tank and the leakage and overflow quantities returned to the tank are united in the low pressure circuit. Figure 5-27 shows the configuration in principle. The basic components are the:
- fuel tank,
- primary fuel filter with hand primer (optional) and main fuel filter,
- cooler for the control unit (optional),
- presupply pump and
- (optional) fuel cooler.

Electric fuel pumps (EFP, Fig. 5-28) or gear pumps (GP) are used as presupply pumps. Systems with EFP are used solely for cars and light commercial vehicles. EFP are usually installed in fuel tanks (in-tank pump) or, optionally, in the feed line to the high pressure pump (inline pump). The EFP switches on when the starting process begins. This ensures that the necessary pressure exists in the low pressure circuit when the engine starts. The fuel is delivered continuously and independently of engine speed. Excess fuel flows back to the tank through an overflow valve. Roller cell pumps (see the scheme in Fig. 5-28) are usually used for diesel applications. The fuel cools the electric motor. This attains high engine power density. A non-return valve that prevents the fuel lines from running empty once the pump is shut off is integrated in the connection cover. EFP have advantages over mechanically driven presupply pumps in terms of starting performance when fuel is hot, when first started and after engine servicing (e.g. filter change).

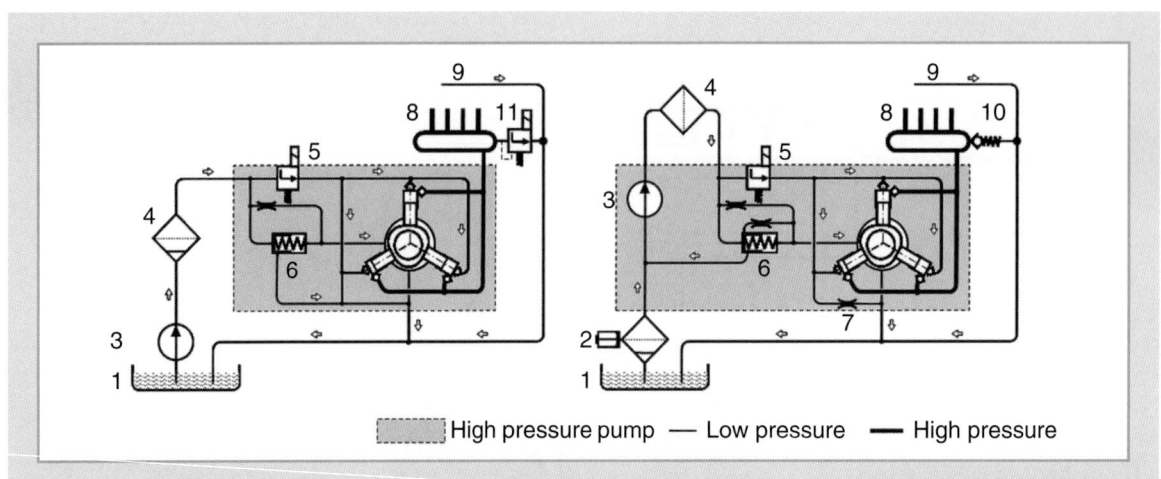

Fig. 5-27 Low pressure circuit for cars (*left*, suction and high pressure side control) and commercial vehicles (*right*, suction side pressure control). *1* Fuel tank; *2* Pre-filter with water separator and hand primer; *3* Electric /mechanical presupply pump; *4* Fuel filter with/without water separator; *5* Metering unit; *6* Overflow valve; *7* Zero delivery throttle; *8* Rail; *9* Injector return line; *10* Pressure limiting valve; *11* Pressure control valve

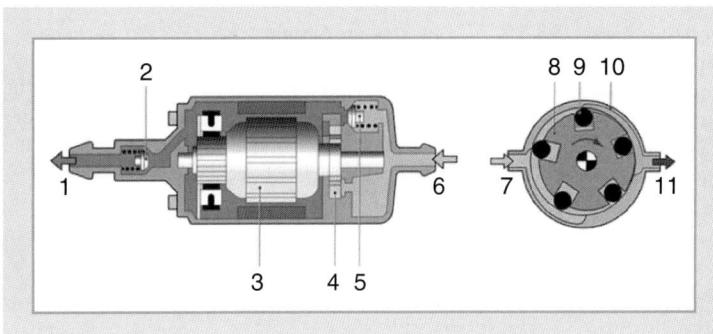

Fig. 5-28
Single-stage electric fuel pump (*left*) and roller cell pump (*right*).
1 Pressure side; *2* Non-return valve; *3* Motor armature; *4* Pump element; *5* Pressure limiting valve; *6* Suction-side; *7* Inlet; *8* Slotted rotor; *9* Roller; *10* Base plate; *11* Pressure side

GP are utilized in car and commercial vehicles systems as presupply pumps. Only GP are used for heavy commercial vehicles. The GP is usually integrated in the high pressure pump and driven by its drive shaft. Thus, the GP delivers fuel only when the engine is running, i.e. it must be designed so that pressure is generated sufficiently faster during the start. This necessitates quantity limiting for high speeds (the delivery rate being approximately proportional to engine speed). As a rule, this is done by throttling on the GP's suction side.

Protecting the injection system from impurities in the fuel (solid particles and water) and thus ensuring the requisite service life requires the use of a fuel filter matched to the particular operating conditions. Pre-filters with integrated water separators are primarily used for commercial vehicles in countries with poor fuel quality and for industrial engine applications. Their separator characteristic is adapted to the main filter. The main filter is normally placed on the pressure side between the presupply pump and high pressure pump.

An infinitely variable solenoid valve, the metering unit (only in systems with suction side fuel delivery control) as well as the overflow valve and the zero delivery throttle are located in the high pressure pump's low pressure stage. The metering unit adjusts the quantity that reaches the high pressure pump so that only the quantity of fuel required by the system on the high pressure side is compressed to high pressure. The excess quantity of fuel delivered is conducted through the overflow valve into the tank or before the presupply pump. In fuel lubricated pumps, throttles in the overflow valve serve to bleed or guarantee a sufficient lubrication quantity. The zero delivery throttle removes leakage quantities that appear when the metering unit closes. This prevents an undesired increase in rail pressure and ensures pressure decays rapidly.

5.3.5.3 High Pressure System

A common rail system's high pressure region is subdivided into three regions, pressure generation, pressure accumulation and fuel metering with the following components:
- high pressure pump,
- rail with pressure sensor and pressure control valve or pressure limiting valve,
- high pressure lines and
- injectors.

The engine drives the high pressure pump. The transfer ratio has to be selected so that the delivery rate is sufficient to satisfy the system's mass balance. Moreover, delivery should be synchronous with injection to obtain largely identical pressure conditions at the time of injection. The fuel compressed by the high pressure pump is delivered through the high pressure line(s) to the rail whence it is distributed to the connected injectors. Along with its accumulator function, the rail has the function of limiting the maximum pressure oscillations generated by pulsating pump delivery or the injectors' extraction of fuel in order to ensure injection metering is precise. On the one hand, the rail volume ought to be as large as possible to satisfy this requirement. On the other hand, it has to be small enough to ensure pressure is generated rapidly during starting. The accumulator volume must be optimized to this effect in the design phase.

The rail pressure sensor's signal with which the current fuel pressure in the rail is ascertained serves as an input variable for pressure control. Various modes are applied to control pressure (Fig. 5-29):

High pressure side control: A pressure control valve (a proportional solenoid valve controlled by the control unit) controls the desired rail pressure on the high pressure side. In this case, the high pressure pump delivers the maximum delivery rate independent of the fuel requirement. Excess fuel flows through the pressure control valve back into the low pressure circuit. While this control allows rapid adjustment of the rail pressure when the operating point changes, the constant maximum delivery and the discharge of the fuel under high pressure are disadvantageous from an energetic perspective. Its poor energetic performance limits the application of such a system to low pressure regions (1,400 bar maximum). This type of control was utilized for the first common rail systems for cars. The pressure control valve is usually in the rail but also installed in the high pressure pump in individual applications.

Suction side control: This method controls the rail pressure on the low pressure side by a metering unit that is flange-mounted on the high pressure pump. Suction side delivery control only delivers the quantity of fuel to the rail with which the required rail pressure is maintained. Thus, less fuel must be compressed to high pressure than in high pressure side

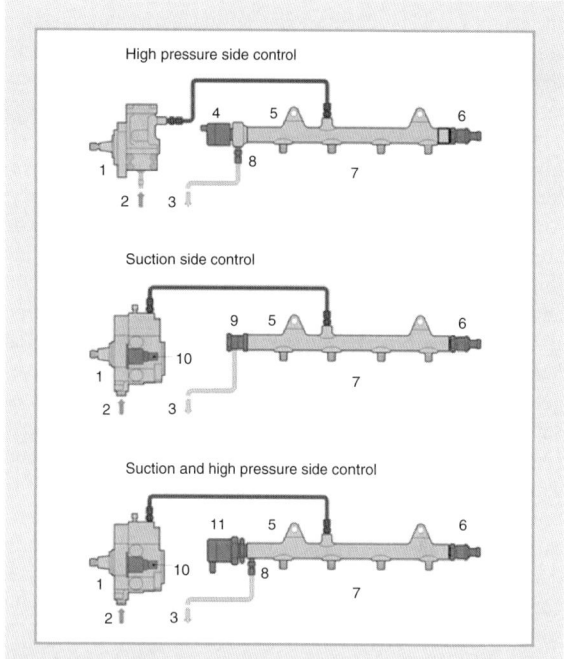

Fig. 5-29 Common rail system high pressure control. *1* High pressure pump; *2* Fuel inlet; *3* Fuel return; *4* Pressure control valve; *5* Rail; *6* Rail pressure sensor; *7* Injectors connections; *8* Connector fuel return; *9* Pressure limiting valve; *10* Metering unit; *11* Pressure control valve

control. Consequently, the pump's power consumption is lower. On the one hand, this affects fuel consumption positively. On the other hand, the temperature of the fuel returning into the tank is lower. This type of pressure control is utilized in all commercial vehicle systems.

A pressure limiting valve is mounted on the rail to prevent an unacceptable rise in pressure in a fault scenario (e.g. a metering unit malfunction). If the pressure exceeds a defined value, a moving piston enables a drainage port. It is designed so that, unrestricted by any engine speed, a rail pressure is reached, which lies significantly below the maximum system pressure. This limp-home function enables limited continued driving to the next service station, a feature that is extraordinarily important in the transportation industry in particular.

Suction and high pressure side control: When the pressure can only be set on the low pressure side, the pressure decay in the rail may last too long during negative load cycles. This particularly pertains to injectors with little internal leakage, e.g. piezo injectors. A pressure control valve mounted on the rail is additionally utilized to step up the dynamics to adjust pressure to changed load conditions. This dual actuator system combines the advantages of low pressure side control with the favorable dynamic performance of high pressure side control.

The possibility of control only on the high pressure side when the engine is cold produces another advantage over low pressure side control alone. The high pressure pump delivers more fuel than is injected. Thus, the excess fuel is heated significantly faster and the need for separate fuel heating is eliminated.

High pressure lines connect the high pressure pump and injectors with the rail. They must withstand maximum system pressure and pressure variations of very high frequency. They consist of seamless precision steel tubes that may also be autofrettaged for very high strength requirements. Given the throttling losses and compression effects, the cross section and line length influence the injection pressure and quantity. Hence, the lines between the rail and injector must be of equal length and kept as short as possible. The pressure waves generated by injection spread in the lines at the speed of sound and are reflected at the ends. As a result, close consecutive injections (e.g. pilot and main injection) interact. This can affect metering accuracy adversely. In addition, the pressure waves cause increased injector stress. Installing optimized throttles in the port to the rail can significantly decrease pressure waves. The effect on metering accuracy is compensated when specifying the maps or by an appropriate software function (see Sect. 5.3.5.7).

The high pressure lines are attached to the engine with clamps mounted at defined distances. Thus, no or only dampened vibrations (engine vibration, delivery pulse) are transmitted to high pressure lines and connected components.

The injectors are attached in the cylinder head by clamping elements and sealed toward the combustion chamber by copper gaskets. Various model types adapted to particular engine concepts are available, which connect the injector with the rail and low pressure circuit (fuel return) (Fig. 5-30). For cars and light duty applications, high pressure is connected by an integrated high pressure connector (a sealing cone on the high pressure line and a union nut). It returns through a slip joint in the injector's head or a threaded socket.

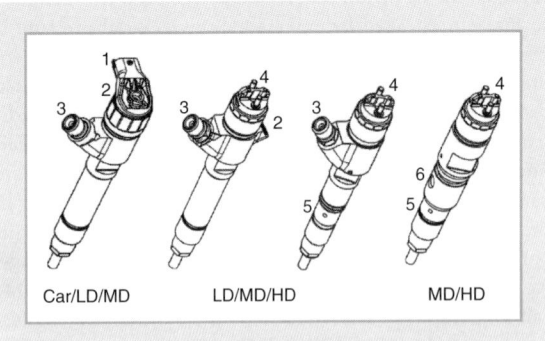

Fig. 5-30 Common rail injector designs (LD: light duty, MD: medium duty, HD: heavy duty). *1* Electric connection; *2* External fuel return; *3* External high pressure connection; *4* Electric screw connection; *5* Internal fuel return; *6* Internal high pressure connection

Internal ports establish the appropriate connections in engines for heavy commercial vehicles. A separate high pressure connector for the high pressure port is implemented as the connecting element between the high pressure line and injector. A bolted connection in the engine block presses the high pressure connector into the injector's conical inlet passage. It is sealed by the sealing cone at the tip of the pressure pipe. At the other end, it is connected with the high pressure line by a conventional pressure port with a sealing cone and union nut. The maintenance-free edge filter installed in the high pressure connector traps coarse impurities in the fuel. The injector's electrical contact is established by a slip joint or bolted connection.

The control unit specifies the injection timing and injected fuel quantity. The duration of control by the actuators mounted in the injector determine the fuel quantity. An electronic diesel control's angle-time system controls injection timing (EDC, see Sect. 6.2). Electromagnetic and piezoelectric actuators are used. The use of piezoelectric actuators is now exclusively limited to injectors for car applications.

5.3.5.4 High Pressure Pumps

The high pressure pump is the interface between a common rail system's low and high pressure stage. Its function is to hold the quantity of fuel required by the system ready at the operating point-dependent pressure level desired. This not only encompasses the injected fuel quantity the engine requires at that moment but also additionally includes reserve fuel quantities for a rapid start and a rapid rise in pressure in the rail as well as leakage and control quantities for other system components including their wear-related drift throughout a vehicle's entire service life.

Design and Function

First generation car common rail systems predominantly utilize high pressure pumps with eccentric shaft drives and three radially arranged pistons (see Fig. 5-31). This design serves as an example below to explain the function of a common rail high pressure pump.

The central drive component is the eccentric shaft (1). The pump elements, i.e. the functional groups of plunger (3), barrel (8), related valves (5, 7) and fuel inlet and outlet (4, 6), are positioned radially to it and are each offset by 120° on the circumference of the pump. A 120° thrust ring, a so-called polygon (2) transmits the eccentric's stroke to the pump plungers. The plunger's foot plate (9) slides back and forth on the polygon. When forced to move upward by the elastic force, the plunger suctions fuel out of the pump's intake port (4) through a suction valve (5) designed as a non-return valve. Depending on the type of high pressure pump, a mechanical pump integrated in the pump or an external electric presupply pump takes over the delivery of the fuel from the tank to the pump and the generation of suction pressure in the intake port. The suction valve closes shortly after the bottom dead center of plunger motion and, during the plunger's subsequent upward motion, the fuel in the barrel is compressed until it reaches the opening pressure of the high pressure valve (7), likewise designed as a non-return valve. This approximately corresponds to the pressure in the rail. After the high pressure valve opens, the fuel flows from the pump to the rail through the connecting high pressure line (6). The end of the delivery stroke is reached at the plunger's top dead center and the pressure in the cylinder drops again during the following upward movement. This closes the high pressure valve. The process of cylinder charging then begins anew.

Combustion engines drive high pressure pumps with a fixed transfer ratio. Only certain values are expedient depending on an engine's number of cylinders and the pump. Transfer ratios of 1:2 and 2:3 relative to engine speed are widespread in four-cylinder engines connected with three-piston pumps. Were the transfer ratios smaller, the pump would have to be designed unreasonably large to compensate for the geometric delivery volume. Larger transfer ratios on the other hand make greater demands on a pump's speed stability. Pump delivery synchronous with injection serves to attain constant pressure conditions in the rail and injector at the time of injection.

The number of pump delivery strokes per camshaft rotation corresponds to the number of an engine's cylinders. In four-cylinder engines with three-piston pumps, this would be provided with a transfer ratio of 2:3. The ideal state of delivery synchronous with pump elements is obtained when the same pump element is always assigned to each single injector furing synchronous injection. In principle, this is only feasible for four-cylinder engines with one or two-piston pumps and an appropriately adapted transfer ratio.

Such coupling of the pump element and injector is necessary for fuel compensation control to reduce the differences in the injected fuel quantity among the cylinders and can also be provided during asynchronous delivery when the pump strokes' phase position relative to the time of injection is retained after every camshaft revolution. Setting an exact value for the phase position of the pump delivery strokes to the injections in degree cam angle by defining the angle of camshaft and pump drive shaft rotation can further enhance the precision of the injected fuel quantity when a pump is assembled.

By virtue of their principle and because of the larger control quantities for the injector, high pressure pumps for pressure amplified common rail systems require a higher delivery rate than pumps for unamplified systems with the same injected fuel quantity. However, the pump is able to deliver fuel at a low level of pressure since the pressure in the injector increases. The increased delivery rate partially compensates the greater component loading.

Fuel Delivery Control

Given the aforementioned design criteria, a high pressure pump usually delivers significantly more fuel than required by the engine or system, especially in part load engine operation. Without regulating measures, this leads to an unnecessarily high expenditure of work when high pressure is generated. This results in heating of the fuel system, which, among other things, is detrimental because it diminishes the hot fuel's lubricating action.

A suction throttle control is implemented in modern high pressure pumps to adjust the delivery rate to the engine requirement. An electrically adjustable throttle, the metering unit (see Fig. 5-32 for the design) is mounted in the inlet passage of the pump elements. A solenoid valve's plunger (10) enables a flow cross section based on its position. It is actuated by means of a pulse width modulated electric signal. Its pulse duty factor is converted into a corresponding intake cross section. In part load engine operation, the restricted inlet passage does not completely charge the pump barrels. Fuel vapor is produced in the latter in certain operating states and the pump's delivery rate decreases altogether. The vapor pocket generated in the pump barrel only collapses during the pump plunger's upward movement before the start of part stroke pressure generation and fuel delivery to the rail. The instantaneous generation of pressure after the collapse of the vapor pocket in the pump barrels causes more loading of the crankshaft assembly than in pumps without suction throttle control.

Main Types for Cars

Without exception, radial piston high pressure pumps used for cars are fuel lubricated. Fuel's lower lubricity than engine

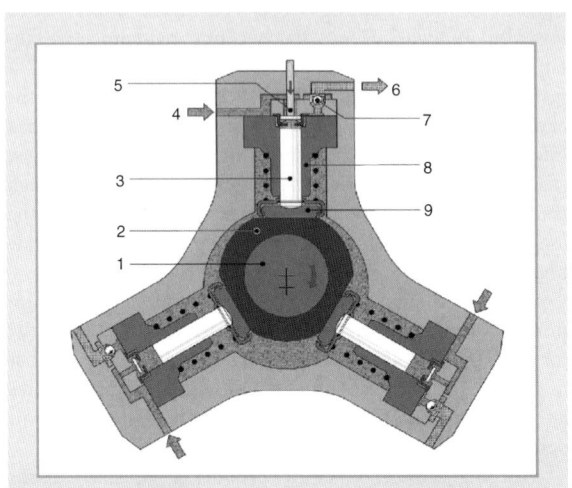

Fig. 5-31 Common rail radial piston high pressure pump (schema, radial section). *1* Eccentric shaft; *2* Polygon; *3* Pump plunger; *4* Intake passage; *5* Inlet valve; *6* High pressure port to the rail; *7* Outlet valve; *8* High pressure cylinder; *9* Plunger foot plate

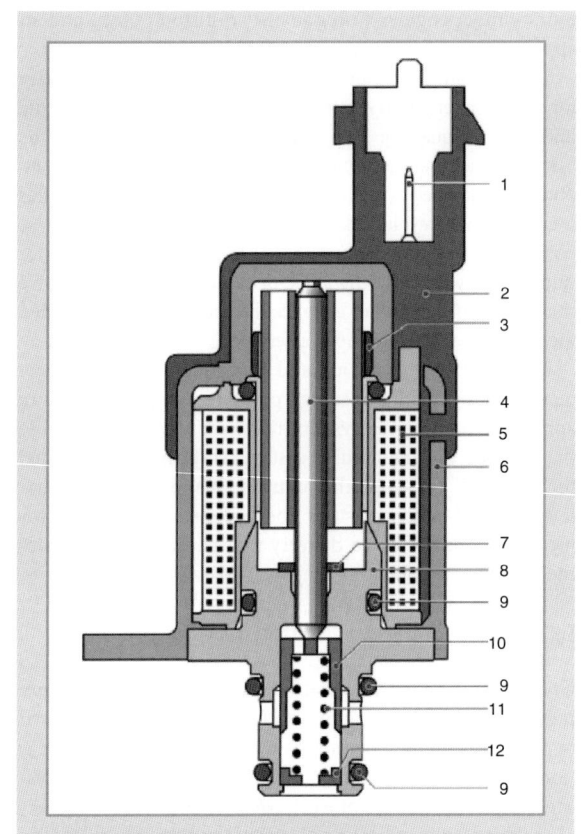

Fig. 5-32 Metering unit design. *1* Plug with electrical interface; *2* Solenoid housing; *3* Bearing; *4* Armature with tappet; *5* Winding with coil body; *6* Cup; *7* Residual air gap; *8* Magnetic core; *9* O-ring; *10* Piston with control slots; *11* Spring; *12* Safety element

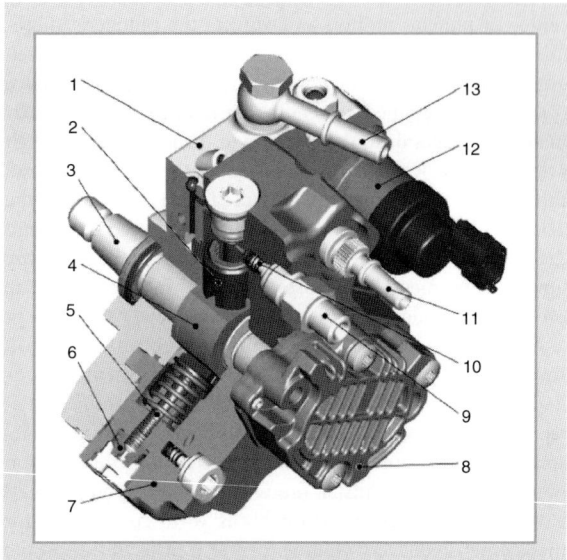

Fig. 5-33 Common rail three-piston radial-high pressure pump, CP3, Robert Bosch GmbH. *1* Pump flange; *2* Bucket tappet; *3* Eccentric shaft; *4* Polygon on eccentric; *5* Pump plunger; *6* Intake non-return valve; *7* Monoblock housing; *8* Gear presupply pump; *9* High pressure port to the rail; *10* High pressure non-return valve; *11* Return port to tank; *12* Metering unit; *13* Low pressure port

oil's imposes high demands on the surface quality of the components involved in the generation of high pressure. Fuel lubrication prevents any intermixing of the fluids fuel and engine oil, which is undesired because of the risk of oil dilution and nozzle coking by oil fractions in the injected fuel.

Common rail three-piston radial high pressure pump: Already described above, this pump principle is characterized by very uniform fuel delivery with three sinusoidal delivery processes offset at eccentric angles of 120°, which yields an extremely constant characteristic for the pump's driving torque. Torque peaks are lower in common rail pumps than in distributor pumps by a factor of five to eight or than in unit injectors with their strongly swelling, pulse-like torque. As a result, a pump drive can be designed more cost effectively. However, the reserve delivery rates necessary in common rail systems produce a comparatively higher mean torque.

The Bosch CP3 pictured in Fig. 5-33 is a typical three-piston radial high pressure pump with suction throttle control (12) and an external gear pump for fuel delivery, which is flange-mounted directly on the pump housing (8). Unlike the CP1 (without suction throttle control), the polygon's movement is not directly transferred to the piston but rather to the bucket tappet (2) placed in between. This keeps the lateral forces induced by friction away from the piston (5) and conducts them to the pump housing (7). As a result, the piston may be loaded more highly. Hence, thusly configured pumps are suitable for higher pressure ranges and larger delivery rates.

The low pressure ports in the CP3 are chiefly located in the aluminum pump flange (1), which represents the customized interface component to the engine and is bolted with the forged steel housing in a monoblock design (7). This attains high compressive strength yet requires extensive machining for the long high pressure bores inside the very hard to machine housing material.

Common rail one and two-piston radial high pressure pump: Newer high pressure pump developments, especially for small and medium-sized car engines, reduce the number of pump elements to two and even one to cut costs. Measures are implemented (and also combined if necessary) to compensate for the resultant reduced delivery rate:

– enlarging cylinder volumes (producing efficiency disadvantages when the piston diameter is enlarged),
– increasing speed (by adapting the drive transfer ratio) and
– implementing a drive shaft with a double cam instead of an eccentric shaft with a polygon (which doubles the piston strokes per drive shaft revolution).

Depending on the drive system concept, the pump elements of two piston pumps have a throw of 90 or 180° to obtain a uniform delivery flow.

The CP4 pump (Fig. 5-34) has pistons arrayed at 90° and a drive shaft with a double cam. A further transmission element must be inserted between the cam (6) and bucket tappet (8) to prevent point contact between them. This type has a roller (7) supported in the tappet and running on the cam.

This pump is also produced as a one piston variant, which can be advantageously driven with a transfer ratio of 1. This produces delivery synchronous with the pump elements in four-cylinder engines and, with regard to the delivery rate, additionally compensates well for the low number of pistons.

Main Types for Commercial Vehicles

The pumps described thus far are also used in the commercial vehicle sector, especially in light and medium duty applications. Allowing for the large loads caused by their high delivery rates and the required service life, the pumps utilized in the heavy duty sector are often designed to be oil lubricated. This is possible because the larger nozzle hole diameters lessen the effect of potential spray hole coking by oil fractions in the fuel.

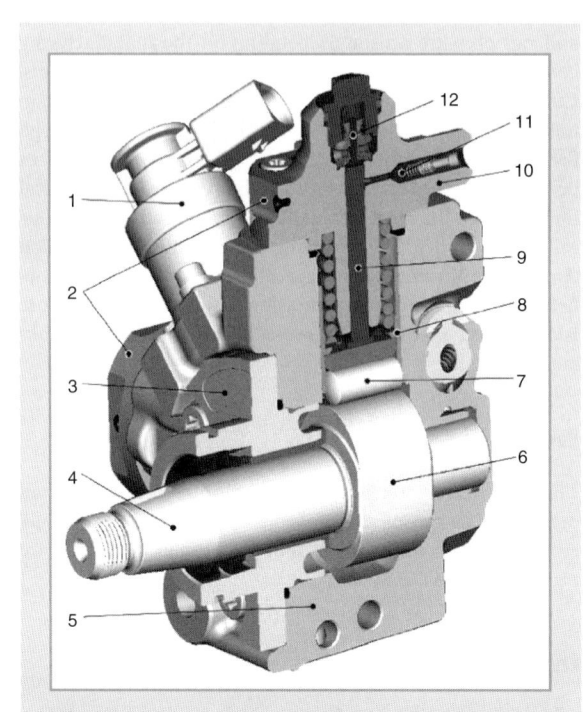

Fig. 5-34 Common rail two-piston radial high pressure pump, CP4, Robert Bosch GmbH. *1* Metering unit; *2* Cylinder head; *3* Pump flange; *4* Drive shaft; *5* Aluminum housing; *6* Double cam; *7* Roller; *8* Bucket tappet; *9* Pump plunger; *10* High pressure port to the rail; *11* High pressure non-return valve; *12* Intake non-return valve

Common rail inline high pressure pump: Pumps such as the Bosch CP2 in Fig. 5-35 are used for very large commercial vehicle engines, which often require installation compatibility with conventional inline pumps (see Sect. 5.3.2). It is an inline two-piston pump with the pump elements arranged side by side. A metering unit (2) placed between the presupply pump (5) and intake valves in the manner described above takes over the fuel delivery control. The pump feeds the fuel to the compression chambers and further conducts it to the rail through a combined intake/high pressure valve (7).

5.3.5.5 Rail and Add-on Components

Rail Function

Accumulator injection systems have a high pressure accumulator also called a (common) rail. The rail's main functions are
- accumulating fuel under high pressure and
- distributing fuel to the injectors.

These main functions also include damping the pressure fluctuations when fuel is supplied to and discharged from the rail. The permissible rail pressure fluctuations represent a design criterion for the rail. Moreover, the rail also performs secondary functions as:
- an add-on point for sensors and actuators in the high pressure circuit,
- throttle elements to damp line pressure oscillations between the high pressure pump and rail as well as the injectors and rail and
- a connecting element for the components in the common rail system's high pressure circuit, e.g. high pressure pump and injectors through the high pressure lines.

The fuel compressed by the high pressure pump reaches the rail through a high pressure line, is stored there and distributed to the injectors through other high pressure lines connected with them. In conjunction with the fuel's compressibility, the volumes stored in the rail enable the rail to damp pressure fluctuation caused by extracting fuel from and supplying it to the rail. Thus, the pressure in the rail depends on the consumers and the pump connected to the rail and the performance of the rail's accumulator itself. A rail pressure sensor measures the current pressure in the rail. Not only the pump and the injectors but also the pressure control valve that may be mounted on the rail itself or on the high pressure pump are actuating variables that influence rail pressure.

Rail Design

Rail design pursues the target compromise of producing maximum storage capacity and thus damping by a large volume to keep rail pressure constant and, on the other hand, to react as dynamically as possible to changes in rail pressure setpoints, e.g. when pressure is generated during starting or the engine

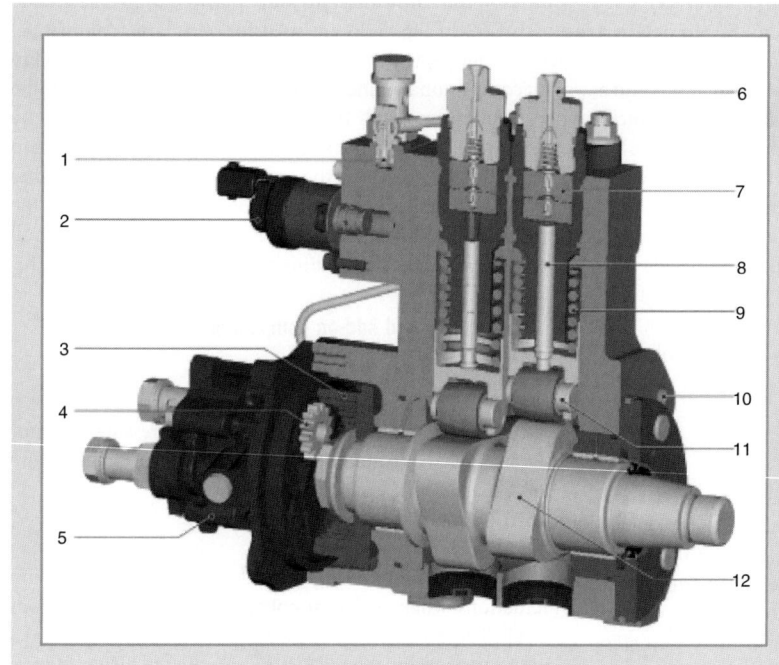

Fig. 5-35
Common rail two-piston inline high pressure pump for commercial vehicles, CP2, Robert Bosch GmbH. *1* Zero delivery throttle; *2* Metering unit; *3* Internal gear; *4* Pinion; *5* Gear presupply pump; *6* High pressure port; *7* Two-part intake/high pressure valve; *8* C-coated plunger; *9* Plunger spring; *10* Oil inlet bore; *11* C-coated roller bolt; *12* Concave cam

load changes dynamically. Large pressure generation and decay rates may be required depending on the load change. A minimum high pressure volume would be optimal. Simulations of the overall system in representative load points and verification on hydraulic test benches are employed to determine the minimum rail volume required as a function of the main injection quantity for a given engine configuration. Table 5-2 presents typical configurations for the rail volume of production applications. The boundary condition of equal line lengths in the high pressure system to prevent variances from cylinder to cylinder can, for example, specify the length of the rail in the engine. The structural space specified in the vehicle and aspects of rail manufacturing represent other factors. Thus, the rail volume actually selected is often larger than the functionally specified minimum volume without noticeably undershooting the requirements of dynamics.

The damping bores placed at the rail outlets are configured as a compromise between the minimum pressure drop and maximum damping of the reflection waves between the rail and the consumers. Functionally, the throttle elements serve to reduce pump and injector loading and to damp line pressure fluctuations that can diminish metering quality during multiple injections.

Rail Types

The rail design selected significantly depends on the engine characteristics and the design of the common rail system itself. Figure 5-36 shows a typical four cylinder rail for a car common rail system with a mounted pressure control valve and rail pressure sensor. Depending on the manufacturing concept, rails are made of forging blanks or tube preforms. Cuts made during machining are normally rounded to obtain the required strength. The damping passages in the high pressure outlet to the injector and pump may be bored or press fit as separate components. Inline engines have one rail implemented the system, while V engines usually employ one rail per cylinder bank. Again, the specific configuration depends on the engine and may include compensating lines between the rails or even connecting rails that ensure pressure is distributed as equally as possible between the engine banks and cylinders.

Rail Pressure Sensor

The rail pressure sensor serves to capture the current rail pressure. The sensor is installed in the rail and electrically connected with the control unit. Other sensor designs are presented in Sect. 6.3.

Pressure Control Valve

A high pressure side actuator in the high pressure control loop, the pressure control valve's function is to set the rail pressure. This is accomplished by altering a cross section in the pressure control valve through which, depending on the pressure and electrical power, more or less fuel is depressurized from high to

Table 5-2 Production designs for typical applications

	QE, max. [mm³/H]	Throttle Ø [mm]	V RAK [ccm]	Total VHD [ccm]	Number of rails	Connecting rail/line
Car engine						
R4	~80	0.85	~25	~20	1	–
R6	~80	0.85	~35	~40	1	–
V6	~80	0.85	~20	~50	2	YES
V8	~90	0.85	~25	~60	2	YES
Commercial vehicle engine						
R4	~200	0.85	14...20	20...30	1	–
R6	~50	0.85...1.3	20...40	35...65	1	–

Fig. 5-36 Typical rail with attached components for a four-cylinder application for cars

low pressure. The valve is primarily mounted on the rail and feeds its depressurized quantity to the common rail system's low pressure circuit. Figure 5-37 presents its configuration and the functionally determinative components.

The valve body houses a valve seat that is in the flow passing through a throttle cross section. The inflow and the elastic and magnetic force exerted by the solenoid valve bolt on the valve ball place it in an equilibrium of hydraulic forces. Larger flow rates through the valve cross section increase the hydraulic force, displacing the ball and thus the solenoid valve bolt even more. This generates an increase of the elastic force and thus a proportionally negative feedback. When a larger mean pressure has to be withstood, the control unit impresses a higher mean current on the magnet by pulse width modulation, which increases the magnetic force. In terms of control system engineering, the valve is designed as a PI element with a slowly integrative reference variable and a rapidly proportional feedforward control. This proportionally equalizes highly dynamic pressure fluctuations and integrators in the cascaded control loops bring the steady state deviations to zero.

A dither frequency that keeps the solenoid pin in constant motion is superimposed on the current signal to eliminate undesired hysteresis effects. The frequency is selected so that it does not affect the current rail pressure adversely.

Dependent on the pressure control valve's operating point, the flow values in typical four cylinder car applications are between 0 and 120 l/h and the mean electrical currents are < 1.8 A at pressures between 250 and 1,800 bar.

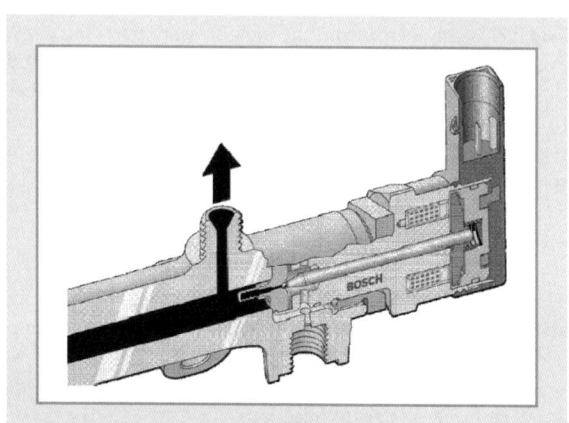

Fig. 5-37 Sectional view of a pressure control valve

Pressure Limiting Valve

Pressure limiting valves (Fig. 5-38) are primarily used for commercial vehicle applications that, on the one hand, do not have any high pressure side actuator in the pressure control loop and, on the other hand, require dry running properties of the engine and thus limited modes of injection system operation. This gives rise to the pressure limiting valve's main functions in the high pressure control loop fault scenario:
- limiting system pressure to a maximum value and
- assuring rail pressure is controlled in the restricted range.

A pressure limiting valve is bolted on the rail and uses a spring loaded valve bolt to maintain contact with the highly compressed fuel through a sealing seat. On the back of the sealing seat, a line connects the valve with the common rail system's low pressure return. When the rail pressure varies within the allowable range, the applied elastic force keeps the valve closed and sealed against the return flow. If the maximum allowable rail pressure is exceeded in the fault scenario, the valve bolt opens, limits the system pressure and uses the working motion of a second valve plunger to control the passage of the high pressure flow. The control edge of the control piston aligned coaxially to the valve plunger on the low pressure side produces the pressure-flow characteristic. Error detection in the control unit makes it possible to maintain the high pressure pump's delivery rate as a function of engine speed so that emergency operation pressure in the rail follows the pressure limiting valve's flow characteristics. The emergency operation characteristic is shaped so that emergency operation pressure remains within limits expedient for the engine in order to operate a commercial vehicle in a restricted load range.

5.3.5.6 CR Injectors

Common rail injectors with identical basic functions are employed in car and commercial vehicle systems. An injector primarily consists of an injection nozzle (see Sect. 5.2. Injection Nozzles), injector body, control valve and control chamber. The control valve has a solenoid or piezo actuator. Both actuators allow multiple injections. The advantage of the piezo actuator's large actuating force and short switching time can only be exploited when injector design has been optimized to do so.

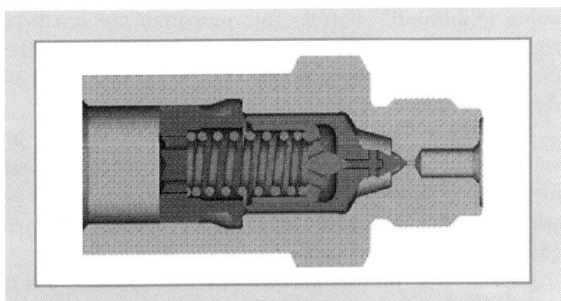

Injectors in a common rail diesel injection system are connected with the rail by short high pressure fuel lines. A copper gasket seals the injectors from the combustion chamber. Clamping elements attach the injectors in the cylinder head. Common rail injectors are suited for straight or oblique installation in direct injection diesel engines, depending on the design of the injection nozzles.

The system characteristically generates injection pressure independent of the engine speed and the injected fuel quantity. The electrically controllable injector controls the start of injection and the injected fuel quantity. The electronic diesel control's (EDC) angle-time function controls the injection timing. It requires two sensors on the crankshaft and on the camshaft for cylinder recognition (phase detection).

Various types of injectors are currently standard:
- solenoid valve (SV) injectors with a one or two-piece armature (Bosch),
- inline SV injectors (Delphi),
- tophead piezo injectors (Siemens) and
- inline piezo injectors (Bosch, Denso).

Solenoid Valve Injector

Configuration

An injector can be broken down into different functional groups:
- the hole-type nozzle (see Sect. 5.2),
- the hydraulic servo system and
- the solenoid valve.

The fuel is conducted from the high pressure port (Fig. 5-39, Pos. 13) through an inlet passage to the injection nozzle and through the inlet throttle (14) into the valve control chamber (6). The valve control chamber is connected with the fuel return (1) by an outlet throttle (12) that can be opened by a solenoid valve.

Function

The injector's function can be subdivided into four operating states when the engine is running and the high pressure pump is delivering fuel:
- injector closed (with adjacent high pressure),
- injector opens (start of injection),
- injector opened and
- injector closes (end of injection).

Fig. 5-38 Sectional view of a pressure limiting valve with limp-home functionality

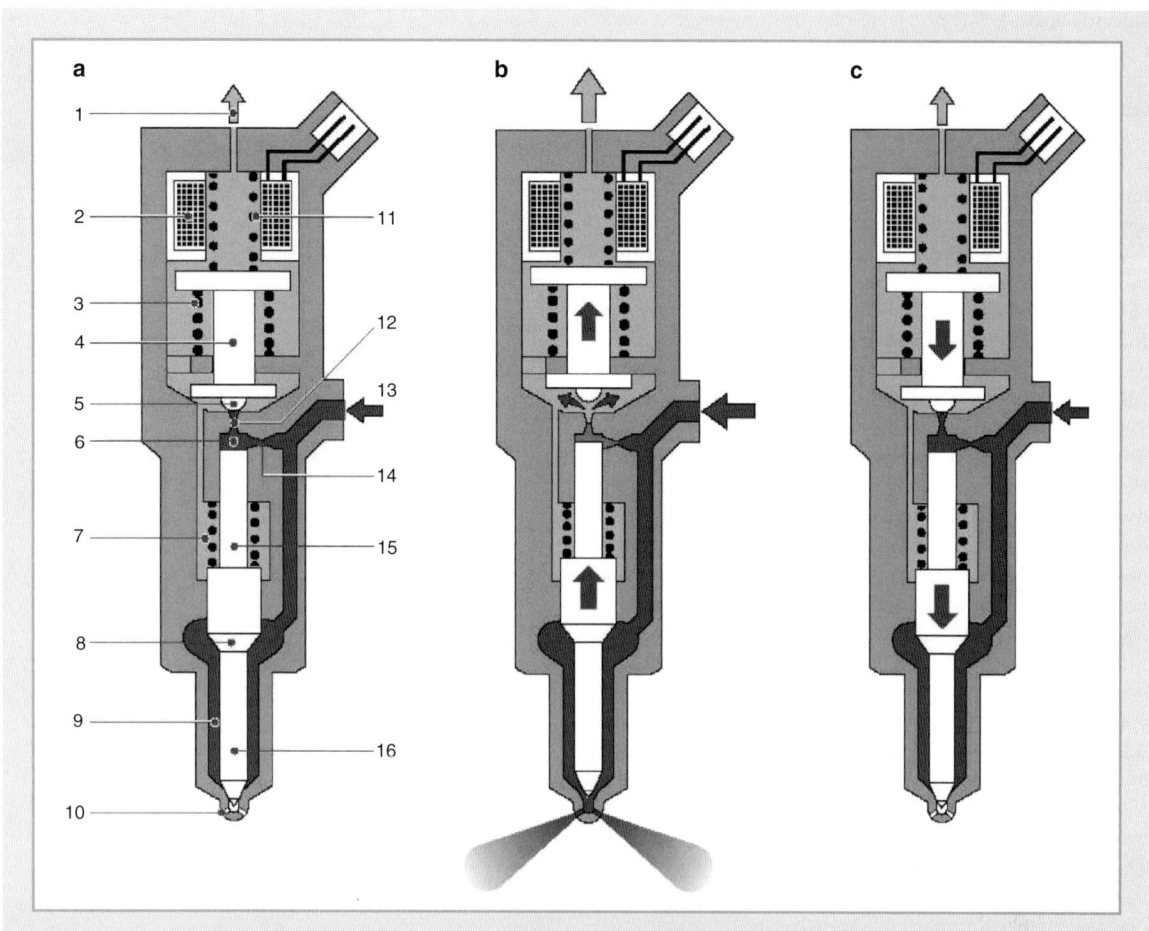

Fig. 5-39 Solenoid valve injector (functional principle). (**a**) Resting state; (**b**) Injector opens; (**c**) Injector closes; *1* Fuel return; *2* Solenoid coil; *3* Overlift spring; *4* Solenoid armature; *5* Valve ball; *6* Valve control chamber; *7* Nozzle spring; *8* Nozzle needle pressure shoulder; *9* Chamber volume; *10* Spray hole; *11* Solenoid valve spring; *12* Outlet throttle; *13* High pressure port; *14* Inlet throttle; *15* Valve piston (control piston); *16* Nozzle needle

These operating states are regulated by the distribution of forces to the injector's components. The nozzle spring closes the injector when the engine is not running and there is no pressure in the rail.

Injector closed (resting state): The injector is not actuated in its resting state (Fig. 5-39a). The solenoid valve spring (11) presses the valve ball (5) into the seat of the outlet throttle (12). The rail's high pressure is generated in the valve control chamber. The same pressure also exists in the nozzle's chamber volume (9). The forces applied to the lateral face of the valve piston (15) by the rail pressure and the force from the nozzle spring (7) hold the nozzle needle closed against the opening force acting on its pressure shoulder (8).

Injector opens (start of injection): The injector is in its neutral position. The solenoid valve is actuated with the "pickup current", which serves to open the solenoid valve quickly (Fig. 5-39b). The short switching times required may be obtained by appropriately designing the energization of the solenoid valves in the control unit with high voltages and currents.

The actuated electromagnet's magnetic force exceeds the valve spring's elastic force. The armature elevates the valve ball from the valve seat and opens the outlet throttle. After a brief time, the increased pickup current is reduced to a lower holding current of the electromagnet. When the outlet throttle opens, fuel is able to flow from the valve control chamber into the cavity located above it and to the fuel tank through the return. The inlet throttle (14) prevents the pressure from fully equalizing. Thus, the pressure in the valve control chamber drops. This causes the pressure in the valve control chamber to be lower than the pressure in the nozzle's chamber volume, which always continues to have the pressure level of the rail. The reduced pressure in the valve control chamber decreases the force on the control piston and causes the nozzle needle to open. Injection begins.

Injector opened: The nozzle needle's opening speed is determined by the differential flow between the inlet and outlet throttles. The control piston reaches its top position and remains there on a fuel cushion (hydraulic stop). The cushion is generated by the fuel flow produced between the inlet and outlet throttles. The injector nozzle is then fully opened. The fuel is injected into the combustion chamber at a pressure approximating the pressure in the rail.

At the given pressure, the injected fuel quantity is proportional to the solenoid valve's operating time and independent of the engine and pump speed (time controlled injection).

Injector closes (end of injection): When the solenoid valve is deenergized, the valve spring pushes the armature downward, whereupon the valve ball closes the outlet throttle (Fig. 5-39c). The closing of the outlet throttle causes the inlet throttle to build up rail pressure in the control chamber again. This pressure exerts increased force on the control piston. The force from the valve control chamber and the force from the nozzle spring then exceed the force acting on the nozzle needle from below and the nozzle needle closes. The flow from the inlet throttle determines the nozzle needle's closing speed. Injection ends when the nozzle needle reaches the nozzle body seat again and thus closes the spray holes.

This indirect control of the nozzle needle by a hydraulic force boost system is employed because the forces needed to quickly open the nozzle needle cannot be generated directly with the solenoid valve. The "control quantity" required in addition to the injected fuel quantity reaches the fuel return through the control chamber's throttles. In addition to the control quantity, leakage quantities are in the nozzle needle and the valve piston guide. The control and leakage quantities are returned to the fuel tank by the return with a manifold to which the overflow valve, high pressure pump and pressure control valve are also connected.

Piezo Injectors

There are two types of piezo injectors:
– Tophead CR injectors (Siemens) and
– inline CR injectors (Bosch, Denso).

The Tophead CR injector functions like Bosch's CR injector with a servo valve. The different temperature correlation of the ceramic actuator and the housing must be compensated in a CR injector with a piezo actuator. The actuator housing assumes this function in a Tophead injector, the hydraulic coupler assumes it in a piezo inline injector. The following describes the inline injector's function in detail.

Design and Function of a Piezo Inline Injector

The body of a piezo inline injector is composed of the following assemblies (see Fig. 5-40):
– actuator module (3),
– hydraulic coupler or amplifier (4),
– control or servo valve (5) and
– nozzle module (6).

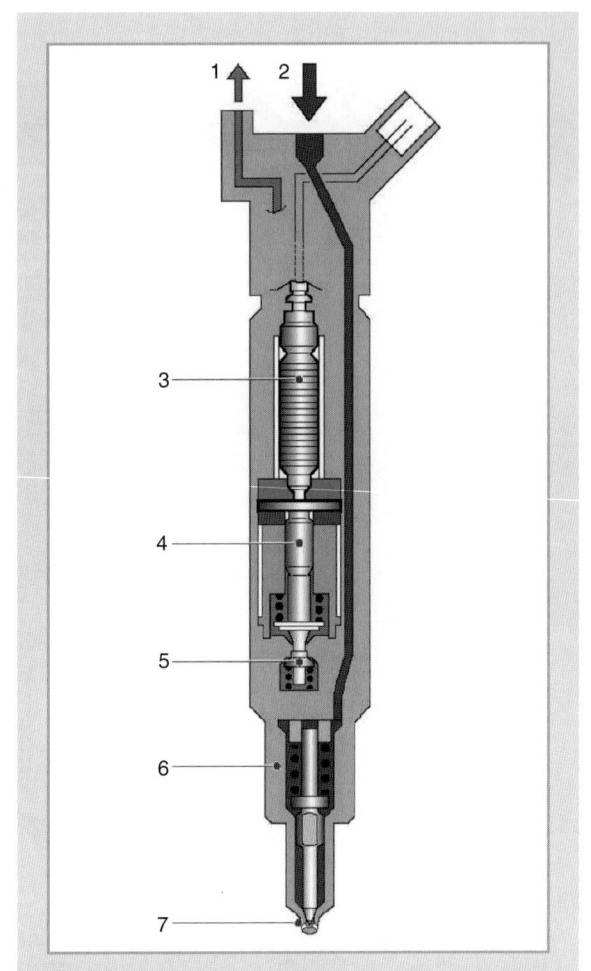

Fig. 5-40 Bosch piezo inline injector design. *1* Fuel return; *2* High pressure port; *3* Piezo actuator module; *4* Hydraulic coupler (amplifier); *5* Servo valve (control valve); *6* Nozzle module with nozzle needle; *7* Spray hole

Attaining a high overall stiffness within the actuator chain of the actuator, hydraulic coupler and control valve was a priority when this injector was designed. Another distinctive design feature is the elimination of the mechanical forces on the nozzle needle, which a push rod may generate in Tophead injectors (solenoid or piezo). All in all, it was possible to greatly reduce the moving masses and friction and thus improve the injector's stability and drift compared with other systems.

The injection system additionally makes it possible to implement very short intervals between injections. At present, the amount and configuration of fuel metering can cover up to seven injections per injection cycle and thus be adapted to the requirements in the engine operating points.

Since the servo valve (5) is closely connected with the nozzle needle, the needle directly reacts to actuator operation. The delay time between the beginning of electric control and the nozzle needle's hydraulic reaction is approximately 150 microseconds. This satisfies the conflicting requirements of simultaneously high needle speeds and minimal reproducible injected fuel quantities.

What is more, by virtue of its principle, the injector does not contain any direct leakage from the high pressure region to the low pressure circuit. This enhances the entire system's hydraulic efficiency.

Piezo Inline Injector Control

The injector is controlled by an engine control unit, the output stage of which was specially designed for this injector. A setpoint for the actuator's voltage is assigned as a function of the rail pressure of the operating point. It is energized intermittently (Fig. 5-41) until a minimum deviation between the setpoint and control voltage has been obtained. The voltage rise is proportionally converted into the piezo actuator's stroke. Hydraulic amplification causes the actuator stroke to generate a pressure rise in the coupler until the force equilibrium is exceeded in the solenoid valve and the valve opens. As soon as the solenoid valve has reached its final position, the pressure in the control chamber above the needle begins to drop and injection takes place.

Their principle gives piezo inline injectors advantages over SV injectors, namely:
- multiple injection with flexible start of injection and intervals between the individual injections,
- production of very low injected fuel quantities for pilot injection and
- low manufactured injector size and weight (270 g instead of 490 g).

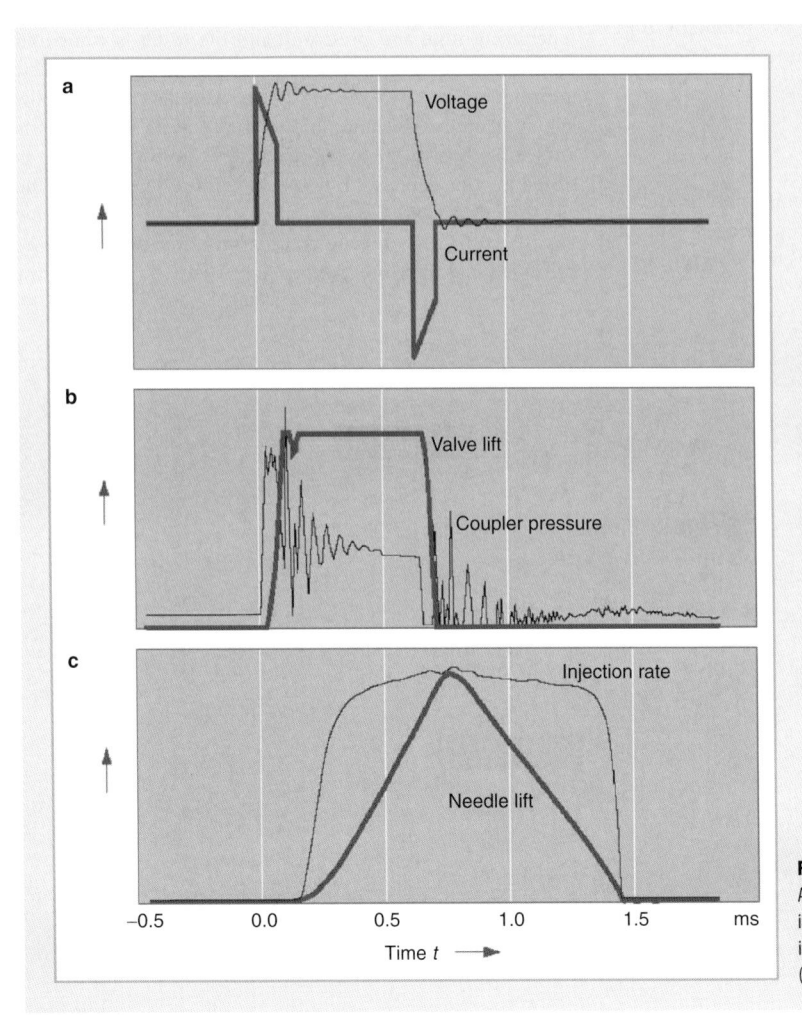

Fig. 5-41
Actuating sequences of the piezo inline injector for one injection. (**a**) Current and voltage curve for the actuated injector; (**b**) Valve lift and the coupler pressure curve; (**c**) Valve lift and injection rate

5.3.5.7 Metering Functions

Definition and Objectives

Metering functions encompass open and closed loop control structures in the electric control unit, which, in conjunction with monotonous injector action, ensure the requisite metering accuracy during injection. The functions utilize a specifically developed action of the injection hydraulics and apply signals from existing sensors as auxiliary quantities as well as model-based approaches based on the physical laws of conservation to precisely meter fuel.

The use of such functions is driven by the increasingly more strictly formulated performance specifications for fuel metering, which, in turn, stem from diesel engine development targets, specifically ever lower raw emissions with simultaneously higher demands for comfort and performance with unrivaled low consumption. Regulating the required accuracies in fuel metering in the hydraulic components themselves and endeavoring to maintain these throughout their service life has proven to be an uneconomical approach.

Overview

Figure 5-42 presents an overview of the four most important metering functions that achieve the metering tolerances required for EU4 applications and future emissions standards worldwide.

While the function of **injector quality adaptation** adjusts injectors' manufacturing tolerances when they are new, **pressure wave correction** corrects the pressure waves' quantitative influence during multiple injections on the basis of models. Both functions operate controlled and require results from measurements of the hydraulic system as input variables.

Zero fuel quantity calibration utilizes the auxiliary quantity of speed change to learn an injector's minimum quantity in situ throughout its period of operation.

Fuel mean value adaptation calculates the air mass of the average injected fuel quantity based on the λ signal. The latter two functions are adaptive injection quantity control functions that utilize auxiliary quantities available from sensor signals.

Injector Quality Adaptation (Fig. 5-43)

This function employs results of measurements from factory wet tests after the final assembly of the injectors. Every injector's injected fuel quantity is measured, e.g. in four test points (energizing time and pressure), and the result is compared with the setpoint for the particular measuring point. This information is stored in the injector as a data matrix code. When an individual injector's tolerance at its test points is known relative to the average injector, the map of which is stored in the control unit, the individual injector can be adapted to the average injector's map. The prerequisite for this is a sufficiently precise description of the entire map by the individual injector's performance based on the four test

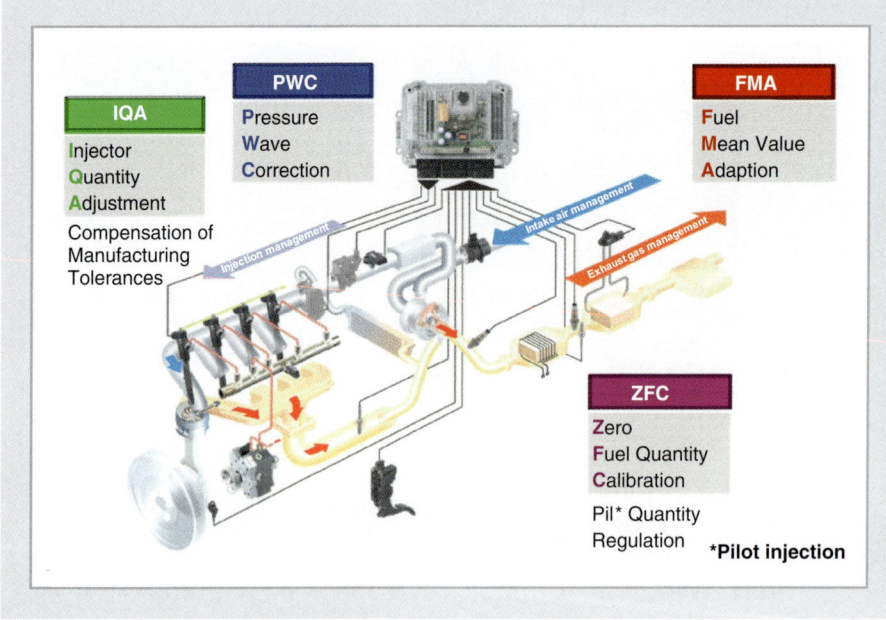

Fig. 5-42 Metering functions for passenger car applications starting with EU4

Fig. 5-43 Injector quality adaptation process chain

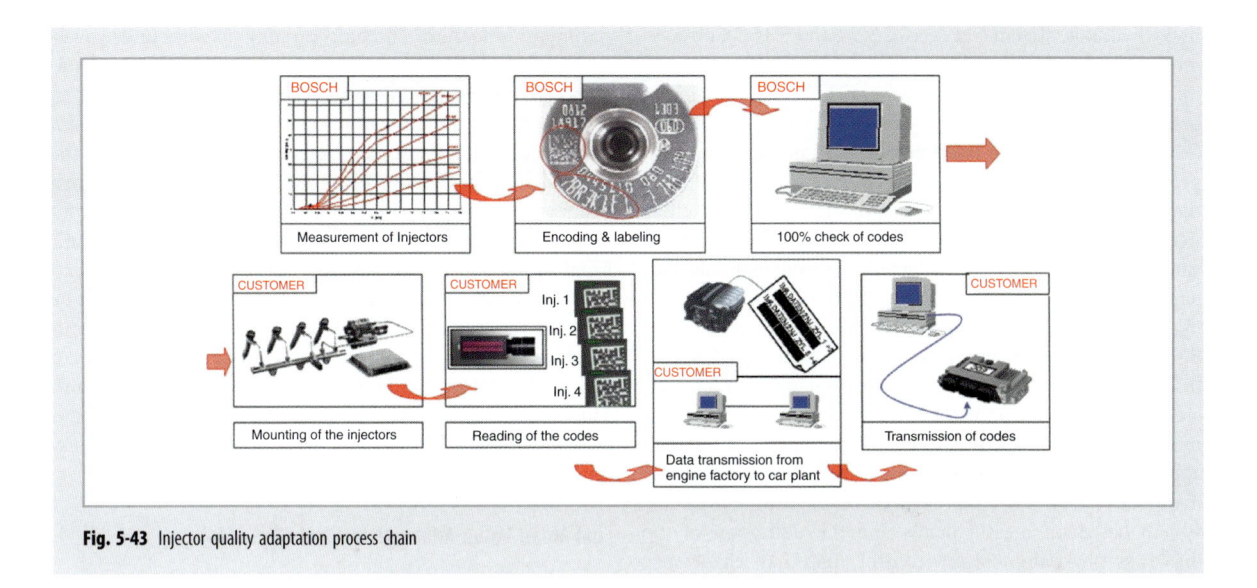

points, e.g. by correlation factors. The correlation factors for all injectors of the same type are stored in the control unit together with the average fuel quantity map. The individual information on an individual injector's performance is entered in the control unit's writeable memory by reading in the data matrix codes. In vehicle production, this is done at the end of the assembly line.

Injector quality adaptation represents an efficient method to equalize quantities of fuel that injectors inject based on their test values. All in all, the method allows a "win–win" strategy by increasing metering accuracy as well as production volume while extending adjustment tolerances. This has a positive impact on costs.

Pressure Wave Correction (PWC) (Fig. 5-44)

Variably positioned and applied to different engine load points and thus, in turn, having different pressures and injecting different masses, multiple injections are indispensible to achieving the emission and comfort goals for diesel engines. Given diesel fuel's compressibility, pressure waves are always triggered during injections in the system, which can in turn influence fuel metering in the combustion chamber as soon as multiple successive injections are initiated. The thusly triggered "quantity waves" expand the tolerance for one or several injections. This is undesired and may influence emission and noise behavior extremely adversely. Along with damping measures

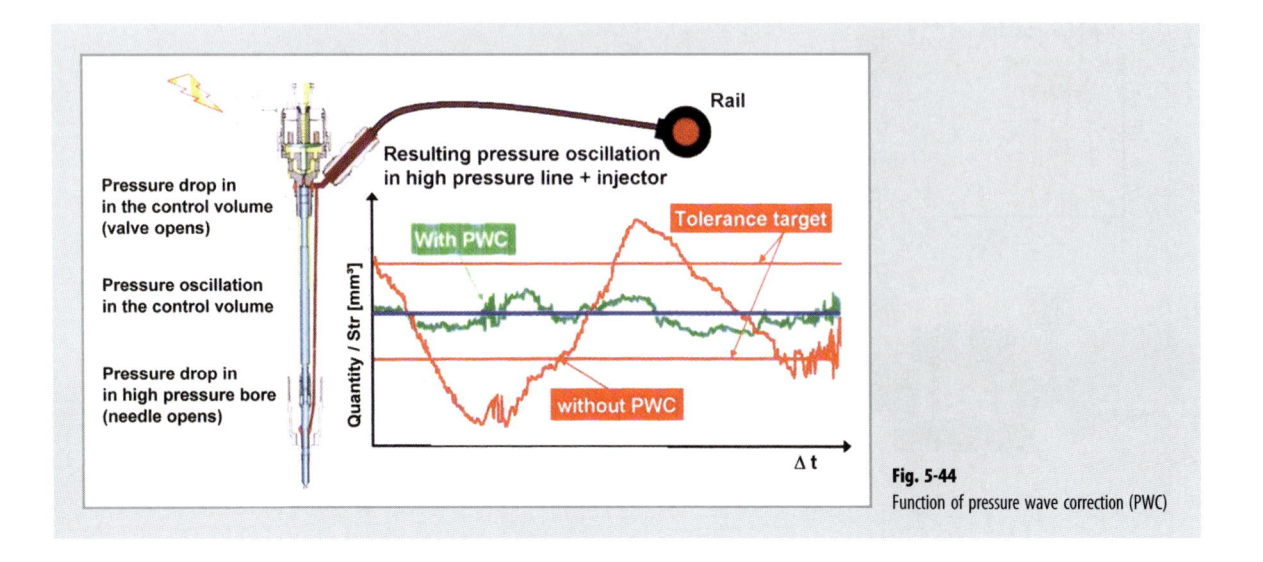

Fig. 5-44 Function of pressure wave correction (PWC)

in the hydraulic system (see Sects. 5.3.5.4 and 5.3.5.5), physical models may serve as the basis to correct the quantitative influences induced by pressure waves. This is the function of pressure wave correction. The quantities of fuel injected in the instantaneous and prior injection, the fuel's pressure and temperature and the injection intervals are variables that influence the changes in fuel quantity. Pressure wave correction effectively corrects quantitative influences on the basis of these variables and the response characteristic of the hydraulic system itself. Pressure wave correction enables incorporating hydraulic injection intervals that are optimal for an engine's thermodynamic application and equalizing the fuel temperature's influence on the fuel waves even when injection intervals are constant. Since pressure wave correction constitutes a model-based control, it is of utmost importance that the fault applied in the model also corresponds to the fault in the hydraulic system, i.e. the reproduction of the existing hydraulic system based on measurements and the utilization of the injectors' physically correct control map has elementary importance for the function of pressure wave correction.

Zero Fuel Quantity Calibration (ZFC) (Fig. 5-45)

From the perspective of the engine, the inherent conflict in the application of pilot injection to lower combustion noise on the one hand and only minimally increase particulate emission on the other hand has to be resolved. Hence, the adaptive function of zero fuel quantity calibration is to ensure that minimum pilot injection quantities are stable over the lifetime. Zero fuel quantity calibration utilizes a highly resolved speed signal from the engine as an auxiliary quantity. It delivers information on cylinder-selective torque generated during the combustion of minimum quantities. The function only operates under overrun conditions so as not to disrupt normal driving. The modulation of every single injector's energizing time allows successively detecting the time an injector has injected a minimum quantity of fuel, the so-called "zero quantity", by analyzing the speed signal. The duration of control is applied to detect and, if necessary, correct changes in the hydraulic system throughout its period of operation. The response characteristic of minimum injected fuel quantities and the speed response depend on the drive train of the vehicle applied and must therefore be calculated specifically. Notably, this function operates without additional sensors and is able to detect minimum quantities of fuel with certainty within an accuracy of less than 0.4 mm^3. Thus, smaller quantities of pilot injected fuel than 1 mm^3 can be represented in the map region as a function of the combustion limit required by the engine.

Fuel Mean Value Adaptation (FMA) (Fig. 5-46)

To reliably obtain the particulate and NO$_X$ emission limits on all certified vehicles, all the tolerances in the air and injection system's sensor and actuator systems must ensure that the right air mass is always available to burn the fuel mass supplied. This is done by controlling the exhaust gas recirculation rate. If the injected mass deviates from the presumed mass in the control unit, the original application shifts along the exhaust gas recirculation rate's particulate-NO$_X$ tradeoff. If somewhat less fuel is injected in a specific load point than presumed by the control unit, then the exhaust gas recirculation will be too low to provide the higher amount of oxygen assumed to be required for combustion. Unlike the nominal application, this produces a rise in NO$_X$ emissions that facilitates particulate emissions.

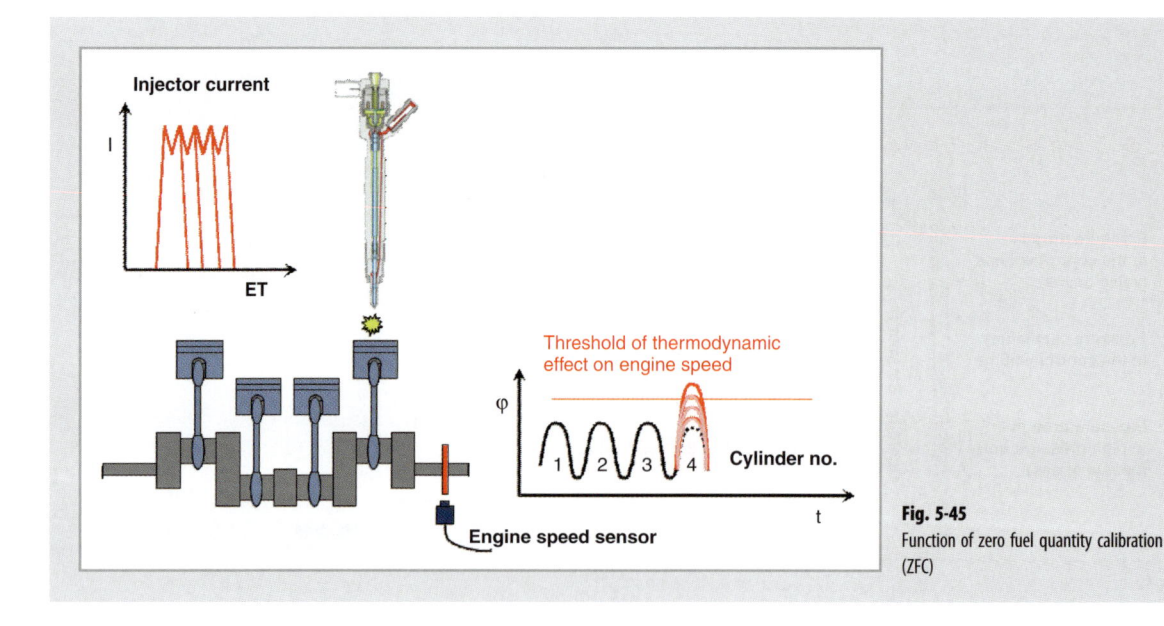

Fig. 5-45 Function of zero fuel quantity calibration (ZFC)

Fig. 5-46 Structure and benefit of fuel mean value adaptation (FMA)

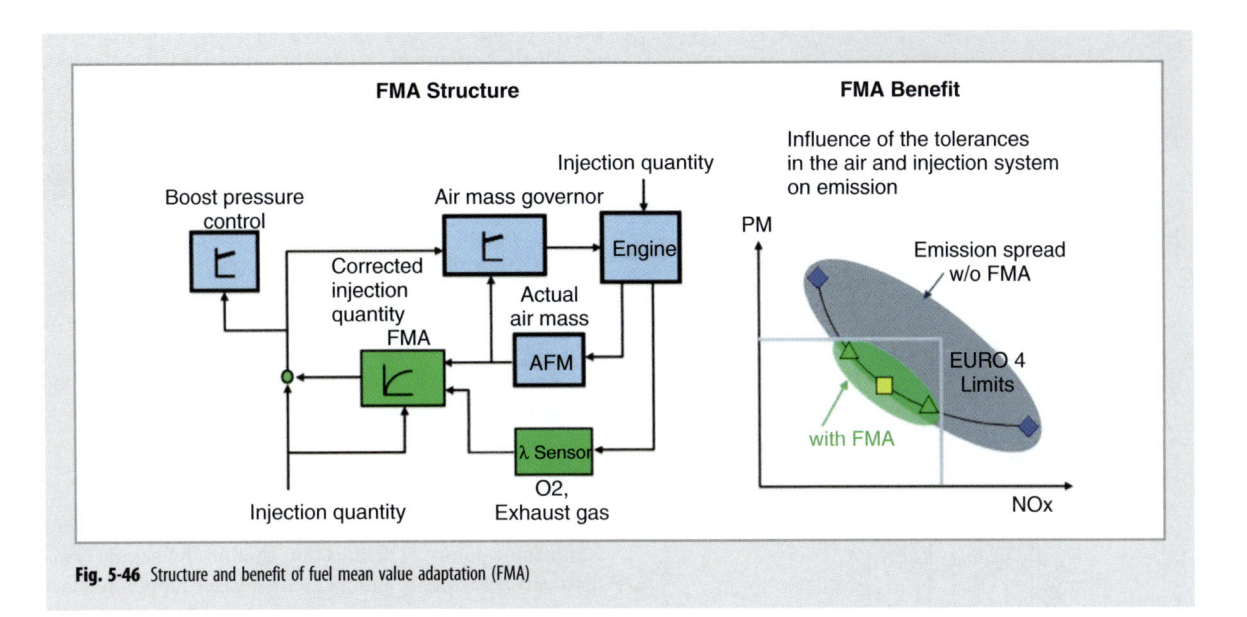

Excess fuel during injection produces the converse and is commensurate with the soot-NO$_X$ tradeoff. The function of fuel mean value adaptation is to ascertain the mass actually injected based on the λ signals and then adjust the air mass flow to obtain the original compromise in the particulate-NO$_X$ concept. Fuel mean value adaptation's response to errors in the injected fuel quantity as well as its compensation for the air mass sensor's errors based on the λ signal is a desired side effect. Thus, tolerances in the air path and engine tolerances may also be compensated. This ultimately manifests itself in a considerably smaller safety margin for emission limits of particulates and NO$_X$ for the corresponding application. When the level of emissions is based on typical distributions of the tolerances of all the components involved, then fuel mean value adaptation can, at the least, halve the safety margin for a Euro 4 application over the original margin of emission limit values without fuel mean value adaptation.

5.3.6 Injection Systems for Large Diesel Engines

5.3.6.1 Field of Application

The range of injection system applications for large diesel engines covers
- cylinder outputs of 70 to 2,000 kW (in large low speed two-stroke diesel engines up to 4,500 kW),
- engine speeds from 60 to 1,800 rpm,
- full load injected fuel quantities from 180 to 20,000 mm³/injection,
- numbers of engine cylinders from 1 to 20 and
- a range of fuel from standard diesel fuel through heavy fuel with a viscosity of up to 700 cSt at 50°C.

Fundamental aspects of the development, engineering and manufacturing of injection systems for large diesel engines include:
- high reliability and long service life,
- large proportion of full load operation,
- easy interchangeability with existing injection systems,
- applicability to any cylinder configurations,
- ease of servicing,
- fuel compatibility,
- competitive costs when manufacturing small quantities and
- controllability of all injection parameters to comply with emission standards.

5.3.6.2 Conventional Injection Systems

Inline pumps with pump plunger diameters of up to 20 mm and pump plunger strokes of up to 15 mm are still used in large engines with cylinder outputs of up to approximately 160 kW. Flexible adaptation to differing numbers of cylinders is only possible with unit pumps. They allow the implementation of short and standardized injection lines.

The rotating injection cam mounted on the engine camshaft moves the pump plunger by a roller tappet (see Fig. 5-47) or valve lever kinematics.

Direct placement of the single cylinder pumps at their related cylinders allows extremely short injection lines. This improves system efficiency because the hydraulic losses are low. Identical components simplify adaptation to engine design and the number of cylinders as well as spare parts management.

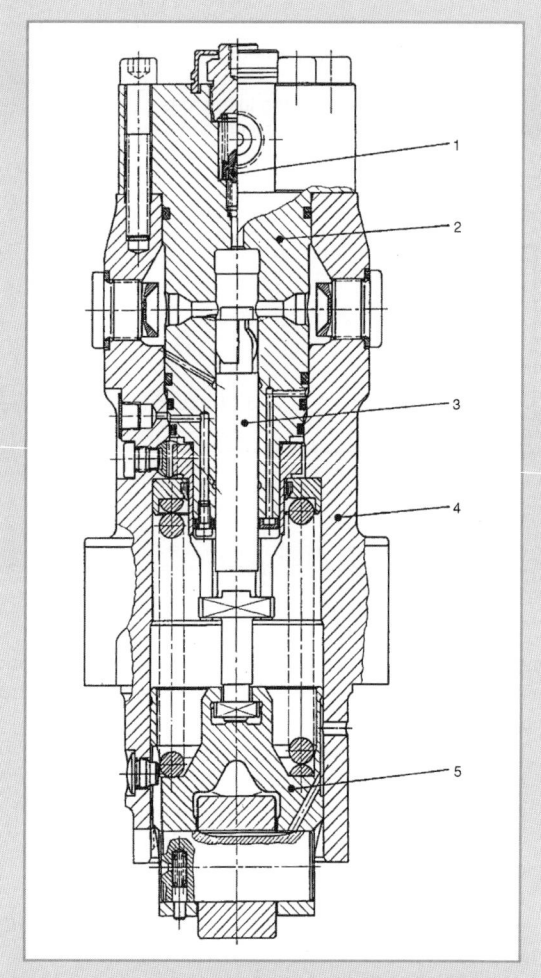

Fig. 5-47 Single cylinder pump with integrated roller tappet (Bosch PFR1CY). *1* Constant pressure valve; *2* Barrel; *3* Plunger; *4* Housing; *5* Roller tappet (= 4–35, 2nd Ed.)

The unit pump illustrated in Fig. 5-47 has a pump barrel with a sac hole design, which supports pump side peak pressures of up to 1,500 bar. A constant pressure valve is mounted in the deformation-resistant flange section of the pump barrel. It prevents the nozzle from post-injecting by rapidly dissipating the pressure after the end of delivery and also keeps the pressure in the injection line high to prevent cavitation in the system. This design avoids large-area high pressure sealing.

Unit pumps for heavy fuel operation usually have three ring grooves incorporated in the pump barrel below the spill port, which perform different functions. As in all diesel applications, the upper groove returns leak oil to the inlet chamber. Lubricating oil from the engine oil circuit, which acts as sealing oil and prevents the fuel from diluting the engine oil, is fed to the bottommost groove through a fine filter. A groove for pressureless removal of compound oil consisting of fuel and lubricating oil is placed in between.

The drive cams for single cylinder pumps are mounted on the same camshaft as the cams for the engine's valve gear. Therefore, cam phasing shared with the drive gear may not be used to shift injection timing. Adjusting an intermediate element, e.g. a roller rocker mounted eccentrically between the cam and roller tappet, can produce an advance angle of a few degrees. Thus, consumption and emission can be optimized or even adapted to the differing ignition quality of various types of fuel. The high structural complexity of this solution led to the development of solenoid controlled pumps.

5.3.6.3 Solenoid Controlled Pumps

Figure 5-48 presents a solenoid controlled single cylinder pump (unit pump). The injection cam on the engine camshaft drives the pump plunger with a roller tappet. A short high pressure line establishes the connection to the nozzle holder assembly in the engine cylinder. These pumps are utilized in high speed ($> 1,500$ rpm) and medium speed engines (rated speed $< 1,200$ rpm) and have pump plunger diameters of 18–22 mm and piston strokes of between 20 and 28 mm.

The pump plunger has neither helices nor grooves; the pump barrel has no spill ports. A pressure valve is installed instead of a solenoid valve with a control piston.

This system's advantages are its:
- smaller installation space,
- good suitability for conventional cylinder head designs,
- stiff drive,
- rapid switching times and precise fuel metering with freely selectable start of delivery,
- easy replacement during repair and servicing and
- option of cylinder shutoff in part load operation.

5.3.6.4 Common Rail Systems

Until recently, speed and load-dependent injection timing was inessential in large high speed and medium speed engines. Marine engines have a fixed speed and load correlation. Generator engines run at constant speed with variable load. Thus, both applications allow employing an upper helix in the pump plunger of conventional pumps to adjust load-dependent start of delivery.

More stringent exhaust emission regulations partly in force in 2006 require the use of new injection concepts even in large diesel engines. The common rail system established in the commercial vehicle sector has advantages that current large diesel engine applications cannot do without. However, the requirements summarized below have to be observed even more strictly than in systems for vehicle engines:

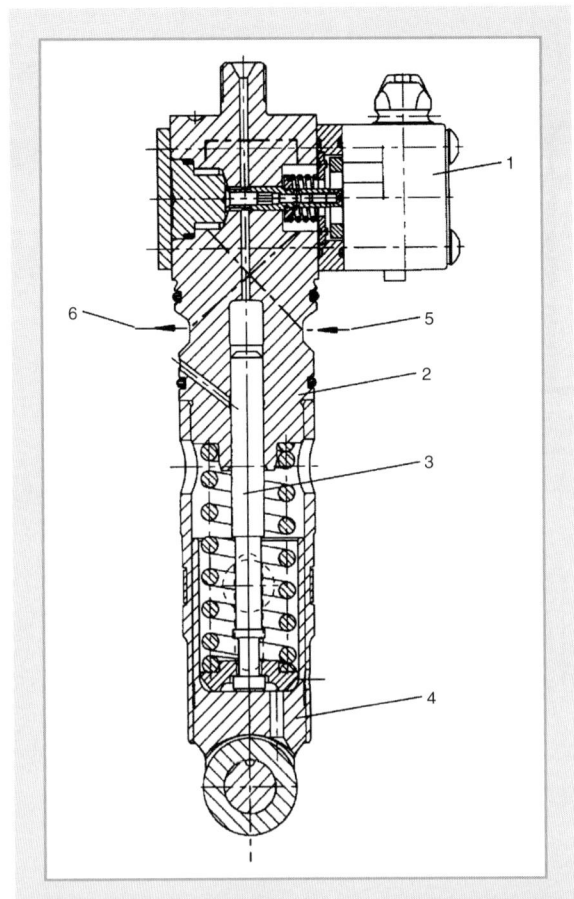

Fig. 5-48 Single cylinder pump with solenoid valve (Bosch PFR1Z). *1* Solenoid valve; *2* Barrel; *3* Plunger; *4* Roller tappet; *5* Fuel feed; *6* Fuel discharge (= 4–36, 2nd Ed.)

– speed-independent, freely controllable system pressure of up to 2,000 bar,
– higher efficiency and thus low pump drive power at low peak torque,
– higher pressure before the spray hole through low pressure losses in high pressure lines, valves and in the nozzle,
– option of multiple injection,
– injection rate shaping,
– low variation from injector to injector, low injected fuel quantity drift throughout the service life,
– high reliability and long service life for life cycles with a high proportion of full load,
– stability with variable fuel quality and contamination,
– easy interchangeability with conventional systems replaced in especially long-lived large engines,
– easy servicing and repair options and
– particularly reliable electronics with redundance and limp-home functions.

System pressures in current applications are 1,400–1,600 bar and 1,800 bar for special applications such as recreational boats. Measures that reduce nitrogen oxides, e.g. exhaust gas recirculation or injection rate shaping, require high injection pressures to compensate for the disadvantages in terms of particulate emission and fuel consumption. The better atomization through increased fuel pressure definitely reduces particulate emission. The pressure increase is connected with higher energy consumption to drive the high pressure pump. Thus, leakage losses, throttling losses and heat generation must also remain low to prevent high losses. Should this not function, the higher drive energy would eliminate the consumption advantages from optimized combustion.

Increasing the system pressure only has the desired effect when it also increases the mean injection pressure and the maximum spray hole pressure. In order to accomplish this, the following aspects have to be observed:
– The injector's design has to be optimized for low throttling resistances and to eliminate regions critical for cavitation.
– Throttling in the region of the nozzle seat has to be kept low during the opening and closing of the nozzle needle.

Thus, the criterion of a high quality injection system is not maximum nominal pressure delivered by the high pressure pump to the rail but rather uniform and high pressure directly at the spray hole, which alone is responsible for appropriately atomizing the fuel for combustion.

Multiple injections are an effective means to reduce emissions. Post-injection reduces particulate emission. Pilot injection moderately increases cylinder pressure with low noise emission. Injectors with multiple injections must be able to inject consistently small quantities of fuel without introducing pressure peaks in the high pressure system.

Pressure controlled injection systems are only able to shape the injection rate by influencing the speed of needle opening. Injection rate shaping largely independent of the injection pressure requires a complex multiple actuator injector design.

Manufacturing costs are a fundamental criterion for the selection of a concept, even for injection systems for large diesel engines. Systems that allow retrofitting long-lived engines to comply with more stringent emissions specifications have cost advantages because larger quantities are manufactured.

5.3.6.5 Common Rail Systems for Diesel Fuel

Pressure amplified systems with one or two actuators are in development for large high speed diesel engines with cylinder outputs of up to 150 kW. They offer advantages for injection rate shaping and multiple injection. Their variance, fuel quantity drift, efficiency and, in particular, cost are disadvantages. The increase in the length of the pressure line from the rail to the injector as engine dimensions increase has an adverse effect. This is the reason for the development of the

Fig. 5-49 Modular common rail system

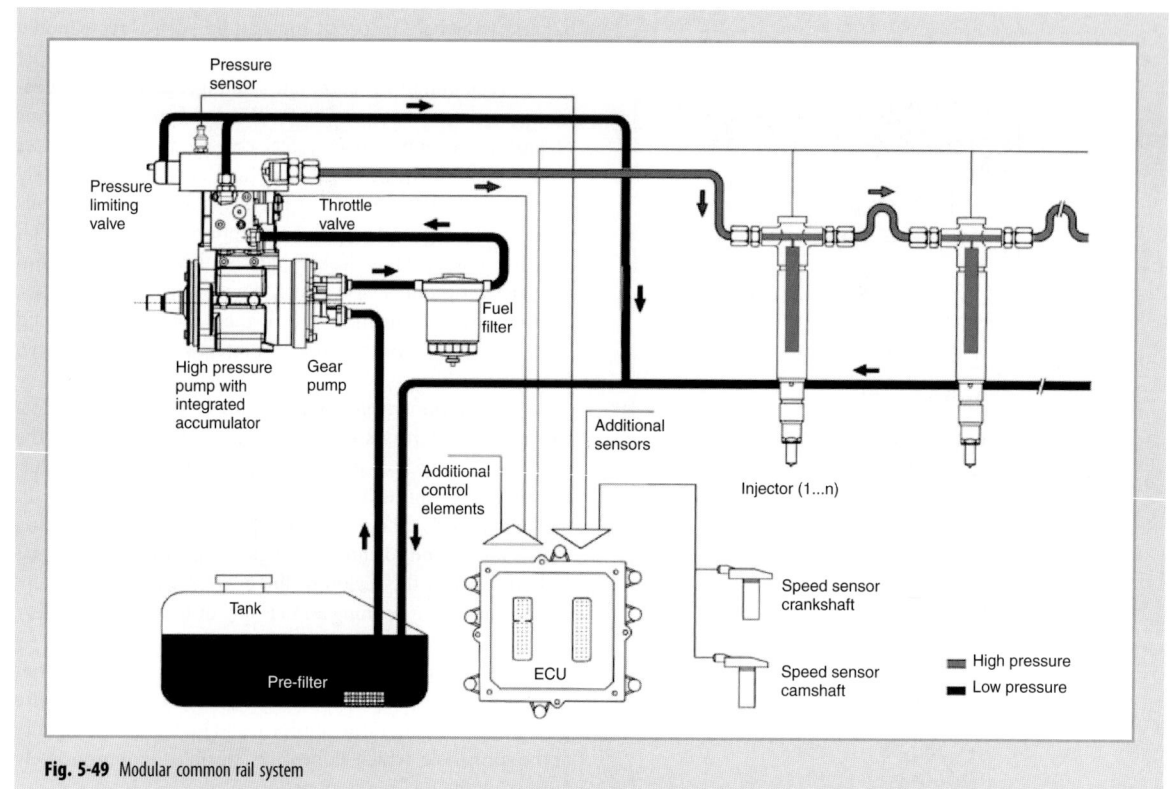

concept of a modular common rail system (see Fig. 5-49) with an injector design based on Fig. 5-50.

A high pressure pump with a multiple cylinder design is driven by the engine and contains a relatively small accumulator volume (see Fig. 5-51).

Delivery is controlled by any inlet metering valve controlled by an ECU. A pressure limiting valve and the pressure sensor are integrated in the high pressure pump. A high pressure line leads from the high pressure pump's accumulator to the individual injectors, each of which has an integrated accumulator chamber. The short distance between the accumulator chambers and the solenoid valve allows multiple injections of minimum quantities of fuel. A rugged design with low wear at the solenoid valve and nozzle seat reduces the fuel quantity drift throughout the period of operation. Large sealing lengths in the pump elements and low throttling losses in the valves deliver high efficiency during pressure generation. Pumps and plungers may be retrofit since the installation space is available.

Conventional systems in medium speed engines with cylinder outputs of up to 500 kW may be replaced by common rail systems. The modular common rail system pictured in Fig. 5-52 is advantageous and cuts costs.

Single plunger pumps generate the high pressure. The pressure peaks from low frequency pressure generation are damped in a larger accumulator, which has a pressure sensor and a pressure limiting valve. A shared controlled suction throttle valve controls the single plunger pumps' delivery. Connected in a series, the individual injectors' accumulator volumes allow equalizing the pressure between the injectors. This system can significantly improve particulate and noise emissions. Retrofitting costs are low.

A broad range of fuel, including heavy fuel of up to 700 cSt at 50°C, may be employed in medium speed engines, especially marine engines. Demands for low fuel consumption, reduced exhaust emissions and improved running smoothness are not only made by users but also appear in IMO emission regulations. Additional requirements include the usability of different grades of fuel with high fractions of impurities and high temperatures of up to 180°C, the option of cylinder shutoff and cylinder balancing control based on fuel temperature or the injector. Obviously, a common rail system for heavy fuel would fulfill these requirements outstandingly. Its trouble-free adaptation to different engine sizes and numbers and configurations of cylinders is an important aspect.

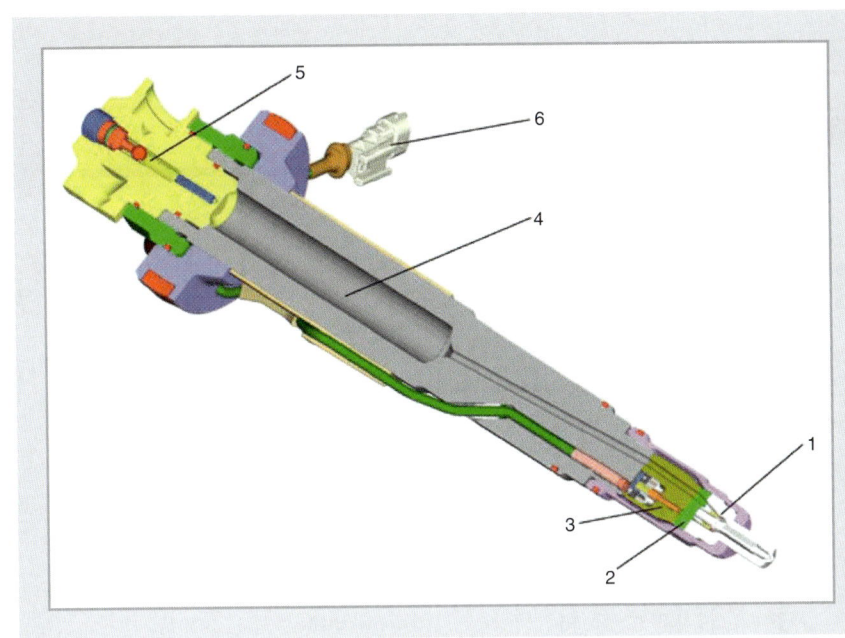

Fig. 5-50
Injector for a modular common rail system.
1 Nozzle; *2* Orifice plate; *3* Solenoid valve; *4* Injector body with accumulator; *5* Flow limiter; *6* Electrical connector

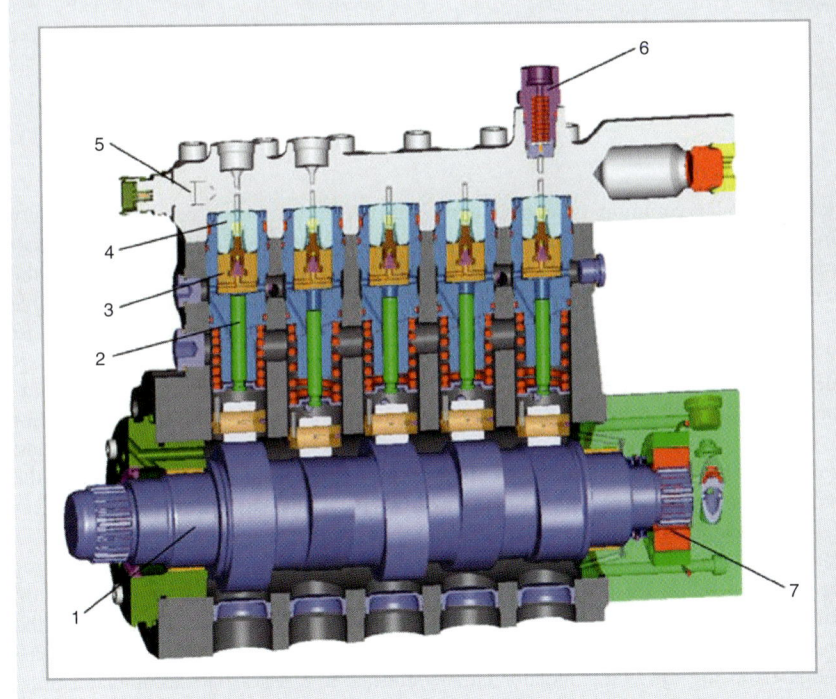

Fig. 5-51
High pressure pump for a modular common rail system. *1* Camshaft; *2* Pump element; *3* Suction valve; *4* Pressure valve; *5* Accumulator; *6* Pressure limiting valve; *7* Low pressure pump

Close interaction between the manufacturers of engines and the manufacturers of injection systems is indispensible for the development of accumulator injection systems for heavy fuel. It is the only way to develop and extensively test an effective concept. High fuel temperatures and impurities require new materials and structural concepts with which even the requisite long service life can be attained.

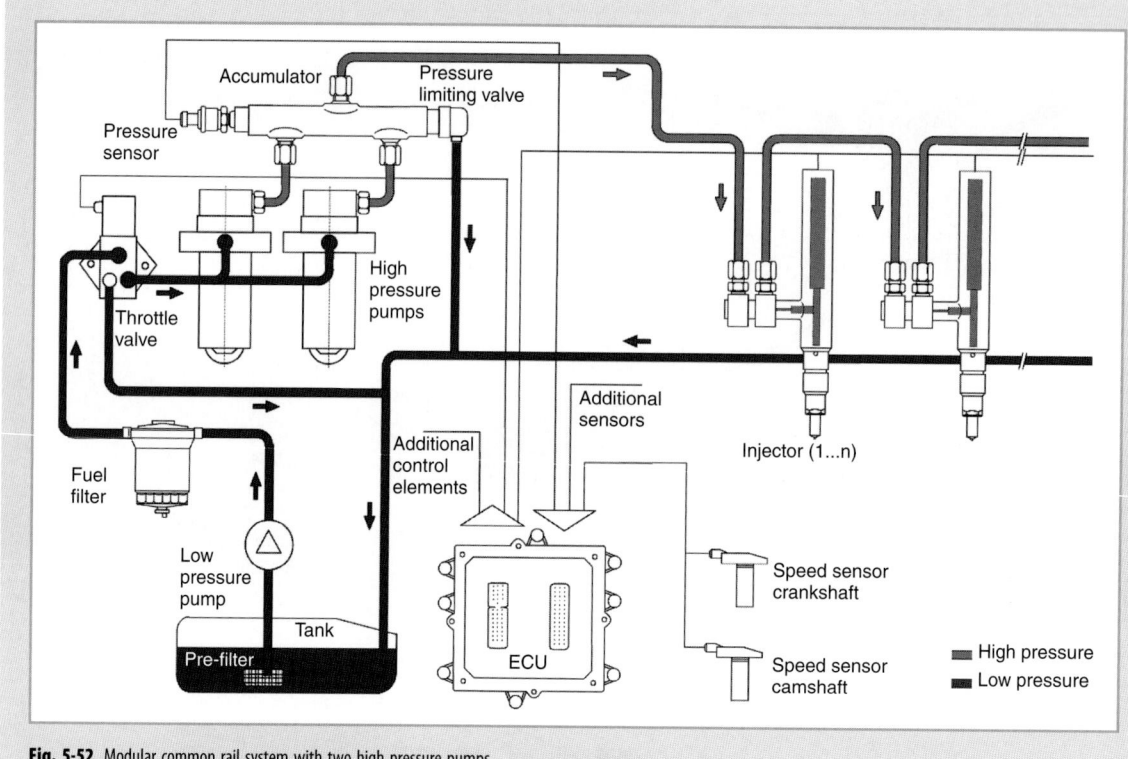

Fig. 5-52 Modular common rail system with two high pressure pumps

Figure 5-53 presents a retrofittable modular common rail system for heavy fuel, the distinctive features of which are described below. The fuel supply system is provided with a heating system that preheats the fuel to up to 160 °C. The injection system in the engine is initially charged to ensure that the diesel fuel is free of bubbles. A special scavenging system charges the fuel supply system with heavy fuel in another sequence before the engine commences operation.

Circulating preheated heavy fuel continually heats the high pressure pumps, pressure accumulators and solenoid valves during operation and during normal short engine shutoffs. However, whenever the engine is shut off longer, e.g. for maintenance, it is switched over to diesel operation for a short time beforehand. This consumes the heavy fuel in the injection system and then charges the system with low viscosity diesel fuel. After an emergency engine stop, the injection system is purged with diesel fuel and only the high pressure lines to the injectors and the injectors themselves contain a charge of heavy fuel. This heavy fuel plug can be purged later when the engine is started.

The considerable dimensions of a large diesel engine make the installation of a single pressure accumulator along the entire length of the engine problematic. Since it is virtually impossible to obtain identical injection conditions for every engine cylinder, excessive pressure oscillations in the system have to be prevented. Thus, it is more practical to divide the accumulator unit into several physical units with a suitable accumulator volume and to divide the supply among at least two high pressure pumps.

One advantage of such a configuration is the greater flexibility when assembling an injection system for different engine configurations. This is of interest for retrofit solutions. More compact physical units enable better utilization of the structural space available in the engine and yield advantages for assembly and spare parts management.

A low pressure fuel pump delivers the fuel through electromagnetically controlled throttle valves to two high pressure pumps, which force the fuel through pressure valves into the pump accumulator whence the fuel is conducted to the accumulator units connected in a series. The accumulator units consist of a massive section of pipe with an accumulator cover attached to each of the two end faces, sealing them. The accumulator cover contains radial ports for the high pressure lines leading to the injectors and for the connecting line to the next accumulator unit.

Interposing the pump accumulator supplied by two to four high pressure pumps can keep dynamic pressure oscillations

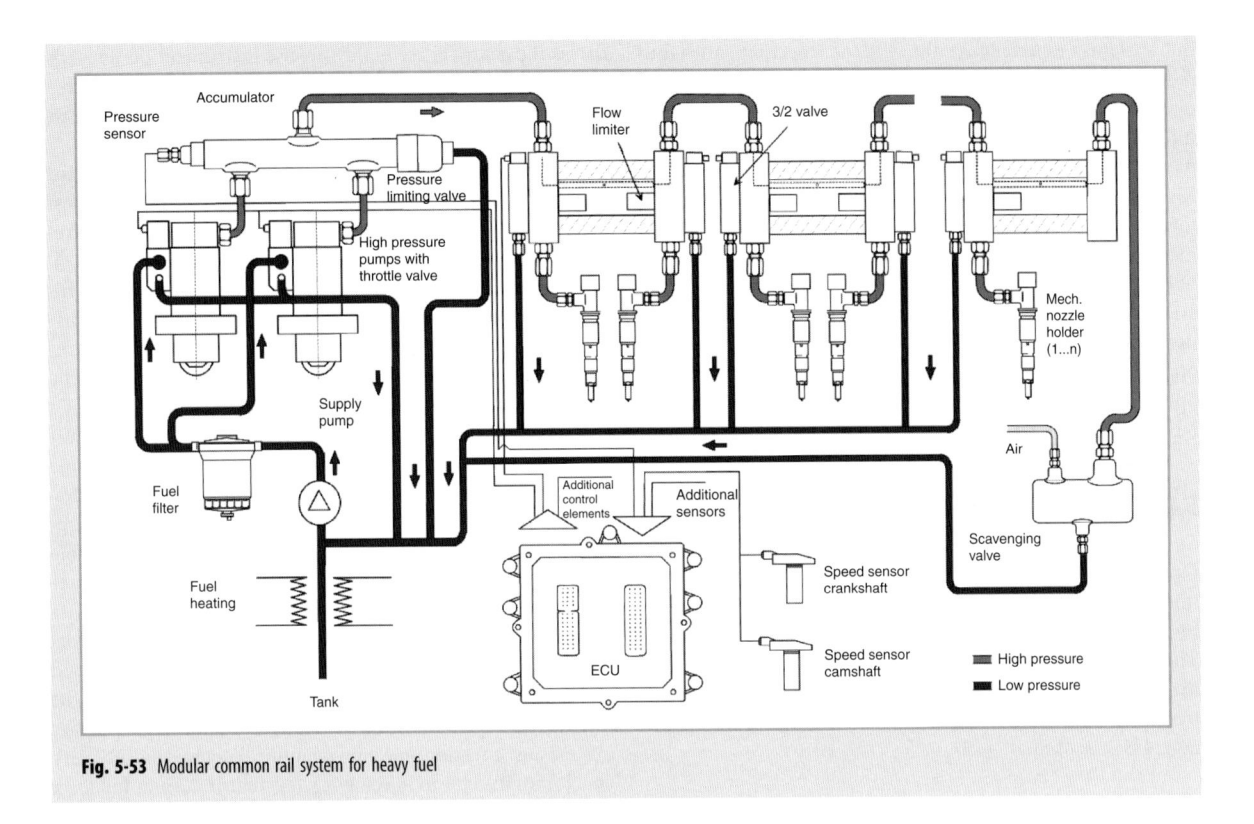

Fig. 5-53 Modular common rail system for heavy fuel

low. The electronic control unit calculates the high pressure pumps' delivery rate by analyzing the fuel pressure reported by the rail pressure sensor and the engine's particular operating state. The electromagnetically controlled throttle valve in the low pressure line measures the quantity of fuel supplied to the high pressure pumps.

Each accumulator cover contains components and ports, which facilitate the supply and transport of fuel and the control of fuel injection in the injectors. The fuel is conducted through a flow limiter on its way from the interior of the accumulator to the 3/2 way valve and from there to the injector. A spring loaded piston in this component completes a stroke proportional to the injected fuel quantity for every injection and returns to its original position when there is a pause in injections. However, when the injected fuel quantity ought to exceed a defined limit value, then the piston is pressed against a sealing seat on the outlet side at the end of its stroke, thus preventing continuous injection in the injector.

Each accumulator cover also contains a 3/2 way valve actuated by a 2/2 way valve that is electromagnetically controlled by the control unit and thus clears the path for the high pressure fuel from the accumulator unit to the injector through the flow limiter. Activating the 3/2 way valve multiple times in the course of the injection process produces pilot and post-injections.

A pressure limiting valve that opens when a defined pressure is exceeded and thus protects the high pressure system from overloading is mounted on the pump accumulator.

The high pressure lines and accumulator units are constructed with double walls so that no fuel can escape when connections typically leak, break or lose their seal. Float switches warn operators in such a case.

The fuel supply system is equipped with a heating system to preheat the heavy fuel. To start a cold combustion engine with heavy fuel, the injection system's high pressure stage is heated with hot circulating heavy fuel. To depressurize it, a pneumatic control opens the scavenging valve placed at the end of the last of the accumulator units connected in a series. Once the injection system has been sufficiently heated, the scavenging valve is closed and the engine started.

The scavenging valve also serves to relieve the pressure of the injection system's high pressure stage for maintenance or repairs. Delivering the high pressure fuel to the pump accumulator through two high pressure pumps is advantageous because it enables an engine to continue operating at part load if one of the two pumps fails.

The arrangement of the 3/2 way valve in the accumulator cover and the utilization of a conventional injector substantially simplify retrofitting of an existing type of engine. This eliminates the pressure oscillations that occur in the high pressure line between the rail and injector in other common rail systems, especially at the end of injection, and thus reduces the stress on the pressurized components. The modularization of the physical units and their allocation to the individual engine cylinders reduces assembly and maintenance work and allows short line lengths to the injectors. Engine customers already value the use of mechanical injectors. Evidently, it will take a few years until ship engineers and mechanics fully place their trust in the electronics.

Low cost retrofittable heavy fuel common rail injection systems for medium speed large diesel engines have been in use in lab engines since 2003 and several field engines with operating times of several thousand hours. The first steps of optimization have already significantly improved smoke and NO_X emissions over mechanical systems.

5.4   Injection System Metrology

Optimization and assessment of the quality of modern injection systems and system components requires highly developed measurement and testing systems. The following overview describes the basic aspects of the technology utilized to test an injection system's hydraulic function. Standard test oil compliant with ISO 4113 is used.

5.4.1   Measurement Principles and Their Application

Flow measurement: Measuring instruments operated in a closed line system based on the gear principle that continuously measure the flow in injection system components are widespread. Detected by a sensor coupled with the gear pair, the gear speed correlates with the fluid's volumetric flow. This enables determining the quantity of fuel the high pressure pumps return in the low pressure circuit. Common flow measurements are 5–150 l/h. Instrument designs that operate between the gear pair's inlet and outlet without differential pressure are also in widespread use. A controlled servomotor drives the gears. These highly precise instrument designs can be applied to detect both the smallest leaks and flows with a minimum value of 0.01 l/h and (in other designs) large flows of 350 l/h.

Another flow measurement system in use is based on the Coriolis principle. A measuring tube connected with a test bench's line system is excited to vibrate at its natural frequency. The Coriolis force's effect on the flowing test oil causes a phase shift of the two tube ends' vibrations. This constitutes a quantity for the instantaneous flow. The distinguishing feature of this continuously operating measurement system is its particularly simple design with little susceptibility to faults. Its measuring time is in a range of a few seconds. The Coriolis measurement system is predominantly employed in production testing, e.g. to determine high pressure pumps' delivery rates and injection nozzles' throughputs. At ± 0.1% of the measured value, the measurement uncertainty of individual commercially available instruments based on the flow measurement principles presented is already very small.

Measurement of the injected fuel quantity: The most important hydraulic quantity measured in an injector is the injected fuel quantity per injection cycle. Since irregularities in the injected fuel quantity directly influence output engine power and exhaust emission, injection quantity tolerances must be very small from injector to injector, e.g. less than ± 2.5% at full load. Therefore, particularly high demands are made on the absolute accuracy and resolution of measuring equipment used to measure the injected fuel quantity. Fuel quantity indicators that employ a measuring principle consisting of a chamber with a sliding piston have proven their value to measure flows in product development and quality testing [5-22]. The injected fuel quantity, which may be dispersed among several partial injections per injection cycle, is injected into the measuring chamber filled with test oil and displaces a moving piston, the stroke of which is determined with an inductive displacement measuring device (Fig. 5-54). The measuring chamber is pressurized with a defined back pressure to suppress piston

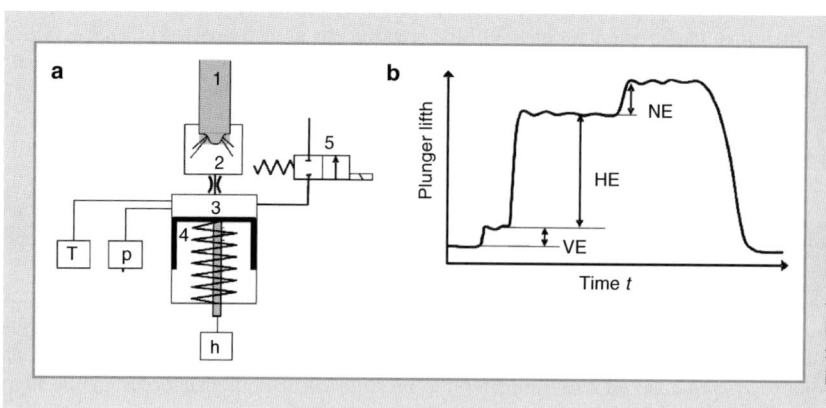

Fig. 5-54
(a) Diagram of the piston displacement principle. *1* Injection system; *2* Spray damper; *3* Measuring chamber; *4* Plunger; *5* High pressure valve; (b) Graph of the plunger stroke signal during pilot (VE), main (HE) and post-injection (NE)

oscillations and prevent the release of air. At the end of the injection cycle, the measuring chamber is evacuated by means of a valve. The injected fuel quantity $m(\rho, h)$ can be easily calculated:

$$m(\rho, h) = \rho(T, p) \cdot h \cdot A_{\text{Kolben}} \tag{5-12}$$

The test oil density $\rho(T, p)$ in (5.12) is a function of the chamber temperature T and the back pressure p in the measuring chambers. h is the piston stroke and A_{Kolben} the piston area.

State-of-the-art measuring instruments that operate based on the principle of piston displacement have a measurement uncertainty of $\pm\,0.1\%$ of the measured value. These devices' measuring range is matched to car and commercial vehicle applications and lies between 0.2 and 600 mm³/injection with a resolution of 0.01 mm³. The specified minimum injection time interval in which two consecutive partial injections can be detected separately is 0.25 ms. Up to ten partial injections can be measured per injection cycle.

Measurement of the injection characteristic: The injection characteristic in the injector nozzle is obtained by differentiating the injected fuel quantity according to time during the injection cycle. In principle, this can be done with the injection quantity measuring systems described above. However, given an oscillating piston's dynamic properties, this method may only be applied conditionally to take temporally highly resolved measurements of the injection characteristic. A measuring instrument based on the principle of the tube indicator is better suited to determine the injection characteristic [5-23]. The quantity of test oil the injector nozzles inject into a tube over time generates a pressure wave that propagates at the speed of sound in the already filled tube. A pressure sensor captures the dynamic increase of pressure $p(t)$ in the tube. The injection characteristic dm/dt is calculated with the tube's cross sectional area A_{Rohr} and the temperature and pressure-dependent speed of sound c in the test oil:

$$\frac{dm}{dt} = \frac{A_{\text{Rohr}}}{c} p(t) \tag{5-13}$$

The injected fuel quantity is calculated by integrating (5.13). Current instrument designs based on the tube indicator principle specify a flow measurement range of nearly 0–150 mg/injection with a resolution of the injected fuel quantity of 0.01 mg/injection. Up to five partial injections are standardly detectable. Instruments based on the tube indicator principle are used in injector development.

Unlike the tube indicator, which injects fuel into a long tube loop to measure the injection characteristic, the measuring principle of the hydraulic pressurization system is based on injection into one chamber with a constant measurement volume. This causes the pressure in the chamber to rise. The characteristic of the injection rate and the injected quantity can be calculated from the change of pressure over time, [5-22]. In current instrument designs, the measuring chamber is filled with test oil and pressurized to prevent the release of dissolved air and cavitation. The injection characteristic dm/dt is calculated with the aid of the chamber volume V_{Kammer}, the speed of sound c and the absolute pressure change in the chamber dp/dt (5-14):

$$\frac{dm}{dt} = \frac{V_{\text{Kammer}}}{c^2} \frac{dp}{dt} \tag{5-14}$$

Distinguished by its very short response time, a highly accurate piezo pressure sensor measures the pressure in the chamber. An ultrasonic transducer is employed to calculate the speed of sound from the echo time of one sonic pulse in the measuring chamber. Such hydraulic pressurization systems are applied in injector development to measure the injection characteristic and quantity simultaneously.

5.4.2 Demands on Metrology

The demands on injection system measurement and test equipment must be considered separately for their two fields of application: development and production. As a rule, development places high demands on both the temporal and local resolution of the test signals and the scope of the test results. Test bench tests serve to experimentally verify the theoretical potentials of new injection systems. To this end, any measuring device used in development must be as flexible and versatile as possible. In contrast to its flexible use in development, test equipment in production requires extensive automation and standardization of both the measuring devices and the measuring processes and sequences. This is necessary to ensure that the manufactured quality of injection system components is consistently high (see Sect. 5.4.3).

Measuring process capability: Regular testing of a measuring process's capability and monitoring of its stability is intended to ensure that a measuring device is able to dependably take measurements of a characteristic of a product, e.g. injected fuel quantity, at its place of operation over a long time with a sufficiently small scattering of measured values relative to the tolerance. The automotive industry formulated criteria [5-24] that help evaluate measuring process capability. Several measuring devices implemented in mass production to inspect products must have similar structural designs and testing sequences to ensure that they all have identical test boundary conditions for a product.

5.4.3 Injector Testing

Measuring device for injector testing: During manufacturing, special measuring devices test the precise injection and return quantities required of a completely assembled injector. In and of themselves, such measuring devices already constitute a complete common rail system. This guarantees maximum transferability of an injector's functional performance on a test bench to its later performance in an engine. While an injection system is usually adapted to a specific engine, the measuring devices' common rail components are always identical for one generation of injectors (see Fig. 5-55).

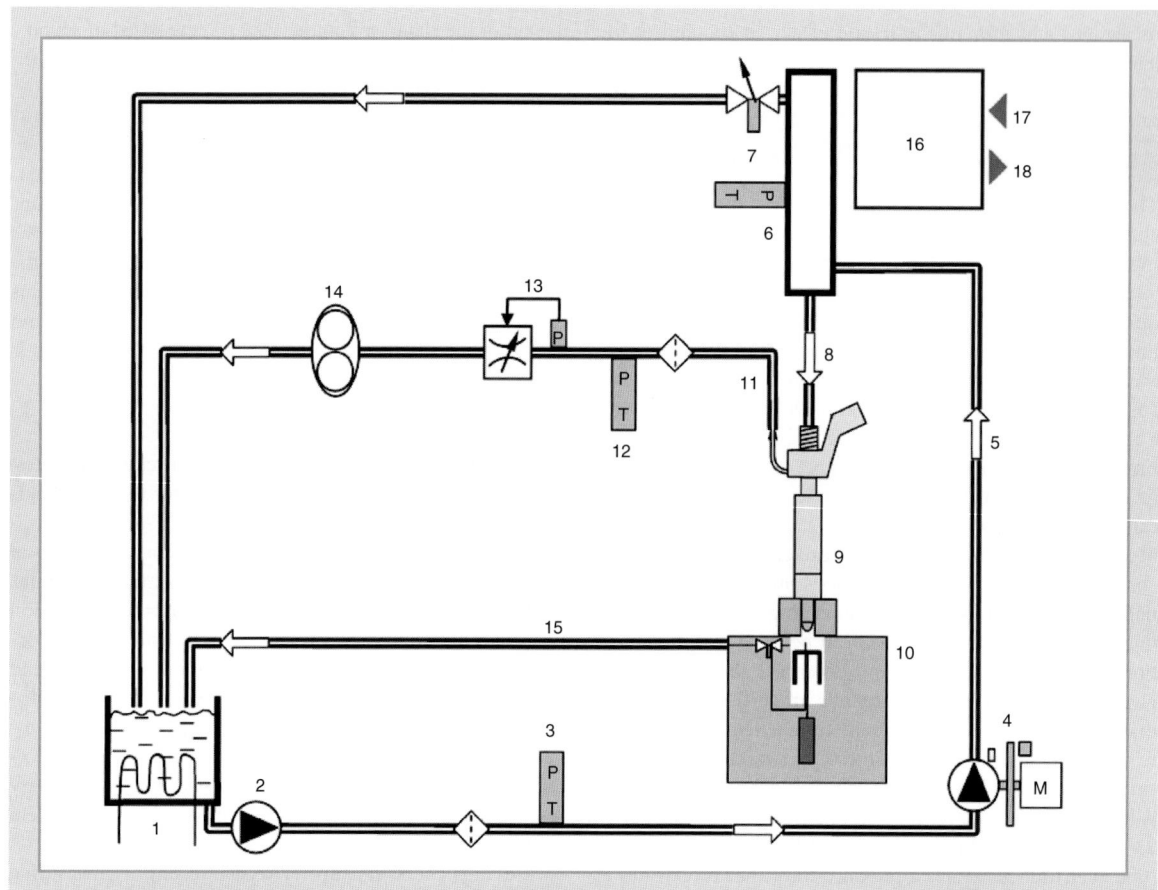

Fig. 5-55 Schematic measuring device for injector testing. *1* Test oil tank with temperature control; *2* Low pressure supply pump; *3* Sensors for inlet pressure and temperature; *4* CR high pressure pump with drive and synchronizer; *5* Flexible high pressure line to the rail; *6* Rail with sensors for pressure and temperature; *7* Pressure control valve with return line; *8* High pressure line to the injector; *9* Common rail injector (here, a piezo CRI); *10* Injected fuel quantity measuring device based on the principle of plunger displacement; *11* Low pressure return from the injector; *12* Sensors for return pressure and temperature; *13* Pressure regulator for low pressure return; *14* Flowmeter based on the principle of a gear; *15* Return of the injected quantity of fuel; *16* Test bench control; *17* Sensors and data input; *18* Control and data output

Injector test sequence: An injector's delivery of the injected fuel quantity per operating point is tested according to a defined test sequence. The required mechanical and electrical contact of the injector with the measuring device is established fully automatically to ensure that installation conditions are always identical and to obtain short cycle times. The test sequence following this has to be optimized so that, on the one hand, the specified features can be measured with sufficient accuracy in short time and, on the other hand, stable test boundary conditions are assured. Injection and return quantities are usually measured in several characteristic injector operating points, e.g. idle, part load, full load and pilot injection. The absolute pressure in the rail and the duration of the injector's electric control in the operating point tested must be matched to the particular engine type.

The low pressure circuit in the injector must be bled completely at the beginning of the test sequence. Afterward, the entire system warms up until it reaches a stable state. The test sequence specifies the time sequences of the subsequent measurements, which are identical for all injectors. A function of the operating point, the thermal energy input to the injector and the flowmeter is in part considerable. At high inlet pressures, temperatures of up to 150°C develop inside the injector and the measuring device. As a rule, the times required for the requisite buildup of temperatures are many times longer than the actual measuring time, which is only a few seconds for each operating point. Apart from stable temperature boundary conditions for the measuring device, stable pressure conditions at the inlet, i.e. in the rail, are essential for accurate and reproducible measurements of the injected fuel quantity. The inlet pressure is controlled with the aid of highly accurate pressure sensors and signal amplifiers. When combined, they achieve an accuracy of ± 1 bar at an absolute pressure of 2,000 bar.

Literature

5-1 Bessières, D.; Saint-Guirons, H.; Daridon, J.: Thermophysical properties of n tridecane from 313.15 to 373.15 K and up to 100 MPa from heat capacity and density data. Journal of Thermal Analysis and Calorimetry 62 (2000), pp. 621–632

5-2 Bosnjakovic, F.; Knoche, K.F.: Technische Thermodynamik Part I, 8th Ed. Darmstadt: Steinkopff-Verlag (1999)

5-3 Davis, L.A.; Gordon, R.B.: Compression of mercury at high pressure. Journal of Chemical Physics 46 (1967) 7, pp. 2650–2660

5-4 Jungemann, M.: 1D-Modellierung und Simulation des Durchflussverhaltens von Hydraulikkomponenten bei sehr hohen Drücken unter Beachtung der thermodynamischen Zustandsgrössen von Mineralöl. Düsseldorf: VDI Fortschrittsberichte 473 (2005) 7

5-5 Cellier, F.E.; Kofman, E.: Continuous System Simulation. Berlin/Heidelberg/New York: Springer (2006)

5-6 Bianchi G.M.; Falfari S.; Parotto M.; Osbat G.: Advanced modeling of common rail injector dynamics and comparison with experiments. SAE-SP Band 1740 (2003) pp. 67–84

5-7 Chiavola O.; Giulianelli P.: Modeling and simulation of common rail systems. SAE-P Vol. P-368 (2001) pp. 17–23

5-8 Spurk, J.; Aksel, N.; Strömungslehre, 6th Ed. Berlin/Heidelberg/New York: Springer (2006)

5-9 Schmitt, T.; Untersuchungen zur stationären und instationären Strömung durch Drosselquerschnitte in Kraftstoffeinspritzsystemen von Dieselmotoren. Diss. Technische Hochschule München 1966, Forschungsberichte Verbrennungskraftmaschinen No. 58

5-10 International Symposium on Cavitation (CAV2006). Wagening: The Netherlands, September 11–15, 2006

5-11 Robert Bosch GmbH (Ed.): Technische Unterrichtung. Diesel-Verteilereinspritzpumpe VE. (1998)

5-12 Bauer, H.-P. et al.: Weiterentwicklung des elektronisch geregelten Verteilereinspritzpumpen-Systems. MTZ 53 (1992) 5, pp. 240–245

5-13 Eblen, E.; Tschöke, H.: Magnetventilgesteuerte Diesel-Verteiler-Einspritzpumpen. 14th Internationales Wiener Motorsymposium May 6–7, 1993. Düsseldorf: VDI Fortschrittberichte, Vol. 12, No. 182, Pt. 2. Düsseldorf: VDI-Verlag (1993)

5-14 Tschöke, H.; Walz, L.: Bosch Diesel Distributor Injection Pump Systems – Modular Concept and Further Development. SAE-Paper 945015, 25. FISITA-Congress, Beijing (1994)

5-15 Hames, R.J.; Straub, R.D.; Amann, R.W.: DDEC Detroit Diesel Electronic Control. SAE-Paper 850542 (1985)

5-16 Frankl, G.; Barker, B.G.; Timms, C.T.: Electronic Unit Injector for Direct Injection Engines. SIA Kongress, Lyon (France), June 1990

5-17 Lauvin, P. et al.: Electronically Controlled High Pressure Unit Injector System for Diesel Engines. SAE-Paper 911819 (1991)

5-18 Kronberger, M.; Maier, R.; Krieger, K.: Pumpe-Düse-Einspritzsysteme für Pkw-Dieselmotoren. Zukunftsweisende Lösungen für hohe Leistungsdichte, geringen Verbrauch und niedrige Emissionen. 7th Aachen Colloquium Automobile and Engine Technology (1998)

5-19 Maier, R.; Kronberger, M.; Sassen, K.: Unit Injector for Passenger Car Application. I. Mech. E., London (1999)

5-20 Egger, K.; Lauvin, P.: Magnetventilgesteuertes Steckpumpensystem für Nfz-Systemvergleich und konstruktive Ausführung. 13th Internationales Wiener Motorensymposium. VDI Fortschrittberichte 167, Vol. 12. Düsseldorf: VDI-Verlag (1992)

5-21 Maier, R. et al.: Unit Injector/Unit Pump – Effiziente Einzelpumpensysteme mit hohem Potential für künftige Emissionsforderungen. Dresdener Motorensymposium (1999)

5-22 Zeuch, W.: Neue Verfahren zur Messung des Einspritzgesetzes und der Einspritzregelmässigkeit von Diesel-Einspritzpumpen. MTZ (1961) 22/9, pp. 344–349

5-23 Bosch, W.: Der Einspritzgesetz-Indikator, ein neues Messgerät zur direkten Bestimmung des Einspritzgesetzes von Einzeleinspritzungen. MTZ (1964) 25/7, pp. 268–282

5-24 DaimlerChrysler Corporation, Ford Motor Company, General Motors Corporation: Measurement System Analysis (MSA) Reference Manual. 3rd Ed. (2002)

Further Literature

Tschöke H. et al. (Ed.): Diesel- und Benzindirekteinspritzung. Renningen: Expert Verlag (2001)

Tschöke H. et al. (Ed.): Diesel- und Benzindirekteinspritzung II. Renningen: Expert Verlag (2003)

Tschöke, H. et al. (Ed.): Diesel- und Benzindirekteinspritzung III. Renningen: Expert Verlag (2005)

Tschöke, H. et al. (Ed.): Diesel- und Benzindirekteinspritzung IV. Renningen: Expert Verlag (2007)

Tschöke, H. et al. (Ed.): Diesel- und Benzindirekteinspritzung V. Renningen: Expert Verlag (2009)

Robert Bosch GmbH (Ed.): Kraftfahrtechnisches Taschenbuch. 26th Ed. Wiesbaden: Vieweg (2007)

Robert Bosch GmbH (Ed.): Dieselmotor-Management. 4th Ed. Wiesbaden: Vieweg (2004)

van Basshuysen; Schäfer (Ed.): Handbuch Verbrennungsmotor. 4rd Ed. Wiesbaden: Vieweg (2007)

Dohle, U.; Hammer, J.; Kampmann, S.; Boecking, F.: PKW Common Rail Systeme für künftige Emissionsanforderungen. MTZ (2005) 67-7/8, pp. 552 ff.

Egger K.; Klügl, W.; Warga, J.W.: Neues Common Rail Einspritzsystem mit Piezo-Aktorik für Pkw-Dieselmotoren. MTZ (2002) 9, pp. 696 ff.

Dohle, U.; Boecking, F.; Gross, J.; Hummel, K.; Stein, J.O.: 3. Generation Pkw Common Rail von Bosch mit Piezo-Inline-Injektoren. MTZ (2004) 3, pp. 180 ff.

Bartsch C.: Common Rail oder Pumpedüse? Dieseleinspritzung auf neuen Wegen. MTZ (2005) 4 pp. 255 ff.

Kronberger, M.; Voigt, P.; Jovovic, D.; Pirkl, R.: Pumpe-Düse-Einspritzelemente mit Piezo-Aktor. MTZ (2005) 5, pp. 354 ff.

Further Literature on Section 5.2

Bonse, B. et al.: Innovationen Dieseleinspritzdüsen – Chancen für Emissionen, Verbrauch und Geräusch. 5th Internationales Stuttgarter Symposium February 2003, Renningen: Expert Verlag (2003)

Gonzales, F.P., Desantes, J.M. (Ed.): Thermofluiddynamic processes in Diesel Engines. Valencia: Thiesel (2000)

Potz, D.; Christ, W.; Dittus, B.; Teschner, W.: Dieseldüse – die entscheidende Schnittstelle zwischen Einspritzsystem und Motor. In: Dieselmotorentechnik. Renningen: Expert Verlag (2002)

Blessing, M.; König, G.; Krüger, C.; Michels, U.; Schwarz, V.: Analysis of Flow and Cavitation Phenomena in Diesel Injection Nozzles and its Effects on Spray and Mixture Formation. SAE-Paper 2003-01-1358

Urzua, G.; Dütsch, H.; Mittwollen, N.: Hydro-erosives Schleifen von Diesel-Einspritzdüsen. In: Tschöke; Leyh: Diesel- und Benzindirekteinspritzung II. Renningen: Expert Verlag (2003)

Winter, J. et al.: Nozzle Hole Geometry – a Powerful Instrument for Advanced Spray Design. Konferenzband Valencia: Thiesel (2004)

Harndorf, H.; Bittlinger, G.; Drewes, V.; Kunzi, U.: Analyse düsenseitiger Massnahmen zur Beeinflussung von Gemischbildung und Verbrennung heutiger und zukünftiger Diesel-Brennverfahren. Konferenzband Internationales Symposium für Verbrennungsdiagnostik, Baden-Baden (2002)

Bittlinger, G.; Heinold, O.; Hertlein, D.; Kunz, T.; Weberbauer, F.: Die Einspritzdüsenkonfiguration als Mittel zur gezielten Beeinflussung der motorischen Gemischbildung und Verbrennung. Konferenzband Motorische Verbrennung. Essen: Haus der Technik (2003)

Robert Bosch GmbH (Ed.): Bosch Technical Instruction: Distributor-Type Diesel Fuel-Injection. Stuttgart: Robert Bosch GmbH (2004)

ETAS – Development tools that move you forward

ETAS Group
Automotive LifeCycle Solutions

ETAS Group

Whether you're looking to shorten your development times or reduce your warranty costs – the ETAS Group has an answer.

We supply a comprehensive portfolio of standardized development and diagnostic tools that cover the entire life cycle of electronic control units, from development to operational use and service.

For many years customers have relied on ETAS as a trusted partner for embedded ECU development.

All over the world ETAS offers tools for model-based design, prototyping, test, validation and calibration, as well as workshop diagnostic tools. INCA, ASCET, or LABCAR are well known examples. Custom solutions provided by ETAS engineering services complete the portfolio.

Depend on tools and expertise by a strong partner – ETAS

ETAS GmbH
Borsigstraße 14
70469 Stuttgart, Germany

Phone +49 711 89661-0
Fax +49 711 89661-106
sales.de@etas.com

www.etas.com

ETAS

6 Fuel Injection System Control Systems

Ulrich Projahn, Helmut Randoll, Erich Biermann, Jörg Brückner, Karsten Funk, Thomas Küttner, Walter Lehle, and Joachim Zuern

6.1 Mechanical Control

6.1.1 Functions of a Mechanical Speed Governor

Rugged and easily serviceable, mechanical governors continue to be used all over the world, especially in off-highway applications and nonroad engines. An inline pump illustrates the basic functions of mechanical control.

The hallmark of closed loop control is *feedback* of a *controlled variable* to the *actuated variable*, e.g. the injected fuel quantity specified by the setting of the inline pump's control rack. Increasing the injected fuel mass at a constant load causes the speed to increase. In turn, the centrifugal force acting on the control device also increases and reduces the amount of rack travel. This creates a closed loop control circuit.

The basic function of every governor is to limit the maximum speed to prevent a diesel engine from exceeding its allowable maximum speed. Other functions of a mechanical speed governor are:
- providing the starting fuel quantity,
- controlling idle speed,
- maintaining a specified speed at various engine loads and
- adjusting the torque characteristic with torque control or auxiliary units.

6.1.1.1 Proportional Degree

When a diesel engine's load is reduced without varying the accelerator pedal's position, the speed in the control range may only increase by a quantity permitted by the engine manufacturer. The increase in speed is proportional to the change in load, i.e. the more the engine load is reduced, the more the speed changes. This is called proportional behavior or proportional degree. The proportional degree is defined as the difference between maximum no load speed n_{Lo} and maximum full load speed n_{Vo} divided by maximum full load speed n_{Vo} (rated speed):

$$\delta = 100 \frac{n_{LO} - n_{Vo}}{n_{Vo}} \text{ in } \%.$$

Common proportional responses are approximately 0–5% for generator engines and approximately 6–15% for vehicle engines.

6.1.2 Design and Function of a Mechanical Speed Governor

An articulated rod connection establishes a connection with the injection pump's control rack. RQ and RQV(K) governors (Fig. 6-1) have two flyweights that act directly on the governor springs integrated in the control device, which are designed for the desired rated speed, proportional degree and idle speed. Increasing quadratically with the speed, the centrifugal forces of the rotating centrifugal weights are counteracted by elastic forces from the governor springs. The control rack's setting corresponds to the particular deflection of the flyweights produced by the biasing of the governor springs resulting from the centrifugal forces.

6.1.2.1 Governor Designs

An RQ idle speed-maximum speed governor is described as an example of governor design. Usually, diesel engines for vehicle applications do not require any control in the speed range between idle and maximum speed. In this range, the driver uses the accelerator pedal to directly control the equilibrium between the vehicle torque demand and the engine torque, i.e. the fuel demand. The governor controls the idle speed and the maximum speed. The following governor functions are identifiable from the schematic governor map (Fig. 6-2):
- The cold engine is started with the starting fuel quantity (A) when the accelerator pedal is fully depressed.
- Once the engine has started and the accelerator pedal is released, the speed adjusts in the idle position (B).
- Once the engine has warmed up, the idle speed (L) adjusts along the idle curve.
- Fully depressing the accelerator pedal when the engine is running but the vehicle is standing still increases the delivery rate to the full load value and the speed from n_{Lu} to n_1.

U. Projahn (✉)
Robert Bosch GmbH, Diesel Systems, Stuttgart, Germany
e-mail: ulrich.projahn@de.bosch.com

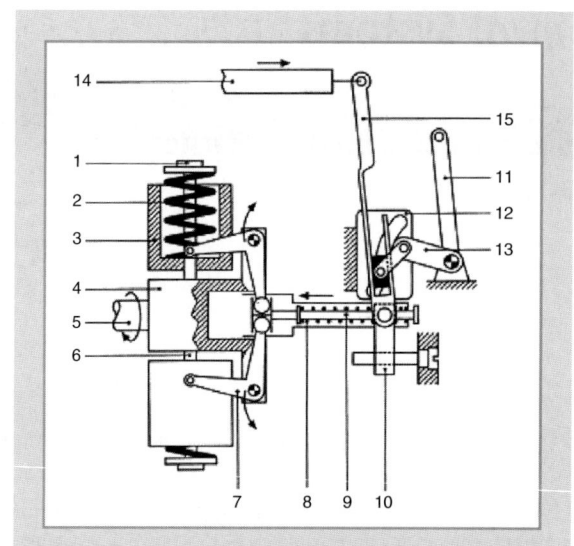

Fig. 6-1 Control device with internally integrated governor springs *1* Adjusting nut; *2* Governor spring; *3* Flyweight; *4* Joint element; *5* Camshaft; *6* Spring loaded bolt; *7* Bell crank; *8* Drag spring; *9* Sliding bolt; *10* Sliding block; *11* Control lever; *12* Cam plate; *13* Linkage lever; *14* Control rack; *15* Fulcrum lever

Then, the torque control activates. The delivery rate is reduced marginally and the speed increases to n_2, the end of torque control. The speed increases again until the maximum full load speed n_{Vo} is reached. Then, corresponding to the proportional degree designed in, the regulation of full load speed begins. The delivery rate is reduced until maximum no load speed n_{Lo} is reached.
– During in-use driving, the driver uses the accelerator pedal to control the current driving condition by establishing equilibrium between the engine torque released and the current torque demand. Based on the accelerator pedal position and the coupled control rack, the injection pump delivers the particular quantity of fuel required.

Readers are referred to [6-1] for other governor designs and auxiliary units/add-on equipment.

6.2 Electronic Control

Since 1986, diesel injection systems have been increasingly equipped with digital electronic control systems. In the beginning, mechanically governed distributor pumps were used. As of 1987, inline pumps with electronic controllers were used to recirculate exhaust gas and control the fuel quantity.[1] With the changeover to advanced direct injection systems, i.e. common rail systems as of 1997 and unit injector systems as of 1998, all automatic control functions were located in one electronic control unit (electronic diesel control EDC). The most important features of electronic engine control are its high availability throughout a vehicle's entire life, full functionality even under extreme environmental conditions and real-time operation in every operating state and at every engine speed.

6.2.1 Control Unit System Overview

Unit injector systems and, even more so, *common rail* direct injection systems allow setting one injection to divide into several single injections (pilot, main and post-injections). The common rail system also allows setting the injection pressure in the control unit as a function of a very large number of engine, vehicle and ambient parameters. Up to 10,000 parameters (i.e. characteristic values and characteristic curves or maps) must be set in advanced control systems and the systems' computing power, memory and functions have grown considerably in complexity in the last ten years to make this possible (Fig. 6-3). Thus, engine control is following Gordon Moore's rule, which is familiar from digital electronics and states that the complexity of integrated circuits increases exponentially [6-2].

6.2.1.1 Functional System Description

Electronic engine control systems can be divided into three groups: the input circuitry consisting of sensors and setpoint generators, the control unit itself and the output circuitry consisting of actuators and displays. In addition, bidirectional communication data buses also exchange information with other control units. The control unit itself is functionally subdivided into an input circuit (signal conditioning), a central processing unit and output stages with the power electronics that control the actuators (Fig. 6-4).

Factoring in a multitude of other input signals such as engine speed, vehicle speed, air and engine temperature, air mass, etc. the control unit ascertains the torque desired by a driver from the accelerator pedal signal (driver demand). This information is matched with other torque requirements, e.g. the additional torque for the generator or the torque desired by the ESP (electronic stability program). Factoring in the torque efficiency in the various parts of the injection (pilot, main and post-injections), the resulting torque is split into the various parts and transmitted to the injectors by the output stages.

Along with the actual engine control, a large number of other functions have been integrated over the course of time, e.g. electronic immobilizers, alternator controls and air conditioning compressor controls for cars or variable speed governors for commercial vehicles. In the latter case, the engine may be utilized to drive auxiliary units even when the vehicle is standing.

---
[1] Electronic governors that regulate the start of injection/delivery already existed in the early 1980s.

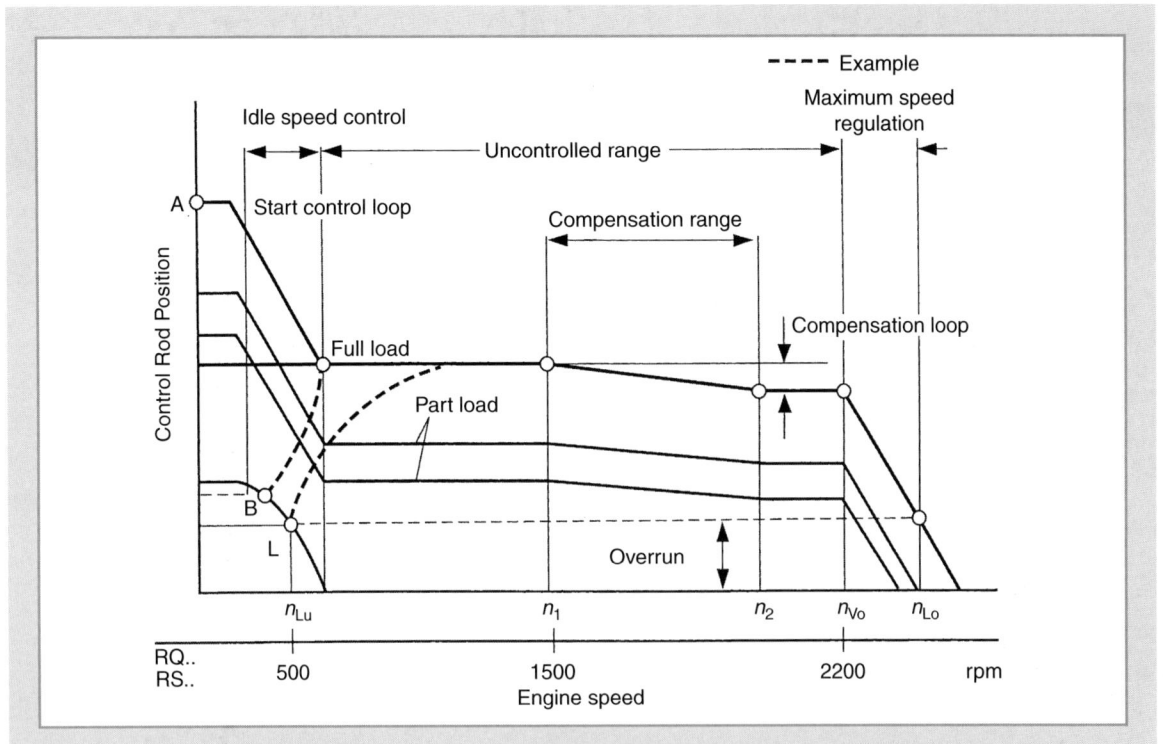

Fig. 6-2 Governor map of an idle speed-maximum speed governor with positive torque control; A Control start position; B Cold engine idle point; L Warm engine idle point; n_{Lu} minimum idle speed; n_1 Start of torque control; n_2 End of torque control; n_{Vo} maximum full load speed; n_{Lo} maximum no load speed

6.2.1.2 System Requirements from Ambient Conditions

The electronics in motor vehicles must be designed for special environmental stress because of the particular installation and operating conditions. In principle, the electronics are designed for a vehicle's lifetime. Thus, their service in cars is typically designed for ten years and 250,000 km. Table 6-1 lists more typical ambient conditions they frequently encounter when installed in a car's engine compartment.

6.2.2 Assembly and Interconnection Techniques

Engine control units are typically designed as printed circuit boards with four to six layers of printed circuit board material. The housing consists of a base/cover combination of die cast aluminum and deep-drawn aluminum plate bolted and bonded together. A diaphragm impervious to water is provided to equalize the air pressure. Thus, the control unit may be equipped with an atmospheric pressure sensor used to correct injected fuel quantities based on the air pressure, i.e. based on the altitude. A typical commercial plug and socket connection has 154 contacts divided among two chambers (Fig. 6-5).

6.2.3 Digital Central Processing Units

Their design includes a 32 bit central processing unit with clock frequencies of up to 150 MHz. Including the calibration parameters, the software fills a 2–4 Mb flash memory. In addition, it has a RAM memory of 32 Kb to execute programs. An EEPROM memory of 2–8 Kb for computed adaptation values, characteristic numbers for an individual vehicle and a fault memory are provided so that the control unit is able to turn off completely without current when switched off. A monitoring circuit checks whether the microcontroller is active and computing with the correct clock frequency.

6.2.4 Input and Output Circuits

The input circuitry converts incoming signals (switching level of switches, analog voltages of sensors and messages through serial interfaces such as CAN [controller area network]) into digital values and supplies them processed to the microcontroller. The output side turns the computed values into electric signals for the injection valves and actuators or messages for serial interfaces. In addition, the input and output circuitry must protect against electromagnetic irradiation and, conversely, prevent interfering radiation.

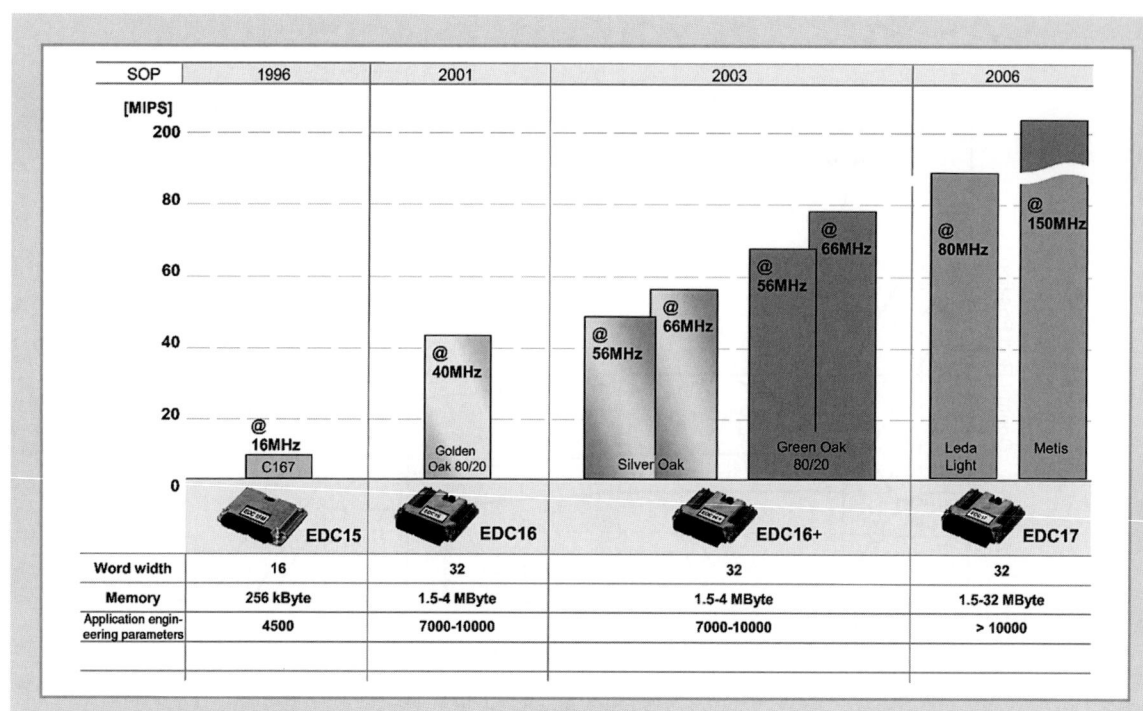

Fig. 6-3 Growth of EDC complexity (BOSCH)

Table 6-1 Typical ambient conditions for engine compartment installation in cars

Ambient temperature	−40°C − +85°C	In moving ambient air
Tropicalization	85°C and 85% relative humidity	
Dustproofness	IP 5 k xx	
Waterproofness	IP xx 9 k	
Anticorrosiveness	Salt spray	
Chemical resistance	Fuel, engine oil, degreaser, etc.	
Acceleration resistance	∼3 g	In all spatial axes

Special monitoring circuits in the output stages detect short circuits to ground, battery voltage and cable faults. The sensors on the input side are also specified so that short circuits and cable faults generate implausibilities in the input circuitry and are detected.

6.2.5 Functions and Software

A control unit usually operates out of sight. Thus, a vehicle's driver is only aware of the software. Programmed in C, the software is increasingly also generated directly from the system specification by autogenerating the executable code. A control unit's functions can be divided very roughly into engine functions and vehicle functions, both of which are implemented in a common architecture.

Serial bus systems connect a control unit with other control units in a vehicle. One or more CAN buses are common. Faster bus systems with defined real-time performance, e.g. FlexRay, will be added in the future. For instance, transmission communication controls the optimal operating point as well as engine torque limiting in the switch point so that the power train cannot overload. Other examples of serial data exchange are communication with the glow control unit, the immobilizer, the alternator or the air conditioner.

6.2.5.1 Software Architecture

Software architecture constitutes a control framework for interaction between an extremely large number of widely varying functions. One important aspect of a highly complex

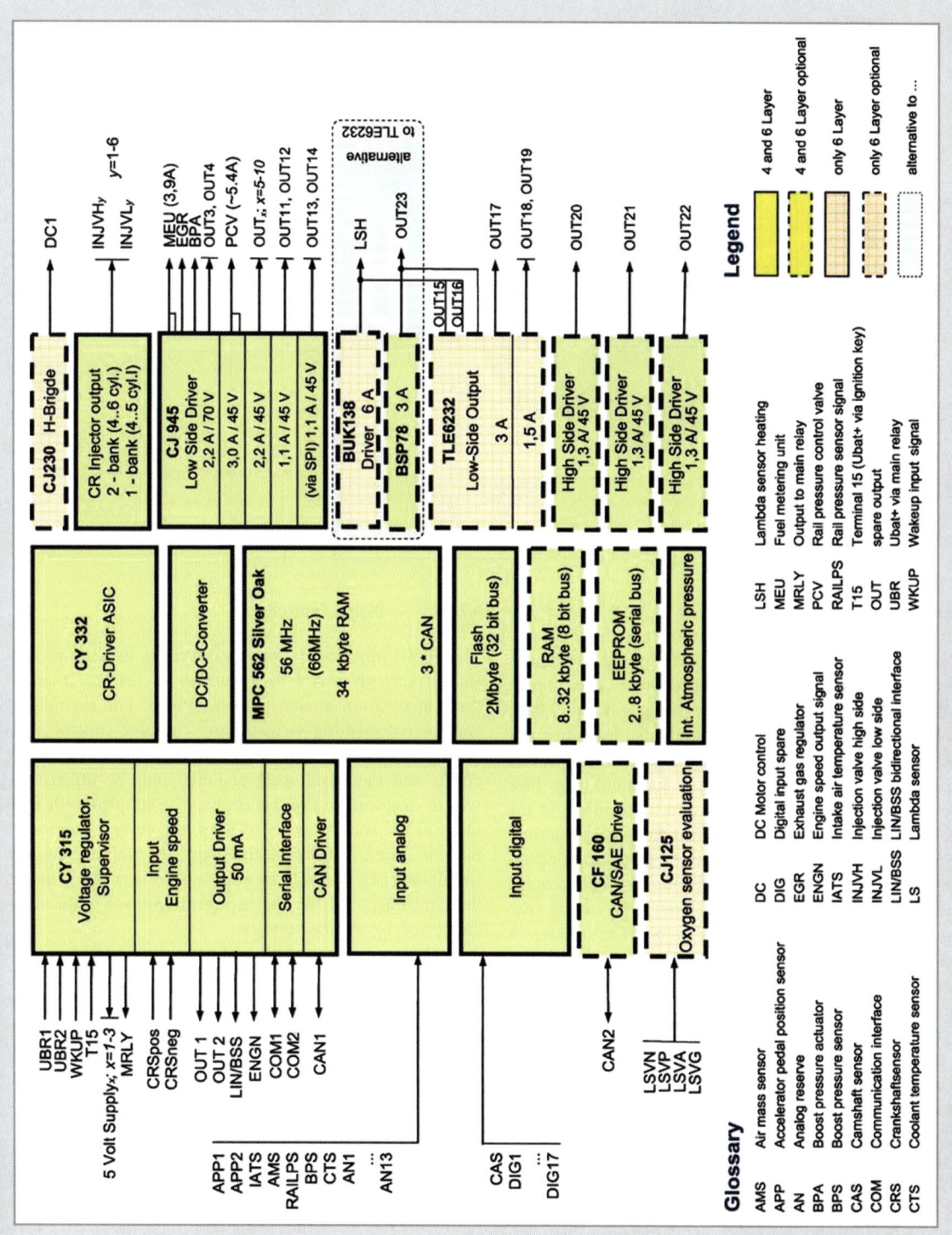

Fig. 6-4 System overview of a control unit: *Left*: signal conditioning; *center*: digital central processing unit; *right*: output stages to control the actuators (BOSCH)

Glossary

AMS	Air mass sensor
APP	Accelerator pedal position sensor
AN	Analog reserve
BPA	Boost pressure actuator
BPS	Boost pressure sensor
CAS	Camshaft sensor
COM	Communication interface
CRS	Crankshaftsensor
CTS	Coolant temperature sensor
DC	DC Motor control
DIG	Digital input spare
EGR	Exhaust gas regulator
ENGN	Engine speed output signal
IATS	Intake air temperature sensor
INJVH	Injection valve high side
INJVL	Injection valve low side
LIN/BSS	LIN/BSS bidirectional interface
LS	Lambda sensor
LSH	Lambda sensor heating
MEU	Fuel metering unit
MRLY	Output to main relay
PCV	Rail pressure control valve
RAILPS	Rail pressure sensor signal
T15	Terminal 15 (Ubat+ via ignition key)
OUT	spare output
UBR	Ubat+ via main relay
WKUP	Wakeup input signal

Fig. 6-5 Cutaway of an engine control unit (BOSCH)

system's control is its modularity, i.e. its subdivision into controllable individual functions and its reusability for economic reasons. To start with, the interfaces of the functions in an engine control unit must be defined according to standardized criteria. Physical quantities (e.g. torque) have established themselves over measured quantities (e.g. air mass), which were employed earlier. The architecture also defines the communication between functions, e.g. how information is provided, requested and transmitted. In the same way, the architecture must incorporate the desired real-time performance: Some functions (e.g. the injection system) must be computed synchronously with engine speed, others, in turn, proceed in fixed time rasters (e.g. communication with other control units in a vehicle). A third software category is event-controlled, i.e. the software reacts to an externally incoming signal. This requires realtime operation systems suitable for engine control, which are compliant with the OSEK standard (see Open Systems and the Corresponding Interfaces for Automotive Electronics at www.osek-vdx.org).

Since its scope is increasing tremendously, software will have to be integrated in different control units across manufacturers in the future. To this end, AUTOSAR is being standardized as a cross-manufacturer architecture (see www.autosar.org). The goal is modularity, scalability, reusability, portability and interface standardization that will enable controlling significantly more complex software systems at reasonable costs.

6.2.5.2 Digital Controllers

The digital implementation of controllers in microprocessor systems gives rise to a number of features that set them apart from analog representations. For example, it enables representing virtually any complex control algorithms. The controllers are not subject to any aging effects and the monitoring of limits may be utilized for system diagnostics. Controllers may be coupled with each other in the widest variety of ways, e.g. several controllers may be time synchronizable and the input variables simultaneously available to many controllers even though the controllers are located in several control units interconnected by a data network.

Analog sensor signals are usually quantized by 8 bit analog/digital converters (ADC). Ten bit ADC are also used when greater resolution is required. This makes it possible to map the sensors' measuring range and accuracy within their physical resolution without any loss of information. Digital signal handling filters the signal to suppress EMC irradiation, micro-interruptions and other disturbances. Moreover, the signal is linearized based on the sensor's characteristic curve, thus making the physical value of the measured quantity available to subsequent controllers. Current microcontroller systems process this physical value as a 32 bit numerical representation. Since the resolution is high, the control algorithms themselves do not have any perceptible influence on the accuracy of the result.

6.2.5.3 Engine Functions

From the measurement of the air mass to the control of the injection valves, all the functions that control the combustion cycle are designated as engine functions. Exhaust gas aftertreatment functions utilized to control combustion and catalysts utilized to optimize exhaust gas quality and stay under legal limits constitute a special category of engine functions.

Table 6-2 provides an overview of engine functions, Table 6-3 a summary of important exhaust gas functions.

Torque Management

Torque is the dominant physically conserved quantity in the power train. It is identical for gasoline and diesel engines. Thus, functional structures are independent of the type of

Table 6-2 Engine functions

- Glow plug indicator
- Main relay control
- Start system control
- Injection output system
  - Division into pilot, main and post-injections
  - Real-time output synchronous with the engine
- Engine coordinator
  - Engine status
  - Servo control
  - Cutoff coordinator
  - Engine torque calculation
  - Torque limiting
  - Torque gradient limiting
  - Fuel consumption calculation
  - Coordinator for engine overrun condition
- Idle controller
  - Surge damper
  - Load reversal damper
- Injection control
  - Injected fuel quantity coordinator
  - Quantity limiter
  - Torque → fuel mass conversion
  - Smoke limiting amount
- Engine speed and angle measurement
  - Overspeed protection
  - Misfire detection
- Engine cooling
  - Fan control
  - Water and oil temperature monitoring
- Air system
  - Exhaust gas recirculation control
  - Boost pressure control
  - Air intake swirl valve control
  - Air butterfly valve control
  - Air mass measurement (by hot film air mass sensor)
- Immobilizer
- Diagnostic system
- Communication through a serial BUS (CAN)

Table 6-3 Exhaust functions

- Diesel particle filter (DPF)
- NOx storage catalyst (NSC)
- Lambda closed-loop control
- Combustion detection using cylinder pressure
- Selective catalytic reduction (SCR)
- Exhaust temperature model

combustion engine selected and torque is therefore used to coordinate the power train. A driver demands a particular torque to the wheels by operating the accelerator pedal. This is back computed using the differential, i.e. the transmission (with transmission losses) for the torque at the clutch. Since various auxiliary units (air conditioner compressor, alternator, etc.) require different torques for themselves at any moment, these torques must be known and included in the initial engine torque. Once the frictional losses of the engine itself have been incorporated, the engine's inner torque is delivered. It is the basis for the injected fuel quantity required. However, modern vehicles do not convert a driver's desire this directly: Traction control systems, cruise control systems or brake assistants may additionally influence wheel torque. Particular advantages of torque management are protection of the power train components by limiting torque, smooth gear shifting in automatic transmissions at constant torque and ease of integration of additional auxiliary drives.

Model-based Air Management

Optimal mixture formation in the cylinder presupposes that the air mass in the cylinder is determined with high accuracy. To this end, a physical model of the air movement in the intake tract and the exhaust gas admixed from exhaust gas recirculation is employed to compute the gas parameters directly at the intake valve with high dynamic accuracy (Fig. 6-6). A chamber model is applied to model the geometric conditions in the intake tract and, in particular, to include the storage and delay effects at high dynamics. A throttle valve model includes a physical model of the flow rate. An air mixing model simulates the mixture cycle of fresh air and exhaust. Finally, a combustion model simulates the combustion cycle itself and the exhaust temperature.

The significantly reduced complexity required for its calibration, especially for various operating modes (normal operation, regeneration operation for exhaust gas aftertreatment systems), give this method an advantage over map specifications of the intake air mass. Further, rapid control makes the transitions between the various operating modes imperceptible to a driver. It is particularly important that the quantities of air and exhaust are always regulated synchronously. This reduces emission tolerances during operation with exhaust gas recirculation, even in normal operation.

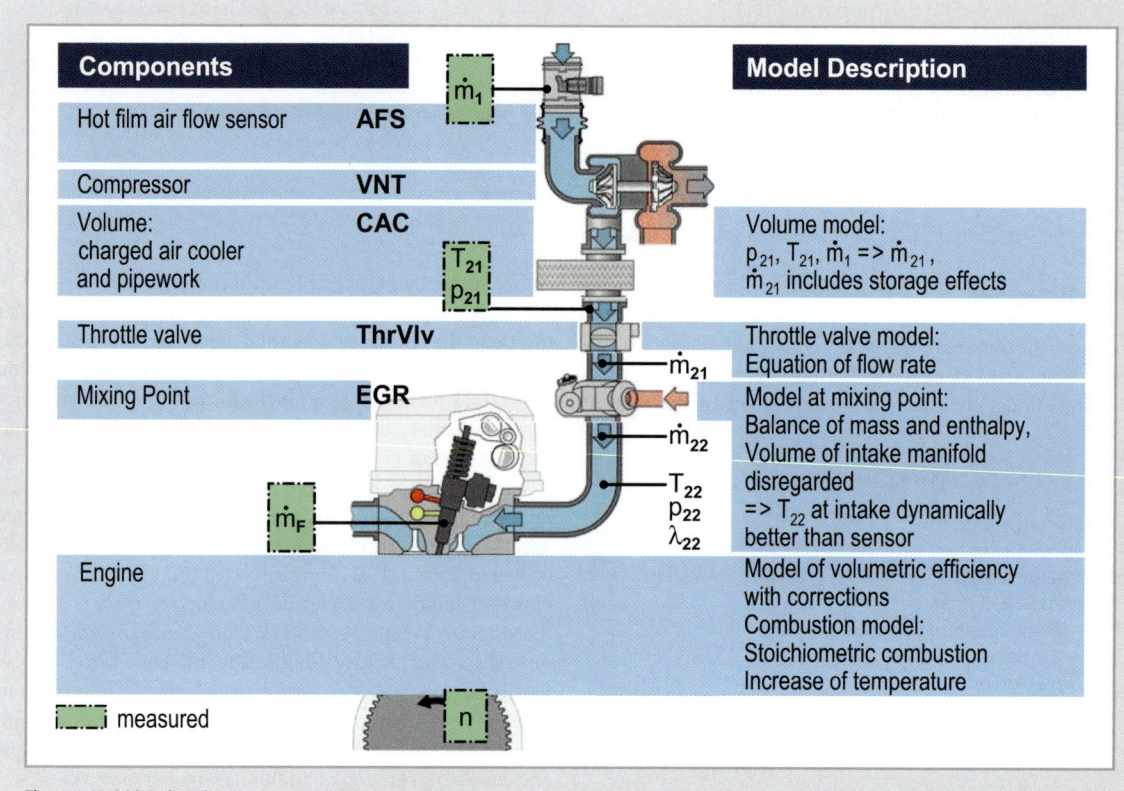

Fig. 6-6 Model-based combustion air management in a diesel engine

Model-based Exhaust Gas Management

Emission control legislation for vehicles is growing increasingly more stringent throughout the world. Taking Euro 4 (2005) as its point of departure, the EU has the reduced the exhaust limits in Euro 5 (2008) by approximately 30% for NO_X and 80% for diesel particulates. Compliance with these standards necessitates complex strategies.

The *exhaust temperature model* simulates the temperature in every point of the exhaust system. This is important for strategies for optimal catalyst use. Consequently, the number of temperature sensors that would be necessary otherwise can be reduced and the points in the exhaust system where water from the combustion reaction condenses can be ascertained by means of a dew point analysis. The exhaust temperature model includes a thermodynamic simulation of the exhaust system. Important submodels are the geometry and the thermodynamic model of the exhaust system, i.e. a thermodynamic model of the turbocharger and the catalysts.

Lambda closed loop control captures the oxygen content of the combustion emissions with the aid of a lambda oxygen sensor in the exhaust. First, a model of the control system simulates the oxygen content at the lambda sensor's position and then compares it with the measured value. Then, the difference is entered into the air model as a corrective value where its serves to correct the air mass, quantity of fuel and exhaust gas return rate. This enables keeping the exhaust constant within narrow tolerances, even during non-stationary operation and despite aging of the mechanical injection components.

Diesel particulate filters are utilized to reduce diesel soot. Particulates deposit on their reactive surface. This would clog a catalyst after a while. Therefore, the soot must be burned off cyclically in so-called regenerative operation.

Various measures increase the temperature for regeneration. The most important are attached and retarded post-injection with combustion in the oxidation catalyst.

The goal is to increase the temperature to 600 °C at which soot burns. A loading model simulates the filter's loading state from the engine's operating parameters and the duration of its operation and differential pressure sensors above the particulate filter measure it. The system is able to identify malfunctions by checking the plausibility of the information from both. The control unit switches to regenerative operation when the appropriate loading state has been reached. Both regenerative operation itself and the transitions to and from regenerative operation must proceed torque neutrally so that a driver does not notice anything.

Since diesel engines run with excess oxygen in normal operation, they emit more *nitrogen oxide* than gasoline engines. At present, it is reduced by two alternative measures: Both NO_X storage catalysts (NSC) and selective catalytic reduction (SCR) may be used in cars. Only SCR is employed in commercial vehicles.

During in-use driving, the NSC initially stores NO_X and SO_2 from combustion emissions. When the storage catalyst is full, excess fuel is generated in the exhaust mixture by means of very retarded additional injections. This is employed to initially generate CO in the oxidation catalyst, which reduces the NO_X to N_2 in the NSC. Finally, the exhaust temperature in the catalyst is increased to over $650°C$ in an approximately stoichiometric air/fuel mixture. This burns off the sulfur molecules that would otherwise clog the catalyst over the course of time.

An SCR on the other hand employs a significantly less expensive catalyst material in conjunction with a separate injection of an aqueous urea solution before the SCR catalyst. This solution (with the brand name of AdBlue®) does not require any engine modifications. Rather, it may be additionally mounted separately in the exhaust tract. Along with fuel, AdBlue® also has to be refilled after several thousand kilometers though.

Injection Management

Injection management computes the exact course of the injection cycle from the torque requirement, the additional requirements of exhaust gas aftertreatment and the current operating point. The injection cycle in a common rail system may consist of several pilot injections, one or more main injections and several post-injections. The start of injection relative to the cylinder's top dead center and the duration of the injection must be computed for every single injection of an injection cycle. In addition, the single injections necessary for the injection cycle must be defined. The control unit's ability to likewise control injection pressure (rail pressure) as a function of the operating point provides an additional degree of freedom. The single injections have extremely different functions: Very early pilot injections serve to precondition the cylinder for the subsequent combustion cycle, retarded pilot injections optimize engine noise, main injections serve to generate torque, attached post-injections increase the exhaust's temperature for exhaust gas aftertreatment and retarded post-injections deliver fuel, e.g. as a reductive agent in the exhaust gas aftertreatment system.

Furthermore, the injection system contains four functions that correct the injected fuel quantity (see also Sect. 5.3.5.7): *Injector quality adaptation (IQA)* compensates for manufacturing tolerances during the production of injectors and thus improves emissions. An injector is measured when it is manufactured and labeled with its characteristic parameters. In engine and vehicle manufacturing, this data is transmitted to the control unit as corrective values dependent on the operating points.

Fuel quantity adaptation (FQA) ascertains differences in the individual cylinders' torque contribution based on the irregularity of the crankshaft revolution and corrects them. This improves the smoothness of the entire engine's operation. *Zero fuel quantity calibration (ZFC)* compensates for the pilot injections' injected fuel quantity drift. This serves to reduce emissions and noise. The energizing time is varied in an engine overrun condition until a change of engine speed is observable. This yields the smallest energizing time effective on torque.

Finally, *pressure wave correction (PWC)* allows for the fact that the fuel pressure does not remain constant but rather drops when the injection valve opens. This generates pressure waves between the injector and the rail, which influence the injected fuel quantities of subsequent injections. This effect is corrected by corrective values dependent on the operating points, thus narrowing the fuel tolerances for multiple injections.

Monitoring Concept

The safety of individuals and, above all, a vehicle's passengers is top priority for the design of electronic systems. Therefore, the control unit includes a monitoring concept with three independent levels, which enables safely controlling a vehicle even in the case of a fault. Should another controllable system reaction no longer be possible, the engine is cut off.

Level 1 of the monitoring concept is the driving functions with their plausibility checks and input and output circuit monitoring. Plausibility checks may include reciprocal checks of the identicalness of the values of the boost pressure in the turbocharger outlet and the atmospheric pressure at engine start. Special circuits in the input and output circuits enable directly detecting short circuits or cable faults, storing them in the fault memory and initiating alternate responses.

At Level 2 of the monitoring concept, continuous torque monitoring is computed from the raw sensor signals independent of the software program. On the one hand, the maximum allowable torque is computed from a redundant signal evaluation and, on the other hand, the current torque is computed from the measured energizing time of the injection output stages. In addition, a check is run in engine overrun condition, i.e. when a vehicle reduces torque when braking for instance, to determine whether the energizing time of the injection valves' output stage is also actually zero in this operating state in which the accelerator pedal is not operated.

Level 3 of the monitoring concept finally checks the correct functioning of the microcontroller itself, both in terms of its computing and its time response, thus making it possible to detect if the microcontroller gets caught in an endless program loop. To this end, the control unit has a second circuit unit that may be another microcontroller or an application-specific integrated circuit (ASIC) furnished with a time base (crystal oscillator) independent of the main microcontroller.

6.3 Sensors

The use of sensors in a motor vehicle requires a high level of sensitivity toward mechanical, climatic, chemical and electromagnetic influences. Not only high reliability and long service life but also high accuracy is demanded. The most important sensors and their functions for diesel engine control are described below (Fig. 6-7).

An *accelerator pedal module* (APM, Figs. 6-7, 7) detects the driver's desired vehicle acceleration. Its signal serves as an input variable to compute the injected fuel quantity in the engine control system. An accelerator pedal module consists of the pedal specific to the vehicle, a mounting bracket and an angle of rotation sensor. The sensor is designed redundantly as a reference and monitoring sensor and contains either a double potentiometer or a contactless pair of Hall effect elements. The potentiometer is screen printed. The Hall effect elements are integrated in two independent Hall IC, which also contain the control and decoding electronics. The sensor output signal is proportional to the pedal position.

Temperature sensors (TS) measure the temperature of the air (ATS), coolant (CTS), oil (OTS) and fuel (FTS). As sensor elements, they normally contain an NTC resistor with a negative temperature coefficient, i.e. their resistance decreases logarithmically as the temperature increases. A temperature sensor's time constant is dependent on the installation of the NTC in the sensor (heat accumulator), the type of medium being measured (thermal capacity) and its flow rate (quantity of heat transferred). Temperature sensors have an extremely large *time constant* for air because of the plastic housing usually employed. An exposed NTC may be utilized for high *dynamic requirements*. This variant is also used in hot film air mass sensors and boost pressure sensors.

A *speed sensor* (SS, Fig. 6-7, 6) measures engine speed with the aid of a pulse generator wheel that is connected and meshed with the crankshaft. This inductive sensor contains a soft iron core enclosed by a coil with a permanent magnet and is positioned directly opposite the pulse generator wheel. The permanent magnet's magnetic field closes above the soft iron core and pulse generator wheel. The magnetic flow through the coil is dependent on the presence of either a pulse generator wheel space with a large air gap or a tooth with a small air gap opposite the core. A running engine's tooth-space alternation causes sinusoidal flow changes, which induce a sinusoidal output voltage. A speed sensor may be

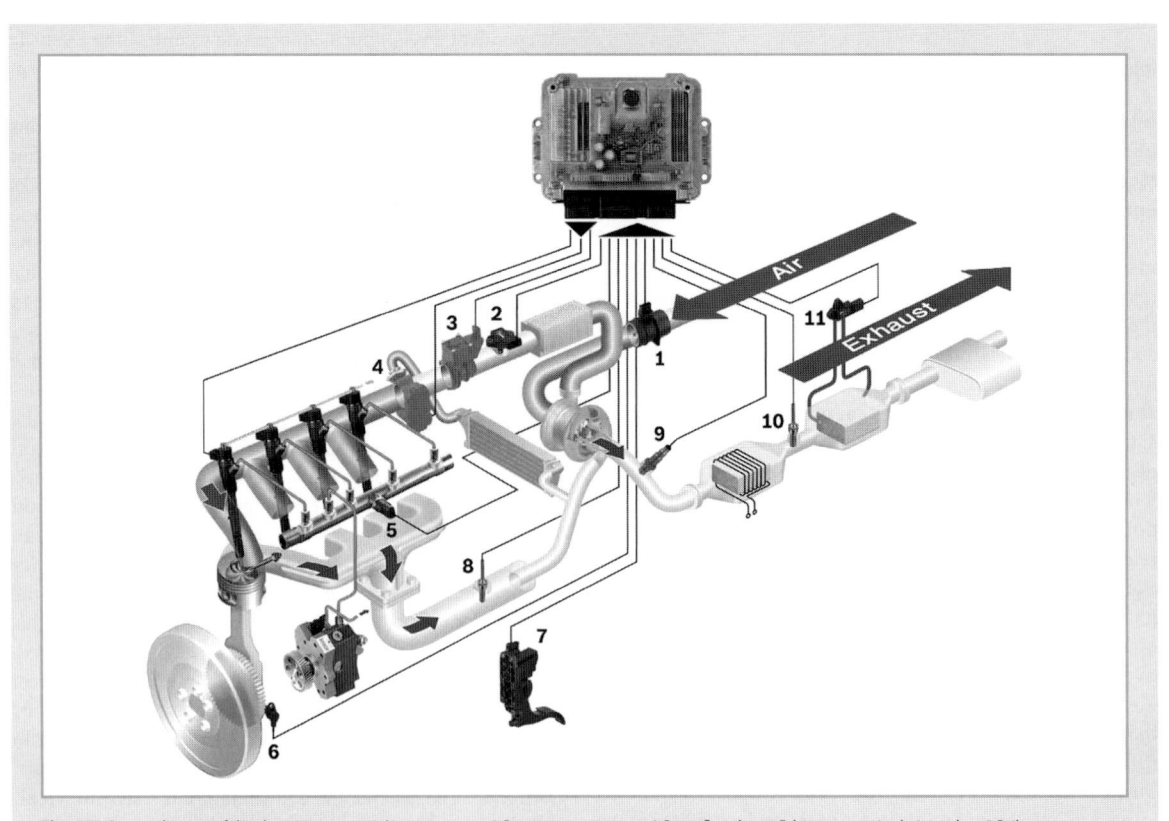

Fig. 6-7 System diagram of diesel engine sensors *1* Air mass sensor; *2* Boost pressure sensor; *3* Butterfly valve; *4* Exhaust gas recirculation valve; *5* Rail pressure sensor; *6* Speed sensor; *7* Accelerator pedal module; *8* Exhaust temperature sensor; *9* Lambda sensor; *10* Exhaust temperature sensor; *11* Differential pressure sensor

manufactured with a digital output signal for use in smaller pulse generator wheels with Hall IC.

Beyond purely measuring engine speed, a speed sensor facilitates the implementation of engine functions that can, for example, correct the *injected fuel quantity* relative to a cylinder on the basis of measurements of smaller speed changes. Thus, the minutest differences in an individual cylinder's particular injected fuel quantities can be corrected (fuel quantity adaptation) or longitudinal vehicle vibrations can be prevented (active damping of engine bucking).

A *phase sensor* (PS) detects the position of the camshaft with the aid of a pulse generator wheel that is connected and meshed with the shaft. The pulse generator wheel's small diameter precludes the use of an inductive speed sensor. Therefore, the phase sensor contains a Hall sensor, which, similar to a speed sensor, uses the pulse generator wheel to detect changes in the magnetic field and outputs them as processed digital signals. The crankshaft's absolute angular position at the first engine cylinder's top dead center (TDC) is computed from phase and speed sensor signals. This makes it possible to reduce emissions and implement comfort functions, e.g. quick start. Should the speed sensor fail, the phase sensor can partially take over its function.

An electric *butterfly valve* (BV, Fig. 6-7, 3) is mounted in the intake manifold on the cold side of the engine in the direction of flow before the exhaust gas recirculation line. It enables reducing the intake cross section in the part load range and thus controlling the pressure level after the valve. Closing the valve increases the flow rate of the intake air and consequently also the vacuum after the valve. The increase of the vacuum increases the exhaust gas return rate by up to 60%. Such exhaust gas return rates make it possible to reduce emissions considerably.

Other functions of a butterfly valve are:
- increasing exhaust temperature to regenerate the particulate filter by throttling and opening the bypass to the intercooler,
- acting as an air controlled governor with an air/fuel ratio < 1 during regeneration of the NO_X storage catalyst,
- providing safety and comfort shutoff and
- throttling in idle to optimize noise by lowering the peak cylinder pressure.

Independent of temperature and density, a *hot film air mass sensor* (HFM, Fig. 6-7, 1) measures the air mass aspirated by an engine and thus the oxygen available for combustion. The hot film air mass sensor is installed between the air filter and compressor. The air mass flow is measured by conducting it through a thin and heated silicon diaphragm. Four temperature-dependent resistors measure the temperature distribution in the diaphragm. The air mass flow changes the diaphragm's temperature distribution and this causes a resistance difference between the upstream and downstream resistors. Since the resistance difference is a function of direction and magnitude, the *hot film air mass sensor* is able to measure pulsating air mass flows with high dynamics and correct the sign. This enables the engine control unit to precisely compute the mean value of the air mass. Optionally, a temperature sensor may be additionally implemented in the hot film air mass sensor to measure the intake air temperature.

Since the measured air mass signal serves as the actual value to control exhaust gas recirculation and to limit the injected fuel quantity (smoke limiting), high accuracy of the hot film air mass sensor is especially important for compliance with emission limits.

A *boost pressure sensor* (BPS, Fig. 6-7, 2) measures the pressure of the aspirated air in the intake manifold, which the compressor supercharges to a higher pressure. Thus, more air mass reaches the individual cylinders. The boost pressure supports the boost pressure control. The current actual pressure and engine speed may be used to compute the air mass introduced. Exact knowledge of the air mass is important to prevent the injection of an excess of fuel into the engine (smoke limiting function).

The boost pressure sensor's sensor element is a silicon chip onto which a diaphragm has been etched. Pressurization causes the sensor diaphragm to bend and the measuring resistors located in the diaphragm to change their resistance. Boost pressure sensors are designed as absolute pressure sensors, i.e. the measured pressure is generated on the one side of the diaphragm and a small insulated reference vacuum is located on the other. The signal evaluation circuits integrated on the chip amplify the changes of resistance and convert them into a voltage signal that is fed to the control unit.

The function of a *differential pressure sensor* (DPS, Fig. 6-7, 11) is to measure the pressure drop in the particulate filter in the exhaust system. This may be used to determine the degree of filter loading with soot. When it is sufficiently loaded, a regeneration cycle is initiated.

The differential pressure sensor's sensor element is a silicon chip similar to that of a boost pressure sensor. However, each of the two measured pressures – pressure before and after the DPF – acts on one side of the pressure diaphragm. The differential pressure signal is generated by the pressure measured on the front of the diaphragm minus the pressure measured on the back of the diaphragm. Pressure changes on both sides of the diaphragm cause the sensor diaphragm's bend to change and thus the measuring resistors' resistances to change accordingly. The signal evaluation circuits on the chip amplify and convert the changes in resistance into a voltage signal that is fed to the control unit.

A *rail pressure sensor* (RPS, Fig. 6-7, 5) measures the fuel pressure in a common rail injection system's distributor rail. The current rail pressure determines the quantity of fuel injected and serves as the actual value for the rail pressure control system.

The rail pressure sensor's sensor element is a stainless steel diaphragm on which a resistance bridge made of thin film resistors is located. Diaphragm deflection at maximum pressure is only roughly 1 µm in this design. Nevertheless, although small (10 mV), the measurement signal delivered by the resistance bridge is very stable and precise throughout the sensor's life. It is amplified in an electronic circuit and fed to the control unit.

The sensor's pressure port securely seals the very high rail pressure (180...200 MPa maximum) from the rail with a so-called knife-edge seal (metal on metal).

A *lambda oxygen sensor* (LS, Fig. 6-7, 9) measures the O_2 concentration in exhaust. This may be used to determine the air/fuel ratio λ in the engine's entire operating range. The exhaust passes through a diffusion barrier to a measuring chamber from which the oxygen can be pumped back into the exhaust pipe through the sensor ceramic. A control circuit keeps the O_2 concentration in the chamber constant at a very low value (ideally zero). The pump current is a measure of the O_2 concentration in the exhaust.

Today, the dominant application for LS (in conjunction with the air mass signal from the HFM) is the correction of the injected fuel quantity for compliance with a vehicle's exhaust emission values throughout its lifetime. A new addition will be the use of LS to monitor and control the regeneration of the NO_X storage catalyst (NSC).

Exhaust temperature sensors (ETS, Fig. 6-7, 8 and 10) have the function of measuring the exhaust temperature at different points in the exhaust pipe. The basic difference between exhaust temperature sensors and other temperature sensors is their measuring range and dynamics. Their most important installation positions are after the oxidation catalyst (diesel oxidation catalyst DOC) and in front of and behind the diesel particulate filter (DPF). An ETS serves both to monitor and control the DPF and to protect components should the particulates in the filter burn off uncontrolled.

Depending on their case of application and measuring range, exhaust temperature sensors consist of ceramic materials with a negative temperature coefficient (NTC ceramics), platinum with a positive temperature coefficient (PTC) or thermocouples, i.e. mechanically connected metallic wires that – based on the Seebeck effect – supply an electrical voltage when subjected to a temperature gradient.

An NO_X *sensor* is important for the monitoring and control of NO_X catalysts. Its principle not only enables an NO_X sensor to determine the nitrogen oxide concentration in exhaust but also the values of the air/fuel ratio. It consists of a two-chamber sensor element. The exhaust passes through a first diffusion barrier to the first chamber in which, comparable to a lambda oxygen sensor, the oxygen content is determined. The oxygen content of the exhaust in the chamber is reduced to nearly zero. The thusly processed exhaust passes through a second diffusion barrier to a second chamber in which the oxygen is extracted from the nitrogen oxide by catalytic reduction and measured. This principle of oxygen measurement also corresponds to that of the lambda oxygen sensor.

This sensor may be used both to control NO_X catalysts (minimum NO_X concentration control) and monitor assemblies important for compliance with exhaust emission limits (NO_X concentration threshold value monitoring).

Future exhaust gas aftertreatment concepts, e.g. particulate filters and NO_X catalysts, will necessitate the use of other sensors both to operate and monitor systems. Along with exhaust temperature and exhaust pressure, specific pollutant fractions in the exhaust flow, e.g. hydrocarbons, nitrogen oxides and soot, will be particularly important.

6.4 Diagnostics

The continually mounting requirements and increasing complexity of the systems utilized for operation have steadily increased the importance of diesel engine diagnostics. Diagnostics development is now an integral part of engine, system and component development and is incorporated in the entire vehicle life cycle from development to production up through service.

The basic functions of monitoring and diagnostics during in-use driving are to ensure reliability and compliance with legal regulations. In a damage scenario, the rapid localization of the defective component has priority (Fig. 6-8).

Driving monitoring and diagnostics during in-use driving are based on the continuous monitoring of functions in the engine control unit and also includes the legally mandated on-board diagnostics (OBD) of components and systems relevant to emissions.

In addition to the results of diagnostics during in-use driving, garages are able to retrieve special diagnostic functions, which are either available in the engine control unit or a diagnostic tester. Moreover, additional testing and measuring instruments support guided troubleshooting in the case of damage.

6.4.1 Diagnostics During In-Use Driving

Monitoring and diagnostics are performed during in-use driving without auxiliary equipment and belong to the basic scope of electronic engine control systems.

It is divided into the OBD system required by law to monitor systems and components relevant to emissions and, where applicable, other manufacturer-specific tests of parameters irrelevant to emissions. Both units take advantage of the control unit's self-monitoring, electrical signal monitoring, plausibility checks of system parameters and tests of system and component functionalities. Detected faults are stored in the control unit's fault memory and can be exported through a serial interface, which is usually specific to the manufacturer.

Part of monitoring and diagnostics during in-use driving is ensuring that the system state is controllable to prevent consequential damage in a fault scenario. When necessary, such default responses employ default functions/values to control limp-home operation or cut off an engine in serious cases.

Diagnostic system management (DSM) is the central management element of diagnostics during in-use driving.

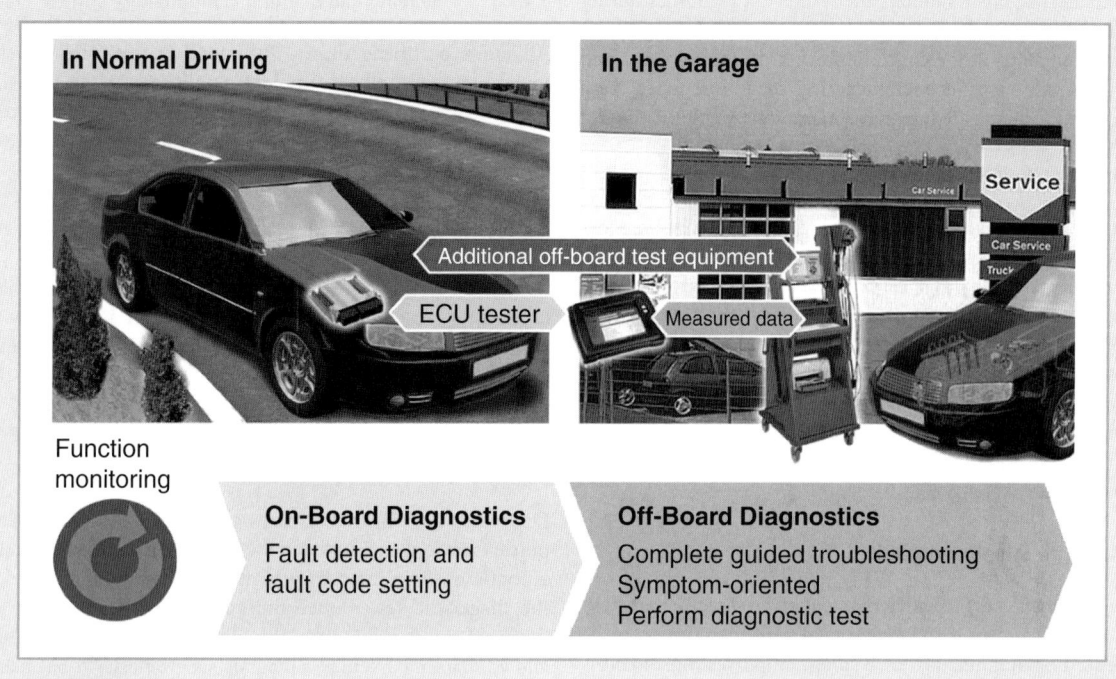

Fig. 6-8 Diagnostics and monitoring in in-use driving and garage diagnostics

DSM consists of the actual storage of faults in the memory, algorithms for fault debouncing and remedying, test cycle monitoring, an exclusivity matrix to prevent entries of secondary faults and the management of default responses.

6.4.2 OBD (On-Board Diagnostics)

As mandated by law, the engine control unit must monitor all components and systems relevant to emissions during in-use driving. The malfunction indicator lamp (MIL) in the instrument cluster signals the driver when an OBD emission limit value has been exceeded.

The activation of diagnostic functions may be linked to certain switch-on and exclusivity conditions. The frequency of activation (In Use Monitor Performance Ratio IUMPR) must also be monitored.

If a fault disappears again ("remedying"), then input usually still remains entered in the fault memory for forty driving cycles (warm up cycles), the MIL usually being shut off again after three fault-free driving cycles.

OBD legislation requires the standardization of both fault memory information and its accessibility (plug and session protocol). Fault memory information relevant to emission is accessed through a diagnostic plug that is easily accessible in every vehicle with the aid of the appropriate commercially available scan tools from various manufacturers.

Legislators may order recall actions at the cost of the vehicle manufacturer when a vehicle fails to fulfill the legal OBD requirements.

California and four other US states enacted the second stage of OBD II (CARB) legislation for diesel engines in 1996(following the first stage in 1988). It requires the monitoring of all systems and components relevant to emissions. Approval requires the demonstration of a monitoring frequency (IUMPR) of diagnostic functions that is adequate for the particular in-use driving. OBD II limits are defined relative to the legal emission limits.

Environmental Protection Agency (EPA) laws have been in effect in the other US states since 1994. The current scope of these diagnostics basically corresponds to the CARB laws (OBD II).

OBD adapted to European conditions is designated EOBD and is functionally based on the EPA's OBD. EOBD for cars and light commercial vehicles with diesel engines has been binding since 2003. Absolute EOBD emission limits are in force in Europe.

6.4.3 Service Diagnostics (Garage Diagnostics)

The function of garage diagnostics is to rapidly and positively localize the smallest replaceable unit. The use of a diagnostic tester, which is usually computerized, is absolutely essential in advanced diesel engines.

Garage diagnostics utilizes the results of diagnostics during in-use driving (fault memory entries) and special garage diagnostic functions in the control unit or a diagnostic tester and testing and measuring equipment. In a diagnostic tester, these diagnostic options are integrated in guided troubleshooting.

6.4.3.1 Guided Troubleshooting

Guided troubleshooting is an integral part of garage diagnostics. Taking the symptom or the fault memory entry as the starting point, a garage employee is guided through fault diagnostics with the aid of a result-driven routine. Guided troubleshooting combines every diagnostic option in a purposeful diagnostic routine, i.e. symptom analysis (fault memory entry and/or vehicle symptom), garage diagnostic functions in the engine control unit (ECU) and garage diagnostic functions in the diagnostic tester and additional test equipment/sensor systems (Fig. 6-9).

6.4.3.2 Symptom Analysis

Faulty vehicle performance may either be directly perceived by a driver and/or documented by a fault memory entry. At the start of fault diagnostics, a garage employee must identify the symptom as the starting point for guided troubleshooting.

6.4.3.3 Reading and Deleting Fault Memory Entries

All faults occurring during in-use driving are collectively saved with the defined ambient conditions at the time of their occurrence and may be read through an interface protocol, which is usually customer-specific. The diagnostic tester may also delete the fault memory.

6.4.3.4 Additional Test Equipment and Sensor Systems

Additional sensor systems, test equipment and external analyzing units extend the diagnostic options in a garage. In a fault scenario, the units are adapted to the vehicle in the garage. The test results are generally evaluated in a diagnostic tester.

6.4.3.5 Control Unit-based Garage Diagnostic Functions

Once they have been started by the diagnostic tester, diagnostic functions integrated in the control unit run completely autonomously in the control unit and report the outcome to the diagnostic tester once finished. A diagnostic tester is designed to parameterize the diagnostic functions. This allows adapting a vehicle's diagnostics even after it has been launched on the market.

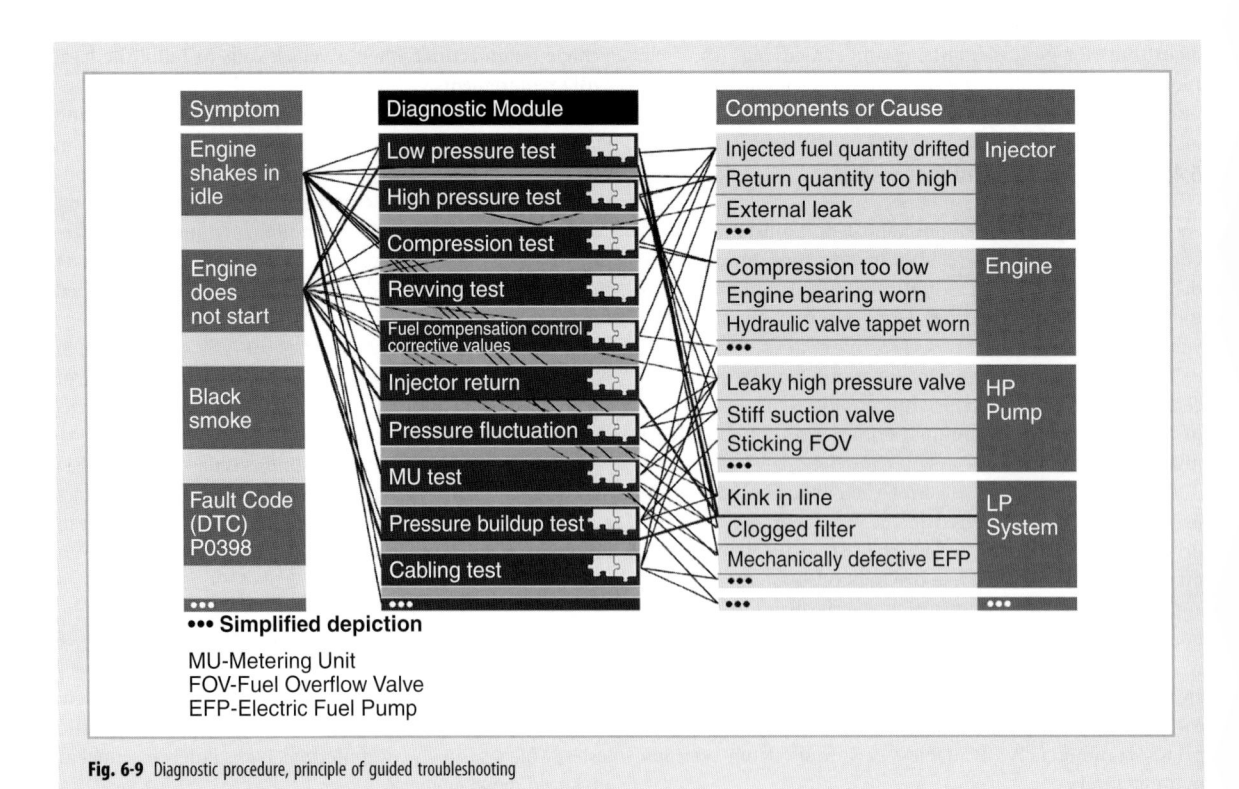

Fig. 6-9 Diagnostic procedure, principle of guided troubleshooting

6.4.3.6 Diagnostic Tester-based Garage Diagnostic Functions

These diagnostic functions are run, analyzed and evaluated in a tester. The vehicle sensors' ECU and/or additional test sensor systems ascertain the measured data referenced for analysis.

Since their default values enable diagnostic testers to utilize their structure variably, dynamic test modules provide a maximum of flexibility. A test coordinator coordinates this functionality in an ECU. Test results from different sensors may either be transmitted to a diagnostic tester in real time or temporarily stored and evaluated in it.

Any garage diagnostic function can only be employed when a diagnostic tester is connected and, as a rule, only when a vehicle is standing. The operating conditions are monitored in the control unit.

6.5 Application Engineering

6.5.1 Significance of Application Engineering

The electronic engine control unit uses electronic actuators to optimally adjust basic engine tuning, performance and emissions to the particular operating conditions, e.g. during cold start and warm-up and at extreme heat or altitudes.

The term application or calibration describes the function (or process) of adjusting the functionalities programmed in the electronic control unit to the individual hardware according to desired specifications. Ultimately, this means utilizing sensors and electronic and electromechanical actuators to give a vehicle combined with its engine and transmission the appropriate performance (engine power, torque characteristic, combustion noise, smooth operation and response). Moreover, legally mandated emission limits must be met and a system's self-diagnostics capability suitably assured.

A multitude of application engineering data, i.e. characteristic values, characteristic curves and maps, has to be applied (calibrated) to achieve these objectives.

Advanced engine control systems process over 10,000 different application data during the application process. The application system (calibration system) supplies this parameterizable data to engineers. The application is produced in labs, at engine and vehicle test benches and on test tracks under real ambient conditions. Once this process has been completed, the data obtained is extensively verified for use in production and stored write-protected in read only memories, e.g. EPROM or Flash.

Understanding the correlations during the entire engineering process is becoming ever more demanding and challenging. Systems are increasing in complexity because of:

– several intercommunicating control units in a vehicle network,
– higher numbers of injections per working cycle (pilot, main and post-injections),
– exhaust gas aftertreatment systems with new requirements for basic engine tuning,
– higher requirements from emission laws,
– interactions of the application engineering data and
– extensions of diagnosability.

As a result, the number of characteristic values, characteristic curves and maps that have to be applied is continually increasing.

6.5.2 Application Engineering Systems

The job of application engineering is performed by tools that enable application engineers to measure internal signals through electronic interfaces to the engine control unit and simultaneously adjust the free application engineering data. The user interfaces to the tools use description files to represent and modify the measured signals and application data on the implementation level or the physical level or even graphically. Additional instrumentation supplements and references the internal variables of engine control to complete the application tasks.

Generally, a distinction is made between offline and online calibration systems. When parameter values are being modified or adjusted, offline calibration interrupts the software program's execution of open loop control, closed loop control and monitoring functions. The need to interrupt these functions adversely affects the calibration process, particularly when they are being used at test benches or in vehicle tests.

The open loop control, closed loop control and monitoring functions are active during the online calibration process. Thus, the effects of modifications made may be evaluated directly in driving mode and the calibration process is completed more effectively. However, online calibration imposes greater demands on the calibration system and the stability of the open loop control, closed loop control and monitoring functions employed since the software program must be designed more reliably for potential emergency situations, e.g. caused by discrepancies in the distribution of interpolation nodes, while the tool is making adjustments.

6.5.3 Application Engineering Process and Methods

The application engineering process entails the completion of the widest variety of tasks. The methods and tools employed are developed for the type and complexity of tasks as well as the frequency of their usage as a project progresses.

The first task is to calibrate the *sensors* and *actuators* so that the physical input and output signals computed in the engine control unit correspond with the real values measured in the engine or vehicle as accurately as possible. The distinctive features of the sensors and actuators, the effects caused by installation conditions, the signal evaluation circuits and the sampling frequency in the electronic control unit are factored in. The application engineering is validated by reference measurements with external sensor systems.

Control functions consist of setpoint values, computed actual values, the actual energization of the governor or controller and the computed on/off ratio for the actuators. Boost pressure, rail pressure, driving speed or idle speed control are some examples of tasks in this domain. Setpoint values are adjusted according to the particular operating state, e.g. cold/warm engine, ambient conditions and driver demand. The controller is designed by identifying the system (e.g. determining the frequency response), modeling the controlled system and applying methods to design controllers (e.g. Bode diagram, Ziegler-Nichols, pole assignment). The control loop's stability and control quality is subsequently checked in real operation and under extreme conditions. Stability is especially important for different aging processes. Extreme conditions are simulated to determine and improve the durability of already optimized systems.

Statistical design of experiments (DOE) has replaced complete raster measurement in *basic engine tuning*. The reason for this is the drastically increasing complexity of optimization tasks caused by the

- continually increasing number of freely selectable parameters (e.g. time and quantity metering of pilot, main and post-injections, rail pressure, EGR rate and boost pressure) and
- the increasing number of complex engine optimization criteria and operating states (e.g. rich operation, lean operation, particulate filter regeneration and homogeneous operation).

The goal of the statistical design of experiments is to greatly reduce the number of operating points measured and to determine the respective local models for representative operating states on the basis of a minimum of parameter variations. The most influential parameters are varied in the representative operating points and their significance for emissions, consumption, power, exhaust temperature, noise, rise in combustion pressure and cylinder pressure is measured.

Repetitive measurements and "filling points" and measuring points in the border operation area provide a basis to compute local engine models, evaluate the statistical value of the results, evaluate the model error and better safeguard against measuring errors.

The reproducibility of the results of measurements and certain detection of outliers is also crucial for the quality of the models. The variations of parameters may not exceed the limits for the engine and specific components, e.g. allowable cylinder pressure or maximum exhaust temperature, during the entire process of measurement.

The optimal parameters for the calibration of the map structures in the engine control unit are determined stationarily on the basis of the computed models. Since the data has only been determined in a rough raster based on a few selected operating points, the requisite intermediate ranges must be computed by interpolation routines. Calibration is normally performed offline. The result obtained must still be validated and documented in stationary operation on an engine test bench and on the vehicle in dynamic operation.

Application engineering takes components with average values as its starting point. In addition, the influence of the manufacturing tolerances must be evaluated to validate the calibrated data for use in production. This may either be simulated by adjusting sensor and actuator characteristic curves or experimentally tested on an appropriately prepared test prototype. Another option is the application of intelligent corrective functions to adjust component tolerances or to correct drift to make the system employed against component aging even more durable, thus preventing adverse effects on performance and emissions. Cylinder-specific correction of the injected fuel quantity to obtain extremely smooth engine operation is one example of this approach.

Engine and transmission protection functions ensure, for example, that the maximum torque and maximum allowable engine operating temperature are not exceeded even under extreme conditions (heat, cold, altitude and high load operation). These functions are applied and validated in road tests under the aforementioned conditions or on air conditioned vehicle test benches.

The *scope of engine and vehicle diagnostics in application engineering* is assuming even greater significance. Specially developed functions make it possible to identify defective components for maintenance and to perform guided troubleshooting. The wealth of knowledge and experience acquired in the entire development process is referenced for diagnostics application engineering and entered into the calibration data. Diagnostics application engineering also defines default reactions that have to be executed when a component is defective. Design criteria include the safety of the vehicle, protection of components and the prevention of secondary faults or undesired secondary reactions.

In light of the large quantity of application engineering data, the division of work among vehicle manufacturers, suppliers and application engineering service providers, the large number of variants of datasets (vehicle, engine, transmission and emission combinations) and the simultaneous engineering of components, software functions and application engineering itself, the defined objectives are only achievable with efficient and knowledge-based *data management*. Various tools (database systems) provide application engineers support to manage data efficiently and satisfy the stringent quality requirements.

Increasingly shorter development times and increasing complexity are making the development of *improved methodological approaches* to application engineering more and more important. The development of methods for automatic system calibration, the use of hardware-in-the loop systems in application engineering and more efficient computation of parameterizing models will become established in application engineering.

Literature

6-1 Robert Bosch GmbH: Gelbe Reihe. Technische Unterrichtung. Motorsteuerung für Dieselmotoren. 2nd Ed. pp. 52–97
6-2 Moore, G. E.: Cramming more components onto integrated circuits. Electronics 38 (1965) 8

Further Literature

Robert Bosch GmbH: Dieselmotor-Management. 4th Ed. Wiesbaden: Vieweg (2004)

Part II Diesel Engine Engineering

7 Engine Component Loading 195

8 Crankshaft Assembly Design, Mechanics and Loading 221

9 Engine Cooling 291

10 Materials and Their Selection 339

7 Engine Component Loading

Dietmar Pinkernell and Michael Bargende

7.1 Mechanical and Thermal Loading of Components

7.1.1 Mechanical Loading of Components

The determination of the effective loads in a diesel engine is crucially important for the design of its individual engine components and assemblies. The determination of stress is an important prerequisite for component sizing. It forms the foundation for determining the geometric dimensions, the material employed or even the manufacturing process applied. Thus, load analysis plays an important role in cost and capacity estimates in the development process and fundamentally determines a diesel engine's reliability.

A stress analysis must differentiate between different *types of loading* since they have different effects. Basically, three types of loading are distinguished.

7.1.1.1 Static Loading

Since a diesel engine's individual assemblies are normally joined by bolted joints following the engineered design, *bolt tightening* produces a strikingly large static preload in engines. Related problems with sizing are normally handled on the basis of VDI Guideline 2230 [7-1]. Interference fits (e.g. in thermally or hydraulically joined components) also cause static stresses.

Some stress concepts additionally also interpret *internal stresses* as a static load. They can be generated by the manufacturing process (e.g. casting or forging, welding and machining) or introduced by a specifically applied mechanical or chemothermal surface treatment such as shot peening, rolling, nitriding, carburizing or the like. However, the determination of these stress variables is difficult and not always feasible. In addition, potential stress redistributions in later engine operation complicate the evaluation of component reliability.

7.1.1.2 Thermal Loading

Thermal loading is relevant for components adjacent to the combustion chamber [7-2] and for diverse pipe systems and the exhaust system. Fluctuating greatly during the combustion cycle, the *gas temperature* heats these components. Figure 7-1 indicates that high peak temperatures are reached briefly in diesel engines. However, the inertia of the heat transfer on the component surface renders this temperature fluctuation practically ineffective, the insulating effect of layers of soot playing a major role here. Accordingly, the temperature fields that appear in components may frequently be regarded as *quasistationary*, i.e. when the engine load does not change, the temperature fields do not change either. This is explicitly untrue for thermal loads produced by switching on and off.

Thermal conduction causes temperature fields to appear in all combustion chamber components, which exhibit the largest *temperature gradients* from their heated surface toward their cooled surface. The resultant temperature differences generate corresponding thermal expansions and thus also thermal stresses. The absolute level of the temperature influences the tolerable stresses (temperature-dependent strengths of materials) as well as the level of the stresses occurring.

The thermal load may be evaluated by interpreting it as a quasi-static load and is then frequently factored into the mean stress. However, other cases may require evaluating the thermal load as a temporally varying load. This especially holds true whenever stresses that cause lasting deformation (local yielding) occur in components because their expansion is impeded. Then, appropriate low cycle concepts are referenced for an evaluation.

7.1.1.3 Dynamic Loading

One basic cause of temporally varying mechanical loading is the variable *gas pressure* in the cylinder (Fig. 7-1), which directly loads such combustion chamber components as the piston, cylinder head or cylinder liner. Since connecting components transfer the gas forces further, loads generated by the gas pressure are ultimately demonstrable in every one of a diesel engine's components. Determining the loads acting

---

M. Bargende (✉)
Universität Stuttgart, Stuttgart, Germany
e-mail: michael.bargende@ivk.uni-stuttgart.de

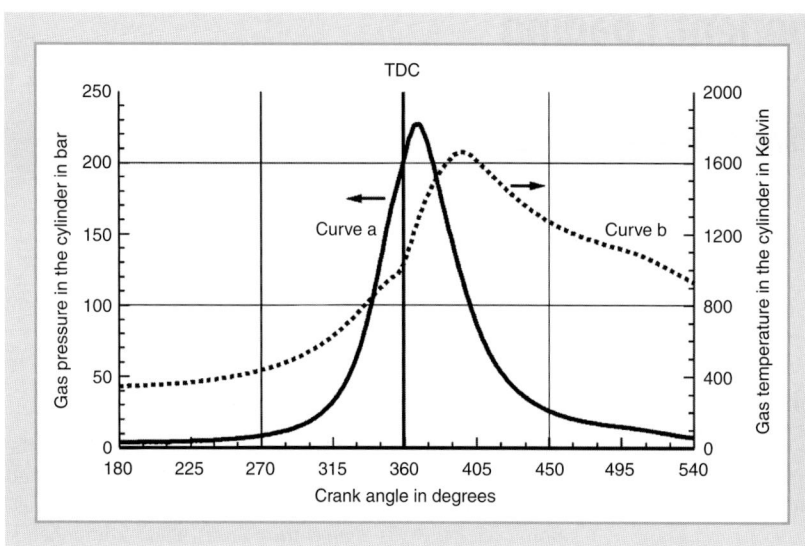

Fig. 7-1 Time characteristic of gas pressure (*curve a*) and the averaged gas temperature (*curve b*) for a medium speed four-stroke engine

on a discrete part requires analyzing the transfer and equilibrium of forces inside the engine.

A significant sizing parameter for many components, the maximum *gas force* ensues from the maximum gas pressure (firing pressure). This simplifying quasi-static approach no longer has validity when elastic components are excited to vibrate, which is the case for instance when components exhibit natural frequencies excited by the gas forces' strong harmonic components, i.e. when a *resonance case* exists. Resonant modes relevant to stress are principally to be feared in components with low natural frequencies since the damaging resonant amplitudes rapidly attenuate toward higher natural frequencies. Add-on parts such as pipe systems or cowlings as well as turbocharger groups or pumps are frequently affected. Along with the resonant range, the respective intrinsic damping, which can assume a magnitude of 1–10%, also plays a role.

Other dynamic loading of components results from the inertial forces caused by engine dynamics [7-3]. A distinction is made between the rotating masses that induce *rotating inertial forces* (centrifugal forces) and masses oscillating with piston motion that generate *oscillating inertial forces*. These can excite vibrations too (see Sect. 8.2). At constant speed, centrifugal forces produce static loading of a component. However, oscillating forces may arise in components that must absorb reactive forces resulting from a centrifugal load. Centrifugal forces in the crankshaft, for example, cause unbalance effects and structural deformations that induce vibratory stresses in the mount and thus in the engine casing. Component stresses caused by inertial forces are much the same as those caused by gas forces, i.e. the maximum inertial forces make quasi-static sizing sufficient for certain components, while the excitation of additional vibrations particularly has to be factored in for others, especially for resonance cases.

Impact dynamics appear, for instance, when valves set down on their valve seats or when the injection nozzle needle closes and may induce high stresses with extremely detrimental effects on component strength or wear.

Total vibration stress not only has to allow for the individual cylinders' gas and inertial force characteristics but also their interaction. This causes a phase shift during one combustion cycle dependent on the firing sequence with corresponding effects, e.g. on the crankshaft's torsional stress or the complete engine's performance.

7.1.2 Component Stress

7.1.2.1 General Background

In principle, the methods to determine stress in a component or assembly can be divided into *methods of calculation* and *methods of measurement*.

A common *method of calculation* to determine stresses is the classic theory of strength of shafts, bolts, pipes, etc. In many cases, such methods obtain results sufficient for a stress analysis while requiring relatively little effort. The use of analytical and/or empirical methods of calculation is generally widespread in the domain of component design and particularly well suited for draft designs, simulations of alternatives and preoptimization. However, the output quantities and their accuracy frequently have limited validity since the incorporation of important input data, e.g. structural elasticity, nonlinear material and contact behavior, is usually impossible. The *finite element method* (FEM) [7-4] is primarily employed to integrate the full complexity of boundary conditions and component geometries in stress calculations. It obtains the best quality and diversity of results. However, the application of this method in the design process is comparatively involved and thus time consuming.

Multibody simulation (MBS), computational fluid dynamics (CFD) or bearing calculations (HD/EHD) are being combined with increasing frequency to properly capture the boundary conditions for such calculations. This makes it possible to simulate an entire functional group, e.g. the crank mechanism, shaft assembly or combustion chamber. The ultimate goal of this development is integrated modeling of the overall "diesel engine" system.

Measurement with *strain gauges* is the most important *method of measurement* for stress analysis in engine design [7-5]. Measurements are taken of components' local expansion proportional to stress, which is generated by the static and dynamic load of the component being measured. Applied in many cases, the three-wire method of measurement can measure both expansion amplitudes and mean strains. Thus, not only dynamic components but also quasi-static stresses, e.g. stresses induced in a component by thermal expansion, can be measured. This method has reached a state that even allows taking measurements under difficult conditions, e.g. when surfaces are washed over by water or subjected to high temperatures or measured values are transmitted from moving components (Fig. 7-2).

Ensuring a maximum of certainty when determining component stress necessitates combining the methods of calculation and measurement (inasmuch as accessibility and proportions allow). Thus, theory and practice are united in one holistic analysis and complement one another outstandingly.

7.1.2.2 Stress of Selected Engine Components

Connecting Rod of a Medium Speed Diesel Engine

Typically, only *mechanical loads* have relevance for the strength of an engine's connecting rod. The component temperatures are in the range of the oil temperature and have only small gradients. Thus, they neither play a role in the ultimate strength factors nor the stress. The main stresses in a connecting rod are induced by the maximum gas force in ignition TDC (compressive stress) and by the inertial force of the piston or connecting rod itself in gas exchange TDC (tensile stress). In some evaluation points, the maximum stresses cannot be allocated to a TDC position. Thus, for example, *transverse bending* may be determinative for some points (e.g. the connecting rod bolts). Even *natural frequencies* may play a role in some connecting rod designs. When they are excited by corresponding excitation frequencies (e.g. from the gas pressure characteristic), then this alone can cause the component to fail. More and more frequently, the *relative motions* between individual components (e.g. between the bearing body and bearing bushing) must also be tested exhaustively. In conjunction with compressive components, an unacceptable degree of relative motion can cause tremendous damage to the area of contact and thus a drop in fatigue strength. Consequently, the determination of a connecting rod's stress is a substantial task that ought to be completed by applying every mathematical and experimental method available.

Fig. 7-2
Typical strain gauge application with wireless inductive transmission, here on a water pump shaft (torque measurement)

Components Adjacent to the Combustion Chamber

Gas pressure and *heating* highly load all the components surrounding a combustion chamber. Moreover, additional loads caused by *inertial forces* have to be allowed for in the piston and the gas exchange valves. What is more, surfaces on the side of the combustion chamber must be kept at a temperature level that does not unduly affect the material's strength.

When thermal loads resulting from combustion are large, thin walls help lower the surface temperature on the side of the combustion chamber and likewise – as a result of the low wall temperature gradient – reduce the thermal stresses. However, this could eliminate the design's resistance to high combustion pressures.

All these circumstance ultimately make combustion chamber components more difficult for engine developers to control than other components. Intensive preliminary tests, e.g. of the combustion process or the coolant flow, are already needed during the selection of the proper boundary conditions. This means investing large, frequently unfeasible amounts of time and money. In addition, some processes, e.g. oil spray cooling, can hardly be calculated in advance. Hence, a multitude of assumptions are already required to define the boundary conditions. Difficult to measure contact behavior (sliding processes and thermal conduction) in a thermally loaded assembly may adversely affect the quality of results.

Figure 7-3 presents the FEM model needed to calculate a cylinder liner's temperature field and the structural analysis that builds upon it. Stresses from the prestressing force of the cylinder head bolts and from the gas pressure are other stress variables that must be factored in.

Presented in Fig. 7-4, the temperature distribution in the steel top section of a composite steel piston, the piston head, clearly shows that the greatest temperature stress and largest temperature gradient appear here. Hence, great importance is attached to the determination of the thermal stresses occurring and the influence on the ultimate strength factors. The mechanical loads from the gas and inertial forces are dominant in the bottom section. Secondary piston motion also

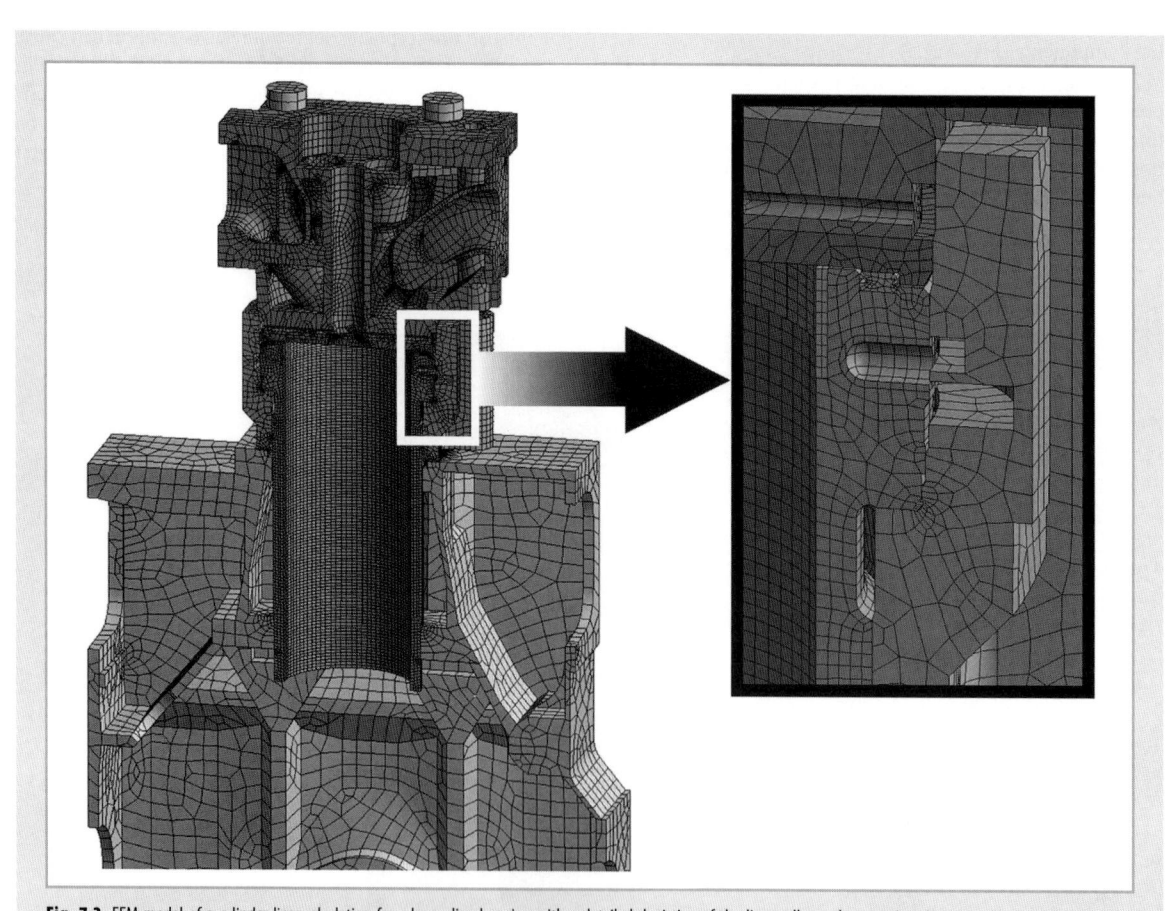

Fig. 7-3 FEM model of a cylinder liner calculation for a large diesel engine with a detailed depiction of the liner collar and supporting ring

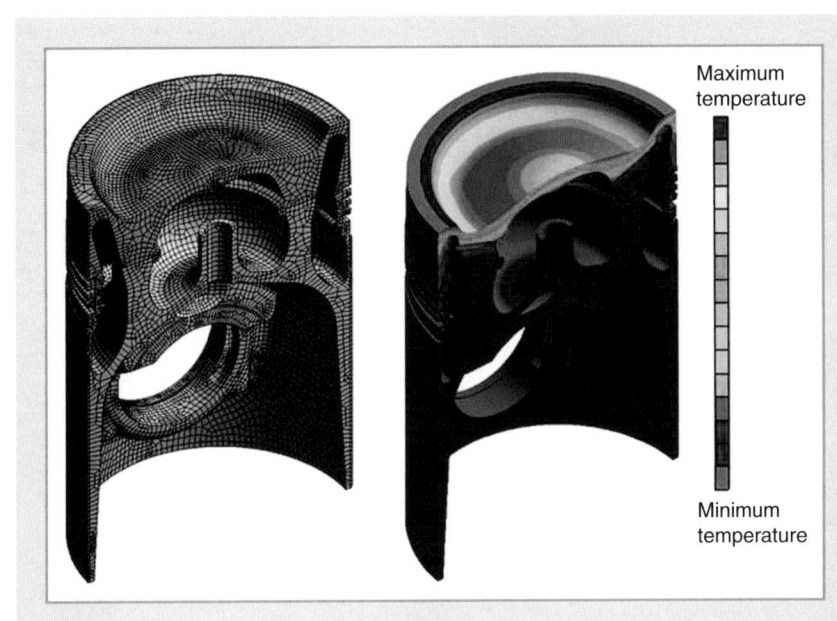

Fig. 7-4 FEM model of a composite piston with mathematically determined temperature distribution

plays an important role in the stress of the piston skirt, particularly in larger diesel engines.

Gas Exchange Valve with Drive

In addition to *gas forces* and *thermal loads* in the valve cone, *inertial forces* particularly play a crucial role in this assembly. The cam defines the curve of the valve stroke and the resulting accelerations generate considerable inertial forces in every component (tappet, push rod, roller rocker, valve spring and valve). The most important prerequisite for a perfectly functioning valve gear is a valve spring force greater than the inertial forces counteracting it at any time. Only this prevents contact losses between the parts of the drive, e.g. when the cam lifts the tappet, and thus high impact forces in the entire system.

In an initial step, analytic multimass models may be employed to calculate valve gear kinematics. Basically, this is based only on the stroke characteristic, acceleration, component masses and valve spring forces. This approach is unable to capture valve gear dynamics that are frequently determinative for a system.

Therefore, multibody simulations, which additionally allow for structural elasticity, system damping, contact stiffness and clearances, are employed as required [7-6]. Such modeling is essential, for example, when the intention is to improve the gas exchange through steeper stroke curve flanks and larger valve diameters. This affects the level of the inertial forces in two ways, i.e. by increasing the accelerations and by increasing the valve mass. Figure 7-5 presents an example of a thusly calculated force characteristic of a push rod, which corresponds to the measured characteristic very closely.

When designing a valve gear, a stress analysis is necessary to ascertain the stresses occurring in the valve stem and valve head. Suitable simulations can do this. The gas pressure stress when the valve is closed and when it is setting on the seat are important boundary conditions. Here too, metrological validation of the results is strongly recommended (provided it is feasible) since the stress is subject to strong variations (Fig. 7-6) caused by free bending vibrations during the opening phase, play in the valve guide and potential valve rotation. As a result, the setting conditions on the valve seat are never clearly defined and are thus subject to stoichiometric fluctuations. This produces a band for the stress curve that varies greatly from combustion cycle to combustion cycle. Manifestations of wear and contamination, which magnify these effects, may additionally appear in the seat and guide region in long-time engine operation.

7.1.3 Component Stress Analysis

Merely determining the stresses (effective stresses and actual stresses) that occur during engine operation does not suffice for an evaluation of component reliability. Determining the tolerable stresses, i.e. component strength, is just as important. The ratio of tolerable to maximum stress is crucial. This ratio is equivalent to a component's safety factor.

The gas and inertial forces furnish the maximum effective stresses for very many diesel engine components. A component's number of load cycles reaches values in the range of over 10^6 alternations of load after relatively short operating times, thus making it necessary to *design in fatigue strength* [7-7–7-9]. Assuming a truck engine has a service life of

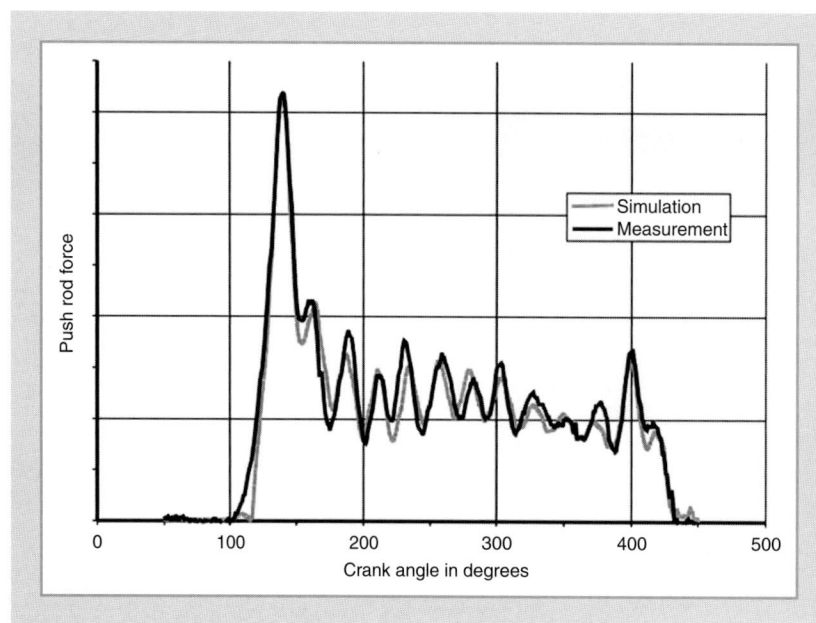

Fig. 7-5 Comparison of measurement/simulation of a typical push rod force characteristic

20,000 h, then the engine casing must tolerate approximately 10^9 load cycles caused by the maximum gas force in this time without fatigue failure. Static stresses in a component (e.g. caused by bolting forces) as well as thermal stresses in combustion chamber components are superimposed on the high frequency dynamic stresses. Thermal stresses vary with engine power and, strictly speaking, thus also ought to be interpreted as "alternating loading". However, since the "stress amplitudes" from thermal stresses have extremely low frequencies, thermal stresses are normally assumed to be quasi-static in engine design. Appropriate concepts that allow for low cycle fatigue (LCF) have to be applied in some cases, e.g. when evaluating local plastic deformations.

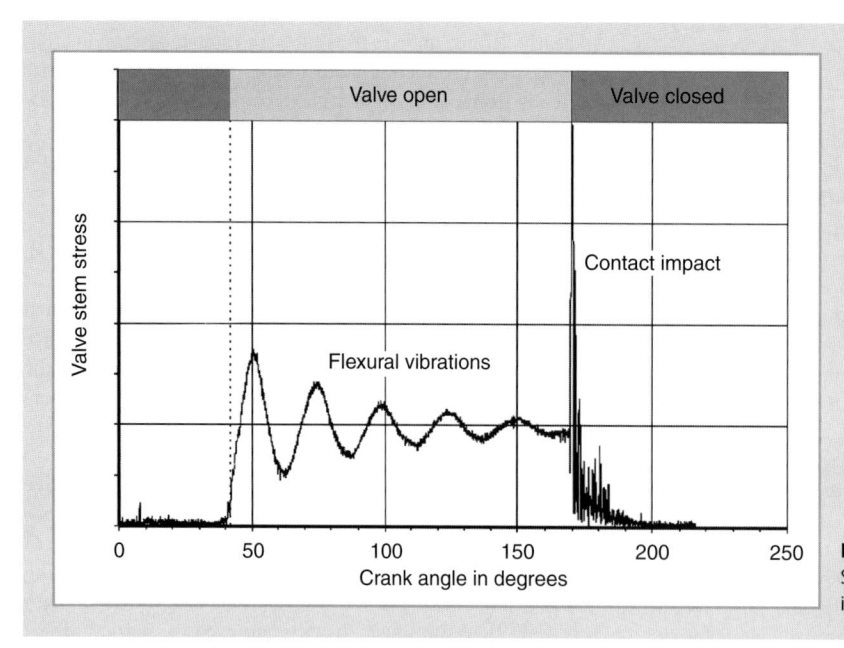

Fig. 7-6 Strain gauge measurement to determine load in a gas exchange valve's stem

High cycle fatigue (HCF) is normally evaluated in a *fatigue strength diagram*, which expresses the correlation of the tolerable stress amplitude (fatigue strength) to the effective static stress (mean stress).

Formerly common practice, solely allowing for material behavior no longer suffices for the materials now utilized in diesel engine design. Hence, it is advisable to employ *component-based fatigue strength diagrams* in which specific boundary conditions of concrete components can be allowed for in the assessment point. Important input parameters are the surface finish (roughness), the supporting effect in the assessment point, the technological size factor, the local component temperature or even a potential surface treatment (surface layer influence). A stress analysis performed on the basis of local component stresses is outlined in the FKM Guideline "Computerized Stress Analysis/Proof of Strength in Mechanical Components" (Fig. 7-7), [7-10].

According to the FKM Guideline, the *safety factors* differ depending on the consequences of damage, the probability of the occurrence of the maximum load, options for component inspection and prior quality assurance measures (specifically for components made of cast iron). Any uncertainties in the determination of stress are not included and must be additionally factored in during the stress analysis. Even when superior methods to determine the effective stresses are available and the fatigue strength can be calculated specifically for a component, experience has demonstrated that an *overall safety factor* of 1.5–2 or (even up to 3 for cast materials) must be planned in to cover remaining uncertainties.

A stress analysis proves to be particularly difficult when stress conditions are complex, e.g. rotating stress tensors or non-proportional principle stress curves. Such cases necessitate the incorporation of correspondingly complex concepts. Normally, the analysis is then performed by special structural integrity software.

Since traditional stress analyses assume material is flawless, another aspect when performing stress analyses is the testing of a component's *fracture mechanical aspects*. Hence, the application of fracture mechanics [7-11] supplements the evaluation of material defects and is thus an important instrument for diesel engine quality assurance.

Component tests with real components and *long-time engine operation* on test benches are other options for analyses of operational reliability. Although extreme load sequences may be used to a limited extent to obtain a time lapse effect, the substantial work required only justifies this approach for smaller components and vehicle engines.

7.1.4 Typical Component Damage in Diesel Engines

As in other domains of mechanical engineering, damage is caused by *product defects* (design, material and manufacturing) and *operating errors* (maintenance and operation). Typical examples in both categories exist for diesel engines. Naturally, they greatly depend on the load situation of the components. Influences that can only be factored in insufficiently or not at all when sizing components but nevertheless

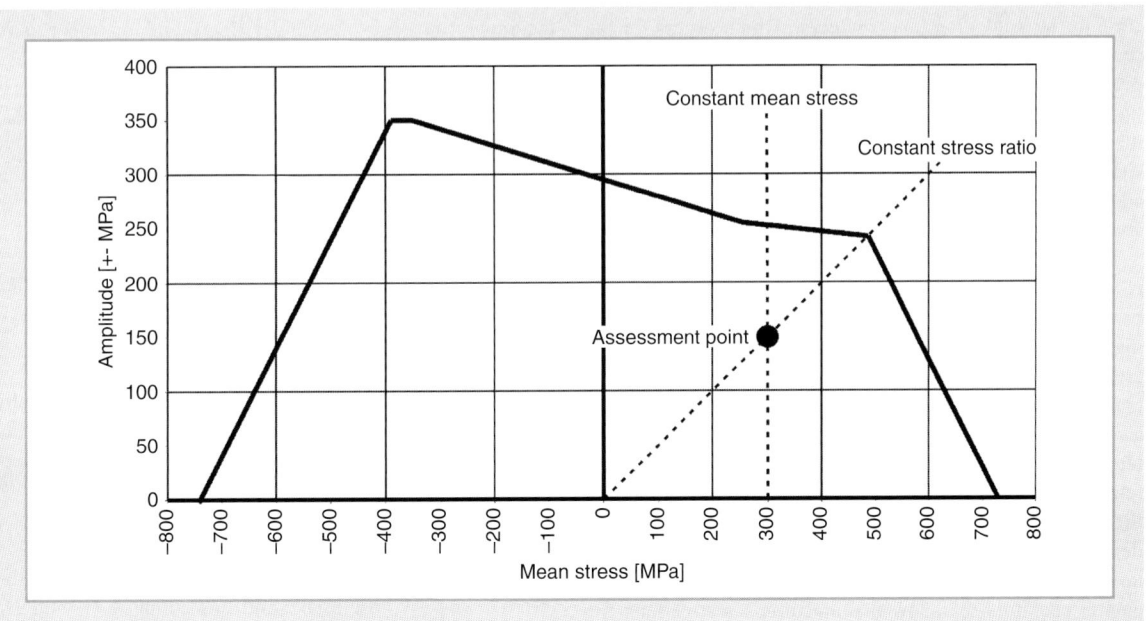

Fig. 7-7 Component-based fatigue strength diagram based on HAIGH

manifest themselves and then cause engine failure are equally important for engine developers and operators.

Failure caused by a fatigue fracture *(fatigue failure)* is the most frequent type of damage in diesel engines. It ensues from the dominant alternating loading of most of the components caused by gas and inertial forces. Fatigue fractures are relatively easy to detect in steel components because of their surface structure. A fatigue fracture's surface is usually smooth and fine-grained, while a residual fracture (overload fracture surface) exhibits a rough, fissured surface structure. So-called beach marks generated by the uninterrupted progression of a crack are frequently found concentric to a fracture's point of origin. As in other machines, incorrect sizing or undetected material defects may be *causes of fatigue fractures*.

Fatigue fractures induced by *fretting* (fretting corrosion) in areas of contact between components appear particularly unexpectedly and often only after a long period of operation. Fretting reduces fatigue strength to approximately 20% of its initial value. A fracture is often detectable in a small projection in the start of a fracture caused by the shear stresses that trigger the first incipient cracks. Among other things, mounting bores for sliding bearing shells (e.g. in connecting rod and main bearings), fitted bolts or press fits on shafts are endangered.

Cavitation can be another cause of fatigue failures. It affects components with coolant or lubricating oil conduits, the media of which are subject to strongly fluctuating pressures or high deflection rates. Crankshaft assemblies or injection components are prime examples. It is always caused by a drop below the particular fluid's vapor pressure, which causes vapor locks to form and break up. This causes the appearance of high accelerations, temperatures and pressure peaks in the medium, which can cause tremendous damage to the surface close to the wall. *Vibration cavitation*, which occurs on the coolant side of cylinder liners for example, is particularly a problem. It can be triggered by high frequency bending vibrations of the cylinder liner, which may be excited by secondary motion of the piston. If the coolant is chemically active as well, then a destructive, local corrosive attack also appears in addition to the cavitation [7-12]. This especially strikes high speed high performance diesel engines.

The extent of the damage very much depends on the type of corrosion. While the damage is relatively slight when a corrosive attack (surface corrosion) is uniform, it increases significantly when an attack is not (e.g. pitting) because of greater notching in the corroded spots. *Vibration corrosion cracking* is particularly damaging and becomes possible when a corrosive attack and dynamic tensile stresses occur simultaneously. In such a case, a component no longer has any fatigue strength because its service life is only determined by the rate at which the crack progresses and this, in turn depends, on the material, the corrosive medium and the level of the stress amplitude. Hence, it is very important for engine operators to follow manufacturer's instructions for coolant care using appropriate anticorrosion agents (see Sect. 9.2.6).

Wet or low temperature corrosion caused by the condensation of steam when sulfur dioxide is present is another type of corrosion, especially in large diesel engines operating with heavy fuel. It allows a very aggressive sulfurous acid to form. The process is defined by the interaction of the temperature and pressure level and, apart from the exhaust lines, primarily affects the cylinder liner. Wet corrosion that occurs during an engine's light load accelerates wear (see Sect. 4.3).

Surfaces on the side of the combustion chamber experience types of damage connected with locally high temperature.

The fuel's combustion products disintegrate protective oxide layers. While the material matrix partially dissolves, the material corrodes with commensurate material attrition. High thermal loading of a piston, e.g. in larger two stroke engines, may cause the piston crown to weaken locally and thus limits service life. Known as *high temperature corrosion*, this phenomenon also occurs in high pressure valves [7-13]. Countermeasures include limiting the valve temperature in the valve seat (preventing deposits of molten slag and thus accompanying corrosive attacks) and using special, corrosion resistant materials with high temperature stability values.

Thermal compression cracks that result from thermal overloading are another type of damage. They arise in spots where a component is strongly heated locally and the volume of surrounding colder components greatly impedes free thermal expansion. This generates compressive stresses that induce plastic buckling in the material. High internal tensile stresses appear in these spots once an engine has been stopped and the temperature equalized. The high "low cycle" stress amplitude produced when this process repeats ultimately causes incipient cracks. Such incipient cracks are common in cylinder heads in the piston lands between the valve openings, on the inner top edge of cylinder liners and on the rim of combustion chamber bowls. Root causes of such damage may be a disrupted combustion process, increased heat transfer during knocking or detonating combustion (e.g. in dual fuel engines) and insufficient cooling.

Readers are referred to the literature for more information on damage analysis and prevention [7-12, 7-14].

7.2 Heat Transfer and Thermal Loads in Engines

7.2.1 Introduction

In addition to the exhaust gas enthalpy H_A and the usually negligible leakage enthalpy H_L, the wall heat loss Q_W is the most important internal loss in a combustion engine's internal energy balance (Eq. (7-1)).

$$Q_b + Q_w + H_A + H_E + H_B + H_L + W_i = 0 \qquad (7\text{-}1)$$

Unlike the other components of the energy balance, the internal work W_i, the intake and fuel enthalpy H_E and H_B and the released fuel energy Q_B, it can be only be measured with difficulty.

However, since its effect on the process flow, efficiency and the formation of pollutants is so substantial, wall heat loss has been intensively researched virtually right from the start of combustion engine development and no end is in sight today. Hence, the literature on this subject is correspondingly extensive.

A search query of the Society of Automotive Engineers website (www.sae.org) with the keywords "wall heat transfer" delivers approximately 2,000 different SAE papers published in the last 5 years.

Although published in 1977, the standard work *Der Wärmeübergang in der Verbrennungskraftmaschine* by Pflaum and Mollenhauer [7-2] is a highly recommendable introductory treatment of the subject.

Therefore, this relatively short section only presents the basics and measurement systems, heat transfer equations and some cases of application, which are most important for understanding wall heat transfer. In addition, it addresses the special problem of modeling heat transfer in 3D CFD (computational fluid dynamics).

7.2.2 Basics of Heat Exchange in Combustion Engines

Wall heat loss (Q_w) in an engine combustion chamber is usually calculated in an engine process simulation by utilizing Newtonian heat transfer as an integral over one combustion cycle (ASP):

$$Q_w = \frac{1}{\omega} \cdot \int_{ASP} h \cdot A \cdot (T_w - T_z) \cdot d\varphi \qquad (7\text{-}2)$$

The temperature difference of the wall-gas temperature ($T_w - T_z$) is defined to produce a negative sign for the numerical value of the wall heat when energy leaves the "combustion chamber" system. Other values in Eq. (7.2) are the crank angle φ, angular frequency ω, instantaneous combustion chamber surface area A and heat transfer coefficient h.

Applying the Newtonian statement already implies that the mechanism of heat transfer in engine combustion chambers is essentially forced convection and the two other heat exchange mechanisms (radiation and thermal conduction) are negligibly small in comparison. The heat exchange in direct proximity to the wall must indeed be caused by thermal conduction since the flow there must be laminar. However, the mechanism of forced convection clearly dominates outside the boundary layer's viscous bottom layer.

The proportion of heat exchanged by radiation greatly depends on how much soot forms in the combustion chamber during combustion since only solid body radiation emanating from the soot is energetically relevant (cf. [7-15, 7-16]). Potential gas radiation on the other hand may be disregarded completely because gases are selective radiation sources and therefore only radiate in narrow wavelength bands. In turn, the proportion of soot radiation loss not only depends on the radiating soot mass and its temperature but also the amount of convective heat transfer. Realistically, radiation loss may only be assumed to attain a significant rate in large low speed diesel engines [7-17]. Therefore, the most frequently applied heat transfer equations dispense with an explicit radiation term (see Sect. 7.2.4).

All state variables (pressure, temperature and gas composition) in an engine combustion chamber may fundamentally be assumed to differ at any time and in any place. This also applies to the wall surface temperature and turbulence that strongly influence heat transfer. Moreover, an absence of developed flow conditions may be expected. Instead, stagnation flows and transient initial flows may be assumed, making this situation virtually unfathomable.

If, however, the conditions in the proximity of the combustion chamber wall are analyzed entirely in principle (Fig. 7-8), it becomes clear that, as in many other technical heat transfer problems, applying similarity theory to analyze the phenomenon of "heat transfer" is conducive to the desired objectives. Dimensionless coefficients describe the phenomena and their relationship. In the case of forced convective heat transfer, this is the Nusselt number $\left(\text{Nu} = \frac{h \cdot d}{\lambda}\right)$ for the temperature and the Reynolds number $\left(\text{Re} = \frac{w \cdot d \cdot \rho}{\eta}\right)$ for the flow boundary layer. The Prandtl number $\left(\text{Pr} = \frac{v}{a}\right)$ describes the interaction between the two boundary layers. Both boundary layers are equally thick in Pr = 1, for example. Pr is always < 1 for air and exhaust gas and thus the flow boundary layer is always thinner than the temperature boundary layer. Strictly speaking, all these analyses only apply to stationary and quasistationary cases, i.e. changes must occur so slowly that no transient effects, e.g. inertia effects, appear. Heat transfer in the combustion chamber of an internal combustion engine is certainly not "quasistationary". Rather, it changes very rapidly temporally and locally, especially in high speed engines. Nevertheless, a quantitatively significant influence of transient effects on heat transfer has not yet been measurably demonstrable (see Sect. 7.2.8).

The heat transfer coefficient h may be calculated by linking the dimensionless coefficients by means of a general power law model:

$$\text{Nu} = C \cdot \text{Re}^n \cdot \text{Pr}^m \qquad (7\text{-}3)$$

Extensive tests [7-18] have demonstrated that the numerical values $n = 0.78$ and $m = 0.33$ for the exponents m and n known from the turbulent pipe flow are valid for combustion engines.

Consequently, within the relevant temperature range, $\text{Pr}^{0.33}$ is constant within a tolerance of $\pm 1\%$ and may be added to the constant C. This produces a general heat transfer equation for engine combustion chambers based on similarity theory:

$$h = C \cdot d^{-0.22} \cdot \lambda \cdot \left(\frac{w \cdot \rho}{\eta}\right)^{0.78} \qquad (7\text{-}4)$$

Pflaum [7-2] includes empirical polynomials for thermal conductivity λ and dynamic viscosity η dependent on the temperature and gas composition. Thermal equations of state may be used to formulate density ρ as $\rho = const \cdot \frac{p}{T}$.

Fig. 7-8 Schematic of heat transfer by forced convection to a combustion chamber wall

Nusselt number
$$Nu = \frac{h \cdot d}{\lambda}$$

Reynolds number
$$Re = \frac{w \cdot d \cdot \rho}{\eta}$$

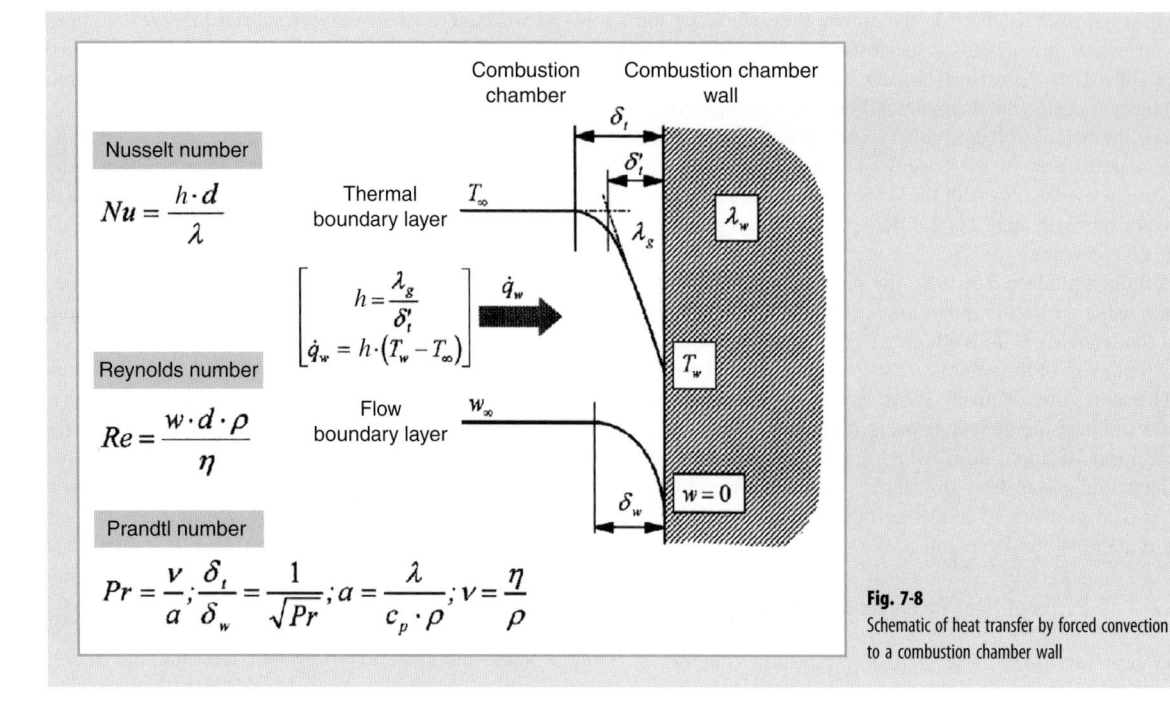

Prandtl number
$$Pr = \frac{v}{a} \cdot \frac{\delta_t}{\delta_w} = \frac{1}{\sqrt{Pr}}; a = \frac{\lambda}{c_p \cdot \rho}; v = \frac{\eta}{\rho}$$

$$h = \frac{\lambda_g}{\delta'_t}$$
$$\dot{q}_w = h \cdot (T_w - T_\infty)$$

Thus, the explicit setup of a heat transfer equation requires a proper interpretation of the characteristic length d and rate w relevant to heat transfer based on measured values and "scaling" by means of the constant C.

"Based on measured values" means that, in absence of a numerical or even analytical solution to the problem, suitable measurements of designed engines must be utilized to create a base of data that allows semi-empirical modeling intended to be as universally valid as possible to ensure that it will also be applicable to future combustion system developments.

7.2.3 Methods of Heat Flux Measurement

Based on the current state of knowledge, three methods are essentially suited to create a base of data to mathematically model a heat transfer equation.

The "internal heat balance" (Fig. 7-9) allows calculating the temporal and local mean wall heat loss based on the combustion chamber's internal energy balance (Eq. (7-1)). Intake, exhaust gas, leakage and fuel enthalpy must be measured. The heat of combustion Q_b is produced by multiplying the mass of injected fuel by the net calorific value H_u ($Q_b = m_b \cdot H_u$). In turn, the internal work is delivered by the ring integral over one cycle of the pressure characteristic p_z measured as a function of the crank angle multiplied by the volumetric change $dV/d\varphi$ (see Sect. 1.2).

Assuming the measurements have been taken meticulously, the "internal" heat balance is a necessary but inadequate method to determine wall heat losses [7-19]. Since only the total energy balance is evaluated, it is impossible to provide information on the development of the local mean wall heat flux over time, e.g. for an analysis of the thermodynamic pressure characteristic (see Sect. 1.3).

Direct methods of measurement to determine wall heat losses are

(a) a heat flux sensor that determines local temporal mean wall heat flux density q_w and
(b) the surface temperature method that measures local temporally variable wall heat flux density (see Fig. 7-10).

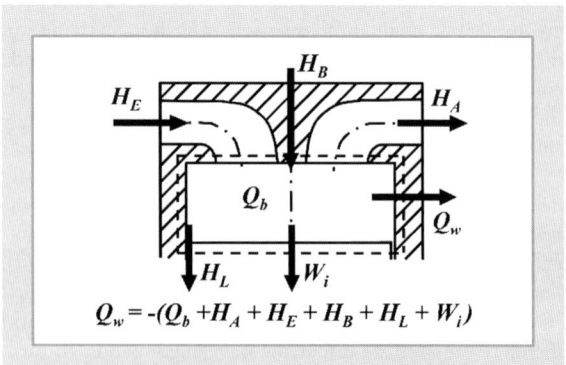

$$Q_w = -(Q_b + H_A + H_E + H_B + H_L + W_i)$$

Fig. 7-9 Principle of the internal heat balance to determine the temporal and local mean wall heat loss Q_w

Fig. 7-10 Principle of probes to directly measure (a) the local, temporal mean and (b) the local, temporally variable wall heat flux density $\dot{q}_w$ on the combustion chamber wall

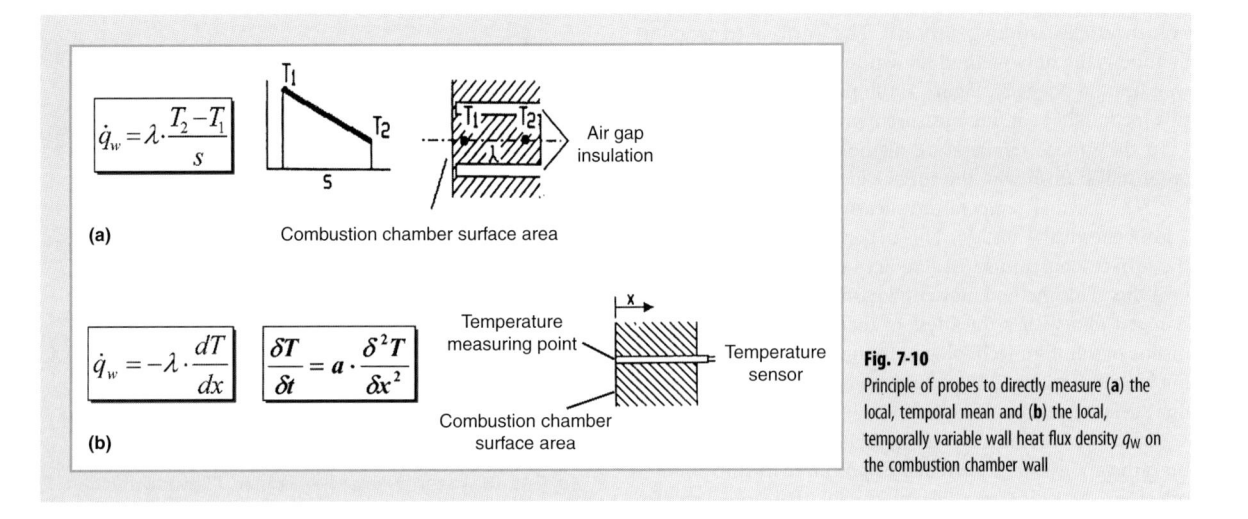

Heat flux sensors should be installed as flush with the combustion chamber surface as possible. Insulation created by an air gap establishes a defined, one-dimensional heat conduction path inside the sensor. When the thermal conductivity of the material of the heat conduction path is known, measurements of the two temperatures T_1 and T_2 in the defined interval s may be used to easily calculate the wall heat flux density with the equation for one-dimensional stationary thermal conduction.

$$\dot{q}_w = \lambda \cdot \frac{T_2 - T_1}{s} \qquad (7\text{-}5)$$

Figure 7-11 presents a heat flux sensor of the type effectively implemented in [7-19] and [7-2]. While the method's principle is very simple, its practical application necessitates relatively complex sensor designs whenever high accuracy is required.

Nevertheless, the relatively easy application of this method of heat flux measurement and the high accuracy achievable are advantageous since it only requires measuring a few component temperatures highly precisely and does not require forming the differences of measured values from several different methods of measurement. As in the method of "internal heat balance", their absolute value in the normal case would be even greater than the differential value being determined.

However, there is one clear disadvantage. Normally a heat flux sensor may only be installed in one or at the most two positions in a combustion chamber since it must be mounted absolutely flush to the surface. Otherwise, the convective heat transfer would be unduly altered by edges that are not present otherwise. Thus, the results of measurement merely represent the local conditions at the point of installation and cannot provide information on the development over time at all.

The temporal mean wall heat flux ascertained with the methods described varies with the gas temperature that constantly changes during a combustion cycle and thus also

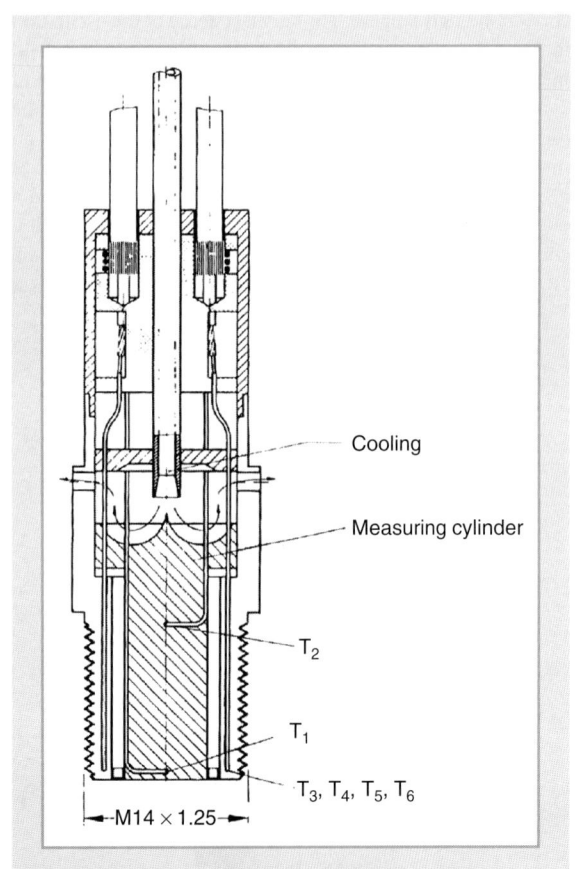

Fig. 7-11 Heat flux sensor with cooling, compensation measuring points and a pure iron graduated cylinder

influences the working process. The surface temperature method is the only method known to measure local variable wall heat flux density. Figure 7-10b presents the principle of this method of heat flux measurement.

In the surface temperature method, temperature sensors are installed flush with the surface of the combustion chamber. Their actual temperature measuring point is less than 2 μm beneath the surface. This makes it possible to measure the surface temperature oscillation caused by the variable wall heat flux. This method may employ sheathed thermocouples. A metal film with a thickness of merely 0.3 μm applied by means of thin film technology produces their actual measuring junction. Figure 7-12 pictures an example of the design of one such thermocouple [7-19].

Figure 7-13 presents typical calculated temperature oscillations in various component depths produced during powered operation of a commercial vehicle DI diesel engine. The slightness of the temperature oscillations relative to changes of the gas temperature is immediately conspicuous and is caused by the widely varying heat penetration coefficient b (Eq. (7.7)), which is larger for metals than gases roughly by a factor of 500. The amplitude of the temperature oscillation is correspondingly smaller. The very fast decay of the oscillation in the component is additionally evident. A temperature field may already be assumed to be purely stationary at a depth of 2 mm. Thus, the "true" surface temperature can only be measured when the temperature measuring junction is no more than 2 μm away from the surface.

Taking the periodicity of the temperature oscillation and a one-dimensional transient temperature field on the combustion chamber wall that expands infinitely in one direction as the starting point, the wall heat flux densities that cause the temperature characteristics may be calculated with Laplace's differential equation. Eichelberg was the first to formulate Laplace's differential equation in the form of a Fourier series [7-21]

$$\frac{\delta T}{\delta t} = a \cdot \frac{\delta^2 T}{\delta x^2}$$

$$T(t,x) = T_m - \frac{\dot{q}_m}{\lambda} \cdot x + \sum_{i=1}^{\infty} e^{-x\sqrt{\frac{i\omega}{2a}}} \cdot$$

$$\left[A_i \cdot \cos\left(i\omega t - x\sqrt{\frac{i\omega}{2a}}\right) + B_i \cdot \sin\left(i\omega t - x\sqrt{\frac{i\omega}{2a}}\right) \right]$$

$$\dot{q} = -\lambda \cdot \left(\frac{\delta T}{\delta x}\right)_{x=0}$$

$$\dot{q} = \dot{q}_m + \lambda \cdot \sum_{i=1}^{\infty} \sqrt{\frac{i\omega}{2a}} \cdot$$

$$[(A_i + B_i) \cdot \cos(i\omega t) + (B_i + A_i) \cdot \sin(i\omega t)]$$

(7-6)

The accuracy achievable with the surface temperature method greatly depends on the meticulousness of the installation of the surface thermocouples in the combustion chamber wall, i.e. both their thermal connection to the surrounding materials and their absolute flushness with the surface are crucially important.

In addition, the heat penetration coefficient

$$b = \frac{\lambda}{\sqrt{a}} = \sqrt{\rho \cdot c \cdot \lambda} \qquad (7\text{-}7)$$

valid for the surface thermocouples including the surrounding materials must be calibrated individually based on the temperature. This may be done mathematically [7-19] or with a special experimental method [7-22].

Ultimately, how to ascertain the temporal mean wall heat flux density $\dot{q}_m$ when employing the surface temperature method remains a question. Metrologically, this might be possible by integrating surface thermocouples in heat flux

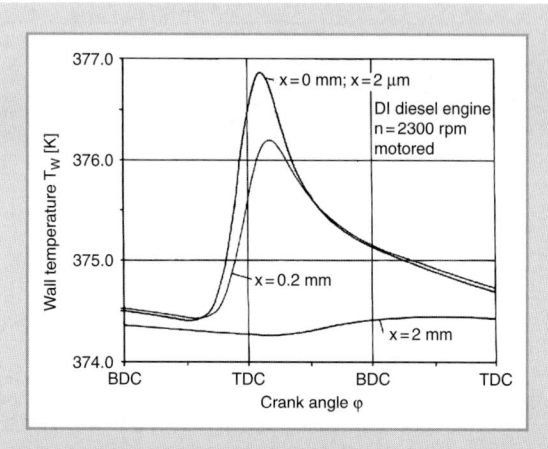

Fig. 7-13 Calculated surface temperature oscillations. Commercial vehicle DI-diesel, over-run condition, $n = 2,300$ rpm [7-20]

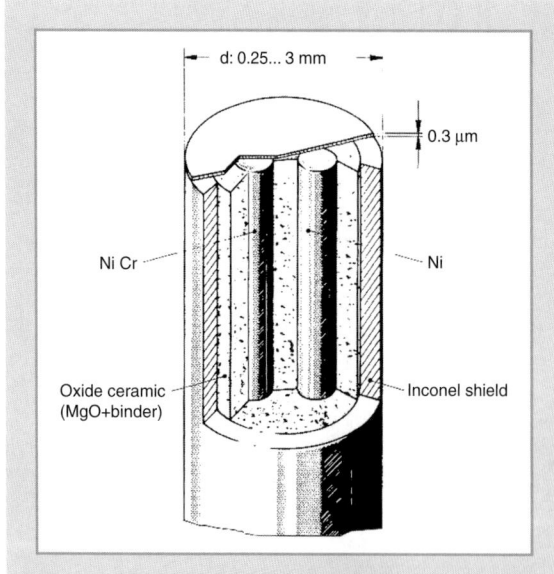

Fig. 7-12 Surface thermocouples based on [7-19]

sensors. A maximum number of thermocouples would have to be installed for the surface temperature method to be able to form a representative local mean value. This will still have to be demonstrated.

One alternative is the "zero crossing method", which utilizes the temperature difference's passage through zero $(T_w - T_g) = 0$. This usually occurs during the compression stroke because, disregarding potential transient effects, the instantaneous wall heat flux density must also be $\dot{q} = 0$ according to Eq. (7.2). Thus, Eq. (7.6) becomes:

$$\dot{q}_m = -b \cdot \sqrt{\frac{\omega}{2}} \cdot \sum_{i=1}^{\infty} \sqrt{i} \cdot$$
$$[(A_i + B_i) \cdot \cos(i\omega t_0) + (B_i + A_i) \cdot \sin(i\omega t_0)]$$
$$t_0 = t_{(T_g = T_w)}$$
(7-8)

In particular, Eq. (7.8) can be used with sufficient accuracy, even locally, for individual surface temperature measuring points since the very great local gas temperature differences that exist during combustion do not exist during compression. Thus, the thermal equation of state calculates the gas temperature T_g as the mass average temperature T_z.

$$T_g = T_z = \frac{p \cdot V}{m_z \cdot R_g} \quad (7\text{-}9)$$

The formation of deposits (soot, carbon, etc.) in the measuring points during operation is a problem that arises in diesel engines in particular but also in gasoline engines too. As a result, the actual temperature measuring point is no longer directly at the surface and, thus, the measuring signal is damped. The temperature characteristics then take on the appearance in Fig. 7-13. Methods to correct such deposits, provided they have not grown too thick and an evaluable vibration is no longer measurable, are specified in [7-19] and [7-22]. In principle, operation with soot-free model fuels is very advisable to prevent this problem entirely. However, this is seldom done since, on the one hand, it is feared that the influence of non-standard fuels on wall heat transfer will not be transferrable and, on the other hand, certain phenomena can only be represented with difficulty with model fuels, e.g. identical mixture formation properties during interaction with the wall (droplet formation, penetration, wall coat, vaporization and ignition quality). Instead, the ability to only measure a few operating points before the cylinder head and/or the piston has to be removed again, cleaned, recoated and remounted is tolerated.

Usually, just a few commercially available surface thermocouples [7-23] are used to minimize work and costs (see [7-24] through [7-25]). However, a very large number of surface thermocouples were used in [7-19]. Although this was a gasoline engine, fundamentally important conclusions are nevertheless inferable. Figure 7-14a presents 182 wall heat flux density characteristics for powered engine operation. As a pure charge port, the intake port's shape caused the flow field to solely consist of non-directional turbulence. Thus, all the measuring points averaged over 100 combustion cycles display nearly identical characteristics. The conditions during combustion are entirely different (Fig. 7-14b).

The wall heat flux densities rise steeply after the ignition point (IP) as soon as the flame front has reached the surface temperature measuring point. The flame front loses intensity (the temperature drops in the burned fractions after peak pressure) as the distance to the spark plug increases, i.e. as the time of combustion is increasingly retarded. The steep rise of the wall heat flux density visibly levels out as a result. While the sequences when the flame front overruns the measuring points are easily interpreted, a convincing explanation of the drop in wall heat flux density after reaching its maxima without any discernible system is challenging. Individual behavior characterized by chaotic turbulent states in the burned fractions when the piston expands appears at every measuring point.

This example impressively demonstrates the need to employ a large number of surface temperature measuring points in order to be able to form a genuinely representative local mean value. Though found repeatedly in the literature, comparisons of wall heat flux density characteristics measured at single measuring points with the familiar heat transfer equations that describe a local mean value are invalid and meaningless.

Since the locally highly variable gas temperatures outside the temperature boundary layer are unknown during combustion and the driving gas-wall temperature gradient consequently has to be determined from the mass average temperature T_z calculated by the thermal equations of state, the heat transfer coefficient h can only be calculated from the representative local wall heat flux density averaged for the entire combustion chamber.

Figure 7-15 presents an example of such an analysis. The discontinuity that appears in the heat transfer coefficient in the range of 110°CA stems from the determination of the temporal mean wall heat flux density $\dot{q}_m$ by means of the "zero crossing method". Thus, this is by no means proof of a "transient effect". Instead, it is more the product of minor phase shifts (the measuring chain's electrical delay time) between the measurement of combustion chamber pressure from which the mass average temperature T_z was calculated and the surface temperature measurements from which the wall heat flux densities were calculated. The formation of the quotients $\dot{q}/(T_z - T_w) = \dot{q}/\Delta T = h$ then produces the effect displayed in the range of $\dot{q} = 0$. Ideally, a pole would have to be produced mathematically.

No method is known to directly allow measuring the local mean and temporally variable characteristic of wall heat losses. Each of the known methods is only able to either deliver local and temporal mean values or local time curves. Even though the latter, the surface temperature method, is very complex, it is the only method that meets every requirement when an adequate number of measuring points are employed.

Used most, the heat transfer equations presented below were either set up with the aid of one or more of the three methods of measurement or at least verified with them many times over.

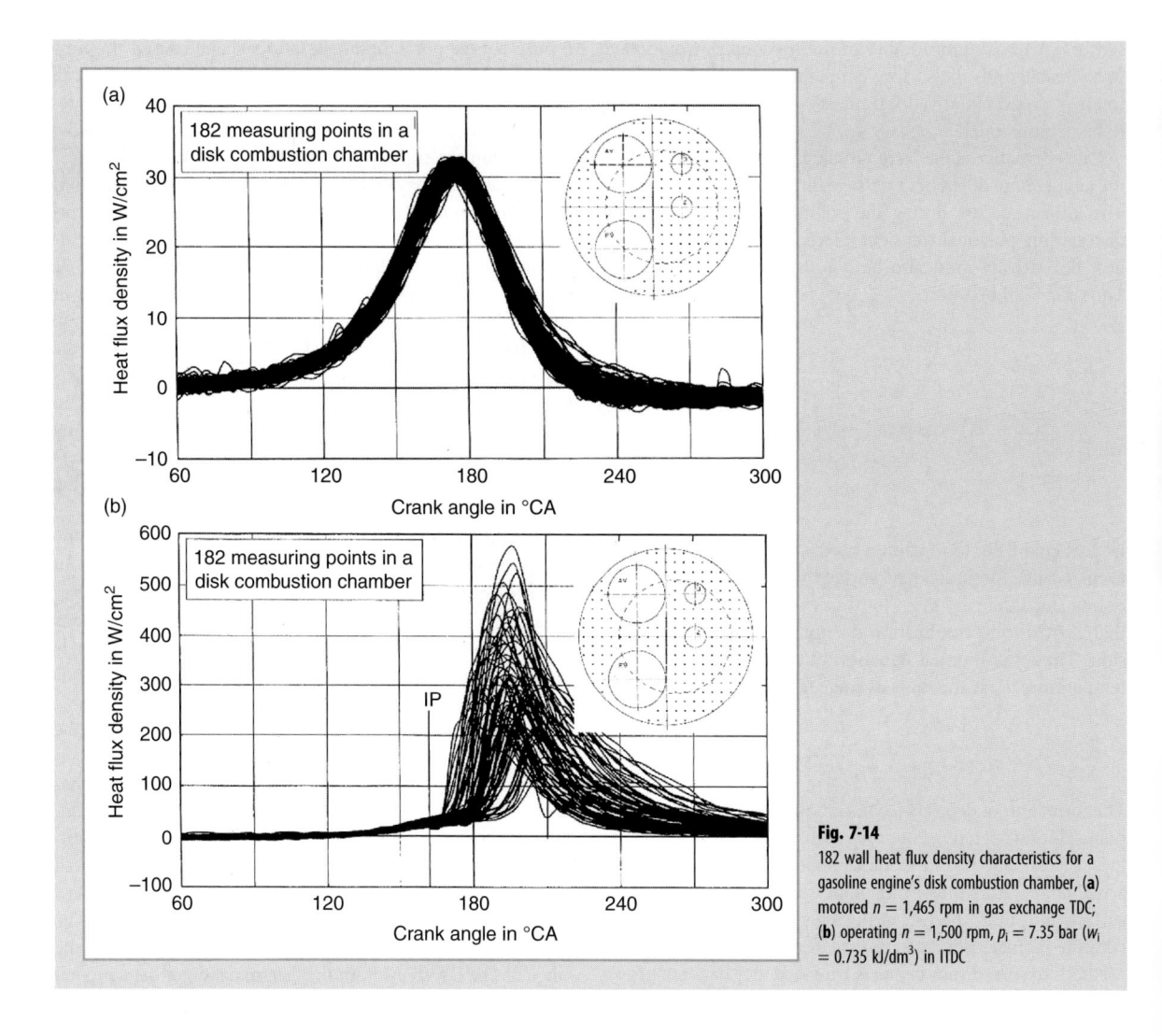

Fig. 7-14 182 wall heat flux density characteristics for a gasoline engine's disk combustion chamber, (a) motored $n = 1{,}465$ rpm in gas exchange TDC; (b) operating $n = 1{,}500$ rpm, $p_i = 7.35$ bar ($w_i = 0.735$ kJ/dm^3) in ITDC

7.2.4 Heat Transfer Equations for Engine Process Simulation

The history of published heat transfer equations that calculate the local mean but temporally variable wall heat losses as part of an engine process simulation began with *Nusselt* at the start of the last century. Although he formulated his Nusselt number as an essentially dimensionless coefficient to calculate forced convective heat transfer, the heat transfer equation he published in 1923 [7-26] was formulated purely empirically and was not based on similarity theory. In 1928 [7-21], *Eichelberg* merely modified the constants and exponents of the "Nusselt equation" without any fundamental changes. Other empirical equations based on the empirical "Nusselt equation" were produced in German and English speaking regions. Given their purely empirical character, all these equations are virtually nontransferable to any other engines than those for which they were modified and consequently lack any universal validity.

In 1954, *Elser* was the first to apply an equation based on similarity theory [7-27]. His equation received relatively little attention though.

It was left to *Woschni* to formulate the first heat transfer equation for combustion engines based on similarity theory. It is still in use today. Initially developed for engine process simulation as part of the implementation of EDP at MAN, it was already published in "raw form" in 1965 [7-28]. Since 1970, the final equation has been [7-18]:

$$h = 130 \cdot D^{-0.2} \cdot T_z^{-0.53}$$

$$\cdot p_z^{0.8} \left(C_1 \cdot c_m + \underbrace{C_2 \cdot \frac{V_h \cdot T_1}{p_1 \cdot V_1} \cdot (p_z - p_0)}_{\text{Combustion term}} \right)^{0.8} \quad (7\text{-}10)$$

$$\underbrace{\phantom{p_z^{0.8} \left(C_1 \cdot c_m + C_2 \cdot \frac{V_h \cdot T_1}{p_1 \cdot V_1} \cdot (p_z - p_0) \right)^{0.8}}}_{w}$$

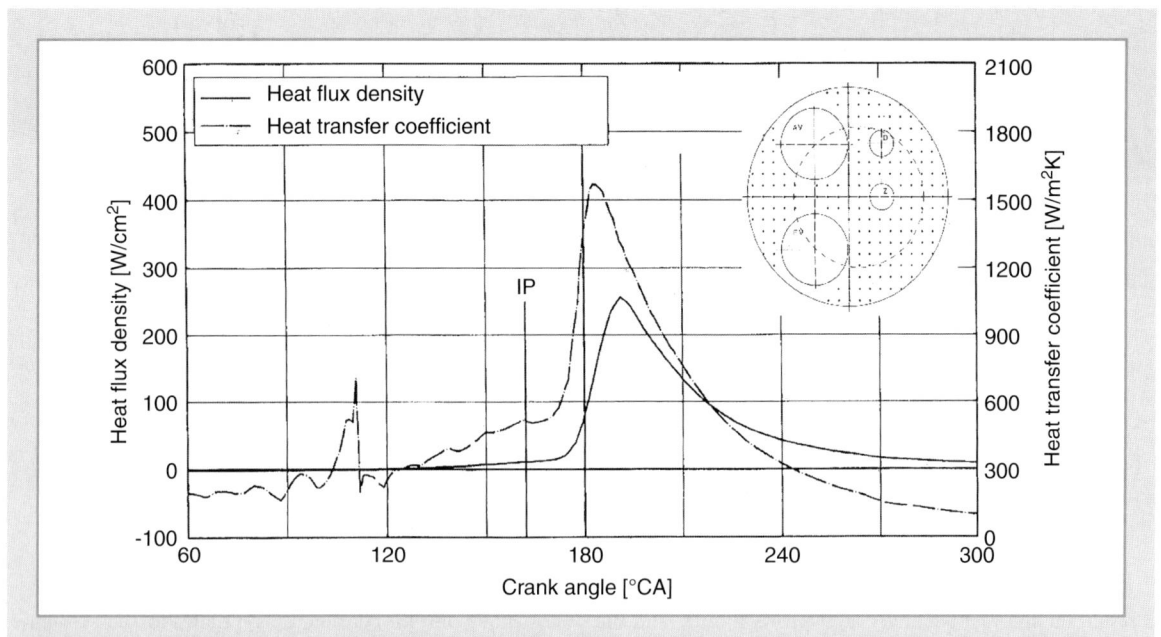

Fig. 7-15 Area-related local mean values and local mean heat transfer coefficient determined from 182 wall heat flux density characteristics, gasoline engine, disk combustion chamber, operating, $n = 1{,}500$ rpm, $p_i = 7.35$ bar [7-19]

Where

$C_1 = 6.18 + 0.417\, c_u/c_m$ from exhaust opens to intake closes

$C_1 = 2.28 + 0.308\, c_u/c_m$ from intake closes to exhaust opens

and

$C_2 = 0.00324$ m/(s K)   for engines with undivided combustion chambers

$C_2 = 0.0062$ m/(s K)   for engines with divided combustion chambers (into main and secondary combustion chambers)

$C_2 = 0$ for compression and gas exchange

In addition to rounding the Re number's exponent n (see Eq. (7.4)) from 0.78 to 0.8, *Woschni* selected a cylinder diameter D as the characteristic length. Applying the mean piston velocity c_m, he formulated a flux term w with the integrated combustion term while dispensing with heat radiation. Accordingly, combustion generates turbulences that are modeled by the difference between "pressure with combustion p_Z and pressure without combustion p_0 (engine overrun condition)" and scaled by the quantity C_2 dependent on the combustion process and by the swirl-dependent quantity $C_1 = f(c_u/c_m)$ for the flow velocity (to determine the swirl number and the peripheral velocity c_u [see Sect. 2.1.2.4]).

Different constants apply to the gas exchange phase than to the high pressure phase. Changing the constants and "switching off" the combustion term when exhaust opens yields a somewhat less attractive "bend" in the wall heat flux curves (Fig. 7-16).

The calculated mass average temperature T_z (Eq. (7.9)) is used to simulate the wall heat flux for the driving temperature difference to the wall. Different "wall temperature ranges" (e.g. cylinder head, piston and cylinder liner) may be incorporated. However, this is not intended to produce a correlation of the heat transfer coefficient. In addition, the equation is adjusted so that the surface of the piston top land is disregarded as if it were not part of the combustion chamber. When the piston top land is explicitly calculated in the process simulation, then the wall heat loss produced there has to be reincorporated into the combustion chamber in a suitable form with an inverted sign. Otherwise, the energy balance cannot be correct.

Equation (7.10) is by far the heat transfer equation most frequently employed throughout the world for over 35 years. Under *Woschni's* direction, the equation has been repeatedly verified (c.f. [7-24] and [7-29]), improved with additions and adapted to specific problems.

Kolesa [7-30] studied the influence of high wall temperatures on wall heat loss, his finding being that the heat transfer coefficient increases significantly as of a certain wall temperature ($T_w > 600$ K) because the flame burns closer to the wall. Thus, the flame quenches in the proximity of the wall later. As

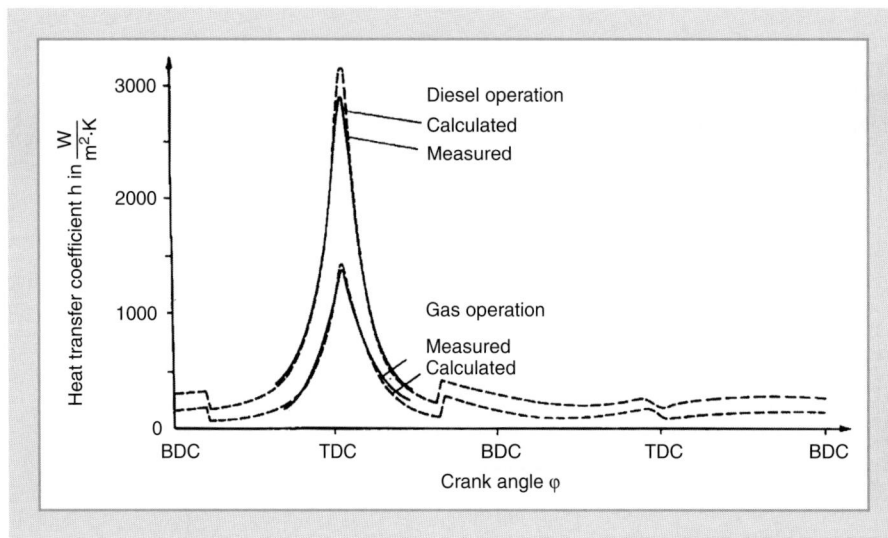

Fig. 7-16 Comparison of the results of Eq. (7.10) with surface temperature measurements for a diesel engine [7-29] and gasoline engine [7-24]

a result, the thermal boundary layer grows thinner and the temperature gradient increases within the boundary layer. This explains the increase of the heat transfer coefficient. This increase counteracts the decline of the wall heat flux caused by the driving temperature difference that grows smaller as the wall temperature rises. Thus, the wall heat loss even increases at first and only begins to decrease at higher wall temperatures until the state of an adiabatic engine is reached when the temperature is identical with the energetic mass average temperature (see Sect. 7.2.7).

The modification developed by *Kolesa* only affects the constant C_2 of the combustion term and was only verified for engines with undivided combustion chambers:

$$T_w \leq 600\,K$$
$$C_2 = 0.00324$$
$$T_w > 600\,K \tag{7-11}$$
$$C_2 = 2.3 \cdot 10^{-5} \cdot (T_w - 600) + 0.005$$

Schwarz eliminated the undesirable distinction of cases in 1993 by converting Eq. (7.11) into a constant form [7-31]:

$$T_w < 525\,K$$
$$C_2^* = C_2 = 0.00324$$
$$T_w \geq 525\,K \tag{7-12}$$
$$C_2^* = C_2 + 2.3 \cdot 10^{-6} \cdot (T_w - 525)$$

More frequent use of *Woschni's* heat transfer equation revealed that the established equation's calculation of heat transfer was too low for low loads and powered engine operation. In addition, wall heat loss was discovered to be a function of the sooting of the combustion chamber surface.

Therefore, *Huber* [7-22] and *Vogel* [7-32] supplemented Eq. (7.10) with a variable term that yields a modified velocity element w. Only the constant C_3 is specified for diesel engines. For other combustion systems and fuels, $C_3 = 0.8$ applies to gasoline and $C_3 = 1.0$ to methanol.

If:

$$2 \cdot C_1 \cdot c_m \cdot \left(\frac{V_c}{V_\varphi}\right)^2 \cdot C_3 \geq C_2 \cdot \frac{V_h \cdot T_1}{p_1 \cdot V_1} \cdot (p_z - p_0)$$

then:

$$h = 130 \cdot D^{-0.2} \cdot T_z^{-0.53} \cdot p_z^{0.8} \cdot w_{\text{mod}}^{0.8} \tag{7-13}$$

where:

$$w_{\text{mod}} = C_1 \cdot c_m \cdot \left(1 + 2 \cdot \left(\frac{V_c}{V_\varphi}\right)^2 \cdot C_3\right)$$
$$C_3 = 1 - 1,2 \cdot e^{-0.65\lambda_v}$$

Equation (7.10) is also being developed further in the USA too. In 2004, Assanis et al. [7-25] adapted the basic *Woschni* equation to HCCI combustion systems.

Nonetheless, experience has shown that none of the modifications mentioned is widely applied. Rather, mainly the original form of Eq. (7.10) is applied despite its known weaknesses. Whenever this now standard form is deviated from, another equation is usually selected for the heat transfer coefficient *HTC*.

In 1980, *Hohenberg* published an equation for heat transfer in diesel engines, which is also based on similarity theory [7-33]:

$$h = 130 \cdot V^{-0.06} \cdot T_z^{-0.53} \cdot p_z^{0.8} \cdot \left(T_z^{0.163}(c_m + 1.4)\right)^{0.8} \tag{7-14}$$

Arranged relatively simply, it is frequently referenced for comparison with the *Woschni* equation.

Hohenberg employed the radius of a sphere with a volume corresponding to the instantaneous volume of the combustion chamber as the characteristic length. He chose a sphere because it is the only geometric body that can be described by specifying a geometric quantity. Engines with differing stroke/bore ratios are reproduced better as a result. Notably, incorporating the exponent "3" ($V = \pi r^3$) produces a very small exponent (–0.06), thus demonstrating that the heat transfer coefficient is largely independent of an engine's geometric dimensions.

For the influence of velocity, *Hohenberg* likewise selected mean piston velocity c_m supplemented by a constant (1.4) for the influence of combustion and a slight temperature correlation of this term: $1.4 \, T_z^{0.163}$. While *Hohenberg* described in [7-33] how he also experimentally detected a slight influence of pressure on velocity relevant to heat transfer ($p_z^{0.25 \cdot 0.8}$), he nevertheless also reduced the pure pressure exponent from 0.8 to 0.6 so that pressure does not exert any influence on velocity mathematically in the strict sense of similarity theory ($\text{Re}^{0.8}$) and, all in all, the pressure exponent remains at 0.8.

Unlike the *Woschni* equation, Eq. (7.2) also allows for the surface of the piston top land in the calculation of the instantaneous total surface of the combustion chamber A:

$$A = A_{\text{Combustion chamber}} + A_{\text{Piston top land}} \cdot 0.3 \quad (7\text{-}15)$$

The factor 0.3 takes into account that the heat transfer in the piston top land constitutes merely 30% of the heat transfer in the combustion chamber. The piston top land surface $A_{\text{Piston top land}}$ ensues from the size of the combustion chamber multiplied by the height of the piston top land times two ($A_{\text{Piston top land}} = D \cdot \pi \cdot 2 \cdot h_{\text{Piston top land}}$).

Equation (7.14) is the outcome of extremely extensive experimental tests on a larger number of very different engines with the widest variety of methods of measurement in [7-33] and [7-34].

In 1991, *Bargende* published [7-19], [7-13] another relationship for the *HTC* initially developed for gasoline engines, which however is also applied to diesel engines [7-35] and [7-36]. It too is based on similarity theory.

$$h = 253.5 \cdot V^{-0.073} \cdot \left(\frac{T_z + T_w}{2}\right)^{-0.477} \cdot p_z^{0.78} \cdot w^{0.78} \cdot \Delta \quad (7\text{-}16)$$

Instead of a rounded value (0.8), the exact value is used for the exponent $n = 0.78$. *Hohenberg*'s formulation was adopted as the characteristic length ($d^{-0.22} \cong \cdot V^{-0.073}$). A temperature averaged from the mass average temperature and wall temperature ($T_m \frac{T_z + T_w}{2}$) is utilized as the temperature relevant for the heat transfer coefficient since the gas to wall temperature is balanced within the thermal boundary layer, thus requiring the calculation of the physical characteristics (λ, η) and density ρ with an average temperature.

A correlation of the gas composition as a function of the air content r may also be incorporated to satisfy greater demands. The air content r is defined as

$$r = \left(\frac{\lambda - 1}{\lambda + \frac{1}{L_{\min}}}\right)_{\lambda \geq 1} \quad (7\text{-}17)$$

and varies in numerical value between $r = 0$ for a stoichiometric air/fuel ratio ($\lambda = 1$) and $r = 1$ for pure air ($\lambda \to \infty$). The material variable term is then:

$$\lambda \cdot \left(\frac{\rho}{\eta}\right)^{0.78} = 10^{1.46} \cdot \frac{1.15 \cdot r + 2.02}{[R \cdot (2.57 \cdot r + 3.55)]^{0.78}} \cdot \left(\frac{T_z + T_w}{2}\right)^{-0.477} \cdot p_z^{0.78} \quad (7\text{-}18)$$

The velocity w relevant for heat transfer is described with a global k-ε turbulence model:

$$w = \frac{\sqrt{\frac{8}{3} \cdot k + c_k^2}}{2} \quad (7\text{-}19)$$

where c_k is the instantaneous piston velocity. The following applies to the change of the specific kinetic energy:

$$\frac{dk}{dt} = \left[-\frac{2}{3} \cdot \frac{k}{V} \cdot \frac{dv}{dt} - \varepsilon \cdot \frac{k^{1.5}}{L} + \left(\varepsilon_q \cdot \frac{k_q^{1.5}}{L}\right)_{\varphi > \text{ITDC}}\right]_{IC \leq \varphi \leq EO} \quad (7\text{-}20)$$

with $\varepsilon = \varepsilon_q = 2.184$ and the characteristic vortex length $L = \sqrt[3]{6/(\pi \cdot V)}$. To calculate the specific kinetic energy of the squish flow k_q, a cup-shaped bowl must be defined, which optimally reproduces the real conditions that usually deviate from the ideal case [7-19].

Most recently, a very similar form of this k-ε model was applied in [7-17] to model the convection relevant for heat release.

Unlike *Woschni* and *Hohenberg*'s models, it does not employ a formulation integrated in the flux term as the combustion term. Rather, the multiplicative combustion term Δ models the different driving temperature gradients of unburned fractions to the combustion chamber walls with a temperature T_{uv} and of burned fractions with a temperature T_v [7-19].

Bargende's heat transfer equation is distinctly less clear than *Woschni* and *Hohenberg*'s older equations with regard to the influence of changed engine parameters on wall heat loss. The influence of a speed increase on wall heat losses can be gathered immediately from *Woschni* and *Hohenberg*'s equations. The *Bargende* equation obscures such simple interpretation with the k-ε model employed to model the velocity term.

This clearly demonstrates that today's exclusively programmed application of such models allows dispensing with clarity in exchange for higher accuracy. As the recent studies of heat transfer modeling [7-17] and [7-16] demonstrate, this trend will continue in the future. However, modeling may not be allowed to mistakenly give an impression a "false gain of accuracy". This occurs whenever experimentally unverifiable phenomena are incorporated in modeling. Such equation systems only appear to be "physical" modeling. In actuality, they are purely empirical adjustments with a correspondingly limited range of validity.

7.2.5 Examples of Application

The different results obtained with the heat transfer equations described in the preceding section are compared relatively frequently, usually either to classify a new equation formulation or to compare results of measurement and computation.

Therefore, the three heat transfer equations are also compared here to make it easier for users to select one that matches their problem.

The diagram on the left in Fig. 7-17 presents the measured pressure characteristic and the characteristic of the mass average temperature analyzed from it thermodynamically and the combustion characteristic of a typical, modern direct injection (DI) common rail (CR) car diesel engine in part load in conventional heterogeneous operation with applied pilot injection. The diagram on the right in Fig. 7-17 presents the heat transfer coefficient according to *Woschni* (Eq. (7.10)), *Hohenberg* (Eq. (7.14)) and *Bargende* (Eq. (7.16)) and the wall heat losses in the high pressure section (HD) relative to the converted heat of combustion.

Accordingly, Fig. 7-18 presents the results of an analysis of homogeneous combustion (HCCI, see Sect. 3.3) for an identical engine. Similar to precombustion in appearance, "cold combustion" (cool flame) directly before the main combustion (hot flame) is clearly recognizable in the combustion characteristic.

The altogether earlier and shorter HCCI combustion causes the peak pressure to rise significantly higher than heterogeneous combustion and to be reached at an earlier crank angle position relative to ignition TDC. Since the pressures are higher, all three equations basically yield slightly higher heat transfer coefficients for HCCI operation. In combination with the mass average temperatures and especially because little fuel energy Q_b is burned, relative wall heat losses produced in the high pressure section (HD) are noticeably higher when combustion is homogeneous than when it is heterogeneous.

The differences between the relative wall heat losses calculated by the three equations are remarkably large and stem from the strong differences in the characteristics of the heat transfer coefficient calculated over time.

Such discrepancies also appear in the energy balances (Fig. 7-19) as a quotient of the maximum of the integrated combustion characteristic $Q_{b_{Max}}$ and the supplied fuel energy Q_{Krst}, reduced by the incompletely (CO) and imperfectly (HC) burned energy fractions $Q_{HC,CO}$. The latter is particularly

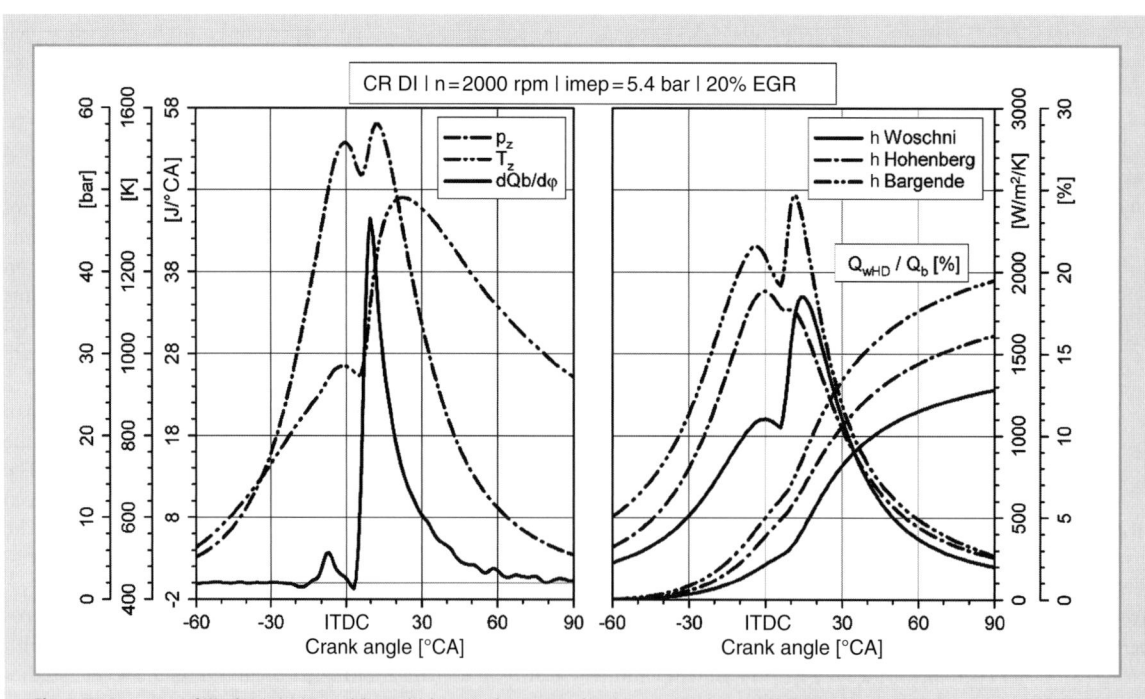

Fig. 7-17 Comparison of *Woschni*, *Hohenberg* and *Bargende's* heat transfer equations for a heterogeneously operated CR DI car diesel engine in part load with applied pilot injection and 20% cooled external EGR

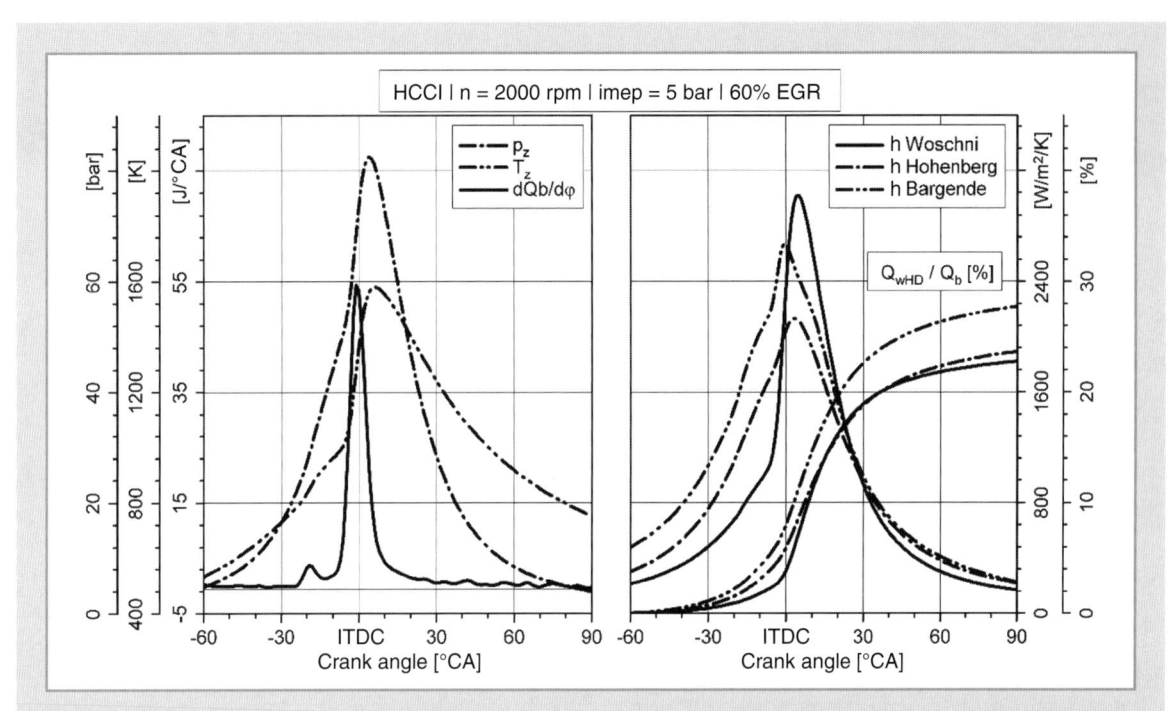

Fig. 7-18 Comparison of *Woschni*, *Hohenberg* and *Bargende's* heat transfer equations of a homogenously (HCCI) operated CR DI car diesel engine in part load with 60% cooled external EGR

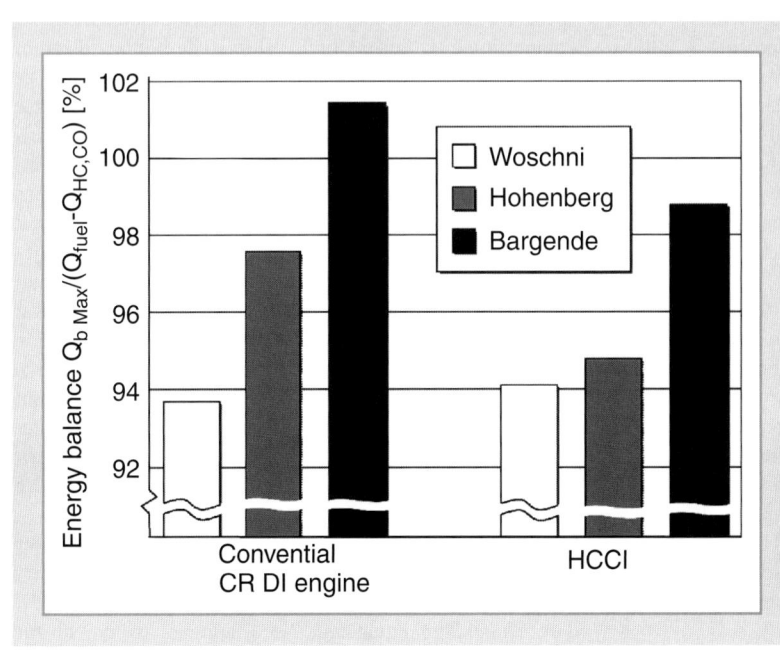

Fig. 7-19
Comparison of the energy balances for *Woschni*, *Hohenberg* and *Bargende's* calculation of the wall heat losses during heterogeneous (conventional CR DI) and homogeneous (HCCI) combustion (CR DI car diesel engine, $n = 2{,}000$ rpm, part load, cooled external EGR

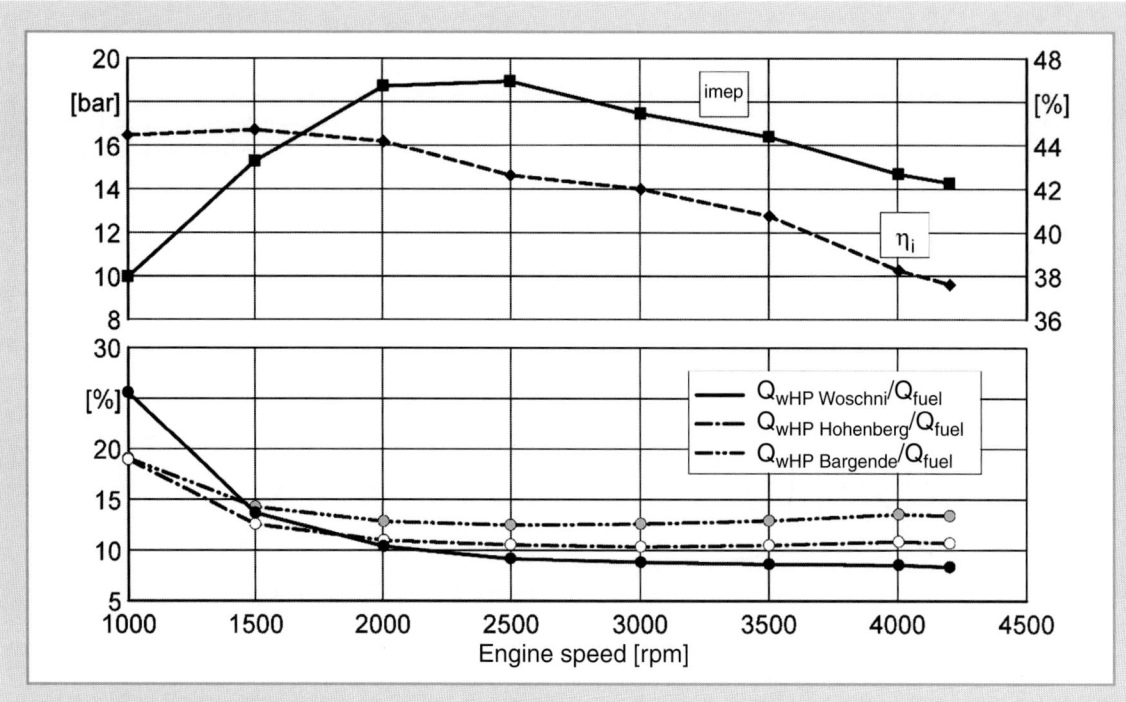

Fig. 7-20 Full load comparison of the *Woschni*, *Hohenberg* and *Bargende* heat transfer equations (CR DI car diesel engine)

necessary for HCCI operation since significant unburned and partially burned energy fractions are present in the exhaust gas.

For some time, the *Woschni* equation has been known to deliver unduly low balance values for heterogeneous operation ([7-22] and [7-32]). Whether the *Hohenberg* or the *Bargende* equation delivers the more accurate result for heterogeneous operation is unclear since energy balances analyzed from measured combustion chamber pressure characteristics have a confidence interval of at least ± 2% even when the measuring equipment has been meticulously calibrated and applied. This relativizes the issue of accuracy somewhat since the deviations for heterogeneous combustion based on *Hohenberg* and plotted in Fig. 7-19 are less than ± 4%.

All in all, it is remarkable that results with any plausibility can be calculated for HCCI operation since none of the three equations was developed or verified for homogeneous autoignited diesel combustion. This is clear evidence that heat transfer equations rigorously and meticulously developed on the basis of similarity theory actually achieve a largely universal validity.

However, as expected, the *Bargende* equation delivers the result with the best energy balance since HCCI operation is energetically relatively similar to premixed gasoline engine combustion with turbulent flame propagation emanating from a single point of ignition.

Significant differences also result when the three equations are applied to the full load of DI diesel engines. The top diagram in Fig. 7-20 presents the mean indicated pressure imep and the indicated efficiency η_i as a function of engine speed for full load operation of a car common rail diesel engine tuned to Euro 4 limit values. The bottom diagram presents the relative wall heat losses in the high pressure section (HP). All three equations have identical qualitative characteristics but their quantitative heat losses differ significantly, above all in the *Woschni* equation. The high percentage loss at $n = 1,000$ rpm is particularly conspicuous, although the indicated efficiency (calculated from internal work and injected fuel mass) does not display any irregularity when compared with $n = 1,500$ rpm. Calculated as less than 10% of the energy with the *Woschni* equation, the wall heat losses at higher speed also appear to be somewhat too low.

A more exact analysis of the heat transfer coefficients invites the conclusion (Fig. 7-21) that the Woschni equation reproduces the influence of combustion on the heat transfer unduly strongly when $n = 1,000$ rpm and unduly weakly when $n = 4,000$ rpm. This explains the differences in the relative wall heat loss.

A comparison of the three relationships for heat transfer produces varying results. The *Woschni* equation appears to deliver the best results for large diesel engines [7-37]. The *Hohenberg* equation [7-33] appears to do so for

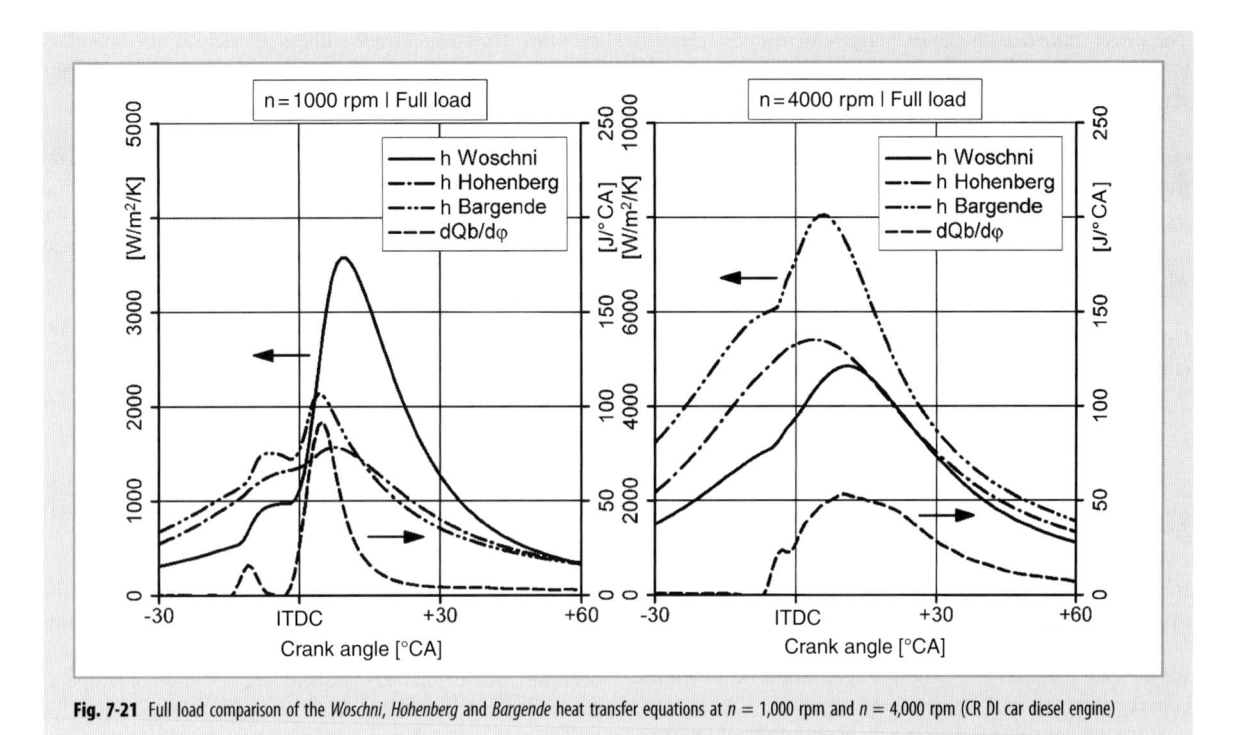

Fig. 7-21 Full load comparison of the *Woschni*, *Hohenberg* and *Bargende* heat transfer equations at $n = 1,000$ rpm and $n = 4,000$ rpm (CR DI car diesel engine)

commercial vehicle diesel engines in particular. The *Bargende* equation likely does so for homogeneous and semi-homogeneous combustion processes, which more strongly resemble gasoline engine process flows, particularly as regards crank angle position and the duration of heat release ([7-38, 7-39]).

However, the *Bargende* equation is now more than fifteen years old, the *Hohenberg* equation is more than twenty-five years old and the *Woschni* equation was published over thirty-five years ago.

The diesel engine process flow has experienced distinct changes since then. The common rail system's flexibilization of the design of injection provides a representative example. A multitude of publications have discussed the effects on the modeling of wall heat losses. Nevertheless, despite all its known weaknesses including those presented in this comparison, the oldest of all the heat transfer equations based on similarity theory, the *Woschni* equation is still employed most frequently.

"Heat transfer from the combustion chamber gas to the combustion chamber walls" is far from finished as a subject of research.

Hence, given the current state, the only viable recommendation is to always critically verify each of the heat transfer equations employed. However, this should be done within the scope of physical plausibility. A balance error of significantly more than 5% for example is certainly not caused by an "incorrect" heat transfer equation. Parallel use of equations is always helpful because it makes a plausibility check significantly easier. Ultimately, nothing can replace a good deal of experience with their application. Moreover, a universally correct solution does not yet exist.

7.2.6 Heat Transfer for Gas Exchange and Intake and Exhaust Ports

By definition, gas exchange extends from exhaust opens to intake closes. The heat transfer in the cylinder during this phase of the process flow has not been researched anywhere near as intensively as the high pressure section of the processes from intake closes to exhaust opens.

Ever higher demands are being made on the accuracy of engine process simulation. Thus, the wall heat loss during gas exchange is also assuming importance. It significantly influences:

- the exhaust gas enthalpy and thus the energization of an exhaust gas turbocharger turbine,
- the exhaust gas temperature and, thus indirectly, the temperature of external exhaust gas recirculation (EGR) and, even more intensely, the temperature of the internal exhaust gas that returns through the intake port during valve overlap or remains in the cylinder and
- the fresh charge and its temperature by heating it more or less strongly during the induction phase, which in turn significantly influences the formation of pollutants (nitrogen oxides) indirectly by changing the temperature level of the entire process.

The direct influence on gas exchange efficiency is relatively low. Rather, the aforementioned influences generate secondary influences on overall process efficiency.

As in the high pressure section, the *Woschni* equation is most frequently used to calculate cylinder side heat transfer in gas exchange:

$$h = 130 \cdot D^{-0.2} \cdot T_z^{-0.53} \cdot p_z^{0.8} \cdot (C_1 \cdot c_m)^{0.8} \quad (7\text{-}21)$$

where $C_1 = 6.18 + 0.417\, c_u/c_m$ for gas exchange.

The constant C_1 has also been revised, for instance by *Gerstle* in 1999 [7-40], for use for medium speed diesel engines to capture exhaust gas enthalpy more accurately:

$$C_1 = f = \left(2.28 + 0.308 \cdot \frac{c_u}{c_m}\right)_{\varphi_{IC} \leq \varphi \leq \varphi_{IO}}$$

$$C_1 = (k \cdot f)_{\varphi_{IO} < \varphi < \varphi_{IC}} \quad \text{mit} \ \ k \ = \ 6.5 \ \text{bis} \ 7.2$$

As a result, the wall heat loss from exhaust opens to intake opens is lowered significantly and the "heating up" of the fresh charge during intake is increased significantly over the original version.

According to [7-33], the *Hohenberg* equation may be used for the high pressure and gas exchange section with identical constants.

Since the *Bargende* equation only applies to the high pressure section of IC to EO, gas exchange requires switching to the *Woschni* equation.

Along with the heat transfer in the cylinder, a gas exchange simulation or analysis also requires appropriate models for the intake and exhaust port as boundary conditions.

In 1969, Zapf [7-41] also published two equations for the ports based on the similarity theory for pipes with turbulent flows:

$$\begin{aligned}Nu_{EK} &= 0.216 \cdot Re^{0.68} \cdot \left(1 - 0.785 \cdot \frac{h_v}{d_i}\right) \\ Nu_{AK} &= 2.58 \cdot Re^{0.5} \cdot \left(1 - 0.797 \cdot \frac{h_v}{d_i}\right)\end{aligned} \quad (7\text{-}22)$$

Their use requires resolving the Nusselt numbers for the intake port Nu_{EK} and the exhaust port Nu_{AK} based on the heat transfer coefficient h. The term h_v/d_i describes the quotient from the valve lift and inner valve diameter.

Equation (7-22) was researched very extensively and systematically in [7-42] and proved to be very well suited to calculate the heat transfer in ports with good accuracy.

7.2.7 Energetic Mean Gas Temperature for the Calculation of Thermal Loading of Components

Calculating the temperature fields in combustion chamber walls does not require temporal resolution of the combustion cycle on the scale of a crank angle. Implementing both the heat transfer coefficient and the gas temperature as mean values over one combustion cycle suffices to calculate the boundary condition of "wall heat flux density on the surface of the combustion chamber".

An even lower temporal resolution is recommended for the calculation of cold and warm starts as well as general changes of the operating point to keep calculating times limited and ensure a good convergence [7-43].

The following applies to the mean heat transfer coefficient h_m:

$$h_m = \frac{1}{ASP} \cdot \int_{ASP} h \cdot d\varphi \quad (7\text{-}23)$$

The following applies to the energetic mean mass average temperature T_{zm}:

$$T_{zm} = \frac{\int_{ASP} (h \cdot T_z) \cdot d\varphi}{\int_{ASP} h \cdot d\varphi} \quad (7\text{-}24)$$

This weighting of the mass average temperatures T_z with the instantaneous heat transfer coefficient h produces significantly higher temperatures than mathematical averaging over the combustion cycle (ASP). The calculation of temperature fields with a mathematically average temperature yields wall temperatures that are utterly incorrect and far too low. The wall heat loss and the thermal loads of the components are significantly underestimated.

However, this also means obeying the condition:

$$\int_{ASP} dQ_w/d\varphi \cdot d\varphi = 0 \rightarrow \Delta T = 0 \rightarrow T_w = T_{zm}$$

$$\int_{ASP} (h \cdot T_z) \cdot d\varphi \Big/ \int_{ASP} h \cdot d\varphi$$

to obtain an adiabatic engine. Accordingly, the wall temperature at full load would have to reach significantly more than $T_w = 1{,}000$ K. Apart from this difficulty, relevant studies [7-30] have demonstrated that such measures are unable to achieve any improvements in efficiency. On the contrary, fuel consumption even increases.

7.2.8 Heat Transfer Modeling in 3D CFD Simulation

As fast computers became sufficiently available, transient three-dimensional simulation programs were developed to design the engine process flow and combustion in particular. Initially, these simulation tools were expected to eliminate the need for heat transfer equations based on similarity theory and to make significantly better and, above all, locally resolved calculations of wall heat fluxes possible.

Classic turbulent and logarithmic laws of the wall stemming from turbulence modeling were employed just as they are very effectively applied in other CFD applications. The basic correlations are presented very clearly in [7-44] and [7-45].

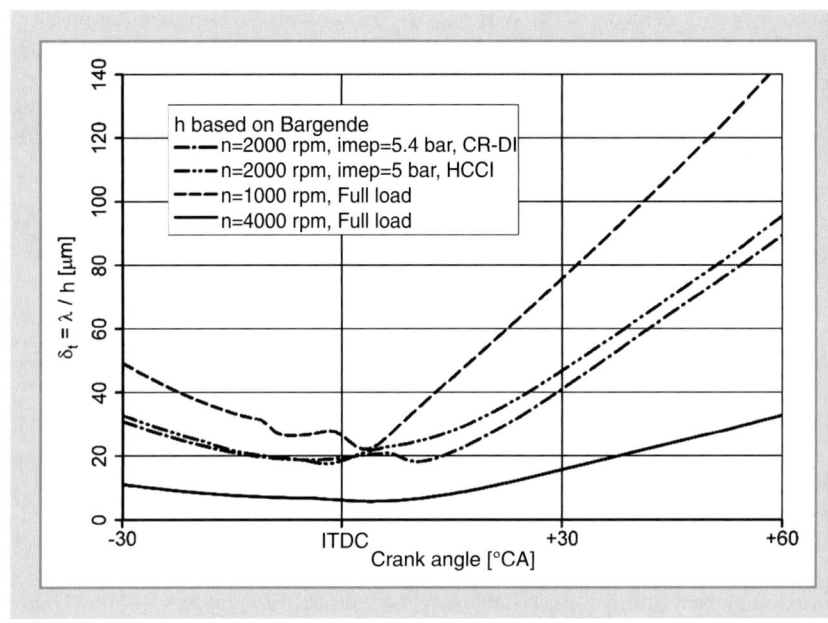

Fig. 7-22 Characteristic of the thickness of the thermal boundary layer's viscous bottom layer. Heat transfer coefficient according to *Bargende* (CR DI car diesel engine)

However, compared with balance measurements and engine process simulations, application of the logarithmic law of the wall in a combustion engine was quickly proven to underestimate the global integral wall heat loss by as much as a factor of 5. Such a large discrepancy made it impossible to expect that correct results could be obtained for all the other parameters of interest.

The reason for these differences can be found in the extremely thin boundary layers in combustion engines. The laminar viscous bottom layers are in fact extremely thin as an estimate using the following simple equation for the thickness of the viscous bottom layer δ'_t corresponding to Sect. 7.2.2 demonstrates.

$$\delta'_t = \frac{\lambda}{h} \qquad (7\text{-}25)$$

Figure 7-22 presents the characteristics of the local mean thicknesses of the thermal boundary layer's viscous bottom layer calculated with Eq. (7.25) for the four operating points discussed in detail in Sect. 7.2.5. Although the simple relationship can only be an estimate, a numerical discretization of a bottom layer with a maximum thickness of 20 μm on the combustion chamber wall surface is clearly impossible and therefore approximations must be worked with.

Despite the advances in the localized formulation of a problem of the law of the wall suitable for 3D CFD simulations, developed by Reitz for instance (c.f. [7-15]), 3D CFD simulation still very frequently employs one of the three heat transfer equations discussed in Sect. 7.2.5. When implemented intelligently, this yields the invaluable advantage of an ongoing check (or, if necessary, a correction) of the 3D CFD simulation's energy balance by a simultaneously executed engine process simulation [7-46].

Literature

7-1 Verein Deutscher Ingenieure (Ed.): Systematische Berechnung hochbeanspruchter Schraubverbindungen. Düsseldorf: VDI 2230 (2003) 2

7-2 Pflaum, W.; Mollenhauer, K.: Der Wärmeübergang in der Verbrennungskraftmaschine. Vienna/New York: Springer (1977)

7-3 Maas, H.; Klier, H.: Kräfte, Momente und deren Ausgleich in der Verbrennungskraftmaschine. In: Die Verbrennungskraftmaschine. Vol. 2 Vienna/New York: Springer (1981)

7-4 Bathe, K.-J.: Finite-Element-Methoden. Berlin/Heidelberg/New York: Springer (2002)

7-5 Keil, S.: Beanspruchungsermittlung mit Dehnungsmessstreifen. Lippstadt: Cuneus-Verlag (1995)

7-6 Robertson, R.-E.; Schwertassek, R.: Dynamics of Multibody Systems. Berlin/Heidelberg/New York: Springer (1988)

7-7 Issler, L.; Ruoss, H.; Häfele, P.: Festigkeitslehre Grundlagen. Berlin/Heidelberg/New York: Springer (2003)

7-8 Radaj, D.: Ermüdungsfestigkeit. Berlin/Heidelberg/New York: Springer (2003)

7-9 Naubereit, H.; Weiher, J.: Einführung in die Ermüdungsfestigkeit. Munich/Vienna: Hanser (1999)

7-10 Forschungskuratorium Maschinenbau (FKM): Rechnerischer Festigkeitsnachweis für Maschinenbauteile. 4th Ed. Frankfurt: VDMA-Verlag (2002)

7-11 Forschungskuratorium Maschinenbau (FKM): Bruchmechanischer Festigkeitsnachweis. 3rd Ed. Frankfurt: VDMA-Verlag (2006)

7-12 Broichhausen, J.: Schadenskunde – Analyse und Vermeidung von Schäden in Konstruktion, Fertigung und Betrieb. Munich/Vienna: Hanser (1985)

7-13 Bargende, M.; Hohenberg, G.; Woschni, G.: Ein Gleichungsansatz zur Berechnung der instationären Wandwärmeverluste im Hochdruckteil von Ottomotoren. 3rd Conference: Der Arbeitsprozess des Verbrennungsmotors. Graz (1991)

7-14 Zima, S.; Greuter, E.: Motorschäden. Würzburg: Vogel (2006)

7-15 Wiedenhoefer, J.; Reitz, R.D.: Multidimensional Modelling of the Effects of Radiation and Soot Deposition in Heavy-Duty Diesel Engines. SAE Paper 2003-01-0560

7-16 Eiglmeier, C.; Lettmann, H.; Stiesch, G.; Merker, G.P.: A Detailed Phenomenological Model for Wall Heat Transfer Prediction in Diesel Engines. SAE Paper 2001-01-3265

7-17 Schubert, C.; Wimmer, A.; Chmela, F.: Advanced Heat Transfer Model for CI Engines. SAE Paper 0695 (2005) 1

7-18 Woschni, G.: Die Berechnung der Wandverluste und der thermischen Belastung von Dieselmotoren. MTZ 31 (1970) 12

7-19 Bargende, M.: Ein Gleichungsansatz zur Berechnung der instationären Wandwärmeverluste im Hochdruckteil von Ottomotoren. Dissertation TU Darmstadt (1991)

7-20 Bargende, M.: Berechnung und Analyse innermotorischer Vorgänge. Vorlesungsmanuskript Universität Stuttgart (2006)

7-21 Eichelberg, G.: Zeitlicher Verlauf der Wärmeübertragung im Dieselmotor. Z. VDI 72, Issue 463 (1928)

7-22 Woschni, G.; Zeilinger, K.; Huber, K.: Wärmeübergang im Verbrennungsmotor bei niedrigen Lasten. FVV-Vorhaben R452 (1989)

7-23 Bendersky, D.: A Special Thermocouple for Measuring Transient Temperatures. ASME-Paper (1953)

7-24 Fieger, J.: Experimentelle Untersuchung des Wärmeüberganges bei Ottomotoren. Dissertation TU München (1980)

7-25 Filipi, Z.S. et al.: New Heat Transfer Correlation for An HCCI Engine Derived From Measurements of Instantaneous Surface Heat Flux. SAE Paper 2004-01-2996

7-26 Nusselt, W.: Der Wärmeübergang in der Verbrennungskraftmaschine. Forschungsarbeiten auf dem Gebiet des Ingenieurwesens. 264 (1923)

7-27 Elser, K.: Der instationäre Wärmeübergang im Dieselmotor. Dissertation ETH Zürich (1954)

7-28 Woschni, G.: Beitrag zum Problem des Wärmeüberganges im Verbrennungsmotor. MTZ 26 (1965) 11, p. 439

7-29 Sihling, K.: Beitrag zur experimentellen Bestimmung des instationären, gasseitigen Wärmeübergangskoeffizienten in Dieselmotoren. Dissertation TU Braunschweig (1976)

7-30 Kolesa, K.: Einfluss hoher Wandtemperaturen auf das Brennverhalten und insbesondere auf den Wärmeübergang direkteinspritzender Dieselmotoren. Dissertation TU München (1987)

7-31 Schwarz, C.: Simulation des transienten Betriebsverhaltens von aufgeladenen Dieselmotoren. Dissertation TU München (1993)

7-32 Vogel, C.: Einfluss von Wandablagerungen auf den Wärmeübergang im Verbrennungsmotor. Dissertation TU München (1995)

7-33 Hohenberg, G.: Experimentelle Erfassung der Wandwärme von Kolbenmotoren. Habilitation Thesis Graz (1980)

7-34 Hohenberg, G.: Advanced Approaches for Heat Transfer Calculations. SAE Paper 790825 (1979)

7-35 Barba, C.; Burkhardt, C.; Boulouchos, K.; Bargende, M.: A Phenomenological Combustion Model for Heat Release Rate Prediction in High-Speed DI Diesel Engines With Common Rail Injection. SAE-Paper 2933 (2000) 1

7-36 Kozuch, P.: Ein phänomenologisches Modell zur kombinierten Stickoxid- und Russberechnung bei direkteinspritzenden Dieselmotoren. Dissertation Uni Stuttgart (2004)

7-37 Woschni, G.: Gedanken zur Berechnung der Innenvorgänge im Verbrennungsmotor. 7th Conference: Der Arbeitsprozess des Verbrennungsmotors. Graz 1999

7-38 Haas, S.; Berner, H.-J.; Bargende, M.: Potenzial alternativer Dieselbrennverfahren. Motortechnische Konferenz Stuttgart, June 2006

7-39 Haas, S.; Berner, H.-J.; Bargende, M.: Entwicklung und Analyse von homogenen und teilhomogenen Dieselbrennverfahren. Tagung Dieselmotorentechnik TAE Esslingen (2006)

7-40 Gerstle, M.: Simulation des instationären Betriebsverhaltens hoch aufgeladener Vier- und Zweitakt-Dieselmotoren. Dissertation Uni Hannover (1999)

7-41 Zapf, H.: Beitrag zur Untersuchung des Wärmeüberganges während des Ladungswechsels im Viertakt-Dieselmotor. MTZ 30 (1969) pp. 461 ff.

7-42 Wimmer, A.; Pivec, R.; Sams, T.: Heat Transfer to the Combustion Chamber and Port Walls of IC Engines – Measurement and Prediction. SAE Paper 2000-01-0568

7-43 Sargenti, R.; Bargende, M.: Entwicklung eines allgemeingültigen Modells zur Berechnung der Brennraumwandtemperaturen bei Verbrennungsmotoren. 13th

Aachen Colloquium Automobile and Engine Technology (2004)

7-44 Merker, G.; Schwarz, C.; Stiesch, G.; Otto, F.: Verbrennungsmotoren, Simulation der Verbrennung und Schadstoffbildung. 2nd Ed. Wiesbaden: Teubner (2004)

7-45 Pischinger, R.; Klell, M.; Sams, T.: Thermodynamik der Verbrennungskraftmaschine. 2nd Ed. Vienna /New York: Springer 2002

7-46 Chiodi, M.; Bargende, M.: Improvement of Engine Heat Transfer Calculation in the three dimensional Simulation using a Phenomenological Heat Transfer Model. SAE-Paper 2001-01-3601

Further Literature

Enomoto, Y.; Furuhama, S.; Takai, M.: Heat Transfer to Wall of Ceramic Combustion Chamber of Internal Combustion Engine. SAE Paper 865022 (1986)

Enomoto, Y.; Furuhama, S.: Measurement of the instantaneous surface temperature and heat loss of gasoline engine combustion chamber. SAE Paper 845007 (1984)

Klell, M., Wimmer, A.: Die Entwicklung eines neuartigen Oberflächentemperaturaufnehmers und seine Anwendung bei Wärmeübergangsuntersuchungen an Verbrennungsmotoren. Tagung: Der Arbeitsprozess der Verbrennungsmotors. Graz (1989)

8 Crankshaft Assembly Design, Mechanics and Loading

Eduard Köhler, Eckhart Schopf, and Uwe Mohr

8.1 Designs and Mechanical Properties of Crankshaft Assemblies

8.1.1 Function and Requirements of Crankshaft Assemblies

Together with both the piston pins and the crankshaft's crank pins, the connecting rod in reciprocating piston engines converts oscillating piston motion into rotary crankshaft motion. Running smoothness is a universally important criterion for the design of crankshaft assemblies. High speed has priority in gasoline engines, thus making a minimum of moving masses an absolute imperative. The emphasis shifts somewhat for diesel engines. Firing pressures can be twice as high as in gasoline engines and continue to increase as the size increases. Thus, controlling the effects of the gas force is the primary challenge.

In conjunction with direct injection and one or two-stage exhaust gas turbocharging and intercooling, car diesel engines now attain the volumetric power outputs of gasoline engines. Moreover, cuts in fuel consumption (reduction of CO_2 emission), strict emission laws and lightweight and increasingly compact designs that do not sacrifice reliability are currently driving engine development. However, in principle, steadily increasing ignition and injection pressures result in "harsher" combustion. This inevitably creates more problems with acoustics and vibrations as demands for comfort increase. Now common in diesel cars, multiple injections, optimized vibration damping, camshaft drives shifted to the flywheel-side, dual-mass flywheels and partial encapsulations serve to improve modern diesel engines' acoustic and vibration performance. Not least, the steadily increasing amplitudes of the gas torque characteristic make using the engine mounting and the entire power train to control crankshaft assembly vibrations, improve mass balancing and reduce the excitation of vibrations even more important. In Europe, the diesel engine has evolved from merely being the primary commercial vehicle engine to also being a frequently used car engine.

A modern diesel engine's tremendous mechanical stress on the crankshaft assembly also has to be accommodated. Crankshaft assembly components require a structural design optimized for structural strength, stiffness and mass. Knowledge of locally present fatigue limits of materials that affect components has not quite kept pace with the simulation of loading conditions, which has become quite precise in the meantime. This reveals a potential weak point in fatigue strength simulations in the limit range. Thus, even more precise measurement of the practical impact of technological influences and quality fluctuations in manufacturing will be essential in the future.

8.1.2 Crankshaft Assembly Forces

The literature includes numerous studies devoted to the crankshaft assembly of a reciprocating piston engine (c.f. [8-1, 8-2]). Varying with the crank angle φ, the piston force $F_K(\varphi)$ acts on the crankshaft assembly on the piston side. According to Fig. 8-1, this ensues from the superimposition of the oscillating inertial force $F_{mosz}(\varphi)$ on the gas force $F_{Gas}(\varphi)$:

$$F_K = F_{Gas} + F_{mosz} \tag{8-1}$$

The gas force $F_{Gas}(\varphi)$ is the product of the cylinder pressure $p_Z(\varphi)$ and the area of the piston A_K. The cylinder pressure characteristic can be measured by *cylinder pressure indication* or calculated with the aid of *engine process simulation*. The oscillating inertial force can be measured according to Eq. (8-30). When ψ is the pivoting angle of the connecting rod, the connecting rod force F_{Pl} in a housing-related reference system follows from the piston force F_K:

$$F_{Pl} = \frac{F_K}{\cos \psi} \tag{8-2}$$

The connecting rod's oblique position causes the piston side thrust F_{KN} to act on the area of the cylinder bore surface:

$$F_{KN} = F_K \tan \psi \approx F_K \lambda_{Pl} \sin \varphi \tag{8-3}$$

where λ_{Pl} is the stroke/connecting rod ratio (the quotient of the crank radius r and connecting rod length l_{Pl}). In a

---

E. Köhler (✉)
KS Aluminium Technologie GmbH, Neckarsulm, Germany
e-mail: eduard.koehler@de.kspg.com

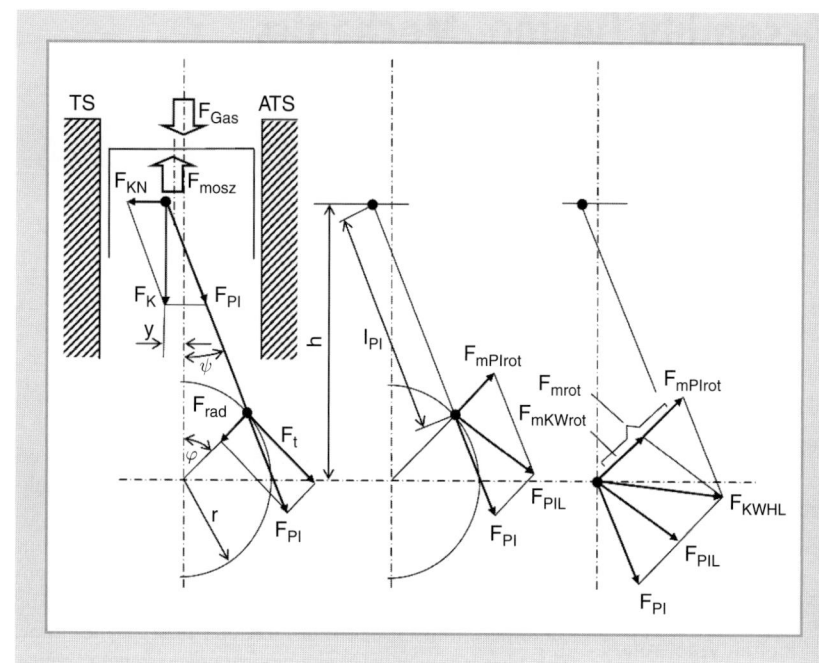

Fig. 8-1 Forces acting in a reciprocating piston engine, offset piston pin (TS – Thrust side, ATS – Antithrust side)

crankshaft-related reference system, the tangential or torsional force F_t acts on the crank's crank pin:

$$F_t = F_{Pl} \sin(\varphi + \psi) = F_K \frac{\sin(\varphi + \psi)}{\cos \psi} \quad (8\text{-}4)$$

The radial force F_{rad} acts on the crank pin radially:

$$F_{rad} = F_{Pl} \cos(\varphi + \psi) = F_K \frac{\cos(\varphi + \psi)}{\cos \psi} \quad (8\text{-}5)$$

With the crank radius r as a lever arm, the tangential force F_t generates the crank torque M. Dependent on the piston side thrust F_{KN} and the instantaneous distance h between the piston pin and crankshaft axis, it can also be defined as a reaction torque in the cylinder crankcase:

$$M = F_t r = -F_{KN} h \quad h = r \cos \varphi + l_{Pl} \cos \psi \quad (8\text{-}6)$$

The bearing forces F_{PIL} or F_{KWHL} in the connecting rod and the crankshaft bearings are obtained by adding the vectors of the connecting rod force F_{Pl} with the related rotating inertial forces. In the case of the connecting rod bearing, this is the inertial force of the assumed mass component of the connecting rod F_{mPlrot} rotating with the crank pin. In the case of the crankshaft bearings, it is the entire inertial force F_{mrot} of the masses rotating with the crank. The latter are the total of the rotating inertial forces of the connecting rod F_{mPlrot} and the crank F_{mKWrot}:

$$\vec{F}_{PIL} = \vec{F}_{Pl} + \vec{F}_{mPlrot} \quad \vec{F}_{KWHL} = \vec{F}_{PIL} + \vec{F}_{mKWrot} = \vec{F}_{Pl} + \vec{F}_{mrot} \quad (8\text{-}7)$$

The transmission of forces through adjacent cylinders – as a function of their phase relation and distribution on the main bearings of the crankshaft throw concerned – has to be allowed for in the statically indeterminate crankshaft supports of multiple cylinder engines. Realistically, capturing every effect (elastic deformation of the bearing bulkhead and crankshaft, different bearing clearances, hydrodynamic lubricating film formation in dynamically deformed bearings and misalignment of the main bearing axis caused by manufacturing) proves to be commensurately difficult and complex (see [8-3–8-8]).

At constant engine speed (angular frequency ω), one crank angle φ run through in the time t is $\varphi = \omega t$. The following relationships deliver the correlation between the angles φ and ψ:

$$r \sin \varphi = l_{Pl} \sin \psi$$

$$\sin \psi = \frac{r}{l_{Pl}} \sin \varphi = \lambda_{Pl} \sin \varphi \quad (8\text{-}8)$$

$$\sin \varphi = \frac{l_{Pl}}{r} \sin \psi = \frac{1}{\lambda_{Pl}} \sin \psi$$

$$\sin^2 \psi + \cos^2 \psi = 1$$

$$\cos \psi = \sqrt{1 - \sin^2 \psi} = \sqrt{1 - \lambda_{Pl}^2 \sin^2 \varphi} \quad (8\text{-}9)$$

Thus, the equations containing the pivoting angle of the connecting rod ψ may also be represented as a function of the crank angle φ.

Equation (8-9) only exactly applies to *non-crankshaft offset* reciprocating piston engines, which additionally do not have

Fig. 8-2 Offset crankshaft assembly

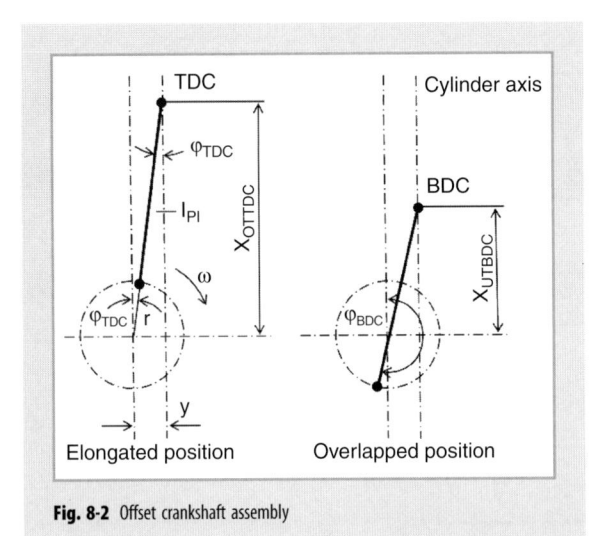

any piston pin *offset*. The offset is usually slight and hence may be disregarded when the calculation is sufficiently accurate. Otherwise, Eq. (8-8, left) is supplemented by the lateral offset y or e (Fig. 8-2):

$$r \sin \varphi \pm y = l_{Pl} \sin \psi \qquad e = \frac{y}{l_{Pl}} \qquad (8\text{-}10)$$

A distinction is made between the offset on the thrust side for reasons of noise (to control the contact change) and the "thermal" offset on the anti-thrust side (to prevent one-sided carbon buildup) [8-9]. Both offsets enlarge the stroke and move the top and the bottom dead center from its defined position. The dead centers are reached in the so-called *elongated* and *overlapped positions*, i.e. when the crank and connecting rod have the same direction geometrically. By accepting interference with the mass balancing and the control of the excitation amplitudes, a certain offset can be contemplated for inline engines to reduce the piston side thrust induced by the gas force and thus the frictional losses.

8.1.3 Engine Design and Crankshaft Assembly Configuration

8.1.3.1 Influencing Variables

The actions of the free gas and inertial forces determine a crankshaft assembly's mechanical and dynamic properties. In this context, the

- number of cylinders z and the
- engine design as the configuration of the cylinders

take on fundamental importance.

Basic influencing variables are the

- cylinder pressure characteristic,
- crankshaft assembly masses,
- kinematic parameters of the crankshaft assembly,
- cylinder bore diameter and distance between cylinder bore center lines and
- crankshaft throw configuration (ignition interval and firing sequence).

8.1.3.2 Common Designs

Gasoline engines and high speed diesel engines for cars, commercial vehicles, locomotives and small ships are constructed with trunk-pistons to cut space and weight. Low speed engines, i.e. large two-stroke diesel engines, on the other hand are designed with crosshead crankshaft assemblies. A crosshead is indispensible for the large stroke relative to the cylinder bore, which is typical for these engines. Otherwise, the limited pivoting angle would only make it possible to design trunk-piston engines with extremely long connecting rods. Another useful advantage is the complete structural and functional separation between the piston, the head side of which is part of the combustion chamber and thus thermally highly stressed, and the linear guidance [8-10].

Inline (abbreviated with I) and V designs are common for diesel engines. The opposed cylinder design plays a niche role for gasoline engines. The cylinder banks or rows in V engines, including VR engines [8-11], are V-shaped. They are arranged fan-like in the special case of the W engine (only gasoline engines [8-12]). A V angle α_V or angles between the cylinder banks is characteristic of the latter.

8.1.3.3 Number of Cylinders

The actions of the free gas and inertial forces are reduced considerably as the number of cylinders z increases in a suitable configuration. In particular, the torque delivered at the flywheel is equalized (cf. [8-2]) as illustrated in Fig. 8-3. On the other hand, a crankshaft's susceptibility to torsional vibrations (see Sect. 8.3) intensifies as its overall length increases.

The installation situation and maintenance as well as the desire to keep vibration damping measures to a minimum limits the number of cylinders in vehicle engines. Hence, gasoline engines and car and commercial vehicle diesel engines are limited to six inline cylinders. Larger numbers of cylinders require a V design with eight, ten and twelve cylinders.

The running smoothness of higher numbers of cylinders proves beneficial when large diesel engines are applied as marine propulsion since it minimizes the vibration excitation of a ship's hull and superstructures. Likewise, the smoothest possible torque characteristic is beneficial for generator operation. The damping of torsional vibrations is particularly important. Medium speed engines with uneven numbers of cylinders are also available in increments of I6, I7, I8, I9 and I10 (e.g. MAN diesel SE, model L32/40). A low speed

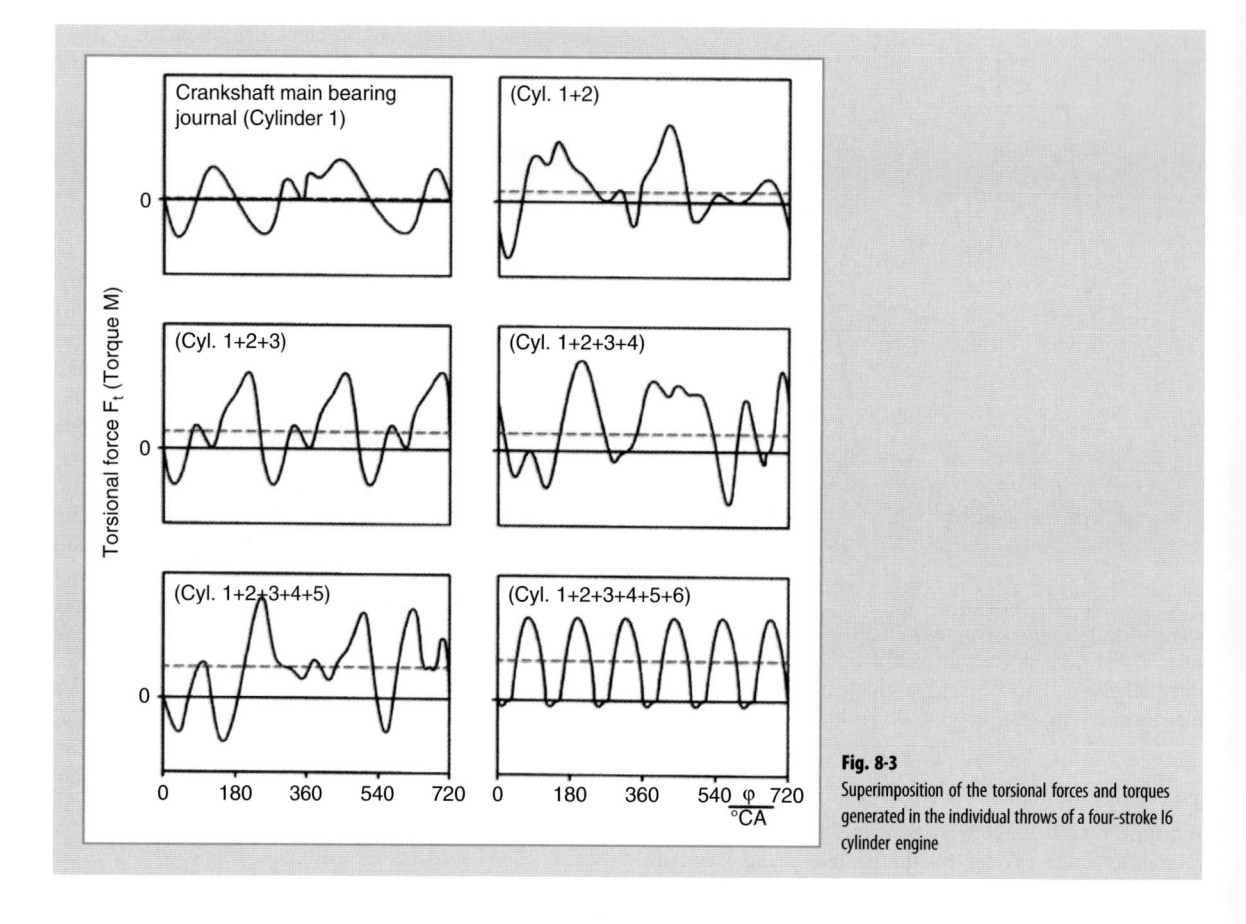

Fig. 8-3 Superimposition of the torsional forces and torques generated in the individual throws of a four-stroke I6 cylinder engine

two-stroke engine series can also range incrementally from an I6 engine up through and including I11 to I12 engines and even an I14 variant (e.g. MAN Diesel SE, model K98MC MK7). Medium speed diesel engines are manufactured in V designs with different increments of up to twenty cylinders.

8.1.3.4 Throw Configuration and Free Mass Actions

Appropriate configuration of the throws suffices to obtain quite good dynamic performance from a crankshaft assembly in multiple cylinder engines. The first order free inertial forces are fully balanced in inline engines when the crankshaft is centrosymmetric (see Sect. 8.3.3.2 for a centrosymmetric *first order star diagram*). This is the case when the crankshaft's throw angle φ_K corresponds to one uniform ignition interval φ_z:

$$\varphi_z = \frac{4\pi}{z} \text{(Four-stroke engine)} \quad \varphi_z = \frac{2\pi}{z} \text{(Two-stroke engine)} \tag{8-11}$$

The firing sequence does not play a role at all. If the *second order star diagram* (a fictitious doubling of every cylinder's throw angle) remains centrosymmetric, then the second order free inertial forces also disappear. Longitudinally symmetrical crankshafts of four-stroke inline engines with an even number of cylinders are also free of all orders of moments of inertia.

In a V engine, the forces of each single V cylinder pair act on one double crankshaft throw. A "connecting rod next to connecting rod" configuration, i.e. two connecting rods that are offset axially but connected to the same crank pin, has become common. Axial connecting rod offset generates a bending moment that is dependent on the gas and inertial forces and transmitted by the crankshaft to the bearing bulkhead walls. This has given rise to significantly more structurally complex connecting rod designs such as a forked connecting rod and a master connecting rod with a connected auxiliary connecting rod. Although they eliminate the drawbacks, they are however no longer employed for reasons of cost (Fig. 8-4).

As Sect. 8.3.3.2 explains in more detail, the conditions in V engines appear more complex. A V crankshaft assembly's first order free inertial forces may be fully balanced by counterweights in the crankshaft when the following condition is met:

$$\delta = \pi - 2\alpha_v \tag{8-12}$$

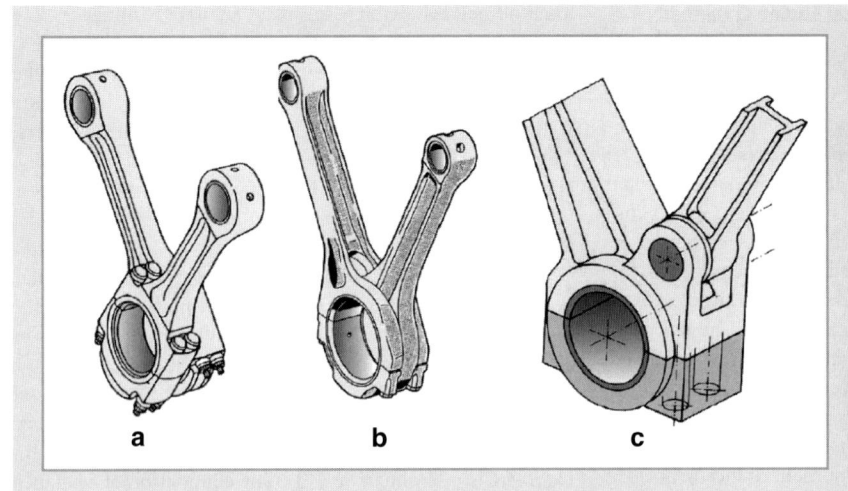

Fig. 8-4
Different connecting rod arrangements for V engines: **a** Connecting rod next to connecting rod, **b** Forked connecting rod, **c** Master and auxiliary connecting rod

Along with the V angle α_V, δ denotes the crank pin offset, more precisely, the angle of connecting rod offset (the angle by which a double throw's crank pins may be offset, i.e. counterrotated). At 90°, this angle would be $\delta = 0°$ for a V8 engine, $\delta = 180°$ for an opposed cylinder engine ($\alpha = 180°$) and $\delta = +60°$ for a 60°/V6 engine. The + sign signifies an offset counter to the crankshaft's direction of rotation. When the angle of connecting rod offset is large, an intermediate web between the offset crank pins is essential for reasons of strength. This entails an increase of overall length though since cylinder bank offset increases by its width at the least (Fig. 8-5).

In conjunction with measures to enhance fatigue strength, an intermediate web may be dispensed with entirely when the crank pin offset (angle of connecting rod offset) is small. (This is called a *split-pin* design.) Small crank pin offsets are increasingly being applied to equalize non-uniform ignition intervals in modular designs. This is primarily found in V6 and V10 engines at a V angle of 90° aligned for a V8 engine.

Assuming the engine is a four-stroke engine, the angle of connecting rod offset is then calculated with the number of cylinders z as follows:

$$\delta = \frac{4\pi}{z} - \alpha_v \qquad (8\text{-}13)$$

It follows that $\delta = +30°$ for the 90°/V6 engine and $\delta = -18°$ for the 90°/V10 engine. (The signs are to be interpreted as explained above.) A uniform ignition interval may have more importance depending on the free torsional forces. A balance shaft (see Sect. 8.3.6) additionally has to be incorporated to fully balance the first order free moment of inertia. Given their regular, i.e. even, numbers of cylinders, V engines only have longitudinally symmetrical crankshafts in exceptional cases. Then, first order free moments of inertia usually do not occur. The crank pin offset in a centrosymmetric first order star diagram neither produces a first order free inertial force nor a first order free moment of inertia.

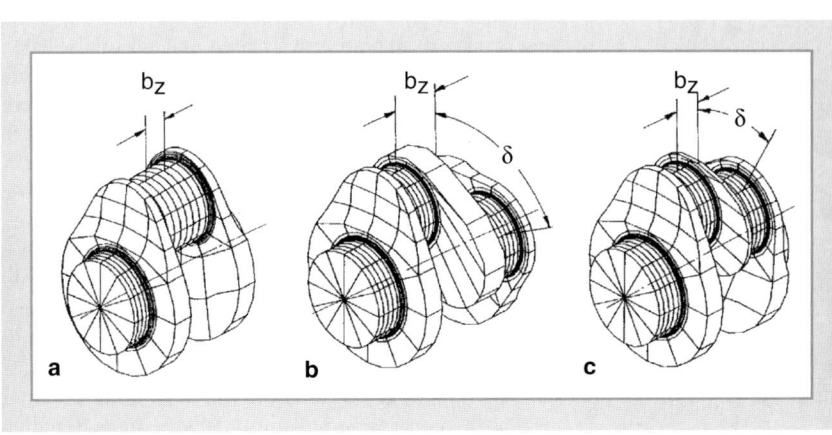

Fig. 8-5
Different crankshaft throw designs for V engines: **a** Common crank pin of two axially offset connecting rods, **b** Double throw with intermediate web with large crank pin offset (angle of connecting rod offset), **c** Double throw with small crank pin offset (special case without intermediate web (split-pin design))

8.1.3.5 Characterization of Diesel Engine Crankshaft Assemblies

The following characterization of crankshaft assemblies is based on the predominant working principle of the four-stroke engine. Single cylinder diesel engines (see Sect. 8.3.3.1 on the balancing of masses of a single cylinder crankshaft assembly) are important drives for small power sets. Unconventional solutions with combined measures that improve running smoothness are also applied in small engines [8-13].

Conflicting goals exist for four-stroke I2 engines. At a uniform ignition interval of 360°, the first and higher order free inertial forces add up unfavorably. On the other hand, free moments of inertia do not occur. A crankshaft with a throw angle of 180° (corresponding to a two-stroke I2 crankshaft) reverses the conditions. First order free inertial forces do not occur but a first order free moment of inertia does. In light of the shorter overall length, a uniform ignition interval is less critical for the excitation of torsional vibrations. The I2 diesel engine is insignificant as a car engine. The VW Eco Polo [8-14] with an ignition interval of 360° formerly represented this design (Fig. 8-6). A laterally mounted balance shaft negatively (counter-) rotating at crankshaft speed fully balances the first order inertial force. Single-sided configuration generates an additional moment around an engine's longitudinal axis, which however favorably influences the free inertial torque.

Disregarding four-stroke radial aircraft engines and the more recent VW VI5 car engine [8-15], only an inline engine is fundamentally suited for uneven numbers of cylinders. The aforementioned engine types are designed as gasoline engines. The VR design is less suited for the diesel process because of the high loads (piston side thrust and bearing load).

The four-stroke I3 engine is not particularly popular because of its poor comfort due to the free moments of inertia, particularly those of the first order. When the numbers of cylinders are uneven, counterweights in the crankshaft cannot manage the balancing alone. Nonetheless, I3 diesel engines have occupied the niche of particularly fuel economizing small car engines for years (e.g. VW Lupo 3L [8-16], DaimlerChrysler Smart [8-17]). The elimination of a balance shaft is considered better for these small engines. In terms of perceptibility in the passenger compartment, the 1.5 order vibrations induced by the gas force become more critically apparent than the first order mass actions [8-17]. Nonetheless, a balance shaft (counter-) rotating at crankshaft speed is now used in Smart/Mitsubishi I3 diesel engines [8-18].

The majority of car diesel engines are inline four-cylinder engines. The I4 engine does not have any first order free inertial forces or any first and higher order free moments of inertia. However, the inertial forces of all four cylinders add up in the second order. What is more, ignition frequency and the second order coincide poorly. Without proper reinforcing measures, an engine-transmission unit's lowest natural bending frequency may be too close to the corresponding excitation. Unpleasant humming noises in the passenger compartment are the consequence. In light of the increased demands on comfort, two balance shafts counterrotating at double speed are increasingly being employed to improve running smoothness (see [8-19]).

Reservations about the I5 engine are in turn based on the conflicting goals when balancing masses. It is designed at the expense of an unfavorably large first or second order tilting moment (see Sect. 8.3.4) [8-20, 8-21]. The former cannot be fully balanced by counterweights in the crankshaft alone. A large crankshaft is comparatively heavy and, since it oscillates, the latter requires a pair of balance shafts rotating in opposite directions at double speed. Despite the recent development of new R5 diesel engines (Fig. 8-7 [8-22]), this design appears to be losing significance again.

I4 diesel engines' high torque and increased volumetric power output make it possible to utilize them as basic motors for cars above medium size. Moreover, a trend toward V6 diesel engines is discernible. *Vehicle package*, *crash length* and the option of transverse engine installation (an increasing diversity of variants) are generally influencing this development in all cars. While the I6 engine is built longer, its outstanding running smoothness (no free mass actions even of the fourth order) is impressive. It enjoys an outstanding

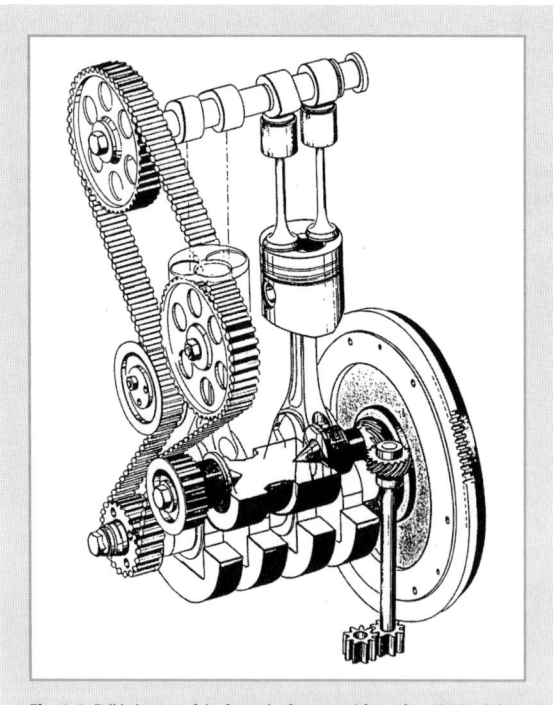

Fig. 8-6 Full balancing of the first order free inertial forces for a VW Eco Polo I2 diesel engine [8-14]

Fig. 8-7 VW I5 TDI® diesel engine crankshaft (with vibration damper integrated in the engine front side counterweight [8-22])

reputation in commercial vehicles too, especially from the perspective of cost and benefit.

V6 engines' first and second order free moments of inertia prevent them from attaining this quality of running smoothness (the special case of $\alpha_V = 180°$ not being considered here). However, a compromise, so-called *normal balancing* (see Sect. 8.3.3) makes it possible to completely balance the first order free moments of inertia for V angles of 60° – the crank pin offset likewise being +60° – and 90°. However, 90° is well known to translate into unequal ignition intervals at the throw angle of 120° typical for V6 engines. Thus, based on Eq. (8-13) already mentioned in Sect. 8.1.3.4, equalization in the AUDI V6TDI® engine [8-23] requires a crank pin offset of +30°. It is +48° in DaimlerChrysler's V6 diesel engine with an atypical V angle of 72° [8-24]. This produces a negatively (counter-) rotating first order moment of inertia. A balance shaft likewise rotating negatively at crankshaft speed provides complete balancing.

The V8 design with a 90° V angle is widespread in larger diesel engines. Neither first and second order free inertial forces nor second order free moments of inertia occur in this configuration. "Normal balancing" (see Sect. 8.3.3) can in turn fully balance the remaining first order free moments of inertia. The non-uniform ignition intervals per cylinder bank (without continuous alternate ignition of both cylinder banks) are acoustically perceptible together with the gas exchange. This can only be prevented with a "flat" crankshaft, i.e. a crankshaft offset by 180° instead of 90° (I4 crankshaft), while accepting second order free mass actions.

A V8 engine only deviates from the V angle of 90° for valid reasons, e.g. 75° in DaimlerChrysler's OM629 V8 car diesel engine [8-25]. A uniform ignition interval is in turn given precedence with a crank pin offset of +15°. Hence, the aforementioned first order balance shaft is used. It is housed in the main oil gallery to save space. SKL's 8VDS24/24AL diesel engine with a V angle of 45° was designed for narrower engine width [8-26]. A *cross shaft* with a +90° crank pin offset balances the first order masses. A thick intermediate web is essential for reasons of strength.

High engine powers not only require large displacement but also large numbers of cylinders. V10 engines have attained importance between V8 and V12 engines. Larger diesel engines are usually based on cost cutting modular engine concepts. The balancing of V8 engine masses normally dictates a uniform V angle of 90°. First order free moments of inertia no longer occur in "normal balancing" (see Sect. 8.3.3). Appropriate throw configuration (firing sequence) can additionally minimize second order free moments of inertia.

A 90° V series can be designed more broadly for four-stroke medium speed engines and not only include V8, V10 and V12 but also V16 and V20 variants (e.g. MTU 2000 (CR), DEUTZ-MWM 604 [8-27], 616 and 620, MTU 396 and 4000). Figure 8-8 presents an MTU 4000 series V16 engine. The best standard V angle for other numbers and increments of cylinders has to be determined allowing for the particular boundary conditions. While the free mass actions are increasingly less significant as of 12 cylinders and upward, non-uniform ignition intervals on the other hand require effective vibration damping. A V angle of 90° still creates sufficient space to house auxiliary units between cylinder banks. However, increasing numbers of cylinders reduce the V angle produced purely mathematically. An engine grows increasingly top heavy whenever auxiliary units continue to be mounted. This aspect may also influence the decision to select a proportionately larger V angle for V12 and V16 engines as in the MTU 595 series [8-28] (72° instead of 60° or 45°).

Fig. 8-8 MTU 4000 series V16 diesel engine (source: MTU Friedrichshafen GmbH)

8.2 Crankshaft Assembly Loading

8.2.1 Preliminary Remarks on Component Loading in the Crankshaft Assembly

Pistons, piston pins, the connecting rod and the crankshaft (*crankshaft assembly*) as well as the flywheel all form a reciprocating piston engine's crankshaft assembly. The components of the crankshaft assembly are not only subject to high gas forces but also tremendous acceleration (mass actions). Conflicting demands such as minimum mass on the one hand and high stiffness and endurance strength on the other present a challenge to component design. Since this handbook handles the components of the piston and connecting rod elsewhere (see Sect. 8.6), these remarks about their loading are kept brief.

8.2.2 Remarks on Piston and Connecting Rod Loading

Piston mass contributes substantially to the oscillating mass and is thus subject to strict criteria of lightweight construction. Combustion chamber pressure and temperature highly stress a piston thermomechanically. Highly thermally conductive, aluminum piston alloys combine low material density with high thermal load relief. However, their use in diesel engines reaches its physical limits at firing pressures above 200 bar.

The piston force (Eq. (8-1)) is supported in the piston pin boss. The inertial force counteracts the gas force, decreasing the load as speed increases. The piston skirt acts on the cylinder bore surfaces with the kinematically induced side thrust (Eq. (8-3)). Gas force predominantly stresses diesel engine pistons. In high speed gasoline engines, the stress of inertial force may outweigh the stress of the gas force. Preventing the piston pin boss from overstressing necessitates limiting the contact pressure and minimizing pin deformation (bending deflection and oval deformation). Diesel engine piston pins have to be engineered particularly solidly even though they contribute to the oscillating mass.

The connecting element between the piston pin and crank pin, the connecting rod is divided into an oscillating and a rotating mass component (Eqs. (8-24) and (8-25)). Lightweight aluminum connecting rods only play a role in very small engines. Titanium alloys are out of the question for reasons of cost and primarily remain reserved for racing. Just as for pistons, the reduction of mass and the optimization of structural strength are inseparably interconnected.

The gas force (ITDC) and oscillating inertial force (ITDC and GETDC) load a diesel engine's connecting rod shank with pulsating compressive stress. Above all, it must be adequately protected against buckling. Alternating bending stress of a connecting rod shank caused by transversal acceleration is less important at diesel engines' level of speed. Inertial force generates pulsating tensile and bending stress in the connecting rod small end and the eye. Bending, normal and radial stresses act in the curved eye cross section. The oscillating inertial force from the piston and connecting rod minus that of the connecting rod bearing cap acts on the connecting rod small end; only the oscillating inertial force from the piston acts on the small connecting rod eye. The bolted connecting rod joint produces static compressive prestress in the clamped area. The press fit of the small end bushing and the connecting rod bearing bushing halves causes static contraction stresses. The connecting rod eyes should only deform slightly to prevent adverse effects on the lubricating film including "bearing jamming". Eccentrically acting bolt and motive forces cause bending moments in the parting line. The one-sided gaping this facilitates in the parting line must be prevented. A form fit (a serration or currently a *fracture-split connecting rod*) prevents dislocation caused by transverse forces, particularly in larger diesel engines' connecting rod small ends, which are split obliquely for reasons of assembly.

8.2.3 Crankshaft Design, Materials and Manufacturing

8.2.3.1 Crankshaft Design

A crankshaft's design and outer dimensions are determined by the crank spacing (distance between cylinder bore center lines) a_Z, the stroke s, the number of crankshaft throws and the throw angle φ_K between them or, optionally, the crank pin offset (angle of connecting rod offset δ) and the number, size (limited by free wheeling in the cylinder crankcase) and arrangement of counterweights. A crankshaft throw's "inner" dimensions are the main pin width and crank pin width, the related journal diameters and the crankshaft web thickness and width (Fig. 8-9). The flywheel flange is located on the output end with its bolt hole circle and centering. The shaft's free end is constructed as a shaft journal for the attachment of the belt pulley, vibration damper and so forth. Powered by the crankshaft, the camshaft drive may be mounted on the front end or, frequently in diesel engines, on the flywheel end for reasons of vibrations.

The number of throws depends on the number of cylinders z and the number of main crankshaft bearings ensuing from the design (I engine: z throws, $z + 1$ main pins, V engine: $z/2$ double throws, $z/2 +1$ main pins).

The distance between cylinder bore center lines a_Z – the cylinder bore diameter D_Z and the width of the wall between cylinders Δa_Z – defines the crank spacing in inline engines. Conversely, unlike corresponding inline engines, "inner" dimensions may be relevant for cylinder clearance in V engines when, for instance, enlarged bearing width, reinforced crankshaft webs and a double throw with crank pin offset and intermediate web become necessary. "Inner" dimensions determine the crank spacing anyway when there are $z +1$ main bearings, i.e. a double throw is dispensed with, particularly in the so-called "Boxer" design with

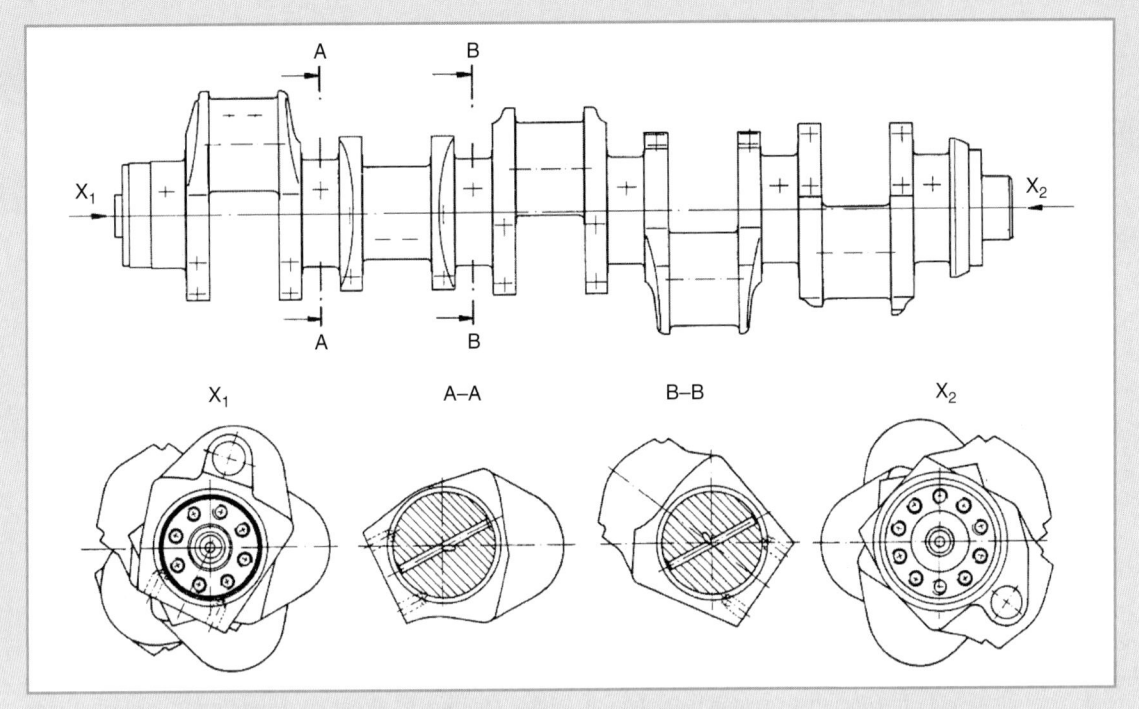

Fig. 8-9 Example of crankshaft design: steel crankshaft of a MAN V10 commercial vehicle diesel engine [8-2]

(horizontally) opposing cylinder pairs with crankshaft throws offset by 180°.

8.2.3.2 Crankshaft Materials and Manufacturing

Forged crankshafts made of high grade heat-treated steels are best able to meet the high requirements off dynamic strength and, in particular, stiffness too. Less expensive *microalloyed* steels heat-treated by controlled cooling from the forging heat (designated "BY") are increasingly being used. In less loaded car engines (primarily naturally aspirated gasoline engines), they may also be cast from nodular graphite cast iron (*nodular cast iron* of the highest grades GJS-700-2 and GJS-800-2) [8-29, 8-30]. This reduces the costs of both manufacturing and machining a blank. 8 to 10% less material density than steel and the option of a hollow design additionally benefit crankshaft mass. A significantly lower Young's modulus, lower dynamic strength values and less elongation at fracture than steel have to be accepted (the label "-2" stands for a guaranteed 2%).

Cost effective machining limits the materials' tensile strength to approximately 1,000 MPa. Hence, measures that enhance the fatigue strength of the transition and the concave fillet radii, which are critical to loading, between the journal and web are indispensible. Mechanical, thermal and thermochemical processes are employed. Pressure forming, roller burnishing [8-31] and shot peening, inductive and case hardening and nitriding build up intrinsic compressive stresses and strengthen the surface areas of materials (Fig. 8-10).

Each process enhances fatigue strength with varying quality. Nitriding's penetration is comparatively slight, thus making it impossible to entirely rule out fatigue failures in the core structure near the surface. The journals in vehicle engines are also hardened. Inductive hardening is quite a low cost process [8-32, 8-33]. The expense connected with case hardening on the other hand limits its cost effective application to larger crankshafts. The sequence of concave fillets must be followed when they are hardened because of distortion. The high heat output required during a very brief heating-up period, makes thin intermediate webs or oil bores in a shallow depth beneath the concave fillets critical during tempering.

Casting and forging necessitate designing blanks adapted to the manufacturing process. Sand (green sand or bonded sand cores), shell mold, evaporative pattern or full mold casting are used for cast crankshafts. Large lots are drop forged (the fiber flow facilitating fatigue strength). Large crankshafts on the other hand are hammer forged (poorer fiber flow). Fiber flow forging is employed for larger crankshafts and smaller quantities. The shaft is cranked

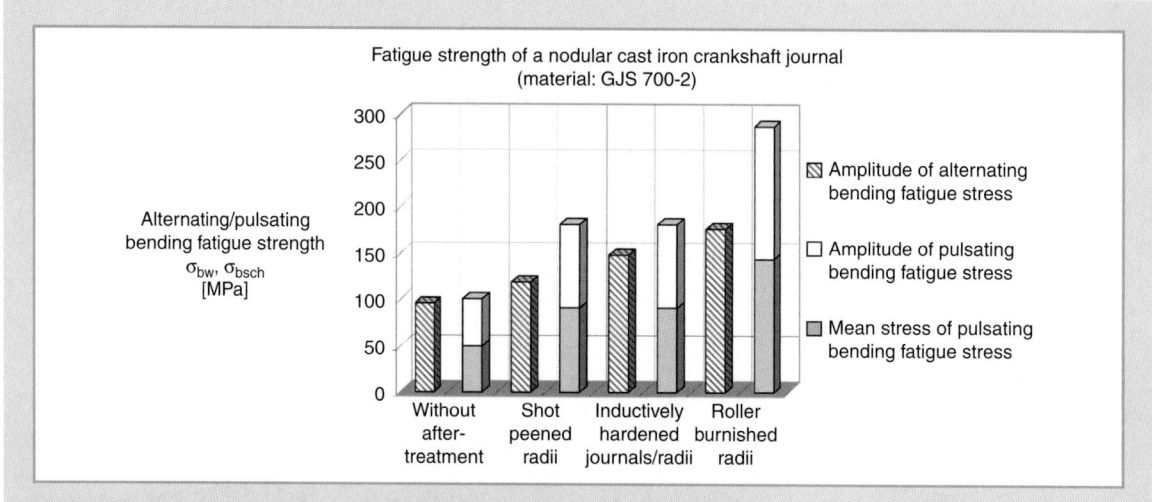

Fig. 8-10 Example of increased fatigue limit in nodular cast iron (GJS) crankshafts for different methods to strengthen the surface area (based on a lecture by H. Pucher, 'Grundlagen der Verbrennungskraftmaschinen', TU Berlin)

individually under a hammer or by hammering the throws in partial dies. Crankshafts are increasingly also being hammer forged in one plane and subsequently twisted in the main pin areas. Low-speed two-stroke diesel engines have "full" or "half" composite crankshafts (shrink joint or submerged-arc narrow gap welding) [8-34]. Machining of crankshafts is limited to the pins, thrust collars and counterweight radii. Only high stress justifies all-around machining. Larger crankshafts have bolted on counterweights (high-strength bolted joints).

8.2.4 Crankshaft Loading

8.2.4.1 Loading Conditions

A crankshaft's loading conditions are quite complex. The directional components of the connecting rod force, the tangential or torsional force and a component of the radial force (Eqs. (8-4) and (8-5)) are based on the gas and oscillating inertial force, which varies periodically with the crank angle. An additional inertial force component of the radial force only varies with the speed. The centrifugal force of the rotating component of the assumed connecting rod mass also acts on the crank pin. The crank and the counterweights additionally exert centrifugal forces.

Circumferentially, the gas and mass torque characteristic produces torque that varies with the crank angle but is nevertheless often denoted as "static". The superimposition of the individual crankshaft throws' torque contribution (Fig. 8-3) generates widely differing torque amplitudes in the individual main/crank pins. The torsional force characteristic's harmonics (see Sect. 8.3.8) additionally excite rotational vibrations (torsional vibrations) of the crankshaft assembly. Torque rises occur when the "static" torque characteristics are superimposed on the "dynamic" torque characteristics. The radial force's harmonics also excite bending and axial vibrations, all modes of vibration being interconnected. In addition, weight force and the flywheel or vibration damper's lever arm (flywheel wobble), belt forces or even forced deformation in misaligned main bearings make gyroscopic effects unavoidable (precession in synchronous and counter rotation). The resonance case additionally has a bending moment that gyrates with precession frequency. Finally, a rapid rise in firing pressure $dp_Z/d\varphi$ in conjunction with bearing clearances causes mass actions of the crankshaft, which likewise cause a dynamic increase in crankshaft loading [8-35].

8.2.4.2 Equivalent Models for Crankshaft Simulation

Crankshafts are now usually designed in two steps, a concept and a layout phase [8-8]. The concept phase involves defining the main dimensions, elementarily designing the bearings, simulating strength on the basis of analytical methods (see Sects. 8.2.4.3 and 8.2.4.4) and simulating 1D torsional vibration. The detailed analysis is then performed in the layout phase where 3D multibody dynamics (MBS) and the finite element method (FEM) are employed. All significant phenomena are reproduced as realistically as possible:
- stress on the multiply supported (statically indeterminate) crankshaft under all spatially distributed and temporally shifted external and internal forces and bending moments in the entire working cycle,
- incorporation of additional stress resulting from coupled vibrations including their damping (of the elastic dynamic deformation of the rotating crankshaft),

- incorporation of the reactions (including mass action) of the only finitely rigid cylinder crankcase with elastically flexible bearing bulkheads (bearing supporting action) and
- elastically deforming main bearing bores, including their clearances, with nonlinear hydrodynamic reactions of the lubricating film (EHD).

Suitable approaches to simulation and proposals to reduce computing time exist [8-8, 8-36, 8-37]. The permissible simplification of equivalent models, also termed model compression (Fig. 8-11) is extremely important. The basis is a finite element or boundary element model (FEM, BEM) of a crankshaft (stress simulation) discretized from a 3D CAD model. It is used to derive a dynamically equivalent 3D structural model (equivalent beam/mass model) and dynamically equivalent 1D torsional vibration model.

Only up-to-date mathematical simulation techniques reliably manage to obtain higher stiffness with less mass and satisfying dynamic behavior as well as uncover hidden risks to endurance strength and prevent oversizing, specifically for vehicle engines. A discussion of the pros and cons of common models [8-36] and a treatment of dynamic simulations in elastic multibody systems [8-37] would exceed the scope of this handbook. Hence the following remarks are limited to conventional, greatly simplified *rough sizing* of the crankshaft in the concept phase.

Multiply supported statically indeterminate bearings and temporally shifted influences of adjacent throws on the load in multiple cylinder engines are disregarded. Thus, the mathematical model is reduced to the statically determined model of a throw. According to the argumentation common earlier, disregarding the main bearings' fixed end moments is justifiable provided they counteract the moment of load and consequently reduce the stress.

The radial force only subjects a throw to bending stress; the tangential force subjects it to both bending (disregarded here) and torsional stress (Fig. 8-12). Simulation concentrates on the locations of maximum stress. These are the transitions from the journals to the webs (e.g. [8-1]) where stress peaks occur because of force deflection and the notch effect. Older technical literature deals exclusively with the question of whether the main pin or crank pin transition is more endangered. Modern methods of simulation that deliver a very exact stress distribution throughout an entire component have relegated this issue to the background. The inconsistency of some statements is related to the varying dimensions. While thin crankshaft webs and sizeable journal overlap characterize vehicle engine crankshafts, the opposite applies to large engines. Tensile stresses occur in the crank pin transition under gas force. The compressive stresses induced by the transverse (normal) force act to relieve stress. Tensile stresses in the crank pin have to be evaluated more critically than compressive stresses in the main pin occurring under the same conditions. However, they increase there by the

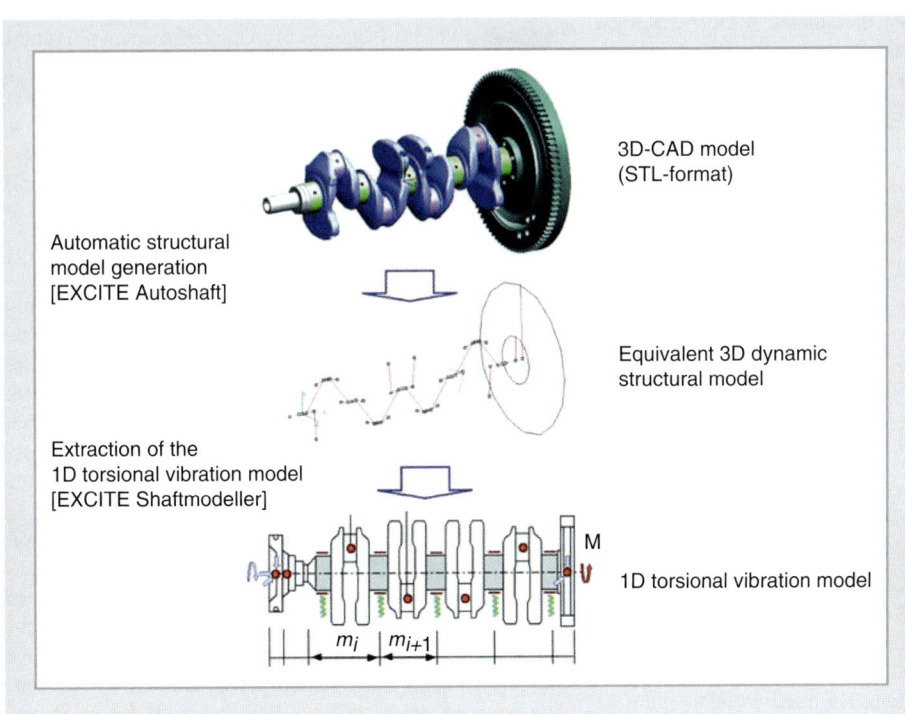

Fig. 8-11 Computerized crankshaft simulation; equivalent crankshaft models satisfying specific requirements [8-8]

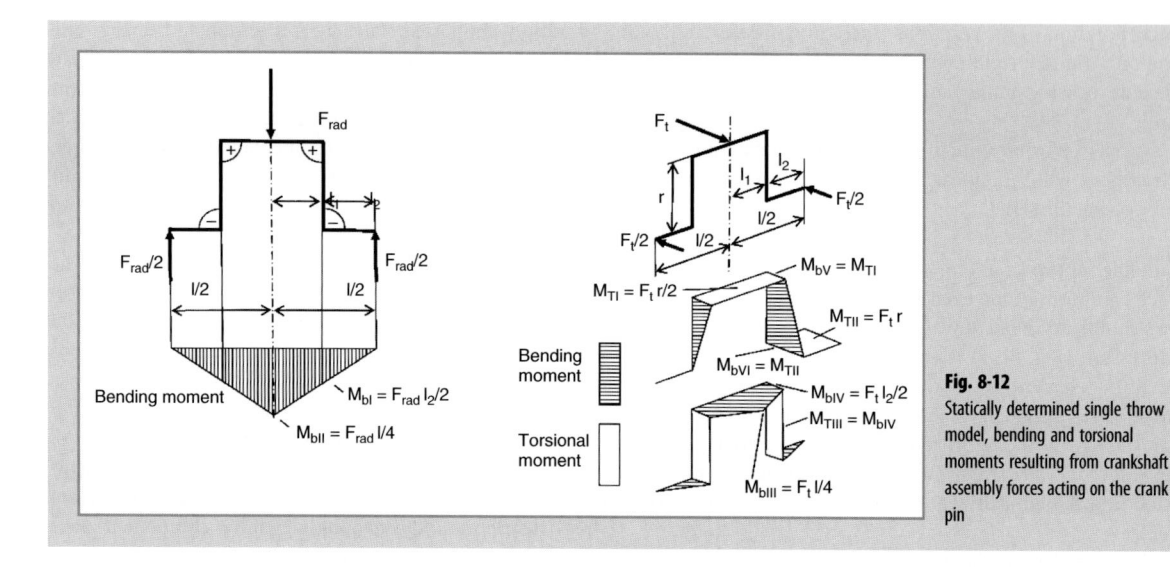

Fig. 8-12 Statically determined single throw model, bending and torsional moments resulting from crankshaft assembly forces acting on the crank pin

component of the transverse force. Test setups that load a crankshaft throw with a constant rather than a triangular bending moment disregard this condition.

8.2.4.3 Bending and Torsional Stresses

The simplified simulation of stresses is based on defined nominal stresses resulting from bending, transverse force and torsion σ_{bn}, σ_{Nn} und τ_{Tn}. Bending and normal stress relate to the crankshaft web cross section A_{KWW} [8-38–8-40] with an axial planar moment of inertia I_b or moment of resistance W_b; torsional stress relates to the particular journal cross section being analyzed with a polar planar moment of inertia I_b or moment of resistance W_T:

$$\sigma_{bn} = \frac{M_{bI}}{W_b} \quad M_{bI} = F_{rad}\frac{l_2}{2}$$

$$W_b = \frac{2I_b}{h_{KWW}} = \frac{b_{KWW} h_{KWW}^2}{6} \tag{8-14}$$

$$\sigma_{Nn} = \frac{F_{rad}}{A_{KWW}} \quad A_{KWW} = b_{KWW} h_{KWW} \tag{8-15}$$

$$\tau_{Tn} = \frac{M_{TI,II}}{W_T} \quad \begin{aligned} M_{TI} &= F_t \frac{r}{2} \\ M_{TII} &= F_t r \end{aligned} \tag{8-16}$$

$$W_T = \frac{2I_P}{d_{KWG,H}} = \frac{\pi(d_{KWG,H}^4 - d_{KWGi,Hi}^4)}{16 d_{KWG,H}}$$

The bending moment M_{bI} in Eq. (8-14) and torsional moments $M_{TI,\ II}$ in Eq. (8-16) stem from Figs. 8-12 and 8-13 (see also the definition of the geometric parameters there). Forces, moments and resultant stresses are quantities that vary with the crank angle φ. The quasistationary nominal bending stress σ_{bKWrot} of the rotating inertial forces relative to the center crankshaft web, i.e. the point $x = l_2$ in Fig. 8-13, is calculated separately:

$$\sigma_{bKWrot} = \frac{l_2}{W_b} \sum_i \vec{m}_{KWroti}\, r_i \omega^2 \left(1 - \frac{l_i}{l}\right) \tag{8-17}$$

m_{KWroti} being the rotating mass components. The arrow indicates that only the directional components of the inertial forces lying in the throw plane have to be added for the counterweights' centers of mass. r_i are the radii of the centers of gravity and l_i the distances from the center main pin (point $x = 0$ in Fig. 8-13).

The analytical method employs experimentally determined stress concentration factors for bending, transverse force and torsion α_b, α_q and α_T to allow for the local stress peaks $\sigma_{b\,max}$, $\sigma_{N\,max}$ and $\tau_{T\,max}$ in the transition radii and concave fillets:

$$\sigma_{bmax} = \alpha_b \sigma_{bn} \quad \sigma_{Nmax} = \alpha_q \sigma_{Nn} \quad \tau_{Tmax} = \alpha_T \tau_{Tn} \tag{8-18}$$

Strictly speaking, the stress concentration factors only apply to the range of the throw parameters covered by tests. The German Research Association for Combustion Engines (FVV) is representative of numerous efforts here [8-38, 8-39]. Supplemented by the influence of the recesses still common in large crankshafts, the stress concentration factors have now been adopted by every classification body in the IACS Unified Requirements M53 [8-40].

The calculation of stress concentration factors is based on empirical formulas. According to [8-38, 8-39] a specific constant is multiplied by values f_i calculated with the aid of polynomials. The polynomials include powers of the geometric throw parameters (with the exception of the concave fillet radii), always in relation to the crank pin diameter. The bending stress concentration factor α_b for crank pin

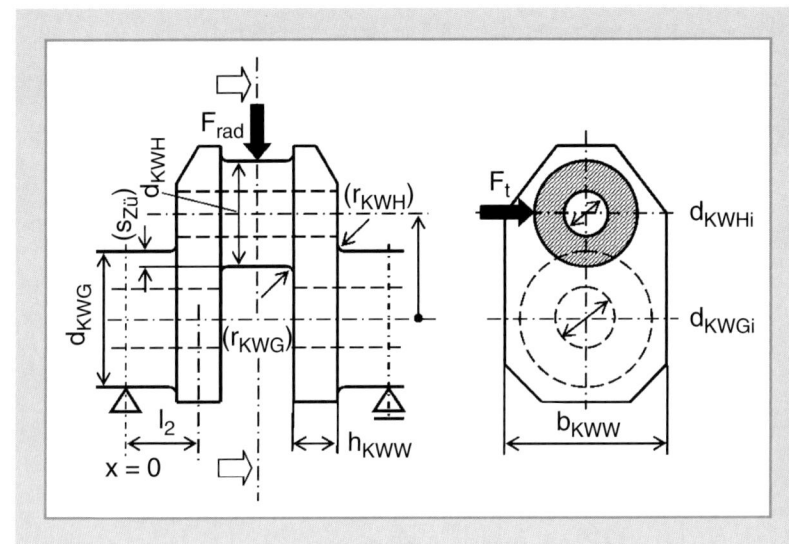

Fig. 8-13
Geometric crankshaft throw parameters for the simulation of nominal stresses and maximum stresses in the journal-web transition (parameters additionally necessary for the latter are indicated parenthetically)

transition, for example, is based on the following mathematical approach:

$$\alpha_b = 2.6914 \times$$
$$\times f_1\left(\frac{d_{KWHi}}{d_{KWH}}\right) f_2\left(\frac{d_{KWGi}}{d_{KWH}}\right) f_3\left(\frac{d_{KWW}}{d_{KWH}}\right) f_4\left(\frac{d_{KWW}}{d_{KWH}}\right) \times$$
$$\times f_5\left(\frac{s_{Z\ddot{u}}}{d_{KWH}}, \frac{h_{KWW}}{d_{KWH}}\right) f_6\left(\frac{r_{KWH}}{d_{KWH}}\right)$$

(8-19)

with: $s_{Z\ddot{u}}$ – journal overlap
r_{KWH} – radius of the crank pin's concave fillet

The other parameters are already known. Since its importance is diminishing, a complete presentation of the analytical method is foregone here (e.g. [8-2]). Certain geometric conditions may have deviations of up to 20% [8-36].

Strictly speaking, a stress concentration factor α_K only applies to static stress and must be converted by a fatigue notch sensitivity factor η_K into a fatigue notch factor β_K for dynamic stress (cf. [8-41]). This has repeatedly been a source of heated debate in the case of crankshafts. Reservations about this approach are rooted in the related uncertainties compared with α_K values based on reliable measurements.

8.2.4.4 Reference Stresses

Allowing for the torsional vibrations, the journal with the largest alternating torque is crucial for dynamic stress (see Sect. 8.4). The hypothesis of energy of deformation allows simulating a reference stress that results from bending and torsional stress. Great importance is attached to the temporal and spatial relation of both stress maxima. Strictly speaking, relating them to a fatigue strength diagram would require their simultaneous occurrence in the same place and synchronous alternation, all of which does not hold true. Moreover, the sudden phase shift of 180° when passing through a resonance (transient operation) additionally compounds the problems of in-phase superimposition of bending and torsion.

Nonetheless, the rough formulation allows the following approach with a temporally exact allocation of bending and torsional stress according to Table 8-1 (Eq. (8-20)): First, the extrema of the bending and torsional stress characteristics during one working cycle – the maximum and minimum stresses σ_o and σ_u or τ_o and τ_u for the related crank angles φ_1 and φ_2 or φ_3 and φ_4 – are ascertained. The inertial force acts against the gas force around ITDC; only the inertial force acts around GETDC. The related reference stresses follow from the mean and alternating bending and torsional stresses that then have to be simulated. The alternating torsional stress is of prime importance. It has a significantly higher amplitude and frequency than alternating bending stress.

Figure 8-14 reproduces the principle of the characteristic of the torsional moment in an I4 engine's flywheel side main bearing at full load based on [8-36]. Quantitatively, the simple equivalent 1D torsional vibration model employed here corresponds well with reality. Typical characteristics of the bending moment are presented next to this (in the same engine but at marginally higher speed). The dynamic characteristic deviates distinctly from the purely kinematic characteristic. While the need to capture every dynamic phenomenon is evident, this requires a rotating 3D beam/mass equivalent model. Deformation follows from the external load as node shifts. The simulation of stress based on this can, for example, be performed so that the simulated node shifts are applied to the volume model (of the finite element

Table 8-1 Calculation of the mean reference stresses and reference alternating stresses in the journal-crankshaft web transition (Eq. (8-20))

Maximum positive and negative bending stress for the crank angles φ_1 and φ_2		Maximum positive and negative torsional stress for the crank angles φ_3 and φ_4	
$\sigma_o = \sigma_b(\varphi_1)$ (at ITDC)		$\tau_o = \tau_T(\varphi_3)$	
$\sigma_u = \sigma_b(\varphi_2)$ (GETDC, only inertial force)		$\tau_u = \tau_T(\varphi_4)$	
Related torsional stresses		Related bending stresses	
$\tau_o = \tau_T(\varphi_1)$		$\sigma_o = \sigma_b(\varphi_3)$	
$\tau_u = \tau_T(\varphi_2)$		$\sigma_u = \sigma_b(\varphi_4)$	
Mean bending and torsional stresses		Alternating bending and torsional stresses	
$\sigma_{bm} = \frac{1}{2}(\sigma_o + \sigma_u)$	(8.21)	$\sigma_{ba} = \left\| \frac{1}{2}(\sigma_o - \sigma_u) \right\|$	
$\tau_{Tm} = \frac{1}{2}(\tau_o + \tau_u)$	(8.22)	$\tau_{Tm} = \left\| \frac{1}{2}(\tau_o - \tau_u) \right\|$	
Mean reference stress		Reference alternating stress	
$\sigma_{vm} = \sqrt{\sigma_{bm}^2 + 3t_{Tm}^2}$	(8.25)	$\sigma_{va} = \sqrt{\sigma_{ba}^2 + 3t_{Ta}^2}$	

structure) incrementally. To simulate strength, this has to be reasonably adjusted in the range of large stress gradients to discretize and select the type of element. Figure 8-15 [8-42] visualizes the distribution of the maximum main normal stresses resulting from bending in the flywheel side throw for a BMW V8 diesel engine.

8.2.4.5 Local Fatigue Strength in the Journal-Crankshaft Web Transition

Exact knowledge about local fatigue strength constitutes a particular challenge especially in the case of highly dynamically stressed crankshafts since the influencing factors are many and diverse:

– technological influences:
  – casting quality: microstructure, pores, liquations and oxide inclusions undetected during production monitoring,
  – forging quality: fiber flow, forging reduction, slag lines not detected during production monitoring,
  – internal stresses,
  – heat treatment, quenching and drawing,
– the influence of aftertreatment (polishing, cold-work hardening, tempering, nitriding),
– the influence of the surface: surface roughness,
– the influence of component size (fatigue strength decreases as component size increases)

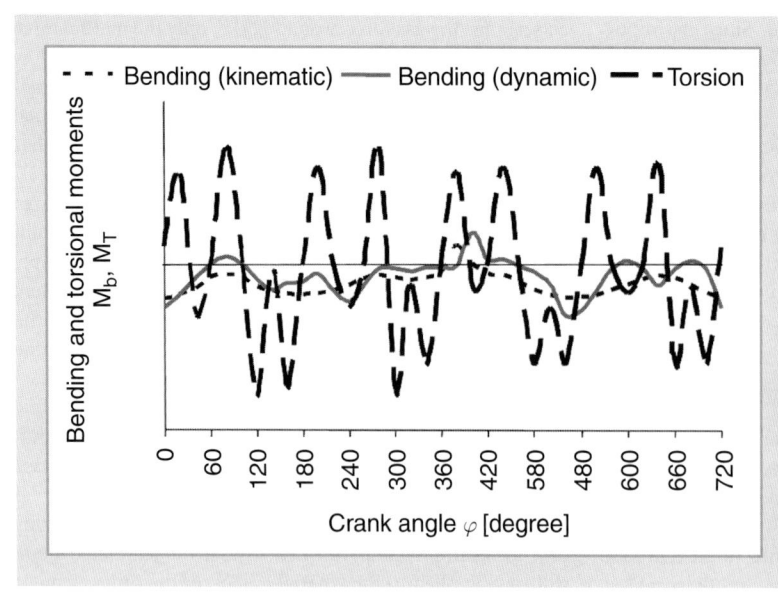

Fig. 8-14
Simulated bending and torsional moment characteristic in the flywheel-side crankshaft web and in the flywheel-side main pin of a I4 crankshaft (merging of two conditionally comparable representations based on [8-36] in one)

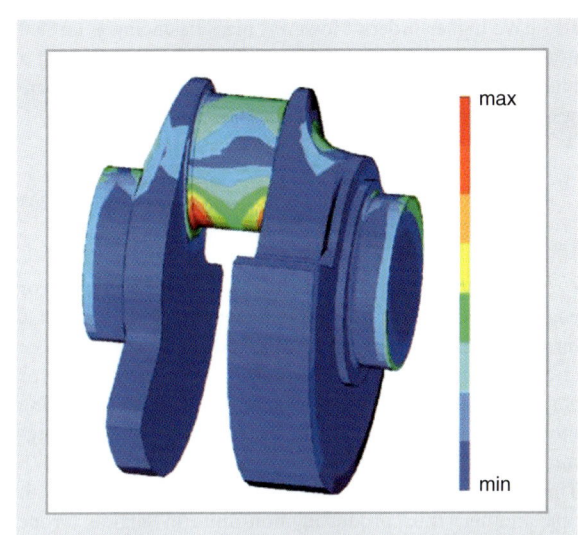

Fig. 8-15 Maximum principal normal stress distribution of the flywheel-side crankshaft throw of a BMW V8 car diesel engine crankshaft [8-42]

- the relative stress gradient in a component's cross section decreases as component size increases

$$\chi^* = \frac{1}{\sigma_{max}} \left(\frac{d\sigma}{dx}\right)_{max} \quad (8\text{-}21)$$

- forging reduction decreases as component size increases.

The size of the sample renders local sampling unfeasible for the dynamic material test. On the other hand, a cast-on or forged-on sampling rod does not deliver the local resolution. Consequently, the local fatigue limit can only be inferred indirectly. The outcome of several FVV projects, a formula for the fatigue properties σ_w of the crankshaft material [8-43–8-45] is based on a familiar approach [8-46]. This value allows a direct comparison with the reference alternating stress σ_{va} (and other proposals are also based on this approach: CIMAC [8-40] and [8-47] for example):

$$\sigma_w = 0.9009(0.476R_m - 42)(1 + \sqrt{0.05\chi^*}) \times \\ \times \left[1 - \sqrt{(1-o_k)^2 + (1-i_k)^2}\right] \quad (8\text{-}22)$$

influencing factors:
internal notch effect: $i_k = 1 - 0.2305 \cdot 10^{-3} R_m$
surface quality:

$$o_k = 1.0041 + 0.0421 \cdot lgR_t - 10^{-6} R_m (31.9 + 143 \cdot lgR_t)$$

relative stress gradient (bending):

$$\chi^* = \frac{2}{r_{KWH}} + \frac{2}{d_{KWH}}$$

The tensile strength R_m (in alternative formulas, the alternating tensile-compression fatigue limit σ_{zdw}) of the smooth forged-on sampling rod is inserted in MPa, the depth of roughness R_t in μm and the crank pin diameter d_{KWH} and the related concave fillet radius r_{KWH} in mm (here). The mean reference stress is initially ignored. While, as already noted in Sect. 8.1.1, there are various proposals for auxiliary constructions of suitable fatigue strength diagrams, exact knowledge of materials' properties is nevertheless likely continue to lag behind the presently quite exact knowledge of their stress despite engines manufacturers' extensive experience. The usual reference to a bending fatigue strength diagram is justified with the stress tensor present at the maximally stressed location and stress gradients (cf. [8-37]). The specified fatigue strength values usually refer to a 90% probability of survival. According to [8-37], a value of 99.99% means a safety factor of 1.33 (only valid for a variation coefficient σ/x̄ = 0.09). Assuming the simulation model is suitable, estimating the uncertainty of the simulated load of the crankshaft with the factor of 1.05 (accordingly low) suffices according to [8-37]. Multiplication then produces a required safety factor of 1.4.

Figure 8-16 provides an impression of the relationship of the reference alternating stress to the mean reference stress in

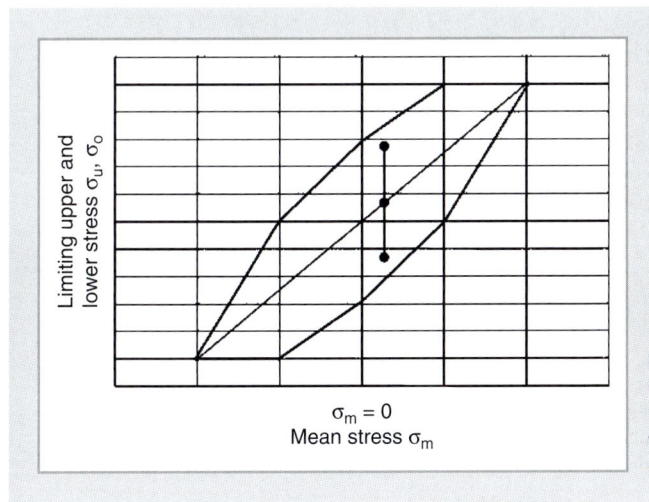

Fig. 8-16
Fatigue strength diagram based on [8-37] (representation based on Smith);Example of the ratio of alternating stress amplitudes to mean stress in the crank pin-web transition in a medium speed engine crankshaft, safety clearance against fatigue failure

the fatigue strength diagram for a simulation of a medium speed engine's crankshaft [8-37]. Without generalization, the influence of the mean stress does not yet decisively diminish the tolerable stress amplitude.

8.3 Balancing of Crankshaft Assembly Masses

8.3.1 Preliminary Remarks on Mass Balancing

A crankshaft assembly's free inertial forces are based on the acceleration of crankshaft assembly masses resulting from the action of irregular motion even when engine speed and centrifugal force are constant. The distances between the cylinders in multi-cylinder engines and the crank pin offset in V engines are lever arms that generate free moments of inertia. Mass balancing facilitates vibrational comfort by minimizing such mass actions through primary measures in the crankshaft assembly. Secondary measures that isolate vibrations usually reach their physical limits quickly. The following remarks are limited to the trunk-piston engines predominantly in use. Detailed presentations are contained in [8-1, 8-6] and summaries in [8-9, 8-48].

Mass balancing is largely related to constant engine speed. At variable speed, additional harmonic components of the free inertial forces and moments of inertia can be responsible for the excitation of structure-borne noise in the drive train. Counterweights in the crankshaft webs are standard balancing measures. Depending on the engine concept and comfort requirement, up to two balance shafts are additionally implemented according to the particular needs. Oscillating crankshaft assembly masses also influence the characteristic of the engine torque through their torsional force component. In a good case, torque peaks and thus the crankshaft's alternating torsional stress can be reduced. Counterweight masses increase a crankshaft's mass moment of inertia. Depending on their number, size and configuration, they exert influence on a crankshaft's bending stress as well as the loading of the main bearing. The quality of crankshaft balancing must be satisfactory and there are recommendations for the permissible residual imbalance. (The Association of German Engineers VDI recommends DIN ISO 1940-1(1993-12) for the minimum quality level.) *Master rings* and possibly not yet extant balancing masses must be simulated when a blank is being mass centered. Unlike simulation, the somewhat complicated utilization of master rings during final balancing also helps allow for unavoidable manufacturing tolerances and thus improve the quality of balancing.

8.3.2 Crankshaft Assembly Masses, Inertial Forces and Moments of Inertia

Mass balancing distinguishes between oscillating and rotating inertial forces. The latter rotate with the crankshaft and therefore only appear in the first order. Oscillating inertial forces also have higher orders. However, mass balancing is limited to the second order at most. Low residual imbalances of a higher order may be virtually ignored. A Cartesian coordinate system is beneficial for the calculation:

$$\vec{F} = \begin{bmatrix} F_x \\ F_y \end{bmatrix} \quad \vec{M} = \begin{bmatrix} M_x \\ M_y \\ M_z \end{bmatrix} \qquad (8\text{-}23)$$

Longitudinal forces F_{xi} act in the direction of a cylinder or engine's vertical axis, transverse forces F_{yi} in the transverse direction, i.e. viewed along the longitudinal axis of an engine (z-axis) perpendicular to the vertical axis of the cylinder or engine. The total longitudinal forces generate a tilting moment M_y around the y-axis, the transverse forces a longitudinal yawing moment M_x around the x-axis (Fig. 8-17). The torques generated by gas and mass forces interfere with one another, thus producing non-uniform engine torque around the z-axis. In addition, the *rotational moments of inertia* also act around this axis. They are generated by the connecting rod's non-uniform pivoting even at constant speed and by the rotating masses' angular acceleration at variable speed (see Sect. 8.3.5).

As in Fig. 8-18, the parameters of a crankshaft assembly's free inertial forces are:
- total piston mass m_{Kges}, crankshaft mass without counterweights m_{KW} and connecting rod mass m_{Pl},
- crank radius r and radius r_1 of the crank's center of gravity,
- the connecting rod eyes' center distance l_{Pl} and connecting rod's center of gravity distance l_{Pl} from the big end eye's center and the
- stroke/connecting rod ratio $\lambda_{Pl} = r/\lambda_{Pl}$.

The rotating crankshaft mass m_{KWrot} is reduced to the crank radius r. The connecting rod mass is divided into a component oscillating with the piston (m_{Plosz}) and a component

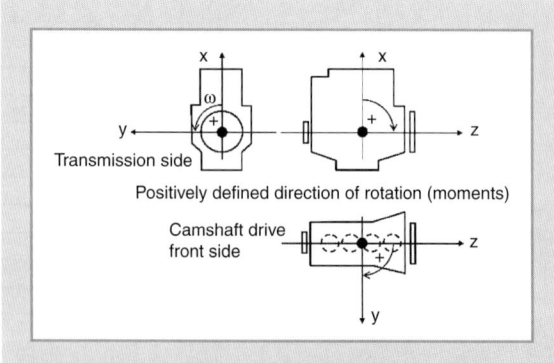

Fig. 8-17 Engine coordinate system including the definition of an assumed positive direction of rotation (x axis: vertical engine axis, y axis: diagonal engine axis, z axis: longitudinal engine axis, coordinate origin: e.g. crankshaft center)

series of the stroke/connecting rod ratio λ_{pl}. Thus, they are reduced to the coefficients subsequently denoted with A_i, and uneven orders disappear:

$$A_2 = \lambda_{Pl} + \frac{1}{4}\lambda_{Pl}^3 + \frac{15}{128}\lambda_{Pl}^5 + \ldots$$
$$A_4 = -\frac{1}{4}\lambda_{pl}^3 - \frac{3}{16}\lambda_{pl}^5 - \ldots \quad (8\text{-}28)$$

The following simple formula may be applied, usually with an error of $< 1\%$:

$$\ddot{x}_k = r\omega^2(\cos\varphi + \lambda_{Pl}\cos 2\varphi) \quad (8\text{-}29)$$

Thus, when rotary motion ($\omega = const.$) is idealized and assumed to be regular, the first and second order oscillating inertial forces $F^{(1)}_{mosz}$ and $F^{(2)}_{mosz}$ and the rotating inertial force F_{mrot} can be defined as follows:

$$F^{(1)}_{mosz} = \hat{F}^{(1)}_{mosz}\cos\varphi = m_{osz}r\omega^2\cos\varphi$$
$$F^{(2)}_{mosz} = \hat{F}^{(2)}_{mosz}\cos\varphi = m_{osz}r\omega^2\lambda_{Pl}\cos 2\varphi \quad (8\text{-}30)$$
$$F_{mrot} = m_{rot}r\omega^2$$

Only longitudinal forces still appear in higher orders. The first order splits into a longitudinal force and transverse force component $F_x^{(1)}$ and F_y:

$$F_x^{(1)} = \left(\hat{F}^{(1)}_{mosz} + F_{mrot}\right)\cos\varphi = r\omega^2(m_{osz} + m_{rot})\cos\varphi$$
$$F_y = F_{mrot}\sin\varphi = r\omega^2 m_{rot}\sin\varphi \quad (8\text{-}31)$$

8.3.3 Balancing of Free Inertial Forces by Counterweights

8.3.3.1 Counterweights for Single-cylinder Engines

Balancing of the rotating inertial force is considered the minimum requirement for single cylinder engines. Incorporating Eqs. (8-24), (8-25) and (8-31), this is done by *balancing the transverse forces*:

$$F_y = 0 \rightarrow F_{mrot} = 0: \ m_{KW}r_1 + m_{Pl}r\left(1 - \frac{l_{Pl1}}{l_{Pl}}\right) = 0 \quad (8\text{-}32)$$

Fulfilling Eq. (8-32) means a negative sign for the crank's radius of the center of gravity r_1, which is impracticable. This problem is resolved by applying a counterweight mass m_{Gg} with zero-torque (symmetrical). Its center of mass lies at the distance r_{Gg} in the reverse extension of the crank. This shifts the shared center of mass to the crankshaft's axis of rotation.

Fig. 8-18 Equivalent crankshaft assembly mass system

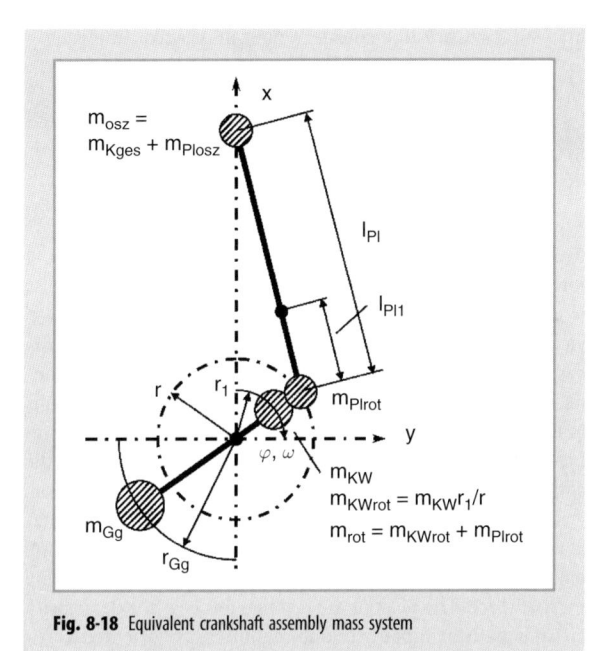

rotating with the crankshaft (m_{Plrot}). Thus, the oscillating (m_{osz}) and rotating masses (m_{rot}) are defined as follows:

$$m_{KWrot} = m_{KW}\frac{r_1}{r} \quad m_{Plosz} = m_{Pl}\frac{l_{Pl1}}{l_{Pl}} \quad m_{Plrot} = m_{Pl}\left(1 - \frac{l_{Pl1}}{l_{Pl}}\right) \quad (8\text{-}24)$$

$$m_{osz} = m_{Kges} + m_{Plosz} \quad m_{rot} = m_{KWrot} + m_{Plrot} \quad (8\text{-}25)$$

The acceleration of the mass m_{osz} follows from the piston movement x_K (viewed from the TDC position here), which can be easily derived with the Eqs. (8-6, right) and (8-9):

$$x_K = r + l_{Pl} - (r\cos\varphi + l_{Pl}\cos\psi) =$$
$$r\left[1 + \frac{1}{\lambda_{Pl}} - \cos\varphi - \frac{1}{\lambda_{Pl}}\sqrt{1 - \lambda_{Pl}^2\sin^2\varphi}\right] \quad (8\text{-}26)$$

Equation (8-26) neither allows for piston pin nor crankshaft offset (piston pin offset or lateral crank dislocation). An alternative, initially still general representation by series is expediently employed. Double differentiation based on time t yields the acceleration:

$$\ddot{x}_k = r\omega^2 \times$$
$$\times (\cos\varphi + B_1\sin\varphi + B_2\cos 2\varphi + B_3\sin 3\varphi + B_4\cos 4\varphi + \ldots) \quad (8\text{-}27)$$

Without piston pin or crankshaft offset (piston pin offset or lateral crank dislocation), the coefficients B_i are only power

Appropriately modified, Eq. (8-32) may be used to calculate the necessary counterweight mass:

$$m'_{KW} r'_1 = m_{KW} r_1 - m_{Gg} r_{Gg}$$

$$m_{KW} r_1 - m_{Gg} r_{Gg} + m_{Pl} r \left(1 - \frac{l_{Pl1}}{l_{Pl}}\right) = 0 \qquad (8\text{-}33)$$

$$m_{Gg} = \frac{1}{r_{Gg}} \left[m_{KW} r_1 + m_{Pl} r \left(1 - \frac{l_{Pl1}}{l_{Pl}}\right) \right]$$

Solving Eq. (8-32) based on l_{Pl1} requires $l_{Pl1} > l_{Pl}$, i.e. shifting the connecting rod's center of mass beyond the small connecting rod eye's axis of rotation. The minimized piston *compression height* and low connecting rod clearance make this solution impracticable.

When Eqs. (8-24), (8-25) and (8-31) are incorporated, the inclusion of the first order oscillating inertial force leads to the *balancing of first order longitudinal forces*:

$$F_x^{(1)} = 0 \quad F_{\text{mosz}}^{(1)} + F_{\text{mrot}} = 0 :$$

$$\left(m_{Kges} + m_{Pl} \frac{l_{Pl1}}{l_{Pl}} \right) r + m_{KW} r_1 + m_{Pl} r \left(1 - \frac{l_{Pl1}}{l_{Pl}} \right) = 0 \qquad (8\text{-}34)$$

Analogous to Eq. (8-33, 1st line), counterweight masses are required in the crankshaft, which shift the common center of gravity to the crankshaft's axis of rotation:

$$(m_{Kges} + m_{Pl}) r + m_{KW} r_1 - m_{Gg} r_{Gg} = 0$$
$$m_{Gg} = \frac{1}{r_{Gg}} \left[m_{KW} r_1 + (m_{Kges} + m_{Pl}) r \right] \qquad (8\text{-}35)$$

The result obviously contradicts the balancing of transverse forces (Eq. (8-33)). Although the oscillating inertial force is compensated longitudinally, it appears transversely. The necessary compromise – *normal balancing* – fully balances the rotating inertial force and 50% of the first order oscillating inertial force in addition. Incorporating these facts into Eq. (8-34) reduces the counterweight mass accordingly:

$$\left[\frac{m_{Kges}}{2} + m_{Pl} \left(1 - \frac{l_{Pl1}}{2 l_{Pl}} \right) \right] r + m_{KW} r_1 - m_{Gg} r_{Gg} = 0$$

$$m_{Gg} = \frac{1}{r_{Gg}} \left\{ m_{KW} r_1 + \left[\frac{m_{Kges}}{2} + m_{Pl} \left(1 - \frac{l_{Pl1}}{2 l_{Pl}} \right) \right] r \right\}$$
(8-36)

Complete balancing of all the orders of the oscillating inertial force is theoretically possible. The common center of mass would have to shift to the crank pin. The condition $F_{\text{mosz}} = 0$ yields a negative center of gravity distance l_{Pl} of the connecting rod. Concretely, a larger supplementary mass below the connecting rod small end fails because of the restricted conditions in the crank chamber. Foot balancing delivers partial improvements [8-49, 8-50], limiting the supplementary mass in the connecting rod small end to a structurally justifiable dimension.

The free "residual force vector" in *normal balancing* rotates at a constant magnitude. Underbalancing or overbalancing balances less or more than 50% of the first order oscillating inertial force. The peak of the residual force vector describes a circular path or a path resembling an ellipse oriented vertically or horizontally as a function of the degree of balance (Fig. 8-19). The same holds for the perceptible vibration amplitude of a single-cylinder engine.

The free inertial force or the residual force as a function of the degree of balance can each be visualized as the addition of the two vectors rotating in opposite directions at crankshaft speed (Fig. 8-20). An inverted counterbalancing force vector can only balance the vector rotating positively – in the crankshaft's direction of rotation. Its value corresponds to half the oscillating inertial force. Provided its peak does not describe a circular path, the inertial force vector's angular velocity varies periodically when rotary motion is uniform. The literature treats the method of adding positively and negatively rotating vectors in detail (cf. [8-1, 8-6]). The analytical approach favored here is beneficial for all problems and programming.

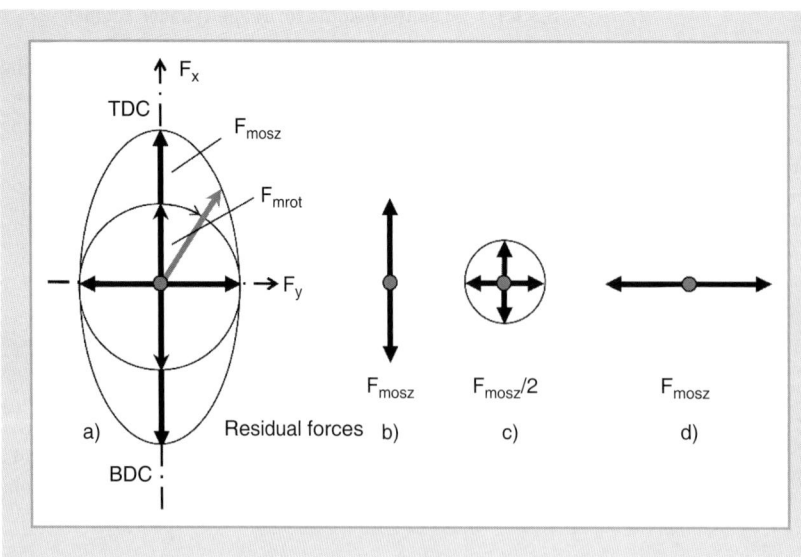

Fig. 8-19
First order free inertial force and residual force vector as a function of the degree of balance Ω: **a** Starting state, **b** $\Omega = 0$: Balancing of transverse forces (rotating masses), **c** $\Omega = 0.5$: 'normal balancing' (100% rotating + 50% oscillating masses), **d** $\Omega = 1$ Balancing of longitudinal forces

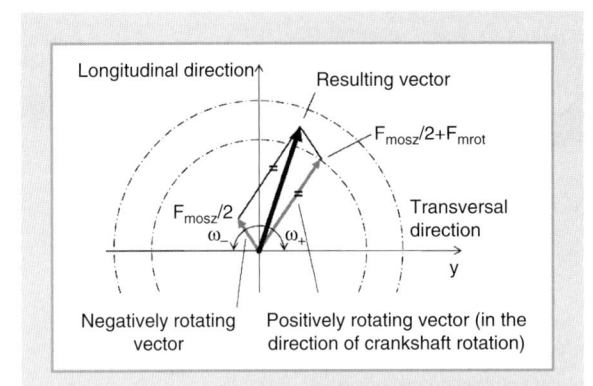

Fig. 8-20 Free inertial force of a single-cylinder engine represented with positively and negatively rotating vectors

8.3.3.2 Counterweights in Multi-cylinder Engines

Inline engines shall be considered first. The free inertial forces in multi-cylinder engines compensate one another when their star diagrams are centrosymmetric. The firing sequence plays no role at all. *Star diagrams* of the ith order are produced by taking cylinder 1 in TDC position as the starting point and applying the throw angle φ_{Kk} ($k = 1, 2, \ldots, z$) allocable to the individual cylinders or their multiples $\varphi_{Kk}^{(i)}$ ($k = 1, 2, \ldots, z; i = 1, 2, 4, 6$) counter to the direction of crankshaft rotation (Fig. 8-21).

In higher orders, the calculation is reduced to the longitudinal forces. The following generally applies to unbalanced orders:

$$\sum_{k=1}^{z} \cos \varphi_{Kk}^{(i)} = c_F^{(i)} = z \qquad (8\text{-}37)$$

The values $c_F^{(i)}$ are denoted as influence coefficients of the first order inertial forces. In inline engines, they correspond to the number of cylinders z, i.e. the oscillating inertial forces of all cylinders are added for the pertinent order.

Conditions for V engines are more complicated. Therefore, a V cylinder pair (V2 engine) shall be analyzed first (Fig. 8-22), i.e. the general case with any V angle α_V and crank pin offset (angle of connecting rod offset δ). The longitudinal and transverse forces of both single cylinder engines in cylinder-specific coordinates x_1, y_1 and x_2, y_2 do not require further explanation (Eq. (8-30) and (8-31)). The bank-specific crank angles φ_1 and φ_2 are measured starting from both cylinders' respective TDC positions. Once the crankshaft assembly forces have been broken down into their directional components in the superordinate x-y coordinate system and added up and one of the crank angles φ measured from the x-axis has been inserted, the V2 engine's first order inertial forces can be specified with the transformation equations:

$$\varphi_1 = \varphi + \frac{\alpha_v}{2} \qquad \varphi_2 = \varphi - \frac{\alpha_v}{2} - \delta \qquad (8\text{-}38)$$

Higher orders then appear in both x and y direction:

$$F_x^{(1)} = \left\{ F_{\text{mosz}}^{(1)} \left[\cos\frac{\delta}{2} + \cos\left(\alpha_v + \frac{\delta}{2}\right) \right] + 2F_{\text{mrot}} \cos\frac{\delta}{2} \right\} \times$$
$$\times \cos\left(\varphi - \frac{\delta}{2}\right)$$

$$F_y^{(1)} = \left\{ F_{\text{mosz}}^{(1)} \left[\cos\frac{\delta}{2} - \cos\left(\alpha_v + \frac{\delta}{2}\right) \right] + 2F_{\text{mrot}} \cos\frac{\delta}{2} \right\} \times$$
$$\times \sin\left(\varphi - \frac{\delta}{2}\right)$$

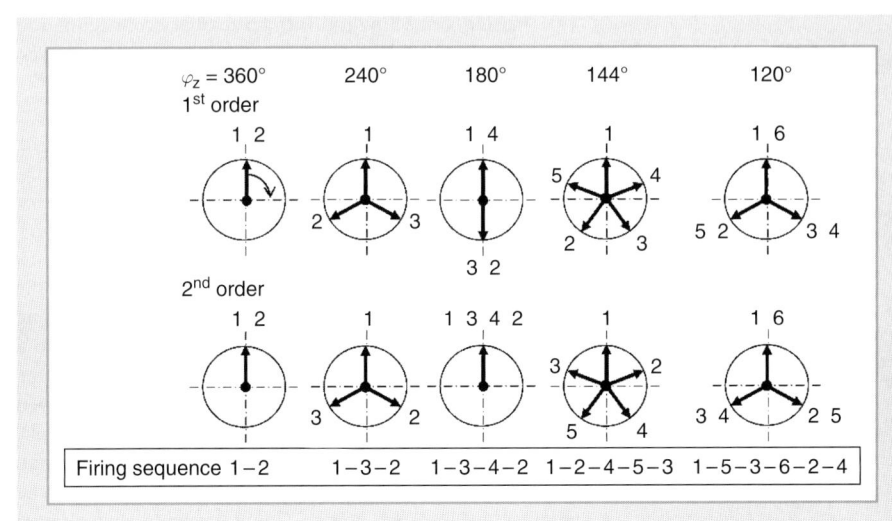

$$(8\text{-}39)$$

Fig. 8-21 Ignition interval and first and second order star diagrams for four stroke inline engines

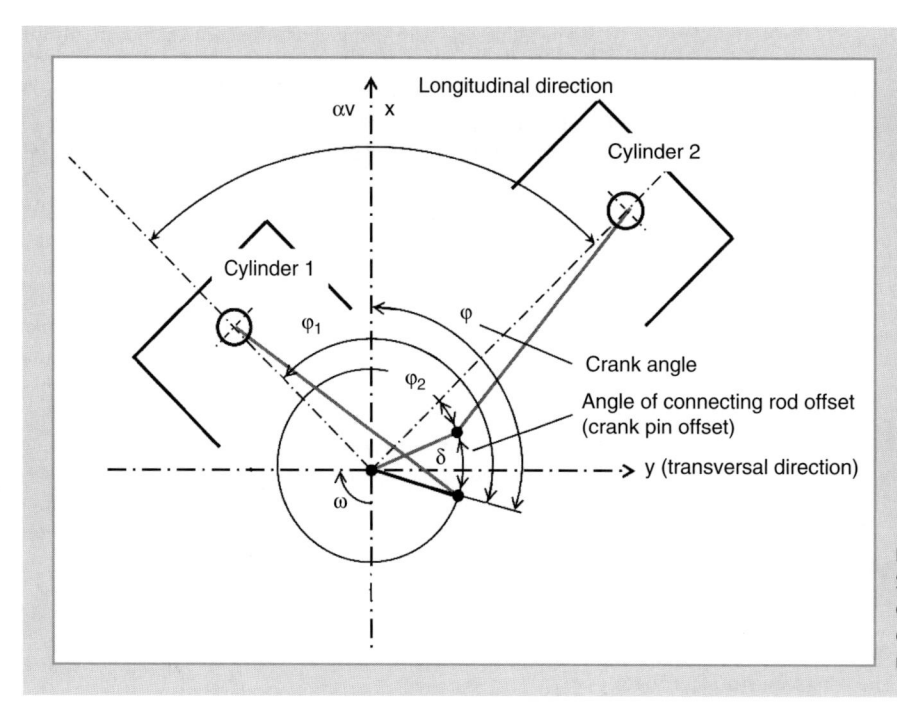

Fig. 8-22 Schematic crankshaft assembly of a V cylinder pair with any V angle α_V and crank pin offset (angle of connecting rod offset δ)

Counterweights alone can fully balance the mass in the crankshaft under the following circumstances:

$$\cos\left(\alpha_v + \frac{\delta}{2}\right) = 0 \quad \begin{array}{l} 1.)\ \alpha_v = 90°\ \delta = 0° \\ 2.)\ \delta = 180° - 2\alpha_v \end{array} \quad (8\text{-}40)$$

The following condition then applies:

$$F_{\text{mosz}}^{(1)} + 2F_{\text{mrot}} = 0. \quad (8\text{-}41)$$

This corresponds to *normal balancing*. Since only one crank is present, only half of one cylinder's mass is incorporated per cylinder. Already employed several times, the statement for counterweight mass yields the following result:

$$m_{Gg} = \frac{1}{r_{Gg}}\left[m_{KW}r_1 + m_{Kges}r + m_{Pl}r\left(2 - \frac{l_{Pl1}}{l_{Pl}}\right)\right] \text{ respectively.}$$

$$m_{Gg} = \frac{r}{r_{Gg}}\left[m_{KW}\frac{r_1}{r} + m_{Kges} + m_{Plasz} + 2m_{Plrot}\right]$$

$$(8\text{-}42)$$

V2 engines have only attained importance for motorcycle engines. The first order inertial forces for V engines also compensate one another when the star diagram is centrosymmetric. When the number of cylinders z is even and the crankshaft with its throws is centrosymmetrically offset by the angle $\Delta\varphi_K$, the *i*th order inertial forces can still be calculated as follows:

$$F_x^{(i)} = 2F_{\text{mosz}}^{(i)} \sum_{m=1}^{z/2} \cos\left[i\left(\frac{\alpha_v}{2} + \frac{\delta}{2}\right)\right] \cos\frac{\alpha_v}{2} \times \\ \times \cos\left\{i\left[\varphi - \frac{\delta}{2} - 2(m-1)\Delta\varphi_K\right]\right\}, \quad (8\text{-}43)$$

$$F_y^{(i)} = 2F_{\text{mosz}}^{(i)} \sum_{m=1}^{z/2} \sin\left[i\left(\frac{\alpha_v}{2} + \frac{\delta}{2}\right)\right] \sin\frac{\alpha_v}{2} \times \\ \times \sin\left\{i\left[\varphi - \frac{\delta}{2} - 2(m-1)\Delta\varphi_K\right]\right\}$$

Alternatives to a "connecting rod next to connecting rod" configuration (Fig. 8-4) have little practical significance. Consequently, their effects on mass balancing are not examined any more closely here:
(a) forked connecting rod with an inner connecting rod directly hinged on it (unequal connecting rod masses)
(b) master connecting rod with an auxiliary connecting rod directly hinged on it (unequal connecting rod masses, altered crankshaft assembly kinematics).

8.3.4 Balancing Free Moments of Inertia with Counterweights

Free moments of inertia result from the longitudinal distribution of the free inertial forces from the lever arms (see the definition of the longitudinal tilting moment and tilting moment in Sect. 8.3.2). Free moments of inertia relate to a reference point in the crankshaft assembly, by convention, the center of the crankshaft. Four-stroke inline engines with even numbers of cylinders have longitudinally symmetrical crankshafts and, consequently, all orders of free moments of inertia. First and higher order free moments of inertia occur when the number of cylinders is uneven. Counterweights can no longer

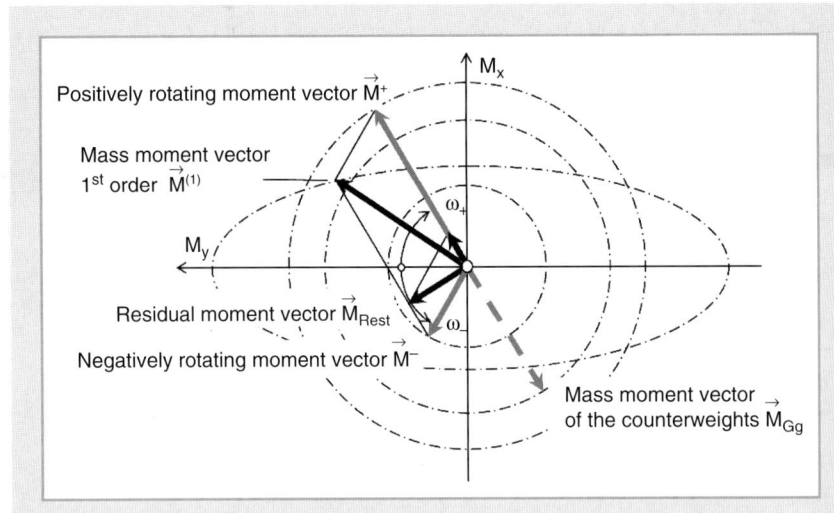

Fig. 8-23 Free mass moment and its partial balancing by counterweights represented with positively and negatively rotating vectors

fully balance a first order free moment of inertia under these circumstances. A free moment of inertia can also be visualized by adding two moment vectors rotating in opposite directions (Fig. 8-23).

V engines likewise profit from longitudinal symmetry, at least in the first order. This is a special case however. In certain cases, counterweights in the crankshaft can nonetheless fully balance a first order free moment of inertia without longitudinal symmetry through *normal balancing* (cf. [8-1]). This requires a centrosymmetric star diagram, V angles suitable to the throw angle, an appropriate throw configuration and the requisite crank pin offset when necessary. The latter does not produce any first order moments of inertia when the crankshaft is centrosymmetric.

On the one hand, the balancing of moments of inertia should not generate any free inertial forces. An I5 engine serves as an example to explain the first order free moment of inertia and its partial balancing for the firing sequence 1 – 2 – 4 – 5 – 3 (one of the $(z-1)!/2 = 12$ "dynamically" varying firing sequences). The ignition interval for a four-stroke engine is $\varphi_Z = 4\pi/5 = 144°$. The firing sequence determines the following phase relationships:

$$\varphi_1 = \varphi$$
$$\varphi_2 = \varphi - 144°$$
$$\varphi_4 = \varphi - 2 \cdot 144° = \varphi - 288° \qquad (8\text{-}44)$$
$$\varphi_5 = \varphi - 3 \cdot 144° = \varphi - 432°$$
$$\varphi_3 = \varphi - 4 \cdot 144° = \varphi - 576°$$

Incorporating the phase relationships, the first order tilting moment $M_y^{(1)}$ and the longitudinal tilting moment M_x can be derived with the aid of the *crank throw diagram* (Fig. 8-24) (the reference point being the center of the crankshaft, assuming a clockwise moment with a positive sign):

$$M_y^{(1)} = \left(F_{\text{mosz}}^{(1)} + F_{\text{mrot}}\right)a_z \times$$
$$\times \begin{bmatrix} 2\cos\varphi + \cos(\varphi - 144°) - \cos(\varphi - 288°) \\ -2\cos(\varphi - 432°) \end{bmatrix}$$
$$M_y^{(1)} = \left(F_{\text{mosz}}^{(1)} + F_{\text{mrot}}\right)a_z(-0.3633\sin\varphi + 0.2640\cos\varphi)$$
$$\qquad (8\text{-}45)$$

$$M_x = F_{\text{mrot}}a_z \begin{bmatrix} -2\sin\varphi - \sin(\varphi - 144°) + \sin(\varphi - 288°) \\ +2\sin(\varphi - 432°) \end{bmatrix}$$
$$M_x = -F_{\text{mrot}}a_z(0.2640\sin\varphi + 0.3633\cos\varphi)$$
$$\qquad (8\text{-}46)$$

Experience requires at least balancing the rotating masses' moment of inertia. This is also called *longitudinal and partial tilting moment compensation*. Consisting of the components M_y (component of the rotating inertial forces of $M_y^{(1)}$) and M_x, the vector of the moment of inertia rotates at a constant rate in the crankshaft's direction of rotation and is consequently fully balanceable with counterweights:

$$|\vec{M}| = \sqrt{M_y^2 + M_x^2} = 0.449\, F_{\text{mrot}} a_z = const. \qquad (8\text{-}47)$$

While the alternative firing sequence 1 –5 – 2 – 3 – 4 reduces the second order influence coefficient from 4.980 to 0.449, it increases the first order influence coefficient to 4.98, thus creating a conflict of objectives [8-51]. Such results may also be obtained with graphic vector addition using first and second order star diagrams (Fig. 8-25). The moment vector's directional components yield its phase angle β relative to the

Fig. 8-24 Longitudinally unsymmetrical crankshaft, star diagram first order and crank throw distance diagram of an I5 engine with the firing sequence 1−2−4−5−3

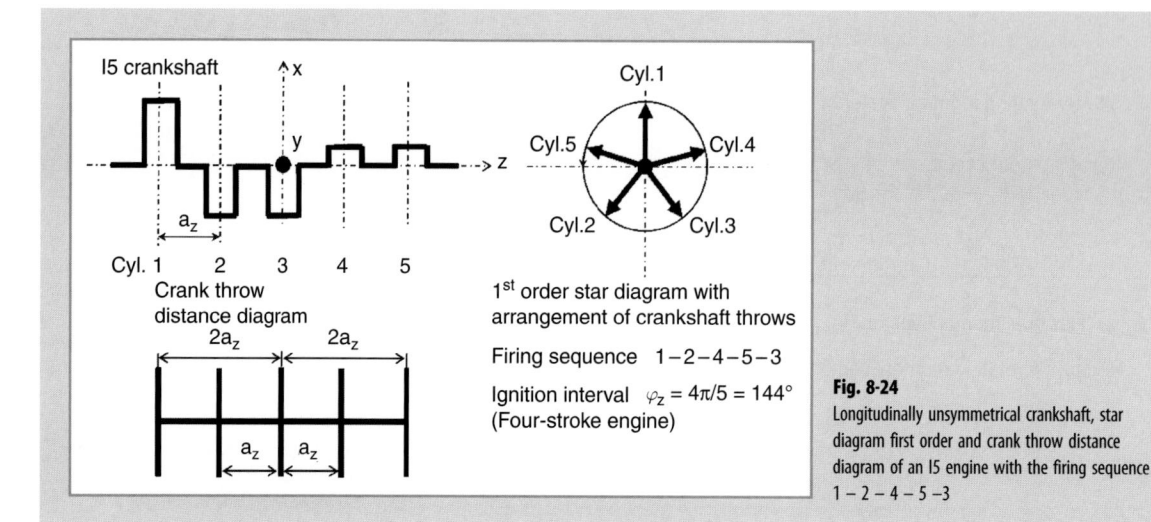

TDC of cylinder 1 ($\varphi_1 = \varphi = 0$):

$$\tan \beta = \frac{M_y}{M_x} = -\frac{0.2640}{0.3633} = -0.7267 \tag{8-48}$$

$$\beta = \arctan(-0.7267) = 144°$$

Dynamic balancing attaches two counterweight masses m_{Gg1}, m_{Gg2} to both ends of the crankshaft spaced at a_{Gg1}, a_{Gg2} at the angles β_1, β_2 in such a way that free forces and moments do not occur. If the counterbalancing masses m_{Gg} and lever arms a_{Gg} are identical and a centrosymmetric crankshaft causes the inertial forces to balance each other, the mathematical statement to ascertain the counterbalancing force F_{Gg} and the angle β_1, β_2 is reduced as follows:

$$\sum_i F_{xi} = 0 \quad \sum_i F_{yi} = 0 \quad \cos \beta_1 + \cos \beta_2 = 0$$

$$\sin \beta_1 + \sin \beta_2 = 0$$

$$\sum_i M_{yi} = 0 \quad F_{Gg} a_{Gg}(\cos \beta_1 - \cos \beta_2) + M_y = 0$$

$$\sum_i M_{xi} = 0 \quad F_{Gg} a_{Gg}(-\sin \beta_1 + \sin \beta_2) + M_x = 0$$

$$\tag{8-49}$$

Incorporating Eqs. (8-45) and (8-46) and the relation to the TDC of cylinder 1 ($\varphi_1 = \varphi = 0$), the equation system (8-49)

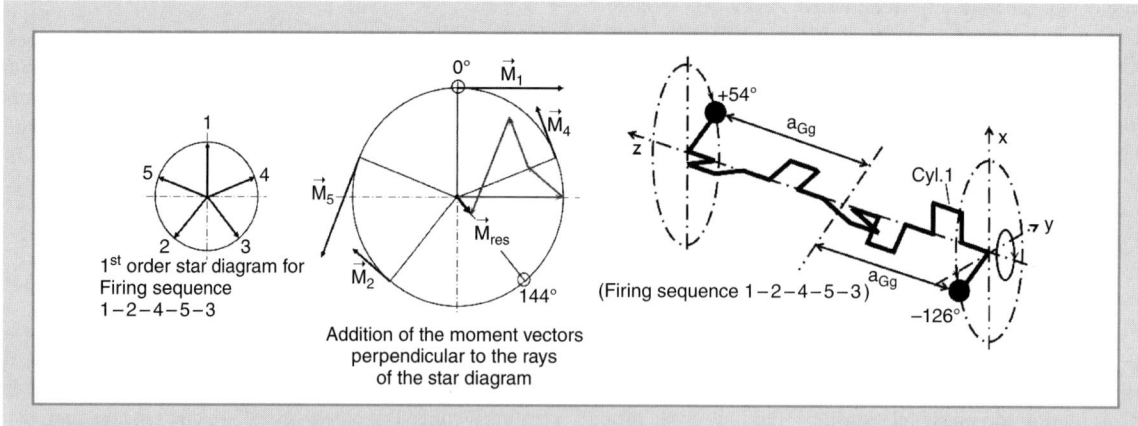

Fig. 8-25 Graphic determination of the resulting mass moment vector – here the rotating inertial forces – and arrangement of the counterweights for the longitudinal and partial tilting moment compensation in the crankshaft of an I5 engine

yields the following results:

$$\tan \beta_1 \frac{\sin \beta_1}{\cos \beta_1} = -\frac{M_x}{M_y} = \frac{0.3633}{0.2640} = 1.3761 \quad (8\text{-}50)$$

$$\beta_1 = 54° \quad bzw. \quad 234°(-126°)$$

β_1 must be selected so that the counterbalancing force becomes positive. Accordingly, $\beta_1 = 234°$ (see Eq. (8-51)) and $\beta_2 = 54°$ (Fig. 8-25). The counterweight mass m_{Gg} still has to be determined by means of Eq. (8-49), e.g. with the moment of inertia M_y, one counterweight's inertial force F_{Gg}, the crank radius r and radius of the counterweights' center of gravity r_{Gg}:

$$F_{Gg} = \frac{-M_y}{2a_{Gg}\cos\beta_1} = \frac{-0.2640 F_{mrot} a_z}{2a_{Gg}\cos 234°} \quad (8\text{-}51)$$

$$m_{Gg} = 0.2246 \frac{r a_z}{r_{Gg} a_{Gg}} m_{rot}$$

Not yet considered, the oscillating inertial forces generate a residual vector of the oscillating tilting moment. *Normal balancing* provides further improvement, i.e. halves the vector magnitude. Then, only a negative residual torque vector rotating at a constant magnitude remains, which can again be fully compensated by an appropriate balance shaft. Using the oscillating inertial force's degree of balance to control the primary direction of the excitation of vibrations is common practice. A rotating residual torque vector causes engine lurching. Unless it has a constant rate, its angular velocity is likewise subject to periodic variations.

Further, uneven cylinder spacing can compensate for the tilting moment [8-1]. It is seldom employed because it entails lengthening the engine. Table 8-2 summarizes the features of engines with practical importance.

8.3.5 Moments of Rotational Inertia

Moments of rotational inertia act around the z-axis, the crankshaft's axis of rotation. According to the *principle of angular momentum of mechanics*, when a mass moment of inertia exists, non-uniform rotary motion generates torque. The connecting rod's angular momentum $T_{Pl} = T_{Plosz} + T_{Plrot}$ may not be disregarded:

$$T_{Plosz} = -\Theta_{Plosz}\dot{\psi} = -m_{Plosz}k_{Plosz}^2 \dot{\psi} \quad (8\text{-}52)$$

$$T_{Plrot} = -\Theta_{Plrot}\dot{\psi} + m_{Plrot}r^2\omega = m_{Plrot}\left(-k_{Plrot}^2\dot{\psi} + r^2\omega\right) \quad (8\text{-}53)$$

where:
Θ_{Plosz} and Θ_{Plrot} – mass moments of inertia of the masses m_{Plosz} and m_{Plrot},
k_{Plosz} and k_{Plrot} – related radii of gyration (the reference axis being the piston pin or crank pin axis.),
$\dot{\psi}$ – angular velocity of pivoting,

ω – angular velocity of the crankshaft and
r – crank radius.

Inserting the connecting rod's radius of gyration k_{Pl} relative to its center of mass and the connecting rod mass m_{Pl} and incorporating Eq. (8-24), the *parallel axis theorem* delivers the following result:

$$m_{Plosz}\left[k_{Plosz}^2 + (l_{Pl} - l_{Pl1})^2\right] + m_{Plrot}\left(k_{Plrot}^2 + l_{Pl1}^2\right) = m_{Pl}k_{Pl}^2$$

$$m_{Plosz}k_{Plosz}^2 + m_{Plrot}k_{Plrot}^2 = m_{Pl}\left[k_{Pl}^2 - l_{Pl1}(l_{Pl} - l_{Pl1})\right] \quad (8\text{-}54)$$

Only relevant when speed is variable, the crankshaft's angular momentum can be defined accordingly:

$$T_{KW} = \Theta_{KW}\omega = m_{KW}k_{KW}^2\omega \quad (8\text{-}55)$$

Finally, with the aid of the "reduced" mass moments of inertia Θ_{red1} and Θ_{red2}, the total angular momentum of the "rotational inertia" can be specified in the following simplified form:

$$T_{ges} = \dot{\psi}\,\Theta_{red1} + \omega\,\Theta_{red2}$$
$$\Theta_{red1} = m_{Pl}\left[l_{Pl1}(l_{Pl} - l_{Pl1}) - k_{Pl}^2\right] \quad (8\text{-}56)$$
$$\Theta_{red2} = m_{Pl}r^2\left(1 - \frac{l_{Pl1}}{l_{Pl}}\right) + m_{KW}k_{KW}^2$$

The change of the angular momentum produces the moment of rotational inertia:

$$M_z = -\frac{dT_{ges}}{dt} = -\left(\ddot{\psi}\,\Theta_{red1} + \dot{\omega}\,\Theta_{red2}\right) \quad (8\text{-}57)$$

$\ddot{\psi}$ being the angular acceleration of the connecting rod's pivoting, $\dot{\omega}$ that of the crankshaft at variable speed (otherwise, $\dot{\omega} = 0$ applies). Based on expansion in a series and double differentiation based on the time t, the pivoting angle $\psi = \arcsin(\lambda_{Pl}\sin\varphi)$ (Eq. (8-8)) delivers the following angular acceleration of pivoting:

$$\ddot{\psi} = -\lambda_{Pl}\omega^2(C_1\sin\varphi + C_3\sin 3\varphi + C_5\sin 5\varphi + \ldots) +$$
$$+ \lambda_{Pl}\dot{\omega}\left(C_1\cos\varphi + \frac{C_3}{3}\cos 3\varphi + \frac{C_5}{5}\cos 5\varphi + \ldots\right) \quad (8\text{-}58)$$

The constants C_i are polynomials of the stroke/connecting rod ratio λ_{Pl}:

$$C_1 = 1 + \frac{1}{8}\lambda_{Pl}^2 + \frac{3}{64}\lambda_{Pl}^4 + \ldots \approx 1$$
$$C_3 = -\frac{3}{8}\lambda_{Pl}^2 - \frac{27}{128}\lambda_{Pl}^4 - \ldots \approx -\frac{3}{8}\lambda_{Pl}^2 \quad (8\text{-}59)$$
$$C_5 = \frac{15}{128}\lambda_{Pl}^4 + \frac{125}{1024}\lambda_{Pl}^6 + \ldots \approx \frac{1}{8}\lambda_{Pl}^4$$

The moment of rotational inertia is balanced when the reduced mass moments of inertia (Eq. (8-56), second and third line)) become zero. The connecting rod's radius of gyration must satisfy the requirement $k^2_{Pl} = l_{Pl1}(l_{Pl} - l_{Pl1})$. It can be satisfied

Table 8-2 Overview of the dynamic properties of engines with great practical significances, influence coefficients of the free inertial forces und moments of inertia, orders of excitation

Engine	Working principle	α_v [a]	φ_K [b]	δ [c]	Common firing sequences [d]	Influence coefficients of mass force without balancing [e] $c_F^{(1)}$ and $c_F^{(2)}$	Influence coefficients of mass force with normal balancing [e] $c_F^{(1)}$ and $c_F^{(2)}$	Influence coefficients of the mass moment without balancing [f] $c_M^{(1)}$ bzw. $c_M^{(2)}$	Influence coefficients of the mass moment with normal balancing [f] $c_M^{(1)}$ and $c_M^{(2)}$	Most important orders of excitation [g]
I2	Two-stroke	–	180°	–	1–2	(1): 0 (2): 2	(1): 0 (2): 2	(1): 1 (2): 0	(1): 0.5 (2): 0	1; 2; 3; … 0.5; 1.5; 2; 2.5; … [h]
I2	Four-stroke	–	360°	–	1–2	(1): 2 (2): 2	(1): 1 (2): 2	(1): 0 (2): 0	(1): 0 (2): 0	1; 2; 3; …
I3	Two-stroke	–	120°	–	1–3–2	(1): 0 (2): 0	(1): 0 (2): 0	(1): 1.732 (2): 1.732	(1): 0.866 (2): 1.732	3; 6; 9; …
I3	Four-stroke	–	240°	–	1–3–2	(1): 0 (2): 0	(1): 0 (2): 0	(1): 1.732 (2): 1.732	(1): 0.866 (2): 1.732	1.5; 3; 4.5; 6; …
I4	Two-stroke	–	90°	–	1–4–2–3 1–3–2–4 or 1–3–4–2 1–2–4–3	(1): 0 (2): 0	(1): 0 (2): 0	(1): 1.414 (2): 4 (1): 3.162 (2): 0	(1): 0.707 (2): 4 (1): 1.581 (2): 0	4; 6; 8; …
I4	Four-stroke	–	180°	–	1–3–4–2 1–4–3–2	(1): 0 (2): 4	(1): 0 (2): 4	(1): 0 (2): 0	(1): 0 (2): 0	2; 4; 6; …
I5	Four-stroke	–	144°	–	1–2–4–5–3 or 1–5–2–3–4	(1): 0 (2): 0 (1): 0 (2): 0	(1): 0 (2): 0 (1): 0 (2): 0	(1): 0.449 (2): 4.98 (1): 4.98 (2): 0.449	(1): 0.225 (2): 4.98 (1): 0.225 (2): 0.449	2.5; 5; 7.5; …
I6	Two-stroke	–	60°	–	1–6–2–4–3–5 1–5–3–4–2–6	(1): 0 (2): 0	(1): 0 (2): 0	(1): 0 (2): 3.464	(1): 0 (2): 3.464	3; 6; 9; …
I6	Four-stroke	–	120°	–	1–5–3–6–2–4 1–4–2–6–3–5 1–3–5–6–4–2	(1): 0 (2): 0	(1): 0 (2): 0	(1): 0 (2): 0	(1): 0 (2): 0	3; 6; 9; …
V6	Four-stroke	60°	120°	+60°	1–6–3–5–2–4 1–4–2–5–3–6	(1): 0 (2): 0	(1): 0 (2): 0	(1): 1.5 (2): 15	(1): 0 (2): 1.5	3; 6; 9; …
V6	Four-stroke	90°	120°	– [i]	1–4–3–6–2–5	(1): 0 (2): 0	(1): 0 (2): 0	(1): 1.731 (2): 2.449	(1): 0 (2): 2.449	1.5; 3; 4.5; 6; …
V6	Four-stroke	90°	120°	+30° [j]	1–4–3–6–2–5	(1): 0 (2): 0	(1): 0 (2): 0	(1): 2.121 (2): 2.121	(1): 0.448 (2): 2.121	3; 6; 9; …
V8	Four-stroke	90°	90°	–	1–5–4–8–6–3–7–2 1–5–4–2–6–3–7–8	(1): 0 (2): 0	(1): 0 (2): 0	(1): 3.162 (2): 0	(1): 0 (2): 0	4; 8; 12; …

Table 8-2 (Continued)

Engine	Working principle	$\alpha_v{}^a$	$\varphi_K{}^b$	δ^c	Common firing sequences^d	Influence coefficients of mass force without balancing^e $c_F{}^{(1)}$ and $c_F{}^{(2)}$	Influence coefficients of mass force with normal balancing^e $c_F{}^{(1)}$ and $c_F{}^{(2)}$	Influence coefficients of the mass moment without balancing^f $c_M{}^{(1)}$ bzw. $c_M{}^{(2)}$	Influence coefficients of the mass moment with normal balancing^f $c_M{}^{(1)}$ and $c_M{}^{(2)}$	Most important orders of excitation^g
V10	Four-stroke	90°	72°	$-^k$	1–6–5–10–2–7–3–8–4–9 1–10–9–4–3–6–5–8–7–2	(1): **0** (2): **0**	(1): **0** (2): **0**	(1): **4.98** (2): **0.634**	(1): **0** (2): **0.634**	2.5; 5; 7.5; ...
V12	Four-stroke	60°	120°	—	1–7–5–11–3–9–6–12–2–8–4–10 1–12–5–8–3–10–6–7–2–11–4–9	(1): **0** (2): **0**	(1): **0** (2): **0**	(1): **0** (2): **0**	(1): **0** (2): **0**	6; 12; 18; ...

^{a} V angle of the cylinder banks;
^{b} Throw angle of the crankshaft;
^{c} Angle of crank pin offset;
^{d} Different methods of counting cylinders for I, VR and V engines are important: I or VR: Continuously from the end of the free shaft to the output side, V: For cylinder bank A beginning at the end of the free shaft continuously to the output end, continuing accordingly for cylinder bank B;
^{e} $c_F{}^{(1)} = F{}^{(1)}{}_{res(osz)}/F{}^{(1)}{}_{mosz}$, $c_F{}^{(2)} = F{}^{(2)}{}_{res(osz)}/F{}^{(2)}{}_{mosz}$;
^{f} $c_M{}^{(1)} = M{}^{(1)}{}_{res(osz)}/(F{}^{(1)}{}_{mosz} \cdot a_z)$, $c_M{}^{(2)} = M{}^{(2)}{}_{res(osz)}/(F{}^{(2)}{}_{mosz} \cdot a_z)$ (factors influencing the inertial forces relative to the oscillating inertial forces, which multiply the moments of inertia to the latter with the cylinder clearance; moment of inertia of the rotating inertial forces completely balanceable by counterweights);
^{g} Excitation of the engine's torsional vibrations;
^{h} Unequal ignition interval 90°–150°–90°–150°;
^{i} Equal ignition interval 120°;
^{j} Ignition interval 90°–54°–90°–54°;
^{k} Unequal ignition interval 180°–540°–180°–540° and half orders in four-stroke operation.

with small supplementary masses on the connecting rod small end and small connecting rod eye but obviously not with a connecting rod's center of gravity distance $l_{Pl1} > l_{Pl}$.

8.3.6 Mass Balancing with the Aid of Balance Shafts

Balance shafts generate additional work for engine design, development and manufacturing. Hence, their use is usually reduced to compensating the free residual forces and moments that cannot be balanced by crankshaft counterweights. They are integrated in the cylinder crankcase and power takeoff concept to save space and driven by a chain, gear or toothed belt. They must have long-lasting phase fidelity, be durable, have low friction and be acoustically unobtrusive. Balance shafts must be securely mounted on bearings. Doubled engine speed imposes increased requirements for second order balancing. Plain bearing mounting requires integration in the lubricating oil circuit. As a rule, their positioning should neither produce free forces nor moments.

8.3.6.1 Balancing Inertial Force with the Aid of Balance Shafts

Balancing a free residual force vector requires two eccentric shafts aligned symmetrically to the crankshaft's axis of rotation with (unilaterally) eccentric masses. Their resultant inertial force vector counteracts the free residual force vector. The following cases are differentiated (Fig. 8-26):
– Negatively (counter-) rotating first order residual force vectors → both eccentric shafts symmetrically aligned to prevent a moment of inertia also rotate negatively with crankshaft speed.
– Oscillating residual force vectors → mirror symmetrically aligned eccentric shafts rotate in opposite directions at crankshaft (first order) or double speed (second order). The Lanchester system – shaft height offset – enables controlling second order free torque to a limited extent but only in a particular operating point (Fig. 8-27).

8.3.6.2 Balancing Moments of Inertia with the Aid of Balance Shafts

One balance shaft suffices to balance a rotating free residual torque vector. It has two eccentric masses twisted 180° and mounted at the maximum distance. The thusly generated balancing moment vector counteracts the residual torque vector. The balance shaft (counter-) rotates negatively at crankshaft speed. (A rotating second order moment of inertia corresponding to doubled speed also appears in certain V engines in second order balancing.) This already implies that counterweights fully balance the positively rotating first order moment vector (see Sect. 8.3.4). The balancing of an oscillating moment of inertia on the other hand requires two mirror symmetrically aligned counterrotating balance shafts. The vector components parallel to the residual torque vector but acting in opposite directions add up. Those perpendicular to it compensate each other. Consequently, the resulting vector oscillates in phase opposition.

8.3.7 "Internal" Balancing of Crankshaft Assembly Masses

The following aspects remain to be treated:
– inertial force's stress of a crankshafts' main bearings and
– excitation of vibrations resulting from internal bending moments transmitted from the main bearings to the cylinder crankcase.

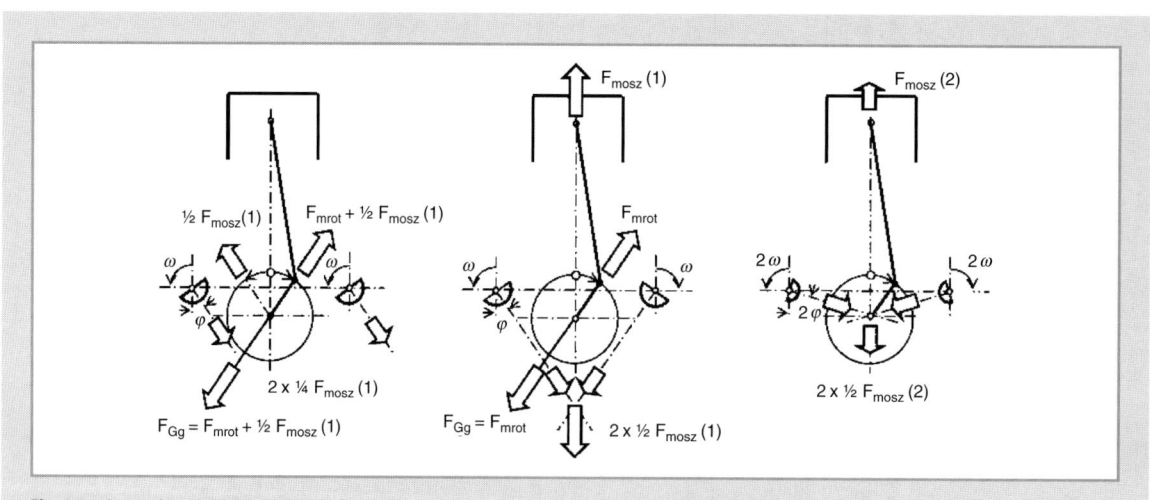

Fig. 8-26 Options for completely balancing the first order free inertial force for a single-cylinder crankshaft assembly by means of normal balancing with two balance shafts (*left*), balancing of the rotating inertial forces by counterweights (*center*); balancing of the second order inertial force (*right*)

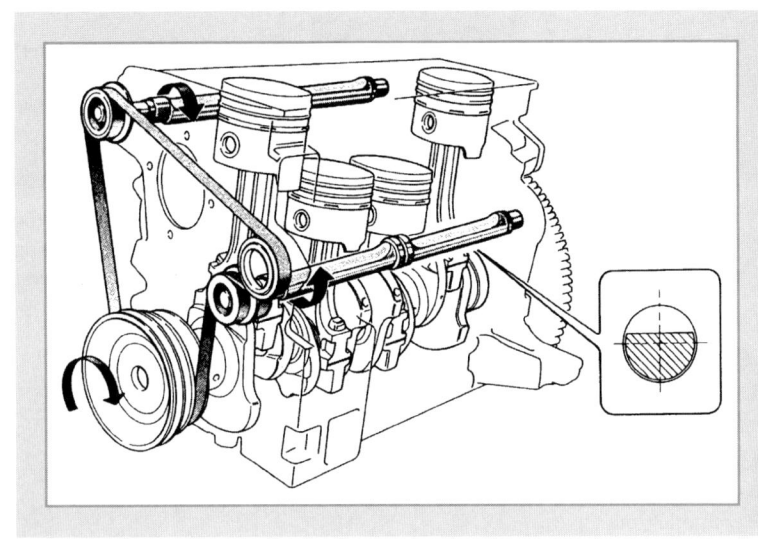

Fig. 8-27
Arrangement of the balance shafts in the Lanchester system (along with balancing the second order free inertial forces, influence is exerted on the second order free moment of inertia in a certain operating point)

Counterweights are also employed to limit crankshaft deformation and its interaction with the housing whenever outwardly balanced mass actions do not require this. For its part, the "internal" mass balancing should not generate any free mass actions (centrosymmetry and longitudinal symmetry). Otherwise, there are not any universally valid rules. Appropriate counterweight masses are expediently provided on every crankshaft web. The correction of the crankshaft's deflection curve obtained in the process is crucially important.

The "internal" moment denotes the longitudinal distribution of the bending moments of an unsupported crankshaft, i.e. "freely floating in space" (Fig. 8-28). Usually, two mutually perpendicular planes of action must be considered. The crankshaft's free deflection under the "internal" moment is a fictive reference parameter to assess the quality of "internal" mass balancing. The "bending moment of the housing" without the support of the inner main bearing is also of interest. Figure 8-28 illustrates this with a simple example of a four-stroke I4 engine. The throws only allow analysis of one plane. It also illustrates the reduction of the bending moment of the housing when differently sized counterweight masses are attached.

8.3.8 Torsional Force Characteristics

8.3.8.1 Inertial Force Induced Torsional Force Characteristic

Varying with the crank angle, the torsional forces $F_t(\varphi)$ excite torsional oscillations of the crankshaft assembly (see Sect. 8.4). Equation (8-4) calculates the torsional force F_t from the piston force F_K. In the case of the mass torque F_{tmosz}, the piston force is reduced to the oscillating inertial force

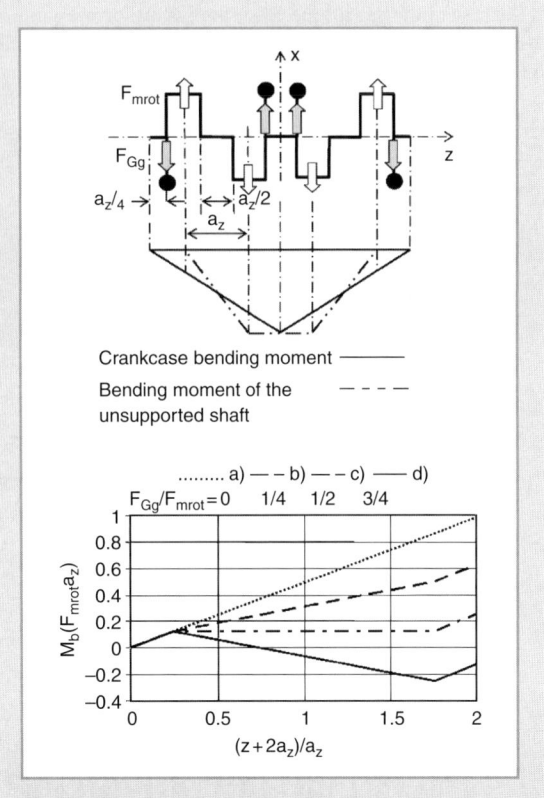

Fig. 8-28 Longitudinal distribution of the bending moment of an unsupported crankshaft and the housing bending moment in an I4 four-stroke engine (*top*); housing bending moment for differing internal mass balancing (*bottom*)

F_{mosz} (Eq. 8-1):

$$F_t \approx F_K \left(\sin \varphi + \frac{\lambda_{Pl}}{2} \sin 2\varphi \right)$$

$$F_{\text{tmosz}} \approx \ddot{x}_K m_{\text{osz}} \left(\sin \varphi + \frac{\lambda_{Pl}}{2} \sin 2\varphi \right)$$

(8-60)

Inserting the acceleration of the oscillating mass $\ddot{x}$ (Eq. (8-29)), i.e. incorporating the first and second order terms, ultimately yields the following result:

$$F_{\text{tmosz}} \approx m_{\text{osz}} r \omega^2 \times$$
$$\times \left(\frac{\lambda_{Pl}}{4} \sin \varphi - \frac{1}{2} \sin 2\varphi + \frac{3}{4} \lambda_{Pl} \sin 3\varphi - \frac{\lambda_{Pl}}{4} \sin 4\varphi \right)$$

(8-61)

The relation to the piston area A_K delivers the oscillating inertial force's *tangential pressure* $p_{\text{tmosz}} = F_{\text{tmosz}}/A_K$. In multi-cylinder engines, certain orders cancel each other out as a result of superimposition during phase shift and do not contribute to the resulting torque [8-1].

8.3.8.2 Gas Force Induced Torsional Force Characteristic

In the case of the gas torque F_{tGas}, the piston force F_K in Eq. (8-60, top) is replaced by the gas force $F_{\text{tGas}} = p_Z(\varphi) A_K$ acting in the cylinder. Since the curve of the cylinder pressure $p_Z(\varphi)$ is periodic, the gas torque can be broken down into its harmonic components by means of *harmonic analysis* [8-52] and thus represented as a Fourier series (Figs. 8-29 and 8-30) [8-53]:

$$F_{\text{tGas}}(t) = F_{\text{tGas0}} + \sum_{m=1}^{m=n, \, i=m\kappa} \check{F}_{\text{tGasi}} \sin(i\omega t + \delta_i) \quad (8\text{-}62)$$

two-stroke: $\kappa = 1$
four-stroke: $\kappa = 1/2$
$m = 1, 2, 3, \ldots, n$

$\hat{F}_{\text{tGasi}}$ is the ith amplitude of harmonic excitation of the torsional force, δ_i of the related phase angle ($i = 1/2$ being the lowest order for a four-stroke engine, $i = 1$ the lowest

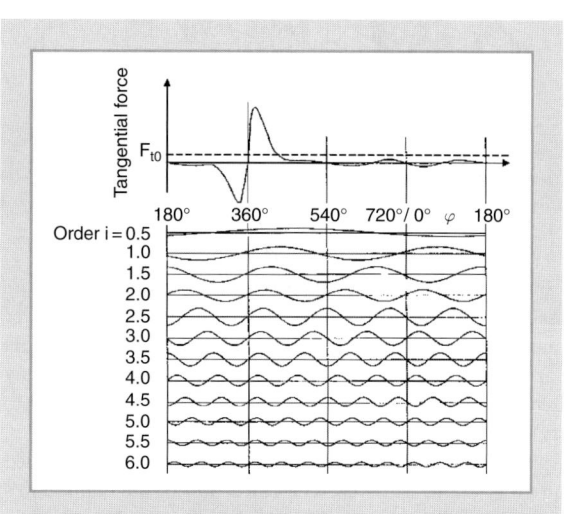

Fig. 8-29 Harmonic components of the torsional force (tangential force) F_t in a four-stroke reciprocating piston engine

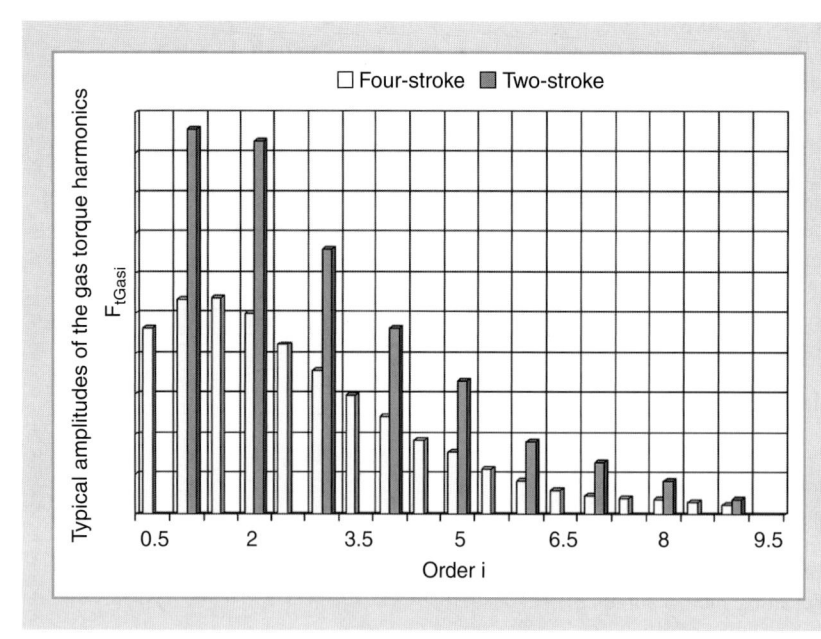

Fig. 8-30 Amplitudes of the harmonics of the gas torque F_{tGasi} for a two and four-stroke engine based on [8-53]

order for a two-stroke engine). A tangential pressure ($p_{tGas} = F_{tGas}/A_K$) can also be defined for the gas force. In addition, $p_{mi} = 2\pi F_{tGas0}/A_K$ applies to the mean indicated operating pressure. In multiple cylinder engines, the ignition intervals produce the phase shifts of the torsional force characteristics. In turn, superimposition produces the resulting torque. In V engines, it is expedient to first determine a cylinder pair's torsional force characteristic. Given their phase relationships, the gas force and mass torque's harmonics are superimposed by adding the vectors (cf. [8-2]). Mass torque is negligibly small above the fourth order.

8.3.8.3 Main and Secondary Exciter Orders

Torsional vibrations are excited by higher orders of gas torque. Analogous to the star diagrams, there are also vector diagrams corresponding to the harmonics of the torsional forces. The vectors in a main exciter order i_H are unidirectional (Fig. 8-31). The vectors are first order secondary exciters i_N when they lie on one line but point in opposite directions. Unlike the amplitudes of the secondary exciter torques, the amplitudes of the resulting main exciter torques are independent of the firing sequence [8-2], which has to be selected to be appropriately advantageous (for rules on this, see for example [8-1]).

8.3.9 Irregularity of Rotary Motion and Flywheel Rating

A reciprocating piston engine's irregular torsional force characteristic generates a speed fluctuation $\Delta\omega$ during one working cycle. Relative to the mean speed $\bar{\omega}$, this speed fluctuation is the *degree of cyclic irregularity* $\delta_U \leq 10\%$:

$$\Delta\omega = \omega_{max} - \omega_{min} \quad \bar{\omega} = \frac{1}{2}(\omega_{max} + \omega_{min}) \quad \delta_U = \frac{\Delta\omega}{\bar{\omega}} \quad (8\text{-}63)$$

The internal work W_i produced during one working cycle (assuming a four-stroke engine here) requires the following equivalence for the mean tangential pressure $\bar{p}_t$ and the mean indicated working pressure p_{mi}:

$$W_i = 4\pi\bar{p}_t A_K r = 2p_{mi}A_K zr \quad \bar{p}_t = \frac{z}{2\pi}p_{mi} \quad (8\text{-}64)$$

A_K is the piston area, r the crank radius and z the number of cylinders. Incorporating the crankshaft assembly's mass moment of inertia Θ_{ges} relative to the crankshaft, the change of the kinetic energy of rotation during a phase of acceleration or deceleration ($\omega_1 \to \omega_2$) in the crank angle range of φ_1 to φ_2 delivers an energy analysis:

$$\frac{1}{2}(\omega_2^2 - \omega_1^2)\Theta_{ges} = A_K r \int_{\varphi_2}^{\varphi_1} [p_t(\varphi) - \bar{p}_t]d\varphi \quad (8\text{-}65)$$

Consequently, "excess work" $W_{\ddot{U}}$ – by definition, the difference between the largest positive and negative amount of energy appearing – can be determined and the degree of cyclic irregularity δ_U specified:

$$\omega_{max}^2 - \omega_{min}^2 = \frac{2W_{\ddot{U}}}{\Theta_{ges}} \to \delta_U = \frac{W_{\ddot{U}}}{\bar{\omega}^2 \Theta_{ges}} \quad (8\text{-}66)$$

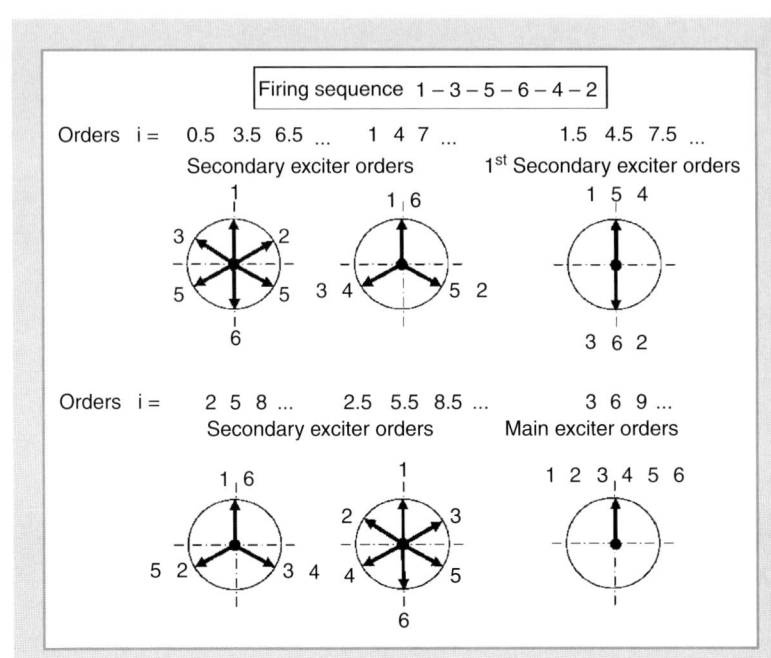

Fig. 8-31
Vector diagrams (star diagram) of the excitation torques; main exciter order; 1st and other secondary exciter orders of a I6 four-stroke engine with the firing sequence 1 – 3 – 5 – 6 – 4 – 2

Not least, acceptable idle quality requires a sufficient flywheel mass moment of inertia Θ_{Schw}. This can be correctly rated in correlation with the crankshaft's mass moment of inertia Θ_{KWges} including the related masses (cf. [8-2]):

$$\Theta_{Schw} = \frac{W_{\ddot{U}}}{\bar{\omega}^2 \delta_U} - \Theta_{KWges} \qquad (8\text{-}67)$$

8.4 Torsional Crankshaft Assembly Vibrations

8.4.1 Preliminary Remarks on Crankshaft Assembly Vibrations

A torsionally, bending and longitudinally elastic crankshaft with its crankshaft assembly masses constitutes a vibration system, which is excited to vibrations by the alternating forces, alternating torques and alternating bending moments. These cause additional dynamic stress of the crankshaft and its bearings (see Sect. 8.2.4). Unduly large vibration amplitudes can cause fatigue failure in the resonance case. In addition, vibrations of the crankshaft assembly contribute quite substantially to structure-borne noise excitation. The intermediate bearing support now common for every crankshaft throw results in crankshafts with short main bearing spacing. Bending-critical speeds increase beneficially as a result. While all the vibrational phenomena are interconnected in reality, flexural vibrations are initially not viewed with the same importance as torsional vibrations in the phase of crankshaft assembly design (for connected crankshaft assembly vibrations, see Sect. 8.2.4.2; cf. [8-37]).

8.4.2 Dynamic Equivalent Models of the Crankshaft Assembly

8.4.2.1 Equivalent Models for the Engine and Power Train

The irregular gas force and mass torque characteristics (Eqs. (8-61) and (8-62)) as well as the variable load moments in the power train excite reciprocating piston engines' crankshaft assemblies to torsional vibration. The vibration amplitudes, i.e. elastic, dynamic torsional deformations of the crankshaft assembly, are superimposed on the rotary motion. The simulation of torsional vibrations may apply to the engine itself (crankshaft assembly including the flywheel) or include the complete power train with or without branching (engines, machine, interposed transmission, drive shafts, etc.) (Fig. 8-32).

An isolated analysis of an engine is valid when, a conventional flywheel – unlike a dual-mass flywheel – equals an extensive decoupling from the rest of the power train in the analyzed frequency range because of its large mass inertia. Common equivalent torsional vibration models correspond to the elastically coupled multi-degree of freedom system (cf. [8-54]). Equivalent rotating masses are mounted on a

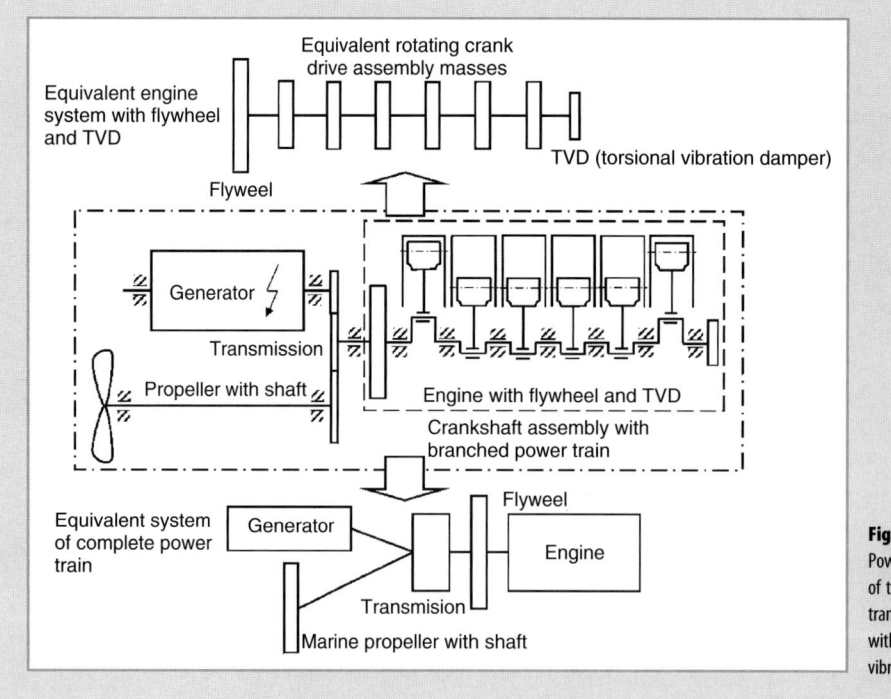

Fig. 8-32
Power train with branching consisting of the engine with the flywheel, transmission, generator and drive shaft with a screw here; equivalent rotational vibration systems

straight, massless equivalent shaft. A crankshaft assembly has one equivalent rotating mass per throw. The flywheel is added to this. Simplified equivalent models for the entire power train are common.

8.4.2.2 Mass and Length Reduction

The equivalent lengths l_{red} of shaft sections and the equivalent rotating masses Θ_{red} mounted in between are determined by reducing the length and mass [8-1, 8-6]. Mass reduction is based on the principle of the equivalent model's equivalent kinetic energy:

$$\frac{1}{2}\Theta_{red}\omega^2 = \frac{1}{2}m_{rot}r^2\omega^2 + \frac{1}{2}m_{osz}\dot{x}_K^2 \qquad (8\text{-}68)$$

(see Sect. 8.3.2 for the formula symbols). Unlike a rotating mass, the kinetic energy of an oscillating mass and thus the equivalent rotating mass (the mass moment of inertia) Θ_{red} changes in a reciprocating piston engine with the crank angle φ:

$$\Theta_{red}(\varphi) = r^2 \left[m_{rot} + m_{osz} \left(\sin\varphi + \frac{\lambda_{Pl}\sin\varphi\cos\varphi}{\sqrt{1-\lambda_{Pl}^2\sin^2\varphi}} \right)^2 \right] \qquad (8\text{-}69)$$

Hence, a representative value is required, which is delivered by *Frahm's formula* for instance:

$$\bar{\Theta}_{red} = r^2 \left(m_{rot} + \frac{1}{2} m_{osz} \right) \qquad (8\text{-}70)$$

The length is reduced so that the torsionally elastic deformation of an equivalent shaft segment corresponds to the length l_{red} of one crankshaft throw that, in reality, consists of two components of the main shaft pin, the crank pin and the two crankshaft webs. This inevitably establishes a relationship to the main pin's polar planar moment of inertia $I_{red} = I_{KWG}$. The equivalent torsional spring stiffness c of the following equation suffices for a homogeneous crankshaft:

$$c = G \frac{I_{red}}{l_{red}} \qquad (8\text{-}71)$$

G is the shear modulus of the equivalent shaft's material consisting of the reduced lengths $l_i = l_{red}$. Neither the pure torsion caused by external torque nor the additional deformation resulting from the application of tangential force to the crank pin may be ignored (see Fig. 8-12). The procedure to reduce length must allow for this. The finite element method (FEM) is particularly suited for this. On the other hand, various reduction formulas may be used (rules of thumb for large low speed diesel engines follow Geiger or Seelmann, those for high-speed vehicle and aircraft engines the British Internal Combustion Engine Research Association BICERA, Tuplin or Carter (cf. [8-53, 8-55, 8-56])). Carter is clearest:

$$l_{red} = (l_{KWG} + 0.8\,h_{KWW}) \frac{I_{red}}{I_{KWG}} + 1.274\,r \frac{I_{red}}{I^*_{KWW}}$$
$$+ 0.75\,l_{KWH} \frac{I_{red}}{I_{KWH}} \qquad (8\text{-}72)$$

where:

I_{red} – polar planar moment of inertia of the reference cross section,

I_{KWG}, I_{KWH} – corresponding value of the main pin and crank pin cross section,

l_{KWG}, l_{KWH} – related journal lengths,

$I^*_{KWW} = \dfrac{h_{KWW}\,b_{KWW}}{12}$ (!) – equatorial planar moment of inertia of a representative (mean) crankshaft web cross section ($I^*_{KWW} \perp I_{KWW}$) and

r – crank radius (see also Fig. 8-13).

8.4.2.3 Transmission Reduction

A power train with its transmission can be reduced to an equivalent system with a continuous shaft and individual rotating masses. The prerequisite is equivalence of the potential and kinetic energy. The reference shaft – normally the crankshaft – corresponds to the real system. The transmission ratio i (not to be confused with the index i) is utilized to reduce the rotating masses Θ_i and torsional stiffness c_i to the reference shaft as follows (Fig. 8-33):

$$\frac{1}{2}\Theta_i\omega_2^2 = \frac{1}{2}\Theta_{ired}\omega_1^2 \quad i = \frac{\omega_2}{\omega_1} = \frac{n_2}{n_1} = \frac{r_1}{r_2} \quad \Theta_{ired} = i^2 \Theta_i \qquad (8\text{-}73)$$

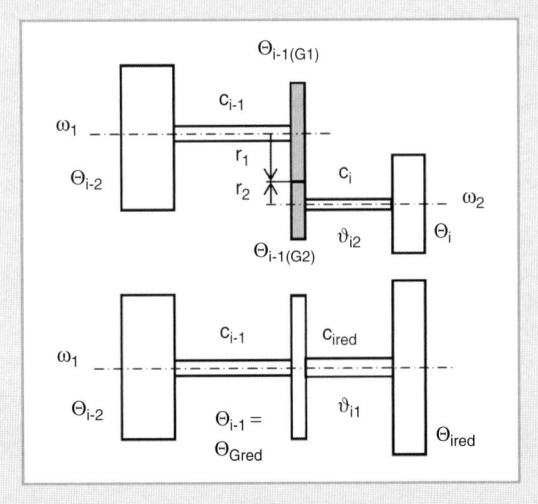

Fig. 8-33 Reduction of a real rotational vibration system with speed transforming gear to an equivalent system with a continuous shaft

$$\frac{1}{2}c_i\vartheta_{i2}^2 = \frac{1}{2}c_{ired}\vartheta_{i2}^2 \quad i = \frac{\vartheta_{i2}}{\vartheta_{i1}} \quad c_{ired} = i^2 c_i \qquad (8\text{-}74)$$

Figure 8-33 contains the formula symbols not explained in the text. A transmission's rotating masses must be merged into an equivalent rotating mass Θ_{Gred} according to the following principle, the index "1" denoting the reference shaft, the index "2" the geared shaft:

$$\Theta_{Gred} = \Theta_{G1} + i^2 \Theta_{G2} \qquad (8\text{-}75)$$

8.4.2.4 Reciprocating Piston Engine Damping

Damping influences on a crankshaft assembly's torsional vibrations are extremely varied. Frictional and marginal oil and splash oil losses damp externally and material properties as well as joints in composite crankshafts damp internally. The common approaches are linear and proportional to speed. The insertion of realistic coefficients in the damping matrix $\mathbf{K}$ (see Sect. 8.4.2.5) proves to be difficult because the external friction changes with an engine's operating state (lubricating oil temperature) and operating time (wear of crankshaft assembly components, aging of the lubricating oil, etc.). Strictly speaking, internal damping is also a function of the torsional angle. Further, both components are a function of frequency and thus speed. Hence, practice is often oriented toward empirical values from a comparison of measurement and simulation. A common method is closely linked to decoupling the vibration equation system (see Sect. 8.4.3.3). The following statement applies to the external and internal damping components $\mathbf{K_a}$, $\mathbf{K_i}$ of the damping matrix $\mathbf{K}$ [8-54]:

$$\mathbf{K} = \mathbf{K_a} + \mathbf{K_i} = \alpha \Theta + \beta \mathbf{C} \qquad (8\text{-}76)$$

It contains the proportionality factors α for external damping relative to the matrix of the rotating masses Θ and β for internal damping relative to the stiffness matrix $\mathbf{C}$.

8.4.2.5 Vibration Equation System, System Parameters

An equivalent rotational vibration model is a system of n rotating masses and thus $n - 1$ natural frequencies and natural vibration modes (Fig. 8-34). The number of nodes corresponds to the level of the natural vibration mode. Since the other critical speeds are above the operating speed range, solely allowing for the mode of first degree vibration with only one node is often sufficient.

The angular acceleration $\ddot{\vartheta}_i$ of the individual rotating mass Θ_i follows from the *principle of angular momentum*:

$$\Theta_i \ddot{\vartheta}_i = M_{Ti+1} - M_{Ti} \qquad (8\text{-}77)$$

$$M_{Ti+1} = c_{i+1}(\vartheta_{i+1} - \vartheta_i) \qquad (8\text{-}78)$$
$$i = 0, 1, \ldots, n-1 \quad M_{T0} = M_{Tn} = 0 \quad c_n = 0$$

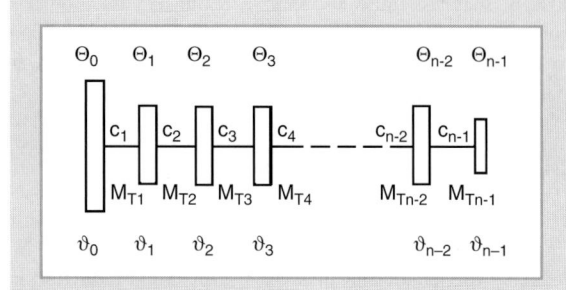

Fig. 8-34 Torsional vibration equivalent model of a crankshaft assembly not incorporating damping

the latter being the torsional moment resulting from the torsional moments acting on both sides of a rotating mass. The torsional moment M_{Ti+1} in the analyzed shaft segment is proportional to the difference of the torsional angles $\vartheta_{i+1} - \vartheta_i$ of the adjacent equivalent rotating masses. The constant c_{i+1} is the torsional spring stiffness. As in Fig. 8-34, a system of differential equations can be set up on the basis of Eqs. (8-77) and (8-78):

$$\Theta \ddot{\vartheta} + \mathbf{C}\vartheta = 0 \qquad (8\text{-}79)$$

Thus, disregarding the vibration system's relatively low damping, the modal variables can be determined. ϑ or $\dot{\vartheta}$ should be interpreted as vectors, Θ as a matrix of rotating masses and $\mathbf{C}$ as the stiffness matrix.

The simulation of forced torsional vibrations requires allowing for damping in the form of the damping matrix $\mathbf{K}$ and the vector of the excitation torques $\mathbf{M_E(t)}$, i.e. the contributions of the individual cylinders in a multiple cylinder engine (Fig. 8-35). Attention must be paid to the phase relationships (ignition interval and firing sequence). The damping moments are proportional to the vibration velocities $\dot{\vartheta}$, also interpreted as a vector here:

$$\Theta \ddot{\vartheta} + \mathbf{K}\dot{\vartheta} + \mathbf{C}\vartheta = \mathbf{M_E}(t) \qquad (8\text{-}80)$$

The differential equations of motion for an equivalent model with four still manageable equivalent rotating masses assumes the following form, extended to n equivalent rotating masses on the basis of the system recognizable here:

$$\begin{bmatrix} \Theta_0 & 0 & 0 & 0 \\ 0 & \Theta_1 & 0 & 0 \\ 0 & 0 & \Theta_2 & 0 \\ 0 & 0 & 0 & \Theta_3 \end{bmatrix} \begin{bmatrix} \ddot{\vartheta}_0 \\ \ddot{\vartheta}_1 \\ \ddot{\vartheta}_2 \\ \ddot{\vartheta}_3 \end{bmatrix} + \begin{bmatrix} k_{00} & k_{01} & 0 & 0 \\ k_{10} & k_{11} & k_{12} & 0 \\ 0 & k_{21} & k_{22} & k_{23} \\ 0 & 0 & k_{32} & k_{33} \end{bmatrix} \begin{bmatrix} \dot{\vartheta}_0 \\ \dot{\vartheta}_1 \\ \dot{\vartheta}_2 \\ \dot{\vartheta}_3 \end{bmatrix} +$$

$$+ \begin{bmatrix} c_{00} & c_{01} & 0 & 0 \\ c_{10} & c_{11} & c_{12} & 0 \\ 0 & c_{21} & c_{22} & c_{23} \\ 0 & 0 & c_{32} & c_{33} \end{bmatrix} \begin{bmatrix} \vartheta_0 \\ \vartheta_1 \\ \vartheta_2 \\ \vartheta_3 \end{bmatrix} = \begin{bmatrix} M_{E0}(t) \\ M_{E1}(t) \\ M_{E2}(t) \\ M_{E3}(t) \end{bmatrix}$$

$$(8\text{-}81)$$

The excitation torques follow from the superimposition of the torques related to gas and mass forces (see Sect. 8.3.8).

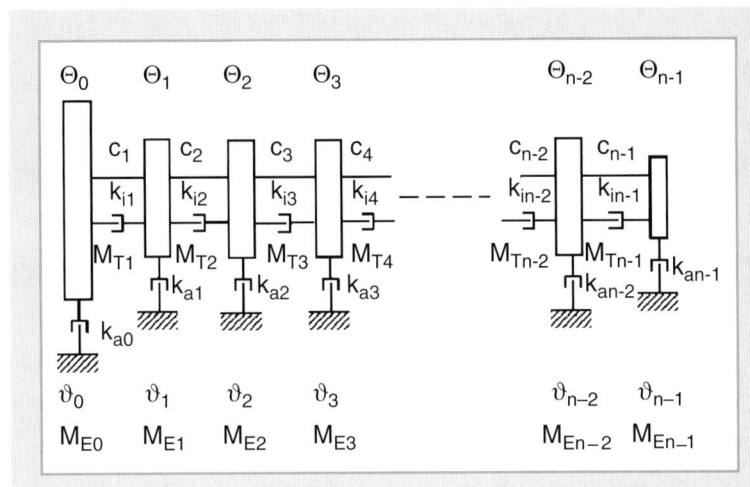

Fig. 8-35 Torsional vibration equivalent model of a crankshaft assembly incorporating the internal and external damping

When the complete power train is analyzed for a multiple cylinder engine, only a resulting excitation torque with its relevant harmonics is formed as a component of the vector of the excitation torques. This produces the same work of excitation as the phase shifted excitation torques of the individual cylinders (cf. [8-2]). Further, the particular load and frictional torques of the individual system components are elements of the pertinent vector. Section 8.4.2.4 already differentiates between internal and external damping (Eq. (8-76) being the common statement for this). Allowing for the relationships represented in Fig. 8-35, the following calculation rules apply to the coefficients k_{ij} of the damping matrix and c_{ij} the stiffness matrix:

$$\begin{aligned}
k_{00} &= k_{a0} + k_{i1} & c_{00} &= c_1 \\
k_{11} &= k_{i1} + k_{a1} + k_{i2} & c_{11} &= c_1 + c_2 \\
k_{22} &= k_{i2} + k_{a2} + k_{i3} & c_{22} &= c_2 + c_3 \\
k_{33} &= k_{i3} + k_{a3} & c_{33} &= c_3 \\
k_{01} &= k_{10} = -k_{i1} & c_{01} &= c_{10} = -c_1 \\
k_{12} &= k_{21} = -k_{i2} & c_{12} &= c_{21} = -c_2 \\
k_{23} &= k_{32} = -k_{i3} & c_{23} &= c_{32} = -c_3
\end{aligned} \quad (8\text{-}82)$$

8.4.3 Simulation of Torsional Vibrations

8.4.3.1 Solution to the Differential Equation Systems of Motion

The solution to the system of homogeneous differential Eqs. (8-79) delivers the vibration system's modal variables:
- natural angular frequencies ω_{0k} and
- related natural vibration modes $\hat{\vartheta}_{ek}$ or $\hat{M}_{Tek}$, i.e. amplitudes of the torsional angle and moments.

The natural vibration modes can be depicted in the form of relative (standardized) torsional vibration amplitudes of the equivalent rotating masses relative to Θ_0 or the greatest value of Θ and their number and position of nodes. The system of inhomogeneous differential Eqs. (8-81) is solved in two steps by:
- determining the amplitude frequency responses $\alpha_{ij}(\omega)$ with the related phase relationships $\beta_{ij}(\omega)$ and transmission factors γ_{ij} and
- determining the torsional angle $\vartheta_i(t)$ and torsional moments $M_{Ti}(t)$ in the individual shaft segments forced by the excitation torques.

Unbranched as well as branched systems are characterized by coefficients of the stiffness and damping matrix aligned symmetrically to the main diagonals. The band-shaped structure of unbranched systems is additionally striking (Eq. 8-81). The selection of suitable algorithms is geared toward developing the aforementioned matrices. The modal variables are provided as solutions to the *eigenvalue problem* explained below. In particular, the following algorithms are available to solve the equation system:
- determinant method,
- *Holzer-Tolle* method and
- transfer matrix method.

The only difference between the latter two methods is their mathematical representation. They ought to only be applied to unbranched systems. The first two methods are explained briefly below. The finite element method (FEM) may employed to perform a *modal analysis* of an appropriately discretized crankshaft assembly structure.

Determinant Method

For the equation system (8.85), the method of resolution

$$\vartheta = \hat{\vartheta} \cos \omega t \tag{8-83}$$

produces the eigenvalue problem

$$(\mathbf{C} - \omega^2 \mathbf{\Theta})\hat{\vartheta} = 0. \tag{8-84}$$

This equation system only has non-trivial solutions in the following case:

$$\det(\mathbf{C} - \omega^2 \mathbf{\Theta}) = 0 \tag{8-85}$$

The calculation of the determinants produces a characteristic polynomial of nth degree $P_n(\omega)$, the zeros of which represent the natural angular frequencies ω_{0k}, $k = 0, 1, 2, \ldots, n - 1$, ($\omega_{00} = 0$ being part of the *rigid body motion*, the trivial natural vibration mode of which has equally large vibration amplitudes for every degree of freedom). Once a particular natural angular frequency ω_{0k} has been inserted, the solutions of the equation system (8-85) form the related eigenvector of kth degree $\hat{\vartheta}_{ek}$. Its n components, i.e. torsional vibration amplitudes of the equivalent rotating masses, are represented in standard form since the solutions are linearly dependent (Fig. 8-36). The corresponding $n - 1$ standardized torsional moment amplitudes of the vector $\hat{M}_{Tek}$ can be calculated according to Eq. (8-78). Common mathematics software contains methods to solve such equation systems.

Holzer-Tolle Method

Once the statements

$$M_{Ti} = \hat{M}_{Ti} \cos \omega t \quad \vartheta_i = \hat{\vartheta}_i \cos \omega t \quad i = 0, 1, \ldots, n \tag{8-86}$$

have been inserted, the Eqs. (8-77) and (8-78) are rendered in the following form:

$$\begin{aligned}\hat{M}_{Ti+1} &= -\Theta_i \omega^2 \; \hat{\vartheta}_i + \hat{M}_{Ti} \\ \hat{\vartheta}_{i+1} &= \hat{\vartheta}_i + \frac{1}{c_{i+1}} \hat{M}_{Ti+1}\end{aligned} \tag{8-87}$$

The initial values are $\hat{M}_{T0} = 0$ and any $\hat{\vartheta}_0 = 1$ for example. The recursive method is excellently suited for programming. The starting value of the angular frequency is, for example, $\omega = 0$. Then, the angular frequency is increased from computational loop to computational loop by a suitable increment of $\Delta \omega$. If the "residual torsional moment" satisfies the condition $\hat{M}_{Tn} = 0$, then $\omega = \omega_{0k}$. The program terminates the calculation when $k = n - 1$ natural angular frequencies have finally been found when "automatically ramping" from ω. The calculation also

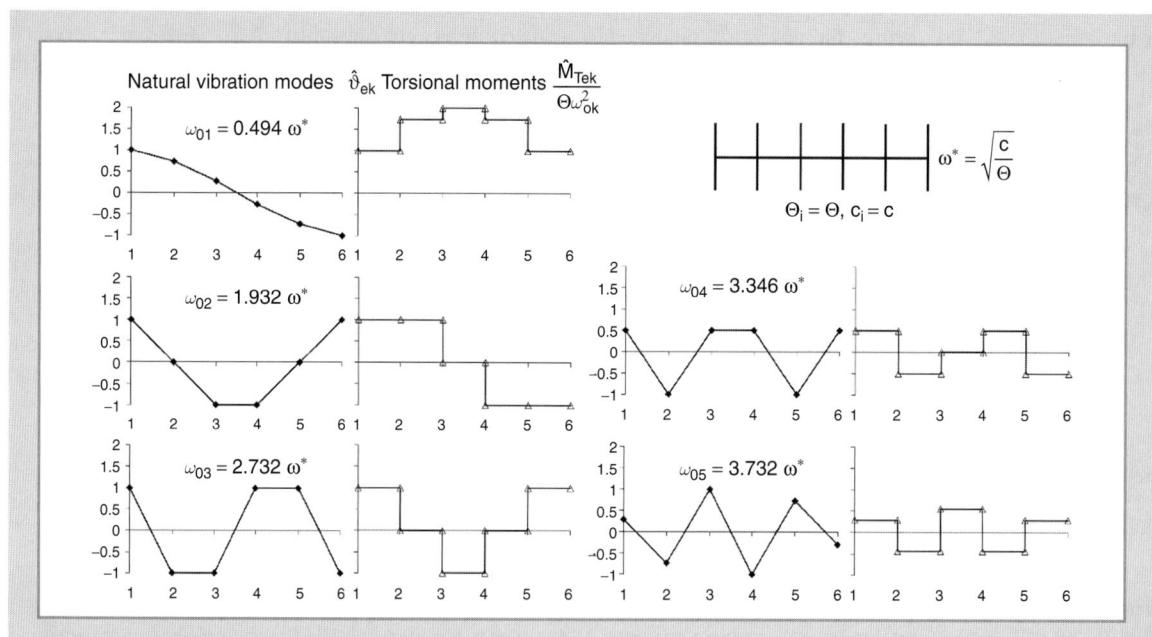

Fig. 8-36 Natural vibration modes (standardized vibration amplitudes) of a torsional vibration system with six degrees of freedom and related standardized torsional moments in the shaft segments

inevitably delivers the components of the eigenvector $\hat{\vartheta}_{ek}$ and $\hat{M}_{Tek}$.

8.4.3.2 Determination of the Frequency Responses

The transfer function $H(\omega)$ is a quotient that describes the relationship between a system's dynamic input and output variable that varies with the angular frequency ω. The torsional vibration amplitudes $\hat{\vartheta}$ are related to the amplitudes of the excitation torques $\hat{M}_E$ as a system response. It is expedient to use mathematically complex notation and *Euler's formula* for any harmonic excitation torque $M_E(t)$ and to select an analog statement for the torsional angle $\vartheta(t)$:

$$M_E(t) = \hat{M}_E(\cos\omega t + j\sin\omega t) = \hat{M}_E e^{j\omega t} \quad \vartheta(t) = \hat{\vartheta} e^{j\omega t}$$
(8-88)

Then, the system of inhomogeneous differential Eqs. (8-80) assumes the following form:

$$\hat{\vartheta}(-\Theta\omega^2 + j\omega K + C) = \hat{\vartheta} B = \hat{M}_E$$
$$\hat{\vartheta} = B^{-1}\hat{M}_E \quad B^{-1} = H(j\omega)$$
(8-89)

$B^{-1} = H(j\omega)$, i.e. the inverted matrix B, is the matrix of the transfer functions with the complex elements $H_{ij}(j\omega)$ (index j must be distinguished from the imaginary unit $j = \sqrt{-1}$ represented with the same symbol):

$$H_{ij}(j\omega) = \gamma_{ij}\alpha_{ij}(\omega) e^{-j\beta_{ij}(\omega)} \quad i,j = 0,1,\ldots,n-1 \quad (8\text{-}90)$$

These transfer functions can be used to calculate the amplitude frequency responses $\alpha_{ij}(\omega)$ and the related phase shifts $\beta_{ij}(\omega)$ of the equivalent rotating masses in the points i resulting from the excitation torques in the points j:

$$\alpha_{ij}(\omega) = \frac{1}{\gamma_{ij}}\left|H_{ij}(j\omega)\right| = \frac{1}{\gamma_{ij}}H_{ij}(\omega)$$
$$\beta_{ij}(\omega) = -\arctan\frac{\text{Im}\{H_{ij}(j\omega)\}}{\text{Re}\{H_{ij}(j\omega)\}}$$
(8-91)

The variables γ_{ij} denote the transmission factors. Figure 8-37 presents an example and schematic of a two degree of freedom system's amplitude frequency responses $\alpha_{ij}(\omega)$ with and without a particular damping.

8.4.3.3 Determination of the Forced Torsional Vibration Amplitudes

Taking a statement analogous to Eq. (8-62) as the starting point

$$M_{Ej}(t) = \sum_{q=1}^{q=m, n=qk} \hat{M}_{Ejn}\cos(n\omega t + \delta_{jn}) \quad \begin{array}{l}\text{two-stroke:} \kappa = 1 \\ \text{four-stroke:} \kappa = 1/2\end{array}$$

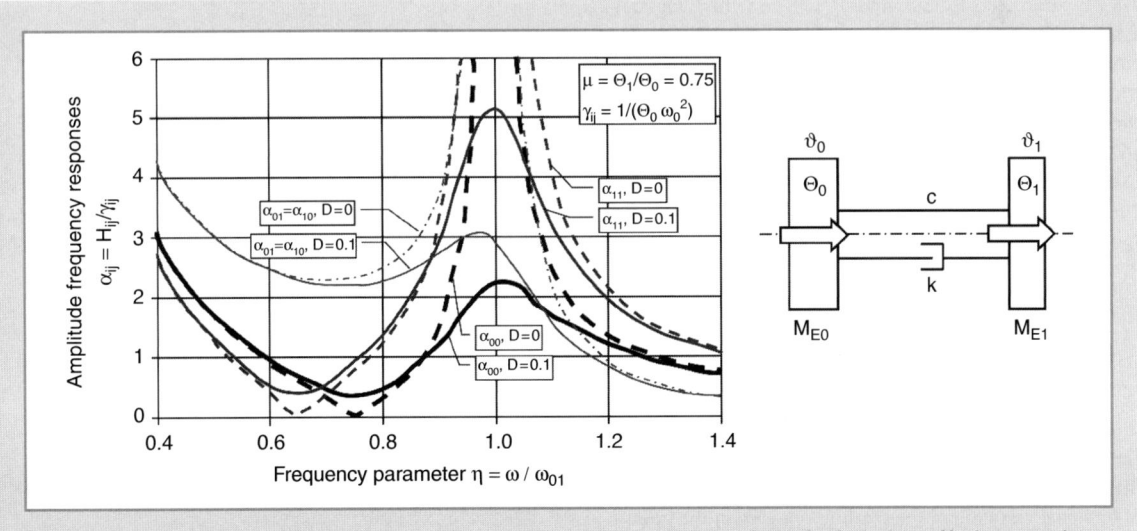

$$q = 1, 2, 3, \ldots, m$$
(8-92)

Fig. 8-37 Schematic of the amplitude frequency responses $\alpha_{00}(\omega)$, $\alpha_{11}(\omega)$ and $\alpha_{01}(\omega)$ for the rotating masses Θ_0 and Θ_1 of a two degree of freedom system relative to the excitation torques $M_E0(t)$ and $M_E1(t)$ with and without damping

the torsional angles of the equivalent rotating masses in the points i can be calculated for the harmonic excitation torques for the engine order n acting simultaneously in the points j (see Sect. 8.3.8) by superposition (superimposition of the individual components):

$$\vartheta_i = \sum_{j=0}^{l-1} \vartheta_{ij} = \tag{8-93}$$

$$\sum_{j=0}^{l-1} \sum_{q=1}^{q=m,n=q\kappa} H_{ij}(n\omega) \hat{M}_{Ejn} \cos\left[n\omega t + \delta_{jn} - \beta_{ij}(n\omega)\right]$$

When another random harmonic excitation is in the power train, the exciting angular frequencies $n\omega$ in the Eqs. (8-92) and (8-93) have to be replaced by $n\omega_j$. Then, $\kappa = 1$ applies. Alternatively, the torsional angle components ϑ_{ij} may also be easily calculated with a Fourier transformation:

$$\vartheta_{ij}(j\omega) = H_{ij}(j\omega) M_{Ej}(j\omega) \tag{8-94}$$

Instead of complex transfer functions, fully decoupled equations may also be referenced to describe dynamic system properties. With the aid of modal variables, the Eq. system (8-80) is transformed with the matrix of the eigenvectors ϑ_e into *modal coordinates* ξ (called main coordinates) and multiplied by the transposed vector ϑ_e^T:

$$\vartheta = \vartheta_e \xi$$
$$(\dot{\vartheta} = \vartheta_e \dot{\xi} \quad \ddot{\vartheta} = \vartheta_e \ddot{\xi}) \tag{8-95}$$

$$\vartheta_e^T \xi \vartheta_e \ddot{\xi} + \vartheta_e^T K \vartheta_e \dot{\xi} + \vartheta_e^T C \vartheta_e \xi = \vartheta_e^T M_E(t) \tag{8-96}$$

The transformation (Eq. (8-95)) that decouples the equations only leads to the goal for $\mathbf{K} = 0$ [8-54]. The matrices of the system parameters $\vartheta_e^T \Theta \vartheta_e$ and $\vartheta_e^T C \vartheta_e$ become "diagonal" as a result. They only still contain modal (generalized) coefficients in the main diagonals. A statement following Eq. (8-76) must be used for the damping matrix $\vartheta_e^T K \vartheta_e$ to ensure that there is decoupling with damping. Then, independent, easy to solve equations of a modal single degree of freedom system exist for the $k = 1$ to $n - 1$ natural vibration modes:

$$\Theta_k^* \ddot{\xi}_k + k_k^* \dot{\xi}_k + c_k^* \xi_k = M_{Ek}^*(t) \tag{8-97}$$

$$\Theta_k^* = \vartheta_{ek}^T \Theta \vartheta_{ek} \quad k_k^* = \vartheta_{ek}^T K \vartheta_{ek}$$
$$c_k^* = \vartheta_{ek}^T C \vartheta_{ek} \quad M_{Ek}^*(t) = \vartheta_{ek}^T M_{Ek}(t) \tag{8-98}$$

It is then possible to employ modal damping $D_k = k_k^*/(2\Theta_k^* \omega_k)$ for the resonant amplitudes. The inverse transformation according to Eq. (8-95, first line) delivers the amplitude of torsional vibration of the rotating mass Θ_i in the point i:

$$\vartheta_i(t) = \sum_{k=1}^{n-1} \vartheta_{eik} \xi_k(t) \tag{8-99}$$

The related torsional moments in the shaft segments are in turn calculated according to Eq. (8-78).

Figure 8-38 presents another significant example of the influence of the firing sequence on the excitation of torsional vibration. An I6 engine provides an example of the important orders of equivalent excitation torques already mentioned in Sect. 8.4.2.5. When the calculation is simplified, these are intended as equivalents of n order excitation torques $\hat{M}_{Ejn}$ in the points j and act fictitiously on the free shaft end. The relative torsional vibration amplitudes of the analyzed natural vibration mode must be plotted in the direction of the rays of the star diagram of the analyzed order. The sum of the vectors delivers a resulting vector. Finally, its value must still be multiplied by the amplitude of the corresponding harmonic of the torsional force and the crank radius. Figure 8-38 compares the relative magnitudes of such equivalent excitation torques as well as the relevant critical speeds (based on [8-2]).

8.4.4 Torsional Vibration Damping and Absorption

8.4.4.1 Torsional Vibration Dampers and Absorbers

In principle, appropriately tuning the system makes resonance-free operation possible at a fixed engine operating speed. However, rotational vibration resonances must be run through

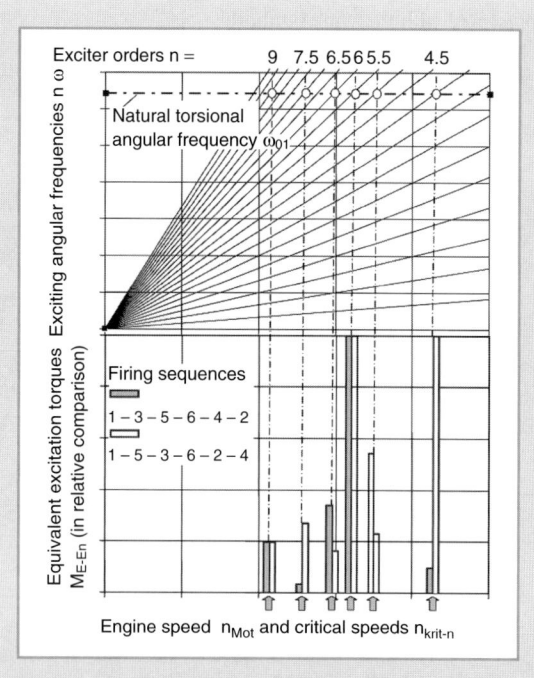

Fig. 8-38 Equivalent excitation torques of the relevant orders n (achieve the same excitability as the phase shifted excitation torques M_{Ejn} of the orders n in the locations j) in a comparison of relative size and corresponding critical speeds of an I6 engine for different firing sequences (based on [8-2])

when accelerating and decelerating. Repeatedly running through and temporarily remaining in resonant ranges can hardly be prevented in transient operation. Hence, torsional vibration dampers and absorbers are usually mounted on a crankshaft's free end where large vibration amplitudes occur. They reduce vibration amplitudes in the pertinent resonant frequency range, thus limiting vibration stress, act to reduce wear and improve comfort. Figure 8-39 presents the operating principle of vibration absorbers and dampers as well as the operating principle for a combination of both concepts.

- *Torsional vibration dampers* convert kinetic energy into heat. They consist of a supplementary rotating mass Θ_D connected with the shaft by a theoretically purely damping (k_D) but in reality elastic element ($c_D > 0$). The damping effect is produced by shear stresses in high viscosity silicon oil, friction surfaces in conjunction with steel springs, oil displacement or material damping by elastomers. The unavoidable elasticity generates an additional resonant frequency with the coupled damper mass that must be below the excitation frequency range. The original resonant frequency is shifted toward higher frequencies. Sufficient heat dissipation is the prerequisite for the function and service life of damping primarily based on energy dissipation. Appropriate materials' spring stiffness and damping are more or less a function of amplitude and temperature. The additional effect of vibration absorption is determined by the design. The farther away a damper is installed from a node, the more effective it is.
- *Torsional vibration absorbers* generate an inertial torque directed counter to the excitation torques. Rather than eliminating vibrational energy, they deflect it for the most part to the connected vibration absorber system. They also consist of a supplementary rotating mass Θ_T and a torsion spring with the stiffness c_T. The unavoidable damping properties ($k_D > 0$) of the torsion spring as well as the entire vibration system reduce the vibration absorber's effectiveness in the frequency range for which it has been designed. This vibration absorber natural frequency $\omega_{T0} = \sqrt{c_T/\Theta_T}$ is tuned to the exciting frequency (resonance tuning). Large vibration amplitudes can occur in processes of acceleration and deceleration as a result of low damping when the resonant frequency additionally caused by the vibration absorber is run through – the original point of resonance being split in turn into two neighboring points of resonance. This also applies to the absorption mass itself. Hence, the acceptability of a vibration absorber's torsional vibration amplitudes has to be verified when it is designed/adjusted. High torsion spring stress limits a vibration absorber's life. This is a practical reason for combining vibration absorption with damping.

"Internal vibration absorption" entails an adjustment of the system parameters – usually only possible in a narrow frequency range afterward. Generally desirable, a "speed-adaptive" vibration absorber can suppress the influence of the excitation order to which a vibration absorber is tuned in the entire speed range (cf. *Sarazin-Tilger* [8-1, 8-53]). Its natural frequency responds proportionally to the speed. The function is based on the principle of a *centrifugal pendulum*.

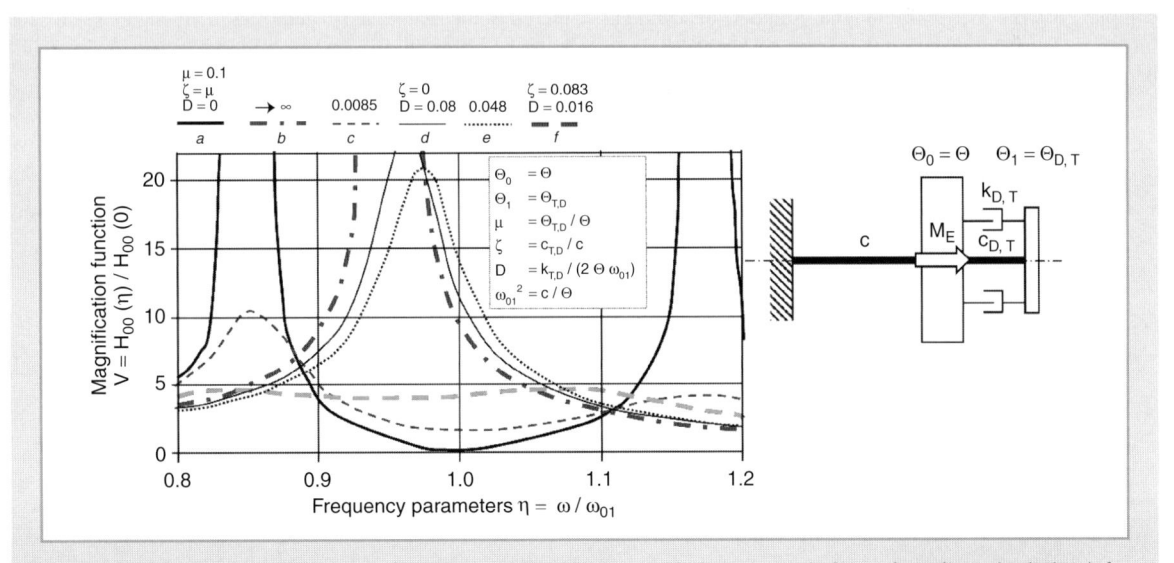

Fig. 8-39 Schematic of a torsional vibration damper, absorber and a combination of both concepts for the simple example of a magnification function (amplitude rise) of an undamped one degree of freedom system; a), b) and c) resonance tuning; **a** Pure vibration absorption without damping; **b** Rigid coupling of the absorption mass; **c** Vibration absorption with damping; **d** and **e** Damping with vibration absorption; **e** Optimal design based on [8-2]; **f** Optimal vibration absorption with damping based on [8-2]

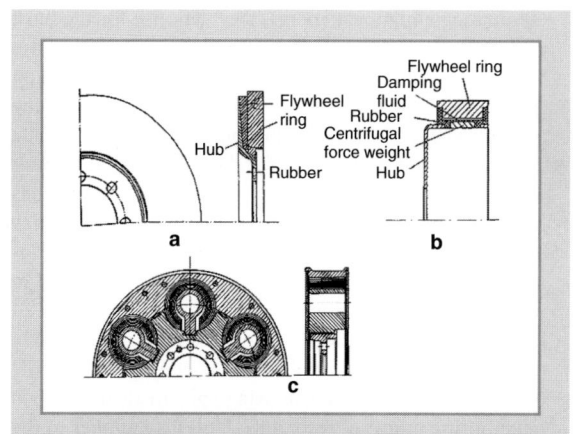

Fig. 8-40 Designs of rotational vibration dampers/absorbers, **a** Rubber torsional vibration absorber, **b** Viscoelastic torsional vibration absorber, **c** Hydrodynamically damped torsional vibration absorber (sleeve spring damper, MAN B&W Diesel AG)

Equivalent systems for vibration dampers and absorbers are identical for the aforementioned reasons. Only their field of application differs to a certain extent. Depending on their particular layout (Fig. 8-40), they more or less satisfy both requirements. Intrinsic damping by displacing lubricating oil in the main and connecting rod bearings [8-56] is usually no longer sufficient for more than four cylinders. A common, simple "rubber damper with a flywheel ring" quickly reaches its limits because of space limitations, thus making more complex designs necessary as the number of cylinders increases. Torsional vibration dampers and vibration absorbers are essential, especially at the very high gas torque amplitudes of supercharged engines. A belt pulley and a damper or absorber combination is also frequently found in vehicle engines. Decoupled belt pulleys, additional crankshaft flexural vibration dampers and camshaft and balance shaft vibration absorbers and dampers are also being used increasingly.

8.4.4.2 Dual Mass Flywheel

Current developments in the motor vehicle industry – high dynamic torques, reduced mass in the drive train, increased numbers of transmission gears, powershift transmission, low viscosity lubricating oils and especially fuel economizing, low speed driving – intensify the problem with vibrations in the entire power train. Hence, a dual-mass flywheel (DMF) serves to decouple the crankshaft assembly from the power train in vehicle engines (Fig. 8-41). Its principle is based on the separation of the flywheel mass into a primary engine side and a secondary drive side mass by interposing torsionally elastic or viscoelastic elements.

A DMF generates an additional, extremely low natural frequency in the vibration system of the crankshaft assembly and power train. Accordingly, it functions as a mechanical low-pass filter. Its response changes with the engine speed though. This has to be factored into design [8-57]. The engine's excitation of the transmission and the rest of the power train is reduced considerably. The consequently nearly uniform rotary motion of the transmission input shaft has an extremely beneficial effect on the familiar noise phenomena of a power train (transmission chatter, rear axle hum, etc.). However, since the primary flywheel mass decreases, a DMF increases the rotational irregularities of an engine itself. This must be borne in mind when tuning a camshaft drive and power takeoffs.

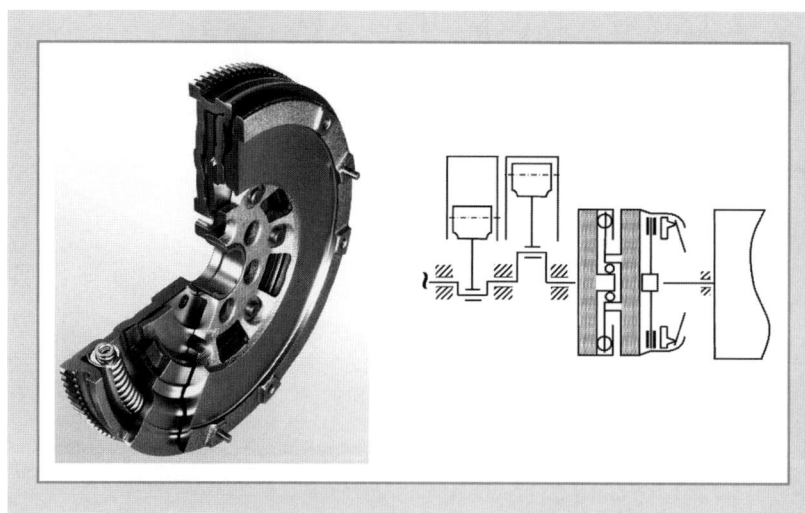

Fig. 8-41
Dual-mass flywheel, schematic and example of design (Photo: LuK GmbH & Co. oHG, Bühl)

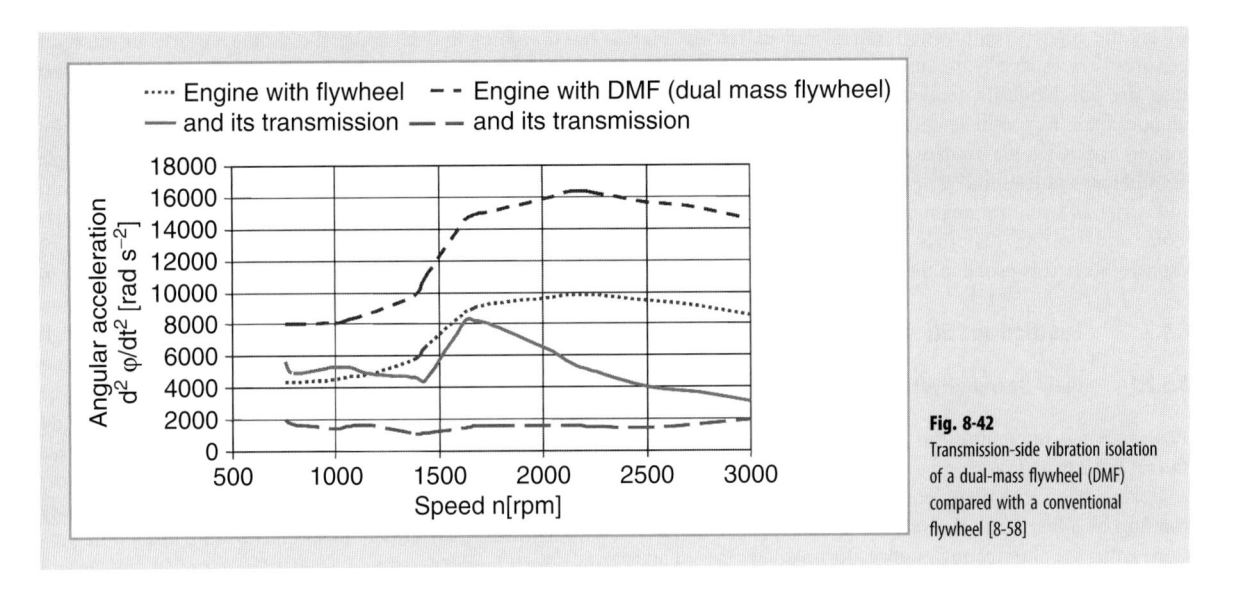

Fig. 8-42 Transmission-side vibration isolation of a dual-mass flywheel (DMF) compared with a conventional flywheel [8-58]

All in all, a DMF has a rather beneficial effect on a crankshaft assembly's torsional and flexural vibrations because the smaller primary side flywheel mass generates fewer crankshaft reactions (Fig. 8-42). A damper/absorber can even be dispensed with under good conditions. The very slight rotational irregularities on the drive side eliminate high frequency torque amplitudes from the transmission. In diesel engines, this allows transmitting a static torque that is up to 10% higher [8-58].

A sufficiently large secondary flywheel mass suppresses the resonant frequency below idle speed. Nevertheless, running through a resonance with large torque amplitudes constitutes a certain problem when starting an engine. Diesel engines with three and four cylinders constitute a particular challenge in this respect. Auxiliary damping is partly necessary, yet limits a DMF's efficacy. High starting torque and high starter speed as well as optimal tuning of the system parameters can keep this problem within limits. Annoying bucking may occur when starting an engine and rattling when stopping it. "Wide angle designs" with a large angle of torsion at low torsional resistance have proven themselves in diesel engines.

A torsional vibration damper in the clutch disc also fulfills its function in the conventional power train of motor vehicles, yet without appreciably isolating vibrations at low speeds. Hydraulic torsional vibration dampers with a spring-mass system are used as a link between the engine and transmission in commercial vehicles. New developments (e.g. Voith's Hydrodamp®) are resolving the conflict of objectives between damping and isolation. *Integrated starter alternator dampers* (ISAD) also have an additional damper/absorber function in the power train. Other applications also employ highly flexible shaft couplings that function as a torsional vibration damper and overload couplings in the power train.

8.5 Bearings and Bearing Materials

8.5.1 Bearing Locations in Diesel Engines' Crankshaft Assemblies

Just as in gasoline engines, *plain bearings* have proven to be the best solution for engine crankshaft assemblies in diesel engines. The primary reasons for this are:

– their capability to absorb strong shock-like loads because the lubricant film between the bearing and shaft constitutes a highly loadable bearing and damping element,
– their suitability for high speeds and their long operating life that usually lasts an engine's entire life span,
– their simple configuration as low-mass, thin-walled bearing shells that can be easily split as necessary for lighter weight assembly in combination with crankshafts and
– their potential for cost effective manufacturing by coil coating, thus ensuring uniformly high quality.

Customized in shape and material for the particular *case of application*, plain bearings are used as:

– crankshaft bearings in the form of half shells,
– connecting rod bearings in the form of half shells in the connecting rod big end,
– connecting rod bearings in the form of bushings in the connecting rod small end,
– piston bearings in the form of bushings,
– camshaft bearings in the form of half shells or bushings,
– rocker arm bearings in the form of bushings,
– idler gear bearings in the form of bushings in timing gears,
– balance shaft bearings in the form half shells or bushings,
– crosshead bearings in the form of half shells and
– crosshead guides in the form of guideways.

Given the higher compression ratios and extremely high gas pressures, especially because of turbocharging, plain bearings are mechanically loaded far more greatly in diesel engines than in gasoline engines. In addition, a considerably longer operating life is demanded from diesel engines in commercial vehicles and an even longer operating life from industrial and marine engines. Both larger bearing dimensions and bearing materials with a higher load carrying capacity solve this problem for plain bearings.

8.5.2 Function and Stresses

8.5.2.1 Plain Bearing Hydrodynamics

Plain bearings in combustion engines function according to the *principle of hydrodynamics*. The lubricating oil supplied to a bearing at an expedient point travels in the direction of rotation by adhering to the surface of the shaft. In conjunction with the lubricating wedge formed by the shaft's eccentric shift relative to the bearing, the thusly produced drag flow builds up pressure in the lubricating oil. The pressure field produced acts as a *spring force*. In addition to its rotary motion, the load causes the shaft to execute movements with a radial component. This squeezes the lubricant out of the decreasing lubricating gap in both circumferential directions and in both axial directions. The pressure field this process produces acts as a *damping force*. The *pressure fields from rotation and displacement* superimpose on each other, thus producing a pressure field that generates the bearing reaction force that separates the sliding surfaces of the shaft and bearing. Figure 8-43 presents the two pressure fields and the related bearing reaction forces.

The basis for the simulation of hydrodynamics in plain bearings is the *Reynolds differential equation* (see for example [8-59] for its derivation), which describes the lubricant flow in the lubricating gap by simulating motion (*Navier-Stoke's* equation) linked with the condition of continuity. For the sake of simplification, numerous *idealized assumptions and preconditions* are usually made in the only numerically possible solution of the Reynolds differential equation (cf. [8-60] through [8-64]). The acceptability of most of these assumptions has been sufficiently validated in experiments and in the field.

The assumption that bearing geometry is completely rigid constitutes an exception however. In earlier engines, lower loads and sizing provided with reserves rendered this still approximately correct. Boosted substantially by enhanced performance and lightweight construction, specific bearing loads now produce such large operating deformations, especially in car and commercial vehicle diesel engines, that they must be taken into account for mathematical bearing design to be reliable.

Such applications' simulations of the hydrodynamics (HD) of plain bearings with rigid bearing geometries now only serve as initial rough estimates in the design stage. Once exact component geometries exist (e.g. connecting rod, engine block and crankshaft), simulations of plain bearings' elastohydrodynamics (EHD) normally follow, incorporating mechanical-elastic component deformations and, if necessary, thermal-elastic deformations (TEHD).

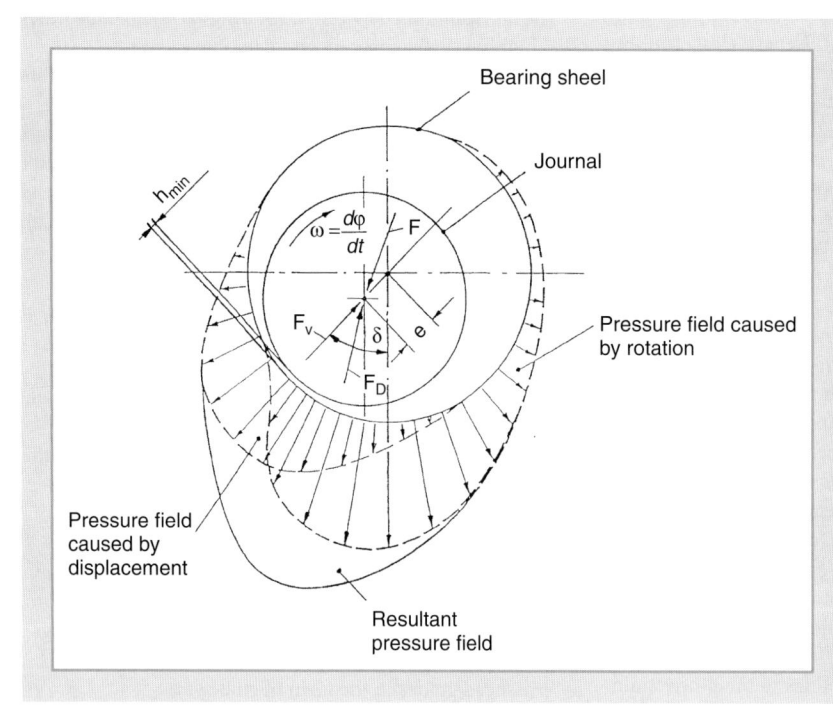

Fig. 8-43
Pressure buildup in the lubricant of transiently loaded hydrodynamic plain bearings. F bearing load force; F_D bearing reaction force from the pressure field caused by rotation; F_V bearing reaction force from the pressure field caused by displacement; e journal eccentricity; δ position angle of journal eccentricity; h_{min} minimum lubricating film thickness; ω angular velocity of the journal

8.5.2.2 Bearing Stress Simulation

Bearing simulation determines the operating parameters of a plain bearing's tribosystem consisting of the elements of bearing housing, bearing shells, lubricant and shaft journal. The *operating parameters* determined for every relevant operating state have to be compared with *standard operating values* (limit values from tests and experience) and their validity checked to verify operational reliability.

The mathematical check of plain bearings' function chiefly consists of determining the mechanical load and the journal's distance from the bearing (minimum lubricating film thickness) in the operating state. Other parameters of evaluation are the friction losses, lubricating oil flow rate and the resultant bearing temperature.

The simulation of the operation of an engine's plain bearings begins by determining the *load forces* (see Sect. 8.2.4).

Figure 8-44 presents the typical characteristic of the bearing load of a connecting rod in a commercial vehicle diesel engine in the form of a polar diagram. The connecting rod's axis proceeds vertically and the shank above and the cover below hare incorporated.

To simulate the *displacement orbit of the journal* in the connecting rod bearings and main bearings, the bearing force F is reduced to two components in the direction of and perpendicular to the minimum lubricating gap h_{min} (see Fig. 8-43), which are then set in equilibrium with the corresponding components of the bearing reaction force F_D and F_V (index D: development of pressure by rotation; index V: development of pressure by displacement). The Sommerfeld numbers then formed with F_D and F_V (see for example [8-59] for an explanation of the dimensionless bearing operating number) are used to obtain the changes of journal eccentricity Δe and its position angle $\Delta \delta$ when the crank angle is altered by the increment $\Delta \varphi$. Adding these changes to the respective current values simulates the journal's displacement orbit for one or more combustion cycles until a periodic and closed curve, i.e. a convergence, is obtained.

Figure 8-45 presents the thusly simulated characteristic of shaft journal displacement inside the bearing belonging to Fig. 8-44. The distance between the displacement curve and the outer circle is a measure of the distance between the shaft journal and the bearing (minimum lubricating film thickness h_{min}).

Conclusions about good positioning of structurally necessary lubricant supply elements (lubrication holes, lubricating grooves and lubricant recesses) in the bearings, which will minimally affect their load carrying capacity, can also be drawn for the load characteristics (Fig. 8-44) and the journal displacement (Fig. 8-45) from polar diagrams with the bearing shell reference frame. Polar diagrams with the journal reference frame provide a corresponding aid to position the lubricant supply elements in the journal (lubrication holes) [8-65].

Figure 8-46 presents an example of the strong influence of the deformations of the connecting rod big end on bearing

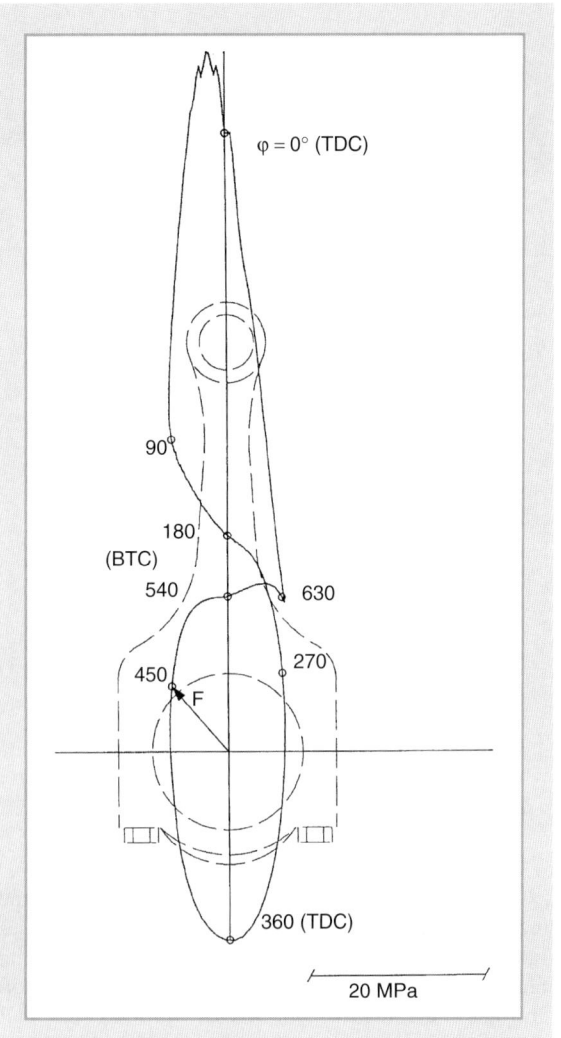

Fig. 8-44 Typical load characteristic in a commercial vehicle engine's connecting rod bearing specifying the crank angle φ (bearing reference frame)

performance. It not only affects the magnitude, position and distribution of lubricating film pressures and the related stress mechanisms in the bearing (with potential material fatigue) and connecting rod but also the magnitude and position of the lubricating film thicknesses and the related risk of wear.

8.5.2.3 Operating Parameters of Present Day Bearings

The evaluation of the level of stress of plain bearings in combustion engines is primarily based on the *maximum values of the specific bearing load* and the *lubricating film pressure* as well as the *minimum values of the lubricating film thickness*. The extreme values of these evaluation criteria for diesel engines have changed greatly in recent years.

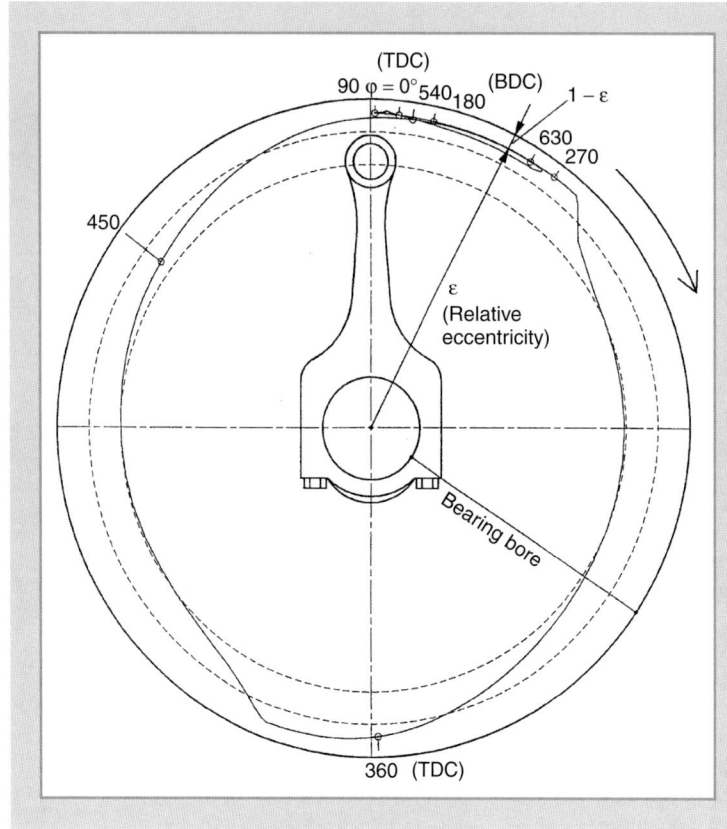

Fig. 8-45
Typical journal displacement orbit of the crank pin in a commercial vehicle engine's connecting rod bearing specifying the crank angle φ (bearing reference frame). $\varepsilon = 2e/C$ relative journal eccentricity; e absolute journal eccentricity; C absolute bearing clearance (difference of diameters of bearing bore and journal); $1 - \varepsilon = h_{min} / (C/2)$

In principle, the most extreme stresses occur in the smaller types of plain bearings in turbocharged car engines and tend toward somewhat more moderate operating values as engine size increases. The most convincing reason for this is the demand for longer *operating life* that increases with engine size. In and of itself desirable and demanded by the engine industry, a theoretically reliable projection of plain bearings' operating life is however just as impossible today as a projection of a combustion engine's exact service life. Hence, since bearings are greatly influenced by the different operating conditions during engine use, especially speeds and loads, and any faults caused by servicing errors for instance, any projection of operating life must for the most part revert to tests, practical experience and statistics.

Despite increases in operating loads, it has proven possible to meet the the demand for longer operating life in recent years. Ultimately, this can be attributed to a whole cluster of technical improvements.

Advances in engine plain bearings include:
- bearing materials,
- bearing shell and shaft journal finishing,
- low-deformation bearing design (e.g. by means of the finite element method),
- lubricating oils,
- filter technology,
- lubricating oil circulation and
- precision and cleanliness during engine assembly.

The magnitude of the mathematical operating values may differ depending on the method of calculation. All the values specified in Table 8-3 were determined with the HD method of calculation mentioned in Sects. 8.5.2.1 and 8.5.2.2. All the data are extreme values from individual production diesel engines. Further increases may be expected in coming years. The majority of engines exhibit strongly moderated conditions though. On average, the maximum loads are approximately 65% of the values specified in Table 8-3 and the minimum lubricating film thicknesses approximately 150%–200%.

8.5.3 Structural Designs

8.5.3.1 Basic Design

Plain bearings for diesel engines are now almost exclusively made of *composite materials*. Steel strips are coated with bearing material in different, usually continuous processes

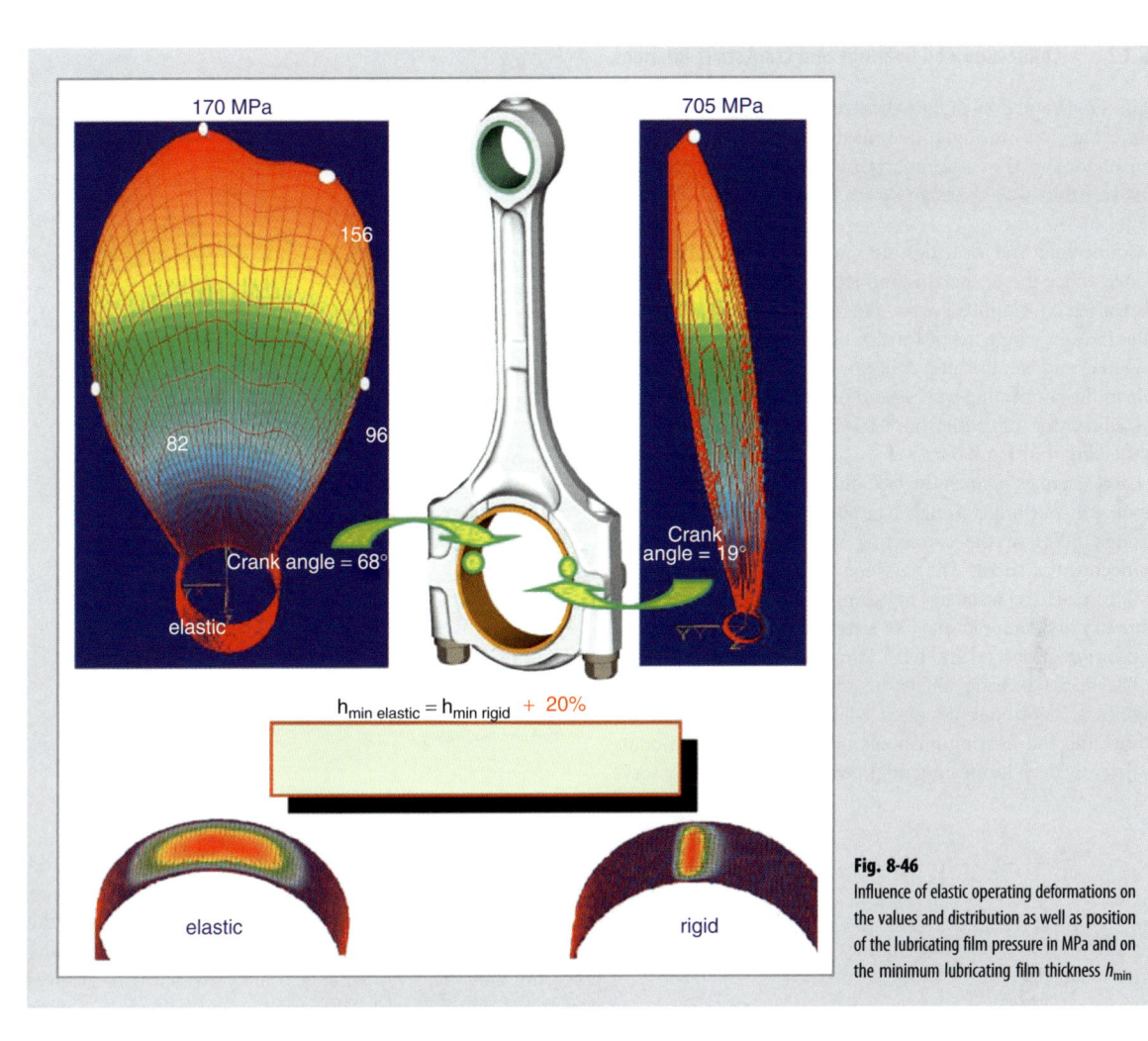

Fig. 8-46 Influence of elastic operating deformations on the values and distribution as well as position of the lubricating film pressure in MPa and on the minimum lubricating film thickness h_{min}

treated in more detail in Sect. 8.5.4. Blanks are punched out of the bimetal strips, bent into half shells and subsequently machined. In certain cases, additional extremely thin sliding layers are also applied to the actual bearing material to improve function.

When the dimensions are large (roughly diameters of 200 mm and up), steel tubes instead of steel strips are also used as the base material, the bearing material being centrifugally cast onto the inner surface of the tube. Afterward, the tubes are split into two half shells.

Table 8-3 Compilation of current operating extremes (computational values) of connecting rod bearings and main bearings of European mass produced car, commercial vehicle and large diesel engines

Operating values	Car diesel engines Shaft diameter ≤ 75 mm Expected service life 3,000 h		Commercial vehicle diesel engines 75 mm ≤ shaft diameter ≤ 150 mm Expected service life 15,000 h		Large diesel engines Shaft diameter ≥ 350 mm Expected service life 50,000 h	
	Connecting rod bearing	Main bearing	Connecting rod bearing	Main bearing	Connecting rod bearing	Main bearing
Maximum specific bearing load in MPa	130	60	100	60	55	40
Minimum lubricating film thickness in µm	0.15	0.25	0.30	0.60	2	3

8.5.3.2 Connecting Rod Bearings and Crankshaft Bearings

Figures 8-47 and 8-48 picture the structural design of a typical connecting rod bearing and crankshaft main bearing (main bearing for short) of a commercial vehicle diesel engine. The relatively thin-walled design serves to reduce space requirements and weight.

Connecting rod bearings do not need any lubricating grooves since the connecting rod journal supplies them with lubricating oil. Connecting rod bearings have appropriate oil holes in cases where the piston pin bushing in the connecting rod small end is lubricated either by oil spray through a split hole in the connecting rod big end or by pressurized oil from the lubricating oil outlet bore in the crank pin through a connecting rod shank bore.

Large engines frequently feed the pressurized oil to the piston pin bushing through a groove in the cap shell and through bores in the connecting rod big end and in the connecting rod shank. The varying wall thickness in Fig. 8-47 (wall eccentricity) facilitates matching the connecting rod bore geometry to obtain the shape of a regular cylinder optimal for load carrying capacity and lubricating film thickness.

The main bearing shells in small and medium-sized engines are normally designed with thicker walls than the connecting rod bearing bushings in order to accommodate sufficiently deep lubricating oil grooves. These not only serve

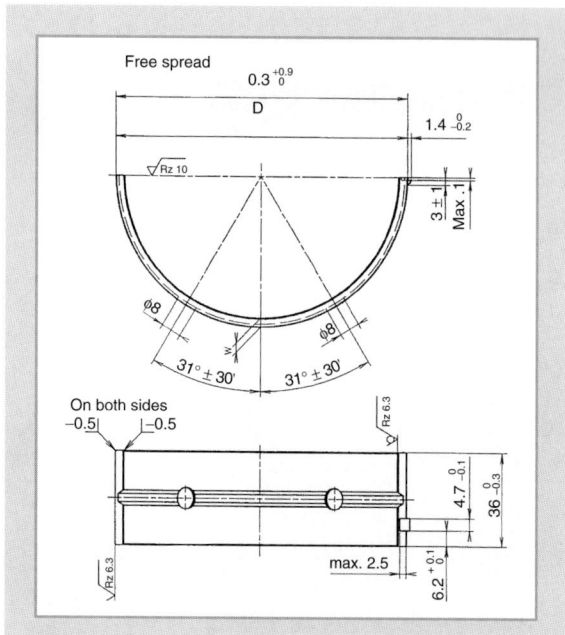

Fig. 8-48 Crankshaft main bearing housing shell of a commercial vehicle engine with an a outer diameter of $D=115.022$ mm and a bearing shell wall thickness $w=3.466+0.012$ mm

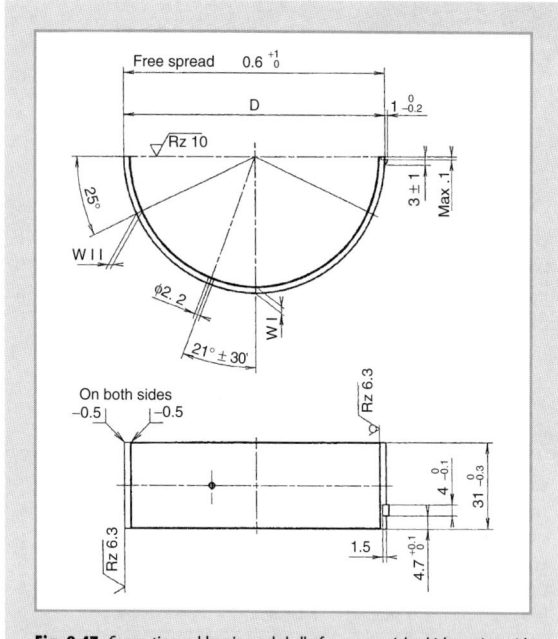

Fig. 8-47 Connecting rod bearing rod shell of a commercial vehicle engine with an outer diameter of $D=98.022$ mm, a bearing shell wall thickness of $wI=2.463 + 0.012$ mm at its crown and a bearing shell wall thickness $25°$ from the joint face of $wII=0.010 + 0.010$ mm thinner than the actual dimension of wI

to supply oil to the main bearing but also the connecting rod bearing. The oil flows from the grooves through crankshaft bores of the main journal to the crank pin. Given the improvement in load carrying capacity, only the housing bearing half shells (upper shells) are provided with grooves in diesel engines, while the cap bearing half shells (lower shells) either have no grooves at all or only partial grooves running from the groove root toward the sliding surfaces. The lubricant is fed from the main oil gallery into the main bearing either through round holes or slots, which benefit any potential angular offset between the bores in the housing and in the bearing shell.

The lugs in the joint face solely serve as assembly aids for correctly positioned installation (positioning aid). A properly designed press fit must prevent the bearing shell from moving relative to the insertion bore.

The trend in all combustion engine and even diesel engine design is geared toward steadily reducing the space requirements and cutting weight. Ever smaller diameters, widths and shell wall thicknesses are being striven for in engine bearings. Bearing-holding components are being constructed increasingly lighter too. This means that particular attention will have to be devoted in the future to obtaining just as adequate a press fit of plain bearings in diesel engines as has already been done in extremely lightweight gasoline engines for some time.

Common mean bearing clearances have a magnitude of 1/1000 of the inner bearing diameter. The smallest possible minimum clearances are targeted for reasons of better load carrying capacity. The limits are established by the irregularities of the elements forming a plain bearing (housing, bearing shell and shaft) induced by manufacturing and assembly and a continued sufficient cooling effect from the oil flowing through the bearing.

8.5.3.3 Thrust Bearings

One of a crankshaft's main bearing points is fitted with a *bilaterally acting thrust bearing* to axially guide the crankshaft and absorb the clutch pressure (and also to partly absorb constant axial loads when automatic transmission is employed). Earlier, vehicle engines often utilized flange bearings in which radial bearings (shells) and thrust bearings (two collars) consisted of one piece. Today, state-of-the-art flange bearings are primarily constructed in a composite form. Bearing half shells are connected with half thrust washers (clinched or welded) on both sides. The advantage over one-piece flange bearings is the possibility of making half thrust washers and shells out of various materials adapted to different tasks.

However, separate half thrust washers are often used, which are inserted on both sides of a radial bearing in recesses in the housing or in the bearing cap (Fig. 8-49). The present trend in smaller engine sizes is mainly toward this lower cost solution with loose thrust washers, which has always been standard in large engines for reasons of manufacturing.

8.5.3.4 Piston Pin, Rocker Arm and Camshaft Bearings

Piston pin bearings und rocker arm bearings are basically designed as *bushings*, the inner bores of which are normally only finish machined after installation to comply with smaller bearing bore tolerances. Where possible, bushings are also used for camshaft bearings. However, bearing shells are necessary in some cases for structural reasons.

8.5.4 Bearing Materials [8-66]

8.5.4.1 Stresses and Function of Plain Bearings in Engine Operation

Plain bearings in combustion engines are *mechanically* loaded by the gas forces and inertial forces that appear there. Due to the dynamic load, the hydrodynamic oil film pressure between the surfaces of the sliding partners, i.e. the bearing and journal, cause *pulsating pressure of the sliding surfaces.* The frictional heat inevitably produced and thermal incidence from the combustion chambers that may occur generate the *thermal* stress.

Wear stress results from the inability to obtain the desired virtually wear-free fluid friction with a complete separation of the sliding surfaces by the lubricating film in every operating point and the unavoidability of temporary mixed friction conditions as a result. In addition, inaccuracies in manufacturing and assembly cause mating surfaces to adapt to each other.

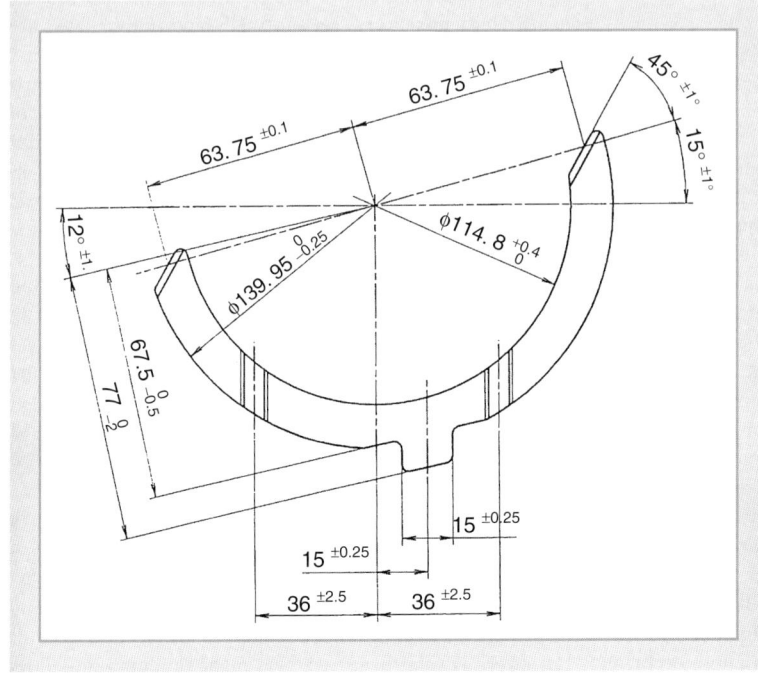

Fig. 8-49
Half thrust washer of a commercial vehicle engine with a wall thickness of *3.360 + 0.05* mm

This also elicits additional compression and wear stresses of the sliding surfaces as well as mechanically and thermally induced deformations of the sliding surfaces in operation.

Ultimately, the sliding surfaces may also be subjected to *corrosive* stresses if either externally or internally induced changes of the lubricant cause chemical reactions with their materials.

The following preconditions must be met in order to ensure the tribological "plain bearing" system functions to the extent the interaction of these stresses requires,:
– Undue mechanical or corrosive wear should not occur!
– Undue bearing temperatures should not occur!
– Material fatigue should not occur!

This catalog of basic requirements not only presupposes that the conditions for hydrodynamic lubrication are fulfilled but also demands quite a number of specific properties from the plain bearings' materials, which are described in the following.

8.5.4.2 Bearing Material Requirements

Compatibility in operation with the material of its sliding partner, e.g. a shaft, is most likely the most important property that makes a certain material a "plain bearing material". ISO 4378/1 defines compatibility and other particulars required of the performance of plain bearing materials. These are:
– adaptability (compensation for geometric imperfections),
– embeddability (embedding of hard particles from the lubricating oil),
– running-in ability (reduction of friction during run-in) and
– wear resistance (abrasion resistance).

Additional properties not in the list are:
– emergency running ability (maintaining bearing operation during poor lubrication through good seizure resistance) and
– fatigue resistance (against pulsating pressure).

Only relatively soft materials, usually with a quite low melting range, facilitate *adaptability and running-in ability* as well as the equally desired *low tendency to seize*. This imposes a limit on further demand for *fatigue resistance against pulsating pressure* as well as *maximum wear resistance* since such properties require rather relatively hard materials.

Good thermal conductivity and *adequate corrosion resistance*, e.g. against aggressive components in the lubricating oil, are additionally desirable. Technically and economically good *manufacturability* is also of major importance. Bearing shells and bushings that are solely joined by a press fit with the insertion bore simplify replacement and are advantageous in the case of repair.

The properties cited are prioritized according the particular use of a plain bearing and the operating conditions. If, for example, disturbances of the fluid friction occur in hydrodynamically lubricated plain bearings with low sliding speeds, experience has shown that the resulting *mixed friction* usually only produces abrasive wear. It increases with the bearing load. While it shortens a bearing's service life, damage caused by spontaneous bearing seizing mostly occurs when bearing loads become extremely high. If *poor lubrication* cannot quite be prevented at high specific loads and low sliding speeds (e.g. in piston pin bushings and rocker arm bushings), priority is given to plain bearing materials with maximum wear resistance and fatigue resistance values over those with high adaptability. Relatively hard bearing materials are options.

Extremely high and diverse requirements are made on bearing material when high sliding speeds occur at the same time as strong, transient specific loads (i.e. in the connecting rod bearings and main bearings of vehicle diesel engines). Along with high fatigue resistance, the bearing material must also possess maximum wear resistance and compatibility.

These diverse, partly contradictory requirements are best met by *heterogeneously structured bearing materials*. Heterogeneity may be obtained both by the microstructure of the bearing alloy itself (mixed crystals or insolubilities of the alloying components) and by layering the bearing materials.

Hence, plain bearings consisting of *multilayer compound materials* have been established in combustion engine design for decades as half shells or bushings with a steel back that produces the strength necessary for the press fit in the insertion bore.

8.5.4.3 Basic Plain Bearing Materials and their Compositions

Lead or tin bearing alloys were used earlier for *low bearing loads*. Hard mixed crystals (normally antimony alloys) are incorporated in a soft matrix with good emergency running ability to improve the load carrying capacity and wear resistance. These materials have as good as disappeared because of their excessively low load capacity, the ban on lead in vehicle combustion engines imposed by the European Directive on End of Life Vehicles and the worldwide ban on the use of toxic substances in combustion engines observed by most engine manufacturers.

Aluminum bearing alloys are frequently used at *medium loads* and *copper* bearing alloys at *high loads*.

Inverting the principle of lead and tin alloys, softer, lower melting components (e.g. tin) are incorporated in a harder matrix with a higher solidus point. The advantages are high fatigue strength and good wear resistance, thus attaining the requisite emergency running properties.

8.5.4.4 One-layer Bearings

One-layer plain bearings are used in some cases in combustion engines as bushings in the connecting rod small end and as piston pin bushings made of copper alloys in the pistons but rarely as solid thrust washers made of aluminum alloys for crankshafts. In the vast majority of cases, multi-layer bearing shells and bushings are used in both diesel and gasoline engines because of their greater strength for the operating load and press fit.

8.5.4.5 Two-layer Bearings

Two-layer bearings are usually manufactured by continuously coating steel strips with the sliding layer by *cast coating, sinter coating or roller coating*.

One of the most important applications are heavy-duty, wrapped bushings in the connecting rod small end or in rocker arms and in the piston bosses in highly loaded pistons. They are made of bronze bimetallic strips manufactured by cast coating or sinter coating. Formerly standard, the alloy CuPb10Sn10 has been replaced by lead-free alloys because of the ban on lead. The base is normally a CuSn material with different additional alloy elements. CuSn10Bi3 is one example of the numerous new lead-free material developments.

Other applications of two-layer compound materials in bushings include camshaft bearings and transmission bearings. In light of the higher sliding speeds of the latter, somewhat softer adaptable alloys are usually employed.

Figure 8-50 presents micrographs of the lead-containing cast alloy CuPb10Sn10 used earlier and the lead-free sintered alloy CuSn10Bi3.

Roller bonded two-layer bearings are employed particularly frequently. The bearing alloys are *rolled on* a steel strip. This method has proven itself particularly well for aluminum bearing materials.

AlSn20Cu is widespread. This bearing material is utilized in crankshaft main bearings of both gasoline and diesel car engines and also used for connecting rod bearings and crankshaft bearings in some large engines because of its good corrosion resistance.

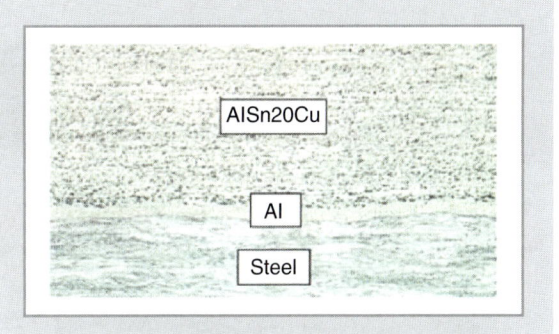

Fig. 8-51 Micrograph of the two-layer compound material steel/AlSn20Cu

Figure 8-51 presents a micrograph of the material structure. Virtually indissoluble in the aluminum, the tin is finely distributed by annealing, thus achieving satisfactory fatigue resistance.

Lead-free without exception because of the lead ban, numerous AlSn alloys with different additional alloy elements have been developed to obtain even higher fatigue resistances approaching bronze. AlSn alloys with silicon have proven advantageous when nodular cast iron crankshafts are employed, a trend in car diesel engines. Thin additional layers of various compositions between the aluminum alloy and steel increase fatigue resistance.

8.5.4.6 Three-layer Bearings

Together with the high sliding speeds, the high specific loads that occur in combustion engines' connecting rod bearings, which have been boosted extremely in recent years by steady increases in specific power and the reduction of dimensions, make maximum demands on plain bearing materials.

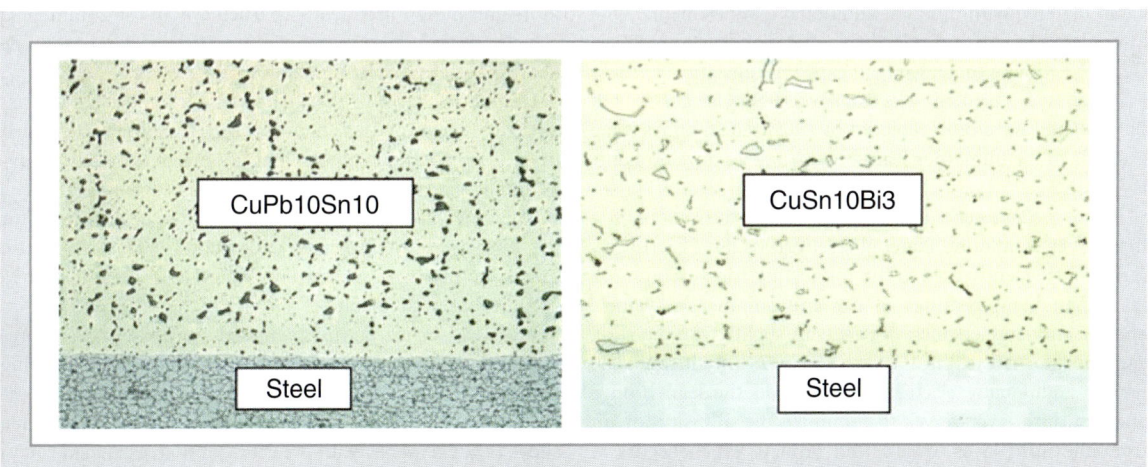

Fig. 8-50 Micrographs of both two-layer composite materials steel/CuPb10Sn10 and steel/CuSn10Bi3

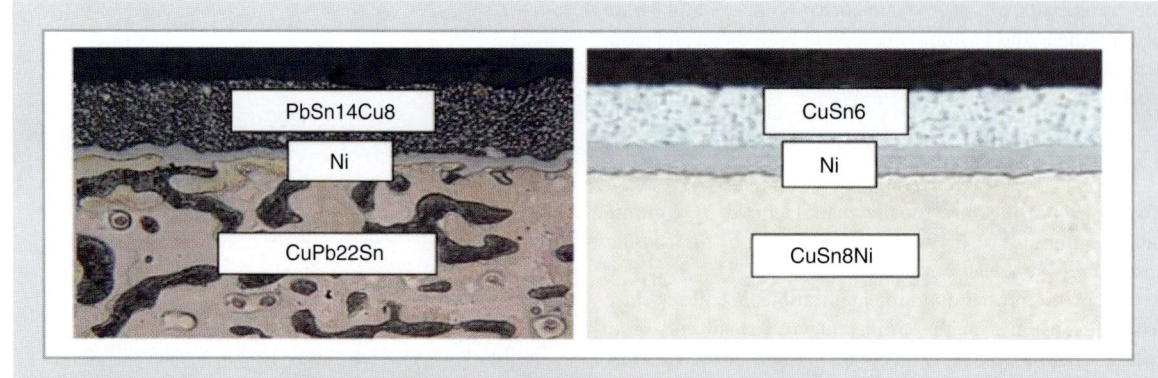

Fig. 8-52 Micrographs of both three-layer compound materials steel/CuPb22Sn/PbSn14Cu8 and steel/CuSn8Ni/CuSn6

Three-layer bearings are employed for these demanding applications.

A *third layer* (sliding layer) that improves the emergency running properties and the adaptability to and embeddability of foreign particles in the lubricating oil is additionally *electroplated* or *vapor deposited* onto relatively hard two-layer compound materials, the second layer of which normally has a thickness of 0.2–0.7 mm.

The very soft third layers in electroplated bearings should only be approximately 0.010–0.040 mm thick to prevent them from lowering a bearing's loading capacity too greatly. Casting cannot achieve this.

The second layer is responsible for ensuring the requisite load carrying capacity. However, the second layer must also sufficiently resist seizing to ensure the bearing continues to function in case the third layer wears locally.

Earlier second layer bronzes of three-layer CuPbSn (e.g. CuPb22Sn) bearing shells have been replaced by lead-free alloys because of the ban on lead. The base is normally a CuSn material with different additional alloy elements. One example of the numerous lead-free material developments is CuSn8Ni with an increased load carrying capacity.

As a rule, formerly electroplated PbSnCu coats (e.g. PbSn14Cu8) have been replaced by lead-free Sn alloys. One example of the numerous lead-free third layer developments is SnCu6.

Figure 8-52 presents a micrograph of the lead-containing three-layer compound material steel/CuPb22Sn/PbSn14Cu8 formerly frequently used in the connecting rod bearings and crankshaft main bearings of car and commercial vehicle engines and large engines, in part with slightly varying alloy components in the third layer. The 1–2 µm thick nickel barrier additionally electroplated between the lead bronze and the third layer serves to minimize the diffusion of tin from the third layer toward lead bronze, which occurs at operating temperature. Otherwise, this would greatly diminish the lead-containing third layer's corrosion resistance.

Figure 8-52 presents a micrograph of the lead free three-layer compound material steel/CuSn8Ni/SnCu6.

By maintaining the principle structure of lead-containing bearing materials even in lead-free materials, combinations with various novel intermediate layers significantly increase fatigue resistance and wear resistance.

Given its good corrosion resistance, a third layer consisting of SnSb7 is also used in heavy fuel diesel engines and gas engines.

An earlier approach to improving the electroplated third layer's frequently inadequate fatigue resistance and wear resistance entailed filling the *recesses* introduced in the surface of a two-layer compound material by only electroplating them rather than completely electroplating the sliding surface. The harder regions increase load carrying capacity and wear resistance. The softer regions still provide sufficient emergency running ability. Once used as a connecting rod and crankshaft bearing in commercial vehicles and medium speed diesel engines, this type of bearing is now being used more and more infrequently (particularly in the commercial vehicle sector) because of its unsatisfactory service life because the electroplating material washes out.

Despite increases in the wear resistance and fatigue resistance by increasing their hardness (e.g. CuSn6), electroplated third layers are frequently no longer a match for today's maximum demands in highly supercharged diesel engines.

Therefore, even *harder alloys* have had to be developed for the third layer. Physical vapor deposition (PVD) coating, also called "sputtering", is utilized. Material is deposited in a vacuum in the vapor phase.

Cast lead bronzes with increased tin content were previously used as a second layer for maximum load carrying capacity. However, just like electroplated three-layer compound materials, they have now been replaced by lead-free alloys (e.g. CuSn8Ni) with even further increased load carrying capacity. The relatively hard PVD third layer normally consists of AlSn20 (Fig. 8-53).

Fig. 8-53 Micrograph of a sputter bearing

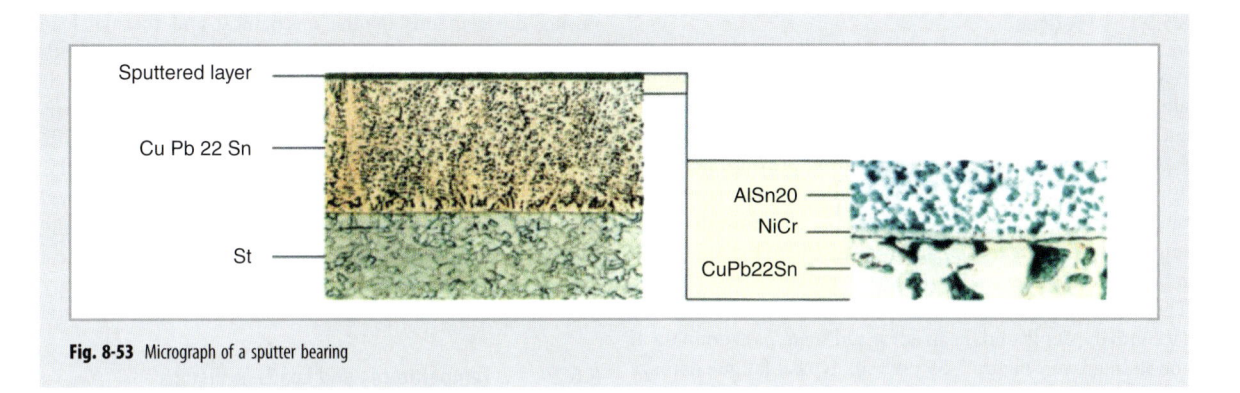

Examples of application are connecting rod bearings in highly supercharged car diesel engines and connecting rod and crankshaft bearings in highly supercharged commercial vehicle diesel engines, especially those with state-of-the-art direct injection and intercooling.

8.5.5 Bearing Damage and Its Causes [8-67, 8-68]

8.5.5.1 Impairment of Operation

Various disturbances can impair a bearing's operation, i.e. any causes that prevent hydrodynamic operation under complete fluid lubrication despite proper design. This may include poor lubrication, contaminated, diluted or air-foamed lubricant, overloading and faulty geometry of the sliding partners due to manufacturing or assembly errors. Disturbances can cause wear, overheating, material fatigue or corrosion. A bearing normally retains its functionality when such impairments are only marginally present. However, in an advanced stage, they can ultimately cause a bearing to fail.

8.5.5.2 Wear

If the thickness of the lubricating film between two sliding partners is insufficient to fully separate their sliding surfaces from one another, then *mixed friction wear* occurs instead of wear-free complete fluid lubrication with pure fluid friction. Surfaces roughened by manufacturing cause the harder of the two sliding partners to score the softer partner and the peaks of roughness to shear. Such microcutting processes are called *abrasive wear*. It may be considered acceptable when the rate of bearing abrasion is so low that it does not reduce the targeted service life of the plain bearing. Planar abrasion is less harmful than scoring abrasion since it impairs the buildup of the hydrodynamic bearing pressure less. Laminar *break-in wear* may also be viewed positively since the thusly generated smoothing of the sliding surfaces improves hydrodynamics. Figure 8-54 pictures surface abrasive wear.

Bearing wear produced by hard foreign particles in the lubricating oil is called *erosive wear*. Many small particles tend to produce laminar wear, a few large particles scoring wear.

Unduly high rates of wear can cause bearing material to overheat and melt. This produces local connecting bridges with the sliding partners, which cause *adhesive wear* during shearing. When the adhesive bridges are no longer separable or this entails substantial destruction, this is called bearing seizure and normally means total bearing failure.

Corrosive wear can arise as a result of the bearing material's chemical reaction with aggressive media in the lubricating oil. Especially when the thermal load is high and the lubricant has deteriorated, media such as acids may form when stipulated lubricant change intervals are exceeded. Corrosive media can also reach the lubricating oil from the fuel (heavy fuel, landfill gas). Corrosive wear is a chemical dissolution of bearing material, which alters its properties and can frequently cause material loss in the form of pitting and even irreparable damage with bearing seizing.

Fig. 8-54 Abrasive wear in a three-layer bearing shell

8.5.5.3 Fatigue

According to [8-59] and [8-69], excessively high dynamic alternating peripheral stresses induced by the pulsating lubricating film pressure cause fatigue cracks in bearings.

At variable bearing load, the pulsating pressure field that appears in the lubricant between the sliding surfaces produces tangential alternating normal and shear stresses in the bearing material. *Fatigue damage* occurs when the material's fatigue strength is exceeded. Fatigue is purely a stress process. The damage manifests itself as cracks running from the surface into the depth and, in an advanced stage, as crumbling in the sliding layer. Figure 8-55 presents typical fatigue damage of electroplated plain bearings. According to [8-59] and [8-69], not only the absolute level of the lubricating film pressure but, above all, its local gradient also influences a bearing material's fatigue. The destruction of a bearing's sliding surface diminishes its load carrying capacity because the hydrodynamic pressure buildup is disrupted. Additional wear, overheating and, ultimately, irreparable damage caused by seizing usually follow.

8.6 Piston, Piston Rings and Piston Pins

8.6.1 Function of a Piston

The piston is the first link in a combustion engine's chain of force transmitting elements. A moveable force transmitting wall, it, together with the piston rings, must reliably seal the combustion chamber against the passage of gas and the inflow of lubricating oil in every operating and load state. Specifically designed as an integral part of the combustion chamber, the piston crown influences flow conditions during gas exchange, mixture formation and combustion during various operating modes. Increases in speed and brake mean effective pressure by supercharging are constantly raising the demands on the mechanical and thermal load capacity of diesel engine pistons. However, increased loads have made it increasingly difficult to meet the demands made on modern piston designs. e.g. adaptability to changing operating conditions, seizure resistance, high structural strength, NVH performance, low oil consumption and long service life. Moreover, the limitations on the use of some conventional piston designs and common materials have become apparent. Hence, every option for material and structural design must be explored when designing pistons for high performance diesel engines.

Fig. 8-55 Fatigue of the third layer of a three-layer bearing shell

8.6.2 Temperatures and Loads in Pistons

The extremely rapid conversion of fuel-based energy into heat causes the temperature and pressure to increase considerably during combustion. The various working principles require different compression and air/fuel ratios. Thus, the gases in a combustion chamber have peak temperatures between 1,800 and 2,600°C. Although the temperatures drop greatly during the combustion cycle, the exhaust gases exiting a combustion chamber can still have temperatures between 500 and 1,000°C. The heat from the hot combustible gases is primarily transmitted to the combustion chamber walls and thus the piston crown by convection and only in small part by radiation (see Sect. 7.2).

The *strong periodic fluctuations of temperature* in a combustion chamber cause temperatures to vary in the uppermost layer of the piston crown. The amplitudes of these fluctuations have a magnitude of about 10–40°C on the surface and subside inwardly within a few millimeters according to an exponential function [8-70].

A large part of the heat absorbed by the piston crown during the expansion stroke is released to the coolant by the piston ring zone and by the cylinder wall. Depending on the design of the engine and the piston and influenced by the operating mode and the piston velocity, between 20 and 60% of the amount of heat that accumulates in a piston crown is primarily released by the piston rings. A small portion of the heat is transferred to the fresh gas during gas exchange. The lubricating and cooling oil that reaches a piston's inner wall absorbs the rest of the dissipated heat.

The *three-dimensional temperature field* produced in a piston can be simulated with finite element programs utilizing appropriate boundary values. Figure 8-56 presents characteristic surface temperatures in pistons of gasoline and diesel engines.

They are measured in an engine with non-electrical methods of measurement (fusible plugs, Templugs and residual hardness) or sophisticated electrical methods of measurement (thermocouples, NTC thermistors and telemetry). Cooling (spray cooling and forced oil cooling) can influence piston temperature significantly.

An equilibrium of gas, inertial and supporting forces (connecting rod and side thrust) prevails in a piston. The *maximum*

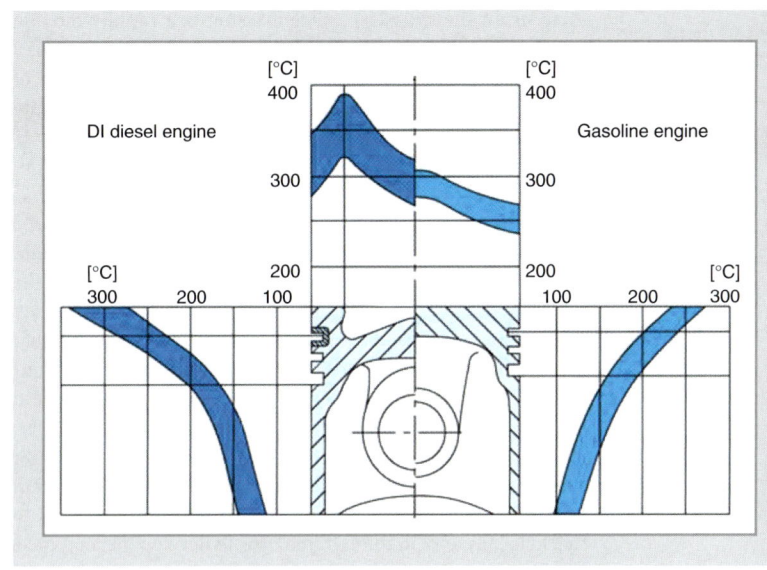

Fig. 8-56 Operating temperatures in aluminum pistons of vehicle engines at full load (schematic)

gas pressure is crucially important for a piston's mechanical stress. It is 80–110 bar in naturally aspirated diesel engines and 160–250 bar in supercharged diesel engines. The pressure increases at a rate of 3–8 bar/°CA and can exceed 20 bar/°CA when combustion is disrupted [8-71–8-73].

8.6.3 Piston Design and Stress

8.6.3.1 Main Piston Dimensions

A piston's functional elements include the piston crown, ring zone with piston top land, piston pin boss and skirt. Additional functional elements such as the cooling gallery and ring carrier are characteristic of piston design. The piston assembly also includes the piston rings and the piston pins and pin retainers.

The main piston dimensions (Fig. 8-57, Table 8-4 [8-74]) are closely interrelated to the main engine dimensions and the dimensions of other components (crankcase, crankshaft andconnecting rod). Apart from the *cylinder diameter*, the most important piston dimension is the *compression height*, i.e. the distance between the center of the pin and the upper edge of the piston top land. Piston mass plays a major role, especially in high speed engines.

8.6.3.2 Loading Conditions of Pistons

General Description

The gas, inertial and skirt side forces that act on a piston generate deformations and thus stress in a piston. The pressure from the combustion gases acts on the piston crown. The piston transmits the resulting force to the crankshaft via the piston pin and the connecting rod. Since the piston pin bosses in the area of the skirt are supported on the piston pin, the *open end* of the piston skirt *deforms ovally*.

Deformation is additionally influenced by the *deformation of the pin* (ovalization and bending deflection) and by the contact forces in the cylinder, which likewise promote piston deformation.

Deformations produced by the temperature field in the piston are superimposed on the piston deformations

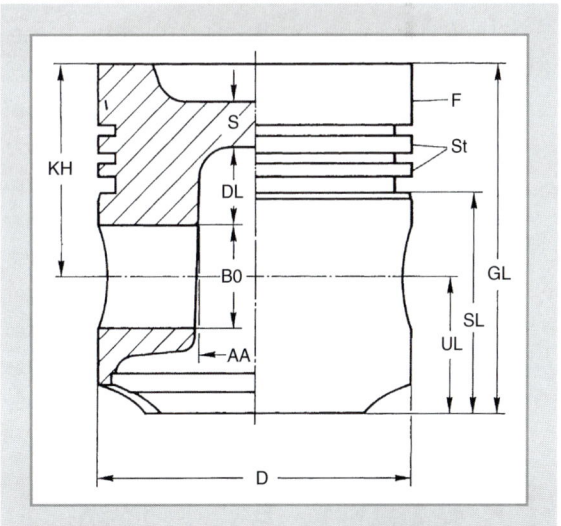

Fig. 8-57 Important piston dimensions and terms. *F* Piston top land; *St* Ring land; *s* Crown thickness; *KH* Compression height; *DL* Length of elongation; *GL* Total length; *BO* Pin bore diameter (pin diameter); *SL* Skirt length; *UL* Bottom length; *AA* Boss spacing; *D* Piston diameter

Table 8-4 Main dimensions of light alloy pistons for diesel engines

Four-stroke diesel engines		
Diameter D in mm	75...100	>100
Total length GL/D	0.8...1.3	1.1...1.6
Compression height KH/D	0.50...0.80	0.70...1,00
Pin diameter BO/D	0.35[a]...0.40	0.36...0,45
Piston top land height F/D	0.10...0.20	0.10...0.22
1st ring land St/D[b]	0.07...0.09	0.07...0.12
Groove height for 1st ring in mm	1.5...3.0	3.0...8.0
Skirt length SL/D	0.50...0.90	0.70...1.10
Boss spacing AA/D	0.27...0.40	0.25...0.40
Crown thickness s/D	0.10...0.15[c]	0.13...0.20
Weight ratio G_N/D in g/cm^3	0.8...1.1	1.1...1.6

[a] Lower value for car diesel.
[b] Values apply to ring carrier pistons.
[c] For DI engines ~ 0.2 x recess diameter.

stemming from the mechanical forces. They cause *the crown to warp* in pistons at operating temperature and *diameters to enlarge* from the lower skirt end to the piston top land.

Piston design aims for wall thicknesses that are sufficient to prevent fractures and deformations, even caused by side forces, which, however, are elastic enough in certain regions that they are able to yield to externally applied deformations (e.g. from the cylinder). Since a piston is a rapidly moving engine part, its mass should always be kept to a minimum.

Deformations and stresses resulting from the action of forces and temperature fields may be simulated with the finite element method (Fig. 8-58) [8-75].

Piston Crown Loading

A diesel piston crown is subjected to *extreme temperature loading* with temperatures far above 300°C at which the fatigue strength of the aluminum materials employed already diminishes considerably. As far as possible, its design and that of the bowl rim must prevent the occurrence of local temperature peaks and weakening notch effects, which could facilitate the initiation of cracks [8-76–8-78].

Engine operation produces significant *thermal cycling* at the bowl rim. Local heating and the inhibition of deformation by surrounding colder regions can cause piston materials to yield locally and, in the extreme case, to develop thermal cycle fatigue. Hard anodizing a piston crown produces a hard aluminum oxide layer that reduces its *propensity to incipient thermal cracking*.

The optimization of exhaust emission for direct injection engines frequently necessitates designing combustion bowls with the smallest possible radius and often with an undercut. A pronounced temperature maximum inevitably arises at the edge of the bowl. Aluminum oxide ceramic fibers incorporated

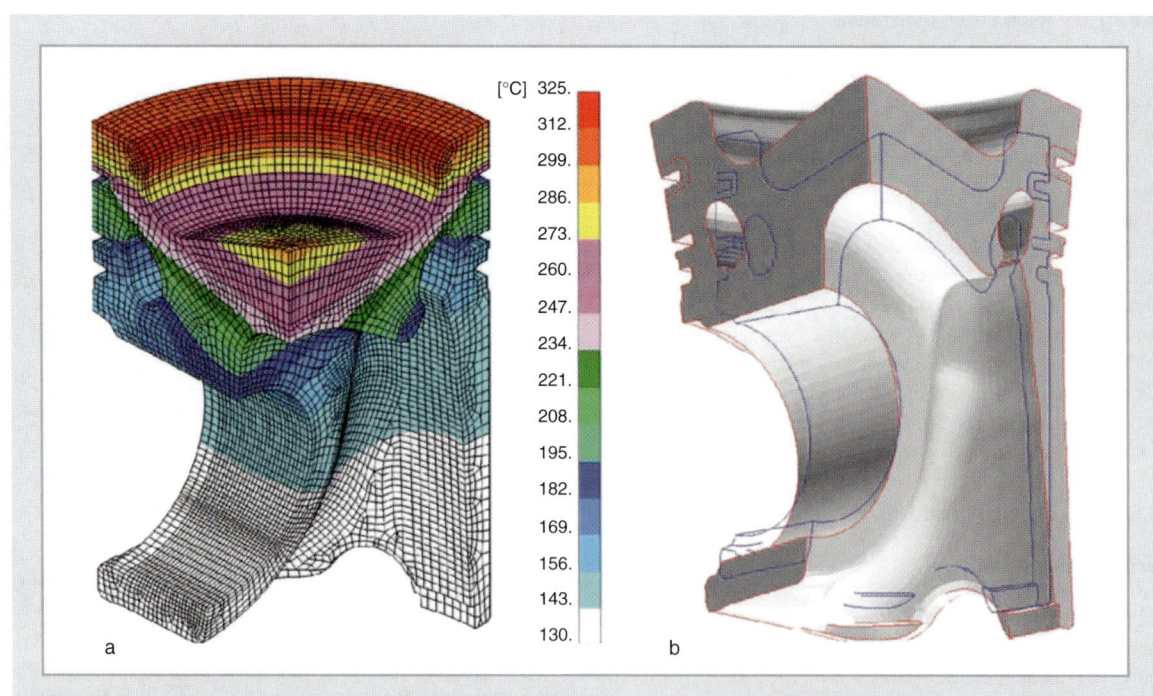

Fig. 8-58 Finite element simulation of a piston. **a** FE mesh and temperature field in °C; **b** Deformation under temperature load

Fig. 8-59 Ceramic fiber reinforced piston with cooled ring carrier and pin bore bushing

in the piston alloy significantly increase the strength of the bowl rim. However, a special casting method is required to be able to manufacture such pistons with fiber reinforced bowl rims (Fig. 8-59) or locally fiber reinforced crown zones since the piston alloy only penetrates the porous fiber package under increased pressure in its liquid state [8-79].

Ring Zone Loading

Since their first (in some cases also their second) ring groove is thermally and mechanically loaded, aluminum diesel engine pistons require additional measures to attain the required service life. The rings' radial motion caused by the piston's rocking motion in the cylinder is the primary cause of *ring groove wear* (flank wear). The axial motion resulting from gas, mass and frictional forces and the rings' rotation also contribute to ring groove wear.

On the one hand, the high thermal stress of the first ring groove particularly intensifies mechanical wear. On the other hand, it can also cause the *ring groove to coke* and the first ring in diesel engines to *stick*.

The first ring groove of aluminum diesel pistons is normally reinforced by casting in a *ring carrier* usually made of "Ni-resist". This is austenitic cast iron with approximately the same thermal expansion as aluminum. The Alfin process produces an intermetallic phase between the ring carrier and the piston material, which prevents gas and inertial forces from loosening the ring carrier and facilitates better heat transfer.

Other measures that *reinforce the ring groove* such as steel lamella in the upper groove flank or alloyed-in wear resistant materials usually provide less wear protection than a ring carrier.

Boss and Support Zone

The forces acting on a piston are transmitted by the piston pin bore to the piston pin and by the connecting rod to the crankshaft. The piston pin boss is one of a piston's *highly stressed* zones.

Various bore geometries (block, trapezoidal or stepped support designs) are employed to properly support the piston crown to transfer load. The pin bore contact pressure in larger diesel engines for rail and marine drives is frequently reduced by a stepped pin boss (stepped connecting rod). Center bores that hold the piston pin are located in the boss supports. The pin bores must be machined with a high quality finish since they function as bearings.

The dimension of the piston pin and the width of the small connecting rod eye significantly influence a pin bore's load capacity. When engine performance has been tuned, the resulting operating conditions (firing pressure, temperature and service life) may make it necessary to engineer the pin bore's fine geometry to deviate from the cylindrical contour to ensure reliable operation in the pin bore without cracking [8-80, 8-81]. A *slight ovality* of the pin bore or an optimized conical shape toward the connecting rod can increase component strength in the zone of the piston pin boss by approximately 5–15%. Side reliefs in the pin bore or piston pins with profiled outer contours (see Sect. 8.6.5 for more on profiled piston pins) furnish other options. However, every one of these measures increases the mechanical stresses in the piston crown.

Piston Cooling

A piston's thermal load may only be increased to a limited extent because AlSi alloys' fatigue strength strongly diminishes as temperatures increase above approximately 150°C. Hence, diesel engines need systematic *piston cooling* in most cases [8-82]. Spray cooling the interior of a piston with oil from stationary jets (reducing the temperature in the first ring groove by 10–30°C) suffices in many cases. An annular cavity, a cooling gallery is frequently needed behind the ring zone.

Cooling gallery pistons are manufactured with water-soluble salt cores or as "cooled ring carriers" with a sheet metal cooling gallery attached to the ring carrier. A meticulously adjusted jet with a fixed housing injects the cooling oil into the annular cooling gallery through an inlet orifice in the piston. Maintaining the correct quantity of oil decisively influences heat dissipation. The outlet is formed by one or more bores inside a piston, preferably positioned on the side of the cooling gallery approximately opposite the inlet.

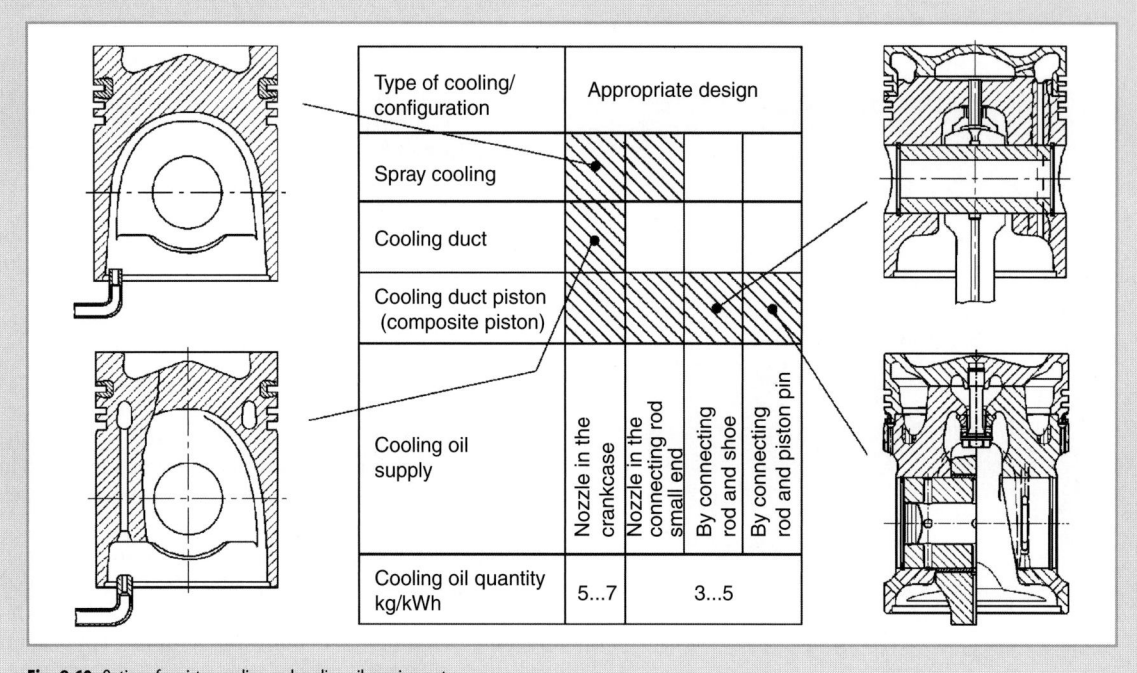

Fig. 8-60 Options for piston cooling and cooling oil requirement

The temperature at the bowl rim and in the ring, support and boss zone can be dropped significantly depending on the cooling gallery's position. Temperatures in the zone of the first groove can be lowered by 25–50°C (Fig. 8-60, Table 8-5).

Cooling ducts, which facilitate effective cooling into the region of the ring zone in particular, may be engineered in the steel crowns of composite pistons. Usually, the cooling oil is fed into the outer annular cooling gallery and returns from the inner cooling gallery. The temperature in the region of the first compression ring's groove should not fall below 150°C in the part load range because acids form when the temperature drops below the dew point of the SO_2 and SO_3 contained in the combustion gases. Thus, it may become necessary to modify cooling oil distribution in a piston so that the piston crown is cooled before the region of the ring zone.

Table 8-5 Influence of engine operating conditions and cooling on vehicle engines' piston temperatures

Operating parameter	Change of the operating parameter	Temperature change in the 1st ring groove
Speed n (p_e = const.)	100/min	2…4°C
Load p_e (w_e) (n = const.)	1 bar (0.1 kJ/dm^3)	≈10°C (bowl rim ≈20°C)
	1 bar for cooling channel pistons	5…10°C (bowl rim 15…20°C)
Start of injection (advanced)	1° CA	+1…3°C (bowl rim <4.5°C)
Compression ratio	By 1 unit	4…12°C
Coolant temperature (water)	10°C	4…8°C
Coolant composition (water cooling)	50% glycol	+5…+10°C
Lubricating oil temperature (oil sump)	10°C	1…3°C
Piston cooling by oil	Spray jet in connecting rod big end	−8…−15°C locally
	Fixed oil jets	−10…−30°C
	Cooling gallery	−25…−50°C
	Cooling oil temperature 10°C for cooling channel pistons	4…8°C (also at the bowl rim)

8.6.3.3 Piston Designs

Car Diesel Engine Pistons

By virtue of their process, diesel engines not only experience high thermal but also high mechanical loads. Hypereutectic aluminum material with a silicon content of 18% is only sufficient in relatively lightly loaded car diesel engines. The ring carrier has established itself as very effective protection against wear in the first groove in more highly loaded engines. Such pistons are usually made of *eutectic AlSi alloys* with 11–13% silicon.

Piston cooling is essential in charged engines with higher thermal loads. It may take the form of *oil spray cooling* of the inner piston contour or *forced oil cooling* in pistons with cooling galleries.

In addition to conventional salt core cooling gallery pistons (Fig. 8-61), cast pistons with cooled ring carriers (Fig. 8-62) are utilized, which create considerable potential to decrease temperatures in critical component points. An effective reduction of the temperature at the bowl rim by 15°C more than doubles service life.

Fig. 8-62 Cooling gallery piston with cooled ring carrier

Heavy Duty Diesel Engine Pistons

Gravity die cast *ring carrier pistons* have become the standard design for commercial vehicle diesel engines. Cast pistons without ring carriers only continue to be used in engines with low specific power or limited service life.

Forged pistons were used in many vehicle diesel engines because their increased strength set them apart from cast pistons. However, as power increased, the first groove's wear performance revealed limits on the use of hypereutectic alloys without a reinforced ring groove. The purely mechanical anchoring between the ring carrier and piston material in forged pistons has never been satisfactory.

Unlike pistons with oil only sprayed onto the piston underside locally, pistons with cooling galleries with a forced oil flow generally allow lowering the temperature level. The cooling channel is filled only partially to achieve the desired "shaker" effect with good heat transfer [8-70].

Aluminum cooling gallery pistons are now used for highly loaded turbocharged diesel engines with intercooling and brake mean effective pressures above 20 bar and maximum cylinder pressures of up to more than 200 bar.

Higher cylinder pressures and combustion bowls with modified geometries and exposed edges in the piston crown further reduce the soot content in exhaust gas and increase thermal efficiency. This increases the thermal load on pistons, especially in the region of the combustion chamber.

Articulated pistons (Ferrotherm® pistons, Fig. 8-63) are designed for extremely high mechanical and thermal stresses. They consist of two parts, a steel piston crown and an aluminum piston skirt flexibly interconnected by the piston pin. Together with the piston rings, the piston crown assumes the function of forming a seal against hot combustion gases and transferring the gas force to the crankshaft assembly. The piston skirt absorbs the side forces induced by the crankshaft assembly and the resultant guideway forces. Another variant is the one-piece steel piston (*Monotherm®* pistons, Fig. 8-64), which is finish machined from an unmachined forged steel part. Its skirt is only connected to the pin bosses or, additionally, to the piston crown. Since the iron material has poorer thermal conductivity, the piston requires more intense cooling than an aluminum piston to prevent oil aging. Spray jets feed

Fig. 8-61 Cooling gallery piston with ring carrier (salt core piston)

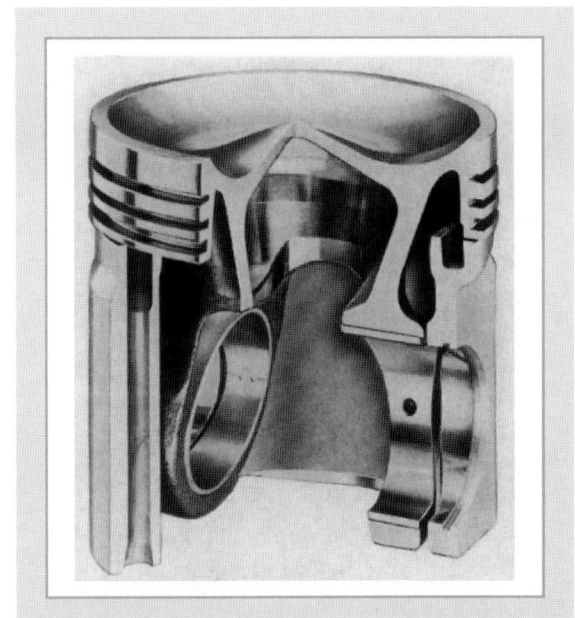

Fig. 8-63 Articulated piston (Ferrotherm piston) with steel crown and aluminum skirt

the cooling oil into the piston's cooling duct. Cover plates on the underside of the cooling chamber in the piston head can turn the open cooling duct into a closed cooling gallery much like a cooling gallery piston ("shaker" effect of the cooling oil).

Unlike an aluminum piston, the first ring groove of a steel piston crown normally needs no additional reinforcement [8-83].

Locomotive, Stationary and Marine Diesel Engine Pistons

The more stable operating conditions in power plants, marine propulsions and railcar and locomotive engines formerly allowed the utilization of *aluminum full skirt pistons* in many cases.

The crown and ring zone in such pistons is constructed sturdily and furnishes sufficient cross sections for heat flux. Either such pistons with ring carriers and cooling coils or cooling galleries are cast or the forged piston body and the cast ring band incorporating the ring carrier and cooling channel are joined by electron beam welding (Fig. 8-65).

The limit on the usability of lightweight alloys proved to be in medium-speed engines powered with heavy fuel. They no longer sufficiently resisted mechanical abrasion in the piston top land caused by solid combustion residues.

Nodular cast iron's better properties, especially when fuels are highly sulfurous, led to one-piece lightweight iron pistons, which, however, are very demanding to cast (*monoblock*, Fig. 8-66).

Composite pistons on the other hand unite the properties of various materials in one piston. Three generations of composite pistons are in use today. All share a piston design in which the piston crown with the ring zone forms the one main element of the piston and the piston skirt with the piston pin bosses forms the other. Both parts are interconnected by appropriate elements.

The classic composite piston design with a forged axisymmetric steel piston crown with outer and inner cooling chambers and a forged aluminum skirt (Fig. 8-67) has been complemented by a design with a bore-cooled piston crown (Fig. 8-68). Despite their thin walls, the feature that sets steel

Fig. 8-64 One piece forged steel piston (Monotherm piston)

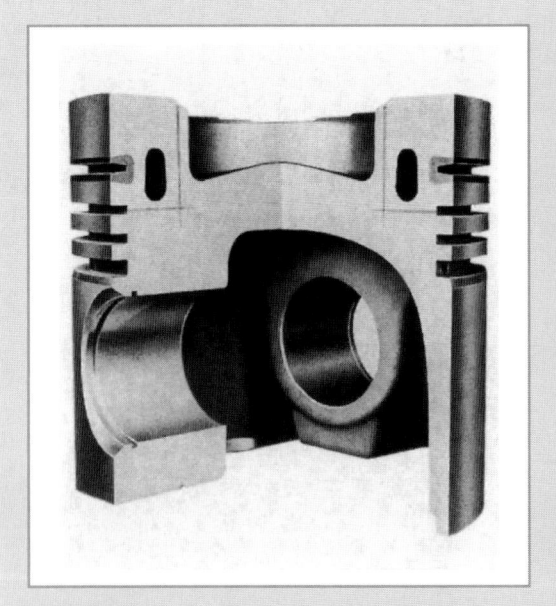

Fig. 8-65 Electron beam welded piston with cooling channel and ring carrier

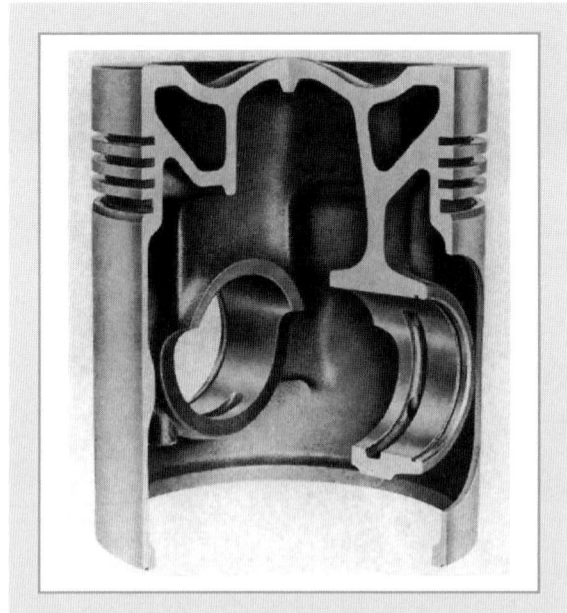

Fig. 8-66 One piece nodular cast iron piston (monoblock)

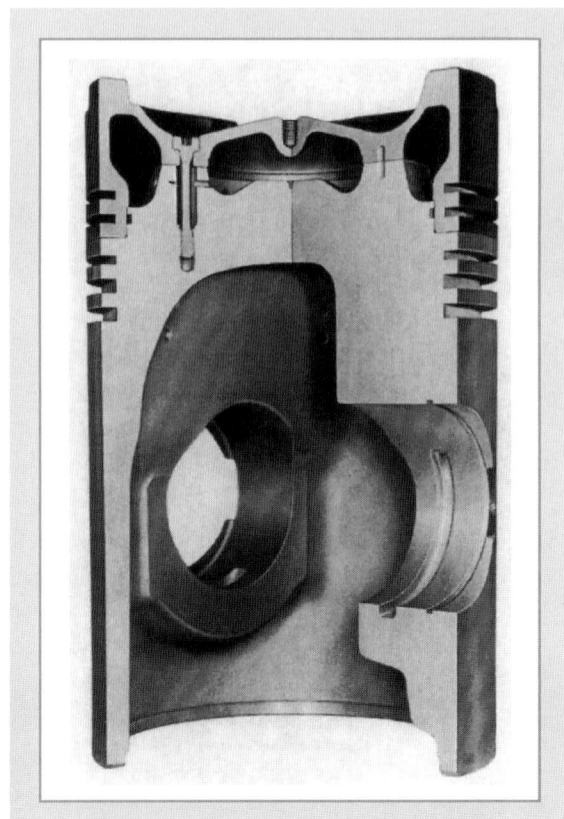

Fig. 8-67 Conventionally cooled composite piston with aluminum skirt

Fig. 8-68 Bore cooled steel crown of a composite piston

piston crowns with radially aligned cooling oil bores apart from conventional steel crowns is their increased bolting stiffness while the effective heat dissipation remains unchanged.

The piston ring grooves of steel crowns are often inductively hardened or chrome plated to reduce wear.

The advantage of aluminum skirts is the option of designing them with thicker and stiffer walls without a weight penalty than would be possible with other materials.

Low piston cold clearance and thus low secondary piston motion, high seizure resistance and low corrosion are the foremost considerations for composite pistons with nodular cast iron instead of forged aluminum skirts. A classic forged, axisymmetric steel piston crown with outer and inner cooling galleries is employed (Fig. 8-69).

Composite pistons intended for use in the cylinder pressure range exceeding 200 bar are designed with nodular cast iron or forged steel skirts and steel crowns. Low deformations, high structural strength and small cold clearances are characteristic properties of this design [8-84].

The use of materials with higher density increases the total mass over an aluminum alloy piston by approximately 20–50%.

Pistons for two-stroke diesel engines (crosshead engines) usually consist of a cup-shaped, axisymmetric steel part (e.g. made of 34CrMo4) bolted to the piston rod directly or connected by an insert or adapter (Fig. 8-70). The application of bore cooling (a principle developed by Sulzer) leads to simple, rugged designs with high reliability with regard to thermal and mechanical loads. The mechanical stresses and the unavoidable deformations of the piston crown can be absorbed without appreciable increases in stress [8-85, 8-86].

Figure 8-71 presents fields of application for large bore piston designs in medium-speed four-stroke diesel engines.

Fig. 8-69 Conventionally cooled composite piston with thin-walled nodular cast iron skirt

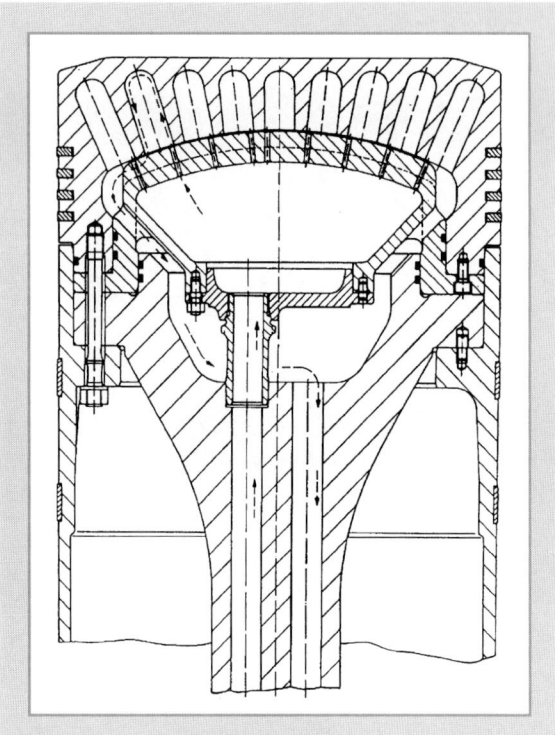

Fig. 8-70 Bore cooled piston for a two-stroke crosshead diesel engine

8.6.3.4 Materials

Piston Materials

The use of dedicated alloys not used in other components for pistons over the course of combustion engine development is a manifestation of the unusual demands that have to be met for this component. In fact, common piston materials constitute a compromise between a whole series of in part contradictory demands.

Aluminum silicon alloys (predominantly eutectic) are overwhelmingly employed as materials to manufacture pistons (Fig. 8-72). High thermal conductivity, low density, good castability and workability as well as good machinability and high high temperature strength are only a few of these lightweight alloys' properties (Table 8-6).

The noticeable decline in the material properties including hardness starting at 150–200 °C is a drawback. Modifications of the alloy composition, especially a higher copper content, have significantly increased AlSi alloys' fatigue strength in the range of temperatures above 250 °C but at the expense of increased casting complexity. AlSi alloys are utilized for cast and forged components (Table 8-7).

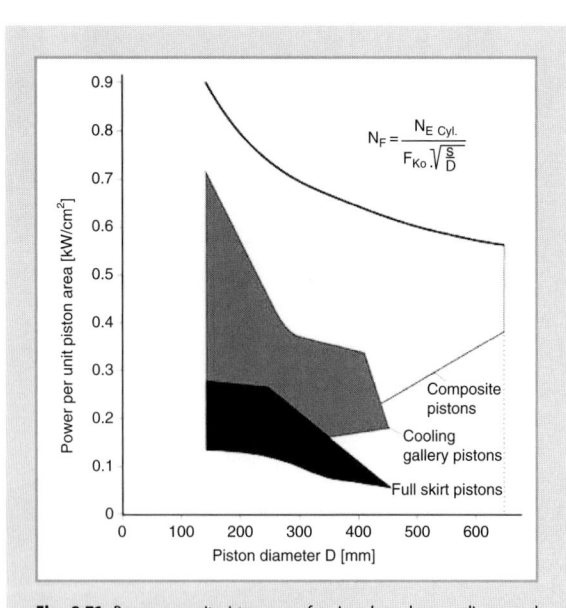

$$N_F = \frac{N_{E\,Cyl.}}{F_{Ko.} \sqrt{\frac{s}{D}}}$$

Fig. 8-71 Power per unit piston area of various large bore medium-speed four-stroke diesel engines

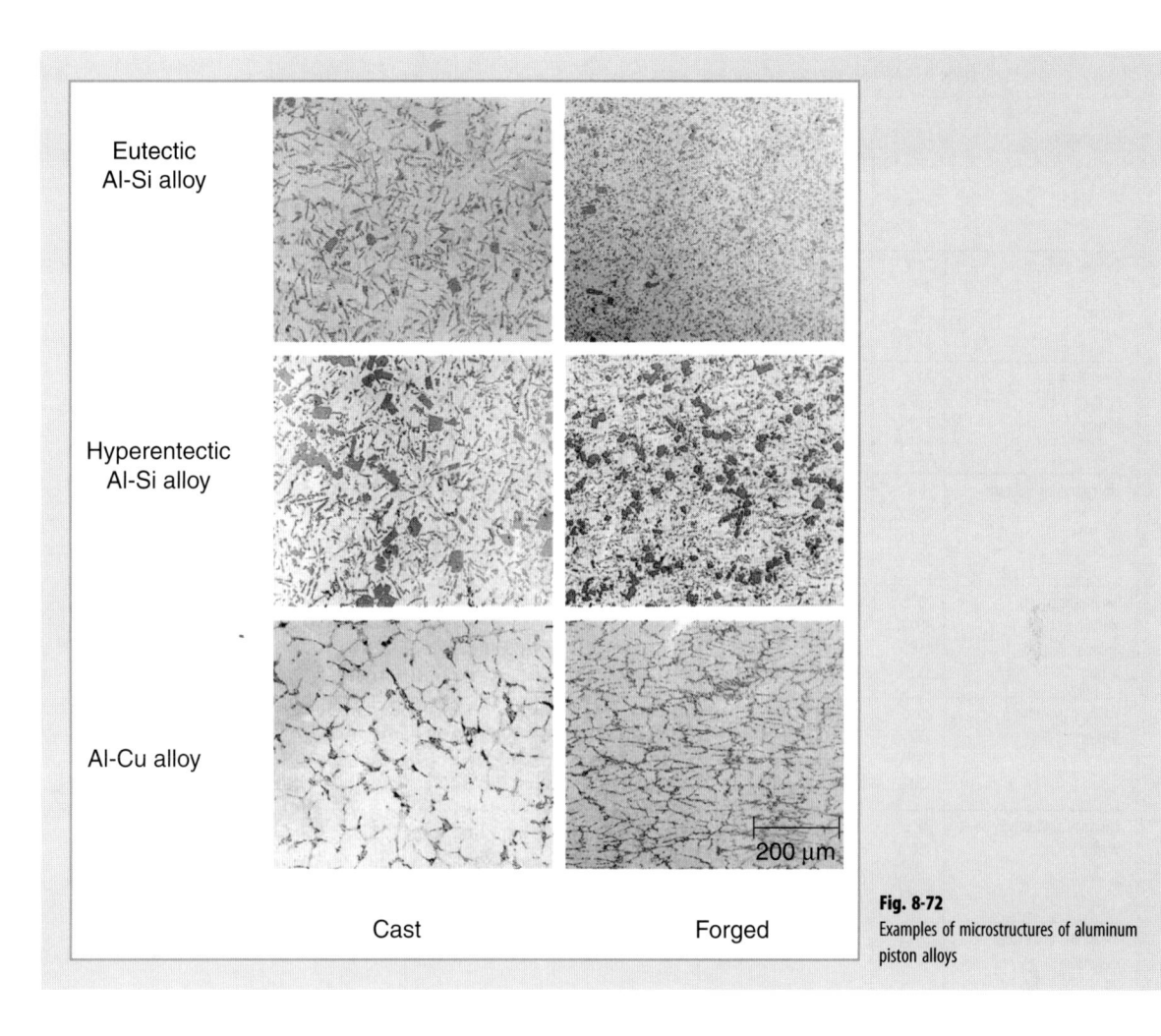

Fig. 8-72 Examples of microstructures of aluminum piston alloys

Table 8-6 Chemical compositions of aluminum piston alloys

Alloying elements %	AlSi alloys			
	Eutectic		Hypereutectic	
	AlSi 12 CuMgNi	AlSi 12 Cu4Ni2Mg	AlSi 18 CuMgNi	AlSi 25 CuMgNi
Si	11...13	11...13	17...19	23...26
Cu	0.8...1.5	3...5	0.8...1.5	0.8...1.5
Mg	0.8...1.3	0.5...1.2	0.8...1.3	0.8...1.3
Ni	0.8...1.3	1...3	0.8...1.3	0.8...1.3
Fe	≤ 0.7	≤ 0.7	≤ 0.7	≤ 0.7
Mn	≤ 0.3	≤ 0.3	≤ 0.2	≤ 0.2
Ti	≤ 0.2	≤ 0.2	≤ 0.2	≤ 0.2
Zn	≤ 0.3	≤ 0.3	≤ 0.3	≤ 0.2
Cr	–	–	–	≤ 0.6
Al	Residual	Residual	Residual	Residual

Table 8-7 Material properties of aluminum piston alloys (strength values determined from bars produced separately)

Parameter		Eutectic			Hypereutectic	
		AlSi 12 CuNiMg permanent mold casting	AlSi 12 CuNiMg forged	AlSi 12 Cu4Ni2Mg permanent mold casting	AlSi 18 CuNiMg permanent mold casting	AlSi 25 CuNiMg forged
Tensile strength R_m N/mm^2	20°C 50°C 250°C 350°C	200...250 180...200 90...110 35...55	300...370 250...300 80...140 50...100	200...280 180...240 100...120 45...65	180...220 170...210 100...140 60...80	230...300 210...260 100...160 60...80
Yield limit R_p 0,2 N/mm^2	20°C 150°C 250°C 350°C	190...230 170...200 70...100 20...30	280...340 220...280 60...120 30...70	190...260 170...220 80...110 35...60	170...210 150...190 100...140 20...40	220...260 200...250 80...120 30...40
Elongation at fracture A %	20°C 150°C 250°C 350°C	0.1...1.5 1.0...1.5 2...4 9...15	1...3 2.5...4.5 10...20 30...35	<1 <1 1.5...2 7...9	0.2...1.0 0.3...1.2 1.0...2.2 5...7	0.5...1.5 1...2 3...5 10...15
Fatigue strength (rotating bending fatigue) σ_{bw} N/mm^2	20°C 150°C 250°C 350°C	90...110 75...85 45...50 20...25	110...140 90...120 45...55 30...40	100...110 80...90 50...55 35...40	80...110 60...90 40...60 15...30	90...120 70...110 50...70 20...30
Young's modulus E N/mm^2	20°C 150°C 250°C 350°C	80,000 77,000 72,000 65,000	80,000 77,000 72,000 69,000	84,000 79,000 75,000 70,000	83,000 79,000 75,000 70,500	84,000 79,000 76,000 70,000
Thermal conductivity λ W/(mK)	20°C 150°C 250°C 350°C	155 156 159 164	158 162 166 168	125 130 135 140	143 147 150 156	157 160 163 –
Mean linear thermal expansion 20...200°C (1/K)x10^{-6}		20.6	20.6	20.0	19.9	20.3
Density ρ (g/cm^3)		2.68	2.68	2.77	2.68	2.68
Relative wear		1	1	~0.9	~0.8	~0.8
Brinell hardness HB 2.5/62.5			90...125			

These alloys are no longer sufficiently wear resistant when larger amounts of dust form or operating temperatures are high. Thus, ring carriers become necessary as mentioned in Sect. 8.6.3.2.

Nodular cast iron (GGG 70) is used for the skirts of composite pistons or for entire pistons (monoblock) for marine diesel and stationary engines, especially when they are operated with heavy fuel (Table 8-8). *Steels* (42CrMo4) are used for the crown of composite pistons and for the crowns of Ferrotherm pistons for commercial vehicle engines and for Monotherm pistons. Their higher thermal stability, improved wear performance, greater stiffness and altogether higher strength are advantageous but their mass is usually greater and their production more complex (Table 8-9).

Surface Coating

Different surface coatings are applied to pistons and can be divided into two groups according to their function. The first group includes coatings that particularly *protect* against thermal overstress. The coatings in the second group are intended to *improve running properties*.

Protective Coatings

Hard anodizing produces an aluminum oxide layer distinguished by its intimate bond with the base material. These layers ensure a piston crown is protected against the thermal and mechanical attack of hot combustion gases and increase

Table 8-8 Cast iron materials, recommended values

	GGG 70 Nodular cast iron	Austenitic cast iron with lamellar graphite	Austenitic cast iron with nodular graphite
Alloying elements %			
C	3.5...4.1	2.4...2.8	2.4...2.8
Si	2.0...2.4	1.8...2.4	2.9...3.1
Mn	0.3...0.5	1.0...1.4	0.6...0.8
Ni	0.6...0.8	13.5...17.0	19.5...20.5
Cr	–	1.0...1.6	0.9...1.1
Cu	≤ 0.1	5.0...7.0	–
Mo	–	–	–
Mg	0.04...0.06	–	0.03...0.05
Tensile strength R_m in N/mm^2			
20°C	≥ 700	≥ 190	≥ 380
100°C	640	170	–
200°C	600	160	–
300°C	590	160	–
400°C	530	150	–
Yield limit R_p 0.2 in N/mm^2			
20°C	≥ 420	150	≥ 210
100°C	390	150	–
200°C	360	140	–
300°C	350	140	–
400°C	340	130	–
Elongation at fracture A in %			
20 °C	≥ 2	≥ 2	≥ 8
Brinell hardness HB 30			
20°C	240...300	120...160	140...180
Fatigue strength (rotating bending fatigue) σ_{bw} in N/mm^2 20°C	≥ 250	≥ 80	–
Young's modulus E in N/mm^2			
20°C	177,000	100,000	120,000
200°C	171,000	–	–
Thermal conductivity λ in W/(mK)			
20°C	27	32	13
Mean linear thermal expansion 20...200°C (1/K)x10^{-6}	12	18	18
Density ρ in (g/cm^3)	7.2	7.45	7.4
Special properties and use	Highly stressed pistons, piston crowns and skirts for composite pistons	High thermal expansion; for ring carriers	High thermal expansion and strength; for ring carriers

resistance against thermally and mechanically induced cracking in the bowl rim and the piston crown.

Since hard oxide layers do not have good dry-running properties, a piston's skirt and ring zone are masked during the coating process.

Skirt Coating

Above all, the mating of piston and cylinder materials and their surface roughness determine a piston skirt's operating performance. Skirt roughness strongly influences the formation and adhesion of a lubricating film capable of supporting a load even when the oil supply is scant. In addition, thin soft metal or graphite coatings can guarantee good sliding properties, at least temporarily, even in the case of boundary lubrication.

Tin plating a piston produces good operating properties. The coating process is based on the principle of ion exchange. Aluminum pistons are submerged in solutions of tin salts. Since it is nobler than aluminum in the electrochemical series, tin is deposited on a piston's surface. At the same time, aluminum is dissolved during the process until a closed surface of tin forms. The 1–2 µm thick metal coating produced is

Table 8-9 Steel materials for forged piston parts, recommended values

	AFP steel 38 Mn VS6	42 CrMo 4 V
Alloying elements %		
C	0.34...0.41	0.38...0.45
Si	0.15...0.80	0.15...0.40
Mn	1.2...1.6	0.60...0.90
Cr	≤ 0.3	0.90...1.20
Mo	≤ 0.08	0.15...0.30
V	0.08...0.2	–
(P, S)	(≤ 0.025, 0.02...0.06)	(≤ 0.02)
Tensile strength R_m N/mm²;		
20°C	≥ 910	≥ 920
100°C	–	–
200°C	–	–
300°C	840	850...930
400°C	–	–
450°C	610	630...690
Yield limit R_p N/mm²		
20°C	≥ 610	≥ 740
100°C	–	–
200°C	–	–
300°C	540	680...750
400°C	–	–
450°C	450	520...580
Elongation at fracture A %		
20°C	≥ 14	12...15
200°C	–	–
300°C	11	10...13
400°C	–	14
450°C	15	15...16
Brinell hardness HB 30	240–310	265...330
Fatigue strength (rotating bending fatigue) σ_{bw} in N/mm²		
20°C	≥ 370	≥ 370
200°C	–	–
300°C	320	340–400
400°C	–	–
450°C	290	280–400
Young's modulus E N/mm²		
20°C	210,000	210,000
200°C	189,000	193,000
400°C	–	–
450°C	176,000	180,000
Density ρ (g/cm3)	7.8	7.8
Thermal conductivity λ W/(mK)		
20°C	44	38
200°C	–	–
300°C	40	39
400°C	–	–
450°C	37	37
Mean linear thermal expansion (1/K)x10⁻⁶		
20...200°C	–	–
20...300°C	13.2	13.1
20...400°C	–	13.2
20...450°C	13.7	13.7
Special properties and use	BY steel for pistons, crowns and skirts of composite pistons	High temperature quenched and tempered steel for pistons, piston crowns and bolts of composite pistons

still applied to pistons for commercial vehicle and car engines to a limited extent because of its good dry-running properties.

Graphite facilitates lubricant adhesion and itself develops a lubricating effect should the lubricating oil film fail. Producing strongly adhesive, protective graphite-filled polymer coatings (solid lubricant) on a piston is important. To do so, the metallic surface is coated in alkaline baths with a metal phosphate layer approximately 3–5 µm thick. This is a good primer for synthetic resin graphite coating, which consists of a fine graphite powder mixed with a polyamide-imide (PAI) resin. Once it has been applied, the approximately 10–20 µm thick coating is cured at higher temperature (polymerization). Such coatings are used for larger pistons and pistons for gasoline and diesel car engines. Graphite protective coatings may be applied both to aluminum and iron pistons. Their "oil-friendly" surface has very good dry-running properties.

8.6.4 Piston Rings

8.6.4.1 General Description

Piston rings in internal combustion engines have the function of sealing off pistons, as moving parts of the combustion chamber, from the crankcase chamber as completely as possible. Further, they facilitate the dissipation of heat from pistons to the cylinder wall and regulate the oil supply, distributing and scraping oil from the liner surface. Therefore, the different rings are divided into compression rings and oil control rings according to their function.

The rings are given the form of an open ring expander (Fig. 8-73) to generate the required *pressure* against the cylinder wall. Once a section (gap) corresponding to the free gap has been cut out, double cam turning, i.e. simultaneously copy turning the inside and outside of the flanks on the cut blank, gives a ring the radial pressure distribution desired for its function in a cylinder. The ring forms a tight seal and applies the specified radial pressure to the cylinder wall.

Once installed, the gas pressure behind the compression ring intensifies the elastic force acting radially during engine operation. The gas pressure applied to the ring flank essentially generates the axial contact on the piston groove flank (Fig. 8-74 [8-87]).

8.6.4.2 Compression Rings

Compression rings take over the function of sealing and dissipating heat to the cylinder wall. In addition, they are involved in regulating the lubricating oil supply. Their sealing function primarily entails preventing combustion gases from passing from the combustion chamber into the crankcase. Increased blow-by endangers pistons and rings by overheating, disrupts the lubrication on the cylinder wall and adversely influences the lubricating oil in the crank chamber.

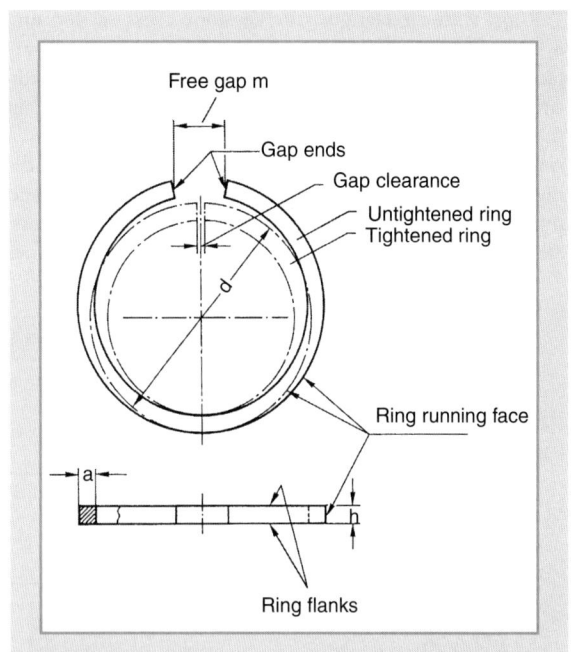

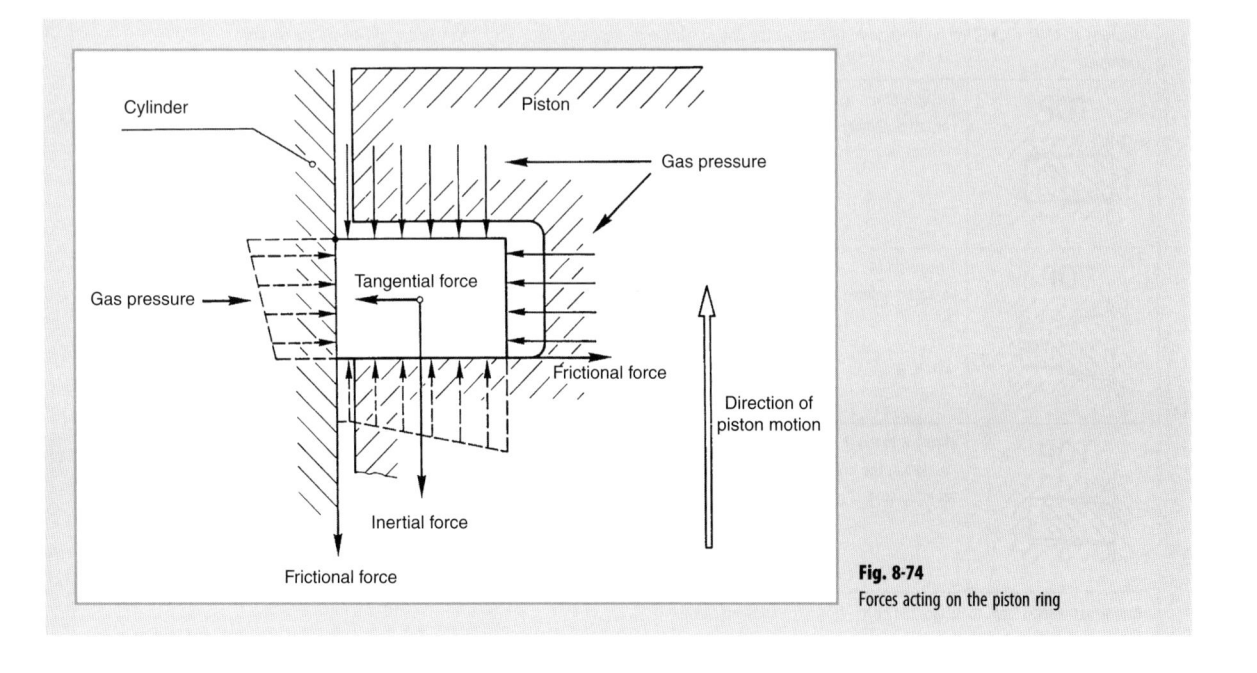

Fig. 8-73 Piston ring denominations. *a* (radial) wall thickness; *h* (axial) ring height; *d* nominal diameter

Fig. 8-74 Forces acting on the piston ring

Rings with a rectangular or keystone cross section and a convex, asymmetrically convex or conical contact surface have proven to be the most efficient compression rings. When the contact surface is designed conically in the range of a few angular minutes (*tapered compression rings*), then the linear area of contact initially produces high unit loading between the ring contact surface and the cylinder wall, which shortens the run-in process. Structural asymmetry in the ring cross section (inner angle and inner bevel) also achieves an identical effect. Then, the cross section altered on one side twists the ring into a dish shape when it is installed. It forms or intensifies a conical surface of contact to the cylinder liner ("twist" effect). Under gas pressure, the ring is pressed flat against the ring groove. This produces additional dynamic stress in operation. The "twist" angle is likewise in the range of only a few angular minutes (Table 8-10 [8-88]).

Table 8-10 Main types of compression rings

	Rectangular ring with *cylindrical* contact surface (R ring). For normal operating conditions
	Rectangular ring with *convex* contact surface (R ring B). Preferred for coated rings.
TOP	Rectangular ring with *conical* contact surface = taper face ring (M ring). Short run-in phase with scraping effect.
TOP	Half keystone ring (ET ring). Against coking and sticking in the groove.
	Full keystone ring (r ring) with cylindrical, conical or convex contact surface. Against coking and sticking in the groove, especially for diesel engines.
Applying an inner edge (IE) or an inner angle (IA) causes the ring to twist in its tightened state. Fits in the cylinder and groove are obtained. Examples:	
TOP	R ring with IE For faster running-in by fitting in the lower edge of the contact surface.
TOP	T ring with IA For faster running-in by fitting in the lower edge of the contact surface.
TOP	M ring with lower IE (reversed torsion) Positioning the ring on the cylinder underside and on the upper inner groove flank causes negative twisting. Preferably installed in the second groove.
Ring flanks labeled "top" must face the piston crown (combustion chamber).	

The first groove's piston rings are usually convex on the contact surface to achieve rapid run-in. When they are asymmetrically convex, the ring floats far more during the upward stroke and allows a larger quantity of oil in the top dead center, the upper part of the cylinder liner being endangered by liner wear from the mixed lubrication conditions.

8.6.4.3 Oil Scraper Rings

Oil scraper or oil control rings regulate and limit the oil supply. They scrape excess lubricating oil off the cylinder wall and return it to the crankcase. Insufficient scraping can cause oil coking or overflow into the combustion chamber and thus increased oil consumption. Oil scraper rings usually have two separated contact surfaces. Spiral expanders define the contact pressure.

Taper faced rings or taper faced Napier rings occupy a hybrid position. They function both as compression and oil scraper rings (Table 8-11).

Even though a compression ring's primary function is to form a seal against combustion gases and an oil control ring's primary function is to distribute and scrape oil, their materials and the surface coating and fine surface structure play an important role in the tribosystem of the piston, piston ring and cylinder. Piston rings usually consist of high-grade cast iron with lamellar or nodular graphite. In addition, various special materials designed for strength and low wear and various types of steel are used for compression rings and oil control rings as well as for spiral expanders that increase tangential stress.

To prolong the service life of piston rings, their contact surfaces are chrome plated, filled or completely coated with molybdenum spray coated with metallic, metal-ceramic and ceramic mixed coats in flame spraying, plasma spraying or

Table 8-11 Main types of oil scraper rings

Ring	Description
	Napier ring with *cylindrical* contact surface (N ring). The simplest oil scraper ring.
	Napier ring with *conical* contact surface (NM ring). The conical shape shortens the run-in phase and reinforces the scraping effect.
	Oil control ring with spiral expander (SSF ring). For normal operating and run-in conditions.
	Double-beveled spiral expander ring (DSF ring), preferably with chrome plated bearing ridges. Greater specific contact pressure causes shorter run-in phase and increased oil scraping effect.
	Spiral expander top beveled oil control ring (GSF ring). Accelerate run-in process and strong oil scraping effect.
	Slotted oil control ring (S ring). Profiles include: Beveled edge (D ring) and Double beveled (G ring) Little used for lightly loaded engines.

Ring flanks labeled "top" must face the piston crown (combustion chamber).

high velocity oxygen fuel (HVOF) processes or coated in a PVD process.

The surface of the entire ring is treated to improve run-in or reduce wear on the flanks and the contact surface. Methods such as phosphating, nitriding, ferro-oxidation, copper plating, tin plating and others are used.

8.6.5 Piston Pins

The piston pin establishes the link that transfers force between the piston and connecting rod. The piston's oscillating motion and the superimposition of gas and inertial forces subject it to high loads from varying directions. The *lubricating conditions are poor* (Table 8-12) because of the slow rotary motion in the bearing surfaces between the piston, pin and connecting rod.

The tubular piston pin has established itself as the standard design for most applications (Fig. 8-75 (a)) and optimally meets demands for simple technology and economical manufacturing.

Circlip rings secure piston pins against laterally drifting out of the center bore and running up against the cylinder wall. Spring steel expanding snap rings inserted in grooves in the outer edge of the center bores are employed for this almost exclusively. Wound flat wire rings or punched expanding snap rings are mainly used for commercial vehicle diesel engines and large engines. A pin "floating" in the piston and connecting rod is preferred for diesel engines. Profiled piston pins in which grinding has slightly recessed the piston pin boss area of the outer pin diameter in the region of the faces of the inner bore edges are partially used to adapt the pin and boss more gently (Fig. 8-75 (b)).

The cooling oil is frequently conducted from the connecting rod to the piston through the pin, especially in large pistons. Hence, pins are provided with longitudinal and transverse bores and closed bore ends (Fig. 8-75 (c) and (d)).

Table 8-12 Typical piston pin layout, recommended values

Application		Ratio of outer pin diameter to piston diameter	Ratio of inner pin diameter to outer pin diameter
Diesel engines	Car engines	0.30...0.42	0.48...0.52
	Commercial vehicle engines	0.36...0.45	0.40...0.52
	Medium-sized engines (160...240 mm piston diameter)	0.36...0.45	0.45
	Large bore engines (>240 mm piston diameter)	0.39...0.48	0.30...0.40

A piston pin's function generates the following requirements:
– low mass,
– maximum stiffness,
– adequate strength and ductility and
– high quality of finish and accuracy of shape.

Figure 8-76 is a schematic of the distribution of stress in a piston pin loaded by the gas force. Pins are case hardened or nitrided to increase the wear resistance in the contact surface and the strength in the surface of the bore.

According to DIN 73126, case hardening steels (17Cr3, 16MnCr5) or, for more highly loaded engines, nitriding steels (31CrMoV9) are used as pin materials.

Once they have been heat treated, these steels have the required hard, wear resistant surface and a ductile core. Piston pins for large bore engines are made of ESR materials. Electroslag remelting (ESR) attains a higher degree of purity.

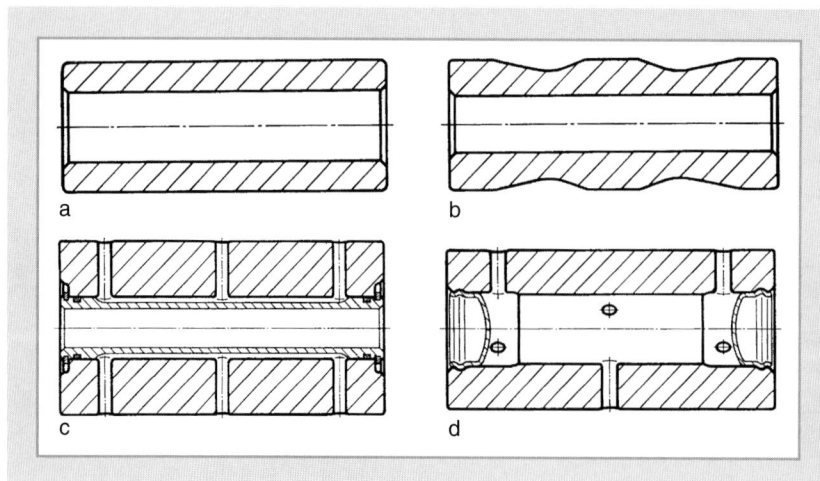

Fig. 8-75
Piston pin designs. **a** Standard piston pin; **b** Profiled piston pin (schematic); **c** Pin with oil duct for piston with oil cooling; **d** Pin with closing covers for piston with oil cooling

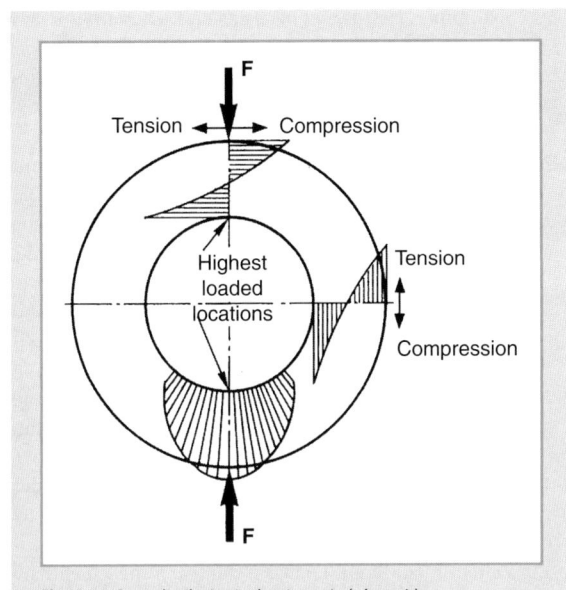

Fig. 8-76 Stress distribution in the piston pin (schematic)

Fig. 8-77 Piston for highly loaded diesel engine with pin bore bushing

Piston pins' bending stress, ovalization stress and overall stress may be approximately simulated according to [8-89]. Extended approaches to simulation have been developed to calculate stress more precisely [8-90]. The three-dimensional finite element method can simulate the entire assembly of the piston, piston pin and small connecting rod eye and their stresses and deformations.

8.6.6 Trends in Development

Objectives of future diesel engine development will be:
- increasing volumetric power output,
- lowering specific fuel consumption,
- increasing service life and
- especially reducing exhaust emissions (NO_X, soot/particulates, CO, HC) as required by restrictions in emission control legislation.

Other aspects will be noise, mass and cost reduction.

Engine-related goals of development affect piston development in many ways. The pistons' mechanical load (maximum cylinder pressure) is increasing just like their thermal load. Hence, AlSi alloy pistons require a more sturdily shaped interior geometry (wall thickness and dome), keystone or stepped machined eye supports and enlarged outer pin diameters. Given the increase in the stresses in the piston crown – depending on the bowl geometry – very carefully considered structural modifications are being made in the pin bore to increase pin bore strength. Shrunk bushings (e.g. brass or aluminum bronze) are increasingly being used (Fig. 8-77). Apart from designing the geometry of the bowl with undercuts in direct injection engines, positioning the first ring groove higher satisfies the demand for minimum dead volumes in the region of the piston top land to reduce emissions. This increases the thermal load of the first ring and the ring groove and the mechanical stresses at the bowl rim in the pin axis. This may necessitate local reinforcement through fiber inserts. The percentage of fibers by volume selected is usually 10–20%. At the temperatures occurring at the bowl rim, this can increase the fatigue strength (rotating bending fatigue) by approximately 30%. Steel pistons might also be an option for high speed diesel engines if loads continue to increase.

Literature

8-1 Lang, O.R.: Triebwerke schnellaufender Verbrennungsmotoren. Konstruktionsbücher Vol. 22 Berlin/Heidelberg/New York: Springer (1966)

8-2 Urlaub, A.: Verbrennungsmotoren. Vol. 3: Konstruktion. Berlin/Heidelberg/New York: Springer (1989)

8-3 Gross, W.: Beitrag zur Berechnung der Kräfte in den Grundlagern bei mehrfach gelagerten Kurbelwellen. Diss. TU Munich (1966)

8-4 Schnurbein, E.v.: Beitrag zur Berechnung der Kräfte und Verlagerungen in den Gleitlagern statisch unbestimmt gelagerter Kurbelwellen. Diss. TU Munich (1969)

8-5 Maass, H.: Calculation of Crankshaft Plain Bearings. CIMAC Congress Stockholm. Paper A-23 (1971)

8-6 Maass, H.; Klier, H.: Kräfte, Momente und deren Ausgleich in der Verbrennungskraftmaschine. Vienna/New York: Springer (1981)

8-7 Salm, J.; Zech, H.; Czerny, L.: Strukturdynamik der Kurbelwelle eines schnellaufenden Dieselmotors: Dynamische Beanspruchung und Vergleich mit Messungen. MTU FOCUS 2 (1993) p. 5

8-8 Resch, T. et al.: Verwendung von Mehrkörperdynamik zur Kurbelwellenauslegung in der Konzeptphase. MTZ 65 (2004) 11, p. 896

8-9 van Basshuysen, R.; Schäfer, S. (Ed.): Handbuch Verbrennungsmotor: Grundlagen, Komponenten, Systeme, Perspektiven. ATZ-MTZ-Fachbuch. Wiesbaden: Vieweg (2002)

8-10 Zima, S.: Kurbeltriebe: Konstruktion, Berechnung und Erprobung von den Anfängen bis heute. ATZ-MTZ-Fachbuch. Wiesbaden: Vieweg (1998)

8-11 Krüger, H.: Sechszylindermotoren mit kleinem V-Winkel. MTZ 51 (1990) 10, p. 410

8-12 Metzner, F.T. et al.: Die neuen W-Motoren von Volkswagen mit 8 und 12 Zylindern. MTZ 62 (2001) 4, p. 280

8-13 Kochanowski, H.A.: Performance and Noise Emission of a New Single-Cylinder Diesel Engine – with and without Encapsulation. 2nd Conf. of Small Combustion Engines, Inst. Mech. Eng., England 4-5 April (1989)

8-14 Wiedemann, R. et al.: Das Öko-Polo-Antriebskonzept. MTZ 52 (1991) 2, p. 60

8-15 Ebel, B.; Kirsch, U.; Metzner, F.T.: Der neue Fünfzylindermotor von Volkswagen – Part 1: Konstruktion und Motormechanik. MTZ 59 (1998) 1

8-16 Piech, F.: 3 Liter / 100 km im Jahr 2000. ATZ 94 (1992) 1

8-17 Thiemann, W.; Finkbeiner, H.; Brüggemann, H.: Der neue Common Rail Dieselmotor mit Direkteinspritzung für den Smart – Part 1. MTZ 60 (1999) 11

8-18 Digeser, S. et al.: Der neue Dreizylinder-Dieselmotor von Mercedes-Benz für Smart und Mitsubishi. MTZ 66 (2005) 1, p. 6

8-19 Suzuki, M.; Tsuzuki, N.; Teramachi, Y.: Der neue Toyota 4-Zylinder Diesel Direkteinspritzmotor – das Toyota D-4D Clean Power Konzept. 26th International Vienna Engine Symposium 28–29 April (2005), p. 75

8-20 Fünfzylinder-Dieselmotor im Mercedes-Benz 240 D. MTZ 35 (1974) p. 338

8-21 Hauk, F.; Dommes, W.: Der erste serienmässige Reihen-Fünfzylinder-Ottomotor für Personenwagen: Eine Entwicklung der AUDI NSU. MTZ 78 (1976) 10

8-22 Hadler, J. et al.: Der neue 5-Zylinder 2,5 l-TDI-Pumpedüse-Dieselmotor von VOLKSWAGEN. Aachen Colloquium Automobile and Engine Technology, Aachen 7–9 October (2002), p. 65

8-23 Bach, M. et al.: Der neue V6-TDI-Motor von Audi mit Vierventiltechnik – Part 1: Konstruktion. MTZ 58 (1997) 7/8

8-24 Doll, G. et al.: Der Motor OM 642 – Ein kompaktes, leichtes und universelles Hochleistungsaggregat von Mercedes-Benz. 26th International Vienna Engine Symposium 28–29 April (2005), p. 195

8-25 Schommer, J. et al.: Der neue 4,0-l-V8-Dieselmotor von Mercedes-Benz. MTZ 67 (2006) 1, p. 8

8-26 Frost, F. et al.: Zur Auslegung der Kurbelwelle des SKL-Dieselmotors 8VDS24/24AL. MTZ 51 (1990) 9, p. 354

8-27 DEUTZ-MWM Motoren-Baureihe 604B – jetzt mit Zylinderleistung bis 140 kW. MTZ 52/11, p. 545

8-28 Rudert, W.; Wolters, G.M.: Baureihe 595 – Die neue Motorengeneration von MTU. MTZ 52 (1991) 6, pp. 52 and 274 (1991) 11, p. 38

8-29 Krause, R.: Gegossene Kurbelwellen – konstruktive und werkstoffliche Möglichkeiten für den Einsatz in modernen Pkw-Motoren. MTZ 38 (1977) 1, p. 16

8-30 Albrecht, K.-H.; Emanuel, H.; Junk, H.: Optimierung von Kurbelwellen aus Gusseisen mit Kugelgraphit. MTZ 47 (1986) 7/8

8-31 Finkelnburg, H.H.: Spannungszustände in der festgewalzten Oberfläche von Kurbelwellen. MTZ 37 (1976) 9

8-32 Velten, K.H.; Rauh, L.: Das induktive Randschichthärten in der betrieblichen Anwendung im Nutzfahrzeugmotorenbau. Tagung: Induktives Randschichthärten. Darmstadt: Arbeitsgemeinschaft Wärmebehandlung und Werkstofftechnik (AWT) (1988)

8-33 Conradt, G.: Randschichthärten von Kurbelwellen: Neue Lösungen und offene Fragen. MTZ 64 (2003) 9, p. 746

8-34 Metz, N.H.: Unterpulver-Engstspaltschweiss-Verfahren für grosse Kurbelwellen. MTZ 48 (1987) 4, p. 147

8-35 Maass, H.: Gesichtspunkte zur Berechnung von Kurbelwellen. MTZ 30 (1969) 4

8-36 Fiedler, A.G.; Gschweitl, E.: Neues Berechnungsverfahren zeigt vorhandene Reserven bei der Kurbelwellenfestigkeit. MTZ 59 (1998) 3, p. 166

8-37 Rasser, W.; Resch, T.; Priebsch, H.H.: Berechnung der gekoppelten Axial-, Biege- und Torsionsschwingungen von Kurbelwellen und der auftretenden Spannungen. MTZ 61 (2000) 10, p. 694

8-38 Eberhard, A.: Einfluss der Formgebung auf die Spannungsverteilung von Kurbelkröpfungen, insbesondere von solchen mit Längsbohrungen. Forschungsvereinigung Verbrennungskraftmaschinen e.V., FVV-Forschungsbericht 130 (1972)

8-39 Eberhard, A.: Einfluss der Formgebung auf die Spannungsverteilung von Kurbelkröpfungen mit Längsbohrungen. MTZ 34 (1973) 7/9

8-40 IACS (International Association of Classifications Societies): M53 Calculation of Crankshafts of I.C. Engines. IACS Requirements (1986)

8-41 Wellinger, K.; Dietmann, H.: Festigkeitsberechnung: Grundlagen und technische Anwendungen. Stuttgart: Alfred Körner (1969)

8-42 Nefischer, P. et al.: Verkürzter Entwicklungsablauf beim neuen Achtzylinder-Dieselmotor von BMW. MTZ 60 (1999) 10, p. 664

8-43 Forschungsvereinigung Verbrennungskraftmaschinen e.V.: Kurbelwellen III – Studie über den Einfluss der Baugrösse auf die Dauerfestigkeit von Kurbelwellen. FVV-Forschungsbericht 199 (1976)

8-44 Zenner, H.; Donath, G.: Dauerfestigkeit von Kurbelwellen. MTZ 38 (1977) 2, p. 75

8-45 Forschungsvereinigung Verbrennungskraftmaschinen e.V.: Kurbelwellen IV – Dauerfestigkeit grosser Kurbelwellen, Part 1 and 2. FVV-Forschungsbericht 362 (1985)

8-46 Petersen, C.: Gestaltungsfestigkeit von Bauteilen. VDI-Z. (1952) 14, p. 973

8-47 Lang, O.R.: Moderne Berechnungsverfahren bei der Auslegung von Dieselmotoren. Symposium Dieselmotorentechnik. Esslingen: Technische Akademie (1986)

8-48 Köhler, E.: Verbrennungsmotoren: Motormechanik, Berechnung und Auslegung des Hubkolbenmotors. 3rd Ed. Braunschweig/Wiesbaden: Vieweg (2002)

8-49 Ebbinghaus, W.; Müller, E.; Neyer, D.: Der neue 1,9-Liter-Dieselmotor von VW. MTZ 50 (1989) 12

8-50 Krüger, H.: Massenausgleich durch Pleuelgegenmassen. MTZ 53 (1992) 12

8-51 Arnold, O.; Dittmar, A.; Kiesel, A.: Die Entwicklung von Massenausgleichseinrichtungen für Pkw-Motoren. MTZ 64 (2003) 5, p. 2

8-52 Beitz, W.; Küttner, K.-H. (Ed.): Taschenbuch für den Maschinenbau. 14th Ed. Berlin/Heidelberg/New York: Springer (1981)

8-53 Haug, K.: Die Drehschwingungen in Kolbenmaschinen. Konstruktionsbücher Vol. 8-9. Berlin/Göttingen/Heidelberg: Springer (1952)

8-54 Krämer, E.: Maschinendynamik. Berlin/Heidelberg/New York/Tokyo: Springer (1984)

8-55 Nestrorides, E.J.: A Handbook on Torsional Vibration. Cambridge: British Internal Combustion Engine Research Association (B.I.C.E.R.A) (1958)

8-56 Hafner, K.E.; Maass, H.: Die Verbrennungskraftmaschine. Bd. 4: Torsionsschwingungen in der Verbrennungskraftmaschine. Wien: Springer (1985)

8-57 Nicola, A.; Sauer, B.: Experimentelle Ermittlung des Dynamikverhaltens torsionselastischer Antriebselemente. ATZ 108 (2006) 2

8-58 Reik, W.; Seebacher, R. Kooy, A.: Das Zweimassenschwungrad. 6th LuK-Kolloquium (1998), p. 69

8-59 Lang, O.R.; Steinhilper, W.: Gleitlager. Konstruktionsbücher, Bd. 31 Berlin/Heidelberg/New York: Springer (1978)

8-60 Sassenfeld, W.; Walther, A.: Gleitlagerberechnungen. VDI-Forsch.-Heft 441 (1954)

8-61 Butenschön, H.-J.: Das hydrodynamische, zylindrische Gleitlager endlicher Breite unter instationärer Belastung. Diss. TU Karlsruhe (1957)

8-62 Fränkel, A.: Berechnung von zylindrischen Gleitlagern. Diss. ETH Zurich (1944)

8-63 Holland, J.: Beitrag zur Erfassung der Schmierverhältnisse in Verbrennungskraftmaschinen. VDI-Forsch-Heft 475 (1959)

8-64 Hahn, H.W.: Das zylindrische Gleitlager endlicher Breite unter zeitlich veränderlicher Belastung. Diss. TH Karlsruhe (1957)

8-65 Schopf, E.: Ziel und Aussage der Zapfenverlagerungsbahnberechnungen von Gleitlagern. Tribologie und Schmierungstechnik 30 (1983)

8-66 Steeg, M.; Engel, U.; Roemer, E.: Hochleistungsfähige metallische Mehrschichtverbundwerkstoffe für Gleitlager. GLYCO-METALL-WERKE Wiesbaden: GLYCO-Ingenieur-Berichte (1991) 1

8-67 Engel, U.: Schäden an Gleitlager in Kolbenmaschinen. GLYCO-METALL-WERKE Wiesbaden: GLYCO-Ingenieur-Berichte (1987) 8

8-68 DIN 31661: Plain bearings; terms, characteristics and causes of changes and damages. (1983)

8-69 Lang, O.R.: Gleitlager-Ermüdung. Tribologie und Schmierungstechnik 37 (1990) pp. 82–87

8-70 Pflaum, W.; Mollenhauer, K.: Wärmeübergang in der Verbrennungskraftmaschine. Wien/New York : Springer (1977)

8-71 MAHLE GmbH: MAHLE-Kolbenkunde. Stuttgart-Bad Cannstatt (1984)

8-72 Kolbenschmidt AG: KS-Technisches Handbuch Part 1 and 2 Neckarsulm (1988)

8-73 ALCAN Aluminiumwerk Nürnberg GmbH: NÜRAL-Kolben-Handbuch. (1992)

8-74 MAHLE GmbH: MAHLE-Kleine Kolbenkunde. Stuttgart-Bad Cannstatt April (1992)

8-75 Röhrle, M.D.: Ermittlung von Spannungen und Deformationen am Kolben. MTZ 31 (1970) 10, pp. 414–422

8-76 Röhrle, M.D.: Thermische Ermüdung an Kolbenböden. MAHLE Kolloquium (1969)

8-77 Lipp, S.; Issler, W.: Dieselkolben für Pkw und Nkw – Stand der Technik und Entwicklungstendenzen. Symposium Dieselmotorentechnik. Esslingen: Technische Akademie 6 December (1991)

8-78 Rösch, F.: Beitrag zum Muldenrandrissproblem an Kolben aus Aluminium-Legierungen für hochbelastete Dieselmotoren. MTZ 37 (1976) 12, pp. 507–514

8-79 Müller-Schwelling, D.: Kurzfaserverstärkte Aluminiumkolben prüfen. Materialprüfung 33 (1991) 5, pp. 122–125

8-80 Seybold, T.; Dallef, J.: Bauteilfestigkeit von Kolben für hochbelastete Dieselmotoren. MAHLE-Kolloquium (1977)

8-81 Sander, W.; Keim, W.: Formgedrehte Bohrungen zur Bolzenlagerung hochbelasteter Kolben. MTZ 42 (1981) 10, pp. 409–412

8-82 Röhrle, M.D.: Thermische Beanspruchung von Kolben für Nutzfahrzeug-Dieselmotoren. MTZ 42 (1981) 3, pp. 85–88 and MTZ 42 (1982) 5, pp. 189–192

8-83 Röhrle, M.D.: Ferrotherm® und faserverstärkte Kolben für Nutzfahrzeugmotoren. MTZ 52 (1991) 7/8, pp. 369 f.

8-84 Röhrle, M.D.; Jacobi, D.: Gebaute Kolben für Schwerölbetrieb. MTZ 51 (1990) 9, pp. 366–373

8-85 Böhm, F.: Die M.A.N.-Zweitaktmotoren des Konstruktionsstandes B/BL. MTZ 39 (1978) 1, pp. 5–14

8-86 Aeberli, K.; Lustgarten, G.A.: Verbessertes Kolbenlaufverhalten bei langsam laufenden Sulzer-Dieselmotoren. Part 1: MTZ 50 (1989) 5, pp. 197–204; Part 2: MTZ 50 (1989) 12, pp. 576–580

8-87 Goetze AG: Goetze Kolbenring-Handbuch. Burscheid (1989)

8-88 Kolbenringreibung I. Frankfurt: FVV-Forschungsbericht 502 (1992)

8-89 Schlaefke, K.: Zur Berechnung von Kolbenbolzen. MTZ 1 (1940) 4 pp. 117–120

8-90 Maass, K.: Der Kolbenbolzen, ein einfaches Maschinenelement. VDI-Fortschritts-Berichte 41 (1976) 1

Further Literature

Affenzeller, J.; Gläser, H.: Lagerung und Schmierung von Verbrennungsmotoren. Vienna/New York: Springer (1996)

Schopf, E.: Untersuchung des Werkstoffeinflusses auf die Fressempfindlichkeit von Gleitlagern. Diss. Universität Karlsruhe (1980)

9 Engine Cooling

Klaus Mollenhauer and Jochen Eitel

9.1 Internal Engine Cooling

9.1.1 The Function of Engine Cooling

9.1.1.1 Heat Balance and Heat Transport

Depending on its size, principle of operation and combustion system, a diesel engine converts up to 30–50% of the fuel energy supplied into effective brake work. Apart from conversion losses during combustion, the remaining percentage is released into the environment as heat (Fig. 9-1), predominantly with the exhaust and by the cooling system. Only a relatively small percentage reaches the environment by convection and radiation through the surface of the engine. In addition to the component heat transferred to the coolant, the heat dissipated by a cooling system also includes the heat dissipated in the lubricating oil cooler and intercooler.

Utilizing the energy loss for heating purposes and the like (see Sect. 14) requires a detailed analysis of the enthalpy content of the individual kinds of heat as well as the engine's use and type. The external cooling system (Sect. 9.2) also has to be incorporated in the analysis. Internal engine cooling essentially covers the wall heat losses that occur when energy is converted in the combustion chamber (see Sects. 1.3 and 7.2) and reaches the coolant by heat transmission. Other engine components, e.g. injection nozzles, exhaust gas turbochargers and exhaust manifolds, are often directly cooled too.

Analyzed from the perspective of energy conversion alone, engine cooling appears to waste energy. This raises the question of whether an uncooled *adiabatic engine* might not represent a worthwhile goal of development. The belief that high temperature strength and heat insulating materials had been discovered in newly developed *ceramic materials* and an adiabatic engine, one of Rudolf Diesel's basic ideas, was one step closer was widespread in the early 1980s.

The rise of engine component temperatures to approximately 1,200 °C when cooling is discontinued was already pointed out in 1970. Even today, this remains an uncontrollable temperature level for reciprocating piston engines [9-1, 9-2] and is compounded by the decrease of the cylinder charge and thus the specific power at such wall temperatures when the charge loss is not compensated by supercharging or increased boost pressure. Experimental tests on an engine with an insulated combustion chamber detected a noticeable deterioration of fuel consumption instead of the expected improvement in consumption [9-3]. A strong rise of the gas-side heat transfer coefficient in the first part of combustion, thus causing more rather than less heat to reach the coolant, was demonstrated to be the reason for this (see Sect. 7.2). Engine process simulations ultimately revealed [9-4] that effective engine cooling that prevents component temperatures from rising above the level common today is one of the basic prerequisites for low nitrogen oxide emission.

Thus, one essential function of engine cooling is to lower the temperatures of the components that form the combustion chamber (piston, cylinder head and cylinder liner) far enough that they retain their strength. Limited thermal expansion of a piston must ensure that the running clearance is sufficient to prevent frictional wear between the cylinder liner and piston. Moreover, the lubricating oil must have the requisite viscosity and may not be thermally overloaded. High temperature corrosion establishes an additional temperature limit for highly thermally loaded exhaust valves (see Sect. 4.3).

In addition, engine cooling also serves to

– improve performance by better charging,
– reduce fuel consumption and exhaust emission,
– improve turbocharger compressor efficiency and, finally,
– keep engines safe and protect operators.

Engine cooling is basically divided into *liquid cooling* and *air cooling*. The type of cooling given preference depends on the engine's level of power and type of use, climatic conditions and often buyers' attitudes as well. Market demands and rationales related to use have resulted in certain concentrations of use of liquid and air-cooled engines in the past: Low and medium power high speed diesel engines for construction, agricultural and auxiliary equipment in particular, are frequently air-cooled. Air-cooled engines are limited to only just a few makes in the commercial vehicle sector and have

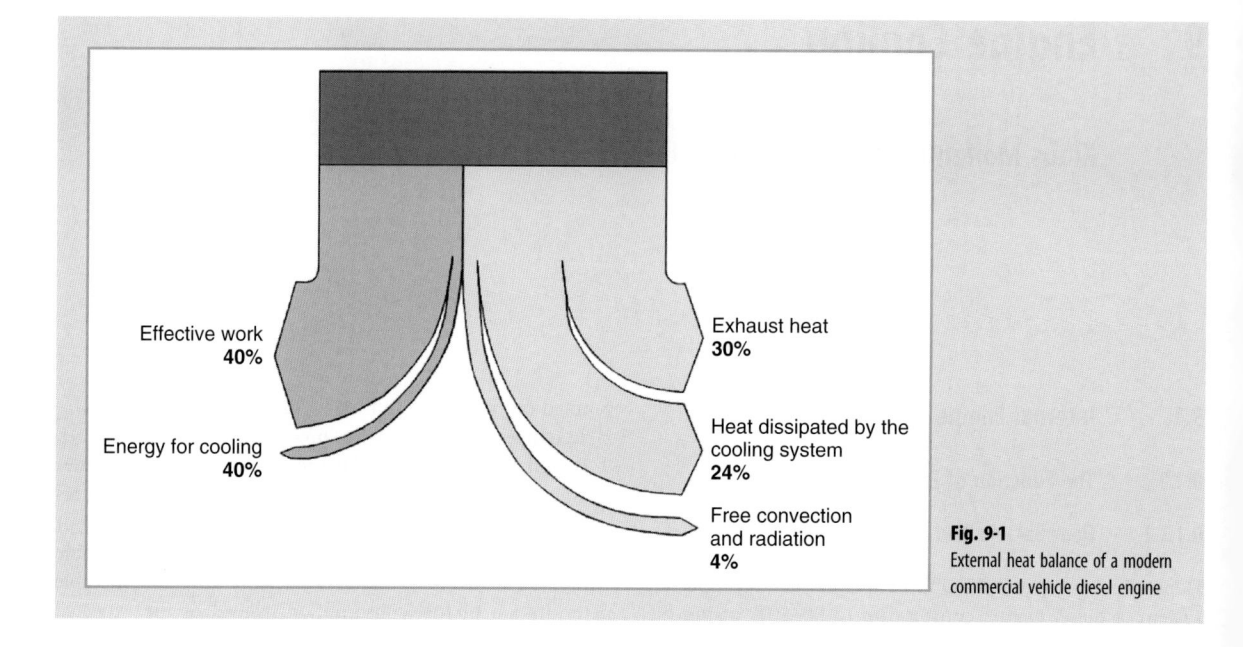

Fig. 9-1 External heat balance of a modern commercial vehicle diesel engine

- Effective work 40%
- Energy for cooling 40%
- Exhaust heat 30%
- Heat dissipated by the cooling system 24%
- Free convection and radiation 4%

largely been edged out of the car sector by liquid-cooled diesel engines in recent years.

Unlike air-cooled engines, liquid-cooled engines dissipate component heat to the environment *indirectly* rather than directly. A closed, separate cooling circuit transports the heat absorbed by the coolant in the heated zones of the cylinder head and cylinder to a coolant/air or coolant/water heat exchanger where it is released to the environment as depicted schematically in Fig. 9-2. Mathematical modeling of this heat transport process in its entirety, which consists of several processes of heat transfer and heat conduction, is extremely involved. Engine engineers normally restrict themselves to the heat transport from the working gas to the coolant, which, in simplified terms, corresponds to the transmission of heat by a flat wall. Section 7.2 already described the transfer of heat from the working gas to the surrounding combustion chamber walls.

A liquid-cooled engine's coolant is normally water or a water/ethylene glycol blend with anti-corrosion and anti-cavitation additives. The addition of ethylene glycol can lower a coolant's freezing point to −50°C. While the cooling effect of

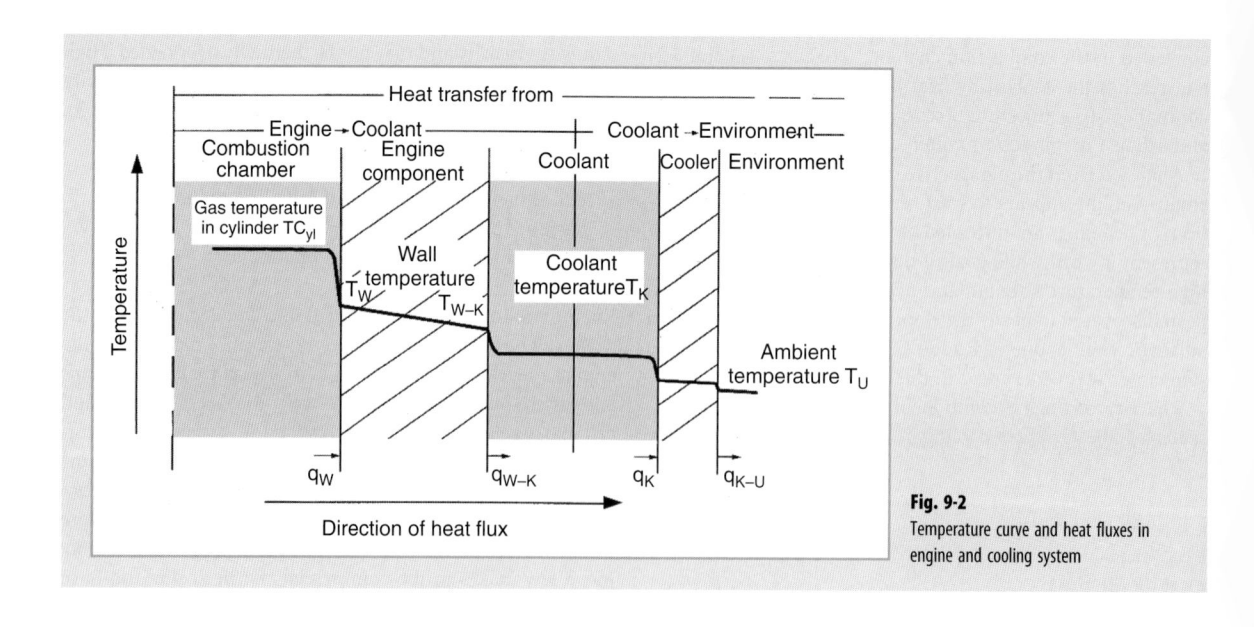

Fig. 9-2 Temperature curve and heat fluxes in engine and cooling system

water without added ethylene glycol is superior to any other coolant, pure water cooling is limited to only medium and low speed marine diesel engines and a few auxiliary equipment engines in which either the aid of external power or the engine's location make it impossible for the cooling water to freeze (For the sake of simplicity, water/ethylene glycol cooling is referred to as water cooling from here on.)

Utilized for some lower power high speed engines, *oil cooling* is a special case of liquid cooling (but is not the same as utilizing lubricating oil for piston cooling; see Sect. 8.6). The cylinder heads and cylinders in oil-cooled engines are cooled with the engine's lubricating oil so that only two resources (lubricating oil and fuel) are needed and additional coolant can be dispensed with. However, the achievable cooling capacity is limited since lubricating oil has poorer properties for cooling purposes (see Sect. 9.1.3).

9.1.1.2 Mathematical Analysis of Heat Transport

Along with model tests and experimental investigations on prototypes, a variety of CAE techniques can now be drawn on to analyze cooling water transport in an engine's heated components. Given the filigreed geometry of engine components and their cooling chambers evident in the engines presented in Sect. 17.1, the available programs based on the *finite element method* are truly optimally suited to simulate flow models as well as temperature fields and heat fluxes. They may be used to already analyze the effects of varying boundary conditions, e.g. geometry changes or parameter variations, during the design stage without consuming much time or money.

Such simulations presuppose the generation of an FEM system for an engine block's cooling water jacket as pictured in Fig. 9-3. Thus, for example, a *flow model* can serve as the basis to analyze the velocity and pressure distributions in an engine's entire coolant jacket, localize existing dead zones and optimize the coolant flow rate in the area of the valve bridge. Further, it is also now possible to use the value of the coolant-side heat transfer coefficient to obtain guide values.

Employing these values and utilizing engine process simulation, the temperature distributions inside components can be presimulated on the basis of structural models of the cylinder head and the cylinder crankcase. The temperature distribution in the cylinder head bottom of an oil-cooled engine presented in Fig. 9-4 is the result of such a simulation.

9.1.2 Water Cooling

9.1.2.1 Water Cooling Heat Transfer

Sufficient cooling of cylinder heads and cylinder liners *with coolant* is the prerequisite for effective component cooling. Dead zones must be prevented in the cooling chambers. The coolant flow must purge vapor locks that form in the highly thermally loaded zones of components. Figure 9-2 contains values that describe the heat transfer from a component to the coolant. Assuming that the heat transferred from the component wall to the coolant corresponds to the heat absorbed by the cooling system and inserting the heat transfer coefficient and the wall temperature on the coolant side, the following applies to the *heat flux density*

$$q_{WK} = q_K = h_K(T_{WK} - T_K).$$

The coolant temperature and local wall temperatures can be measured relatively easily. Determining the heat transfer

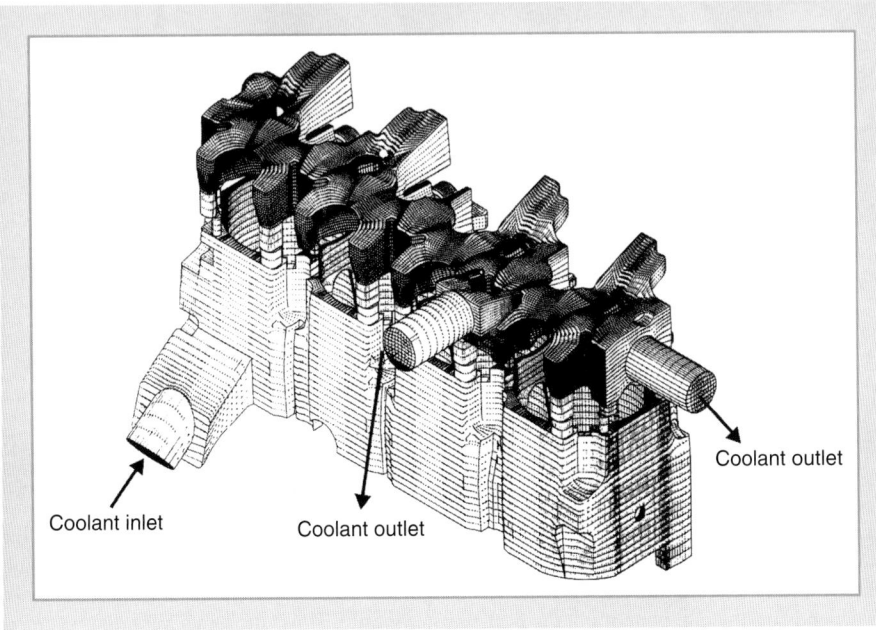

Fig. 9-3 FEM mesh of a car diesel engine's coolant jacket (V_H=1.9 l; VW)

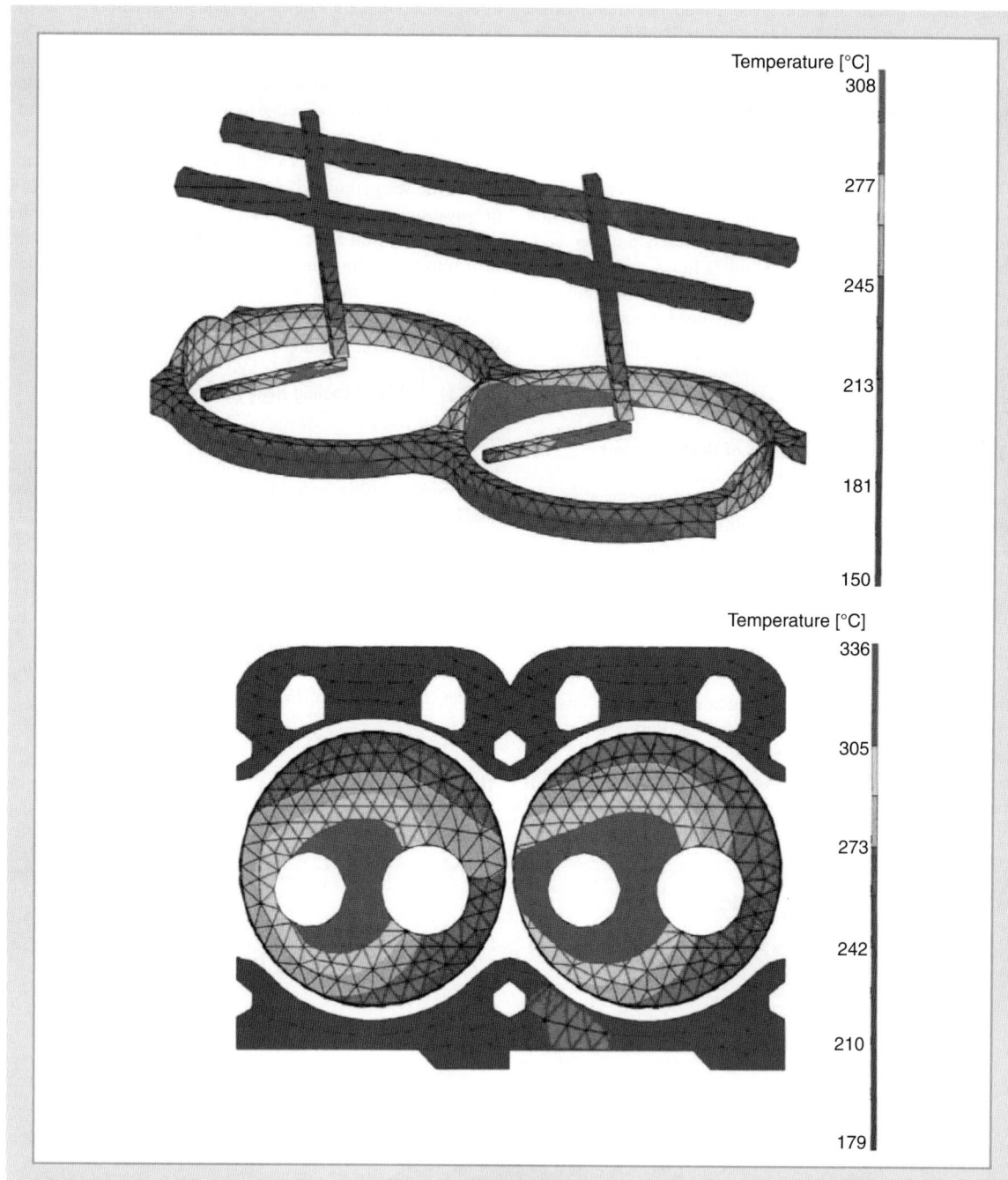

Fig. 9-4 Temperature distribution in a Deutz BF4M 2011 oil-cooled diesel engine's cylinder head bottom (*top*) and cylinder head base plate (*bottom*); see Fig. 9-10 for the cooling oil flow

coefficient for the heat transfer on the coolant side proves to be significantly more difficult. Along with the cooling water's flow velocity, density, specific heat capacity, thermal conductivity and composition, it also depends on the component's thermal load and shape as well as the flow conditions at the component. In addition, nucleate boiling and cavitation caused by engine vibrations sometimes strongly influence the heat transfer locally. Thus, altogether *very different local heat transfer conditions* may exist in one engine. The laws of heat exchange derived from similarity theory with its parameters [9-5, 9-6] and the relationships developed from them mainly apply to geometric bodies and defined flow conditions. Hence, they can only be transferred to engines conditionally [9-7, 9-8].

Cylinder liner. Cooling water chambers of a larger engines' cylinder liners (Fig. 9-5) predominantly have a vertical flow with low flow velocities far below 1 m/s. Free convection has a more dominant influence than forced convection below 0.5 m/s. As the temperature gradient $T_{WK} - T_K$ increases and the length L within the flow decreases, free laminar convection turns into free turbulent convection and heat transfer increases. In addition, rather than the smooth tubes usually assumed, the roughness on the outside of the cylinder liners also has to be allowed for. Thus, the empirical values for the heat transfer coefficients (HTC) α_K specified in Table 9-1 and in the literature [9-9] often have to suffice.

Depending on the construction of the engine block and the configuration of the cylinders (Fig. 9-6), the heat transfer to the coolant in smaller (car) diesel engines (Sect. 17.1) may be modeled as *aligned tube bundles in a cross flow* [9-10]. This was demonstrated by comparisons of engine measurements and simulations of both real engines and engine component models and with water and oil as coolant. In addition, taking the coolant velocity c_0 in the empty port cross section and the proportion of the flow port that is a cavity ψ as the starting point and employing the values specified in Fig. 9-6, the velocity c_ψ is calculated as follows:

$$c_\psi = c_0/\psi = c_0/[1 - \pi D'/(4H)].$$

Where $a = S_1/D'$ and $b = S_2/D'$, the arrangement factor f_A is determined as

$$f_A = 1 + \{[0.7(b/a) - 0.21]/\psi^{1.5}[(b/a) + 0.7]^2\}$$

With the outer diameter D' of the cylinder in the flow and the characteristic "overflow length" $L = \pi \cdot D'/2$ and using c_ψ as the velocity term, the following ensues for the Reynolds number:

$$\text{Re}_\psi = c_\psi \cdot L/\nu = c_\psi \cdot \pi D'/2\nu$$

Thus, when the Prandtl number Pr is known, the independent Nusselt number Nu^* can be determined (Fig. 9-7) as

$$Nu^* = Nu/f_R,$$

or when the tube bank coefficient

$$f_R = [1 + (z-1)f_A]/z,$$

incorporates the number of cylinder barrels z in the flow, the Nusselt number $Nu = h_K L/\lambda$ and thus the *mean heat transfer* h_K can be determined for the heat transfer in the cylinder barrels in the flow (see Table 9-2 for the physical properties).

In highly loaded water-cooled engines, the piston's impact against the cylinder liner when it changes contact with the cylinder liner can cause cavitation in the cylinder liner on the coolant side [9-11]. This increases the heat transfer coefficient locally by as much as tenfold and speeds up the destruction of materials induced by the cavitation attack (see Sect. 7.1).

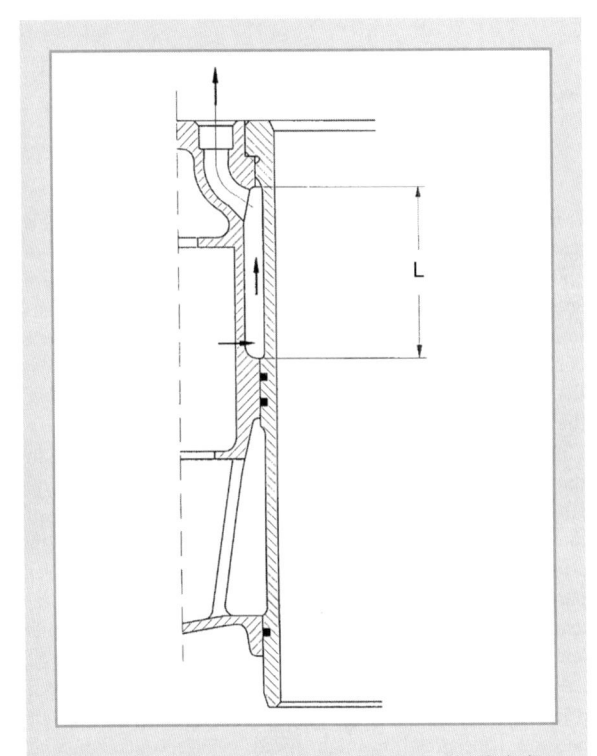

Fig. 9-5 Flow conditions in a larger water-cooled engine's wet cylinder liner

Table 9-1 Guide values for coolant-side heat transfer coefficients

Type of heat exchange	h_K in W/(m² · K)		
	Water	Water/ethylene glycol 50%/50%	Oil
Free convection	400...2,000	300...1,500	
Forced convection	1,000...4,000	750...3,000	300...1,000
Nucleate boiling	2,000...10,000	1,500...7,500	

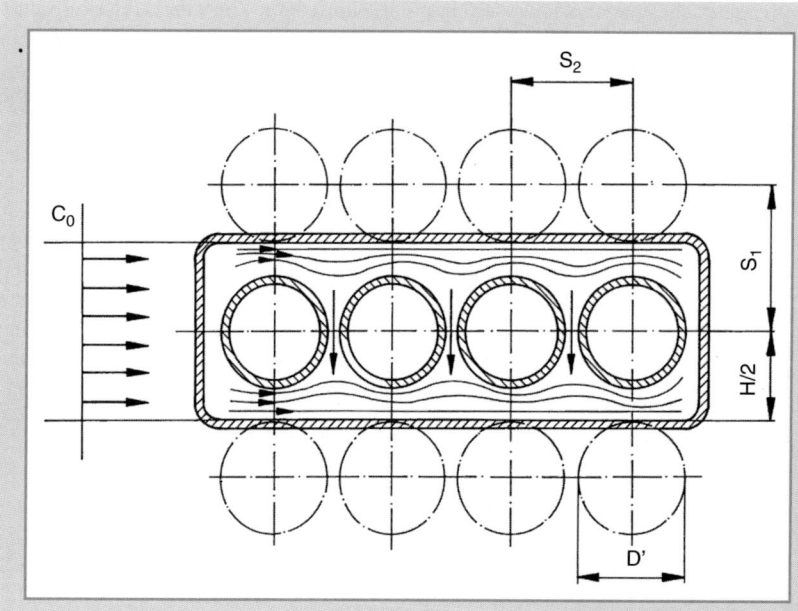

Fig. 9-6 Flow conditions and cylinder configuration for vehicle engines

Cylinder head. Given their complex arrangement, every law-like approach to the HTC, e.g. relationships for "plates in a transverse flow" for cylinder head bottoms, falls short for most cylinder heads too, particularly because a proper flow rarely develops. Moreover, since there are only low flow velocities in a cylinder head and the thermal load is higher, the heat transfer is largely determined by intensive nucleate boiling [9-8] with local heat transfer coefficients h_K of up to 20,000 W/(m² K) (see Table 9-1 and Sect. 9.1.2.3).

Accordingly, clear flow conditions only exist for bore-cooled engine components (Fig. 9-8) where there is a turbulent pipe flow with relatively large flow velocities induced by the forced flow in narrow pipes. Nearly every surface temperature desired may be obtained by varying the bores' spacing and diameters and their distance to the surface and the coolant flow rate. Hence, the thickness of the main walls decisive for mechanical stress may be freely selected [9-12]. Thus, *bore cooling* takes up the design principle of separating thermal and mechanical loading (see Sect. 7.1).

9.1.2.2 High Temperature Cooling

As the name indicates, the significantly *higher temperature level of the coolant* distinguishes high temperature cooling from a conventional cooling system. High temperature cooling aims for temperatures of up to 150°C on the engine's coolant outlet side in the part load range. Naturally, this results in a corresponding rise of component temperatures. However, suitable control of the coolant temperature must ensure that the maximum permissible component temperatures are not exceeded in any of the engine's load points. A component temperature-regulated cooling concept is described in [9-13]. Since the vapor pressure of water at 150°C is just under 5 bar, engines with high temperature cooling cannot use a conventional coolant unless the intention is to put the entire cooling system under high pressure corresponding to the boiling pressure. Experience has shown that the hermetic sealing this requires is unachievable. Thus, the minutest leaks cause an engine to "mist".

Aircraft engines operated with high temperature cooling in the past used pure ethylene glycol as coolant. However, pure ethylene glycol reduces the heat transfer coefficient to a fifth of that of pure water. The *advantage of high temperature cooling* is a perceptible drop in fuel consumption in the engine's lower part load region. The comparatively higher lubricating oil temperature and thus lower oil viscosity is the reason for this and reduces an engine's hydrodynamic frictional losses [9-14]. Despite the advantages in consumption and intensive research primarily in the 1970s, high temperature cooling has still not become widespread in high speed vehicle and industrial engines because a suitable cooling medium that meets the diverse requirements imposed on a coolant is lacking. Described in Sect. 9.1.3, oil cooling in which the maximum coolant temperature at the engine outlet is 130°C represents an important step toward high temperature cooling.

9.1.2.3 Evaporation Cooling

Basics

Evaporation cooling is based on the physical principle of cooling by latent heat and the inherent regulation of temperature

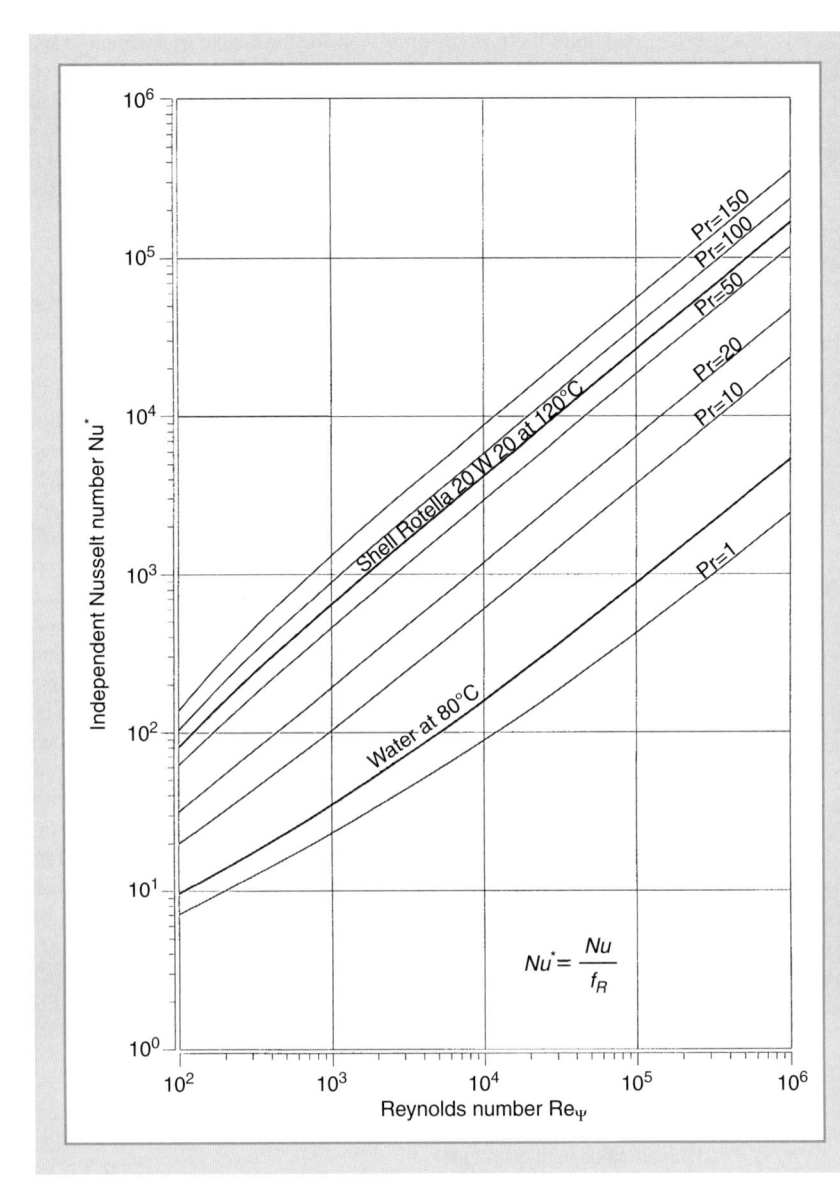

Fig. 9-7
Nusselt number as a function of the Reynolds number and Prandtl number for coolant-side heat transfer in integrated cylinder liners in vehicle engines

$$Nu^* = \frac{Nu}{f_R}$$

Table 9-2 Physical properties of air and water

			Air at 60°C	Water at 95°C	Ratio of air to water
Density	ρ	kg/m³	1.045	961.70	1 : 920
Spec. heat capacity	c_p	kJ/(kg · K)	1.009	4.21	1 : 4.2
Thermal conductivity	λ	W/(m · K)	$28.94 \cdot 10^{-3}$	$675.30 \cdot 10^{-3}$	1 : 23
Kinematic viscosity	ν	m²/s	$18.90 \cdot 10^{-6}$	$0.31 \cdot 10^{-6}$	61 : 1
Thermal diffusivity	a	m²/s	$27.40 \cdot 10^{-6}$	$0.17 \cdot 10^{-6}$	161 : 1

connected with it. The cooling liquid in the components being cooled is heated to boiling temperature so that *nucleate boiling* causes heat to transfer to the coolant without having to force the flow through the engine's cooling chambers: When the thermal load of the component is higher, the temperatures on the surface on the cooling side rise so that an overheated state is reached in the boundary layer near the wall although the cooling liquid's mean temperature is below its saturation

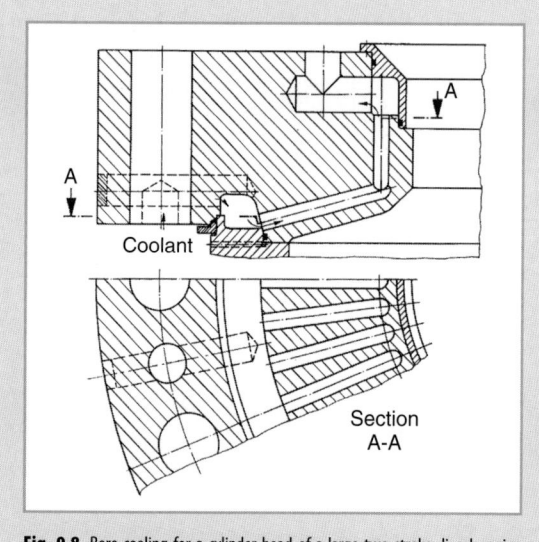

Fig. 9-8 Bore cooling for a cylinder head of a large two-stroke diesel engine (MAN B&W) based on [9-8]

into thermal self-protection for the engine to a certain extent since the increased heat transfer only causes the wall temperature on the coolant side to increase by approximately 10–20 K. Nonetheless, the larger temperature gradient in a component when nucleate boiling is local can cause stronger deformations and stresses, especially during dynamic load operation.

Closed System Evaporation Cooling

The advantages of this type of cooling over forced circulation cooling include a substantially more even temperature distribution in the component while temperature fluctuations in the coolant as a function of load are only slight. Since the coolant's boiling temperature depends on the vapor pressure and thus the system pressure, programmed pressure control can keep the components' temperature level roughly constant at differing engine loads. Thus, as with high temperature cooling (see Sect. 9.1.2.2), lower fuel consumption is attained in part load operation as a result of the higher component temperatures that persist in the normal range even at full load. The vaporized coolant liquefies in the schematic of the closed circuit evaporation system presented in Fig. 9-9. An electrically driven coolant pump constantly circulates the coolant, which exists in both a liquid and gaseous state, through the engine cooling chambers, through the bypass line when the engine is cold and through the heat exchanger when the engine is warm. The fluid/vapor separator ensures that primarily vapor is conducted to the heat exchanger. Given the water-based coolant's high heat of evaporation, its volumetric flow is only approximately 1–3% of its value in forced circulation cooling. Thus, relatively small pumps with greatly reduced power consumption may be used to facilitate lower fuel consumption.

temperature. The vapor locks produced by this are swept along by the coolant flow, thus causing implosions in the proximity of the wall and an intensive pulsation flow as a result of the continuing vapor lock. This increases as the thermal load increases and thus the heat transfer coefficient HTC increases too. An experimentally corroborated relationship for the HTC published in [9-10, 9-15] incorporates the simultaneous occurrence of nucleate boiling and convection.

Nucleate boiling occurs more or less strongly in every highly thermally loaded component, even in the top region of cylinder liners. The strong rise of the HTC connected with this translates

To prevent cavitation in the pump inlet, the coolant suctioned in must be sufficiently undercooled by designing the

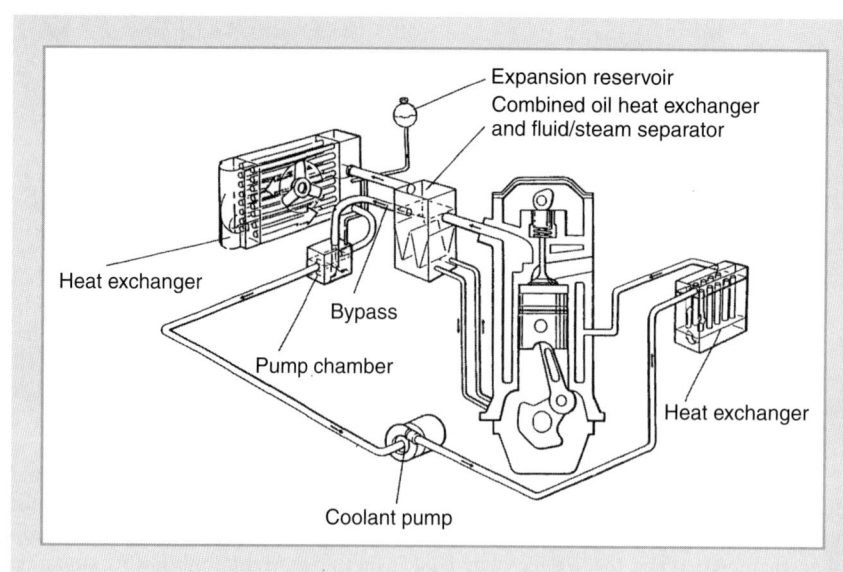

Fig. 9-9 Schematic of evaporation cooling (VW)

vehicle heat exchanger and the coolant flow appropriately. The flow of vapor into the heat exchanger is produced by the partial pressure gradient opposite the engine cooling chambers.

In thermal equilibrium, a common ethylene glycol/water blend with a content of 50/50% by volume produces system-independent boiling temperatures of 105–120 °C at a differential pressure in the system of 100–300 mbar, i.e. significantly lower than the liquid circulation cooling common in cars. This reduces the requirements on components' compressive strength. The achievable cuts in consumption are as much as 5% for gasoline engines but no more than 3% for diesel engines because of their lack of thermal dethrottling [9-16, 9-17].

It is problematic that the usually additized ethylene glycol/water blend is not azeotropic, i.e. the low boiling water fraction in the engine's cooling chamber distills and thus causes the glycol concentration to increase. This entails a steady rise of the boiling temperature and a constant feed of undercooled cooling liquid to compensate. On the other hand, the glycol fraction in the heat exchanger and thus the antifreeze protection are reduced. Suitable alternatives for the cooling liquid have not been found yet.

Moreover, the continued lack of solutions to challenges in the design of cooling chambers, e.g. additional space requirements for the water separator, expansion reservoir, larger pipe cross sections, etc., explains the failure thus far to develop a state-of-the-art engine with evaporation cooling to the stage of production despite its advantages, particularly since it does not generate any cost advantages over conventional cooling in terms of the overall technical work required.

Open System Evaporation Cooling

Used earlier even in Europe for small, rugged single-cylinder diesel engines of low power, this cooling system has been implemented as a horizontal single-cylinder diesel engine millions of times, e.g. in China, to motorize so-called *walking tractors*, which, among other things, relieve farmers or construction workers from heavy physical labor. The evaporated cooling water can be replenished with simple untreated water.

9.1.3 Oil Cooling

Oil cooling alters both the thermal stress of components and the running properties of an engine. Oil neither freezes nor boils in the relevant temperature range of −50 °C to 150 °C. Oil-cooled engines (Fig. 9-10) do not experience corrosion

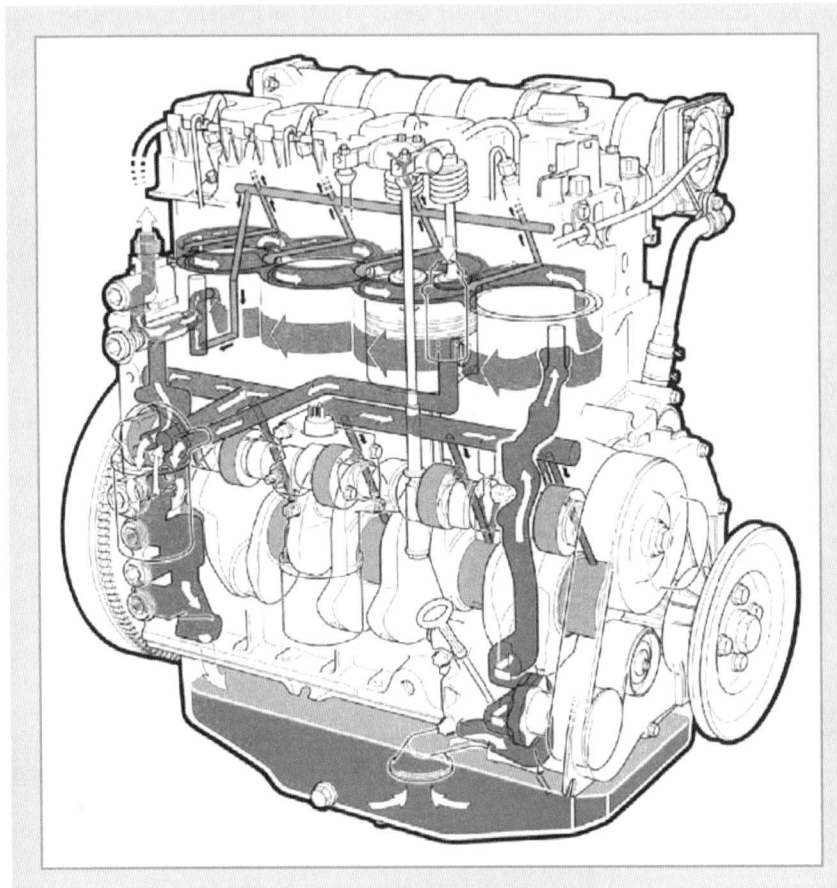

Fig. 9-10
Schematic of a Deutz BF-4 M oil-cooled diesel engine's lubricating and cooling oil flow [9-19]

and cavitation problems. Cooling oil's working temperature of 100–130°C is significantly higher than the usual cooling water temperatures. This results in a correspondingly higher component temperature level, which, in turn, leads to slightly less component heat being eliminated by the coolant and the heat capacity of the exhaust gas flow being somewhat larger than in conventional water-cooled engines. With its low specific fuel consumption in the lower part load region, an oil-cooled engine comes quite close to a "high temperature cooled" engine.

Oil heated to 130°C has a density approximately 15% lower, a specific heat capacity roughly only half as large, thermal conductivity approximately only a fifth as high and a dynamic viscosity 10–30 times greater than water at 95°C. Given the coolant's greater viscosity, particular attention must be devoted to the design of the cross section of cooling chambers and cooling oil bores for oil-cooled engines.

The heat transfer coefficient in oil cooling reaches only 25–30% of the value in water-cooled engines (see Table 9-1). The exponent of the velocity in the correlation $h_K \sim c^n$ averages 0.3. Given the moderate heat transfer coefficient on the coolant-side, an oil-cooled engine naturally cannot attain the same specific power as a water-cooled engine. A supercharged, oil-cooled engine's maximum specific power is 21 kW/l. However, the less intensive heat transfer and the higher coolant temperatures have the advantage of lower temperature gradients in the component and thus significantly lower material stress. Compared to water cooling, oil cooling is *gentle cooling*.

The new 2011 series of oil-cooled diesel engines [9-18] has been presented as a further development of the 1011 series [9-19] (see also Sect. 18.2). Primarily designed for light tractors, construction equipment and auxiliary equipment, these direct injection engines with a power of up to 65 kW at a nominal speed of 2,800 rpm are distinguished by an "open deck" oil-cooled cylinder crankcase with cast-in cylinder liners and a cylinder head with block bore cooling, which is also oil-cooled. The cylinder's oil flows in the longitudinal direction of the engine. All the coolant delivered by the oil pump initially flows around the cylinder walls (Fig. 9-10). The flow cross sections in the crankcase were varied from cylinder to cylinder to ensure that the flow through the intermediate cylinder walls is also adequate. The cylinder head bottom's cooling chamber geometry is a mirror image of the cylinder crankcase and cooling oil flows through it in the opposite direction (see Fig. 9-4).

The cylinder head gasket separates the crankcase and cylinder head cooling chambers. The cooling oil from the crankcase flows through bores located in it into the cylinder head's cooling gallery. Upwardly inclined and interconnected bores meter every cylinder unit's outflow of oil. Two V-shaped bores arranged in the cylinder head above the last cylinder on the side of the fan connect the cooling gallery in the cylinder head bottom with the return gallery. All in all, this reduces the resistance to flow in the cylinder head, increases the flow rate and improves the cylinder head's cooling by uniformly distributing the cooling liquid. The experiences acquired with oil-cooled engines have demonstrated that deposits do not form in the cooling chambers, viscosity properties do not deteriorate and oil oxidation does not increase.

9.1.4 Air Cooling

9.1.4.1 Historical Review

The idea of air cooling – dissipating the heat of an engine's components directly to the ambient air – is as old as the internal combustion engine itself. The Frenchman de Bisschop already introduced an air-cooled internal combustion engine in 1871. Lenoir's single cylinder gas engine, which operated according to the atmospheric principle [9-20], had longitudinal fins cast on the working cylinder, which conducted the cooling energy to the environment by free convection.

The meteoric development of the aviation industry after Bleriot's flight across the English Channel in 1909 also included the development of an air-cooled aircraft engine and was marked by the following milestones: The development of aluminum alloys (1915), the introduction of light alloy cylinder heads (1920) and research on the physical correlations of heat dissipated by cooling fins [9-21], their optimal design [9-22, 9-23] and the influence of the cooling airflow routing.

Apart from the aircraft gasoline engines that attained powers per unit piston area of approximately 0.5 kW/cm^2 in 1944, many firms throughout the world saw themselves induced in the 1930s to work on the development of aircraft diesel engines to counter the danger of altitude induced misfires and achieve higher efficiencies. Out of a total of twenty-five projects, twelve engines were air-cooled, yet only Junkers' Jumo 205 water-cooled opposed piston engine was produced in larger quantities. Air-cooled vehicle diesel engines were launched on the market at approximately the same time. In 1927, Austro-Daimler introduced the first high speed diesel engine, a four-cylinder inline engine with 11 kW of power.

Water-cooled engine systems' susceptibility to relatively major breakdowns under extreme climatic conditions brought the industry orders for developments of air-cooled commercial vehicle and tank diesel engines during the Second World War. Building upon the results obtained, air-cooled diesel engines were developed and implemented in commercial vehicles and agricultural and construction equipment as of the 1950s [9-24–9-26]. In those days, a few engine firms deliberately only offered air-cooled diesel engines: Klöckner-Humboldt-Deutz (now Deutz AG) has been the global market leader since the 1950s. Around 1980, air-cooled diesel engines could be found in 80% of the construction equipment made in Germany.

Initially, it was possible to counter the increase in thermal component loading with a trend toward higher power density by switching to direct injection and employing aluminum alloys with higher high temperature strength. Yet, the number of air-cooled diesel engines produced has been declining since the mid 1980s. Small industrial diesel engines with approximately 15 kW of power are still almost solely air-cooled because of the cost advantage of the integrated cooling system. Even today, air-cooled diesel engines with up to approximately 100 kW of power are still preferred for construction equipment and auxiliary equipment. Both types of cooling exist side by side in the higher power range of up to approximately 400 kW [9-27, 9-28].

9.1.4.2 Heat Exchange from Components to the Cooling Air

Heat Transfer and Cooling Surface Design

The heat transfer coefficient h, temperature gradient ΔT_K and heat exchanging surface A determine the transferrable heat (see Sect. 9.1.2). The heat transfer coefficient (HTC) is a function of the flow velocity c and the material properties (coefficient of thermal conductivity λ, kinematic viscosity v and thermal diffusivity a). The comparison of physical properties in Table 9.2 indicates the different heat transfer conditions for air and liquid cooling.

With the aid of similarity theory and its parameters, i.e. the Nusselt number $Nu = h \cdot D/\lambda$, Reynolds number $Re = c \cdot D/v$ and Prandtl number $Pr = v/a$, the heat transfer in a turbulent flow tube with an internal diameter D can be described by the following power equation ($10^4 < Re < 10^5$):

$$N_u = 0.024 \cdot Re^{4/5} \cdot Pr^{1/3}.$$

Retaining the tube diameters and flow velocities, the following ensues for the relationship of the heat transfer coefficients for air and water (index "Lu" and "Wa"):

$$h_{Lu}/h_{Wa} = (\lambda_{Lu}/\lambda_{Wa}) \cdot (v_{Lu}/v_{Wa})^{-7/15} \cdot (h_{La}/h_{Wa})^{-1/3},$$

and the following after inserting the physical properties (Table 9.2):

$$h_{Lu}/h_{Wa} = 1870.$$

Thus, under identical conditions, the heat transfer coefficient for water cooling is approximately 900 times greater than for air cooling. However, its significantly lower density than water's allows considerably greater air velocities and thus an eight to tenfold increase of the heat transfer. Despite the improvement this makes possible, according to a relation of

$$h_{Lu}/h_{Wa} = 1/(110...60)$$

the same cooling capacity as for water cooling can only be dissipated by larger temperature gradients between a component and coolant and by giving the components fins and thus enlarged surfaces. With the same cooling capacity and an empirical ratio of temperature differences of

$$(T_{WK} - T_K)_{Lu}/(T_{WK} - T_K)_{Wa} \approx 2...4$$

the heat dissipating surface on the cooling side must be enlarged to

$$A_{Lu}/A_{Wa} \approx 15...55.$$

For *straight fins with a rectangular cross section*, a fin height h, width b and space width s and a coefficient of thermal conductivity λ_R, the following applies to the cooling energy flow q_K at a wall temperature T_{WK} in the fin base and a cooling air temperature T_K:

$$q_K = h_K(T_{WK} - T_K)[(2h + s)/(b + s)] \cdot \eta_R$$

with the fin efficiency factor

$$\eta_R = \tanh(h\sqrt{2h_K/\lambda_R b})/(h\sqrt{2h_K/\lambda_R b}),$$

which compares the heat flux actually released by one cooling fin with that a fin with constant surface temperature would transfer, i.e. with infinitely large thermal conductivity. As indicated in Fig. 9-11, the fin height necessary for heat exchange and thus the importance of finning decreases rapidly as the heat transfer coefficient grows larger. Therefore, the cooling chamber walls of water-cooled engines remained unfinned. For the limit case where fin height and fin thickness approach zero, the equation for the heat flux density is simplified to the form that applies to a flat smooth wall $q_K = h_K (T_{WK} - T_K)$.

Knowledge of not only the local component temperature but also the value of the heat transfer coefficient is the prerequisite for a sufficiently precise determination of the cooling energy flow. In principle, the different equations for the heat transfer coefficient can be traced back to one of the approaches of [9-8]

- heat transfer in a turbulent flow channel or
- heat transfer in a body within a flow.

For the finned channel of a fully encased finned tube, the first approach to channel flow can revert to a multitude of validated heat transfer equations for turbulent pipe flow inserting the hydraulic diameter [9-29]. The second approach is based on *Krischer* and *Kast's* studies [9-30, 9-31] and corresponds to an unencased finned tube. The mathematical treatment of heat transfer applied to an air-cooled engine is described in [9-27]. Experimentally obtained heat transfer coefficients must be reverted to in cases when the flow conditions in the finning are not known precisely enough [9-32].

The *enlargement of the exothermic surface* by means of finning can be achieved both by greater fin height and a larger number of fins. However, the two options do not inevitably lead to the same increase of heat flux density when the enlargement of their surfaces is identical. Beyond a certain extent, the increase of fin height does not cause any heat dissipation since the temperature of the fin tip already approaches that of the cooling air. As the comparison of aluminum finning with gray cast iron finning in Fig. 9-12 reveals, the thermal conductivity of the fin material decisively determines the temperature curve between the fin base and fin tip.

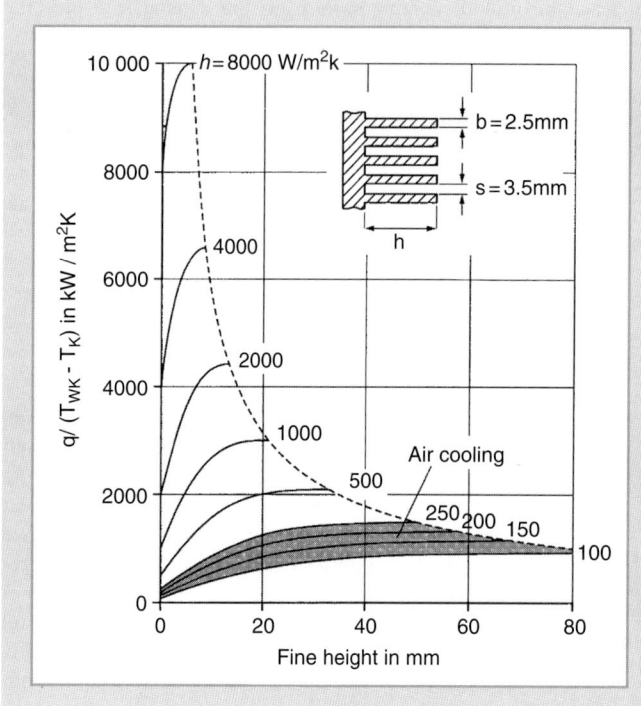

Fig. 9-11 Influence of fin height and the heat transfer coefficient on the heat flux relative to the temperature gradient for gray cast iron finning with $\lambda_R = 58$ W/(m K)

Likewise, the surface can only be enlarged to a certain extent by increasing the number of fins and reducing the space width. This is defined by the space width at which the flow turns turbulent immediately before the adjacent boundary films coalesce. The minimum, still economical space width is approximately 1.2 mm [9-33]. Such small space widths are at the limit of manufacturability and only found in the machined steel cylinder barrels of high performance air-cooled engines.

Fin heights of up to 70 mm with minimum space widths of 3.5 mm at the fin base and fin thicknesses of 3–2 mm from base to tip are cost effective and also perfectly feasible for aluminum cylinder heads produced by permanent mold casting. Given their lower thermal conductivity, gray cast iron cylinder barrels are produced by sand casting with fin heights of not more than 35 mm and space widths of 3 mm at the fin base and fin thicknesses of 2.5–1.5 mm from base to tip. Machining of the type in aircraft engine manufacturing must be turned to for smaller fin thicknesses and space widths.

When the mean heat transfer coefficient is $h_K = 250$ W/(m²K), the maximum heat dissipated in a cylinder barrel with gray cast iron fins is estimated to be approximately $0.5 \cdot 10^3$ kW/m; in a cylinder barrel with machined Alfin fins it is estimated to be approximately $0.2 \cdot 10^3$ kW/m². By comparison, water cooled highly thermally loaded components can dissipate up to $5 \cdot 10^3$ kW/m² of heat by locally occurring nucleate boiling.

Cooling Airflow Routing and Heat Transfer

Cooling airflow routing influences the heat transfer and thus the temperature distribution at a component. The cooling air flowing through a finned cylinder barrel is initially baffled at the point of incidence and then flows around the cylinder barrel until it breaks away laterally in the height of the meridian section so that a dead zone forms behind the cylinder. This causes the highest wall temperature to appear on the back of the cylinder and the lowest on the inflow side (Fig. 9-13). Air cowling by encasing the finned cylinder barrel is required to prevent different thermal expansions from warping the cylinder (Fig. 9-14). This may only be dispensed with in engines with small bore diameters and low power outputs per displacement, e.g. motorcycle engines.

The guide plates are intended to ensure cooling air passes through the fin channels without dead zones. By virtue of the principle, the absorption of the heat from the cooling air in the finned channel makes it impossible to attain a fully uniform temperature distribution in the circumference when the inflow is transverse. Experience has shown that, as a function of the heating of the cooling air by ΔT_{KL}, the following applies to the maximum temperature deviation in the circumference ΔT_{Umfang} [9-34];

$$\Delta T_{\text{Umfang}} / \Delta T_{KL} \approx 0.8.$$

Consequently, the heating of the cooling air is limited to 50 K for a maximum permissible temperature deviation of 40 K.

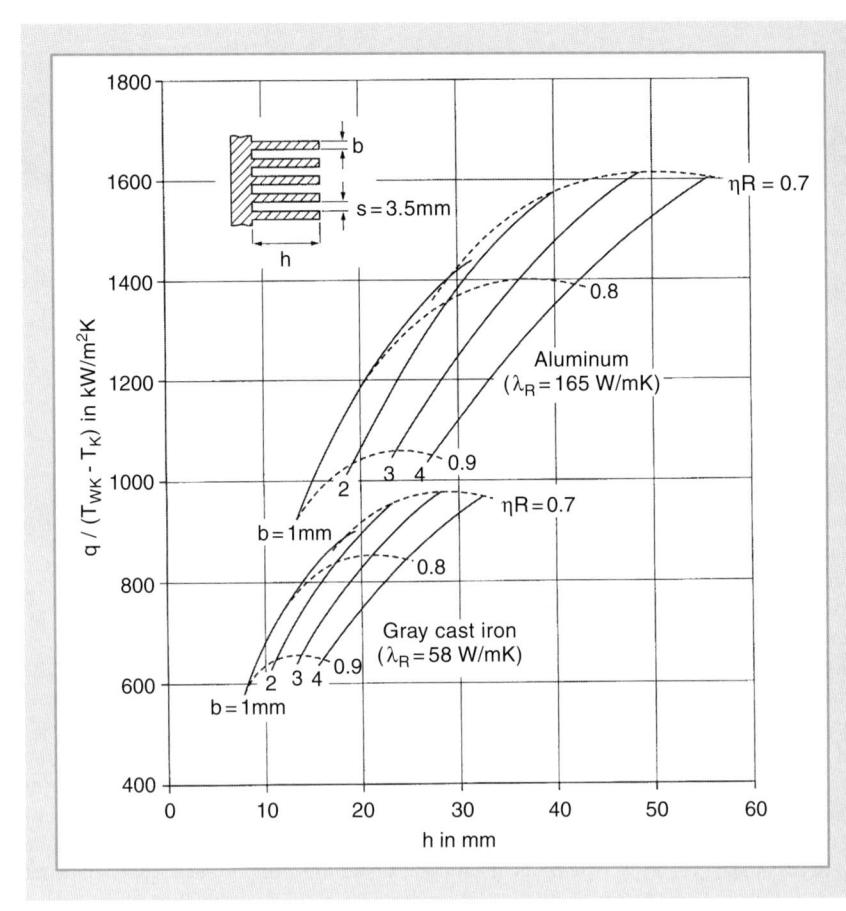

Fig. 9-12
Influence of fin height h and the fin's thermal conductivity λ_R on the heat flux relative to the temperature gradient with a constant heat transfer coefficient $h_K=150$ W/(m² K)

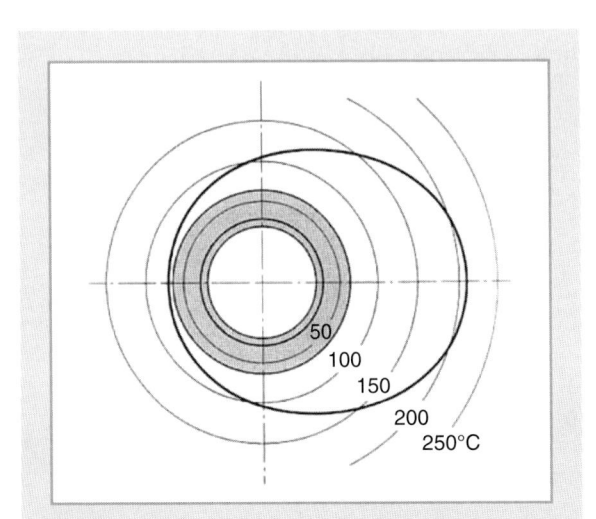

Fig. 9-13 Temperature curve for a cross-flow, unenclosed cylinder barrel (flow from the *left*).

9.1.4.3 Design Features of Air-cooled Engines

Overall Engine Design

The most striking design features of air-cooled engines are their single cylinders and integrated cooling systems. Since aluminum and cast iron have very different thermal expansion, the virtually universal use of aluminum cylinder heads (more uniform temperature distribution in the cylinder head bottom and maximum dissipation of heat to the cooling fins) and the requirement of unimpeded thermal expansion of components inevitably necessitate *single cylinder construction*. Hence, air-cooled engines are downright predestined for the modular principle with a large number of identical parts and resultant cost advantages for manufacturing and spare parts management, particularly for small and medium quantities [9-36]. Further, a modular engine allows easy maintenance around the cylinder head, cylinder barrel and piston without having to remove the engine and disassemble the oil pan.

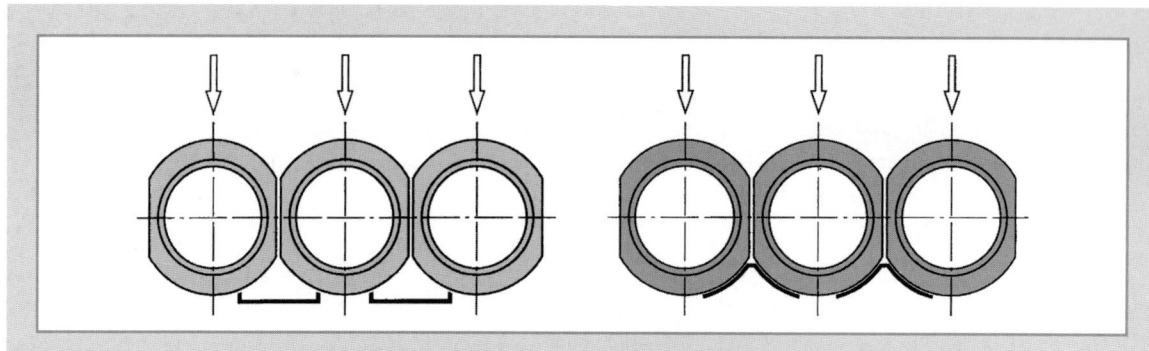

Fig. 9-14 Cooling airflow routing: Argus shroud (*left*); shroud abutting the finned tube (*right*) [9-35]

In most cases, an air-cooled diesel engine's cylinder unit is equipped with *four tension bolts* that connect the cylinder head with the crankcase, brace the cylinder barrel between the two and transfer the gas forces directly to the crankcase. Tension bolts are slender, highly elastic expansion bolts that reduce the alternating loads in the tensile stress areas of the threads and limit the increase of force resulting from the greater thermal expansion of the cylinder head and cylinder barrel than of the significantly cooler bolts.

Significant for the overall length of an engine, the design of the cooling fins and not the spacing of the crankshaft bearings determines the cylinder clearance in air-cooled inline engines. Relative to the cylinder bore, this is normally 1.35–1.45 and only 1.275 in the extreme case (KHD engine model B/FL 913).

A comparison of the installation dimensions of air and water-cooled engines including the latter's external cooling system reveals that an air-cooled engine with an integrated cooling system is an exceptionally *compact engine* despite its greater cylinder clearance. This is especially true of the V engine in which the cooling fan is mounted between the cylinder banks, thus saving space.

Crankcase

Together with the cylinder barrel, the crankcase absorbs the load transfer from the cylinder head to the crankshaft assembly with the crankshaft bearing and hence requires high inherent stability for smooth piston travel, even when the crankcase is simultaneously one of the vehicle's bearing elements, e.g. in tractors. A comparison of the cross sections of the crankcases of an air-cooled and a water-cooled engine with identical main dimensions important for their flexural and torsional rigidity in Fig. 9-15 reveals that an air-cooled engine's crankcase must be designed with greater care because of the significantly smaller overall height due to the absence of a water reservoir and the resultant poorer

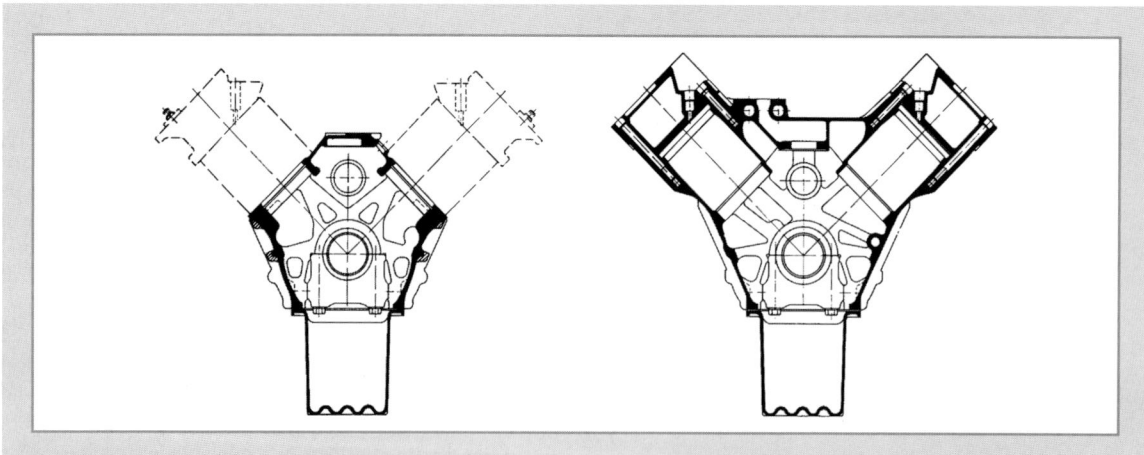

Fig. 9-15 Cross section crucial for the inherent stability of an air-cooled engine with a cylinder barrel and single cylinder head (*left*) and a water-cooled engine with a block cylinder head (*right*)

conditions for strength. Effective and proven measures that provide an air-cooled engine's crankcase high inherent stability are:
- side housing walls that are dropped far under the crankshaft axis and, if possible, curved,
- continuous bracing ribs in the side and cross housing walls and a sturdy, broad oil pan flange,
- a relatively thick, only slightly perforated supporting cylinder base deck,
- a cast oil pan instead of a sheet metal oil pan and
- cross bolting of the main bearing cover with the housing walls in V engines [9-37].

Air-cooled engines are also frequently constructed in a tunnel design for reasons of rigidity [9-27].

Cylinder Barrel

Constructed as a cylinder liner with cooling fins, the cylinder barrel is usually made of gray cast iron in one piece. Sand casting remains economical even for larger quantities. The walls in the upper and lower barrel region are usually designed somewhat thicker and the uppermost cylinder barrel zone provided with cooling fins closed along the circumference to *minimize liner deformations* caused by the bolting forces and the in-cylinder pressure. For reasons of strength and cooling, engines with high power outputs per liter displacement must switch to cast steel barrels and the filigreed body of the cooling fins must be machined.

The composite casting process (Fig. 9-16) known as the Alfin process [9-38] in which an approximately 0.03 mm thick intermediate metal layer ensures a gap-free connection between a steel barrel and its finned aluminum jacket is and has only been used in aircraft engine manufacturing with the exception of a few military applications. It is also employed for diesel engines in the commercial sector, utilizing gray cast iron instead of steel. Light alloy cylinder barrels with appropriately finished barrel surfaces are only used for gasoline engines [9-39, 9-40].

Cylinder Head

Its diversity of functions and alternating mechanical and thermal stresses make the cylinder head the most complex component of an engine. A cylinder head must have *high inherent stability* to transfer the gas forces acting on it to the crankcase and simultaneously ensure that the connection to the cylinder barrel is gastight. An air-cooled cylinder head not only has to house the gas ports, the injection nozzle, possibly a prechamber and the cylinder head bolts but also the cooling fins and flow cross sections necessary for the cooling air. This is a difficult task considering that a direct injection diesel engine, for example, requires a fin area 30–35 times larger than the piston area to cool the cylinder head bottom and the region of the exhaust port. In addition, the cooling fins must be designed so that the maximum cylinder head temperature between the valves remains limited and large temperature differences are avoided to prevent high thermal stresses in the cylinder head bottom. Normally, only *aluminum* cylinder heads can satisfy these requirements. Their high thermal conductivity facilitates the distribution of heat in the cylinder head bottom and allows cost effective manufacturing of thin, high cooling fins.

Such cylinder heads fundamentally require *valve seat inserts*, which are usually centrifugally cast and shrink fit. The transverse arrangement of the valves to the crankshaft axis allows a better cooling fin design in the valve region than a parallel arrangement as well as larger flow cross sections for the cooling air. However, the combustion system must allow the valves to incline strongly (gasoline engines). When they are arranged parallel to the crankshaft, the valves should only incline toward the cylinder axis slightly. However, provided the combustion system permits a weakly curved cylinder head bottom, this allows the cooling fins an even somewhat larger area above the valve bridge. Accommodating the maximum of cooling fins in the highly thermally stressed area of the valve bridge necessitates designing the gas ports' cross section to be relatively narrow and high. While four valve cylinder heads are virtually standard in air-cooled motorcycle gasoline engines, the extremely limited space above the cylinder head bottom precludes the implementation of a cylinder with four valves in air-cooled diesel engines.

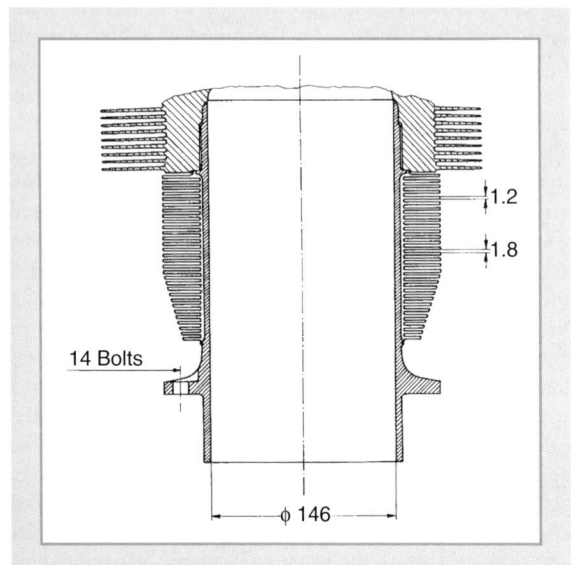

Fig. 9-16 Alfin cylinder barrel of an air-cooled diesel engine (Teledyne Continental engine) [9-28]

Table 9-3 Cast aluminum alloys for air-cooled cylinder heads [9-44]

Alloy type	Alloy name	Alloy elements in % by weight								
		Cu	Ni	Si	Mg	Mn	Ti	Co	Zr	Sb
AlMgSiMn	Hydronalium, Ho 411, 511, Hy 418, 511, Hy 51, Hy 71	0 – 1.0	0 – 1.5	0.7 – 1.8	3.5 – 6.5	0.1 – 1.0	0.1 – 0.2	– – –	– – –	– – –
AlCuNiMg	Y alloy, A-U4NT	3.5 – 4.5	1.7 – 2.3	0.2 – 0.6	1.2 – 1.7	0.02 – 0.6	0.07 – 0.2	– – –	– – –	– – –
AlCuNiCoMnTiZrSb	RR 350, A-U5NZr	4.5 – 5.5	1.3 – 1.8	0 – 0.3	0 – 0.5	0.2 – 0.3	0 – 0.25	0.1 – 0.4	0.1 – 0.3	0.1 – 0.4

The distinctive feature of the cast aluminum alloys of air-cooled cylinder heads is their particularly high high temperature strength and resistance to cyclic temperature. Both of these material properties important for a cylinder head's dimensional stability are obtained by complex alloy compositions (see Table 9-3), controlled cooling in the permanent mold and special postheat treatment. The multi-alloyed material RR350 high temperature strength of 230 N/mm² at 200°C is the highest by far and even its high temperature strength of 200 N/mm² at 250°C is good. Aluminum alloys' greater thermal expansion and ductility make the *danger of valve bridge cracks* disproportionately greater in air-cooled cylinder heads than in water-cooled cylinder heads: Two cast-in steel plates that function as expansion joints can keep the region of the cylinder head base between the valves largely free of tensile stresses when the cylinder head cools down and thus prevent valve bridge cracks (Fig. 9-17). *Thermal shock tests* [9-41] in which the valve bridge is warmed to 300°C and cooled to 100°C in approximately 2 min intervals are an important aid for the development of air-cooled cylinder heads.

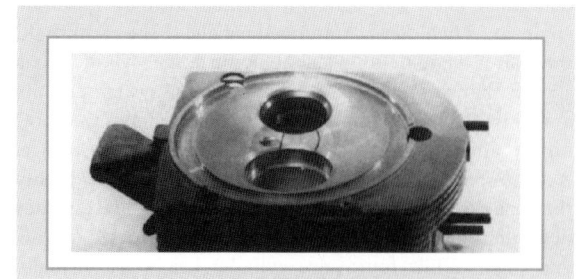

Fig. 9-17 Steel plates cast in valve bridges to prevent valve bridge cracks in the cylinder head of an air-cooled diesel engine (Deutz AG FL 513)

9.1.4.4 Engine-integrated Cooling Systems

Comparison of the External Cooling Systems of Liquid-cooled Engines

By virtue of the principle, cooling systems in air-cooled engines are *engine-integrated* since the component heat dissipates directly into the ambient air. The engine and cooling system constitute a unit. Cooling fans with cowling and integrated lubricating oil coolers as well as equipment or vehicle coolers, e.g. the hydraulic oil coolers of construction equipment or transmission oil coolers of vehicles, are add-on engine parts. Air-cooled engines require smaller quantities of cooling air since they utilize it better (greater temperature increase). However, the narrower flow cross sections cause relatively high air velocities in the finning and thus relatively high cooling air resistances (Fig. 9-18). The adaptation of cooling systems to these conditions results in varying sizes and designs of components in water-cooled engines. Thus, air-cooled engines' fans are only approximately half as large in diameter as fans of comparable liquid-cooled engines. They are however operated at speeds two to three times higher and constructed somewhat longer because of the guide vane required. The radiators are also substantially more compact, their end faces on the cooling air side being up to 60% smaller than conventional radiators. They are usually mounted on an engine without elastic intermediate elements. This causes high mechanical stresses and requires aluminum radiators with low inertial forces and higher strength and rigidity. Efficient intercooling based on the air/air principle has been applied to air-cooled engines from the start. Thus, the charge air can be cooled far below the temperature of a liquid-cooled engine's coolant. Depending on the type of intercooler arrangement (before the cooling fan or in the parallel flow to the other cooling air consumers), charge air temperatures

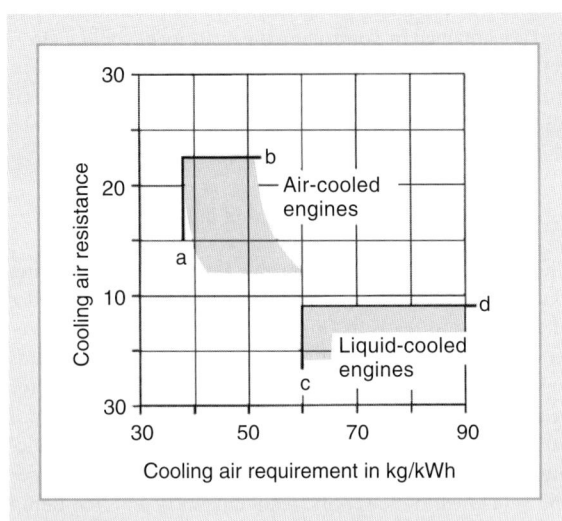

Fig. 9-18 Cooling systems of direct injection diesel engines: Operating ranges for cooling air resistance as a function of the cooling air requirement: **a** Maximum permissible cooling air heating, **b** Economically still justifiable cooling air resistance, **c** Maximum cooling air heating for conventional systems, **d** Maximum pressure increase of conventional fans

are reached, which are only 25–45 K above the respective ambient temperature.

Cooling Fan Designs and Design Criteria

An air-cooled engine's relatively high cooling air resistance and its requirement of a fan with the smallest possible dimensions and low speeds leads to axial cooling fans with aerodynamically highly loaded flow cascades. Basically accommodated in the flywheel, radial fans are reserved for small one and two-cylinder engines.

The axial fan's compact design facilitates simple cooling airflow routing. High efficiencies with low noise emission can be achieved when the flow cascade is designed and manufactured more meticulously. Two designs are distinguished based on the arrangement of the guide vane: Both bladings in fans with a guide vane downstream from the impeller (outlet guide vane fan) are deceleration cascades in which pressure builds up. When the guide vane is upstream (inlet guide vane fan), it is an acceleration cascade and lowers the static pressure so that the impeller alone must produce the pressure increase, compensating for the preceding pressure drop.

Given the multitude of constraints, a decision on the suitability of one of the two fan designs can only be made after weighing their respective features. Table 9-4 lists the most important features [9-42]. Both designs have proven themselves in the field. Overall efficiencies of 80–84% are achievable when certain values for the flow and pressure coefficient and the limits for a flow without separation in the axial cascade are observed [9-43]. The total power required for engine-integrated cooling is 2.5–4.5% of the rated engine power depending on the fan efficiency, cooling air requirement and resistance to flow.

Along with good aerodynamic design, the noise a fan generates is acquiring ever greater significance. It should not appreciably influence overall engine noise and must be free of tonal components. In light of mounting environmental regulations (see Sect. 16), this may often necessitate higher development costs. A fan's aerodynamic noise consists of three different components: The strongest noise source, turbulence and vortex noise extends over the audible frequency range. On the other hand, the tonal noise generated by the impeller with a multiple of the fundamental frequency of the human ear (number of blades times speed) is perceived as many times louder than an equally strong broadband noise. Uneven arrangement of the impeller blades provides a remedy.

The *sound power* emitted by the cooling fan can be described by the empirically grounded law

$$L = L_{sp} + 10 \log[(\dot{V}/\dot{V}_0)(\Delta p_g / \Delta p_{g0})^2]$$

Table 9-4 Differences between the outlet guide vane and inlet guide vane fan

	Outlet guide vane fan	Inlet guide vane fan
Maximum efficiency	84%	80%
Minimum specific sound power level	31 dB(A)	33 dB(A)
Tool requirements for pressure die casting	Only two mold halves (bladings are free of overlap and allow axial demolding)	Two mold halves plus radial slide (bladings are not free of overlap)
Engine installation	High	Low
	Acoustic sensitivity during disturbances upstream from the fan (obstacle, flow constrictions)	

where the parameters of the operating point are $\dot{V}$ = volumetric flow and Δp_g = overall pressure increase and the reference variables are $\dot{V}_0 = 1\,m^3/s$ and $\Delta p_{g0} = 1\,mbar$. A constant characteristic for every fan design, L_{sp} is the specific sound power; the second term corresponds to the sound power of the operating point. Fan loudness can only be reduced in a defined operating point (cooling air requirement and the cooling system's resistance to flow) by decreasing the specific sound power. In addition to the flow coefficient and pressure coefficient, the magnitude of which may only vary in a certain range of values for optimal design, it is significantly influenced by the:
– type of guide vane arrangement,
– aerodynamic quality of the flow cascade and the fan inlet,
– position of the operating point to the design point,
– radial gap between the impeller and housing wall,
– axial distance between both bladings,
– type of impeller blade shape (radial or sickle),
– type of blade arrangement on the impeller circumference and
– number of blades and type of pairing of the number of blades.

The high quality axial fans of air-cooled engines now reach specific sound power levels of 31 dB(A) and are distinguished by low broadband noise.

A uniform temperature level of components and lubricating oil and thus optimal conditions for engine operation (consumption, exhaust gas quality, noise emission and service life) are obtained by controlling the quantity of cooling air. A control in which the cylinder head temperature is the controlled variable and, allowing for the lubricating oil temperature, remains constant is especially advantageous. A hydraulic clutch installed in the fan hub can control fan speed.

9.1.4.5 Examples of Implemented Diesel Engines

The range of commercially available air-cooled diesel engines extends from universally implementable small single cylinder direct injection diesel engines with their typical design preferred for construction equipment and power and pump units (see Sect. 18.1) up through powerful V12 diesel engines for heavy commercial vehicles (Fig. 9-19). The high performance variant with exhaust gas turbocharging and intercooling pictured here is part of a line of six, eight and ten cylinder engines and is tailored to the specific requirements of a 38 ton dump truck used at large construction sites and strip mines. It exemplarily demonstrates the advantages of an integrated engine cooling system, which not only encompasses the actual engine cooling but also the heat exchanger that cools the charge air, the engine oil and the oil utilized in the transmission and the retarder. Thus, the engine merely has to be connected to the fuel supply and the exhaust manifold when it is installed. In normal driving, the specific air requirement for engine cooling is 41 kg/kWh. The fan's power consumption requires

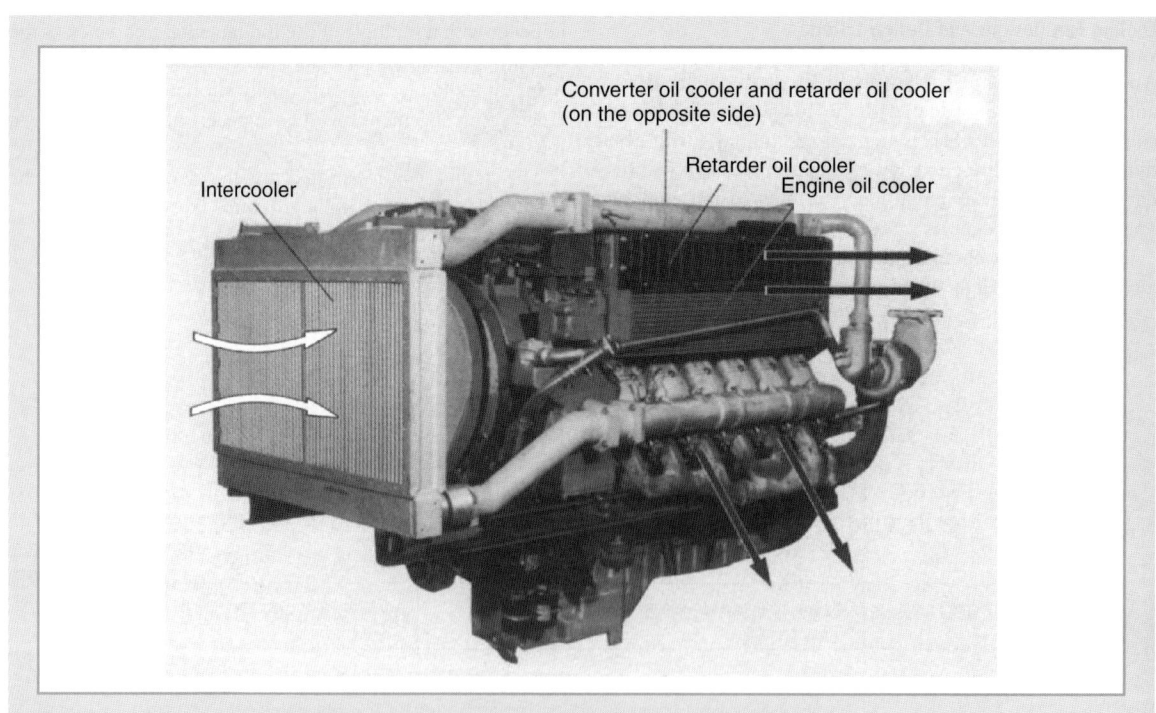

Fig. 9-19 Deutz AG BF12L523CP air-cooled twelve cylinder diesel engine, V_H=19.144 dm^3; P_e= 441 kW at 2,300 rpm

approximately 11.6 kW. Air-cooled industrial diesel engines are found in the mid and lower power range in between (see Sect. 18.2) and are employed for many purposes of installation including soundproofed diesel engines equipped with encapsulation (see Sect. 16.5).

9.1.4.6 Limits of Air-cooled Engines

The air-cooled diesel engines available in the power range up to 440 kW have reached a high level of maturity and are utilized preferably as industrial engines because of their simplicity and ruggedness (see Sect. 18.2). Air-cooled diesel engines remain little accepted in the motor vehicle sector because of their insufficient comfort, particularly their low supply of heating heat.

As supercharging rates increase, air cooling is being called into question as an equivalent, alternative cooling process more and more frequently since high mean effective pressures cause substantial mechanical loading of the cylinder unit and crankshaft assembly and a significant increase of the thermal loading of the piston, cylinder barrel and cylinder head. The limits of thermal loading ensue from the high temperature strength of the aluminum alloy for the cylinder head and the maximum dissipatable cooling energy. The disadvantage of smaller heat transfer coefficients h_K on the cooling side may be compensated by a larger heat gradient between a component and cooling air and by enlarging the exothermic component surface through thinner and closer fins on which there are also limits however. There are limits on a larger temperature gradient too since it inevitably increases component temperatures: Factoring in sufficient lubrication, the temperature ought to be a maximum of 190 °C in the slide face of the liner, 240 °C on average in the cylinder head bottom for dimensional stability and not above 280 °C in the valve bridge.

Aluminum alloys' lower high temperature strength than gray cast iron's makes the cylinder head the weakest component and, thus, the component of an air-cooled engine that determines its performance. The cylinder head's dimensional stability determines the quality of the cylinder head seal. Given the cylinder barrel's increased temperature and the aluminum alloys' greater thermal expansion, thermally induced component deformations have proven to substantially exceed the mechanical deformations caused by forces and pressure [9-44]. In addition to a shift of the maximum sealing pressure on the inner sealing area resulting from the curved cylinder head bottom, the exhaust port's plastic deformation also observable in highly supercharged engines influences the sealing pressure distribution.

However, the higher temperature level of air-cooled engine components also causes greater warming of the aspirated combustion air and thus a lower cylinder charge. When their smoke emission is identical, air-cooled diesel engines' rated power in suction mode is approximately 2.5% lower than water-cooled engines' and 3.5% lower in the point of the maximum torque. A simulation determined that the hotter intake port causes approximately 50% of this, the higher wall temperatures of the cylinder liner cause approximately 30% and the hotter cylinder head and piston top each cause 10%. Since the temperature differences between the components and the combustion air is relatively small, no appreciable power loss occurs in supercharged engines without intercooling; it can be compensated by a marginally higher charge air pressure in engines with intercooling.

The higher temperature level of components that form the combustion chamber together with the higher temperature of the charge determines the final compression temperature and thus the maximum combustion temperature. As basic tests on cooling's influence on the formation of NO_X confirmed [9-45], this is decisive for the formation of NO_X, (see Sect. 15.3) and it complicates efforts to lower NO_X emissions as emission limits are continually tightened and causes considerable problems for air-cooled engines [9-46].

While the measures that
- intensify internal cooling (intercoolers and cooling gallery pistons),
- elevate component temperature limits necessitating the use of expensive materials, lubricants and more complex machining (cylinder head material with extremely high high temperature strength, fully formulated lubricating oils, cold worked cylinder barrel surfaces, Alfin bonded piston ring carriers and dual keystone rings with molybdenum coating) and
- insulate components against heat (thermal insulation of the exhaust port from the cylinder head)

could significantly shift the limits set by the restricted external heat dissipation in air-cooled diesel engines, they entail a higher level of technology that comparable water-cooled engines do not require at all or not to the same extent.

9.2 External Engine Cooling Systems

9.2.1 The Function of Engine Cooling Systems

9.2.1.1 Definitions

As explained in Sect. 9.1, the cooling system establishes an important prerequisite for trouble-free engine operation by dissipating heat from thermally critical points in engine components (cylinder head, piston, cylinder liner, etc.) and agents (engine oil, fuel, charge air, etc.) as a *cooling load* to the environment either directly (e.g. in air cooling, see Sect. 9.4) or, usually, by a closed coolant circuit and a radiator to comply with functional limit temperatures.

The cooling of agents is referred to as direct cooling when a heat exchanger releases the heat directly to the cooling air and as indirect cooling when the heat is released to a closed coolant circuit.

The arrangement of the components that dissipate heat to one another and their control constitute a cooling system.

From the perspective of thermal engineering, engine cooling for components consists of "heat exchangers" with usually small heat exchanging surfaces connected in parallel and/or a series. This, in turn, requires large heat transfer coefficients [9-47] (see Sects. 7.2 and 9.1).

Along with the radiator for the coolant, heat exchangers for
- engine oil,
- piston cooling oil or water,
- nozzle cooling with fuel, cooling water or cooling oil,
- charge air,
- exhaust gas recirculation and
- transmission oil

are also integrated in the coolant circuit.

Compromises often have to be made when designing a cooling system and its components. While cooling used to be primarily specified for rated and extra power, factoring in unfavorable boundary conditions (e.g. climate and seasons), the ever stricter requirements for fuel consumption, pollutant emission and idle and part load performance have made *control of the temperature of the components* and agents rather than their cooling the primary function of the cooling system. This not only necessitates providing sufficient cooling capacity but also using adjustable actuators such as thermostats, valves, pumps or fans.

9.2.1.2 Engine Cooling

Depending on the engine, the heat dissipated from engine cooling with the coolant as the cooling load is 40–100% of the rated power or 20–40% of the supplied fuel energy (see Table 9-5). High pressure supercharging's increase of the power density has led to a shift in heat balances as effective efficiencies have increased. When an engine's release of heat to the coolant decreases, the heat absorbed in the oil cooler and intercooler increases so that the overall cooling load remains approximately constant. On the other hand, the exhaust gas energy decreases as the effective brake work increases.

In turn, the cooled exhaust gas recirculation implemented to reduce NO_X introduces additional quantities of heat into the coolant as a function of the recirculation rates. This often increases the complexity of a cooling system considerably.

Regarded as energy loss, the release of the cooling load to the environment may be expediently utilized ecologically and economically by cogeneration to increase system efficiency (see Sect. 14.1).

9.2.1.3 Cooling Agents

The heat loss from transmissions and brakes in power train systems and vehicles must be dissipated too and the cooling system designed appropriately.

The additional cooling requirement relative to the input power to the transmission is
- 1–3% for mechanical transmissions
- up to approximately 5% for automatic transmissions with a torque converter
- 5% for hydraulic transmissions for locomotives when oil/water cooled (40% when air/oil cooled) and
- 25% for hydraulic transmissions for rail coaches (when the allowable temperature of the transmission oil is T = 80–100°C and only briefly 125°C to a maximum of 130°C).

Table 9-5 Dissipated heat fluxes in % of rated power

Engine type/speed range	Engine cooling[a]	Engine oil	Intercooling	Coolant heat	EGR cooling
Low speed engines 60...200 rpm	14...20	6...15.3[b]	20...35	40...70	–
Medium speed engines 400...1,000 rpm	12...25	10...15	20...40	40...80	–
High performance engines 1000...2,000 rpm	30...50	5...15	10...20	10...20	–
Commercial vehicle engines 1800...3,000 rpm with exhaust gas turbocharging and intercooling	30...50	30...50	15...30	45...80	10...20[c]
Naturally aspirated engines	50...70	50...70	n. a.	50...70	–

[a] Cylinder, cylinder head and exhaust gas turbocharging cooling.
[b] Lubricating oil and piston cooling oil cooling.
[c] Lubricating oil cooling (piston cooling by water).
Note: These and the following figures for quantities of heat, volumetric flows and temperatures are guide values. The diversity of engine design, power ranges and operating conditions explain the large range.

Table 9-6 Cooling water temperatures in °C

	Low speed two-stroke engines	Medium speed four-stroke engines	High speed four-stroke engines
Engine cooling water			
Inlet into engine	65–75	70–80 (82)	76–87
Outlet from engine	75–80	80–90	80–95 (110)
Temperature difference in the engine	5–10	5–10	4–8
Preheating to	50	40–50	40
Preheating during heavy fuel oil operation to	60–70	60–70	n. a.
Fresh water (sea water)			
Inlet into radiator, maximum[a]	32–38	32–38	32–38
Outlet from radiator, maximal	≤ 50	≤ 50	≤ 50

[a] Tropical operation.

Table 9-7 Volumetric coolant flows relative to engine power in l/kWh

	Low speed two-stroke engines	Medium speed four-stroke engines	High speed four-stroke engines	Commercial vehicle engines
Engine cooling water	6...15	30...40	50...80	50...90
Fresh water	30...40	30...50	30...50	N.A.

Hydrodynamic brakes in commercial vehicles (e.g. retarders) can introduce larger quantities of heat into the cooling system than the engine releases in normal operation.

The temperatures for the agents water, oil and charge air are a function of the engine's size, type and mode of operation. Engine cooling water temperatures are kept somewhat lower in large low and medium speed engines than in high speed engines. Making allowances for a bearing's construction, materials and design, lubricating oil temperatures are often significantly lower (Table 9-6). The dissipated heat fluxes and the desired or allowable temperature differences produce the volumetric coolant flows (see Table 9-7).

9.2.2 Cooling System Design

9.2.2.1 Cooling Systems with Direct Heat Dissipation

Of the two options to dissipate engine heat to the environment [9-48], direct cooling dissipates the cooling load to the environment in an open coolant circuit. This is part of
- air-cooled engines,
- engines with circulating water cooler,
- engines with cooling tower cooling and
- engines with evaporation cooling.

With a few exceptions, direct cooling is now synonymous with air cooling (see Sect. 9.1.4).

9.2.2.2 Cooling Systems with Indirect Heat Dissipation

Heat Dissipation

During indirect heat dissipation or cooling (Fig. 9-20), an engine initially releases its cooling energy to a coolant in a closed circuit (main circuit). A heat exchanger in the cooling system then transfers this heat to another coolant (secondary circuit).

Water-cooled engines are divided into:
- Engines with fan cooling (air/water cooling): These are used wherever cooling water is unavailable for the secondary circuit (autonomous cooling systems), i.e. primarily vehicle engines but also stationary engines. The advantage of zero water consumption must be obtained with the fan's consumption of power. High fan powers must be installed especially when the installation conditions in a vehicle are poor (< 1% of the rated power for cars, up to approximately 5% for medium commercial vehicle engines and approximately 10% for large commercial vehicle engines). So as not to unduly increase fuel consumption, such fans are regulated so that the maximum power consumption is only needed in critical cooling conditions. Air cooling systems do not have any components or lines that conduct

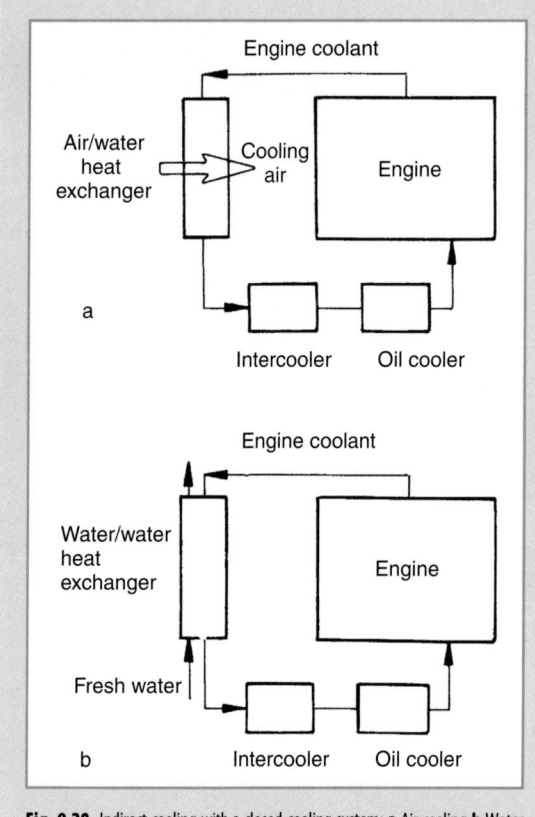

Fig. 9-20 Indirect cooling with a closed cooling system; **a** Air cooling **b** Water cooling

fresh or sea water. However, the higher system costs and larger space requirements are drawbacks.
- Engines with water/water cooling in a closed secondary circuit: The secondary circuit's fresh water (external water) is conducted from above into a cooling tower where it is distributed over a large area or atomized in a natural or fan generated counterflow air draft and releases its heat to the air by evaporation and cooling. The cooling capacity depends on the air temperature, flow rate and humidity. The water loss is approximately 3%. The greater complexity of equipment, problematic antifreeze protection and plume formation are disadvantages. This type of cooling only makes sense for large plants.
- Engines with water/water cooling in an open secondary circuit: The fresh or raw water circuit (secondary circuit) is open. The water is freshwater (river or lake water) or sea water or brackish water. Fresh open water, especially sea water, constitutes a virtually infinite heat sink. However, this cooling has a number of problems in practice, which must be taken into account when designing cooling circuits. Hull-mounted and keel cooling are special forms of this type of cooling (Fig. 9-21).

Cooling System Design

Main and secondary circuit cooling: The heat exchangers are connected in parallel on the secondary side, i.e. the external water side (Fig. 9-22).

Single, double and multiple circuit cooling: The heat exchangers in these systems are connected in a series on the secondary side. Several configurations are possible:
- *Single circuit system:* In a single circuit system, the components being cooled to the desired temperatures (cooling

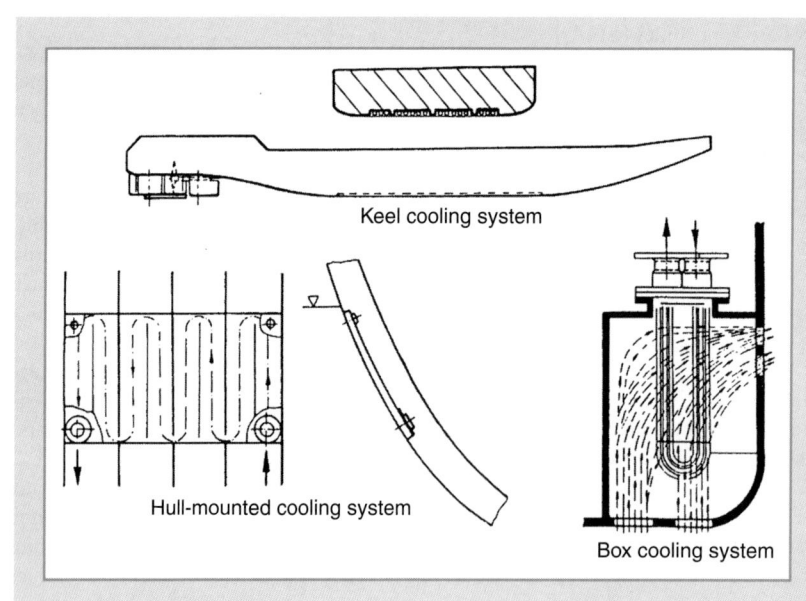

Fig. 9-21 Water cooling for inland ships

– *Double circuit system:* A better adaptation of the cooling system to the variables such as engine load, heat requirement or fresh water temperature is obtained by thermally separating the circuits into a high and low temperature circuit (HTC and LTC). LTC and HTC heat exchangers are connected in a series on the fresh water side (on the cooling air side in vehicle engines) (Fig. 9-24). This also has the advantage of functioning with smaller volumetric fresh water flows. The allocation of the heat flux to the subcircuits depends on the design of the system: Which heat exchangers are in which branches [9-48]? Thus, for example, vehicle engines and marine and generator engines often have a mixed double circuit system in which the air intercooler is placed upstream from the air/water heat exchanger.

– *Mixed double circuit system:* In this cooling system, also known as an "integrated cooling system", the HTC and LTC are hydraulically connected (Fig. 9-25). The HT cooling water is cooled by mixing it with the LT cooling water. This eliminates the HTC heat exchanger. However, this requires meticulous design work to ensure that the desired coolant flow is present in every branch of the circuit. In compact engines with fully integrated accessories, one branch of the HT and one of the LT circuit are united (Fig. 9-26). Approximately two thirds of the coolant flow passes through the high temperature circuit (HTC) and the rest through the low temperature circuit (LTC). The HTC cooling water is cooled by mixing it with the LTC cooling water and not by a heat exchanger. At low and part load, the LTC is controlled so that the heated engine cooling water reaches the intercooler directly through a bypass line. The subcircuit remains uncooled and heats the charge air. The engine cooling water is increasingly sent through the heat exchanger and the intercooler as the engine temperature rises [9-50].

Combining these features and placing the heat exchangers in the individual branches of the coolant circuits furnishes many options to design and optimize a cooling system based on the particular engine design and the placement of the cooler in the secondary circuit. Sea water operation in particular must allow for contamination and deposits [9-51].

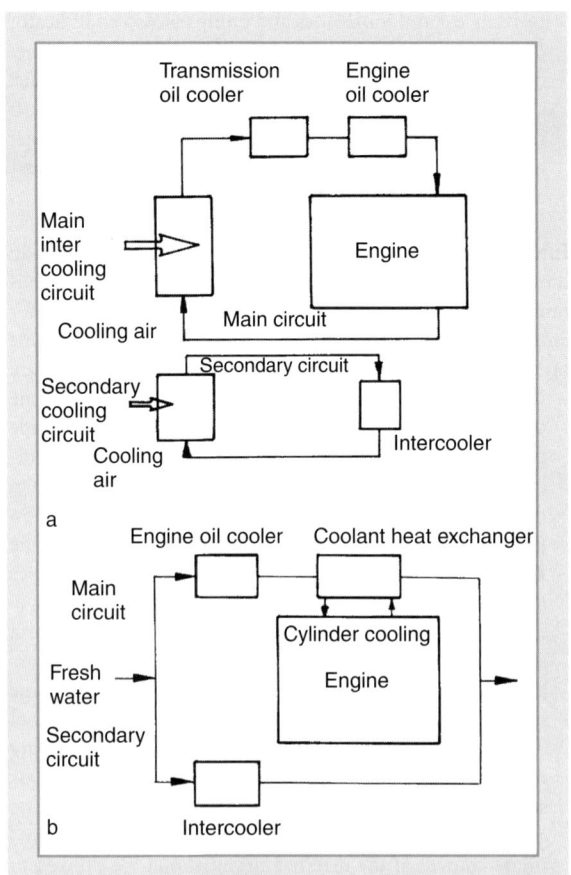

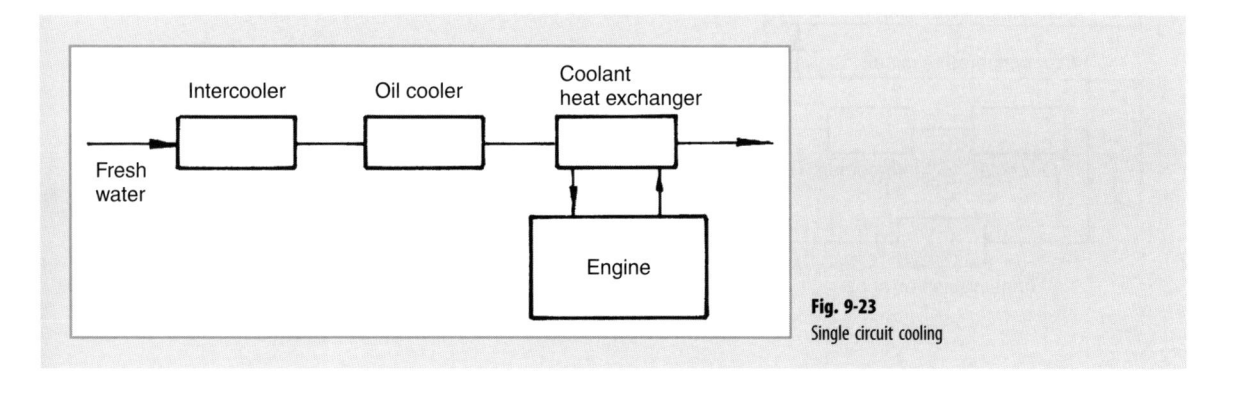

Fig. 9-22 Main and secondary circuit cooling: The cooling system is connected in parallel on the coolant side (secondary side). The greater part of the heat is dissipated in the main circuit, the lesser part in the secondary circuit: **a** Cooling with air; **b** Cooling with fresh water

priority) are connected in a series and/or parallel in the cooling circuit (Fig. 9-23). The heat exchangers in the cooling circuit interact. When configured appropriately, this is exploited for the purpose of "self-regulation" [9-48, 9-49].

Fig. 9-23 Single circuit cooling

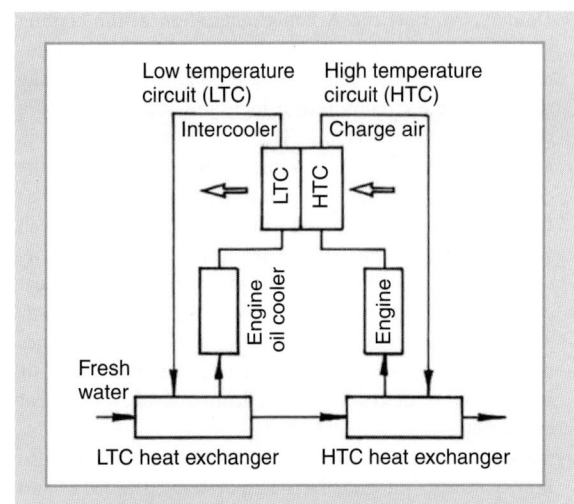

Fig. 9-24 Double circuit cooling with a high (HTC) and low temperature circuit (LTC) and with fresh water cooling

Cooling systems for sea water operation: These can be divided into:
- "Conventional" sea water cooling (Fig. 9-27): The technically simple thermal engineering advantage of greater inlet temperature differences and lower pump outputs resulting from smaller volumetric sea water flows entails accepting the disadvantages of sea water operation.
- Central cooling (Fig. 9-28): In light of the disadvantages of conventional cooling, central cooling in which a large central cooler cools the fresh water of all other heat exchangers is usually preferred. The high technical complexity generated by more as well as larger coolers (lower inlet temperature differences and additional heat transfer resistances) and more lines and pumps is also an indicator of the problems of indirect sea water cooling. On the other hand, fewer lines and components are needed to conduct sea water. Since the temperature in the fresh water circuits remains largely constant, the cooling system proves easier to control. Therefore, central cooling is advantageous, above all, for multiple engine systems.

Intercooler placement: The placement of the intercooler in the cooling circuit is determined by the demand for a minimum charge air temperature at full load and by the need for an acceptable charge air temperature at the particular engine operating points. There are several solutions to choose from:
- *Internal intercooling:* Intercoolers integrated in the engine cooling water circuit (main circuit) on the coolant side produce a certain self-regulation of the charge air temperature (Fig. 9-29). While this can only cool the charge air to the level of the engine cooling water temperature at best (power losses), the engine load adjusts (improved engine running properties). The charge air is cooled in the upper load range and heated in the lower.
- *External intercooling:* The intercooler/s is/are placed in a secondary circuit and coolant flows through it/them parallel to the other heat exchangers (Fig. 9-23). Thus, a large inlet temperature difference is on hand, which can cool down charge air even more. Efforts have also been being made for roughly twenty years to exploit this advantage for smaller engines (commercial vehicle engines). Since the charge air is thermally decoupled from the engine, appropriate control of the external circuit should be utilized to adjust the temperature to the engine's load level.
- *Two and multistage intercooling:* The intercooler is divided into one subcooler connected to the high temperature circuit and one subcooler connected to the low temperature circuit (two-stage intercooling in Fig. 9-24). When the

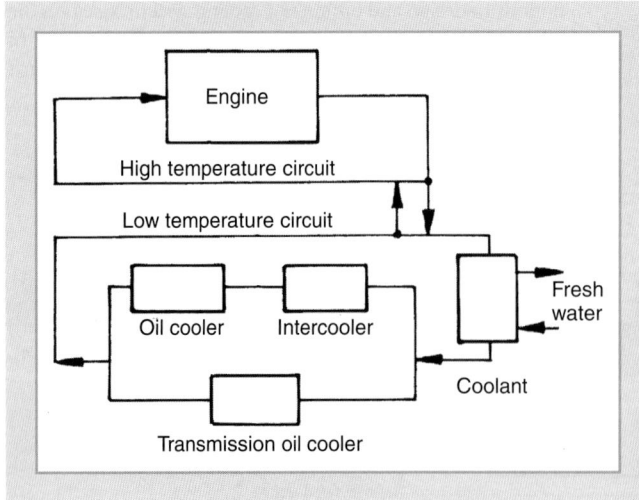

Fig. 9-25
Mixed double circuit cooling: High and low temperature circuit are hydraulically connected

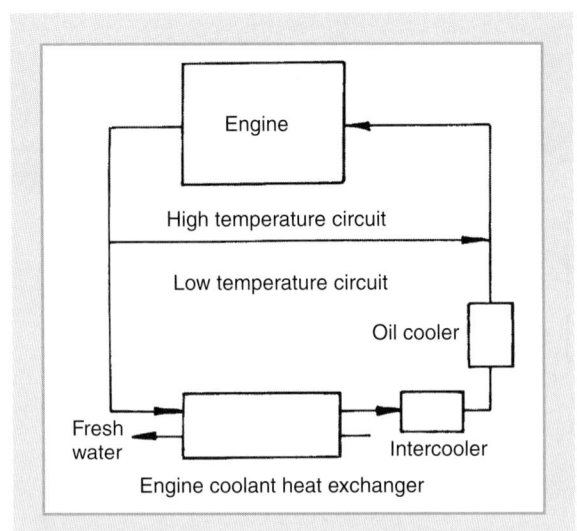

Fig. 9-26 Mixed double circuit cooling with combined branches of the high and low temperature circuit

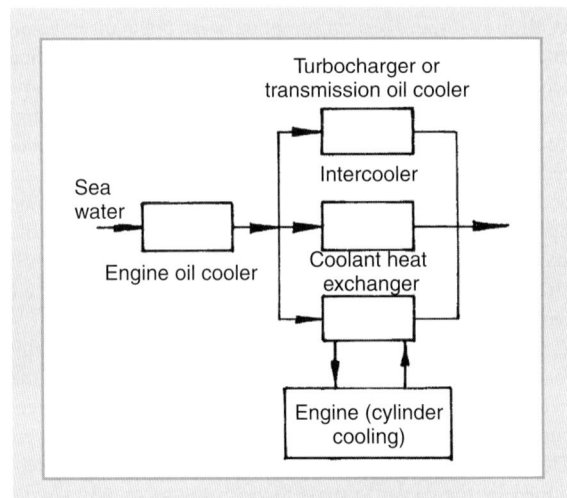

Fig. 9-27 "Conventional" sea water cooling (single circuit system)

Such complexly configured cooling systems are implemented with numerous flow branches and unions and components with different flow resistances (see Sect. 9.2.4). The pressure and volumetric flows are modeled for individual branches of the circuit and verified in a test of the system or when it is installed. Where necessary, the individual partial flows may be corrected with adjustable restrictors [9-52].

Cooling Systems in Vehicle and Compact Engines

Engines for vehicles, mobile or transportable generator units and all types of fast boats assume a special position. Since compact design is required, the cooling system components also have to be integrated in the engine assembly or situated as closely to the engine as possible [9-53, 9-54].

The actual engine cooling water cycle in vehicle engines or engines with similarly compact designs includes the engine itself, radiator, lines, expansion tank, temperature controller and circulation pump. Powered directly by the engine, the engine cooling water pump forces the cooling water through the heat exchangers (oil cooler and intercooler), which are each connected in a series or in parallel, into the engine's cooling chambers (cylinders and cylinder heads) and through other components being cooled, e.g. exhaust gas turbochargers, exhaust manifolds, etc., or through other heat exchangers, e.g. oil coolers, exhaust gas recirculation coolers, etc.. The cooling water flow is divided according to the cooling capacity requirement. A temperature controller ensures that the cooling water in a cold engine reaches the suction line to the engine cooling water pump completely or partially through a line that bypasses the radiator (Fig. 9-29). This ensures that the operating temperature is reached quickly and the desired temperature maintained even when engine load varies.

Positioned at the highest point of the cooling circuit, the cooling water expansion tank collects the changes in cooling water volumes when temperatures fluctuate and removes exhaust gas (e.g. during cylinder head gasket blow-by) and air (e.g. inlet through water pump gasket) from the cooling circuit. Air bleeding and venting of the cooling circuit is so important because air or gas in the cooling water reduces throughput. The cycle becomes unstable at an air content of 12% and collapses at 15% [9-55]. Moreover, heat transfer is diminished or a corrosion attack facilitated. In addition, the expansion tank contains a coolant reserve for smaller leaks and creates a certain pressure buffer.

A combined pressure and suction valve provides the pressure equalization necessary for excess pressure produced when the cooling water is heated and when a vacuum is produced when it is cooled. Continuous bleed lines run from the engine – as well as from the air/water radiator in vehicle engines – to the expansion tank. In addition, a connecting line runs from the expansion tank to the suction side of the cooling water pump.

The cooling water's flow through the cylinder and cylinder head is laid out meticulously. In engines with a water chamber shared by all cylinders, defined *flow conditions* must ensure

charge air enters the cooler in the HT circuit, its high temperature can increase the coolant temperature for heat consumers (e.g. fresh water producers) to 98 °C and correspondingly increase the useful heat gradient. The majority of the charge air heat is dissipated in the HT circuit (1.5–2.6:1). The low temperature stage is switched off in part load operation ($\leq 40\%$, particularly at full speed) or in arctic operation and the charge air is heated by the engine cooling water in the HT circuit. This translates into better combustion, lower firing pressures and less smoke emission. Engine cooling water heating may be switched in when the engine load drops below 15% [9-48].

Fig. 9-28 Central cooling

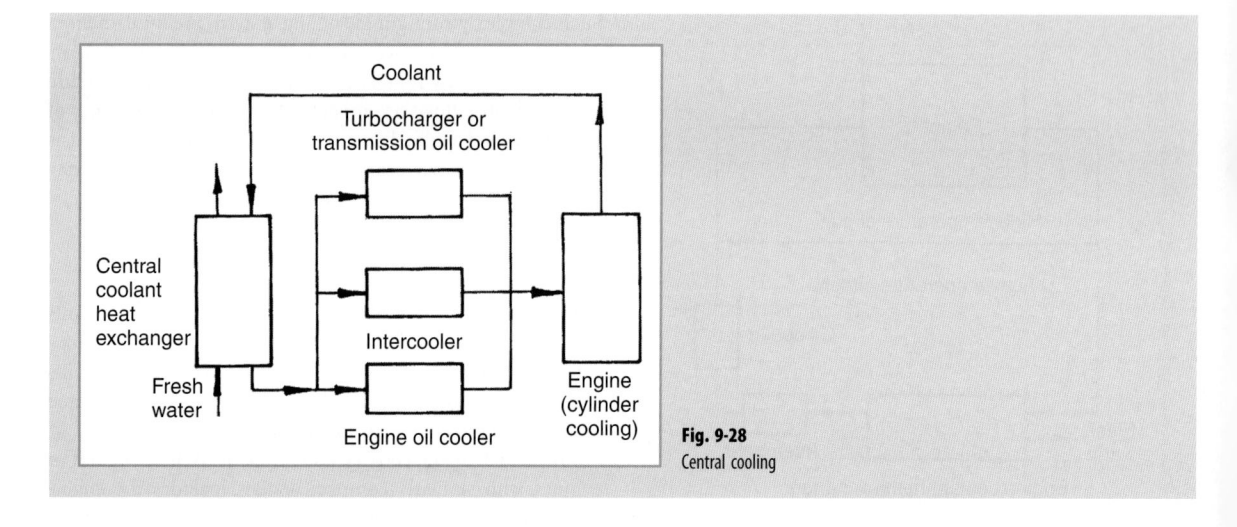

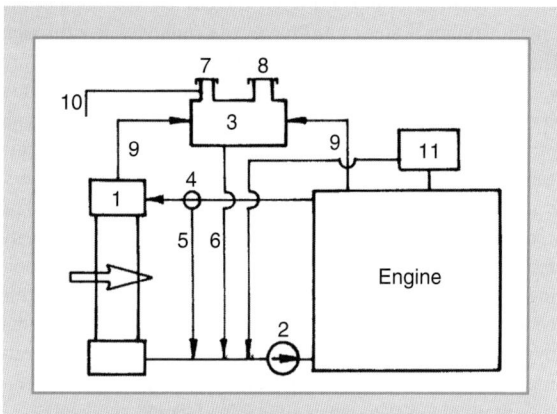

Fig. 9-29 Commercial vehicle engine cooling. *1* Radiator; *2* Water pump; *3* Expansion tank; *4* Thermostat; *5* Bypass line; *6* Filling line; *7* Working valve; *8* Fill cap with safety valve; *9* Bleed lines; *10* Overflow line; *11* Heat exchanger for cab heating

that the cooling water flows uniformly around the thermally loaded zones (cross flow). On the one hand, the flow velocity must be high enough for heat exchange. On the other hand, cavitation should not occur. Pressure losses should remain low as well. Conditions prove to be simpler in large single cylinder engines because the cylinders are individually supplied from a manifold. Since cylinder heads have a complicated structure in which aspects of thermal and mechanical strength and castability are prominent, experience is required to design the cooling water paths in them so that temperatures that are as uniform as possible set in. Neither leakage nor the venting of water when they are heating up should cause air pockets to form. Local boiling should not cause any aggregation of vapor bubbles. The cooling water flow is optimized with flow simulations (CFD) and can be tested in flow tests on a Plexiglas model.

9.2.3 Cooling System Control

9.2.3.1 Control Requirements

With an eye toward fuel consumption, wear, pollutant emission and noise, efforts are being made to reach the engine operating temperature quickly, to maintain desired temperatures regardless of engine load (preventing undercooling and overheating) and to lose a minimum of power through additional assemblies (e.g. cooling fan) in order to sustain cooling. Moreover, the coolant circuit must be able to adjust to the demands of engine operation and additional boundary conditions. In addition, the coolant circuit must not only adapt to the cooling load, which changes as the engine load changes, but also different air temperatures (e.g. seasonal and geographic conditions).

Engines are also utilized to different capacities, e.g. full load, part load and idle. In an initial approximation, the heat that accumulates from the engine cooling water, intercooler or engine oil cooler is a function of the engine power. The heat exchangers' cooling capacity is primarily determined by the coolant's mass flows in the secondary circuit and the fluid being cooled. Thus, the cooling capacity requirement and supply diverge whenever the coolant flow changes disproportionately to the engine power.

Since the heat exchangers and the entire cooling system are designed corresponding to an engine's rated power for the maximum cooling capacity required (cooling load), it must be assured that the desired (intended) temperatures are obtained even when the engine operating points diverge. Thus, control of coolant temperatures is indispensible. The temperature of the coolant is regulated in the main circuit.

Return temperature: The coolant temperature in the engine outlet serves as an *index for the thermal state of the engine*. The actuating element is a thermostatic control valve equipped with a temperature sensor. The desired temperature is set by mixing cooled cooling water conducted past the radiator in the bypass with uncooled cooling water in the control valve. The process is similar for oil coolers.

Flow temperature: The coolant temperature is regulated at the inlet. Implemented similarly, mixed temperature control is primarily applied to multiple engine systems with a central cooling system. Larger engines have electrically or pneumatically actuated PI controllers; smaller engines function with controllers without auxiliary power (expansion element). The engine cooling water's thermal inertia may be disregarded when the engine load changes; the cooling water reaches the operating temperature after 50–70 seconds, even in large medium speed engines. When their region of travel changes, marine engines use water pumps with pole reversing electric motorss to adapt the flow rate of sea water to the actual cooling requirements.

Preheating: Engines from which full power is demanded immediately after starting (especially quick-starting and no-break standby generating sets) must be preheated or kept heated, to be precise, by heating the engine cooling water to 40 °C by means of electric preheating equipment or thermal heat exchangers. Larger and large engines are usually preheated (see Table 9-6).

9.2.3.2 Fan-Cooler Combinations

Fan-cooler combinations, i.e. primarily for vehicle engines and stationary units, use the fan speed to control the flow of cooling air to establish the desired cooling water temperature. This can effectively reduce energy consumption and noise emission. The temperature set by the fan is higher than the thermostat's operating range. The fan may be powered:

- *Mechanically:* The rigid connection of fan speed to engine speed does not allow any control intervention and is thus disadvantageous. When bimetal viscous clutches are employed, the speed can be regulated as a function of the cooling air temperature (based on the vehicle radiator). Modern electrically controlled clutches can control fan speed as a function of any reference variable.
- *Electrically:* Fan speed is independent of engine speed. The fan can continue running when the engine is stopped. This provides flexibility in the arrangement of the fan and radiator.
- *Hydraulically:* A hydraulic variable speed coupling produces the drive. The control range is limited. The fan must be placed directly at the engine. All in all, this is a complex solution primarily used for larger vehicle engines when the power input for the fan can no longer be transmitted by a belt without problems and vibrations in the drive train need to be damped.
- *Hydrostatically:* Higher power inputs can also be transmitted over greater distances. Thus, the radiator can be positioned flexibly. The control range and the power loss are large, unit volume and mass on the other hand are small.

There are various options for the design of mechanically powered fans. Axial fans with small gaps (approx. 8 mm) are used for maximum demands on efficiency and air mass flow. Given the engine motion, such a small gap necessitates firmly installing the frame ring around the fan on the engine. Were the frame not firmly on the engine, the relative motion would have to be compensated by a large gap (20–30 mm). Shrouded fans are used to prevent a strong reverse flow through the gap and flow dispersion. In addition to higher fan gap tolerances, this arrangement has advantages for acoustics and flow stability [9-56].

9.2.4 Implemented Cooling Systems

9.2.4.1 Cooling Systems for Commercial Vehicle Diesel Engines

Figure 9-29 illustrates the standard cooling system described here. The heat exchanger output (approximately 6–10 kW) needed to heat the cab is not incorporated into the design of the radiator circuit. However, to prevent the engine cooling water temperature from dropping too greatly, no more than 20% of the heat may be drawn from the cooling circuit. The heat from the engine cooling water in highly boosted engines may be insufficient to cover the heat requirement, particularly in raised roof cabs (approx. 20 kW). In such cases, an additional heating unit must be designed in. Other heat exchangers (for retarders and transmission oil) are connected in line with the engine between the water outlet out of the engine and the thermostat.

Extreme requirements are imposed on the radiators of armored vehicles, which must dissipate sizeable heat fluxes from the engine and transmission under extreme ambient temperatures (−30 °C to +45 °C). This requires a large cooling air mass flow and a large gradient from the cooling water temperature to the outside temperature, e.g. 95 °C (110 °C) at an outside temperature of 20 °C (45 °C). The cramped space in a tank requires an extremely compact radiator design. In addition, the fan must overcome high intake and pressure losses at correspondingly higher fan output (e.g. 13.5% of the engine power of 1,100 kW in a Leopard II tank).

9.2.4.2 Diesel Locomotive Cooling Systems

The cooling system of the Deutsche Bahn's 215, 218 and 210 series diesel locomotives is constructed as a two circuit system with a high and low temperature circuit (HTC and LTC) (Fig. 9-30). The actual cooling system consists of two radiator cores configured in a V-shape through which a hydrostatically or hydrodynamically powered axial fan aspirates the cooling air. The fan speed is regulated so that the coolant temperature remains nearly constant as a function of the engine load and outside temperature. A mixing valve (with a response temperature of 30 °C) hydraulically connects the LTC and HTC to reduce undercooling of the LTC when outside temperatures are low. A second mixing valve (approximately 60 °C) enables preheating both cooling circuits and keeping them warm by a heater and a circulation pump. The quantity of

Fig. 9-30 Radiator of Deutsche Bahn's model 215, 218 and 210 diesel locomotives (double circuit cooling with LTC and HTC; see Fig. 9-24): **a** schematic of the cooling circuit; **b** radiator in V configuration

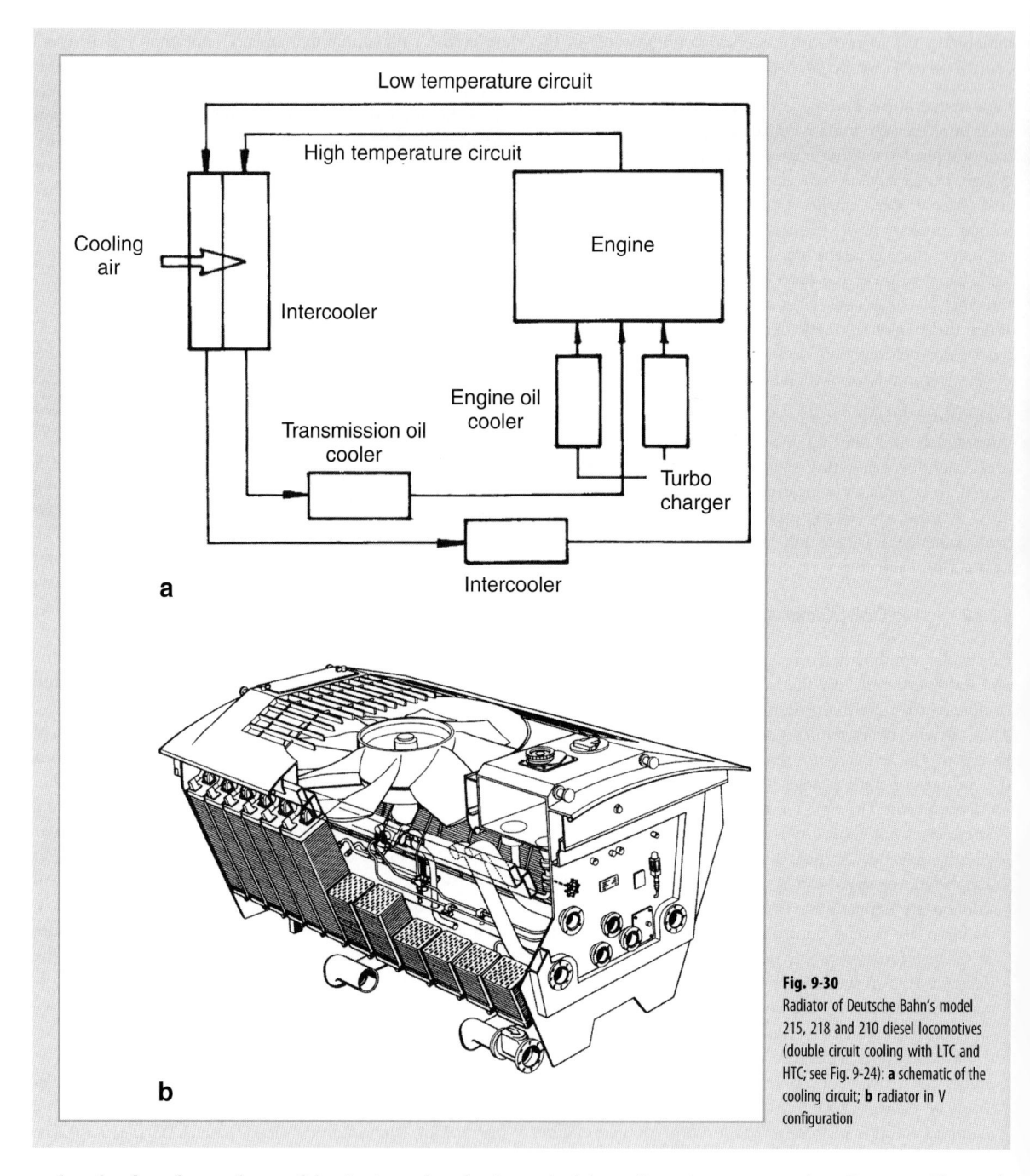

coolant that flows from each one of the circuits to the other is returned to the expansion tank [9-57, 9-58].

9.2.4.3 Cooling Systems for Marine Propulsion Systems

Figure 9-31 is a detailed schematic of a marine engine system's central cooling system with a hydraulically coupled HTC and LTC (integrated cooling system) including pumps, expansion tanks and other accessories. The principle configuration corresponds to Fig. 9-26. This cooling system is the basis for the MAN B&W 40/54 to 58/64 engine series.

9.2.4.4 Vehicle Cooling System Cooling Modules

In vehicles, the basic components necessary for cooling are united in a cooling module that normally consists of a radiator, intercooler, air conditioner condenser and fan

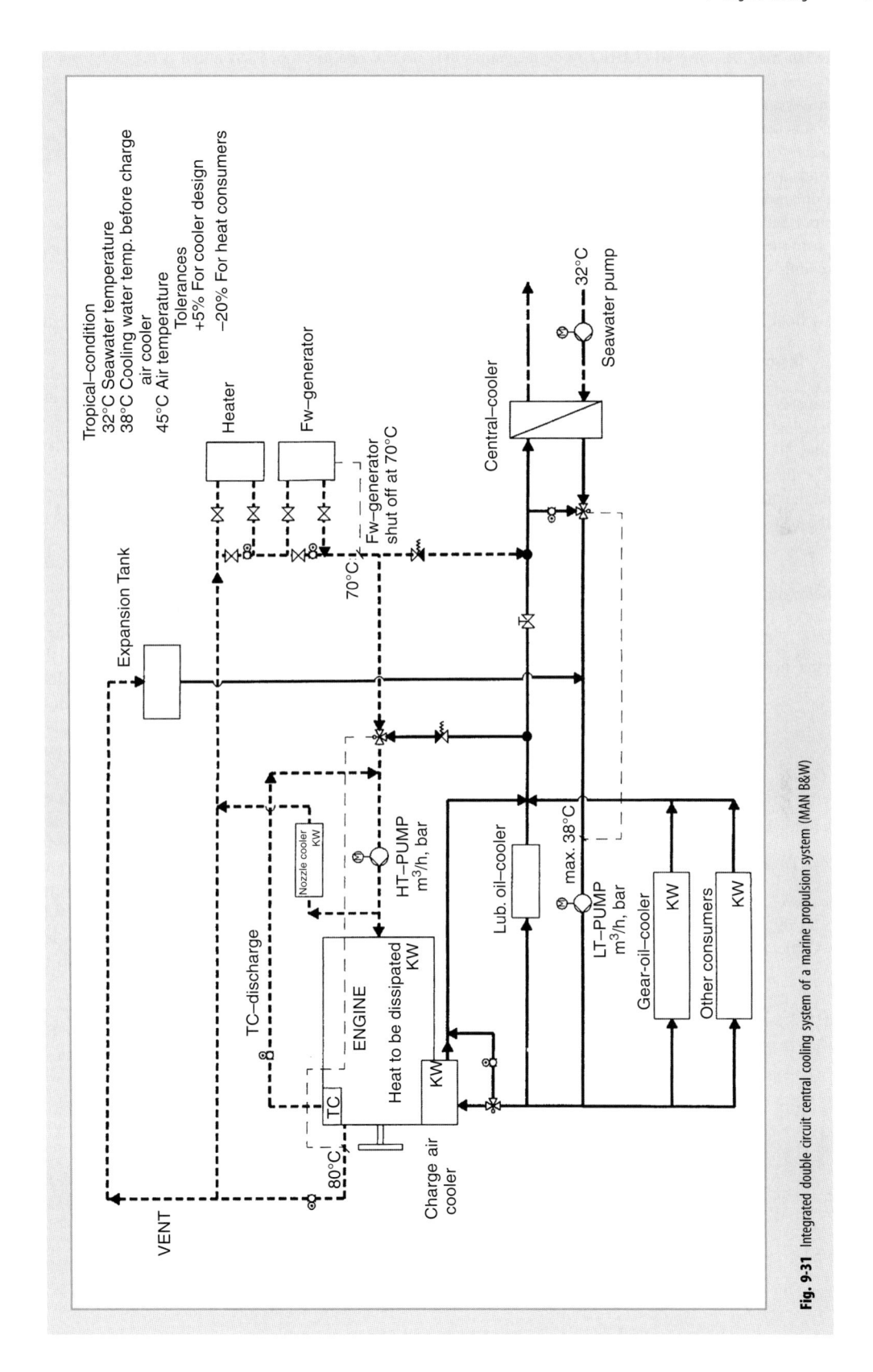

Fig. 9-31 Integrated double circuit central cooling system of a marine propulsion system (MAN B&W)

cowl. The fan may be powered electrically or mechanically directly by the engine. Hydrostatic fan drives are frequently used in busses and special vehicles.

The arrangement of the heat exchangers in the cooling air flow ensues from the particular inlet and target temperatures of the fluids being cooled. Normally, this results in the sequence of condenser, intercooler, radiator and fan.

It is important that air guides and sealings prevent heated air from being recirculated in order to optimally utilize the cooling module's installed cooling capacity.

9.2.5 Heat Exchangers

9.2.5.1 Thermal Engineering

Only stationarily operated heat exchangers with fixed baffles between two fluids, so-called *recuperators*, are treated here [9-59].

The index "1" always designates the exothermic fluid and the index "2" the heat absorbing fluid. The index "e" stands for "inlet" and the index "a" for "outlet".

Thus, the following applies to the cooling of the mass flow $\dot{m}_1$:

$$\Delta T_1 = T_{1e} - T_{1a} \geq 0 \qquad (9\text{-}1)$$

and the following to the heating of the mass flow $\dot{m}_2$:

$$\Delta T_2 = T_{2a} - T_{2e} \geq 0 \qquad (9\text{-}2)$$

Inserting the heat capacity flows

$$\dot{W}_1 = \dot{m}_1 c_{p1} \qquad (9\text{-}3)$$

$$\dot{W}_2 = \dot{m}_2 c_{p2} \qquad (9\text{-}4)$$

yields the following for the heat transfer capacity

$$\dot{Q} = \dot{W}_1 \Delta T_1 = \dot{W}_2 \Delta T_2 \qquad (9\text{-}5)$$

When the power is related to the largest possible temperature difference in the heat transfer, i.e. the difference of the two mass flows' inlet temperatures

$$\Delta T_e = T_{1e} - T_{2e}, \qquad (9\text{-}6)$$

then, with $\dot{W}_{min}$ as the smaller of the two heat capacity flows, the following is obtained

$$\frac{\dot{Q}}{\Delta T_e} = \frac{\dot{W}_1 \Delta T_1}{\Delta T_e} = \frac{\dot{W}_2 \Delta T_2}{\Delta T_e} = \dot{W}_{min}\varepsilon \qquad (9\text{-}7)$$

with

$$\varepsilon = \frac{\Delta T_1}{\Delta T_e} \quad \text{for } \dot{W}_{min} = \dot{W}_1 \qquad (9\text{-}8)$$

or

$$\varepsilon = \frac{\Delta T_2}{\Delta T_e} \quad \text{for } \dot{W}_{min} = \dot{W}_2 \qquad (9\text{-}9)$$

In this relationship, ε is denoted as the *exchanger efficiency* or *operating characteristic* [9-60, 9-61].

It can be demonstrated that ε is a function of the number of heat transfer units

$$N = \frac{kA}{\dot{W}_{min}} \qquad (9\text{-}10)$$

the ratio of the heat capacity flows

$$R = \frac{\dot{W}_{min}}{\dot{W}_{max}} \qquad (9\text{-}11)$$

and the flow configuration in the heat exchanger [9-61, 9-62]. In Eq. (9-10), kA is the product of the coefficient of heat transmission k and the area of heat exchange A, which cannot be separated expediently in finned surfaces – as usually exist here.

The VDI Heat Atlas [9-63] contains formulas and diagrams for different flow configurations (Fig. 9-32). It and [9-62] and [9-64] specify a mathematical method, the cell method, which may be used to calculate exchanger efficiency for any flow configurations.

The task when designing heat exchangers is to calculate the necessary exchanger efficiency ε from Eq. (9-7) at heat capacity flows specified on the basis of Eqs. (9-3) and (9-4) and the relative heat transfer capacity $\dot{Q}_c/\Delta T_c$ to be obtained. Depending on the intended flow configuration, the number of heat transfer units N and thus the kA value required according to Eq. (9-10) can be determined from the related formula or the relevant diagram. In many cases, N cannot be calculated explicitly from the specified formulas. Thus, it must be determined iteratively.

Since the magnitude of the heat capacity flows is often dependent on the selected size and design of the heat exchanger, e.g. a vehicle radiator dynamically pressurized with cooling air, an assumed size of the heat exchanger is postulated in practice and iterated by the mean fluid temperature until a satisfactory correspondence has been obtained. When necessary, the thusly calculated heat exchanger output serves as the basis to adjust the size of the heat exchanger to the performance requirements.

The heat transfer coefficients h_1 and h_2 and the coefficients of thermal conductivity of the baffle λ_w and the fins λ_{r1} and λ_{r2} are needed to actually calculate the kA value. Then, the Péclet equation applies:

$$\frac{1}{kA} = \frac{1}{h_1 A_1} + \frac{\delta_W}{A_W \lambda_W} + \frac{1}{h_2 A_2} \qquad (9\text{-}12)$$

A_w being the area of the flat baffle, δ_w its thickness and A_1 and A_2 the heat exchanging surfaces, rated in part with a fin efficiency η_r [9-65]. The heat transfer coefficients are normally specified as Nusselt numbers

$$Nu = h \cdot l/\lambda = Nu(\text{Re}, \text{Pr}) \qquad (9\text{-}13)$$

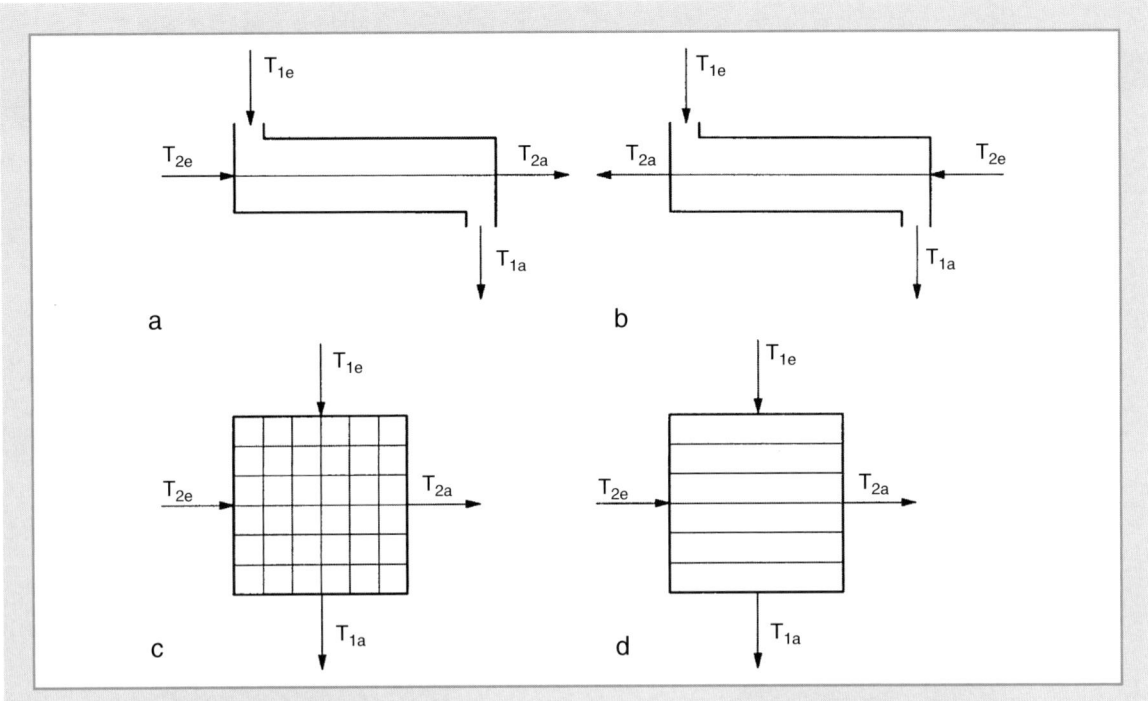

Fig. 9-32 Some basic forms of flow configuration in heat exchangers: **a** pure parallel flow; **b** pure counter flow; **c** pure cross flow; **d** cross flow cross-mixed on one side

The following

$$\mathrm{Re} = v \cdot l / \nu \tag{9-14}$$

is the Reynolds number and

$$\mathrm{Pr} = \rho \cdot v \cdot c_p / \lambda \tag{9-15}$$

is the Prandtl number, l being a characteristic dimension of the fixed body, λ the coefficient of thermal conductivity of the flowing medium, v its flow velocity, ν its kinematic viscosity, ρ its density and c_p its isobaric mass-specific heat capacity.

Heat exchangers for motor vehicle engine cooling are characterized by very compact designs. The transfer area A_1 or A_2 relative to the unit volume V of the heat exchanger matrix is typical. Values for this are generally available:

$$\frac{A}{V} > 700 \frac{m^2}{m^3} \tag{9-16}$$

However, the kA values relative to the matrix volume V are even more informative. Table 9-8 specifies some recommended values.

9.2.5.2 Radiators

A distinction is made between air-cooled and liquid-cooled radiators. Air-cooled heat exchangers used for vehicle diesel engines should be designed very compactly and lightweight but the drops in pressure allowable on the air and coolant side limit their compactness. Moreover, the finned tube core's sensitivity to contamination increases as the surface density on the air side increases in accordance with Eq. (9-16) and thus also establishes limits, especially for tractors and construction vehicles for instance. Finned tube cores consist of combinations of fins and tubes, which are differently shaped depending on the manufacturing process. A brazed radiator design that consists of flat tubes and corrugated fins brazed together is pictured in Fig. 9-33.

Mechanically joined radiators (Fig. 9-34) have round or oval tubes stuck through precisely dimensioned holes in louvered fins. These are flared mechanically to produce lasting contact pressure between the tube and hole to conduct heat from the tube to the fin. In both cases, the fins are provided with a multitude of gill-like notches to improve the heat transfer [9-66]. Made of glass fiber reinforced polyamide, the dispensing and receiving tank is interconnected with the particular header by an elastic seal. Solutions for brazed radiators with brazed-on aluminum receiving tanks also exist. The tubes may be provided with turbulence inserts to prevent laminar flow that might occur when coolant throughputs are small.

Mechanically joined radiators cost less to manufacture than brazed radiators. However, their heat transfer capacity relative to the air side end face or the air side drop in pressure

Table 9-8 Common maximum mass flows for specific cross sections and attendant kA/V values for a high performance heat exchanger. Fluid 1 exothermic, Fluid 2 heat absorbing

Fluid 1	Fluid 2	Specific mass flows		
		$\dot{m}^*_1$ kg/(m² · s)	$\dot{m}^*_2$ kg/(m² · s)	(kA/V) · 10⁻³ W/(m³ · K)
Coolant[a]	Air	250	12	370
Lubricating oil[b]	Air	150	6	200
Lubricating oil[b]	Coolant[a]	150	200	1,300
Charge air	Coolant[c]	12	40	280
Charge air	Air	20	10	100

[a] Mixture of water and glycol with a volumetric content of 70/30% at approximately 100 °C.
[b] Essolub HDX 30 at approximately 130 °C.
[c] Same as [a] but at approximately 70 °C.

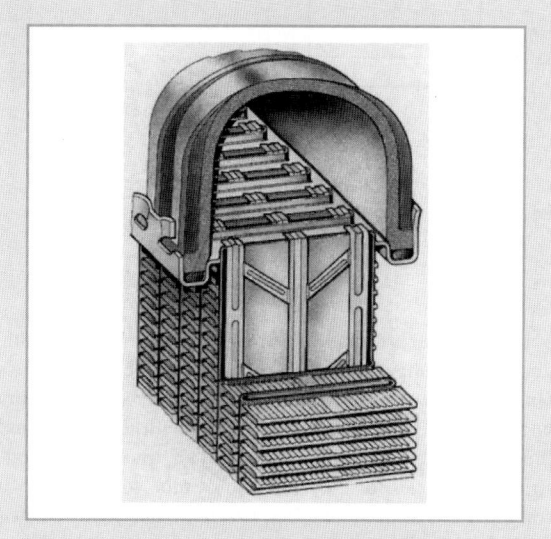

Fig. 9-33 Design of an aluminum brazed radiator for use in motor vehicles (source: Behr)

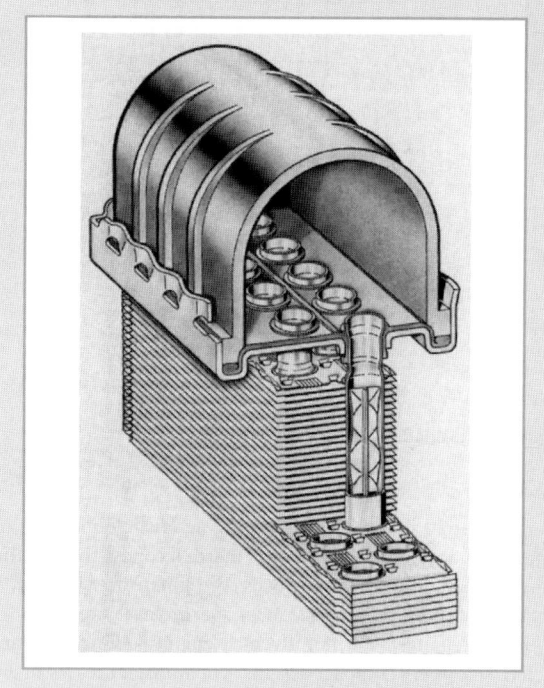

Fig. 9-34 Design of an aluminum mechanically joined radiator for use in motor vehicles (source: Behr)

is normally poorer. This makes larger radiators or higher fan powers necessary [9-67, 9-68].

Motor vehicle applications in European cars and commercial vehicles now almost exclusively use aluminum radiators.

Another air-cooled radiator design is the radiator core section pictured in Fig. 9-35 for a cooling system with a main and secondary circuit (see Sect. 9.2.2). A partial flow coming from the main circuit is cooled further in the secondary circuit in order to use it obtain a low charge air outlet temperature in an intercooler for instance. The core consists of louvered, usually smooth copper fins with holes for the flat copper tubes. A layer of lead-tin solder on the tubes connects tubes and fins. The headers, the receiving and dispensing tank and the cast connectors brazed to it consist of brass.

The tube-header is connected and the tank is brazed on with soft solder. Very rugged and insusceptible to contamination, such radiator elements are preferred in rail vehicles but also employed in stationary engines. Figure 9-36 illustrates the use of simple turbulent flow radiator core sections in an underfloor cooling system for the Deutsche Bahn.

Particularly vibrostable designs have to be selected for radiators mounted directly on engines. Deutz AG utilizes an air-cooled radiator in a shell or plate design for its new

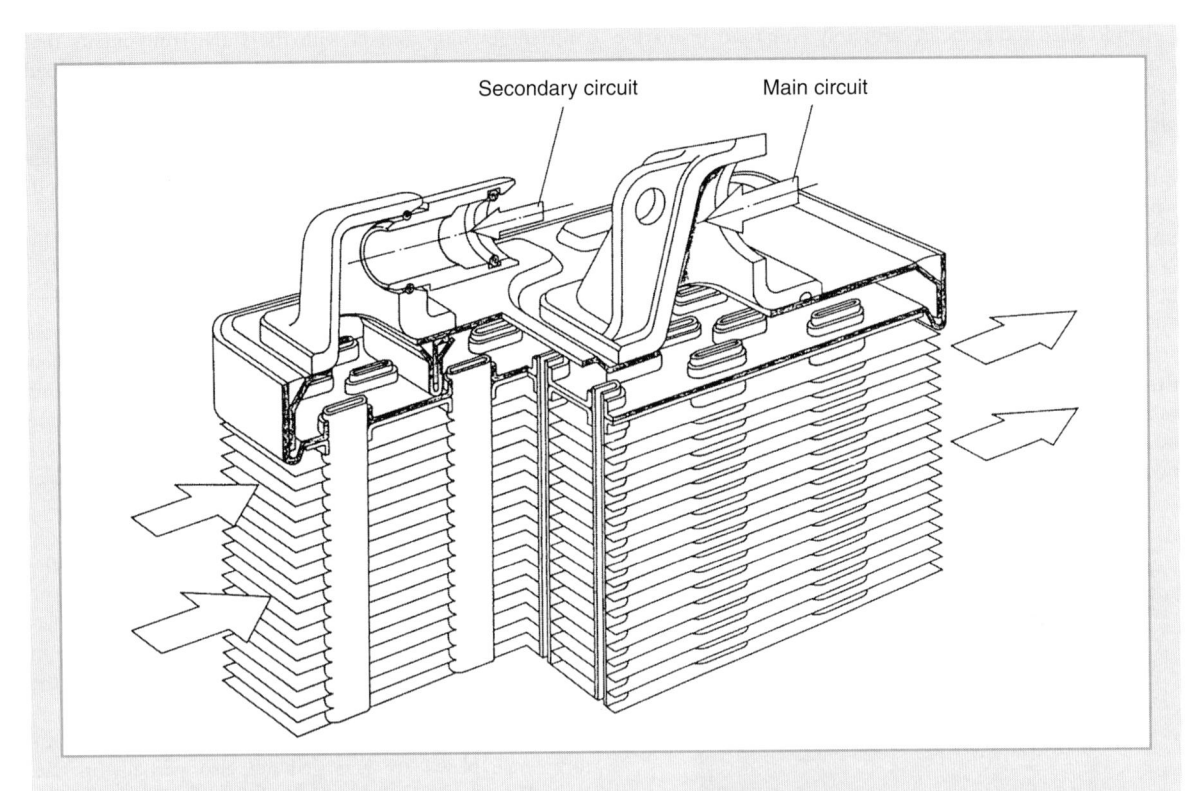

Fig. 9-35 Design of a nonferrous metal radiator core section with a main and secondary circuit in rail vehicles (source: Behr)

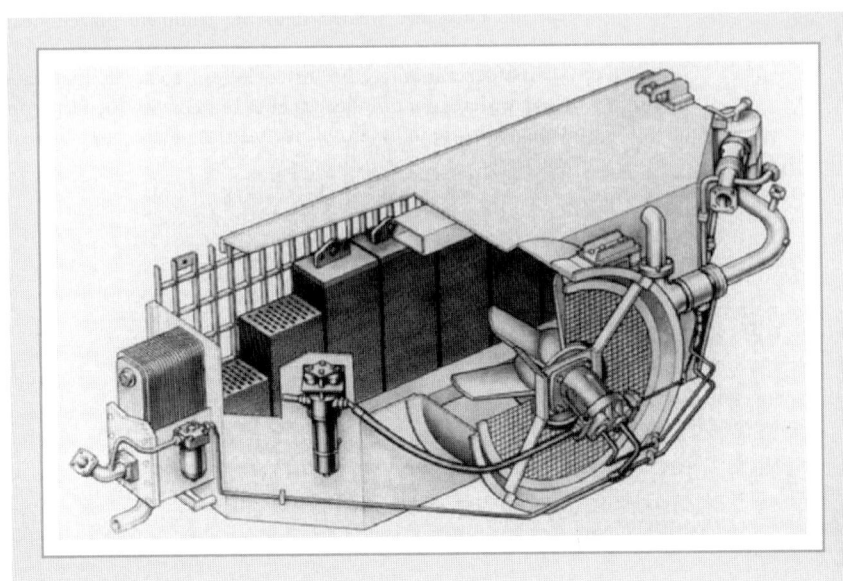

Fig. 9-36
Underfloor cooling system for a Deutsche Bahn railcar, consisting of ten radiator core sections with lengths of 700 mm and widths of 200 mm and a liquid-cooled transmission oil cooler in plate design and a hydrostatically driven fan. 405 kW are transmitted to the cooling air flow at an inlet temperature difference of 58 K (source: Behr)

generation of engines with an engine-integrated cooling system (see Sect. 18.2). Each of the elements that conduct coolant are formed of two aluminum half plates. Aluminum corrugated fins similar to those pictured in Fig. 9-33 are arranged between the individual elements. Outfitted with appropriate end plates and connectors, the entire radiator is brazed. Performance can be easily modified by the number of elements.

Liquid filled radiators are primarily employed in marine engines but also in part in stationary engines. Unfinned systems are normally employed since the heat transfer coefficients are similarly high on both sides of a heat exchanger. What is more, low sensitivity to contamination and good options for cleaning and resistance to corrosion are important when sea water is used. In addition to the classic tube bundle units [9-59], plate heat exchangers made of CuNi or titanium are used. They are usually installed separately from the engine but may also be integrated in the engine as in MTU's newer 595 engine series [9-69].

Figure 9-37 presents another sea water filled radiator design, which is also integrated in the engine. A flat tubular cooling unit is incorporated in a cast aluminum housing through which engine coolant flows. The sea water flows around the flat tubes in a register configuration and comes into contact with the CuNi components that are brazed together. The individual flat tubes are stamped with dimples, which are brazed to the dimples on the opposite side of the tube. This produces the mechanical strength required for the operating pressure of 6 bar gauge pressure as well as better heat transfer than in smooth tubes.

9.2.5.3 Oil Coolers

Oil coolers must be designed for higher operating pressures than radiators, i.e. between 10 and 20 bar. In addition, their heat transfer from the oil to the wall surfaces is far lower. This always requires having to work with turbulence generating structures on the oil side. Soldering them to the wall surfaces also achieves the requisite increase of internal mechanical strength.

Air-cooled oil coolers are made of aluminum virtually without exception. Figure 9-38 pictures an oil cooler in the flat tubular design used for motor vehicle diesel engines. The design of the finned tube core resembles that of brazed radiators. However, the tanks are made of aluminum because of the increased operating pressures. As Fig. 9-39 shows, the turbulence plates are formed as offset fins.

Instead of flat tubes, aluminum plates between which the turbulence plates are arranged are also partially in use. Brazed-in as a boundary, plates with a rectangular profile produce channels that conduct air and oil. Such pack construction has the advantage of making it possible to manufacture any radiator designed without special tools. The type-specific tanks are welded onto the complete brazed heat exchanger block.

Another variant of an air-cooled oil cooler is the plate design. Here too, turbulence plates brazed with the plates on the oil side are indispensible for reasons of strength and heat exchange.

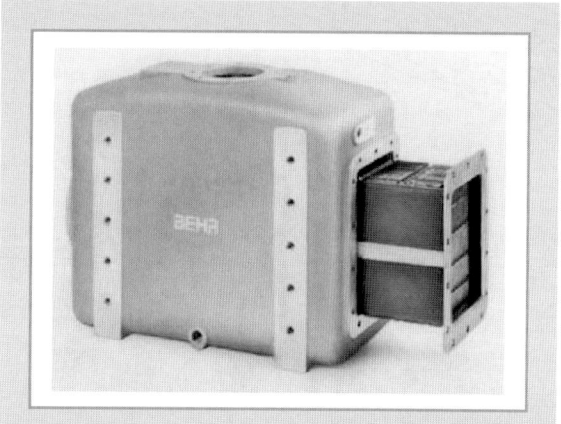

Fig. 9-37 Sea water filled radiator in flat tubular design for mounting on an engine (source: Behr)

Fig. 9-38 Air-cooled oil cooler in flat aluminum design for use in motor vehicles (source: Behr)

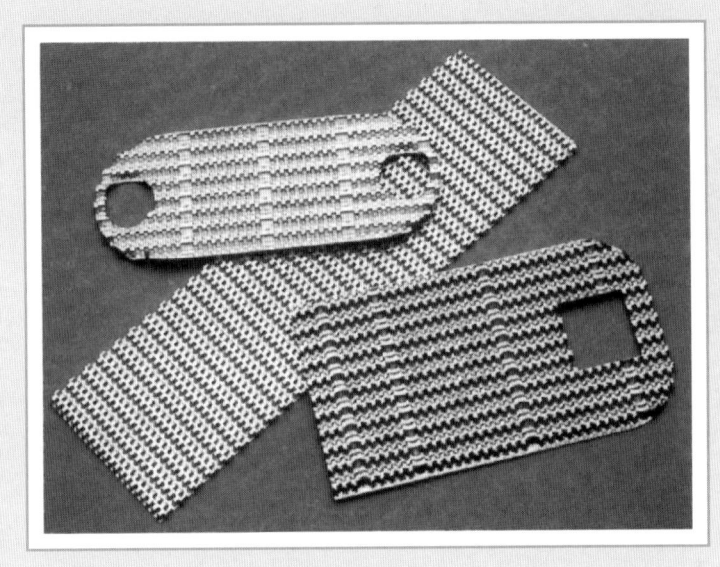

Fig. 9-39
Different turbulence plates for use in air and liquid-cooled oil coolers. Constructed of aluminum or steel (Behr)

Pictured in Fig. 9-40, aluminum stacked plate oil coolers are now only used for liquid filled oil coolers in car diesel engines. Thanks to the elevated edges of the plates, each of which is brazed with the next plate, these radiators do not need a separate housing. Turbulence inserts similar to the plates pictured in Fig. 9-39 are located on the coolant side and on the oil side or the plates already have a distinct structure that assumes this function. Such radiators are frequently bolted to the filter module with an oil filter as well.

The plate design illustrated on the right in Fig. 9-41 is frequently employed for heavy duty engines, e.g. for commercial vehicle applications. It has a longitudinal plate oil cooling unit in a special housing as pictured in Fig. 9-36 or incorporated directly in the engine block. The oil flows parallel through the individual plates equipped with turbulence plates and the coolant flows around the plate packs in the counter or cross flow.

The cut open round stack plate oil cooler pictured on the left in Fig. 9-41 can be installed under the oil filter and thus

Fig. 9-40
Stacked plate oil cooler for car applications (source: Behr)

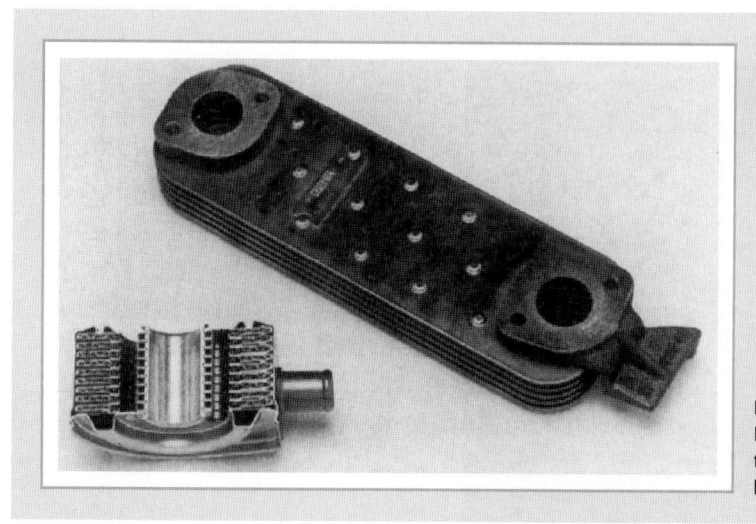

Fig. 9-41 Liquid-cooled oil cooler in plate design for installation under the oil filter (split, left) and in the engine block or a separate housing (right) (source: Behr)

integrated in the oil flow. Since the use of steel as a material entails higher weight, this product is now being used more and more rarely. Plate oil coolers are also occasionally manufactured entirely of aluminum or a CuNi alloy when sea water is used.

Not only plate oil coolers but also tube bundle heat exchangers are often implemented in marine and larger industrial engines. Allowing for oil's poorer heat exchange properties, the tube bundles are provided with intensive louvered finning over which oil flows based on the arrangement of the baffles. Another design utilizes smooth tube bundles and conducts the oil through tubes provided with special packing. Engine-integrated plate heat exchangers are also utilized for oil cooling [9-69]. However, since they are only furnished with a primary area of heat exchange, such units only have a limited capability to counteract the imbalanced heat exchange properties of oil and engine coolant or sea water.

9.2.5.4 Intercoolers

Intercooling increases the air density by reducing the air temperature. Thus, a larger mass of air is introduced into the combustion process. This results in higher engine power. Hence, the pressure drop Δp_1 on the charge air side has to be kept as low as possible when an intercooler is being developed. Otherwise, the pressure drop can overcompensate for the increase in air density induced by cooling and the intercooler causes a reduction in engine power. An efficiency factor

$$\eta_\rho = \frac{\dfrac{T_{1e}}{T_{1a}}\left(1 - \dfrac{\Delta p_1}{p_{1e}}\right) - 1}{\dfrac{T_{1e}}{T_{2e}} - 1} \qquad (9\text{-}17)$$

is introduced to evaluate the density recovery obtained in the intercooler [9-70]. The temperatures T_{1e}, T_{1a}, T_{2e} and T_{2a} have to be inserted in K. p_{1e} is the absolute pressure of the charge air at the inlet into the radiator. $\eta_\rho=1$ in an ideal, infinitely large heat exchanger without any drop in pressure. However, η_ρ can also assume negative values, i.e. the density decreases, when high pressure drops occur at low boost pressures.

For the most part, air-cooled intercoolers are implemented in motor vehicles. They are placed upstream from the radiator and, in part, in other positions in passenger cars. The construction of an aluminum radiator core resembles a brazed radiator. As in radiators, glass fiber reinforced polyamide tanks are used when charge air temperatures are below 190°C. Particularly durable plastics can even be used at up to approximately 220°C. At even higher temperatures, permanent mold cast aluminum tanks must be welded on as pictured in Fig. 9-39. The flat tubes contain inner finning, which, making allowance for the pressure drop, are only intended to moderately increase the surface area.

Unless they are used in other vehicle or stationary engines, air-cooled intercoolers are normally constructed as an aluminum pack as described for oil coolers in Sect. 9.2.5.3. The fins on the charge and cooling air side have a design similar to that in Fig. 9-42. In turn, the advantage of pack construction is the ability to cost effectively manufacture small quantities of compact coolers with individual dimensions. Flat tubular systems made of steel or nonferrous metal are also used wherever space and weight play a subordinate role. To facilitate a low pressure drop on the charge air side, the flat tubes often do not contain any inner finning. The cooling capacity required must then be achieved by designing the radiator end face large enough.

Liquid-cooled intercoolers are preferable for charge airflows that facilitate flow and thus prevent pressure losses. They can be harmoniously integrated in an engine's charge air line [9-69, 9-71]. In addition, they are built even more

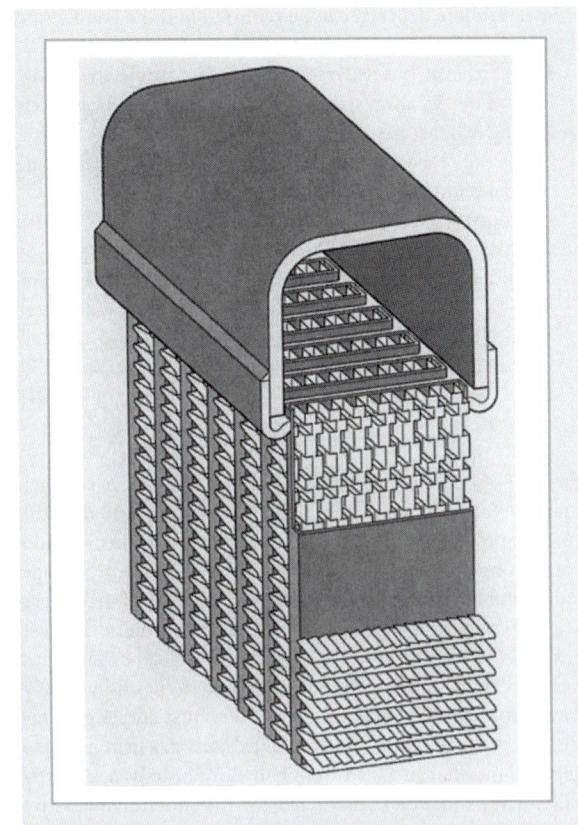

Fig. 9-42 Design of an aluminum brazed intercooler for use in motor vehicles (source: Behr)

Fig. 9-43 Sea water filled, round tube intercooler for installation in the charge air line of marine diesel engines (Source: Behr)

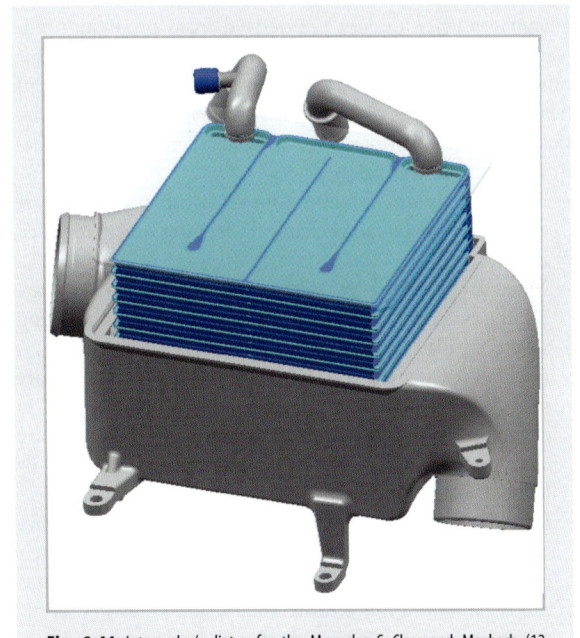

Fig. 9-44 Intercooler/radiator for the Mercedes S Class and Maybach (12 cylinder bi-turbo, M275)

compactly than air-cooled designs because their kA/V values have to be larger (see Table 9-8) and the ratio R of the heat capacity flows is better. Moreover, a cross-counter flow configuration is often implemented to obtain high exchanger efficiencies.

With the exception of motor vehicle engines, larger boosted diesel engines and marine engines in particular are now the main field of application for liquid-cooled intercoolers. This technology is also already being increasingly implemented in motor vehicles. Figure 9-43 pictures such a cooler with a heat exchanger matrix that intercools a marine diesel engine. It consists of round tubes with an outer diameter of 8 mm and lightly corrugated, louvered copper fins with a thickness of 0.12 mm. The fins and tubes are connected by flaring the tubes as in mechanically joined radiators (see Fig. 9-34). A CrNi or CuNi alloy is employed for the tubes depending on the individual case. The steel or special brass tube plates are connected by rolling in the tubes. Gray cast iron or an AlSi or CuAl alloy is selected for the receiving and dispensing tanks depending on the coolant used. Since the charge air and coolant temperatures can differ considerably, one of the plates is designed as a sliding plate, partly to prevent thermal stresses. Aluminum can then be utilized for the cast side sections. Otherwise, gray cast iron is preferred to obtain ratios of expansion similar to those in the tubes.

A liquid-cooled intercooler Mercedes Benz has been using for several years in its twelve cylinder biturbo engine that powers the Mercedes S Class and the Maybach is illustrated in Fig. 9-44. This intercooler is made completely of aluminum. The heat exchanger matrix consists of plates, each plate pair forming a narrow channel for the coolant routing. Fins that absorb the heat of the charge air and release it to the coolant are arranged between these plate pairs. The housing is made of cast aluminum.

Other design variants exist for commercial vehicles.

Generally, liquid filled intercoolers are also suitable as *charge air preheaters* and are used as such with the identical design.

9.2.5.5 Exhaust Gas Heat Exchangers

Designs, Forms of Construction

Exhaust gas heat exchangers are gas/water (oil) heat exchangers. So far, every form has been designed with varying success, i.e. smooth and finned tubular heat exchangers with the gas flow configured in the tubes (gas tube design) and the external gas flow oriented vertically and horizontally. The diversity of designs arises from the cramped installation conditions (vertical and horizontal designs or the use of finned tubes from air cooler production), from manufacturing advantages when designs are derived from production air/water coolers or from particular price constraints. On the other hand, there are clear rules for engineering exhaust gas heat exchangers for diesel engines, which stem from regulations [9-72] and various tests. Specialist companies automatically observe these rules:

- Basically only straight tubes with inner exhaust routing are used, thus enabling mechanical cleaning and simplified repair.
- Tubes are attached by welding them in the tube plates. Rolling is inadequate when local overheating occurs and weakens rolled fits.
- Possibilities for the gas tubes to thermally expand are provided. For example, the tubes are given freedom to deflect from straight lines and directions of deflection are specified by small bends or an adjustable tube plate or flexible housing is designed in.
- Designs are primarily vertical, i.e. as far as possible, outlet chambers under the tubes with the exhaust flow directed downward are employed so that agglomerated particulates can separate out more easily and, where necessary, accumulating condensation can discharge. Since greater heights for maneuvering are required above the heat exchanger, disadvantages for mechanical cleaning may have to be accepted.
- Tube diameters not smaller than 12 mm Ø are selected.
- The design compels the cooling water to flow reliably at the exhaust gas inlet tube plate and in the inlet region of the gas tubes.
- The cooling water flow is always designed from bottom to top and good air bleeding is provided even in operation.
- Using circular flanges, exhaust inlet and outlet chambers are designed so both sides open easily or sufficient openings are designed in so that soot or potentially present solid metal sulfate deposits can be removed and the condensate outlet unblocked.
- Since exhaust gas valves never close completely, a low water flow is always incorporated to absorb heat from leaking exhaust gas.
- Given the reserve of wall strength usually present in commercial semi-finished tubes, small gas tubes (12–30 mm Ø) are constructed of high-grade steel X10CrNi MoTi 18.10, material no. 1.4571, and large tubes even of mild steel, St 37.10. Tube plates are usually produced from mild steel for reasons of cost and manufacturing. High-grade steel X10CrNiMoTi 18.10, material no. 1.4571, is also selected for exhaust inlet and outlet chambers. The materials 1.4539 and 1.4404 are also used for exhaust tubes in vehicle applications.

Figure 9-45 presents a vertical exhaust gas heat exchanger manufactured in small lots [9-73]. Horizontal models for smaller powers are also constructed for special installation conditions, low machine shops or spaces without package volume above the exhaust gas heat exchanger. Since there are still constraints on the recovery of diesel engine exhaust gas and the quantities are usually small, special exhaust gas heat exchangers are generally manufactured individually for each system in job production. Exhaust steam generators on the other hand are special exhaust gas heat exchanger designs that are also designed horizontally and vertically. They must be constructed for the individual case by special firms and equipped with considerably expanded safeguards.

Problems with Hot Side Fouling

The exhaust gas heat exchangers (exhaust gas/water heat exchangers) employed in diesel engines often register clogging of their gas passages with dry and loose to slightly moist deposits of soot particulates and liquefiable components. Therefore, common gas/water heat exchanger designs have only been effectively employed with additional measures.

Since soot formation in diesel engines is system induced, not even large engines with partly very low soot emission values are fully soot-free despite all the advances in combustion technology (not even after a soot filter!).

Research studies [9-74–9-76] have largely been able to explain the deposition mechanism in systematic experiments: Accordingly, the particles entrained in exhaust gas are subject to different forces (Fig. 9-46). Thermophoresis, i.e. the attraction of particles moving in the direction of the temperature gradient, and the adhesion of condensing hydrocarbons, water and sulfuric or sulfurous acids to the wall (adsorption) cause particles to accrete.

Exhaust pulsations and turbulence intensity in the exhaust gas cause the particles to be transported away again with the exhaust flow. In addition, drying of the wall layer holds promise as "cleaning".

- Without outside influence, fouling can gradually cause clogging.
- Fouling extremely impedes or prevents heat exchange from the exhaust gas to the heat recovery medium, e.g. water.
- Rather than leaving the heat exchanger steadily, particles agglomerate to larger particulates that are expelled aperiodically and pollute the environment.

Not only soot particulates but also metal sulfates (especially $Fe_2(SO_4)_3 \cdot H_2O$), which deposit in exhaust gas heat exchangers as voluminous solids in temperature ranges of the dew point of sulfuric acid/water, may contaminate the heating surfaces. These metal sulfates are formed by gradually detaching – especially from iron – from the engine's exhaust system for instance. The more sulfur is contained in the fuel and, in particular, the more SO_3 is contained in the exhaust, the more intensely they appear. This process increasingly manifested itself as oxidation catalytic converters were used in diesel engines in which increased SO_3 forms from SO_2.

Solutions

Both research studies and experience with the manufacturing of diesel engine plants deliver principle measures to prevent or decrease soot deposits. Moreover, periodic mechanical or chemical cleaning for fully or partially opened exhaust gas heat exchangers must be scheduled during plant downtimes.

Water Injection

Water injection is a mechanical process that was first implemented in combined heat and power stations that have diesel engines as refitting and retrofitting measures. It is based on the soot blower effect familiar from steam boiler operation. One example of a successful installation is the peak load diesel powered CHPS with $2 \times 1,000$ kW$_{el}$ in operation in Lülsfeld since 1979. By 1991, it had been in operation 18,000 hours and there has been good experience with injection cycles of approximately 10 minutes with spray times of only 5–6 s.

However, this method cannot be considered a universally applicable and reliable measure. Water injection must be applied periodically and increases adverse impacts on the environment. In addition, condensate that contains soot must be disposed of.

Scrubbing with Exhaust Gas Condensate

Cooling exhaust gases to very low temperatures, e.g. a coolant temperature of merely 15°C, causes a large quantity of condensate to separate, which is able to wash out the deposits. The low coolant temperatures required limit this measure's effectiveness to the low temperature range of a heat exchanger.

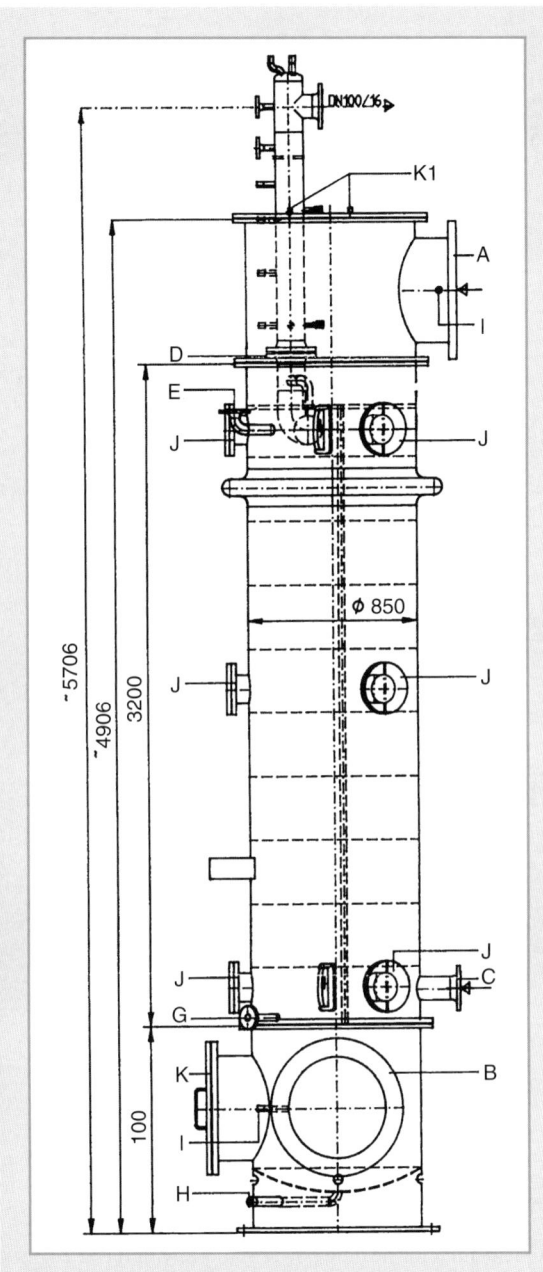

Fig. 9-45 Vertical exhaust gas heat exchanger (MBN, Neustadt/Wied). Heat output 718 kW to heat water from 100 °C to 120 °C at TA1=475 °C; TA2=180 °C; $\dot{m}_A/\dot{m}_W$ = 8475/3087 kg/h/kg/h; **A** Exhaust inlet and **B** Exhaust outlet; **C** Water inlet and **D** Water outlet; **E** Safety valve; **J** Inspection holes; **K, L** Water spray nozzles

While findings acquired in systematic experiments reveal potentials for improvements, not all difficulties have been eliminated. Exhaust heat recovery from diesel engines still entails the following problems:

Fig. 9-46 Deposit formation process in exhaust gas heat exchangers

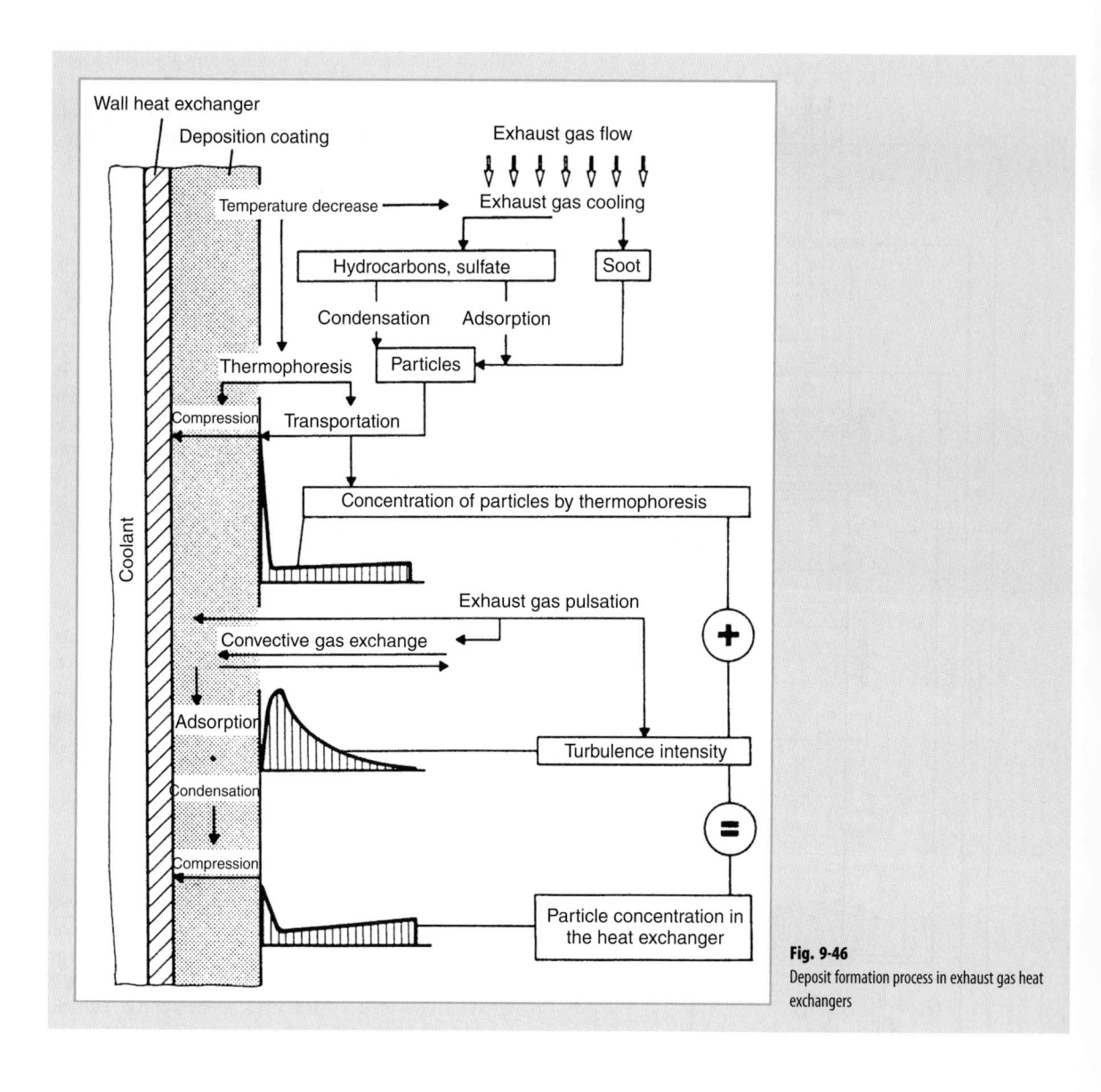

Mechanical Cleaning

Shot cleaning equipment [9-77] familiar from boiler manufacturing has been implemented effectively in nearly forty large diesel heating power stations, especially for dry deposits. Roughly every 60–100 min but sometimes only just once a week, soft iron shot with a diameter of 3 mm is flung against the heating surfaces and strikes the dirt particles that have deposited there. The shot is recollected and returned to the heat exchanger with a driving air flow. Soot and dirt particulates are discharged with the exhaust flow. The increased technical complexity makes it only worthwhile for large plants though.

Thermal Regeneration of the Heat Exchanger

Implemented for heating power stations with diesel engines, [9-74, 9-75, 9-78], this method employs two identically constructed exhaust gas heat exchangers configured in a series in which a valve can redirect the exhaust gas flow. The cooling water flow can be shut off in both parts as Fig. 9-47 illustrates for the left part of the equipment filled with exhaust gas. The residual water contained in it evaporates, reaches a steam separator and then a holding tank. The dried particles detach from the walls heated by the exhaust gas and are partly separated by the cyclone installed between the two heat exchangers or reach the atmosphere with the exhaust gas. If

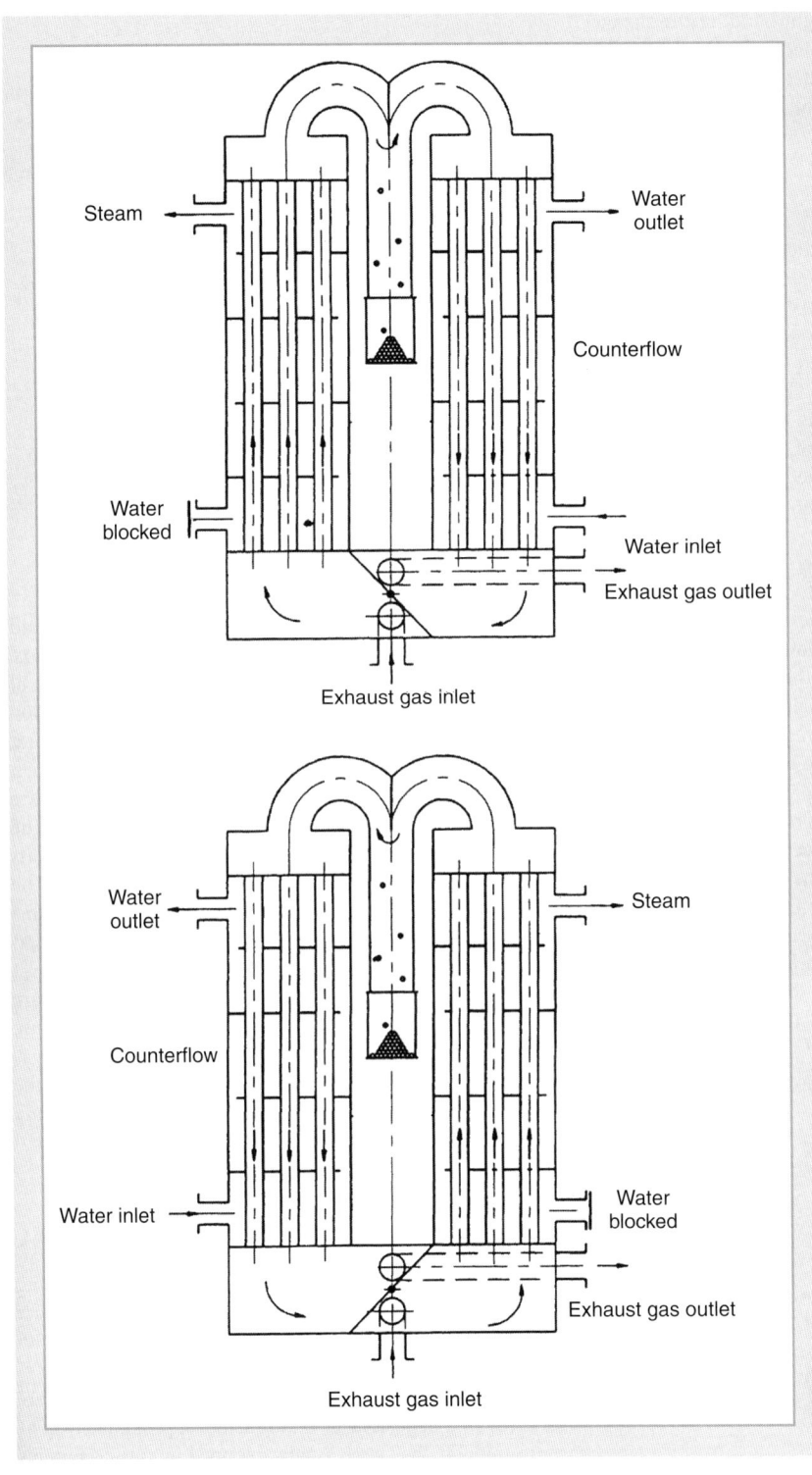

Fig. 9-47
Self-cleaning exhaust gas heat exchanger by thermal regeneration

the exhaust back pressure in the right hand heat exchanger that serves to heat the cooling water increases, the exhaust gas valve is reversed, the right hand water inlet blocked and the left hand heat exchanger reopened and the process repeats for the redirected flow of exhaust.

Turbulence Boost

Boosting the flow turbulence in the exhaust flow in the proximity of the heat transferring wall enables particles surging to the wall to return to the main exhaust flow as a result of the thermophoresis. In addition to the given exhaust pressure pulsations, pulsations generated by Helmholtz resonators may also be employed [9-75]. The propensity to form deposits turns out to diminish as pulsation amplitudes increase and as the temperature of the cooling medium increases, while the pulsation frequency apparently has no influence. These findings have still not been applied industrially though.

Exhaust gas heat exchangers (coolers) have been developed for cooled exhaust gas recirculation to meet the Euro 4 emission standard for vehicle diesel engines [9-79] (see Sect. 15.4). Built-in turbulence generating components, so-called "winglets", are mounted on the surface of the tubes to their reduce fouling [9-80]. They generate exhaust gas turbulence at acceptable pressure losses and thus reduce the propensity to foul and simultaneously improve the heat transfer. Winglets are welded on or pressure formed (Fig. 9-48). The water spaces are arranged around the exhaust ports. The port matrix is laser welded from thin high-grade steel plate (Fig. 9-49). The heat exchanger is fabricated in lengths of 100–800 mm.

An exhaust gas recirculation or EGR cooler is usually implemented in a branch line that runs from the engine's exhaust manifold to the intake manifold back to the intercooler. The flow is maintained by the exhaust gas excess pressure that is slightly above the suction pressure and occasionally by an evacuating venturi tube in the intake system as well as a non-return valve to take advantage of brief pressure waves. The exhaust flow is cooled to 100–200 °C to effectively reduce NO_X emissions. The EGR cooler is connected to the coolant circuit. Apart from the high pressure application, which recirculates the exhaust to the high pressure side, concepts are being tested in which the exhaust is first extracted downstream from the exhaust turbine and downstream from the particulate filter. The exhaust mixed with the fresh air is inducted by the compressor and conducted compressed through the intercooler. The advantage of this solution is its elimination of the limit imposed on the recirculation rate by the differential pressure between the engine's exhaust manifold and inlet connection. Since the exhaust gas pressure in low pressure exhaust gas recirculation is only slightly above the ambient pressure, the EGR cooler on the exhaust side must have a significantly lower pressure loss.

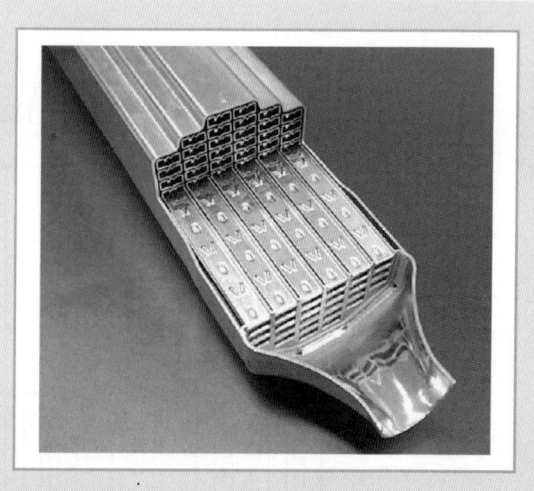

Fig. 9-49 Cross section through the tube matrix of an EGR cooler (source: Behr)

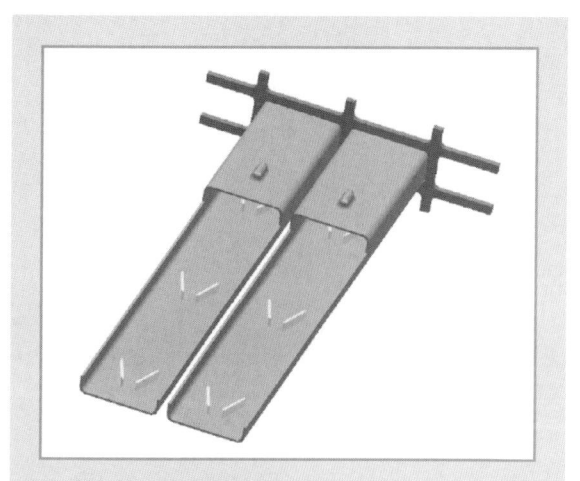

Fig. 9-48 Exhaust flow channels with winglets and tube plate (source: Behr)

High Temperature Operation

The operation of exhaust gas heat exchangers at high temperatures, i.e. >240 °C, prevents components condensing from the exhaust from bonding. Such heat exchangers are implemented with thermal oil as the cooling medium or with pressurized water to generate steam. This equipment functions satisfactorily because the deposits and erosion are balanced in a deposit layer that remains approximately constant. Initially, they were usually designed in marine systems as two-pass boilers with counter flow (superheaters) and parallel flow (evaporators). While they experience fewer difficulties with fouling, they only utilize the high temperature fraction of the exhaust heat.

In principle, the use of low sulfur fuels reduces the deposition of sulfates and the formation of particulates (see Sect. 4.3). Exhaust gas heat exchangers should be designed so that the gas passages are easily accessible for cleaning purposes.

9.2.6 Coolant

9.2.6.1 Cooling Water: Properties and Requirements

Liquid cooling predominantly employs water with refining additives (engine coolant, chemicals, anticorrosion oil, etc.). The use of fuel to cool injector nozzles and sodium to cool the insides of valve stems are special cases.

An ideal coolant because of its thermal properties, water also has disadvantages with regard to engine operation:
- The freezing and boiling point limit the maximum usable temperature range to 100 K. However, nucleate boiling can intensively locally cool highly thermally loaded components (see Sect. 9.1.2 on thermal self-protection).
- Substances dissolved in water can act corrosively and/or interfere with the heat transfer through deposits.
- Malfunctions and damage in the cooling water system can cause severe damage. 15 to 20% of all engine damage can be traced back directly or indirectly to malfunctions and defects in the cooling water system.

Tables 9-9 and 9-10 compare the most important physical properties for heat exchange.

Like fuel and lubricants, cooling water is an *automotive fluid*. An engine's reliability and service life depend on its composition. Its suitability as a coolant is determined by the qualities described below.

Water Hardness

Hardness indicates the water's content of calcium ions and magnesium ions (DIN 38409 Part 6). The total alkaline earths are referred to as the total hardness, which consists of:
- Temporary hardness (carbonate hardness): Carbonates separate out of hydrogen carbonates as the temperature increases during warming and deposit as hardness scale primarily in hot spots where they inhibit heat exchange with the coolant. Since the carbon dioxide released acts corrosively, only a small percentage of carbonate hardness ought to be present when it forms.
- Permanent hardness (noncarbonate hardness): The content of calcium and magnesium chlorides and sulfates does not change with the temperature. However, it influences electrical conductivity and thus facilitates corrosion.

Water hardness is specified as the concentration of quantities of hardness ions in mmol/l. However, outmoded specifications in mval/l or German hardness (°d) are common in engine engineering. Table 9-11 contains the correlation between the quantity of dissolved hardness constituents and degrees of hardness [9-81].

Table 9-9 Physical properties of the heat transfer mediums important for engine cooling (at 80 °C and 1 bar)

Variable		Unit	Air	Water	Engine oil	Coolant[a]
Density	ρ	kg/m^3	0.986	972.0	843.6	1035
Spec. heat capacity	c_p	kJ/(kg · K)	1.010	4.194	2.154	3.59
Kinematic viscosity	ν	10^{-6} m^2/s	21.2	0.366	23.34	1.01
Thermal conductivity	λ	10^{-3} · W/(m · K)	29.93	666.6	127	429
Prandtl number	Pr	–	0.70	2.24	333.1	8.73

[a] Coolant: Mixture of 50% water by volume + 50 % glycol based engine coolant by volume.

Table 9-10 Physical properties of heat transfer mediums relative to air (at 80°C and 1 bar)

Physical variables of a heat transfer medium	Air	Water	Engine oil	Coolant (see Table 9-2)
Density	1	986	856	1,050
Specific heat capacity	1	4.15	2.13	3.55
Kinematic viscosity	1	0.017	1.10	0.048
Thermal conductivity	1	22.27	4.24	14.33
Prandtl number	1	3.20	475.9	12.47

Table 9-11 Overview of degrees of hardness

Ranking	Total hardness [mmol/l]	German degrees of hardness [°d]	Former German classification [°dH]	Parts per million (USA) [ppm]
Very soft	0 – 1	0 – 5.6	0 – 4	0 – 70
Soft	1 – 2	5.6 – 11.2	4 – 8	70 – 140
Moderately hard	2 – 3	11.2 – 16.8	8 – 18	140 – 320
hard	3 – 4	16.8 – 22.4	18 – 30	320 – 530
Hard Very hard	> 4	>22.4	> 30	> 530

Table 9-12 Required properties of engine cooling water

Total hardness [°d]	pH value [–]	Chloride [mg/l]	Sulfate [mg/l]	Hydrogen carbonate [mg/l]
max. 20	6.5...8.5	max. 100	max. 100	max. 100

pH Value

The pH value is a measure of the hydrogen ion concentration. Neutral solutions have a pH value of 7. Basic solutions have a higher pH value (> 7), acidic solutions a lower pH value (< 7). Engine cooling water should have a pH value between 6.5 and 8.5 (see Table 9-12).

Chloride Content

Chlorides strongly facilitate corrosion (particularly in aluminum materials and high grade steel radiators) and sludge formation. Thus, the chloride content should be minimal or less than 100 mg/l (see Table 9-12).

Corrosion, Cavitation and Erosion

Corrosion of materials and their resultant disintegration is caused by an electrochemical reaction between metals as galvanic elements and cooling water as an electrolyte. Gases dissolved in it, e.g. oxygen, carbon and sulfur dioxide, and pH values that diverge from specified recommended values intensify corrosion [9-82]. The same applies to the occurrence of flow or vibration cavitation excited by piston contact change. Vapor lock that occurs when surfaces are overheated is a special case, which, comparable to the process during cavitation, damages the material surface and its protective layers by imploding near the wall. This is hot corrosion. Erosion is to a heat exchanger what cavitation is to an engine cooling water tank. Erosion, the material attrition caused by mechanical friction between the coolant and the surface of material, is dependent on the coolant's flow velocity and the solids and gases contained in it.

9.2.6.2 The Influence of Cooling Water Routing

The danger of engine damage up through complete failure caused by unsuitable cooling water makes its preparation and care essential. Design measures can additionally decrease the potential hazards [9-81]. These include:
- air bleeding and venting the cooling circuit,
- designing lines and cross sections to facilitate the flow,
- increasing the system pressure in the cooling circuit and
- selecting optimal flow velocities.

A cooling circuit can be operated with coolant temperatures of up to approximately 115°C, an excess system pressure of up to 2 bar being necessary. The system is protected by a pressure relief valve.

Low flow velocities facilitate the formation of deposits and increase material attrition induced by corrosion depending on the material. Recommended values are 0.2 m/s$<c_{Grenz}<3.0$ m/s for aluminum, $c_{Grenz}<2.2$ m/s for CuZn20Al for sea water cooled components, <2.7 m/s for CuNi10Fe and <4.5 m/s for CuNi30Fe. The electrochemical compatibility of materials is important when coolants conduct electricity well. This holds particularly true when one cooling system is designed by different enterprises (engine manufacturers, shipyard, etc.). Finally, cooling water losses have to be prevented or minimized because refilling allows additional oxygen and carbon dioxide to reach the cooling circuit. Moreover, active ingredients can concentrate (*accumulation of minerals*).

9.2.6.3 Cooling Water Care

Cooling Water Requirements

Cooling water care begins with the selection of the water. Engine manufacturers specify clean, clear water free of impurities. However, condensate or completely deionized water may also be employed. Sea water, brackish or river water and rainwater are fundamentally unsuitable. Soft water prevents scale deposits (see Table 9-11), yet must have a minimum hardness of $2°d$ to prevent increased metal ion solubility (see Table 9-12).

Engine Coolant

Antifreezes

An engine coolant must be added to the engine cooling water when an engine is operated at ambient temperatures at or below the freezing point,. Its basis is *ethylene glycol* (ethane-1,2-diol) or less often *propylene glycol* (propane-1,2-diol). Its effectiveness depends on the blending ratio with the water. A glycol content of approximately 50% by volume lowers the mixture's freezing point to approximately $-50°C$. The effect diminishes again when percentages are greater (Fig. 9-50).

The percentage of glycol also changes a coolant's physical properties (density, heat capacity, viscosity, thermal conductivity and boiling point). This must be factored into radiator design (Fig. 9-51) [9-53].

Since a water/glycol blend provides insufficient protection against corrosion, anticorrosion agents (*inhibitors*) are additionally blended into commercial engine coolants.

Siliceous Engine Coolants

Siliceous engine coolants predominantly contain inorganic inhibitors, e.g. silicates, nitrites, nitrates or molybdates, as

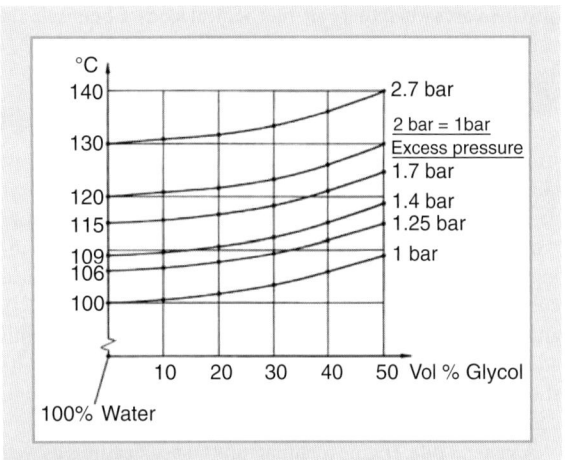

Fig. 9-51 Boiling point curves of water/ethylene glycol mixtures as a function of pressure

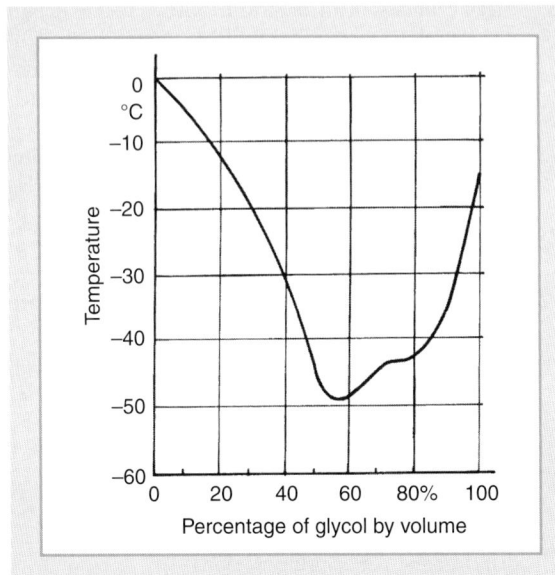

Fig. 9-50 Change in the freezing point of water/ethylene glycol as a function of glycol content by volume

well as organic inhibitors, e.g. tolyl or, to a small extent, benzotriazole. Additives, e.g. borate, phosphate, benzoate or imidazole, serve as buffer substances to assure the desired, mildly alkaline pH value throughout the entire period of operation. In addition, detergents (e.g. sulfonates), antifoaming agents and pigments are also added.

These engine coolants have proven themselves in the field over decades. Their silicate content counteracts dangerous hot corrosion in aluminum engines by forming thin protective coats. However, the low solubility of some additives is disadvantageous. Thus the concentration of inhibitors cannot be increased randomly. Inorganic additives also degrade in normal driving, thus making a new dosage or replacement necessary if the minimum concentration is fallen below.

OAT (Organic Acid Technology) Engine Coolant

OAT engine coolants have been available for several years and contain a combination of organic rather than inorganic inhibitors. Combinations of aliphatic monocarboxylic and dicarboxylic acids, azelaic and aromatic carboxylic acid are typical. Blends of carboxylic acid with inorganic inhibitors are also employed but silicates are usually never employed. The advantage of OAT products is their slower degradation in normal driving. This increases their working life and thus reduces the maintenance required.

Chemicals

Chemicals are only used when there is no danger of freezing and engine manufacturers' specifications allow them. Additives used do not contain any glycol and thus only protect against corrosion. However, this increases evaporation losses and thus the maintenance intervals. The danger of corrosion also increases when there are mechanical loads. Chemicals are usually added in a concentration of 5–10%.

Anticorrosion Oils

Infrequently employed, anticorrosion oils' protection against corrosion and cavitation is based on their prevention of oxygen and other gases dissolved in the cooling water from reaching the walls in engine cooling water chambers by forming a protective film and on their stoppage of electrochemical corrosion. Added in a blending ratio of 1:200 to 1:100 (70), anticorrosion oils consist of emulsifiable mineral oils with additives to protect against corrosion or sludge deposits should the emulsion break, e.g. when heated to >95°C or in contact with copper materials.

A general rule, engine manufacturers' specifications as well as compatibility for humans and the environment during use and disposal have to be observed for every additive.

Literature

9-1 Zinner, K.: Einige Ergebnisse realer Kreisprozessrechnungen über die Beeinflussungsmöglichkeiten des Wirkungsgrades von Dieselmotoren. MTZ 31 (1970), pp. 249–254

9-2 Zapf, H.: Einfluss der Kühlmittel- und Zylinderraumoberflächentemperatur auf die Leistung und den Wirkungsgrad von Dieselmotoren. MTZ 31 (1970), pp. 499–505

9-3 Woschni, G.; Kolesa, K.; Spindler, W.: Isolierung der Brennraumwände – ein lohnendes Entwicklungsziel bei Verbrennungsmotoren? MTZ 47 (1986), pp. 495–500

9-4 Kleinschmidt, W.: Einflussparameter auf den Wirkungsgrad und auf die NO-Emission von aufgeladenen Dieselmotoren. VDI-Berichte No. 910. Düsseldorf: VDI-Verlag (1991)

9-5 Eckert, E.: Einführung in den Wärme- und Stoffaustausch. 3rd Ed. Berlin/Göttingen/Heidelberg: Springer (1966)

9-6 Schack, A.: Der industrielle Wärmeübergang. 7th Ed. Düsseldorf: VDI-Verlag (1969)

9-7 VDI Heat Atlas, 5th Ed. Düsseldorf: VDI-Verlag (1988)

9-8 Pflaum, W.; Mollenhauer, K.: Wärmeübergang in der Verbrennungskraftmaschine. Vienna/New York: Springer (1977)

9-9 Wanscheidt, W.A.: Theorie der Dieselmotoren. 2nd Ed. Berlin: Verlag Technik (1968)

9-10 Willumeit, H.-P.; Steinberg, P.: Der Wärmeübergang im Verbrennungsmotor. MTZ 47 (1986), pp. 9–12

9-11 Pflaum, W.; Haselmann, L.: Wärmeübertragung bei Kavitation MTZ 38 (1977), pp. 25–30

9-12 Wolf, G.; Bitterli, A.; Marti, A.: Bohrungsgekühlte Brennraumteile an Sulzer-Dieselmotoren. Techn. Rundschau Sulzer 61 (1979) pp. 25–33

9-13 Willumeit, H.-P.; Steinberg, P.; Ötting, H.; Scheibner, B.: Bauteiltemperaturgeregeltes Kühlkonzept und seine Auswirkungen auf Verbrauch und Emission. VDI-Berichte 578, pp. 291–306. Düsseldorf: VDI-Verlag (1985)

9-14 Mühlberg, E.; Besslein, W.: Variable Heisskühlung beim Fahrzeug-Dieselmotor. MTZ 44 (1983), pp. 403–407

9-15 Norris, P.M.; Hastings, M.C.; Wepfer, W.J.: An Experimental Investigation of Liquid Coolant Heat Transfer in a Diesel Engine. International Off-Highway & Powerplant Congress and Exposition, Milwaukee/Wisconsin, September (1989), SAE-Paper 891898

9-16 Held, W.: Untersuchungen über die Verdampfungskühlung an Nutzfahrzeugdieselmotoren. Diss. TU Braunschweig (1986)

9-17 Schäfer, H.J.: Verdampfungskühlungssysteme für Pkw-Motoren. Diss. TH Darmstadt (1992)

9-18 Schmitt, F.; Münch, K.-U.; Rechberg, R.: Optimierte Zylinderkopfkühlung der luft-/ölgekühlten Deutz-Dieselmotorenbaureihe 1011F. In: Pischinger, St.; Wallentowitz, H. (Ed.): 8th Aachen Colloquium Automobile and Engine Technology (1999) pp. 1003–1015

9-19 Garthe, H.; Wahnschaffe, J.: Konstruktion und Entwicklung der Deutz-Dieselmotorenbaureihe FL 1011. MTZ 53 (1992) pp. 148–155

9-20 Sass, F.: Geschichte des deutschen Verbrennungsmotorenbaues. Berlin/Göttingen/Heidelberg: Springer (1962)

9-21 Schmidt, E.: Die Wärmeübertragung durch Rippen. VDI-Z. (1926), pp. 885–888 and pp. 947–951

9-22 Löhner, K.: Leistungsaufwand zur Kühlung von Rippenzylindern bei geführtem Luftstrom. Diss. TH Berlin (1932)

9-23 Biermann, A.E.; Pinkel, B.: Heat Transfer from finned metal cylinders in an air stream. NACA Techn. Report 488 (1936)

9-24 Berndorfer, H.; Keidel, W.: Luftkühlungsuntersuchungen an geheizten Rippenzylindern. Deutsche Luftfahrtforschung FB 981 (1938)

9-25 von Gersdorff, K.; Grasmann, K.: Flugmotoren und Strahltriebwerke. Munich: Bernard & Graefe Verlag (1981)

9-26 Flatz, E.: Der neue luftgekühlte Deutz-Fahrzeug-Dieselmotor. MTZ 8 (1946), pp. 33–38

9-27 Mackerle, J.: Luftgekühlte Fahrzeugmotoren. Stuttgart: Franckh'sche Verlagshandlung (1964)

9-28 Scheiterlein, A.: Der Aufbau der raschlaufenden Verbrennungskraftmaschine. Vienna: Springer (1964)

9-29 Gröber, H.; Erk, S.; Grigull, O.: Grundgesetze der Wärmeübertragung. Berlin/Göttingen/Heidelberg: Springer (1963)

9-30 Kast, W.; Krischer, O.; Reinicke, H.; Wintermantel, K.: Konvektive Wärme- und Stoffübertragung. Berlin/Heidelberg/New York: Springer (1974)

9-31 Krischer, O.; Kast, W.: Wärmeübertragung und Wärmespannungen bei Rippenrohren. VDI-Forschungsheft 474. Düsseldorf: VDI-Verlag (1959)

9-32 Biermann, A. E.; Ellerbrock, H.: The design of fins for air cooled cylinders. NACA Techn. Report 726 (1937)

9-33 Biermann, A.: Heat transfer from cylinders having closely spaced fins. NACA Technical Note 602 (1937)

9-34 Lemmon, A.W.; Colburn, A.P.; Nottage, H.B.: Heat transfer from a baffled finned cylinder to air. Transactions ASME 67 (1945), pp. 601–612

9-35 Berndorfer, H.: Die Beeinflussung der Kühlungsverhältnisse an luftgekühlten Motorenzylindern durch die Gestaltung der Luftführung. MTZ 15 (1954), pp. 291–298

9-36 Mai, O.; Mettig, H.: Das Baukastensystem als Konstruktionsprinzip im Motorenbau. ATZ 67 (1965), pp. 77–85

9-37 Mettig, H.: Die Konstruktion schnellaufender Verbrennungsmotoren. Berlin/New York: Walter de Gruyter (1973)

9-38 Bertram, E.: Das Alfin-Verfahren. Giesserei 44 (1957) 20

9-39 Maier, A.: Verchromte Aluminiumzylinder in luftgekühlten Motoren. MTZ 14 (1953), pp. 60–62

9-40 Hübner, H.; Osterman, A.E.: Galvanisch und chemisch abgeschiedene funktionelle Schichten. Metalloberfläche 33 (1979), pp. 456–458

9-41 Howe, H.U.: Der luftgekühlte Deutz-Dieselmotor FL 413. SAE-Paper 700028

9-42 Esche, D.: Beitrag zur Entwicklung von Kühlgebläsen für Verbrennungsmotoren. MTZ 37 (1976), pp. 399–403

9-43 Esche, D.; Lichtblau, L.; Garthe, H.: Cooling Fans of Air Cooled Deutz-Diesel Engines and their Noise Generation. SAE-Paper 900907

9-44 Nitsche, J.: Entwicklung und Einsatz von warmfesten Aluminium-Gusslegierungen für luftgekühlte Zylinderköpfe. Presentation at the österreichische Giesserei-Tagung, Leoben, Austria, May 8 (1980)

9-45 Schöppe, D.: Einfluss der Kühlbedingungen und des Ladeluftzustandes auf die Stickoxidemission bei direkteinspritzenden Dieselmotoren. Diss. RWTH Aachen (1991)

9-46 Moser, F.X.; Haas, E.; Sauerteig, J.E.: Optimization of the DEUTZ aircooled FL 912/913 engine family for compliance with future exhaust emission requirements. SAE-Paper 951047

9-47 Pflaum, W.; Mollenhauer, K.: Wärmeübergang in der Verbrennungskraftmaschine. Vienna/New York: Springer (1977)

9-48 Brun, R.: Science et technique du moteur diesel industriel et de transport, Tome 3. Paris: Éditions Technique

9-49 Mau, G.: Handbuch Dieselmotoren im Kraftwerks- und Schiffsbetrieb. Braunschweig: Vieweg Verlag (1984)

9-50 Rudert, W.; Wolters, G.-M.: Baureihe 595 – die neue Motorengeneration von MTU. MTZ 52 (1991) 6

9-51 Lilly, L.R.C.: Diesel Engine Reference Book. London: Butterworth (1986)

9-52 Küntscher, V. (Ed.): Kraftfahrzeugmotoren. Berlin: Verlag Technik (1987)

9-53 Süddeutsche Kühlerfabrik Julius Fr. Behr (Ed.): Die Kühlung der Brennkraftmaschine.

9-54 Wilken, H.: Derzeitiger Stand der Motorkühlung. ATZ 82 (1980) 12

9-55 Rüger, F.: Luftabscheidung aus Kühlkreisläufen von Fahrzeugmotoren. ATZ 73 (1971) 8

9-56 Süddeutsche Kühlerfabrik Julius Fr. Behr (Ed.): Lüfter und Lüfterantriebe für Nutzfahrzeug- Kühlanlagen

9-57 Feulner, A.: Entwicklung der Kühlanlagen in den Diesellokomotiven der Deutschen Bundesbahn. ZEV/Glasers Annalen (1974) 3

9-58 Feulner, A.: Neue Kühlanlage für die Diesellokomotiv-Baureihe 218 der DB. ZEV/Glasers Annalen (1975) 12

9-59 Gelbe, H.: Komponenten des thermischen Apparatebaus. In: Dubbel – Taschenbuch für den Maschinenbau. 19th Ed. Berlin/Heidelberg/New York: Springer (1997)

9-60 Kays, W.M.; London, A.L.: Compact heat exchangers. 3rd Ed. New York: Krieger Publishing (1984)

9-61 Bošnjaković, F.; Viličić, M.; Slipčević, B.: Einheitliche Berechnung von Rekuperatoren. VDI Forschungsheft 432. Düsseldorf: VDI-Verlag (1951)

9-62 Martin, H.: Wärmeübertrager. Stuttgart/New York: Thieme-Verlag (1988)

9-63 Roetzel, W.; Spang, B.: Design of Heat Exchangers. In: VDI Heat Atlas. 8th Ed. Berlin/Heidelberg/New York: Springer (1997)

9-64 Gaddis, E.S.; Schlünder, E.-U.: Temperaturverlauf und übertragbare Wärmemenge in Röhrenkesselapparaten mit Umlenkblechen. Verfahrenstechnik 9 (1975) 12, p. 617ff

9-65 Grigull, U.; Sandner, H.: Wärmeleitung. 2nd Ed. Berlin/Heidelberg/New York: Springer (1990)

9-66 Webb, R.L.: The flow structure in the louvered fin heat exchanger geometry. SAE-Paper 900722 (1990)

9-67 Illg, M.; Kern, J.: Aluminium-Wärmeaustauscher für Kraftfahrzeug-Motorkühlung. METALL 42 (1988) 3, pp. 275–278

9-68 Pietzcker, D.; Fuhrmann, E.: A mechanically bonded AKG-ML3-aluminum radiator as a direct replacement for brazed units. SAE-Paper 910198 (1991)

9-69 Rudert, W.; Wolters, G.-M.: Baureihe 595 – Die neue Motorengeneration von MTU. MTZ 52 (1991) 6, pp. 274–282 (Part 1) and 11, pp. 538–544 (Part 2)

9-70 Jenz, S.; Wallner, R.; Wilken, H.: Die Ladeluftkühlung im Kraftfahrzeug. ATZ 83 (1981) 9, pp. 449–454

9-71 Reimold, H.W.: Bauarten und Berechnung von Ladeluftkühlern für Otto- und Dieselmotoren. MTZ 47 (1986) 4, p. 151–157

9-72 DIN 4751, Part 2: Wasserheizungsanlagen bis 120°C mit TRD (Technische Regeln für Dampfkessel) 702 sowie für Wassertemperaturen > 120°C TRD 604

9-73 Konstruktionsskizzen von Maschinen und Behälterbau. Köln: Neuwied GmbH, MBN

9-74 Abgaswärmeübertragerverschmutzung; Verminderung der Ablagerungen in Abgaswärmeübertragern von Verbrennungsmotoren. FVV-Forschungsbericht, Issue 429, Project No. 374 (1989)

9-75 Abgaswärmeübertrager; Entwicklung kompakter Wärmeübertrager mit geringer Verschmutzungsneigung für Verbrennungsmotoren. FVV-Forschungsbericht, Issue 474, Project No. 425 (1991)

9-76 Mechanism of Deposit Formation in Internal Combustion Engines and Heat Exchangers. SAE 931032, Detroit (1993)

9-77 Kugelregen-Anlagen. Firmenunterlagen von FEG Heizflächen Reinigungsanlagen, Oer Erkenschick

9-78 Selbstreinigende Abgaswärmetauscher. Firmenunterlagen Kramb Mothermik. Patentschrift DE 3243114C1, Simmern

9-79 Lösing, K.H.; Lutz, R.: Einhaltung zukünftiger Emissionsvorschriften durch gekühlte Abgasrückführung. MTZ 60 (1999)

9-80 Lutz, R.; Kern, J.: Ein Wärmeübertrager für gekühlte Abgasrückführung. 6th Aachen Colloquium Automobile and Engine Technology (1997)

9-81 Hütter, L.A.: Wasser und Wasseruntersuchung. 3rd Ed. Frankfurt: Diesterweg (1988)

9-82 Pflug, E.; Piltz, H.H.: Korrosion und Kavitation in den Kühlwasserräumen von Hochleistungs-Dieselmotoren. MTZ 26 (1965) 3

10 Materials and Their Selection

Johannes Betz

10.1 The Importance of Materials for Diesel Engines

Materials engineering is one of the key technologies that have been closely linked with the diesel engine development from its beginnings to the present day. Not only the high firing pressures and temperatures already connected with the basic idea but also corrosive and tribological influences stress engine components very complexly and affect the selection of materials, which is primarily oriented toward the objectives of

– reliability by employing diverse high performance materials and
– cost effectiveness by limiting labor and costs.

Typically, as little as one tenth of the material required for the weight-to-power ratios of 25–50 kg/kW at the start of the twentieth century is used in today's engines [10-1, 10-2]. Substantially greater demands for operating life and reliability are now satisfied despite higher stresses.

Metals are the material group best qualified for industrial application. Given their orderly crystalline structure, they not only provide *good basic properties* but also particularly numerous and attractive *options to influence properties* and therefore have been produced and utilized industrially for a long time. As Fig. 10-1 indicates, roughly 90% of a modern high performance diesel engine consists of approximately equal parts of *cast iron alloys*, *steel* and *aluminum*. Other metals and nonmetals only constitute a mass fraction of 10%. The proportions of the main elements may vary depending on an engine's design, size and application. Thus, for example, the requirements of lightweight vehicle engine manufacturing distinctly reduce the percentage of iron in favor of lightweight aluminum or magnesium alloys or unreinforced and reinforced plastics. However, since a sweeping replacement of technically and economically highly developed metal alloys by other materials is unlikely in the foreseeable future, metals will continue to constitute the fundamental basis of engines.

Fatigue strength serves as an indicator for the overwhelming number of engine components at rapidly alternating cyclic loads (high cycle fatigue HCF). However, additional properties, which are sometimes required and in part conflicting, have led to the use of nearly all technical metals and their alloys (Table 10-1).

The following provides an overview of the state-of-the-art and anticipated future developments. Not only DIN standards but also EN standards [10-3], which have been expanded greatly in recent years, are considered. Readers are referred to the standard literature for more exact information on material properties [10-4–10-7].

10.2 Technical Materials for Engine Components

10.2.1 Materials for Crankshaft Assembly Components

10.2.1.1 Crankshafts

With the exception of pistons, the components of crankshaft assemblies were originally made of *forged steel*. Especially in smaller engines, it has been replaced by *spheroidal graphite cast iron* since the early 1970s for economic reasons. Economic competition between the shaping processes of casting and forging gave rise to the development of pearlitic microalloys that have been controlled cooled out of the forging heat in BY (best yield) state. Small vanadium or niobium additions normally replace expensive alloy elements in this material group, also called *AFP steel* for its (precipitation hardened, ferritic-pearlitic) structural constitution. Moreover, energy costs are cut since heat treatment is eliminated [10-8]. Compared with the standard forging process, controlled cooling in this forging process causes the formation of special carbides of the microalloy elements V or Nb as well as increased pearlite fractions. Both structural measures increase the strength properties.

Table 10-2 outlines the present state of materials technology for crankshafts. Forged types of steel are best suited for high stresses [10-9]. Since steel has nonmetallic inclusions

J. Betz (✉)
MTU Friedrichshafen GmbH, Friedrichshafen, Germany
e-mail: johannes.betz@mtu-online.com

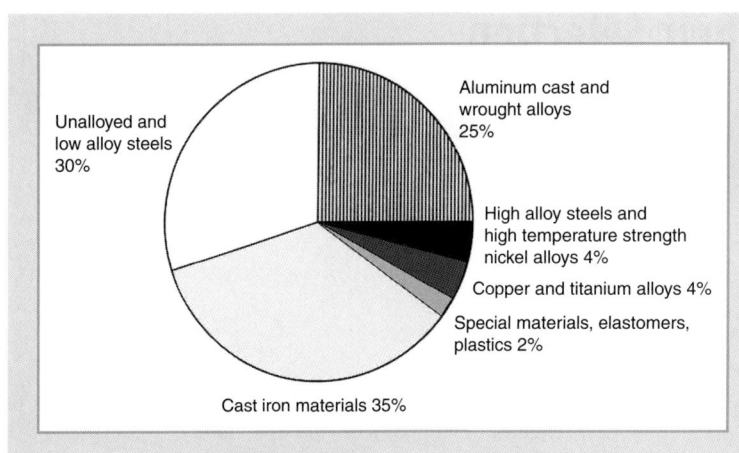

Fig. 10-1
Distribution of the mass of a high performance diesel engine for marine propulsion based on materials

- Aluminum cast and wrought alloys 25%
- Unalloyed and low alloy steels 30%
- High alloy steels and high temperature strength nickel alloys 4%
- Copper and titanium alloys 4%
- Special materials, elastomers, plastics 2%
- Cast iron materials 35%

and inhomogeneities due to melting and rolling, the rules of *structural strength* must already be observed during forging. Unlike hammer forging, *drop forging* produces a continuous flow of fibers adapted to the shape of the shaft, (see Fig. 10-2).

Since the strongest changes in shape occur in a die's parting plane and the contaminated core zone is pressed outwardly, nonmetallic inclusions and liquations normally weaken the region of the flash line most. Shifting the flash line from the highly stressed center plane of the throw can be an expedient means to increase reliability against fatigue failures caused by torsional stress. Throw-by-throw or *fiber flow forging* is one option. Each single throw is upset into a die with moving

Table 10-1 Typical materials for engine components

Component	Density	Thermal expansion	Heat conductivity	Young's modulus	Ductility	Static strength	High cycle fatigue	Low cycle fatigue	High temperature strength	Wear resistance	Corrosion resistance	Material
Crankshaft assembly components	(▽)			△	△	△	▲			△		Quenched and tempered steels and cast iron and aluminum alloys
Control components and injection equipment							▲	▲		▲		Hardenable steels, chill cast iron
Bolts						▲	▲	▲			△	Quenched and tempered steels
Supporting structures	▽	▽		△	△	△	▲					Cast iron and aluminum alloys
Hot parts		▽	△	▽		▲		▲	▲	△		High and very high temperature strength alloys
Bearings							▲	▲		▲	▲	Laminated metallic composites
Radiators, coolers		▽	▲				△	△			▲	Aluminum copper and titanium alloys
Seals, filters, insulation			(▽)				△		△	▲		Special materials, elastomers and plastics

Symbols: ▲ high, required   △ high, desirable   ▽ low, desirable

Table 10-2 Materials for crankshafts

	Nodular graphite cast iron	Pearlitic iron/micro alloyed steels	Unalloyed steel and steel alloys
Types	EN-GJS-600-3... EN-GJS-700-2	C38mod. 30MnVS 6 46MnVS 3	C 35 E...42CrMo4 mod. 34CrNiMo6
Tensile strength R_m MPa	700–850	700–1000	700–1150
Shaping	Casting	Drop forging	Drop, throw-by-throw and hammer forging
Heat treatment		Controlled cooling out of the forging heat (BY)	Quenching and tempering (V) Normalizing (N)
Use	Car engines	Car and commercial vehicle engines	Car, commercial vehicle and large engines

parts. Since the entire shaft must be reheated in the furnace to forge each end throw, this method is only suitable for smaller quantities of larger shafts (with a *mean range of length between 4...9 m*) for economic reasons. Depending on the quantities and dimensions, the shafts are forged on press lines or in forging hammers based on the counterblow principle or in large hydraulic presses.

According to Hooke's law, $\varepsilon = \sigma/E$ applies to the elastic deformation of a metallic material. Regardless of its fraction of alloy elements, steel's Young's modulus of approximately 210,000 MPa at room temperature is the highest among the metals used industrially. Thus, variations of material cannot improve the stiffness of the crankshaft structure at the given overall dimensions. Therefore, great value is attached to

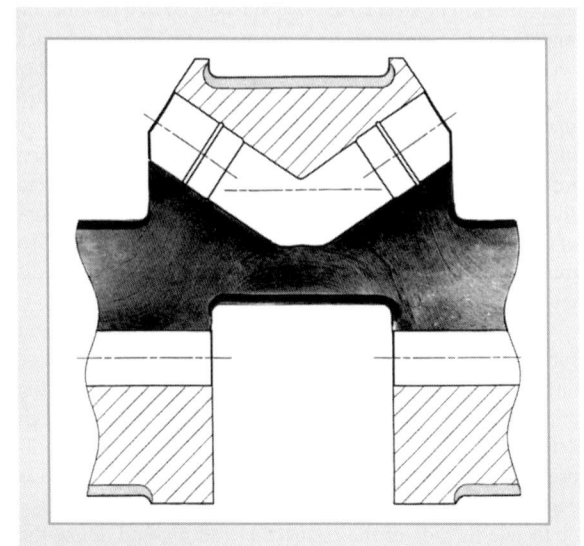

Fig. 10-2 Fiber flow of a drop forged crankshaft throw with an inductively hardened journal and fillet surface

the *design of the support* since a crankshaft cannot be self-supporting at high firing pressures. However, even then, *measures that increase fatigue strength* must often be applied to highly stressed regions such as fillets to assure their durability. To this end, journals and fillets are frequently surface hardened. Based on hardening the contact surfaces of the main and connecting rod bearings because of their wear, the technology of fillet hardening gradually established itself after 1950. The flame hardening originally employed was replaced by *induction hardening*. Improved methods of heat treatment and manufacturing have made it possible to harden all of one shaft's fillets (see Fig. 10-2). Since they had to be flattened on a press or stamped with a die over unhardened regions, only part of the fillets could be hardened earlier because they would warp. The intrinsic compressive stresses generated by the martensitic structural transformation and the increase in strength in the surface of hardened fillets make it possible to *increase fatigue strength at bending stress* by 50–100% (Fig. 10-3).

Hardening normally increases torsion fatigue strength less than bending fatigue strength since other weak points with high stress peaks can occur, e.g. in oil bores. Other thermal or mechanical processes, e.g. nitriding, peening or rolling, or combinations of different measures also increase the fatigue strength of crankshafts (see Sect. 10.5.2). The technical options and cost effectiveness are determinative in each individual case.

Advantages from forming and mass can compensate for the fundamentally lower strength values of the *spheroidal graphite cast irons* EN-GJS-600 and EN-GJS-700 (GGG-60 and GGG-70), which are frequently used for car and commercial vehicle engines. The aforementioned processes that increase service life are similarly applicable and effective. Surface treatment is only uncommon in large engines because they require refinishing. Since it reduces "chafing" by spheroidal graphite inclusions, roller burnishing the surfaces of crankshaft journals can reduce the demonstrably higher rates of wear in the main and connecting rod bearings of

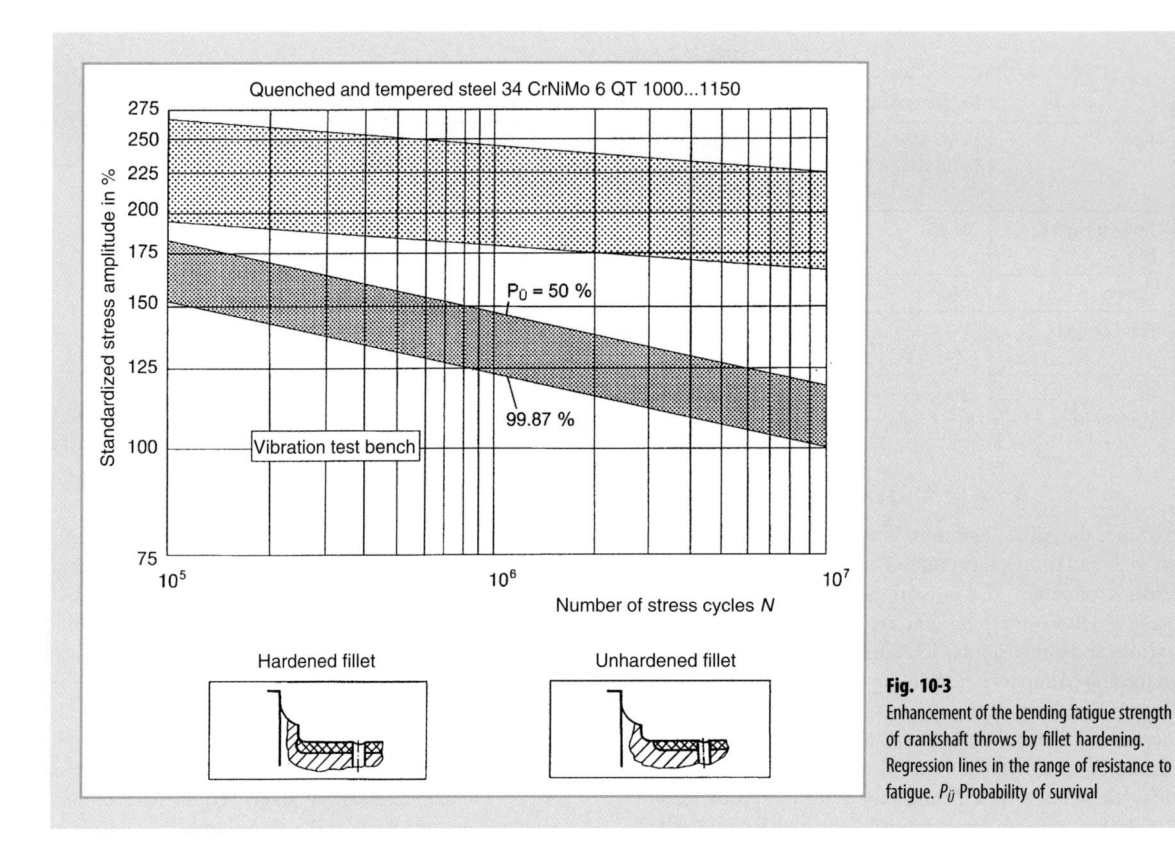

Fig. 10-3
Enhancement of the bending fatigue strength of crankshaft throws by fillet hardening. Regression lines in the range of resistance to fatigue. $P_{ü}$ Probability of survival

nodular cast iron crankshafts. In addition to all the types of cast iron and unalloyed and low alloy steels, sintered tungsten-based heavy metal inserts with a density of 17–19 g/cm^3 are employed for bolted on counterweights in special cases.

10.2.1.2 Connecting Rod

The material groups listed in Table 10-2 are also used for connecting rods [10-10]. Instead of spheroidal graphite cast iron, malleable cast iron of the strength classes EN-GJMB-650 to EN-GJMB-700 (GTS-65 to GTS-70) is also employed for smaller, cast parts. However, *heat treatable steels*, which may be unalloyed or low Cr, CrMo, CrMoV or CrNiMo alloys, are generally dominant. *AFP grades* of pearlitic iron also established themselves in the mid 1980s for cost reasons. At identical tensile strength values, these grades have a fatigue strength identical to quenched and tempered steels despite their lower yield point. Given the demand for ever higher firing pressures, the comparatively low yield strength and compressive yield point of AFP steels must already be factored into connecting rod design. Since the influence of surface defects with depths of a few tenths of a millimeter make up for their differences in static strength, the same fatigue strength can be expected from all unfinished forged

steels regardless of their basic strength. The development of fracture splitting technology for connecting rods began in the early 1990s. The steel C70S6 was introduced as a production material for fracture split connecting rods in the car and truck sector in the mid 1990s. While the material C70S6 had somewhat lower strength properties than C38mod, the standard material in this vehicle segment, fracture splitting technology brought significant economic advantages when it was first introduced in mass production by ending saw-cutting and shortening the process chain. Systematic further development of fracture splitting technology has elevated the yield point so that the strength properties of C70S6 now satisfy most requirements in this segment.

Strength properties can be exploited best by machining and subsequently *shot peening* surfaces. This, in turn, can increase fatigue strength by up to 100% depending on the alloy and the tensile strength. Certain regions of connecting rods with high deformation or strong notching are normally critical to failure and must be improved by design, material or process engineering measures. Examples are toothed parting planes or female threads of the connecting rod big end. Once the design options have been exhausted, the fatigue strength in these points can also be significantly enhanced by shot peening the surfaces or shaping (rolling) the thread without

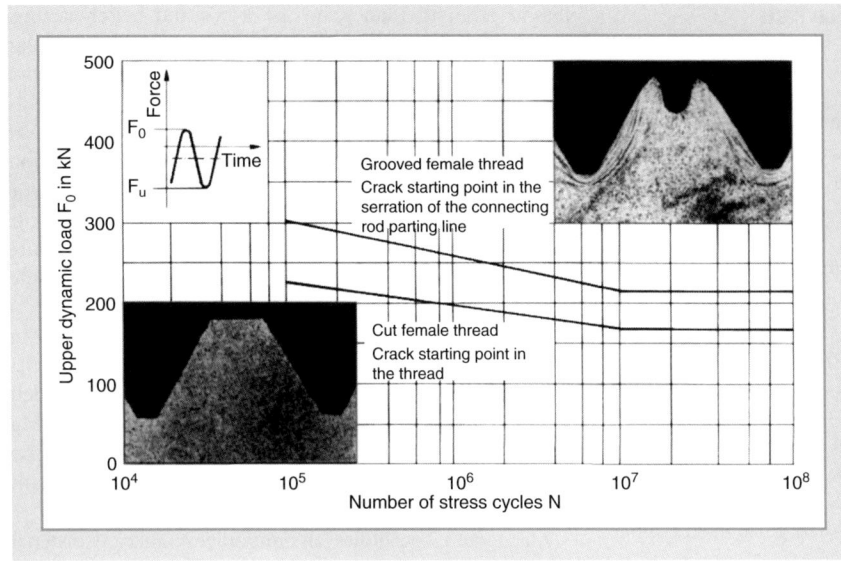

Fig. 10-4
Pulsator tests on obliquely split connecting rods. Increase in fatigue strength by rolling the female threads

cutting. Figure 10-4 presents the results of corresponding endurance tests on obliquely split connecting rods with increases in fatigue strength of over 30%. In addition to mechanical methods, thermal (surface hardening) or thermochemical (nitriding) methods of finishing are also occasionally applied to connecting rods to locally or fully *finish the surface*. However, they have not attained the same importance for connecting rods as for crankshafts.

10.2.1.3 Pistons

Piston and piston ring materials are especially important for diesel engines and are treated in Sect. 8.6.

10.2.2 Materials for Control Components and Injection Equipment

Steels with surface hardening of around 60 HRC are preferred for camshafts and gears. These include *case hardening steels* and *quenched and tempered steels* suitable for inductive surface hardening or nitriding. Camshafts and flat followers for smaller engines are usually made of ledeburitic solidified chill cast iron; the aforementioned hard steels and fully hardened steels of the 100Cr6 type are employed for roller tappets. Highly quenched and tempered and shot peened CrV or CrSi spring steels alloys are typical for valve springs.

Case hardening and nitriding steels are also employed for injection nozzles and high pressure pump components. High alloy cold, hot or high speed tool steels with carbide forming additions such as Cr, Mo, W and V are also employed in individual cases, e.g. for pump plungers. The demand for ever lower emission values and thus higher system pressures necessitates the use of coatings of classic resistant materials for tribologically stressed common rail crankshaft assemblies or injector components, e.g. titanium nitride or diamond-like, thin, amorphous carbon coatings that reduce the coefficient of friction [10-11]. With internal pressures of up to 6,000 bar, autofrettaging reinforces internally pressurized components such as housings to increase fatigue strength and thus service life [10-12, 10-13].

10.2.3 Bolt Materials

Materials and properties of bolts have been standardized worldwide in ISO 898 and DIN EN 20898. Primarily the *high strength classes* of 8.8, 10.9 and 12.9 are important for durable engine design. Quenched and tempered steels with low hardness and strength enhancing alloying additions of Cr, Mo, Ni and B are employed for such bolts [10-14]. The threads are rolled after quenching and tempering and thus acquire increased fatigue strength. The materials become susceptible to embrittlement when their strength properties are above 1,000 MPa, usually because hydrogen has been atomically absorbed. Therefore, very high strength classes with limited toughness are avoided and only such protective surface coatings are applied in which the supply of hydrogen derived from the process is low and the hydrogen can easily diffuse out again (inorganic zinc or manganese phosphate coats, metallic zinc coatings). The European Union's Directive on End of Life Vehicles requires coatings free of chromium (VI) as of 2007. In recent years, this has advanced the development and application of so-called duplex coatings (paint coats with zinc lamella) as a replacement for chromating, especially for bolts.

10.2.4 Materials for Core Engine Parts

10.2.4.1 Crankcases

Cast materials are employed with diverse manufacturing technologies for complex crankcases and cylinder heads, the design of which is made additionally more complicated by coolant chambers that conduct cooling water [10-15].

While cast steel and cast light alloys were already employed early on, gray cast iron such as EN-GJL-250 (GG-25) with low alloying additions of Cr, Mo or Ni and spheroidal graphite cast iron alloys such as EN-GJS-500 (GGG-50) are now prominent. Table 10-3 is a compilation of typical materials with their characteristic structural constitutions and properties. Since only cast steel is suitable for structural welds, the remaining alloys are machined as monoblock. A highly advanced hot welding technology that touches up unfinished parts makes spheroidal graphite cast iron cost effective, especially for crankcases.

Since combining the good properties of gray and nodular cast iron suggested itself, "vermicular graphite cast iron EN-GJV" was developed; the graphite's "worm-like" formation is its distinctive feature. This takes advantage of vermicular graphite cast iron's higher stiffness and fatigue strength than gray cast iron's and better casting, damping and heat conductivity properties than nodular cast iron's. However, EN-GJV is metallurgically more complex than gray and nodular cast iron; this holds particularly true when wall strengths vary greatly. Novel casting technology appears to be making its application for larger lots possible (see Sect. 10.6.1). New engine developments employ vermicular graphite cast iron, e.g. EN-GJV-450 (GGV-45), as a crankcase material. Fracture splitting technology is also being employed in part in the bearing cover area.

Artificially aged cast aluminum materials of the AlSiMg type are excellently suited for designs that aim for a higher power density and weight-to-power ratio. Such *silumin alloys* (e.g. EN AC-ALSi7Mg, EN AC-ALSi9Mg and EN AC-ALSi10Mg) combine low specific weight and good casting properties with good strength properties.

Magnesium constitutes an innovative crankcase material [10-16, 10-17] and its specific weight delivers an exceptional advantage: Magnesium is approximately 30% lighter than aluminum (magnesium's density is 1.81 g/cm^3, aluminum's 2.68 g/cm^3 and cast iron's 7.2 g/cm^3). The use of crankcases made of aluminum-zinc-magnesium alloys, e.g. AZ-91, can lower the weight-to-power ratio from 1.19 (aluminum crankcase) to 1.02 [10-16]. Combining magnesium and aluminum or cast iron

Table 10-3 Crankcase and cylinder head materials

Property	Gray cast iron	Vermicular graphite cast iron	Nodular graphite cast iron	Cast steel	Cast aluminum
Structure					
Density in g/cm^3	7.2	7.2	7.2	7.7	2.7
Tensile strength in MPa	250–400	300–500	400–800	450–800	160–320
Young's modulus in GPa	100–135	130–160	160–185	210	75
Castability	Excellent	Good	Good	Poor	Excellent
Weldability	Poor	Poor	Good	Excellent	Good
Form	Monoblock	Monoblock	Monoblock	Composite welded with steel plate	Monoblock
Application	Car and commercial vehicle engines	Large engines	Commercial vehicle and large engines	Large engines	Special engines

materials eliminates magnesium's drawbacks, e.g. low stiffness, propensity to creep and corrosion and wear resistance. Thus, for example, ENAC-AlSi17Cu4 inserts are employed for cylinder liners and their cooling water chambers or vermicular graphite cast iron inserts for the basic bearing support. Only such a combination of a magnesium core and aluminum or cast iron insert meets the requirements for high stiffness, acoustics and durability in modern car diesel engines.

10.2.4.2 Oil Pans

Depending on the implementation of the design, the oil pan and the crankcase lower half can contribute to the stiffness of the overall engine structure. Parts cast of AlSi, AlSiMg and AlSiCu alloys or FeP03/04 (St13/14) or steel sheet constructions of S235JR/S355J2 (St37/52) are employed. For the first time, plastics (glass fiber reinforced polymers) are also being used to a greater extent for vehicle engine oil pans [10-18].

10.2.4.3 Cylinder Heads

The *cast alloys* contained in Table 10-3 also represent the range of materials common for cylinder heads. The temperature load on the combustion chamber side intensifies a cylinder head's loading conditions, thus generating low frequency thermal stresses triggered by the engine's load changes as a consequence of impeded thermal expansion (low cycle fatigue LCF). Conventional gray cast iron is still preferred because of its better physical properties, good castability and lower costs; spheroidal graphite cast iron is also used at high stresses. *Vermicular graphite cast iron* (EN-GJV) in the strength classes 400 and 500 has generally not been able to establish itself apart from individual applications.

Highly stressed by temperature and wear, the valve seats in cylinder heads are in part inductively or laser hardened. When this is insufficient, *centrifugally cast seat inserts* with a Brinell hardness of 35 to approximately 50 HRC are provided. Medium and high alloying additions of Cr, Mo and V give graphitic and graphite-free grades the required temperature and wear resistance. For maximum firing pressure and temperature load, cobalt alloys with hard intermetallic phases are used as materials for valve seat inserts with a hardness of 42–58 HRC and identical surface properties over a broad temperature range. This material group guarantees high resistance against valve seat wear and hot corrosion, particularly when lubricating conditions are poor too.

The gray cast iron EN-GJL-250 (GG-25) has also proven itself for valve guides. Precipitation hardenable copper materials (e.g. CuNi2Si) are additionally used. Sintered iron alloys with wear inhibiting interstitial carbide and bronze have been developed for the mass production of vehicle engines.

10.2.4.4 Cylinder Liners

The material's *wear resistance* is top priority for cylinder liners. Materials have remained fundamentally unchanged since the beginning of engine development. The dominant type of material was and is *gray cast iron* of the class EN-GJL-250 (GG-25) with Cr and Mo additions and, in Europe, an increased phosphorous content that facilitates the formation of a wear inhibiting phosphide network in the structure. Large liners are sand cast, smaller centrifugally cast. Cylinder liners are in part inductively or, in individual cases, laser hardened to increase their wear resistance (Fig. 10-5). Low-distortion nitriding is another measure that improves wear.

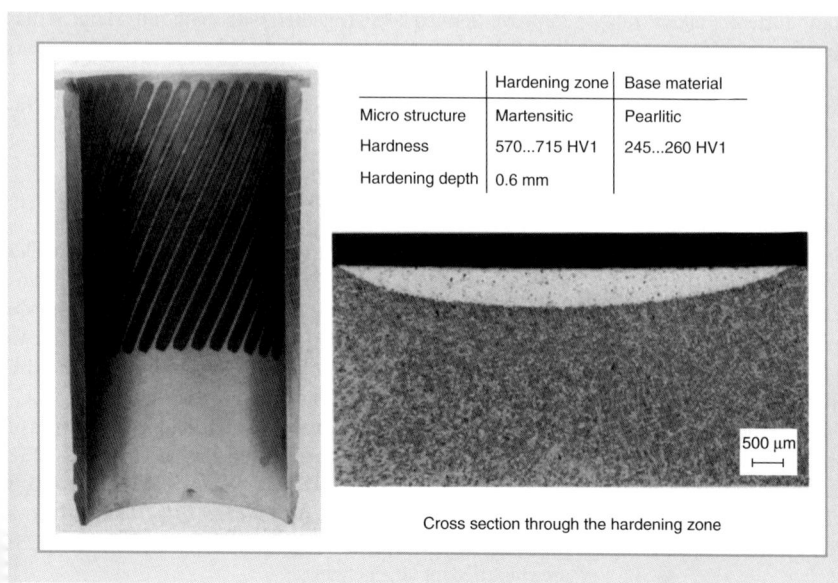

	Hardening zone	Base material
Micro structure	Martensitic	Pearlitic
Hardness	570...715 HV1	245...260 HV1
Hardening depth	0.6 mm	

Cross section through the hardening zone

Fig. 10-5
Laser hardened cylinder liner

As explained in the section on "crankcases", car diesel engines with aluminum crankcases employ cast-in gray cast iron or hypereutectic, highly siliceous cast aluminum alloy cylinder liners as the contact surface for the piston rings.

10.2.5 Materials for Hot Parts

10.2.5.1 Fundamental Problems

Since they must withstand mean exhaust gas temperatures that can exceed 750°C, valves, parts conducting exhaust and exhaust gas turbocharger components require the use of *very high temperature strength alloys*. In recent years, diesel engine engineering has been able to take advantage of material developments for gas turbines, e.g. nickel-based alloys with strength improving intermetallic phases (γ'-phase $Ni_3(Ti, Al)$).

One particular problem occurs when engines are run with heavy fuel oil or blended fuel. Their high sulfur and vanadium contents substantially intensify hot corrosion attacks on parts that conduct gas. Nickel sulfide or vanadium pentoxide form when limiting temperatures are exceeded, thus destroying the material. Therefore, *hard face welding* is also used to plate critical points such as hot exhaust valve seat faces in cylinder heads and valves with stellites or similar alloys (see Sect. 7.1.4).

10.2.5.2 Valves and Prechambers

Given their chemical, tribological and thermomechanical stresses, diesel engine valves have a complex requirement profile, which the selection of materials must satisfy. While high temperature strength quenched and tempered steels with CrMoV additives were long used for intake valves, high alloy hardenable CrSi steels or special steels such as austenite with CrNiW alloying elements were developed for exhaust valves early on. Heat resistant austenite alloys' relatively low high temperature strength necessitated limiting the maximum temperatures in the valve head region and reinforcing highly loaded zones. This is done by filling the body of hollow stem valves with sodium and surface welding or plating the valve seat and stem end. Such complex designs are largely foregone today because of their additional potential for defects and failure. *Monometal valves* are state-of-the-art for the intake side and friction or butt welded *bimetallic valves* state-of-the-art for the exhaust side (Fig. 10-6). According to DIN EN 10090, the range of materials extends from the heat treatable high temperature strength steel X45CrSi9-3 to precipitation hardenable nitrogen alloyed austenite up through the nickel alloy NiCr20TiAl (Nimonic 80A) [10-19]. New developments are also dealing with pressure nitridated steels as intake valve material [10-20]. This new material group links the good properties of the conventional intake valve material X45CrSi9-3 such as hardness and ductility with higher high temperature strength and corrosion resistance.

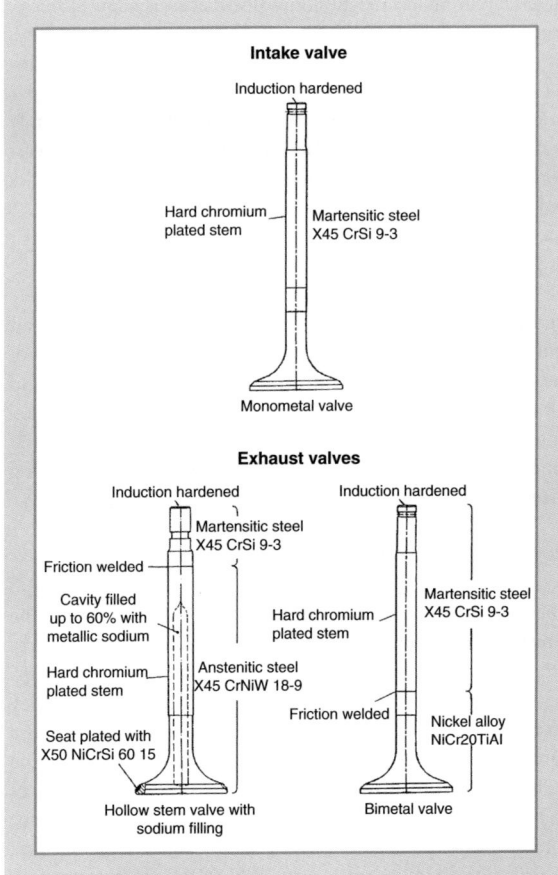

Fig. 10-6 Valve designs

Chrome plating the valve stems is one of the measures that improve service life. Together with bath nitriding, it has established itself as a standard process for these regions stressed by wear and corrosion.

In addition to high alloy steels, primarily nickel and frequently cobalt alloys are employed for prechambers.

10.2.5.3 Exhaust Lines and Housings

For reasons of cost effectiveness, components with smaller dimensions are produced from gray cast iron with high temperature strength enhancing alloying additions of Cr and Mo or ferritic nodular cast iron EN-GJS-400-15. In addition to more oxidation resistant spheroidal graphite cast iron with higher Si and Mo contents (EN-GJS-X SiMo5-1 and GGG-SiMo51), the high alloy austenite EN-GJL and EN-GJS types of "Ni-Resist" are available for thermally highly loaded parts [10-15]. Encapsulated, cooled constructions with inner shells made of high temperature strength steels (e.g. X15CrNiSi20-12) or nickel alloys up through mixed crystal hardening NiCr22Mo9Nb (Inconel 625) are employed for larger engines.

10.2.5.4 Exhaust Gas Turbochargers

A number of high performance materials are used to rigorously take advantage of exhaust gas turbocharging. The very high temperature strength, precipitation hardened superalloy G-NiCr13MoAl (Inconel 713), in part with a carbon content specifically lowered below 0.08%, is preferred as a material for turbine wheels or blades. The range of materials for extremely highly loaded compressor wheels is substantially broader [10-7] and ranges from cast aluminum parts of AlSi and AlCu types for car and commercial vehicle engines to higher strength, forged types of the composition of AlCuMgNi through the titanium alloy TiAl6V4 for high pressure ratios and circumferential speeds (Fig. 10-7).

10.2.6 Bearing Materials

Laminated composite materials are employed for highly loaded plain bearings (see Sect. 8.5.4). Copper alloys, e.g. tin bronze (CuSn8) or brass (CuZn31Si1), as well as aluminum materials, e.g. the piston alloy AlSi12CuMgNi, have proven themselves for the numerous other bearing positions in engines. Bushings made of sintered iron, steel or bronze alloys can fulfill low stress bearing functions.

The antifriction bearings sporadically employed for crankshafts are based on fully hardenable rolling bearing steels such as 100Cr6 and 100CrMn6 and case hardenable chromium steels with high percentage purity.

10.2.7 Materials for Corrosive Stress

The range of types of corrosive stress on engine parts is extremely large. Thus, many materials that have no natural *corrosion protection* because of their composition must first be protected for support and operation. Where treatment with anticorrosion oils and greases or paints is unable to do this, surfaces, e.g. of bolts, are coated with electroplated zinc layers, zinc lamella coatings, or bonder coats of zinc or manganese phosphate.

Every surface wetted with cooling water exhibits one particular problem of complex corrosive stress. Even though every engine manufacturer only permits purified water of suitable quality for cooling, the energy support in thermally and mechanically stressed, components that conduct cooling water causes corrosion attacks, frequently in the form of particularly critical *pitting*. Remedial material engineering measures that employ coatings and corrosion resistant special materials have proven to be unsuitable for components in the internal circuit or economically unfeasible. The work by the Research Association for Combustion Engines in the early 1960s delivered fundamental knowledge on *inhibiting corrosion* of cooling liquids, which has made the unproblematic use of aluminum for components and radiators that conduct cooling water possible (AlMn alloys).

The selection of materials for sea water cooling systems in the secondary circuit of marine engines is extremely limited. Only *special materials* such as copper alloys (red brass/CuSnZnPb-C, multicomponent aluminum bronze/CuAL10Ni-C and copper-nickel/CuNi10/30Fe) or pure titanium exhibit satisfactory to good resistance. Austenitic steels only have sufficient resistance to chloride ions when nitrogen or appropriately high molybdenum contents have been alloyed. Since contamination with liquids and solids strongly increases the aggressiveness of water in harbor areas, pipes and cooler inserts are partly protected by *plastic coatings*. Parts of the (sea) water routing are also constructed completely of plastic.

Turbine wheel
G-NiCr13MoAl (Inconel 713 C / LC)

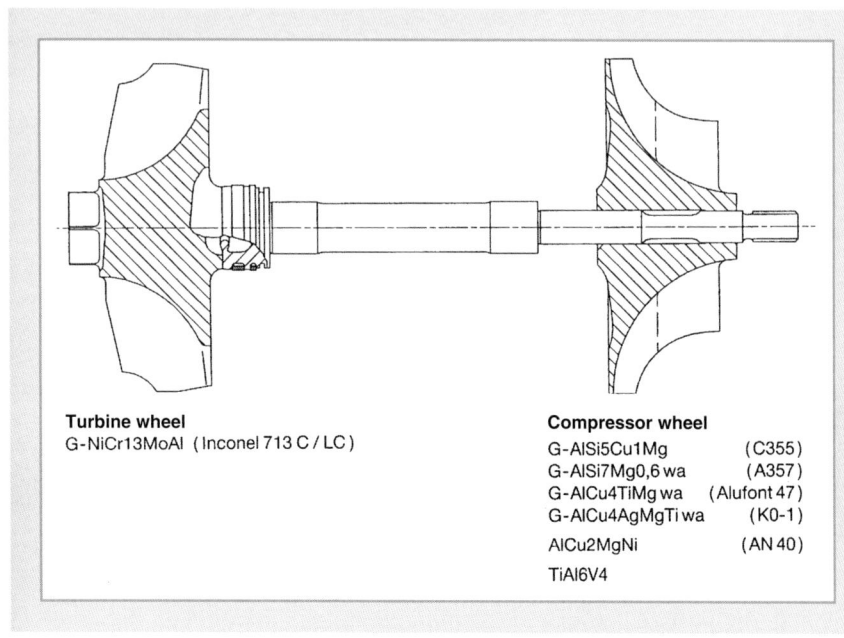

Compressor wheel
G-AlSi5Cu1Mg   (C355)
G-AlSi7Mg0,6 wa   (A357)
G-AlCu4TiMg wa   (Alufont 47)
G-AlCu4AgMgTi wa   (K0-1)
AlCu2MgNi   (AN 40)
TiAl6V4

Fig. 10-7
Rotor assembly with radial blade wheels as an example of materials for exhaust gas turbochargers

10.2.8 Special Materials for Functional Parts

Seals, filters and other functional elements of an engine are usually made of *nonmetallic organic materials* [10-21]. A few problems and trends are mentioned here only to highlight the multitude of different parts and materials:
- the replacement of asbestos by aramide fibers or entirely new material combinations for thermally stressed flat gaskets,
- increasing use of more highly resistant fluorine rubber (FPM) instead of nitrile rubber (NBR) and other types of elastomers for coolant and oil seals,
- the use of polytetrafluoroethylene (PTFE) for parts stressed by wear, e.g. radial shaft sealing rings,
- the use of rings made of Al_2O_3 or SiC ceramic instead of special carbon for the axial seal ring seals of water pumps,
- the use of synthetic fiber fleeces with higher fineness instead of impregnated papers for intake air and lubricating oil filters and
- disposable or recyclable oil filter concepts.

Formed part elastomer seals mountable at many different points continue to be indispensable for engine function. However, parting planes in mass produced vehicle engines, e.g. between crankcases and oil pans, are now starting to be sealed with pasty silicone applied by dispensing robots.

10.3 Factors for Material Selection

An evaluation of materials for their suitability for diesel engine components should not be limited to composition and properties. Rather, a *holistic view* that incorporates design, manufacturing, machining and finishing is essential. Aspects of environmental protection and disposal have also been added in recent years. In addition to the aforementioned factors, material costs are extraordinarily important (Fig. 10-8). Not only expensive basic materials or alloy elements but, above all, complex shaping and machining methods may also have a dominant influence on costs.

Figure 10-8 lists the material costs of some material groups utilized in engine manufacturing. The variations in the costs of cast alloys chiefly result from the complexity of casting induced by the geometry. Material costs are normally compared on the basis of weight. Since the performance of materials is proportional to the fully utilized volume, volume prices would actually facilitate better comparisons. Plastics frequently rate well. However, since extremely different materials usually do not allow similar geometric designs, a component must be compared with a component in particular cases [10-22].

10.4 Service Life Concepts and Material Data

As long as only a few physical and static properties were available to describe material properties, the suitability and reliability of materials could only be tested in component or engine tests. Other rating concepts were developed to supplement the experimental investigation of endurance strength [10-23]. Above all, the long established *concept of nominal stress* and, more recently, the *concept of local stresses* (local concept and notch base concept) are employed in engine manufacturing (see Sects. 7.1.3 and 8.6). While tests on original components are essential to the concept of nominal stress (Fig. 10-9), the concept of local stresses builds directly

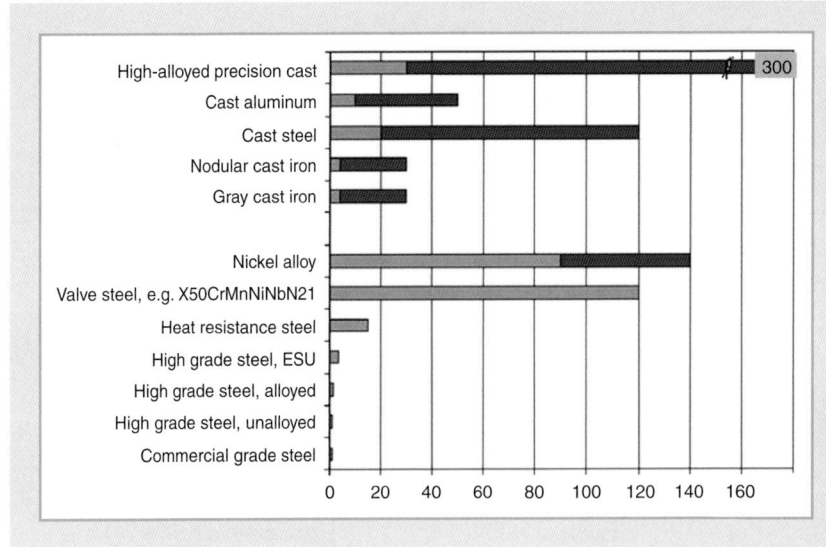

Fig. 10-8
Reference values for material prices. Basis of comparison: Commercial grade steel with a price factor of 1 (as of December 2005)

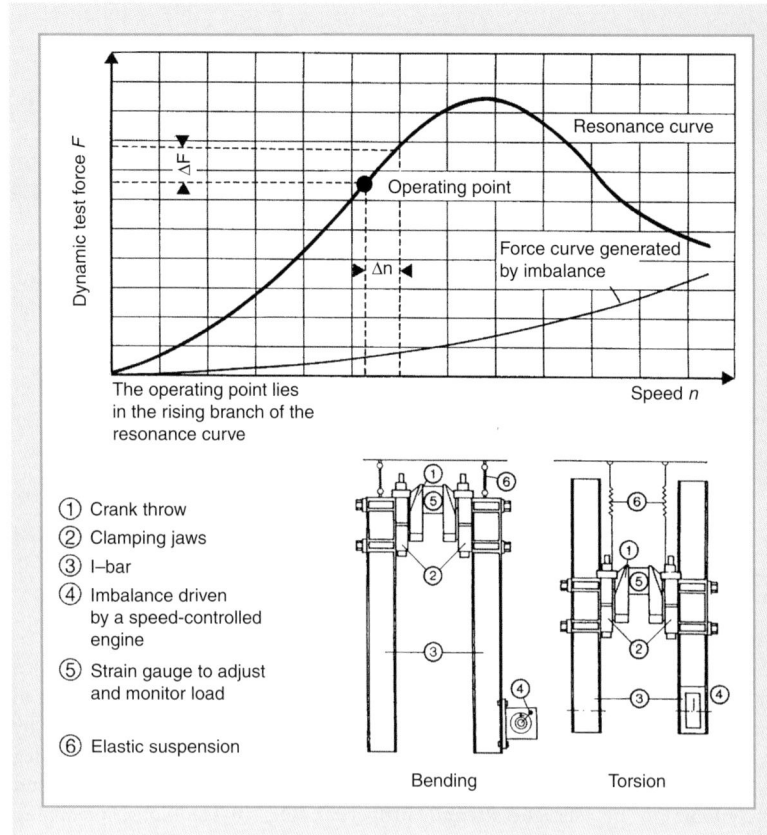

Fig. 10-9 Alternating cyclic stress tests of crankshaft cranks based on the resonance principle

① Crank throw
② Clamping jaws
③ I-bar
④ Imbalance driven by a speed-controlled engine
⑤ Strain gauge to adjust and monitor load
⑥ Elastic suspension

upon a stress analysis using FE calculation or DMS measurement and only needs material characteristics determined on a solid rod.

The solid rod's properties are assumed to be transferrable to the properties of the maximally stressed material element in the component's notch base. Plastic deformations can be factored in. Since they still have numerous gaps, the concepts are frequently combined in practice [10-24]. A need for *reliable material properties* to precisely predict endurance strength and service life is common to both. Values of cyclic material properties taken from publications, standard works and material databases are usually inadequate for assured component design. Experimental investigations under known boundary conditions are indispensible in particular cases. The test procedure and number of samples should be selected to allow statistical analysis, e.g. by the staircase method. Further improvement of predictions of service life will also necessitate *comprehensive experimental and theoretical testing* of the different materials' stress and failure properties in the future.

10.5 Service Life Enhancing Processes

10.5.1 Creative Forming

The better a technically utilizable material's properties may be adapted to widely varying stresses through a component's cross section and volume according to the *principle of task sharing*, the more valuable it is (Fig. 10-10).

Above all, the different methods for treating the surfaces of metallic materials, in particular cast iron materials and steels, have acquired fundamental importance because the maximum stresses caused by temperature, stress and wear usually appear at a component's surface or in areas near the surface with a steep drop in stress toward the center of the material. Goals of development such as lowering fuel consumption by increasing the maximum cylinder pressure or lowering the power-to-weight ratio by raising the brake mean effective pressure while retaining the same crankshaft assembly dimensions are only realizable with their assistance. The different coating

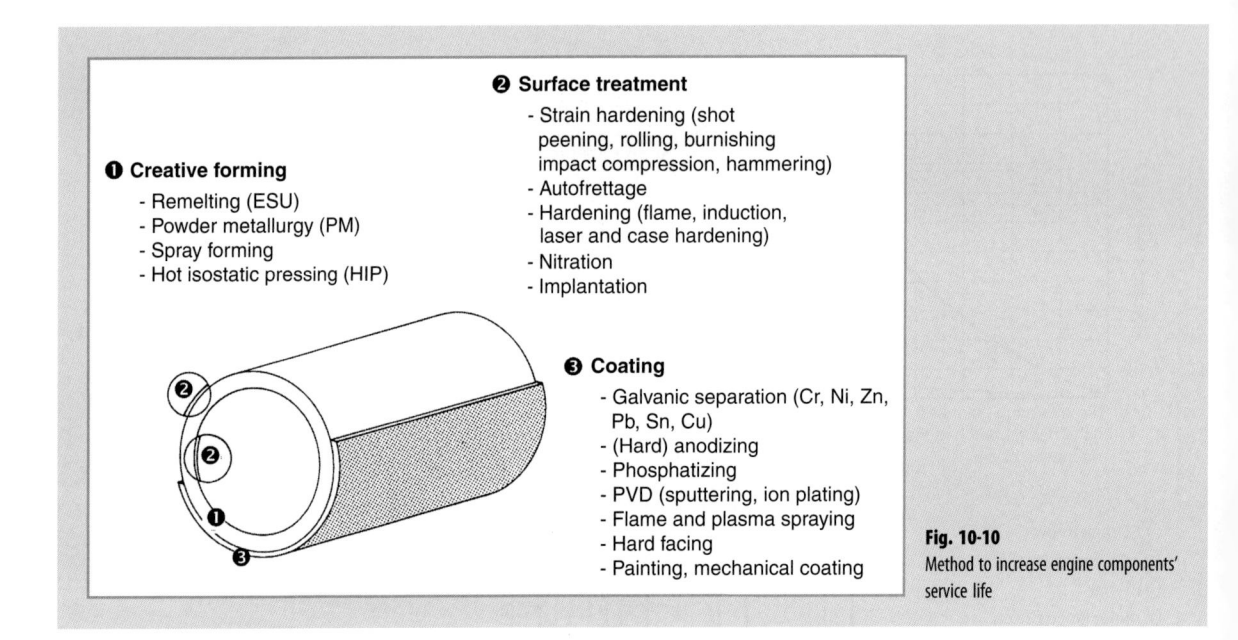

❶ Creative forming
- Remelting (ESU)
- Powder metallurgy (PM)
- Spray forming
- Hot isostatic pressing (HIP)

❷ Surface treatment
- Strain hardening (shot peening, rolling, burnishing impact compression, hammering)
- Autofrettage
- Hardening (flame, induction, laser and case hardening)
- Nitration
- Implantation

❸ Coating
- Galvanic separation (Cr, Ni, Zn, Pb, Sn, Cu)
- (Hard) anodizing
- Phosphatizing
- PVD (sputtering, ion plating)
- Flame and plasma spraying
- Hard facing
- Painting, mechanical coating

Fig. 10-10 Method to increase engine components' service life

technologies have attained the same importance for such tribologically stressed components as plain bearings and piston rings, while the further potentials of creative forming are only applied sporadically for economic reasons.

Liquations and *nonmetallic* inclusions continue to be problems for steel. The risk of fatigue failure they can cause has been substantially lowered by the change from ingot to continuous casting, by improved testing technology and by vacuum melting and remelting [10-25], e.g. electroslag remelting (ESU). ESU significantly boosts toughness and fatigue strength properties (Fig. 10-11). Thus, quenching and tempering, case hardening and nitriding steels of

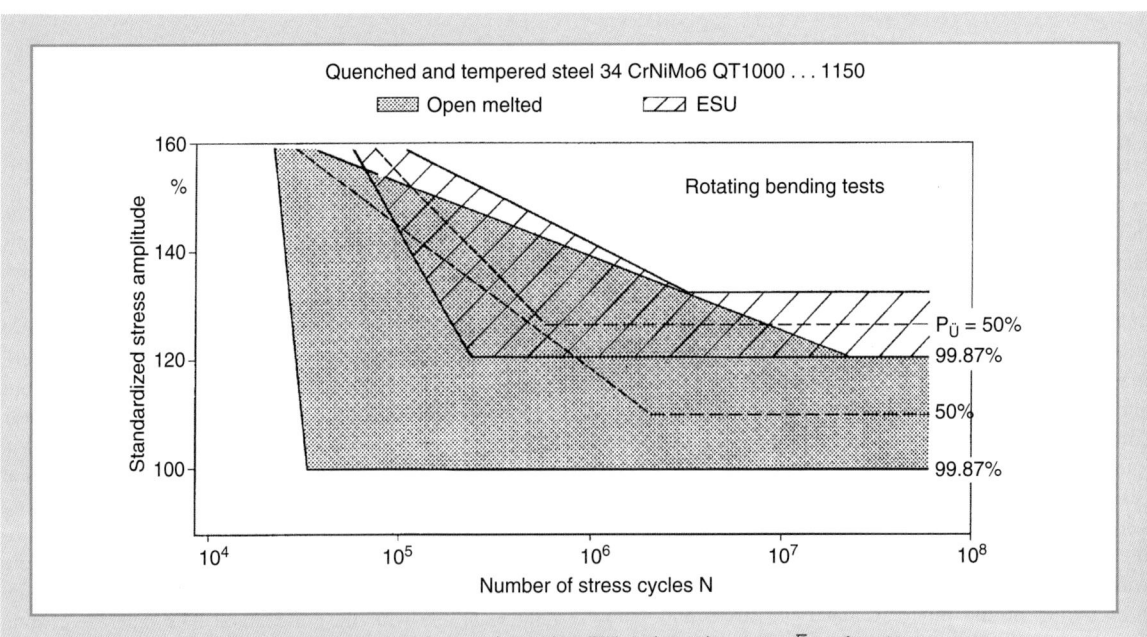

Fig. 10-11 Enhancement of steel's bending fatigue strength by electroslag remelting (ESU). Analysis with an arcsin $\sqrt{P}$-transformation

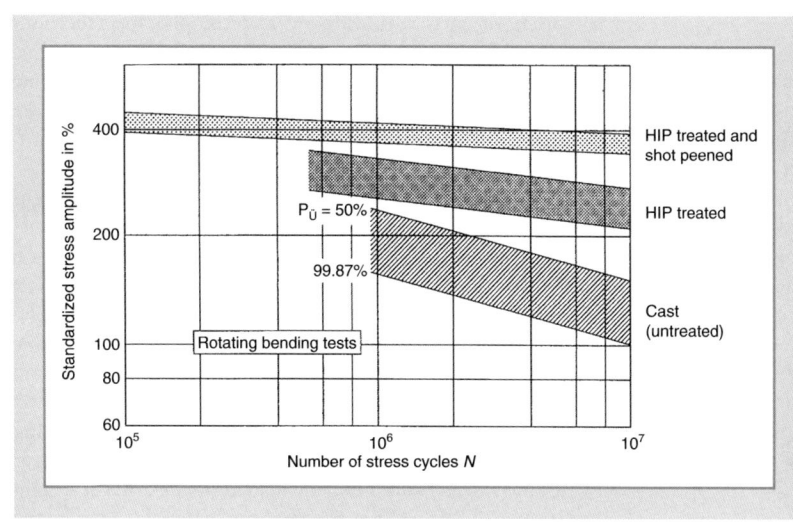

Fig. 10-12 Enhancement of cast aluminum's bending fatigue strength by hot isostatic pressing (HIP). Regression lines in the range of resistance to fatigue

an ESU grade can be employed for highly stressed components such as crankshafts, piston pins and injection pump parts.

Gas pores and shrink holes frequently form internal notches in castings, which shift the points of failure inside the material volume despite low stress. *Hot isostatic pressing* (HIP) can provide a remedy [10-26]. The pressures and temperatures present in the HIP process are utilized to weld together cavities in specific castings, e.g. in high nickel fine casting alloys (turbine wheels) and cast light alloys (crankcases), and thus increase fatigue strength (Fig. 10-12). However, it is unable to eliminate surface porosities and structural defects.

Powder metallurgy (PM) delivers virtually homogeneous materials with isotropic properties. It not only facilitates cost effective mass production (e.g. oil pump gears and valve guides) but also allows sinter forging of connecting rods and even manufacturing of effective piston alloys.

Different special methods of powder metallurgy, thermal spraying based on the Osprey process, are still in development [10-25]. Powder metallurgy applications require certain minimum quantities and are limited by unit weight and shape.

10.5.2 Surface Treatment

The importance of different methods of treating surface zones in iron materials cannot be appreciated enough. The thusly achievable enhancements of properties far exceed the possibilities of alloying techniques, particularly in terms of wear and fatigue load. Various thermal methods of *quenching and tempering* and *hardening* have long been established and in use. The hardening of a steel part's surface is connected with a measurable enlargement of volume due to the formation of martensite. This produces a state of superimposed internal stress with high intrinsic compressive stresses in the hardened area. In conjunction with the increased hardness and strength, they are the basic cause for the increase in fatigue strength. For reasons of equilibrium, the intrinsic compressive stresses must be compensated by corresponding internal tensile stresses. When the hardening zone is positioned poorly, failure under low stress can occur in another spot.

The thermochemical process of *nitriding*, including the related methods of nitrocarbonizing in a salt bath, gas or plasma (ion nitriding), provides similar possibilities to improve properties. Attention must be paid to the degree of porosity of any compound nitride layers and the nitrided surfaces' generally higher sensitivity to damage and mean stress. Therefore, thread zones should not be nitrided at all.

Although they hold the greatest potential to increase the strength of metals, *mechanical strain hardening methods* were somewhat neglected for a long time. Simple blast cleaning of castings and wrought alloys can already produce perceptible increases in fatigue strength. The range of options extends from shot peening, impact compression and rolling to various methods of roller burnishing [10-27]. Depending on the method, smoothing of the microgeometry and plastic deformations of the surface zones improve properties while increasing strength and developing suitable intrinsic compressive stresses much like thermal methods.

Autofrettaging (self-shrinkage) of internally pressurized components represents a special method in which inside surfaces are plastified by nonrecurring stress at very high internal pressure and subjected to compressive prestresses. This method is suitable to reliably control pressures above 1,500 bar in seamlessly drawn St30Al injection lines.

10.5.3 Coating

While coating techniques have facilitated significant leaps in the development of sliding and wear stressed engine components, their role in other component groups is still comparatively minor (Table 10-4). The interruption of the manufacturing process normally necessary and the additional, frequently considerable costs are impediments. Nevertheless, a multitude of methods and coatings are being developed and tested on lab scale since demand continues to exist for customized solutions for technically sophisticated areas of engines, e.g. the piston–piston rings–cylinder liner system. In addition to the established methods of chemical and electrochemical coating, metal injection molding and hard face welding, solutions are primarily expected from techniques of physical and chemical vapor deposition (PVD and CVD).

10.6 Trends in Development

10.6.1 Materials

Once the potentials of engineering design, state-of-the-art manufacturing and machining and treatment technology are consistently availed of, it will be possible to take advantage of the potentials of the properties of conventional, established materials to achieve a large number of the goals of modern engine developments. However, the great variety of parts with different and complex load mechanisms and new engine manufacturing requirements continue to dictate demand for *innovations in materials engineering* (Fig. 10-13). Requirements of lightweight construction are continuing to grow in importance for vehicle engines [10-22, 10-8].

Extremely great hopes were placed in substituting conventional metals with *structural ceramics*. In engine manufacturing, these are *nonmetallic*, inorganic materials based on nitrides, carbides and metal oxides. Since the physical approach of a heat-tight engine with an insulated combustion chamber with better thermal efficiency has proven to be unfeasible (see Sect. 7.2), potential future applications are primarily envisioned in lessening friction and wear (piston rings, cylinder liners and valve guides), reducing moving masses (valves, piston pins and exhaust gas turbocharger turbine wheels) and thermally insulating the exhaust system [10-28, 10-29]. Although ceramic materials have many properties superior to steel (Table 10-5), their substitution is not anticipated in most engine parts for reasons of cost alone. Moreover the deficits of their deformation and failure properties have to be allowed for in design engineering and minimized in development. In addition to individual, already common structural ceramic parts, e.g. highly stressed axial water pump seal rings and tappet rollers, advanced ceramic coatings will most likely be employed as standard, e.g. for piston rings. Thermal insulation by coatings and composite structures, exhaust gas turbocharger

Table 10-4 Potential applications for coatings of engine components

Coating method	Type of coating	Application
Electrochemical (electroplated and chemical deposition)	– Cr	Cylinder liners, cylinder heads, valve stems, Piston top parts, Piston ring grooves
	– Cr, Cr-Al$_2$O$_3$	Piston rings
	– Ni-SiC/NiKasil	Cylinder liners
	– PbSnCu/CuPb	Plain bearing wear layers
	– Zn	Threaded parts, bolts, lines
	– Zn/Mn-phosphate	Low alloy steel components, bolts
	– (Hard) Anodized coatings	Aluminum components, pistons
Physical vapor deposition (PVD)	– TiN	Injection pump plungers, control parts,
	– WC/C	Injection pump plungers, control parts, axial seal ring seals
	– AlSn	Plain bearing wear layers
Thermal spraying process	– Mo	Piston rings, contact surfaces for radial shaft sealing rings
	– Mixed metallic/ceramic coatings	Piston rings
	– ZrO$_2$	Parts conducting exhaust and adjacent to the combustion chamber
Hard face welding	– Hard alloys	Valves, valve seats
Painting/mechanical coating	– Graphiting	Pistons
	– Antifriction lacquer	Bolts
	– Plastic coatings	Sea water conducting cooler components, springs
	– Baked coatings containing metal	Bolts, springs

Fig. 10-13 Future goals of material development

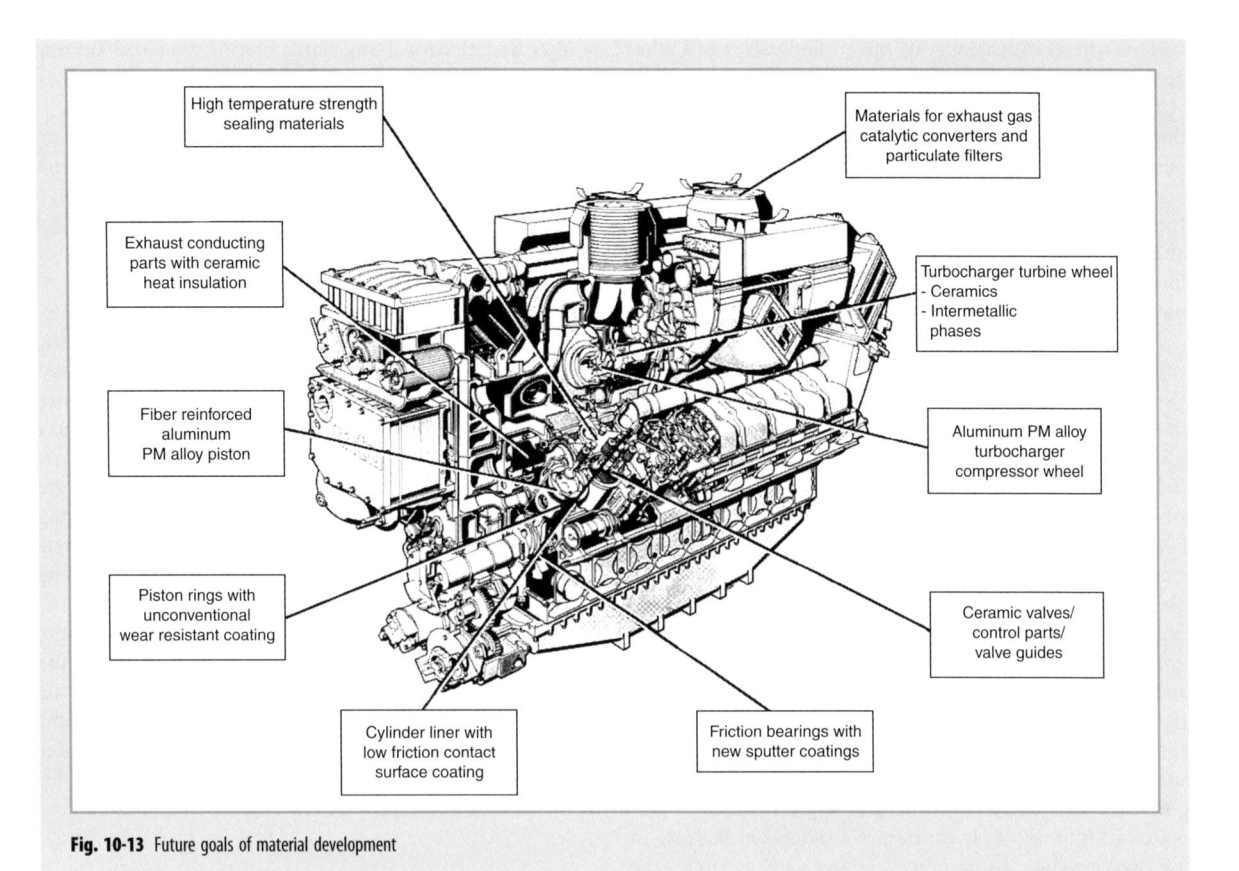

Table 10-5 Ceramic materials for engine manufacturing. Properties compared with steel

Properties materials	Density p g/cm^3	Young's modulus E GPa	Coefficient of elongation α 10^{-6} m/(m · K)	Thermal conductivity λ W/(m · K)	Tensile and flexural strength R MPa	Maximum operating temperature T_{max} °C	Properties Potential applications
Silicon nitride Si$_3$N$_4$	2.5...3.2	180...320	3.0...3.5	11...35	220...700	>1,400	Piston, cylinder liner insert, prechamber, valve guide
Silicon carbide SiC	2.5...3.2	350...410	4.4...4.8	70...90	300...450	1,600	Exhaust gas turbocharger turbine wheel
Aluminum oxide Al$_2$O$_3$	3.9	360	4.0	30	300	1,200...1,900	Axial seal ring, structural reinforcement
Zirconium oxide ZrO$_2$	5.8	200	9.5	2.5	>500	900	Piston, cylinder liner insert, cylinder head plate, valve, port liner
Aluminum titanate Al$_2$TiO$_5$	3.2	20	2.0	2.0	40	900	Piston bowl, port liner
Quenched and tempered steel 34CrNiMo6	7.8	210	11.0	50	1000	500	Crankshaft assembly components (crankshaft, connecting rod, etc.)

turbine wheels with smaller diameters and valves and other engine control components are already technically feasible.

Fiber reinforced or advanced, powder metallurgically produced *aluminum alloys* have potentials for use for high performance requirements just like ceramic materials in pistons or exhaust gas turbocharger compressor wheels. Interest in employing polymer materials in lightly loaded areas of the engine periphery (covers, cowlings, etc.) is increasing because of their potential to reduce weight, noise and costs.

Intermetallic phases, e.g. nickel or titanium aluminide, are coming to be regarded as an alternative to structural ceramics. Only slightly denser (titanium aluminide $\rho = 3.8$ g/cm^3) yet more ductile than structural ceramics (silicon nitride $\rho = 2.5\ldots3.2$ g/cm^3), intermetallic phases are an interesting material group for components of internal combustion engines, which are subjected to high accelerations and temperatures [10-30]. Initial applications for turbine wheels or exhaust valves are being tested at present. In addition to intermetallic phases, high nitrogen austenitic steels will also be used as valve materials.

AFP steels are being developed further for higher strength and yield strength properties. Costly alloy elements are being replaced by boron or heavy nitrogen additives. The range of cast iron types can be supplemented by types of higher strength bainitic nodular cast iron, which have been quenched and tempered in the interstage. Vermicular graphite cast iron will likely experience a significant increase in use once a stable casting process (sinter cast) generates the microstructure in all areas of the engine.

Among light alloys, the use of *magnesium alloys* continues to be pushed for such components as crankcases, cylinder head covers and oil pans to further improve the weight-to-power ratios for vehicle engines.

More work is being done on unconventional materials as substitutes for very high temperature strength alloys and on their optimization, e.g. fine grain carbon for pistons and intermetallic titanium or nickel aluminides [10-31].

10.6.2 Methods

Fundamental importance will be attached to further developing and perfecting process engineering and technology for cost effective manufacturing. The methods described in Sect. 10.5 are being optimized and augmented by new developments. Examples of the wide variety of methods are:
- hard turning,
- water jet cutting and strain hardening,
- fracture splitting of connecting rods and main bearing covers,
- remelting and additional alloying of surface coatings,
- laser alloying of aluminum cylinder liners,
- laser and electron beam welding of the turbine wheel/shaft connection,
- laser and electron beam hardening of the radial bearing contact surfaces of turbine blades and
- ion beam technology (implantation).

The future potential of individual methods remains difficult to assess however [10-10, 10-27, 10-32, 10-33, 10-22, 10-8].

10.6.3 Environment

A number of environmentally hazardous and polluting materials have had to be replaced over the past years. In some cases, this was connected with problems, e.g. replacing cadmium coatings with phosphate and zinc coatings and asbestos with other fiber materials. Further developments and changes may be expected in the wake of new findings and requirements. At present, protective anticorrosion coatings free of Cr(VI) are being qualified for bolts (see Sect. 10.2.3.) and lead-free bearing metals for plain bearings. Strategies to systematically recycle components and materials will also play a role in the future. Above all, logistical issues still have to be resolved.

Challenging problems related to the reduction of pollutants in exhaust gas will require altogether new solutions in the future. Above all, there is a need for suitable catalyst materials that are effective even at low temperatures and materials and elements that filter soot. Innovative contributions that safeguard and advance the competitiveness of diesel engines on the market may be expected in this field.

Literature

10-1 Sperber, R. (Ed.): Technisches Handbuch Dieselmotoren. Sect. 9, Berlin: Verlag Technik (1990)

10-2 Schmidt, R.M.; Trebs, J.: Werkstoffe im Hochleistungsdieselmotor. Ingenieur-Werkstoffe 4 (1992) 1/2, pp. 56–59

10-3 Klein, M.: Einführung in die DIN-Normen. 13th Ed. Berlin: Beuth (2001)

10-4 Bargel, H.J.; Schulze, G. (Ed.): Werkstoffkunde. 9th Ed. Berlin/Heidelberg/New York: Springer (2005)

10-5 Metals Handbook. American Society for Metals. Metals Park, Ohio 44073

10-6 Stahlschlüssel. 20th Ed. Marbach: Verlag Stahlschlüssel (2004)

10-7 Aluminium-Zentrale (Ed.): Aluminium Taschenbuch. Vol. 1, 16th Ed.: Grundlagen und Werkstoffe (2002); Vol. 2, 13th Ed.: Umformen, Giessen, Oberflächenbehandlung (1999); Vol. 3, 16th Ed.: Weiterverarbeitung und Anwendung (2003). Düsseldorf: Aluminium Verlag

10-8 Werkstoffe im Automobilbau 1999/2000. Special Issue ATZ/MTZ (1999)

10-9 Coenen, H.P.; Wetter, E.: Das Pressenschmieden von Kurbelwellen. Thyssen Edelstahlwerke AG: Technische Berichte (1988) 2

10-10 Pischel, H.: Stand und Entwicklungstendenzen beim Herstellen von Pleueln. Werkstatt und Betrieb (1990) 12, pp. 949–955

10-11 Benz, G.: Harte Schichten für hohe Beanspruchungen. Bosch Research Info (2001) 3

10-12 Greuling, S., et al.: Dauerfestigkeitssteigerung durch Autofrettage. FVV e.V., Issue R 506, Frankfurt (2000), pp. 29–46

10-13 Jorach, R.; Doppler, H.; Altmann, O.: Heavy Fuel Common Rail Systems for Large Engines. MTZ Motortechnische Zeitschrift 61 (2000)

10-14 Kayser, K.: Hochfeste Schraubenverbindungen. Die Bibliothek der Technik, Vol. 52. Landsberg/Lech: Verlag Moderne Industrie (1991)

10-15 Fussgänger, A.: Eisengusswerkstoffe im Fahrzeugbau gestern, heute – morgen? Konstruktion 44 (1992), pp. 193–204

10-16 Krebs, R. et al.: Magnesium Hybrid Turbomotor von Audi. MTZ (2005) 4

10-17 Schöffmann, W. et al.: Magnesium Kurbelgehäuse am Leichtbau-Dieselmotor. 3. Tagung Giesstechnik im Motorenbau, Magdeburg, 2005. VDI-Bericht 1830. Düsseldorf: VDI Verlag (2005)

10-18 Weibrecht, A.: Ölwannen aus SMC. Kunststoffe 88 (1998), pp. 318–322

10-19 TRW Thompson, Barsinghausen: TRW-Motorenteile. 7th Ed. (1991)

10-20 Escher, C.: Druckaufgestickte Stähle für Verbrennungsmotoren – Fortschr.-Berichte VDI 5/569 Düsseldorf: VDI Verlag

10-21 Walter, G.; Eyerer, P. (Eds.): Kunststoffe und Elastomere in Kraftfahrzeugen. Stuttgart: Verlag Kohlhammer (1985)

10-22 Weber, A. (Ed.): Neue Werkstoffe. Düsseldorf: VDI-Verlag (1989)

10-23 Haibach, E.: Betriebsfestigkeit. Verfahren und Daten zur Bauteilberechnung. Berlin/Heidelberg/New York: Springer (2002)

10-24 Sonsino, C.M.: Zur Bewertung des Schwingfestigkeitsverhaltens von Bauteilen mit Hilfe örtlicher Beanspruchungen. Konstruktion 45 (1993), pp. 25–33

10-25 Gräfen, H. (Ed.): VDI-Lexikon Werkstofftechnik. Düsseldorf: VDI-Verlag (1991)

10-26 Hofer, B.W.: Nachverdichten statt Nachgiessen – Verbesserung der Eigenschaften durch heissisostatisches Pressen. Giesserei 75 (1988) 2/3, pp. 45–49

10-27 DVM-Arbeitskreis Betriebsfestigkeit: Moderne Fertigungstechnologien zur Lebensdauersteigerung. 17. Vortrags- und Diskussionsveranstaltung GF/Schaffhausen Schweiz. 16. u. 17.10.1991

10-28 Tietz H.D. (Ed.): Technische Keramik – Aufbau, Herstellung, Bearbeitung, Prüfung. Düsseldorf: VDI-Verlag (1994)

10-29 Eisfeld, F. (Ed.): Keramik-Bauteile in Verbrennungsmotoren. Referate der Fachtagung am 15. u. 16.3.1988 in Essen. Wiesbaden: Vieweg (1988)

10-30 Knippscheer, S.; Frommeyer, G.: Intermetalic TiAl(Cr,Mo,Si) Alloys for lightweight engine parts – Structure, Properties and Applicationstpdel 99. Advanced Engineering Materials 1 (1999) $^3/_4$

10-31 Dowling, W.E. Jr.; Allison, J.E.; Swank, L.R.; Sherman, A.M.: TiAl-Based Alloys for Exhaust Valve Applications. SAE-Paper 930620. International Congress and Exposition Detroit/Michigan (1993)

10-32 Georg Fischer Formtech AG Schaffhausen, Schweiz: Pleuel Cracken (1993)

10-33 Haefer, R.A.: Oberflächen und Dünnschicht-Technologie Part II. Berlin/Heidelberg: Springer (1991)

Further Literature

Adlof, W.: Neuere Entwicklungen bei geschmiedeten Kraftfahrzeug-Kurbelwellen. Schmiede-Journal (2001) 9

Adlof, W.: Fortschritte beim Schmieden grosser Kurbelwellen. Schmiede-Journal Report 1999

Technische Keramik. Jahrbuch Ausgabe 1. Essen: Vulkan-Verlag (1988)

Part III Diesel Engine Operation

11 Lubricants and the Lubrication System 359

12 Start and Ignition Assist Systems 377

13 Intake and Exhaust Systems 387

14 Exhaust Heat Recovery 401

Part III Diesel Engine Operation

11 Lubricants and the Lubrication System

Hubert Schwarze

11.1 Lubricants

11.1.1 Requirements for Engine Oils for Diesel Engines

Diesel engines not only impose maximum demands on the loading capacity of every component but also on the lubricant, i.e. the engine oil, which is therefore a technically complex operating agent [11-1–11-3]. Since the operating conditions of car and commercial vehicle diesel engines and large diesel engines differ considerably, differently optimized lubricating oils are employed for the intended purposes.

Engine oil is not only a lubricant but also an important *design element* that is crucial and *integral* to an engine's function and service life and must therefore be matched to an engine's standard of technical quality. Tribologically optimized layers, which determine an engine's friction and wear performance, only form when component surfaces and lubricants interact under mechanical and thermal loads (particularly during running in). Different engine concepts impose fundamentally different requirements on engine oil, which a manufacturer has to specify, factoring in the lubrication system, the maintenance concept, the metallurgy and the design of the engine components. What is more, initiated by the Kyoto Protocol and various national exhaust standards, new technologies that cut the emission of climatically relevant gases and pollutants make very specific additional demands, which have necessitated the development of fuel economy engine oils and catalyst compatible formulations.

Engine oil is required to fulfill all of its functions, which far exceed lubricating oil's functions, under every operating condition at least until the scheduled oil change.

Characterized by chemical, physical and technological properties, the requirements for engine oil are derived from the *main functions* required by engine operation, namely the
- separation of sliding surfaces,
- transmission of forces,
- neutralization of undesired products,
- wear protection,
- corrosion protection,
- sealing and
- cooling.

Table 11-1 compiles the functions of engine oils together with their effective engine ranges and operating states and the complexes of requisite properties. On the one hand, certain functions must be viewed collectively rather than treated separately from one another. On the other hand, different and in part conflicting properties must exist to fulfill certain functions simultaneously to obtain optimal conditions.

The *properties* of engine oils can be subdivided into the following somewhat simplified *complexes*:
- viscosity and flow characteristics,
- surface activity and
- corrosion protection.

Surface activity is differentiated according to whether it affects on engine surfaces or impurities. Corrosion protection refers to the capability to both neutralize combustion products and stabilize the lubricant itself against oxidative degradation.

In addition to its main functions, the following *additional requirements* are imposed on engine oil:
- neutrality toward sealing materials,
- low foaming tendency,
- long service life and large oil change intervals,
- low oil consumption,
- low fuel consumption and
- low loading of emission control systems.

11.1.2 Makeup and Composition

Like other lubricating oils, engine oils consist of a *base oil* or *base oil blend* and *additives* or *agents*. Pure mineral oils, blends of mineral and synthetic oils or purely synthetic oils are employed as base oils. In addition to conventional mineral oils, hydrocrack oils and certain synthetic fluids such as polyalphaolefins or diesters, polyesters, etc. have become

Table 11-1 Functions of engine oils and resultant requirements

Function	Engine range and operating state	Requisite property
Separate sliding surfaces and		
a) Transfer forces a 1) Through load carrying film under pressure a 2) Through chemophysical reaction film	Friction points with fluid friction Friction points with mixed friction	High viscosity Surface activity/high viscosity (EP and AW properties)[a]
b) Neutralization of undesired products (against contamination) neutralizing properties and detergent-dispersive properties b1) Neutralization of liquid impurities b2) Suspension of solid impurities	Low and high operating temperatures	Neutralizing properties Detergent-dispersive properties
c) Wear protection c 1) see a 1) c 2) see a 2)	Friction points with – Fluid friction – Mixed friction	High viscosity Capillary activity/high viscosity (EP and AW properties)
d) Corrosion protection d 1) see a 2) d 2) see b 1)	Low and high operating temperatures Engine stoppage and operation	Surface activity Neutralizing properties
e) Sealing	Piston-piston ring zone/cylinder wall	High viscosity
f) Cooling	Areas with fluid and mixed friction	Heat dissipation properties/low viscosity

[a] Suitable for maximum pressures (extreme pressure) with low wear (anti-wear).

extremely important, especially for the production of multigrade oils.

Synthetic oils are employed as base oil components because they have higher oxidative stability, lower evaporative losses, fewer carbon residues and better resistance to viscosity-temperature change than mineral oils,. However, they are less soluble and their compatibility with sealing materials is more problematic. Only synthetic oils satisfy the most stringent requirements.

Present day engine oils invariably contain additives to perform the functions required of them and are therefore called *fully formulated oils*. The additive-free engine oils used formerly are no longer employed. Advanced high performance engine oils may be assumed to contain the types of additives listed in Table 11-2, which includes examples of the chemical compounds common in various types of additives and some information on their functions.

Detergents and *dispersants* attach their polar "heads" to oil-insoluble combustion and oxidation products, hold them in suspension with the aid of their oleophilic hydrocarbon chains and thus inhibit deposits on metal surfaces, oil thickening and sludge formation in an engine. Basic (overbased) detergents contain a hyperstoichiometric quantity of metal ions (calcium and magnesium) and thus additionally function as corrosion protection. Their alkaline reserve neutralizes acidic combustion products that have reached the oil. Specified by the TBN (total base number), an engine oil's alkalinity is particularly important, especially for large diesel engines powered with highly sulfurous heavy fuel oil. Specifically for the cylinder lubrication of such engines, highly alkaline, oil-soluble additives form homogeneous single phase oils with a TBN of up to 100 (corresponding to 100 mg KOH/g oil) and diminish the danger of chemically corrosive wear. Additives are additionally required to produce a minimum ash with only a soft structure during combustion.

The polar head groups of ashless dispersants only contain functional groups with oxygen and nitrogen atoms and thus no ash-forming metal ions. Dispersants with a fraction of macromolecular oleophilic molecules additionally function as viscosity index (VI) improvers.

Antioxidants are intended to inhibit the degradation of the lubricant by the action of oxygen. While inhibited phenols interrupt the chain reaction of hydrocarbon oxidation by trapping free radicals, zinc dithiophosphates intervene by interacting with the peroxides also formed during oxidation.

VI improvers are long-chain hydrocarbon polymers, which congregate in tight balls (with predominantly intramolecular interactions) at lower temperatures and unravel again at high temperatures and cause thickening with an increase of

Table 11-2 Engine oil additives. Types, chemical compounds and function

Type	Examples	Function
1. Basic detergents	Calcium or magnesium sulfonates, phenolates or salicylates	1. Neutralization of acids 2. Inhibition of varnish formation
2. Ashless dispersants	Polyisobutene succinimides	1. Dispersion of soot and oxidation products 2. Inhibition of the deposition of impurities and varnish
3. Antioxidants	Zinc dithiophosphates, inhibited phenols, phosphor-sulfurized olefins, metal salicylates, amins	Inhibition of oil oxidation and oil thickening
4. High pressure (EP) additives	Zinc dithiophosphates, organic phosphates, organic sulfur compounds	Inhibition of wear
5. Corrosion/rust protection additives	Calcium or sodium sulfonates, amine phosphates, zinc dithiophosphates	Inhibition of corrosion
6. Viscosity index improvers	Polymethacrylates, ethylene propylene copolymers, styrene butadien copolymers	Reduction of the drop in viscosity that occurs as the temperature rises
7. Antifoaming agents	Silicon compounds, acrylates	Inhibition of foaming when intensely circulated
8. Friction modifiers	Fatty acids, fatty acid derivatives, organic amines, amine phosphates, generally milder EP additives	Reduction of friction force losses

viscosity (as intermolecular interactions increase). A formulation of oil with VI improvers must allow for a potential decrease of efficacy by mechanical shearing (shear stability) and a temporary reduction of viscosity in the bearing gap at high temperatures and high shear rates (HTHS viscosity). In this respect, the more complex formulation of high performance oils with base oils with good resistance to viscosity-temperature change (e.g. synthetic oils) and a correspondingly low content of select VI improvers is the more expensive but technically more advantageous method. Permanent and temporary viscosity losses are increasingly playing a role, especially in fuel economy engine oils. So-called stay-in-grade oils maintain the viscosity limits of their specified SAE classification during their service life.

The addition of *pour point improvers* retards the crystallization of paraffins (waxes) contained in mineral base oils and thus improves the oils' low temperature performance.

Friction reducing additives, *friction modifiers* (FM) are increasingly acquiring importance in conjunction with measures to cut fuel consumption. They act in a region of mixed friction to reduce adhesion by solid body friction by forming an adsorption layer on the surface, which separates metal surfaces from one another. Ashless, purely organic FM have extremely polar head groups with a high affinity to surfaces and olefinic hydrocarbon chains that separate surfaces under mixed friction conditions. Molybdenum organic FM act on friction points by forming friction reducing molybdenum sulfide.

The combustion products of ash-forming and sulfurous and phosphorus containing additives are potential catalyst poisons that reduce the service life of three-way catalytic converters and diesel particulate filters. Therefore, catalytic converter compatible high performance oils were developed with low limits for sulfur, phosphorus and sulfate ash contents.

11.1.3 Characterization of Engine Oils

11.1.3.1 General Characteristics

The requirements cited in Sect. 11.1.1 are usually characterized by a combination of the following criteria:

– chemophysical data, e.g. density, flash point, viscosity, alkaline reserve, sulfate ash, etc. (of both fresh and used oil),
– measured values from standardized as well as non-standardized single and multiple cylinder engine test runs (e.g. MWM B, OM 602 A, OM 411 LA),
– results from more or less closely controlled road tests and
– empirical values from the field.

Table 11-3 contains a selection of physical, chemical and technological test methods that may be drawn on to specify new and used engine oils.

11.1.3.2 SAE Engine Oil Viscosity Grades

The Society of Automotive Engineers (SAE) introduced the internationally applied *viscosity classification system*. Table 11-4 presents an excerpt from the SAE J300 classification of

Table 11-3 Physical, chemical and technological test methods to characterize new and used engine oils and base oil components

Parameters	Standard	Parameters	Standard
Density	DIN 51 757	Sealing material performance	
Flash and fire point		Swelling	DIN 53 521
Fresh	DIN EN ISO 2592	Shore hardness	DIN 53 505
Used	DIN EN 22719	Ball indentation hardness	DIN 53 519
Viscosity		Reference elastomer nitrile rubber (Tensile strength test)	DIN 53 504
Ubbelohde	DIN 51 562	Wear test:	
Cannon-Fenske	DIN 51 366	FZG-Test	DIN 51 354
Viscosity index	DIN ISO 2909	VKA-Test	DIN 51 350
m value	DIN 51 563	Iron content	DIN 51397
Cold cranking simulator	ASTM D 5293	IR analysis	DIN 51451
Mini rotary viscosimeter	ASTM D 4684	IR soot content	DIN 51452
Cold viscosity	DIN 51 377	X-ray fluorescence analysis (RFA)	DIN 51396-2
SAE grades for engine oils	DIN 51 511	Carbon distribution	DIN 51 378
Neutralization number	DIN 51 558	Metal contents (Ba, Ca ,Zn)	DIN 51 391
Saporification number	DIN 51 559	Magnesium content	DIN 51 431
Total base number	DIN ISO 3771	Chlorine content	DIN 51 577
Coking tendency		Phosphorus content	ASTM D 1091
Conradson	DIN 51 551	Sulfur content	DIN 51 768
Ramsbottom	ASTM D 524		DIN 51 450
Ash content		Lead content	ASTM D 810
Oxide	DIN ISO 6245	Other metals (Sn, Si, Al etc.)	ASTM D 811
Sulfate	DIN 51 575	Undissolved matter	
Color (ASTM)	DIN ISO 2049	Diaphragm filter methods	DIN 51 592
Evaporation	DIN 51 581	Water content - Distillation method	DIN ISO 3733
Foaming tendency		Ethylene glycol content	DIN 51 375
Seq. 1–3	ASTM D 892	Lubricating oil content in two-stroke mixture	DIN 51 784
Seq. 4	ASTM D 6082	Sampling	DIN 51 750
Air separation properties	DIN 51381	Test error	DIN 51 848
Aging stability	DIN 51 352		
	IP 48		
Oxidation stability	ASTM D 2272		
Thermostability	MIL-II-27601A		
Shear stability: mechanical	DIN 51 381		
Sonication	ASTM D 2603		
Corrosion protection test			
Seawater	DIN 51 358		
HBr	DIN 51 357		

engine oil viscosities. It covers "winter" oils from SAE 0 W to SAE 25 W and "summer" oils from SAE 20 to SAE 60. Winter oils have to satisfy requirements of maximum viscosity at low temperatures and minimum viscosities at high temperatures. By contrast, only a minimum of high temperature viscosity is required of summer oils. A minimum viscosity at high temperature and high shear rate was also introduced in 1994.

Provided expected outside temperatures allow it, both single grade and multigrade engine oils are employed for engine lubrication. Leading manufacturers are increasingly using fuel economy engine oils such as 0 W-30. The observance of manufacturer specifications is imperative in modern high performance engines. Single grade oils are no longer common in Central Europe.

11.1.3.3 Classifications, Specifications and Tests

Military Specifications

Specifications stipulating the subjects as well as the precise sequence of test runs have been defined to facilitate the evaluation of the performance characteristics of engine oils customized for operating conditions. The US military's field specifications, MIL-L-2104 (the different letters indicating different versions) were once tremendously important throughout the world, even in the civilian sector, but this is no longer the case.

Civil Specifications

A distinction is made between general and manufacturers' civil classifications and specifications of engine oils and their

Table 11-4 Viscosity grades for engine oils according to SAE J 300 (2007)[*], excerpt

SAE viscosity grade[a]	Low-temperature cranking (CCS) viscosity[b] max.		Low-temperature pumping viscosity[c] max. with no yield stress		Low-shear-rate Kinematic viscosity[d] (mm²/s) at 100°C		High-shear-rate viscosity[e] (mPa s) at 150°C and 10^6 s^{-1} min.
	mPa s	at °C	mPa s	at °C	min.	max.	
0 W	< 6200	−35	< 60000	−40	3.8		–
5 W	< 6600	−30	< 60000	−35	3.8		–
10 W	< 7000	−25	< 60000	−30	4.1		–
15 W	< 7000	−20	< 60000	−25	5.6		–
20 W	< 9500	−15	< 60000	−20	5.6		–
25 W	< 13000	−10	< 60000	−15	9.3		–
20	–	–	–	–	5.6	< 9.3	2.6
30	–	–	–	–	9.3	< 12.5	2.9
40	–	–	–	–	12.5	< 16.3	3.5[f]
40	–	–	–	–	12.5	< 16.3	3.7[g]
50	–	–	–	–	16.3	< 21.9	3.7
60	–	–	–	–	21.9	< 26.1	3.7

Notes: [a] Requirement in accordance with ASTM D3244; [b] CCS - Cold Cranking Simulator: ASTM D5293 or DIN 51 377; [c] MRV - Mini rotary viscometer: ASTM D4684; [d] ASTM D445 or DIN 51 562; [e] ASTM D4683 or CEC L-36-A-90 (ASTM D4741); [f] For 0 W-40, 5 W-40 and 10 W-40 oils; [g] For 15 W-40, 20 W-40, 25 W-40 and 40 oils
[*] SAE J300 Engine Oil Viscosity Classification, revised in November 2007, effective May 2009

Table 11-5 API classification for diesel engine oils

Category	Status	Service
CJ-4	Current	Introduced in 2006. For high-speed, four-stroke engines designed to meet 2007 model year on-highway exhaust emission standards for all applications with diesel fuels with up to 500 ppm sulfur effective at sustaining emission control system durability where particulate filters and other advanced aftertreatment systems are used. Optimum protection for control of catalyst poisoning, particulate filter blocking, engine wear, piston deposits, low- and high-temperature stability, soot handling properties, oxidative thickening, foaming, and viscosity loss due to shear. May be used in place of API CI-4 with CI-4 PLUS, CI-4, CH-4, CG-4 and CF-4. Using CJ-4 oil with higher than 15 ppm sulfur fuel may impact exhaust aftertreatment systems.
CI-4	Current	Introduced in 2002. For high-speed, four-stroke engines designed to meet 2004 exhaust emission standards implemented in 2002. CI-4 oils are formulated to sustain engine durability where exhaust gas recirculation (EGR) is used and are intended for use with diesel fuels ranging in sulfur content up to 0.5% weight. Can be used in place of CD, CE, CF-4, CG-4, and CH-4 oils. Some CI-4 oils may also qualify for the CI-4 PLUS designation.
CH-4	Current	Introduced in 1998. For high-speed, four-stroke engines designed to meet 1998 exhaust emission standards. CH-4 oils are specifically compounded for use with diesel fuels ranging in sulfur content up to 0.5% weight. May be used in place of CD, CE, CF-4, and CG-4 oils.
CG-4	Current	Introduced in 1995. For severe duty, high-speed, four-stroke engines using fuel with less than 0.5% weight sulfur. CG-4 oils are required for engines meeting 1994 emission standards. May be used in place of CD, CE, and CF-4 oils.
CF-4	Obsolete	Introduced in 1990. For high-speed, four-stroke, naturally aspirated and turbocharged engines. May be used in place of CD and CE oils.
CF-2	Current	Introduced in 1994. For severe duty, two-stroke-cycle engines. May be used in place of CD-II oils.
CF	Current	Introduced in 1994. For off-road, indirect-injected and other diesel engines including those using fuel with over 0.5% weight sulfur. May be used in place of CD oils.
CE	Obsolete	Introduced in 1985. For high-speed, four-stroke, naturally aspirated and turbocharged engines. May be used in place of CC and CD oils.
CD-II	Obsolete	Introduced in 1985. For two-stroke cycle engines.
CD	Obsolete	Introduced in 1955. For certain naturally aspirated and turbocharged engines.
CC	Obsolete	CAUTION: Unsuitable for use in diesel-powered engines built after 1990.
CB	Obsolete	CAUTION: Unsuitable for use in diesel-powered engines built after 1961.
CA	Obsolete	CAUTION: Unsuitable for use in diesel-powered engines built after 1959.

Table 11-6 ACEA specifications for car diesel engine oils

ACEA-grade	Status	Range of applications/requirements
B1	Current	Category for fuel economy engine oils with particularly low HTHS viscosity (corresponding to A1)
B2	Withdrawn	Category for conventional and high lubricity engine oils
B3	Current	Category for conventional and high lubricity engine oils; exceed ACEA B2 in terms of cam wear, piston cleanliness, viscosity stability at soot load
B4	Current	New category for direct injection diesel engines (TDI)
B5	Current	Corresponds to ACEA B4, however with reduced HTHS viscosity. Fuel economizing $\geq 2.5\%$ compared to a 15 W-40 reference oil must be demonstrated in a test engine.
C1	Current	As of 2004 for car diesel engines with diesel particulate filters, maximum sulfate ash content of 0.5%, with reduced HTHS (Ford).
C2	Current	As of 2004 for car diesel engines with diesel particulate filters, maximum sulfate ash content of 0.8%, with HTHS > 2.9 mPas (Peugeot).
C3	Current	As of 2004 for car diesel engines with particulate filters, maximum sulfate ash content of 0.8%, with HTHS > 3.5 mPas. (Mercedes Benz and BMW)

performance characteristics. Generally valid civil engine oil specifications and classifications include those from the API (American Petroleum Institute) and the ACEA (Association des Constructeurs d'Automobiles), which superseded the classifications of the CCMC (Comite des Constructeurs d'Automobiles du Marche Commun).

The following API grades are particularly important:
– SM, SL and SJ for gasoline engine oils and
– CJ-4, CI-4, CH-4, CG-4 CF-2, and CF for diesel engine oils (Table 11-5).

The ACEA employs the classifications AxBx for car gasoline and diesel engines (Table 11-6), Ex for commercial vehicle diesel engines (Table 11-7) and the new classification Cx for catalyst compatible engine oils intended to guarantee longer service life of three-way catalytic converters and diesel particulate filters and a sustainable reduction of fuel consumption. There are ACEA test sequences for:
– car gasoline and diesel engines: A1/B1-08, A3/B3-08, A3/B4-08 and A5/B5-08,
– catalyst compatible oils: C1-08, C2-08 and C3-08 (Table 11-8) and
– commercial vehicle diesel engines: E4-08, E6-08 and E7-08 (Table 11-9).

The ACEA [11-4] has expanded the scope of engine test runs. Limits on sulfate ash and sulfur and phosphorus content were introduced or tightened for catalyst compatible Cx oils.

The ACEA has more stringent viscosity requirements than the SAE. The following engine tests have been designated for the aforementioned requirements:

Table 11-7 ACEA Specifications for truck diesel engine oils

ACEA-grade	Status	Range of applications / requirements
E1	Withdrawn	Largely corresponds to previous CCMC D 4.
E2	Current	Largely based on MB 228.1, a Mack T8 test is additionally required
E3	Withdrawn	Largely based on MB 228.3, a Mack T8 test is additionally required
E4	Current	Largely based on MB 228.5; no engine test OM 364 A but a Mack T8 & T8E test, longest oil change, suited for Euro III engines.
E5	Withdrawn	Category for Euro III engines, reduced ash content over E4. Quality level is between ACEA E3 and E4.
E6	Current	For EGR engines with and without diesel particulate filters and SCR NO_x engines; recommended for engines with diesel particulate filters in combination with sulfur-free fuel; sulfate ash content $< 1\%$ (wt.)
E7	Current	For most EGR and most SCR NO_x engines without diesel particulate filters; sulfate ash content max. 2% (wt.).

Table 11-8 ACEA test sequences for gasoline and diesel engine oils and for catalyst compatible oils

Requirement	Test procedure	Properties	Unit	Limit values								
				Oils for car gasoline and diesel engines					Catalyst compatible oils			
				A1/B1-08	A3/B3-08	A3/B4-08	A5/B5-04	C1-08	C2-08	C3-08	C4-08	
Lab tests												
Viscosity grades		SAE J 300		No restrictions except for requirements for shear stability and HTHS. Manufacturers can indicate specific requirements for viscosity relative to the ambient temperature								
Shear stability	CEC L-14-A-93 or ASTM D6278	100°C viscosity after 30 cycles	mm²/s	xW-20 s.i.g. xW-30 ≥ 9.3 xW-40 ≥12.0	The viscosity of all grades must stay in grade (s.i.g.)							
HT/HS viscosity	CEC L-36-A-90 (2nd Ed.) (Ravenfield)	Viscosity at 150°C and 10^6 s^{-1} shear rate	mPa s	≥ 2.9 and ≤ 3.5 xW-20: 2.6 min.	≥ 3.5	≥ 3.5	≥ 2.9 and ≤ 3.5	≥ 2.9	≥ 2.9	≥ 3.5	≥ 3.5	
Evaporative losses	CEC L-40-A-93 (Noack)	Maximum weight loss after 1 h at 250°C	%	≤15	≤13	≤13	≤13	≤13	≤13	≤13	≤11	
Sulfated ash	ASTM D874		% m/m	≤1.3	≤1.5	≤1.6	≤1.6	≤0.5	≤0.8	≤0.8	≤0.5	
Sulfur	ASTM D5185		% m/m	Report				≤0.2	≤0.3	≤0.3	≤0.2	
Phosphorus	ASTM D5185		% m/m	Report				≤0.05	≤0.090	≤0.090	≤0.090	
Chlorine	ASTM D6443		ppm m/m	Report								
TBN	ASTM D2896		mg KOH/g	≥ 8						≥ 6	≥ 6	
Foaming tendency	ASTM D892	Tendency-stability	ml	Sequence I (24°C) 10 – nil Sequence II (94°C) 50 – nil Sequence III (24°C) 10 – nil								
High temperature foaming tendency	ASTM D6082	Tendency-stability	ml	Sequence IV (150°C) 100 - nil								
Engine tests												
High temperature deposits Ring sticking Oil thickening	CEC L-088-T-02 (TU5JP-L4) 72 h Test	Ring sticking Piston varnish Increase in viscosity Oil consumption	Merit Merit mm²/s kg/test	≥ 9.0 ≥ RL 216 ≤ 0.8 x RL 216 Report								
Low-temperature sludge	ASTM D6593-00	Avg. engine sludge	Merit	≥ 7.8								
		Rocker cover sludge	Merit	≥ 8.0								
		Avg. piston skirt varnish	Merit	≥ 7.5								
		Avg. engine varnish	Merit	≥ 8.9								
		Compression ring (hot stuck)		none								
		Oil screen clogging	%	≤ 20								
Valve train scuffing wear	CEC L-038-94 (TU3M)	Cam wear, avg. Cam wear, max. Pad merit	μm μm Merit	≤ 10 ≤ 15 ≥ 7.5								

Table 11-8 (Continued)

Requirement	Test procedure	Properties	Unit	Limit values								
				Oils for car gasoline and diesel engines					Catalyst compatible oils			
				A1/B1-08	A3/B3-08	A3/B4-08	A5/B5-04	C1-08	C2-08	C3-08	C4-08
Black sludge	CEC L-53-95 (M111)	Sludge in engine (Average)	Merit	$\geq$ RL 140	$\geq$ RL 140 + 4σ or $\geq$ 9.0						
Fuel economy	CEC L-54-96 (M111)	Improvement over RL 191 (15 W-40)	%	$\geq$ 2.5			$\geq$ 2.5		$\geq$ 3.0	$\geq$ 2.5	$\geq$ 1.0 (for xW-30)
Medium temperature dispersivity	CEC L-093-04 (DV4TD)	Increase in viscosity Piston merit	mm/s^2 Merit	$\leq$ 0.6 RL 223 result $\geq$ RL 223 - 2.5 pts.							
Wear	CEC L-099-08 (OM646LA)	Avg. cam wear outlet Avg. cam wear inlet Avg. cylinder wear Bore polishing Avg. tappet wear inlet Avg. tappet wear outlet Avg. piston cleanliness Avg. engine sludge	μm μm μm % μm μm Merit Merit	$\leq$ 140 $\leq$ 110 $\leq$ 5.0 $\leq$ 3.5 Report Report Report Report	$\leq$ 140 $\leq$ 110 $\leq$ 5.0 $\leq$ 3.5 Report Report Report Report	$\leq$ 120 $\leq$ 100 $\leq$ 5.0 $\leq$ 3.0 Report Report Report Report	$\leq$ 120 $\leq$ 100 $\leq$ 5.0 $\leq$ 3.0 Report Report Report Report	$\leq$ 120 $\leq$ 100 $\leq$ 5.0 $\leq$ 3.0 Report Report Report Report	$\leq$ 120 Report $\leq$ 5.0 $\leq$ 3.0 Report Report Report Report	$\leq$ 120 $\leq$ 100 $\leq$ 5.0 $\leq$ 3.0 Report Report Report Report	$\leq$ 120 $\leq$ 100 $\leq$ 5.0 $\leq$ 3.0 Report Report Report Report
DI diesel piston cleanliness and ring sticking	CEC L-078-99 (VW TDI)	Piston cleanliness Ring sticking (rings 1 & 2) Average Max for 1st ring Max for 2nd ring EOT TBN (ISO 3771) EOT TAN (ASTM D 664)	Merit ASF ASF ASF mgKOH/g mgKOH/g	$\geq$ RL 206 - 4 pts. $\leq$ 1.2 $\leq$ 2.5 $\leq$ 0.0 $\geq$ 4.0 Report	$\geq$ RL 206 - 4 pts. $\leq$ 1.2 $\leq$ 2.5 $\leq$ 0.0 $\geq$ 4.0 Report	$\geq$ RL 206 $\leq$ 1.0 $\leq$ 1.0 $\leq$ 0.0 $\geq$ 4.0 Report	$\geq$ RL 206 $\leq$ 1.0 $\leq$ 1.0 $\leq$ 0.0 $\geq$ 4.0 Report	$\geq$ RL 206 $\leq$ 1.0 $\leq$ 1.0 $\leq$ 0.0 Report Report	$\geq$ RL 206 $\leq$ 1.2 $\leq$ 2.5 $\leq$ 0.0 Report Report	$\geq$ RL 206 $\leq$ 1.0 $\leq$ 1.0 $\leq$ 0.0 Report Report	$\geq$ RL 206 $\leq$ 1.0 $\leq$ 1.0 $\leq$ 0.0 Report Report

Table 11-9 ACEA test sequences for oils for commercial vehicle diesel engines

Requirement	Test procedure	Properties	Unit	Limit values			
				Oils for commercial vehicle diesel engines			
				E4-08	E6-08	E7-08	E9-08
			Lab tests				
Viscosity grade		SAE J 300		No restrictions except for requirements for shear stability and HTHS. Manufacturers can indicate specific requirements for viscosity relative to the ambient temperature.			
Shear stability	CEC-L-14-A-93 or ASTM D6278	Viscosity after 30 cycles 100°C	mm²/s	stay in grade			
	ASTM D6278	Viscosity after 90 cycles 100°C	mm²/s			stay in grade	
HT/HS viscosity	CEC-L-36-A-90 (2nd Ed.) (Ravenfield)	Viscosity at 150°C and 10⁶ s⁻¹Shear rate	mPa s		≥ 3.5		
Evaporative losses	CEC-L-40-A-93 (Noack)	Maximum weight loss after 1 h at 250°C	%		≤13		
Sulfated ash	ASTM D874		% m/m	≤2.0	≤1.0	≤2.0	≤1.0
Sulfur	ASTM D5185		% m/m		≤0.3		≤0.4
Phosphorus	ASTM D5185		% m/m		≤0.08		≤0.12
Foaming tendency	ASTM D892	Tendency - stability	ml		Sequence I (24°C) 10 – nil Sequence II (94°C) 50 – nil Sequence III (24°C) 10 – nil		Seq I 10/0 Seq II 20/0 Seq III 10/0
High temperature foaming stability	ASTM D6082	Tendency - stability	ml		Sequence IV (150°C) 200-50		
Oxidation	CEC-L-085-99 (PDSC)	Induction time	min	R&R	R&R	≥ 65	≥ 65
Corrosion	ASTM D 6594	Copper increase Lead increase Copper strip rating	ppm ppm max	R&R R&R R&R	R&R R&R R&R	R&R ≤100 R&R	≤20 ≤100 3
TBN	ASTM D 2896		mg KOH/g	≥ 12	≥ 7	≥ 9	≥ 7
			Engine tests				
Wear	CEC L-099-08 (OM646LA)	Cam wear outlet (avg.)	μm	≤ 140	≤ 140	≤ 155	≤ 155
Soot in oil	ASTM D 5967 (Mack T-8E)	Test duration 300 h Rel. viscosity at 4.8% soot 1 test/2 test/3 test average			≤ 2.1/2.2/2.3		
Soot in oil	Mack T11	Min TGA soot @ 4.0 cSt (100°C) Min TGA soot @ 12.0 cSt (100°C) Min TGA soot @ 15.0 cSt (100°C)	%				3.5/3.4/3.3 6.0/5.9/5.9 6.7/6.6/6.5
Bore polishing Piston cleanliness	CEC L-101-08 (OM501LA)	Bore polishing, avg. Piston cleanliness, avg. Oil consumption	% Merit kg/test	≤ 1.0 ≥ 26 ≤ 9		≤ 2.0 ≥ 17 ≤ 9	

Table 11-9 (Continued)

Requirement	Test procedure	Properties	Unit	Limit values Oils for commercial vehicle diesel engines			
				E4-08	E6-08	E7-08	E9-08
Soot induced wear	Cummins ISM	Engine sludge, avg.	Merit	$\leq$ R&R		$\leq$ R&R	
		Merit				$\leq 7.5/7.8/7.9$	≥ 1000
		Rocker pad average weight loss at 3.9% soot	mg			$\leq 7.5/7.8/7.9$	≤ 7.1
		1 test/2 test/3 test average Oil filter diff. press @ 150 h	kPa			$\leq 55/67/74$	≤ 19
		1 test/ 2 test/3 test average Engine sludge	Merit			$\geq 8.1/8.0/8.0$	≥ 8.7
		1 test/2 test/3 test average Adj. screw weight loss	mg				≤ 49
Wear (liner-ring-bearings)	Mack T12	Merit			≥ 1000	≥ 1000	≥ 1000
		Avg. liner wear	μm		≤ 26	≤ 26	≤ 24
		Avg. top ring wt. loss	mg		≤ 117	≤ 117	≤ 105
		Lead content (end of test)	ppm		≤ 42	≤ 42	≤ 35
		Delta lead 250–300 h	ppm		≤ 18	≤ 18	≤ 15
		Oil consumption (Phase II)	g/h		≤ 95	≤ 95	≤ 85

- Ring sticking and piston cleanliness (CEC-L-78-T-99): VW-1.9-1-DI, which superseded VW 1.6 TC D, requires a demonstration of oils' performance in the specified disciplines in comparison with RL 206.
- Bore polishing and piston cleanliness: The OM-501LA-Test (CEC-L-101-08) is the determinative quality criterion for E4 to E7 oils. A minimum of potential deposits in piston ring grooves is critical.
- Oil thickening: The Mack T-8E (Mack T11 for E9-08) was selected with a limit on the increase in viscosity per hour.
- Wear: The OM 646 LA CEC L-099-08 engine was selected for wear and other criteria.
- Piston cleanliness and increase in viscosity are measured in a Peugeot DV4TD CECL-093-04 engine.

Naturally, some engine manufacturers have developed their own classifications and specifications. Some examples are:
- MAN 270 and 271,
- QC 13-017,
- Mercedes Benz 228.1 to 229.5,
- MTU MTL 5044,
- Volvo Drain Specification VDS and VDS-2 and
- VW 505.00 to 507.00.

Requirements that exceed the ACEA specifications are imposed on certain technologies (diesel particulate filters and unit injectors) and servicing (prolongation of the servicing interval).

No international specifications exist for the alkaline engine oils used in large diesel engines that run on heavy fuel oil. Thus, they are only approved after longer test runs at a manufacturer's facilities. However, apart from their TBN, they must at least fulfill the specification MIL-L-2104C or API CD [11-5] (Table 11-4).

11.1.4 Changes in Engine Oil Caused by Operation

11.1.4.1 Predominantly Caused by Evaporation, VI Improvers and the Infiltration of Solid Impurities

Evaporative Losses

Changes in engine oil during engine operation are partly caused by physical effects resulting from mechanical and thermal stresses and partly by chemical reactions. Physically induced changes include evaporative losses. At a given temperature level, they depend on the parameters of the:
- molecular weight and viscosity of the base oil and type of base oil (mineral oil and synthetic oil),
- base oil formulation (the core fraction being better than a blend of high and low viscosity base oils) and
- molecular weight and distribution of VI improvers.

The consequences of evaporative loss are increased oil consumption and oil thickening, i.e. increased viscosity.

Reduction of Viscosity by Shearing

Mechanical, thermal and oxidative degradation of the high molecular polymers employed as VI improvers lower the viscosity level and diminish engine oil's viscosity-temperature characteristics.

The extent of the losses from shearing depends on the operating conditions, the chemical structure and the molecular weight and concentration of the polymer additives. Oils are designated stay-in-grade oils when they remain in their initial viscosity grade even after shearing.

Temporary Reduction of Viscosity

At a high shear rate, oils formulated with viscosity index improvers can temporarily lose viscosity in the oil film between sliding surfaces. This can be explained by the alignment of the additives' long-chain molecules in the direction of flow, which especially constitutes a problem for low viscosity fuel economy engine oils. Therefore, a viscosity measurement at 150°C under a high shear rate (HTHS viscosity) was introduced in addition to the still common Ubbelohde method of viscosity measurement in a capillary at 100°C. Despite their lower nominal viscosity (SAE grade), fuel economy engine oils must meet appropriate HTHS limits to ensure sufficient certainty of lubrication at high load.

Reduction of Viscosity by Fuel

The condensation of unburned fuel on cool cylinder walls and its infiltration of engine oil cannot be ruled out, above all in short distance driving below the normal operating temperature. This reduces viscosity.

In practice, up to 2% fuel in engine oil is regarded as normal, larger values as critical. Any evaluation of these figures should allow for the capability of a slight percentage of fuel to drop an engine oil's viscosity into the next lower viscosity grade (danger of wear).

Increase in Viscosity by Solid Impurities

Solid impurities in used engine oil include dirt particulates that enter from the outside with the air, oil-insoluble reaction products of oil aging and residues from fuel combustion. Primarily the soot produced by the latter process is important in diesel engines.

In practice, 1–2% soot by mass in oil is considered harmless. The percentage increase of engine oil viscosity as a function of the content of solid impurities demonstrates that less than 1% soot already suffices to increase the viscosity of high viscosity oils to the next higher viscosity grade. Less soot infiltrates engine oil in direct injection diesel engines than in engines with chamber combustion systems.

11.1.4.2 Predominantly by Additive Depletion

Acidification

Engine oil acidification is caused by the oxidation of the oil, which produces organic oil-soluble acids. Even the compounds produced during the combustion of sulfurous fuels – particularly pronounced in heavy fuel oil operation of large diesel engines – contribute to oil acidification. Effects of engine oil acidification are chemical wear in cylinder barrels and corrosion, especially in bearings.

The use of oxidation and corrosion inhibitors in an engine oil are some of the most important measures against engine oil acidification and its effects. The oil must always remain alkaline and be changed early enough before it acidifies.

Residue Formation

The thermal decomposition and oxidation of oil produce high molecular, organic, hydrogen-deficient and oil-insoluble compounds and are causes for the formation of residues. These processes act to separate the compounds from the oil and deposit them inside an engine on the pistons and in the oil bores and lines.

The use of detergent/dispersant additives in engine oil is the most important measure against residue formation.

Sludge Formation

The bonding of oil residues with solid impurities, water and acids causes sludgy masses to form. Two types of sludge are distinguished:
– Low-temperature sludge is typical for stop-and-go driving in an engine's undercooled operating state. If the normal operating temperature is not reached, water and unburned fuel produced during combustion can condense and bond with oil residues to form sludge.
– High-temperature sludge primarily forms at high operating temperatures and consists of oil-insoluble reaction products from blow-by gases, nitrogen oxides and engine oil. In an engine, this type of deposit is particularly found on the valve cover (cylinder head) where it cannot be eliminated by an oil change (danger to the crankcase ventilation system). Along with engine characteristics and operating conditions, engine oil and fuel are factors that influence the formation of high temperature sludge.

Appropriately fully formulating engine oil with detergents/dispersants generally counteracts sludge formation.

11.2 Lubrication Systems

11.2.1 Functions and Requirements

Optimal oil circuit design significantly determines the service life of diesel engines.

In addition to its main function of lubrication, the oil circuit, consisting of the main components of the oil reservoir, oil pump, oil cooler and oil filter, must also ensure that engine components are cooled and protected against corrosion [11-6–11-10].
– An oil supply system must supply the requisite volumes of oil for every component in every engine operating point and under every operating condition, e.g. at extremely high and low operating temperatures. Among other things, this entails measuring and monitoring the oil pressure in the main oil gallery.
– A lubrication system must be designed so that routinely changing the oil and filter and monitoring the oil volume (e.g. with an oil dipstick) and the oil pressure gauge are the only work required during an engine's entire service life.

11.2.2 Design

11.2.2.1 Oil Circuit Sizing

Figure 11-1 illustrates the lubricating oil circuit of an OM 906 LA commercial vehicle engine with the real arrangement of the individual components, the oil galleries and feed lines.

It reveals that adjusting the supply and cooling the components with the available oil volume and providing every component sufficient oil is extremely involved. Oil distribution may be assumed to require approximately 25–30 l of oil per KWh. This volume is divided into thirds among the assemblies of the engine suspension, the piston cooling system and the control system with the turbocharger. The total volume of oil in a motor vehicle engine is approximately 2 l/dm^3 of displacement.

The consumers in an oil circuit should have the lowest flow resistances possible to minimize pressure losses. This leads to compact design, i.e. external lines are eliminated the oil cooler is often integrated in the crankcase and the oil filter is fitted in the engine in order to produce short oil paths with low pressure losses.

The sequence of the components in a circuit is particularly important. The oil filter should be placed before the main oil gallery to be able to trap manufacturing debris and any impurities. The oil cooler is located after the oil pump to keep the temperature level in the crankshaft assembly low.

Generally, the crankshaft bearings must receive a higher volume of oil to provide the bearing shells hydrodynamic lubrication as well as cooling. However, increased oil flow causes increased flow resistance and thus bearing heating. Yet again, this underscores the need to match oil volumes exactly.

Fig. 11-1 OM 906 LA oil circuit

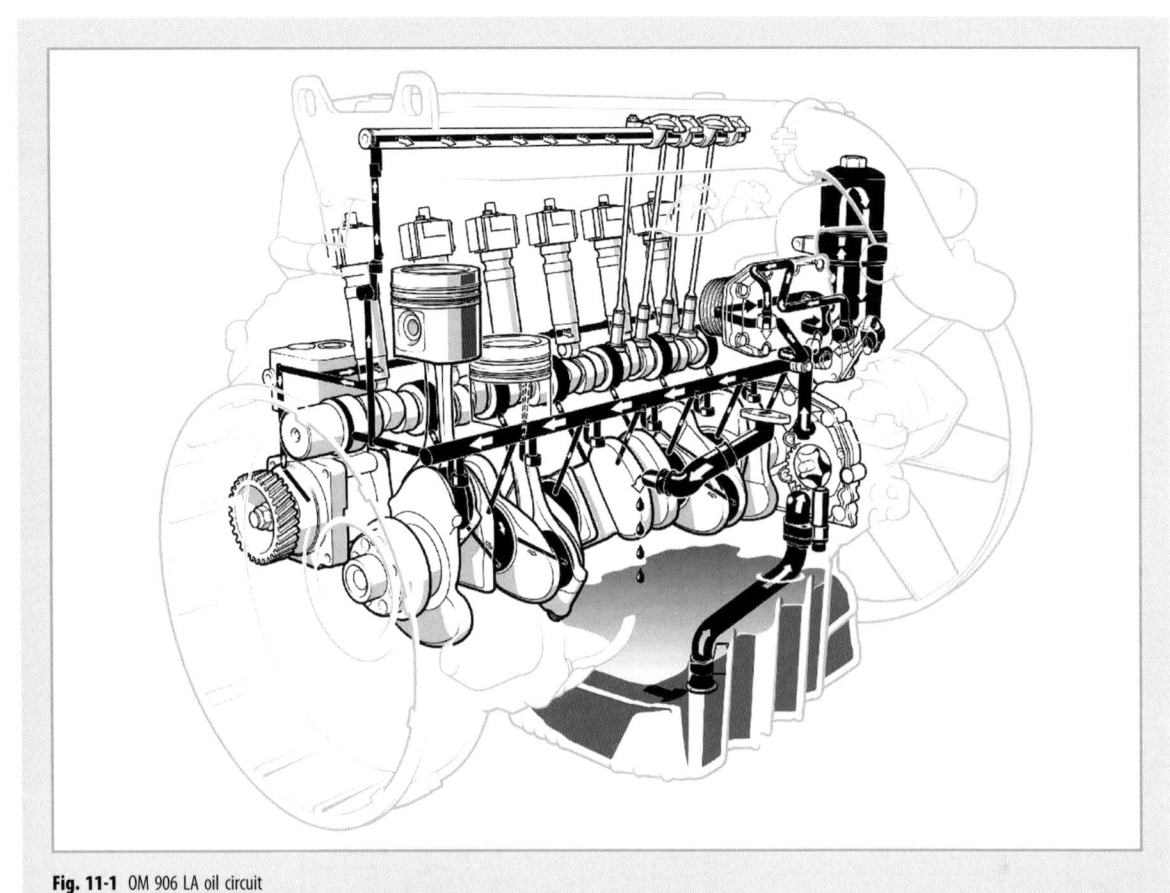

Control components such as the camshaft and roller rocker bearings are supplied oil by oil galleries. Cams and tappets receive thrown oil spray or are supplied by oil spray nozzles.

Oil circuits are now designed with experimental methods, e.g. by means of flow measurement (flowmeter), drip collecting pans, particle image velocimetry (PIV), laser Doppler anemometry (LDA) and temperature measurements as well as high-speed filming.

In the simplest case of mathematical design, simulation programs are employed to simulate one-dimensional flows. CFD flow simulations are performed to capture phenomenological flow effects. Measurement and simulation must always factor in every operating condition from cold starting to extremely high temperatures.

11.2.2.2 Lubrication Systems

Figure 11-2 presents a schematic of a commercial vehicle engine's oil circuit.

The oil pump sucks oil out of the reservoir and transports it through the ports to the consumers. Since the oil volume circulated over time increases with the engine speed, a valve limits the maximum oil pressure to 5 bar in order to prevent any damage to the oil cooler, oil filter and seals.

The flow of oil from the oil pump initially proceeds through the cooler and filter to the main oil gallery with an oil pressure gauge point. All branches to the other lubricating points emanate from here. The exhaust gas turbocharger and injection pump are also lubricated from the main oil gallery.

The oil reaches the basic bearings in the area of the crankshaft assembly whence it passes through the crankshaft to the connecting rod bearings and through bores in the connecting rod to the piston pins. Alternatively, the piston pin and the connecting rod small end may be supplied with oil spray.

Oil reaches the camshaft, tappets and roller rockers through ports to lubricate the control components. Once they have been lubricated, the oil flows unpressurized through returns and ports back into the oil pan.

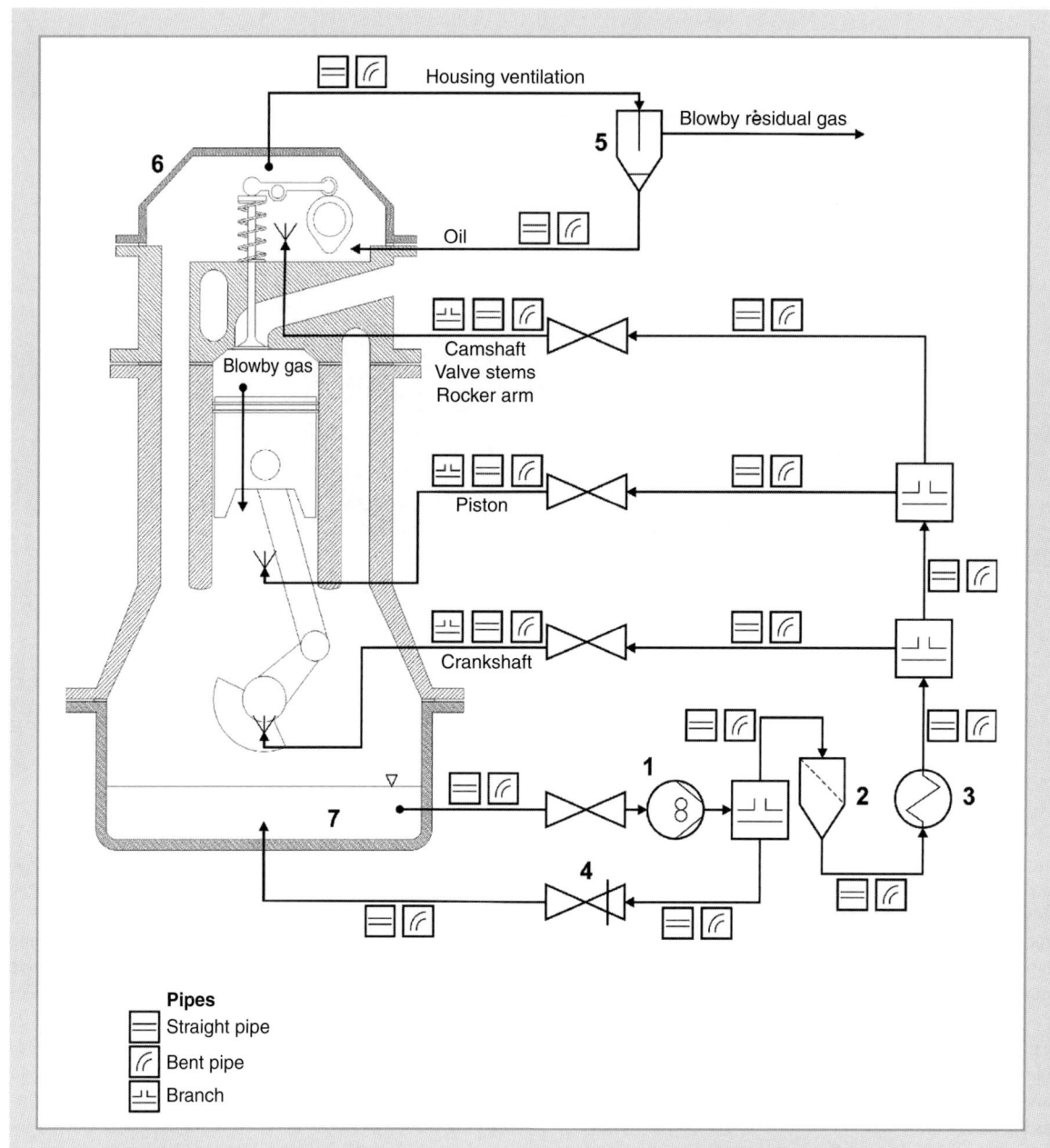

Fig. 11-2 Oil circuit schematic *1* pump, *2* oil filter, *3* oil cooler, *4* pressure control valve, *5* oil separator, *6* engine housing, *7* oil pan

11.2.3 Lubrication System Components

11.2.3.1 Oil Pumps

The function of an oil pump is to provide the oil volume and pressure required in an engine's speed range. In most cases, gear pumps (Fig. 11-3), crescent pumps and vane pumps directly driven by the crankshaft are used. Larger engines often have several oil pumps connected in parallel to provide sufficiently large volumetric flows, remain within space limits and create redundancy. Some engines have additional prelubrication or permanent lubrication to keep the crankshaft assembly ready for operation before starting. Engine design must ensure that none of the permissible tilt angles aspirates any air. However,

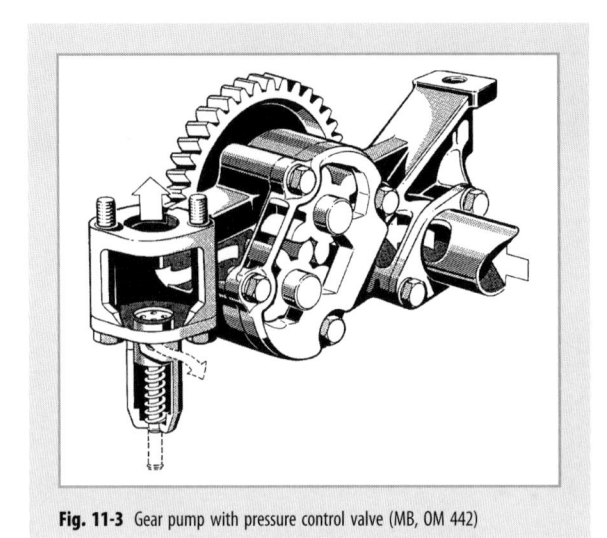

Fig. 11-3 Gear pump with pressure control valve (MB, OM 442)

Pressure oil from oil pump
Filtered engine oil
Oil filter bypass
Coolant

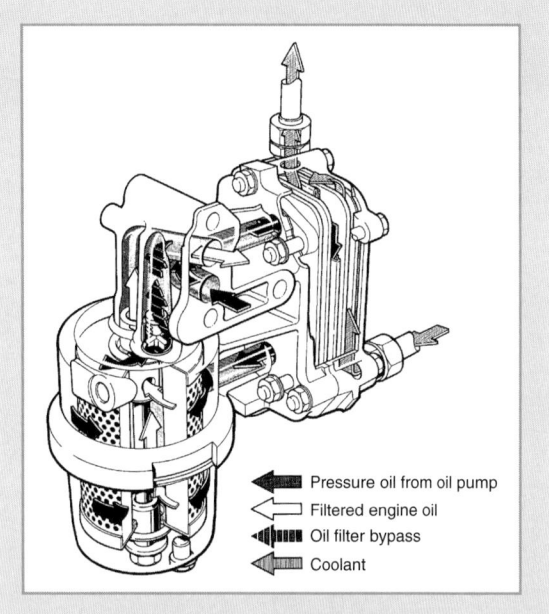

Fig. 11-4 Oil filter unit, consisting of the oil filter and oil cooler

the crankshaft must be prevented from immersing in the oil so that splashing does not cause any frictional work.

11.2.3.2 Oil Coolers

Either tube or plate heat exchangers are employed, which are usually intercooled by engine cooling water (oil-water heat exchange). A cooler must be sized so that unduly high oil temperatures cannot occur even at maximum water temperatures. Aluminum heat exchangers achieve optimal heat transfer and the best heat output but with the drawback of lower mechanical stability and higher danger of corrosion than more common stainless steel heat exchangers. Mechanical stability is essential since an oil pump generates pressure peaks during cold starting and even in normal engine operation.

11.2.3.3 Oil Filters

An oil filter normally consists of an oil filter housing and an oil filter cartridge containing the filter element. The oil filter housing contains the inlet and outlet passages and the oil filter bypass valve. Sometimes, the oil cooler is also integrated (Fig. 11-4). The entire oil filter unit is usually flange mounted on the engine block.

Figure 11-5 illustrates the flow of oil through a filter. Unfiltered oil is conducted through the inlet passage in the oil filter housing into the filter cartridge where it flows through the filter element outside-in and dirt particles are separated. The inlet and outlet passage is connected with the oil filter bypass valve to properly maintain the oil circulation in the engine even when the oil filter element is clogged.

11.2.3.4 Oil separators

Combustion products in engine oil seldom only take the form of solid particulates. Rather, blow-by gases consisting of oil particulates, fuel vapor and steam also reach the crankcase chamber through the piston rings. They are drawn off through the separation system into the engine intake system together with oil vapors at high temperatures. If the blow-by gases in an intake system cool down, then the oil particulates usually separate onto the walls. Figure 11-6 illustrates an oil separation system that further conducts the ventilation gases into the air intake. In the process, the oil vapors initially arrive at a steel wool mesh in which larger oil droplets are separated, while the excess pressure in the intake system feeds drawn off oil vapors to the intake air through a diaphragm. This serves to maintain a constant vacuum in the crankcase, thus reducing oil leaks.

11.2.4 Lubricating Oil Care

A multitude of dirt particulates form during engine operation, which cause sedimentation and wear or chemical reactions in oil and thus accelerate oil decomposition, The particulates have different sizes, are only partly soluble in oil and can be divided into two groups:
- inorganic products, primarily wear products from cylinder barrels, piston rings and bearings or silicates (casting sand, dust borne by combustion air and manufacturing debris) and

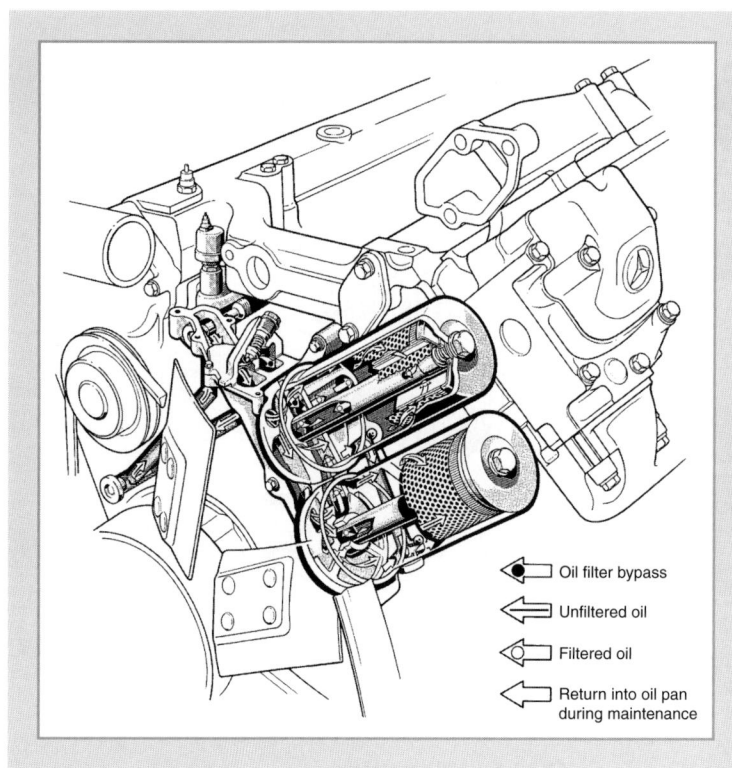

Fig. 11-5 Oil flow through the oil filter (MB, model 400, NG 90)

Legend:
- Oil filter bypass
- Unfiltered oil
- Filtered oil
- Return into oil pan during maintenance

- organic products such as soot products from oil aging, fuel consumption products or impurities caused by leaks in the coolant circuit.

Various wear mechanisms manifest themselves as a function of particulate size and distribution. Particulates > 20 μm are now generally filtered out since they cause unduly high wear in basic and connecting rod bearings, piston rings, cylinder barrels and gears. Figure 11-7 demonstrates that even smaller particulates < 20 μm influence component wear. Particulates of 2.5–5 μm alone cause more than 50% as much wear as larger particulates.

The following oil filter systems are employed to effectively keep impurities out of the oil circuit:
– Full flow filters: The entire volume of oil is filtered before it enters the main oil gallery.
– Full flow and bypass filter systems: A small portion of the oil is conducted through a finer bypass filter to trap finer particulates.
– Long life oil filter systems: The bypass filter cartridge is housed in a casing located separately from the engine and the quantity of engine oil is additionally increased (Fig. 11-8). Better filtering and a larger volume of oil make it possible to prolong the oil change intervals.
– Secondary filter elements: These are placed after the actual filter element and, in the case the main filter element is damaged, protect an engine against dirt particulates with the aid of a filter with 50 μm pores.
– Oil treatment systems: Central oil treatment systems in which all of an engine's oil is collected and purified are often used in large diesel engines and multi-engine plants.

Fig. 11-6 Oil separator

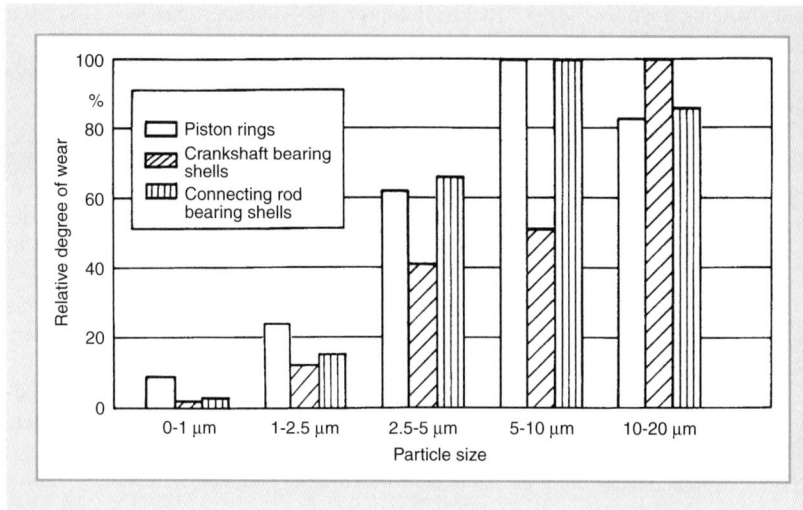

Fig. 11-7 Relative wear as a function of the particulate size of the impurities

Paper (allowing exact definition of particulate size), fiber materials (better penetration than paper), cotton or stacked paper disks are employed as filter media.

Oil centrifuges are a special form of bypass filter for vehicle diesel engines. A centrifuge (impulse centrifuge) driven by the oil jet attains extremely fine filtration.

Along with the actual filter size, the total volume of oil available and the required oil change intervals are crucial for the sizing of oil filter systems.

An oil filter must be designed with a pore size and oil particulate retention capacity that filters out all engine damaging particulates until the next oil or filter change. As a general rule, a filter may be changed without changing the

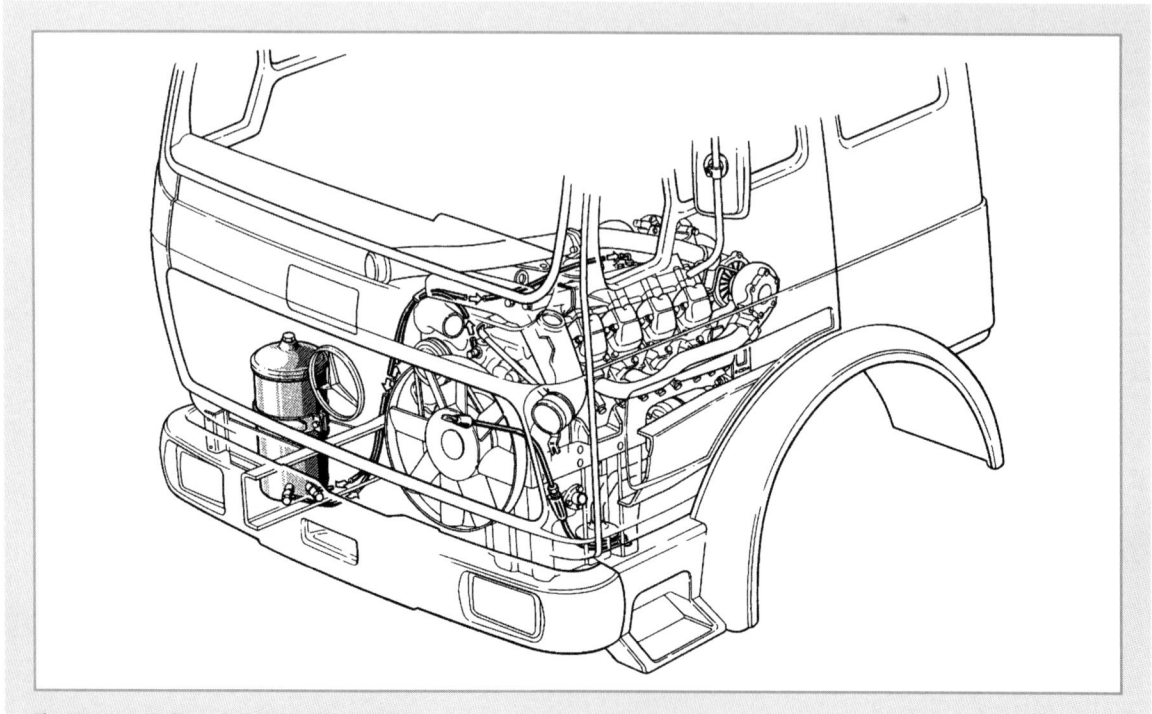

Fig. 11-8 Long life filter and oil extractor (MB, OM 442 A)

oil but oil may never be changed without changing the filter because fresh additives in new oil potentially elute filtrate from used filters. What is more, filter resistance should not be high enough at the end of a filter change interval that the oil filter bypass valve activates and unfiltered oil reaches the engine. When a bypass valve is not implemented, unduly high filter resistance can cause oil pressure decay in an engine or the destruction of the oil filter element.

Literature

11-1 Schilling, A.: Automobile Engine Lubrication Vol. I and II. Broseley, Shropshire: Scientific Publications (1972)
11-2 Reinhardt, G. P.: Schmierung von Verbrennungskraftmaschinen. Ehningen: expert (1992)
11-3 Bartz, W. J.: Aufgaben von Motorölen. Mineralöltechnik 25 (1981) 2, pp. 1–21
11-4 ACEA: European Oil Sequences 2004 Rev. 1. Brussels 2004, pp. 1–15
11-5 Groth, K. et al.: Brennstoffe für Dieselmotoren heute und morgen. Ehningen: expert (1989)
11-6 Treutlein, W.: Schmiersysteme. In: Handbuch Dieselmotoren. Berlin/Heidelberg: Springer (1997)
11-7 Gläser, H.: Schmiersystem in Kraftfahrzeugmotoren. In: Küntscher, V. (Ed.): Kraftfahrzeugmotoren: Auslegung und Konstruktion. 3rd Ed. Berlin: Verlag Technik (1995)
11-8 Affenzeller, J.; Gläser, H.: Lagerung und Schmierung von Verbrennungskraftmaschinen Bd. 8. Wien: Springer (1996)
11-9 Zima, S.: Schmierung und Schmiersysteme. In: Handbuch Verbrennungsmotor. 2nd Ed. Wiesbaden: Vieweg (2002)
11-10 Zima, S.: Kurbeltriebe. 2nd Ed. Wiesbaden: Vieweg (1999)

12 Start and Ignition Assist Systems

Wolfgang Dressler and Stephan Ernst

Early vehicles with diesel engines could only be started at low temperatures with intense smoke formation accompanied by loud knocking noises. Glow and engine starting times of several minutes were common. Therefore, startability was and is an important goal of engine manufacturers, and suppliers' development activities. The refinement of suitable start assist systems and their adaptation to engine characteristics has led to crucial advances in diesel engine starting and cold running performance. Sheathed-element glow plugs (GLP) are used for cars (typically with displacements of less than 1 l/cylinder) to assist engine starting and running at low outside temperatures. Intake air glow systems such as glow plugs, intake air heaters or flame starting systems are used for commercial vehicles with larger displacements.

Ongoing development of car diesel engines to increase driving satisfaction by enhancing specific power output and improve environmental compatibility by further reducing exhaust emissions (soot/NO_X) is leading to engine concepts with lowered compression ratios. Present day direct injection (DI) car engines have compression ratios of 16–18:1. The currently lowest compression ratio (ε) in a production car diesel engine is 15.8:1 [12-1]. Early (prechamber) diesel engines typically had compression ratios of approximately 21:1. Without additional measures, lowering compression ratios from 18:1 to 16:1 impairs cold starting and cold idle performance. This, in conjunction with the desire for cold start and cold idle performance similar to gasoline engines, increases demands on future glow systems (glow control unit, GCU and GLP) for cars:

- fastest heating rates (1,000°C in less than 2 s) for cold starts down to −28°C similar to gasoline engines,
- flexible adaptation of the glow temperature to engine demands,
- maximum glow temperatures of up to 1,300°C and continuous glow temperatures of up to 1,150°C to reduce exhaust emissions for low compression engines (17:1 and lower),
- extended afterglow capability in the minute range for low emission cold idle with increased running smoothness even in the warm-up phase of low compression engines,
- intermediate glow capability, e.g. to facilitate particulate filter regeneration,
- high glow system service life of up to 200,000 km,
- constant glow characteristics (temperature and heating rate) throughout the service life of sheathed-element glow plugs,
- OBDII and EOBD capability and
- support of advanced communication interfaces, e.g. CAN or LIN.

Typical compression ratios for commercial vehicle engines are presently between 17:1 and 20.5:1. Commercial vehicle engines with two-stage supercharging will be introduced in the future. This will increase peak pressure without additional measures in the cylinder. Lowering the compression ratio to 16–16.5:1 will allow maintaining peak pressure at the present level despite higher supercharging.

A distinction is made between diesel engines with auto-ignition and ignition assisted by a start aid.

12.1 Conditions for the Auto-Ignition of Fuel

Diesel engines auto-ignite fuel once they are sufficiently warm. For an engine to self-operate, the temperature of the air/fuel mixture compressed in the cylinder must exceed the (typically > 250°C). The compression work performed during the compression cycle generates the quantity of heat required to reach this temperature. Whether the auto-ignition temperature is reached depends on the temperature of the intake air and the engine, the compression ratio and the leakage and heat losses during compression.

The compression work and thus the final compression temperature increase as the compression ratio increases. Leakage losses in an engine reduce the effective compression ratio and, together with heat losses, cause the final compression temperature to drop. Increasing the engine speed can reduce leakage and heat losses. The engine speed at which

S. Ernst (✉)
Robert Bosch GmbH, Diesel Systems, Stuttgart, Germany
e-mail: stephan.ernst@de.bosch.com

auto-ignition occurs is called the starting speed and is dependent on other influencing variables such as the cylinder volume, number of cylinders and engine design. Typical values of diesel engine starting speeds are 80–200 rpm.

At identical combustion chamber volumes, DI car engines have a smaller combustion chamber surface area dissipating heat than prechamber and swirl chamber engines. This results in improved cold start performance and a lower starting speed for engines of this type. DI car engines with a compression ratio of 18:1 only need additional starting aids below 0°C. The higher heat losses in swirl chamber and prechamber car engines already make the use of starting aids necessary at around 20–40°C.

The ratio between the surface area and volume of the combustion chamber in large volume engines, e.g. commercial vehicle engines, is significantly lower than in car engines. Today's commercial vehicle engines start at temperatures as low as –20°C without starting aids.

The injected fuel quantity is increased up to the maximum air utilization to overcome the extremely high frictional torque during a cold start.

12.2 Fuel Ignition Aids

Only additional starting aids make complete combustion of the fuel possible at low temperatures (cold start). Two principles of cold start assist are distinguished: On the one hand, sheathed-element glow plugs supply locally limited energy to the air/fuel mixture in a combustion chamber. This is called glow ignition. Glow systems consist of sheathed-element glow plugs and glow control units and are used for cars and light commercial vehicles. On the other hand, flame starting systems, intake air heaters or glow plugs supply heat energy to the intake air so that the compression in the combustion chamber causes the auto-ignition temperature to be reached. Given the very large volume to surface area ratio (low heat losses) in large volume engines, the intake air requires relatively little heating for a cold start.

A glow system usually supplies heat energy to the mixture formed from the fuel spray and aspirated air through a concentrated hot "spot" in its sheathed-element glow plugs. The combustion cycle begins near the glow plug when the combustible mixture there is brought to a sufficiently high temperature (Fig. 12-1). Such locally limited premature ignition causes flame propagation throughout the entire injection spray during pilot injection in the combustion chamber presented here. Under real conditions in a cylinder, the mixture's turbulence (swirl) additionally supports flame propagation. A sheathed-element glow plug's efficiency is determined by its temperature and its geometric arrangement in the combustion chamber.

The use of a sheathed-element glow plug in the cylinder to increase the mean air temperature plays a subordinate role for glow ignition.

Since glow ignition is locally limited, combustion may be incomplete in a cold engine or the air/fuel mixture may fail to ignite at all (misfire). The degree of combustion and the burn rate depend on the particular ignition conditions in the region of the glow plug. A high burn rate causes a rapid increase in pressure. This makes harsh combustion noises perceptible (diesel detonation, see Fig. 12-2, top). Since ignition conditions in cold engines vary, the pressure gradients in the compression cycles do too. Varying the combustion chamber pressure gradient modulates combustion noise.

Increasing a sheathed-element glow plug's surface temperature produces nearly complete and slower combustion.

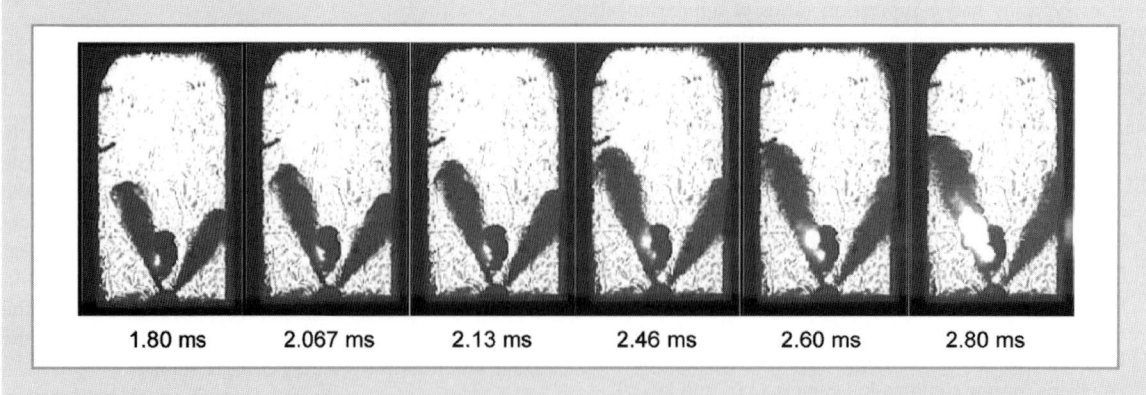

Fig. 12-1 A metal sheathed-element glow plug igniting diesel spray under cold start conditions (in pictures taken in a combustion chamber). The sheathed-element glow plug initially ignites a small region of the injection spray. As the supply of energy increases, several such locally limited flame cores arise and, as mixing continues, ignite the entire injection spray. Source: RWTH Aachen/VKA

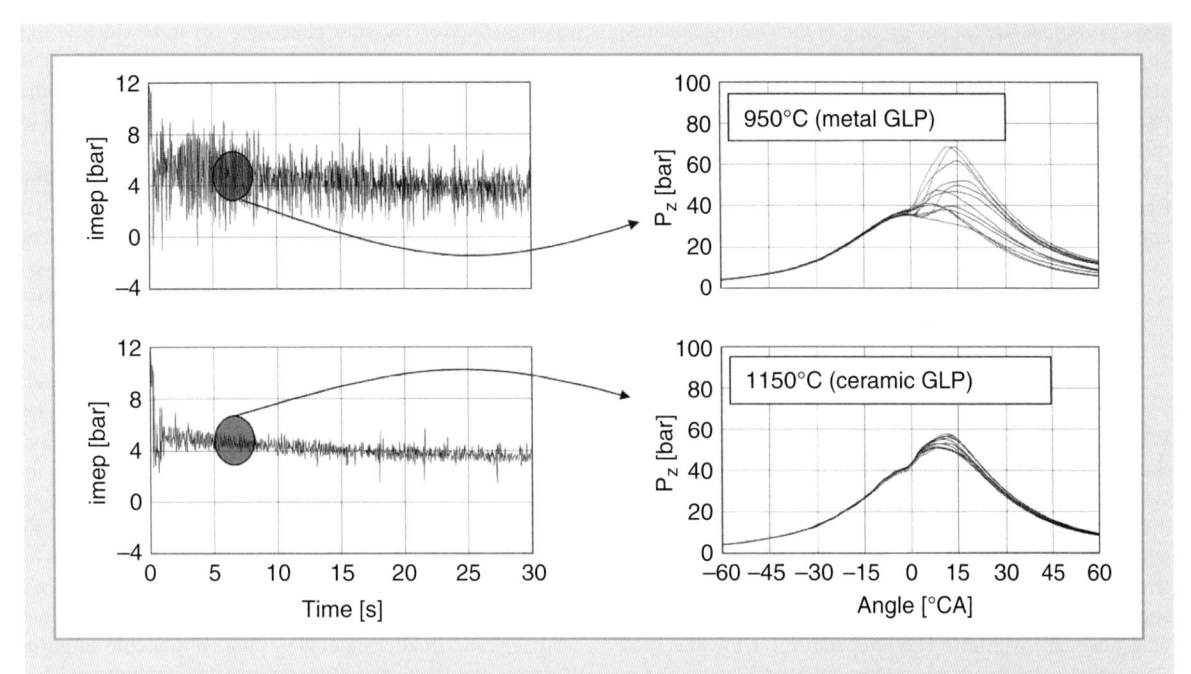

Fig. 12-2 Brake mean indicated pressure and cylinder pressure curves as a function of glow temperature in cold idle with an optimized position of pilot injection at −20°C for a four cylinder common rail engine with a compression ratio of 16:1

The resultant gentler increase in pressure causes softer engine noise. Depending on the particular type of engine, several minutes of glow assist at temperatures of up to over 1,150°C may be necessary for smooth cold idle.

12.3 Start and Ignition Assist Systems

12.3.1 Start and Ignition Assist Systems for Cars

Glow systems consist of sheathed-element glow plugs, a glow control unit and potentially a software module. Conventional glow systems employ sheathed-element glow plugs with a rated voltage of 11 V (voltage at which the rated temperature of the GLP is reached), which are directly controlled with the vehicle system voltage. Low-voltage glow systems require sheathed-element glow plugs with rated voltages below 11 V. An electronic glow control unit adjusts their heat output to the engine requirement.

Glow consists of up to five phases.
- *Preglow* heats the GLP to their operating temperature.
- During *standby glow*, the glow system maintains a GLP temperature required for starting for a defined time.
- *Start glow* is applied when an engine is revving up.
- Once the starter has discharged, the *afterglow phase* begins.
- *Intermediate glow* facilitates diesel particulate filter regeneration or reduces smoke formation in a cooled down engine with alternating loads.

12.3.1.1 Vehicle System Voltage Glow Systems

In conventional vehicle system voltage glow systems, the EDC software module starts and ends the glow process as a function of the activation of the ignition switch and the parameters stored in the software (e.g. cooling water temperature). The GCU controls sheathed-element glow plugs with the vehicle system voltage through a relay based on the settings of the EDC. The sheathed-element glow plugs' rated voltage is 11 V. Thus, the heat output is dependent on the current vehicle system voltage and the GLP's resistance as a function of temperature (PTC). This produces the GLP's self-regulating function. Dependent on the engine load, a cutoff function in the engine control unit's software module safely protects the GLP against overheating. Adjusting the afterglow time to the engine requirement makes long GLP service life possible while keeping cold driveability good.

12.3.1.2 Low-Voltage Glow Systems

Low-voltage systems control the GLP so that the glow temperature is optimally adjusted to the engine requirements. In this phase, the sheathed-element glow plugs are briefly operated with push voltage in order to reach the glow temperature required to start the engine as rapidly as possible during glow. The push voltage is higher than the GLP's rated voltage. The voltage is then lowered to the GLP rated voltage during start standby heating. The voltage is elevated again during start

glow to compensate for the cooling of the sheathed-element glow plugs by the cold charge air. The charge air's cooling effects and combustion's heating effects can be compensated in the afterglow and intermediate glow range. Pulse width modulation (PWM) generates the requisite voltage from the vehicle system voltage. The respective PWM value is taken from a map matched to the particular engine during the application engineering. A map contains important parameters that characterize an engine's operating state such as:
– engine speed,
– injected fuel quantity (load),
– time after starter discharge and
– cooling water temperature.

When engine speed is high and the injected fuel quantity low, sheathed-element glow plugs are cooled during gas exchange. When engine speed is low and the injected fuel quantity high, they are heated by the combustion heat. The voltage is selected as a function of these two parameters so that the temperature of the sheathed-element glow plug necessary for engine operation is reached and the sheathed-element glow plug does not overheat. The temperature of the sheathed-element glow plug necessary for quiet and low-emission engine operation drops continuously once the starter discharges. This can be taken advantage of to increase GLP service life by continually lowering the GLP glow temperature in the afterglow phase. Reducing the GLP temperature and shortening the afterglow time when cooling water is "warm", e.g. upwards of approximately 10°C in DI car engines, further prolongs the service life of sheathed-element glow plugs.

12.3.1.3 Metal Sheathed-Element Glow Plugs

Metal GLP have a glow plug consisting of a tubular heating element press fit gastight in a housing (5 in Fig. 12-3). The tubular heating element consists of a hot gas and corrosion proof Inconel or Nicrofer glow tube (4) into which a coil embedded in compressed magnesium oxide powder (2) is incorporated. This coil normally consists of two metallic resistors connected in a series, the heating coil (1) located in the tip of the glow tube and the control coil (3).

The heating coil's electric resistance is independent of the temperature. The control coil on the other hand has an electric resistance with a positive temperature coefficient (PTC). Its resistance increases as the temperature increases. Advanced 11 V sheathed-element glow plugs reach the surface temperature of 850°C necessary to ignite fuel in approximately 4 seconds. Typically, the GLP can be post-heated for several minutes. The heating coil is welded into the rounded tip of the glow tube to contact it on the ground side. The control coil is contacted at the terminal pin by the connection to the glow control unit.

When voltage is applied to the sheathed-element glow plug, the majority of the electrical power in the heating coil is initially converted into heat. The temperature in the tip of the sheathed-element glow plug rises steeply. The control coil and thus the resistor's temperature increases with a time delay. The sheathed-element glow plug's power consumption decreases and the temperature approaches equilibrium. This produces the heating characteristics presented in Fig. 12-4.

In principle, the design and function of low-voltage metal sheathed-element glow plugs correspond to the 11 V type. The heating and control coil are designed for a lower rated voltage and higher heating rates. The slenderer design is matched to the limited space in four valve engines. The tip area of the glow plug is tapered to reduce the space between the heating coil and glow tube and thus accelerate the transport of heat from the coil to the surface of the sheathed-element glow plug. The "push operation" principle of control applied here (control of the GLP with a voltage above the rated voltage) makes heating rates of up to 330°C/s possible. The maximum heating temperature is above 1,000°C. The

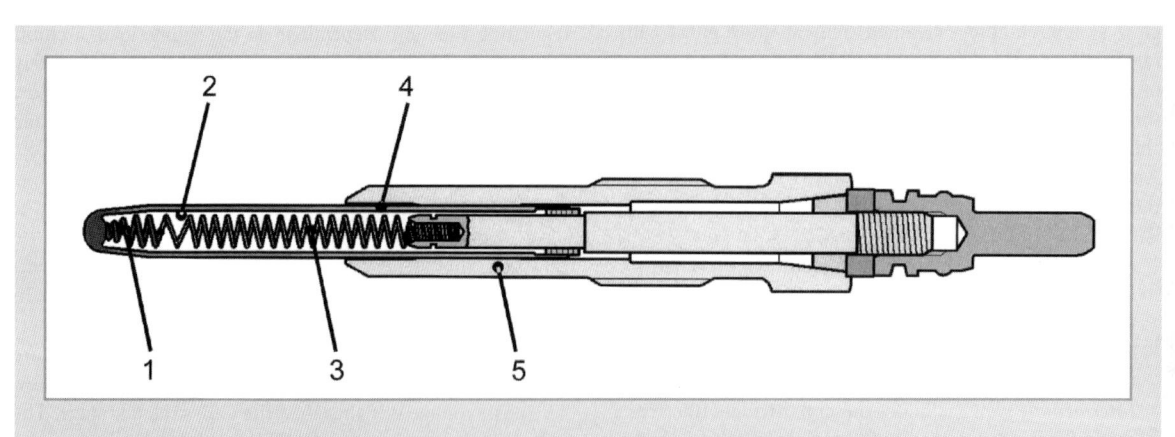

Fig. 12-3 Design of a metal sheathed-element glow plug

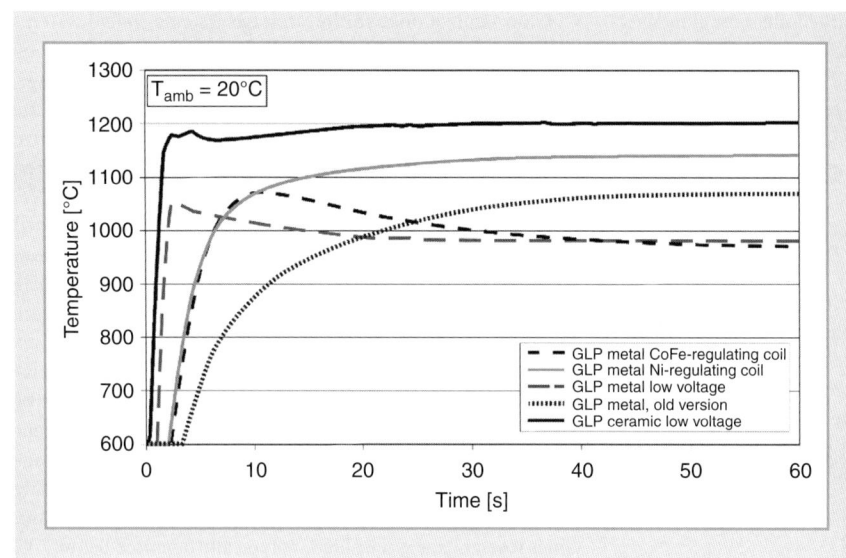

Fig. 12-4
Heating characteristic of various metal sheathed-element glow plugs and a ceramic sheathed-element glow plug

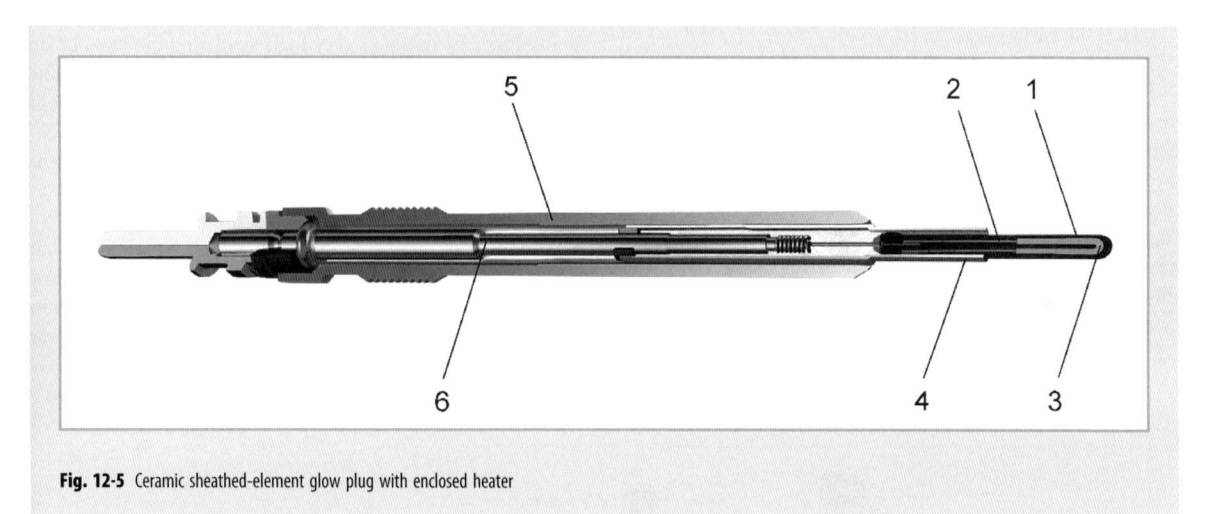

Fig. 12-5 Ceramic sheathed-element glow plug with enclosed heater

temperature during standby glow and afterglow operation is approximately 980°C. The typical afterglow time is up to 3 min for engines with compression ratios greater than 18:1 and extends to up to ten minutes for engines with compression ratios less than 18:1.

12.3.1.4 Ceramic Sheathed-Element Glow Plugs

Modern car engines with low compression ratios require sheathed-element glow plugs that allow maximum temperatures of up to 1,300°C and long heating times at over 1,150 °C without deteriorating. The desire for immediate starting similar to that of gasoline engines even at extremely low temperatures necessitates heating rates of up to 600°C/s (see Fig. 12-4).

Sheathed-element glow plugs with Si_3N_4 ceramic heaters have been developed for the combination of high thermal shock and hot gas corrosion loading (Fig. 12-5).

The ceramic heater (1) consists of electrically insulating Si_3N_4 in which highly electrically conductive supply lines (2) and an enclosed heater (3) with PTC effect are embedded. Unlike metallic sheathed-element glow plugs, the heating and control function is combined in the heater. The ceramic heater is fixed in a metal tube (4) and press fit gastight in a housing (5). A terminal pin (6) electrically contacts the + pole. The metal tube and the housing establish the ground connection to the engine block. Eleven volt and low-voltage variants are produced by adjusting the heater resistance. The glow control unit controls the ceramic sheathed-element glow plugs just like metallic GLP.

Often observable in metallic sheathed-element glow plugs, a drop in heating temperature caused by deterioration and thus the gradual impairment of cold start and cold running performance is rarely observed in ceramic variants. Even at constant extremely high heating temperatures of 1,200°C, the heating temperature typically drops less than 50°C after 3,000 h of operation.

12.3.1.5 Glow Control Units

The function of a glow control unit (GCU) is to control and monitor sheathed-element glow plugs. There are different concepts for this:
- Depending on the operating voltage of the sheathed-element glow plugs, relays (for 11 V systems) or transistors (low-voltage systems) are employed as power switches.
- The sheathed-element glow plugs can be driven by one supply line for all GLP or by supply lines for every single GLP.
- The GCU may be operated dependent on (software module and parameters in the EDC, Fig. 12-6) or independent of (autonomous systems, software and parameters in the GCU) the EDC.

Relays are typically employed as power switches in 11 V systems. A relay solely serves as an on-off switch for the GLP. Thus, the glow system's heat output is determined by the current vehicle system voltage and the sheathed-element glow plug's electrical resistance as a function of the temperature. This can have a particularly adverse effect at very low ambient temperatures since a voltage drop of as low as 7–8 V during starting when a battery is weak significantly reduces a GLP's heating power.

Glow control units for low-voltage systems enable systematically controlling sheathed-element glow plugs' voltage by pulse width modulation (PWM) of the vehicle system voltage. The ratio at which the power switch in the GCU switches the vehicle system voltage on and off allows setting an effective voltage in the sheathed-element glow plug. The required effective value is taken from a map specific to the engine.

As a result, a glow system's heat output can be adjusted perfectly to the dynamic engine requirements. In addition, this type of control also enables compensating for the vehicle system voltage, i.e. the PWM signal's duty factor adjusts to the level of the vehicle system voltage. Requirements on power switches for vehicle system voltage modulation increase relative to operating speed and frequency. Thus, semiconductor switches (Power MOSFET) are necessary. Sheathed-element glow plugs may additionally be controlled temporally staggered. This minimizes the maximum load of the vehicle electrical system during a cold start and the afterglow phase. Since low-voltage glow plugs are typically operated below 7 V, a drop in vehicle system voltage during starting has no appreciable influence on the heating power.

Controlling all the glow plugs in parallel through one supply line can result in very high switching currents of up to 300 A. The malfunction of a single glow plug is impossible to diagnose in this type of wiring. Hence, sheathed-element glow plugs are often controlled with one line apiece to facilitate individual diagnostics. Single glow plug monitoring is the prerequisite for such advanced diagnostics concepts as OBDII that will be mandatory in the future.

Autonomous systems use sensor information that describes an engine's operating state to control the glow process. Once a control unit has been activated by the ignition switch, the cooling water temperature is analyzed. The glow

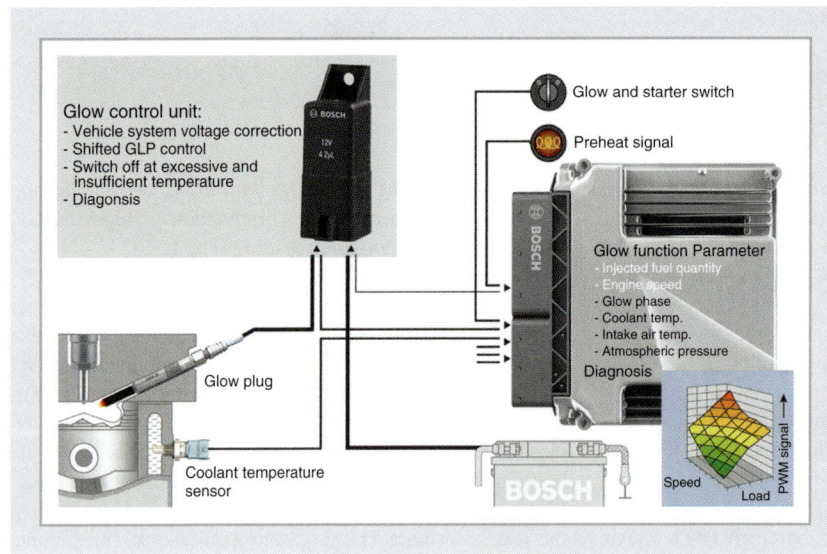

Fig. 12-6
Glow system with non-autonomous GCU controlled by an engine control unit with a stored map

plugs are only activated when the EDC issues a release (master/slave). All information for the control of the glow process is stored in a microprocessor in the GCU or imported through interfaces. When a critical injected fuel quantity is exceeded, the afterglow process is shortened to protect a glow plug from overheating. These units are also able to take over the control of a diagnostic indicator in a vehicle.

Non-autonomous glow control units control sheathed-element glow plugs according to an engine control unit's specifications and return the diagnostic information to it.

12.3.2 Start and Ignition Assist Systems for Commercial Vehicles

Large volume diesel engines are usually started with preheated intake air. When this is insufficient, highly volatile and combustible hydrocarbons (e.g. Start Pilot) may additionally be injected into the air cleaner.

The intake air is preheated with glow plugs, intake air heaters (IAH) or flame starter plugs installed in the intake tract. Glow plugs and intake air heaters operate on the basis of the principle of a blowdryer, i.e. cold air is conducted past maximally large and hot metal surfaces and heated in the process. They lower a diesel engine's natural starting threshold by 3–5°C. Since they not only heat the intake air but also produce reactive radicals that enter a combustion chamber with the intake air, flame starter plugs lower this threshold by 8–12°C.

Flame starter plugs consist of one to three glow plugs connected in parallel. Diesel fuel vaporizes and partially burns on their hot surfaces, which are usually expanded by several ply of a wire mesh screen wrapped around the glow plugs. Protective sleeves surrounding the glow plugs and the ply of the mesh screen prevent the intake air from blowing out the flame. Special integral checks properly matched to the engine displacement supply diesel fuel to the flame starter plugs.

Intake air heaters and glow plugs require substantial electrical power in the range of 1–3 kW.

12.4 Cold Start, Cold Running Performance and Cold Running Emissions for Cars

Without additional measures, cold diesel engines produce increased HC and soot because their fuel burns incompletely. Visible smoke (exhaust gas opacity) makes this apparent. In addition, increased diesel detonation occurs. The goal is to lower emissions and noise generation to the lowest level possible by using ignition aids.

12.4.1 18:1 Compression Ratio

A reduction of exhaust gas opacity (translucence) by sheathed-element glow plugs and an optimization of the position of the injection spray are represented in Fig. 12-7.

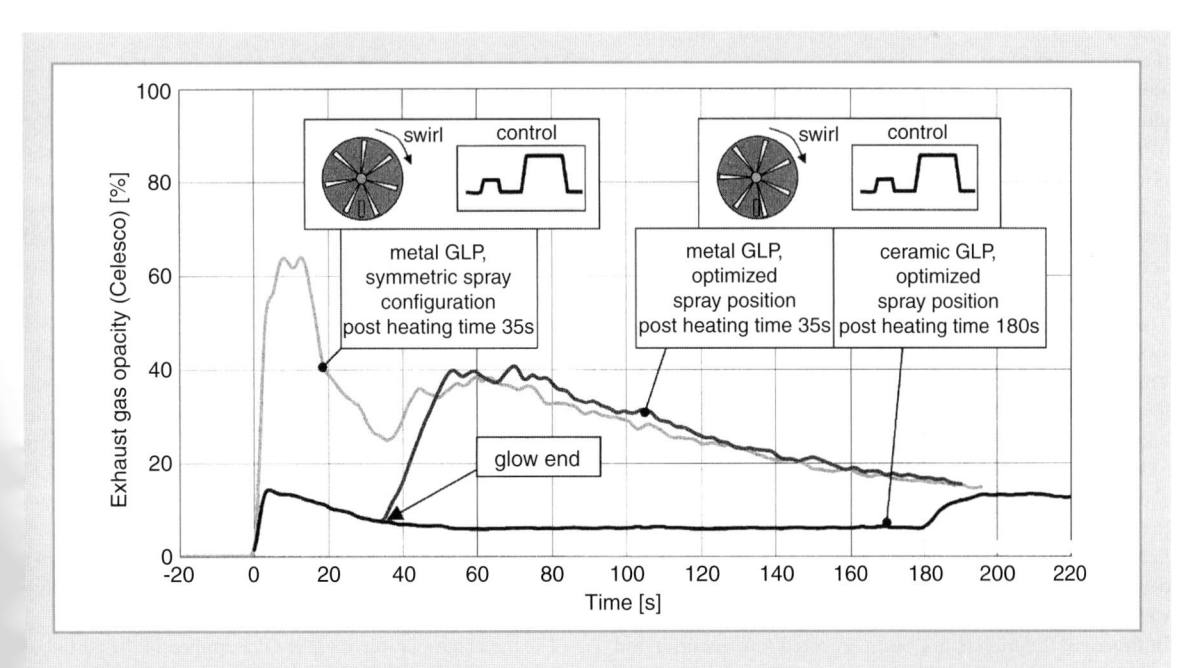

Fig. 12-7 Exhaust gas opacity as a function of the spray configuration and the afterglow time for one pilot injection. 2.2 l four cylinder common rail engine with a compression ratio of 18:1 at −20°C. (source: Bosch)

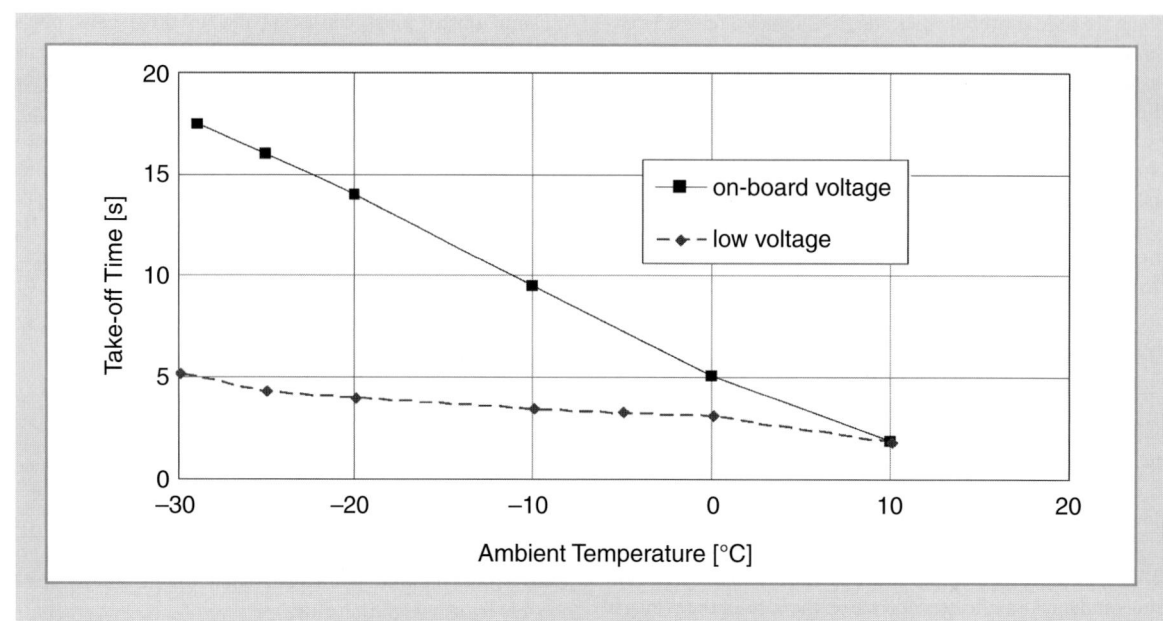

Fig. 12-8 Total engine starting time as a function of GLP heating rates and ambient temperature (source: Bosch)

Switching off the sheathed-element glow plugs significantly increases the exhaust gas opacity in the analyzed period up to 220 seconds after cold start. Long afterglow reduces cold idle emissions. Depending on the engine boundary conditions, the usual afterglow time range for metal GLP is up to 180 seconds. Depending on the engine requirement, the afterglow time of ceramic GLP can be several minutes without loss of service life. The use of ceramic sheathed-element glow plugs can reduce overall exhaust gas opacity.

In the common rail engine with a compression ratio of 18:1 presented here, the optimized geometric configuration of the sheathed-element glow plug and injection spray reduces exhaust gas opacity from more than 60% to less than 20%. The injection spray is directed close to the GLP. The intake air swirl additionally directs the air/fuel mixture toward the GLP. This increases the probability that combustible mixture is located near the GLP. Combustion is stabilized. Exhaust gas opacity decreases.

When the injection spray position is optimized and sheathed-element glow plug temperatures are above 900°C, the engine starting time (the time between an engine's speed increase and reaching 800 rpm) during cold starts is independent of the GLP temperature down to –20°C. The fuel burns well during the first injection and the engine revs up rapidly. In this case, the interaction of the injected fuel quantity and rail pressure (300–600 bar) is most important for brief engine starting times.

Particularly at low temperatures < 0°C, the total starting time, i.e. glow time plus engine starting time, is significantly shorter for low-voltage GLP than for 11 V GLP (Fig. 12-8). This can be attributed to the extremely short heating time of low-voltage GLP.

12.4.2 Compression Ratio of 16:1

Lowering the compression ratio (ε) allows increasing the specific engine power output kW/displacement at identical peak combustion pressure by increasing the boost pressure and the injected fuel quantity. At the same time, soot and NO_X emissions at operating temperature can be reduced significantly as engine tests on steady-state operating points have demonstrated. However, cold start and cold idle in conjunction with a reduction of the compression ratio are problematic. The engine starting temperature at which a cold start is possible without a glow system increases by approximately 10°C because of the low final compression temperature (when 18:1 is compared with 16:1). In addition, lowering the compression ratio increases exhaust gas opacity in cold idle significantly because the low final compression temperature causes poorer complete combustion of the fuel.

At a compression ratio of 16:1, the exhaust gas opacity values during cold idle greatly depend on the glow plug temperature and glow time (Figs. 12-9 and 12-10).

The mean temperatures of the GLP applied in the period of 0–35 seconds after starter discharge are 950°C (11 V metal GLP), 1,010°C (5 V metal GLP) and 1,200°C (ceramic GLP). The low mean glow temperature of the 11 V metal glow plug can be attributed to the drop in the vehicle system voltage

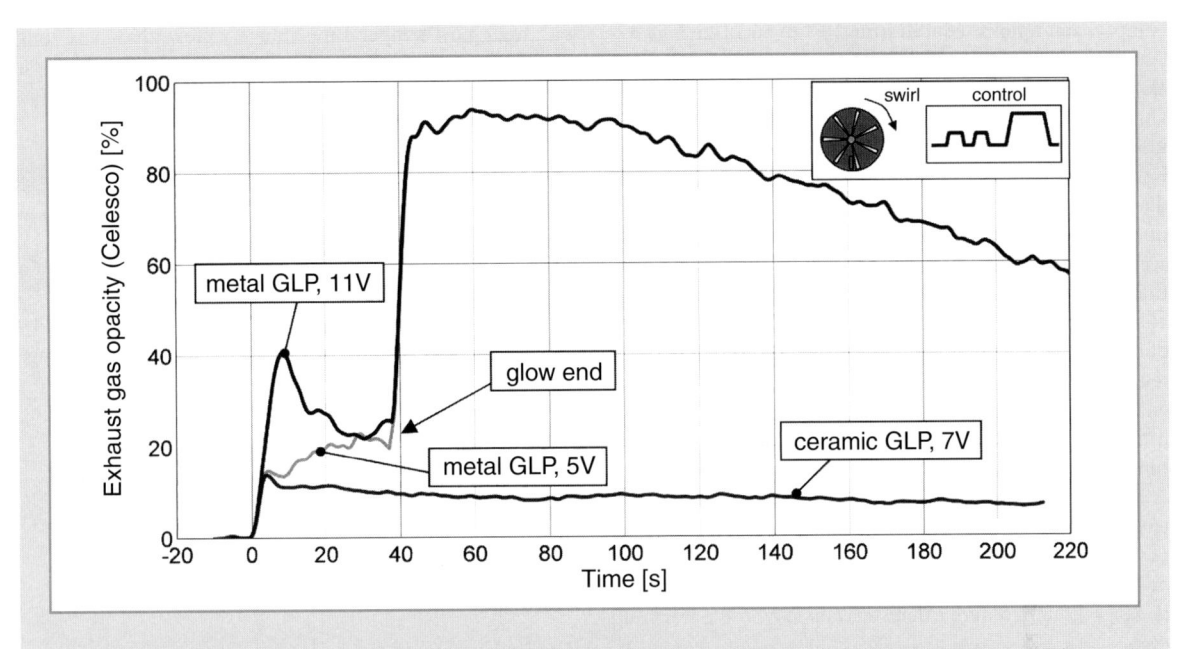

Fig. 12-9 Exhaust gas opacity as a function of the properties of sheathed-element glow plugs and the afterglow time for two pilot injections and optimized spray configuration. 2.2 l four cylinder common rail engine with a compression ratio of 16:1 at −20°C (source: Bosch)

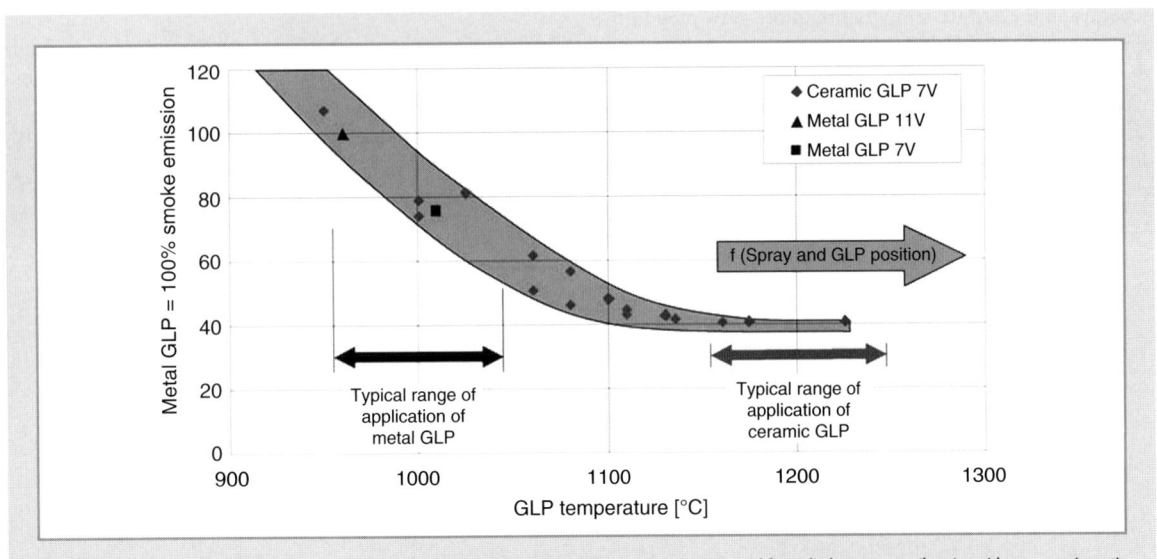

Fig. 12-10 Influence of GLP surface temperature on the relative exhaust gas opacity with two pilot injections. 2.2 l four cylinder common rail engine with a compression ratio of 16:1 at −20°C (source: Bosch)

during starting (down to 7 V here) and the ramp up of the generator normally around 3 seconds after the engine has started. Consequently, the 11 V metal glow plug is operated during the starting process with a voltage below its rated voltage and the heating temperature drops. The cooling down of the 11 V GLP during the starting process causes exhaust gas opacity to increase from less than 20% to approximately 40% right after starter discharge while the compression ratio is reduced from 18:1 to 16:1 (see Figs. 12-7 and 12-9). Low-voltage glow plugs on the other hand can also be controlled during the starting process ($U_{ansteuer} \geq U_{Nenn-GLP}$) so that they do not cool down. Therefore, at a lower compression ratio (= 16) and two pilot injections, low-voltage systems have exhaust gas opacities after starter discharge comparable to 11 V GLP when = 18 and there is one pilot injection. After starting, the exhaust gas opacity curves of metal 11 and 5 V GLP are equal in cold idle since their glow temperatures balance. The increase in vehicle voltage after the starter discharges causes the temperature of the 11 V metal GLP to increase. If the GLP surface temperature is higher than 1150 C° ($T_{Keramik-GLP}$) in the entire measuring interval, exhaust gas opacity remains significantly below 20% and continues decreasing as cold idle time increases because of the heating of the engine.

Exhaust gas opacities increase significantly once glow has ended. Misfires or retarded combustion occur. Ceramic glow plugs have a longer lifetime than metal glow plugs. This can be taken advantage of to substantially prolong afterglow times and significantly reduce exhaust gas opacity in the warm-up phase.

Figure 12-10 plots the influence of the GLP surface temperature averaged over the time interval of 0–35 seconds after engine start on the cumulative relative exhaust gas opacity (11 V metal sheathed-element glow plug = 100%) in the same period. Clearly, exhaust gas opacity greatly depends on the surface temperature of the glow plug in the temperature range below 1,100°C and does not improve at all above 1,150°C.

These findings demonstrate that, even when the compression ratio has been lowered, only optimization of the injection strategy and position of the pilot injection combined with high glow temperatures in every glow phase and long afterglow times reduces exhaust gas opacity to very low values. At the same time, noise generation (diesel detonation) can be reduced.

12.5 Conclusion

Lowering the compression ratio to $\varepsilon = 16.0$ imposes certain requirements on glow and injection systems:
(a) a high surface temperature of the sheathed-element glow plug (1,150°C) even when the vehicle system voltage drops,
(b) long afterglow (approximately 2–15 minutes depending on the warm-up characteristic),
(c) an optimized injection strategy and
(d) optimal position of the injection spray toward the GLP.

Literature

12-1 Reichenbach, M.: Neuer 2,2 l Dieselmotor mit Verdichtungsverhältnis 15,8:1. MTZ 66 (2005) 9, p. 638

Further Literature

Tafel, St.: Entwicklung eines Brennverfahren für kleine, drallfreie Direkteinspritzer-Dieselmotoren. VDI Fortschritts-Berichte 377, Diss. (1998)
Robert Bosch GmbH (Ed.): Dieselmotor-Management. 4th Ed. Wiesbaden: Vieweg (2003)

13 Intake and Exhaust Systems

Oswald Parr, Jan Krüger, and Leonhard Vilser

13.1 Air Cleaners

13.1.1 Requirements

The function of air cleaners is to hinder dust contained in the intake air from entering the engine and thus to prevent premature engine wear. Air cleaners are additionally employed to damp intake noise.

The air's dust content depends on the application and road conditions. Table 13-1 presents an overview of mean dust concentrations and Fig. 13-1 particle size distributions of various real dusts in the field. Two standardized test dusts employed in lab air cleaner tests have also been entered for comparison [13-1]. The dust concentrations and particle size distributions differ from one another by orders of magnitude depending on the applications. This data is important when designing air cleaners for different cases of application and enables estimating their expected service life.

Dust that enters an engine with the intake air causes abrasive wear in sliding systems. Cylinder liners, pistons and piston rings, crankshafts and connecting rod bearings, valves and gaskets are critical points.

Wear in these components affects an engine's service life adversely. Figure 13-2 presents the percentage of wear in three of an engine's different critical points measured with a dust test rig, once without an air cleaner, once with an older type of air cleaner, an oil wetted air bath cleaner, and once with a paper dry type air cleaner. Significantly reducing wear, paper dry type air cleaners provide the most effective wear protection.

Along with basic knowledge of various types of air cleaners' ability to reduce wear, the influence of particle sizes is also important for design.

Taking a piston ring in a diesel engine as an example, Fig. 13-3 presents percentages of wear from different particle size ranges as a ratio of dust wear to basic wear without the influence of dust. The intake air's dust concentration is 2.3 mg/m^3. The 5–10 µm fraction causes the strongest increase of wear over basic wear without the addition of dust, followed by the 10–20 µm fraction. Therefore, an air cleaner must reliably filter these fractions out of the intake air.

In addition to the abrasive wear caused by dust particles, deposits of fine particulate matter can also form on the compressor walls in exhaust gas turbocharged engines. This reduces compressor efficiency [13-4]. Space and cost normally preclude implementing the very fine filtration that is quite possible technically. The injection of fluids during operation constitutes an effective measure to clean compressor surfaces of deposited fine particulate matter.

Yielding good protection against wear and sufficient hours of operation, recommended values for filtration efficiency and service life gathered from years of practical experience are applied as the basis of cleaner design. Failure to adhere to these values results in reduced filtration efficiencies and insufficient dust capacity. This is particularly critical whenever engines are operated in high-dust environments.

Frequently, automotive engineers must additionally adapt a cleaner's size and design to very tightly sized installation spaces. Furthermore, air cleaners require servicing. To this end, good cleaner accessibility has to be designed in with the easiest handling possible.

13.1.2 Filter Materials and Performance Data

Cellulose and synthetic *fiber materials* are predominantly used as filter materials for air cleaners. This type of cleaner usually has replaceable elements and is called a *dry air cleaner*. In addition, *oil wetted air cleaners* and *oil wetted air bath cleaners*, in which oil assumes an important role in the filtration of dust particles, are employed. However, these cleaners have lower filtration efficiencies than dry air cleaners and their filtration efficiencies are dependent on their airflow. Whenever the airflow drops, the filtration efficiency drops as well. While oil wetted air cleaners and oil wetted air bath cleaners have to be cleaned, they do not require any replacement elements when serviced.

Dry air cleaners have a constant high filtration efficiency independent of the airflow. They are easy to service by replacing the filter elements, are independent of their installation position and can be flexibly designed for engine operating conditions.

L. Vilser (✉)
J. Eberspächer GmbH & Co. KG, Esslingen, Germany
e-mail: leonhard.vilser@eberspaecher.com

Table 13-1 Mean dust concentrations

Operating conditions	in mg/m³
Normal European road traffic	0.6
Non-European road traffic	3
Off-road (construction site)	8
Buses with rear intake in normal European road traffic	5
Buses with rear intake in non-European road traffic	30
Construction equipment (wheel loader, tracked vehicles)	35
Farm tractors in Central Europe	5
Farm tractors in non-European regions	15
Combine harvesters in fleets	35

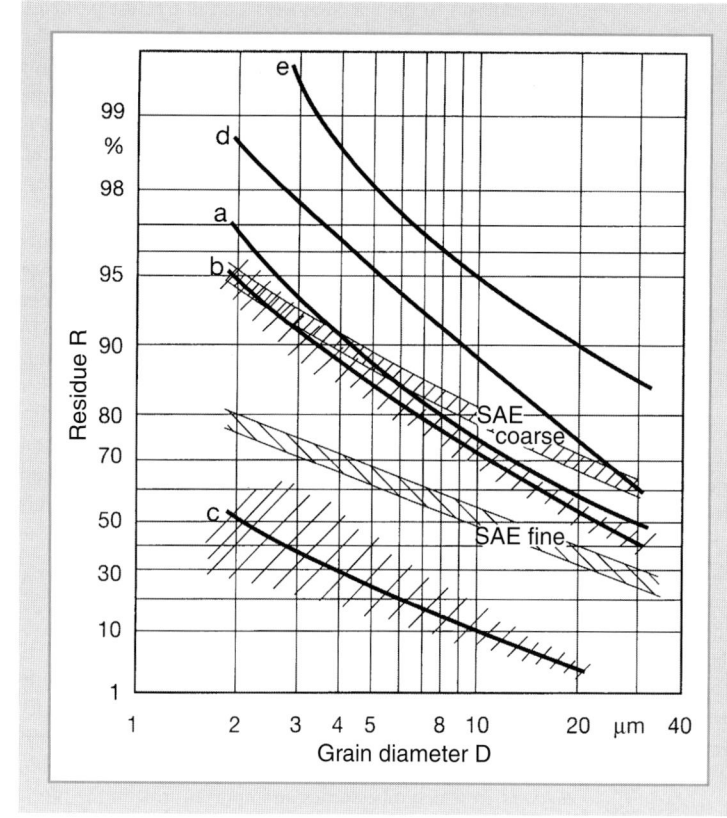

Fig. 13-1 Particle size distributions of real dusts in the field. Curve **a**: at construction sites; curve **b**: on unpaved roads; curve **c**: on paved roads; curve **d**: on non-European dirt roads; curve **e**: during sandstorm conditions

Dry air cleaners' filter elements consist of technical *special papers* or *fleeces* with precisely defined compositions, fiber structures and pore sizes. Worldwide, they have a high and uniform quality standard and a good price/performance ratio. *Filtration efficiency* is defined by the fraction filtration efficiency, overall filtration efficiency, dust capacity and flow restriction. *Fraction filtration efficiency* specifies the percentage share of a particular particle size retained based on a test dust, while overall filtration efficiency specifies the percentage share of all particle sizes retained.

Dust capacity is defined as the volume of dust a cleaner traps until a specified restriction is reached. These terms are defined in the standards [13-5, 13-6] with related test procedures.

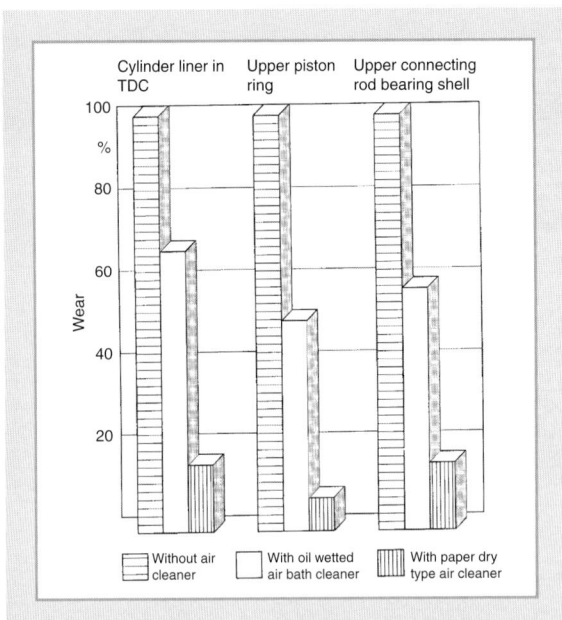

Fig. 13-2 Reduction of wear by air filtration in engine components [13-2]

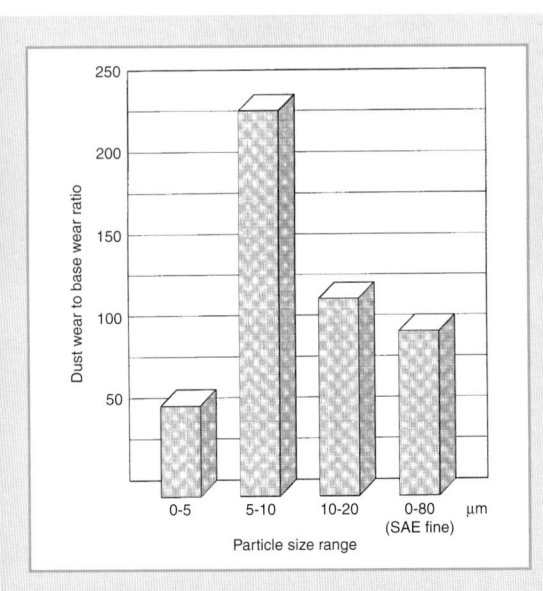

Fig. 13-3 Piston ring wear caused by various dust particle fractions [13-3]

These performance data depend on the structure of the filter material, the air velocity and the allowable increase in restrictions. Naturally, a cleaner's surface area is also decisive for its dust capacity.

Practical experience has yielded the following data as recommended values for the design of paper dry type air cleaners with the established filter materials predominantly in use:

cars: $>2,500$ cm^2/m^3/min or 200 cm^2/kW and
commercial vehicles: $>4,000$ cm^2/m^3/min or 320 cm^2/kW.

These values can vary considerably depending on the field of application, engine type and filter material and thus only represent a mean. The same design criteria apply to large engines as to commercial vehicle engines. (The specification of the specific surface area in cm^2/m^3/min corresponds to the reciprocal value of the inflow velocity.)

Figure 13-4 contains typical functional data for car and commercial vehicle dry air cleaners. The restriction increases quadratically with the volumetric intake flow. It should not exceed an initial value of 10–30 mbar so that it does not influence engine operation and it has sufficient dust capacity. The right hand graph provides data on the total mass filtration efficiency and the cleaner's dust capacity. These values were determined by tests conducted in conformance with standards [13-6] using the specified nominal volumetric flows and SAE coarse test dust. Smaller specific surface areas and higher air velocities lead to lower total mass filtration efficiencies and smaller dust capacities in air cleaners for cars. However, they are sufficient for a car's specific operating conditions. The smaller design of car air cleaners' specific surface area is a good compromise between the requisite filter function and the limited space conditions under car hoods as well as the lower dust concentrations prevalent there on average.

13.1.3 Designs

The central filter element of dry air cleaners is a replaceable filter element made of a pleated filter medium shaped into a flat or round filter with the appropriate seals. Virtually without exception, the filter medium is impregnated with a synthetic resin to resist thermal, mechanical and chemical stresses. Furthermore, spacers or grooves are stamped into the pleats to mechanically stabilize the filter element.

Air cleaners have different designs based on the different installation and mounting conditions in cars and commercial vehicles. The typical car air cleaner pictured in Fig. 13-5 consists of a rectangular plastic (polypropylene or polyamide) housing with a flat rectangular paper element. The cleaner is installed directly at the engine or in the lateral area of the engine. Filter elements are replaced once they have reached the maintenance interval specified by the engine manufacturer. Car air cleaners must be separately developed for and matched to every engine type to optimize power, fuel consumption, the torque characteristic and intake noise damping. Paper dry type air cleaners are predominantly used for commercial vehicles too. Less effective but self-regenerating, oil wetted air bath cleaners are also used in a few cases. Despite their lower protection against wear, they are advantageous when a supply of spare filter elements is uncertain (see Chaps. 17 and 18).

Figure 13-6 presents a schematic of an oil wetted air bath cleaner. The air flows through the center tube to an oil bath in

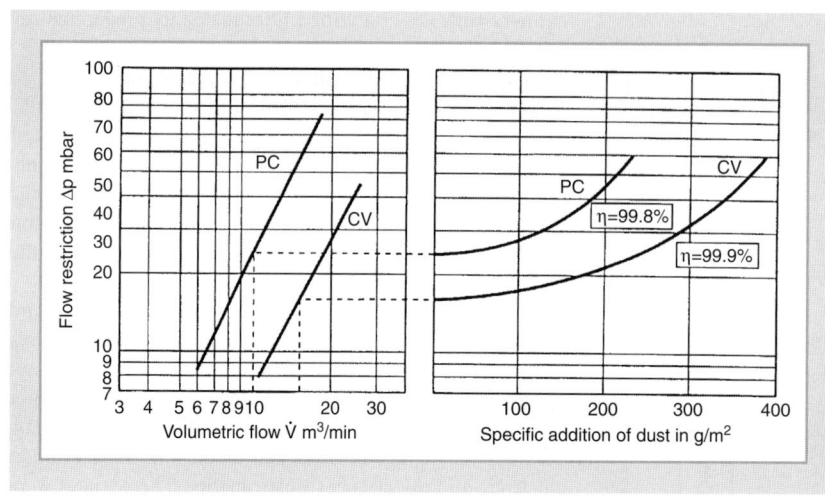

Fig. 13-4 Typical characteristic curves of paper dry type air cleaners for cars and commercial vehicles (specific filter areas: 2,500 and 4,000 cm^2/m^3/min)

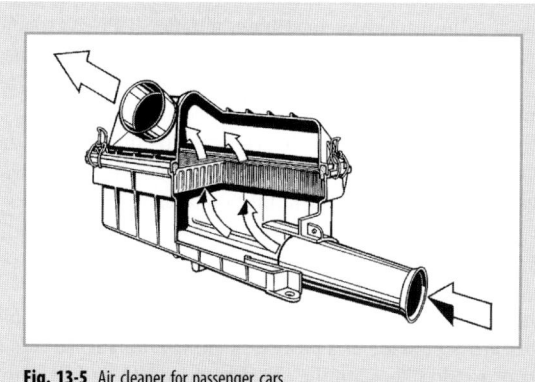

Fig. 13-5 Air cleaner for passenger cars

Fig. 13-6 Oil wetted air bath cleaner for diesel engines

the lower part of the cleaner housing and is conducted upward into the filter packing. The air initially flows through the lower region wetted by oil and then through the upper dry region of the filter packing to the clean air outlet. This achieves a filtration efficiency of 99% at full load, which tends to decrease in part load range. An oil bath air cleaner's design must match the engine's air demand exactly. Oversizing or undersizing results in oil carryover or lower filtration efficiency. Therefore, proper attention must be paid to the air cleaner's design and the manufacturer's installation instructions.

Figure 13-7 pictures a standard dry air cleaner design for commercial vehicles and similar applications. It contains a cylindrical paper element with a cyclone integrated in the housing to increase service life and thus prolong the maintenance interval. A vane ring sets the air rotating and the centrifugal forces separate a majority of the coarse particulate matter. This initially separated dust is released to the atmosphere through a dust discharge valve or collected in a dust bowl. The pulsation of the intake air determines which option is more suitable in each case. Dust discharge valves have proven to be simple and good solutions when the pulsations are strong enough. These cyclones achieve initial filtration efficiencies of 85% based on SAE coarse test dust, which is commensurate with an approximately fourfold prolongation of service life. It can be increased even more by small cyclone cells that may be installed with both dry air cleaners and oil wetted air bath cleaners. This can achieve initial filtration efficiencies of up to 95%. Figure 13-8 presents an example of a simple standard upstream cyclone for commercial vehicle engines. It is mounted on the air cleaner's

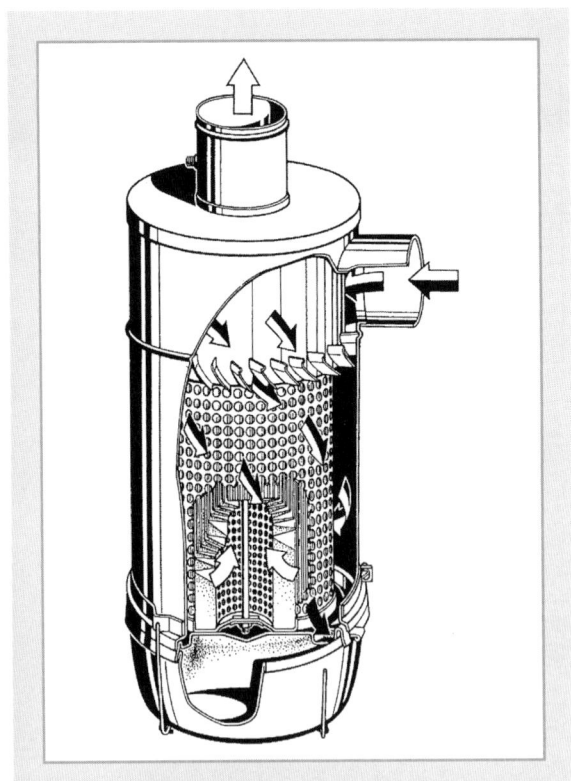

Fig. 13-7 Combination air cleaner for commercial vehicles (MANN & HUMMEL)

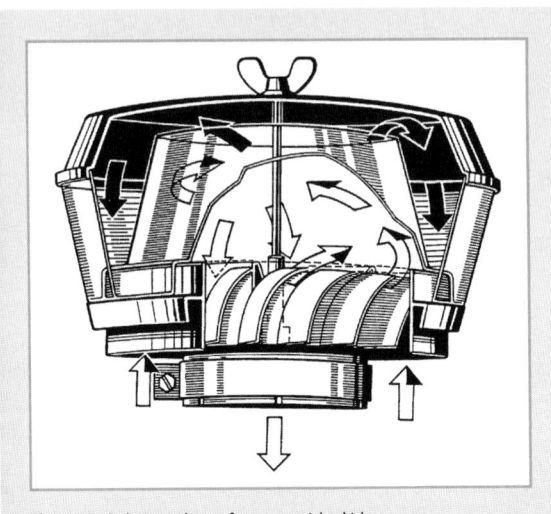

Fig. 13-8 Cyclone precleaner for commercial vehicles

raw air pipe and collects the discharged dust in a surrounding transparent dust bowl.

Larger stationary diesel engines operated in closed rooms or on ships have relatively low dust concentrations. Simple oil wetted air cleaners work satisfactorily under such conditions: A cylindrical filter body made of steel fabric, expanded metal or plastic with an outside-in flow is wetted with oil to increase efficiency. More recently, dry air cleaners with a design similar to cleaners for commercial vehicle engines have also been increasingly being used for stationary diesel engines because of their high filtration efficiency.

13.1.4 Intake Noise Damping

The damping of diesel engine intake noise in cars and commercial vehicles is essential for compliance with the legal requirements for overall vehicle noise. The level of interior noise, which is very strongly influenced by intake noise, must also be minimized for reasons of driving comfort too.

Intake noise is damped by designing the air cleaner as a reflection muffler in the special form of a Helmholtz resonator. A *Helmholtz resonator* consists of a muffler chamber with a connected tube section. Figure 13-9 presents the characteristic progression of a relevant damping curve for a car air cleaner.

Both the theoretical characteristic and two measured characteristics are plotted. According to the relationship

$$f_0 = \frac{c}{2\pi}\sqrt{\frac{s}{l \cdot V}}$$

the resonant frequency f_0 that appears is a function of the speed of sound c of the air, the volume V of the resonator, the length l and the mean cross section S of the intake manifold. Sound intensifies when it is resonant. As of $f = f_0\sqrt{2}$, the damping increases as the frequency increases. f_0 must lie as far below the frequencies occurring in operation as possible to obtain optimal damping. This can be accomplished by enlarging the air cleaner volume, reducing the intake cross section or lengthening the intake manifold [13-8].

An intake muffler volume of at least fifteen or, even better, twenty times the cylinder's displacement is a *recommended value* for four-stroke engines. This normally achieves damping of 10–20 dB(A). Reductions of the intake cross section and lengthening of the intake manifold quickly reach their limits because of the increase in resistance and the space conditions.

A diesel engine's intake noise can reach 100 dB(A). Figure 13-10 compares the undamped intake noise of a commercial vehicle diesel engine with the improvement obtained with a standard air cleaner.

Further improvements can be achieved when the length and placement of the intake line and the dimensions of the cleaner housing are matched to one another. Figure 13-10 also contains the additional damping achieved.

Resonances additionally occurring in the cleaner housing or in the intake manifold can diminish the effect of an air cleaner/muffler considerably (see Fig. 13-9). Additional mufflers are employed when the air cleaner alone no longer produces the desired amount of intake noise damping. Table 13-2 lists the acoustic properties of the most important types of mufflers

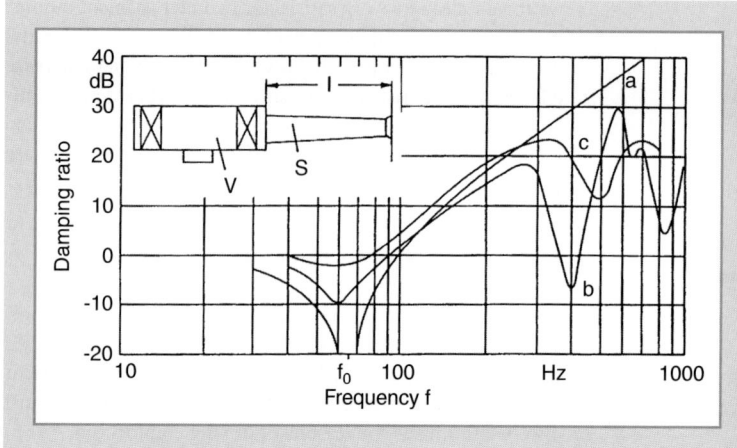

Fig. 13-9
Sound damping of an intake muffler (Helmholtz resonator)[13-7]. Curve **a**: theoretical damping characteristic ($f_0 = 66$ Hz); curve **b**: measured damping characteristic for lower sound energy density without parallel flow (loudspeaker measurement); curve **c**: measured damping characteristic for higher sound energy density with parallel flow (measurement in engine)

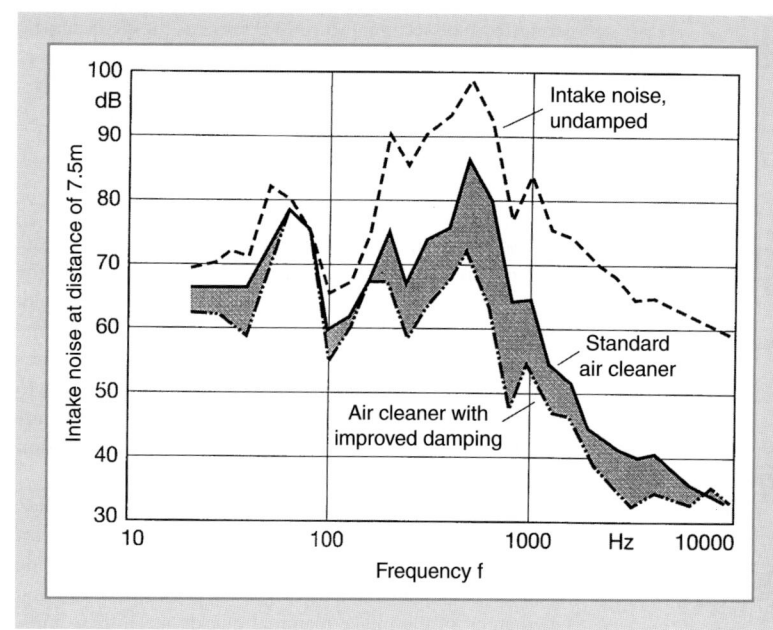

Fig. 13-10
Intake noise of a 147 kW six cylinder commercial vehicle diesel engine [13-9]

Table 13-2 Acoustic properties of the most important muffler types

Type of damping	Properties
Absorption muffler	Broadband; suited for middle and higher frequency range of approximately 300–5,000 Hz
Flow control muffler	Broadband; suited for middle and higher frequency range
Inline resonator	Narrowband damping in the range above the resonant frequency f_0; suitable for frequencies to approximately 500 Hz
Side branch resonator	Narrowband damping in the range above the resonant frequency f_0; suited for low and middle frequencies; variously tuned resonators can be connected in parallel
Whistle resonator	Narrowband damping in the frequencies $f = C(2m+1)/4\,l$; $m = 0, 1, 2, \ldots l =$ length of the whistle
Interference damping through bypass	Ultra narrowband damping in the frequencies $f = C(2m+1)/4\,l$; $m = 0, 1, 2, \ldots l =$ path length difference; allows very high damping values

capable of doing this. Disturbing individual noises can be treated and damped very systematically with such additional mufflers. They are also frequently integrated in the air cleaner housing to produce space-saving and low cost solutions.

13.2 Exhaust Systems

13.2.1 Function and Basic Design

Exhaust systems basically serve three functions in motor vehicles:
– discharging hot exhaust gas produced during the combustion of the air/fuel mixture in the engine to the atmosphere,
– purifying exhaust gas of harmful chemical components and particulate matter to comply with legal requirements and
– damping exhaust noise to the legal minimum [13-10] and additionally shaping it to deliver the sound design desired by customers.

Therefore, the following components have to be designed into exhaust systems:
– exhaust manifolds that aggregate exhaust gases after they exit the valves,
– catalysts and diesel particulate filters (DPF) for exhaust gas aftertreatment (see Sect. 15.5),
– mufflers that reduce and control exhaust noise and
– piping that conducts the exhaust gas until it is released into the environment.

Engines with exhaust gas recirculation (EGR, see Sect. 15.4) have an exhaust gas recirculation valve additionally integrated in the exhaust system. The exhaust manifold in diesel engines with an exhaust gas turbocharger also serves as an inflow passage to the exhaust turbine.

Exhaust and intake systems have considerable influence on engine power and torque through gas exchange. The initial reflections of the pressure waves generated in such upstream pipe installations as exhaust manifolds and catalysts during the exhaust cycle are particularly influential and require meticulous tuning.

The exhaust back pressure in diesel engines with exhaust gas turbocharging should not exceed a certain level (see Sect. 2.2). Thus, low flow resistance is particularly important. It also affects the exhaust muffler. Hence, dual exhaust lines and muffler systems are sometimes implemented in extremely powerful engines with correspondingly strong exhaust gas mass flows. Consequently, the exhaust gas is either routed through two pipe runs starting from the manifold, e.g. in V engines, or the exhaust gas flow is separated in the middle of the exhaust system, e.g. in a central muffler. In such cases, the exhaust gas is usually fed to two rear mufflers.

In turn, the different gas routing has considerable influence on an exhaust system's acoustic transmission properties. In addition, the space conditions in the vehicle underbody complicate sound engineering design in most cases since the location and maximum size of a muffler is frequently already specified when design work begins. Hence, along with emission control, the functional development of an exhaust system is an optimization process typical for development, which is influenced by the following partly conflicting parameters:
– back pressure and thus engine power,
– muffler volume and weight,
– sound pressure level in the outlet and
– system costs.

13.2.2 Tailpipe Noise Damping

13.2.2.1 General Background

Pulsating gas emissions from cylinders are the main cause of exhaust noise. A specific frequency spectrum is produced as a function of the cylinders' firing sequence and the operating time differences in a particular exhaust path. This is called an engine order (EO). Since the air/fuel mixture is ignited during every second crankshaft revolution and subsequently expelled, an EO of 0.5te is dominant in a single cylinder in a four-stroke engine. In multi-cylinder engines, the individual cylinders' orders add up. Thus, the 2nd EO is dominant in four-cylinder engines, the 3rd EO in six-cylinder engines and so on. Each of these respectively dominant EOs is also called ignition frequency. The amplitude ratio of the main engine order (ignition frequency) and secondary order (half, even and uneven EO) fundamentally determines the sound character. This ratio depends on the firing sequence and the acoustic propagation of the individual cylinders' noise contributions up to a common exhaust gas run, i.e. it depends on the manifold in particular. The frequency of alternating pressure fluctuations resulting from the EO is always directly proportional to the engine speed.

Figure 13-11 presents the frequency characteristic of the engine orders of a four-stroke engine. Thus, at the 1st EO or rotational frequency of 50 Hz, a speed of 3,000 rpm in a four-stroke engine as well as in a four-cylinder engine corresponds to an ignition frequency of 100 Hz. All in all, the EO in a typical vehicle diesel engine only substantially contribute to outlet sound in a frequency range of approximately 30–600 Hz.

Furthermore, considerable flow noises that must also be damped are produced when exhaust gas is emitted and routed further. They are characterized by a spectral broadband and, unlike pulsation noises, also extend into the high frequency range of up to approximately 10 kHz. In principle, the strength of flow noises is disproportionately linked to the flow velocity. Hence, preventing unduly high gas velocities in pipes and bypasses and during inflow and discharge can already reduce this noise component at its source.

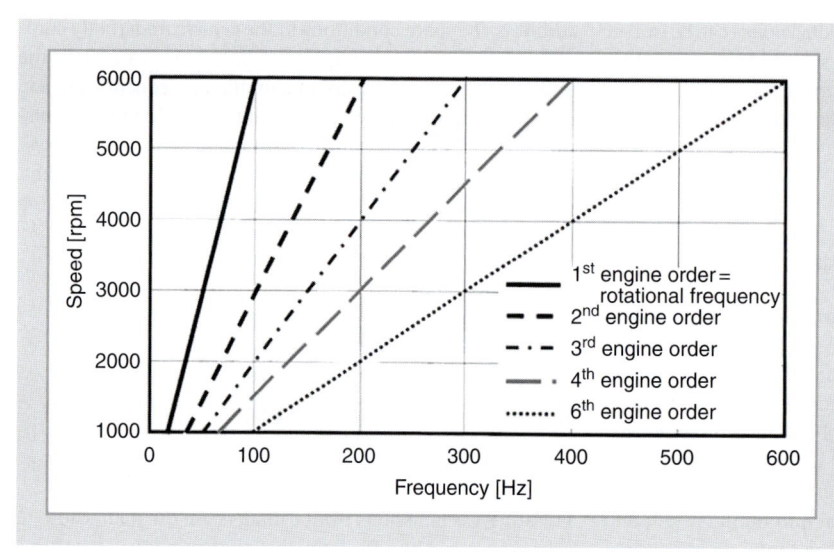

Fig. 13-11 Engine orders for four stroke engines

Together with every pipe and connected muffler and catalyst volume, the exhaust system in a combustion engine collectively constitutes an oscillating system with numerous acoustic and mechanical natural resonances. The geometric position of the mufflers has great significance for the position of the natural frequencies and the amount of damping [13-11]. Frequently, only just the important orders are examined to analyze tailpipe noise and the level of the orders is plotted as a function of the speed. The Campbell diagram in Fig. 13-12 indicates the sound pressure level of a six-cylinder diesel engine for every speed and frequency. The significant EO of 3, 6 and 9 are discernible as dark regions as is a weak resonance excited by flow noises at approximately 1,000 Hz.

Computerized simulations now systematically tune the natural frequencies of entire exhaust systems for cars. Suitable commercially available software packages allow modeling intake systems, combustion engines and exhaust systems as well as simulating gas exchange in one application [13-12]. Among other things, the results yield the temperatures and pressures in an exhaust system and the outlet sound in every engine operating point of interest (speed and load). By rigorously applying statistical methods, e.g. design of experiment (DoE), and linking diverse software tools, automatic acoustic exhaust system design has become the state-of-the-art in development [13-13].

A basic distinction is made in muffler design between the physical principles of absorption and reflection. Furthermore, mufflers are also subdivided according to whether a switching element alters the acoustic effect during operation (semi-active mufflers) or a sound generator directly cancels the sound by wave superimposition (active mufflers). From this perspective, conventional absorption and reflection mufflers are also considered passive mufflers. However, semi-active and active mufflers ultimately function according to the principle of reflection.

13.2.2.2 Absorption Mufflers

In acoustics, the conversion of sound energy caused by gas molecules rubbing against each other or structures into heat is referred to as absorption. Since the friction between gas molecules in air is relatively low, air absorption may usually be disregarded. The larger the surface area of the material, the greater the friction in structures is. Hence, porous and fibrous materials, e.g. wool and foam, absorb particularly well

Fig. 13-12 Campbell diagram of the sound pressure level measured at the outlet of a typical six-cylinder diesel engine

Fig. 13-13 Muffler schematics

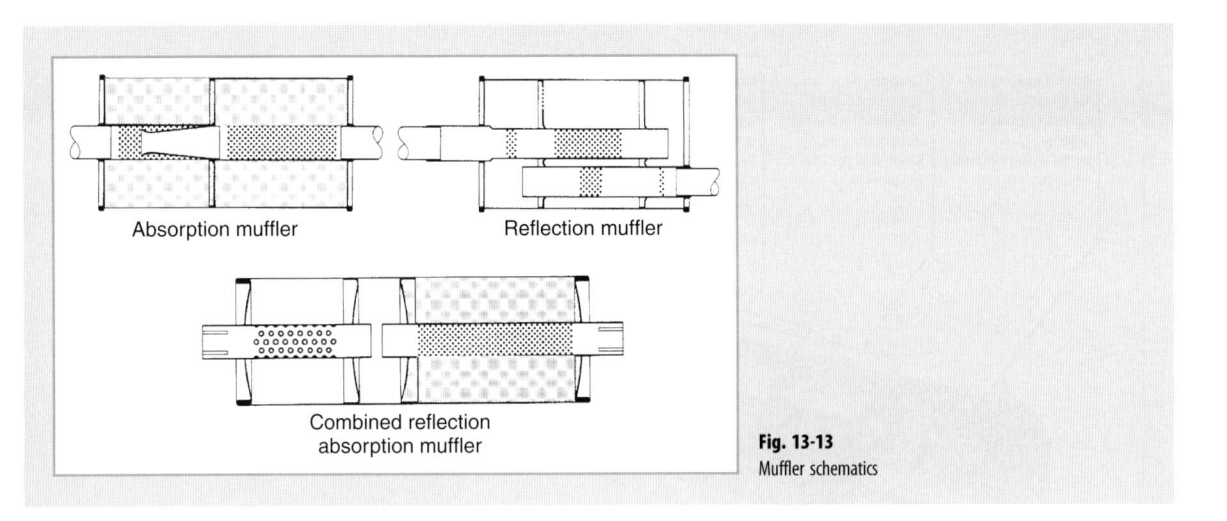

because air molecules penetrate the material easily and rub intensely in the many thin fibers or many small pores. The activity of this acoustic effect is measured with the absorption coefficient defined as the ratio of the sound power absorbed to the sound power occurring.

Just like sound damping, absorption generally increases from low to high frequencies. Glass wool sometimes serves as the absorption material. However, long-fiber basalt or rock wool with a bulk density of approximately 100 g per liter of volume is usually employed instead because it has resistance to heat.

Figure 13-13 contrasts the typical design of a packed absorption muffler through which a perforated pipe runs with a conventional reflection muffler and a combination of both principles. The shaping of the perforations and the routing of the pipe through the wool ensure that the exhaust gas pulsation cannot blow the material out despite high flow velocities. The mineral wool is occasionally protected with a layer of stainless steel wool around the perforated pipe. Absorption mufflers are especially implemented to reduce flow noises because of their broadband effect in the middle and high frequency range.

13.2.2.3 Reflection Mufflers

This muffler design consists of various long chambers interconnected by pipes. The cross sectional jumps between pipes and chambers as well as the exhaust gas bypasses and the resonators formed by the pipes connecting with the chambers produce damping that is particularly effective for low frequencies. However, every bypass and every inflow and discharge in reflection mufflers increases exhaust back pressure and, thus, normally causes a greater loss of power than absorption mufflers with directly routed flows.

Catalysts and particulate filters also should be regarded as acoustic elements and adjusted. Clever shaping of a catalyst's inlet cone including the pipe routing before the element achieves uniform pressurization while simultaneously increasing service life and efficiency.

As in the intake system, branch or intake resonators based on the Helmholtz principle are implemented to eliminate particularly disturbing low frequency resonances in muffler systems' tailpipe noise (e.g. starting humming). A lateral connection through which there is no gas flow at the main pipe conducting gas feeds sound energy into a sealed volume where it is temporarily stored and subsequently added back into the main gas flow with a time delay. At the resonant frequency, the time delay is exactly long enough that the two sound waves cancel each other. However, the flow velocity adversely affects the Helmholtz resonators' effect. Hence, they may only be mounted in locations where the velocity is low. Since a Helmholtz resonator only functions at one frequency, its volume is neither available to damp other frequencies nor at other speeds (see Sect. 13.1). What is more, this principle is only applied in problem cases since its effect usually requires quite extensive sealing. Figure 13-14 pictures a fictitious design of a conventional passive muffler with diverse options for sound damping.

13.2.2.4 Semi-active Mufflers

A good acoustic damping effect can be obtained by partially blocking the path of flow. If one of two muffler tailpipes is sealed, e.g. by an exhaust gas valve (Fig. 13-15), then the low frequency tailpipe noises drop by up to 10 dB compared to a system without an exhaust gas valve. This is commensurate with halving the perceived loudness. Low frequency tailpipe noises chiefly occur during city driving and intensify in engine overrun condition (e.g. at stoplights). However, the reduction of exhaust back pressure has priority at higher speed and load (e.g. fast highway driving) when rolling and driving noises predominate anyway. Then, the flap is opened,

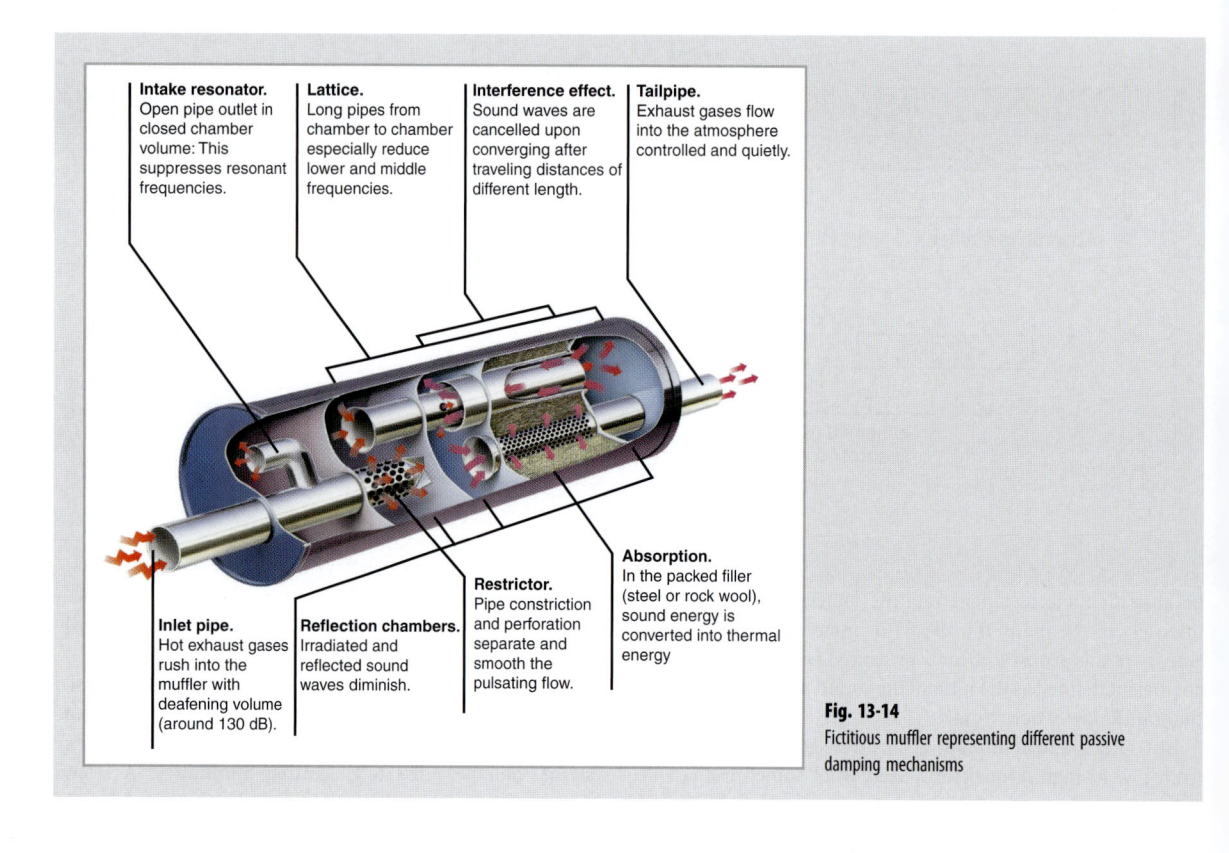

Intake resonator.
Open pipe outlet in closed chamber volume: This suppresses resonant frequencies.

Lattice.
Long pipes from chamber to chamber especially reduce lower and middle frequencies.

Interference effect.
Sound waves are cancelled upon converging after traveling distances of different length.

Tailpipe.
Exhaust gases flow into the atmosphere controlled and quietly.

Inlet pipe.
Hot exhaust gases rush into the muffler with deafening volume (around 130 dB).

Reflection chambers.
Irradiated and reflected sound waves diminish.

Restrictor.
Pipe constriction and perforation separate and smooth the pulsating flow.

Absorption.
In the packed filler (steel or rock wool), sound energy is converted into thermal energy

Fig. 13-14
Fictitious muffler representing different passive damping mechanisms

Fig. 13-15 Actuated exhaust gas valve in a rear muffler tailpipe

the gas flows through both tailpipes, the flow noise is reduced, the exhaust back pressure drops and the engine is able to release its full power. Some flaps are self-controlled by pressure and flow and some are externally controlled by an interface to the engine electronics. Thus, this technology provides significantly more flexibility in the load and speed-dependent selection of the changeover point, which opens up considerably more options for map-dependent sound design. Moreover, the back pressure is usually lower when it is generated by an externally controlled flap than a flap actuated by the flow. Naturally, they are also considerably more technically complex than self-controlled flaps and therefore only employed in sophisticated applications.

13.2.2.5 Active Mufflers (ANC)

The functional principle of active noise control (ANC) is simple: A negative mirror image of disturbing sound waves is systematically generated and both components are superimposed in one point. The sound waves cancel out each other as a result. Since the propagation of the waves can be predicted well and the noises are principally low frequency, this technology is fundamentally suited for use in mufflers in particular. A loudspeaker usually serves as the source of the sound in phase opposition. Reliable and fast electronics must ensure that the anti-noise is synchronized and has the correct volume.

However, some basic problems must be resolved to assure the operability of an anti-noise system. The ambient conditions (heat, moisture and high sound pressure level) prevalent in exhaust systems diminish the service life of the loudspeaker that generates the anti-noise. In addition, the engine's quickly changing speeds and load conditions require a controller

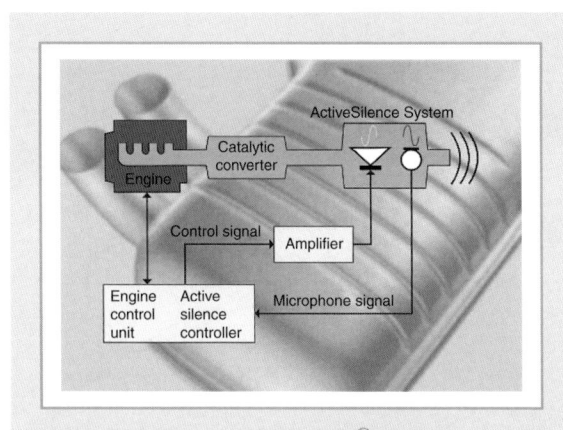

Fig. 13-16 Schematic diagram of Active Silence® technology

supported by a processor. Significant advances for motor vehicles have been achieved with newly developed efficient, low cost sound transducers [13-14]. What is more, this technology is especially suited for diesel engines since they normally have lower exhaust gas temperatures than gasoline engines.

Advances in microelectronics and their massive spread in the automotive and consumer goods industry over the last few decades have resolved another problem. The requisite controller hardware is now so small and efficient that it can be integrated in the engine control units common today or, as a small separate control unit, communicate with them through a standard bus (CAN, MOST) (Fig. 13-16).

Apart from cancelling sound waves and thus purely damping sound, an ANC system can also boost certain frequencies (engine orders), e.g. pleasant frequencies, to obtain a desired acoustic pattern (sound design). Thus, vehicles with unattractive or inconspicuous sound can be acoustically enhanced. Software settings alone accomplish this. As a result, in contrast to conventional exhaust systems with passive mufflers, it will be possible in the future to adapt exhaust noise within certain limits independent of the engine, vehicle type and/or driving situation.

13.2.3 Structure-Borne Noise Emission from Exhaust Systems

In addition to the noise from its tailpipe, an exhaust system also emits noise through its surface. This structure-borne noise emission is caused by vibrations, which for their part may be mechanically excited by the engine or the turbocharger or forced by the pulsating gas column. Upstream structure-borne noise decoupling elements can effectively suppress further transmission of structure-borne noise that reaches the mufflers from the engine or the turbocharger through pipes. Technically, several options exist to reduce the emission of structure-borne noise by using a housing shell (mufflers, catalyst and diesel particulate filter):

– increasing the plate's wall strength,
– using double plate and
– optimizing the outer shape.

A thicker wall usually also reduces structure-borne noise emission since it increases the mass and stiffens the structure. However, this option is employed reluctantly since it always makes an exhaust system heavier and more expensive. Generated by the relative motion between layers of plate during oscillations, friction in double plate also reduces structure-borne noise emission. A fleece layer may also be incorporated to thermally insulate both layers and further enhance their decoupling. However, this solution is technically complex and not as mechanically stable as simple plate. Optimizing the outer shape of mufflers made of half shells contributes to the prevention of audible structural resonances. However, compromises frequently have to be made in terms of the utilization of structural space, durability and tool manufacturability. Consequently, identifying the most preferable solution remains a challenge of development, which must be dealt with on a case to case basis.

13.2.4 Sound Design

Not only technical and economic but also emotional factors play a considerable role in the purchase of an automobile. The sensory experience, e.g. the perceived sound, appeals to emotions. Hence, distinctly sporty premium products particularly also depend on their "sound" setting them positively apart from other manufacturers' vehicles. Along with the intake system, an exhaust system's tailpipe noise significantly influences a vehicle's overall acoustic impression, which affects pedestrians outside a car as much as drivers and passengers inside. Demanded sound quality and legal requirements impose clear limits on boosting loudness simply to emphasize sportiness. Manufacturers of exhaust systems work extensively on producing a particularly distinct sound impression at the same or similar level [13-15–13-17]. Systematic sound design involves gradually optimizing noise from exhaust systems in intensive acoustic tests in a sound studio. Depending on their positioning in the market segment and the type of engine and vehicle, carmakers' requirements vary greatly in response to potential customers' demands. Unlike gasoline engines, narrow limits are imposed on the sound design of diesel engine exhaust systems since the exhaust gas turbochargers frequently employed and the diesel particulate filters (DPF) necessary for emission control remove emotionally interesting noise components from the gas pulsation. This continues to be an obstacle to the implementation of diesel engines in sports cars, which ought not to be underestimated.

13.2.5 Geometric Designs

Exhaust systems in cars are designed quite differently depending on the vehicle type, motorization and underbody. The top half of Fig. 13-17 illustrates a single line system for a four-cylinder diesel engine with a front pipe emanating from the turbocharger, a catalyst, an air gap insulated intermediate pipe, a DPF and a larger rear muffler. The DPF simultaneously assumes the function of the central muffler. The intermediate pipe is insulated to feed the exhaust gas to the DPF with a maximum temperature because this makes it easier to thermally regenerate a filter loaded with soot. The bottom half illustrates a V8 bi-turbocharger engine's dual exhaust line system. Splitting the exhaust system necessitates two catalysts, diesel particulate filters and rear mufflers. Instead of a central muffler, there is solely a small crosstalk point in the center section of the exhaust system.

Figure 13-18 presents a particularly innovative solution for a compact exhaust system for a compact car with a three-cylinder diesel engine where the catalyst, DPF and the mufflers have been integrated in one housing.

In principle, exhaust systems for commercial vehicles are designed similarly to those of cars. Since the engines employed usually have a larger displacement though, the mufflers have larger volumes of up to 1,000 liters. However, space conditions in a vehicle impose limits on the mufflers' dimensions too. The legally mandated noise limits for commercial vehicles are higher than those for passenger cars and the acoustic properties differ considerably from passenger cars. Thus, a large-sized muffler also containing catalysts and, where necessary, particulate filters is normally sufficient in today's commercial vehicles. Since the mufflers have such large surface areas, not only tailpipe noise itself but also structure-borne noise emission plays a role in legally restricted passing noise.

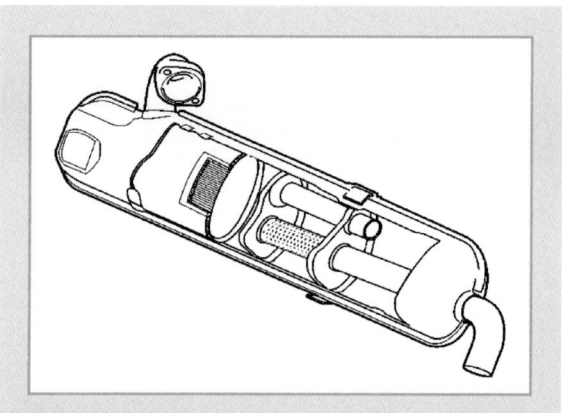

Fig. 13-18 Design of an extremely compact exhaust system for a compact car with a three-cylinder diesel engine

Literature

13-1 Erdmannsdörfer, H.: Trockenluftfilter für Fahrzeugmotoren – Auslegungs- und Leistungsdaten. MTZ 43 (1982) 7/8, pp. 311–318

13-2 James, W.S.; Brown, B.G.; Clark, B.E.: Air cleaner – Oil Filter Protection, Critical Factor in Engine Wear. SAE Journal 1952, pp. 18–26

13-3 Thomas, G.E.; Culbert, R.M.: Ingested Dust, Filters and Diesel Engine Ring Wear. SAE Paper 680536 (1968)

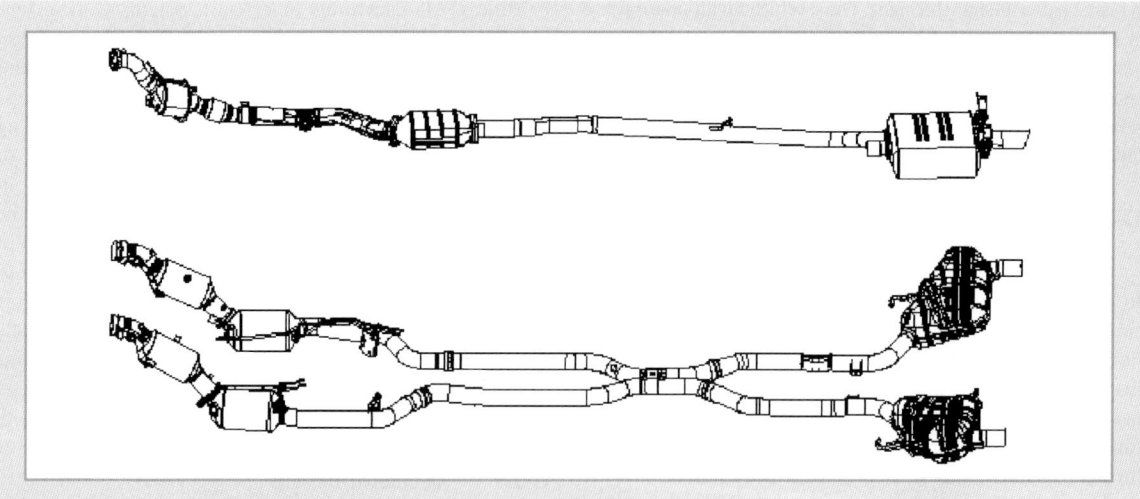

Fig. 13-17 Designs of two exhaust systems for cars with four and eight-cylinder diesel engines

13-4   Schropp, G.: Versuche über Entstehung und Auswirkung der Verschmutzung in Verdichtern. Brown Boveri Mitteilungen Vol. 55, No. 8

13-5   DIN 71450: Filter für Kraftfahrzeuge und Verbrennungsmotoren: Begriffe für Filter und Komponenten. Deutscher Normenausschuss (Ed.), No. (1990) 5

13-6   Entwurf DIN ISO 5011: Luftfilter für Verbrennungsmotoren und Kompressoren; Prüfverfahren. Deutscher Normenausschuss (Ed.), Entwurf (1992) 5

13-7   Blumenstock, K.-U.: Motorenfilter. In: Die Bibliothek der Technik Vol. 31. Munich: Moderne Industrie 1989

13-8   Bach, W.: Beitrag des Luftfilters zur Geräuschdämpfung und Leistungsbeeinflussung von Verbrennungsmotoren. ATZ 78 (1976) 4, pp. 165–168

13-9   Bendig, L.: Ansauggeräuschdämpfung an Nutzfahrzeugen. ATZ 80 (1978) 4, pp.171–173

13-10 ECE Regulation 51: Uniform provisions concerning the approval of motor vehicles having at least four wheels with regard to their sound emissions

13-11 Munjal, M.L.: Acoustics of ducts and mufflers. New York: Wiley Interscience Publication 1987

13-12 Ricardo Software, Bridge Works: WAVE V7 Manuals. Shoreham-by-sea, West Sussex, England 2005

13-13 Jebasinski, R.; Halbei, J.; Rose, T.: Automatisierte Auslegung von Abgasanlagen. MTZ 67 (2006) 3, pp. 180–187

13-14 Krüger, J.; Pommerere, M.; Jebasinski, R.: Active Exhaust Silencers for Internal Combustion Engines. Fortschritte der Akustik – DAGA 2005, in print

13-15 Heil, B.; Enderle, Ch.; Bachschmid, G.; Sartorius, C.; Ermer, H.; Unbehaun, M.; Zintel, G.: Variable Gestaltung des Abgasmündungsgeräusches am Beispiel eines V6-Motors. Motortechnische Zeitschrift MTZ 62 (2001) 10, pp. 787–797

13-16 Krüger, J.; Castor, F.: Zur akustischen Bewertung von Abgasanlagen. Fortschritte der Akustik – DAGA 2002, pp. 188–189

13-17 Krüger, J.; Castor, F.; Müller, A.: Psychoacoustic investigation on sport sound of automotive tailpipe noise. Fortschritte der Akustik – DAGA 2004, pp. 233–234

Further Literature

Kurtze, G.: Physik und Technik der Lärmbekämpfung. Karlsruhe: Verlag G. Braun 1964

Forschungshefte. Forschungskuratorium Maschinenbau e.V. 26 (1974)

14 Exhaust Heat Recovery

Franz Hirschbichler

14.1 Basics of Waste Heat Recovery

14.1.1 Preliminary Remarks

Although public awareness of the finiteness of fossil fuel reserves has receded into the background somewhat after being raised in the 1970s, the impact of pollutant and CO_2 input into the earth's atmosphere is again making the need for a longer range environmentally compatible energy policy with concrete goals evident.

In the future, both challenges – conserving resources and protecting the environment – will increasingly require an approach that endeavors to take full advantage of the ample potentials to save energy and additionally intensify the utilization of renewable, i.e. inexhaustible, energy sources. Both goals will have to be pursued simultaneously, i.e. in parallel, rather than sequentially.

This will necessitate research on the types of waste heat that accumulate during diesel engine combustion as well as expedient recovery methods for the purpose of conserving primary energy and protecting the environment.

14.1.2 Diesel Engine Waste Heat

The following types of waste heat can be distinguished on the basis of their origin:
- waste heat from exhaust gas generated by gas exchange,
- waste heat produced as cooling energy to protect metallic walls, e.g. cylinder cooling, piston cooling and, where applicable, cooling of turbocharger turbine housings and oil cooling of bearings and interior walls,
- waste heat from intercooling, which serves to boost engine power and net efficiency, and
- waste heat emitted from the engine surface to the environment as radiation and convection heat.

While exhaust gas heat is dissipated by gas exchange in the exhaust process, all other waste heat must inevitably be dissipated with the aid of a coolant (water, oil or air).

Heat that accumulates in various points of an engine (Fig. 14-1) is transferred to water as the heat transfer medium for recovery of varying complexity. While cooling energy is transferred to water/water or air/water heat exchangers without any problem, the transfer of exhaust gas heat loaded with particulate matter and soot particulates to a gas/water heat exchanger proves to be somewhat more complicated (see Sect. 9.2.5.5).

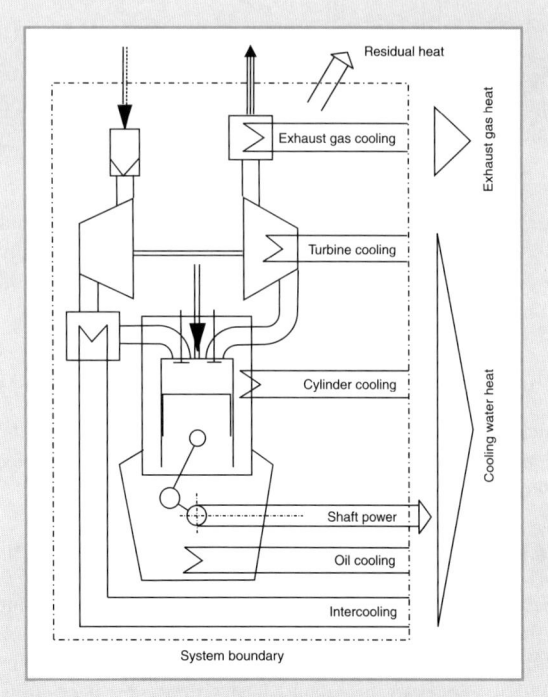

Fig. 14-1 External heat balance and waste heat of a diesel engine

F. Hirschbichler (✉)
München, Germany
e-mail: franz.hirschbichler@gmx.de

Radiation and convection heat emitted by an engine is usually dissipated by aerating and ventilating the underhood environment. In principle, it may also be dissipated with the aid of an air/water heat pump and recovered. This has been implemented in very few cases though.

In addition to being dissipated and exchanged differently, the different types of diesel engine waste heat also have different temperature levels corresponding to their place of origin in an engine. The waste heat with the highest temperature level, exhaust gas heat accumulates in the range of 300–500°C depending on the type and size of the engine. Engine cooling water outlet temperatures are usually in the range of 75–95°C. While water temperatures in an intercooler or low temperature intercooler are 30–40°C during multistage intercooling, water temperatures in a high temperature intercooler can reach the temperature level of engine cooling water. Temperatures of water used to cool lubricating oil heat are also usually in or slightly below the range of the temperature level of the engine cooling water.

14.1.3 Determination of Waste Heat Outputs

14.1.3.1 Diesel Engine Energy Balance

The following relationships may be referenced to determine a diesel engine's waste heat output.

The following applies to a diesel engine's external heat balance:

$$P_B = P_e + \Phi_A + + \Phi_K + \Phi_R.$$

Accordingly, the heat Φ_{zu} supplied as fuel power P_B corresponds to the product from the fuel mass flow $\dot{m}_B$ and calorific value H_u

$$\Phi_{zu} = P_B = \dot{m}_B \cdot H_u,$$

to the sum of the net (mechanical) power P_e, the heat output Φ_A discharged with the exhaust gas and the cooling capacity Φ_K and to the loss to the environment by radiation and convection contained in the remainder Φ_R.

The overall cooling capacity Φ_K

$$\Phi_K = \Phi_{ZK} + \Phi_{ÖK} + \Phi_{LLK} \quad (14\text{-}1)$$

includes the cooling energy Φ_{ZK} emitted by the engine (cylinders) and the heat fluxes that accumulate in the oil cooler ($\Phi_{ÖK}$) and intercooler (Φ_{LLK}).

14.1.3.2 Exhaust Heat Output Φ_A

Relative to the system boundary specified by the ambient condition (p_U, T_U), the following, expressed by the enthalpy difference employing the particular specific enthalpy h in (kJ/kg), applies to the exhaust heat output after it exits the turbocharger (index L: air; index A: exhaust gas):

$$\Phi_A = \dot{m}_A h_A - \dot{m}_L h_L = \dot{m}_A [h_A - (1/\delta_0) h_L].$$

With the minimum air requirement L_{min} and the total air/fuel ratio λ_V of combustion, the mass flow ratio $1/\delta_0 = \dot{m}_L/\dot{m}_A$ follows from:

$$1/\delta_0 = L_{min}/(1 + \lambda_v L_{min}). \quad (14\text{-}2)$$

Of this, the following is utilizable in the exhaust gas heat exchanger (index AK):

$$\Phi_{AK} = \dot{m}_A \eta_{AWT}(h_{A1} = h_{A2}),$$

where $\eta_{AWT} = 0.95\ldots 0.98$ can be employed as the exhaust gas heat exchanger's efficiency factor and $T_{A1} = T_A - 5$ K or $T_{A2} = 160\ldots 180°C$ as the exhaust gas temperatures (to prevent wet corrosion). Values for the net enthalpies h of air and exhaust gas relative to the absolute zero point as a function of the temperature and the air/fuel ratio can be gathered from the graph in Fig. 14-2 based on [14-1].

14.1.3.3 Cooling Energy Output Φ_K

The heat an engine dissipates as cooling energy is normally composed of three components (see Eq. (14-1)). Chapter 11 lists guide values for the distribution of the three components in different engines and the valence of the cylinder and lubricating oil cooling energy.

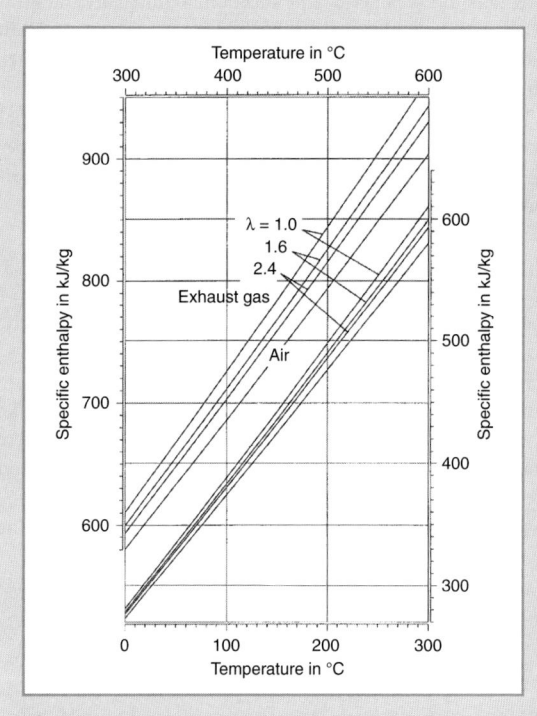

Fig. 14-2 Specific enthalpy of air and exhaust gas as a function of the temperature and the air/fuel ratio

14.1.3.4 Heat Output Φ_{LLK} Dissipated in an Intercooler

Corresponding to the isentropic compression ratio π_L of a "supercharger" compressor, the compression of the air aspirated at ambient temperature T_U increases the temperature of the charge in the compressor outlet to T_{L1} = intercooling inlet temperature. With the compressor's isentropic efficiency η_{SL}, the following ensues for the relative temperature increase T_{L1}/T_U or T_2/T_1, Eq. (2-37):

$$T_{L1}/T_U = [1 - (\pi_L^{\kappa-1/\kappa} - 1)/\eta_{SL}].$$

With the air temperature T_{L2} in the intercooler outlet, the heat output dissipated in an intercooler is then

$$\Phi_{LLK} = \dot{m}_L (h_{L1} - h_{L2}).$$

With the reference temperature T_U = 298 K according to ISO 3046-1 and the temperature of the charge when it enters the engine $T_{L2} > T_L$, Φ_{LLK} can also be determined with the aid of an h-T diagram (Fig. 14-2).

14.1.3.5 Air and Exhaust Gas Mass Flow Guide Values

With the guide values for the specific air flow rate l_e in kg/kWh from Table 14-1, applying the mass flow ratio δ_0, (Eq. (14-2)) as a function of the increase of the air mass during combustion yields the following:

Table 14-1 Specific air consumption l_e, in kg/kWh

Commercial vehicle diesel engine	With turbocharger and intercooler	l_e = 6.0 .. 6.4
High speed high performance diesel engine	With turbocharger and intercooler	l_e = 6.8 .. 7.2
Medium speed diesel engine	With turbocharger and intercooler	l_e = 7.0 .. 7.2
Low speed two-stroke diesel engine	With turbocharger and intercooler	l_e = 9.8 .. 10.5

$$\dot{m}_A \approx l_e P_e \delta_0.$$

The stoichiometric air/fuel ratio L_{min} follows from the calculation of combustion in the common elemental analysis of the fuel, applying the following as guide values:

- diesel fuel (DK) L_{min} = 14.6 kg air/kg fuel or
- heavy fuel oil (HF) L_{min} = 14 kg air/kg fuel.

Table 14-2 lists components measured at full load relative to the diesel engines' power supplied by the fuel as guide values

Table 14-2 Diesel engine parameters

Parameter		Engine type	
		18V 32/40 MAN Diesel	18V 48/60 MAN Diesel
p_e	bar	24.9	23.2
Bore/stroke	mm/mm	320/400	480/600
Power	kW	9,000	18,900
Speed	rpm	750	500
Exhaust gas temperature	°C	310	315
Percentage of fuel power			
HT cooling water circuit[a]	%	14.2	13.8
LT cooling water circuit[b]	%	10.7	9.8
Exhaust gas (180°C)[c]	%	12.5	12.7
Radiation and convection	%	1.9	1.7
Efficiencies[d]			
η_e^d (effect.)	%	46.2	47.7
η_a (therm. usable)[e]	%	37.4	36.3
η_{ges} (effect. + thermal)[e]	%	83.6	84.0

[a] Includes: cylinder cooling + intercooling
[b] Includes: LT percentage of intercooling + oil cooling
[c] Percentage of exhaust gas heat when cooled to 180°C
[d] Including oil pump(s) without cooling water pumps
[e] Including low temperature heat recovery

14.2 Options of Waste Heat Recovery

14.2.1 Recovering Waste Heat as Mechanical Energy

14.2.1.1 Turbocompounding

While the conversion of waste heat into mechanical energy whenever possible would appear to be an obvious extension of a diesel engine's primary purpose, i.e. to emit mechanical energy, narrow limits are imposed in practice, particularly with regard to cost effectiveness. The impacts on specific engine costs resulting from technical complexity and low conversion efficiency make the use of such processes questionable, particularly in smaller engines. Nonetheless, turbocompounding in which exhaust gas performs additional effective brake work in a downstream exhaust gas turbine, which is either transferred to the output shaft or a generator, is employed in commercial vehicle engines and above all in large engines (see Sects. 2.2.4.4 and 18.4.4).

14.2.1.2 Steam Plant (Bottoming Cycle or Organic Rankine Cycle)

Engine waste heat may also be utilized in steam plants. Also called bottoming cycle plants, they are usually based on the Clausius-Rankine process as an ideal process (Fig. 14-3). According to the Carnot process, the maximum effective temperature interval is subject to narrow limits between the steam temperature achievable with the exhaust gas temperatures and the ambient air as the process waste heat sink. Exhaust gas provides the highest temperatures of 300–500°C (see Tables 14-2 and 1-3). The steam temperatures vary in the range of 200–250°C depending on the particular design (exhaust gas cooling interval). With the exception of partial recovery to preheat feed water with commensurate additional complexity (Fig. 14-3), cooling water heat (engine cooling water, charge air heat, oil heat, etc.) is poorly suited for the generation of mechanical or electrical energy because of its low temperatures.

Continuous advances and increases in effective engine efficiencies and the attendant lower exhaust gas temperatures are increasingly diminishing the option of exhaust gas heat recovery in steam plants.

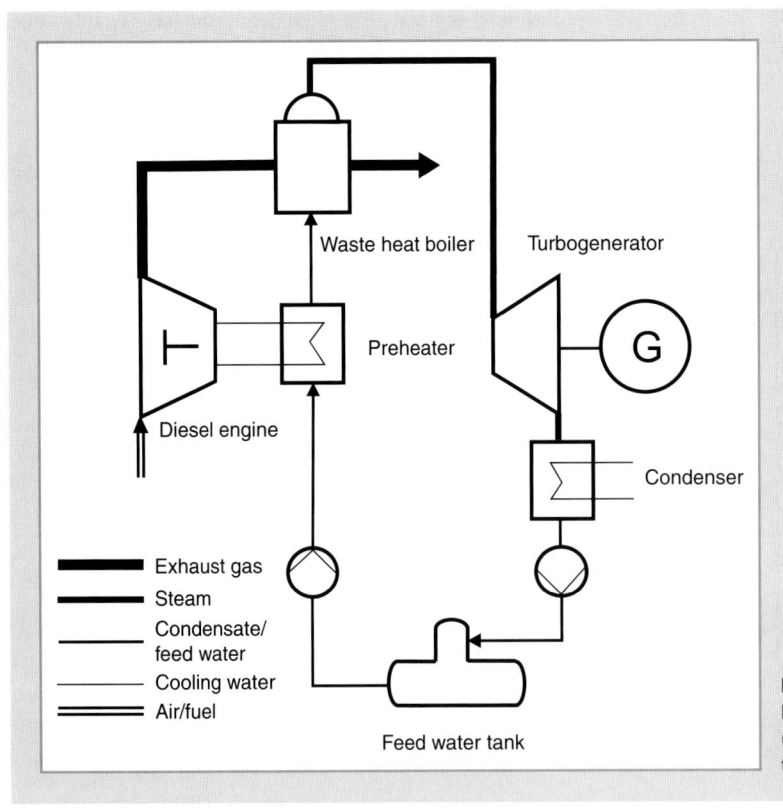

Fig. 14-3 Diesel engine with a downstream steam power process (bottoming cycle) to generate electricity in a turbogenerator

Steam turbines, screw engines and reciprocating steam engines are possible expansion engines. While steam engines with speeds between 750 and 1,500 rpm can directly drive generators, relatively high speed steam turbines and screw engines require gears to adjust speeds to connect to a generator. This additionally diminishes the efficiencies of expansion engines, which are low anyway because of their low outputs.

In addition to steam, fluids with better boiling characteristics than exhaust gas temperatures may also be used as the cycle medium [14-2]. Often called cold vapors or organic vapors, these are common refrigerants that operate in the organic Rankine cycle. Advantages from the better cycle efficiencies anticipated offset disadvantages in terms of toxicity, thermal stability, material compatibility, etc. Steam makes safe operation possible, even with conventional fluorochlorohydrocarbon refrigerants that have a harmful impact on the ozone layer. Detailed tests on a commercial vehicle diesel engine have also corroborated this [14-3]. The maximum increase in output determined was 3%. Although these results demonstrate the limitedness of options for the recovery of waste heat for conversion into mechanical or electrical energy, this process is increasingly attracting interest because of steadily rising fuel prices. Systematic developments would definitely improve the results, above all in large plants [14-4].

In the meantime, the automotive industry is also working on potential applications of this process in passenger cars [14-5].

14.2.2 Recovering Waste Heat as Thermal Energy

14.2.2.1 Heating and Process Heat

Apart from direct utilization in residential heating systems, the technically simplest option for waste heat recovery is the heating of process water. Furthermore, it may be recovered as process heat in manufacturing or to produce fresh water from salt water on ships. Yet, the main application is probably its recovery in motor vehicles to heat the passenger cabin. A vehicle without this is unimaginable today. Classic heating is integrated in the engine coolant circuit. Not only have efficiency-optimized vehicle engines proven to heat passenger compartments poorly in the cold start and warm-up phase because their heat output is inadequate but their pollutant emission, specific fuel consumption and engine wear also surpass that of engines at operating temperature. Hence, so-called auxiliary heaters, fuel-powered heating units that compensate for the heat deficit, are increasingly being implemented in diesel engines in particular [14-6].

14.2.2.2 Cogeneration

General

The simultaneous recovery of mechanical energy and accumulating heat leads to the principle of cogeneration, which aims to recover a maximum of the primary energy supplied by the fuel and thus conserve energy resources as well as take advantage of the attendant reduction of combustion products to reduce the emission of pollutants.

The use of cogeneration plants is both economically and ecologically advantageous.

Combined Heat and Power Stations

According to VDI Guideline 3985 [14-7], combined heat and power stations (CHPS) are cogeneration plants with combustion engines or gas turbines, which generate power and effective heat simultaneously.

A combined heat and power station (Fig. 14-4) consists of one or more CHPS modules with the auxiliary equipment necessary for operation, the related switching and control equipment, noise protection measures, exhaust gas outlets and the appropriate installation space.

The basic unit of a combustion engine CHPS, a CHPS generating set consists of a combustion engine as the generator of mechanical and thermal energy, a generator as the converter of mechanical energy into electrical energy and power transmission and suspension elements. Together with the heat exchange components, the control and monitoring systems, the intake and exhaust system, the lubricating oil and fuel system and the safety systems, it forms a CHPS module.

While, CHPS generating sets for large outputs are usually delivered to construction sites and all other components are custom finished for the systems, compact modules (Fig. 14-5) containing every component including the primary exhaust silencer are usually employed for smaller outputs.

Combined heat and power stations are implemented in municipal facilities such as hospitals, swimming pools and schools and in industries and businesses as well as in office and residential buildings. Their electrical output extends from the low kilowatt to two digit megawatt range. While gas turbines are frequently implemented as drive engines when outputs are large and CHPS heat is recovered to generate steam, internal combustion engines are used when units and outputs are smaller. In addition to diesel engines and dual fuel engines, spark ignited engines are predominantly employed because their exhaust gas emits few pollutants (see Sect. 4.4).

In the context of the use of renewable energy sources, vegetable oils, particularly rape oil or rape oil methyl ester (RME), are available as fuel for diesel engines and dual fuel engines [14-8].

A number of laws passed in the German Bundestag have further improved the cost effectiveness of CHP plants in Germany.

The law introducing the Ecological Tax Reform exempts CHPS with outputs of up to 2 MW from the electricity tax and the mineral oil tax when their annual utilization ratio is at least 70%. The Act on Promoting the Generation of Electricity from Renewable Energy Sources establishes lucrative feed-in

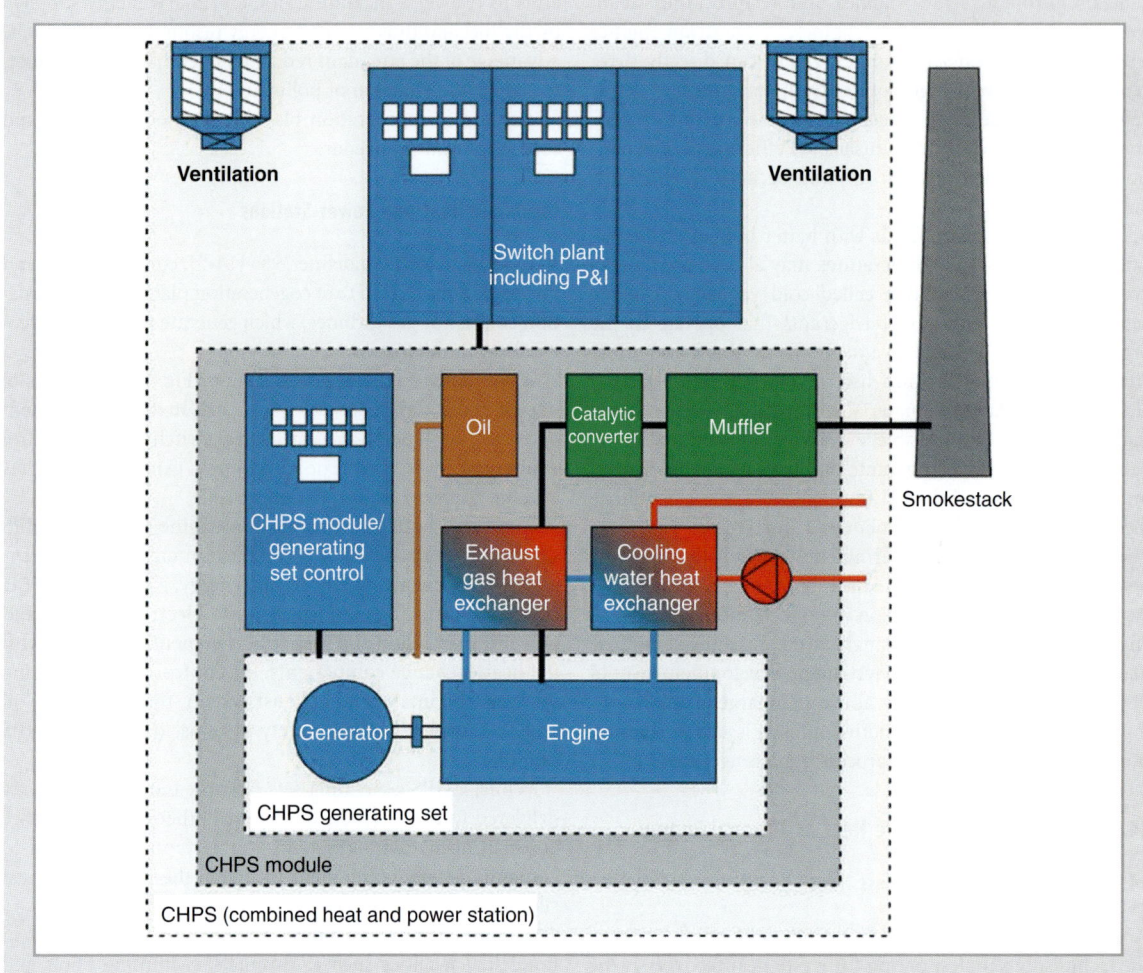

Fig. 14-4 Definition and specification of CHPS components according to DIN 6280

tariffs. A bonus in addition to the feed-in tariff is also awarded when fuels made from renewable raw materials are used.

Combined heat and power generation in CHPS indisputably saves a more substantial share of primary energy than the separate generation of power in a power plant and heat in a boiler. In terms of NO_X and CO_2, the input of pollutants into the atmosphere is lower than the separate generation of power in a thermal power plant and heat in a boiler.

However, the prerequisite for this is the implementation of appropriate measures that reduce NO_X emissions. The limits of the Technical Instructions on Air Quality Control (*TA Luft*) have to be observed when combined heat and power stations with thermal firing capacities above 1 MW are operated in Germany (see Sect. 15.2). While in-engine measures (lean burn systems) help spark ignited engines considerably undershoot the limits that are already lower than for diesel engines, exhaust emission control systems are essential for the operation of diesel and dual fuel engines in the majority of cases.

An economically and ecologically trendsetting concept, a combined heat and power station consisting of four CHPS modules powered with the renewable fuel rape oil methyl ester supplies the Reichstag building in Berlin with power.

Consequently, the potential savings in CO_2 is taken advantage of in two ways, namely by burning a renewable fuel known to have a good CO_2 balance as well as by applying the principle of cogeneration with its inherent potential to save primary energy and thus emit little CO_2.

With 400 kW of electrical power output apiece at electrical efficiencies of 42.5%, the four CHPS modules recover 90% of the primary energy. Heat is decoupled in an HT circuit (engine cooling water – heat + oil cooler heat + exhaust gas heat) at 110°C and an LT circuit (charge air cooling energy) at 40°C.

The units were provided with a system for exhaust gas aftertreatment to – as the client demanded – minimize not only CO_2, which is admittedly nontoxic but damaging to the earth's atmosphere, but also exhaust gas pollutants formed

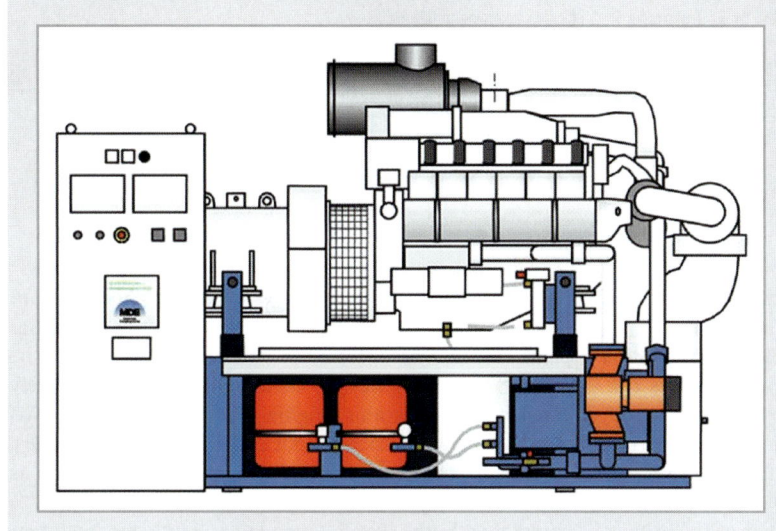

Fig. 14-5
Compact design of a CHPS module

during combustion. The nonvolatile particles are separated out of the exhaust gas in a particulate filter by means of catalytically coated filter cartridges. An SCR catalyst consisting of coated honeycombs reduces the NO_X emission by spraying in urea and an additional oxidation catalyst reduces the emission of carbon monoxide and hydrocarbons.

Thus, the mandated emission values, which are far lower than in *TA Luft* (see Sect. 15.2), are observed:
- soot (particulates) < 10 mg/m^3
- particulate matter < 20 mg/m^3
- nitrogen oxides $NO_X < 100$ mg/m^3
- carbon monoxide CO < 300 mg/m^3
- hydrocarbons HC < 150 mg/m^3

(According to *TA Luft*, based on the standard level (273.15 K; 101.3 kPa) after deducting the moisture content in exhaust gas with an oxygen content of 5%, 4,000 mg/m^3 of NO_X were permissible at the time of construction!). Even though experiences are available from a large number of successfully operating CHP plants, the expedience of CHPS implementation and the most promising concept have to be evaluated in every individual case.

Resource conservation and environmental relief alone rarely suffice to sway an operator to make a positive buying decision. Therefore, like other capital goods, the cost effectiveness of CHP must also be subjected to analysis. However, a plant must be designed before a feasibility study is performed. The current situation of the energy flows, the energy supply and procurement contracts must be reviewed. This requires detailed data on the power and heat demand over time. The operating hours of the CHPS modules can be determined by entering module power based on annual load duration curves and daily and weekly load curves [14-9].

Naturally, the mode of CHPS operation is also entered. Is it operated in heat driven, power driven or alternating mode? Is running it at peak load advantageous based on the power procurement contract? What does the energy demand forecast look like?

Once the concept has been established, a feasibility study can be performed, applying methods of investment mathematics familiar from capital expenditure budgeting. Along with the annuity method, the net present value method has particularly proven itself.

The curve of present value over years of useful life not only delivers the amortization period but also the return at the end of the useful life. Along with information on cost effectiveness, the amortization period, which specifies the time span between the time of investment and the time at which the higher capital investment is recouped through savings of energy costs, represents an important parameter for the assessment of the financial risk. The shorter the amortization period, the lower the investment risk is. Both information on the risk and data on the achievable surplus provide support for a decision on the appropriate concept.

Parameters of CHP Plants

Part 14 of the standard DIN 6280 defined important parameters in order to establish uniform rules for the many terms employed over the years, (see Table 14-3) [14-10]. Issued in August 1997, the standard applies to combined heat and power stations (CHPS) with reciprocating internal combustion engines, which generate alternating current and useful heat.

Part 14 of DIN 6280 defines utilization ratios similarly to efficiencies. The generated energy (electrical, thermal and

Table 14-3 Definition of CHPS efficiency according to DIN 6280

Electrical efficiency	η_{el}	Ratio of the true electrical power output generated to the heat input from the fuel supplied based on the calorific value (Hu)
Thermal efficiency	η_{th}	Ratio of the generated thermal power to the heat input from the fuel supplied based on the calorific value (Hu)
Overall efficiency	η_{ges}	Sum total of the electrical efficiency and thermal efficiency. The overall efficiency does not account for the power for the auxiliary drives.

total) is set in relation to the thermal energy of the quantity of fuel supplied relative to the calorific value (H_u) within a longer period (e.g. one year). Unlike efficiencies, utilization ratios also incorporate the energy of auxiliary drives (e.g. pumps and fans) and downtime losses.

Each of the measured or specified efficiencies (see Table 14-4) depend on the operating state of the CHPS (rated load or part load, speed, cooling water temperature, charge air temperature, cooling temperature of the exhaust gas, etc.).

All parameters, efficiencies and utilization ratios for a particular unit (generating set, module or CHPS) relate to a defined system boundary, e.g. current output at the generator terminals, hot water inlets and outlets in the module and charge air cooling water inlets and outlets in the module. These boundary conditions are indispensible for the classification of the parameter data.

While the measurements or specifications of efficiencies relate to constant operating conditions, startup and shutdown processes, part load operation and downtimes are also incorporated in the data of utilization ratios. Consequently, the planning and design of a complete plant and the mode of operation selected by the operator substantially influence a plant's utilization ratios. Hence, it is impossible to draw any conclusions about a plant's quality from the utilization ratios.

The desire to prolong the operating hours of emergency diesel units by attaching process equipment to supply heat from cogeneration or to even run them in continuous operation has been expressed repeatedly as combined heat and power stations with gas engines have spread. This is absolutely inadvisable since diesel emergency power plants are solely designed for emergency use. This pertains to engines engineered for such operation (higher output at low operating hours) and every other component not intended for continuous operation. An emergency power plant can only be operated in parallel operation with the grid or as an emergency power supply system after relatively extensive modifications (engine, control, etc.) or a reconfiguration of the electrical output and the attachment of every component necessary to supply heat from cogeneration as well as the exhaust gas aftertreatment system needed to comply with the legally mandated limits for pollutant emissions.

Diesel Engine Heat Pump

A diesel engine heat pump is a compression heat pump with a compressor driven by a diesel engine. Figure 14-6 is a schematic of the principle of diesel engine heat pump design. Piston compressors, screw compressors and turbocompressors are the types of compressors predominantly implemented to compress the working medium (refrigerant). The compressed and simultaneously heated working medium flows into a condenser where it emits useful or heating heat at a high temperature level. After the pressure in the throttle drops, the working medium in the evaporator absorbs energy in the form of unusable heat from the ambient air, river water, salt water or other low temperature heat sources.

In accordance with the principle of a heat pump, the addition of compressor work converts low temperature heat into high temperature heat.

A ratio of the heat output $\dot{Q}_c$ emitted in the condenser to the compressor input power P_v where $P_v \equiv P_e$ applies, the heat coefficient of performance

$$\varepsilon = \dot{Q}_c / P_e$$

serves as a parameter for the evaluation of the heat pump process.

Dependent on the temperature of the heating heat and the low temperature heat from the refrigerant fed to the evaporator, the coefficients of performance for heat pumps commonly implemented for residential or process water heating have values between 2 and 4.

However, a diesel engine heat pump also recovers the thermal energy emitted by the engine. It may either be supplied to the heating circuit after the condenser to increase the

Table 14-4 Ranges of common CHPS efficiencies

Electrical efficiency	η_{el}	25–48%
Thermal efficiency	η_{th}	30–56%
Overall efficiency	η_{ges}	65–92%

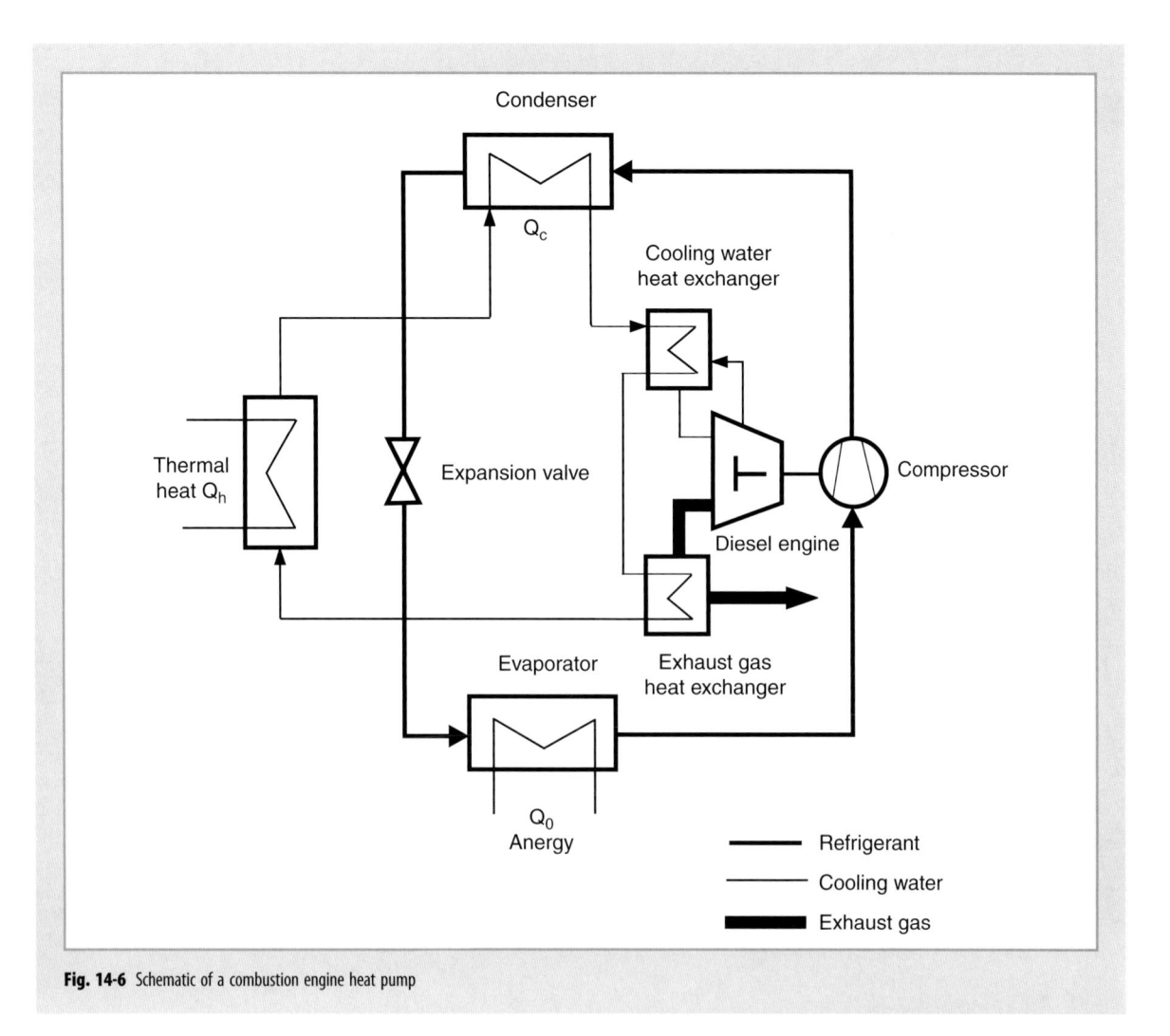

Fig. 14-6 Schematic of a combustion engine heat pump

flow temperature or drawn into a second separate circuit to supply other consumers.

A ratio of the total useful heat output $\dot{Q}_N$ to the energy input of the fuel P_B based on the calorific value (H_u), the heat factor

$$\zeta = \dot{Q}_N/P_B$$

serves as a parameter for the evaluation of the overall process of a diesel engine heat pump.

The heat factor can also be calculated from the relationship

$$\zeta = \varepsilon\eta_e + \eta_a,$$

i.e., from the coefficient of performance, the engine efficiency and the percentage of useful engine waste heat relative to the fuel power supplied.

Dependent on the temperature of the heating heat and the low temperature heat from the refrigerant supplied to the evaporator and the corresponding engine data (see also Table 14-2), the heat factors for heat pumps commonly implemented for residential or process water heating have values between 1.5 and 2.

Figure 14-7 presents the energy balance of a diesel engine heat pump system designed with an engine power of 250 kW and a heat output of 1,085 kW.

Heat pumps utilized for residential and process water heating emit heating heat with temperatures of 70°C/50°C (forward flow/return). Groundwater with a temperature of 10°C is used as the energy source. When the water is cooled down by 4 K, evaporator temperatures of +1°C can be run at full load and 4–5°C at part load operation. Crucial for the

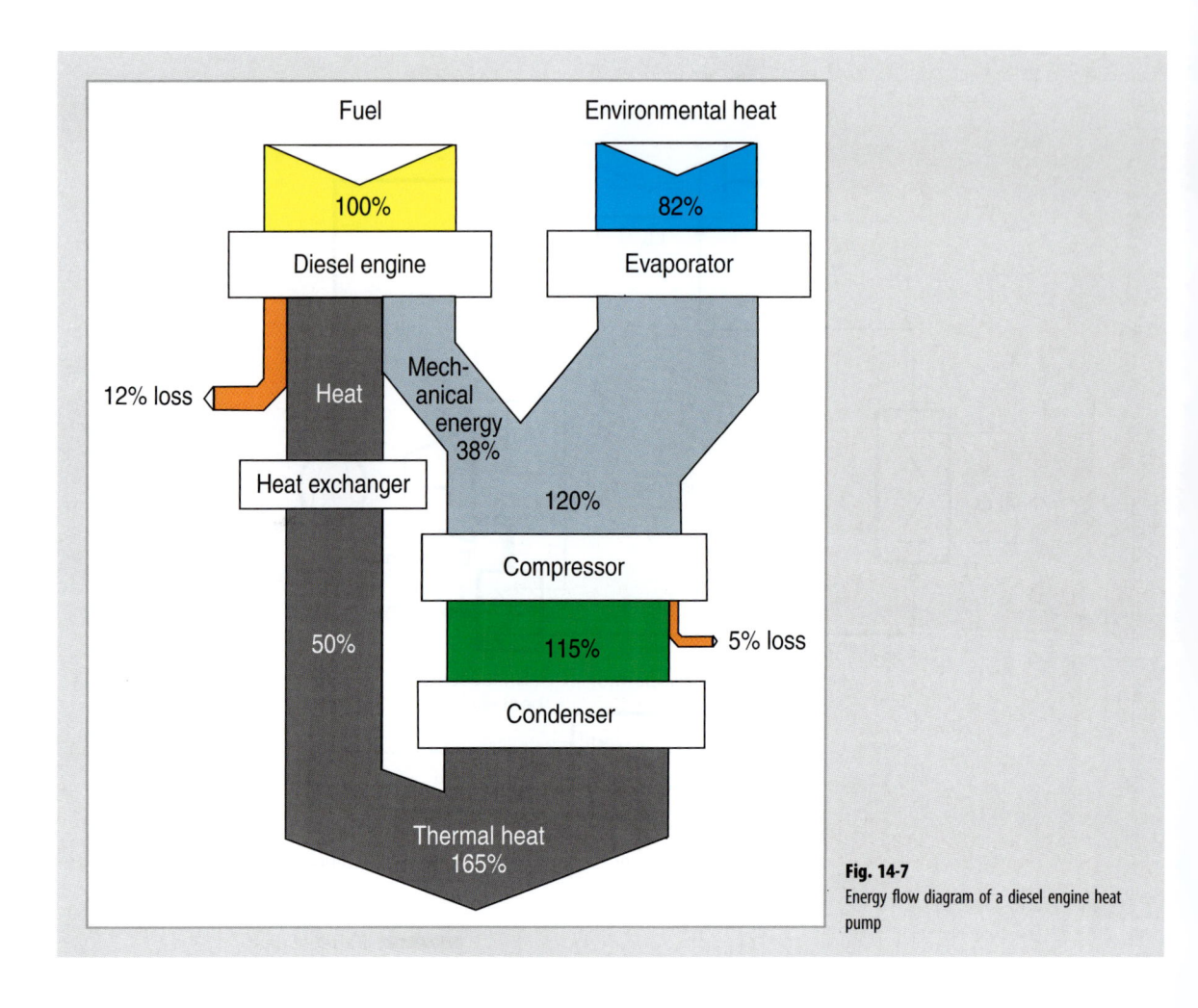

```
                    Fuel              Environmental heat

                   100%                       82%
              ┌──────────┐              ┌──────────┐
              │Diesel engine│           │ Evaporator │
              └──────────┘              └──────────┘
                         Mech-
  12% loss ←   Heat     anical
                         energy
                          38%
              ┌──────────┐
              │Heat exchanger│                120%
              └──────────┘
                           ┌──────────┐
                           │Compressor│
                           └──────────┘
                 50%          115%       → 5% loss
                           ┌──────────┐
                           │ Condenser │
                           └──────────┘

                     Thermal heat
                         165%
```

Fig. 14-7
Energy flow diagram of a diesel engine heat pump

heat pump's coefficient of performance, a temperature lift of 50–59 K takes effect between refrigerant evaporation and condensation at condenser temperatures of 60 °C at full load; values are correspondingly lower at part load operation.

Compressor Module

The operation of compressors driven with internal combustion engines furnishes another option for direct waste heat recovery. The units employed to generate compressed air are predominantly screw compressors and turbocompressors.

The design of a compressor module is comparable to a combined heat and power station module with a compressor instead of a generator.

As the energy flow diagram in Fig. 14-8 indicates, 87% of the primary energy may be utilized with a diesel engine compressor module in the 400 kW performance class.

14.2.2.3 Trigeneration

In the majority of cases of application, CHP plants are designed for heat driven operation, i.e. the output is specified so that CHPS heat can be recovered the entire year if possible. This results in high operating hours and guarantees cost effective operation. In cases of application with heat recovery for residential heating systems, this leads to CHPS output designed more for the lower heat demand arising in summer. Hence, planning should entail verifying that cooling is not required in the summer months instead of heat. Administrative buildings with larger EDP facilities, hospitals, hotels, shopping centers, etc. especially require cooling and air conditioning in many cases. A plant based on the principle of combined heating, cooling and power (CHCP) can definitely be operated cost effectively when the annual demand for electrical energy, thermal heat and cooling energy resembles the curve in Fig. 14-9.

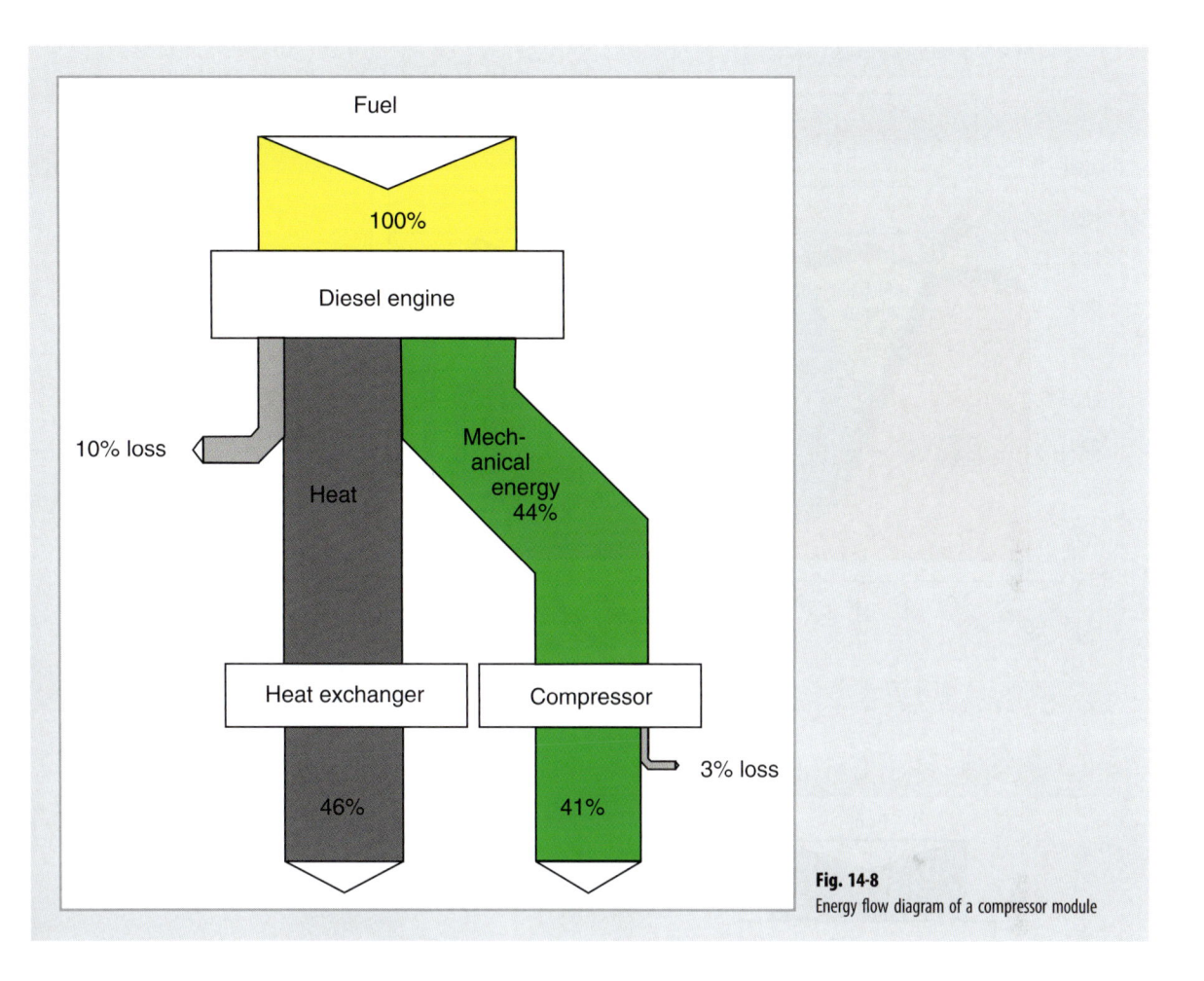

Fig. 14-8 Energy flow diagram of a compressor module

A plant based on the principle of trigeneration consists of one or more combined heat and power station modules coupled with a sorption refrigerator. For the most part, absorption refrigeration units are utilized. Adsorption refrigeration units are also utilized in some cases [14-11].

Mainly water is used as the refrigerant and lithium bromide as the solvent in air conditioning systems (cold water circuit 12°C/6°C). Mainly ammonia is used as the refrigerant and water as the solvent in cooling systems below 0°C.

Figure 14-10 presents the energy balance of a CHCP plant with an absorption refrigeration unit at CHPS heating water temperatures of 95°C/85°C, a cooling tower temperature of 27°C for absorber cooling and cold water temperatures of 6°C/12°C.

The advantage of any sorption unit is its low wear (only a few moving parts) and thus the low maintenance required, its good part load characteristics with infinitely variable load control and its low noise emission. Since they do not require fluorochlorohydrocarbon refrigerants that contribute to the greenhouse effect and adversely impact the ozone layer

sorption units, also have advantages over compression refrigeration units in terms of environmental compatibility.

Examining the expedience of a cogeneration or trigeneration plant and identifying the most promising concept is fundamentally necessary in every individual case.

Ecological benefits constitute a sufficient basis for a decision to build a plant in only the rarest of cases. The investment required makes meticulous plant planning and a detailed evaluation of cost effectiveness essential.

14.2.3 Concluding Remarks

Although established types of mechanical and thermal energy recovery from diesel engine waste heat have often failed to satisfy the criteria of cost effectiveness in the past, these technologies are increasingly attracting attention again as fuel prices continue to rise. Moreover, the high level of fuel prices expected in the future too will also advance the use of renewable energy sources in combustion engines. This will be financially advantageous, not least because of lower taxes. Particularly from the

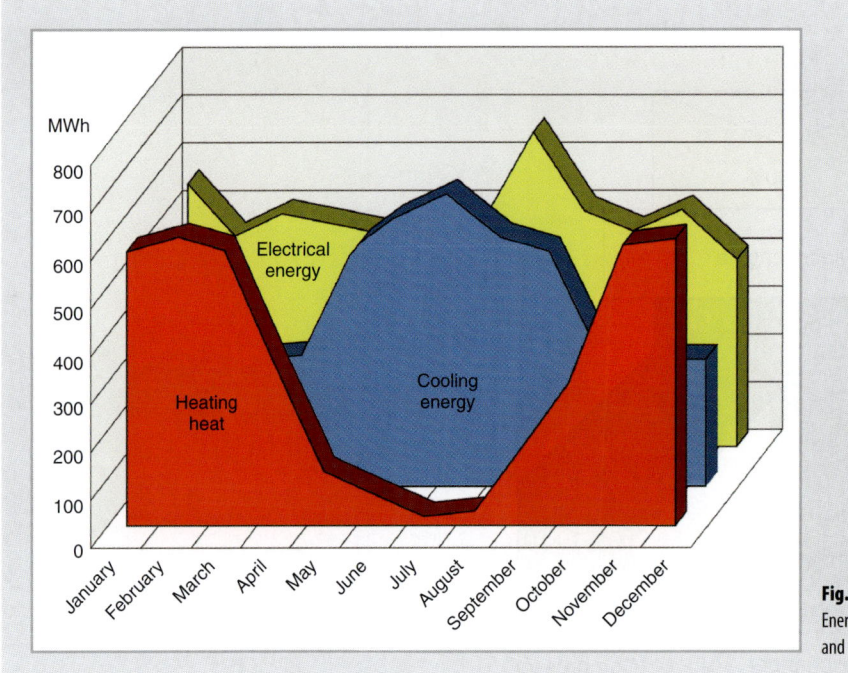

Fig. 14-9
Energy profile of electrical energy, heating heat and cooling energy

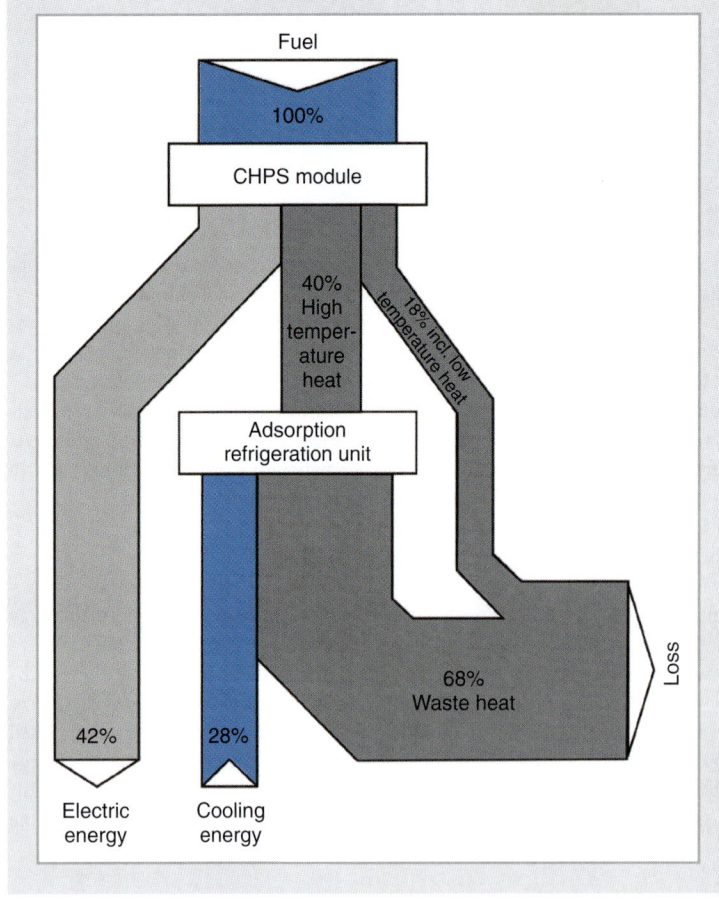

Fig. 14-10
Energy flow diagram of a trigeneration plant

perspective of conserving fossil energy sources and protecting the environment by lowering the input of CO_2 into the atmosphere, the recovery of diesel engine waste heat, especially in combination with the use of biogenic energy sources, will continue to grow in importance in the future.

Literature

14-1   Pflaum, W.: Mollier-Diagramme für Verbrennungsgase Part I and II. 2nd Ed. Düsseldorf: VDI-Verlag 1960, 1974

14-2   Gneuss, G.: Arbeitsmedien im praktischen Einsatz mit Expansionsmaschinen. VDI- Berichte No. 377. Düsseldorf: VDI-Verlag 1980

14-3   Gondro, B.: Forschungsbericht 03 E-5373-A des BMFT. October 1984

14-4   MAN B&W Diesel A/S, Copenhagen: Thermo Efficiency System (TES) for Reduction of Fuel Consumption and CO2 Emission

14-5   Spiegel Online: Turbosteamer Heizkraftwerk im Auto. December 14, 2005

14-6   Lindl, B.: Kraftstoffbetriebene Heizgeräte für das Wärmemanagement in Fahrzeugen. ATZ 105 (2003) 9

14-7   VDI-Richtlinie 3985: Grundsätze für Planung, Ausführung und Abnahme von Kraft-Wärme-Kopplungsanlagen mit Verbrennungskraftmaschinen. (1997) 10

14-8   Ortmaier, E.; Hirschbichler, F.: Regenerative Energieträger als Brennstoff für BHKW. VDI-Berichte 1312, Düsseldorf: VDI-Verlag 1997

14-9   Hirschbichler, F.: Auslegung eines Blockheizkraftwerks. Fachzeitschrift der Deutschen Mineralbrunnen (1998) 2

14-10 DIN 6280: Blockheizkraftwerke (BHKW) mit Hubkolben-Verbrennungsmotor. Part 14: Grundlagen, Anforderungen, Komponenten und Ausführung und Wartung. Part 15: Prüfungen. (1995) 10

14-11 Wärme macht Kälte. Kraft-Wärme-Kopplung mit Absorptionskältemaschinen. ASUE, Arbeitsgemeinschaft für sparsamen und umweltfreundlichen Energieverbrauch e.V. ASUE-Druckschrift No. 190990

Part IV Environmental Pollution by Diesel Engines

15 Diesel Engine Exhaust Emissions 417
16 Diesel Engine Noise Emission 487

15 Diesel Engine Exhaust Emissions

Helmut Tschoeke, Andreas Graf, Jürgen Stein, Michael Krüger, Johannes Schaller, Norbert Breuer, Kurt Engeljehringer, and Wolfgang Schindler

15.1 General Background

The direct release of exhaust gas components from combustion processes into the environment, i.e. *emission,* is the primary and most important process in the chain of emission, *transmission, pollutant input* and *impact*. Naturally, a basic distinction is made between emissions from vegetation, oceans, volcanic activity or biomass decomposition for instance and anthropogenic emissions, i.e. emissions caused or influenced by humans, from power generation, traffic, industry, households and farming for instance. The following deals exclusively with anthropogenic emissions from diesel engine combustion processes. Figure 15-1 depicts the functional chain with the major anthropogenic sources [15-1]. Exhaust gas components may be harmful or harmless as well as gaseous, liquid or solid.

Figure 15-2 is an overview of anthropogenic emissions and their sources in Germany.

Influenced by topography, climatic conditions, temperatures, moisture and air movements, exhaust gas components are diluted and subjected to physicochemical reactions and dispersed over large distances by atmospheric transport (*transmission*).

Pollutant input (air quality) is the concentration that ultimately manifests itself after being transmitted to a particular location where measurements are taken, e.g. at an intersection. Pollutant input is the contamination of humans or nature engendered by exhaust emission and transmission.

Impact denotes the consequences of *pollutant input* on the environment, organisms or goods. The complex processes of emission transport cause pollutant input readings to fluctuate daily and seasonally. Therefore, there are emission limits on exhaust pollutants, e.g. from vehicles or household heating systems, as well as air quality limits on sulfur dioxide, particulates, lead and ozone [15-2].

The exhaust pollutants produced during combustion processes can be divided into harmless exhaust gas components, which are unavoidable natural products of combustion, and harmful exhaust gas components, which may or may not be subject to limits. Figure 15-3 presents the composition of exhaust gas for ideal and complete combustion with pure oxygen, which only produces carbon dioxide and combustion water in addition to the desired heat that the internal combustion engine converts into mechanical energy. Both components are harmless but relevant to the climate. Aside from the aforementioned harmless components, exhaust gas from combustion with air under ideal conditions only contains nitrogen and additionally oxygen in the case of diesel engines with excess air.

However, real combustion produces other components that are harmful and therefore partly subject to limits, e.g. carbon monoxide CO, unburned hydrocarbons HC, nitrogen oxides NO, NO_2 (NO_X), particulates, sulfur compounds, aldehydes, cyanide, ammonia, special hydrocarbons such as benzene and polycyclic aromatic hydrocarbons such as phenatrene, pyrene and fluorene [15-3].

Figure 15-4 presents the raw emission from real combustion and its composition in percent by weight. Diesel engines have substantially smaller fractions of pollutants than gasoline engines. However, diesel engines also produce additional particulates, i.e. solids (predominantly soot) and components present as condensate, since their mixture formation is inhomogeneous. Their particular composition greatly depends on an engine's operating point. Particulate emission also has to be allowed for in gasoline engines with charge stratification, i.e. inhomogeneous mixture fractions.

All in all, harmful exhaust gas components subject to limits have been reduced dramatically over the past years and projections up through 2020 confirm this trend will continue in the future too. Figure 15-5 clearly indicates that vehicles powered by diesel engines are particularly responsible for nitrogen oxide and particulate emission. Reduction measures will have to concentrate on this.

In light of the projected and highly probable increase of the earth's mean temperature by approximately 1.5–5 °C by the end of this century, greenhouse gases will also be limited in the future. Above all, carbon dioxide CO_2 and methane CH_4 (predominantly from agriculture) are responsible for the

M. Krüger (✉)
Robert Bosch GmbH, Diesel Systems, Stuttgart, Germany
e-mail: michael.krueger2@de.bosch.com

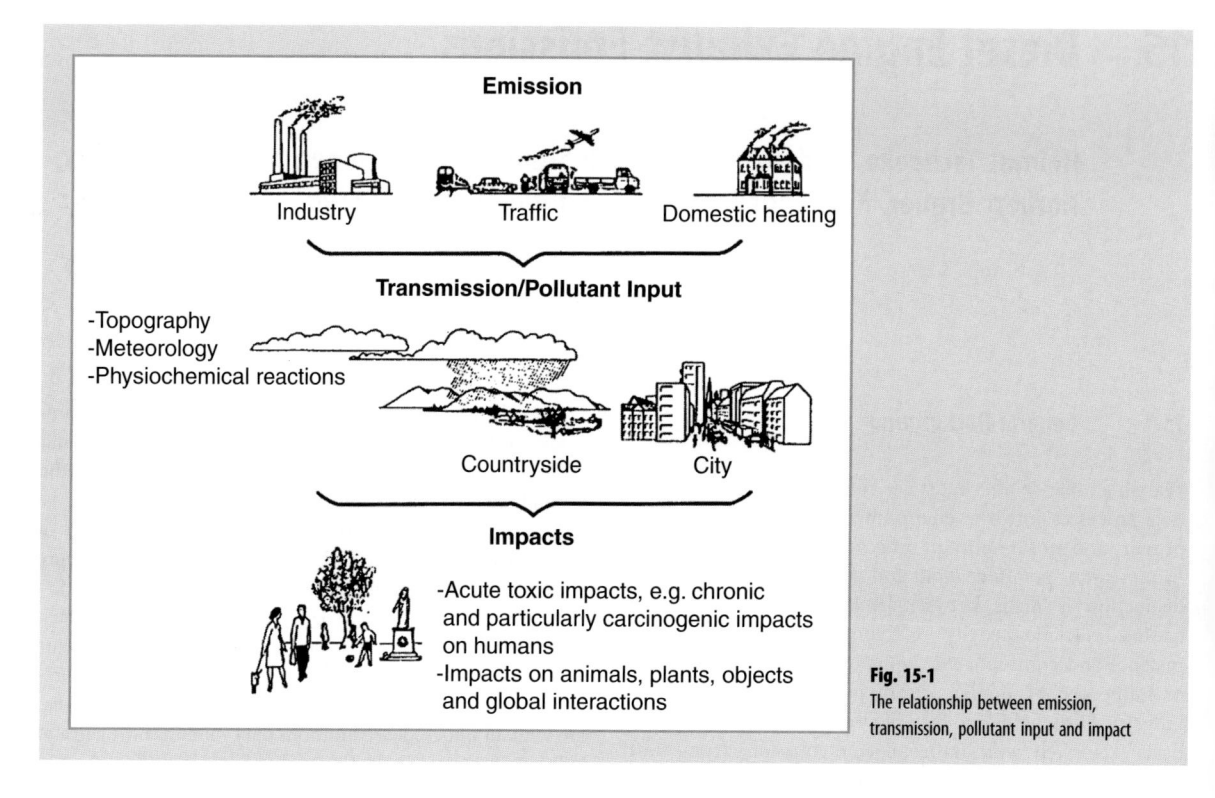

Fig. 15-1 The relationship between emission, transmission, pollutant input and impact

"greenhouse effect". Other greenhouse gases, e.g. fluorohydrocarbons FCKW, nitrogen dioxide N_2O and sulfur compounds, have in part a significantly higher global warming potential (GWP) than CO_2. However, they appear in substantially smaller concentrations and are therefore less relevant.

CO_2 accounts for 65% of the projected temperature increase. Figure 15-6 presents the development of global CO_2 emission of which traffic is estimated to produce between 15 and 20% (see Fig. 15-7).

Along with already existing air quality limits (Climate Conference, Kyoto Protocol and EU laws), intensive political and economic debates and efforts are also establishing limits on the CO_2 emission of motor vehicles. A fleet standard of 120–130 g/km of CO_2 is planned for cars by 2012.

In practice, only an integrated approach, i.e. the interaction of various measures, can effectively reduce harmful or climatically relevant emissions. Such measures may include the:
- provisions of legislation (emission limits in conjunction with the underlying test cycles),
- promotion of environmentally compatible vehicles through incentive systems (tax advantages, increased attractiveness of public transportation, etc.),
- reduction of emissions at their place of origin by improving combustion processes and exhaust gas aftertreatment,
- increase of efficiency of energy conversion,
- use of low emission fuels,
- power train and vehicle improvements (transmission design, road load and aerodynamics),
- power train energy management utilizing optimized drive systems (hybridization of vehicle propulsion),
- dynamic traffic routing (traffic flow control adjusted to the particular volume of traffic) and
- low emission and fuel economizing operation by drivers (driver training).

Transmission and pollutant input interact closely. Influencing factors are:
- local emission,
- road routing,
- development,
- traffic density,
- climatic influences such as
  - wind speed,
  - wind direction,
  - temperature and
  - solar irradiation and
- physicochemical reactions.

The expected air quality in particular geographic locations can be projected on the basis of dispersion models and by incorporating the aforementioned influencing factors. This is important, for instance, when planning new residential areas and routing the connected roads. While carbon monoxide and hydrocarbon emission have now reached a low level, not least because of the introduction of catalytic converter technology in gasoline powered vehicles (three-way catalytic

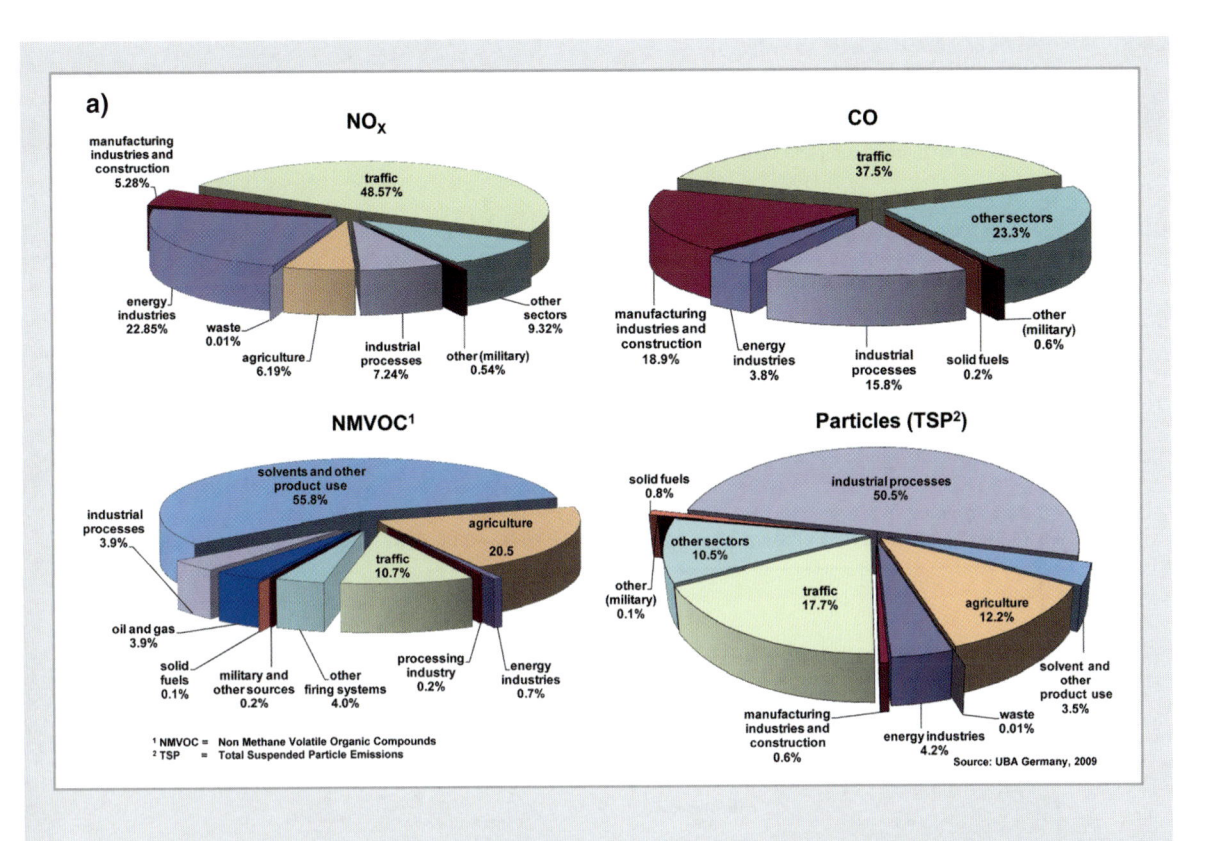

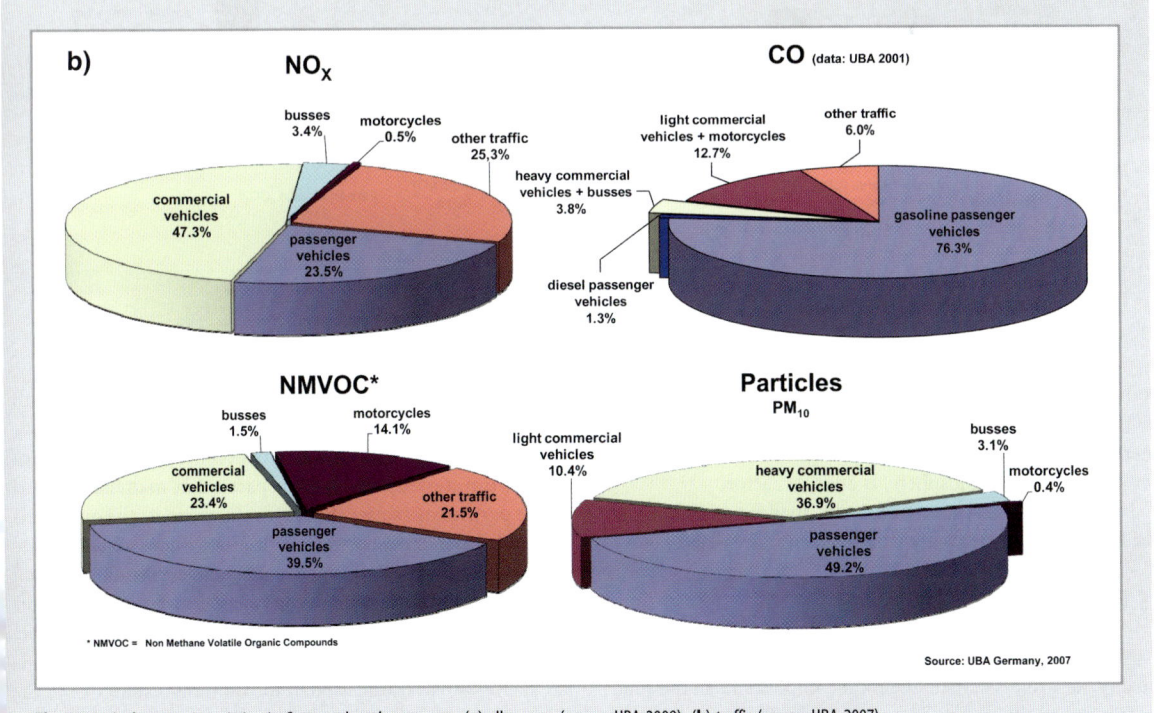

Fig. 15-2 Anthropogenic emission in Germany based on sources: (**a**) all sources (source: UBA 2009); (**b**) traffic (source: UBA 2007)

Composition of Exhaust Gas

Reaction equation when there is complete combustion with pure oxygen:

$$C_nH_{2n+2} + \frac{3n+1}{2}O_2 \Rightarrow nCO_2 + (n+1)\,H_2O + heat$$

Mass balance when there is complete combustion with air:

Gasoline Engine ($\lambda = 1$):

1 kg fuel + 14.7 kg air =
1 kg fuel + 3.4 kg O_2 + 11.3 kg N_2 $\Rightarrow$ 3.1 kg CO_2 + 1.3 kg H_2O + 11.3 kg N_2

Diesel Engine ($\lambda = 3$):

1 kg fuel + 3 · 14.5 kg air =
1 kg fuel + 3 · 3.3 kg O_2 + 11.2 kg N_2 $\Rightarrow$ 3.2 kg CO_2 + 1.2 kg H_2O + 6.6 kg O_2 + 33.6 kg N_2

Fig. 15-3 Exhaust gas components during ideal combustion

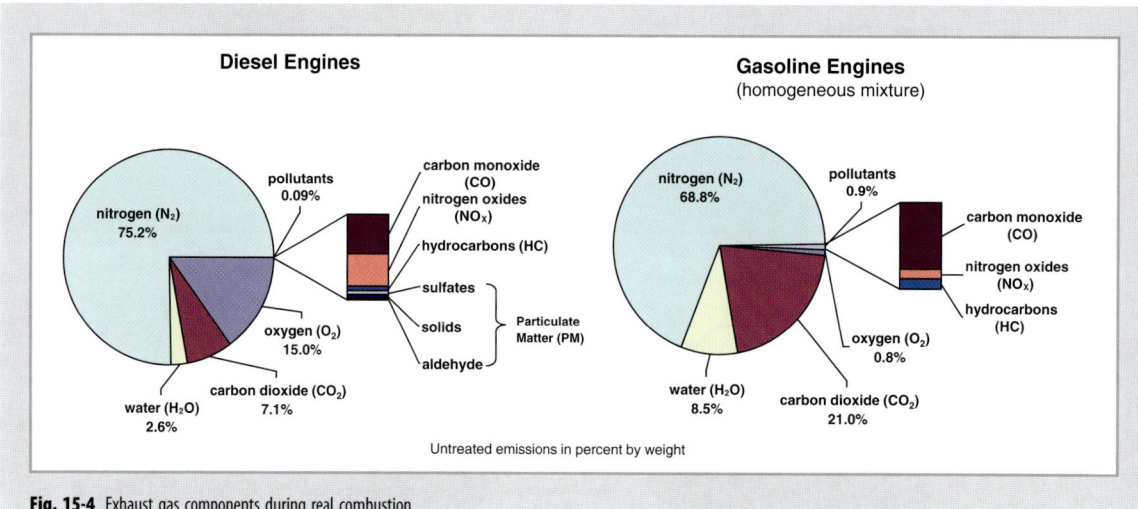

Fig. 15-4 Exhaust gas components during real combustion

converters), particulate emission (fine particulate matter) and NO_X emission remain particularly relevant for cars and commercial vehicles powered by diesel engines. Ozone O_3, which only forms in the atmosphere, is another factor.

Important because they affect air hygiene and humans, these three components are examined in more detail below.

15.1.1 Fine Particulate Matter

Particulate matter is one of the air pollutants that determine air quality. Regardless of their chemical composition, all the solids distributed in the air are subsumed under the terms "particulate matter" or "particulates". Primarily fine particulate matter that is no longer perceptible with the human eye is significant for humans. Such suspended particulates are relevant to health and classified according to their aerodynamic diameter d_{Aero}. The mass of all the particulates with aerodynamic diameters of up to 10 μm contained in the total particulate matter is denoted as fine particulate matter (PM10) and represents the overwhelmingly inhalable proportion of the total mass of particulate matter. Fine particulate matter may have natural (e.g. soil erosion) or anthropogenic origins (i.e. human action). Power and industrial plants, stationary firing plants, metal and steel production and bulk material handling produce fine particulate matter. Road traffic is often the dominant source of particulate matter in urban areas.

Although not applied uniformly in the literature, the following designations are employed:

15 Diesel Engine Exhaust Emissions   421

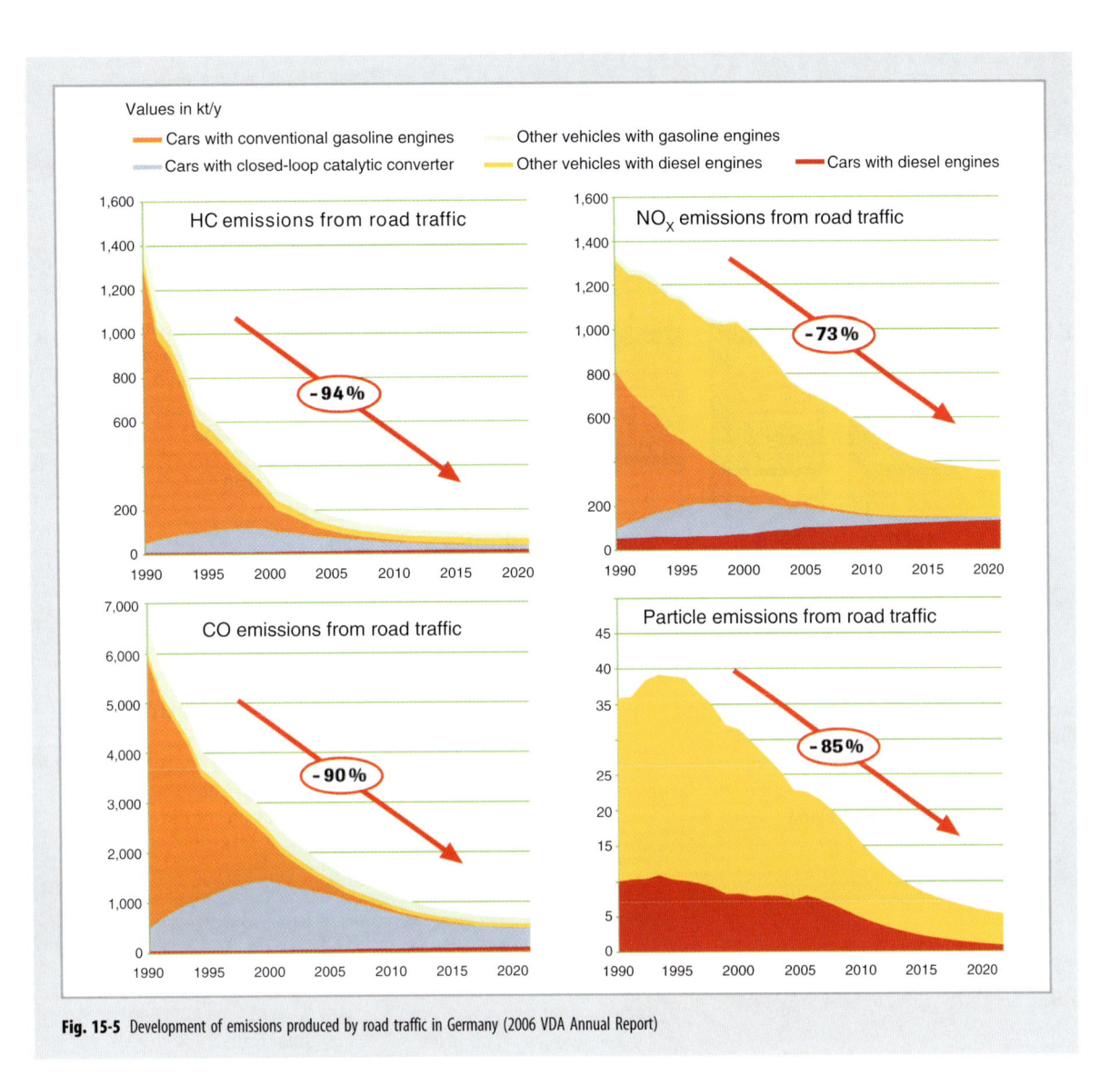

Fig. 15-5 Development of emissions produced by road traffic in Germany (2006 VDA Annual Report)

Legend: Values in kt/y — Cars with conventional gasoline engines; Cars with closed-loop catalytic converter; Other vehicles with gasoline engines; Other vehicles with diesel engines; Cars with diesel engines

Charts: HC emissions from road traffic (−94%); NO$_x$ emissions from road traffic (−73%); CO emissions from road traffic (−90%); Particle emissions from road traffic (−85%)

suspended particulates (total suspended particulates, TSP):	particulates up to a d_{Aero} of around 30 µm (total particulate matter, all airborne particulates)
fine particulate matter (PM10):	particulates that pass through a size-selective air inlet with a 50% filtration rate for $d_{Aero} = 10$ µm
coarse fraction:	particulates larger than 2.5 µm (and smaller than 10 µm)
fine fraction (PM2.5):	particulates that pass through a size-selective air inlet with a 50% filtration rate for $d_{Aero} = 2.5$ µm
ultrafine fraction (PM0.1):	particulates smaller than 100 nm
nanoparticulates:	particulates smaller than 50 nm

The particulate fraction PM2.5 is also designated as respirable fine particulate matter.

The aerodynamic diameter d_{Aero} of a particulate of any shape, chemical composition and density is equal to the diameter of a sphere with the density of 1 g/cm³, which has the same settling velocity in air as the particulate being analyzed.

In the European Union, the air quality framework directive (96/62/EC on ambient air quality assessment and management) concretized by a subsidiary directive (1999/30/EC relating to limit values for sulphur dioxide, nitrogen dioxide and oxides of nitrogen, particulate matter and lead in ambient air) mandates strict limits for air quality. These limits (air quality values) have a legally binding character, i.e. in order to prevent harmful environmental impacts, they may not be

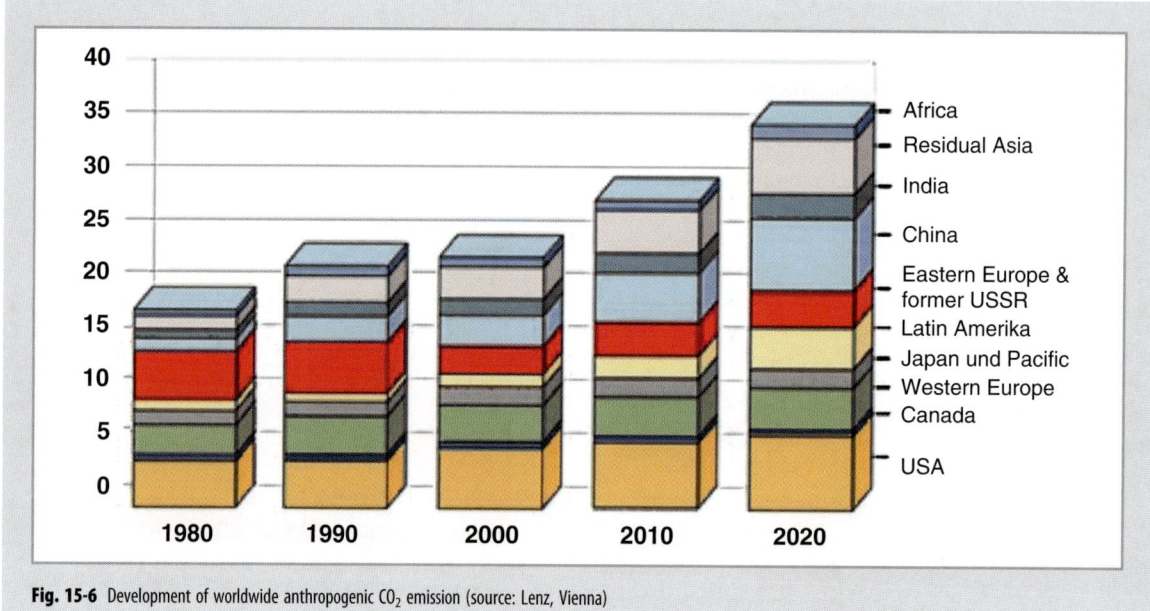

Fig. 15-6 Development of worldwide anthropogenic CO_2 emission (source: Lenz, Vienna)

exceeded. A revision of Directive 96/62/EC and the three subsidiary directives was planned for 2007.

As of January 1, 2005, the limits were:
- 50 µg/m³ for the 24 h mean value of PM10 with 35 excedances allowed per year and

- 40 µg/m³ for the annual mean value of PM10.

Limits as of January 1, 2010 were:
- 50 µg/m³ for the 24 h mean value of PM10 but with only seven excedances allowed per year and
- 20 µg/m³ for the annual mean value of PM10.

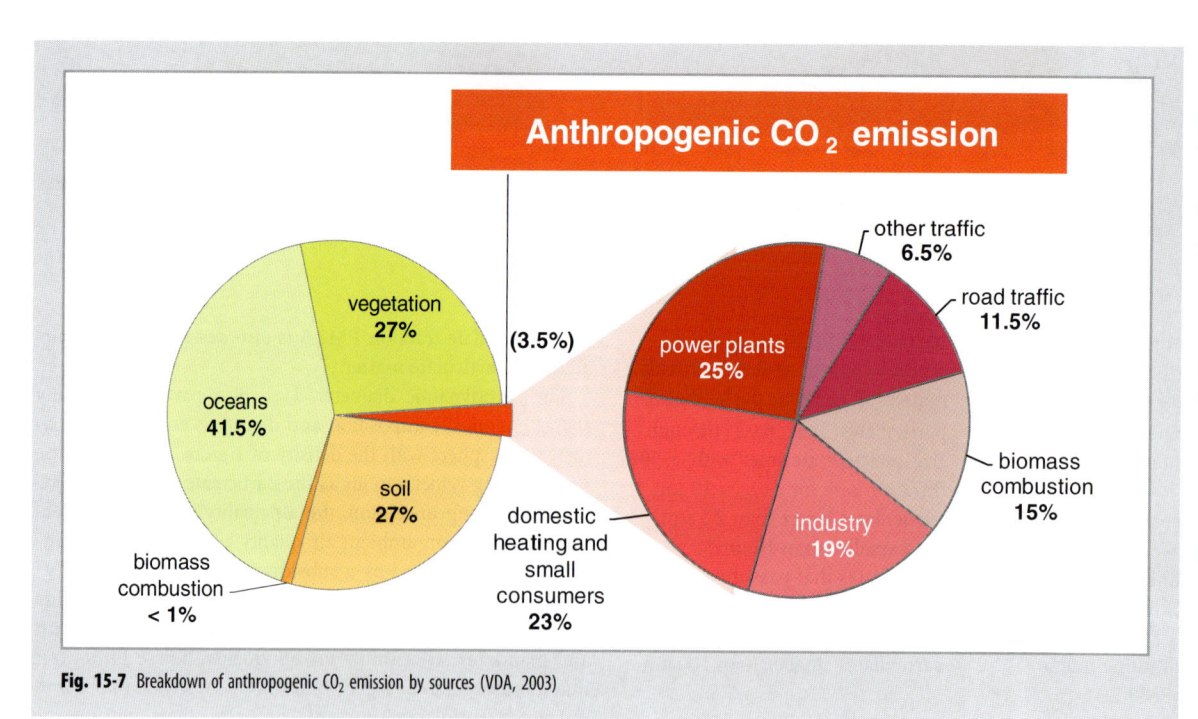

Fig. 15-7 Breakdown of anthropogenic CO_2 emission by sources (VDA, 2003)

The limits mandated in the subsidiary directive are based on WHO studies and are normally below the values of earlier regulations. Thus, the new limits for fine particulate matter (PM10) supersede the previous limits for total suspended particulates (TSP with particulate diameters up to 40 μm). The German 22nd Federal Ambient Pollution Control Act (22. BimSchV) of September 11, 2002 [15-2] implemented the air quality framework directive and the first two subsidiary directives as federal law.

A distinction is usually made between anthropogenic and natural sources. Both sources can be divided into primary and secondary sources.

Primary anthropogenic sources directly produce and release dust particulates and include stationary sources such as combustion plants that supply power (power plants and district heating plants), waste incineration plants, domestic heating (gas, oil, coal and other solid fuels), industrial processes (e.g. metal and steel production, sintering plants), agriculture and bulk material handling.

The dominant sources in urban areas are mobile sources such as road traffic, e.g. diesel commercial vehicles and cars . In addition to soot particulates from exhaust, abraded particles from tires, brakes and clutch linings and stirred up particulate matter have to be factored into road traffic as diffuse emissions. Rail traffic, shipping (with diesel engines) and aviation are other mobile sources with appreciable emissions of particulate matter.

Secondary anthropogenic sources release reactive gases (including sulfur and nitrogen oxides and ammonia), which atmospheric reactions convert into secondary dust particulates. Among others, these include ammonia sulfates and ammonia nitrates, which bind to fine particulates already present in the atmosphere and thus form secondary aerosols.

Natural sources include volcanoes, oceans (salt water aerosols), forest fires and biological organic material (e.g. plant pollen). Such particulates may be transported from their original source region over long distances and thus contribute to long distance transport. The size of the particulates essentially determines their residence time in the atmosphere and potential transport routes. Thus, small particulates may be transported several 1,000 km within a few days. Saharan dust for instance can reach Europe or America depending on the wind direction.

Particulate matter emissions in Germany have dropped dramatically in the last 15 years. However, mainly total particulate matter was measured until 2001. Figure 15-8 plots the development of particulate matter emissions with the following boundary conditions.

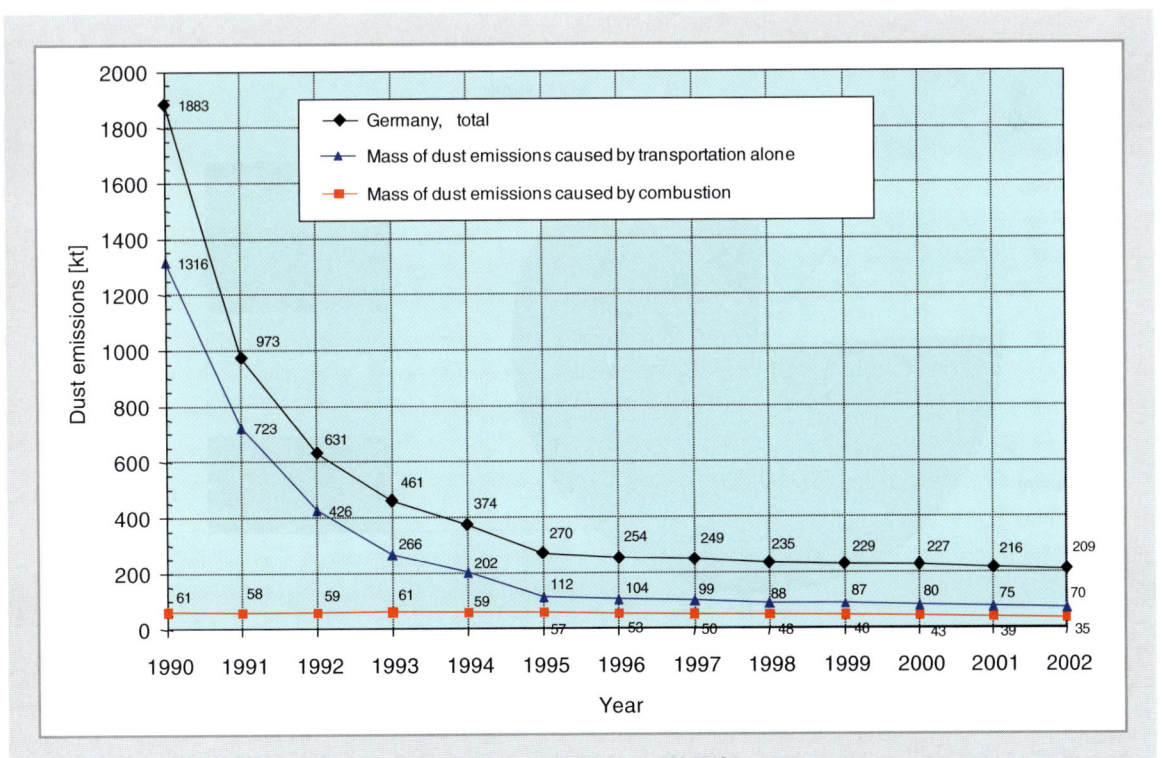

Fig. 15-8 Development of particulate matter emission in Germany from 1990 to 2002 (source: UBA 2004)

- It includes emissions from combustion processes caused and not caused by transportation and from industrial processes and bulk material handling.
- It does not include dust and grit from road surfaces stirred up by road traffic nor the large amount of abraded particles from tires and brakes.

The proportion of diesel soot particulates in fine particulate matter is difficult to estimate. The local percentage can be relatively high in urban regions with a high volume of traffic. Authorities are taking action against increased traffic with high diesel soot emissions. In addition, stricter emission limits (Euro 5 and 6) and tax incentives are stepping up the outfitting of diesel vehicles with diesel particulate filters.

Figure 15-9 presents a source analysis of PM10 in a measuring station in Berlin near heavy traffic. 49% of the total particulate matter load can be attributed to road traffic but only 11% stems from the exhaust of vehicles in local traffic (approximately 8% from trucks and 3% from cars).

Regulations for the reduction of fine particulate matter in conformance with the World Health Organization's (WHO) recommendations were implemented in the wake of various studies that demonstrated they are a health hazard.

Ultrafine particulates constitute only a low percentage of PM by mass but possess a sizeable particulate surface area because of their large numbers (up to 90%). Harmful substances (e.g. heavy metals or organic materials such as polycyclic aromatic hydrocarbons or dioxins) can bind to them. The soot from a diesel vehicle's exhaust system also consists of ultrafine fractions.

Suspended particulate fractions can act as irritants that cause inflammation whenever they deposit in the human body as foreign bodies. The smaller the particulates, the deeper they are able to penetrate the respiratory tract. Particulates that are over 10 μm in size hardly get past the larynx and only a small fraction can reach the bronchial tubes and the alveoli. However, particulates under 10 μm and under 2.5 μm can do so. Ultrafine particulates that are less than 0.1 μm in size can even reach the blood stream through the pulmonary alveolus and disperse in the body through the bloodstream.

Increasing evidence in recent years has indicated that – contrary to earlier assumptions – there is no threshold as of which suspended particulate matter adversely affects health.

Findings from cohort studies indicate that the inhalation of fine particulate matter can be expected to shorten humans' life expectancy [15-5]. Such studies are not undisputed though.

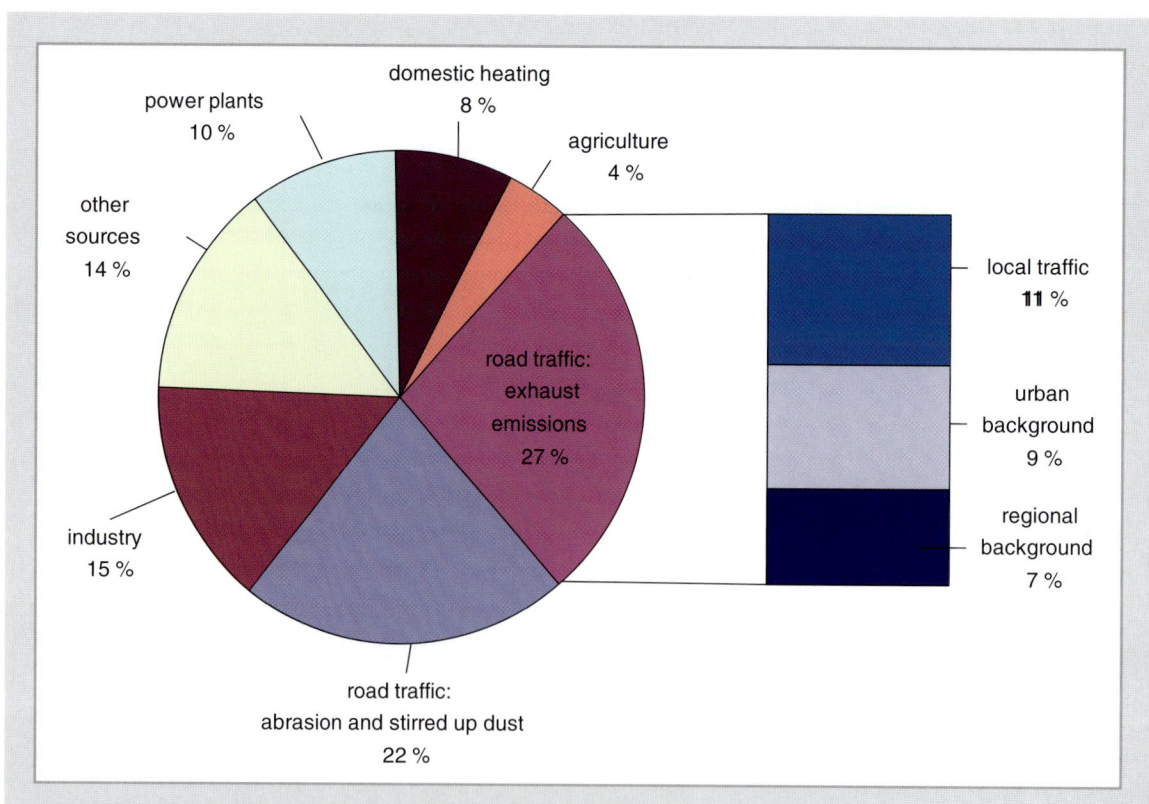

Fig. 15-9 Fine particulate matter loads in a measuring station in Berlin near heavy traffic [15-4]

15.1.2 Nitrogen Oxides NO$_X$

The combustion process mainly produces NO (approximately 60–90%) and little NO$_2$ in the combustion chamber. They are considered a mixture called NO$_X$ but only NO$_2$ is relevant for air hygiene as a pollutant input. Diesel engines systematically produce NO$_2$ from NO in catalytic converters. deNOx systems use it to oxidize and efficiently reduce soot particulates. In air, NO oxidizes to NO$_2$, a gas that irritates mucous membranes and is caustic when combined with moisture (acid rain). It increases asthma sufferers' physical stress, especially when they exert themselves physically. NO$_2$ has a "fertilizing" effect on plants, i.e. it promotes growth.

According to [15-2], a limit concentration of NO$_2$ of 40 µg/m^2 will be in force in the EU in 2010. Figure 15-10 presents the current concentration of NO$_2$, which has remained constant since around 1997 regardless of location. Achieving the aforementioned limits will necessitate considerable efforts in all sectors, i.e. not only in the transportation sector.

15.1.3 Ozone O$_3$

Two domains of ozone activity are distinguished:
a) The ozone layer: Ozone at an altitude of 30–40 km (stratosphere) acts as a filter against harsh UV radiation (UV-C and UV-B) and is therefore essential to life. Increasing thinning of the ozone layer caused by FCKW and halones has been observable over the last 20 years. The O$_3$ concentration over the Antarctic (ozone hole) declines particularly dramatically during southern spring (September and October). Exhaust emissions do not play any part. According to the latest calculations, the ozone hole will have closed again by around 2040–2050.
b) Low level ozone: This is summer smog that forms from atmospheric oxygen influenced by photooxidants (NO$_X$, VOC, CO) when there is strong solar irradiation. It is toxic and, hence, undesired. Since the precursor substances partly stem from motor vehicle exhaust gases, information is passed along to motorists as of certain O$_3$ concentrations and even driving bans are imposed to reduce motor vehicle traffic. On the one hand, nitrogen oxides trigger the formation of O$_3$ but, on the other hand, they also facilitate its disintegration. A higher ozone concentration can actually appear in rural regions than in urban regions with heavy traffic.

Ozone is an irritant gas that alters the transport of oxygen in the human body. It reduces resistance to viral infections, impairs plant growth and affects breathing.

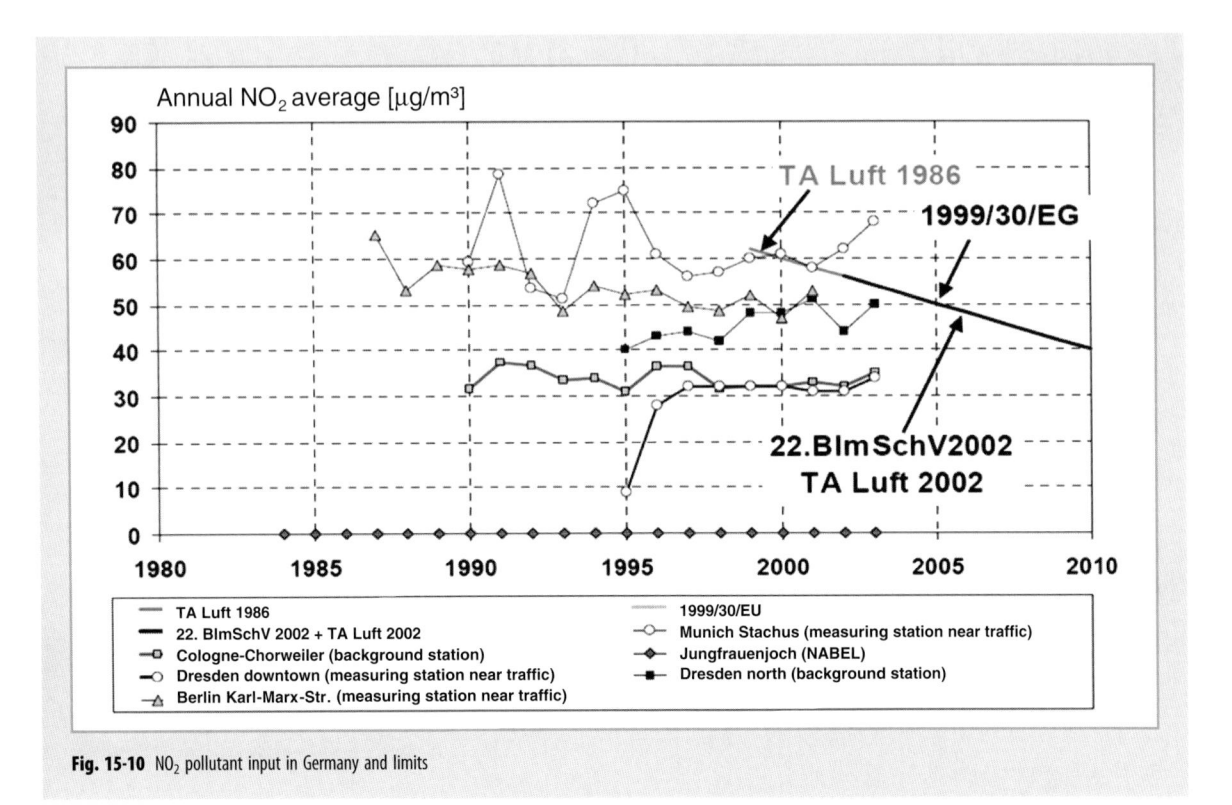

Fig. 15-10 NO$_2$ pollutant input in Germany and limits

15.2 Emission Control Legislation

15.2.1 Emission Control Legislation for Car Engines

The USA issued the first emission control standards for diesel vehicles in the early 1970s. Japan and Europe followed somewhat later. Legislation is still focused on limiting diesel smoke, particulate emission and nitrogen oxide emission. An emission test performed on a chassis dynamometer with standardized driving cycles is intended to deliver quantitative information on vehicle emissions under representative operating conditions. However, different driving cycles for the measurement of exhaust gas and fuel consumption have also been developed in parallel since driving conditions in the USA, Europe and Japan differ. Figure 15-11 presents the major driving cycles.

Although higher nitrogen oxide emissions from diesel vehicles were initially accepted, the development of emission control standards is now heading toward treating gasoline and diesel engines equally, i.e. toward technology-independent limits.

The beginnings of emission control legislation in the USA, Europe and Japan were followed by the development of national emission standards in many other countries and emission control standards of varying stringency now exist in over 30 countries. Since these standards basically rely on American, European and Japanese models, the following remarks concentrate on the standards in these three regions.

15.2.1.1 Diesel Smoke

Initially also meaningful for the evaluation of new vehicles, diesel smoke measurement now only makes sense for tests of vehicles in service since smoke levels have become so low overall.

Basically, two types of diesel smoke tests are employed, full load measurement (4–6 points on the engine's full load curve) and free acceleration (the engine is accelerated to its maximum governed speed). The metrology applied is either light absorption or filter measurement. However, the results of these measurements do not allow direct and clear conclusions

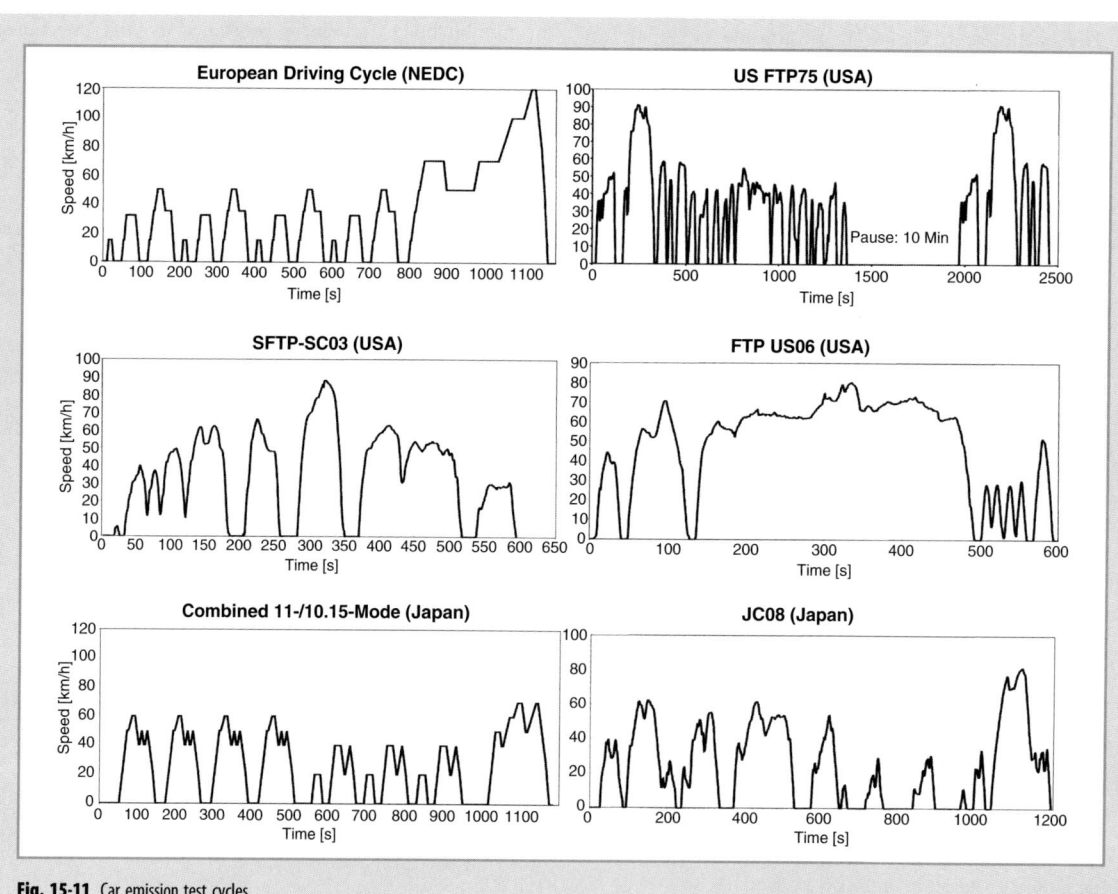

Fig. 15-11 Car emission test cycles

about the level of particulate emission in a chassis dynamometer test. A diesel smoke test is only able to effectively identify high emission vehicles and, in certain cases, also detect tampering with an injection system.

15.2.1.2 Gaseous Emissions and Particulate Emission

New vehicle type approval testing requires the determination of the HC, CO, nitrogen oxide and particulate emissions during the driving cycle on a chassis dynamometer. Moreover, legislation mandates conformity of production testing and, in some countries, the testing of vehicles in use too (vehicles with a kilometrage of up to 100,000 km are currently be subjected to testing in the EU). Type approval testing also requires compliance with the emission limits over a vehicle's useful life. To demonstrate compliance, manufacturers normally have the choice between applying fixed deterioration factors (Table 15-1) or conducting a durability test extrapolating emissions to the useful life by linear regression. In the USA however, the option of fixed deterioration factors only applies to small production runs and must be additionally adapted to the useful life distance. The useful life period is defined differently. Thus, while the USA requires demonstration of proof of compliance over 120,000 miles (193,000 km), Europe requires proof over 160,000 km.

A chassis dynamometer test analyzes the emissions from the diluted exhaust gas flow. Filter disks collect exhaust gas particulates from a partial exhaust gas flow and are weighed on a scale before and after the test. Debate over the potential health hazard of diesel soot particulates has raised questions about the real relevance of particulate mass as a metric when the number of particles may instead be the actual determinant of health risks. Hence, particle number limits will become mandatory for Euro 5 and 6 and thus supplement particulate mass measurement in Europe.

When particulate filter systems were launched on the market, European legislation specified a supplementary test for periodically regenerating particulate filters. Emission and fuel consumption are determined during both the particulate accumulation phase and the regeneration phase. The higher emissions during particulate filter regeneration are weighted in the final result based on the regeneration frequency.

15.2.1.3 American Emission Control Standards

The different emission limit levels in the USA are referred to as tiers, Tier 2 being the most recent. In turn, tiers contain different sets of limits or bins. Manufacturers may choose between several bins but must ensure that the NO_X values of their vehicle fleets do not exceed 0.07 g/mile on average. This greatly limits the selection of bins since bin 5 must be reached on average. Bin 8 is the least stringent. The application of the same emission limits to both diesel and gasoline vehicle is a principal of American legislation. This is a particular challenge for diesel engine emission control (see Table 15-2). In addition, compliance with emission limits in additional test cycles, the *Supplemental Federal Test Procedure* (SFTP), is required.

Given its special situation with regard to climate, population density and traffic density, California has maintained its own emission standards until the present. The NMOG (non-methane organic gases) fleet standard supersedes the NO_X fleet standard in force in the USA. A manufacturer's average NMOG fleet value determines its production volume in the three limit classes, i.e. LEV (low emission vehicle), ULEV (ultra low emission vehicle) and SULEV (super ultra low emission vehicle) (see Table 15-3). California has the most stringent emission limits in the world.

15.2.1.4 European Emission Control Standards

The development of emission control standards in the European Union (EU) only caught up to the stringency of legislation in the USA relatively late with the *Euro 1* emission level. In the meantime, *Euro 4* (see Table 15-4) has been enacted and applied to all new types since January 1, 2005 and all newly approved vehicles since January 1, 2006. Further specifications already exist for Euro 5 (in force as of September 1, 2009) and Euro 6 (as of September 1, 2014). At present, the Euro exhaust levels still apply different limits to gasoline and diesel engines. What is more, cars are differentiated by their allowable total mass. Thus, vehicles with a gross vehicle weight rating of less than 2,500 kg must meet more stringent emission limits than vehicles of 2,500 kg and more. Mainly vans profit from this regulation.

EU emission control standards are proposed by the EU Commission and ratified by its institutions. Components of

Table 15-1 European/Japanese deterioration factors

	CO	HC	NOx	Σ HC+NOx	PM	Useful life
EU (Euro 4)	1.1	–	–	1.0	1.2	80,000 km
EU (Euro 5)	1.5	–	1.1	1.1	1.0	160,000 km
EU (Euro 6)	*	*	*	*	*	160,000 km
Japan	1.2	13	1.0	–	1.2	80,000 km

* To be set later

Table 15-2 Tier 2 bin8/bin5 emission limits for "half useful life" (50,000 miles)

	Emissions in the FTP-75 cycle				
	NMOG [g/mi]	CO [g/mi]	NOx [g/mi]	PM [g/mi]	HCHO [g/mi]
Bin 8	0.100	3.4	0.14	0.02	0.015
Bin 5	0.075	3.4	0.05	0.01	0.015

Table 15-3 Californian emission limits for "half useful life" (50,000 miles)

	Emissions in the FTP-75 cycle				
	NMOG [g/mi]	CO [g/mi]	NOx [g/mi]	PM [g/mi]	HCHO [g/mi]
LEV	0.075	3.4	0.05	0.01	0.015
ULEV	0.040	1.7	0.05	0.01	0.008
SULEV[a]	0.010	1.0	0.02	0.01	0.004

NMOG: non-methane organic gases; HCHO: formaldehyde.
[a] Full useful life (120,000 miles).

Table 15-4 Euro 4/5/6 limits for diesel vehicles

	European vehicle cycle NEDC				
	CO [g/km]	HC + NOx [g/km]	NOx [g/km]	PM [g/kg]	P[a] [#/km]
Euro 4	0.5	0.30	0.25	0.025	–
Euro 5	0.5	0.23	0.18	0.005	6×10^{11}
Euro 6[b]	0.5	0.17	0.08	0.0045	6×10^{11}

[a] Number of particles.
[b] in accordance with Regulation 715/2007/EC.

the legislation, e.g. test procedures, are being adopted from the work of the United Nations' *World Forum for Harmonization* (UNECE) with increasing frequency. One example is the test procedure for vehicles with periodically regenerating exhaust gas aftertreatment systems such as diesel particulate filters. This procedure determines the regeneration factor (k_i) by measuring emissions before, during and after the regeneration of the exhaust gas aftertreatment system (Fig. 15-12). The measured values from the emission test with an unloaded particulate filter are multiplied by the value of k_i.

Recently adopted Euro 5 and future Euro 6 emission control legislation are further tightening the limits on NO_X emissions from diesel engines in particular to close the gap between gasoline and diesel limits. Future levels of EU emission standards may well harmonize them. EU legislative bodies are also contemplating revising the particulate emission limits and there is movement toward not only measuring the quantity of particulate mass but also determining the number of particles emitted.

Any revisions of the test cycle are still open. Statements from legislators to the effect that the current NEDC test cycle is no longer representative seem to suggest they are likely. Such activities may be coordinated with those at the UNECE.

15.2.1.5 Japanese Emission Control Standards

For some time now, major Japanese cities and their metropolitan areas have been suffering from the problem of smog caused by high concentrations of hydrocarbons and nitrogen oxides. Road traffic contributes to this substantially. Hence, the development of emission control legislation has particularly concentrated on reducing NO_X emissions. Along with emission limits for the entire country, the NO_X *Control Law* contains more stringent limits for the approval of diesel vehicles in metropolitan areas. However, revisions of the exhaust emission standard in Japan are also moving toward treating gasoline and diesel vehicles equally. The implementation dates are staggered for Japanese made vehicles and imports. However, this practice may be discontinued in the future. Moreover, two different limit are being defined, one for mass production and another for small lots or low numbers of imports (a maximum of 2,000 vehicles per type and calendar year) (see Table 15-5).

On-Board Diagnostics (OBD)

Legislation requires *on-board diagnostics* that monitors the function of components related to emission during vehicle operation. California required this first with *OBD I* in 1988. The current Californian version is *OBD II*. Euro 3 legislation (*EOBD*) introduced on-board diagnostics in the EU in 2000. The function of OBD is to inform drivers of malfunctions or failures of components relevant to emissions through an optical malfunction indicator. Detected malfunctions are entered in a fault memory and can be retrieved by an external diagnostic tool through a standardized interface. American and European OBD have different monitoring parameters and OBD thresholds at which a malfunction is indicated.

15.2.2 Emission Control Legislation for Commercial Vehicle Engines

15.2.2.1 Test Mode and Emission Limits

The diversity of types of commercial vehicles, their evolution into component trucks and tire and brake wear during chassis

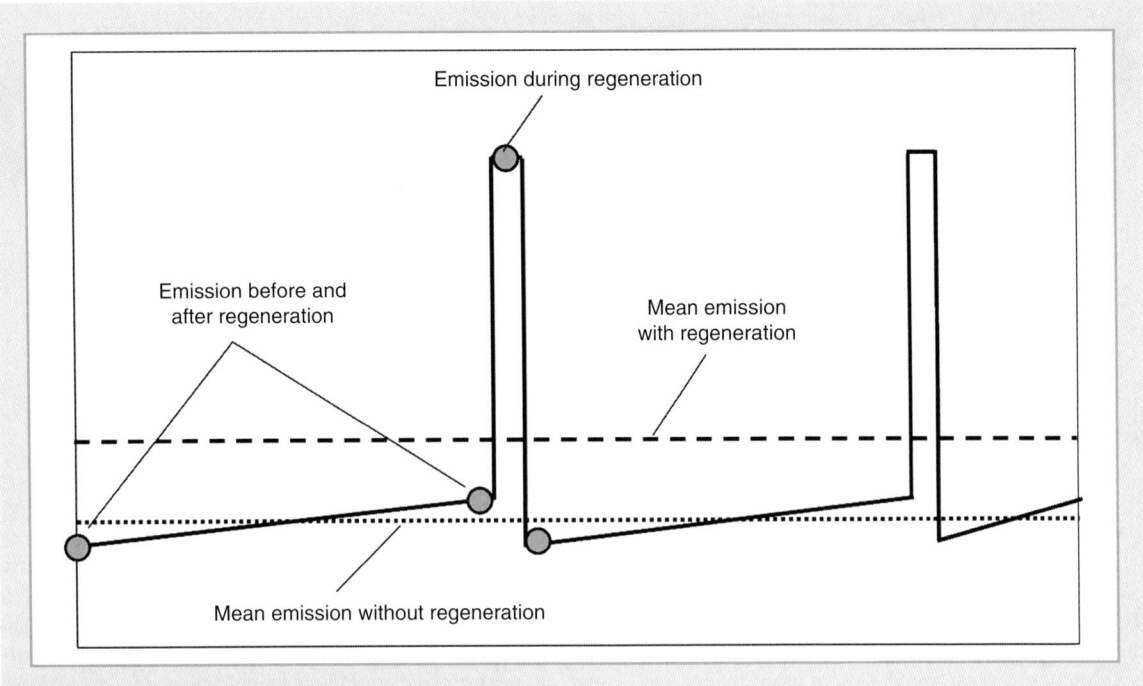

Fig. 15-12 Method for determining emissions from periodically regenerating systems

dynamometer operation gave rise to emission testing for heavy commercial vehicles on engine test benches. The European Union's vehicle classes are listed in Table 15-6.

Since it is impossible to test every engine application profile, an engine must be run through a test cycle that matches the application profile as closely as possible in order to evaluate emissions and keep the time and effort required for engine certification within limits. A test measures the emissions of the following components:

– nitrogen oxides (NO_X),
– hydrocarbons (HC),
– carbon monoxide (CO),

Table 15-5 Japanese diesel car emission limits

		HC [g/km]	CO [g/km]	NO_x [g/km]	PM [g/km]
		Emissions in the 10.15-mode			
Long-term targets	since 2004	0.12 (0.24)	0.63 (0.98)	0.28 (0.43)	0.052 (0.11)
		Emission in the 11/10.15-mode As of 2008: emissions in the JC08/10.15-mode			
New long-term targets	as of Sept. 1, 2007	0.024 (0.032)	0.63 (0.84)	0.14 (0.20)	0.013 (0.017)
		Emissions in the JC08/10.15-mode Emissions in the JC08/JC08-mode (as of 2013)			
Post new long-term targets	as of 2010	0.024 (0.032)	0.63 (0.84)	0.08 (*)	0.005 (*)

Parenthetical values apply to small lots/small volumes of imports up to a maximum of 2000 vehicles per type and calendar year. (*) To be set later.

Table 15-6 Vehicle classes according to Directive 70/156/EEC[1] for diesel-powered vehicles. Test requirements for gaseous emissions and particulates

Transportation of passenger				Transport of goods		
Class	Seats[2]	Gross vehicle weight rating	Test cycle	Class	Gross vehicle weight rating	Test cycle
M_1	≤8	≤3.5 t >3.5 t	NEDC[3] ESC; ELR; ETC	N_1	≤3.5 t	Either NEDC[3] or ESC, ELR; ETC
M_2	>8	≤5 t	ESC; ELR; ETC (NEDC)[4]	N_2	3.5 t < … 12 t	ESC, ELR (NEDC)[4]
M_3	>8	>5 t	ESC; ELR; ETC	N_3	≥12 t	ESC; ELR; ETC

ESC: European Steady State Cycle; ETC: European Transient Cycle; ELR: European Load Response test.
[1] Applicable to vehicles with at least four wheels and a maximum speed of more than 50 (25) km/h (without driven machines).
[2] Without driver.
[3] New chassis dynamometer test cycle up to 120 km/h in compliance with Directive 70/220/EEC.
[4] Chassis dynamometer test can be performed for M_2 and N_2 vehicles when the reference mass of 2,840 kg is not exceeded.

- particulates (PM) and
- visible emission (exhaust gas opacity).

Tables 15-7 and 15-8 present the emission limits applied in the different regions. All over the world, emission legislation is exhibiting a clear trend toward ever tighter emission limits. Since engines must be tuned for the particular test cycle, a direct comparison of the individual regions' limits is impossible.

The most important test procedures are described below.

15.2.2.2 US Test Cycle

A transient test procedure (US-FTP) has been mandatory for commercial vehicle engines in the USA since 1985 (see Fig. 15-13). The US-FTP test cycle is specified in a standardized form (percentage speed and percentage torque), lasts 20 min and is run twice (cold start test and hot start test). The cold start test is weighted as 1/7 and the hot start test as 6/7 of the final result. The operating frequencies of the cycle are concentrated on the higher engine speed range. The range around the maximum torque speed applied in real operation to reduce fuel consumption is only represented very little. Therefore, European legislation has not adopted US-FTP.

The test cycle's failure to capture typical in-use driving conditions has been exploited to design engines that obtain good fuel consumption in the field at the expense of NO_X emission while meeting the NO_X limit in the test cycle. This abuse (cycle beating) led to a significant tightening of American legislation. Along with additionally implementing Euro III ESC tests, a region of the engine map was stipulated in which emissions may not exceed the respective limits multiplied by one of the NTE factors, which are dependent on the

Table 15-7 Development of the limits for gaseous and particulate emissions of commercial vehicles over 3.5 t in the EU

	1992/1993 Euro I	1995/1996 Euro II	2000/2001 Euro III		2005/2006 Euro IV		2008/2009 Euro V		2012/2013 Euro VI	
Test cycle	13-mode	13-mode	ESC	ETC	ESC	ETC	ESC	ETC	WHSC	WHTC
CO g/kWh	4.5 (4.9)	4	2.1	5.45	1.5	4	1.5	4	1.5	4
HC g/kWh	1.1 (1.23)	1.1	0.66		0.46		0.46		0.13	0.16
NMHC g/kWh (gas engines)				0.78		0.55		0.55		0.416
NO_x g/kWh	8.0 (9.0)	7	5	5	3.5	3.5	2	2	0.4	0.44
PM g/kWh	0.61 (0.68) 0.36 (0.4)	0.25 0.15	0.13 0.10	0.21 0.16	0.02	0.03	0.02	0.03	0.01	0.01
CH_4 g/kWh (gas engines)				1.6		1.1		1.1		0.5
ELR smoke m^{-1}			0.8		0.5		0.5			

Table 15-8 Development of the limits for gaseous and particulate emissions of diesel powered commercial vehicles over 3.5 t in the USA and over 2.5 t in Japan

Effective date	Test cycle	USA FTP [g/kWh]	USA ESC [g/kWh]	Japan D 13 [g/kWh]	Japan JE05 [g/kWh]
1998	NO_x HC CO PM	5.4 1.7 20.8 0.13/0.07			
1999	NO_x HC CO PM			4.5 2.9 7.4 0.25	
2004	NO_x HC CO PM	3.35 (NO_x + HC) 20.8 0.13/0.07		3.38 0.87 2.22 0.18	
2005	NO_x HC CO PM				2.0 0.17 2.22 0.027
2007	NO_x HC CO PM	1.5 0.19 20.8 0.02	1.5 0.19 20.8 0.02		
2009	NO_x HC CO PM				0.7 0.17 2.22 0.01
2010	NO_x HC CO PM	0.27 0.19 20.8 0.02	0.27 0.19 20.8 0.02		

level of the emission limits. This is called the NTE (not-to-exceed) zone (Fig. 15-14).

A dynamic test cycle that includes the different motoring, loading and deceleration phases (Fig. 15-15) has been specified to limit visible emissions (exhaust gas opacity).

15.2.2.3 European Test Cycle

Headquartered in Geneva, the UN Economic Commission for Europe (ECE) introduced the first European test cycle (13-mode test) in 1982 as part of its regulation ECE R49, which was in force until Euro III.

The European Union then implemented a new test cycle for Euro III in 1999. Its operating points were determined by taking extensive measurements of in-use driving [15-6]. Engines are tested on the new ESC (European Steady State Cycle) together with the ELR (European Load Response test) at three test speeds and several load points (Fig. 15-16). The test range is determined from the engine's full load curve (Fig. 15-17). NO_X emission is measured in three measuring points randomly selected by the approval authority within the test range to validate the homogeneity of the NO_X map. The ELR limits dynamic particulate emission (Fig. 15-18).

Engines with exhaust gas aftertreatment systems (particulate filters and $deNO_X$ systems), gas engines and all Euro IV and Euro V engines must be additionally tested on the ETC (European Transient Cycle) derived from the same base of data (Fig. 15-19). Like the US transient test, the ETC is specified on the basis of normalized speed and torque values but it is only run as a hot start test and lasts thirty minutes.

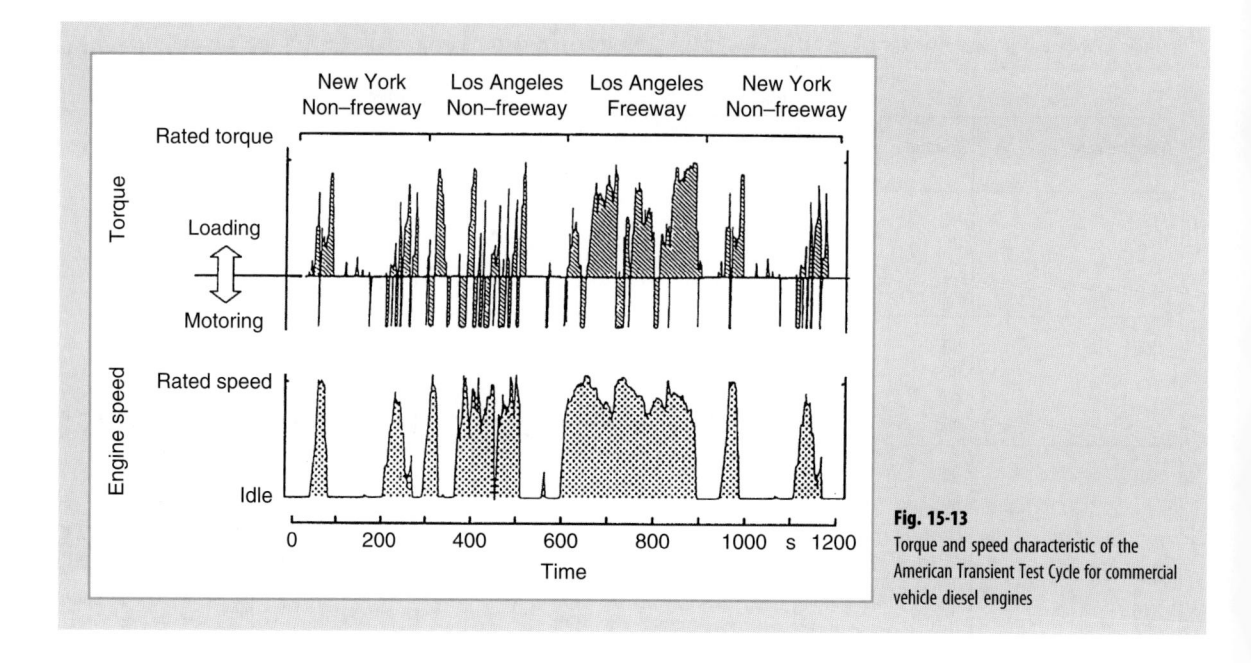

Fig. 15-13 Torque and speed characteristic of the American Transient Test Cycle for commercial vehicle diesel engines

The European regulation ECE-R24 limits visible full load smoke as a function of the theoretical engine air flow rate (Fig. 15-20). Smoke is additionally measured during free engine acceleration from idle to full load to test the effectiveness of the smoke limiter. The certified smoke value is indicated on the engine nameplate and serves as the basis for periodic inspection implemented in some EU countries.

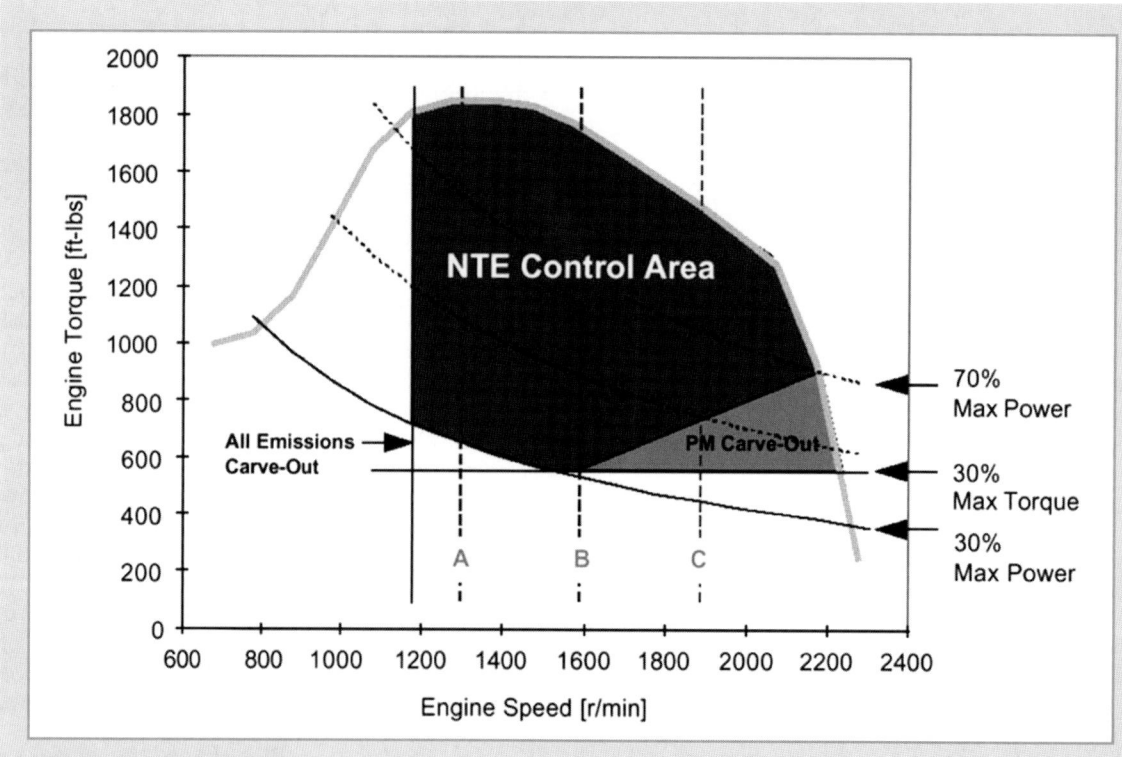

Fig. 15-14 American control area (NTE)

Fig. 15-15 American smoke test

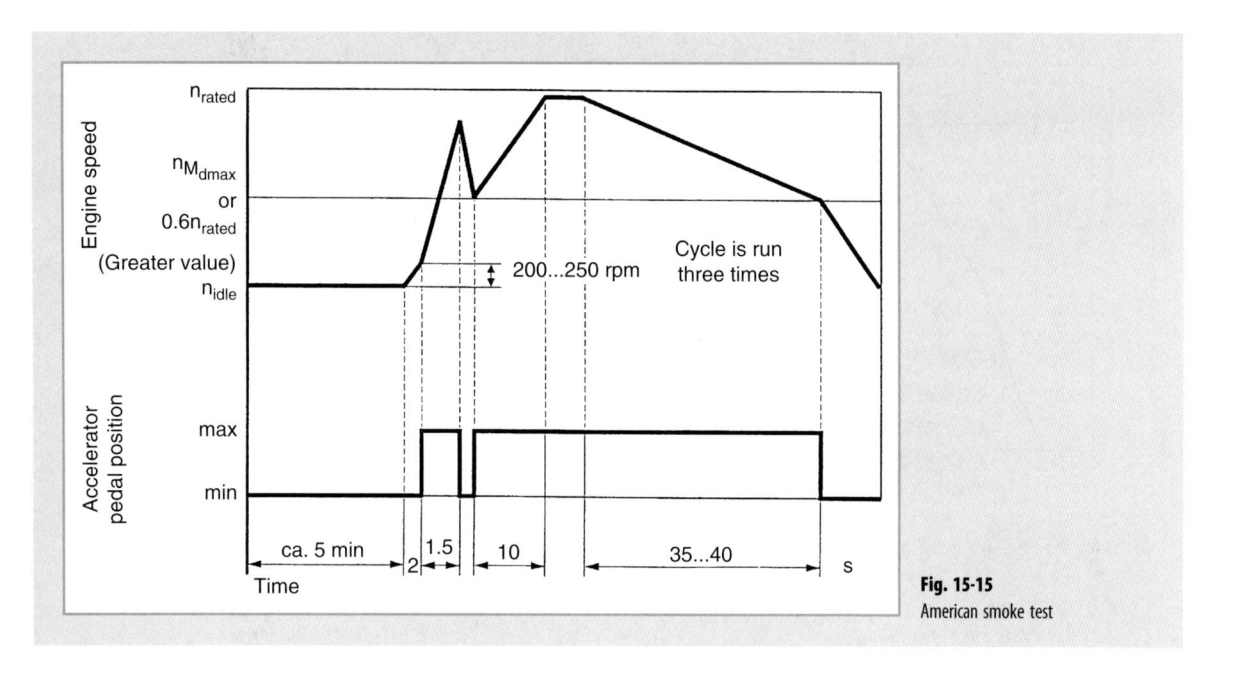

15.2.2.4 Japanese Test Cycle

As in Europe, a steady state 13-mode test (D 13) was originally also implemented in Japan. Its testing points chiefly cover the range of maximum torque and idling (Fig. 15-21). A transient test cycle (JE 05) has been in effect since 2005. Unlike the US-FTP and ETC, it is specified as a vehicle cycle in km/h like that for cars (Fig. 15-22). As in the EU and the USA, the emission test is performed on the engine. Japanese legislation additionally provides a computer program that converts the vehicle cycle into the engine cycle as a function of vehicle type.

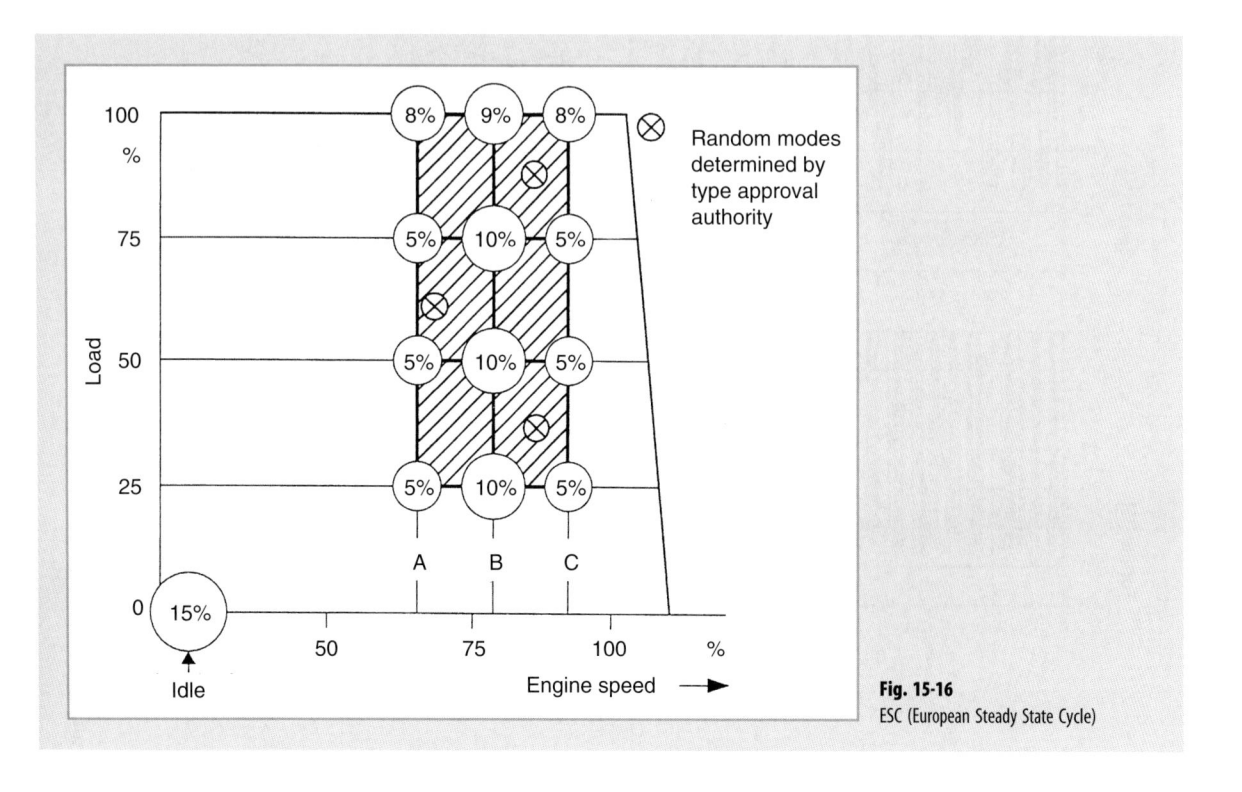

Fig. 15-16 ESC (European Steady State Cycle)

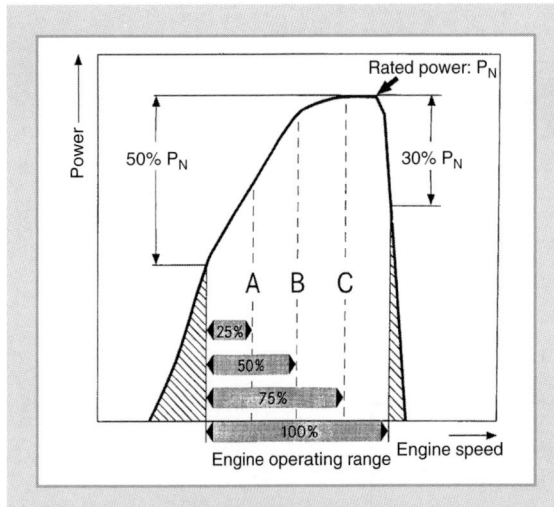

Fig. 15-17 Determination of the ESC (European Steady State Cycle) test range

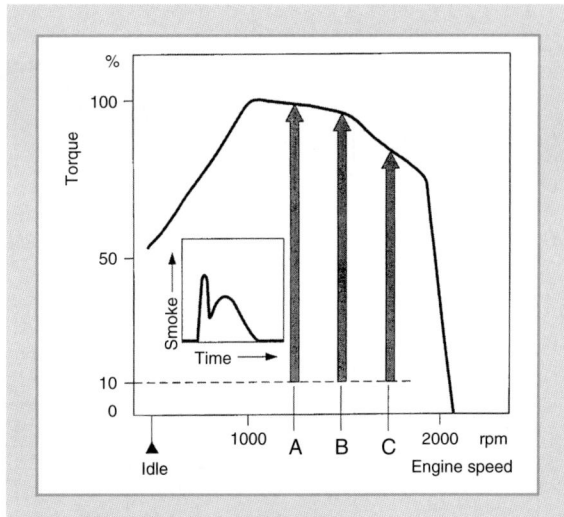

Fig. 15-18 ELR (European Load Response test)

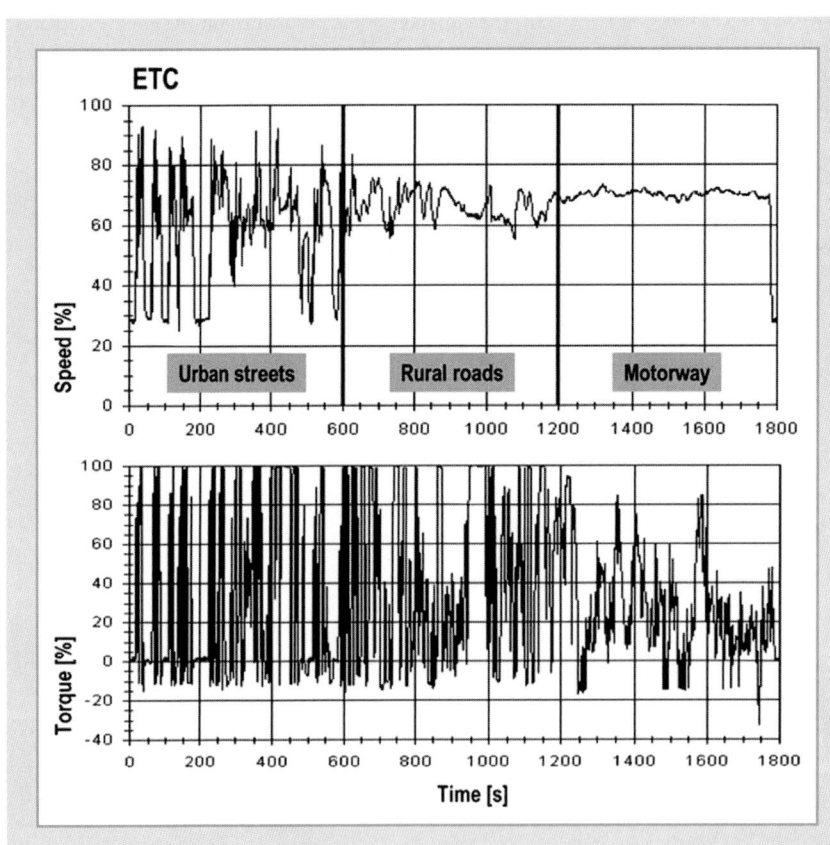

Fig. 15-19 ETC (European Transient Cycle)

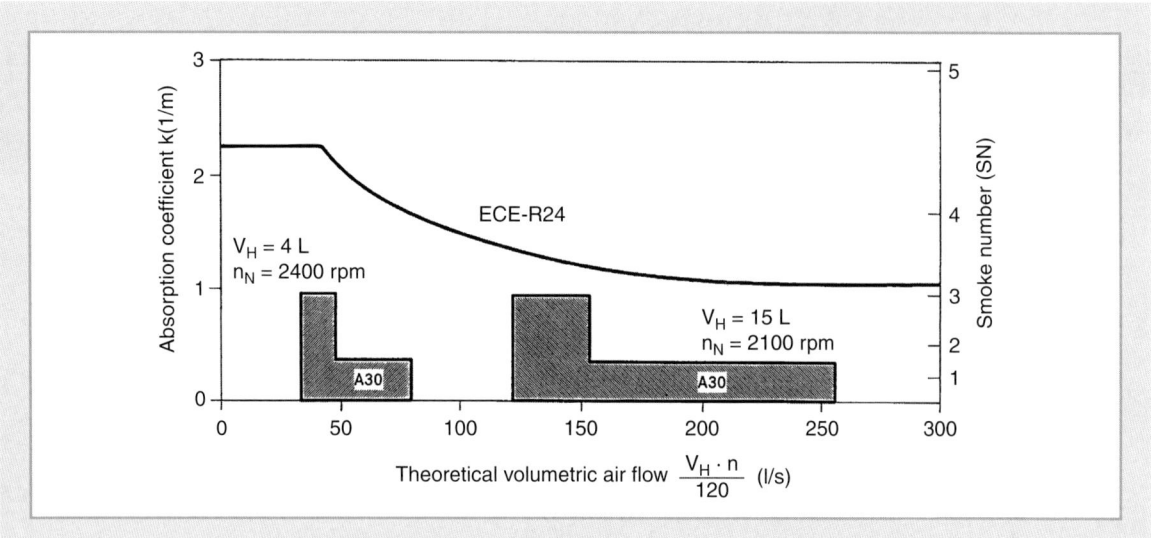

Fig. 15-20 Swedish smoke limit curves based on ECE-R 24 and A30

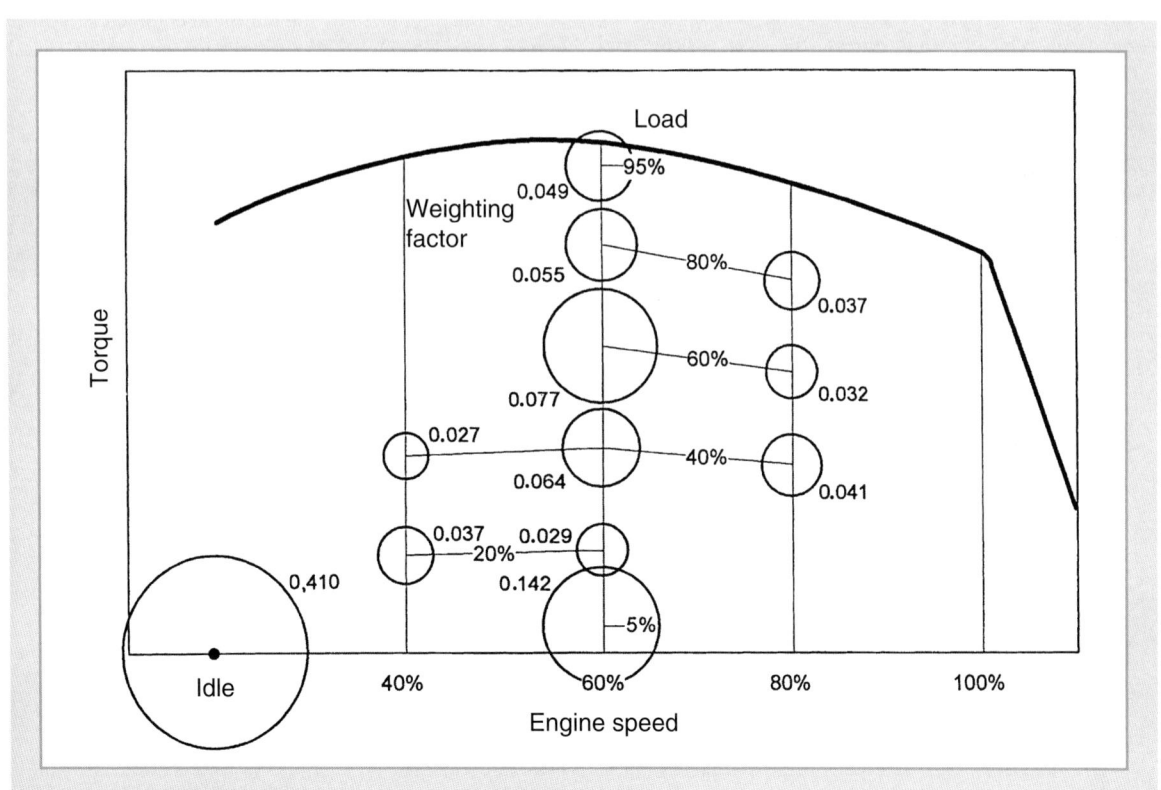

Fig. 15-21 Modes and weighting factors in the Japanese 13-mode cycle (D 13)

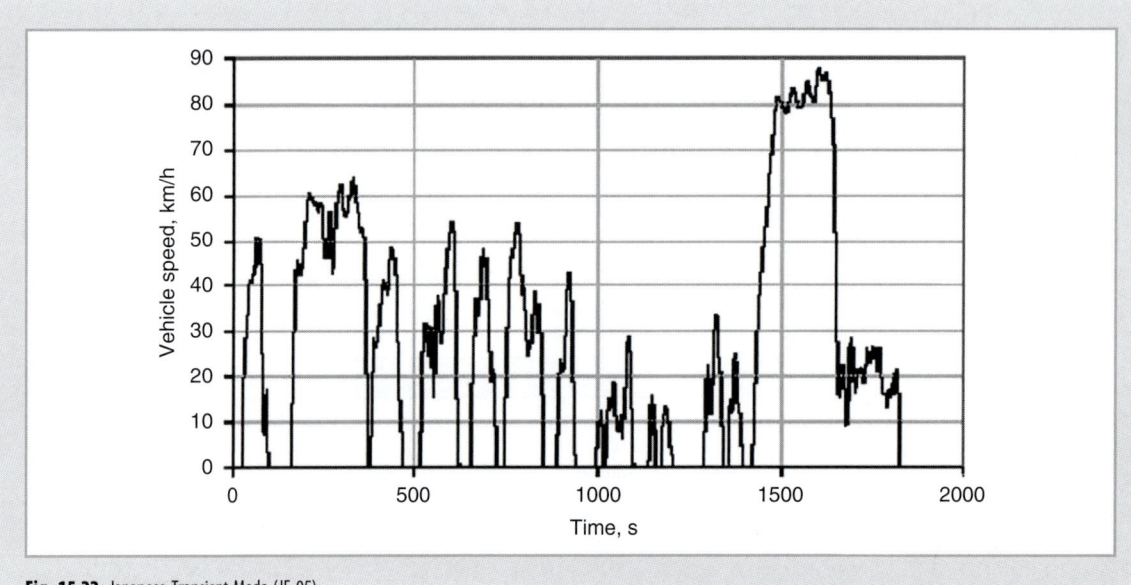

Fig. 15-22 Japanese Transient Mode (JE 05)

15.2.2.5 Worldwide Harmonized Test Cycle (WHDC)

American, European and Japanese test cycles have significantly different operating ranges (Fig. 15-23). For globally operating commercial vehicle manufacturers, the ever stricter and converging limits translate into disproportionately high costs for the development of emission control technology that is identical in principle.

Therefore, prompted by the European commercial vehicle industry, the UN Economic Commission for Europe (ECE)

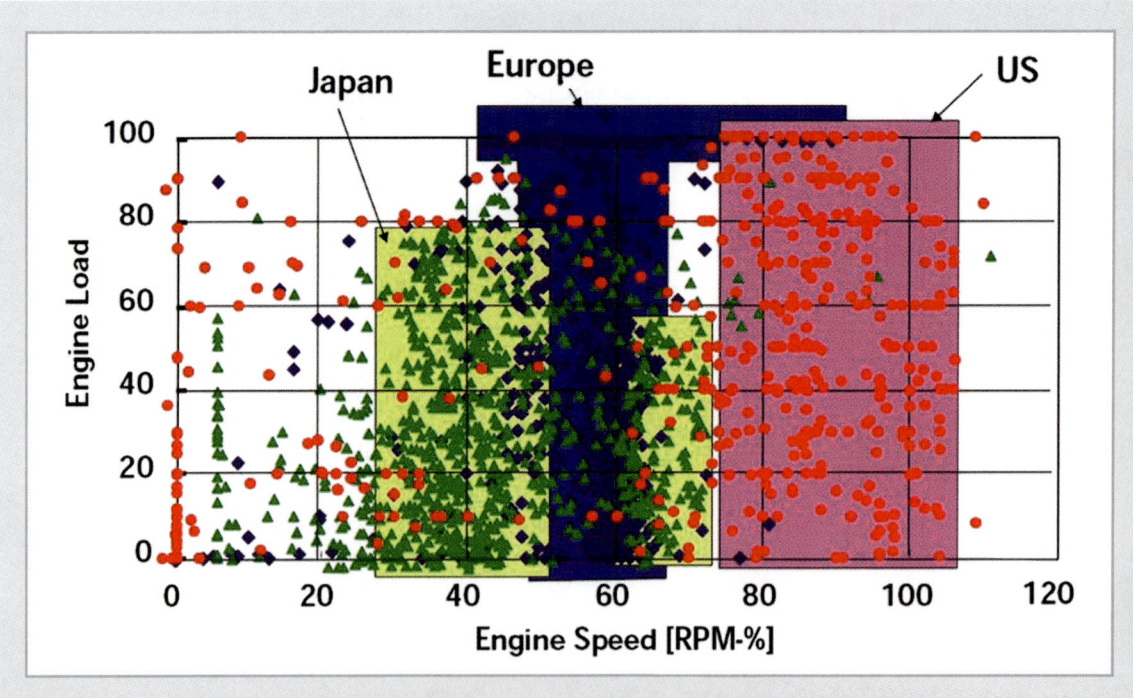

Fig. 15-23 Operating ranges of American, EU and Japanese test cycles

developed a global test procedure in 1997 for the certification of heavy commercial vehicle engines. This WHDC procedure was adopted as a global technical regulation (gtr No. 4) in 2006 and is intended to harmonize test procedures and measuring equipment worldwide. It will be implemented in Europe with Euro VI in 2012–2013. The cycle was developed from driving data from over 80 vehicles worldwide and is a compromise between European, Japanese and American driving profiles (Fig. 15-24). Thus, it is not representative for specific vehicle usage (e.g. long-distance traffic) but facilitates the development and use of efficient exhaust gas control technology across the entire range.

Like the US-FTP and ETC, the transient test cycle (WHTC) is specified in normalized speed and torque (Fig. 15-25). The main engine speed range has clearly been shifted to lower speeds also found in actual in-use driving. Normalization is now substantially more complex than in the US-FTP and ETC, thus minimizing the risk of cycle bypass. In addition, there is also a steady state test cycle WHSC since the combination of transient and steady state testing has established itself worldwide.

15.2.2.6 New Elements of Emission Control Legislation

Until Euro III was phased out, emission legislation basically only consisted of the test cycle(s) and the respective emission limit(s). Already familiar from passenger cars, OBD (on-board diagnostics) systems became mandatory for commercial vehicles in the EU for the first time as of Euro IV (2005). When a malfunction of an emission-related component is detected, a malfunction indicator informs the driver to find a garage for repairs. The ECE has already formulated a global regulation (WWH-OBD) in this field too. It was adopted as gtr No. 5 in 2006 and is expected to be applied worldwide by 2013–2014.

Another key point of future emission control regulations will be emissions monitoring of customer vehicles in the field (in-use compliance and in-service conformity). Commercial vehicles will be breaking new ground here. Since the test cycle is based on the engine but removing an engine from a vehicle for the test is inexpedient, emission will be measured directly under real vehicle operation with a portable emission measurement system (PEMS). While such regulations were already adopted in the USA in 2008, the technicalities of implementation are currently still being sorted out in the EU and adoption for Euro V vehicles is foreseen in 2010.

15.2.3 Emission Control Legislation for Nonroad Engines

Engines not installed in a road vehicle are referred to as non-road engines. Since this category covers an extremely wide range from small lawnmower engines through large marine diesel engines, the following can only address the most relevant regulations. An ISO committee formulated the standard ISO 8178 [15-7] (see Standards and Guidelines for Combustion

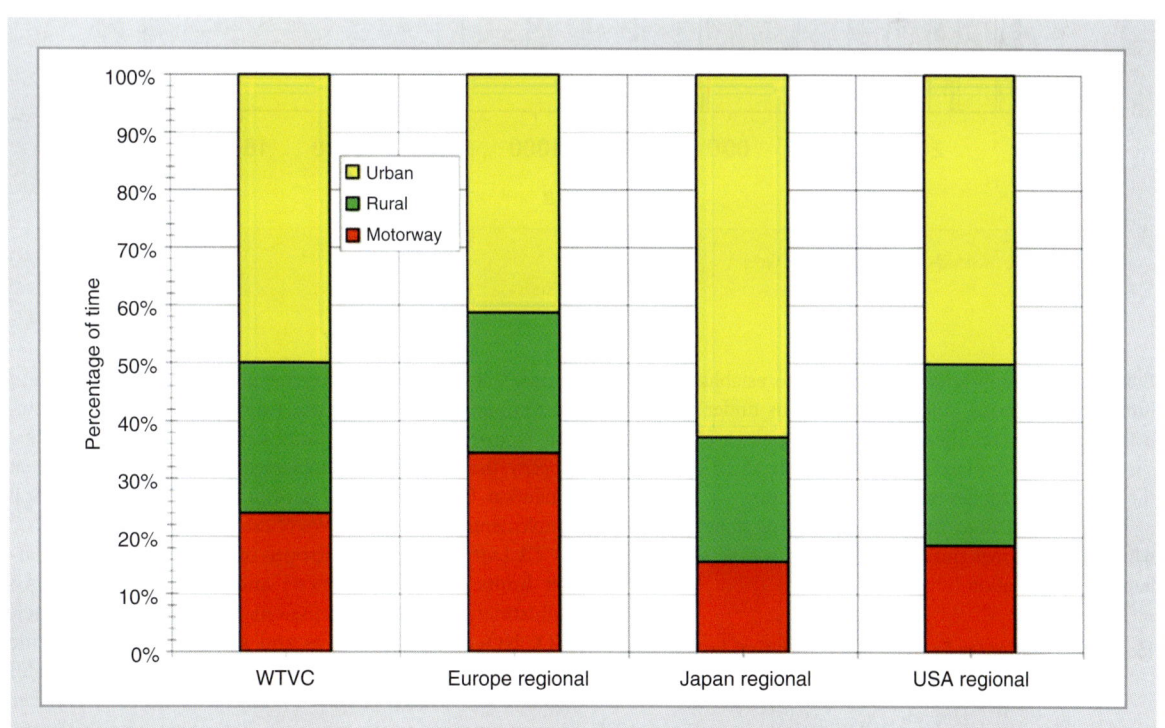

Fig. 15-24 Comparison of the frequency of road categories in the USA, EU and Japan WTVC - World-wide Transient Vehicle Cycle

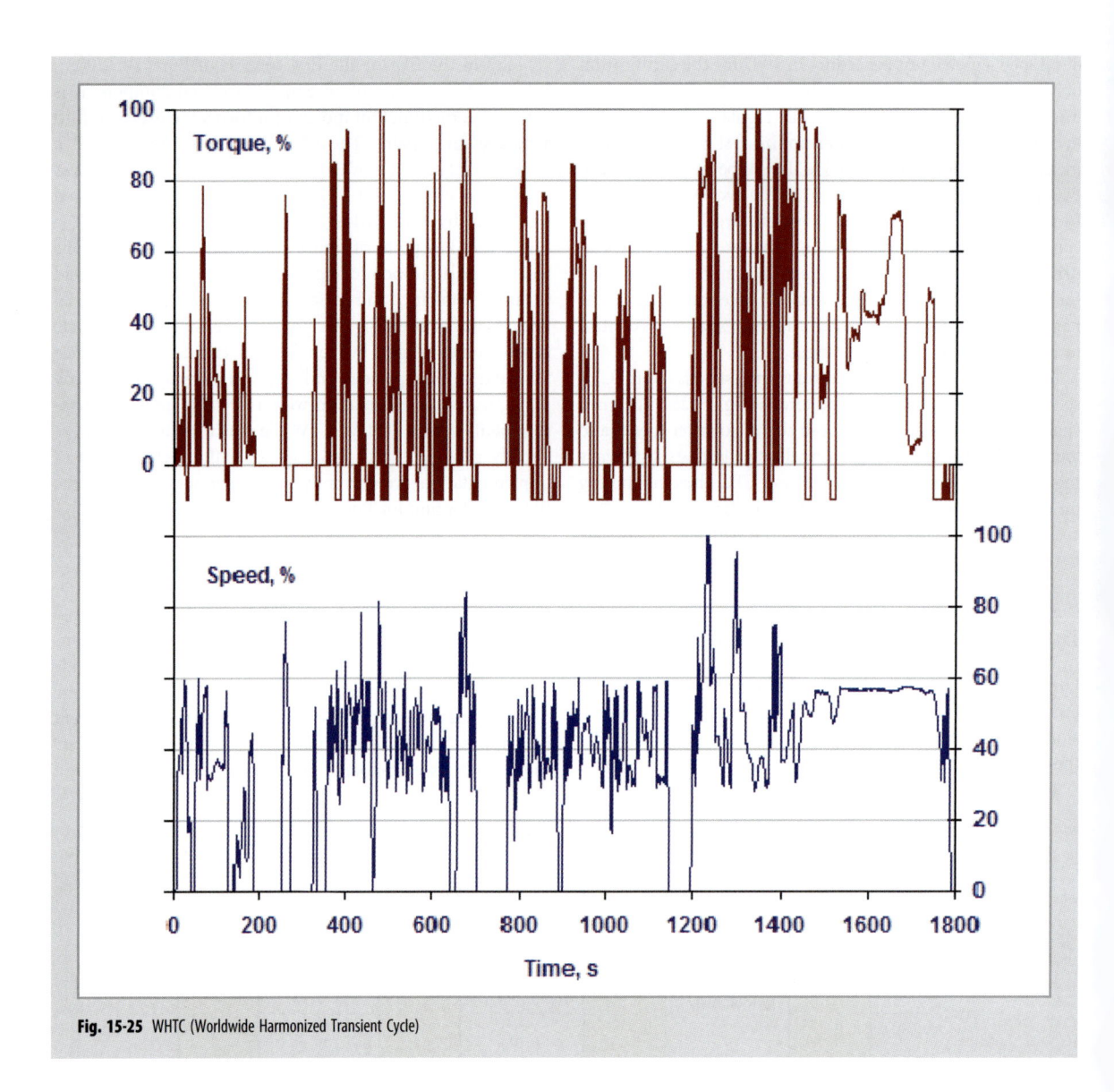

Fig. 15-25 WHTC (Worldwide Harmonized Transient Cycle)

Engines in the appendix) as the basis to establish emission limits for this broad range of applications under the appropriate operating conditions. Specific test cycles are assigned to specific applications (see Table 15-9). Table 15-10 summarizes the relevant test modes and weighting factors.

The standard does not address the emission limits with which these engines must comply. They will be set by regulators and legislation.

15.2.3.1 Stationary Engine Plants: Technical Instructions on Air Quality Control

The German Federal Ambient Pollution Control Act (BImSchV) limits the emissions of stationary combustion engines, e.g. emergency power units, cogeneration plants, combined heat and power stations and the like. Known as *TA Luft*, the Technical Instructions on Air Quality Control from July 24, 2002 specify the measures for implementing the Federal Ambient Pollution Control Act and provide instructions for practical implementation and examples of existing laws (Table 15-11).

TA Luft has different nitrogen oxide and particulate matter limits for gasoline, diesel and dual fuel engines. An adjustment clause for emissions (e.g. particulate matter, CO, NO_X, SO_2 [for biogas and sewage gas] and organic substances) exists for some fuels (sewage, biogas and landfill gas). It demands that every option be exhausted to further reduce emissions with measures that reflect state-of-the-art emission control technology (cf. *TA Luft* 2002). The limits

Table 15-9 Test cycles for nonroad engines: application ranges

Test cycle		Example applications
C1	Off-road diesel engine	Construction equipment, agricultural equipment, material transport
C2	Off-road gasoline engines	Construction equipment, agricultural equipment, material transport
D1	Cruising speed	Power plants, irrigation pumps
D2	Cruising speed	Gas compressors, generators
E1	Marine	Ships with diesel engines less than 24 m in length
E2	Marine	Oceangoing ships with constant speed
E3	Marine	Oceangoing ships with propeller curve
E4	Marine	Sport boats with gasoline engines less than 24 m in length
E5	Marine	Ships with diesel engines less than 24 m in length (propeller curve)
F	Rail	Locomotives, railcars, switch engines
G1	Small engines (lawnmowers and the like)	Intermediate speed applications
G2	Small engines	Rated speed applications
G3	Small engines	Handheld equipment

Table 15-10 Test cycles for nonroad engines: operating modes and weighting factors

	Rated speed					Intermediate speed					Idle
	Torque/power* [%]										
Type	100	75	50	25	10	100	75	50	25	10	0
C1	0.15	0.15	0.15		0.10	0.10	0.10	0.10			0.15
C2				0.06		0.02	0.05	0.32	0.30	0.10	0.15
D1	0.30	0.50	0.20								
D2	0.05	0.25	0.30	0.10							
E1	0.08	0.11					0.19	0.32			0.30
E2	0.20	0.50	0.15	0.15							
*E3	0.20						0.50 / 91%	0.15 / 80%	0.15 / 63%		
*E4	0.06							0.14 / 80%	0.15 / 60%	0.25 / 40%	0.40
*E5	0.08						0.13 / 91%	0.17 / 80%	0.32 / 63%		0.30
F	0.25							0.15			0.60
G1						0.09	0.20	0.29	0.30	0.07	0.05
G2	0.09	0.20	0.29	0.30	0.07						0.05
G3	0.90										0.10

* Values for test cycles E3, E4 and E5 are power values. The values for the other test cycles are torque values.

Table 15-11 TA Luft limits for stationary combustion engines ≥ 1 MW$_{th}$ (no output limit exists for landfill gas) in mg/m³

Particulate matter	20[1]		
SO$_2$	Dependent on fuel, e.g. 350 for biogas or digester gas		
Formaldehyde	60		
Total C	–		
Chlorine, fluorine, halogens	3%		
CO[2,3]	(a) Auto-ignition engines and spark ignition engines with liquid fuels, auto-ignition engines (ignition spray engines) and spark ignition engines with gaseous fuels (excluding bio, digester and natural gas)	300	
	(b) Spark ignition engines with bio or digester gas[4]	< 3 MW$_{th}$ 1,000[3]	≥ 3 MW$_{th}$ 650
	(c) Spark ignition engines with natural gas	650	
NO$_x$[3]	(c) Spark ignition engines with natural gas	< 3 MW$_{th}$ 2,000[3]	≥ 3 MW$_{th}$ 650
	(a) Auto-ignition engines with liquid fuels	< 3 MW$_{th}$ 1,000	≥ 3 MW$_{th}$ 500
	(b) Gas powered auto-ignition engines (ignition spray engines) and spark ignition engines		
	- Ignition spray engines with bio or digester gas	< 3 MW$_{th}$ 1,000	≥ 3 MW$_{th}$ 500
	- Lean gas engines and other four-stroke gasoline engines with bio or digester gas	500	
	- Ignition spray engines and lean gas engines with other gaseous fuels	500	
	(c) Other four-stroke gasoline engines	250	
	(d) Two-stroke engines	800	

values in mg/m³, O$_2$–content 5%
[1] 80 mg/m³ when exclusively for emergency drive or up to 300 h to cover peak load during power generation.
[2] Landfill gas at present generally 650 mg/m³.
[3] Emission values are not applied when exclusively for emergency drive or up to 300 h to cover peak load during power generation.
[4] For spark ignition engines with natural gas 650 mg/m³.

are based on an exhaust gas oxygen content of 5% by volume.

According to VDI 2066, particulate matter emissions must be measured close to the point of emission, i.e. in the hot exhaust gas, while measurement of particulate emission mandated for commercial vehicle engines and nonroad mobile machinery is based on diluting the exhaust gas.

The thermal outputs (P$_{th}$ in MW$_{th}$) specified in Table 15-11 are based on the overall fuel flow rates of the specific systems.

15.2.3.2 Nonroad Mobile Machinery and Agricultural and Forestry Tractors

Nonroad mobile machinery (NRMM) encompasses such applications as front loaders, excavators, forklifts, road construction equipment, etc. The C1 cycle is applied to diesel engines implemented in these applications. In keeping with the aforementioned principle of transient and steady state testing, the USA and the EU intend to introduce a transient cycle (NRTC) for these engines in addition to the C1 cycle as of 2011. The NRTC is also the basis for a newly developed global technical regulation (gtr) set to be adopted in 2009.

The emission limits are specified as a function of engine power. After long negotiations, test cycles and limits have largely been harmonized worldwide. The limits specified in Table 15-12 apply to engines in nonroad mobile machinery and agricultural and forestry tractors [15-8, 15-9].

15.2.3.3 Marine Engines

The Central Commission for Navigation on the Rhine (ZKR) incorporated a new section, Sect. 8a "emissions of gaseous pollutants and air contaminating particulates from diesel engines", in the Rhine Shipping Inspection Directive (RheinSchUO). The first stage went into effect on January 1, 2002 (Table 15-13). Level II followed on July 1, 2007.

The European Commission also extended the scope of its emissions Directive 97/68/EC to inland marine engines in the amendment 2004/26/EC. Since their levels of technology are virtually identical, mutual recognition of EU and ZKR directives is assured.

Table 15-12 Limits for engines in nonroad mobile machinery NRMM

Year	1996	1997	1998	1999	2000	2001	2002	2003	2004	2005	2006	2007	2008	2009	2010	2011	2012	2013	2014	2016
Power (kW) EU																				
>560																				
130 =< 560				NOx: 9.2 / PM: 0.54			NOx: 6.0 / PM: 0.2				NOx + HC: 4.0 / PM: 0.2				NOx: 2.0 / PM: 0.025			NOx: 0.4 / PM: 0.025	
75 =< 130				NOx: 9.2 / PM: 0.70				NOx: 6.0 / PM: 0.3					NOx + HC: 4.0 / PM: 0.3				NOx: 3.3 / PM: 0.025			NOx: 0.4 / PM: 0.025
56 =< 75				Mar. 99	NOx: 9.2 / PM: 0.85					NOx: 7.0 / PM: 0.4			NOx+HC: 4.7 / PM: 0.4				NOx: 3.3 / PM: 0.025			NOx: 0.4 / PM: 0.025
37 =< 56				Mar. 99	NOx: 9.2 / PM: 0.85					NOx: 7.0 / PM: 0.4					NOx+HC: 4.7 / PM: 0.4			NOx+HC: 4.7 / PM: 0.025		
19 =< 37						NOx: 8.0 / PM: 0.8							NOx + HC: 7.5 / PM: 0.6							
< 19																				
USA																				
> 560						NOx: 9.2 / PM: 0.54						NOx + HC: 6.4 / PM: 0.2				NOx: 3.5 / PM: 0.1	NOx: 3.5 / PM: 0.04			
450 =< 560		NOx: 9.2 / PM: 0.54					NOx + HC: 6.4 / PM: 0.2					NOx + HC: 4.0 / PM: 0.2				NOx: 2.0 / PM: 0.02			NOx: 0.4 / PM: 0.02	
225 =< 450		NOx: 9.2 / PM: -					NOx + HC: 6.4 / PM: 0.2					NOx + HC: 4.0 / PM: 0.2				NOx: 2.0 / PM: 0.02			NOx: 0.4 / PM: 0.02	
130 =< 225				NOx: 9.2 / PM: 0.54				NOx + HC: 6.6 / PM: 0.2				NOx + HC: 4.0 / PM: 0.2				NOx: 2.0 / PM: 0.02			NOx: 0.4 / PM: 0.02	
75 =< 130				NOx: 9.2 / PM: -				NOx + HC: 6.6 / PM: 0.3				NOx + HC: 4.0 / PM: 0.3	NOx + HC: 4.0 / PM: 0.3				NOx: 3.4 / PM: 0.02			NOx: 0.4 / PM: 0.02
56 =< 75					NOx: 9.2 / PM: -					NOx + HC: 7.5 / PM: 0.4				NOx + HC: 4.7 / PM: 0.4			NOx: 3.4 / PM: 0.02			NOx: 0.4 / PM: 0.02
37 =< 56					NOx: 9.2 / PM: -					NOx + HC: 7.5 / PM: 0.4			NOx + HC: 4.7 / PM: 0.3					NOx + HC: 4.7 / PM: 0.03		

Table 15-12 (Continued)

Year	1996	1997	1998	1999	2000	2001	2002	2003	2004	2005	2006	2007	2008	2009	2010	2011	2012	2013	2014	2016
19 = < 37					NOx + HC: 9.5 / PM: 0.4					NOx + HC: 7.5 / PM: 0.6			NOx + HC: 7.5 / PM: 0.3							
8= < 19						NOx + HC: 9.5 / PM: 0.8				NOx + HC: 7.5 / PM: 0.8					NOx + HC: 7.5 / PM: 0.40			NOx +HC: 4.7 / PM: 0.03		
< 8						NOx + HC: 10.5 / PM: 1.0				NOx + HC: 7.5 / PM: 0.8					NOx + HC: 7.5 / PM: 0.40					
Japan																				
Power (kW)																				
> 560																				
130 = < 560								NOx : 6.0 / PM : 0.2			Oct 06		NOx : 3.6 / PM : 0.17							
75 = < 130								Oct 2003	NOx : 6.0 / PM : 0.3			Oct 07	NOx : 3.6 / PM : 0.2							
56 = < 75								Oct 2003	NOx : 7.0 / PM : 0.4				Oct 08	NOx : 4.0 / PM : 0.25						
37 = < 56									NOx : 7.0 / PM : 0.4			Oct 07	Oct 08	NOx : 4.0 / PM : 0.3						
19 = < 37								Oct 2003	NOx : 8.0 / PM : 0.8					NOx : 6.0 / PM : 0.4						
< 19																				

Table 15-13 RheinSchUO emission limits for inland waterway engines as a function of rated power P_N and rated speed n_N

Level I

Rated power P_N [kW]	CO [g/kWh]	HC [g/kWh]	NO_x [g/kWh]	PM [g/kWh]
$37 \leq P_N < 75$	6.5	1.3	9.2	0.85
$75 \leq P_N < 130$	5.0	1.3	9.2	0.70
$P_N \geq 130$	5.0	1.3	$n_N \geq 2{,}800$ rpm $= 9.2$ $500 \leq n_N < 2{,}800$ rpm $= 45\, n_N^{-0.2}$	0.54

Level II

P_N [kW]	CO [g/kWh]	HC [g/kWh]	NO_x [g/kWh]	PM [g/kWh]
$18 \leq P_N < 37$	5.5	1.35	8.0	0.8
$37 \leq P_N < 75$	5.0	1.3	7.0	0.4
$75 \leq P_N < 130$	5.0	1.0	6.0	0.3
$130 \leq P_N < 560$	3.5	1.0	6.0	0.2
$P_N \geq 560$	3.5	1.0	$n_N \geq 3{,}150$ rpm $= 6.0$ $343 \leq n < 3{,}150$ rpm $= 45 \times n_N^{(-0.2)} - 3$ $n < 343$ rpm $= 11.0$	0.2

The IMO (International Maritime Organization) imposed limits on NO_X emissions for marine diesel engines with power outputs >130 kW as of January 1, 2000 (Table 15-14 [15-10] and Fig. 15-26). It plans to later adapt the limits determined from the test cycles dependent on use (main propulsion or auxiliary engine) and the mode of operation (constant speed or propeller drive) specified in Fig. 15-27 to its technical program and environmental policies.

Sweden has been charging emission-related harbor fees based on the IMO Code since January 1, 1998 [15-11]. However, the incentive to use emission reducing exhaust gas aftertreatment largely failed because of the imbalance between the additional operating costs and the potential reductions in fees.

Table 15-14 IMO emission limits for marine diesel engines with a power of 130 kW and up

Rated speed n_N in rpm	NO_x emission in g/kWh
$0 < n_N \leq 130$	17
$130 < n_N \leq 2{,}000$	$45\, n_N^{-0.2}$
$n_N > 2{,}000$	9.8

15.3 Pollutants and Their Production

Apart from the desired heat energy, the only products produced by ideal combustion of hydrocarbons (HC) in an engine are water (H_2O) and carbon dioxide (CO_2), their mass ratio being a function of the fuel's H:C ratio. Both diesel fuel and gasoline may be specified with the empirical formula C_xH_y. An ideal stoichiometric combustion process yields:

$$C_xH_y + \left(x + \frac{y}{4}\right) \cdot O_2 \Rightarrow x \cdot CO_2 + \frac{y}{2} \cdot H_2O. \quad (15\text{-}1)$$

The water produced is environmentally harmless. CO_2 is nontoxic but contributes significantly to the greenhouse effect. Lowering the specific fuel consumption (b_e) reduces CO_2 emissions during engine combustion.

Combustion does not proceed ideally in gasoline or diesel engines and it produces other byproducts, which are in part harmful to the environment. Incomplete combustion during homogeneous gasoline engine operation ($\lambda = 1$) not only produces nitrogen oxides (NO_X) and unburned hydrocarbons (HC) but also chiefly carbon monoxide (CO) as pollutants. Combustion is incomplete because the flame front quenches on the cold combustion chamber wall or in squish gaps. Inhomogeneous combustion cycles where $\lambda > 1$, typical diesel engine combustion and stratified operation of direct injection gasoline engines produce another pollutant, soot (Fig. 15-28).

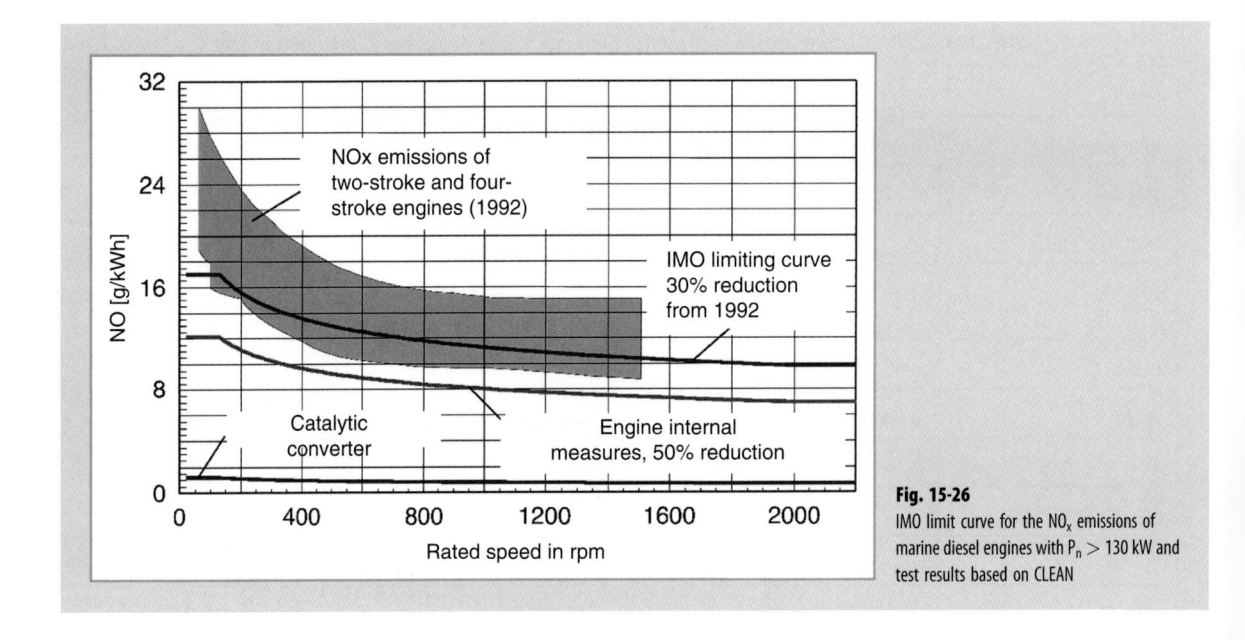

Fig. 15-26
IMO limit curve for the NO$_x$ emissions of marine diesel engines with P$_n$ > 130 kW and test results based on CLEAN

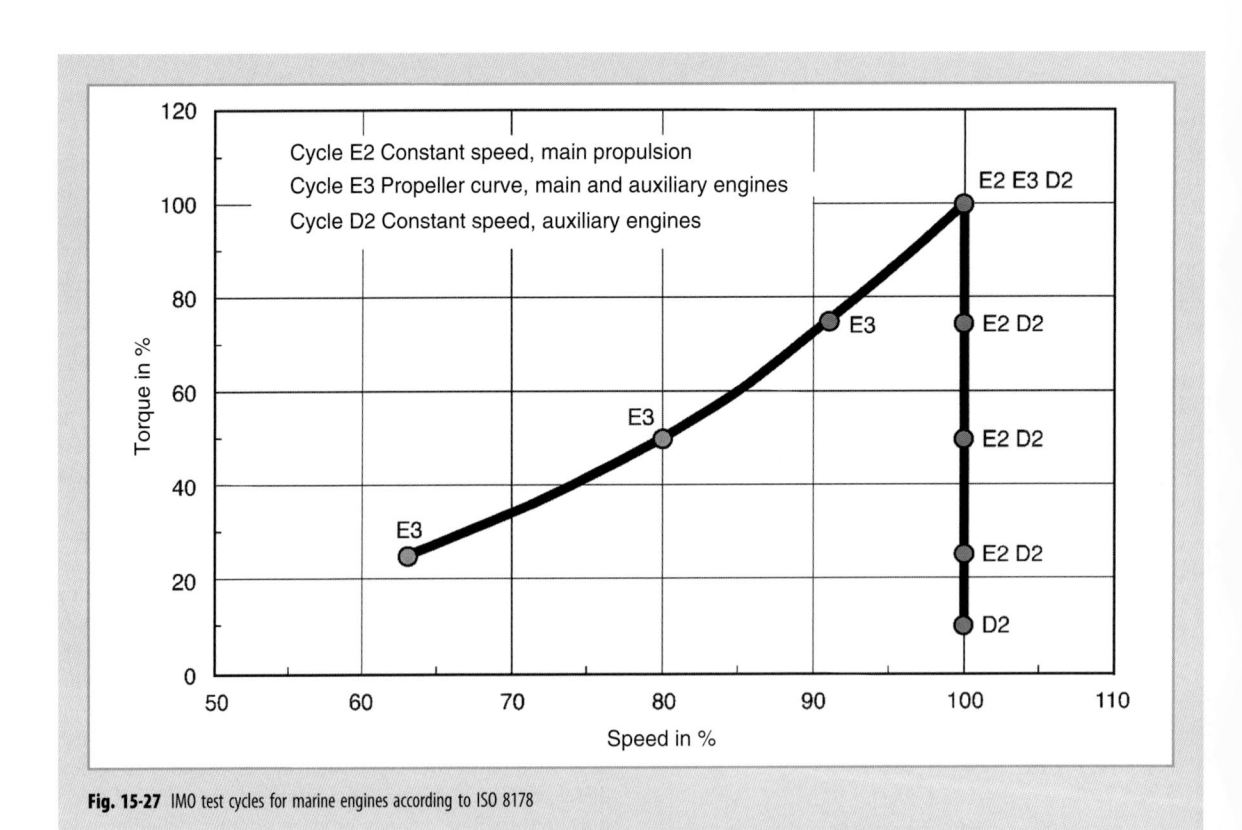

Fig. 15-27 IMO test cycles for marine engines according to ISO 8178

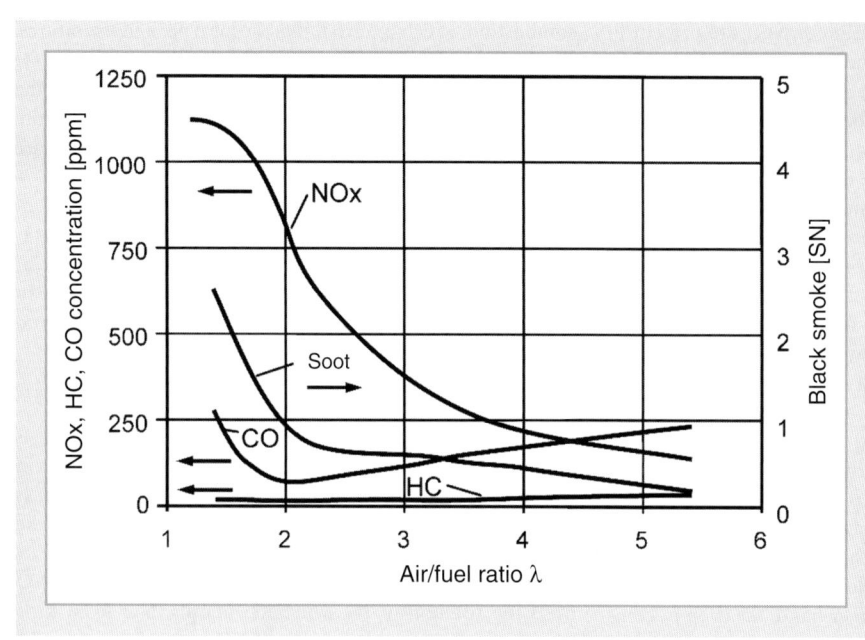

Fig. 15-28
Pollutant concentrations in diesel engine exhaust gas with a varied air/fuel ratio λ (based on [15-14])

The main advantages of diesel engines – lower fuel consumption and high torque at low speed – especially unfold in direct injection turbocharged engines. Their combustion systems are characterized by locally strongly fluctuating air/fuel ratios. There is a deficiency of air ($\lambda \ll 1$) inside the individual flames that form around the injection sprays. There is an excess of air ($\lambda \gg 1$) between injection sprays and on the combustion chamber wall. Soot is produced in the air deficient regions. Nitrogen oxides are chiefly produced directly behind the locally extremely hot flame front. Thus, the production of the two main pollutants during diesel engine combustion is directly connected with the combustion system.

Diesel engine development remains focused on continually reducing these pollutants while continually lowering fuel consumption and optimizing performance at the same time [15-12]. Before emission reduction measures in and downstream from the engine are described in Sects. 15.4 and 15.5 (exhaust gas aftertreatment), the production of the different pollutants is treated in detail below.

15.3.1 Nitrogen Oxides (NO$_X$)

Of the different nitrogen oxides (NO, NO$_2$, N$_2$O, N$_2$O$_3$, N$_2$O$_5$), only the compounds NO and NO$_2$ (nitrogen monoxide and nitrogen dioxide) are produced in appreciable quantities and the designation NO$_X$ (nitrogen oxides) is frequently employed as shorthand for the total NO and NO$_2$.

The most important mechanism for the production of NO$_X$ is the formation of thermal NO. It was described for the first time by Zeldovich in 1946 [15-13]. Specifically, the following elementary reactions occur:

$$O_2 \Leftrightarrow 2 \cdot O \qquad (15\text{-}2)$$

$$N_2 + O \Leftrightarrow NO + N \qquad (15\text{-}3)$$

$$O_2 + N \Leftrightarrow NO + O \qquad (15\text{-}4)$$

$$OH + N \Leftrightarrow NO + H \qquad (15\text{-}5)$$

Equations (15-3) and (15-4) describe the Zeldovich chain reaction: When elementary oxygen (O) is present, N$_2$ produces NO and N. The molecular nitrogen (N) formed reacts with O$_2$ in the following step to become NO and O. This completes the cycle and the reaction chain begins all over again. Equation (15-5) describes the formation of NO in fuel-rich zones like those located behind the flame front.

The presence of atomic oxygen, which forms from molecular oxygen at temperatures above 2,200 K (see Eq. (15.2)), is the basic condition for the start of the Zeldovich reactions that follow Eqs. (15-3) and (15-4). Thus, one prerequisite for the formation of NO is peak temperatures, explicitly, local peak temperatures, not mean combustion chamber temperatures. The second prerequisite for the formation of NO is the presence of excess oxygen, i.e. local excess air [15-13].

Ideal conditions for the formation of NO_X are present in gasoline engines when $\lambda = 1.1$. The maximum concentration of NO_X in diesel engine exhaust gas shifts toward a somewhat higher air/fuel ratio. The concentrations of the different pollutants are plotted as a function of the air/fuel ratio λ in Fig. 15-28 [15-14]. The curve of the NO_X concentration rises continuously as λ drops. This can be attributed to the increasing exhaust gas temperature. Despite the diminishing oxygen content, the increasing process temperature facilitates an increase in the NO_X concentration up to a value of $\lambda = 2$. When the exhaust gas temperature continues to increase, free oxygen is no longer sufficiently available below $\lambda = 2$. The gradient of the NO_X concentration decreases as a function of the air/fuel ratio and a local maximum is produced.

Zeldovich reactions are equilibrium reactions. Their equilibrium parameters are established as a function of temperature. However, engine combustion processes are so rapid that the equilibrium concentrations are usually not obtained and the actual NO_X concentrations are below what would be obtained at thermal equilibrium. On the other hand, the drop in combustion chamber temperature during the expansion phase causes reverse reactions to "freeze" according to Eqs. (15-3) and (15-4). NO no longer reforms significantly below approximately 2,000 K. As a result, the NO_X concentrations in the actual exhaust gas are above the equilibrium concentrations. Consequently, NO_X emissions cannot be predicted by equilibrium reactions alone. They can only be predicted with the aid of the reaction kinetics and by incorporating the time sequence of the combustion that actually occurs.

The process sequence can strongly influence NO_X emissions in diesel engines. Thus, the combustion temperature may be limited by cooling the charge air and recirculated exhaust gas as well as by retarding injection and combustion after TDC. Exhaust gas recirculation lowers the oxygen supply and, thus, directly reduces the formation of NO_X. At the same time, the lower oxygen concentration reduces the rate of combustion, which in turn limits the local peak temperatures. The higher specific heat capacity of the triatomic gases (CO_2 and H_2O) in the recirculated exhaust gas further reduces the local peak temperatures during exhaust gas recirculation. Different engine processes that reduce NO_X emissions are discussed in detail in Sect. 15.4.

NO_2 makes up 1–10% of the total NO_X emissions in gasoline engines and 5–15% in diesel engines. Higher concentrations are also detectable in the lower part load range at correspondingly low exhaust gas temperatures [15-14]. In an engine, NO_2 forms from NO by reacting with HO_2 and OH radicals. The most probable equation is:

$$NO + HO_2 \Leftrightarrow NO_2 + OH \qquad (15\text{-}5a)$$

At ambient temperature, the chemical equilibrium for NO_2 is virtually complete. When there is incident light, NO reacts in the atmosphere with ozone to become NO_2. Equilibrium is established after a few hours to days, depending on the ambient conditions.

Other mechanisms of NO formation such as prompt NO from the reaction of N_2 with fuel radicals, NO from fuel-based nitrogen or the formation of NO from N_2O tend to be of secondary importance for diesel engine combustion.

15.3.2 Particulate Matter (PM)

According to the statutory test specifications, a vehicle's particulate matter emissions are the total mass of solids and attached volatile or soluble constituents. The test conditions are defined precisely: an exhaust gas sample is diluted with filtered ambient air and cooled to a maximum of 52°C [15-15]. The particulate matter is separated onto a defined and conditioned sample holder and the total mass is determined by weighing under defined conditions.

Figure 15-29 presents the typical composition of particulate matter [15-16]. The particulates mainly consist of soot, i.e. of elementary carbon. This is addressed in more detail

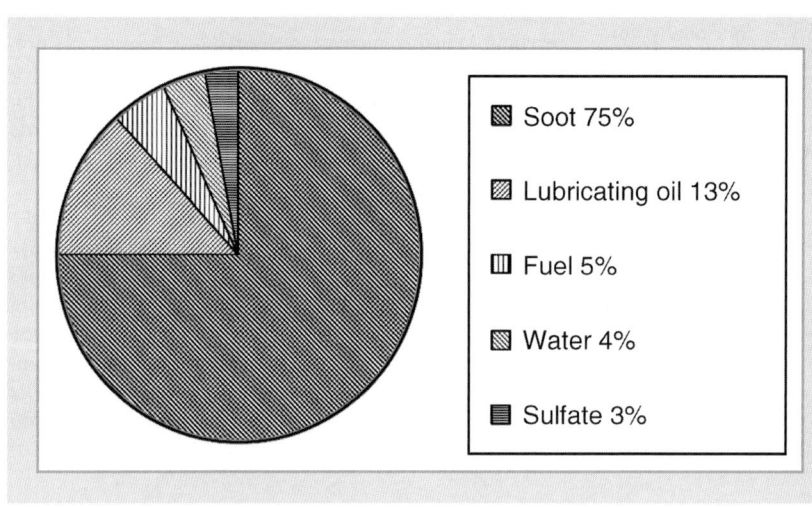

- Soot 75%
- Lubricating oil 13%
- Fuel 5%
- Water 4%
- Sulfate 3%

Fig. 15-29
Typical particulate composition with an standard oxidation catalyst (based on [15-16])

below. Organic compounds consisting of unburned hydrocarbons that may stem from the lubricating oil or the fuel itself constitute the second largest fraction. The dew point of numerous hydrocarbons is fallen below under the aforementioned conditions for particulate matter sampling and weighing: the compounds condense and bond to the solid cores.

The particulates' sulfate fraction is basically determined by the sulfur content of the fuel and the engine oil. During combustion, the sulfur oxidizes into SO_2 and, at exhaust gas temperatures above 450°C, into SO_3. The latter oxidation process can also be facilitated by downstream exhaust gas aftertreatment in an oxidation catalytic converter [15-17]. Interaction with water causes the formation of sulfate ions to produce sulfuric acid (H_2SO_4), which condenses on the particulates in the cooled exhaust gas. Metal oxides are produced as products of lubricating oil or fuel additives and are only present in particulate emissions in traces. These oxides can assume a significant proportion of the particulate mass when an additive is blended into fuel for particulate filter regeneration.

The composition of particulate matter presented in Fig. 15-29 corresponds to the mean value of measurements in various cars and can vary greatly depending on vehicle operation and type. Thus, a commercial vehicle engine run at high load has a larger fraction of elementary carbon. The fraction of hydrocarbons in cars operated at part load can significantly exceed the value presented in Fig. 15-29.

Soot constitutes the largest fraction of particulate matter relative to mass. This fraction can be controlled by engine measures described in more detail below. Soot is generally produced in air deficient zones. In older model prechamber engines, these zones include wall films in which fuel coking partly produces large soot particulates [15-18], which then become visible in the exhaust gas. The particulates in modern direct injection diesel engines are usually significantly smaller and consequently no longer visible in the exhaust gas. The production processes also differ distinctly. Two hypotheses of soot production exist and are presented here in greatly simplified form [15-19]:

15.3.2.1 Elementary Carbon Hypothesis

This hypothesis assumes that the fuel dissociates at high combustion temperatures, i.e. breaks down into its basic elements of carbon and hydrogen. Hydrogen molecules diffuse in the oxygen-containing environment significantly faster than larger carbon atoms. Their quadruple valences enable them to form clusters, primarily hexagonal and pentagonal structures, very quickly when they are deoxygenated. Curved shells form and grow to typical particulate sizes of approximately 10 nm within milliseconds.

15.3.2.2 Polycyclic Hypothesis

This hypothesis assigns critical importance to ethyne (formerly known as acetylene C_2H_2). Ethyne is formed by pyrolysis – the decomposition of fuel under O_2 exclusion – of aliphatics and aromatics while cleaving hydrogen. Assuming the structure is polycyclic, repeated bonding of ethyne molecules can cause a graphitic structure to grow [15-20]. Advanced growth through repeated ethyne bonding is number 1 in Fig. 15-30. The occasionally quintuple rings curve the macromolecules produced. Several of these molecules accrete in layers atop one another, producing primary particulates. Figure 15-30 depicts the formation process. The primary particulates produced typically have a size of 2–10 nm.

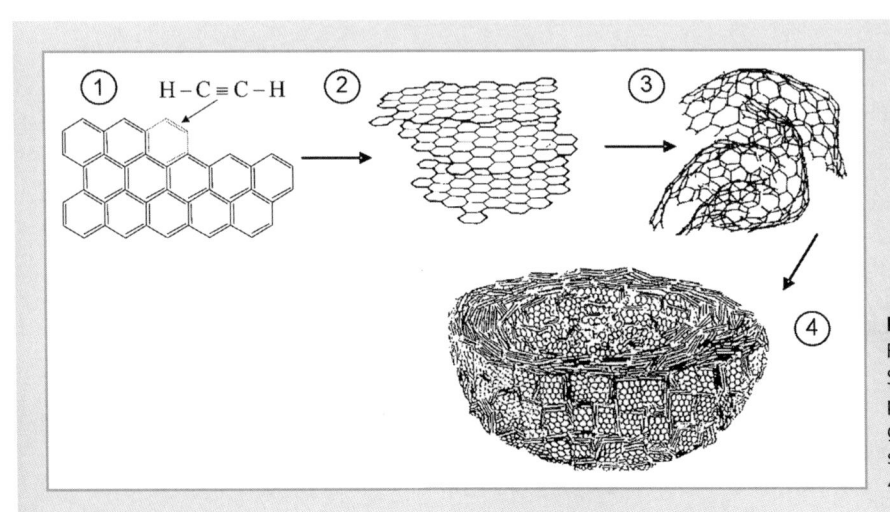

Fig. 15-30 Formation of soot particles according to Siegmann (based on [15-21]). *1* polycyclic (PAK) growth; *2* planar growth of the PAK; *3* formation of soot shaping by the formation of 3D clusters; *4* soot nuclei growth by condensation

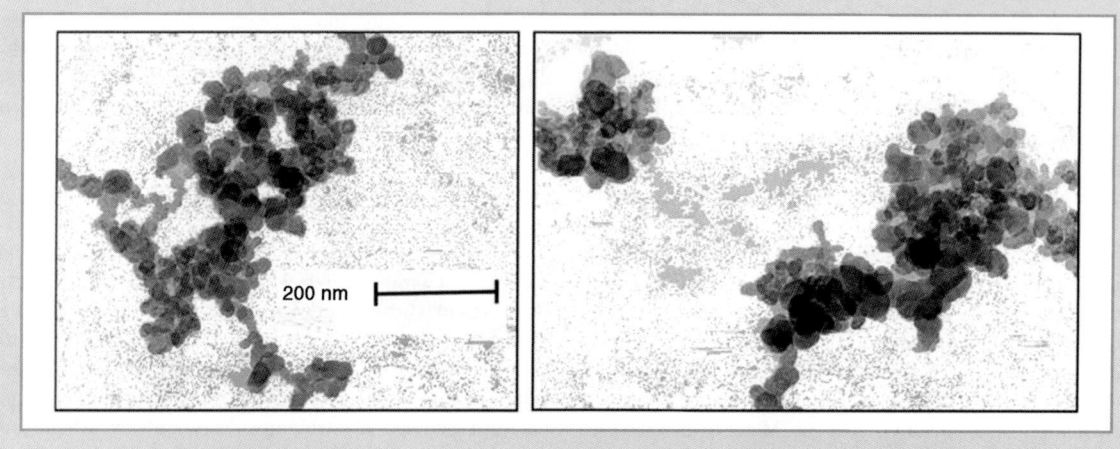

Fig. 15-31 Diesel particle agglomerates

According to both hypotheses of formation, primary particulates with diameters of less than 10 nm form first, are approximately spherical and have a density of 1.8 g/cm^3. These primary particulates subsequently agglomerate into the actual soot particulates, the individual particulates continually adhering to one another. Some SEM micrographs of typical agglomerates are pictured in Fig. 15-31. Very loose agglomerates only have a density of 0.02–0.06 g/cm^3 [15-21].

The bonding of highly mobile primary particulates causes the agglomerates to grow very rapidly at first. However, as the concentration of primary particulates decreases and the mobility of larger agglomerates diminishes [15-22], their growth in size decreases and a typical agglomerate size distribution appears. The size distribution is quite uniform even for different engines and is normally a log-normal distribution with a value of approximately 80–100 nm. Figure 15-32 presents the size distributions of particulate emissions of different vehicles at a constant operating point corresponding to a vehicle speed of 100 km/h [15-23]. The absolute number of particulates corresponds to the different emissions of the vehicles in the operating point and displays fluctuations that are three times as large in the maximum. However, the shape of the distribution and the position of the maximum are virtually independent of the vehicle and considered characteristic of all modern combustion systems.

A majority of the soot produced during combustion oxidizes while still in the combustion chamber. This afterburning takes place at temperatures above 1,000 K as soon as the combustion gases mix with the remaining fresh air, i.e. when

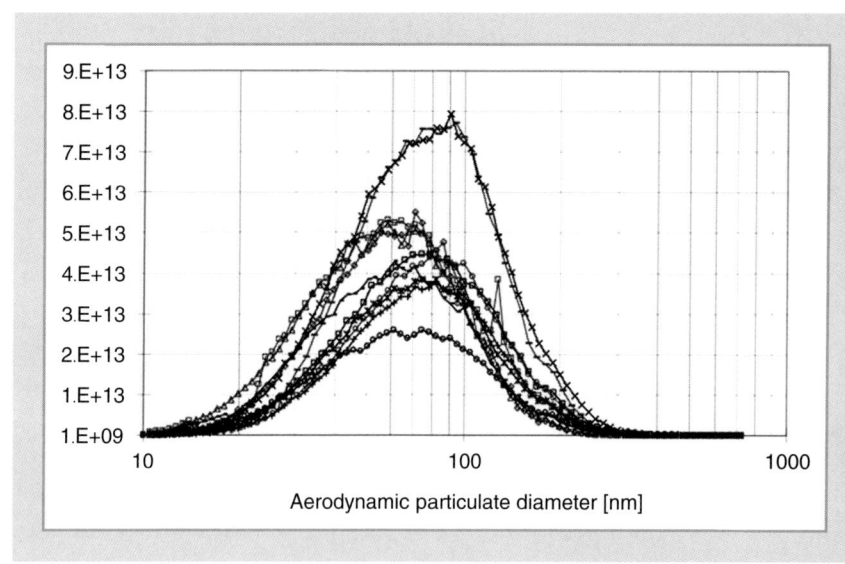

Fig. 15-32 Size distribution of particulate emissions of 11 different cars with state-of-the-art diesel engines (based on [15-23])

sufficient oxygen is available again. Unlike in gasoline engines, the mixture temperature in lean burn diesel engines rapidly drops below a critical value during the expansion phase and post-oxidation is frozen. The remaining particulates are discharged and can only be eliminated by exhaust gas aftertreatment with a particulate filter (see Sect. 15.5) [15-17].

15.3.3 Carbon Monoxide (CO)

Carbon monoxide (CO) emissions from diesel engines are normally very low and only increase during a conventional combustion process when the soot limit is approached. The remaining area of the map contains sufficient oxygen to completely oxidize the fuel. However, the prerequisite for this is good intermixing of the partially burned gases with the remaining fresh air at sufficiently high temperatures. Prechamber engines have particularly low CO emissions because of their intensive mixture turbulence during the flow from the prechamber into the main combustion chamber [15-18]. Modern combustion systems have swirl and/or squish flows, which help support air side mixture formation. Adjusting the combustion chamber air flow, the combustion chamber geometry and the injection geometry to one another minimizes CO emissions. The entire map must be optimized, paying particular attention to the lower part load range since lower temperatures cause the post-oxidation of CO to come to a standstill earlier [15-24]. In some engines, one of the charge ports is closed in this range to promote swirl, i.e. air side mixture formation.

The latest studies [15-24] have demonstrated that the fraction of CO from the rich spray core in the total CO emissions is low. The temperature in this range is high enough to assure that CO completely oxidizes to CO_2 when it is mixed with air after the end of combustion.

15.3.4 Hydrocarbons (HC)

Diesel engines may emit unburned hydrocarbons (HC) when inadequately prepared fuel reaches regions where the temperature no longer suffices for combustion. Such conditions exist in the lower part load range when there is substantial excess air. The function of an injection system is to prepare the fuel by atomizing it well so that it can evaporate completely even at low temperatures.

Locally very rich mixture zones can be other sources of HC emissions, e.g. when fuel sprays strike the combustion chamber wall. Complete evaporation cannot be assured during cold start and HC emissions increase. Misfires similar to the combustion of leaner mixtures in gasoline engines are usually not observed in diesel engines since direct injection always establishes a range with approximately stoichiometric mixing and thus with ideal conditions for auto-ignition.

Fuel contained in the nozzle holes and the injection nozzle's sac hole after the end of injection is a further source of HC emissions. This fuel evaporates during the expansion phase at temperatures far below the limit necessary for oxidation and is forced into the exhaust system unburned. This source of HC emissions has been reduced significantly in recent years by minimizing the volume of the sac hole [15-25].

An oxidation catalytic converter (see Sect. 15.5) can further reduce both hydrocarbons and carbon monoxide [15-17].

15.4 In-Engine Measures for Pollutant Reduction

The optimization of diesel engines generally entails a conflict of objectives between fuel consumption and emission reduction. Only a few additional measures allow simultaneously optimizing both parameters and a solitary measure is seldom able to simultaneously reduce every legally limited pollutant in equal measure. Fine tuning an engine requires a compromise. Other important parameters for engine tuning are comfort (noise) and engine dynamics. Since the impact of some measures is limited to certain ranges of the map and can reverse in other ranges, one challenge is tuning the engine in the entire map – for the entire speed range at different loads.

The numerous optimization parameters are weighted according to the legal requirements for pollutants and according to customer demand when the objectives of comfort and fuel consumption conflict. Thus, for example, combustion noise is increasingly receiving attention as a parameter for cars engines. Fuel consumption is generally the crucial parameter for customer utility in commercial vehicles with high mileage. Of course, the costs must be considered for every engine type in order to be able to offer a competitive product.

Table 15-15 provides an overview of different, currently common methods to optimize engines and their influence on pollutant emissions as well as specific fuel consumption and combustion noise. The principal correlations are identical for all vehicle types. Only the weighting changes according to market requirements. Thus, the different measures always have to be reevaluated on an individual basis. An advanced DI combustion system provides the basis for evaluation.

No one measure affects every parameter positively. Frequently, a combination of several measures is required to compensate a detrimental effect. Thus, for example, an increase in injection pressure only has a positive effect on NO_X emissions when it is combined with exhaust gas recirculation. Selected measures are discussed in detail below.

15.4.1 Start of Injection

The start of injection was the first parameter of diesel engine injection on which influence could be systematically exerted as a function of the operating point. Some distributor injection pumps were already able to control the start of injection,

Table 15-15 Various measures for the optimization of diesel engine combustion and their influence on different parameters

Measure	NOx	HC/CO	Soot	bsfc	Noise
Retarded start of injection	+	−	−	−	+
Exhaust gas recirculation	+	−	−	−	+
Cooled EGR	+	−	+	+	0
Supercharging	−	+	+	+	0
Intercooling	+	−	+	+	0
Pilot injection	0	+	−	0	+
Added post-injection	+	0	+	−	0
Injection pressure increase	0	+	+	+	0
Lower compression ratio	+	−	+	0	−

Symbols: +: reduction; −: increase; 0: no change

initially mechanically and later by an electrically actuated solenoid valve.

Figure 15-33 visualizes the importance of the start of injection for engine emissions. The NO_X and PM emissions are plotted for different starts of injection, the crank angle of 0° (0°CA) denoting top dead center. The steady rise of particulate emissions and the continuous drop of NO_X emissions are clearly discernible. The measurements are based on a commercial vehicle engine at mean load and 1,425 rpm.

Figure 15-34 presents the cylinder pressure curves measured during engine operation for four starts of injection selected from Fig. 15-33. The rise in pressure after top dead center (TDC) indicates the start of combustion. The pressure rises more steeply when the start of injection is advanced than when injection and combustion are long after TDC.

The flatter rise in pressure in retarded combustion can be attributed to the continuing expansion. On the one hand, it directly limits the rise in pressure. On the other hand, the expansion causes combustion chamber temperatures to drop and thus combustion to proceed more slowly. The pressure curves also allow the inference that the peak temperatures occurring in the cylinders will also be lower when the start of injection is retarded since the heat generated by slower combustion has more time to disperse from the zone of direct combustion. As explained in Sect. 15.3.1, not only the oxygen supply but also the local peak temperature is a crucial parameter for the formation of NO_X. Thus, the decreasing NO_X emissions (Fig. 15-33) when the start of injection is retarded can be explained by the cylinder pressure curves in Fig. 15-34.

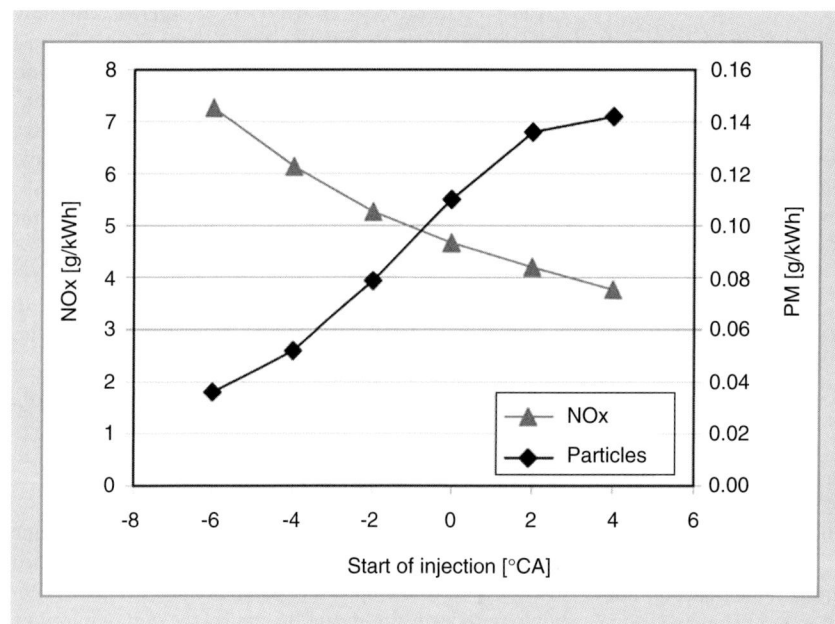

Fig. 15-33
Influence of the start of injection on PM- and NO_X emissions of a commercial vehicle engine at 1,425 rpm and mean load

Fig. 15-34
Cylinder pressure curve for four different starts of injection (cf. Fig. 15-33)

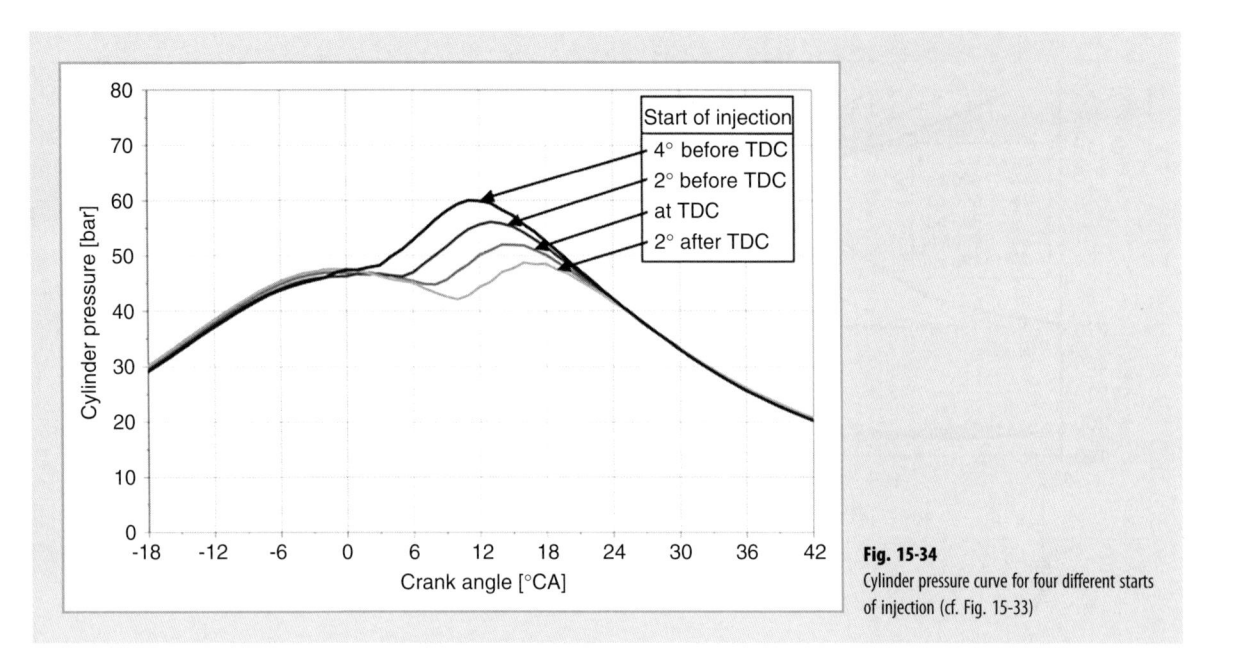

An increase in particulate matter when the start of injection is retarded is discernible in Fig. 15-33. This is a typical example of the aforementioned conflicts of objectives when every emission parameter is minimized. The increase in particulate matter applies to the majority of the map with the exception of the low load ranges and is caused by the decrease in the quality of mixture preparation as mixture density decreases and the reduced post-oxidation of the particulates due to lower temperatures in the combustion chamber.

A decrease in particulate emissions when the start of injection is retarded is also observable in the lower load range because the temperatures in the combustion chamber drop. This prevents particulates from forming at all [15-24]. However, a significant increase in the CO and HC emissions as well as fuel consumption must be accepted in this case. Thus, this strategy to simultaneously reduce NO_X and particulates can only be implemented to a certain extent.

Furthermore, the retarded start of injection affects brake specific fuel consumption adversely (not represented here). The consumption disadvantage can also be inferred from the cylinder pressure curves in Fig. 15-34. The rapid combustion near TDC when the start of injection is advanced resembles isochoric combustion that is optimal for consumption, while a convergence with isobaric combustion when injection is retarded worsens consumption.

15.4.2 Injection Pressure

Technical advances since the introduction of direct injection diesel engines have steadily increased the maximum injection pressure [15-25]. Figure 15-35 presents the influence of injection pressure on NO_X and PM emissions (the parameter black smoke (SN) being specified here) and brake specific fuel consumption b_e for a car engine's part load point at 50% load and $n = 1,400$ rpm.

The rise in brake specific fuel consumption for the retarded start of injection discussed above is recognizable for every injection pressure in the lower portion of the figure. A significantly more retarded start of injection without any disadvantage in fuel consumption can be set when injection pressures are higher than when they are lower. In this example, the minimum brake specific fuel consumption shifts from $-19°CA$ at 500 bar injection pressure to $-12°CA$ at 1,100 bar injection pressure because the duration of injection is shorter – since the injection rate increases with the injection pressure – and the quality of the mixture preparation improves as the injection pressure increases. Thus, the position of the center of combustion, which essentially determines brake specific fuel consumption, can be kept approximately constant.

The crucial advantage of higher injection pressure is evident in the clearly reduced particulate emission – the smoke number SN – in the center section of Fig. 15-35. When the start of injection is constant, soot emissions decrease significantly as injection pressures increase. An increase as injection pressure increases is only observable when the starts of injection are retarded considerably. Once again, the reason is the improved mixture preparation. On the one hand, less soot is produced because atomization improves at a higher injection pressure. On the other hand, the higher energy of mixture formation facilitates post-oxidation.

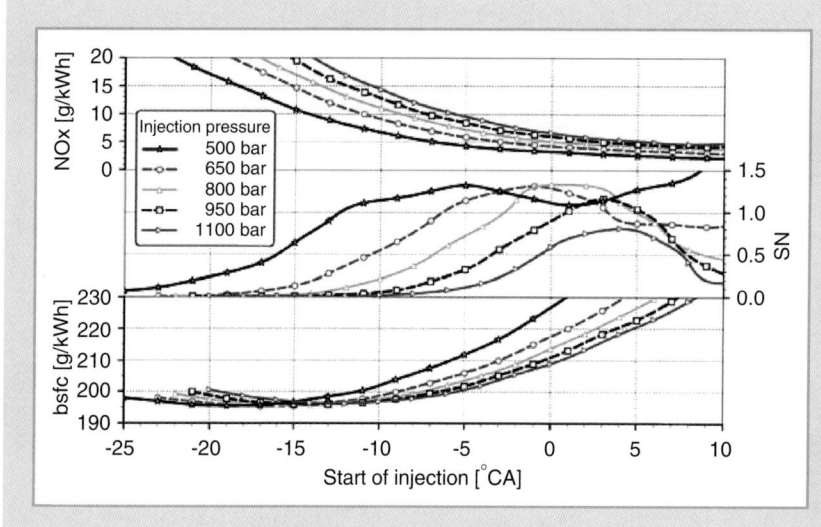

Fig. 15-35 NO$_X$ emissions, black smoke (SN) and specific fuel consumption as a function of the start of injection with the injection pressure as a parameter

However, a higher injection pressure as a function of higher local peak temperatures affects NO$_X$ emissions adversely. When the start of injection is constant, NO$_X$ emissions increase significantly as injection pressures increase. On the other hand, any comparison of NO$_X$ emissions becomes relative when brake specific fuel consumption is constant (e.g. at minimum brake specific fuel consumption, NO$_X$ emissions have roughly constant values of approximately 16 g/kWh for 500 bar injection pressure at $-19°$CA and for 1,100 bar injection pressure at $-12°$CA). Combining increased injection pressure with exhaust gas recirculation (see Sect. 15.4.3) can significantly reduce NO$_X$ emissions. This strategy has its limits though., Further elevation of the pressure fails to generate any further advantages above an upper limit on injection pressure, which is dependent on the combustion system and load point.

15.4.3 Exhaust Gas Recirculation

Exhaust gas recirculation (EGR) is now employed in cars everywhere as an important means to reduce NO$_X$. The exhaust gas recirculation rate x_{AGR} is defined as the ratio of the recirculated exhaust gas mass flow to the total mass flow in the intake tract:

$$x_{AGR} = \frac{\dot{m}_{AGR}}{\dot{m}_{AGR} + \dot{m}_{Luft}} \quad (15\text{-}6)$$

It may be as much as 50% in state-of-the-art combustion systems and is controlled by an electrically or pneumatically activated valve. A hot film air-mass flowmeter determines the mass of the fresh air.

Figure 15-36 presents the influence of exhaust gas recirculation on brake specific fuel consumption, noise and HC and PM emission as a function of the NO$_X$ emissions. The measured NO$_X$ emissions were selected as the abscissa and the other parameters were plotted on the ordinates in the individual graphs as dependent values. The experimentally determined individual points came from varying the EGR rate. The typical PM-NO$_X$ trade-off hyperbolae in the lower right graph are produced as a function of the EGR rate. Two different injection pressures were investigated. The EGR rate rises from right to left in each of the individual graphs, from 0% to the maximum value of 40%.

A continuous reduction of NO$_X$ emissions as the EGR rate increases is identifiable for both injection pressures in all the graphs. Slower combustion, retarded timing of the start of combustion by prolonging the ignition delay time and enlarged specific heat capacity through the higher percentage of triatomic gases (inert gas) in the cylinder charge all reduce the maximum local peak temperature and thus NO$_X$ emissions. The influence of the EGR rate on HC emissions turns out to be quite slight in the engine tested here. On the other hand, an increase in both PM emission and brake specific fuel consumption are recognizable in the two lower graphs. Slower combustion and thus the shift of the position of the center of combustion toward retardation cause the increase in fuel consumption. The limitation of the oxygen necessary for soot oxidation is also a primary reason for the higher soot emissions. The oxygen content reduced by EGR always acts to reduce NO$_X$ emissions and increase soot emissions. The hyperbolae produced in the lower right graph are characteristic of the conflicts of objectives when optimizing diesel engines.

Discussed in Sect. 15.4.2, the increase in NO$_X$ emissions when injection pressure is increased without other measures is also discernible in Fig. 15-36. The circled points denote the

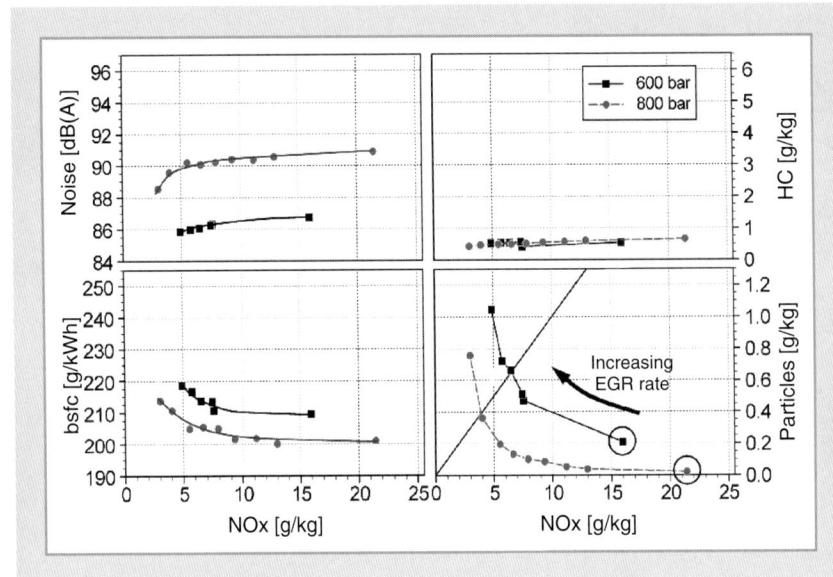

Fig. 15-36 Influence of the exhaust gas recirculation rate on noise, specific fuel consumption and HC and PM emissions as a function of the NO_X emissions for different injection pressures at 2,000 rpm and 50% load

experiments in which emissions without EGR were measured. The increase in NO_X emissions with the injection pressure is clearly recognizable. The PM emissions clearly exhibit lower values.

When the EGR rate is increased, the curve of the PM-NO_X trade-off for 800 bar injection pressure is clearly below the trade-off for 600 bar. This effect is called *enhanced EGR compatibility*. Since the energy of mixture formation is predominantly obtained from the injection spray, increasing the EGR rate can reduce O_2 more substantially without increasing PM emissions too much. However, these approaches are subject to limits that are dependent on the combustion system.

The straight line in the lower right graph in Fig. 15-36 denotes a 10:1 ratio of NO_X emissions to PM emissions. This corresponds to typical Euro 4 car application engineering since the legally mandated emission limits have exactly this ratio. The trade-offs' intersections with the straight line indicate that an increase of injection pressure from 600 to 800 bar reduces NO_X and PM emissions by approximately 35%. A roughly 3% reduction of the specific fuel consumption for this operating point is recognizable in the lower left graph.

An increase in injection pressure solely has an adverse effect on noise (upper left graph). The increase in noise can be attributed to the large mass of fuel injected during injection delay, which burns virtually instantaneously at the start of combustion. When the EGR rate is constant, the chemical ignition delay is constant and the physical ignition delay is slightly reduced by the smaller droplets. However, the effect of the higher injection rate outweighs this at higher injection pressure.

The following discusses the most important in-engine measure to reduce combustion noise, i.e. pilot injection.

15.4.4 Pilot Injection

Pilot injection (PI) has established itself as an effective measure to reduce noise in DI diesel engines. Small quantities of fuel (1–3 mm³ per injection) are injected in a short time interval before the main injection. This small quantity typically starts to combust shortly before TDC, resulting in an increase of temperature and pressure in the combustion chamber before the start of the main injection. Figure 15-37 presents typical cylinder pressure curves for a part load point with different pilot injected fuel quantities at a constant ratio of soot to NO_X of 1:10.

The lower portion of the graph represents the stroke of the injector needle. The pilot injected fuel quantity is the parameter for the different curves. The prolongation of the duration of the pilot injection as the pilot injected fuel quantity increases is evident. The duration of the main injection is shortened accordingly to keep the load set on the test bench constant.

The black curve in the upper portion of the graph represents the pressure curve measured in the cylinder during operation without pilot injection. This pressure curve corresponds to that of a motored engine (motored curve) until the start of combustion of the main injection (approximately 12°CA after TDC). The steep rise in pressure at the start of combustion until approximately 15°CA after TDC causes undesired high noise emission. Pressure oscillations after the end of combustion are artifacts attributable to the method of measurement.

The three curves of differing grays were determined with differently sized pilot injected fuel quantities. The start of the control of the main and pilot injection was kept constant. The

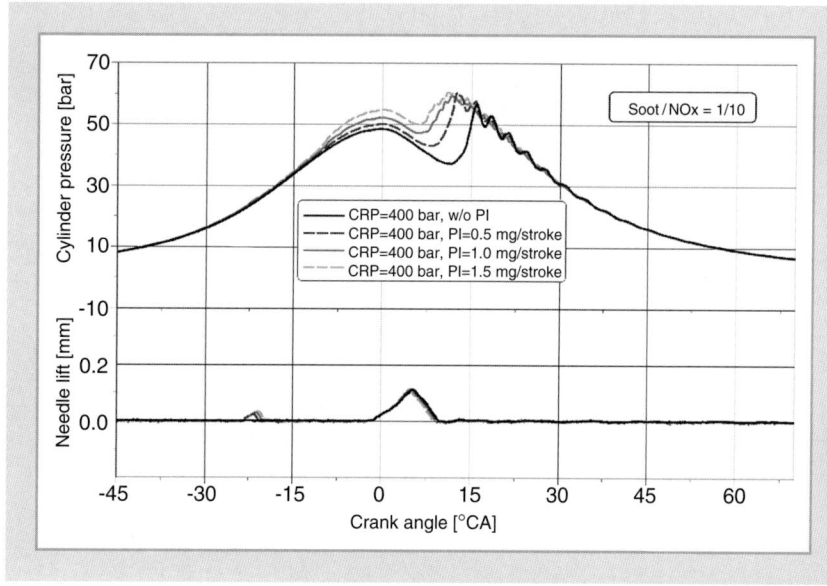

Fig. 15-37 Cylinder pressure curves in part load operation with a varied pilot injected fuel quantity

start of pilot injection combustion is discernible in the increase in pressure at approximately 12°CA before TDC where the pressure curve begins to deviate significantly from the motored curve. A local pressure maximum is reached at approximately TDC in each case. The absolute value is a function of the parameter of the pilot injected fuel quantity and increases with it.

The cylinder pressures and temperatures that increase when the pilot injected fuel quantity increases shorten the chemical ignition delay for the main injected fuel quantity. This is indicated in Fig. 15-37 by the increase in pressure, which characterizes the start of main injection combustion (6–12°CA after TDC), and moves closer and closer after TDC as pilot injected fuel quantities increase. The gradient of the increase in pressure of the main injection always grows smaller as pilot injected fuel quantities increase – at approximately constant maximum pressure in this example. The reasons have already been discussed: a shorter ignition delay decreases the quantity of fuel injected during the ignition delay. The rapid combustion of this fuel quantity at the start of combustion affects the increase in pressure and the combustion noise. Therefore, the cylinder pressure gradient is a measure of the level of combustion noise.

The influence of the pilot injected fuel quantity on noise and emissions as a function of the exhaust gas recirculation rate is represented in Fig. 15-38 as a function of the NO_x emissions.

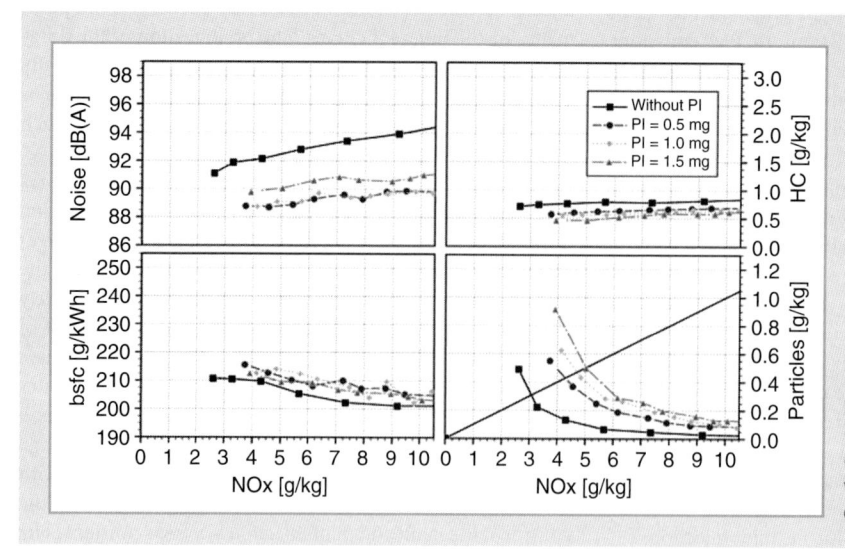

Fig. 15-38 HC and PM emissions, specific fuel consumption and noise in part load operation with and without pilot injection as a function of the NO_x emissions with a varied EGR rate

This type of representation is already familiar from Fig. 15-36: the different parameters have been represented as a function of the NO_X emissions by varying the EGR rate. The upper left graph plots the aforementioned decrease in combustion noise brought about by the pilot injection. A minimum is already obtained at a pilot fuel quantity of 0.5–1 mg per injection. Another increase of the pilot fuel quantity to 1.5 mg causes an increase in noise: The combustion of the pilot injected fuel quantity itself determines the combustion noise. Moreover, an increasing EGR rate slightly reduces noise for every pilot injected fuel quantity. This can be attributed to the slower combustion caused by the combustion air's reduced oxygen content.

As the upper right graph indicates, pilot injection lowers HC emissions. This effect greatly depends on the combustion system and the operating point.

As the lower right graph indicates, the combustion system's significantly reduced EGR compatibility is a negative effect of PI. The larger the pilot fuel quantity, the more detrimental the effect is. This underscores the importance of precise injection of the minutest pilot fuel quantities for the combined optimization of noise and particulates [15-25]. Since the increased combustion noise must be compensated by a pilot injection in some cases of application, part of the advantages for the PM-NO_X trade-off and brake specific fuel consumption obtained by increasing the injection pressure are lost. Thus, optimization of the overall system again requires a weighted analysis of every influencing variable.

Advanced combustion systems apply a second pilot injection and/or a post-injection positioned shortly after main injection for further optimization. While the second pilot injection is applied to further optimize noise, the added post-injection reduces soot emissions. Increased turbulence in the combustion chamber caused by the added post-injection facilitates soot oxidation. At the same time, the temperature at the end of combustion is increased. This also has a positive effect on soot oxidation. The efficacy of the added post-injection greatly depends on the combustion system and the map range being analyzed.

15.4.5 EGR Cooling and Intercooling

A minimum charge air temperature before the intake valves is advantageous for several reasons. On the one hand, charge air density that increases as the temperature drops produces efficient cylinder charging. This is called thermal engine dethrottling. As a consequence, the EGR compatibility increases and NO_X and PM emissions can be lowered further. In addition, thermal dethrottling produces a consumption advantage through the air/fuel ratio. This underscores the positive influence of EGR and intercooling on an engine's thermodynamic performance.

Both the compression of the charge air in the turbocharger and the recirculation of hot exhaust gas increase the air temperature in the cylinder intake. Therefore, processes have been developed to limit the temperature. These include the use of intercoolers virtually all over and the use of processes to cool the recirculated exhaust gas for cars in particular. The latter can entail long line length, line routing through the cylinder head and/or a heat exchanger cooled with cooling water. Numerous applications are able to control the cooling capacity or bypass the refrigeration unit to quickly heat up the engine cooling water to heat the passenger cell during a cold start or to prevent unduly low combustion chamber temperatures from increasing HC and CO emissions during operation in the lower part load range.

Whether the gas temperature in the intake valves is reduced by cooling the recirculated mass of exhaust gas or by cooling the compressed charge air is irrelevant thermodynamically.

Figure 15-39 presents the NO_X and particulate emissions for two different temperature levels after the compressor (T2) as a function of the EGR rate. The upper graph plots the resultant air/fuel ratio. Specific fuel consumption and noise emissions as well as boost pressure and exhaust back pressure were equalized during the measurements. The higher value of λ for an EGR rate of 0% on the far right in the top graph indicates that the lower temperature causes better cylinder charging. The lower graph displays lower NO_X emissions for colder intake air when EGR is disabled. This can be attributed to the low peak temperatures that result. Improved emission performance with exhaust gas recirculation is also recognizable for the lower temperature in Fig. 15-39: significantly higher EGR rates with commensurately lower NO_X emissions can be applied without an unduly high increase in particulate emissions.

15.5 Exhaust Gas Aftertreatment

15.5.1 Introduction

Emission limits for diesel vehicles have dropped steadily in recent years and further reductions for the future have already been agreed upon. Emission reduction refers to every technology and system that is instrumental in lowering vehicle and engine emissions. Emission reduction systems not only have to satisfy legal requirements but also other technical boundary conditions (e.g. structural space and exhaust gas temperatures) and economic requirements.

Emission reduction can be divided into engine measures and downstream measures, which are frequently also combined as exhaust gas aftertreatment.

In the past, it was initially possible to lower emission limits by improving engine combustion with corresponding decreases of raw emissions (see Sect. 15.4). Although raw emissions are continuing to decrease, this will no longer suffice in the future.

The term exhaust gas aftertreatment subsumes the systems located in the exhaust gas system with the primary function of reducing engine emissions. They include catalytic converters,

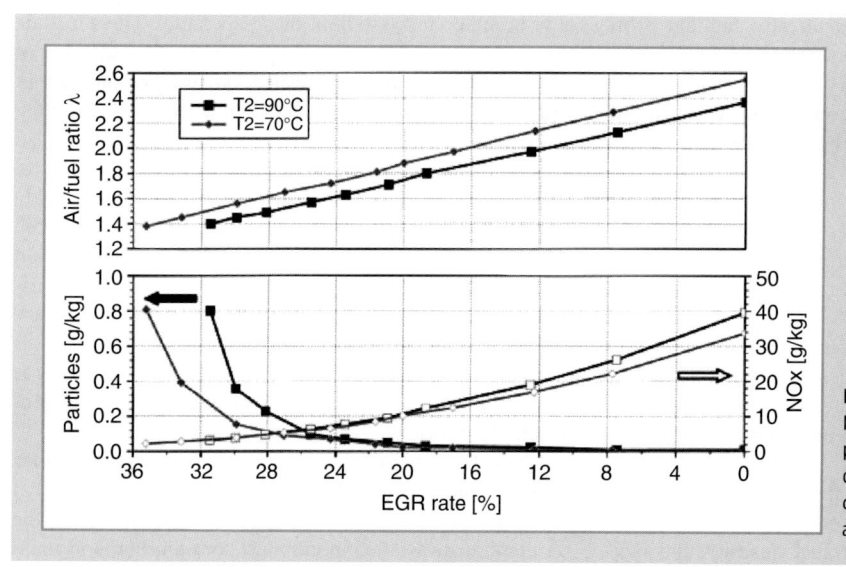

Fig. 15-39 Influence of the EGR rate on NO_x and particulate emissions and resultant λ for different air temperatures after the compressor (T2) for a part load point at approximately 50% load and 2,000 rpm

sensors, particulate filters and auxiliary systems that may introduce a reductant or support particulate filter regeneration. Exhaust gas aftertreatment systems primarily reduce pollutant concentrations through chemical processes. Particulate filters also employ physical separation.

Apart from the desired conversion of pollutants, a number of additional technical requirements must be considered when selecting a system. The boundary conditions are briefly presented below before the individual components are described.

The *exhaust gas temperatures* of diesel engines are now usually so low that the chemical processes are frequently limited kinetically even when high quality catalytic converters are used. In addition, the exhaust gas temperatures greatly depend on the engine operating conditions, especially torque. Temperatures can reach values between the outside temperature (immediately after engine start) and up to over 700°C at full load. Temperatures in the front underbody area of the exhaust gas system are typically between 150 and 250°C, e.g. during the European driving cycle. Therefore, not only the optimization of the chemical reaction rate but also the protection of the components must be considered when designing and positioning components.

The *exhaust gas mass flow* also largely depends on engine operating conditions, particularly engine speed, the supercharging rate and the exhaust gas recirculation rate. These operating conditions can change considerably within a few seconds, thus changing the space velocity (the ratio of the volumetric exhaust gas flow to the volume of the individual catalytic components) to the same extent. Space velocity is a measure of the residence time of exhaust gas in catalytic converters. When residence times are too short, the gas is only able to react inadequately and conversion is diminished.

Therefore, catalytic converters must be sized to ensure sufficiently large conversion even at high exhaust gas mass flows. However, the components' thermal mass increases as their sizes increase. This affects the temperature curve of upstream components. The allowance for such interactions and the optimization of the overall system for all requirements are the object of system development and discussed after the description of the components.

Exhaust gas aftertreatment components represent aerodynamic obstacles that generate *exhaust back pressure* as a function of the volumetric exhaust gas flow. This exhaust back pressure must be overcome by the engine and increases the energy loss of gas exchange. In a bad case, this causes measurably increased fuel consumption, which translates into increased operating costs as well as increased CO_2 emission. Therefore, minimal flow losses are important in the design of exhaust gas aftertreatment systems.

In addition, the flow restriction damps engine noise induced by exhaust gas. Vehicle operation transmits acceleration torque to the exhaust gas system. These mechanical stresses must be factored into the engineering of structural elements and the overall system and are treated elsewhere (see Sect. 13.2).

Engine exhaust gas contains vapor, which condenses while the exhaust gas system cools after the engine has been cut off. This exhaust condensate also contains corrosive constituents formed in reactions with the nitrogen oxides and sulfur dioxide contained in the exhaust gas. Engineering and material selection must include *corrosion protection*.

A suitable exhaust gas aftertreatment system that incorporates the aforementioned restrictions meets the requirements of reducing several pollutants at minimum *cost*. The design must also ensure the system's *service life* is long enough.

Along with the aforementioned corrosive and mechanical stresses, the deterioration of catalytic coats must also be considered.

The multitude of requirements and interactions require complex design optimization customized for each vehicle model.

The individual components of an exhaust gas aftertreatment system are treated below first. Usually, they also have secondary functions in addition to their primary function. Suitable designs of exhaust gas aftertreatment systems with properly adjusted raw engine emission reduce the complexity and cost of overall emission reduction. The following section addresses this aspect of system design.

15.5.2 Exhaust Gas Aftertreatment Components

15.5.2.1 Diesel Oxidation Catalyst (DOC)

Diesel oxidation catalysts (DOC) were the first catalytic converters installed in diesel vehicles as standard. Their primary function is to oxidize engine carbon monoxide (CO) and hydrocarbon (HC) emissions with the residual oxygen of the exhaust gas into the harmless gases H_2O and CO_2. Precious metal coatings are employed for this. Advanced exhaust gas aftertreatment systems also include additional components and the DOC in these systems also assumes other functions:

- It oxidizes the volatile constituents of the particulate matter (adsorbed hydrocarbons), thus reducing particulate mass by up to 30%.
- It improves the ratios of nitrogen dioxide (NO_2) to nitrogen monoxide (NO). This step is conducive to nitrogen oxide reduction, especially for the SCR process.
- It releases heat by oxidizing deliberately supplied hydrocarbons and CO (as a so-called catalytic burner). This increases the temperature of an exhaust gas system after DOC. It is applied to facilitate the increase in temperature necessary for particulate filter regeneration and is additionally utilized for temperature management measures that bring deNOx systems to the operating temperature as quickly as possible after a start. This improves NO_X conversion.
- When suitable coated, it additionally reduces NO_X slightly (approximately 5–10%) by reacting with HC and CO.

All these functions are performed by the same principle made possible by the catalytic converter's basic structure. The converter body consists of a ceramic or metallic honeycomb structure in which the exhaust gas is routed through channels approximately 1 mm in width. The channel walls consist of a ceramic or metal substrate structure covered with a catalyst washcoat containing precious metal. Exhaust gas components diffuse onto this catalyst coat when they flow through the converter body and are oxidized.

The basic variables influencing conversion are the:
- activity of the catalyst coating,
- size and inner geometry of the catalytic converter to which the residence time of the gas and the space velocity are related,
- catalytic converter temperature and
- concentration of the reaction partners.

The washcoat's catalytic activity is basically defined by the type and quantity of its material and the spatial structure of its surface. DOCs utilize precious metals from the platinum group (platinum and palladium), which are dispersed on an oxide washcoat (aluminum oxide, cerium oxide or zirconium oxide) in the form of very small particulates (with a size of a few nm). The washcoat furnishes a very large internal surface area, stabilizes the precious metal particulates against sintering and supports the course of reactions either directly by reacting on the boundary between particulates and the substrate or indirectly by adsorbing catalyst poisons. Also frequently called catalyst loading, the quantity of catalyst utilized is in the range 50–90 g ft^{-3} (1.8–3.2 g l^{-1}).

The external dimensions (diameter and length), the density of the channels (specified in channels per square inch or cpsi) and the strength of the walls between the individual channels are the basic structural properties of a converter body. These properties determine a catalytic converter's mechanical stability, exhaust back pressure and heating performance.

Typical values for the space velocity are between 150,000 and 250,000 h^{-1}. The volumetric exhaust gas flow depends on the engine displacement among other things. Placing the volume of a catalytic converter in relation to displacement yields values of $V_{Kat}/V_{Hub} = 0.4-0.8$.

As already mentioned in the introduction, a catalytic converter's temperature is dependent on the engine's operating state. If the exhaust gas temperature increases after the turbine of the turbocharger, then the catalytic converter temperature also follows with a delay induced by the exhaust gas system's thermal mass. Figure 15-40 presents a typical conversion

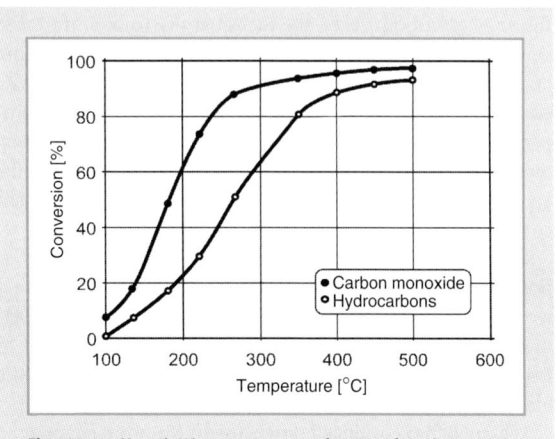

Fig. 15-40 CO and HC conversion as a function of catalytic converter temperature

curve for CO oxidation. A very steep increase in conversion is discernible. The exhaust gas temperature at 50% conversion is referred to as the catalytic converter's light-off temperature and is 150–200°C, depending on the catalyst composition, flow velocity and exhaust gas composition. Over 90% of the CO is then converted at higher temperatures. The oxidation of hydrocarbons proceeds similarly but at somewhat higher temperatures and specifically depends on the composition of the hydrocarbons. Thus, for example, methane is only converted at very high temperatures, while short-chain alkenes already react at low temperatures.

A basic function of a DOC is to improve the ratio of NO_2 to NO. NO_2 is beneficial for a number of exhaust gas aftertreatment systems (DPF, NSC and SCR). In the presence of oxygen, NO and NO_2 are in equilibrium with each other, at low temperatures (<250°C) on the NO_2 side and at high temperatures (>450°C) on the NO side. Depending on the operating point, there is somewhere between 5 and 50% NO_2 in NO_X in the engine exhaust gas. This is far under the equilibrium value effective for the exhaust gas temperatures for most operating conditions. Therefore, a DOC can use catalytic reactions to increase the ratio of NO_2 to NO toward equilibrium as of 180–230°C and above. At high temperatures (>450°C), the NO_2 concentration commensurate with the thermodynamic equilibrium drops as the temperature rises. Aside from the exhaust gas temperature, the HC and CO concentration is also a significant factor that influences NO oxidation. Thus, HC or CO reduction can lower the percentage of NO_2 below the initial value, even in the mean temperature range.

Given its low concentration, the reaction heat released during oxidation does not cause any appreciable increase in the exhaust gas temperature. Additional hydrocarbons must be introduced before the DOC when an increase in temperature is desired, e.g. to initiate particulate filter regeneration. Then, the DOC assumes the function of a catalytic heating component (a catalytic burner). HC may either be introduced by an engine post-injection or by a device downstream from the engine. In both cases, the fuel quantity to be introduced can be calculated from the desired increase in temperature and the exhaust gas mass flow. A 1% increase in the CO concentration for an approximately 90°C increase in temperature is the approximation applied. The energy is released on the catalytic surface area, which transfers the heat to the exhaust gas by convection. The maximum permissible washcoat temperature (e.g. 800°C) limits the heat output.

Engine post-injection must be applied so that, as far as possible, it no longer plays a role in combustion. Otherwise it would cause an undesired increase in torque. However, unduly retarded injection causes part of the diesel fuel to reach as far as the cylinder wall. This would dilute the lubricating oil.

When HC is introduced downstream from the engine, the required quantity must disperse throughout the flow cross section as homogenously as possible and reach the catalytic converter fully vaporized. Various approaches are in development:
- liquid introduction of the diesel sprays with a metering valve,
- the introduction of vaporized diesel fuel and
- the introduction of gases with low light-off temperatures, e.g. from H_2/CO mixtures preferably generated from diesel fuel directly in the vehicle.

Operation can diminish the effectiveness of catalytic converters over the course of time. This is known as catalyst deterioration. Two deterioration mechanisms are fundamental. On the one hand, the precious metal particulates can agglomerate at very high exhaust gas temperatures. This reduces the specific surface area of the precious metals. On the other hand, catalyst poisons can either coat the precious metal surface directly or make them inaccessible to the requisite diffusion processes by forming voluminous layers on the washcoat. The best known catalyst poison is the sulfur contained in the fuel. It forms sulfates on the surface, which inhibit the precious metal's accessibility. Today's fuels are sulfur-free or at least low-sulfur. This reduces the risk of DOC sulfurization.

Part of the processes of deterioration are irreversible. Such processes must be prevented by selecting appropriate exhaust gas temperatures and fuel qualities. However, some catalyst poisoning can be reversed by selecting appropriate operating conditions.

15.5.2.2 Particulate Filters

The function of diesel particulate filters (DPF) is to separate a very large fraction of particulates from the exhaust gas flow. Given the small size of the particulates (the majority being smaller than 100 nm; see Sect. 15.3.2), only filtration provides sufficiently large filtration efficiency with relative ease.

Over time, the increasing quantity of filtrate increases the flow restriction through a filter. This causes higher fuel consumption. Therefore, a filter requires regeneration in certain intervals, i.e. utilizing appropriate operating conditions to oxidize the combustible filtrate constituents (incombustible filtrate constituents remaining behind as ash). Thus, operation can be divided into long phases of particulate filtration interrupted by short phases of regeneration. Therefore, particulate filter operation requires an operation strategy and other components that collectively form the DPF system. The following describes the structure of the DPF, which is important for the filtration phase, then examines the regeneration phase and finally explains the other elements of a DPF system.

The requirements for a particulate filter are:
- a high filtration rate even for extremely small particulates (50–95% depending on the law and raw emission),
- low flow restriction,
- high temperatures strength against temperatures of up to 1,000°C, which occur during regeneration, and

- good structural and aerodynamic tolerance toward nonoxidizable particulate constituents (filter ash).

Four filter types in are in use at present:
(a) ceramic extrudates made of cordierite,
(b) ceramic extrudates made of silicon carbide (SiC),
(c) sintered metal filters (particularly for the retrofit market) and
(d) particle separators with open structures.

The first three filter types are based on the wall-flow filter principle (closed filter systems), which routes the complete quantity of exhaust gas through a porous wall. Every other channel in the ceramic extrudate is closed on the front and back (Fig. 15-41). This makes the extrudate walls a porous filter surface with a very large surface area (approximately 1 m^2 l$_{Filter}^{-1}$). When particulates pass through the porous wall, they initially diffuse onto the inside pores. After a short time, a thin surface filtration layer begins forming on the surface of the walls, which have significantly smaller pores than the substrate structure and consequently trap the majority of the particulates. The thickness of the filtrate layer increases with the loading time. This initially increases the flow restriction and then also increases the flow restriction in the filter's inlet channels as it continues. Flow restriction and filtration efficiency are a function of wall thickness (0.3–0.4 mm) and pore size. Moreover, channel density (100–300 cpsi) is important for flow restriction. While high channel density increases the inner surface area and thus reduces the wall's resistance to penetration, it also results in smaller channel diameters. This increases the flow restriction in the channels, especially when the surface filtrate additionally constricts the cross section of the inlet channels. In a new development, the diameters of the ingoing channels are larger than the diameters of the outgoing channels. This improves the flow loss induced by surface filtrate and the compatibility with ash deposits.

Sintered metal filters are composed of filter cells with large inlet cross sections that increasingly taper in the direction of flow. This geometry also reduces the inlet flow loss and increases ash compatibility.

Wall-flow filters filter the entire exhaust gas, producing filter efficiencies above 95% for the entire relevant size range (10 nm^{-1} μm). When a filter cannot be regenerated in due time, e.g. because a retrofit solution fails to provide every measure to trigger regeneration, then the exhaust back pressure can increase so much that engine operation is impeded. Such filter blockage is impossible in an open particle separator.

Instead of imperatively routing the exhaust gas through a wall, the structure of an open particle separator initially deflects it into the cells protruding into the filter channels. This delivers a pulse to the gas, thus accelerating it toward a porous channel wall. When filter loading is low, the majority of it penetrates the channel wall much like in a wall-flow filter and filtrate builds up. As the resistance to penetration increases, the fraction of gas that penetrates the wall decreases and the fraction in the channel that bypasses a cell increases. Therefore, the filtration efficiency in such structures is a function of loading and between approximately 30 and 70% in real applications.

Similar to a DOC, the filter's manufactured size and thus its surface area are oriented toward displacement (typically $V_{DPF}/V_{Displ.} = 1.2$–2.0).

Depending on the design and the raw emissions, filter regeneration is necessary after 300–800 km. Above approximately 600°C, soot burns with the oxygen contained in the exhaust gas to become CO_2, releasing heat. The design point for regeneration is the accumulated quantity of soot (5–10 g l^{-1} depending on the filter material). If this quantity is too large, then local temperatures can peak during exothermic regeneration and damage the substrate or the catalytic coating as the case may be.

The temperatures necessary for regeneration only exist in an engine's rated power range. Therefore, additional measures must be planned, which facilitate timely filter regeneration even under ordinary in-use driving.

The following regeneration strategies exist:
1. uncatalyzed oxidation by residual oxygen at 550–650°C,
2. additively supported regeneration,
3. regeneration with NO_2 and
4. regeneration with catalyzed diesel particulate filters (CDPF).

Non-catalytic Oxidation

Uncatalyzed oxidation employs various measures to raise the filter temperature to the ignition temperature. In principle, the temperature of the DPF can be increased by degrading the engine efficiency (changing the injection characteristic), reducing the exhaust gas mass flow (lowering boost pressure or throttling) or increasing the temperature downstream from the engine (e.g. by a catalytic burner in the DOC). All these measures must ensure that the residual oxygen content in the DPF is high enough for rapid regeneration (>5%). The measures are a function of the engine operating point and assembled into packages of measures (Fig. 15-42).

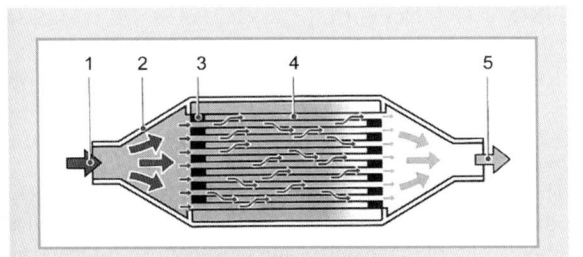

Fig. 15-41 Filter structure. *1* Raw exhaust gas from the engine, *2* Housing, *3* Ceramic plug, *4* Ceramic honeycomb, *5* Effluent exhaust gas

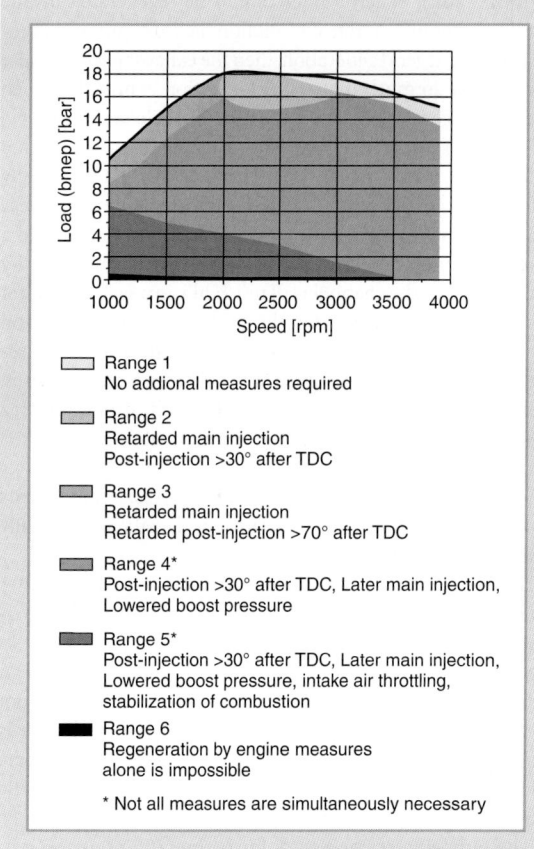

Fig. 15-42 Engine measures that boost exhaust gas temperature

Legend:
- Range 1: No addional measures required
- Range 2: Retarded main injection; Post-injection >30° after TDC
- Range 3: Retarded main injection; Retarded post-injection >70° after TDC
- Range 4*: Post-injection >30° after TDC, Later main injection, Lowered boost pressure
- Range 5*: Post-injection >30° after TDC, Later main injection, Lowered boost pressure, intake air throttling, stabilization of combustion
- Range 6: Regeneration by engine measures alone is impossible

* Not all measures are simultaneously necessary

The engine temperatures in *range 1* (rated power range) are already so high that no other measures must be introduced. This range is very rarely found in car operation.

The very high torques required in *range 2* must be available despite regeneration measures. The main injection is shifted to be somewhat more retarded. This degrades engine efficiency and increases the exhaust gas temperature as desired. In addition, an added (advanced) post-injection takes place, which still plays a part in combustion and delivers another torque contribution. These measures aimed at increasing the engine exhaust gas temperature by degrading efficiency are called engine burners.

The supercharging in *range 3* is low and the air/fuel ratio already below 1.4 when combustion is optimal. In this case, an added post-injection would locally cause very low air/fuel ratios and thus a strong increase of black smoke. Therefore, the post-injection is retarded long enough that it no longer plays a part in combustion (retarded post-injection). The additional HC and CO emissions are converted into heat in the DOC (catalytic burner).

Various measures are combined with one another in *range 4*. However, lowering the boost pressure reduces the exhaust gas mass flow. Retarding the main injection and adding a post-injection further increases the exhaust gas temperature in *range 2*. The proportion of individual measures must be optimized for noise, emissions and consumption and are usually not all required simultaneously.

A sizeable temperature increase of 300–400 °C is required in *range 5*. A throttle valve additionally reduces the air mass in this range. This necessitates further measures to stabilize combustion, e.g. increasing the pilot injected fuel quantity and adjusting the interval between the pilot and main injection.

The very small torques in *range 6* make it impossible to trigger regeneration at temperatures >600 °C.

The scope of the measures in every range significantly depends on the temperature generated. The thermal loss between the engine and filter should be kept to a minimum to fully maximize the effect of the heat generated in the engine in the filter. Therefore, many applications place the filter as close to the engine as possible.

Catalytically supporting soot oxidation is another option to lower the required regeneration temperature.

Additive Systems

Additive systems blend additives (usually cerium or iron compounds) into fuel. The metal bonds to the soot during engine combustion. This produces a surface area of soot in the filter that is doped with mixed oxides that lower the ignition temperature to 450–500 °C, which allows reducing the scope of the aforementioned measures. After the soot oxidizes, the metal oxide remains in the filter as residue and thus increases the ash fraction that thermal regeneration is unable to remove. Therefore, additive regeneration requires the removal and mechanical cleaning of conventional wall-flow filters every 120,000–180,000 km.

NO$_2$ Regeneration

NO$_2$ is an extremely active oxidant that already oxidizes soot as of temperatures of 250–350 °C. Such temperatures are frequently reached in commercial vehicle applications and in car applications during highway driving for instance. NO forms during soot oxidation.

$$2\,NO_2 + C \rightarrow 2\,NO + CO_2 \tag{15-7}$$

$$NO_2 + C \rightarrow NO + CO \tag{15-8}$$

$$CO + NO_2 \rightarrow CO_2 + NO \tag{15-9}$$

$$CO + {}^1/_2 O_2 \rightarrow CO_2 \tag{15-10}$$

The equations indicate that eight times the mass of NO$_2$ must be present to completely oxidize soot. On average, the amount of soot that oxidizes and new soot that is separated is the same when the temperature and mass ratio are high enough ($T > 350$ °C). This is referred to as a CRT® (continuous regenerating trap). The required NO$_2$ forms from

NO in upstream oxidation catalysts. In practice, CRT® steadily oxidizes a certain fraction of the soot, especially in high load phases, thus making it possible to prolong the regeneration intervals. However, CRT® is unable to completely regenerate DPF under every single driving condition, particularly in car applications. Thus, the aforementioned additional active regeneration measures must be designed in.

Catalyzed Diesel Particulate Filter (CDPF)

Catalyzed diesel particulate filters (CDPF) can lower the regeneration temperature slightly. While their catalytic coating has far less of an effect than the use of fuel additives, it does not produce any additive ashes.

Catalytic coating performs several functions:
- oxidizing CO and HC and
- oxidizing NO to NO_2.

Just like a DOC, a CDPF can also oxidize HC and CO while releasing heat. In this case, the temperature lift produced acts directly in the point where high temperatures are required to ignite soot. The heat losses that may occur when an upstream catalytic burner is employed can be prevented. Just as for a catalytic burner, either engine post-injection or a metering unit downstream from the engine supplies the requisite HC and CO to the exhaust gas system.

In addition, NO oxidizes to NO_2 in the catalytic coating. In small quantities, it can support soot oxidation at low temperatures.

A DPF system consists of the particulate filter and other components:
- a *DOC* employed as a catalytic burner and to boost the NO_2 fraction,
- a *temperature sensor before the DOC* that determines the convertability of HC in the DOC (light-off state),
- a *differential pressure sensor* that measures the differential pressure through the particulate filter from which the flow restriction can be calculated with the volumetric exhaust gas flow and
- a *temperature sensor before the DPF* that determines the DPF temperature important to control regeneration.

Moreover, a DPF system must have appropriate control unit functions to control and monitor regeneration. First, a particulate filter's loading state must be measured during the loading phase (loading detection). Various methods are employed to do this. The flow restriction, which increases as the filter load increases, is determined with the aid of a differential pressure sensor. A function of the preceding operating conditions, the influence of the soot coat's morphology (e.g. equipartition of the coat thickness over the filter and filtrate porosity) interferes with the correlation between the quantity of soot and the flow restriction.

Therefore, the quantity of included soot arrived at by integrating the engine soot mass flow is additionally modeled mathematically. Moreover, the removal of soot by the CRT® is also incorporated. The soot burn-off is calculated during the regeneration phase as a function of the filter temperature and the oxygen mass flow.

A so-called coordinator determines the soot mass decisive for the regeneration strategy from the soot mass values calculated with both methods.

The regeneration strategy determines when regeneration is triggered and what measures are initiated. To prevent thermal damage to the substrate during regeneration, a threshold at which regeneration should take place is defined for the soot mass as a function of the material used. Advancing regeneration is expedient when particularly favorable conditions exist (e.g. highway driving). A regeneration strategy defines the regeneration measures to be executed as a function of the soot mass and the operating state of the engine and vehicle. These are transmitted to the other engine control functions as a status value.

The DPF temperature is regulated during regeneration to prevent uncontrolled overheating in a filter or uncontrolled termination of regeneration. The injection characteristic and the air mass are available as manipulated variables.

15.5.2.3 NO_X Reduction Catalysts

Like methods of particulate matter reduction, only some of the theoretically conceivable methods of NO_X reduction are implementable in vehicles.

Effectively installed in gasoline engines, three-way catalytic converters in which NO_X reacts with HC and CO with very high conversion at $\lambda = 1$ to become N_2, H_2O and CO_2 cannot be implemented in lean diesel exhaust gas. The desired reduction of NO_X vies with the reduction of the residual oxygen, which is present in an approximately thousand-fold concentration.

Two systems (SCR and NSC) are being launched on the market at present.

Selective Catalytic Reduction (SCR)

Using ammonia (NH_3) as the reductant, selective catalytic reduction (SCR) reduces NO_X to nitrogen in a suitable catalytic converter. SCR has proven itself in large firing plants and is found in commercial vehicles that have been launched widely on the market (since around 2005). Initial applications in cars are even planned (starting around 2007).

Basically, the actual SCR reaction proceeds according to the following reaction equations:

$$4\,NO + O_2 + 4\,NH_3 \rightarrow 4\,N_2 + 6\,H_2O \quad (15\text{-}11)$$

$$NO + NO_2 + 2\,NH_3 \rightarrow 2\,N_2 + 3\,H_2O \quad (15\text{-}12)$$

$$6\,NO_2 + 8\,NH_3 \rightarrow 7\,N_2 + 12\,H_2O \quad (15\text{-}13)$$

The reductant does not react with oxygen in the SCR catalyst at ordinary vehicle temperatures below 550°C, i.e.

the selectivity for NO_X reduction is 100%. The first two reactions are dominant in most cases. The required quantity of reductant can then be calculated directly from the desired NO_X reduction. Based on one refueling interval, this would produce a substantial NH_3 requirement (approximately 0.3–1% of the fuel quantity, depending on the raw emission). Its safe storage in a vehicle is questionable because NH_3 is toxic.

However vehicles can also produce NH_3 from a number of precursor substances relatively easily. Such precursor substances differ from one another in terms of storage density, toxicity, availability and stability. In the 1990s, the European automotive industry reached an agreement to use a 32.5% aqueous urea solution (with the brand name AdBlue®) in commercial vehicles. Used industrially as fertilizer, urea is sufficiently chemically stable for environmental conditions. Moreover, urea is easily soluble in water and, as a 32.5% solution, forms a eutectic mixture with a freezing point of −11°C.

The urea/water solution is metered in before the SCR catalyst and, an intermediate product, isocyanic acid hydrolyzes the urea to NH_3 in a two stage process in the exhaust gas system at temperatures of approximately 250°C and above.

$$(NH_2)_2CO \rightarrow NH_3 + HNCO \quad \text{(thermolysis)} \quad (15\text{-}14)$$

$$HNCO + H_2O \rightarrow NH_3 + CO_2 \quad \text{(hydrolysis)} \quad (15\text{-}15)$$

Solid deposits (biuret and higher molecular compounds) can be produced from the isocyanic acid in a secondary reaction. Appropriate catalytic converters and sufficiently high temperatures (of approximately 250°C and above) must be selected so that the hydrolysis reaction proceeds with sufficient speed to prevent solid precipitates.

The ammonia produced is adsorbed in the SCR catalyst and then available for SCR reactions. High conversions can be obtained with NH_3 in the temperature range of 180–450°C. Allowing for the upstream hydrolysis reaction, stable high conversion with the urea/water solution is possible only at 250°C and above.

The reaction mainly proceeds through Eq. (15-12) at low temperatures (<300°C). Therefore, an oxidizing catalytic converter that elevates the ratio of NO_2 to NO_X to around 50% is located before the SCR catalyst and the metering point to boost conversion. The oxidation catalytic converter may be either a DOC or a CDPF.

The SCR and hydrolysis reactions directly link the reduction of NO_X to the amount of AdBlue metered in. The mass ratio of required AdBlue to reduced NO_X is 2 g_{AdBlue}/g_{NOx}. The metering ratio α (also called a feed ratio) is defined as the molar ratio of the NH_3 equivalent metered into the NO_X present in the exhaust gas. The theoretically possible maximum reduction of NO_X corresponds to the metering ratio α Theoretically, NO_X can be eliminated completely when $\alpha = 1$. If it is metered at $\alpha > 1$ over a longer time, then the catalytic converter's adsorptive capacity will be exceeded and unconverted NH_3 will leave the SCR catalyst (NH_3 slip). NH_3 has a very low odor threshold (15 ppm in air). Excessive NH_3 would cause an odor nuisance in the environment. Therefore, not only conversion optimized by maximizing the metering ratio but also minimum NH_3 slip is important. In addition to limiting slip by a sufficiently small metering ratio, a downstream oxidation catalytic converter (trap catalyst) can also remove escaping NH_3.

The conversion achieved in existing systems may be less than the theoretical potential for conversion defined by the metering ratio. There are a number of possible reasons for this discrepancy:
– NH_3 slip can occur even at $\alpha < 1$ when insufficient homogenization of the AdBlue solution in the exhaust gas produces an inhomogeneous concentration of reductant in the SCR catalyst inlet. The quantity of NH_3 that passes through reduces the NO_X conversion achieved.
– When hydrolysis is incomplete, the formation of precipitates causes the reductant to be lost for the SCR reaction.
– At high temperatures, part of the NH_3 can react with oxygen by oxidizing.
– When the NO_2 to NO_X ratio is unduly large, part of the NO_2 reacts according to Eq. (15-13). 30% more NH_3 is required for this fraction than for reactions that follow Eqs. (15-11) and (15-12).

Just like a DOC and DPF, an SCR catalyst is dependent on the volume after displacement ($V_{SCR}/V_{Displ.} = 1.0-2.5$). When implemented in the field, 90% NO_X conversion can be achieved when the NH_3 slip is <20 ppm.

The achievable conversion greatly depends on the metering strategy. In the simplest design, the reductant is metered at a constant ratio. The required quantity of reductant yielded by the NO_X mass flow can be determined in engine or vehicle tests. The metered amount is released when the temperature in the SCR catalyst has reached the operating temperature (e.g. 250–450°C). This simple metering strategy is suitable for stationary engine operation. It is not suited for high NO_X conversion at low NH_3 slip under the dynamic operating conditions common in vehicle operation.

SCR catalysts have excellent NH_3 storage capacity (approximately $1\,g\,l^{-1}$), which greatly depends on the temperature (e.g. only 10% of the low temperature value is retained as of 350°C and above). The advantage of such storage capacity is that overmetering of AdBlue does not directly cause NH_3 slip and NO_X conversion can proceed with the stored NH_3 even at temperatures that are too low for a hydrolysis reaction. However, the storage capacity's dependence on the temperature harbors a risk that part of the adsorbed NH_3 desorbs should the temperature increase too rapidly. This causes NH_3 slip. To control this characteristic, an extended metering strategy includes a storage model that incorporates the SCR catalyst's storage capability and storage level. The SCR catalyst's storage level is increased by the reductant metered in and decreased by the NO_X conversion and the NH_3 slip that occur. The goal is to attain high NO_X conversion with an optimal storage level as a function of temperature. Therefore, the metered quantity is reduced by metering with a constant metering ratio in phases of decreasing storage capability and it is increased when the temperature decreases.

Higher conversion with low NH_3 slip is only possible when the calculation of the storage level is correct. In real systems, drifts in reductant metering and deviations in the raw NO_X emissions cause the calculated storage level to deviate from the real state. Insufficient NO_X or excessive metering ratios would cause continuous NH_3 slip. Maximum conversion can be achieved when the NO_X and NH_3 concentrations are measured after the SCR catalyst and the metering is readjusted commensurately.

A complete AdBlue SCR system includes the following components and subsystems (Fig. 15-43)
– a *DOC* or *CDPF* that increases the NO_2 to NO_X ratio,
– a *temperature sensor* downstream from the SCR catalyst, which determines the SCR temperature,
– an *NO_X sensor downstream from the SCR* (optional), which determines the NO_X concentration and, when appropriately cross sensitive, the NH_3 concentration,
– an *NO_X sensor upstream from the SCR* (optional), which improves the control quality,
– a *tank system* that stores the urea/water solution, an integrated level sensor, a heater, a temperature sensor (optional) and a quality sensor (optional),
– a *delivery module* consisting of a pump that conveys the urea/water solution from the tank to the metering module and, when mixture formation is air-supported and, additionally an air control valve and an air pressure sensor to establish an appropriate air mass flow from the air tank (commercial vehicle system) to the metering module,
– a *metering module* in which a solenoid valve regulates the exact quantity of urea/water solution. (In an air-supported system, this quantity together with the compressed air reaches a mixing chamber whence a line transports the aerosol to the metering point in the exhaust pipe; in systems without air support, an appropriate nozzle atomizes and forms the mixture directly at the exhaust pipe.) and
– a *control unit* that reads the sensors and controls the relevant actuators according to the metering strategy, while appropriate diagnostic functions monitor the components; communication with the engine control unit takes place through a CAN bus.

NO_X Storage Catalyst (NSC)

An NO_X storage catalyst (NSC or lean NO_X trap LNT) facilitates the reduction of NO_X without having to replenish any additional agent (Fig. 15-44).

NO_X is decomposed in two steps:
– *a loading phase* in which NO_X is stored in the lean exhaust gas in the catalytic converter's storage components and
– *a regeneration phase* in which the stored NO_X is discharged and reduced to N_2 in the rich exhaust gas.

A function of the operating point, the loading phase lasts approximately 30–300 seconds, the regeneration phase approximately 2–10 seconds. This operating mode initially concentrates the nitrogen oxides to be reduced. In the extremely short regeneration phase, the reductant only reduces the concentrated NO_X together with the residual oxygen. This reduces the fraction of oxygen that acts parasitically during NO_X reduction. Thus, the reductant required can be limited to a 2–4% increase in fuel consumption.

NO_X Storage

An NSC is coated with chemical compounds with a high propensity to enter stable but chemically reversible bonds with nitrogen oxides. Examples are oxides and carbonates of alkali and alkaline earth metals. Barium compounds are employed particularly frequently because of their thermal characteristics.

NO must progressively oxidize to form nitrate. NO is initially oxidized to NO_2 in a catalytic coating. The NO_2 reacts with the storage compounds in the coating and afterward with oxygen (O_2) to become nitrate:

$$BaCO_3 + 2\ NO_2 + 1/2\ O_2 \leftrightarrows Ba(NO_3)_2 + CO_2 \quad (15\text{-}16)$$

Thus, an NO_X storage catalyst stores the nitrogen oxides emitted by an engine. Storage is only optimal in a material-dependent exhaust gas temperature interval between 250 and 450°C. The NO oxidizes to NO_2 very slowly. The nitrate formed is unstable above 450°C and NO_X is discharged thermally.

Apart from its aforementioned capability to store NO_X as nitrate, a catalytic converter also has a limited capability to bind NO_X in surface storage at low temperatures. Such storage systems are sufficient to adequately store nitrogen oxides with low catalytic converter temperatures, e.g. during the start phase.

Nitrates form from carbonates in an equilibrium reaction. The catalyst's ability to bind more nitrogen oxides decreases as the quantity of nitrogen oxides stored increases (loading). Thus, the quantity of NO_X allowed through increases over time. There are two ways to recognize when a catalyst is so loaded that the storage phase must be terminated:
– a model-aided method incorporating the state of the catalyst calculates the quantity of nitrogen oxides stored and, from this, the remaining storage capacity, the storage efficiency and thus the quantity of NO_X allowed through or
– an NO_X sensor after the NO_X storage catalyst measures NO_X in the exhaust gas and thusly determines the current fill level.

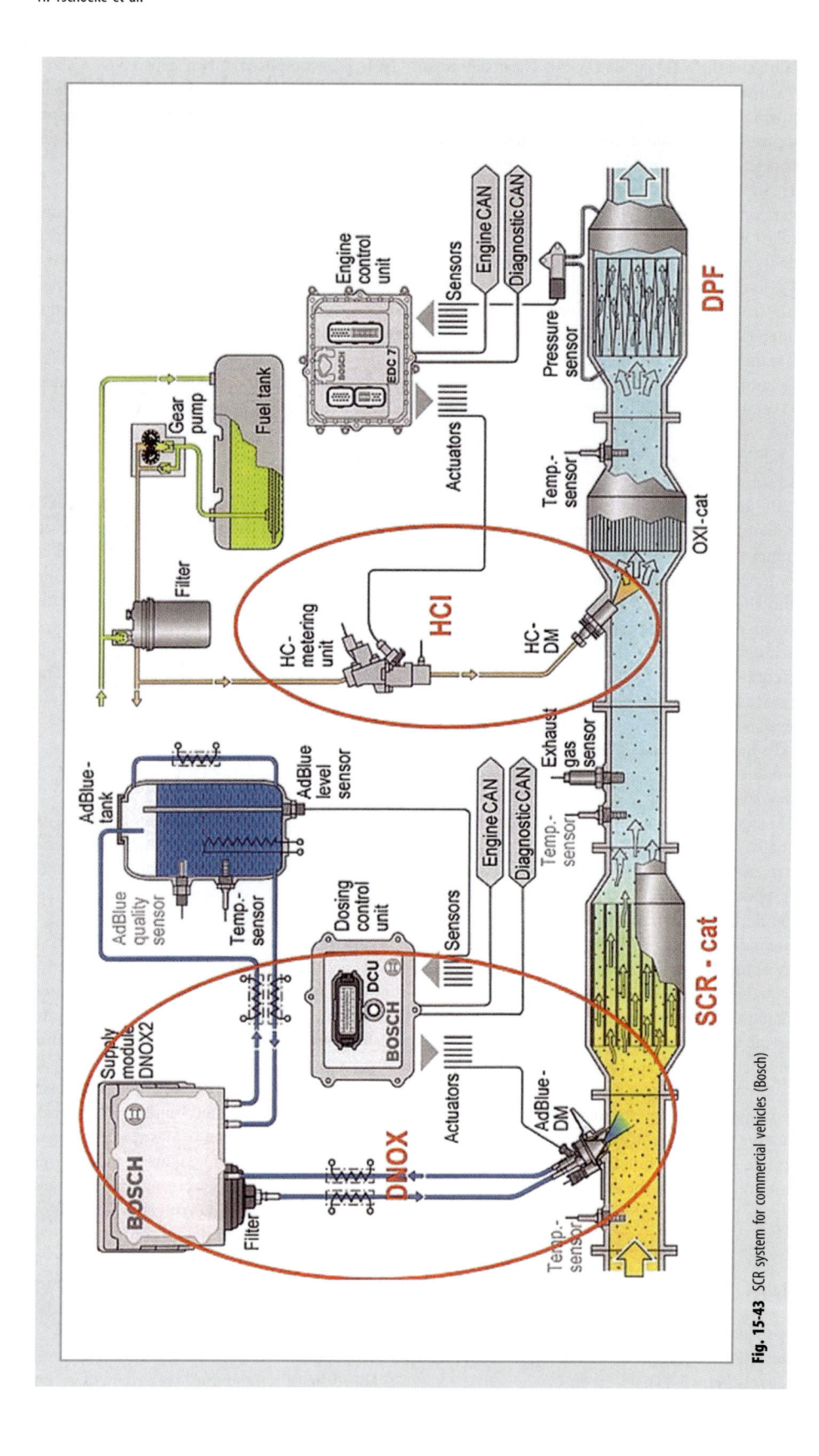

Fig. 15-43 SCR system for commercial vehicles (Bosch)

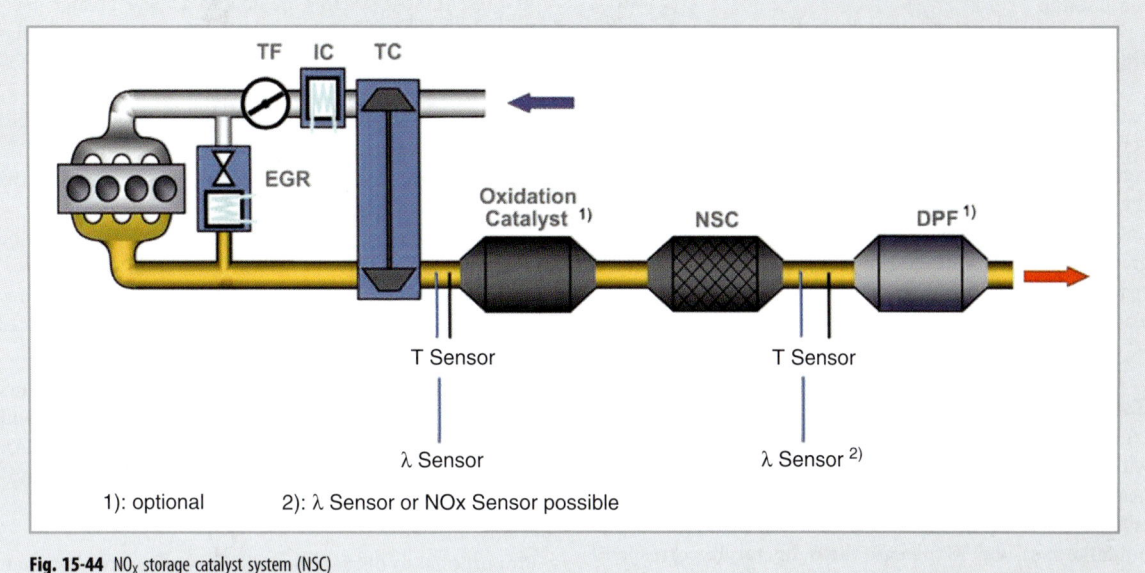

Fig. 15-44 NO$_X$ storage catalyst system (NSC)

A storage catalyst must be regenerated after the storage phase to limit the passage of NO$_X$.

NSC Regeneration

During regeneration, the stored nitrogen oxides are discharged from the storage components and converted into the harmless components of nitrogen and carbon dioxide. The processes of NO$_X$ discharge and conversion proceed separately.

To this end, oxygen deficiency ($\lambda < 1$, also called a rich exhaust gas condition) must be established in the exhaust gas. The components of carbon monoxide (CO) and hydrocarbons (HC) present in the exhaust gas serve as the reductant. Discharge – presented below in an example with CO as the reductant – proceeds so that CO reduces the nitrate (e.g. barium nitrate Ba(NO$_3$)$_2$) to NO and, together with barium, reforms the carbonate originally present.

$$Ba(NO_3)_2 + 3\ CO \rightarrow BaCO_3 + 2\ NO + 2\ CO_2 \quad (15\text{-}17)$$

CO$_2$ and NO are produced in the process. In a method familiar from three-way catalytic converters, a rhodium coating subsequently uses CO to reduce the nitrogen oxides to N$_2$ and CO$_2$:

$$2\ NO + 2\ CO \rightarrow N_2 + 2\ CO_2. \quad (15\text{-}18)$$

Less nitrogen oxide is discharged as regeneration advances and thus less reductant is consumed.

There are two methods to recognize the end of the discharge phase:

- a model-aided method calculates the quantity of nitrogen oxides still present in the NO$_X$ storage catalyst or
- a lambda oxygen sensor after the catalytic converter measures the excess oxygen in the exhaust gas and indicates a change of λ to $\lambda < 1$ when discharge has ended (CO breakthrough).

The rich operating conditions ($\lambda < 1$) required for regeneration can be established in diesel engines by retarding injection and throttling the intake air. The efficiency of engine operation is poor during the regeneration phase because of throttling losses and the suboptimal introduction of fuel. Therefore, the ratio of the duration of the regeneration phase to the duration of the storage phase should be minimized to keep increased fuel consumption low. Unlimited drivability as well as the constancy of torque, response and noise has to be assured when switching from lean to rich operation.

Rich operating conditions may also be established by introducing the reductant downstream from the engine. Similar to the operation of a catalytic burner (see Sect. 15.5.2.1), it may be introduced as diesel spray, vaporized fuel or a highly reactive species, usually a reformate gas (H$_2$/CO mixture). Intervention in engine combustion, in particular application engineering of retarded injection, can be reduced commensurately. However, the air mass should to be reduced, e.g. by lowering the boost pressure or throttling, to reduce the reductant requirement.

Desulfation

One problem with NO_X storage catalysts is their sulfur sensitivity. Sulfur compounds in the fuel and lubricating oil are oxidized to sulfur dioxide (SO_2) during combustion. The compounds utilized in NSC to form nitrate ($BaCO_3$) have very high bonding strength (affinity) to sulfate, which exceeds the bonding strength of the nitrate. Normal regeneration does not remove the sulfates. Thus, the quantity of sulfate stored gradually increases during the operating life. As a result, less space exists for NO_X storage and NO_X conversion decreases. Therefore, storage catalysts require the use of sulfur-free fuel (≤ 10 ppm).

The decreasing NO_X storage capability after a driven distance of between 500 and 2,500 km necessitates sulfur regeneration (desulfation) even during operation with fuel with a sulfur content of 10 ppm. Typically, the catalyst is heated to over 650°C for a period of more than five minutes and pulses of rich exhaust gas ($\lambda < 1$) are applied. The potential measures to increase the temperature correspond to those for DPF regeneration. These conditions reconvert the barium sulfate to barium carbonate. Appropriate process control (e.g. oscillating λ by 1) must be selected to ensure that the deficiency of residual oxygen does not reduce the discharged SO_2 to hydrogen sulfide (H_2S) during desulfation. (Alternatively, an appropriate trap catalyst must be designed in.)

In addition, the conditions established during desulfation must be selected to avoid excessive increase of catalyst deterioration. While, high temperatures (typically >750°C) accelerate desulfation, they also intensify catalyst deterioration. Therefore, desulfation optimized for a catalytic converter must occur in a limited temperature and excess air coefficient window and not interfere with a vehicle's in-use driving in the process.

Above all, the storage process is significant for the design of storage catalysts. Storage efficiency is a function of the catalyst temperature, precious metal loading, space velocity and available amount of storage. The ratio of catalytic converter volume to displacement (V_{NSC}/V_{Hub}) is 0.8:1.5. Efficient oxidation of NO to nitrate by NO_2 and maximum recovery of HC compounds during regeneration require high precious metal loading of approximately 100 g ft^{-3}. An NSC mounted as closely to the engine as possible can assume the functions of the DOC, which may then be dispensed with.

NSC catalysts allow a 50–80% reduction of NO_X throughout their service life.

NO_X Reduction Systems in Comparison

NSC and SCR systems have a multitude of different properties. Any decision on which system to implement in a vehicle greatly depends on the requirements and boundary conditions. The most important differences are examined below.

- At present, SCR systems are able to achieve greater maximum efficiency than NSC systems. This criterion is fundamental when compliance with extremely ambitious NO_X limits is required.
- SCR systems require another agent (e.g. AdBlue®). This has three consequences that must be borne in mind:
  (a) The agent must be approved for NO_X reduction in the region concerned.
  (b) Replenishment in appropriate intervals must be assured.
  (c) The agent must be stored in the vehicle. This requires space and – depending on the agent – necessitates cold flow and defrosting measures.

An AdBlue infrastructure for commercial vehicle systems is currently being set up throughout Europe, which will make replenishment possible in the fueling interval. An approximately 20–25 l tank could also suffice for a servicing interval for cars, thus making replenishment at garages possible.

- Proportional to displacement, SCR systems cost less than NSC systems. However, AdBlue systems generate fixed costs. Therefore, NSC systems cost less for cars with small displacement, while SCR systems are less expensive for commercial vehicles.
- The type and amount of operating costs differ. An SCR system consumes AdBlue. An NSC system increases fuel consumption by 2–4% depending on the NO_X reduction and the system design. When their NO_X reduction is comparable, SCR systems cost less to operate than NSC systems.

15.5.3 Exhaust Gas Aftertreatment Systems

Knowledge of the function of every individual system module is required to completely design an exhaust gas aftertreatment system (Fig. 15-45). Additionally understanding the interaction of the exhaust gas aftertreatment components is also essential. Moreover, dependable, highly efficient pollutant reduction must be obtained with minimum complexity by a good correspondence between engine measures for emission reduction and the exhaust gas aftertreatment system.

The following first explores the interaction of exhaust gas aftertreatment components and then explains the link with engine emission reduction.

15.5.3.1 Interaction of Exhaust Gas Aftertreatment Subsystems

Exhaust gas aftertreatment components are arranged serially in an exhaust gas system, i.e. one component's physicochemical performance has an effect on downstream components. Allowing for a slight delay (with a magnitude of 0.1–1 s), the exhaust gas mass flow is identical for every component.

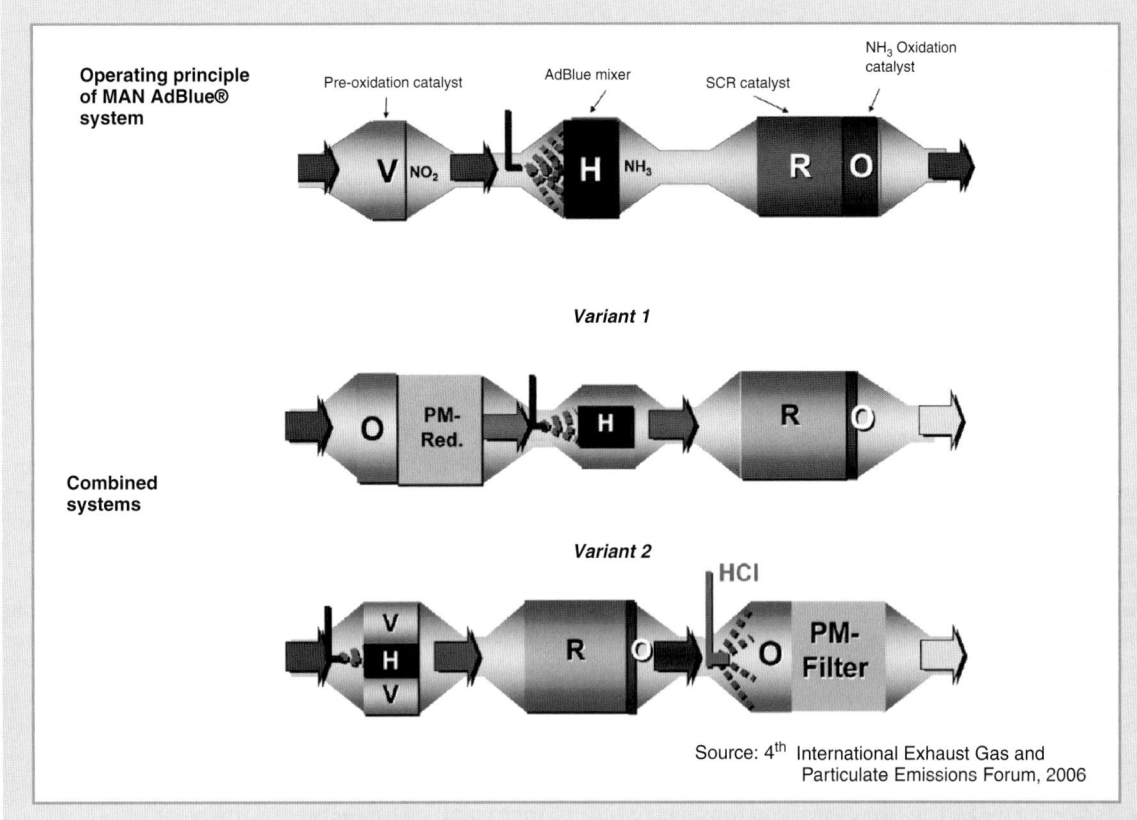

Fig. 15-45 Complete systems for exhaust gas aftertreatment (HCl–HC injection)

Thermal Coupling

Uncoupled from their chemical function, all exhaust gas aftertreatment components including pipes possess heat capacity. A change in the temperature of the exhaust gas before one of the components causes a change in the temperatures of the components and a damped temperature change of the gas downstream from the components (thermal low pass with a time constant in the magnitude of 10–100 seconds). At the same time, outer walls (pipe wall and catalytic converter housing) release heat from the exhaust gas system to the environment. Thus, components further downstream have a lower mean temperature than upstream components.

As described in the preceding section, every chemical process requires a minimum operating temperature of 150 °C (light-off CO oxidation) to 250 °C depending on the components (CRT®). Since conversion cannot take place below these temperatures, high conversion necessitates that the components reach their operating temperature as rapidly as possible after the engine starts. Therefore, the components are mounted as closely to the engine as possible. The use of insulated pipes can reduce their heat losses.

Very rapid starting processes additionally require start assist measures that correspond in principle to the measures that increase the DPF temperature.

Frequently, part of the required temperature increase is reached by an upstream catalytic burner that thermally regenerates the DPF. For this reason, a DOC is always located upstream from a DPF.

Pollutant Coupling

Every active exhaust gas aftertreatment component reduces at least one pollutant component according to its primary function. The other pollutant components can disrupt or facilitate this process or also be partially converted in it. Tables 15-16 presents these interactions.

One particularly important coupling is the required reduction of particulates by NO_X and especially by NO_2 through the action of the CRT®. Furthermore, HC and CO compounds interfere with the storage of NO_X in the NSC and the SCR process. In both cases, NO_2 is a key component that is reduced to NO in the presence of HC and CO.

Table 15-16 Interaction of exhaust gas aftertreatment systems

	Components					
	DOC	CDPF	NSC	SCR	Storage	Regeneration
Pollutant						
HC	X	--	--	-- ↘	- ↗	- ↘
CO	X	--	--	-- ↘	-- ↑	- ↘
Particulates	-	X	-	-	0	-
NO	-	-	- ↗	X	X	X
NO_2	+	+	↑	X	X	X

X: primary purpose of the components; --: pollutant is greatly reduced; -: pollutant is reduced; 0: no change; +: pollutant increases;
↑: pollutant greatly facilitates process; ↗: pollutant facilitates process; ↘: pollutant impedes process.

A decision about the relative arrangement of the components always has to be made whenever an exhaust gas system contains both a particulate filter and a nitrogen oxide reduction system. Along with thermal coupling, the pollutants' influence on the function of the components is important. The following discusses both interactions together.

CRT® can reduce DPF loading when the DPF is located upstream from the NO_X reduction system. Under favorable conditions, this can prolong the regeneration intervals considerably. Moreover, since it dispenses with the thermal low pass of an NSC or SCR catalyst, this arrangement makes it easier to bring the DPF to the regeneration temperature and keep it there. However, the DPF's relatively sizeable heat capacity only allows the NO_X reduction system to reach its operating temperature very late after start. Therefore, start assist measures must be designed in for high NO_X conversion rates. However, when a coated DPF is located downstream from an SCR system, then the DPF can simultaneously assume the function of a trap catalyst for potential NH_3 slip.

The abundance of arguments demonstrates the futility of a universal recommendation for a relative arrangement. The more beneficial arrangement for a particular vehicle frequently only becomes evident in combination with engine and vehicle design.

15.5.3.2 Interaction Between Engine Operating Parameters and Exhaust Gas Aftertreatment

The preceding section revealed the complexity of the interaction between the different components in an exhaust gas aftertreatment system. A number of correlations between engine combustion and exhaust gas aftertreatment also must be optimized for the purpose of overall system design.

Historically, exhaust gas aftertreatment systems were initially operated without intervention in engine combustion. The DOC reaches the light-off temperature necessary for CO and HC oxidation shortly after the engine starts and, with the exception of long motoring or idle phases, remains above this temperature until the engine is cut off. The SCR system is very similar. In the first applications, the metered quantity necessary for the targeted NO_X conversion was determined by tests on engine test benches and in vehicles and metered in production vehicles according to the application engineering data obtained. Only as requirements for NO_X reduction increased have allowances for fluctuations in engine NO_X emission become so important that it is only possible to obtain the conversion values by controlling the metered quantity. However, the engine operating parameters are not adapted to the requirements of exhaust gas aftertreatment during operation here either.

The situation changes when storage catalysts or particulate filters are utilized. Both systems must be regenerated regularly, the operating parameters being set far outside the usual range for diesel engines. In addition, the exhaust back pressure also increases as particulate filters become more loaded. First, this causes the pressure after the turbocharger to rise and, then, the pressure before the turbine. This pressure increase not only increases the energy loss of gas exchange but also influences the pressure gradient through the EGR system. Therefore, a change in exhaust back pressure must be compensated by the air system. Consequently, the operating parameters change with the filter's fill level.

Understandably, component costs as well as the complex interactions of components and their effects on engine operation have prompted attempts to achieve emission goals with a minimum number of components. Engine operating parameters (e.g. EGR rate and injection timing) can be utilized to change the ratio between soot and NO_X emission (the so-called soot-NO_X trade-off). They are employed in two main strategies to reduce complexity in exhaust gas aftertreatment:

(a) combustion optimized for soot with NO_X reduction downstream from the engine and
(b) combustion optimized for NO_X with particulate reduction downstream from the engine.

Very good engine efficiency can be obtained when combustion is optimized for soot (and NO_X rich). What is more, this eliminates a particulate filter's back pressure contribution, which also causes perceptibly increased fuel consumption. Therefore, NO_X reduction (with an SCR system) downstream from the engine is the preferred solution for commercial vehicle applications for which operating costs are extremely important.

However, particulate limits will be lowered considerably in the near future. Implementing a particulate filter with very high filtration efficiency and eliminating the complexity of NO_X reduction downstream from the engine by (particulate rich) combustion optimized for NO_X benefits overall system optimization when the particulate limits are not attainable in the engine or only with considerable complexity. This path is currently being pursued for European car applications.

Finally, some new markets for diesel vehicles, e.g. the USA, have such low particulate and NO_X limits that every measure for emission reduction in and downstream from the engine must be implemented. Although the attendant complexity is considerable, this route may be worthwhile in light of the alternatives. Diesel engines are still significantly superior to gasoline engines in terms of efficiency. In the face of limited global fuel reserves and global CO_2 emissions, this is an increasingly important argument for pushing to increase the number of diesel vehicles.

15.6 Emissions Testing

Exhaust measurement systems consist of exhaust measuring equipment and dilution systems. The determination of environmentally relevant mass emissions requires the determination of their concentrations as well as the engine's exhaust gas volumes.

15.6.1 Exhaust Measurement Systems for Gaseous Emissions

Individual pollutant concentrations are measured by individual exhaust gas analyzers mounted in an exhaust measurement system. Such systems (Fig. 15-46) consist of a base unit that extracts an exhaust gas sample from the engine's exhaust

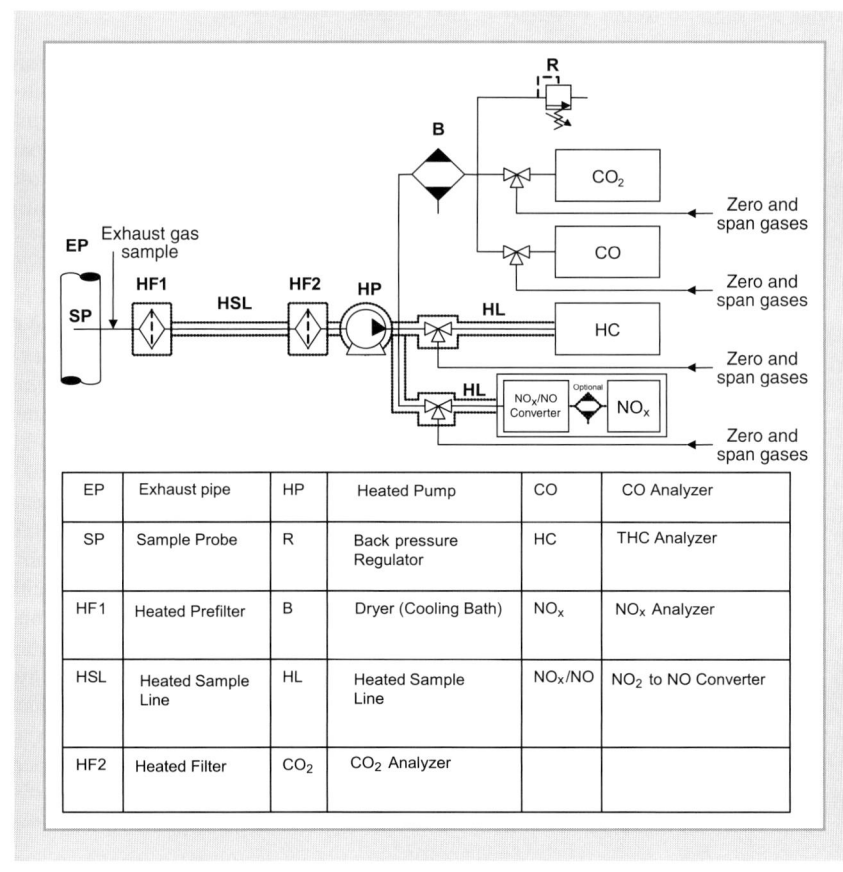

EP	Exhaust pipe	HP	Heated Pump	CO	CO Analyzer
SP	Sample Probe	R	Back pressure Regulator	HC	THC Analyzer
HF1	Heated Prefilter	B	Dryer (Cooling Bath)	NO_x	NO_x Analyzer
HSL	Heated Sample Line	HL	Heated Sample Line	NO_x/NO	NO_2 to NO Converter
HF2	Heated Filter	CO_2	CO_2 Analyzer		

Fig. 15-46
Schematic of an exhaust measurement system (ISO 16183 [15-33])

system, transports the sample gas to the system, conditions it and feeds it to the individual gas analyzers to measure the concentrations. In addition, a multitude of valves that feed the different operating and calibration gases to the analyzers are also necessary. Sample conditioning prevents the composition of the sample from changing on its way to the analyzers, e.g. prevents water from condensing or certain hydrocarbon compounds from depositing. In addition, particulates are filtered out of the sample gas to protect the analyzers.

Zero and Span Point Adjustment (Calibration):
All gas analyzers require regular adjustment of their zero and span calibration point (Fig. 15-47) before every measurement or at least once a day depending on the application. A zero gas and a span calibration gas are supplied to the analyzer from gas cylinders and its measured value is adjusted to the known calibration gas concentration. Normally, calibration is fully automatic. The absolute accuracy of measurement significantly depends on the accuracy of the calibration gases.

Linearization:
Linearization is performed to ensure that the values measured between the analyzer's zero and span points are accurate (Fig. 15-48). To this end, different gas concentrations dispersed throughout the measuring range are supplied to the analyzer and the measured values are compared with the expected concentrations. When the deviations are more substantial (>2%), the measured values are corrected mathematically (linearization curve). Such linearization is usually performed or at least checked in 3-month intervals. The reference concentrations are mixed together from zero and calibration gas in a highly precise multistage gas divider.

Diagnostic Tests:
Depending on the type of analyzer and the application, a multitude of diagnostic tests are performed, which assure the

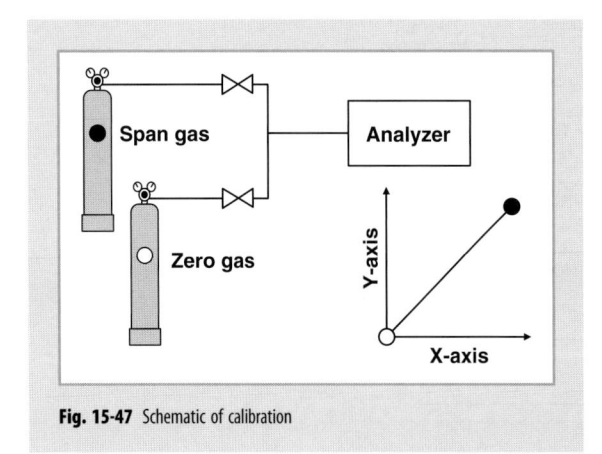

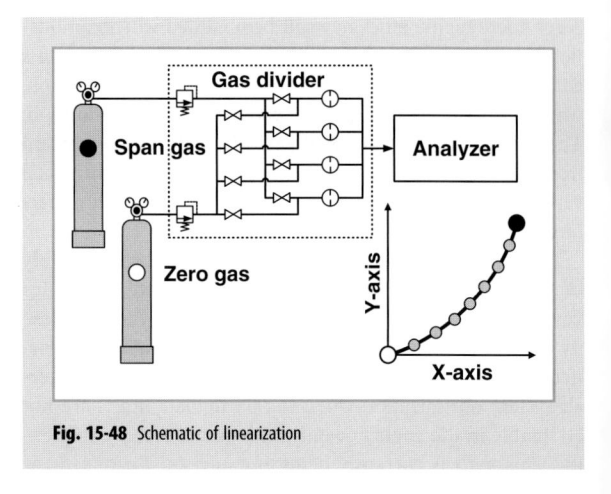

Fig. 15-47 Schematic of calibration

Fig. 15-48 Schematic of linearization

quality of the systems' measurements. Among others, these include cross sensitivity tests, which verify that the measured value of a gas component is not influenced by other gas components at all or only minimally, and leak checks of the system.

15.6.1.1 Hydrocarbon Measurement

Hydrocarbons are measured with a flame ionization detector (FID) (Fig. 15-49). Exhaust gas contains a multitude of hydrocarbon compounds. Normally, it is important to measure their cumulative rather than their individual components. The FID measuring principle measures most different hydrocarbon compounds, thus allowing a cumulative result.

FID Measuring Principle:
A flame generated from a constant flow of synthetic air and a gas mixture of hydrogen and helium burns inside the measuring cell. The flame burns in an electrical field between the cathode and anode. This flame is blended with a constant sample gas flow. Hydrocarbon molecules are cracked and ionized in the process. The ions generated transport a very weak current between the cathode and anode (measurement signal). Ideally, every hydrocarbon molecule would decompose into ionized components, which only contain one carbon atom. Then, the ion flow in the exhaust gas sample would also be proportional to the number of carbon atoms with a bond to hydrogen. However, the cracking and ionization process does not fully function in practice. Nonetheless, its efficiency for the individual hydrocarbon molecules is constant. This is called structural linearity and is identified as response factors. The response factors specify the difference between the value measured by the FID and the real concentration of individual hydrocarbons. Typically, they are between 0.9 and 1.1.

Fig. 15-49 Schematic of an FID (flame ionization detector)

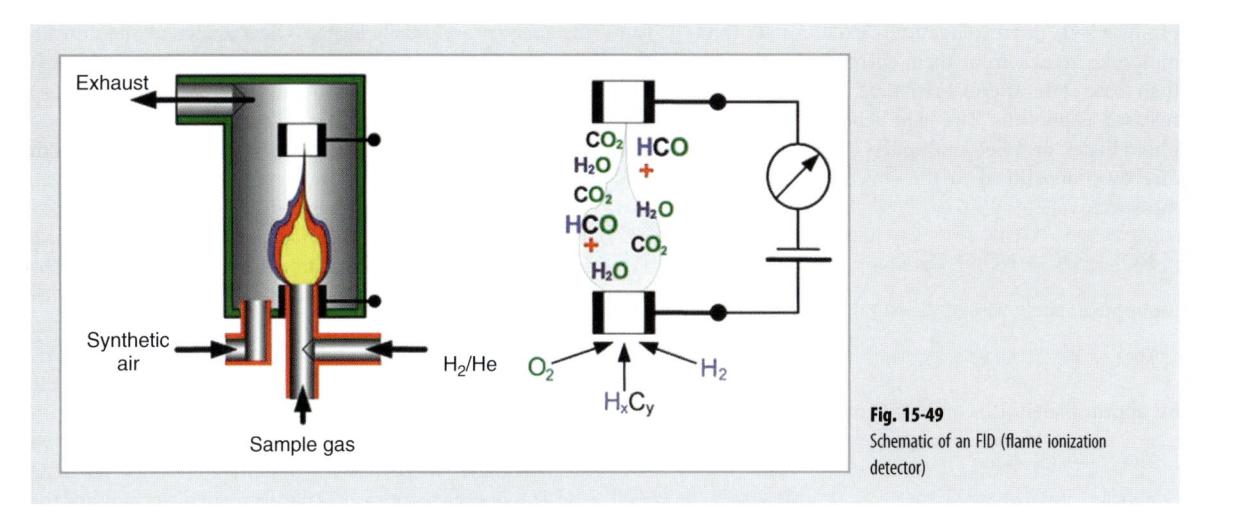

When measuring hydrocarbons in diesel engine exhaust gas, the entire exhaust gas sample must be heated to 190 °C from the sampling point to the FID since diesel engine exhaust gas contains hydrocarbons that would already condense below this temperature. Without heating, the condensed hydrocarbons would not be measured and the gas passages would also be contaminated. This is called HC hang-up.

15.6.1.2 Nitrogen Oxide (NO and NO$_X$) Measurement

Usually, the total content of nitrogen oxides NO and NO$_2$, referred to as NO$_X$, is measured. Typically, a chemiluminescence detector (CLD) is employed to do this (Fig. 15-50).

CLD Measuring Principle:
Measurement is based on the chemiluminescence produced by blending NO and ozone O$_3$. A chemical reaction converts NO and O$_3$ into NO$_2$. Approximately 10% of these reactions

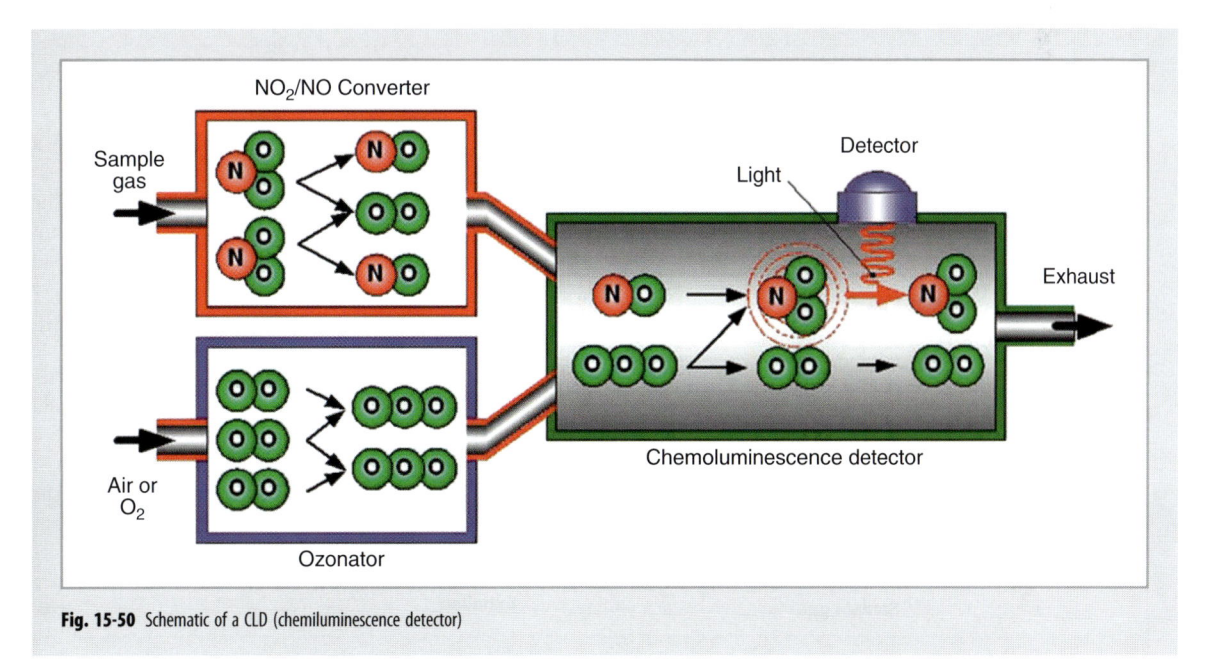

Fig. 15-50 Schematic of a CLD (chemiluminescence detector)

produce NO_2 in an energetically excited state (NO_2^*). The molecules return from their energetically excited state to their base state after a brief time; the excess of energy is released as photons. The light generated is measured with photodiodes or photomultipliers. The light's intensity is directly proportional to the concentration of NO in the measuring cell.

$$NO + O_3 \rightarrow NO_2 + O_2 \qquad (15\text{-}19)$$

for approximately 90% of the NO molecules in the sample

$$NO + O_3 \rightarrow NO_2^* + O_2 \qquad (15\text{-}20)$$

for approximately 10% of the NO molecules in the sample

$$NO_2^* \rightarrow NO_2 + h\nu \qquad (15\text{-}21)$$

with h ... Planck constant $\qquad$ hν ... photons $\qquad$ (15-22)

An ozone generator in the analyzer itself generates the required ozone O_3 from oxygen O_2., which may be pure oxygen, synthetic air or ambient air depending on the type of analyzer.

A CLD is only able to measure NO. Therefore, every NO_2 molecule is converted into NO before the CLD detector so that NO_X ($NO_X = NO + NO_2$) can be measured. An NO_2/NO converter performs this conversion much like a catalytic converter.

Since NO_2 can also react with water, any water condensation must be prevented at least up to the NO_2/NO converter. Otherwise, the NO_2 being measured would be lost and aggressive acids would form. Therefore, the complete sample gas path and the analyzer are usually heated. Older analyzers often use an unheated CLD detector. Then, the sample has to pass through the NO_2/NO converter first (to avoid NO_2 to condensed water) and then proceed to a gas drier (to remove water).

NO_X quenching occurs when NO_2^* molecules collide with other suitable molecules before they have released light. Then, the energy is not released as light but to the other molecules. Thus, less light is generated and the measured value is too small. In exhaust gas, such molecules are primarily H_2O and CO_2. The more molecules are in the measuring cell, the more often such collisions occur and, thus, the greater the quenching is. Hence, most CLD analyzers are operated in a vacuum (at approximately 20–40 mbar absolute). This significantly reduces the number of molecules and thus the potential for quenching.

In addition to analyzer calibration and linearization, two diagnostic tests are particularly important for CLD. The H_2O and CO_2 quench test measures the degree of quenching and the NO_2/NO converter test measures convertibility and efficiency. Typically, the efficiency ought to be over 90%.

15.6.1.3 Carbon Monoxide (CO) and Carbon Dioxide (CO_2) Measurement

Carbon monoxide CO and carbon dioxide CO_2 are measured by nondispersive infrared analyzers (NDIR) (Fig. 15-51).

NDIR Measuring Principle:
An emitter emits a broad infrared spectrum and the radiation is sent through a bipartite measuring cell. One part of the measuring cell is filled with nonabsorbent gas (e.g. nitrogen N_2) and

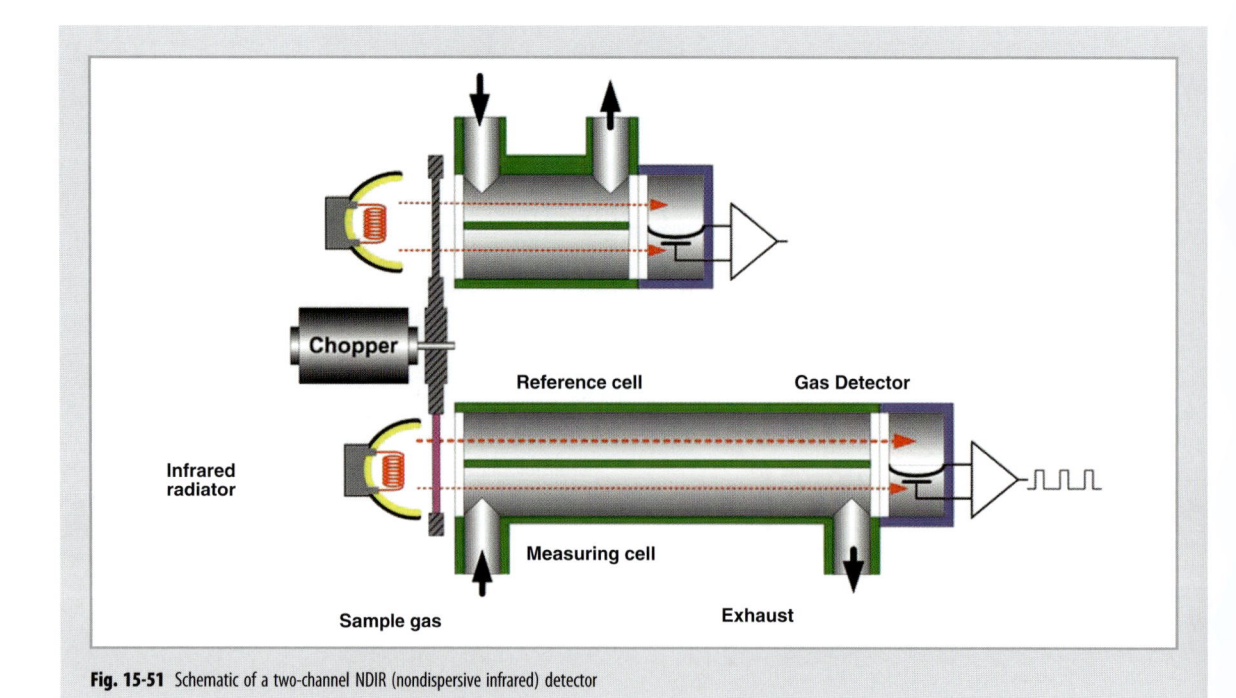

Fig. 15-51 Schematic of a two-channel NDIR (nondispersive infrared) detector

referred to as the reference cell. The sample gas flows through the other part. A chopper (e.g. a rotating perforated disk) intermittently disrupts the infrared radiation. When the sample gas contains gas molecules that are absorbent in the infrared range, e.g. CO or CO_2, part of the radiation from these molecules is absorbed. Thus, the radiation through the measuring cell is lower than the radiation through the reference cell.

A detector measures the difference in the intensity through the two cells. The detector also consists of two chambers. One receives the radiation from the reference cell and the other the radiation from the measuring cell. Both of the detector's chambers are filled with the gas to be measured (e.g. CO or CO_2). Thus, the detector absorbs infrared radiation just like the measuring cell for these gas components. Depending on the concentration of the gas components being measured in the measuring cell, the infrared radiation through the measuring cell becomes weaker than the radiation through the reference cell since radiation has already been absorbed. The remaining radiation is also absorbed in the detector chambers since they are filled with the same gas that is being measured. The absorption increases the energy and thus the pressure in the sealed chamber. The two cells' differing radiation cause a differential pressure in the two detector chambers, which is measured by a flexible diaphragm between them. The diaphragm's movement is measured capacitively (as in Fig. 15-51). Alternatively, flow measurement of a compensating flow between the two chambers may be used.

The larger the concentration of the gas being measured in the measuring cell, the larger the measurement signal is. The correlation between the concentration and measurement signal corresponds to the Beer-Lambert law, which is a nonlinear function. Hence, NDIR detectors must always be linearized.

Many gases absorb in the infrared spectrum and the different gases' absorption spectra also overlap. CO and CO_2 analyzers are particularly cross sensitive to H_2O vapor. CO analyzers are also cross sensitive to CO_2. Such cross sensitivity increases the measured value. Typically, the raw exhaust sample is dried first before it flows to the NDIR analyzers. Thus, cross sensitivity to water is no longer relevant. Wet gas only has to be measured from the bags of a constant volume sampling system (CVS). Such applications require the performance of water cross sensitivity tests (interference checks).

15.6.1.4 Oxygen O_2 Measurement

The oxygen concentration in exhaust gas is measured by a paramagnetic detector (PMD) (Fig. 15-52).

PMD Measuring Principle:

Oxygen is one of the few gases with paramagnetic properties. The sample gas flows through a strong magnetic field in a measuring cell. Their magnetic properties cause the oxygen molecules to move to the center of the magnetic field. However, a quartz sphere without any magnetic properties is in the center. Such detectors are designed

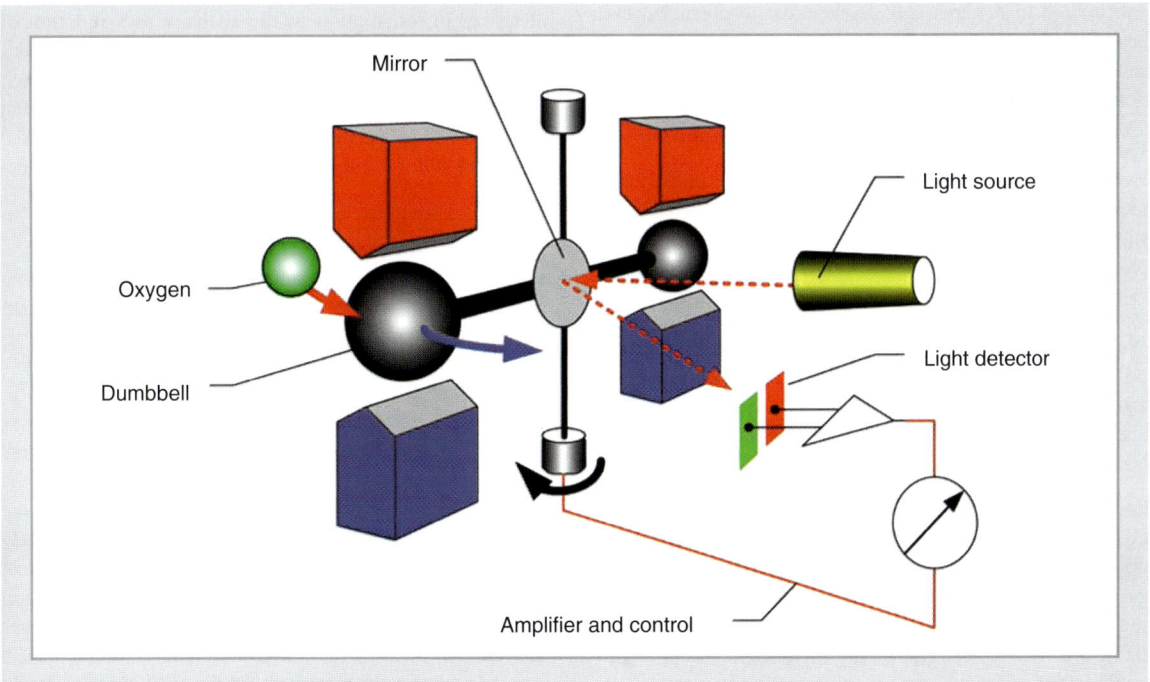

Fig. 15-52 Schematic of a PMD (paramagnetic detector)

symmetrically with two magnetic fields and two quartz spheres. An arm connects both spheres. This is also referred to as a dumbbell. The dumbbell is mounted on a rotating axis. The oxygen molecules surging into the magnetic fields attempt to displace the dumbbell's spheres. The higher the oxygen concentration, the greater its displacement force is. A mirror is mounted on the dumbbell's rotating axis. A beam of light and a light detector measure the deflection. Either the deflection itself is the measurement signal or a controlled magnetic field always keeps the sphere centered and the required electrical current is then the measurement signal. In both cases, the measurement signal is directly proportional to the oxygen concentration in the sample gas.

The measuring principle is also slightly sensitive to NO, NO_2 and CO_2 because these gases are also slightly paramagnetic. Since these gases are measured in every exhaust measurement system, the slight cross sensitivity may be corrected mathematically. Since the cross sensitivity is relatively low, it also only plays a role in gasoline engines that have only a very low O_2 concentration at $\lambda = 1$ and below. Without a correction, errors of a magnitude of 5,000 ppm may occur.

15.6.1.5 Special Measurement Systems

The aforementioned measuring principles constitute the standard methods of measurement, often also referred to as conventional measurement systems, as well as the measuring principles mandated in most emission control legislation [15-26–15-30]. Other methods of measurement are also required and utilized in research and development to obtain an even better picture of exhaust gas composition as well as to measure gas components that are not or are not yet covered by legislation. Some such methods of measurement are presented below.

Fast Response Measurement Systems:
Fast response measurement systems are based on the aforementioned conventional measuring principles but the measurement systems are trimmed to rapid signal response times of only a few milliseconds. However, this results in significantly shorter service lives and maintenance intervals.

Nondispersive Ultraviolet Analyzers (NDUV):
NDUV analyzers measure gas components much like NDIR analyzers. However, unlike an NDIR, ultraviolet light is used. Such analyzers are mainly employed to measure NO, NO_2 and NH_3.

Fourier Transform Infrared Spectroscopy (FTIR):
Fourier transform infrared spectroscopy is a method of optical measurement that measures a multitude of exhaust gas components simultaneously. Measurement is based on the individual gas components' absorption of infrared light. FTIR are widely used, especially in modern diesel engine applications with exhaust gas aftertreatment systems, e.g. NO_X storage and SCR catalysts and diesel particulate filters.

A differentiated measurement of NO, N_2O, NO_2, NH_3 and other exhaust gas components is particularly important in these applications.

FTIR employs a broad infrared waveband to capture all of an exhaust gas sample's spectral information simultaneously. The intensity of the individual infrared wavelengths is varied continuously by means of a Michelson interferometer. A beam splitter divides the light source's infrared radiation into two beams. One of the beams strikes a moveable mirror, the other a stationary mirror. Afterward, the two beams are rejoined into one beam. The mirror's continuous movement produces differential path lengths, which in turn produce interference when the beam is reunited. Individual wavelengths may be cancelled or amplified depending on the moveable mirror's position. The thusly continuously modified infrared beam is conducted through the measuring cell and individual wavelengths are absorbed by the different gas components in the exhaust gas sample. The Fourier transformation applies complex mathematical formulas to calculate an infrared spectrum (intensity as a function of wavelength) from the interferograms and special methods of analysis determine the concentrations of the individual gas components from them.

Mass Spectrometers (MS):
Mass spectrometers make it possible to measure many gas components in exhaust gas. The sample gas is ionized, e.g. by reactant gases or electrical ionization, and the ions are separated in the analyzer, according to their mass. There are several methods for this but all of them are based on the differences in ion motion in the analyzer as a function of their mass. These may be differential transit times or different curve radii during deflection. Given its complexity, such equipment is used more in labs than regular test bench operation. Mass spectrometer systems are especially used to measure sulfur components (SO_2, H_2S, COS) and hydrogen H_2.

Diode Laser Spectroscopy (DIOLA):
Diode laser spectroscopy is similar to the infrared measuring principle employed in NDIR analyzers but it produces very short signal response times. It is especially used for the development of catalytic converter applications.

Exhaust Gas Dilution Systems:
In principle, the masses of environmentally relevant pollutants in exhaust gas are determined from the concentration of the particular exhaust gas components, their density and the engine's volumetric exhaust gas flow. While this is a relatively simple task in stationary engine operation, the method of measurement is relatively complicated in transient operation since it must follow both the rapid changes in the exhaust gas and accurately measure the concentrations over several powers of ten. In addition, the extremely dynamic volumetric exhaust gas flow must also be measured. Since every signal has different

time delays, the measured data must be accurately time aligned before further calculation. Since it was impossible to meet such requirements in the early days of emission control legislation, alternative methods were sought to perform the task with the instruments and computers available at the time. This was achieved by employing full flow dilution. Even though it is now possible to meet these requirements without dilution, full flow dilution is still mandatory in virtually all emission control legislation. The only exceptions are stationary and transient emission tests for commercial vehicles such as EURO IV (2005 [15-32]). The somewhat conservative position in legislation is also rooted in the ease with which the relatively simple full flow dilution method enables approval authorities to check the correctness of the mandatory emission test.

Simplified calculation of a pollutant component's mass in exhaust gas employs the following terms:

Q_{exh} volumetric flow rate of the engine
Q_{CVS} volumetric flow rate of the CVS
V_{CVS} total volume of the diluted exhaust gas over the sample time
q dilution ratio of the CVS
$Conc_{raw}$ concentration of the pollutant component in the undiluted exhaust gas
$Conc_{dil}$ concentration of the pollutant component in the diluted exhaust gas
$Conc_{Bag}$ concentration of the pollutant component in the exhaust sample bag
ρ density of the pollutant component
T sample time (duration of the test cycle)
m mass of the pollutant component

Undiluted exhaust gas (see scheme ①, Fig. 15-53) is

$$m = \int_0^T Conc_{raw} \cdot Q_{exh} \cdot \rho \, dt \qquad (15\text{-}23)$$

With the following, diluted exhaust gas (see scheme ②, Fig. 15-53) from Eq. (15-23) becomes

$$Q_{exh} = \frac{Q_{CVS}}{q}$$
$$m = \int_0^T Conc_{raw} \cdot \frac{Q_{CVS}}{q} \cdot \rho \, dt$$

and the following with

$$Conc_{raw} = Conc_{dil} \cdot q$$
$$m = \int_0^T Conc_{dil} \cdot Q_{CVS} \cdot \rho \, dt$$

and the following with

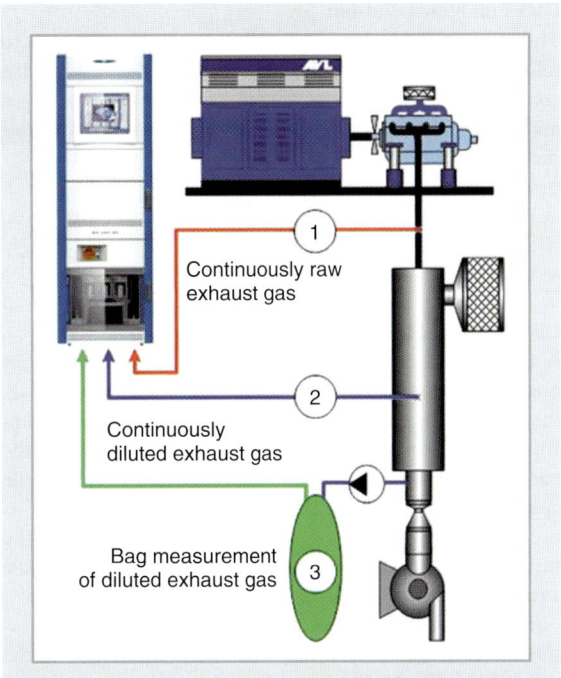

$$Q_{CVS} = const. \text{ and } \rho = const.$$
$$m = Q_{CVS} \cdot \rho \cdot \int_0^T Conc_{dil} \, dt \qquad (15\text{-}24)$$

Fig. 15-53 Schematic of the methods for the calculation of exhaust gas mass emissions (simplified). *1* undiluted exhaust gas; *2* diluted exhaust gas; *3* constant volume sampling

With the following, CVS (diluted exhaust gas, see scheme ③, Fig. 15-53) from Eq. (15-24) becomes

$$\int_0^T Conc_{dil} \, dt = T \cdot (Conc_{dil})_{mean} = T \cdot Conc_{Bag}$$

and becomes the following with

$$V_{CVS} = Q_{CVS} \cdot T \, (da \, Q_{CVS} = const.)$$
$$m = V_{CVS} \cdot \rho \cdot Conc_{Bag.} \qquad (15\text{-}25)$$

Dilution of engine exhaust gas also lowers the water concentration in the diluted exhaust gas so far that no water condenses in the measurement system. Sufficiently high dilution is the prerequisite. Dilution also simulates chemical reactions of the individual exhaust gas components in ambient air (real operation). This is particularly crucial for particulate formation.

Since no exhaust gas is removed, exhaust gas dilution fully retains the entire mass of the pollutants in the exhaust gas. However, the addition of dilution air also introduces small quantities of pollutants that are already contained in the ambient air. To prevent this from skewing the test result, the dilution air is also collected in a sample bag and analyzed. The pollutant mass added by dilution is subtracted when the final result is computed. Pollutants fed to the engine with the intake air are not corrected and are considered to be vehicle emissions.

CVS Full Flow Dilution:

The main function of constant volume sampling (CVS) is to dilute all of the engine's exhaust gas (full flow) and maintain a constant volumetric flow of diluted exhaust gas (exhaust gas and dilution air). This can be accomplished with various methods:
- CFV critical flow venturis,
- PDP positive displacement pumps or Roots blower
- SSV subsonic flow venturis and
- UFM ultrasonic flowmeters.

CFV (Critical Flow Venturi):

A blower sucks the diluted exhaust gas through a venturi nozzle. Constricting the cross section in the nozzle increases the flow velocity. Sonic speed is reached in the narrowest point when the differential pressure between the nozzle inlet and the narrowest point has the factor of approximately two. Since the speed cannot be increased any further, a constant flow is produced, regardless of the strength of the blower after the nozzle. This state is referred to as critical and, hence, the nozzle as a critical venturi nozzle.

The exact flow can be calculated from the nozzle's calibration parameters, the pressure and the temperature in the nozzle inlet (Fig. 15-54):

$$V_s = K_v \cdot \frac{p_v}{\sqrt{T_v}} \quad (15\text{-}26)$$

V_s volumetric flow rate normalized to the standard conditions at 20°C and 1,013 mbar in force for American legislation
or
to normal conditions at 0°C and 1,013 mbar in force for European legislation
K_V venturi calibration factor as a function of the narrowest nozzle cross section
p_V absolute pressure before the venturi nozzle
T_V temperature before the venturi nozzle in Kelvin

Since the pressure downstream from the venturi nozzle does not influence the flow in any way, the blower does not have to be controlled. However, it must be strong enough to generate critical flow conditions in the nozzle.

Depending on the engine and test, usually three or four venturi nozzles are connected in parallel to establish different flows. Ideally, each next larger venturi nozzle has twice the flow than the preceding venturi nozzle. Thus, as in the binary numeral system, a combination of four nozzles produces fifteen different flow rates and three nozzles seven flow rates.

PDP (Positive Displacement Pump):

Also known as a Roots blower, a positive displacement pump PDP is another option to keep the flow through a CVS constant. Two rotary pistons in the housing transport the gas. The volume is not compressed though and the volumetric flow is directly proportional to the speed of the pump.

This advantage of this CVS variant is the adjustability of the flow rate by the pump motor's speed control. CVS systems are no longer constructed with PDP primarily because of the high costs.

Flow Measurement and Active Control:

Some more recent legislation also allows the use of flow measurement and active blower control.

The flow is measured with a subsonic venturi (SSV). Unlike CFV, sonic speed is not reached in this nozzle and the flow is therefore calculated with the differential pressure according to the Bernoulli equation. Alternatively, ultrasonic flowmeters (UFM) are used.

Common CVS Flow Ranges:

CVS with differing flow ranges are implemented depending on the application and engine size. The flow must be large enough to prevent water from condensing in the system and, for diesel engines, to keep the temperature of the diluted exhaust gas below 52°C during particulate measurement:

commercial vehicle engines	120 ... 180 m³/min
passenger cars	10 ... 30 m³/min
motorcycles	1 ... 5 m³/min

CVS Sampling:

Only a small quantity of gas is extracted from a CVS to analyze the diluted concentrations in it. It is filled in sample bags that are evacuated before and analyzed after the test. The exhaust gas concentration in the sample bag yields a mean value over time throughout the test. To measure particulate emissions, the diluted exhaust gas is drawn through analysis filters that trap the particulates from the

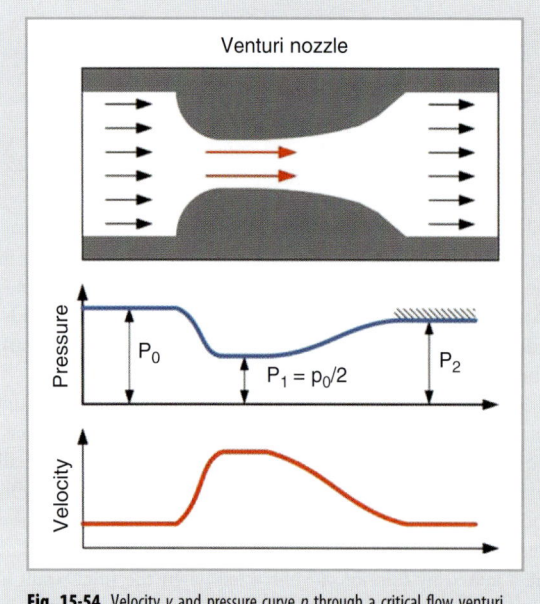

Fig. 15-54 Velocity v and pressure curve p through a critical flow venturi

sample taken. The analysis filters are weighed before and after the measurement to determine the quantity of particulates trapped.

The total masses of emissions from the total volumetric CVS flow during the measurement are calculated from the averaged concentrations in the sample bag and the particulate mass trapped in the filter. Thus, the sample flow into the sample bag and through the particulate filter must be proportional to the flow in the CVS.

The flow in a CVS is not absolutely constant and can vary slightly as a function of pressure and temperature. Therfore, either the sample flow must follow these variations proportionally or the flow through the CVS must be kept sufficiently constant. A CFV-CVS satisfies this requirement for gaseous exhaust gas components by also filling the sample bag by a CFV. Thus, any influence of pressure or temperature is identical for both nozzles (CVS and sample nozzles) and establishes the proportionality required. Sampling nozzles may not be utilized for particulate measurement since the nozzle hole traps particulates and the lower pressures after the nozzle can alter particulate formation. Either the CVS (state-of-the-art particle sampler) actively readjusts the particulate sampling to be proportional to the flow or a heat exchanger before the CVS venturi keeps the temperature constant. Thus, the flow through the CVS venturi is also constant. Since a CVS is always open to the environment through the dilution air inlet, significant changes in pressure that would alter the flow do not occur (Fig. 15-55).

Requirements for Diesel Engine Measurement:
Additional requirements exist for the measurement of particulate and hydrocarbon emissions from diesel engines (see Fig. 15-56).

A dilution tunnel is added to a CVS to measure particulates. Basically, this is simply a long, straight stainless steel pipe, which is intended to form particulates realistically. Its diameter must be selected so that the flow is always turbulent and the Reynolds number (measure of the turbulence) is above 4,000. Its length should be such that the diluted exhaust's residence time in the tunnel is long enough to simulate particulate formation in the environment. The dilution tunnel's length is usually 10 times its diameter. Dilution must also be high enough to maintain the temperature of the diluted exhaust gas in the point of particulate measurement below 52°C. Already very large, CVS systems are usually unable to dilute sufficiently to do this for commercial vehicles. Then, double dilution is performed in which a small flow sampled from the CVS is diluted a second time. This is referred to as secondary dilution.

Particulate measurement is performed at the end of the dilution tunnel by a particle sampler (PTS). It draws diluted exhaust gas through analysis filter disks. The filters are weighed before and after the measurement and the particulate emissions are calculated from the increase in weight and the flows.

Hydrocarbon compounds have a significantly higher boiling point in diesel engine exhaust gas than in gasoline engine exhaust gas. This stems from the fuel production. Put simply,

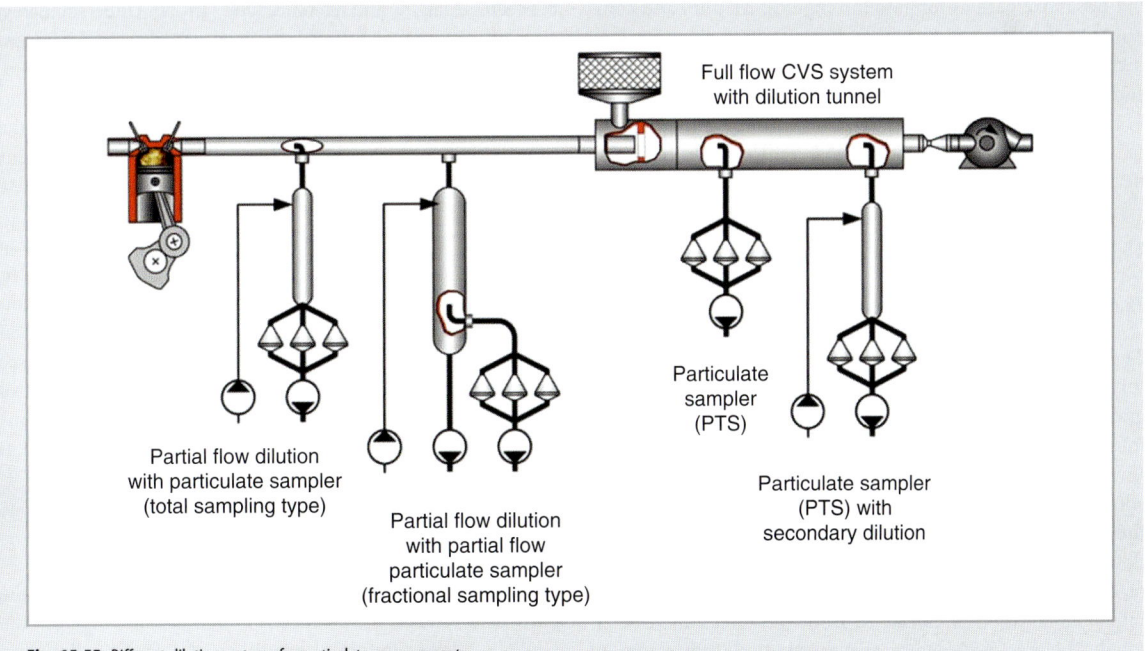

Fig. 15-55 Different dilution systems for particulate measurement

Fig. 15-56 Full flow dilution for certification

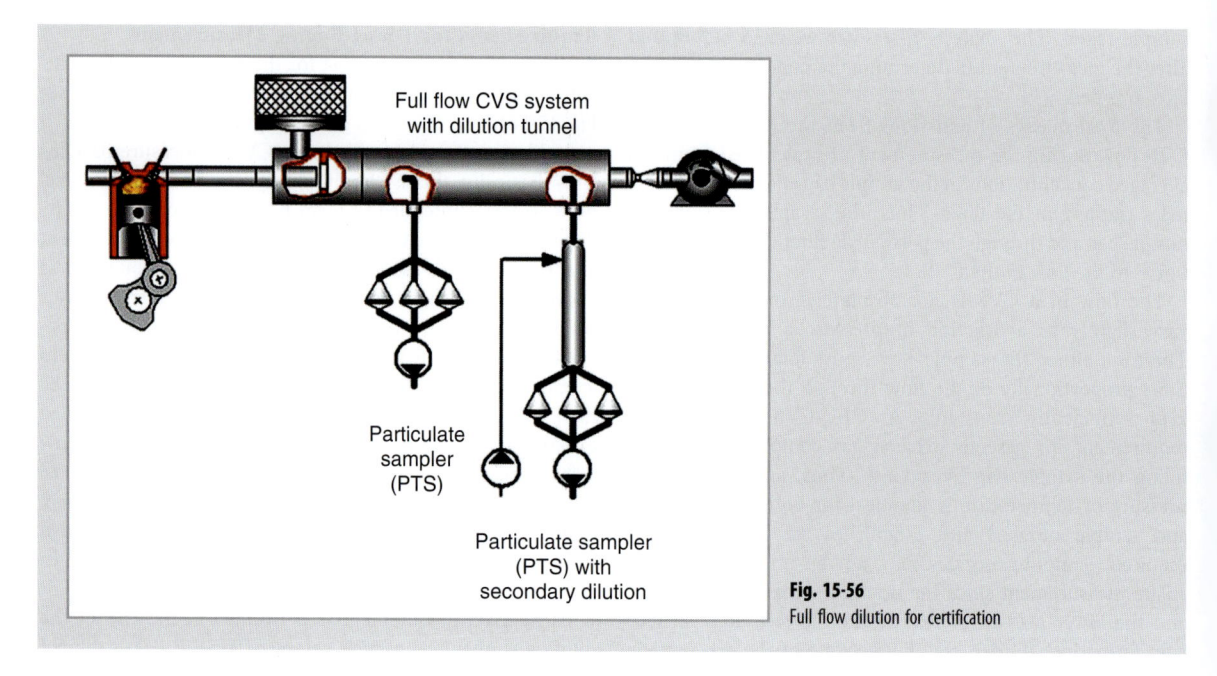

gasoline consists of hydrocarbons that vaporize at up to 200°C and diesel consists of hydrocarbons that vaporize at between 200 and 400°C. Hence, hydrocarbons from diesel exhaust gas can condense at higher temperatures and are thus no longer measured as gaseous components and contaminate the measurement systems. Referred to as HC hang-up, this would affect subsequent measurements. Therefore, the hydrocarbons are measured directly from the CVS dilution tunnel rather than out of sample bags as for gasoline engines. All lines and components that transport gas for hydrocarbon measurement, including the analyzer (FID), are heated to 190°C. This inhibits such condensation.

In addition to hydrocarbon measurement, nitrogen oxide measurement for commercial vehicle engines is also heated and performed directly from the dilution tunnel.

15.6.2 Measurement of Particulate and Dust Emissions

15.6.2.1 Particulate Emission: Dilution Tunnel

The limits on particulate emission specified in every mandatory regulation are based on an integral measurement by gravimetric determination of the particulate mass after dilution in a full flow or partial flow tunnel as first defined by the Environmental Protection Agency (EPA) [15-26] and then adopted worldwide [15-27, 15-28]: Based on the principle of the CVS, exhaust gas is mixed with filtered air and a partial flow of the diluted exhaust gas, which must have a temperature of <52°C, is drawn through inert filters with a >99% filtration rate. The emission is calculated from the filter's increase in weight. Figure 15-56 depicts this schematically for a type of system with secondary dilution usually employed for commercial vehicles. In principle, car emissions are measured identically in a chassis dynamometer but without secondary dilution.

Particulates are composed of soot, adsorbed organic components, condensed and adsorbed sulfuric acid and solid constituents such as grit, ash, etc. Condensed and adsorbed substances basically only form in a dilution tunnel. However, other than initially expected, the soot concentration is not completely stable between the engine and reference filter [15-29]. Understandably, small modifications of the dilution and particle sampler system already influence the particulate mass measured.

The EPA specified the dilution, particulate sampling and weighing system more precisely in 2007 to also increase the repeatability and reproducibility of the measurement method as particulate and, above all, soot emissions decrease [15-30, 15-31].

The EU (and thus nearly every Asian and Latin American country) allows the use of partial flow dilution tunnels for commercial vehicles [15-32], which dilute a constant fraction of the exhaust gas as specified in the standard ISO 16183 [15-33]. The space and cost advantages of these systems (see the schematic in Fig. 15-57) are offset by the complex control of the mass flows. In addition, several boundary conditions must be observed to obtain the same emission as measured in a full flow system [15-34].

15.6.2.2 Particle Number Counting

Since only very sensitive instruments are able to capture the particulate emissions of modern combustion engines, the UNECE GRPE's informal group on the Particle Emission Programme (PMP) is working on new methods of particle

Fig. 15-57 Partial flow dilution for certification

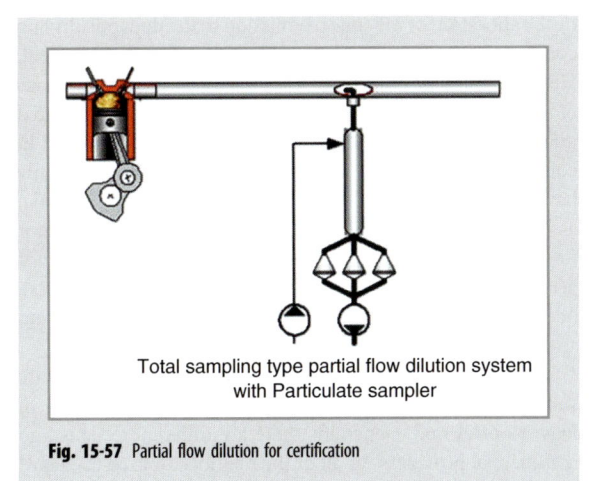

measurement. The group's recommendation for future homologation procedures includes modified EPA particulate measurement (2007) as well as particle number counting [15-35, 15-36].

The PMP group has defined a complex system to condition already diluted exhaust gas, which is presented schematically in Fig. 15-58. First, coarse particles that stem from reintroduced wall deposits rather than directly from combustion are separated. Second, the exhaust gas is diluted and subsequently heated to 400 °C. Third, dilution is performed once more before the particle number counter (PNC) to cool the exhaust gas and lower the particle number further. A particle number is obtained in a condensation particle counter (CPC) and volatile nanoparticles are eliminated. Thus, only nonvolatile particles, i.e. chiefly soot particles, are counted. This requirement is rooted in two factors. On the one hand, nonvolatile particles are toxicologically more relevant to human health. On the other hand, reproducible measurement of volatile particle emissions has proven to be extremely difficult. This is not a problem with the measurement per se since volatile particles can be counted just like solid particles. However, the formation of homogeneously condensed hydrocarbons and sulfates after particulate filters is extremely sensitive to the slightest changes in engine or exhaust gas conditioning.

The conditioned exhaust gas may also be analyzed for such particle properties as size distribution, active surface area, etc. (not required by PMP).

Although not explicitly required by PMP, condensation particle counters (CPC) are the most common and most sensitive systems for particle number counting in the range of submicrons to a few nanometers. Figure 15-59 presents the principle of a CPC. Heterogeneous condensation of supersaturated vapor generates microparticles from nanoparticles, which are subsequently counted by the light scattering method.

15.6.2.3 Dust Measurement

In Germany, stationary diesel engines are subject to the regulations of *TA Luft*. Sampling is performed according to VDI 2066 [15-38]. A sample is taken from the exhaust gas without prior dilution so that, for all intents and purposes, the particulates do not contain any substances condensed and adsorbed onto the soot. Hence, this "dust mass" does not correlate with the particulates measured from diluted exhaust gas. Depending on the load point, the difference can be between 10 and 90%.

15.6.2.4 Alternative Methods

Gravimetric determination of particulate emission has serious disadvantages: it is a tedious, time consuming and integrating method. However, a rapid measurement and/

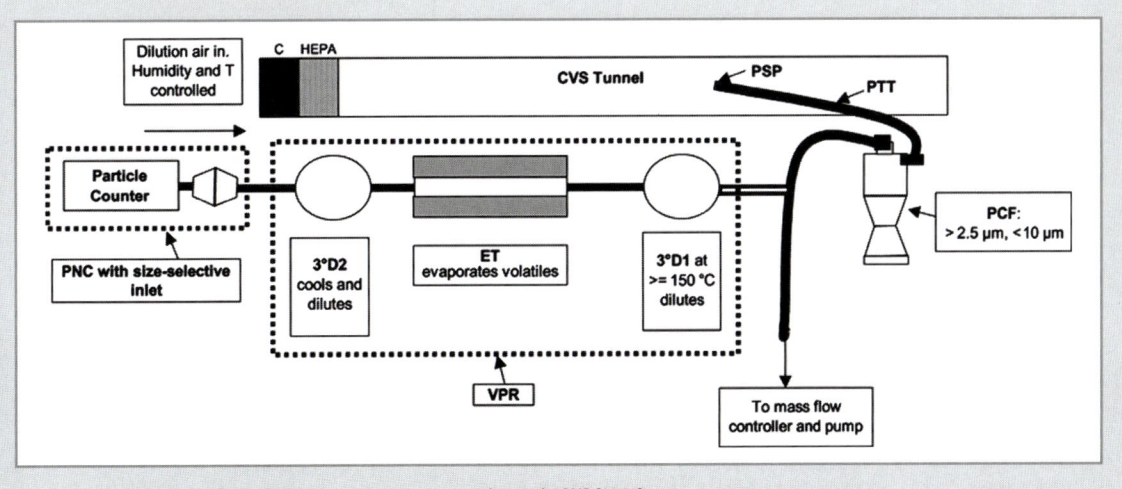

Fig. 15-58 Exhaust gas conditioning for particle number counting according to the PMP [15-35]

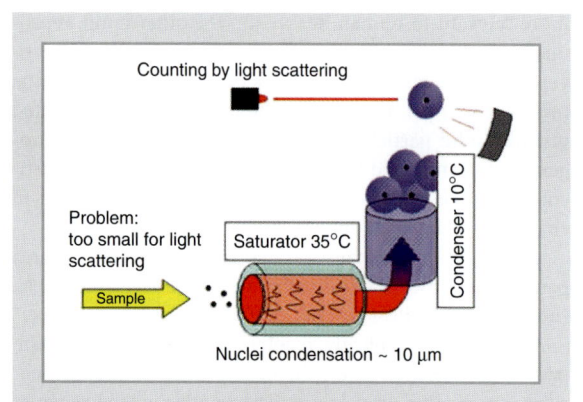

Fig. 15-59 Functional principle of a condensation particle counter (CPC) [15-37]

or the allocation of the time of emission to the dynamic driving conditions are frequently needed for engine development. Hence, a number of simpler and/or more dynamic methods of measurement have been developed. However, the measured quantities generally deviate from the particulates measured in compliance with the law and the established correlations only have limited validity. The measurement of soot emission assumes a special role here since it is an important indicator for the quality of combustion. Several measuring methods generally based on soot's strong absorption of radiation have been developed. New methods have excellent time resolution and/or very high sensitivity.

The most important alternative measurement methods are summarized in Table 15-17 and presented schematically in Figs. 15-60–15-67. Generally, there are different designs and commercial providers for all of the methods of measurement.

Table 15-17 Advantages and disadvantages of alternative methods of particulate/soot measurement

Method	Advantages	Disadvantages
Opacimeter [15-39]	– Mandatory for some certification tests, e.g. ELR – Reliable, cost efficient, established method to measure exhaust gas opacity – Excellent time resolution of 0.1 s – High sensitivity (0.1% opacity, corresponding to approximately 300 $\mu g/m^3$ soot) – Usable with special sample conditioning up to exhaust gas pressures of 400 mbar; higher pressure add-on available – Acceptable soot concentration correlation (mg/m^3) can be found for families of engines	– Sampling systems require sampling flows of up to 40 l/min – High sensitivity necessitates sophisticated system design: long optical path length L, good thermal conditioning – Relatively strong cross sensitivity to NO_2
TEOM (Tapered Element Oscillating Microbalance) [15-40]	– Measurement of particulate (not soot) emission – Result is similar to the statutory method of particulate measurement – Time resolution in the second range	– Replaces the particulate filter but requires exhaust gas dilution – Generally not fully equivalent to the mandatory method – Sensitivity depends on the time resolution, typically 1 mg/m^3 – Expensive
DMM (Dekati Mass Monitor) [15-41]	– Measurement of particulate (not soot) emission – Result is similar to the mandatory method of particulate measurement – Time resolution in the second range – Sensitivity of approximately. 1 $\mu g/m^3$ – Additional estimation of mean particulate size	– Replaces the particulate filter but requires high exhaust gas dilution – Frequently not really equivalent to the mandatory method – Expensive
Smokemeter [15-39]	– Reliable cost efficient – Established method – High sensitivity (0.002 FSN, corresponding approximately to 20 $\mu g/m^3$ soot) during longer sampling times – With special sampling equipment, exhaust gas can be measured before the diesel particulate filter – Good soot concentration correlation (mg/m^3), minimal cross sensitivity to other exhaust gas components	– Integrating method – Time resolution of approximately 1 min.
Photoacoustic soot sensor [15-42–15-44]	– High sensitivity – typically < 5 $\mu g/m^3$ soot – Sensor signal is directly and linearly sensitive to soot concentration, minimal cross sensitivity – Good time resolution, ≈ 1 s	– Requires exhaust gas dilution – Method of calibration is not rigorously established – Measurement upstream from the DPF requires exhaust gas conditioning

Table 15-17 (Continued)

Method	Advantages	Disadvantages
	– Applicable for diesel particulate filter tests – Moderately expensive – High dynamic range (1: 10,000)	– Regular servicing is easy but necessary
Laser induced incandescence [15-45]	– *High sensitivity – typically* < 5 µg/m³ *soot* – Sensor signal is directly and linearly sensitive to soot concentration, minimal cross sensitivity – Good time resolution, ≤ 1 s – Applicable for diesel particulate filter tests	– *Very expensive* – Method of calibration is not established. – High dynamic range is only achievable with optical attenuators (insertion of absorber filters)
Photoelectric aerosol sensor [15-46]	– Compact, cost effective system – High sensitivity – typically < 1 µg/m³ soot – An empirical correlation with the soot emission of diesel engines can be established in most cases	– Time resolution, ≤ 10 s – Strongly influenced by substances with high photoemission (PAH)
Diffusion charging sensor [15-41, 15-46, 15-47]	– Compact, cost effective system – Measures the active particulate surface (Fuchs surface) – High sensitivity – typically < 1 µg/m³ particulates – The signals have been found to empirically correlate with the particulate emission of diesel engines in some cases	– Not proportional to particulate mass – Time resolution of a few seconds

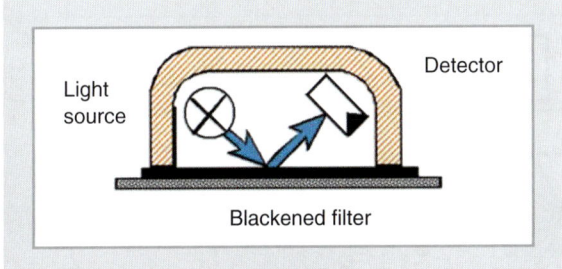

Fig. 15-60 Principle of the smokemeter [15-39]

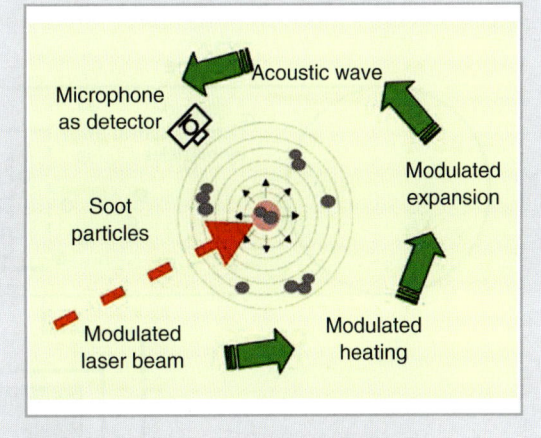

Fig. 15-62 Principle of photoacoustic measurement [15-44]

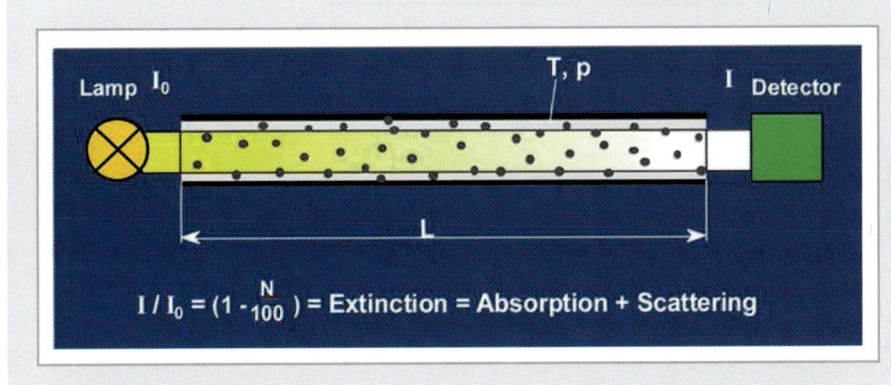

$$I / I_0 = (1 - \tfrac{N}{100}) = \text{Extinction} = \text{Absorption} + \text{Scattering}$$

Fig. 15-61 Principle of the opacimeter [15-39]

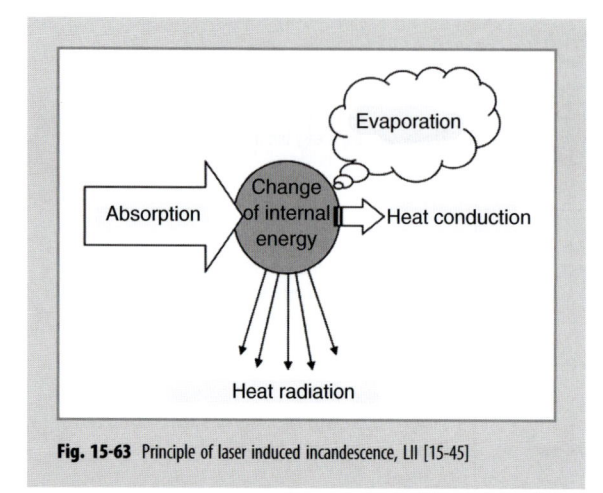

Fig. 15-63 Principle of laser induced incandescence, LII [15-45]

Fig. 15-64 Principle of photoelectric measurement [15-46]

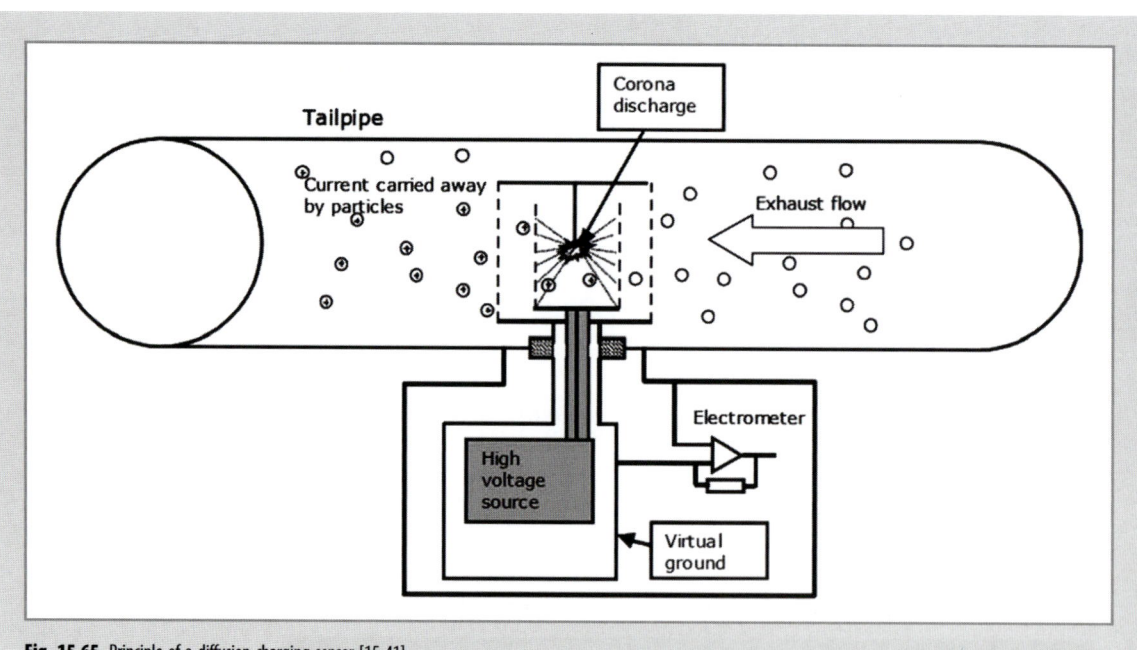

Fig. 15-65 Principle of a diffusion charging sensor [15-41]

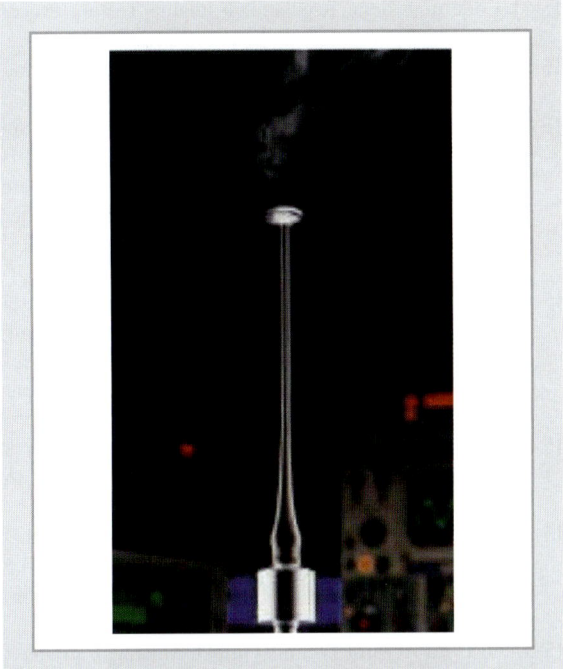

Fig. 15-66 TEOM (Tapered Element Oscillating Microbalance): tapered glass tube with the filter at the tip. The tube's vibration frequency changes with the filter's load [15-40]

The references [15-47–15-50] contain further and/or summarized information on unconventional state-of-the-art methods of diesel particulate measurement.

Literature

15-1 Klingenberg, H.: Automobil-Messtechnik. Vol. C: Abgasmesstechnik. Berlin: Springer (1995)

15-2 22nd Federal Ambient Pollution Control Act of September 18, 2002

15-3 Basshuysen, R.V.; Schäfer, F. (Eds.): Internal Combustion Engine Handbook. Warrendale: SAE International (2004)

15-4 John, A.; Kuhlbusch, T.: Ursachenanalyse von Feinstaub (PM10)-Immissionen in Berlin. Berlin: Senatsverwaltung für Stadtentwicklung (2004)

15-5 Wichmann, E.: Abschätzung positiver gesundheitlicher Auswirkungen durch den Einsatz von Partikelfiltern bei Dieselfahrzeugen in Deutschland. Umweltbundesamt 5/2003

15-6 Fränkle, G.; Havenith, C.; Chmela, F.: Zur Entwicklung des Prüfzyklus EURO 3 für Motoren zum Antrieb von Fahrzeugen über 3,5 t Gesamtgewicht. 4th Aachen Colloquium Automobile and Engine Technology (October 1995)

15-7 European Standard: EN ISO 8178 (1996)

15-8 Directive 97/68/EC of December 16, 1997

15-9 EPA Environmental Protection Agency: Certification Guidance for Heavy Duty On-Highway and Nonroad CI Engines. Code of Federal Regulations 40 CFR 86/89 (1998) 9

15-10 Technical Code on Control of Emission of Nitrogen Oxides from Marine Diesel Engines. Regulations for the Prevention of Air Pollution of Ships. IMO MP/Conf. 3/35, Annex VI to MARPOL 73/78

15-11 SJÖFS: Swedish Maritime Administration Decree (1997) 27

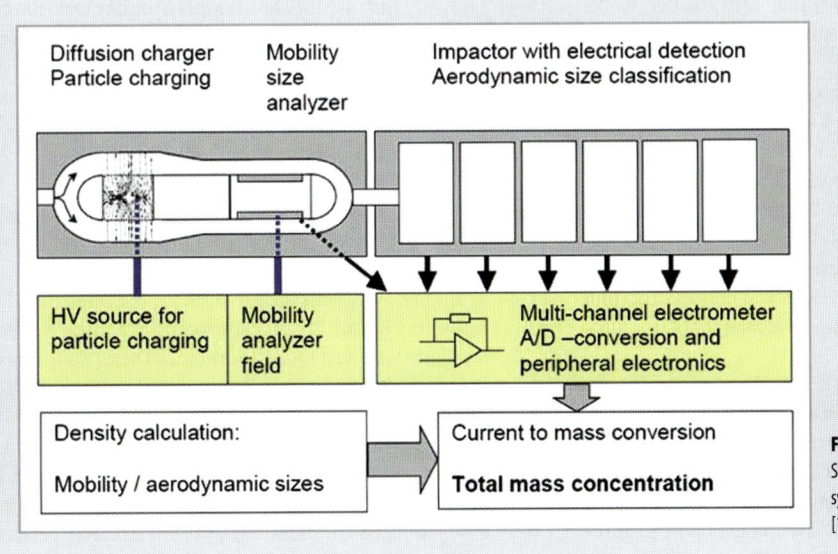

Fig. 15-67 Schematic of the sensor and data processing systems of a Dekati mass monitor DMM [15-41]

15-12 Graf, A.; Obländer, P.; Land, K.: Emissionen, Kraftstoffverbrauch-Vorschriften, Testverfahren und Grenzwerte. Mercedes-Benz Abgasbroschüre (September 2008)

15-13 Zeldovich, Y.B.: Zhur. Tekhn. Fiz. Vol. 19, NACA Tech Memo 1296 (1950) p. 1199

15-14 Pischinger, S.: Verbrennungsmotoren. Vorlesungsumdruck RWTH Aachen (2001)

15-15 Directive 70/220/EEC (Measures to be taken against air pollution by emissions from motor vehicles)

15-16 Hohenberg, G.: Partikelmessverfahren. Abschlussbericht zum Forschungsvorhaben BMWi/AiF 11335 (2000)

15-17 Hagelüken, C.: Autoabgaskatalysatoren. Bd. 612, Reihe Kontakt & Studium. 2nd Ed. Renningen: Expert Verlag (2005)

15-18 Mollenhauer, K.: Handbuch Dieselmotoren. 2nd Ed. Berlin/Heidelberg/New York: Springer (2000)

15-19 Mayer, A.: Partikel (www.akpf.org/pub/lexicon10-3-2000.pdf)

15-20 Siegmann, K.; Siegmann, H.C.: Molekulare Vorstadien des Russes und Gesundheitsrisiko für den Menschen. Physikalische Blatter 54 (1998) p. 149–152

15-21 Siegmann, K.; Siegmann, H.C.: Die Entstehung von Kohlenstoffpartikeln bei der Verbrennung organischer Treibstoffe. Haus der Technik e.V. Veranstaltung 30-811-056-9 (1999)

15-22 Khalek, I.A.; Kittelson, D.B.; Brear, F.: Nanoparticle Growth During Dilution and Cooling of Diesel Exhaust: Experimental Investigation and Theoretical Assessment. SAE Technical Paper Series 2000-01-0515 (2000)

15-23 ACEA report on small particle emissions from passenger cars (1999)

15-24 Pischinger, S. et al.: Reduktionspotential für Russ und Kohlenmonoxid zur Vermeidung des CO-Emissionsanstiegs bei modernen PKW-DI-Dieselmotoren mit flexibler Hochdruckeinspritzung. 13th Aachen Colloquium Automobile and Engine Technology (2004), p. 253

15-25 Robert Bosch GmbH (Ed.): Dieselmotor-Management. 4th Ed. Wiesbaden: Vieweg (2004)

15-26 Control of Air pollution from New Motor Vehicles – Certification and Test Procedures. Code of Federal Regulations 40 CFR 86.110-94

15-27 Directive 91/441/EEC of 26 June 1991

15-28 TRIAS 60-2003, Exhaust Emission Test Procedures for Light and Medium–Duty Motor Vehicles. In: Blue Book. Automobile Type Approval Handbook for Japanese Certification. JASIC (2004)

15-29 Engeljehringer, K.; Schindler, W.: The organic Insoluble Diesel Exhaust Particulates – Differences between diluted and undiluted Measurement. Journal of Aerosol Science 20 (1989) 8, p. 1377

15-30 Code of Federal Regulations: Control of Emissions from new and In-Use Highway Vehicles and Engines. 40 CFR 86.007-11

15-31 Code of Federal Regulations: Engine Testing Procedures and Equipment 40 CFR 1065, (July 2005)

15-32 Directive 2005/55/EC of September 28, 2005 and Directive 2005/78/EC of November 14, 2005

15-33 International Organization for Standardization: Heavy Duty Engines – Measurement of gaseous emissions from raw exhaust gas and of particulate emissions using partial flow dilution systems under transient test conditions. ISO 16183, (December 15, 2002)

15-34 Silvis, W.; Marek, G.; Kreft, N.; Schindler, W.: Diesel Particulate Measurement with Partial Flow Sampling Systems: A new Probe and Tunnel Design that Correlates with Full Flow Tunnels. SAE Technical Paper Series 2002-01-0054 (2002)

15-35 Informal document No GRPE-48-11: Proposal for a Draft Amendment to the 05 Series of Amendments to Regulation No. 83, (2004)

15-36 Dilaria, P.; Anderson, J.: Report on first results from LD Interlab. Working Paper No. GRPE-PMP-15-2 (2005) 5 (www.unece.org/trans/main/wp29/wp29wgs/wp29grpe/pmp15.html)

15-37 GRIMM Aerosol Technik GmbH, Ainring: Datenblatt Nano-Partikelzähler (CPC) Model 5404; TSI Inc., St. Paul, MN: CPC Model 3790 Data Sheet, (2007)

15-38 VDI 2066: Particulate matter measurement – Dust measurement in flowing gases – Gravimetric determination of dust load. In: Air Pollution Prevention. Vol. 4, Berlin: Beuth (1986)

15-39 AVL List GmbH (Ed.): Measurement of Smoke Values with the Filter Paper Method. Application Notes No. AT1007E, 2001; AVL 439 Opacimeter Data Sheet, 2001

15-40 Thermo Electron Co.: TEOM Series 1105 Diesel Particulate Monitor Data Sheet (www.thermo.com)

15-41 DEKATI Ltd.: DMM Dekati Mass Monitor Data Sheet, 2007; Dekati ETaPS Electrical Tailpipe PM Sensor Data Sheet, (2007) (http://dekati.fi)

15-42 Krämer, L.; Bozoki, Z.; Niessner, R.: Characterization of a Mobile Photoacoustic Sensor for Atmospheric Black Carbon Monitoring. Analytical Sciences 17S (2001) p. 563

15-43 Faxvog, F.R.; Roessler, D.M.: Optoacoustic measurements of Diesel particulate Emissions. Journal of Applied Physics 50 (1979) 12, p. 7880

15-44 Schindler, W.; Haisch, C.; Beck, H.A.; Niessner, R.; Jacob, E.; Rothe, D.: A Photoacoustic Sensor System for Time Resolved Quantification of Diesel Soot Emissions. SAE Technical Paper Series 2004-01-0968 (2004)

15-45 Schraml, S.; Heimgärtner, C.; Will, S.; Leipertz, A.; Hemm, A.: Application of a New Soot Sensor for Exhaust Emission Control Based on Time resolved Laser Induced Incandescence (TIRELII). SAE Technical Paper Series 2000-01-2864 (2000)

15-46 Matter Engineering AG: Diffusion Charging Particle Sensor Type LQ1-DC Data Sheet, (2003) (www.matter-engineering.com); EcoChem Analytics: PAS 2000 Photoelectric Aerosol Sensor (www.ecochem.biz)

15-47 Burtscher, H.: Physical characterization of particulate emissions from diesel engines: a review. Journal of Aerosol Science 36 (2005) pp. 896–932

15-48 Burtscher, H.; Majewski, W.A.: Particulate Matter Measurements. (www.dieselnet.com/tech/measure_pm_ins.html)

15-49 Aufdenblatten, S.; Schänzlin, K.; Bertola, A.; Mohr, M.; Przybilla, K.; Lutz, T.: Charakterisierung der Partikelemission von modernen Verbrennungsmotoren. MTZ 63 (2002) 11, p. 962

15-50 Vogt, R.; Scheer, V.; Kirchner, U.; Casati, R.: Partikel im Kraftfahrzeugabgas: Ergebnisse verschiedener Messmethoden. 3. Internationales Forum Abgas- und Partikelemissionen, Sinsheim (2004)

Further Literature

Robert Bosch GmbH (Ed.): Automotive Handbook. 7th Ed. Chichester: John Wiley & Sons Ltd 2007

16 Diesel Engine Noise Emission

Bruno M. Spessert and Hans A. Kochanowski

16.1 Fundamentals of Acoustics

Like many other machines, diesel engines also generate variations in air pressure. These variations disperse in the air as longitudinal vibrations. The human ear is able to perceive such pressure variations as noise within a frequency range of approximately 16 Hz–16 kHz. High and low frequency noise components in this frequency range are perceived to be far less loud than noise in the frequency range of 0.5–5 kHz. This frequency-dependent sensitivity of the human ear can be accommodated in a frequency-dependent evaluation curve (A-weighting).

At a frequency of 1 kHz, the human ear is able to hear sound pressure amplitudes between approximately $2 \cdot 10^{-5}$ and 20 Pa. (Pressure variations above this amplitude range are experienced as pain.) The large amplitude bandwidth of human hearing would make a specification of linear sound (pressure) values extremely unwieldy. Sound (pressure) levels are usually employed instead:

$$L_p = 20 \log p/p_o$$

(where L_p is the sound pressure level, p the sound pressure and $p_o = 2 \cdot 10^{-5}$ Pa the reference sound pressure).

The sound pressure level specifies the level of sound for a measuring point, i.e. the noise exposure (noise pollution) at the measuring point. However, the noise radiated by a component or a machine is either described by specifying the sound pressure level and additionally the measurement distance (!) or, alternatively, the sound power level:

$$L_W = 10 \log P/P_o$$

(where P is the sound power and $P_o = 10^{-12}$ W the reference sound power).

The reference variables p_o and P_o are selected so that the following applies:

$$L_W = L_p + 10 \log A/A_o$$

(where A is the measuring surface and $A_o = 1$ m^2 the reference surface).

The surface enveloping the sound source in which the sound pressure is measured is referred to as the measuring surface. The decibel (dB) is the unit of sound pressure levels and sound power levels. Routinely applied, A-weighting is identified by a suffix (L_{pA} or L_{WA}, in the unit dB$_A$ or dB(A)).

A doubling of the sound energy corresponds to an increase of the sound pressure level or sound power level by 3 dB or 3 dB(A). However, humans do not perceive such a doubling of sound energy as twice as loud. A doubling of noise is not subjectively perceived until the sound energy has roughly decupled, i.e. the sound pressure level has increased by 10 dB(A).

Noise can be measured by microphones that capture air pressure variations through a thin diaphragm, which forms part of a capacitor. Air pressure variations cause diaphragm vibrations, which in turn cause corresponding variations of the capacitor's electrical charge, which are proportional to the air pressure variations and relatively easy to process further.

16.2 Development of Engine Noise Emission

Legislators have been steadily tightening noise limits for practically every vehicle and unit of equipment since the early 1970s. An end to this trend is not foreseeable, particularly since the increasing density of traffic largely offsets noise reductions in individual vehicles or units of equipment. Noise pollution has not yet been reduced sufficiently among the populace, even though this ought to be imperative from the perspective of public health policy (e.g. [16-1, 16-2]).

Mandatory limits on noise emission from combustion engines do not exist. Legislators limit the noise emission of entire vehicles or units of equipment instead of engines. For instance, equipment manufacturers are left to decide whether to use relatively quiet engines or enclosed engines to meet noise limits. Since encapsulation measures incur considerable additional costs for equipment manufacturers, relatively quiet engines have a market advantage. Thus, legislation is also indirectly forcing the reduction of the noise emission of internal combustion engines. Therefore, the development of

---

B.M. Spessert (✉)
FH Jena (University of Applied Sciences), Jena, Germany
e-mail: bruno.spessert@fh-jena.de

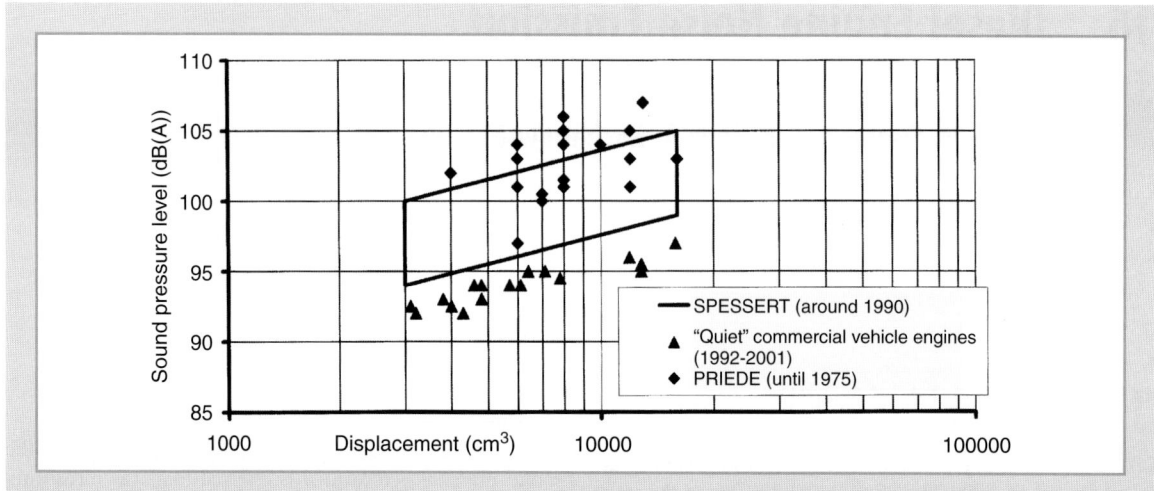

Fig. 16-1 Comparison of noise levels for heavy duty diesel engines before 1975 and around 1990 and for acoustically good new developments launched on the market since 1990 (production engines, rated power, 1 m test distance) [16-3]

ever quieter engines was and continues to be an ongoing task of modern engine design. However, other noise sources, e.g. drive train and tire noise from trucks or hydraulic noise from construction equipment, are also increasingly playing a significant role.

Taking direct injection diesel engines with displacements of 3…16 l used in truck and industrial engines as an example, Fig. 16-1 [16-3, 16-4] illustrates the engine noise reduction achieved in recent decades. Each of the scatter bands of noise emission from different engines are specified for the years 1975 and 1990. In general, the scatter band range is sizeable. However, an average drop of 3 dB(A) is discernible in noise levels between 1975 and 1990.

What is more, a number of newly developed advanced engines have been launched on the market since 1990. Above all, structural optimization rigorously implemented by means of FEM and radical measures to reduce noise excitation (e.g. relocating the timing gear drive to the flywheel

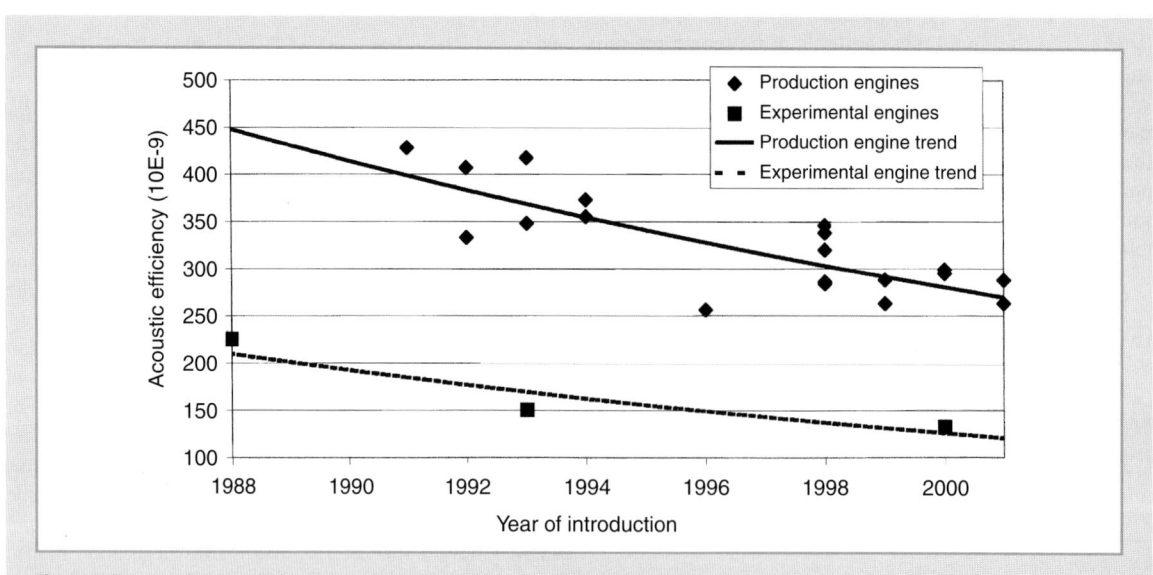

Fig. 16-2 Reduction of acoustic efficiency in acoustically good commercial vehicle diesel engines between 1988 and 2001

side; see Sect. 16.3.1.5) have made a real leap in development possible in these new engine models, which has lowered noise emission by 5 and 8 dB(A) on average (over 1990 or 1975 respectively) (Fig. 16-1).

The existing potential for noise reduction is far from having been exhausted. The "acoustic efficiency", i.e. the ratio of engine sound power to the rated engine power delivered, has been dropping continually since the 1990s (Fig. 16-2, [16-3–16-6]). In addition, a recent study demonstrated that even the noise emission of an already acoustically very good industrial diesel engine can still be reduced by over 6 dB(A). A considerable part of this potential for noise reduction has been verified experimentally in a test engine [16-7, 16-8].

The prerequisite for the reduction of engine noise is knowledge of the noise sources. Engines excite noise by
- engine surface (surface noise) vibrations (structure-borne noise),
- pulsations (aerodynamic noise) generated by intake, exhaust and cooling system(s) and the
- transmission of vibrations by the engine mount to the chassis or foundation.

In most cases, the engine surface plays the biggest role.

16.3 Engine Surface Noise

16.3.1 Structure-Borne Noise Excitation

16.3.1.1 Excitation Mechanisms

The mechanism of engine surface noise excitation is represented schematically in Fig. 16-3 [16-9]. The variables required to describe the mechanism may be treated as functions of the frequency according to Fourier and represented as a spectrum: a force F, the gas force in the combustion chamber in this example, causes an acceleration a of the structure-borne noise. The transfer function $T = a/F$ describes the characteristics of the engine structure's structure-borne noise transmission. The acceleration a is converted on the surface into the sound pressure p. This radiation may be characterized by the radiation factor $A = p/a$. Hence, the following options for noise reduction exist:
- reducing the force F that excites structure-borne noise,
- reducing the structure-borne noise transmission, i.e. the transfer function T, or
- reducing the sound radiation, i.e. the radiation factor A.

However, instead of a single force F, a multitude of dynamic forces between various components are decisive for the excitation of noise in real engines. The different excitation mechanisms are:
- direct combustion noise resulting from structure-borne noise excitation of the combustion chamber walls by the gas force,
- indirect combustion noise resulting from relative movements influenced by the gas force (crank mechanism and spur gear transmission) or influenced by load-dependent forces (injection pump), and
- mechanical noise resulting from relative movements influenced by inertial force (crank mechanism and valve gear).

16.3.1.2 Direct Combustion Noise

A reduction of direct combustion noise can reduce noise
- in the entire operating range of naturally aspirated direct injection diesel engines and
- in the low idle range and transient operation of all diesel engines during cold starts.

Therefore, acoustic optimization of the combustion sequence must be an emphasis in the development of every combustion system. The cylinder pressure curve is crucial to the excitation of direct combustion noise. The pressure curve is normally transformed from the time range into the frequency range in order to evaluate it acoustically. This yields a cylinder pressure excitation spectrum. The excitation spectrum depends on various parameters of cylinder pressure (Fig. 16-4 [16-9]). In practice, the frequency ranges determined by the pressure increase or rate of pressure increase are decisive for the A-weighted level of engine noise. They roughly cover the range of 0.5–3 kHz in which the engine structure is relatively "permeable".

Figure 16-5 compares excitation spectra for different combustion processes [16-10]. The excitation level of DI diesel engines is typically 10 dB higher than that of IDI diesel engines. Moreover, natural vibrations develop in the combustion chamber at even higher frequencies. They cause peaks in the excitation spectrum and particularly increase the noise nuisance (e.g. [16-11, 16-12]). IDI diesel engines specifically experience vibrations of the system of the main combustion chamber-passage-secondary combustion chamber, which normally has a natural frequency of approximately 2 kHz. This induces excitation levels that are often determinative for the level of noise.

Since the engine structure has a very high insulating effect, high excitation levels at low frequencies (less than 0.5 kHz) are normally irrelevant for engine noise. What is more, the radiation factor and the sensitivity of the human ear are both low in this frequency range.

Conflicts of objectives arise during the development of combustion systems. A satisfying compromise that yields a significant potential for noise reduction without appreciable disadvantages can only be found when every complex of requirements for an advanced combustion system – low fuel consumption, low pollutant concentrations in exhaust gas and low combustion noise at acceptable manufacturing costs – is factored in from the beginning of development onward.

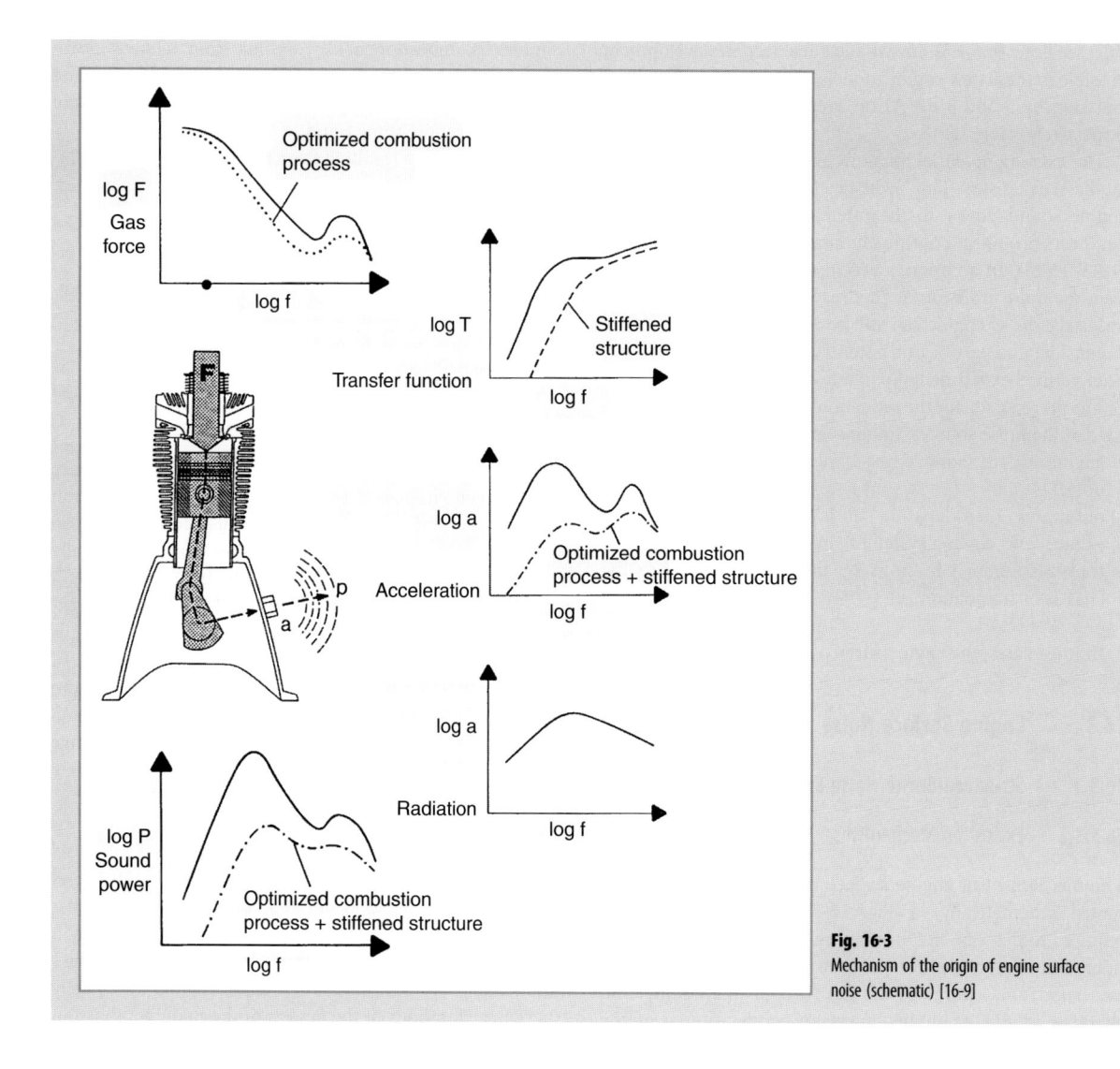

Fig. 16-3 Mechanism of the origin of engine surface noise (schematic) [16-9]

Good combustion excitation, i.e. an excitation spectrum with minimum level values, is obtained by a "gentle" cylinder pressure curve (Fig. 16-4). The conditions that initiate combustion are crucial; direct combustion noise can primarily be reduced by reducing the quantity of combustible fuel at the time of ignition. The most important options for this are
- retarded start of injection,
- advanced start of ignition (e.g. by increasing the compression ratio, supercharging, exhaust gas recirculation and/or high temperature cooling) and
- shaped or split injection.

In addition, a maximally advanced end of injection is striven for to prevent any increase in soot and HC emissions. Therefore, the start of injection can practically always only be retarded in conjunction with a more efficient injection system that reduces the duration of injection. However, this can increase noise excitation in the injection pump itself and in the injection pump drive (see Sect. 16.3.1.3).

Injection can be shaped or split by modifying the injection pump or the nozzle holder for instance. Common rail injection systems with electronically controlled injection nozzles particularly facilitate acoustically good injection rate shaping or splitting in combination with a map-dependent start of delivery and thus enable drastically reducing direct combustion noise in broad operating ranges.

Shaped or split injection can lower direct combustion noise during idling and part load by over 10 dB(A); overall engine noise can definitely be reduced by 3 dB(A) and more (Fig. 16-6 [16-13]). At the same time, the impulsiveness of the noise can be reduced and subjective noise perception improved significantly.

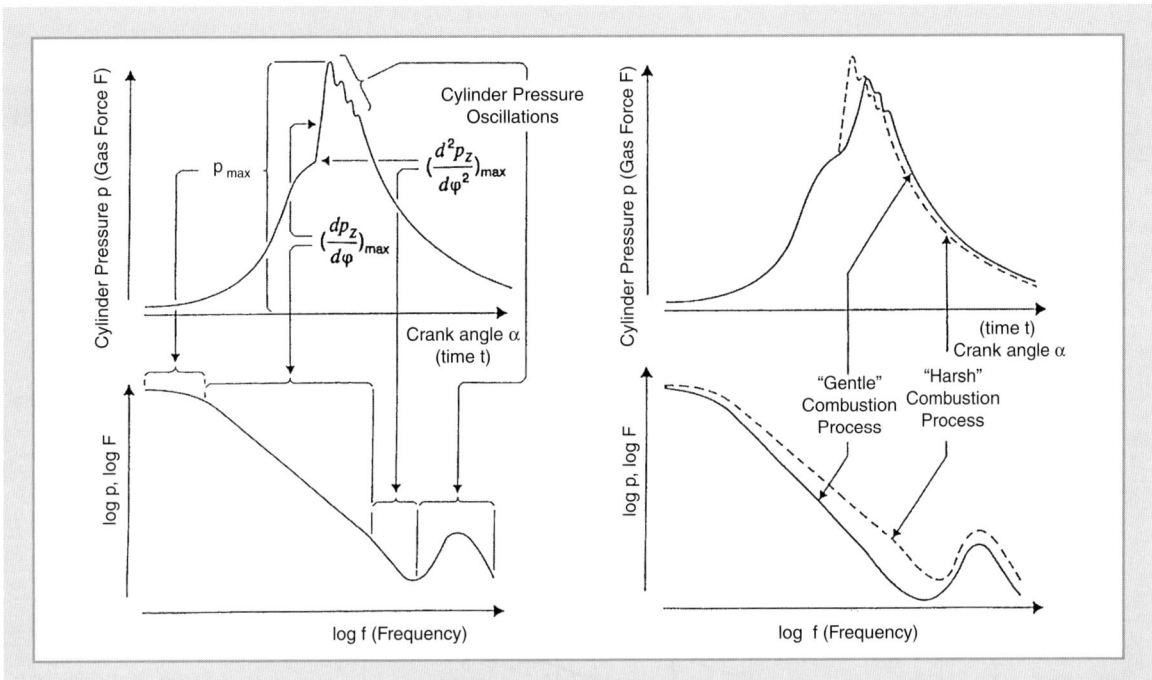

Fig. 16-4 Correlation between the cylinder pressure curve and cylinder pressure excitation spectrum (schematic) [16-9]

Alternative fuels such as RME (biodiesel) or vegetable oils can significantly influence direct combustion noise. Depending on the type of engine and its operating state, the use of alternative fuels can both increase and reduce the noise level significantly [16-5].

16.3.1.3 Indirect Combustion Noise

Indirect combustion noise dominates the full load range in high speed supercharged direct injection diesel engines. Piston noise plays an important role for instance (cf. [16-8,

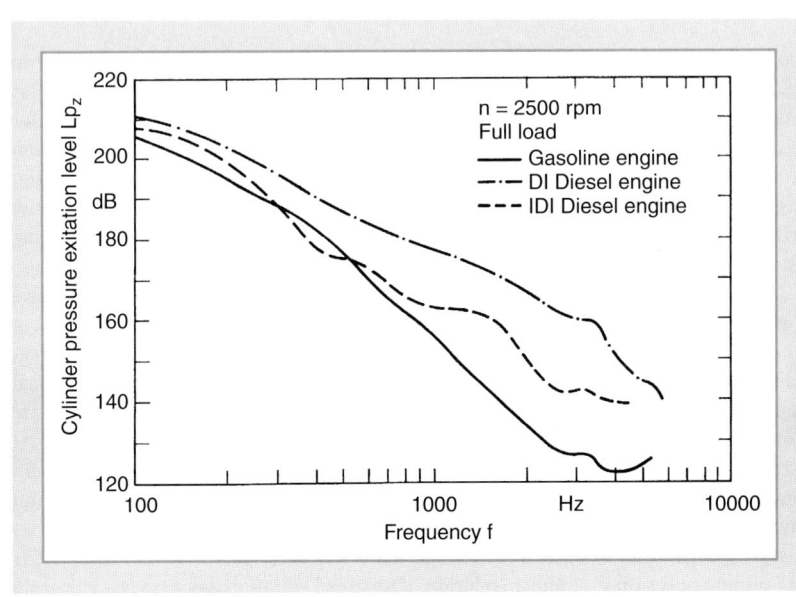

Fig. 16-5 Comparison of gas pressure excitation spectra of different combustion systems [16-10]

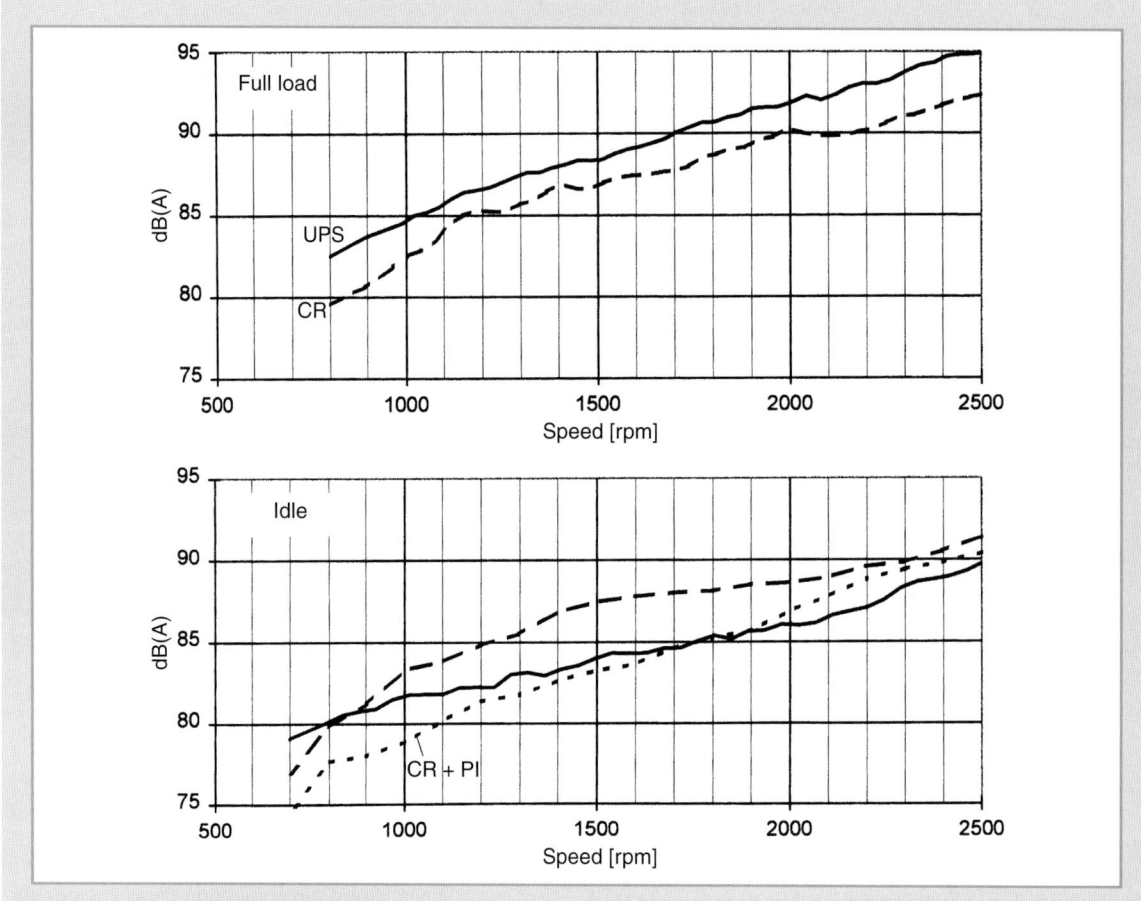

Fig. 16-6 Reduction of overall engine noise by a common rail injection system with pilot injection (supercharged, charge air cooled, heavy duty six cylinder inline engine, $V_H = 5.7$ dm^3) UPS unit pump system, CR common rail injection system, PI pilot injection (split injection) [16-13]

16-14–16-18]). Secondary piston motion produces shock excitations between pistons and cylinders, which are primarily generated by gas forces at high loads. The piston in a cylinder changes position two to ten times per working cycle depending on the operating point. Since the excitation of noise is a function of combustion pressure, it is referred to as "indirect combustion noise". (The inertial force also excites noise, especially at high speeds and during idling to be precise. This is referred to as "mechanical noise".)

Piston noise, i.e. the noise component generated by impact processes between pistons and cylinders, can be systematically reduced by minimizing the shock pulses during piston impact. Clearance, skirt length, shape and stiffness and pin offset are significant parameters that influence piston noise. Thrust side piston pin offset (by 1–2% of the piston diameter) is acoustically beneficial for many industrial and commercial vehicle engines.

Along with piston noise, injection pump and crankshaft noise is also particularly important for supercharged diesel engines (cf. [16-18–16-20]). Injection pump noise has become particularly relevant because of the strong rise in injection pressures and injection pressure gradients as required in advanced combustion systems optimized for exhaust gas and noise. The pump housing's noise emission and the engine block's structure-borne noise excitation caused by the pump as well as the structure-borne noise excitation in the pump drive produced by the alternating torques (see below) can play an important role.

The crankshaft also excites noises in the timing gear drive and the main bearings. This is particularly true at speeds at which the excitation of the lower (n) orders of the gas forces causes a resonance of the crankshaft's first natural torsional frequency. Alternating torques increased by the resonance are triggered in the gear drive where they generate impulses. At the same time, a crankshaft's radial movement in the main bearings coupled with the torsional vibration can excite an engine block to low frequency with the ($n-1$) and ($n+1$) order as well as to high frequency by impulses [16-21, 16-22]. The implementation of torsional vibration dampers can eliminate noise excitations caused by the torsional vibrations of the crankshaft, especially in the resonance.

16.3.1.4 Mechanical Noise

Valve gears, oil pumps, water pumps and, to a certain extent, crankshafts and pistons excite mechanical noise in diesel engines (see Sect. 16.3.1.3).

Noise excitation caused by impact processes when the valves close and (in engines with lower numbers of cylinders) by the triggering of alternating torques in the gear drive is prevalent in valve gears [16-23, 16-24]. Though relatively slight, valve noise is nonetheless perceived as "annoying" because of its high impulsiveness. According to [16-25–16-27], a high natural valve gear frequency, a dynamically optimized cam contour, slightly moving valve gear masses, hydraulic valve clearance compensation and a (slight) angular offset of the cams (phasing) in engines with more than two valves per cylinder are measures that reduce noise excitation caused by the valve gear.

Oil pumps in particular can generate noise by oil pressure pulsations. Noise can be decreased by reducing the pressure peaks (optimizing the gearing in gear pumps, the pressurestat, the oil galleries before and/or after the pump and so on [16-8, 16-28]). Water pumps normally cause negligibly little noise excitation.

16.3.1.5 Primary Drive Noise

Primary drive noise is comprised of injection pump, crankshaft and valve gear noise. Therefore, primary drive noise can neither be fully classified as "indirect combustion noise" nor as "mechanical noise".

Figure 16-7 presents the generation of pulses on the injection pump and intermediate gear as an example of noise excitation in a heavy duty DI diesel engine's gear drive: the pressure increase in the injection pump line (Fig. 16-7, top) and thus also on the injection pump plunger initially retards the injection pump gear. Thus, the injection pump gear teeth and intermediate gear teeth move relatively to each other. A pulse is triggered between the tooth flanks, which is clearly detectable as a pulse in the structure-borne noise and on both gears (Fig. 16-7, center and bottom), and excites the engine structure through the gear bearings. Options for reducing timing gear drive noise include torsional vibration dampers, supplementary masses on the injection pump gear, elimination of clearances by split gears, utilization of so-called "full-depth teeth" with relatively slender teeth and a high contact ratio and relocation of the gear drive to the flywheel side (Fig. 16-8 [16-8, 16-18–16-20, 16-23, 16-24, 16-29–16-32]).

Auxiliary units such as compressors or hydraulic pumps driven by the gear drive's power take-offs can also influence engine surface noise considerably. Piston compressors introduce alternating torques to the gear drive. This causes "chatter" in the gear drive, which increases gear drive noise and is frequently also subjectively perceived as disagreeable, especially during idling and at low speeds [16-8, 16-28]. By contrast, hydraulic pumps driven by camshaft or injection pump gears can even lower noise emission when the quasi static torque they deliver is substantial enough to prevent relative movement in the gear drive.

Chain or toothed belts are an acoustically better alternative than timing gear drives for car and smaller industrial diesel engines. However, chain and toothed belt drives may also be acoustically perceptible. Level peaks that appear in individual speed ranges and can be attributed to resonant vibrations of individual chain or belt segments are particularly disturbing subjectively. Inhomogeneities of the chain or belt, eccentricities of the wheels, oscillating torques and/or torsional vibrations of the pertinent shafts excite natural vibrations. These vibrations increase the chains or belts' impact velocity and thus noise excitation, especially in the resonance case. The tooth meshing order and its multiples stand out in the noise spectrum (Fig. 16-9). Toothed belt noise may be reduced by modifying the wheel and/or belt profile, modifying the contact surface of the belt and wheel, shifting the natural frequencies of the idler pulleys and/or using damping idler rollers [16-33].

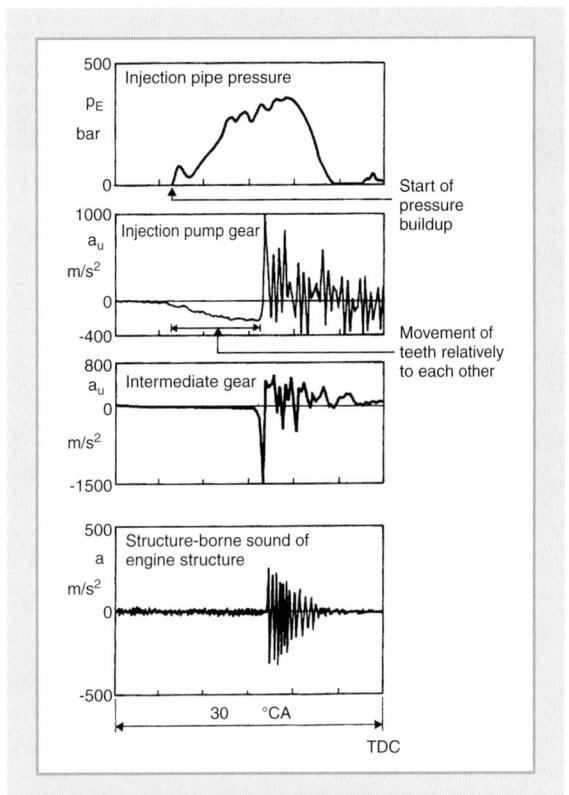

Fig. 16-7 Correlation between the increase in injection line pressure p_E, tangential acceleration a_u of the primary drive gears and structure-borne noise acceleration a in the engine structure [16-24]

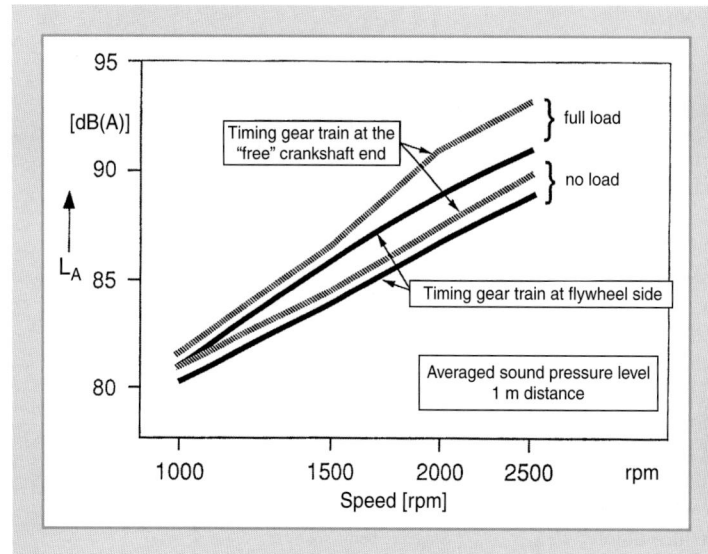

Fig. 16-8
Reduction of overall engine noise L_A by flywheel side placement of the timing gear drive (supercharged heavy duty four cylinder inline engine) [16-7]

Acoustically optimal measures for chain drives partly resemble the aforementioned measures for toothed belt drives. The chains themselves can be optimized acoustically. Gear chains are better than roller chains. The guide influences noise excitation through its rails and clamping elements [16-34]. Absorption of the chain sides' impact on lateral rubber washers increases the damping of the chain meshing [16-35]. Intensive oil injection reduces impacts and noise [16-36]. A duplex chain with offset links reduces whining noises [16-34].

16.3.2 Structure-Borne Noise Transmission in Engines

Along with reducing structure-borne noise excitation, acoustically optimizing the structure that transmits structure-borne noise is essential to fully exploit the existing potential for noise reduction. At least in new designs, the utilizable potential for noise reduction by structural optimization is greater than the utilizable potential by the reduction of noise excitation.

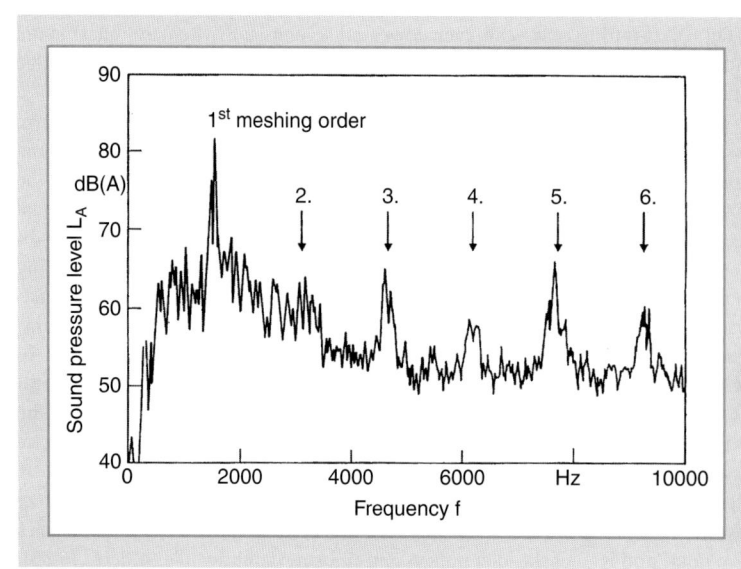

Fig. 16-9
Influence of timing belt meshing on the noise spectrum

16.3.2.1 Structure-Borne Noise Transmitted by Engine Blocks and Cylinder Heads

Simultaneously transmitting force and radiating noise, engine blocks particularly deserve attention since experience has shown that they and the add-on parts base-excited by them cause at least 50% of the overall engine noise in inline engines [16-37]. Therefore (while simultaneously minimizing their total mass and optimizing the distribution of mass) engine blocks have to be constructed rigidly and without material accumulations to thusly shift natural frequencies toward high frequencies or reduce the transmission of structure-borne noise in the frequency range of 0.5–3 kHz, which is particularly important for noise.

Force transmitting structures are normally optimized in a multitude of small steps. The efficacy of every individual step can hardly be tested on operating full engines. This is difficult because of the great effort required to construct numerous engine variants and because of the numerous disturbances (assembly, accuracy of measurement, reproducibility and high damping). Hence, only the efficacy of the sum of measures is usually demonstrated. However, development can be pursued more purposefully, more quickly and more cost effectively by evaluating individual steps and measures on the basis of principle tests, e.g. experimental modal analysis, and on the basis of calculations with the finite element method (cf. [16-7, 16-8, 16-32, 16-38–16-42]).

Figure 16-10 presents the acoustic concept for the engine block of a state-of-the-art, acoustically good commercial vehicle and industrial diesel engine developed with the aid of advanced methods of simulation and analysis:

– Thin but high ribs, a wide oil pan flange and rigid decks attain high horizontal stiffness especially in the area of the (single) injection pump case.
– Linear force distribution minimizes the noise generating motions of the block surface.
– In addition, internal ribbing stiffens the lower block zones in particular.
– The extremely rigid crankcase skirt prevents the main bearing wall and skirt vibrations from coupling, thus rendering the main bearing wall vibrations acoustically irrelevant.

This concept results in engine blocks with good acoustic properties while keeping the weight acceptable and manufacturing costs low.

Figure 16-11 presents engine block structures that can be used for new designs as an alternative to the standard design (a) for reasons of acoustics. An axial main bearing beam (b) shifts the natural frequencies of the bulkheads toward high frequencies and thus reduces skirt vibrations. A ladder frame (c) stiffens the crankcase skirt and, above all, prevents this engine block zone from vibrating in phase opposition. The ladder frame and main bearing beam may also be combined.

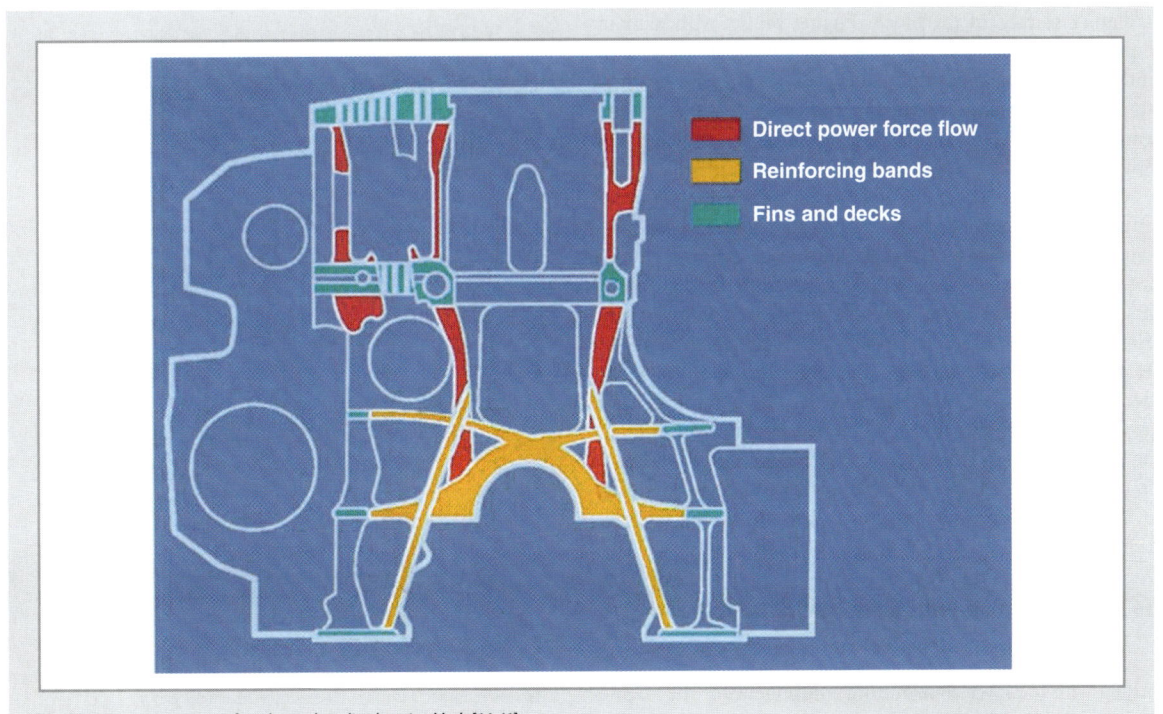

Fig. 16-10 Acoustic concept for a heavy duty diesel engine block [16-41]

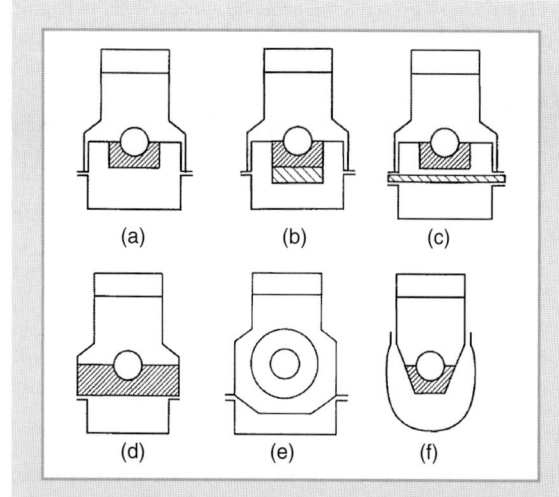

Fig. 16-11 Alternative engine block concepts: (**a**) Standard engine block; (**b**) Standard engine block with main bearing beam; (**c**) Engine block with ladder frame; (**d**) Engine block with bedplate; (**e**) 'Tunnel' engine block; (**f**) Crankframe engine block

The bedplate (d) and tunnel housing (e) make it possible to design extremely rigid engine blocks. A crankframe engine block (f) (cf. [16-43–16-45]) brings the oil pan, solely insulated from the engine block by elastic elements, up to the level of the cylinder block. This can drastically reduce the sound radiating surface of the block. Engine blocks with a main bearing beam, a ladder frame or even a bedplate have been introduced with increasing frequency and gone into production in recent years [16-46] but the other aforementioned engine block designs are still implemented in only very few engine types.

One consequence of the acoustic optimization of other engine components has been a steady increase in the importance of the cylinder head and the add-on parts it base excites for noise emission. The base plate and intermediate deck make the bottom of a cylinder head sufficiently rigid. Hence, the structure-borne sound levels are relatively low. However, high levels appear in the upper region of the cylinder head. Therefore, the structure has to be stiffened by greater wall strengths in this example or (better) by ribbing or (even better) by crowning. An alternative to these stiffening measures is the use of uncoupled or strongly damped cylinder head covers drawn down over the top part of the cylinder head, thus covering it [16-26, 16-46–16-49].

16.3.2.2 Structure-Borne Noise Transmitted by Add-On Parts

Along with the force transmitting components of the engine block and cylinder head, add-on parts also play a major role acoustically. In particular, pressure die cast aluminum add-on

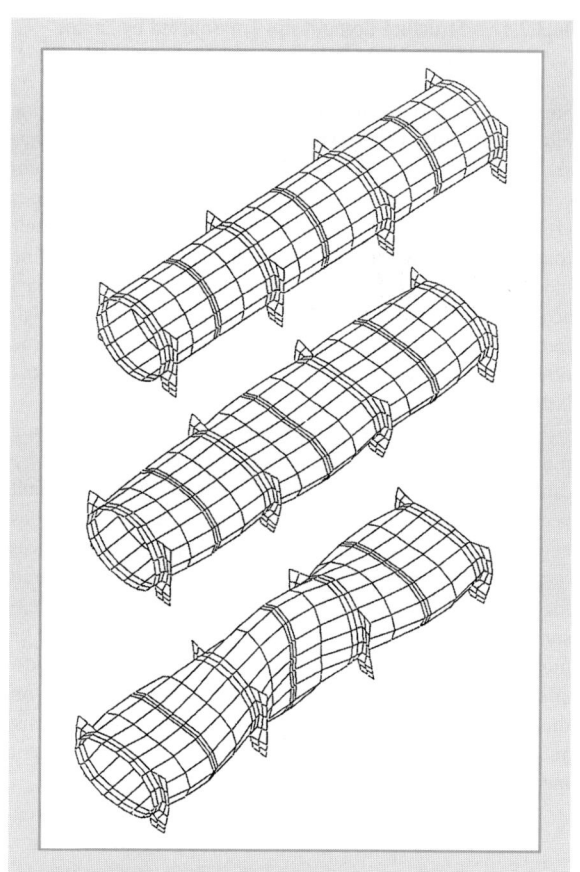

Fig. 16-12 Modes of intake manifold vibration: *top*: unshaped; *center*: 1st mode (FEM: 3,580 Hz, experimental modal analysis: 3766 Hz); *bottom*: 2nd mode (FEM: 4,081 Hz, experimental modal analysis: 3,980 Hz) [16-7]

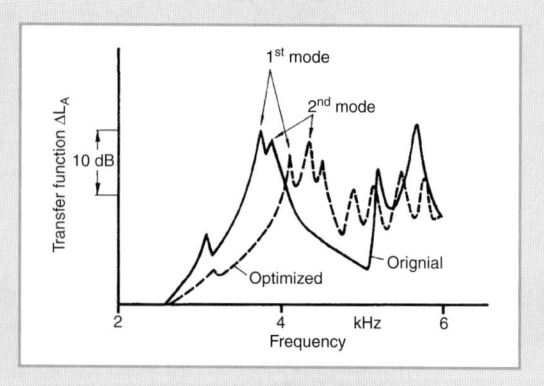

Fig. 16-13 Transfer functions between the cylinder head and pipe surface of two intake manifold variants. The *lower* the amplitudes and the *higher* the frequency of the transfer function's peaks, the better the pipe is acoustically

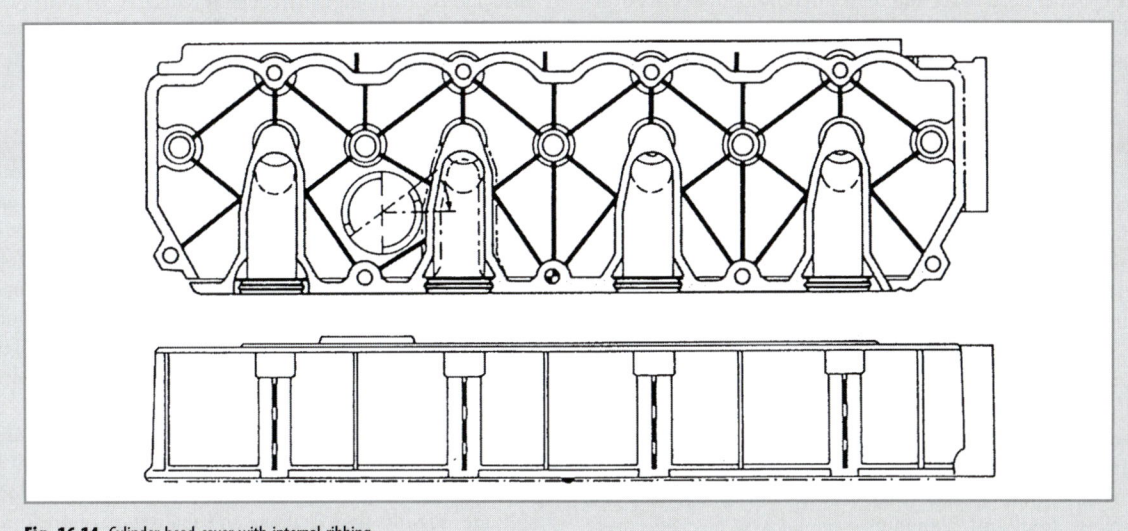

Fig. 16-14 Cylinder head cover with internal ribbing

parts are frequently problematic since they must be light and thin walled for reasons of manufacturing, cost and weight. Figure 16-12 visualizes the modes of vibration of the first two natural vibrations of an intake manifold calculated by means of the finite element method (FEM) and measured by means of modal analysis. Figure 16-13 plots the transfer function between the intake manifold (two variants: before and after optimization) and cylinder head (as the component that base excitates the intake manifold). Circumferential reinforcing bands can shift the natural frequencies toward higher frequencies. Since the spectra of the exciting forces decrease as the frequency increases, the increase of the natural frequencies causes the forces to shift to frequency ranges of smaller exciting forces (Fig. 16-1). At the same time, the amplitudes of the transfer function can be reduced.

A combination of transversal and longitudinal ribs and the creation of closed spaces around the injection nozzles (Fig. 16-14) lowers the component noise of a cylinder head cover by a maximum of 8 dB(A) over a fully unribbed, angular and flat variant. In this example, optimization of

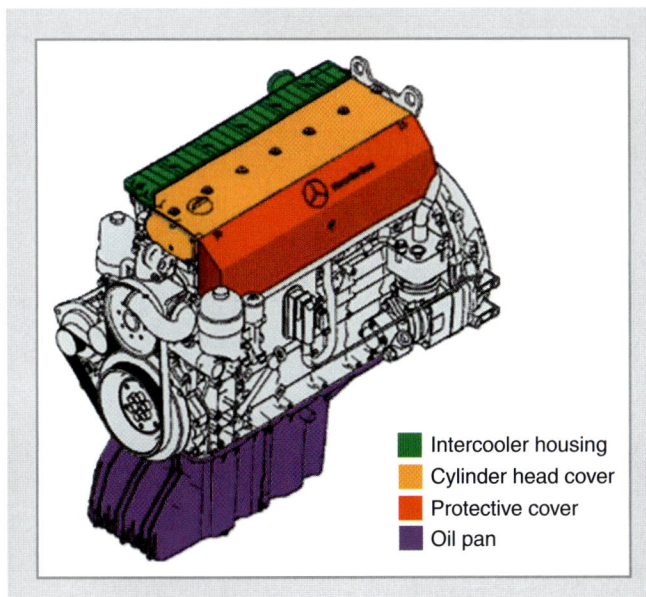

Fig. 16-15 Plastic components of a truck diesel engine [16-51]

the cylinder head cover and intake manifold lowers the overall engine noise emission in the measuring point above the engine by 1.5 dB(A) over a fully unribbed variant and slightly in the lateral measuring points as well [16-7, 16-8].

Structure-borne noise insulation by means of elastic elements between an add-on part and the components that base excite an add-on part can also reduce noise in add-on parts that do not transmit force. Insulating components with relatively low dynamic stiffness can significantly reduce noise (cf. [16-4, 16-5]). Naturally, soundproofing components that have already been structurally optimized beforehand only reduces noise relatively little [16-32].

Strongly damping materials can also reduce the noise emission of add-on parts [16-50]. The use of sandwich sheets for valve covers, oil pans or also air cowling in air-cooled engines is particularly established. Making high damping valve covers, intake manifolds, air cowling and even oil pans out of plastic is acoustically just as effective (Figs. 16-15 and 16-16 [16-51]).

Noise from cast aluminum oil pans is also effectively reduced by coating them with high damping plastic. Another option is to damp vibrating surfaces, preferably in the range of higher vibration amplitudes. Coating a spur gear cover with elastomers with high damping in the area of the center of the cover can drastically lower the vibrations of the cover's diaphragm (which dominate the overall noise of a tested engine) and significantly reduce the engine noise in the measuring point upstream from the engine. However, such coatings not only generate considerable additional costs and additional weight but also problems when they are recycled [16-50].

Insulation or damping can frequently reduce even more noise than increased stiffness. Therefore, such options for noise reductions are being utilized with increasing frequency [16-3].

16.3.3 Noise Radiation

Noise radiation may either be reduced by encapsulation measures (see Sects. 16.5 and 16.6) or the generation of an "acoustic short circuit". An "acoustic short circuit" is generated when the excess and low pressures of noise generated by a vibrating surface cancel each other out. Thus, air pressure no longer vibrates in greater distances from the vibrating surfaces and no more detectable noise radiates in the far field. This effect can be taken advantage of when acoustically optimizing V-belt pulleys for instance.

V-belt pulleys' acoustically relevant modes of vibration are usually their diaphragm vibrations. While openings in a V-belt pulley only alter natural frequencies and vibration amplitudes slightly, they facilitate an equalization of pressure on the front and back of the pulley. Thus, pressure variations are no longer able to propagate and noise no longer radiates into the engine's far field.

16.4 Aerodynamic Engine Noises

16.4.1 Intake and Exhaust System

Pressure pulsates strongly in combustion engines' intake systems and even more strongly in their exhaust systems.

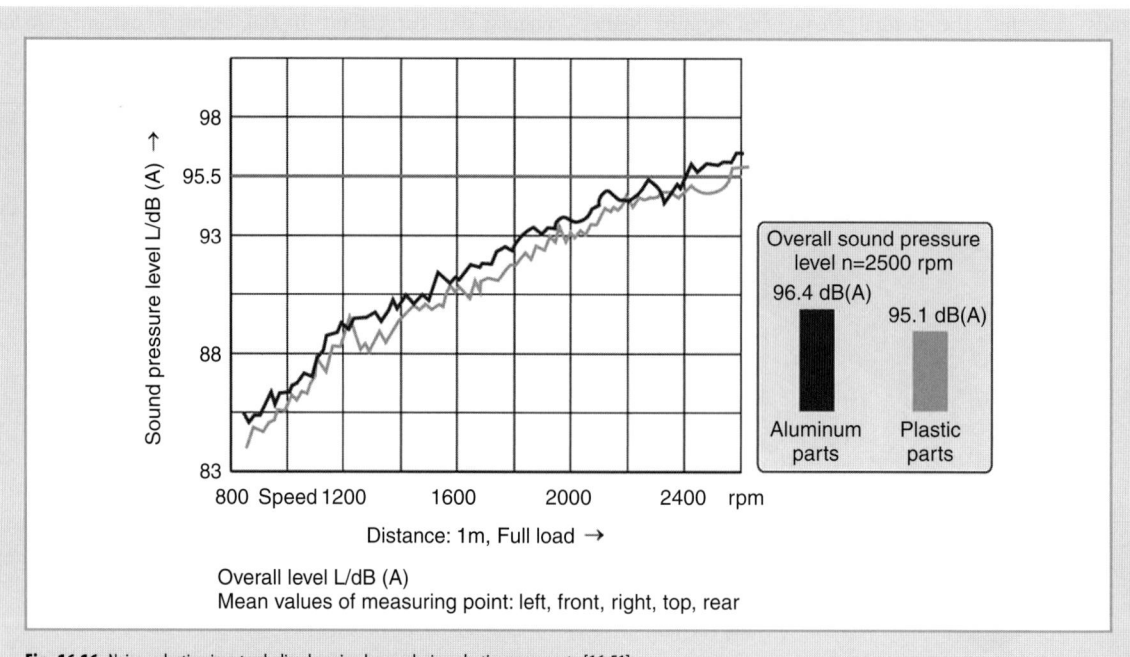

Fig. 16-16 Noise reduction in a truck diesel engine by employing plastic components [16-51]

Undamped pressure pulsations would drown out all other components of engine noise.

At least in supercharged engines, the air cleaner may already generate sufficient damping of the intake noise. Special damping systems in which, for example, venturi tubes primarily reduce the higher frequency intake noise component and/or resonators primarily reduce the low frequency component by reflecting and superimposing the intake noise pulsations (e.g. [16-52]) are implemented for other requirements.

In either case, the reduction of exhaust noise by a muffler (systems) is mandatory. It usually combines the aforementioned principles of sound damping by venturi tubes, bypasses or linings with sound absorbing materials and the cancelation of low frequency noise components by reflection and superimposition. This produces a basic correlation between noise damping capacity, unit volume and damper back pressure (that degrades engine efficiency). The achievable damping capacity grows with both the unit volume and the permissible back pressure (see Sects. 13.1 and 13.2).

Along with conventional "passive" muffler systems, "active" systems will also potentially play a role in the future. Microphones measure sound emission. The measurements are analyzed mathematically and loudspeakers, which generate an amplitude and frequency equal to the sound, are activated accordingly. However, the noise radiated by the loudspeaker is so dephased that the corresponding (low frequency) sound components are cancelled when the measured and additionally generated sound is superimposed. Advantages of "active" muffler systems are a reduction of back pressure and a significant diminution of the required (residual) damper volumes. However, these are offset by a number of disadvantages such as the additional consumption of electrical power or the considerable additional costs for the mechanical and electronic elements required to actively dampen sound.

16.4.2 Cooling System

In the 1970s, the cooling system noise generated by fans or blowers frequently still played a role, even a dominant role at times. Its influence on overall engine noise is now usually rather slight. This was achieved (cf. [16-53–16-57]) by
- designing the entire cooling system and specifically the fan blades or blower blades better aerodynamically,
- dividing blades unequally to prevent tonal noise components and especially
- using control systems to limit fan or blower speeds to the minimum still permissible for cooling in a particular operating point.

The aerodynamic noise generated by alternator fans can dominate engine noise in the upper speed range [16-8, 16-28].

Ever higher electrical outputs have led to ever larger generators and higher generator speeds. Even new generator designs with quieter enclosed fans have been unable to sufficiently lower alternator noise. The use of viscous clutches even between engines and alternators or the use of liquid-cooled alternators is more effective.

16.5 Noise Reduction by Encapsulation

16.5.1 Encapsulation and Enclosed Engines

Every noise reduction measure at the source ought to have been fully exhausted before noise is reduced by additional measures that take the form of partial or complete enclosures. This approach makes sense because measures at the source prove to be more cost effective than encapsulation or commensurate acoustic measures in the engine compartment. From the perspective of weight and structural space too, measures at the source should be preferred initially over encapsulation. However, the effect of measures at the source is limited to only a few dB(A) (see Sects. 16.3 and 16.4).

Complete enclosure of an engine on the other hand reduces noise considerably by 10–13 dB(A) and only necessitates marginally more structural space than an unenclosed engine (Fig. 16-17). A noise reduction of 13 dB(A) means that twenty enclosed engines generate the noise of one unenclosed engine.

The development of noise reduction in diesel engines by partial or complete enclosures has intensified since the end of the 1960s (cf. [16-58–16-60]). The first standard series of enclosed air-cooled two, three and four cylinder engines have been on the market since 1977 [16-61, 16-62]. Encapsulation development must endeavor to reduce noise for every potential operating point and every direction of radiation. Figure 16-18 presents results obtained in mass production.

Enclosed engines are used wherever extreme noise requirements cannot be satisfied in any other way. Unenclosed engines are another sizable area of application. Equipment manufacturers spare the engine hood, engine installation, their own noise reduction measures and cooling measures that may be additionally necessary. This is an especially cost effective solution for special equipment manufactured in smaller quantities.

16.5.2 Partial Sound Sources of Complete Enclosures

An enclosed engine's overall noise (Fig. 16-19) consists of surface noise from the enclosure, noise from the combustion air intake and noise from the exhaust outlet. This is compounded by noise from the cooler/fan systems in water-cooled engines or by noise from the cooling air inlet and outlet in air-cooled engines. Equally effective measures must be implemented against all these partial noises to obtain low overall noise.

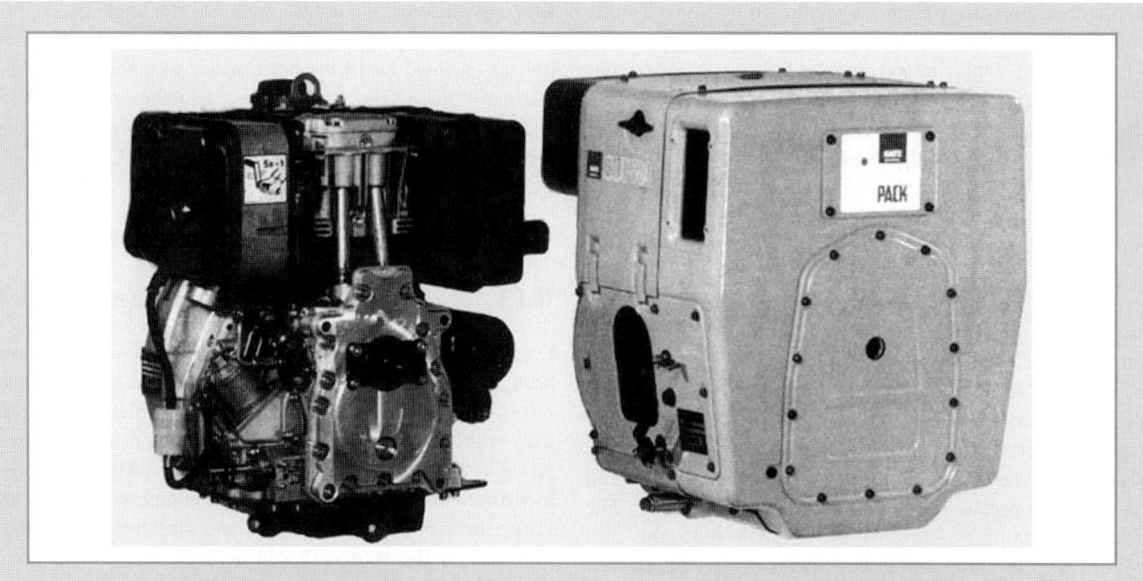

Fig. 16-17 Size comparison of an unenclosed and enclosed air-cooled single cylinder DI diesel engine

When water-cooled engines are enclosed, the cooler/fan system is normally left outside the enclosure, thus leaving noise reduction in these parts to the engine's user. Alternatively and similar to air-cooled engines, liquid-cooled engines are partly furnished with integrated cooling systems.

Air-cooled engines always have a cooling fan mounted directly at the engine itself and thus in the enclosure. Consequently, the engine enclosure also includes noise reduction by inlet and outlet air ducts.

16.5.2.1 Surface Noise

An engine's introduction of structure-borne noise by way of the enclosure mount particularly excites the enclosure surface to vibrate and thus radiate noise. Therefore, meticulously designed elastic enclosure mounts are preferred. They must also be provided for every part abutting the outer enclosure wall, e.g. apex seals that separate hot and cold spaces in an enclosure.

The enclosure wall also reflects sound radiated by the engine surface back to the engine. Therefore, the noise level measured inside an enclosure is approximately 3–5 dB(A) higher than that of an unenclosed engine. Absorption materials inside an enclosure can lessen this noise increase. The noise in the enclosure excites the enclosure walls to vibrate. The enclosure surface radiates these vibrations outwardly in reduced strength.

The extent of this reduction is referred to as wall insulation. The damping factor is a function of frequency and chiefly determined by the weight per unit area and the wall's bending rigidity. 1 mm thick steel plate has proven well suited as wall material for enclosures. Multilayered walls, for example, can further increase sound damping factors [16-63].

Appropriate engineering designs of the enclosure surface must prevent the formation of pronounced natural vibrations or reasons of acoustics and strength.

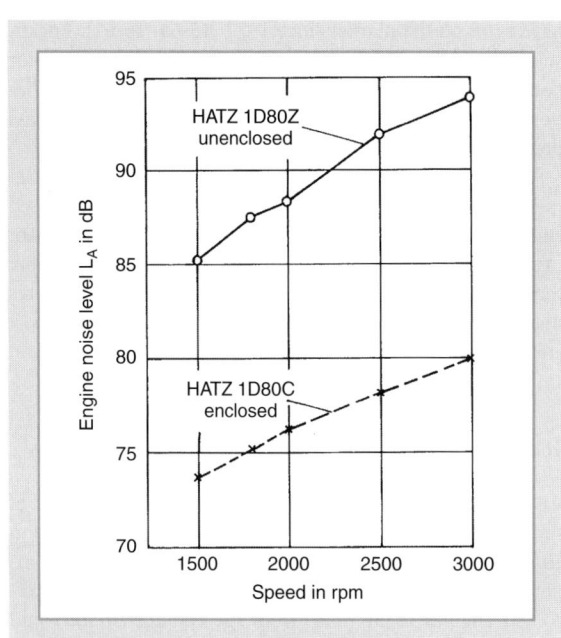

Fig. 16-18 Overall engine noise L_A derived as a function of speed, test distance 1 m, full load, exhaust gas deducted (engines from Fig. 16-17)

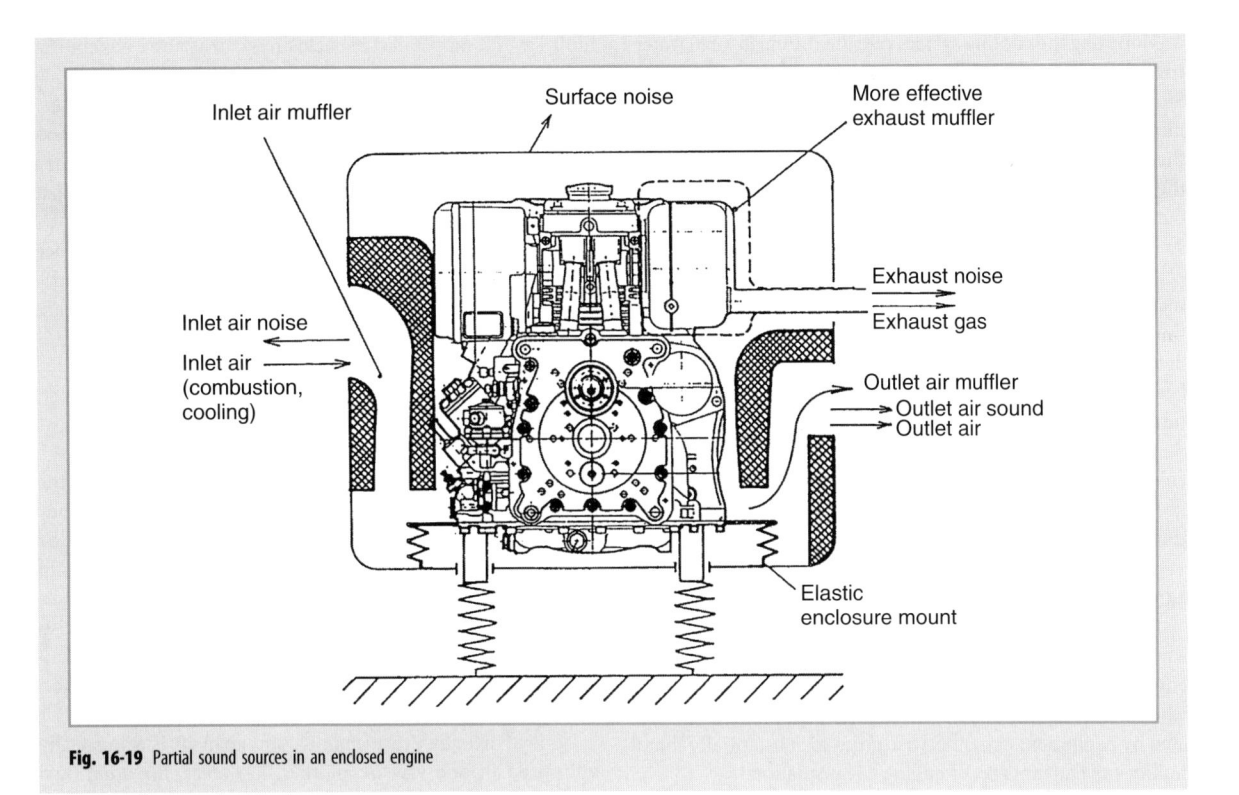

Fig. 16-19 Partial sound sources in an enclosed engine

Leaks in the enclosure surface allow engine noise amplified by enclosure reflection to penetrate outwardly and thus drastically reduce the wall insulation achievable. Therefore, an enclosure should be hermetically sealed.

If the exhaust muffler is placed outside the engine enclosure, then the muffler's surface noise also has to be considered. Measures in the muffler's surface, e.g. double walls with or without an intermediate layer, frequently only yield insufficient improvements. In most cases, the exhaust muffler must also be enclosed.

16.5.2.2 Outlet Noise

Intake and exhaust noise are referred to as outlet noises. The procedure to reduce them in enclosed engines is the same as in unenclosed engines (see Sect. 16.4.1). However, a substantially better acoustical result must be obtained in enclosed engines. The outlet noises must be improved by roughly the same magnitude by which encapsulation is also intended to reduce the overall noise.

The exhaust muffler in an enclosed engine is usually integrated on or in the enclosure structure since the engine presented to customers ought to be complete [16-64]. Thus, enclosed engines forfeit the option of reducing exhaust noise well by tuning volumes and line length, which is common in vehicles. Even today, exhaust muffler design is largely a matter of experience and testing.

Intake noise must also be improved by at least 10 dB(A) over unenclosed engines. Enclosed air-cooled engines are equipped with noise damping inlet air ducts for cooling air. Refraining from aspirating the combustion air until it is inside the enclosure to thus also take advantage of the acoustic effect of the absorption line for the cooling air for the combustion air suggests itself. In this case, the inlet air port cross sections must be adjusted to the total air mass. Combustion air should not heat up substantially inside the enclosure.

16.5.2.3 Noises from Engine Cooling

In addition to basic engine cooling, limiting temperatures also especially have to be observed for rubber parts, seals, V-belts, elastic engine suspensions and add-on parts, e.g. alternators and voltage regulators. The surface temperature of a noise enclosure also has to be kept low. Temperatures may be strongly elevated especially after shutoff from full load. Since users do not always expect this, a touch guard may be necessary for reasons of safety.

The outlet air temperature at the cooling system outlet is lower in water-cooled engines than in air-cooled engines. Since the higher outlet air temperature causes more heat to dissipate per quantity of air that passes through them, air-cooled engines require smaller quantities of air. Thus the inlet and outlet air mufflers required are also smaller than in water-cooled engines.

Mounting the cooling system outside the enclosure (water-cooled engine) necessitates additional enclosure ventilation. Small electrically driven fans are ideal. Inlet and outlet air mufflers are necessary for the air openings – not because of the (low) noise of the fan but to prevent the loud noise inside the enclosure from escaping outside. A cooling system located outside the enclosure likewise requires noise reduction measures since it can be significantly louder than an enclosed engine when the full quantity of heat is dissipated. Absorption materials make inlet and outlet air mufflers effective. Foam and mineral wool are employed. The muffler's effect is intensified by redirections or cross sectional jumps. Mufflers can be simulated precisely in advance when the dimensions, air temperatures and material properties are adequately known [16-63].

An enclosure must be meticulously separated into hot and cold spaces to properly cool an engine. This is necessary to keep exhaust muffler heat from heating combustion air for example.

16.5.3 Engine Installation and Maintenance

The demand for efficient insulation of driven machines from engine structure-borne noise takes on particular importance when enclosed engines are employed since the noise generated by equipment can otherwise grow so substantially that it nullifies the advantages of engine encapsulation.

When encapsulation lowers engine noise by 10 dB(A) or more, it enters ranges in which the noise of the driven machine can no longer be disregarded across the board. Meticulous equipment planning with knowledge of the driven machine's noise level must initially identify the other parts aside from the engine that require noise reduction to obtain the overall result desired. An engine enclosure can be designed so that smaller driven equipment, e.g. hydraulic pumps, can be accommodated inside it.

The maintenance points in an enclosed engine have to be designed and configured so that they are accessible without additional effort. This particularly pertains to frequently needed parts that should be accessible without opening the enclosure, e.g. oil dipsticks. The engine parts projecting from the enclosure and thus affected by structure-borne noise must have minimal sound radiating surfaces so that they radiate little noise.

Other maintenance points that must be accessed less often, e.g. valve covers (valve clearance), are located behind easy to open enclosure covers (Fig. 16-17).

16.5.4 Partial Encapsulation

Partial enclosures are employed to achieve a limited acoustic objective at minimum cost. One frequently encountered example is small power units used at construction sites, which are required to comply with a sound power level of $L_{WA} \leq 100$ dB(A) (electrical output >2 kVA) within the EU. Since diesel engines' noise level only slightly exceeds this limit, several simple partial enclosures can assure a generating set's compliance.

Enclosures of the cylinder head, muffler and air cleaners including the intake manifold are frequently employed. Partial enclosure measures are also often combined with primary noise reduction measures (e.g. oil pans insulated against structure-borne noise).

The application of partial enclosure measures can reduce noise by up to 4 dB(A). Such concepts leave large parts of the engine surface unenclosed. Special cooling measures are unnecessary.

16.6 Engine Soundproofing

Diesel engines are always significant and, frequently, even the most significant sources of noise of the vehicles or units they power. Only in the rarest of cases can the installation space be designed appropriately to satisfy the acoustic requirements in vehicles or units without either engine encapsulation (see Sect. 16.5) or soundproofing. Many options for this exist. They depend on the particular case of application, the structural space available, the amount of the noise level reduction required, etc.

Today, virtually every engine compartment is acoustically optimized in one way or another. However, the additional costs for acoustic measures in engine compartments, which are taken for granted, could often be reduced significantly by using quieter engines [16-8, 16-52]. The more stringent requirements for vehicle, unit or equipment noise emission are, the more substantial the advantages of acoustically relatively good engines become. Therefore, "low noise" vehicles, units or equipment are practically always outfitted with "quiet" engines.

Apart from noise radiated by its surface and aerodynamically excited noise, an internal combustion engine also generates noises through its suspension by exciting the base and thus the connected engine compartment walls, operator cab, etc. to radiate noise. Natural and synthetic rubber engine mount elements are commonly employed. Hydraulically damping bearing elements that combine high damping in the resonance case with low transfer stiffness are frequently even better from the perspective of acoustics and vibrations (cf. [16-65, 16-66]) but used less frequently for reasons of cost and space. Another means to reduce structure-borne noise transmission by the engine suspension is to employ vibration absorbers above the suspension [16-67]. Appropriate placement of an engine mount can also decrease excitation of the body and cab, e.g. by the engine's idle shaking (neutral torque axis).

Literature

16-1 Möse, R.: Sonderstellung des Lärms im Umweltgeschehen. AVL-Tagung Motor und Umwelt Graz (1990)

16-2 Gottlob, D.: Verkehrslärmimmissionen – Gesundheitliche Auswirkungen, Gesetzgebung in Deutschland. AVL-Tagung Motor und Umwelt Graz (1996)

16-3 Spessert, B.: Auf dem Weg zum leisen Motor. 2. Symposium Motor- und Aggregateakustik, Magdeburg: Haus der Technik (2001)

16-4 Spessert, B.: Noise Reduction Potential of Single Cylinder DI Diesel Engines. Small Engines Technologies Conference 03SETC-19 (2003)

16-5 Spessert, B.; Pohl, M.: Akustische Untersuchungen an Einzylinder-Industriedieselmotoren. 4. Symposium Motor- und Aggregateakustik, Magdeburg: Haus der Technik (2005)

16-6 Spessert, B.: Noise Emissions of Engines in Different Vehicle Groups: Historical Review, State of the Art and Outlook. FISITA Congress, Helsinki (2002)

16-7 Spessert, B. et al.: Development of Low Noise Diesel Engines Without Encapsulations. CIMAC Congress, London (1993)

16-8 Moser, F.X.; Spessert, B.; Haller, H.: Möglichkeiten der Geräuschreduzierung an Nutzfahrzeug- und Industriedieselmotoren. AVL-Tagung Motor und Umwelt Graz (1996)

16-9 Flotho, A.; Spessert, B.: Geräuschminderung an direkteinspritzenden Dieselmotoren. Automobilindustrie, (1988) 4 and (1988) 5

16-10 Wolschendorf, J.: Zyklische Schwankungen im Verbrennungsgeräusch von Dieselmotoren und ihre Ursache. Diss. RWTH Aachen (1990)

16-11 Schlünder, W.: Untersuchungen des direkten Verbrennungsgeräusches an einem Einzylinder-Dieselmotor. Diss. RWTH Aachen (1986)

16-12 Schneider, M.: Resonanzschwingungen der Zylinderladung von Dieselmotoren und ihre Bedeutung für das Verbrennungsgeräusch. Diss. RWTH Aachen (1987)

16-13 Miculic, L.: High Power Diesel Engines for Onroad Application. World Engineers Conference, Hannover (June 2000)

16-14 Kamp, H.: Beurteilung der Geräuschanregung durch den Kolbenschlag. Diss. RWTH Aachen (1984)

16-15 Tschöke, H.: Beitrag zur Berechnung der Kolbensekundärbewegung in Verbrennungsmotoren. Diss. Universität Stuttgart (1981)

16-16 Kaiser, H.-J.; Schmillen, K.; Spessert, B.: Acoustical Optimization of the Piston Slap by Combination of Computing and Experiments. SAE 880 100

16-17 Kaiser, H.-J.: Akustische Untersuchungen der Zylinderrohrschwingungen bei Verbrennungsmotoren. Diss. RWTH Aachen (1988)

16-18 Haller, H.; Spessert, B.; Joerres, M.: Möglichkeiten der Geräuschquellenanalyse bei direkteinspritzenden Dieselmotoren. VDI-Tagung (Oct. 1991)

16-19 Spessert, B.; Ponsa, R.: Investigation in the Noise from Main Running Gear, Timing Gears and Injection Pump of DI Diesel Engines. SAE 900012

16-20 Spessert, B.; Haller, H.; Thiesen, U.-P.: Auswirkungen verschärfter Abgasemissionsvorschriften auf die Geräuschemission von DI-Dieselmotoren. 3rd Aachen Colloquium Automobile and Engine Technology (Oct. 1991)

16-21 Ochiai, K.; Nakano, M.: Relations Between Crankshaft Torsional Vibrations and Engine Noise. SAE 790365

16-22 Sheng, H.Y.; Fu, Y.Y.: The Influence of Crankshaft Torsional Vibration on Engine Noise in Diesel Engines. 15th CIMAC Conference (1983)

16-23 Wilhelm, M. et al.: Structure Vibration Excitation by Timing Gear Impacts. SAE 900011

16-24 Wilhelm, M.: Untersuchung des Geräuschverhaltens von Steuerrädertrieben bei Dieselmotoren. Diss. RWTH Aachen (1990)

16-25 Flotho, A.: Mechanisches Geräusch des Ventiltriebs von Fahrzeugmotoren. Diss. RWTH Aachen (1984)

16-26 Kaiser, H.-J. et al.: Geräuschverbesserung an Mehrventilmotoren durch Modifikation der Nockenwelle. 2nd Aachen Colloquium Automobile and Engine Technology (Oct. 1989)

16-27 Kaiser, H.-J.; Schamel. A.: Ventiltriebsgeräusch in Mehrventilmotoren. VDI-Tagung Motorakustik, Essen: Haus der Technik March (1993)

16-28 Haller, H. et al.: Noise Excitation by Auxiliary Units of Internal Combustion Engines. SAE 931293

16-29 Watanabe, Y.; Rouverol, W.S.: Maximum-Conjugacy Gearing. SAE 820508

16-30 Spessert, B. et al.: Noise Excitation by the Timing Gear Train. 19th Congress of CIMAC, Florence (May 1991)

16-31 Wilhelm, M.; Spessert, B.: Vibration and Noise Excitation in the Timing Gear Train of Diesel Engines. IMechE – 5th International Conference of Vibration in Rotating Machinery, Bath (Sept. 1992)

16-32 Spessert, B. et al.: The Exhaust and Noise Emission Concepts of the New DEUTZ B/FM1012/C and BFM1013/C Engine Families. SAE 921697

16-33 Kaiser, H.J.; Querengässer, J.; Bündgens, M.: Zahnriemengeräusche – Grundlagen und Problemlösungen. VDI-Tagung Verbrennungsmotoren-Akustik (1993)

16-34 Gray, M.; Hösterey, J.; Wölfle, M.: Der neue 1,8-l-Endura-DI-Dieselmotor für den Ford Focus. Sonderheft ATZ/MTZ Der neue Ford Focus (1999)

16-35 Bauer, R. et al.: BMW V8-Motoren – Steigerung von Umweltverträglichkeit und Kundennutzen. MTZ 57 (1996)

16-36 Anisits, F. et al.: Der neue BMW Sechszylinder-Dieselmotor. MTZ 59 (1998)

16-37 Spessert, B.: Untersuchungen des akustischen Verhaltens von Kurbelgehäusen und Zylinderblöcken unter besonderer Berücksichtigung des inneren Körperschalleitweges. Diss. RWTH Aachen (1987)

16-38 Spessert, B.; Flotho, A.; Haller, H.: Akustische Gesichtspunkte bei der Entwicklung einer neuen Dieselmotoren-Baureihe. MTZ 51 (1990) 1

16-39 Schmillen, K.; Schwaderlapp, M.; Spessert, B.: Untersuchung des Körperschallübertragungsverhaltens von Motorblöcken. MTZ 53 (1992)

16-40 Spessert, B.; Ponsa, R.: Prediction of Engine Noise – A Combination of Calculation and Experience. London: FISITA Congress (1992)

16-41 Spessert, B. et al.: Neue wassergekühlte Deutz-Dieselmotoren. FM 1012/1013: Rechnerische Bauteiloptimierung MTZ 53 (1992)

16-42 Seils, M.; Spessert, B.: Die neuen wassergekühlten Deutz-Dieselmotoren BFM1015 – Konstruktive Gestaltung und Strukturoptimierung. MTZ 55 (1994)

16-43 Thien, G.E.: A Review of Basic Design Principles for Low-Noise Diesel Engines. SAE 790506

16-44 Priede, T.: In Search of Origins of Engine Noise – An Historical Review. SAE 800534

16-45 Moser, F.X.: Development of a Heavy Duty Diesel Engine with a Full Integrated Noise Encapsulation – the STEYR M3 Engine, Truck and Environment. KIVI-RAI-Seminar, Amsterdam (1990)

16-46 Spessert, B.: Geräusch-Zielwerte für die Fahrzeug-Dieselmotoren des Jahres 2005. Geräuschminderung bei Kraftfahrzeugmotoren. Essen: Haus der Technik März (2000)

16-47 Röpke, P.; Schwaderlapp, M.; Kley, P.: Geräuschoptimierte Auslegung von Zylinderköpfen. MTZ 55 (1994)

16-48 Haiduk, T.; Wagner, T.; Ecker, H.J.: Der Vierventil-DI-Zylinderkopf – eine Herausforderung für die Strukturoptimierung. MTZ 59 (1998)

16-49 Kraus, N. et al.: Cylinder Head Noise Reduction on a 4-Cylinder 4-Valve SI Engine. IMechE C521/036 (1998)

16-50 Spessert, B.: Realisierung ambitionierter Motorgeräusch-Zielwerte mit Motorbauteilen aus Kunststoff. 4. Kunststoff Motorteile Forum (2001)

16-51 Harr, T. et al.: Der neue Sechszylinder-Dieselmotor OM906LA von Daimler-Benz. MTZ 59 (1998)

16-52 Spessert, B.: Geräuschminderung bei Baumaschinen und landwirtschaftlichen Fahrzeugen. AVL-Tagung Motor und Umwelt Graz (1990)

16-53 Esche, D.: Beitrag zur Entwicklung von Kühlgebläsen für Verbrennungsmotoren. MTZ 37 (1976)

16-54 von Hofe, R.; Thien, G.E.: Geräuschoptimierung von Fahrzeugkühlern, Axiallüftern und saugseitig angeordnetem Wärmetauscher. ATZ 4 (1984)

16-55 Esche, D.; Lichtblau, L.; Garthe, H.: Cooling Fans of Air-Cooled DEUTZ-Diesel Engines and their Noise Generations. SAE 900907

16-56 Lichtblau, L.: Aerodynamischer und akustischer Entwicklungsstand von Axialgebläsen für kompakte Motorkühlsysteme. VDI-Tagung Ventilatoren im industriellen Einsatz. Düsseldorf (Febr. 1991)

16-57 Esche, D.: Konzeptmerkmale und Besonderheiten von integrierten Kühlsystemen schnellaufender Dieselmotoren. Tagung Konstruktive Gestaltung von Verbrennungsmotoren. Essen: Haus der Technik März (1991)

16-58 Frietzsche, G.; Krause, P.: Entwicklung von schalldämmenden Motorkapseln. Düsseldorf: VDI Fortschrittsberichte, 26 (1969) 6

16-59 Thien, G.E.; Fachbach, H.A.; Gräbner, W.: Kapseloptimierung. Forschungsbericht der Forschungsvereinigung Verbrennungskraftmaschinen (1979) 262

16-60 Donath, G.; Fackler, M.: Geräuschminderung an mittelschnellaufenden Dieselmotoren durch Teilverschalung. BMFT-FB-HA 82-018 (1982) 9

16-61 N.N.: MTZ 38 (1977) 4, p. 165

16-62 Kunberger, K.: Progress With Quiet Small Diesels. Diesel and Gas Turbine Progress. (May 1977)

16-63 Heckl, M.; Müller, H.A.: Taschenbuch der Technischen Akustik. 3rd Ed. Berlin: Springer (1994)

16-64 Kochanowski, H.A.: Performance and Noise Emission of a New Single-Cylinder Diesel Engine – with and without Encapsulation. Second Conference of Small Internal Combustion Engine C372/023. Institution of Mechanical Engineers (April 1989)

16-65 Härtel, V.; Hoffmann, M.: Optimierung körperschalldämmender Motorlagerungen. Düsseldorf: VDI-Berichte, 437 (1982)

16-66 Holzemer, K.: Theorie der Hydrolager mit hydraulischer Dämpfung. ATZ 87 (1985)

16-67 van Basshuysen, R.; Kuipers, G.; Hollerweger, H.: Akustik des AUDI 100 mit direkteinspritzendem Turbo-Dieselmotor. ATZ 92 (1990)

Part V Implemented Diesel Engines

17 Vehicle Diesel Engines 507

18 Industrial and Marine Engines 559

17 Vehicle Diesel Engines

Fritz Steinparzer, Klaus Blumensaat, Georg Paehr, Wolfgang Held, and Christoph Teetz

17.1 Diesel Engines for Passenger Cars

17.1.1 History

The diesel engine came to be used as a car engine relatively late after the first demonstration of its operation in 1897. The introduction of the first mass produced diesel passenger cars in 1936 finally enabled diesel engines to vie with the dominant gasoline engines of the day as an alternative drive concept in this segment too.

The stage for this had been set by the highly precise fuel metering timing allowed by the injection systems Bosch had started developing and manufacturing (1927) and the control of the processes of mixture formation and combustion at relatively high engine speeds by dividing the combustion chamber into a prechamber and main chamber, an idea that dated back to L'Orange.

Driven by the increasing focus on energy saving propulsion sources that conserve resources and reduce climatically relevant CO_2 emissions toward the end of last century, diesel engines repeatedly experienced reasonable successes in cars but never truly established themselves. The real breakthrough came in the second half of the 1990s when new high pressure injection systems such as unit injector systems and, above all, common rail injection technology became available as standard, thus enabling a changeover to direct injection and the development of innovative exhaust gas turbochargers with variable turbine guide blade systems.

These new technologies made it possible to tremendously improve the performance characteristics of diesel cars, which are relevant to customers. Figure 17-1 highlights this impressive development of power, torque, fuel consumption and emission performance.

17.1.2 Specific Vehicle Requirements

17.1.2.1 Quality Criteria

As car engines, diesel engines are conceptually interrelated to vehicles by various subsystems (transmission, chassis, etc.) in a variety of ways. Therefore, the design requirements for a driving engine must be derived from a vehicle's general quality criteria and aspects of the drive train – essentially the design and characteristics of the transmission employed. In terms of product features, the basic vehicle requirements are based on criteria such as transportation performance, safety, comfort, operational safety and environmental compatibility.

The individual categories can be broken down into secondary aspects from which the criteria relevant for an engine can be derived. The functional requirement of *transportation performance* addresses vehicle performance, energy input and energy conversion.

Vehicle safety requirements also have consequences for the drive train. For instance, they affect:
– engine responsiveness and the controllability of engine power,
– transmission of driving torque on the road or even
– suitable limp-home strategies in a fault scenario.

Fire resistance is also an important aspect of engine design.

The *comfort* requirements are many and diverse. A drive train's vibration characteristics influence driving comfort related to the engine. Ease of operation is based on the force-displacement characteristics (e.g. accelerator pedal) and everything that facilitates operation (e.g. automatic preglow before engine start). Climatic comfort defines the heat output and cooling capacity requirement, which engine design has to provide. Finally, a vehicle's acoustic comfort, which can be significantly influenced by an engine's sound engineering features, is important.

Vehicle *operational safety requirements* can be divided into two categories: long-term quality and usability under special conditions. Hence, the requirements for reliability, service life, functional stability, system diagnostics and serviceability (extent, frequency and accessibility) have to be established with these factors in mind.

Among the aforementioned criteria, passenger cars' *environmental compatibility* is increasingly growing in importance.

G. Paehr (✉)
Volkswagen AG, Wolfsburg, Germany
e-mail: georg.paehr@volkswagen.de

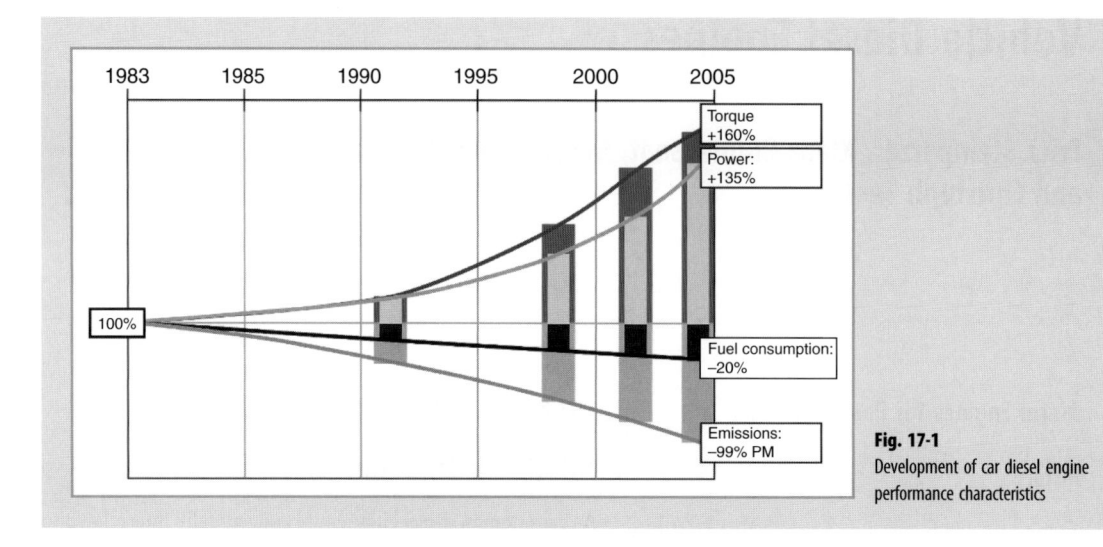

Fig. 17-1 Development of car diesel engine performance characteristics

Secondary aspects are based on exhaust and noise emission, conservation of resources by operation, scrap recycling and disposal and efficient use of the raw materials and energy required for manufacturing. Depending on the level of quality, this basically determines an engine's design and working principle.

The diverse requirements imposed on a vehicle in terms of pleasantness make engine design and the engine compartment important for engine engineering and give rise to independent and customized solutions for both engine placement and component design.

The geometric similarity of car diesel engines and gasoline engines normally employed in the same vehicles is an additional aspect that deserves attention. This particularly affects engines' outer dimensions and their interfaces to the vehicle cooling system, intake airflow, exhaust system and manual and automatic transmissions, which usually originate from the same set of components. Rigorously lightweight construction is also an important element in this context. Regardless of the model of vehicle, standardized chassis components only give diesel engines comparably good driving dynamics when their additional weight is successfully kept very low.

17.1.2.2 Aspects of Drive Train Configuration

The widest variety of operating conditions for passenger car such as:
- starting,
- accelerating,
- uphill driving and
- maintaining constant speeds

impose requirements on a vehicle's tractive force, which an engine must satisfy with its power in a broad speed range.

A motor vehicle's power can be derived from a tractive force-speed diagram (based on the engine, torque being a function of engine speed). An engine's torque characteristic, maximum speed and operating speed range significantly influence the specification of the gear steps, their number and the magnitude of the transmission's individual transmission ratios. The interaction of the engine and transmission makes itself noticeable in a car's hill-climbing performance, acceleration performance and startability.

An engine provides good prerequisites for tuning a car's drive train properly when
- the torque curve has a characteristic that increases to its maximum, preferably at minimum speed (n_{Mmax}/n_{max} approximately 0.4–0.5), as engine speed drops and
- the maximum engine speed or speed range relevant for the main road load is selected to be large enough for optimal transmission design (number of transmission steps and transmission ratio).

17.1.3 Design Features of Car Diesel Engine

17.1.3.1 Engine Size and High Speed Capacity

Engine size and speed level are the prominent distinguishing features of different kinds of diesel engines (low, medium and high speed diesel engines).

Car diesel engines are designed with a cylinder displacement volume of approximately 0.3–0.55 dm^3.

While four, five and six cylinder versions were virtually the only inline designs implemented earlier, three, eight and even ten cylinder engines have significantly expanded the range of products in recent years. Primarily driven by basic vehicle conditions, V designs are also increasingly being employed for six cylinder engines.

From the perspective of thermodynamics, larger displacement is fundamentally desirable because of the small surface-volume ratio and the potential to design combustion chambers compactly. Moreover, engine designs with large displacements are desirable for good start-up performance and low

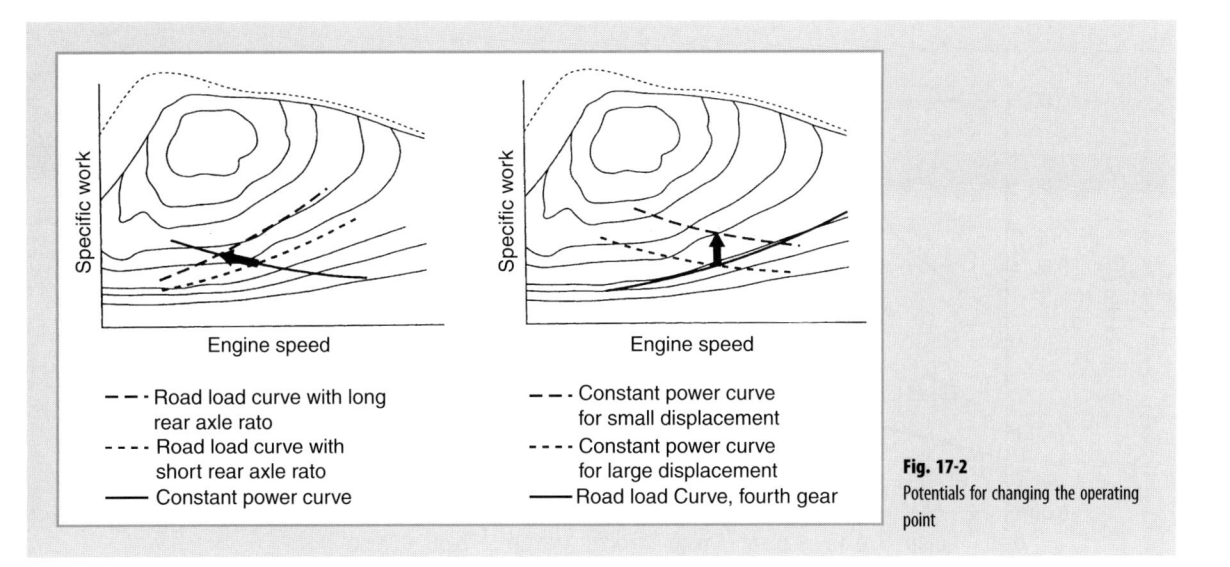

Fig. 17-2 Potentials for changing the operating point

Legend (left): – – – Road load curve with long rear axle rato; - - - - Road load curve with short rear axle rato; —— Constant power curve

Legend (right): – – – Constant power curve for small displacement; - - - - Constant power curve for large displacement; —— Road load Curve, fourth gear

idle speed whenever auxiliary units (power steering pump, air conditioner, etc.) require high power as well as for good starting. Other aspects argue for small cylinder displacements. The most important argument is the "operating point change", which, in conjunction with supercharging, can be utilized to lower a vehicle's fuel consumption and emissions. Supercharging facilitates maintaining power with a low displacement or utilizing a supply of higher torque to select a larger overall transmission ratio. Figure 17-2 illustrates the two principal options to reduce fuel consumption.

Together with a constant engine power line, the two road load lines derived from different rear axle ratios are entered in the left half of the schematic engine map. The overall transmission ratio advances into a range of low specific consumption at reduced speed and increased load. The right half of the map depicts the same for two different displacements. The operating point shifts to a higher load when the displacement decreases, i.e. to a range with better engine efficiency. When engines have equal power but differently sized displacement, the shift can lower consumption considerably as indicated by the transition from a 2.5 dm^3 naturally aspirated engine to a 1.6 dm^3 turbocharged diesel engine. Vehicle consumption is reduced by 16% without impairing vehicle performance.

Since total emission is a product of mass flow and concentration, the reduction of the mass flow resulting from a smaller displacement produces another beneficial effect, namely a reduction of emissions in the low part load range. A measure of *high speed capacity*, the *mean piston velocity* has a range of 13–15 m/s for car diesel engines. It is closely related to an engine's maximum speed. In turn, the selection of the maximum engine speed has far reaching consequences for the design of the drive train.

The dimensionless engine maps for two engines with high and lower rated speed presented in Fig. 17-3 furnish an explanation. The relative reduction of the speed range utilizable for in-use driving at low rated speed necessitates either another transmission step (sixth gear) or extremely widely spread transmission in the low gears.

17.1.3.2 Mixture Formation and Combustion Systems

The mixture formation and combustion system in car diesel engines is of key importance for engine speed and the related extremely brief time interval for the working process. The system selected in each case determines the speed limit, fuel consumption, exhaust gas composition and combustion noise. Systems developed for large and low speed engines are not transferrable to car diesel engines. On the one hand, the valves and the nozzle in smaller cylinder units cannot be arranged according to the scheme of geometric similarity. On the other hand, the low intensity of mixture formation due to ignition delay at high speed does not suffice to end combustion early enough.

Prechamber systems (dual chamber systems) – decentral swirl chambers or central prechambers – long dominated car applications but the picture has changed entirely in the last ten years.

Once advanced high pressure injection systems became available, direct injection rapidly established itself within a few years. In addition to the main motivator of approximately 15% better efficiency over prechamber systems, advanced direct injection combustion systems also have significant advantages in terms of power density and minimum emissions. The development of highly flexible high pressure injection systems with injection pressures of up to 1,800 bar (see Sect. 5.3) prepared the way for the implementation of highly efficient, clean and powerful direct injection car engines. The significantly increased efficiency of electronic engine management was also instrumental in this development.

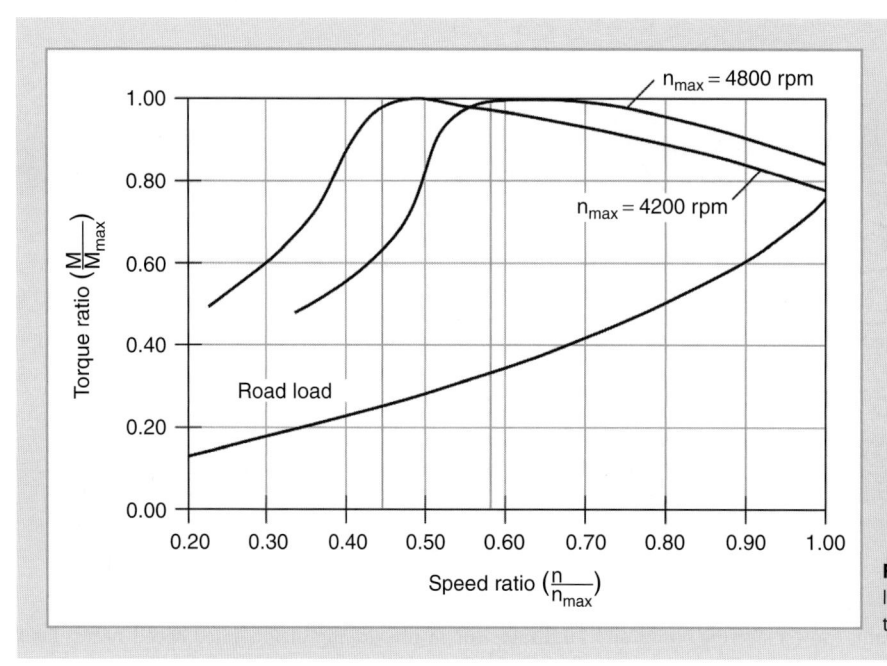

Fig. 17-3 Influence of the usable speed range on transmission design

17.1.4 Engine Design

The following thematic fields constitute the basic emphases of car diesel engine design and development:
- heavy duty, compact basic engines,
- combustion chamber shape and mixture formation elements,
- high pressure fuel injection systems,
- supercharging concepts and gas exchange tuning,
- electronic engine management with interfaces to the vehicle electrical system and
- highly efficient exhaust gas aftertreatment.

The following presents the state-of-the-art and an overview of selected forms of implementation of these thematic fields.

17.1.4.1 Basic Engine Design

Depending on the vehicle segment, car diesel engines are designed with three, four, five, six, and eight or even up to ten cylinders. Different trends in basic engine design based on the number of cylinders and the design are identifiable (see Sects. 8.1 and 8.2).

Along with four-valve engines, a considerable number of two-valve engines and thus asymmetrical combustion chamber configurations are also still in use in three and four cylinder engines, which are implemented in part in extremely cost sensitive vehicles. However, for all intents and purposes, the segment for engines with five and more cylinders is virtually dominated by four-valve engines with symmetrically arranged valves and injection nozzles.

Three, four and five cylinder engines are only produced in inline designs, eight and ten cylinder engines only in V designs. Six cylinder engines include both inline and V designs. The design selected for a six cylinder engine basically depends on two main influences, the installation conditions in the vehicle and the manufacturing strategy. Since they have exceptional vibration characteristics as well as weight and cost advantages, inline six cylinder engines are the technically superior solution. The arrangements of their components that conduct air and exhaust gas also provide significant advantages. However, inline designs have a larger overall engine length than V6 engines, which is one reason many vehicle manufacturers avoid employing them.

The ideal manufacturing network is also crucial in many cases. Manufacturers have to decide whether joint production of inline four and six cylinder engines or V6 and V8 engines is better for the overall cost optimum and the requisite flexibility of production runs.

The desired engine positioning is another crucial factor for basic engine design. While engines for entry-level segments are designed for maximum combustion chamber pressures of up to 150 bar, the requirements in the premium segment are significantly more demanding and they are designed for up to 180 bar. This affects component design as well as the selection of materials and the manufacturing processes.

The core components are the crankcase and the cylinder head. While only one-piece aluminum alloy cylinder heads are generally still employed, the range of technical solutions for crankcases is far broader and extends from aluminum die cast crankcases for small, specifically not very highly stressed engines to normal gray cast iron and heavy duty vermicular gray cast iron crankcases up through high strength special gravity cast aluminum alloy crankcases with subsequent heat treatment, which are increasingly being implemented more in the premium segment. Figure 17-4 pictures such high-end aluminum crankcases for an I6 and a V8 cylinder engine.

For a long time, the seal between the cylinder head and crankcase limited the maximum producible combustion chamber pressures. The replacement of formerly common compressible seals with laminated steel gaskets brought about a quantum leap in the dependability of seals.

High peak pressures also impose high requirements on the components of a crankshaft assembly. Thus, heavy duty forged steel crankshafts are used almost exclusively. Split connecting rods are normally employed in the crankshaft bearing journal. Very often, the connecting rod small end is trapezoidal to optimally integrate it in the piston contour as well as to obtain a maximum bearing surface area in the more highly loaded bottom half bearing. Pistons are generally made of heavy duty aluminum alloys (see Sect. 8.6). Cast-in ring carriers and cooling galleries are also standard. Figure 17-5 presents an example of a piston-connecting rod assembly for a highly loaded application.

Mass balancing systems in inline four cylinder and V6 engines have increasingly established themselves ever since expectations of vibrational comfort for diesel cars reached a level that is just as high as for gasoline cars.

So-called "add-on" systems are implemented in inline four cylinder engines almost universally. Two counterrotating balance shafts driven by the crankshaft by chains or gears are mounted in the oil pan. The advantage of such an add-on solution is the modular engine design. Thus, depending on the requirement for vibrational comfort in different vehicle applications, engines with and without balance shafts can be produced on an assembly line very easily. Figure 17-6 pictures an example of one such balance shaft unit that reduces second order inertial forces. The balance shafts often employed in V6 engines are normally integrated in the crankcase.

Their timing is either driven by maintenance free chain drives or high strength toothed belts. Optionally, a chain drive may be placed at the back of an engine. The location of the camshaft drive near the crankshaft's nodes and, hence, significantly less rotational irregularity than in the front end of the crankshaft is the advantage of this solution, which is already being marketed in its first applications.

Engine height may be reduced additionally in the front to obtain more free space for front end design.

The valves in the majority of car diesel engines are actuated by roller bearing drag levers to minimize frictional power losses (see Fig. 17-8).

Auxiliary units such as water pumps, generators, power steering pumps and air conditioning compressors are usually driven by poly-V-belts mounted on an engine's front end.

A decoupled belt pulley integrated in a rotational oscillation damper does the same for vehicles with elevated comfort requirements. Figure 17-7 presents a cutaway view of an advanced rotational oscillation damper with a decoupled belt pulley for a six cylinder engine.

Figure 17-8 presents a cross section of a current four cylinder direct injection diesel engine.

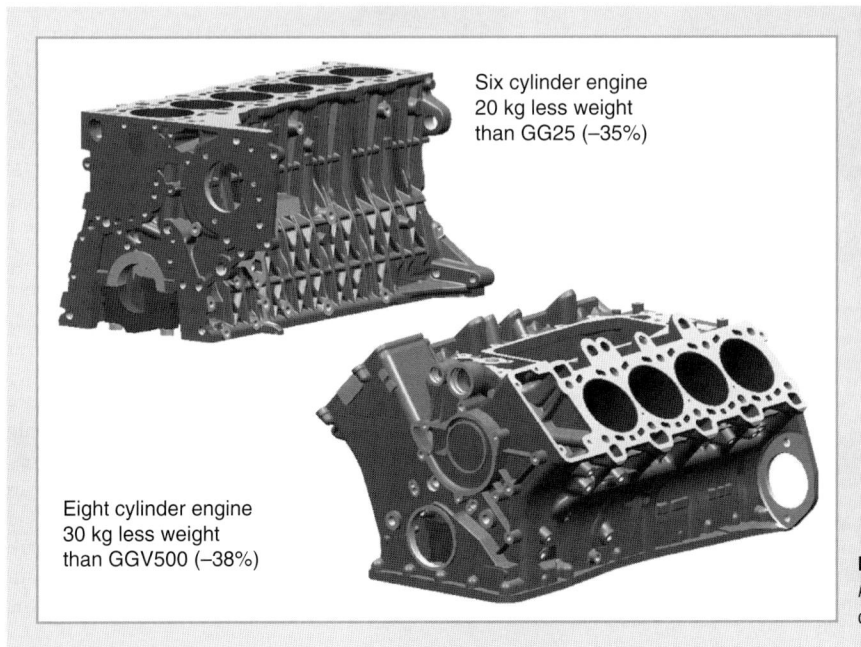

Six cylinder engine
20 kg less weight
than GG25 (−35%)

Eight cylinder engine
30 kg less weight
than GGV500 (−38%)

Fig. 17-4
Aluminum crankcases for highly loaded diesel engines

Fig. 17-5 Engineered design of a piston-connecting rod assembly

- Ring carrier with integrated cooling gallery
- Single-sided vertically oval brass bushing with fitted bore
- Horizontally oval small end bushing with fitted bore
- Trapezoidal small connecting rod eye

Fig. 17-6 Balance shaft unit for a four cylinder engine

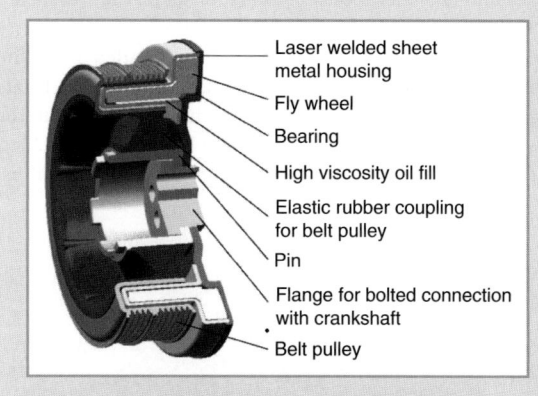

Fig. 17-7 Rotational oscillation damper with integrated decoupled belt pulley

- Laser welded sheet metal housing
- Fly wheel
- Bearing
- High viscosity oil fill
- Elastic rubber coupling for belt pulley
- Pin
- Flange for bolted connection with crankshaft
- Belt pulley

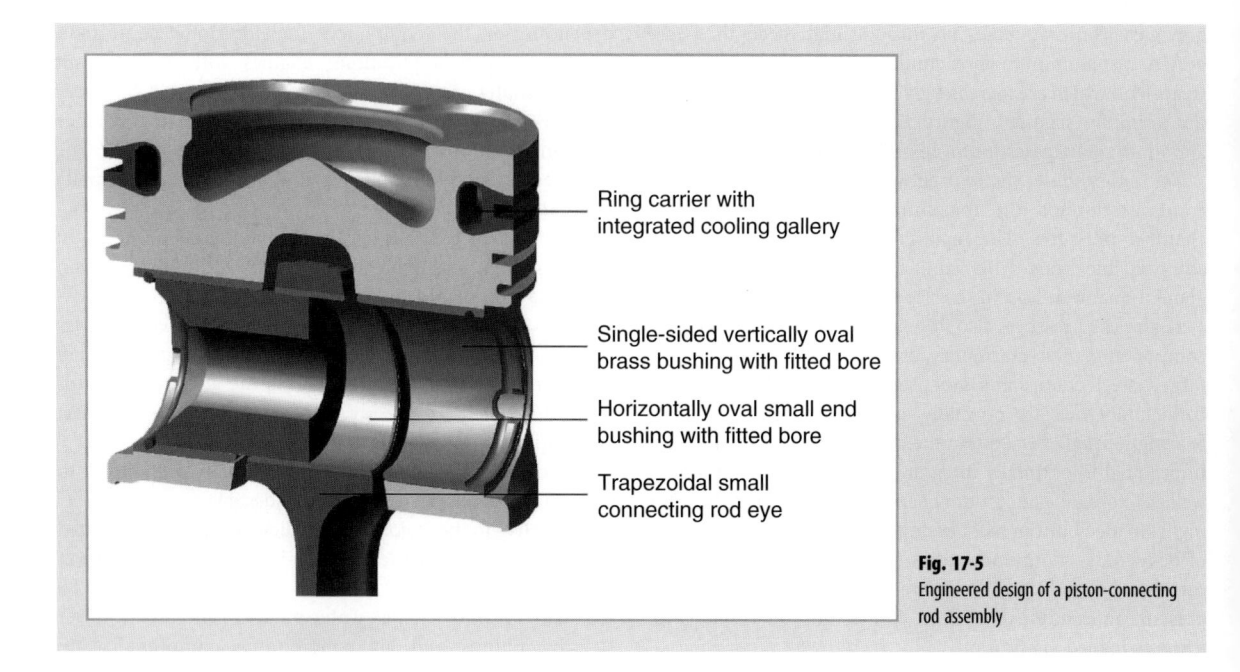

17.1.4.2 Combustion Chamber and Mixture Formation

A completely symmetrical arrangement of the injector and the combustion chamber bowl in the piston is the optimal configuration of a combustion chamber for a direct injection car diesel engine. Naturally, this is only possible geometrically when the design has four valves. The intake ports are routed separately out of the plenum chamber to the cylinder head. This allows the use of a port shut-off, i.e. normally infinitely variable flaps are enabled to completely or partially close a port in certain map ranges. Thus, the directed air movement in the

Fig. 17-8 Engine cross section

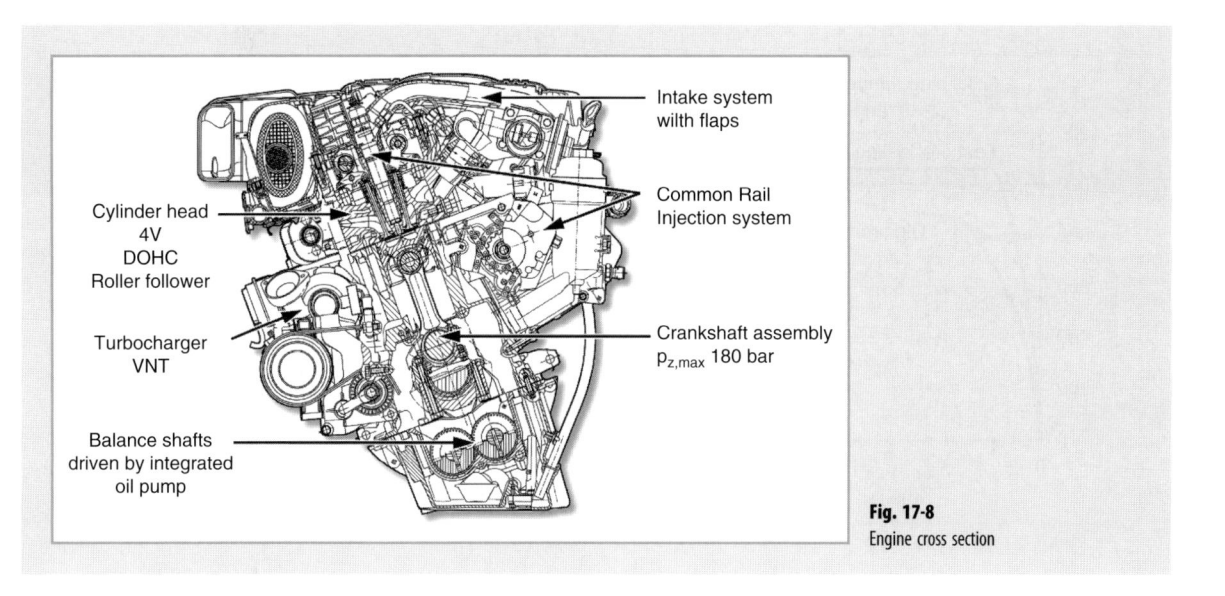

Labels in figure:
- Intake system with flaps
- Common Rail Injection system
- Crankshaft assembly $p_{z,max}$ 180 bar
- Cylinder head 4V DOHC Roller follower
- Turbocharger VNT
- Balance shafts driven by integrated oil pump

combustion chamber, the swirl can be systematically adjusted to the particular operating range. This is extremely important for optimal mixture formation in direct injection diesel engines.

The outlet ports are usually designed as twins, i.e. they already merge inside the cylinder head. Since two-valve engines have significantly smaller free spaces for attainable cylinder charging motion, they now only tend to be implemented in cost sensitive engines in the lower performance classes.

In addition to the valves and the injector, the glow plug (see Chap. 12) is also installed in the combustion chamber. Along with their original function of producing the "hot spot" in a combustion chamber that is required for reliable cold starting, glow plugs are also increasingly being heated during engine operation to positively influence the combustion sequence in terms of noise and emission performance as a function of the operating point. The performance of glow systems has been enhanced tremendously in recent years. While preglow times of twenty seconds during cold starts were formerly not uncommon, the preglow times now required range below five seconds even at very low outside temperatures. Figure 17-9 presents the characteristic of a modern spontaneous glow system as an example.

The injection nozzle injects the fuel with varying injection pressure as a function of the operating point and is currently divided into up to five individual injections per combustion cycle. Injection nozzles with six to eight nozzle orifices are common, depending on the application. Figure 17-10 presents the configuration of a four-valve engine's combustion chamber as an example.

17.1.4.3 High Pressure Injection

Three different high pressure injection systems established themselves once direct injection had also made a breakthrough in the diesel car sector (see Chaps. 5 and 6). In addition to the now dominant common rail system, these included the high pressure distributor pump and the unit injector system. While the distributor pump and unit injector system are cam driven systems in which the generation of pressure is directly coupled to engine speed and crankshaft position, the common rail system allows a fully free selection of the injection pressure and the time of the individual injections independent of speed and crankshaft position.

In light of the mounting requirements on acoustics and vibration characteristics, the required further reduction of exhaust emissions and fuel consumption and the necessity to support smooth operation of complex exhaust gas aftertreatment systems with various injection strategies, the common rail system has gained tremendous importance because of its system flexibility.

For all intents and purposes, the high pressure distributor injection pump and the unit injector system are no longer applied and are becoming obsolete for car applications.

Moreover, the core components of common rail systems have significantly fewer reciprocal effects on basic engine design and may be integrated in different engine designs very flexibly. One example is the arrangement of the high pressure pump driven by the timing chain and the rail with its integrated pressure sensor, pressure control valve, injectors positioned in the cylinder head and respective high pressure

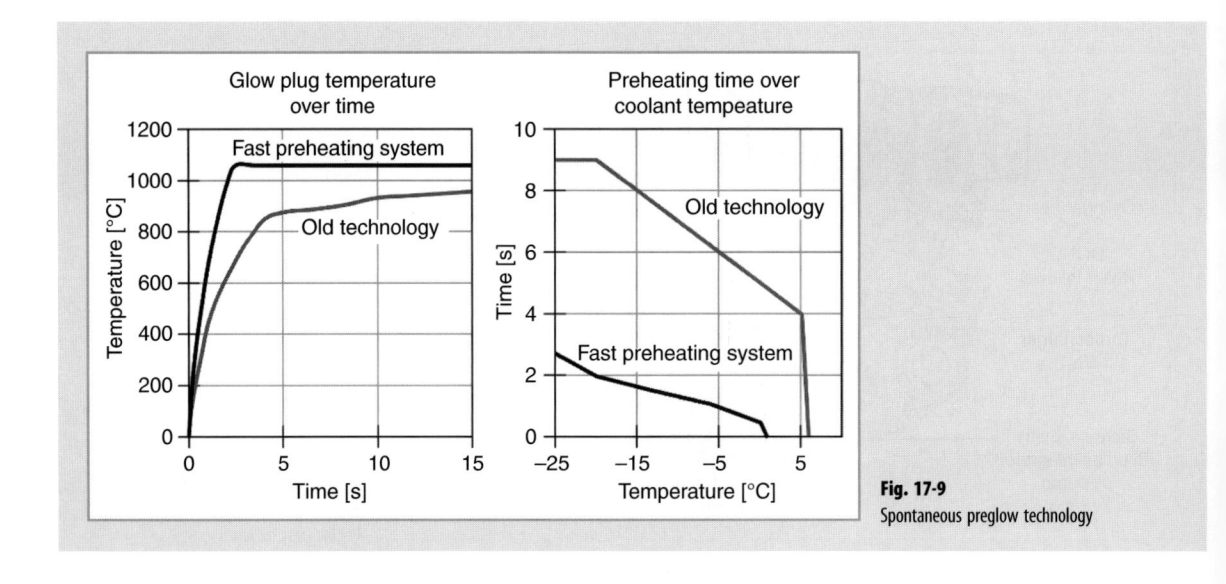

Fig. 17-9 Spontaneous preglow technology

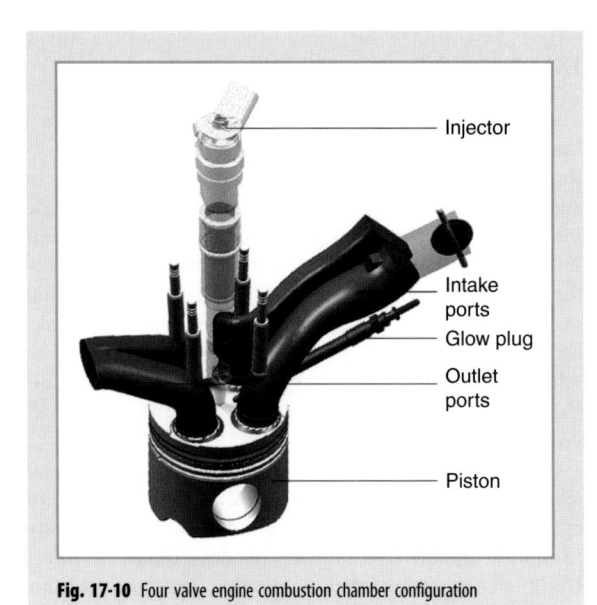

Fig. 17-10 Four valve engine combustion chamber configuration

lines in the inline six cylinder and V8 engines pictured in Fig. 17-11.

The pressure range now employed in car diesel engines extends from 250 bar in the near idle range to up to 1,800 bar in the rated power range. The number of individual injections employed per combustion cycle ranges from one block injection to up to five injections. While pilot injections upstream from the main injection normally serve to lower combustion noise, post-injections are utilized to support any exhaust gas aftertreatment system employed.

17.1.4.4 Supercharging and Gas Exchange

Along with high pressure injection, supercharging has been instrumental in the currently strong market position of car diesel engines (see Chap. 2). Limited high speed capability and the necessity of lean burn operation make a sufficient fresh air supply the basic prerequisite for a suitable power output. In addition, the diesel engine principle is predestined for supercharging. Internal mixture formation and auto-ignition also make high supercharging rates producible without any problem. This has led to the virtual disappearance of naturally aspirated diesel engines from the market.

Variable turbine geometry turbochargers dominate the market. Figure 17-12 presents a cutaway model of an engine with an electric adjuster for the turbine guide blades. Such turbochargers make it possible to produce the large speed range with excellent turbine and compressor efficiencies, which is necessary for car engines. In addition, the guide blades are closed in part load and the thusly increased exhaust back pressure significantly increases the scavenging gradient between the exhaust manifold and fresh air, thus boosting the exhaust gas recirculation rate. Since exhaust gas recirculation is one of the most effective measures to reduce NO_x in diesel engines, this technology also contributes considerably to a diesel engine's environmental compatibility.

Hence, the use of fixed geometry turbines with wastegate control is steadily decreasing and has become limited to the entry-level engine segment.

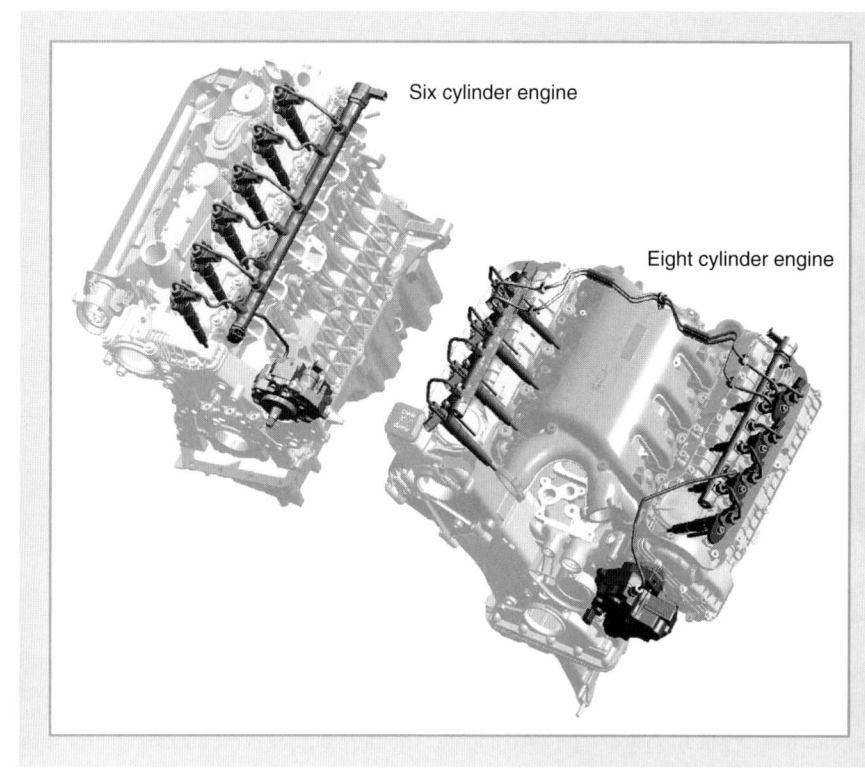

Fig. 17-11 Arrangement of common rail injection components

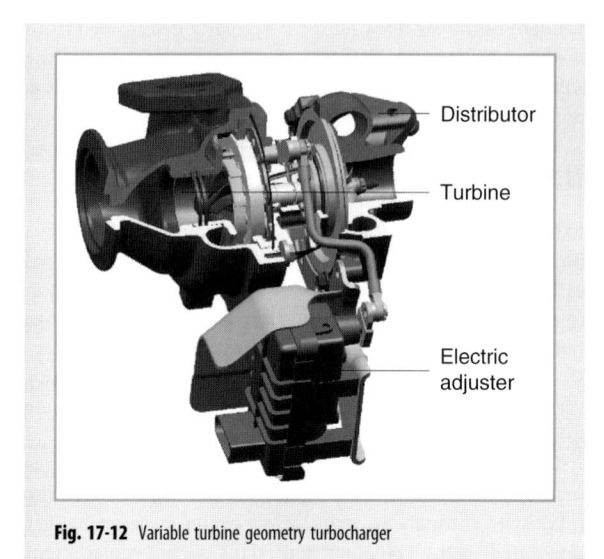

Fig. 17-12 Variable turbine geometry turbocharger

However, attempts are increasingly being made in the premium segment to further increase supercharging rates. The attendant increase of the specific power output generates substantial potentials for downsizing. Thus, it will further contribute to the reduction of fuel consumption and exhaust emissions.

Another promising technology, two-stage exhaust gas turbocharging is already being implemented in the first standard applications. Figure 17-13 presents a cutaway view of one such application in which the two differently sized turbochargers connected in a series simultaneously provide high boost pressures in the lower speed range and high specific power output in the rated power range. Sequential turbocharging connected sequentially with two equally sized turbochargers is also already being marketed in a first standard application.

Intercooling constitutes a basic technology in all supercharged car diesel engines. While air/air cooling is preferred, somewhat more complex water/air intercooling is occasionally reverted too for difficult package conditions. However, it not only requires the usual coolant circuit but also an additional low temperature circuit to reliably obtain the low charge air temperatures aimed for under every driving condition.

17.1.4.5 Electronic Engine Management

Along with the crucial advances achieved in engine load capacity and injection and supercharging technologies, the great advances in electronics also contributed to the diesel engine's breakthrough as a car engine (see Chap. 6).

Fig. 17-13 Two-stage exhaust gas turbocharging

The performance of engine control units and their functions have increased greatly since the first application for fully electronically controlled diesel engines in 1988. In addition to their former core function of metering the correct injected fuel quantity at the right time based on the map, engine control units now assume diverse functions in vehicle electrical systems, which are growing increasingly complex. A car's function architecture is structured hierarchically. A vehicle coordinator that controls vehicle movement, the entire drive train, body functions and the vehicle electrical system, functions as the top level. In many cases, an internal bus system, a power train CAN, is used for the drive field. The engine and transmission control unit and the driving dynamics control systems communicate through the power train CAN. Figure 17-14 is a schematic of a torque-based vehicle electrical system. System components, e.g. the air conditioner or the generator, communicate with system partners through precisely defined physical interfaces. Not only the exact, demand-oriented, torque-oriented control of the drive train mad possible by this but also the modular design of such vehicle electrical system architectures is very advantageous. Individual components may be interchanged for different engine or transmission variants very easily.

Frequently, standardized adaptation functions and, increasingly, even self-learning adaptation are implemented for the significant injection system and air system parameters relevant to combustion. Every injector is precisely measured in defined operating points at the end of its production process and the results are documented on the injector in a data matrix code. The data from every individual injector is read in and assigned a compensation value in the control unit in the course of the "marriage" of the control unit with the finished engine. Another example is the air mass sensor with self-learning adaptation in operation, which measures the air mass introduced to the cylinder. Using the parameters of speed, boost pressure and air temperature, an air mass independent of the air mass sensor is regularly determined from the gas equation in quasi stationary operating states and the software corrects the air mass sensor's signal drift, which is unavoidable throughout a vehicle's service life.

17.1.4.6 Exhaust Gas Aftertreatment

Exhaust gas aftertreatment has also acquired tremendous importance for diesel vehicles as requirements for minimal exhaust emissions have increased, (see Sect. 15.5). In addition to rigorously optimized combustion systems that minimize raw emissions, important technologies include highly effective cooled exhaust gas recirculation, oxidation catalytic converters mounted as close as possible to engines, particulate filters and NO_x reduction systems.

Exhaust gas recirculation diverts part of the exhaust gas before the exhaust gas turbocharger and returns it to the intake air through an exhaust gas/water heat exchanger integrated in the coolant circuit. This partial replacement of the fresh air mass with recirculated exhaust gas can lower nitrogen oxide emission considerably. The quantity of recirculated exhaust gas is normally metered by a pneumatically or electrically actuated recirculation valve controlled by the engine control unit. Strictly speaking, exhaust gas recirculation is an in-engine measure for emission reduction.

Fig. 17-14
Vehicle electrical system network

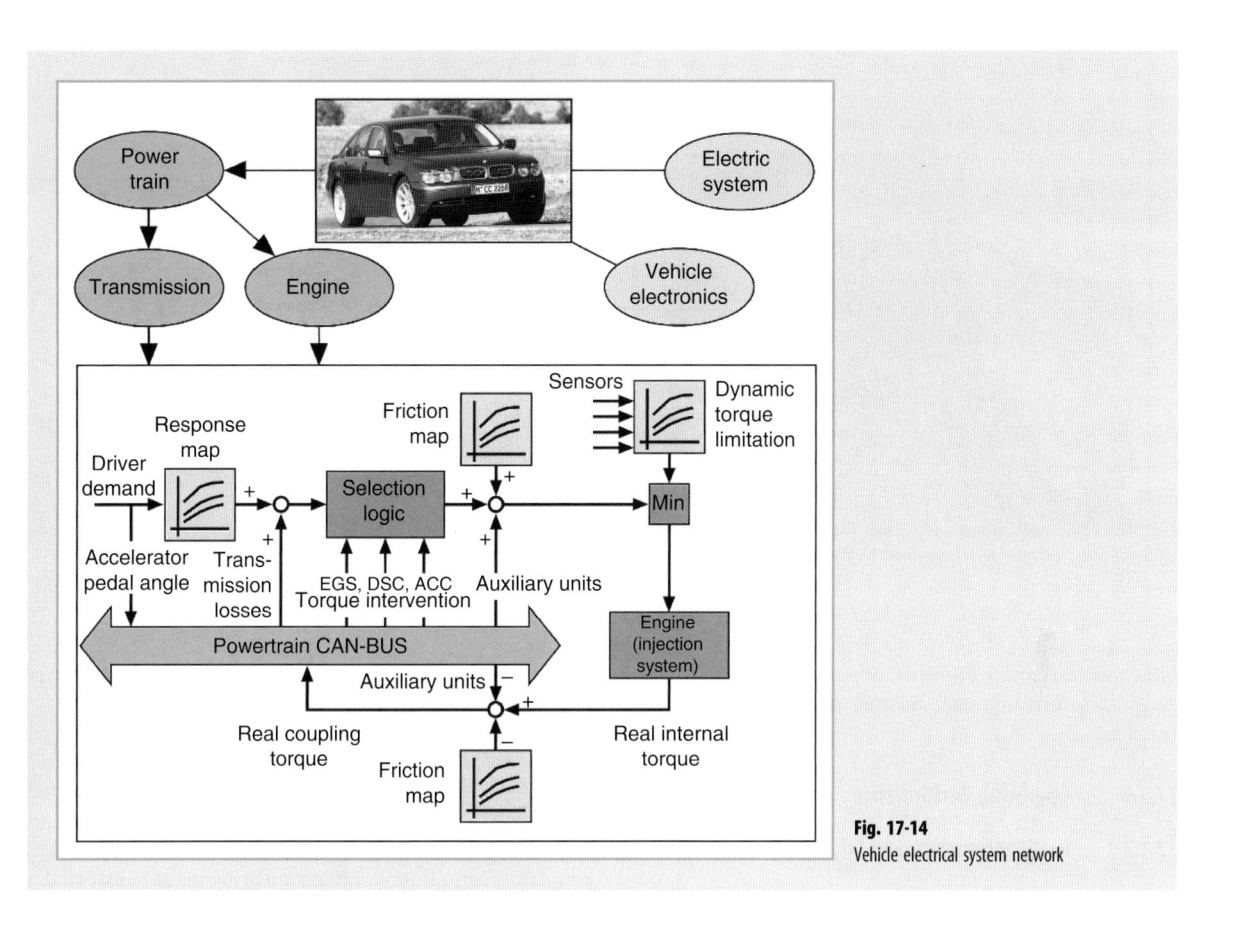

The exhaust gas itself is actually aftertreated either by means of an oxidation catalytic converter alone or by a combination of an oxidation catalytic converter and particulate filter and, where applicable, a deNOx system. Even though many engine/vehicle combinations are capable of undershooting effective EU 4 emission limits even without them, most manufacturers now implement particulate filters.

Platinum coated cordierite or metallic foil substrates are used as oxidation catalytic converters. However, nearly all particulate filters now in use have a substrate of substantially more heat resistant silicon carbide. While oxidation catalytic converters normally operate continuously, particulate filters require a discontinuous operating strategy. Soot particles are separated from the exhaust gas and trapped in the particulate filter in normal engine operation. A filter must be actively regenerated from time to time depending on its loading state, i.e. the mass of soot accumulated in the filter is systematically burned off. In order to regenerate a filter, a modified operating strategy temporarily elevates the engine's exhaust gas temperature above the ignition temperature of soot. This is primarily done by implementing retarded post-injection in combination with intake air throttling. Unrestricted by the engine map, the measures to be taken differ and must be carefully selected and coordinated. In addition, measures are also required to enhance the ignition quality of the stored particulates. First generation particulate filters accomplished this by blending an additive into the fuel but second generation particulate filters dispense with such additivation. The catalytic coating directly on the particulate filter substrate establishes sufficiently good ignition conditions. An optimal loading and regeneration strategy is crucial for reliable operating performance of a particulate filter. The engine control unit employs a loading model to continuously simulate the current mass of soot in the filter during engine operation. When approximately 70% of the loading capacity has been reached, the initiation of active regeneration is triggered, preferably at good temperature conditions (high speed and high engine load). Not only loading but also exhaust back pressure is constantly monitored. Active regeneration is initiated when it exceeds certain limits. Depending on the operating conditions, common regeneration intervals for additive-supported systems are between 300 and 800 km. They are between 500 and 2000 km for catalytically coated particulate filters.

Placement of the particulate filter close to the engine is conducive to obtaining optimal regeneration conditions. Figure 17-15 pictures a closed-coupled arrangement in which the oxidation catalytic converter, the coated particulate

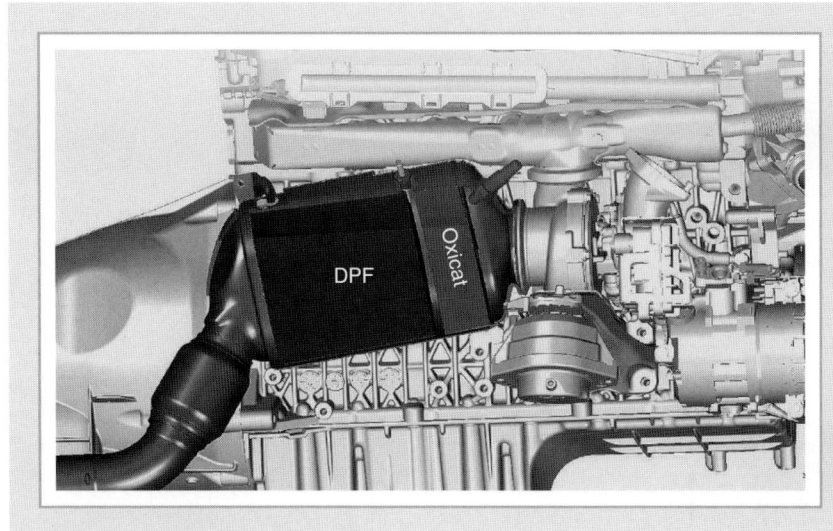

Fig. 17-15 Closed coupled particulate filter

filter and the respective sensors for pressure, temperature and excess air (eight-element sensor) are combined in one housing.

17.1.5 Operating Performance

17.1.5.1 Fuel Consumption and CO_2

Primarily shaped by concerns about the depletion of crude oil reserves, energy policy discussion of more efficient use of fossil fuels is now additionally driven by the dangers of potential global climate change caused by carbon dioxide emissions (CO_2).

Since carbon dioxide is the end product of any combustion of carbon, fuel consumption is directly connected with CO_2 emission and thus – apart from exhaust emission – the most important environmentally relevant factor.

A car's fuel consumption is determined by a multitude of influencing factors, which are not only related to the engine. These include:
- a vehicle's total gross weight, drag and road load,
- the position of the operating point in the engine map produced as a function of the drive train design and displacement,
- the power requirements of a vehicle's auxiliary units (generator, power steering pump, air conditioning compressor and vacuum pump) and
- the driving conditions, traffic routing and individual driving style.

A diesel engine's energy efficiency is determined by the amount of brake specific fuel consumption in the minimum, i.e. it is generally characterized by the marginal effective efficiency in the optimal point of the fuel consumption map. The value for optimal efficiency chiefly depends on the selection and qualitative optimization of the combustion system and the engine friction.

The best consumption values for direct injection car diesel engines are around 200 g/kWh, corresponding to an effective efficiency of up to 44%. Consumption increases during the warm-up phase as a result of poor mixture formation and increased friction. Therefore, the warm-up phase is also factored into optimization. Efforts are being made to extend the ranges that facilitate efficiency with an eye toward good vehicle fuel consumption.

Fuel consumption has been determined in a new European test cycle (MVEG-A) since early 1996. This combined cycle consists of an urban (ECE) and an extra-urban cycle (EUDC). An average value in l/100 km is calculated from total consumption. In the USA, fuel consumption is specified in mpg (miles per gallon, the conversion being 23.5 mpg = 10 l/100 km) and limited as fleet consumption. It is calculated from the ranges per gallon of one manufacturer's vehicles weighted by quantity.

Figure 17-16 compares the CO_2 emission of modern diesel cars with gasoline cars as a function of vehicle weight in the new European test cycle. A diesel car has an advantage of approximately 15–25% over gasoline cars. Since diesel fuel has a higher density, diesel cars have a consumption advantage of between 25 and 35%.

All the energy consuming processes from fuel delivery to final fuel consumption yield an additional CO_2 emission advantage of 6% for diesel fuel.

17.1.5.2 Exhaust Emission

The efforts to improve air quality have also included significant advances in the reduction of toxicologically harmful

Fig. 17-16 CO_2 emissions of cars

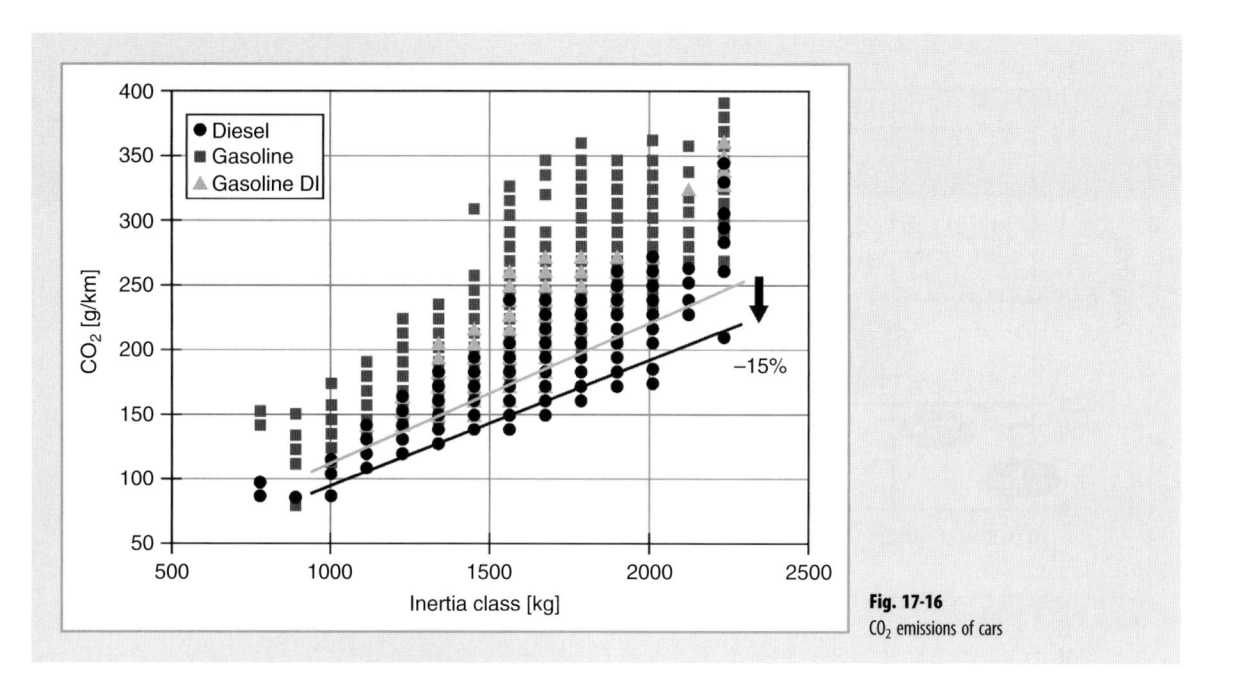

emissions in recent years. Diesel engines inherently have extremely low carbon monoxide (CO), hydrocarbon (HC) and nitrogen oxide (NO_x) emissions (see Chap. 15).

The introduction of state-of-the-art high pressure injection systems, high pressure supercharging and highly efficient exhaust gas recirculation systems has made it possible to lower these emissions substantially in recent years. The additional use of oxidation catalytic converters and particulate filters has put modern diesel cars practically at the detection limit for CO, HC and particulate emission. A diesel engine only has higher NO_x emission than a gasoline engine because of its principle. However, the deNOx systems being implemented will reduce this significantly in the near future.

The development of European limits for particulate and NO_x emissions represented in Fig. 17-17 documents the significant improvements prompted by Euro 1 through Euro 5 limit levels.

17.1.5.3 Performance

Passenger car performance is evaluated on the basis of criteria of quasi stationary and transient operation.

Quasi stationary driveability, e.g. starting acceleration, elasticity, maximum speed and hill-climbing performance, are decisively influenced by the rated power, maximum torque and full load characteristic, i.e. the torque curve as a function of speed.

However, the response to a demanded load and speed change (responsiveness) depends on the inertias contained in the system as mass moments of inertia, mass storage (filling and emptying) and thermal storage.

The introduction of direct injection, high pressure supercharging and electronics has significantly boosted the performance of car diesel engines over the last ten years, enabling them to rate better than car gasoline engines in virtually every vehicle performance criterion. This holds particularly true for the potential for low end torque from low and medium speeds, which is crucial in routine operation.

The index employed as a unit to compare engines' power density is specific power output. This value is the ratio of maximum power to displacement. Today's diesel engines have values ranging from approximately 30 kW/dm^3 in the entry-level segment to up to 70 kW/dm^3 in the top premium segment. The weight-to-power ratio is also a very important index. While the best diesel engines already have values close to 1.0 kg/kW, which is good for gasoline engines, the first car diesel engines to break this "sound barrier" may be expected in the near future.

17.1.5.4 Comfort

Driving, acoustic and climatic comfort and ease of operation are primary vehicle properties, yet decisively influenced by the engine. Since the standards for passenger cars are high, comfort is an aspect emphasized in car diesel engines. Consequently, diesel technology is so advanced in terms of the quality of comfort that it has already become the dominant drive concept in the upper vehicle classes in Europe.

Driving and acoustic comfort is primarily shaped by the engine, the main exciter of vibration and source of noise (see Chap. 16 and Sect. 18.2). Vibrations are excited by the processes of combustion and gas exchange as well as the

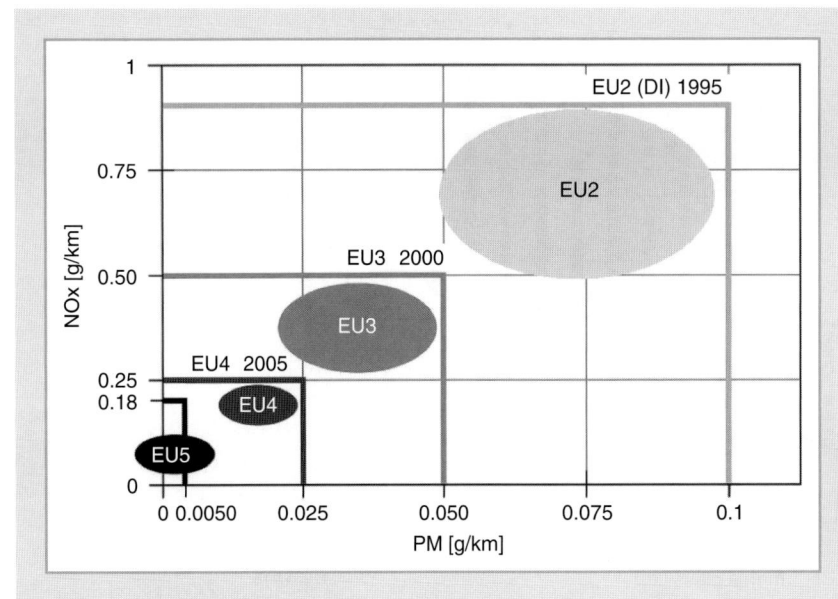

Fig. 17-17 Development of emissions levels in Europe

oscillating motions of reciprocating pistons and the rotation of the crankshaft and camshaft. This may be referred to as combustion noise or mechanical noise depending on the source of vibration. Combustion noise is based on the rapid rise and decay of pressure in the cylinder, which generates impact excitations in the cylinder head, crankcase and crankshaft assembly, which cause vibrations and radiate noise. The alternating forces that periodically occur in a crankshaft assembly are responsible for mechanical engine noise. All other components involved in the drive, e.g. transmission, drive shaft, differential gear and auxiliary units and intake and exhaust system, are considered avenues of structure-borne noise to the vehicle body and/or radiators of noise.

Systematic intervention in the vibration exciter and the transmission mechanism is required to obtain good driving and acoustic comfort.

Some active engine measures are geared toward reducing the oscillating masses (piston, piston pin and connecting rod mass) and the running clearance between the piston and cylinder wall.

Optimized mass balancing, high crankshaft stiffness to counter bending, direct placement of the main bearings in the throw, adequate stiffness of the main bearing centerline and decoupling of the supporting engine structure limit vibration to an acceptable level.

Precautions also have to be taken in the valve gear assembly. The opening and closing of the valves generate impact and inertial forces during gas exchange, which cause a narrowband increase of the level of vibrations both in the lower and upper speed range. The following contribute to a reduction of these forces:
– low valve spring forces,
– lightweight valves,
– cam shapes with low acceleration peaks and
– rigid camshaft carriers with bearing impedance.

Every high mass auxiliary unit attached to a crankcase alters an engine's vibration characteristics and can increase the sound pressure level in the particular frequency range. To counter this, the following rules must be observed:
– the mounting distance to the crankcase should be a minimum,
– the mass of the auxiliary units should be a minimum,
– the holder should be very rigid and
– the eye should be mounted at rigid spots of the crankcase.

Pulsating intake air excites the intake system. Adequate damping can be obtained when the air cleaner housing has a sufficiently large volume (approximately five times the displacement in four cylinder engines). Connecting an additional cavity (Helmholtz resonator) or resonance tube and placing the intake opening in an insensitive point on the body (see Chap. 13) are other measures that improve matters. Passive measures are oriented toward interrupting the paths for the transmission of vibration energy. Decoupling the valve cover, oil pan and beltguards as well as utilizing additional cover panels are effective measures that lower noise emission.

Rattling noises from vehicles with manual transmissions are a special problem that primarily manifests itself in the speed range during part and full load, including idling. A dual-mass flywheel is an effective method to eliminate this problem (see Sect. 8.2).

Constructed of a primary flywheel on the crankshaft, a multistage torsional vibration damper and a turbine with a friction clutch and a rigidly attached driver disk, a dual-mass flywheel responds to rotational irregularities like a low-pass filter. Since resonances only appear below idle speed, noises are only produced when the engine is being started and stopped.

Limiting external noises by covering the engine compartment toward the base or enclosing it to soundproof it is an emerging trend. Depending on the complexity, an enclosed engine compartment can reduce noise by 12 dB during idling and by 5–7 dB in second and third gear over vehicles without enclosures. Additional passive vehicle measures in the crankshaft support, e.g. hydro mounts, and in the exhaust suspension contribute to better insulation of the body against structure-borne noise and, thus, to good vibrational comfort of diesel cars.

17.1.6 Outlook

Modern diesel engines are now fully established in Europe.

The diesel car segment has also already begun to grow significantly in many other markets outside Europe. Other large markets such as the USA have adopted a wait and see attitude. However, the conservation of fossil energy resources and thus the promotion of fuel economizing engine concepts are also gradually becoming important issues.

Figure 17-18 presents the global and European share of diesel car production.

Technologically, numerous potentials to significantly further enhance the performance characteristics of car diesel engines are still identifiable. Further development in materials, manufacturing methods and simulation methods will make it possible to further increase engines' load capacity. Thus, a maximum combustion chamber pressure of 210 bar even in combination with light alloy crankcases now appears to be a thoroughly realistic goal. Moreover, potentially considerable increases in injection pressures and boost pressure levels are also within grasp.

Injection pressures of 2,000 bar will also be achievable in the foreseeable future with CR systems. There are thoroughly realistic approaches to the next innovative steps for injection precision and flexibility, e.g. directly actuated piezo injectors. However, the development and standard implementation of a highly effective NO_x exhaust gas aftertreatment system will be the greatest challenge and simultaneously the greatest opportunity to introduce diesel all over the USA as well. Here too, thoroughly promising approaches and concepts already exist in the form of storage catalysts as well as SCR technology. The technical implementation and commercialization of these technologies will require several years though. In the longer term, it will also enable car diesel engines to shed their last still remaining disadvantage, i.e. slightly higher NO_x.

17.2 Diesel Engines for Light Duty Commercial Vehicles

17.2.1 Definition of Light Duty Commercial Vehicles

Highly reliable and durable, diesel engines have always been the preferred drive for commercial vehicles. Diesel engines are the only drive employed for heavy duty trucks In Europe. Diesel engines have also largely driven gasoline engines out of the European light duty commercial vehicle sector.

Since they are widely used in cities and their environs to transport goods and passengers in the commercial and increasingly in the private sector, light duty commercial vehicles also significantly shape the current traffic pattern.

The European Union's Directive 71/156/EEC classifies self-propelled road vehicles in three different classes based on use and model (Table 17-1). This classification assigns light duty commercial vehicles to the classes M_2, N_1 and N_2.

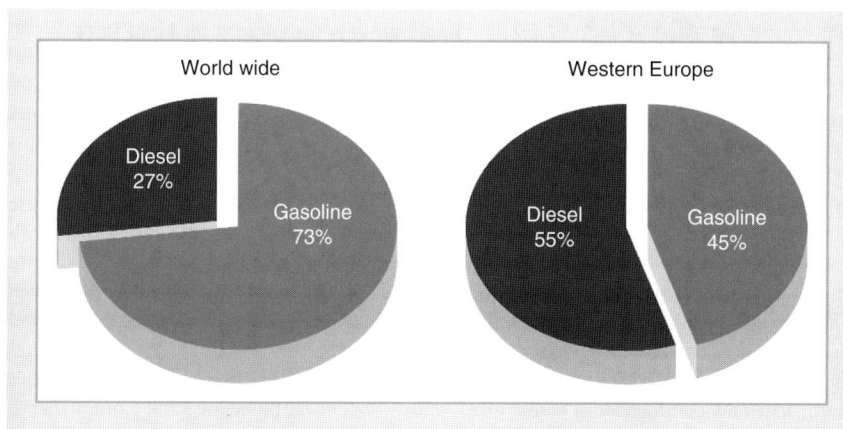

Fig. 17-18 Share of diesel in the car segment

Table 17-1 Classification of motor vehicles according to EC 70/156/EEC

	L Class					M Class			N Class		
Classification	Vehicles with less than four wheels, motorcycles, three-wheelers					Motor vehicles with at least four wheels or three wheels and a total gross weight >1 t designated for the transport of passengers			Motor vehicles with at least four wheels or three wheels and a total gross weight >1 t designated for the transport of freight		
	L1	L2	L3	L4	L5	M1	M2	M3	N1	N2	N3
Model	Two-wheeled	Three-wheeled	Two-wheeled	Three-wheeled (asymmetr.)	Three-wheeled (symmetr.)	–	–	–	–		
Displacement	<50 ccm	<50 ccm	>50 ccm	>50 ccm	>50 ccm						
Maximum speed	<50 km/h	<50 km/h	>50 km/h	>50 km/h	>50 km/h						
Number of seats	–	–	–	–	–	1–5	> 9	> 9			
Gross vehicle weight rating	–	–	–	–	<1 t	>1t	1–5 t	>5 t	1 – 3.5 t	3.5–12 t	>12t

Commercial vehicles for passenger transportation with a gross vehicle weight rating of 5 t are in class M_2, commercial vehicles up to 3.5 t are in class N_1 and commercial vehicles up to 5 t in class N_2.

The German Road Traffic Regulations categorizes trucks with a total gross weight of 2.8–7.5 t as light duty commercial vehicles. However, vehicles with a gross vehicle weight rating below 2.8 t can also be approved as trucks, provided the useful area determined by the design is at least 50% of the total area. The payload of such vehicles is approximately 0.55 t at a total gross weight of 1.7 t, approximately 1 t at 2.8 t and approximately 1.8 t at 3.5 t.

The considerable share of light duty commercial vehicles among trucks reveals their importance. Commercial vehicles with a total gross weight of 1.4–7.5 t accounted for approximately 87% of the new truck registrations in Germany in 2004 [17-1].

While the category of vehicles with a total gross weight of up to 1.5 t is increasingly growing in importance, it is not examined any more closely here since such vehicles are almost always derived from cars. Off-road vehicles' motorization is also not addressed since their engines frequently place them in the luxury car class segment.

The fields of application for light duty commercial vehicles are extremely diverse. Commercial applications of this type of vehicle primarily encompass the transportation of goods to surrounding areas. Moreover, they are employed in trades and small businesses as well as municipal and commercial services (e.g. as taxis, school buses, hospital and special needs transport, street cleaners, fire department vehicles, etc.).

In recent years, this vehicle category has enjoyed growing popularity among private users too. The trend toward organizing active vacations and recreation has led to camping vehicles, motor homes and minibuses becoming a familiar sight on the road.

The requirements imposed on engines for light duty commercial vehicles arise from the aforementioned diverse fields of application. Low operating costs, high torque and high availability are prominent for commercial use. Private users additionally expect high vehicle performance resembling that of cars, e.g. acceleration performance and maximum speed along with comfort and low fuel consumption.

17.2.2 Requirements for Engines for Light Duty Commercial Vehicles

The mandatory emission regulations applied to light duty commercial vehicles are derived from both heavy duty commercial vehicles and regulations for passenger cars.

With the introduction of Euro III limits, a test procedure based on the ESC and ELR test is now applied to vehicles with a gross vehicle weight rating over 3.5 t, which evolved from the 13-mode test in accordance with Directive ECE R49. Vehicles with diesel particulate filters also have to be tested based on the ETC.

Pollutant limits apply to light duty commercial vehicles with a gross vehicle weight rating below 3.5 t. Compliance is determined on a chassis dynamometer and the same chassis dynamometer test mandated for cars is applied to this vehicle class. Different limits apply depending on the reference weight (see Chap. 15). The reference weight class of 1–1,305 kg applies to cars, the classes 2 (<1,760 kg) and 3 (>1,760 kg) apply to vehicles represented by light duty commercial vehicles. The limits for vehicle classes 2 and 3 are set correspondingly higher. Different measures are taken based on these regulations, which are applied based on class.

The continuous tightening of exhaust regulations necessitates adapting the state of engine development to current legislation in correspondingly short cycles. Common rail technology has established itself in recent years virtually across the board. This method of mixture formation furnishes ample degrees of freedom to design injection in terms of quantity, timing, frequency (pilot and post-injection) and pressure. The application of electronics in diesel engines has also made other measures such as cooled, regulated exhaust gas recirculation or boost pressure standard. However, potential engines have cylinder volumes of significantly less than 1 dm^3, which complicates matters. Relatively small compared to that in heavy duty truck engines, this cylinder volume impedes the optimization of combustion because of the short lengths of the free injection sprays. Improvements in oil consumption and, above all, in fuel consumption will be essential in the future.

Exhaust gas aftertreatment is increasingly growing in importance despite the implementation of all the measures intended for engines. Diesel particulate filters have been implemented in light duty commercial vehicles in addition to oxidation catalytic converters in order to meet exhaust limits since the introduction of exhaust level 4 for vehicles with a gross vehicle weight rating >3.5 t based on the thirteen-mode test and for vehicles <3.5 t based on the chassis dynamometer test. The significant reduction of nitrogen oxides with the introduction of Euro 5 limits on October 1, 2008 will presumably also additionally necessitate further exhaust gas aftertreatment systems in this vehicle segment.

Directive 70/157/EEC limits noise emission from commercial vehicles (see Sect. 16.2). Compliance with this directive necessitates measures in the engine and the vehicle. The requirements are satisfied in the engine by designing the cylinder crankcase rigidly, employing a cast oil pan and using a ladder frame for the main bearing centerline. Enclosures can bring secondary improvements. Substantial influence can be exerted on combustion noise by designing the combustion characteristic as one or more pilot injections.

Apart from the statutory constraints, a commercial vehicle's field of application and customer requirements also dictates engine development. High torque at low speeds is desired for local commercial transportation far more than high rated power. Such a torque characteristic is obtained with the aid of variable turbine geometry turbochargers. Another option is two-stage turbocharging, which is already standard in the car sector [17-2]. However, private users demand relatively high engine power, above all in designs that transport passengers. Engines with high specific power are available for vehicles in this category. Exhaust gas turbocharging has become standard in diesel engines and intercooling is implemented without exception. The engine loads induced by high cylinder peak pressure can be controlled by both combustion and design. Gasoline engines are found in the vehicle sector with correspondingly high power.

Factors such as durability, reliability, ruggedness and long maintenance intervals are extremely important in light duty commercial vehicles. The car engines implemented in light duty commercial vehicles of up to 3.5 t are specially modified for these applications. Such engines usually have limited maximum torques, rated powers, maximum boost pressures and rated speeds to meet the aforementioned criteria. In addition, the area of the crankshaft assembly is modified to match the specific commercial vehicle load profile and peripheral measures are taken, e.g. maintenance interval indication and modifications for the installation situation.

Commercial vehicles with a gross vehicle weight rating of more than 5 t are normally equipped with diesel engines with rated speeds of approximately 3,000 rpm or lower to keep engine wear within limits and to ensure long service lives.

17.2.3 Implemented Light Duty Commercial Vehicles Engines

17.2.3.1 Overview

The range of diesel engine designs implemented as drives in motor vehicles extends from small volume 0.8 l engines with 30 kW to 18 l engines with 485 kW.

The engines employed to motorize light duty commercial vehicles with a total gross weight of 2–5 t now cover a power range of approximately 50–170 kW (Fig. 17-19). Both gasoline and supercharged direct injection diesel engines with intercooling and displacements ranging from 1.9 to 3.7 liters are implemented. Along with classic gasoline and diesel engines, natural gas engines are also establishing themselves.

Light duty commercial vehicles with gross vehicle weight ratings of up to 2 t are normally equipped with diesel engines that stem from car engines. They have outputs of between 50 and 90 kW and displacements of approximately 2 l. The engine power in light duty commercial vehicles up to a gross vehicle weight rating of 6 t extends to up to approximately 130 kW with a displacement of 2.5–3 l. Larger volume gasoline engines are partly found in the passenger transport vehicle sector (Fig. 17-20).

Engines obviously derived from the engine families of heavy duty commercial vehicles are designed for trucks with a total gross weight ranging of between 6 and 7.5 t, which are also considered commercial vehicles. Their rated power spans 90 and 160 kW at displacements of between 3 and nearly 5 l.

A comparison of performance data from implemented vehicle diesel engines reveals specific relationships to their fields of application. Figure 17-21 plots the brake mean effective pressure of the engines at rated power as a function of

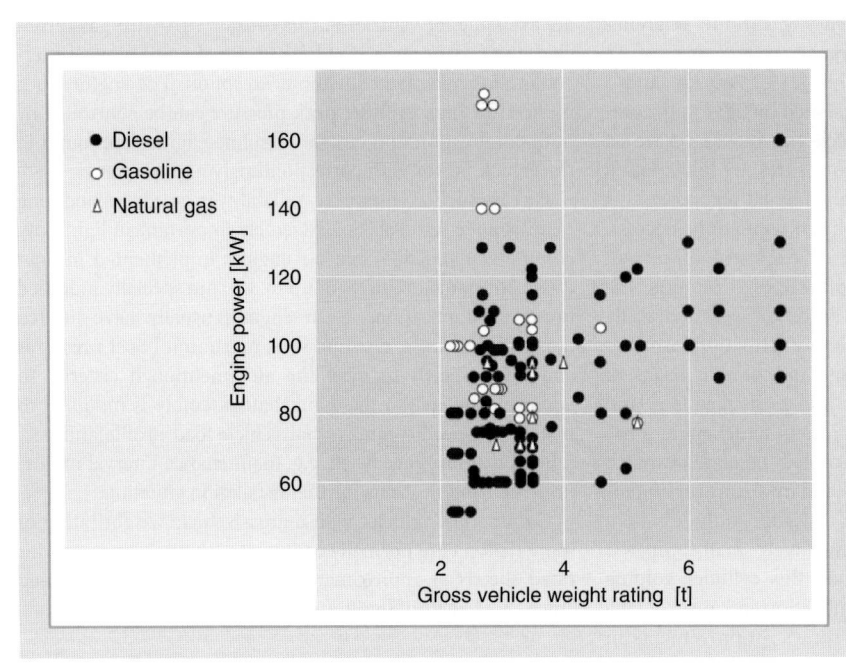

Fig. 17-19 Rated power of combustion engines for light duty commercial vehicles with different total gross weights

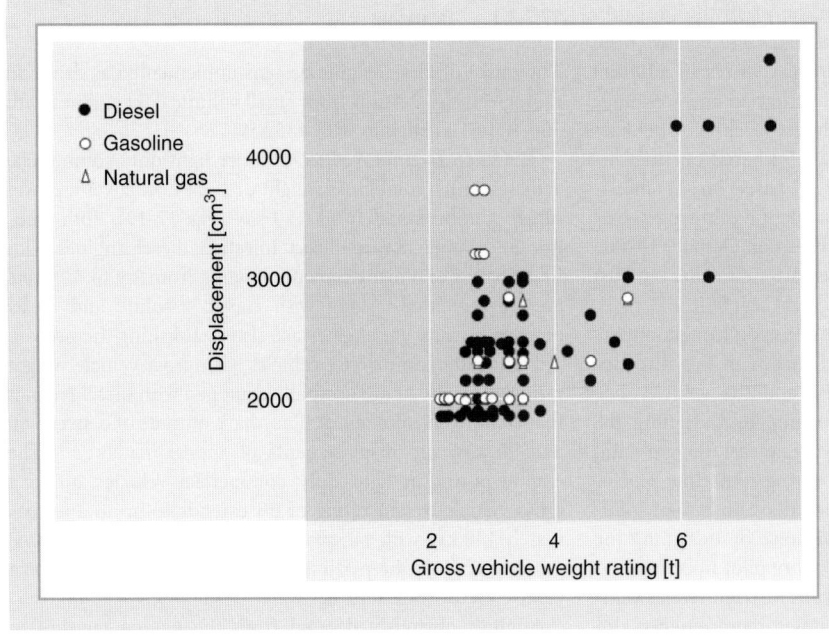

Fig. 17-20 Displacement of combustion engines for light duty commercial vehicles with different total gross weights

their volume-specific power. The graph additionally plots the corresponding rated speeds. As expected, engines for light duty commercial vehicles have design data situated closer to car engines. Power outputs per liter of up to 50 kW/l can be found. The mean is approximately 40 kW/l. Top engines in the diesel car sector reach nearly 70 kW/l. The rated speeds of commercial vehicle engines cover the range of around 3,500 rpm and above and are lower than those of car engines. Their design ensues from the geometric dimensions of the crankshaft assembly on the one hand and is limited by the piston velocities critical for service life for the load profile specific to the commercial vehicle on the other hand.

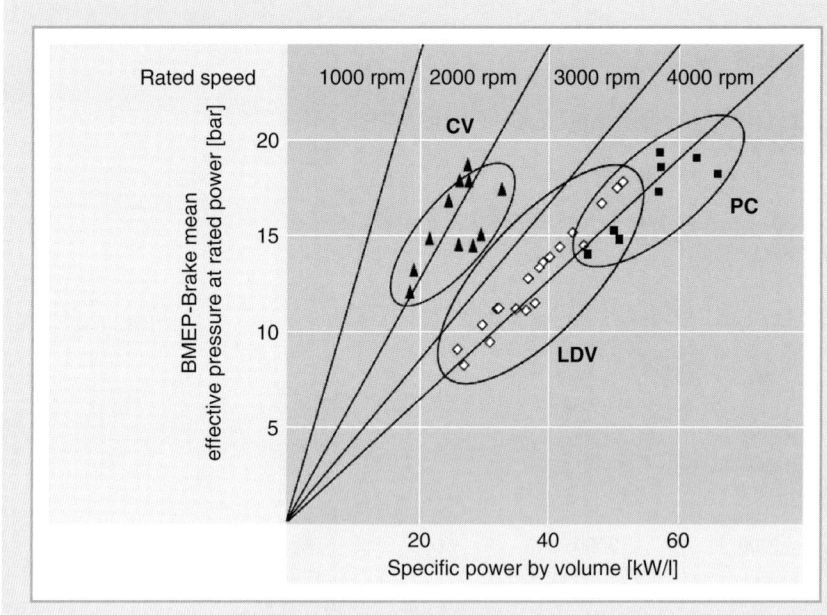

Fig. 17-21 Performance data of selected diesel engines for use in vehicles

Direct injection is the combustion system applied in diesel engines for this category of light duty commercial vehicles. The common rail has now established itself for mixture formation. Chamber combustion systems with one distributor injection pump are no longer applied because emission limits are no longer observable with this type of mixture formation and combustion. Electronic systems have established themselves because of their diverse options for control. This development has been significantly influenced by emission control legislation.

Given their high efficiency and durability, direct injection engines have long since established themselves as the truck drive above the displacement limit of 2.5 liters. Fuel consumption and thus operating costs significantly influenced the development of this concept.

Aspirated engines are no longer avaiable for use in light duty commercial vehicles for the aforementioned reasons. Gasoline engines cover a more modest market share of the upper power range than diesel engines in the retail segment of exclusively equipped vehicles. In addition, a trend toward equipping light duty commercial vehicles with tonnages of up to 5 t with natural gas drives is observable (Fig. 17-22).

Figure 17-23 presents the maximum torques attained by implemented commercial vehicle diesel engines. The lower part of the graph contains natural gas drives followed by gasoline engines, covering a torque range between 100 and 200 Nm. One exception is larger, in part also supercharged gasoline engines. Diesel engines for vehicles with a gross vehicle weight rating of up to 6 t reach torques of 400 Nm. Torques from 400 up to approximately 800 Nm are common in the vehicle class up to 7.5 t.

The engines presented in the preceding figures comply with exhaust level 3 and 4. A further modification of the engines' mixture formation in the diesel range can be expected when exhaust level 5 for the thirteen-mode test is introduced on October 1, 2008. The future generation of high pressure pumps will make it possible to produce injection pressures of 2,000 bar and more. Moreover, the use of an appropriate exhaust gas aftertreatment system to reduce nitrogen oxides can be expected. The extent to which one of these systems will establish itself for light duty commercial vehicles is difficult to judge conclusively at this point.

17.2.3.2 Select Examples

Two engine models selected from different manufacturers, which cover displacements between 2.2 and 2.5 liters, are described as representatives of the range of diesel engines for light duty commercial vehicles. This displacement class constitutes the majority of engines employed in light duty commercial vehicles (Fig. 17-20). Table 17-2 presents the basic data of the two selected engines.

Five Cylinder DI Diesel Engines (Volkswagen)

Along with some gasoline engines, four and five cylinder diesel engines are implemented in Volkswagen's T5 model delivery vans (Fig. 17-24). The successor to the LT2, the *Crafter* is only available with a five cylinder diesel engine [17-3]. This proven engine has its origin in the car engine of the 1980s and has been steadily refined for commercial vehicle application.

The engine was fundamentally redesigned for use in the *Crafter*. An integral element is the use of third generation common rail technology. The injection pressure reaches

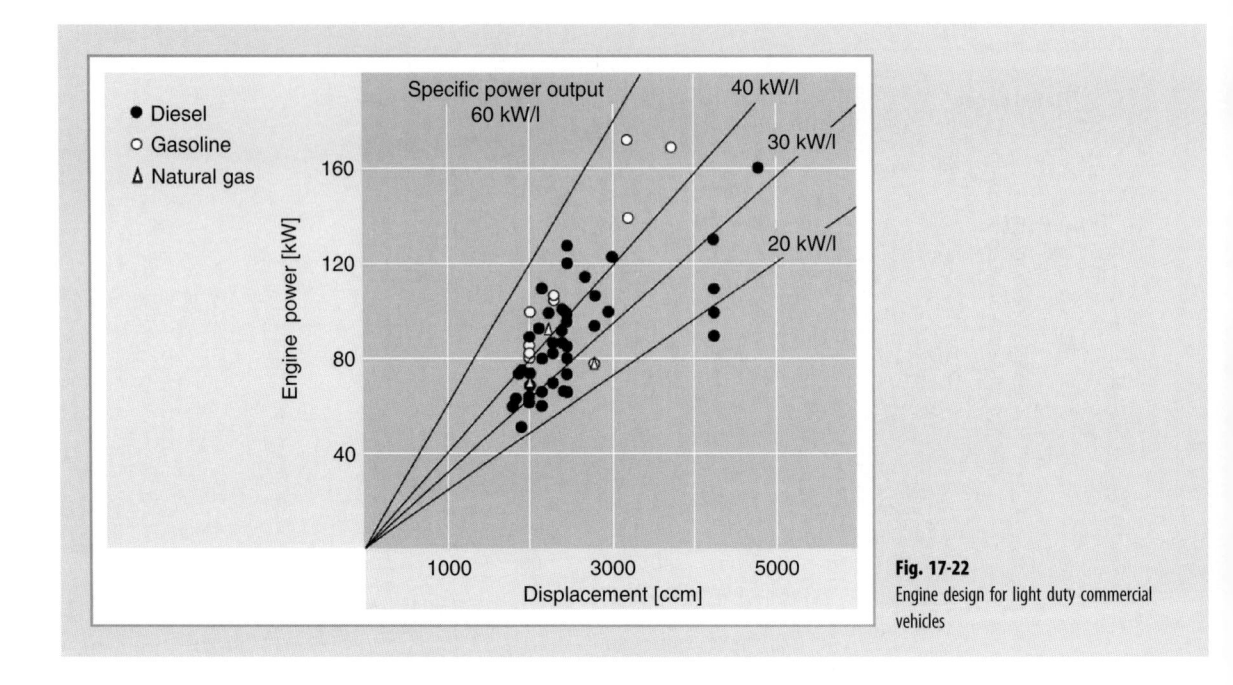

Fig. 17-22 Engine design for light duty commercial vehicles

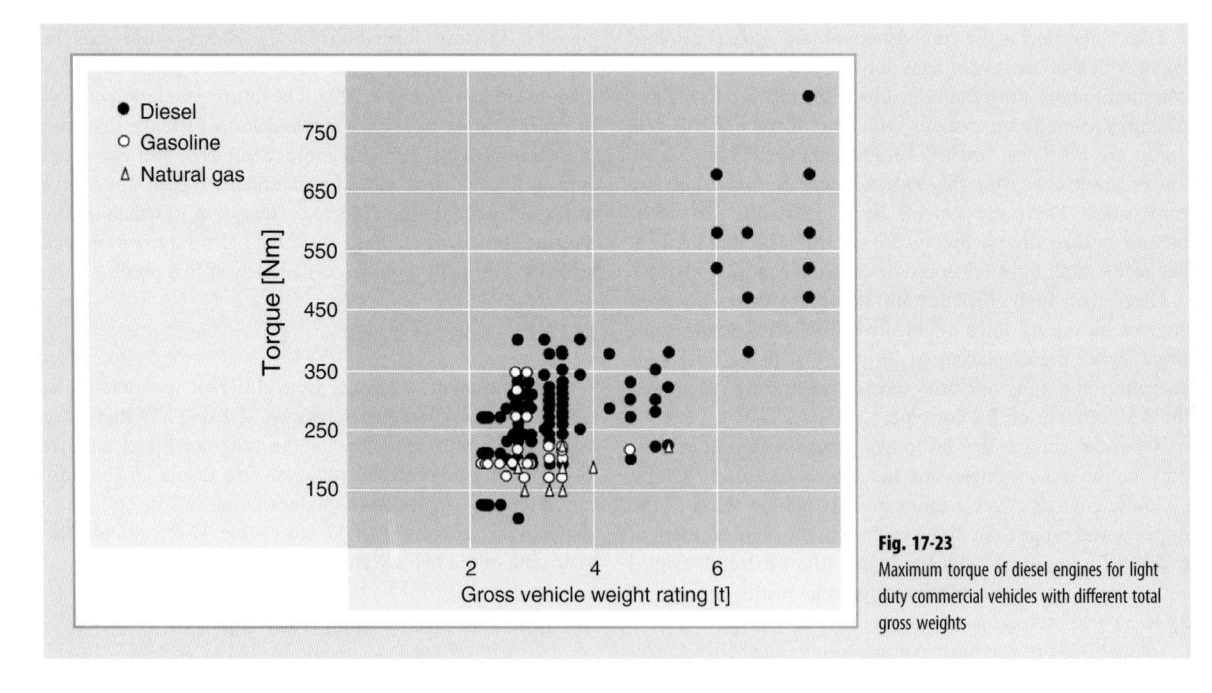

Fig. 17-23 Maximum torque of diesel engines for light duty commercial vehicles with different total gross weights

1,600 bar. The injectors are furnished with piezo actuators. The power is spread from 65 to 120 kW in four classes. The rated speed is generally 3,500 rpm. Maximum torque is reached at a speed of 2,000 rpm. Variable turbine geometry and intercooling have now become standard for this vehicle category. A stroke of 95.5 mm and a cylinder diameter of 81 mm make it a comparatively long-stroke engine. At 2,461 cm^3, the cylinder displacement corresponds to that of four cylinder engines employed for cars and light duty commercial vehicles.

The cylinder crankcase consists of gray cast iron, the cylinder head of an aluminum alloy. Their combination with an aluminum oil pan that also helps form the transmission mounting flange makes a rigid system out of the complete engine. A distance of 88 mm between cylinder bore center

Fig. 17-24 Volkswagen 2.5 l turbo diesel engine

lines and a bore of 81 mm leaves a minimum wall thickness of 7 mm between cylinders, which is controlled by means of a crimped metal layer cylinder head gasket. The seal system helps keep cylinder distortion within narrow limits. This is crucially important for oil consumption. The aluminum-based cylinder head with two valves is designed as a parallel flow head. The valves' hydraulic bucket tappets are actuated by a camshaft.

The camshaft and the common rail pump are driven by an automatically tightened toothed belt. The coolant pump drive is also integrated in this belt drive. The advanced materials for the toothed belts and the moderate loading by the common rail system allow an interval of toothed belt replacement of 200,000 km.

Coated diesel particulate filters are implemented to comply with Level 4 exhaust limits both for the chassis dynamometer test and approval based on the thirteen-mode test.

Four Cylinder 2.2 Liter Diesel Engine (DaimlerChrysler)

DaimlerChrysler also employs four cylinder diesel engines in the light duty commercial vehicle segment, which have their origin in cars [17-4]. The power of the four cylinder engine model in its Sprinter ranges from 65 to 110 kW at a rated speed of 3,800 rpm. The most powerful design has a maximum torque of 330 Nm. A V6 diesel engine tops off the engines.

The cylinder crankcase is constructed of gray cast iron. The side walls of the crankcase assembly are drawn down far to enhance stiffness. A trapezoidal connecting rod accommodates increased loading of the crankcase assembly. Despite the use of a trapezoidal piston, the connecting rod's axial guidance in the piston (piston end guidance) was retained.

The cylinder head is constructed of an aluminum alloy. Two overhead camshafts drive four valves apiece. The CR injector is centered, thus making it possible to center a bowl in the piston. Significant advantages over a two-valve engine with eccentric piston bowls in terms of temperature distribution can be expected as a result.

The engine is equipped with a third generation common rail system. The achievable injection pressure is 1,600 bar. The valve gear's drive is designed as a chain drive in which the common rail pump drive is integrated.

17.2.4 Outlook

Future legislation will also decisively influence the future development of diesel engines intended as drives for light duty commercial vehicles. Further limitation of pollutant and noise emissions, the development of crude oil prices and thus increases in energy costs will constitute important influencing factors for engine developers and lastingly influence their decisions about drive concepts.

Table 17-2 Basic data of selected light duty commercial vehicle engines

	Unit	DaimlerChrysler	Volkswagen
Model		I4	I5
Displacement	ccm	2,148	2,461
Bore	mm	88	81
Stroke	mm	88.3	95.5
Exhaust gas recirculation		Yes	Yes
Intercooling		Yes	Yes
Exhaust gas turbocharging geometry		Variable	Variable
Injection system		Common rail	Common rail
Rated power	kW/min	110/3,800	120/3,500
Max. torque	Nm	330	350
Spec. power	kW/l	51.2	48.8

In the future, gasoline engines will likely be employed only in commercial vehicles that are directly derived from cars and predominantly serve to transport passengers. These will primarily be smaller tonnages equipped with gasoline engines.

More than ever, future bills will be driven by debate on CO_2 in the context of the greenhouse effect. Gas drive is also growing in importance because natural gas has a better H:C ratio than carburetor fuel or diesel oil [17-5]. Some manufacturers of light duty commercial vehicles already offer natural gas engines.

Today, diesel engines are designed solely as direct injection engines; rising fuel prices will intensify this trend for a long time to come. The common rail system has largely established itself for mixture formation. Consequently, the injection characteristic and thus the combustion characteristic can be better shaped. Whether piezo technology will force solenoid valve controlled injectors from the market in the coming years is a question that cannot be answered conclusively at this point. Both systems are employed side by side at present.

Variable turbine geometry has established itself for exhaust gas turbocharging. Two-stage supercharging will increasingly play a role in diesel engines to enhance vehicle performance and comply with emission control legislation. Exhaust gas aftertreatment beyond the use of diesel particulate filters will also be implemented in this vehicle segment to comply with Euro 5 limits.

Given their better fuel consumption, diesel engines will be the dominant drives for light duty commercial vehicles in the coming years; strict mandatory regulations shall be observed, service life and maintenance improved and comfort further adapted to cars.

17.3 Diesel Engines for Heavy Duty Commercial Vehicles and Buses

17.3.1 Definition of Heavy Duty Commercial Vehicles

17.3.1.1 Classification

Legislators classify commercial vehicles for the transportation of goods and passengers according to their gross vehicle weight rating (Fig. 17-25). This classification is referenced for many regulations that pertain to commercial vehicles and their surroundings, including the limitation of emissions, the levying of taxes, the classification of licenses or the setting of speed limits. Beginning at 7.5 t according to the German Road Traffic Regulations, a further subdivision of the heavy duty commercial vehicle sector into light duty, standard and heavy duty classes is common (Fig. 17-25). The limits are fluid. The maximum gross vehicle weight rating for heavy duty commercial vehicles permitted on public roads in Europe is 40 t. Local exceptions to this regulation may be encountered.

17.3.1.2 Fields of Application

Commercial vehicles are the backbone of our modern economy based on the division of labor and a guarantor of growth and prosperity. Roughly two thirds of all commercial transportation services in the European Union are handled with commercial vehicles, the vast majority locally. While semi-trailer trucks dominate European long-distance heavy transport (Fig. 17-26), drawbar combinations are also extensively used in Germany.

Mainly standard and light duty class vehicles are operated in regional and local delivery traffic. These include

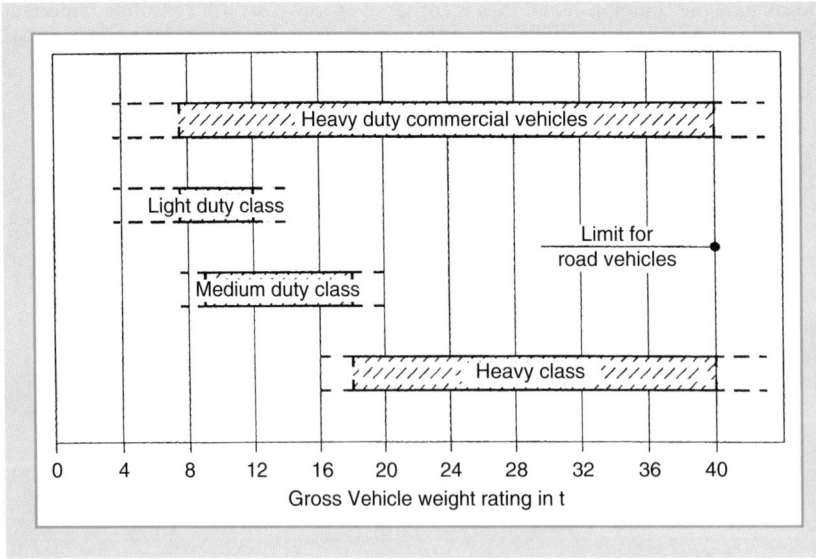

Fig. 17-25
Classification of heavy duty commercial vehicles according to their total gross weight (Europe)

Fig. 17-26 Heavy semitrailer truck, total gross weight 40 t

municipal vehicles for city cleaning and waste disposal as well as fire vehicles. Trailers are rarely used in these fields of application.

Motor coaches are used to transport passengers in long and medium distance tourist travel. They are available in many variants of payload and seating capacity. Buses in short distance public transportation, i.e. municipal and regional buses, are the most numerous representatives of this type of vehicle by far. Numerous variants are found here too, e.g. double decker and articulated buses.

17.3.2 Operator Demands on Engines

17.3.2.1 Economic Factors

Cost Analysis

Commercial vehicles are capital goods. Therefore, economic considerations are uppermost when they are purchased. Once the transport job has been analyzed, a decision is made to buy the vehicle that is most suitable. Primarily the costs during a vehicle's service life are analyzed (cf. [17-6]). Table 17-3 contains an example of long distance vehicles. It presents the different cost types as percentages of the total costs. As expected, the dominant factors are the fuel costs followed by vehicle leasing and highway tolls. Apart from manufacturing costs, vehicle manufacturers are only able to influence maintenance and repair costs. They account for 5.7% of the total costs.

Power, Specific Power

The market demands engine outputs of between 100 and 500 kW for the heavy duty commercial vehicle sector. The most cost effective power output is determined in individual cases based on the given transport job. Once the required power has been defined, specific power comes into play as a competitive criterion. An engine's space requirements and weight should limit a vehicle's freight volume and payload as little as possible.

Fuel Consumption

Apart from reliability and particularly because public debate about the conservation of resources and protection of the environment has also made it prominent, fuel consumption is the most important economic factor for operators because it directly influences costs (see Table 17-3). Therefore, virtually without exception, only drives with the best consumption, i.e. direct injection diesel engines, find acceptance in the heavy duty commercial vehicle market. This holds particularly true in the heavy duty class since long-distance transportation has the highest kilometrage of over 200,000 km/year. Fuel consumption is economically less important in the light duty class sector in delivery traffic. Some such vehicles do not exceed 20,000 km/year.

A vehicle's distance-based consumption values are decisive for operators. The optimization of a vehicle's road load and aerodynamic drag to minimize power loss is also decisive

Table 17-3 Cost breakdown over the useful life of a semitrailer truck suitable for a maximum towing mass of 40 t, useful life of 48 months and total kilometrage of 600,000 km, without costs for drivers, administration and garage (as of 2006)

Variable costs		Boundary conditions		Evaluation	
Diesel costs (€/100 km = ct/km)	32.11	Days of operation (days/year)	240	Fixed costs (€/day)	122.93
AdBlue costs (€/100 km = ct/km)	0.00	Useful life (months)	48	Fixed costs (€/100 km = ct/km)	19.67
Highway tolls (€/100 km = ct/km)	9.60	Vehicle kilometrage (km/year)	150,000	Variable costs (€/100 km = ct/km)	41.71
Variable costs (€/100 km = ct/km)	41.71	Kilometrage on toll roads (km/year)	120,000	Fixed and variable costs (€/100 km = ct/km)	61.38
		Vehicle leasing (€/month)	1,285		
Fixed costs		Servicing and repairs (€/month)	438	Vehicle leasing (€/year)	15,420
Vehicle leasing (€/year)	15,420	Tires (€/month)	75.00	Servicing and repairs (€/year)	5,256
Servicing and repairs (€/year)	5,256	Diesel consumption (l/100 km)	34.90	Tires (€/year)	900
Tires (€/year)	900	AdBlue consumption (l/100 km)	0	Fuel (€/year)	48,162
Insurance, taxes, misc. fixed costs (€/year)	7,926	Highway toll (€/100 km = ct/km)	12	Toll (€/year)	14,400
Fixed costs (€/year)	29,502	Diesel price (€/l)	0.92	Tax and insurance (€/year)	7,926
		AdBlue price (€/l)	0.60	Total (€/year)	92,064

initially. Another goal of vehicle development, particularly for tanker trucks, must be to reduce curb weight. This not only contributes to energy savings but also indirectly to increased transportation performance.

However, attaining the lowest minimum consumption possible is not enough for engine development. The minimum ought to be as broad as possible and in the range in the engine map in which the engine is most frequently operated.

Conversely, the engine consumption map also influences the design of the drive train. Only optimization of the interrelations between the engine, transmission, rear axle and wheel size satisfies operators' demands.

Availability, Service, Repairs and Warrantees

Like all capital goods, a commercial vehicle should be available for its intended use as fully as possible. Ideally, this means that the operator only has to refill automotive fluids. Components with a shorter service life than the vehicle are reconditioned, replaced or readjusted during servicing and maintenance. Since such expenditures are a quite considerable competitive factor, efforts are being made where possible to prolong and synchronize maintenance intervals to reduce necessary downtimes. State-of-the-art servicing and maintenance are examined below from the perspective of operators.

Lubricating oil system. Although oil consumption partially refreshes lubricating oil, its aging makes changing it in certain intervals unavoidable. While this is undesired with respect to emissions and exhaust gas aftertreatment systems, it continues to be unavoidable at a low level. An oil change burdens operators with costs for the oil itself, the oil filter and labor. Hence, the most important goal of development has been to further lower oil consumption and prolong the change interval.

Prolonged oil change intervals necessitate modified oil regeneration systems. Impurities are trapped in the filter system to prevent abrasive wear in tribopartners. While the oil centrifuges used earlier to further prolong oil change intervals are also still available, oil filters in the partial flow are now widespread. Oil change intervals of up to 150,000 km have already been attained on various occasions in long-distance transportation in conjunction with service computers, which allow factoring in different parameters such as oil quality, type of fuel, sulfur content or operating conditions.

Combustion air: Since combustion air must be dust-free to prevent premature wear in cylinders, pistons and piston rings, the servicing of air cleaner systems as required is crucially important for an engine's service life. The intervals are a function of the vehicle's operating conditions. A cyclone separator upstream from the main cleaner is frequently employed when the dust concentration is high. Paper dry type air cleaners in the form of replaceable cartridges have also established themselves in commercial vehicles (see Sect. 13.1).

Vacuum indicators on the pure air side indicate cleaner loading and the need for a cleaner change since engines must be prevented from suffering from air deficiency. Consequences can be incomplete combustion with smoke emission, power loss, increased fuel consumption and even engine damage.

Fuel system: Extraordinarily stringent requirements for the purity of the supplied medium apply to the fuel system where the finest fits exist at high mechanical and hydraulic stresses. Impurities are carried over with the fuel during fueling or can reach the circuit through the required tank ventilation. Only microfilter inserts are able to obtain the requisite purity. Under normal ambient conditions, their service life should also correspond to the oil change interval.

Valve clearance: While the classic design with the camshaft in the crankcase and the transmission of motion to the valves by tappets and rocker arms remained dominant in the past decades, new developments in heavy duty commercial vehicles are also increasingly implementing overhead camshafts that transmit motion by roller rocker arms. One adjuster has to be designed in for each valve pair, which must be checked in certain – maximally large – intervals and corrected if necessary. Hydraulically based automatic clearance compensation is not yet implementable in commercial vehicle engines with acceptable operational safety: automatic clearance compensation must be prevented from becoming active during engine braking operation when the butterfly valve is in the exhaust system and exhaust valves open uncontrolled and thus keeping valves from closing (see Sect. 17.3.3.2).

V-belt drives: Commercial vehicle engines must drive a number of different auxiliary units. Depending on the vehicle and type of application, these include fans, generators, air compressors, power steering pumps, refrigerant compressors, hydraulic pumps and other units. The V-belt was simple and cost effective [17-7]. More efficient ribbed V-belts [17-8] have established themselves for newer developments (Fig. 17-27). These drive elements have a significantly shorter service life than an engine. Therefore they must be replaced in certain intervals.

Their design and stressing subject V-belts to wear and increasing elongation. They only function reliably and attain their expected service life when they are operated with defined preloading. Therefore, the drive geometry must equalize the change in length. Self-adjusting spring-loaded idler pulleys have now established themselves in commercial vehicles too.

Repairs, warrantees: Other engine components should not malfunction before the limit of the engine's service life. Therefore, rather than falling under the term maintenance, such malfunctions count as damage requiring repair. Kilometrages of 1 million km and more are reached in long-distance transportation. Auxiliary units, e.g. water pumps, generators or exhaust gas turbochargers, do not attain such values in some applications.

Warrantees for heavy duty commercial vehicle drive trains have different limits, e.g. 2 years after first approval and/or kilometrage of 200,000 km, depending on which limit is reached first. The warranty covers the replacement of defective components, provided the engine has not been utilized improperly.

Fig. 17-27
Auxiliary units driven by a ribbed V-belt in a diesel engine for buses: (**a**) Crankshaft; (**b**) Idler pulley; (**c**) Generator; (**d**) Water pump; (**e**) Tension pulley; (**f**) Spring-damper element; (**g**) Air conditioner compressor; (**h**) CR high pressure pump

Purchase price, manufacturing costs: Since a maintenance-free engine is impossible to produce or only possible with complex additional measures, finding the right balance between manufacturing costs and servicing and maintenance costs is essential. Moreover, the increasing significance of environmentally sound but expensive measures prompted by legislators either directly through impact-oriented regulations or indirectly through emission-based graduated tax breaks for operators must be taken into account.

17.3.2.2 Drivability, Ease of Operation, Social Acceptance

Power Characteristic, Gear Selection

Figure 17-28 presents the power requirements for a 40 t train for different road inclines as a function of the running speed. An engine power of approximately 100 kW is necessary to overcome road load at a speed of 80 km/h on level road. The final drive ratios in the highest gear step are designed to produce engine speeds of 1,300–1,400 rpm. Since the engine power of this vehicle class is usually between 250 and 400 kW at rated speeds of around 1,900 rpm, this design facilitates a fuel economizing and low noise driving style.

A requirement that engines must already have maximum torque as of approximately 1,000 rpm to enable driving with little shifting even at low to medium upgrades can be inferred from this. Figure 17-29 presents full load characteristics developed on the basis of these requirements.

While manufacturer A favors a combination of constant power and constant torque for drivability, manufacturer B rates steady torque up to the lower full load speeds as better. The resultant torque peaks of approximately 35% produce a low end torque with strongly elastic operating performance and – above all in conjunction with automatic transmissions – relaxed driving. These characteristics have been made possible by applying exhaust gas turbocharging with intercooling, which has been standard for engines for heavy duty commercial vehicles for decades, usually only with a wastegate for higher powers. Unlike in cars or light duty commercial vehicles, variable turbine geometry does not provide engines in heavy duty commercial vehicles any special advantages because the speed band in conjunction with 12 or 16 speed transmissions is smaller.

Starting Performance

Another important parameter is an engine's accelerating performance under load at low speeds. This performance is considered critical, especially for commercial vehicles in urban traffic, e.g. municipal buses, municipal vehicles and vehicles in commercial delivery traffic. Pronounced stop-and-go operation in urban areas already requires high engine torque at low operating speeds. What is more, it must be available promptly when the driver desires. This requires largely preventing somewhat retarded responsiveness (turbo lag) even in turbo engines. Engine measures have had to be developed since additional measures such as mechanically or electrically driven compressors were previously eliminated

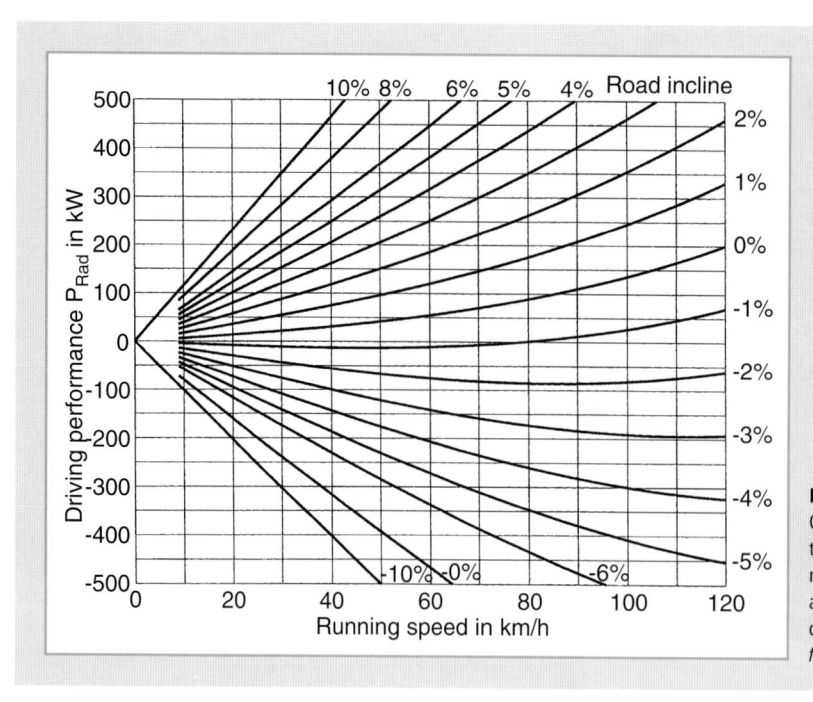

Fig. 17-28
Correlation of vehicle performance requirements in the wheels P_{Rad} to running speed and road incline: maximum vehicle weight $m = 40,000$ kg, frontal areas $A = 8.45$ m^2, drag coefficient $c_W = 0.62$, air density $D = 1,250$ kg/m^3, road load coefficient $f_r = 0.006$

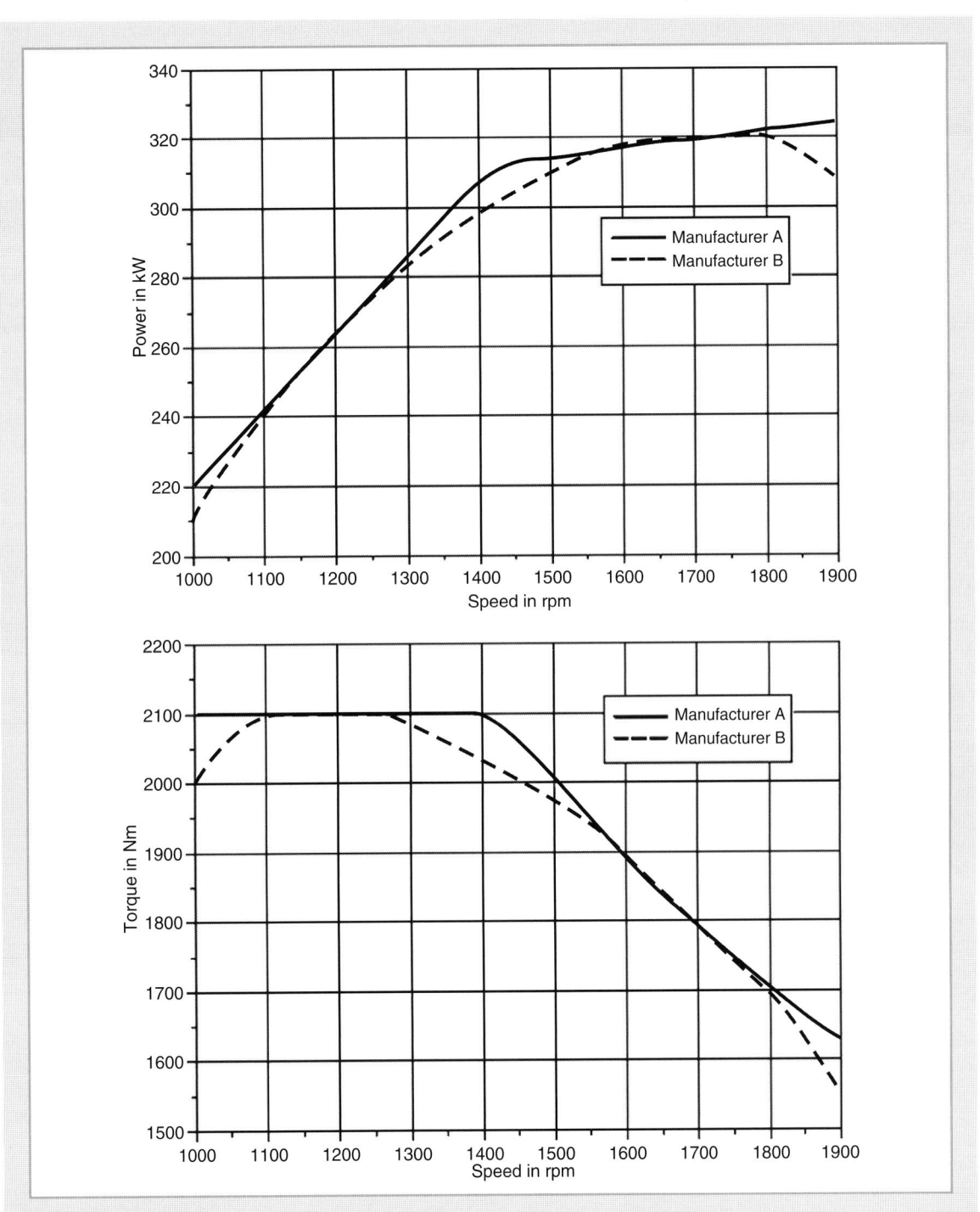

Fig. 17-29 Full load characteristics of commercial vehicle diesel engines with exhaust gas turbocharging und intercooling

for reasons of energy, costs and/or availability. Solutions have taken on the form of electronically controlled injection systems in conjunction with high injection pressure.

Braking and Engine Brake

A vehicle operator's competitiveness depends on the vehicle cruising speed attained, particularly in long-distance transportation. In the ideal case, it should be very close to the particular maximum permissible speed. Reaching this requires driving inclines fast and safely.

According to Fig. 17-28, the required braking power is as high as the power output in the opposite direction, reduced by the vehicle's road load. Since friction brakes would be thermally strained at the wheels during such continuous braking, wear-free additional brake systems have been developed. The engine is utilized as a retarder based on the principle of the air compressor. The exhaust system directly dissipates braking heat with the compressed air. Figure 17-30 presents the power curves of different engine brakes relative to displacement (see also Sect. 17.3.2). The gear steps may be used to adjust braking power to a vehicle's requirements.

Internal Noises, Vibrations

In long-distance transportation, the cab is the workplace of commercial vehicle drivers, which they may never leave during work hours. Development teams give high priority to its design and optimization for functionality and comfort. Standards for cars are attained even though the engine is normally installed directly under the cab. Significant advances have been the development of rigid engine surface structures with low sound radiation or improved damping of vibrations in the engine mounts (see Chap. 16).

Conducted into the vehicle structure by the engine suspension, an engine's outwardly acting, free inertial forces and moments of inertia can interfere with driving comfort. The magnitude and effect of these forces stemming from a piston engine's motion sequence depend on the engine's design and size (see Sect. 8.2). It may be necessary to equip an engine with additional driven shafts, which support counterweights arranged for balance.

Largely six cylinder inline engines but also six and eight cylinder V engines with 90° V angles are implemented as the drive for the heavy duty class. I6 and V8 engines are fully outwardly balanced for inertial forces and moments of inertia. Thus, they represent ideal solutions for high driving comfort. A V8 engine's larger number of cylinders gives it advantages in terms of rotational irregularity. However, this is only perceptible in the field at extremely low speeds under full load.

Other designs have also established themselves in the medium and high load range. These include five cylinder inline engines and the V10 engine in the heavy duty class. The engine suspension for five cylinder inline engines must be meticulously tuned since this structure transmits a pronounced second order free moment of inertia outwardly (see Sect. 8.3). Hence, torque balance shafts are normally necessary in the engine. Figure 17-31 presents an example of a design with moments of inertia internally balanced by two transmissions, each with two opposing shafts with balancing masses running at double crankshaft speed. The transmissions are placed under the second and fifth crankshaft bearings and each is driven by a ring gear on a crankshaft web.

Four cylinder inline engines are frequently encountered at the lower end of the scale of commercial vehicle power in the light duty class range. Such engines with unbalanced second order inertial forces are acceptable when a well tuned engine suspension keeps unacceptable vibrations away from the cab. They are given balance shafts for increased comfort requirements, e.g. in minibuses.

Starting Performance, Low Temperature Performance

Heavy duty commercial vehicle operators expect the engine to start promptly and without conspicuous emissions even at low outside temperatures. Direct fuel injection with high pressure and compression ratios of 17–20:1 make this the norm above an ambient temperature of approximately −15°C during cold starts and even more during hot starts. Low temperatures necessitate cold start assist systems that preheat intake air, e.g. electric heating elements or burners. The preheating time lasts up to 25 seconds depending on the ambient temperature. Rapid sheathed-element glow plugs that project into the combustion chamber like those familiar from cars cannot be implemented in engines for heavy duty commercial vehicles because they affect emissions adversely.

Social Acceptance

Since vehicle traffic on local and through roads is occasionally dense, societal and thus political acceptance of heavy duty commercial vehicle has become very important. Individual road users already view commercial vehicles as a nuisance to their own progress merely by virtue of their presence in traffic. When this is additionally compounded by exhaust and noise pollution, tendencies toward rejecting them manifest themselves even more strongly in political demands and social objectives. Arguments to the effect that commercial vehicles also significantly contribute to individuals' quality of life often fail to find acceptance.

Thus, it is in the business interest of individual commercial vehicle operators and serves the image of the commercial ground transportation industry when polluting emissions do not raise the conspicuousness of commercial vehicle engines in traffic. Therefore, the maxim for engine development is: *An engine should not to be outwardly perceptible above the level of driving noise and its exhaust gases should neither be visible nor smellable.* The challenge is to come as close to this goal as possible at competitive costs [17-9].

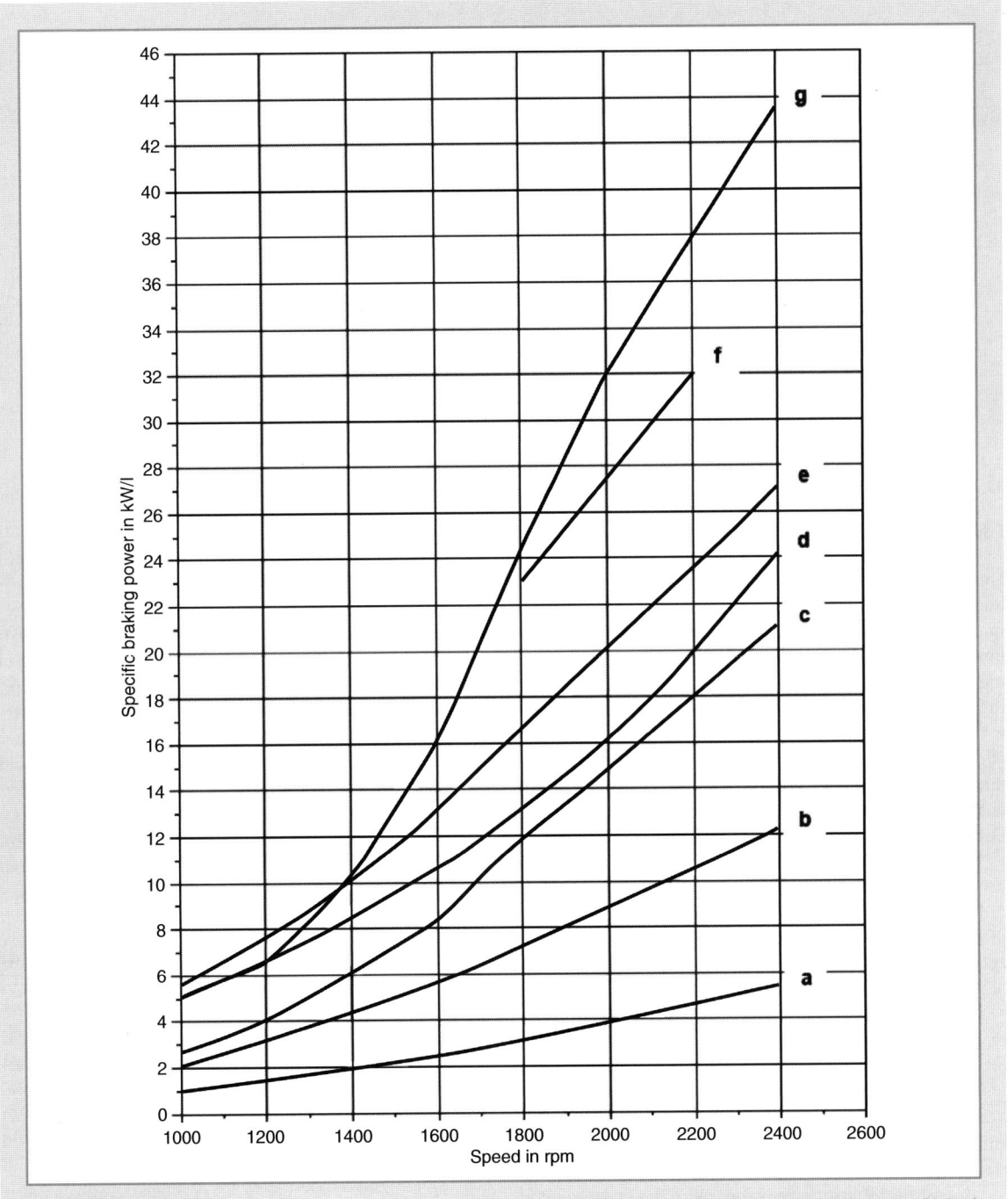

Fig. 17-30 Engine braking power curves of different engine brake designs relative to displacement: (**a**) open engine butterfly valve (drag power), I6 engine; (**b**) closed engine butterfly valve, engine as (**a**); (**c**) EVB (exhaust valve brake), engine as (**a**); (**d**) constant throttle with closed engine brake flap, V8 engine; (**e**) Jake brake, I6 engine; (**f**) intebrake, I6 engine; (**g**) as (**d**) but with turbocharging in brake operation

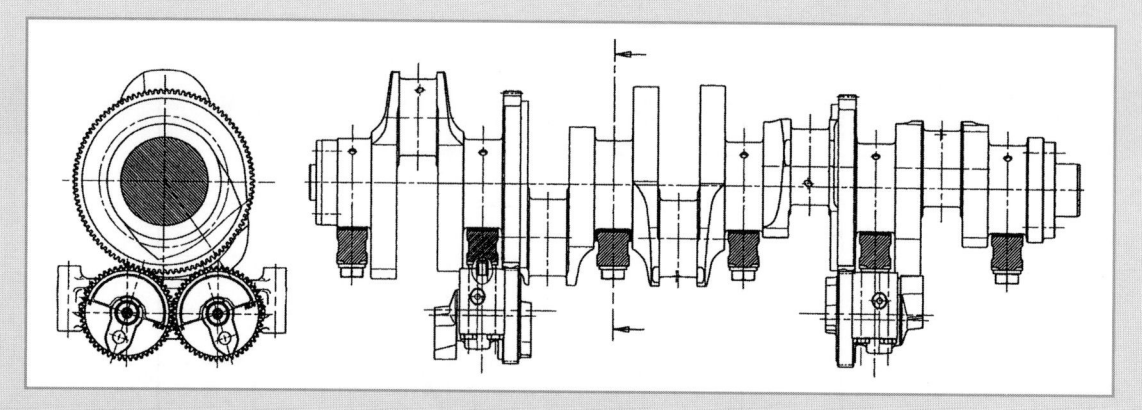

Fig. 17-31 Transmission that balances the second order moments of inertia in a five cylinder inline engine (MAN Nutzfahrzeuge AG)

17.3.2.3 Legislation

Exhaust Emissions

Undesired exhaust emissions, nitrogen oxides (NO_X), carbon monoxide (CO), hydrocarbons (HC) and particulate matter (PM) with solid and liquid constituents are legally limited (see Sect.15.2, which also describes the measurement systems and methods of determination). Unlike cars, the exhaust emissions of diesel engines in heavy duty commercial vehicles are based on the power generated. This allows for a direct correlation between engine power and the transport task. Thus, power-based emissions limits ensure that transportation performance from the light duty class up to a 40 t train is treated appropriately equally.

The most difficult task is resolving the conflict of objectives between nitrogen oxide emission and particulate emission in order to comply with effective and future limits. The influence of the fuel quality is also important (see Sect. 4.1). Hence, engine manufacturers demand the farthest reaching reduction of the content of sulfur and aromatics in fuel [17-10, 17-11].

Another conflict of objectives exists between the emission of nitrogen oxides and carbon dioxide (CO_2). A desired product of complete combustion, carbon dioxide is not a pollutant in the conventional sense but is considered a climatically relevant greenhouse gas. A reduction of nitrogen oxide emission fundamentally degrades an engine's indicated efficiency. This translates into higher fuel consumption and increased carbon dioxide emission. Here too, it is essential to prevent consumption disadvantages when reducing nitrogen oxide emission where possible by refining mixture formation and combustion.

A level has now been reached with the implementation of the Euro IV limit level, which necessitates the implementation of exhaust gas aftertreatment, particularly since the stability of emission must simultaneously be guaranteed over 7 years/500,000 km. Basically, two solutions are suitable (see Sects. 15.4 and Sects. 15.5).

A. In-engine nitrogen oxide minimization and particulate reduction by a particulate filter:
Cooled external exhaust gas recirculation (EGR) is the most effective measure to limit NOx emission in the combustion system. It increases the mass of the cylinder charge and slows the rate of reaction so that lower combustion chamber temperatures occur.

It also slows processes of soot oxidation. This is characterized by an increase in soot emission and is disadvantageous. Further increases of injection pressures (>2,000 bar) and post-injection can improve post-oxidation.

Hence, compliance with the particulate matter limit necessitates a filter. Normally, it is regenerated continuously with the aid of nitrogen dioxide NO_2 generated in a platinum coated catalyst when the exhaust gas temperatures are adequately high.

B. In-engine particulate reduction and nitrogen oxide reduction by SCR technology:
A completely different route is followed when an SCR (selective catalytic reduction) system is employed. The optimization of combustion brings particulates to the required level. A downstream catalyst uses the reactant ammonia to reduce nitrogen oxides back to atmospheric nitrogen N_2. The ammonia is produced from an aqueous urea solution with the brand name of AdBlue. Once sprayed into the exhaust flow, it hydrolyzes when temperatures are adequately high.

Noise

Mandatory limits on heavy duty commercial vehicles' noises are based on their external noise radiation (see Chap. 16). Noise emission is based on the overall vehicle, the engine being one noise source among others. An engine is normally considered the main noise source under the conditions of type approval during accelerated passing based on 70/157/EEC [17-12]. Effective damper systems are available to reduce intake pipe and tailpipe noises (see Chap. 13). The pressure loss and structural space must remain acceptable.

Controlling the noise emission of an engine itself is more difficult. Triggered by the combustion pulse, airborne noise is radiated by structure-borne noise conducted by the engine structure in conjunction with resonance phenomena from the surface of the engine and transmission. This is compounded by noises triggered by the pulses from an engine's mechanics, e.g. by bearing clearances, gearing clearances, valve clearances, etc.

"Gentle" combustion, i.e. a low combustion pressure gradient, can reduce noise during combustion [17-9, 17-10]. The introduction of electronically controlled injection systems with the option of pilot injection have made substantial advances possible without diminishing engine efficiency.

Noise-optimized crankcase structures as well as noise damping elements as a secondary measure, which can be designed to be supported by the engine as well as the vehicle, are effective in new engine developments (see Sect. 17.3.3.2. for partial and full enclosures and Sect. 16.5). Weight increases always have to be accepted and obstacles during necessary engine maintenance often do too.

17.3.3 Engine Designs Implemented for Heavy Duty Commercial Vehicles

17.3.3.1 Overview

Power Density

Emission control legislation is lastingly influencing engine development. In recent years, this has led to the equipping of all heavy duty commercial vehicles only with diesel engines with exhaust gas turbocharging and intercooling and thus high specific powers. Figure 17-32 presents different European engines for heavy duty commercial vehicles with their rated powers. Given the higher rated speeds, maximum values of around 35 kW/l are reached in the lower displacement range.

Another important variable for commercial vehicle engines is their torque relative to displacement. It corresponds to the brake mean effective pressure p_e, (see Sect. 1.2). Figure 17-32 also contains the maximum torque values for the same engines. The maximum specific torque values are approximately 200 Nm/l. This corresponds to a brake mean effective pressure of $p_e \approx 25$ bar and effective brake work of $w_e \approx 2.5$ kJ/dm^3 (see Sect. 1.2.5).

Displacement

Turbocharging and intercooling generally produce an increase in specific power in smaller engines (downsizing). Figure 17-32 provides information on the rated power and displacement values of engines in the different vehicle classes. Engines with displacements of approximately 4 to 16 liters are implemented in European heavy duty commercial vehicles. Four and six cylinder engines are common in the light duty class. Six cylinder engines are implemented almost exclusively in the standard class. Six, eight and ten cylinders are found in the heavy duty class. Their displacement is between 0.7 and 2.2 l/cylinder and the related rated speed is between 2,500 and 1,800 rpm.

Cylinder Pressures

The concentration of power has also necessitated improving the parts adjacent to the combustion chamber as well as the drive train through the connecting rod, crankshaft and flywheel. Figure 17-33 plots the development of combustion pressures since 1990. The significant impact of emission control legislation is evident, especially as of Euro 3.

Fuel Consumption

Diesel engines are implemented in all heavy duty commercial vehicles with the exception of a few special applications for buses or municipal vehicles. Thus, the lowest brake specific fuel consumption is obtained for the drives of road vehicles. The mandatory limits on nitrogen oxide emission as well as particulate emission increase fuel consumption (see Sect. 17.3.2.2). Attempts are being made to keep the adverse effects on fuel consumption to a minimum by systematically refining the combustion sequence for lower exhaust emissions. Nevertheless, values below 190 g/kWh corresponding to an overall engine efficiency of approximately 45% are being obtained in the best point.

A vehicle's fuel consumption relative to kilometrage is more meaningful to vehicle operators. However, it is hardly possible to specify standard values for this because many boundary conditions, e.g. vehicle dimensions, application profile, load, topography, wind conditions and traffic density, influence the results. Professional journals conduct driving and consumption tests with long-distance transportation vehicles usually from the heavy duty class on unvarying routes, which can serve operators as orientation aids (cf. [17-13]).

17.3.3.2 Selected Designs

Combustion Systems, Gas Exchange

Without exception, engines for heavy duty commercial vehicles have direct injection that distributes air (see Sect. 3.1). Support from a concentrated swirl in the combustion chamber helps improve mixture formation and combustion. The amount depends on such boundary conditions as the injection system's pressure potential, the exhaust gas recirculation rate, the supercharging rate, etc. The vortex is built up in a swirl-generating port located before the inlet valves.

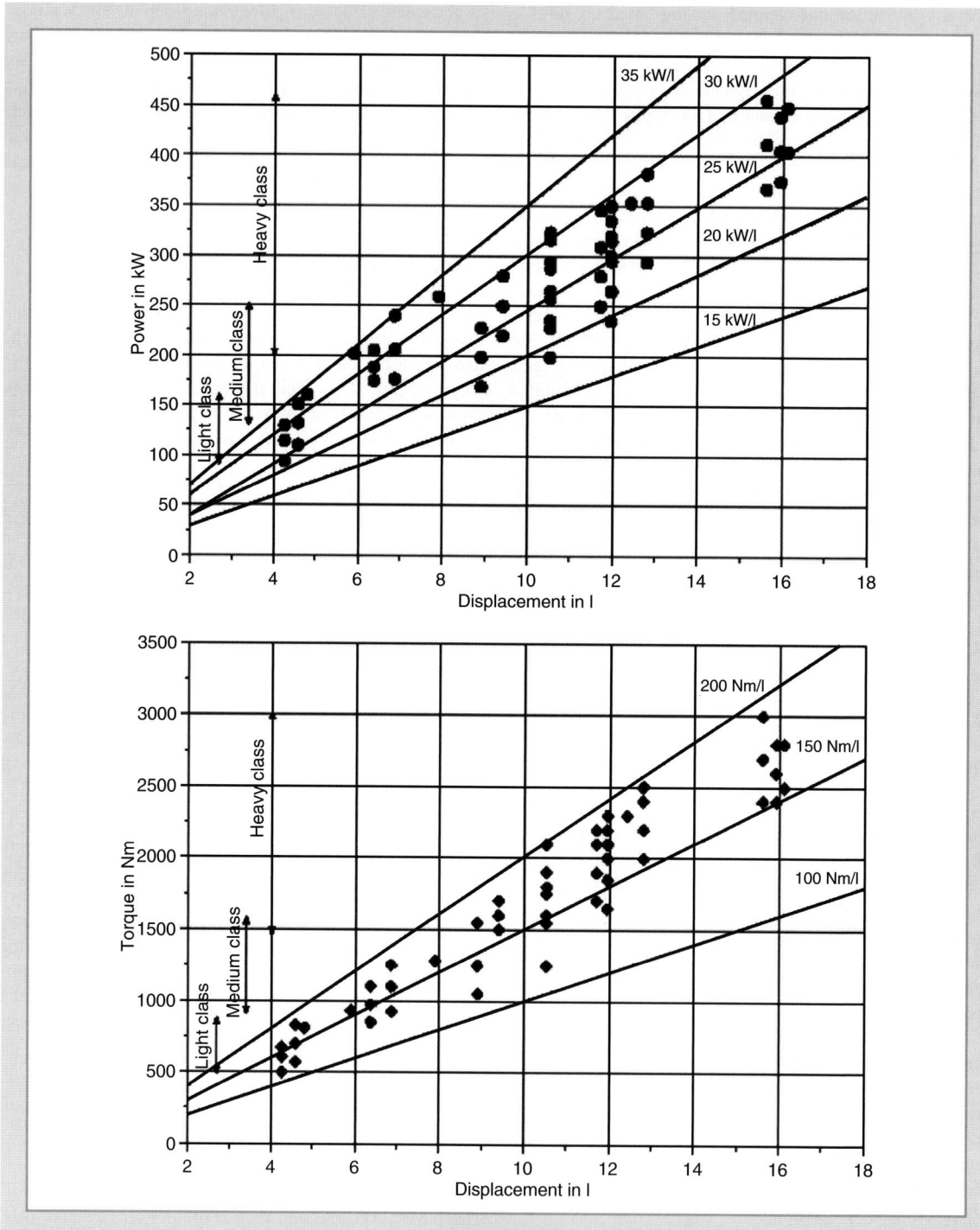

Fig. 17-32 Power and torque of commercial vehicle engines with turbocharging and intercooling as a function of displacement

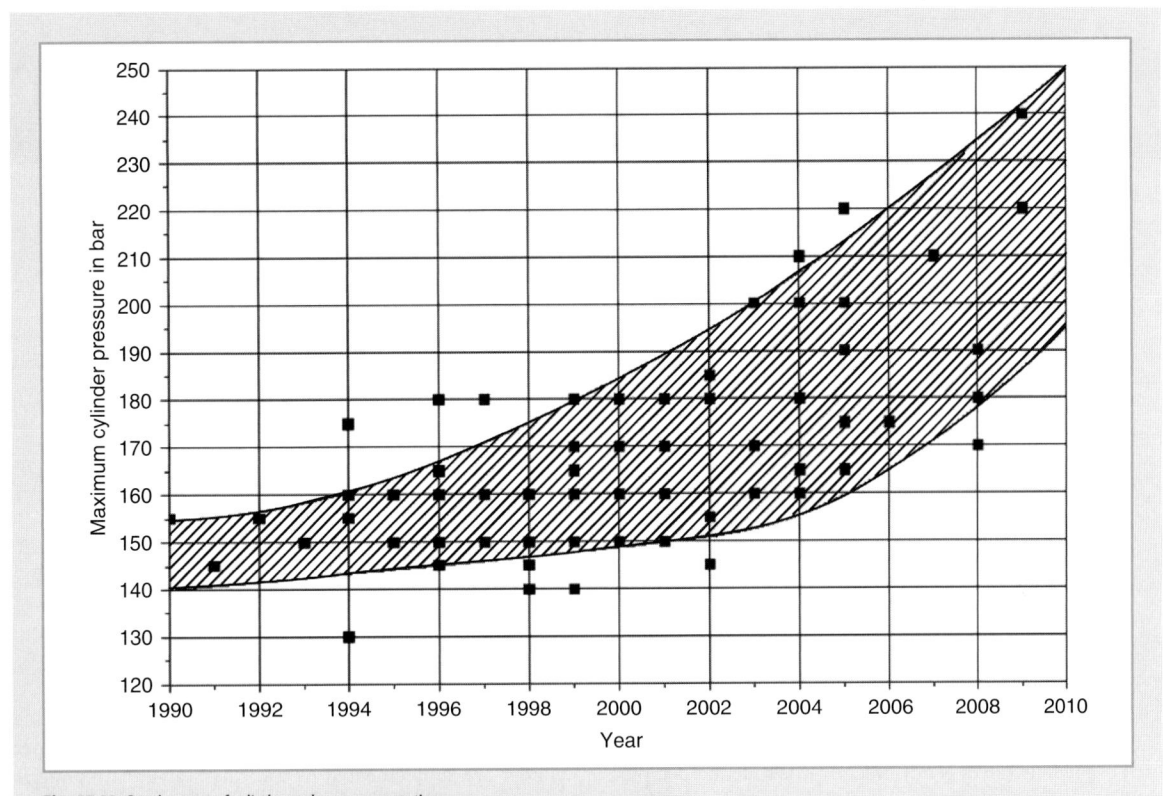

Fig. 17-33 Development of cylinder peak pressure over time

Normally, four valves, i.e. two intake and two exhaust valves apiece, control the gas exchange. Along with enlarging the flow cross sections, this development also brings advantages for combustion because the injection nozzle can be centered in the combustion chamber.

The engine in Fig. 17-34 embodies a variant with an increased swirl requirement. The inline six cylinder engine with a cylinder bore of 108 mm and a stroke of 125 mm corresponding to a displacement of 6.9 l delivers 240 kW at 2,300 rpm. It is provided with externally cooled exhaust gas recirculation (EGR), which alleviates the conflict of objectives between pollutant emission and fuel consumption (see Sect. 17.3.2.2). In conjunction with two-stage supercharging and intercooling, it complies with Euro 4 limit levels for NO_X through in-engine measures. Like virtually all engines in this class, it has two intake and exhaust valves per cylinder. The camshaft is mounted in the cylinder block.

Figure 17-35 [17-14] pictures a representative of low-swirl engines with more energy of mixture formation distributed by fuel. The first representative of a new engine design with a displacement of V_H = 7.8 l, it has also been followed by models with displacements of 10.5 and 13 l with identical design features, [17-15, 17-16]. The inline six cylinder engine has a bore of 115 mm with a stroke of 125 mm and delivers 257 kW at 2,500 rpm. The cylinder head contains two intake and exhaust valves per cylinder. The centers of the intake valves are each aligned diagonally to the crankshaft axis. This applies analogously to the exhaust valves. Among other things, the geometry of the combustion chamber supports mixture formation with turbulences produced at the end of the compression stroke.

Fuel injection contributes the majority of the energy for mechanical mixture formation. A unit injector mounted in the cylinder axis ensures that the fuel sprays are uniformly distributed onto the centric combustion chamber. The maximum injection pressure of 1,900 bar dissolves the fuel into very small droplets. This design uses the slower vortex of the charge to evaporate the fuel and mix it with the air. An injection cam on the camshaft located in the cylinder head and driven by gears drives the unit injector's piston by a rocker arm. A solenoid valve in the unit injector controls the start of injection and the injected fuel quantity. An electronic unit regulates the overall system, which also includes variable turbine geometry VTG in the most powerful design.

Figure 17-36 pictures the cross section of a turbocharged and intercooled V8 engine with 420 kW at 1,800 rpm [17-17].

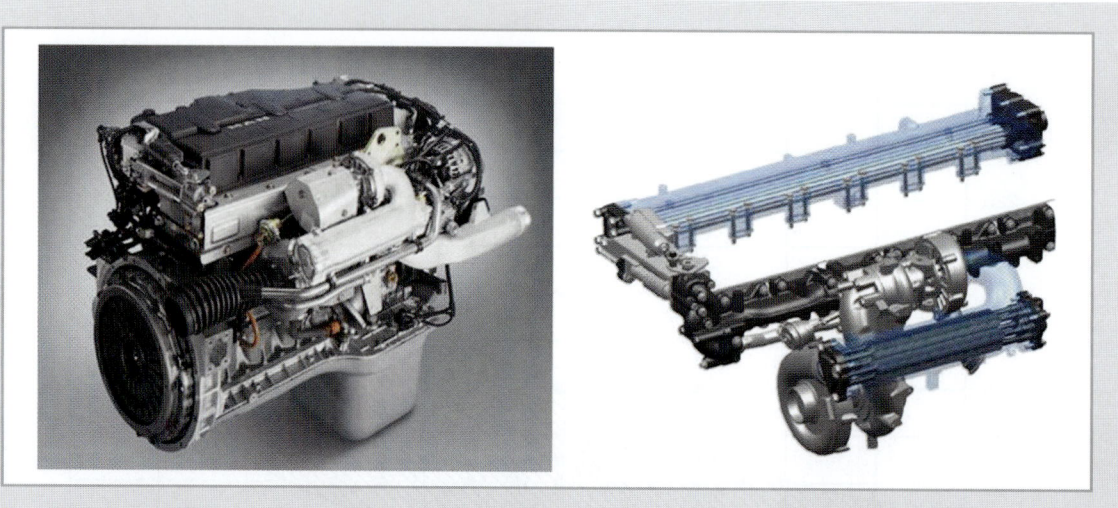

Fig. 17-34 Commercial vehicle diesel engine with controlled exhaust gas recirculation, two-stage supercharging and intercooling and a common rail injection system (MAN D08, MAN Nutzfahrzeuge AG)

The cylinder bore has a diameter of 130 mm and the stroke is 150 mm. A V6 variant is also available. A camshaft actuates each of the four valves per cylinder by one roller tappet, push rod, rocker arm and valve connection per intake and exhaust valve pair.

The camshaft is supported by bearings in the crankcase centered between the two cylinder banks. It is driven directly by the crankshaft on the flywheel end without an intermediate gear. Apart from the cams that actuate the valve, the camshaft supports one more cam per cylinder, which drives the solenoid controlled unit pump for fuel injection implemented in the crankcase (see Sect. 5.3). An electronic unit regulates the injected fuel quantity and start of injection.

Engine Brake

One simple and frequently applied design is a throttle valve in the exhaust manifold, which attains high braking power through the back pressure in the piston's expulsion phase (see Fig. 17-37). However, the back pressure and thus the braking power achievable must be limited to prevent component damage.

Higher braking powers are achieved when not only the piston's expulsion stroke but also the compression stroke absorbs lost work. This is achieved by purging the compressed air toward the end of the compression stroke. Systems that operate with controlled valves are the Jake (Jacobs) brake [17-18], constant throttle [17-19], EVB (exhaust valve brake) [17-20] and intebrake [17-21] (see Sect. 17.3.3.1). Additionally utilizing a VTG turbocharger, turbobrakes further boost braking power (see Sect. 2.2.5.2).

Recognizable in the cylinder bank to the left of the V engine cross section in Fig. 17-36, the four gas exchange valves per cylinder are supplemented by a further valve that connects the compression chamber with the outlet port. This valve is closed in normal operation. In combination with the throttle valve in the exhaust line, it serves to increase engine braking power. Actuated by compressed air, it is constantly opened by a defined gap. The additional negative work is produced in the compression stroke by purging the air through this throttle valve, the so-called constant throttle [17-19] (see Sect. 17.3.2.1).

The exhaust valve in the EVB (exhaust valve brake) engine brake system [17-20] assumes the function of the relief valve (Fig. 17-37). Figure 17-38 illustrates the intervention required in the valve gear in the four-stroke sequence: The throttle valve in the exhaust system is closed to activate the engine brake. Shaped by the pressure wave of an expelling neighboring cylinder, the exhaust pressure flings open the exhaust valve at the end of the intake stroke. An oil-hydro lock in the rocker arm prevents the exhaust valve from closing completely. Part of the compressed air is discharged through the remaining defined gap in the valve seat (1.5 ... 2 mm) during the compression stroke and thus increases the braking power (see Sect. 17.3.2.1). The cam-actuated opening of the exhaust valve beginning at the end of the expansion stroke deactivates the oil-hydro lock.

Controlled opening of the exhaust valve at the end of the compression stroke is another option to blow off a compressed charge and thus increase engine braking power. The Jake brake [17-18] draws on the mechanical actuation of the unit injector in engines with unit injector injection systems to

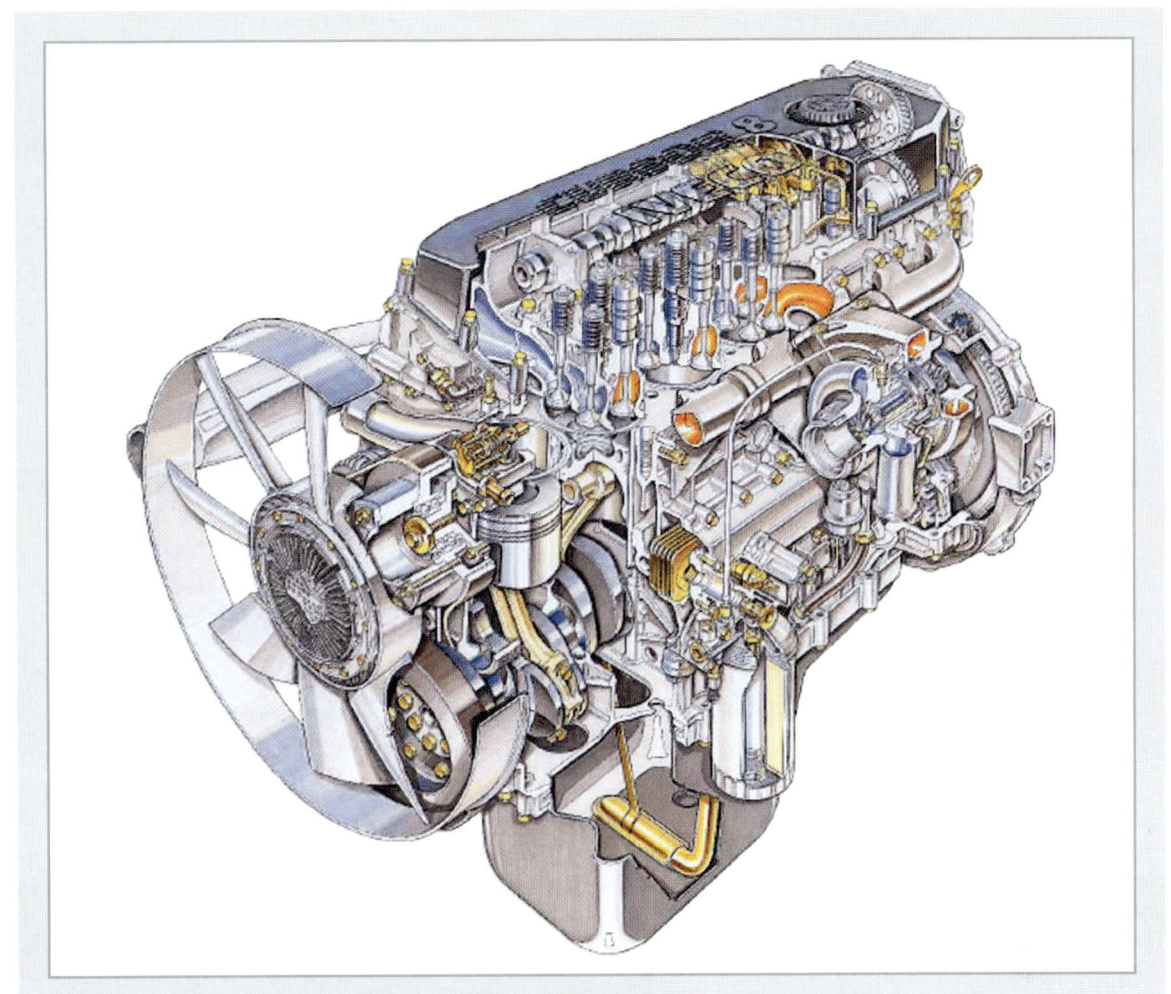

Fig. 17-35 Commercial vehicle diesel engine with a camshaft and unit injection system mounted in the cylinder head (Iveco)

open the exhaust valve in compression TDC with the aid of an additional hydraulic unit.

The intebrake benefits from two overhead camshafts that facilitate higher braking power. The one solely actuates the unit injectors, while the other controls the valves. One additional cam per cylinder, which uses a switchable roller rocker arm to open the exhaust valve to blow off the compressed air, is mounted on the shaft. Thus, the functions of engine braking and fuel injection are independent of one another and optimal valve timing and cam profiles can be produced for both (see Sect. 17.3.2.1).

Crankcase and Crankshaft Assembly

Figure 17-39 pictures the cross section of an inline six cylinder engine with a power of 321 kW. This model has a cylinder diameter of 120 mm and a stroke of 155 mm. For the first time in this engine class, the crankcase was cast of vermicular graphite cast iron EN-GJV-450 and designed for high loads. This material's high strength provides options for compact and lightweight engine design. The crankcase's deck and the cylinder head's base plate are designed very rigidly since the coolant and engine oil overflow through external add-on parts rather than the cylinder head seal. Also a first in a commercial vehicle engine, the cracked main bearing caps are each secured in the crankcase with two high strength thread rolling bolts. When assembled in the correct position, cracked components return exactly to their correct original position. Their rough surface facilitates better absorption of radial loads.

The piston consists of a cast aluminum alloy. A Ni-resist ring carrier with an Alfin bond is cast in for the first piston

Fig. 17-36 Cross section of a V eight cylinder diesel engine with pump-line-nozzle injection system and constant throttle (Daimler Chrysler AG)

ring. The three ring grooves are equipped with a chrome plated double keystone ring, a taper faced ring and a chrome plated double-beveled spiral expander ring as the oil scraper. The piston pin is secured axially by circlips.

The connecting rod is precision drop forged from quenched and tempered steel without weight compensation ribs and divided obliquely by cracking the bearing cap. The bearing cap and rod are secured to one another by the surface macrostructure generated during cracking and bolted with high strength Torx screws. The connecting rod bearings are designed for maximum loads and service life. The rod end shell consists of wear resistant sputter bearing metal. Oil spray from piston cooling is supplied to the connecting rod small end with a press fit bushing.

The crankshaft is drop forged from highly tempered micro alloy steel. Eight forged-on counterweights serve to balance the inertial forces. The crankshaft was designed for high bending and torsional rigidity with the aid of FEM simulation. A viscous torsional vibration damper with cooling fins that effectively dissipate heat is mounted on the front end of the shaft to damp the crankshaft's torsional vibrations. Pre-assembled three material bearings are employed as the main bearing. It is supported axially by half thrust washers inserted in the crankcase.

As Fig. 17-39 indicates, the high pressure common rail injection system (see Sect. 5.3.5) in diesel engines was also developed for heavy duty commercial vehicles. A three-plunger high pressure pump lubricated with fuel is employed as the pressure generator.

Oil additives eliminate the risk of contaminating the exhaust gas aftertreatment system. The fuel pressure desired in the rail is flexibly adjustable in the entire map and is also applied through short lines to the injectors controlled by solenoid valves. It is still limited to 1,600 bar in the second generation system employed here. Further increases are in development.

The injector is perpendicular to and centered over the combustion chamber. A seven-hole nozzle with sprays distributed uniformly throughout the combustion chamber injects the fuel. Depending on the combustion cycle, there is a multiple injection with pilot, main and post-injection. The engine electronics controls their characteristics, i.e. rail pressure, start of injection and injection time, based on stored maps. Thus, a combustion sequence can be obtained, which has a beneficial effect on exhaust emissions, noise emissions and fuel consumption. The combustion pressure in the cylinder reaches a maximum of 200 bar.

17.3.4 Outlook

Functional utility and attendant costs will be the crucial competitive criteria for diesel engines in heavy duty commercial

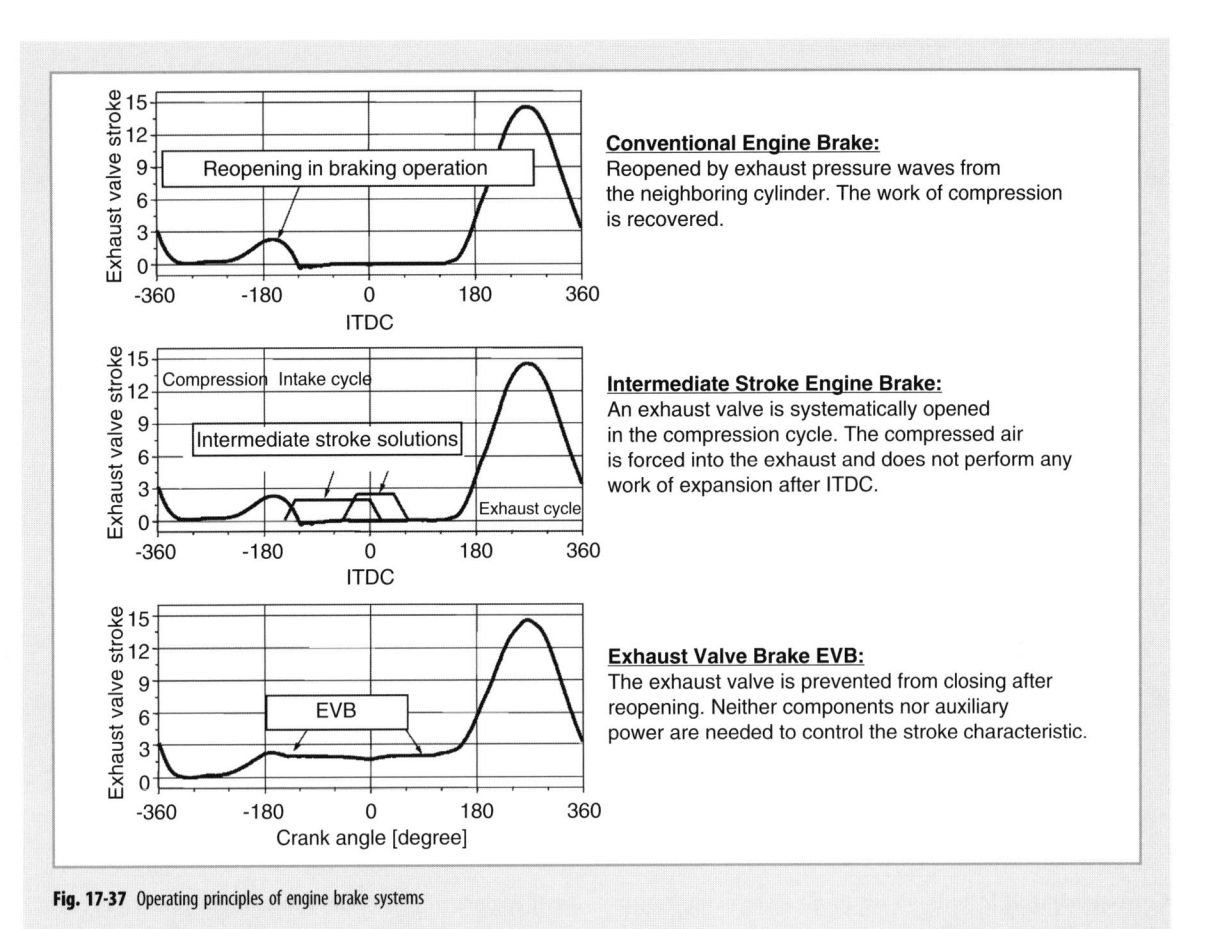

Fig. 17-37 Operating principles of engine brake systems

Conventional Engine Brake:
Reopened by exhaust pressure waves from the neighboring cylinder. The work of compression is recovered.

Intermediate Stroke Engine Brake:
An exhaust valve is systematically opened in the compression cycle. The compressed air is forced into the exhaust and does not perform any work of expansion after ITDC.

Exhaust Valve Brake EVB:
The exhaust valve is prevented from closing after reopening. Neither components nor auxiliary power are needed to control the stroke characteristic.

vehicles in the future too. The legal standards for environmental compatibility will continue to influence development considerably. As a result, in-engine measures alone have already no longer been enough to comply with limit levels for nitrogen oxide emissions and particulate matter since 2005. Efficient and serviceable exhaust gas aftertreatment systems must be developed to market maturity. The commercial vehicle industry has opted for two different strategies, each of which is laden with particular advantages and disadvantages.

Manufacturers of EGR engines monopolize low manufacturing costs, structural space advantages, low weight or independence from an AdBlue infrastructure; SCR engine manufacturers take advantage of lower fuel costs or user benefits from state subsidies for precompliance with future emission standards. Regardless of the path chosen for Euro IV and V, the prediction has been ventured that the Euro VI level, which is still being debated, will require the simultaneous installation of both technologies in the same engine as of around 2012. Individual companies' capability to develop potentials to minimize raw exhaust emissions will lastingly influence the work required for exhaust gas aftertreatment systems and thus their costs. In the interest of customers, functional utility in the form of service life, required maintenance, etc. should not be adversely affected.

At any rate, exhaust gas aftertreatment systems will require diesel fuel with a sulfur content lowered below 10 ppm and an aromatic content lowered to an economically justifiable extent. Longer range, significant contributions can be expected from blends of synthetic fuels such a CTL, GTL or BTL (see Sect. 4.2).

Further increases in specific engine power may also be expected in the future despite the challenges from emission control legislation.

This will necessitate increasing engine components' peak mechanical strength to significantly more than 200 bar on the basis of higher EGR rates and two-stage supercharging with intercooling as well as further developing vehicle cooling systems. A significant push toward larger and thus more powerful engines could entail an increase of the gross vehicle weight rating in normal road use.

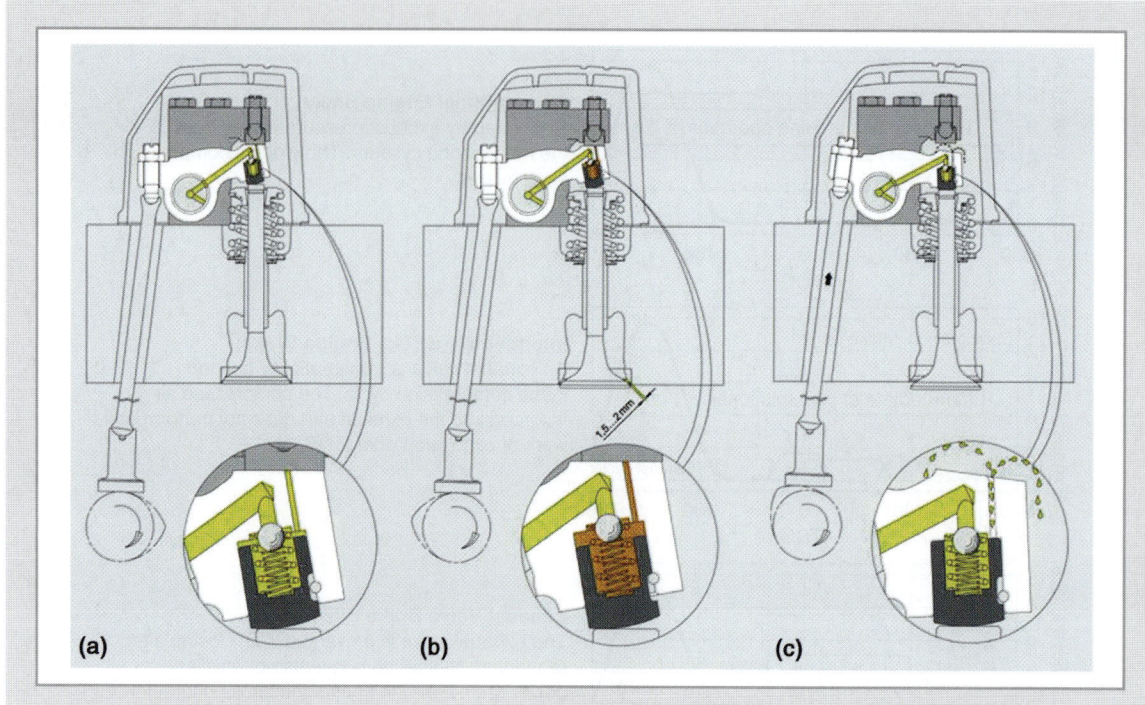

Fig. 17-38 Operating principle of the EVB (exhaust valve brake): (**a**) intake stroke: locking piston abuts valve; (**b**) compression and expansion stroke: an exhaust system pressure wave flips open the valve, locking piston maintains defined valve gap (1.5 ... 2 mm), air blows off; (**c**) exhaust stroke: cam lift deactivates lock (MAN Nutzfahrzeuge AG)

Corresponding field tests on up to 60 t are being run in various European regions.

Whether a power turbine downstream from the supercharger turbine in the exhaust gas flow will establish itself remains to be seen. The outcome will depend on the development of the ratio of complexity to fuel savings for such turbo compound systems (see Sect. 14.1).

In the medium-term, the engine power of long-distance transportation trucks with a gross vehicle weight rating of 40 t will increase to approximately 500 kW. This power will be needed to be able to drive highway downgrades of 4% at the permissible speed of 80 km/h.

17.4 High Speed High Performance Diesel Engines

17.4.1 Classification of High Speed High Performance Diesel Engines

High speed high performance diesel engines (HHD) are generally understood to be diesel engines with high power density. It is impossible to draw clear boundaries between HHDs and other diesel engines since the transitions are fluid. HHDs normally have design speeds above 1,000 rpm. Another distinguishing feature that sets HHDs apart from medium speed diesel engines in particular is their mean piston velocity c_m, which is significantly higher than 10 m/s. A significant parameter for high performance, the specific work w_e (corresponding to the brake mean effective pressure p_e, see Sect. 1.2) in HHDs is more than 2 kJ/dm^3 in the rated power point. The specific power per unit piston area is also characteristic for high performance diesel engines. Its correlation to the bore D is presented for implemented HHDs in Fig. 17-40. Accordingly, the specific power per unit piston area averages values of 5 W/mm^2. The maximum is approximately 10 W/mm^2.

Figure 17-41 presents the mass m_P of implemented high performance diesel engines relative to their power, i.e. the weight-to-power ratio. HHDs are designed with V configurations to produce good weight-to-power ratios.

In principle, at constant specific power per unit piston area P_A, the weight-to-power ratio increases as the bore increases. This corresponds to a general law (see Sect. 17.4.2). When the following are defined as boundary conditions for the high speed capacity of HHDs

$$c_m \leq 13 \text{m/s}, n \geq 1,000 \text{ rpm}$$

then the cylinder volume V_h of high speed high performance diesel engines is given an upper limit by applying

$$s = c_m/2n$$

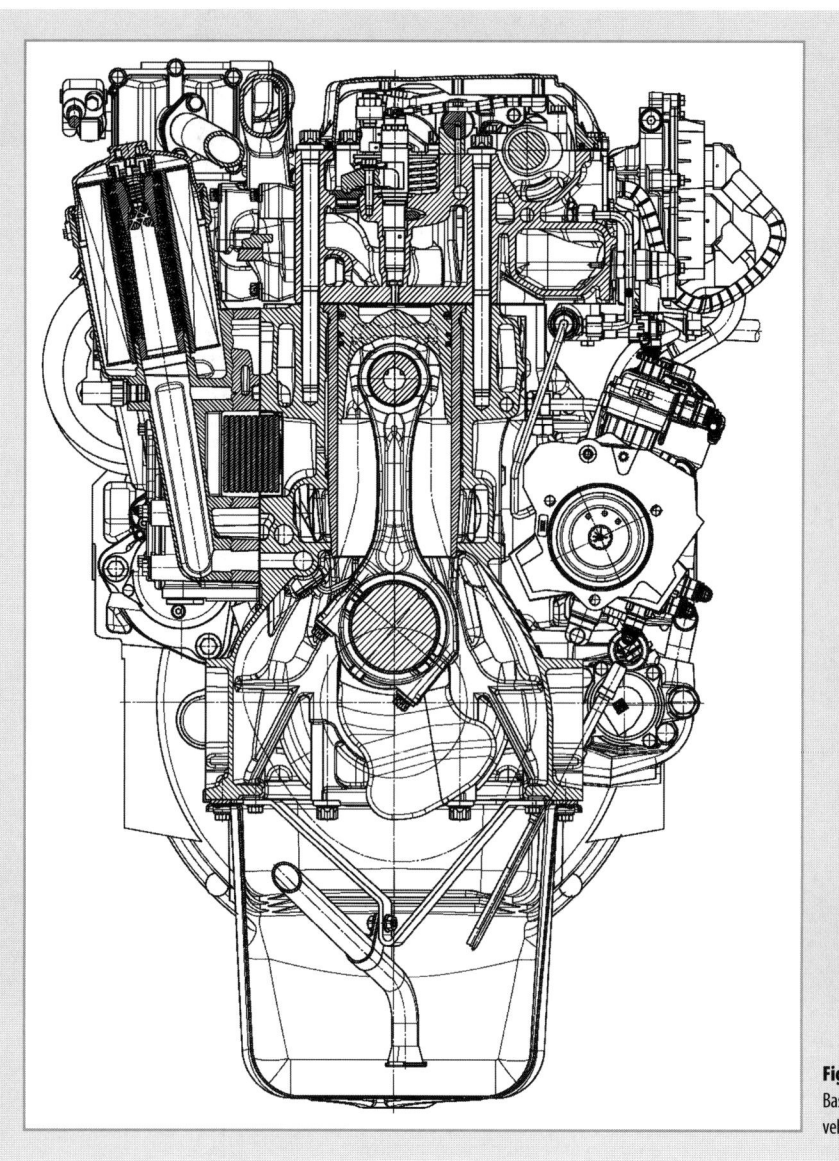

Fig. 17-39
Basic design features of a modern commercial vehicle diesel engine (MAN Nutzfahrzeuge AG)

so that the following ensues for the stroke:

$s < 390$ mm.

The stroke/bore ratio for an effective combustion system should be

$s/D \geq 1.25$

so that the maximum possible bore for a high speed high performance diesel engine can be calculated from it:

$D \leq 300 \, mm$.

When the specific power per unit piston area of $P_A = 10$ W/mm^2 is selected according to Fig. 17-40, then the maximum possible cylinder output is calculated as

$P_Z \approx 700$ kW

with a cylinder displacement of approximately 27 dm^3.

However, this limit range has not yet been ventured into since HHDs of this size do not generate any customer benefits, as explained in more detail below. MTU's 8000 series constitutes the upper limit for high speed high performance engines with $P_e = 9{,}100$ kW (455 kW/cylinder) (Fig. 17-42).

High speed high performance diesel engines are implemented wherever a low weight-to-power ratio and/or low overall volume produce benefits for the overall systems.

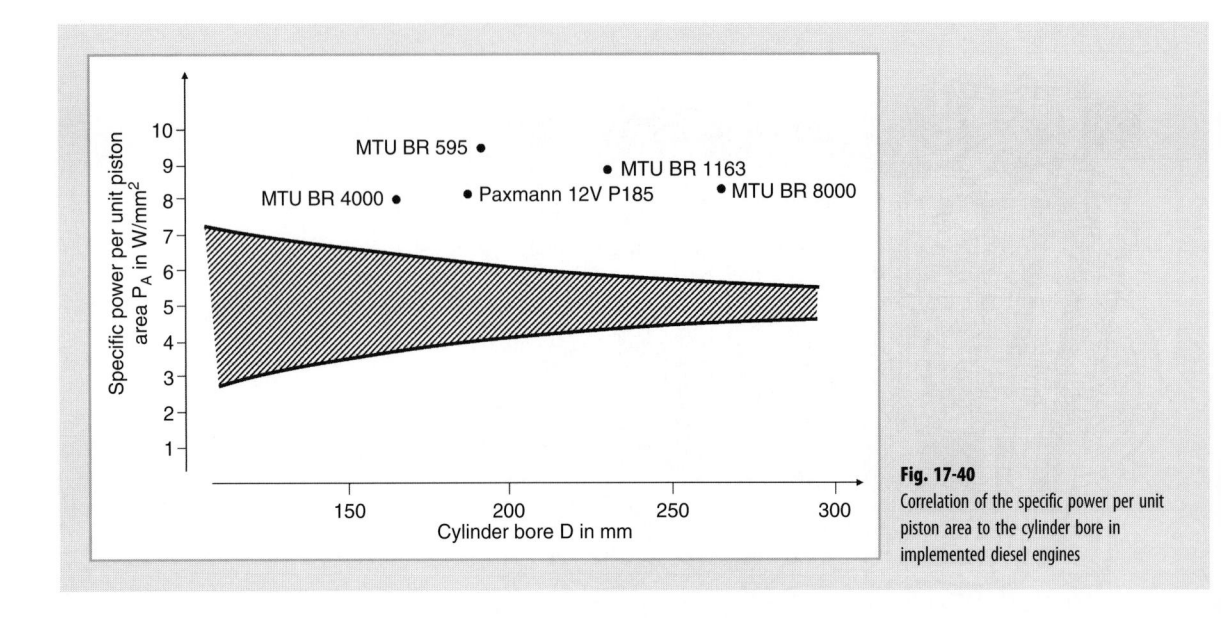

Fig. 17-40
Correlation of the specific power per unit piston area to the cylinder bore in implemented diesel engines

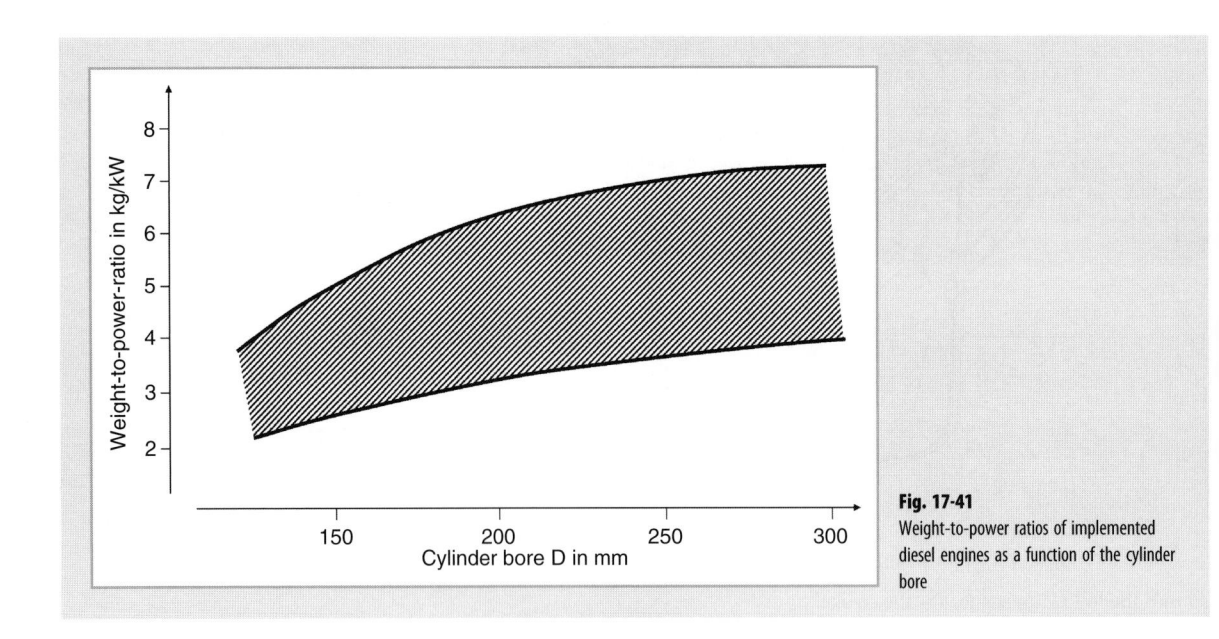

Fig. 17-41
Weight-to-power ratios of implemented diesel engines as a function of the cylinder bore

This may include:
- fast ships,
- power units with special requirements (transportability),
- locomotives and railcars and
- large dump trucks.

HHDs are especially important for fast ships since, much like in aircraft, the payload crucially depends on the weight of the propulsion system. Without intending to treat ship systems in more detail (for more, see [17-22–17-24]), the requirement for a low weight-to-power ratio fundamentally increases as a ship's displacement decreases and its speed increases and vice versa. Thus, as long as extreme demands are not made on speed at the same time, the propulsion system's weight-to-power ratio loses importance as the ship's displacement increases. Exact limits on the use of HHDs in the upper power range are indefinable since they largely depend on the requirements for the drive system. One indicator is the steady decrease of importance of HHDs for the power range between 5,000 and 10,000 kW and speed in the range of ≤30

Fig. 17-42 MTU 20 V 8000 M91, 9,100 kW at 1,150 rpm, bore 265 mm, stroke 315 mm, specific work 2.73 kJ/dm³

kn (kn: knots, i.e. sea miles per hour, 30 kn corresponding to ≈ 55 km/h) and the steady increase of importance of medium speed diesel engines with good weight-to-power ratio ($m_P \leq$ 6 kg/kW). Only gas turbines in conjunction with large marine units are able to supply the required power for high speed requirements.

Low cost derivatives of commercial vehicle diesel engines normally cover the power range up to 1,500 kW. Specially developed HHDs are found above 1,500 kW. Sections 17.4.2 and 17.4.3 specifically treat their design and engineering.

17.4.2 Representing Higher Diesel Engine Power

Based on the common conditional equations for an internal combustion engine's power (see Sect. 1.2), the following correlation for the typical field of application of HHDs, i.e. marine engines, is expediently taken as the starting point because the speed is freely selectable in wide ranges:

$$P_e \sim z \cdot w_e \cdot c_m \cdot D^2 . \quad (17\text{-}1)$$

The number of cylinders z is crucially important for the weight-to-power ratio. The influence of the number of cylinders on the weight-to-power ratio at constant mean piston velocity c_m, constant specific work w_e and constant engine power P_e requires analysis.

In the first approximation, the following applies to the mass of a reciprocating piston engine:

$$m \sim D^3 . \quad (17\text{-}2)$$

Thus, incorporating Eq. (17-1), the following ensues for the engine mass relative to power, i.e. the weight-to-power ratio:

$$m_P = m/P_e \sim D/(z \cdot w_e \cdot c_m) \sim D/(z \cdot P_A). \quad (17\text{-}3)$$

Taking Eq. (17-1) into account, the bore diameter D decreases for P_e, w_e and c_m = const. as the number of cylinders increases. Thus, at otherwise constant boundary conditions, the weight-to-power ratio drops disproportionately at the same time the number of cylinders z increases. Figure 17-43 plots the influence of the number of cylinders on the weight-to-power ratio for constant power engines with a V configuration.

Not only high values for the specific power per unit piston area P_A but also high numbers of cylinders are desirable in HHDs. Twelve, sixteen and twenty cylinders in a V configuration are selected for an HHD series since one series is normally required to cover one power range. A V20 engine is no longer considered for smaller engines with cylinder displacements below 3 dm³ for reasons of crankcase stiffness. Engines with upwards of twelve cylinders in a V configuration can be designed for every V angle so that masses are completely balanceable without additional first or second order mass balancing.

A product of partial efficiencies, a diesel engine's net efficiency (see Sect. 1.2) is essentially influenced by the ratio of maximum cylinder pressure (peak pressure) to specific work p_{Zmax}/w_e, combustion based on quality and time sequence, supercharging efficiency (see Sect. 2.2) and mechanical losses. Neither p_{Zmax}/w_e nor supercharging efficiency is influenced by the number of cylinders.

A comparison of engines with equal power but different numbers of cylinders according to Eq. (17-1) reveals that, at otherwise identical boundary conditions, the engineered cylinder dimensions decrease as the number of cylinders increases. Thus, the efficiency factor and indicated efficiency drop because the wall heat loss increases as a result of the enlarged surface-volume ratios and the gas exchange losses.

However, this measure causes fewer efficiency losses than any other measure that reduces the weight-to-power ratio.

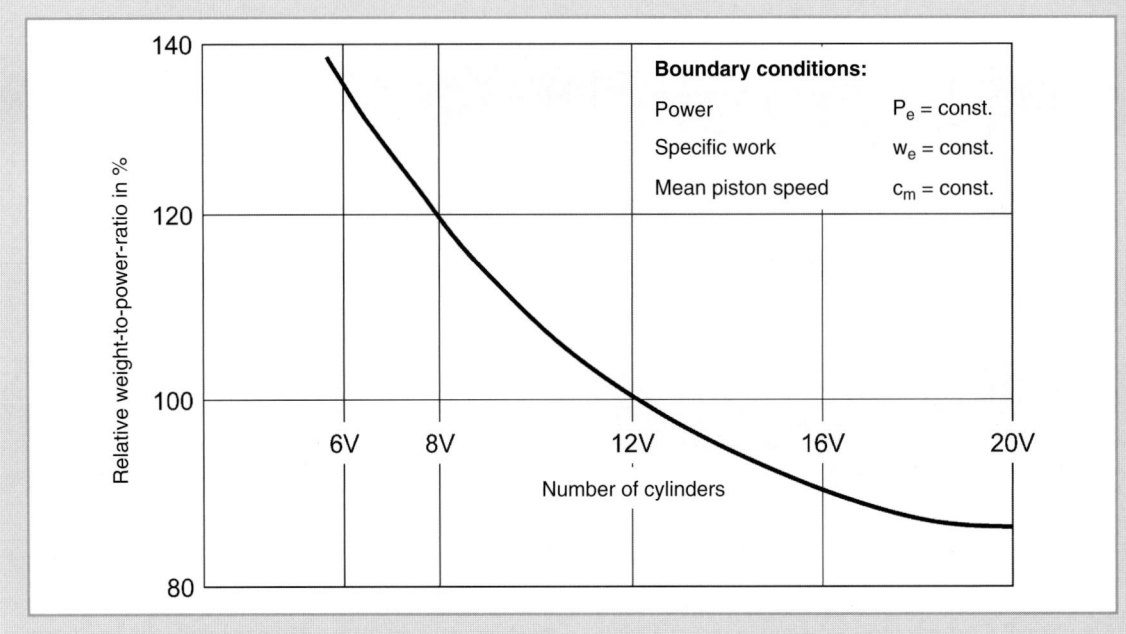

Fig. 17-43 Relative weight-to-power ratio as a function of the number of cylinders for V engines

Since engine power increases proportionally with specific work, engine mass relative to power decreases when dimensions are identical. Viewed relatively, mechanical losses decrease at the same time. Thus, the determination of the upper limit of the specific work is of interest for engine design.

The boost pressure achievable for HHDs with single-stage supercharging is approximately 4.5 bar absolute.

While higher boost pressures are possible in principle, the narrowing compressor map is not enough to produce the engine map necessary for marine engines for instance (see Sect. 17.4.3). 4.5 bar of boost pressure makes a value of up to 2.8 kJ/dm³ achievable for the specific work w_e. Above this value, engines have to be supercharged in multiple stages. This yields the maximum possible compression ratios as a function of permissible peak pressure and the possible injection time. Higher peak pressure also allows high compression ratios. At a constant injection pressure, the start of injection can be shifted later by shortening the duration of injection. In turn, higher compression ratios are possible at a specified peak pressure.

Since a minimum compression ratio is necessary for starting and idle operation free of white smoke, the permissible boost pressure has an upper limit for a specified engine design and thus, in turn, the maximum potential specific work w_e is also below the specified boundary conditions (η_{emax}, λ_{Vmin}). The minimum required compression ratio is 14–15 for small engines ($V_h < 3$ dm³) and 13–14 for larger engines. In sum,
- high peak pressure,
- high efficiency,
- low air/fuel ratio and
- short injection time

facilitate high boost pressures and thus high specific work.

The peak pressure greatly influences both engine mass and engine efficiency. A conflict of objectives exists, which must be carefully weighed.

Figure 17-44 plots the influence of the specific work on a diesel engine's weight-to-power ratio under different boundary conditions. The cylinder output P_Z and the mean piston velocity are constant for both curves. The positioning of the power must first be defined when the engine is designed. Each of the curves specifies the weight-to-power ratios expected for a certain design principle as a function of the specific work w_e, i.e. the bore also varies with w_e along the curves with P_Z = const.

In addition, the ratio p_{Zmax}/w_e is constant for the solid curves, i.e., efficiency and thus brake specific fuel consumption are constant in the first approximation. p_{Zmax} = const. applies to the dashed curve. As specific work increases respective to brake mean effective pressure, the ratio p_{Zmax}/w_e and thus efficiency decrease.

The points of instability at $w_e = 2.8$ kJ/dm³ represent the transition from single-stage to two-stage supercharging. The engine's mass at the components of a concrete engineering concept was calculated by similarity analyses. The potential for reducing the weight-to-power ratio decreases when the specific work is higher. It is necessary to increase the specific work in order to compensate for the additional complexity of

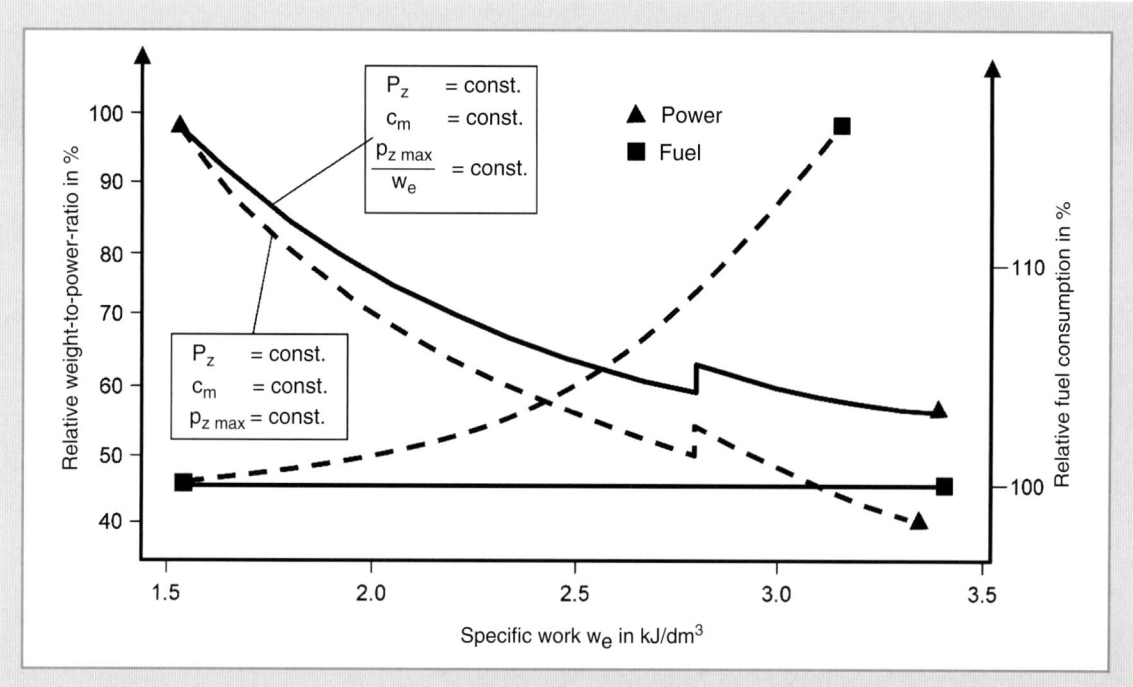

Fig. 17-44 Relative weight-to-power ratio and relative fuel consumption of a diesel engine as a function of the specific work

two-stage supercharging. High specific work only generates perceptible advantages for the weight-to-power ratio when the peak pressure in the cylinder and thus the necessary component strength is limited and concessions to efficiency are possible.

The bore is another parameter that crucially influences a diesel engine's power. The bore is limited geometrically by the distance between cylinder bore center lines on the one hand and, in combination with the desired peak pressure, by the permissible bearing load on the other hand.

Figure 17-45 presents the cylinder bore D as a function of specific work w_e assuming that both the distance between cylinder bore center lines x and the mean piston velocity c_m are constant in the entire graph. At a constant distance between cylinder bore center lines x, the bore has an upper limit that conforms to $D \leq x/1.3$ for reasons of geometry. At a constant distance between cylinder bore center lines, every bore is assigned a maximum combustion pressure that conforms to $D^2 \cdot p_{Zmax}$ = const. The maximum peak pressure has an upper limit of 220 bar. This is a variable limit, which shifts upwardly depending on technical progress. Furthermore, lines of constant cylinder output P_Z and lines with p_{Zmax}/w_e = const. are plotted. Both proceed parallel since power and engine load are both a quadratic function of the bore. In addition, an approximately equal net efficiency can be assumed for lines p_{Zmax}/w_e = const.

The field is limited by a minimum potential $p_{Zmax}/w_e \geq 50$. Problems with starting and part load may be expected below this value. $w_e = 2.8$ kJ/dm³ is the limit of single-stage supercharging, $w_e = 4.0$ kJ/dm³ of two-stage supercharging. Under the given boundary conditions, it would only be possible to increase specific work to values above 4.0 kJ/dm³ by increasing the peak pressure or reducing the permissible ratio p_{Zmax}/w_e. When lines with constant power run through the range of both single-stage and two-stage supercharging, i.e. at high ratios p_{Zmax}/w_e, then high specific work when bores are small and low specific work when bores are larger are equivalent means to produce the desired power. When good fuel consumption is desired, i.e. $p_{Zmax}/w_e \geq 70$, the permissible power can be produced with single-stage supercharging.

Boosting power by increasing the specific work w_e at a given peak pressure p_{Zmax} worsens brake specific fuel consumption because the ratio p_{Zmax}/w_e decreases.

The power for the range of $50 \leq p_{Zmax}/w_e \leq 60$ can only be produced with two-stage supercharging.

Advances in basic engine loading capacity, i.e. increases in peak pressure at a constant distance between cylinder bore center lines, can be expected to make values for p_{Zmax}/w_e above 60 possible with two-stage supercharging and a geometrically maximum bore.

Hence, when the bore cannot be enlarged for reasons of geometry and the limits of single-stage supercharging have

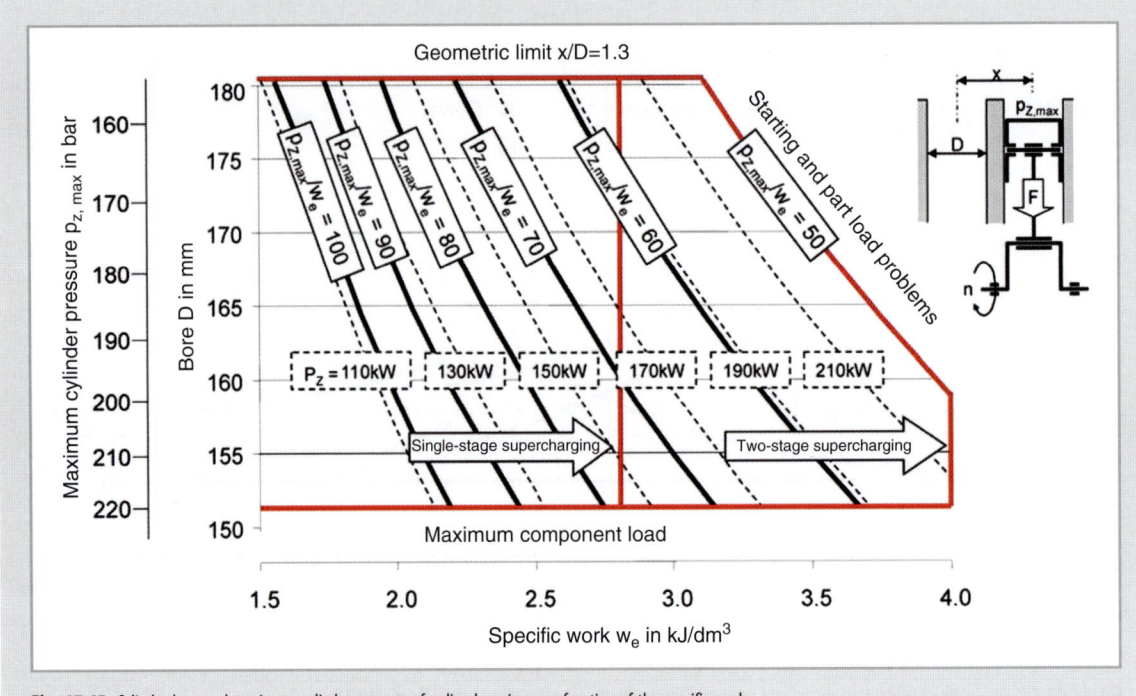

Fig. 17-45 Cylinder bore and maximum cylinder pressure of a diesel engine as a function of the specific work

been reached, two-stage supercharging should be applied in new designs aiming for extreme power density. Once the potential of single-stage supercharging has been exhausted, two-stage supercharging can help boost power even for existing models.

Two-stage supercharging with intercooling can achieve higher supercharging efficiencies than single-stage supercharging. Thus, brake specific fuel consumption enjoys a bonus of 3–4 g/kWh (see Sect. 2.2).

Finally, mean piston velocity is another parameter that influences power. According to Eq. (17-1), power is directly proportional to the mean piston velocity, which can be enhanced by increasing the stroke and/or increasing the engine speed. Both measures affect the weight-to-power ratio and engine efficiency differently.

The correlations are extremely complex and, hence, only discussed qualitatively here. An exact grasp of the effects requires an investigation of their influence on the design of the crankshaft assembly including their influence on the overall vibration system, friction and gas exchange losses for a concrete case.

Thus, for example, a crankshaft's journal overlap grows smaller when the stroke is enlarged. Therefore, the journal diameter must be enlarged to retain stiffness. This makes a crankshaft assembly heavier and increases bearing friction. Gas exchange losses increase quadratically with mean piston velocity. Thus, these two correlations cause engine efficiency to decrease. When specific work has to be maintained, this can only be compensated by increasing the boost pressure, which normally entails reducing supercharging efficiency.

This is considered particularly problematic when the limits of supercharging have already been reached. In addition, higher required injected fuel quantities for given injection equipment lead to a prolongation of the duration of injection, which in turn leads to a degradation of thermal efficiency.

Lengthening the stroke increases the engine mass in every case.

The second measure discussed here, i.e. increasing speed at a constant stroke, has little influence on an engine's mass but causes higher gas exchange losses. At constant mean piston velocities, long-stroke engines have a lower speed than short-stroke engines. Lower speeds result in lower gas exchange losses, lower speed making more time absolutely available for gas exchange since the time cross sections are larger (see Sect. 2.1). In sum, the design of an HHD with specified power must:
– aim for high numbers of cylinders and a
– maximum bore for the specified distance between cylinder bore center lines,
– adapt the ratio of maximum cylinder pressure to specific work to the efficiency desired and the required starting and part load performance and
– aim for high mean piston velocity, yet limit it to 13 m/s.

17.4.3 High Performance Diesel Engine Design

17.4.3.1 Preliminary Remarks

One hundred years of diesel engine manufacturing have produced a highly developed combustion engine as well as many very interesting design solutions. Complex basic concepts are no longer pursued today, chiefly because of the costs. Thus, the basic features of newly developed diesel engines and basic engines differ only little. Design approaches to accommodate displacements in little structural space, e.g. by configuring more cylinder banks in delta, W or H form, will no longer be pursued as exhaust gas turbocharging is developed further.

Consequently, the following remarks solely treat the current state-of-the-art for four-stroke diesel engines. Readers interested in the history are referred to the literature, e.g. [17-25, 17-26].

The design and implementation of high speed high performance diesel engines (HHD) largely depends on an individual manufacturer's know-how and know-why. Effective new and further development of advanced HHDs makes the application of state-of-the-art aids such as

- CAD (computer aided design),
- new findings in mechanics, thermodynamics and aerodynamics,
- process simulation of complete diesel engine systems and
- the latest advances in metrology and analytics

indispensible. Only such aids make it possible to optimize the properties of complete systems, subsystems and components. The basic engine and supercharging are treated here specifically.

17.4.3.2 Basic Engine

The basic engine essentially includes the crankcase, the components of the crankshaft assembly, the cylinder head, the valve gear, the injection system and the gear drive or drives. Crankcases are now normally made of lamellar cast iron or globular cast iron when requirements are more stringent. The crankcase has the function of closing the load transfer between the cylinder head and crankshaft assembly. High dynamic loads also occur. Hence, the use of little material to produce high structural strength is extremely important for HHDs with their good weight-to-power ratios. To satisfy this demand, the crankcase is dropped far below the level of the crankshaft bearings (Fig. 17-46).

Bearing caps are laterally preloaded to increase stiffness. Minute movements in the parting plane of the bearing cap–crankcase, which can cause friction rust, are prevented either by providing a suitable form fit or using tensioning bolts to apply appropriate pretensioning forces in the level of the parting plane. The finite element method is employed to optimize the structure's stiffness. This not only relieves the crankshaft and the crankshaft bearing but also positively influences the transmission of structure-borne noise. An appropriately

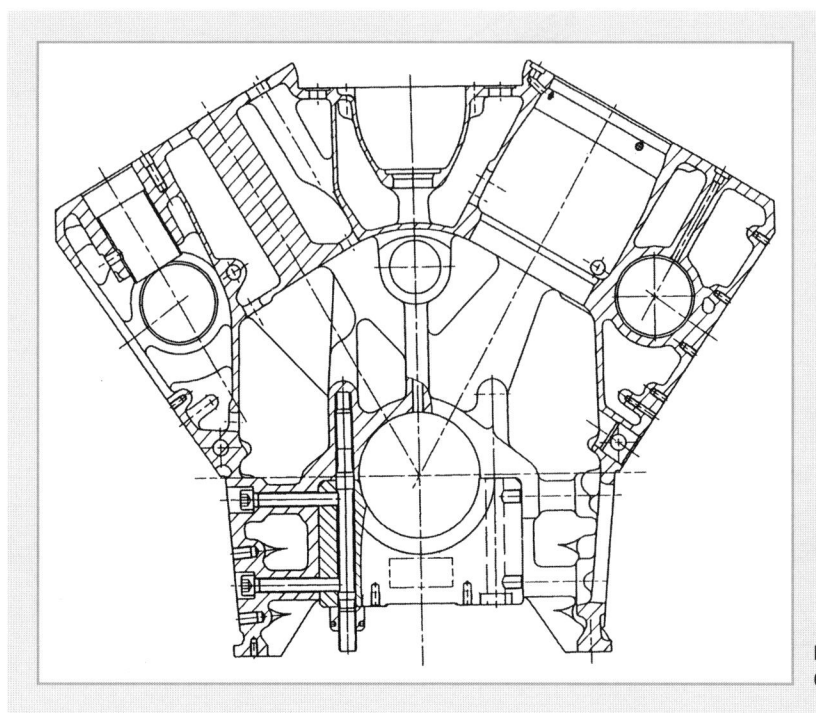

Fig. 17-46
Crankcase with high structural strength

designed oil pan can contribute to increasing the structure's stiffness. Since the crankcase is the most expensive single component of the entire engine, efforts must be made to not load it with secondary functions, which design can also resolve differently. The design of the upper zone of the crankcase greatly depends on such components as the injection system and the exhaust system. Thus, little in this area is universally valid. Arranging the threads for the cylinder head bolts deeply in the crankcase to obtain a good pressure distribution in the seal group of the cylinder liner–cylinder head is particularly important in this zone.

The crankshaft assembly has the function of converting a linear motion into a rotary motion and consists of the components of crankshaft, connecting rod bearing, connecting rod, piston pin and piston. Counterweights are mounted on the crankshaft to reduce loading of the main bearings and the housing structure and to eliminate or reduce outwardly active free forces and torques (as a function of the number of cylinders).

The crankshaft is forged from quenched and tempered steel. The radii at the crank pin may be hardened to increase fatigue properties. When sizing the crankshaft and arranging the crank pins, a compromise must be made between torsional vibration performance for different numbers of cylinders, the firing sequence, mass balancing and bearing loads. The aforementioned influences have to be factored into the sizing of the main bearing journal, the web and the crank pin. Since both the required distance between cylinder bore center lines and the required counterweight radius and thus the engine's weight and structural space depend on them, more extensive structural-mechanical calculations have to be performed in advance. The connecting rods in newly developed engines are only arranged side by side on one crank pin for reasons of cost (Fig. 17-47).

The skirt of a forged connecting rod is designed to buckle. The connecting rod is designed rigidly in the region of the connecting rod bearing to reduce any additional bearing loading by deformation and to prevent relative motions in the bearing back.

The connecting rod has to be fully machined to obtain a mass-optimized connecting rod and additionally reduce the required counterweight radius and thus mass. The connecting rod small end must be given a stepped or trapezoidal shape to enable it to absorb high pressures (Fig. 17-48).

The piston pin is supported by bearings in the piston in a bronze liner to prevent the piston boss from breaking at high loads.

Electron beam welded full skirt pistons are employed for peak pressures of up to 180 bar. Constraints when designing the cooling gallery additionally establish limits for combustion chamber design. Composite pistons must be provided

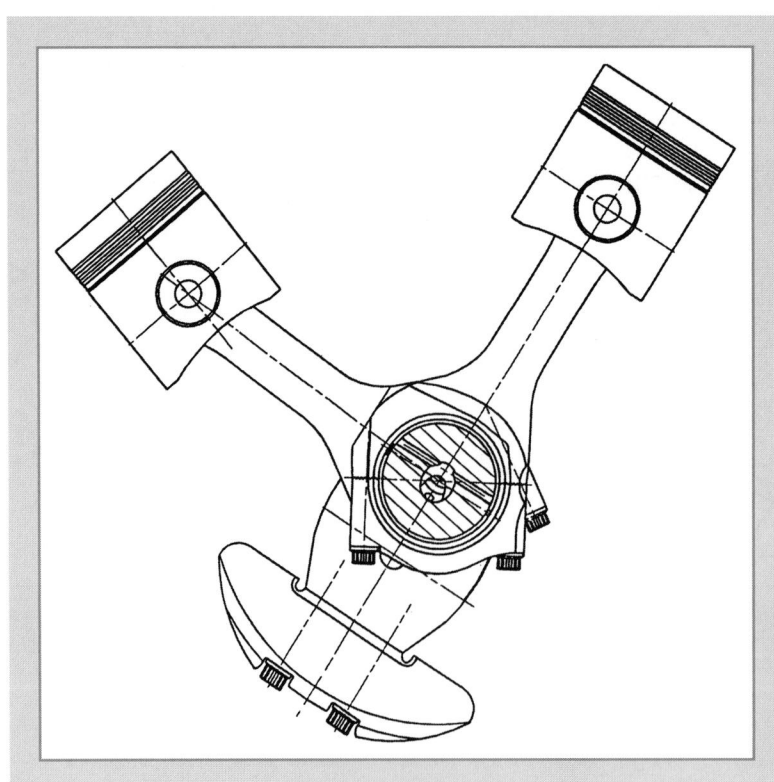

Fig. 17-47
Crankshaft assembly with parallel arrangement of connecting rods

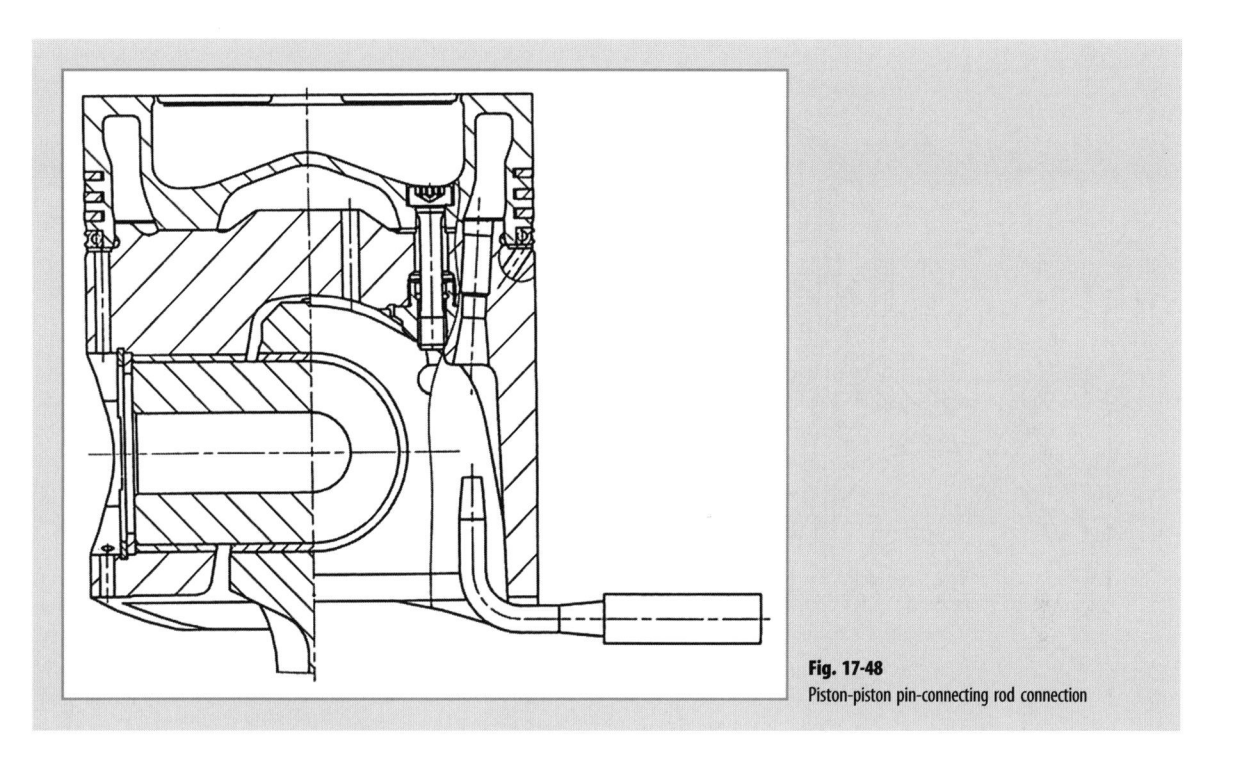

Fig. 17-48
Piston-piston pin-connecting rod connection

with aluminum skirts and steel tops for peak pressures >180 bar and deep combustion chamber bowls (see Fig. 17-48 and Sect. 8.6 for more).

At the given high power densities, the piston must be intensively cooled with oil. This is normally done by a spray jet that plunges into the piston in bottom dead center.

The cylinder head has the function of both imposing an upper limit on and sealing the combustion chamber and holding the gas exchange elements and the injection nozzle. Particular attention has to be devoted to the design of the cylinder head by:
- optimally designing the intake and outlet ports (two apiece in HHDs),
- sufficiently cooling the region of the cylinder head bottom and the outlet port,
- making the structure rigid to obtain a uniform distribution of pressure in the seal group of the cylinder liner-cylinder head,
- arranging the cylinder head bolts to trigger gas forces with the load transfer in the crankcase bearing walls and
- satisfying additional requirements, e.g. decompression and compressed air starting systems.

Extensive structural mechanical and fluid mechanical tests are necessary to harmonize the aforementioned requirements. The cylinder head is normally designed as a cross-flow head, i.e. the inlet of the intake port and the outlet of the outlet port are located opposite one another. The secondary ports are located next to one another in low-swirl combustion systems or are in series in high-swirl combustion systems. Intermediate positions are also possible but make control more complicated. With displacements of more than 3 dm^3, the cylinder heads are solely designed as single cylinder heads since thermal expansion is otherwise able to generate unduly high relative motions in operation. This additionally makes maintenance easier. The gas exchange elements are normally controlled by an underhead camshaft, tappet, push rod and rocker arm (Fig. 17-49).

Depending on the arrangement of the valves, forked rocker arm or bridges are employed on both valves to transmit actuating forces.

The arrangement of an underhead camshaft facilitates simple disassembly of individual cylinder heads during maintenance. Overhead camshafts cause considerable additional design work, which is disproportionate to any benefit.

Injection equipment has the function of supplying the fuel to the combustion chamber in the required quantity at the right time for the appropriate duration. A maximum injection time of 25°CA at full load is aimed for to obtain high thermal efficiencies.

This necessitates maximum injection pressures above 1,500 bar as a function of engine size and the combustion system.

Established injection systems are the (see Sect. 5.3):
- single pump-line-nozzle system,
- unit injector system and
- high pressure accumulator line injector (common rail).

The common rail injection system with full electronic control using solenoid valves is normally selected for new

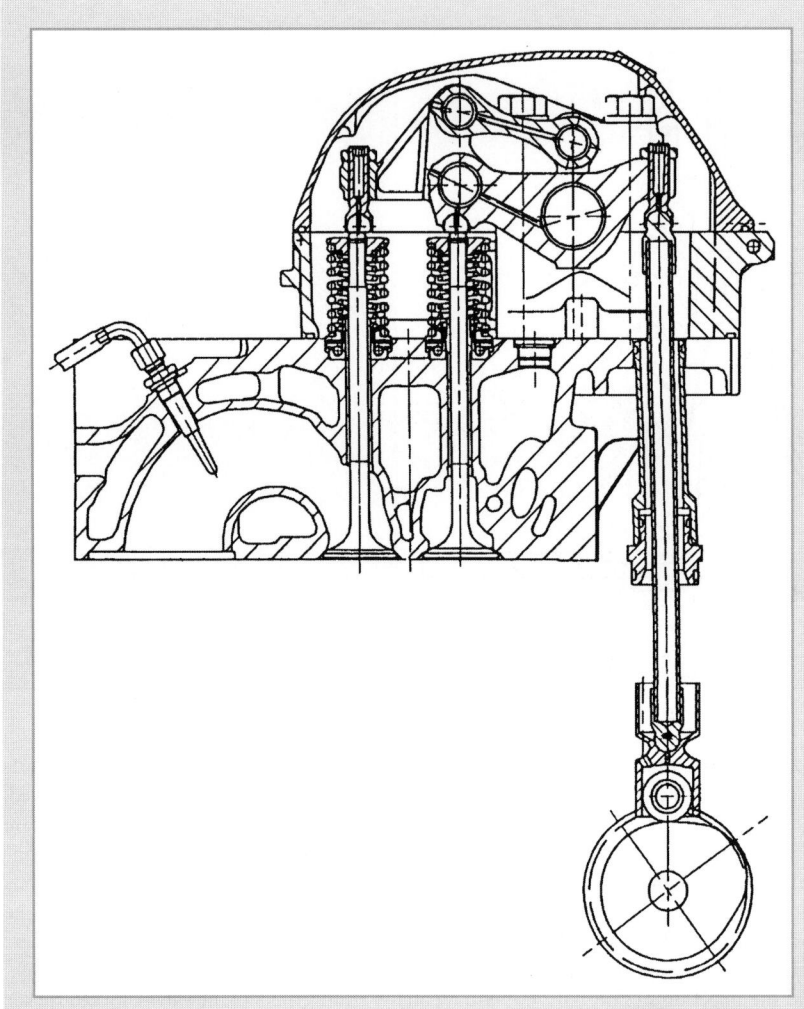

Fig. 17-49 Control of the gas exchange elements

developments. The injection pressure is no longer a function of engine speed and the start of injection is freely selectable. Moreover, the injection characteristic may be varied very easily by software parameters, e.g. by pilot or post-injection. Thus, a common rail injection system makes it possible to decisively influence emissions and fuel consumption. It also enables substantially simplifying the basic engine. A second gear drive on the drive end can be dispensed with. The valve gear assembly can be operated by a central camshaft and one of the usual two camshafts spared. Complex control linkage is eliminated completely. MTU's 4000 series is equipped with this advanced injection system (Fig. 17-50).

17.4.3.3 Supercharging

Supercharging serves to enhance a diesel engine's performance and thermodynamic efficiency. (See Sect. 2.2 for more background on thermodynamics.) Since it furnishes more options for unrestricted engineering, exhaust gas turbocharging has established itself for four-stroke diesel engines. The components of a high speed high performance diesel engine's (HHD) exhaust gas turbocharging system are the:

– exhaust manifold system,
– exhaust gas turbochargers,
– intercooler and
– charge air distribution system.

Since the displacement performance of a reciprocating piston machine, i.e. the diesel engine, and the delivery performance of a turbomachine, i.e. the exhaust gas turbocharger, differ starkly from one another, diesel engines and exhaust gas turbochargers can only be optimally tuned in one operating point in the map. This necessitates adjustments in the operating map as a function of the desired specific effective brake work, the required engine map and the permissible limits of the turbocharger and diesel engine.

Fig. 17-50
Cylinder head with common rail injection system

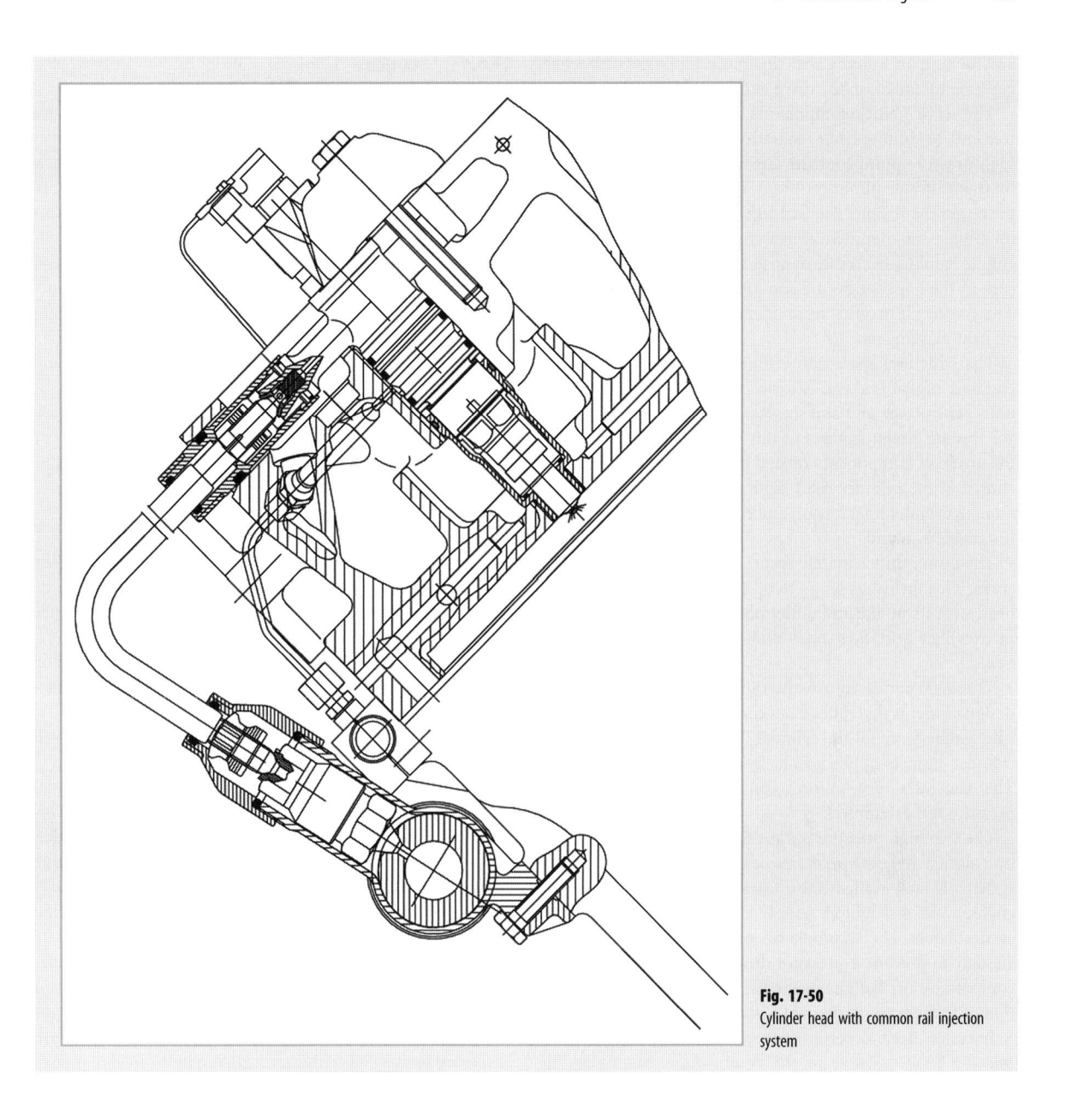

Every supercharging system has to be evaluated in terms of its thermodynamic properties and the design work necessary. Thus, a basic distinction is made between pulse and constant pressure supercharging in HHDs. Intermediate solutions are also possible and are definitely implemented. Section 2.2 describes supercharging in detail.

17.4.3.4 Implemented Designs

Situated in the limit range between high speed high performance diesel engines and medium speed four-stroke diesel engines (see Sect. 17.3), MTU's 8000 series introduced in 2000 is described (Fig. 17-42) as an example of an implemented HHD.

This series was initially available as a V20 version for marine applications. Its cylinder displacement is 17.37 dm^3. A maximum speed of 1,150 rpm produces a mean piston velocity of 12.1 m/s, a respectable value for this engine's size. Single-stage multistage supercharging produces the specific work $w_e = 2.7$ kJ/dm^3, a value which allowed shifting the limit of single-stage supercharging further upward.

The engineering design of the 8000 series engine is clearly structured. The supercharger components, i.e. the so-called junction box with intercoolers and air inlet in the crankcase and the support housing with the four exhaust gas turbochargers, are mounted on the drive end.

The service block components are located on the engine free end, beginning from the bottom with the oil pump, two high pressure pumps for the CR injection system, followed by the saltwater pump on the A side and the engine water pump on the B side. The fuel filter, automatic oil filter and oil cooler are mounted above the two water pumps. The engine cooling water thermostat and centrifuges for oil care in the bypass flow are mounted outside on the A side. The engine electronics components are centered above the oil heat exchangers.

The structural space above the engine between the drive end and engine free end is easily accessible to install power units. The power unit and the common rail injection system are specific design features of this engine. The power unit is the group of components related to the cylinder station and consist of the cylinder head (with control and injection elements), cylinder liner, piston and connecting rod, collectively installable as a unit.

The power unit is merely secured in the crankcase by four studs. This furnishes scope to optimally design the cylinder head with its intake and outlet ports, which manifests itself in excellent port flow rates and rugged, durable cylinder heads.

The 8000 series absorbs holding and sealing force with design elements that differ from conventional design variants. The sealing force for the cylinder liner's connection with the cylinder head is applied from below by twenty-four bolts. This generates very uniform contact pressures in the high pressure seal assembly.

The two high pressure pumps for the common rail injection system generate pressures of up to 1,800 bar in the fuel system. Double-walled lines, installed longitudinally in the engine, and distributors supply fuel to rails or individual accumulators. The accumulators have a volume that is large enough to prevent a pressure drop before the injector and compressive oscillations in the fuel system during the injection of the maximum mass of fuel injected.

Injection itself is performed by injectors centered in the cylinder head (relative to the combustion chamber). The engine management system controls the start and duration of injection by an electric control. A number of sensors provide the electronic engine management system information on the engine's operating state. This information is linked with stored maps and delivers data not only for the timing but also for the pressure in the fuel system as a function of the operating state.

The advantage of the common rail system is the ability to freely select the injection pressure as a function of the engine's operating state. This means that – unlike engines with conventional injection systems – high injection pressures, which also enable efficient and low emission combustion even in the part load range, are also available at low engine speeds.

17.4.4 Outlook

High speed high performance diesel engines (HHD) have now reached an advanced state of development, i.e. customer benefit elements such as weight-to-power ratio, structural space, service life and net efficiency have a high standard. In the future, diesel engines will increasingly be judged in terms of their environmental compatibility.

An analysis of material and energy input during diesel engine production and operation reveals that diesel engines are relatively environmentally compatible machines throughout their service life. Typically, this entails:
- the processing of safe materials such as cast iron, steel and aluminum,
- conventional manufacturing methods,
- controllable environmental concerns,
- a high percentage of recyclable material,
- a long service life and
- high thermal efficiency.

Exhaust emissions are currently considered problematic (see Chap. 15). NO_x constituents and particulate matter are particularly important and CO_2 is also gaining importance because of the greenhouse effect.

Compliance with future limits will necessitate designing diesel engines so that the conditions for low exhaust emissions are good.

In principle, engine internal and external measures will enable this (see Chap. 15).

Engine external measures entail considerable additional work in the form of costs, structural space and weight. This sacrifices specific properties of HHDs, i.e. low mass and smaller structural space. Thus, the goal will be to obtain good exhaust emission values for HHDs with engine internal measures.

Conventional methods such as modifying the combustion chamber, the swirl and the injection system will have to be exhausted first. New developments with low-swirl combustion systems will have to be designed. This will make efficient injection systems imperative.

Additional measures such as exhaust gas recirculation will be needed to meet NO_x limits <5 g/kWh in the field of mobile applications.

Literature

17-1 Kraftfahrt-Bundesamt: Statistische Mitteilungen. Vol. 1, (2005) 12
17-2 Steinparzer, F.; Stütz, W.; Kratochwill, H.; Mattes, W.: Der neue BMW-Sechszylinder-Dieselmotor mit Stufenaufladung. MTZ 66 (2005) p. 334–345
17-3 Krebs, R.; Hadler, J.; Blumensaat, K.; Franke, J.-E.; Paehr, G.; Vollmers, E.: Die neue 5-Zylinder-Dieselmotoren-Generation für leichte Nutzfahrzeuge von Volkswagen. Wiener Motorensymposium (2006)

17-4 Brüggemann, H.; Klingmann, R.; Fick, W.; Naber, D.; Hoffmann, K.-H.; Binz, R.: Dieselmotoren für die neue E-Klasse. MTZ 63 (2002) p. 240–253

17-5 Bach, C.; Rütter, J.; Soltic, P.: Diesel- und Erdgasmotoren für schwere Nutzfahrzeuge, Emissionen, Verbrauch und Wirkungsgrad. MTZ 66 (2005) p. 395–403

17-6 DVZ: Deutsche Verkehrszeitung. 17 (2006) 1, p. 9 ff.

17-7 DIN 7753-3: Endless narrow V-belts for the automotive industry. Berlin: Deutsches Institut für Normung (Ed.) (1986) 2

17-8 DIN 7867: V-ribbed belts and corresponding pulleys. Berlin: Deutsches Institut für Normung (Ed.) (1986) 6

17-9 Neitz, A.; Held, W.; D'Alfonso, N.: Schwerpunkte der Weiterentwicklung des Nutzfahrzeug-Dieselmotors. Düsseldorf: VDI-Z. Special (1990) 1, p. 17

17-10 Held, W.: Die MAN-Strategien für Euro 4, 5 und in der Zukunft, MAN Nutzfahrzeuge. Ambience & Safety Conference (2005)

17-11 Lange, W. et al.: Einfluss der Kraftstoffqualität auf das motorische Verhalten und die Abgasemissionen von Nutzfahrzeug-Dieselmotoren. MTZ 53 (1992) 10, p. 466

17-12 The Council of the European Communities (Ed.): Council Directive 70/157/EEC of 6 February 1970 on the approximation of the laws of the Member States relating to the permissible sound level and the exhaust system of motor vehicles (1990) 3

17-13 TRUCKER. Munich: Verlag Heinrich Vogel GmbH Fachverlag

17-14 Biaggini, G.; Buzio, V.; Ellensohn, R.; Knecht, W.: Der neue Dieselmotor Cursor 8 von Iveco. MTZ (1999) 10

17-15 Im Aufwind. lastauto omnibus (1999) 11, p. 16 ff.

17-16 TEST&TECHNIK. lastauto omnibus. (2001) 1, p. 34

17-17 Schittler, M. et al.: Die Baureihe 500 von Mercedes-Benz. MTZ (1996) 9, p. 460, (1996) 10, p. 558, (1996) 11, p. 612

17-18 Price, R.B.; Meistrick, Z.S.A.: New Breed of Engine Brake for the Cummins L10 Engine. SAE-Paper 831780 (1983)

17-19 Körner, W.-D.; Bergmann, H.; Weiss, E.: Die Motorbremse von Nutzfahrzeugen – Grenzen und Möglichkeiten zur Weiterentwicklung. Automobiltechnische Zeitschrift 90 (1988) 12, p. 671

17-20 Haas, E.; Schlögl, H.; Rammer, F.: Ein neues Motorbremssystem für Nutzfahrzeuge. Dusseldorf: VDI Fortschrittberichte Vol. 12, No. 306, (1997) p. 279–298

17-21 Cummins Engine Company Inc.: Signature Engine. Bulletin 3606151 (1997) 7

17-22 Théremin, H.; Röbke, H. (Ed.): Schiffsmaschinenbetrieb. 3rd Ed. Berlin: Verlag Technik (1978)

17-23 Holden, K.O.; Faltinsen, O.; Moan, T. (Ed.): Fast '91 First International Conference on Fast Sea Transportation. Trondheim: TAPIR Publishers 1/2 (1991) 6

17-24 Jewell, D.A.: Possible Naval Vehicles. Naval Research Reviews, Oct. 1976. Office of Naval Research Arlington (VA), USA

17-25 Zima, S.: Hochleistungsmotoren – Karl Maybach und sein Werk. Düsseldorf: VDI-Verlag (1992)

17-26 Reuss, H.J.: Hundert Jahre Dieselmotor. Stuttgart: Franckh-Kosmos-Verlag (1993)

Further Literature

Rudert, W.; Wolters, G.-M.: Baureihe 595 – Die neue Motorengeneration von MTU, Teil 1. MTZ 52 (1991) p. 274–282

List, H.: Das Triebwerk schnellaufender Verbrennungskraftmaschinen. Berlin/Heidelberg: Springer (1949)

Kraemer, G.: Bau und Berechnung der Verbrennungsmotoren. Berlin/Heidelberg: Springer (1963)

Scheiterlein, A.: Der Aufbau der raschlaufenden Verbrennungskraftmaschinen. Berlin/Heidelberg: Springer (1964)

Pischinger, A.; List, H.: Die Steuerung der Verbrennungskraftmaschine. Berlin/Heidelberg: Springer (1948)

Maass, H.; Klier, H.: Kräfte, Momente und deren Ausgleich in der Verbrennungskraftmaschine. Berlin/Heidelberg: Springer (1981)

18 Industrial and Marine Engines

Günter Kampichler, Heiner Bülte, Franz Koch, and Klaus Heim

18.1 Small Single Cylinder Diesel Engines

18.1.1 Introduction

The industrial single cylinder diesel engine has a history steeped in tradition, which extends from the first operable diesel engine in 1897 up to today's versatile, air-cooled, small single cylinder diesel engine. On account of the low manufacturing costs relative to output, lowfuel consumption, good lubrication conditions and better exhaust quality, it is now only built as a four-stroke engine, primarily in the small diesel engine segment.

Despite the high level of development achieved, potentials for improvement continue to be researched. These are chiefly envisioned in the use of new, high grade materials and in the criteria of air/fuel mixture formation and control.

While the service life and reliability of industrial engines will always have top priority, complicated market mechanisms as well as additional parameters, e.g. exhaust and noise emission, increasingly demand consideration (see Chaps. 15 and 16).

Technical advances in related types of product ranges are also spurring on the development of small diesel engines. However, the same standard cannot always be applied: the development of a new engine generation or the introduction of new technologies necessitates meticulous studies, including exact analysis of present engine engineering, which has proven itself in harsh industrial operation for decades. Only a correct assessment of foreseeable stages of development and future requirements will make it possible to integrate new technologies in internal product development at an early stage and successfully market these products with an edge over the competition. Fields of application for engines with a power range of 2–12 kW at 3,000 to a maximum of 3,600 rpm are construction equipment, municipal vehicles, lawn and garden and agricultural equipment and small tractors, electric generators, water pumps and boat engines.

The classic small single cylinder diesel engine. Horizontal designs are only still found in Asia. Typically implemented as gasoline lawnmower and outboard engines earlier, vertical shaft engines also became available as single cylinder four-stroke diesel engines four years ago (Fig. 18-1). Free inertial forces and moments of inertia are less noticeable in devices with a low location of center of gravity, e.g. lawnmowers.

Engine customers' demands are diverse and extremely varied. They all want to see their current case of installation optimally resolved in terms of function and cost. Responding to every customer demand would result in a multiplicity of different engines. This is only worthwhile when backed up by suitable quantities, which is normally not the case. Consequently, all engine manufacturers attempt to develop their own concept strategies, which, taking a basic engine as the point of departure, can provide maximally variable, universally implementable and customizable engines. This presupposes installation engineering in a dialog between engine suppliers and users. Harmonization of requirements specifications and consultation supporting installation up through the start of a device's production and including acceptance certification have become part of engine manufacturers' standard programs.

18.1.2 Performance Specifications and Basic Data of Single Cylinder Diesel Engines

18.1.2.1 Power Range and Combustion Process

Engine Power

The market demands four-stroke single cylinder diesel engines with powers of 2–12 kW and cylinder volumes of V_h = 200–850 cm^3. Annual worldwide demand for such engines is 1.2 million and growing. Two cylinder engines already make sense above 12 kW of power. Allowing for costs, a lower limit of 8 kW at 3,600 rpm is justifiable depending on the torque requirement. Low free inertial forces and thus gentler, lower vibration engine operation are crucial. Using higher speeds to obtain a power increase is detrimental to the application and hardly effective. Maximum rated speeds of 3,000 rpm in the upper displacement range or up to 3,600 rpm in the lower displacement range have proven themselves.

K. Heim (✉)
Wärtsilä Switzerland Ltd., Winterthur, Switzerland
e-mail: klaus.heim@wartsila.com

Combustion Process

The basic requirement for single cylinder diesel engines continues to be manual startability down to at least –6°C (–12°C) without electric auxiliary devices such as glow plugs or coils. Since a swirl chamber engine is unable to start below 0°C without preglow, direct injection (DI) is the only option for small displacements (≤ 0.4 l).

18.1.2.2 Engineering Requirements

Stroke to bore ratio. While a stroke to bore ratio of s/D>1 is fundamentally preferable, larger strokes impede exploiting the advantages of optimal overall height. Occasionally, this may be the decisive factor for the implementation of a motorization project. Thus, even engines with an s/D ratio of up to ≥ 0.6 are available. Vertical shaft engines are gaining popularity, e.g. for installation in lawnmowers, because of their overall height (see Fig. 18-1).

Cooling systems. For all intents and purposes, only direct air cooling (see Sect. 9.1.4) with a radial exhaust fan integrated in the flywheel is employed. This space saving and low cost principle utilizes baffles to systematically route cooling air to temperature-critical components such as the cylinder head and cylinder liner. The larger the engine surface area drawn on for cooling, the larger the reserve is for implementation in countries with high ambient temperatures. Water cooling of the type in small multiple cylinder engines is too complex and hence out of the question.

Continually tightened regulations for noise emission of appliances, equipment, etc., also target the drive, i.e. the combustion engine. Thus, full engine encapsulation is often the only corrective (see Sect. 16.5). The thermal problems that arise in an engine and the required damping of cooling air noises are compelling engine manufacturers to develop new concepts for fully enclosed engines. Hatz is testing liquid cooling with lubricating oil and an external heat exchanger for its B model engines (Fig. 18-2) with a modified cylinder head, oil circulation on the control side and a larger oil supply. The enlarged, oil-filled gap between the slip-fit cylinder liners serves to cool them or acts as a heat bridge to the housing.

Engine mount. When possible, an engine should be mounted elastically. The supplier industry provides

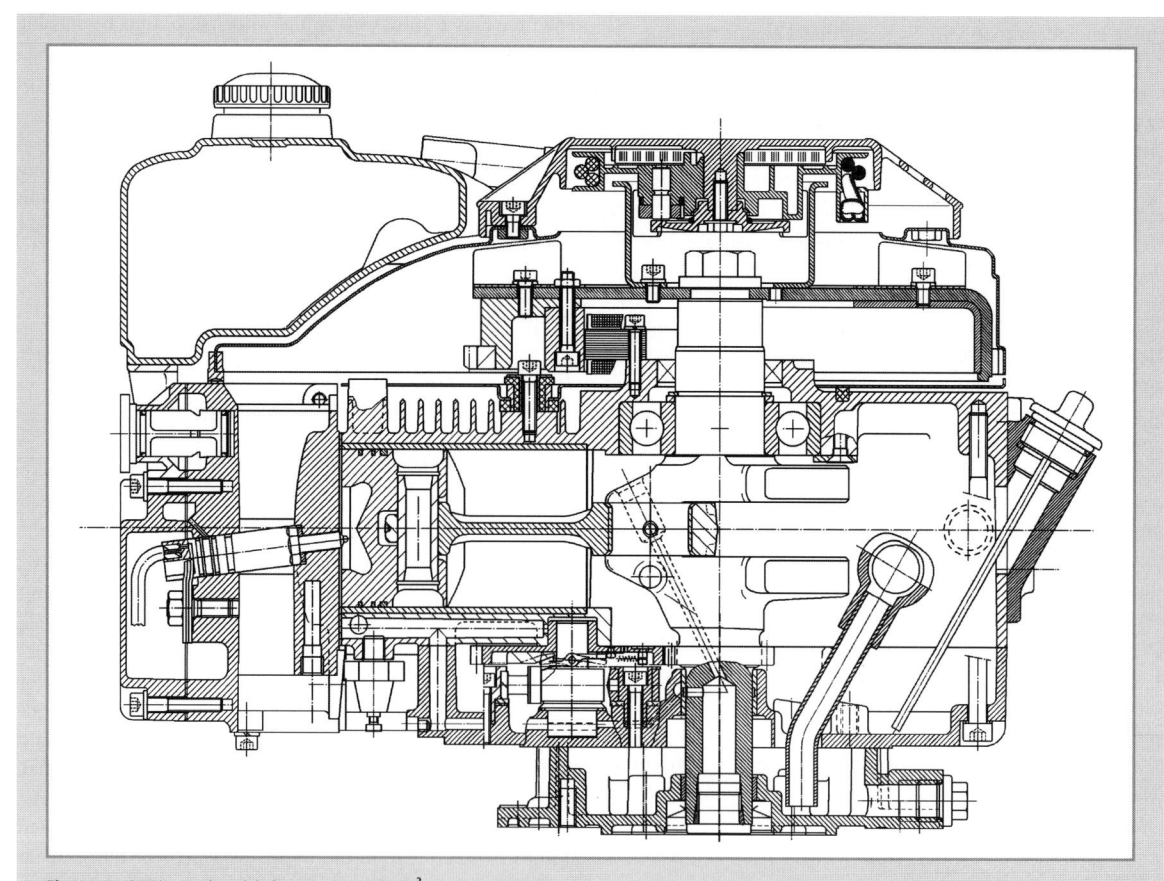

Fig. 18-1 1B20V vertical crankshaft engine V_h = 232 cm³, P_e = 3.8 kW/3,600 rpm

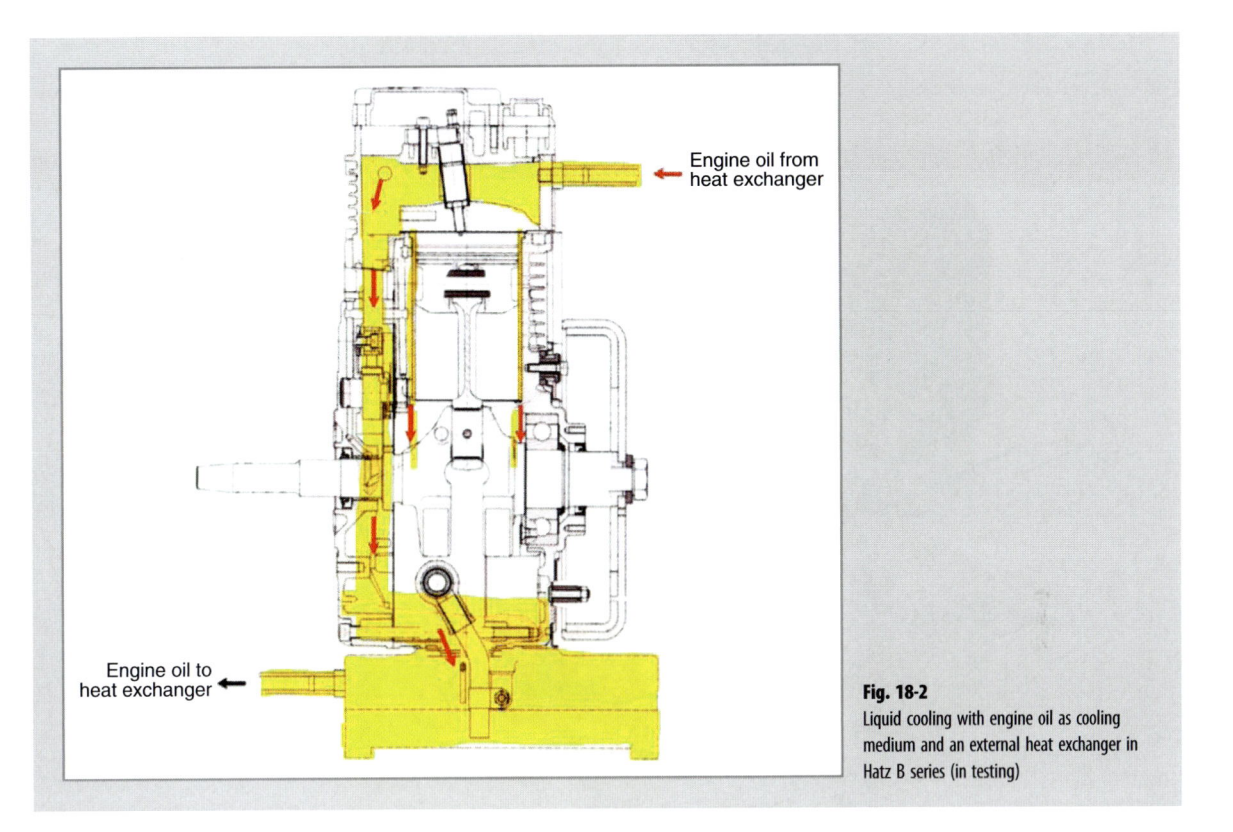

Fig. 18-2 Liquid cooling with engine oil as cooling medium and an external heat exchanger in Hatz B series (in testing)

extensive options. Rubber/polymer elements with hydro mounts have proven to be excellent even when resonance speeds are run through when an engine is revving or coasting when it is stopped.

Belt driven machines may initially be rigidly constructed on an intermediate frame together with the engine. Damping elements for the foundation can decouple the frame. An engine may be designed rigidly in conjunction with very stiff and massive frames and foundations.

Power Take-offs. Universal implementation of an engine requires output on both the flywheel side and the opposite control side. This reverses the direction of rotation. However, the recoil starters increasingly being implemented block power take-off on the flywheel side. Depending on the engine design, dispensing with the simple manual starter and selecting the more convenient but also more vulnerable electric starter produces other power take-offs. For instance, the 1D81 engine has 4 PTOs (see Fig. 18-3).

In addition to the options of counterclockwise (standard) or clockwise rotation, torque may also be decreased:
- 100% axially on the flywheel side by a coupling flange or radially by a belt pulley (1),
- 100% radially and axially on the control side at the crankshaft (2),
- additionally 100% on the control side when hand crank started and at the camshaft by attaching a V-belt pulley (3) and
- to a limited extent at the camshaft that drives small hydraulic pumps (4).

Starter Systems, Starting Option

Hand crank start. A hand crank start always requires a higher mass moment of inertia than an electric start. However, a higher mass moment of inertia means considerably higher engine weight because of the necessarily heavier flywheel. Potential corrosion in electric components and damage to batteries by strong shaking are points against electric starting in harsh construction equipment.

Neglected maintenance, improper operation, etc. typify the construction equipment rental business where "simple", "rugged" and "functionally manageable" are demanded. These are arguments for a hand crank start in larger single cylinder diesel engines upwards of $V_h > 0.5$ dm³ in which several preselectable, decompressed cycles speed the flywheel up enough that its energy allows engine auto-ignition and revving for one or more TDC cycles at full compression. German laws mandate protective measures against dangerous recoil (reversion) when the flywheel action is inadequate and

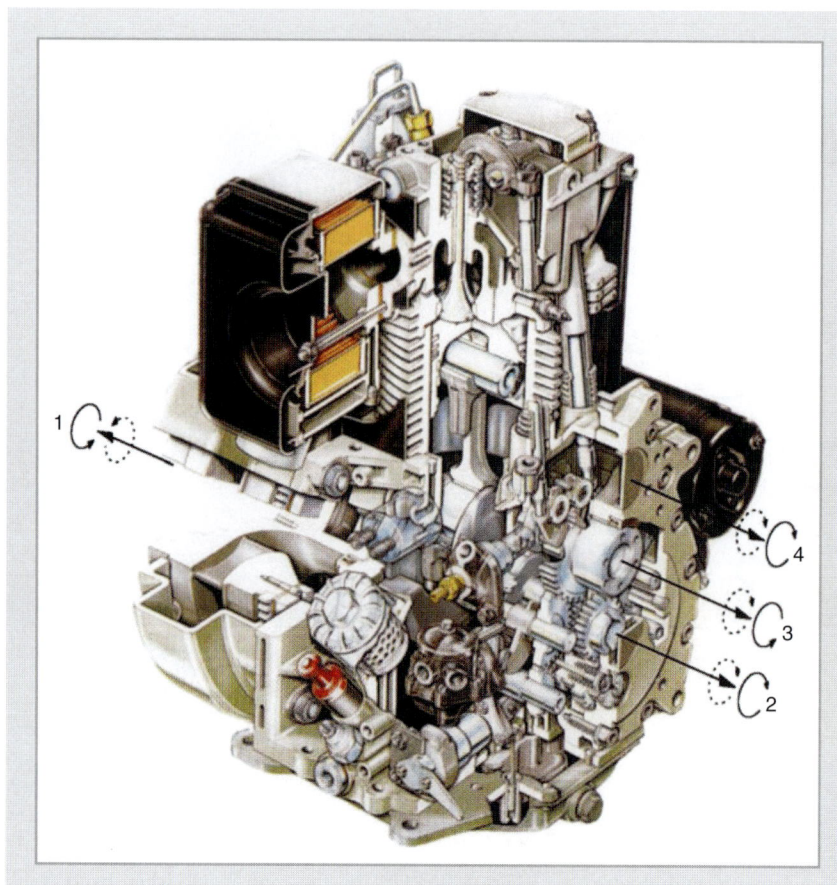

Fig. 18-3
PTOs on a single cylinder air-cooled diesel engine (HATZ SUPRA 1D81). $V_h = 0.667$ dm^3

ignition starts before TDC, e.g. a trip gear installed in the hand crank that stops the transmission of power to the recoil torque after a few angular degrees. Hand crank starting is feasible up to a displacement of 0.8 l when the flywheel and the multiplication of hand crank to crankshaft speed are selected correctly. A hand crank speed of 2.5 revolutions per second with a maximum hand crank radius of 200 mm at full expenditure of energy over a time of 3.5 seconds is barely justifiable ergonomically.

The trend toward using flywheels to decrease engine mass related to power comes at the expense of dependable starting and cold start performance. Figure 18.4 presents the correlations for dependable cold start at –6 °C.

Recoil start. Implemented in stationary *gasoline* engines virtually without exception, recoil starters are increasingly also being employed in *diesel engines*, especially since they present hardly any danger of injury. The energy manually introduced by a cable with a deflection of 0.7–1 m must bring the flywheel to the starting speed within one cycle, i.e. two revolutions. Starting is facilitated by decompression by elevating the exhaust valve by approximately 0.1–0.2 mm by a simple lever system that returns after one cycle or by an automatic centrifugally controlled decompression system. Figure 18-4 also includes the moment of inertia of the flywheel for flicker-free generator operation at 3,000 min^1. However, this already makes cold start by means of recoil start problematic as of a displacement of 300 cm^3 and above. Superior in this case, hand crank start allows higher transmission ratios, which are between 4:1 and 5:1 in IDI engines with $V_h < 0.4$ l. Prechamber engines (IDI) additionally require electric preglow units below –6°C.

Electric start. With the exception of the harsh construction sector, starting with electric starting motors using pinions and a ring gear at the flywheel is increasingly being implemented in air conditioners and elevating and lifting equipment, for example, for which remote control or control electronics have been designed in from the outset. Since single cylinder diesel engines have long settling times (up to 2.5 s), a tooth modulus >2.5 mm has to be selected that prevents teeth from breaking out during inadvertent post-starting. Hence, an electronic start block relay, which is indispensible for remotely controlled engines, is generally recommended. Germany and Europe will continue to prohibit "ignition aids" sprayed into the intake tract during manual starting as well as simple

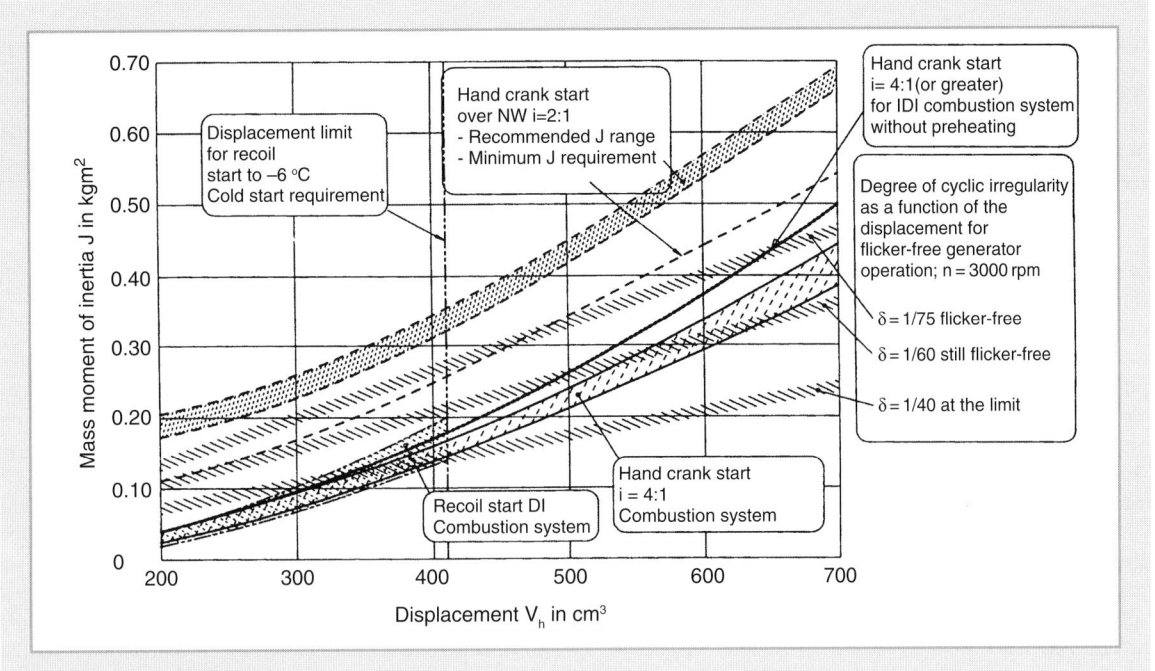

Fig. 18-4 Flywheel moment of inertia as a function of displacements required for single cylinder diesel engines at the cold start limit of −6°C

manual pull starters in the future because of the danger of accidents.

Intake and Exhaust Systems

Air Cleaners. Formerly used for small diesel engines, oil wetted air bath cleaners have poorer filtration efficiencies but are easy to service since they function with the engine oil available at construction sites. Significantly better in terms of their filtration efficiency, dry air cleaners (see Sect. 13.1) not only require the stocking of cleaner cartridges but also maintenance during which cleaner loading is checked. The common cleaner vacuum indicator has problems in single cylinder engines because of the strong intake pulsations. A clogged cleaner cartridge can already initiate severe damage in the cylinder head region after a brief time by overheating due to a deficiency of combustion air. Since increased dust concentrations occur in the construction sector, a basic engine concept should design in both types of cleaners and allow the installation of a precleaner (cyclone) when dust concentrations are extreme. Oil wetted air bath cleaners are preferable when the supply of replacement parts is uncertain, e.g. in "third world countries".

Intake tract. An airtight connection, e.g. elastic hoses, has to be provided from the air cleaner's raw air inlet up to the engine's intake air inlet. When engines are installed in semi or completely enclosed spaces, the air should be supplied externally without pressure losses and temperature increase (recommended values: $\leq$10 mbar (100 mm WS) vacuum and 5 K temperature increase over the external air, measured at the intake port inlet at rated speed).

Exhaust system. The maximum muffler volume should be selected. On the one hand, structural space renders a volume that is ten times the displacement unfeasible even though this is ideal in terms of power loss and tailpipe noise. On the other hand, no effective damping without power losses of up to 10% is achievable below three times the displacement. The mean back pressure at the muffler inlet (= exhaust port outlet) should be <25 mbar (250 mm WS) at rated speed. Apart from being sealed tightly, exhaust manifolds have to be installed rigidly or with flexible compensators depending on the engine mount to decouple vibrations. Engines installed inside enclosed spaces require the shortest insulated exhaust manifolds possible.

Engine compartment ventilation. When engines are installed in tight, enclosed spaces (encapsulation), the cooling air has to be supplied to the impeller inlet with a minimal temperature increase (<3 K). Bellows, hoses or ducts should discharge heated exhaust air to the atmosphere by the shortest path to regulate the heat balance in the engine or installation space. Partial evacuation from the installation space by means of the flywheel fan is normally sufficient when the muffler is mounted outside the installation space and the exhaust manifold is short and insulated. When a muffler is located in an engine compartment, it as well as the following exhaust pipes must be placed in a compartmentalized air duct. External ventilation is only necessary when the exhaust flow is not leak proof.

Fuel Supply

A fuel tank mounted at the engine with a filter, a fuel line to the injection pump and return and leak oil return from the nozzle holder to the tank are standard. Reliable engine operation requires a minimal gradient of ≥ 50 mm from the tank to the injection pump, allowing for potential operation in an inclined position. Horizontal or slightly inclined piping arrangements of $<10°$ have to be avoided as well. This is particularly true for the forward flow from the tank to the injection pump. The discharge of gas or air bubbles in the fuel after engine shut off has to be assured, particularly in the region of the injection pump. When fuel is supplied by a fuel tank positioned lower, a non-return valve installed directly before the diaphragm feed pump prevents the fuel system from discharging when the engine is stopped. Otherwise, protracted ventilation procedures would be required during restarting.

Alternators

Alternators in single cylinder diesel engines are usually designed as space saving flywheel alternators. Magnet segments arranged in a ring on the flywheel bypass star-shaped spool bodies attached to the crankcase with a clearance of approximately 0.4 mm. The voltage regulator is attached to the engine so that it is easily accessible and delivers a rectified charging current of 15 A at 12 V and 8 A at 24 V as a function of speed and thus a charging power of ≈ 200 W. In another concept, permanent magnets mounted on the flywheel interact with a coil bracket attached to the crankcase through an axial air gap. It alone is replaced when there is a malfunction. Thus, the engine or flywheel does not have to be removed. Alternator power can be boosted easily by additional coils.

18.1.3 Engineering Design of Small Single Cylinder Diesel Engines

18.1.3.1 Crankcase

While gray cast iron crankcases provide advantages in terms of noise emission, they are no longer relevant because of their weight just as lightweight sand or gravity die cast aluminum crankcases are no longer relevant because of the expensive machining (Fig. 18-5).

Crankcase or rack designs that can be cast from light alloys are primarily employed for reasons of cost. One technically optimal, low cost solution is a pressure die cast aluminum crankcase design with an integrated, raised holder for the cylinder liner, which is open on one side and thus easy to demold (see Figs. 18-6 and 18-7). As the sealing component, the control cover accommodates one main bearing. However, the development of the force lines reveals the limits of this design principle. Firing pressures that are no longer manageable are produced at displacements that are larger than 0.6 dm³. Here, a one-piece design open on the bottom, which can be pressure die cast, in

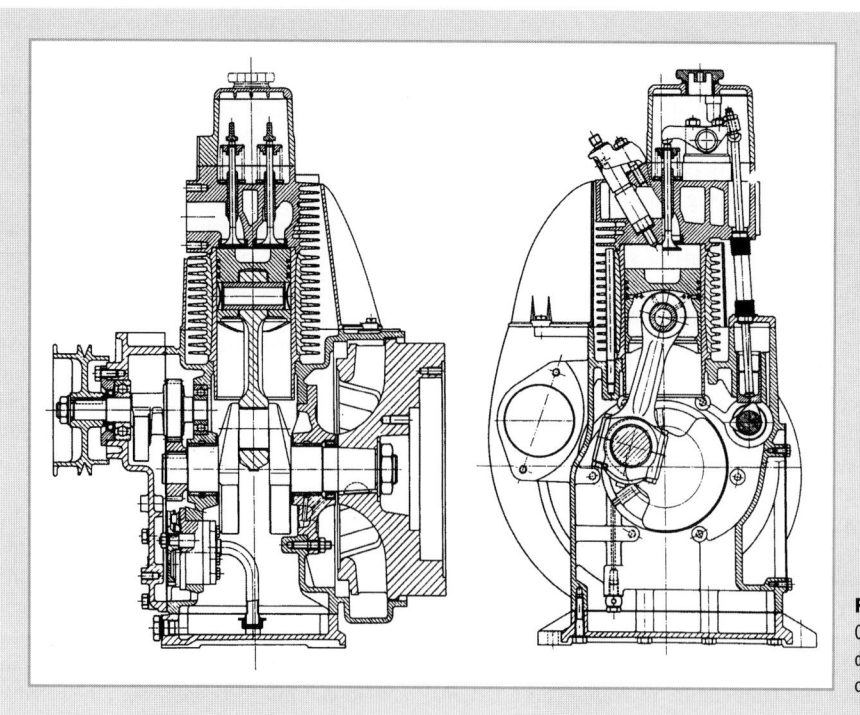

Fig. 18-5
Older model single cylinder air-cooled diesel engine with gray cast iron crankcase. $s/D = 82/82$ (DEUTZ F1L 208)

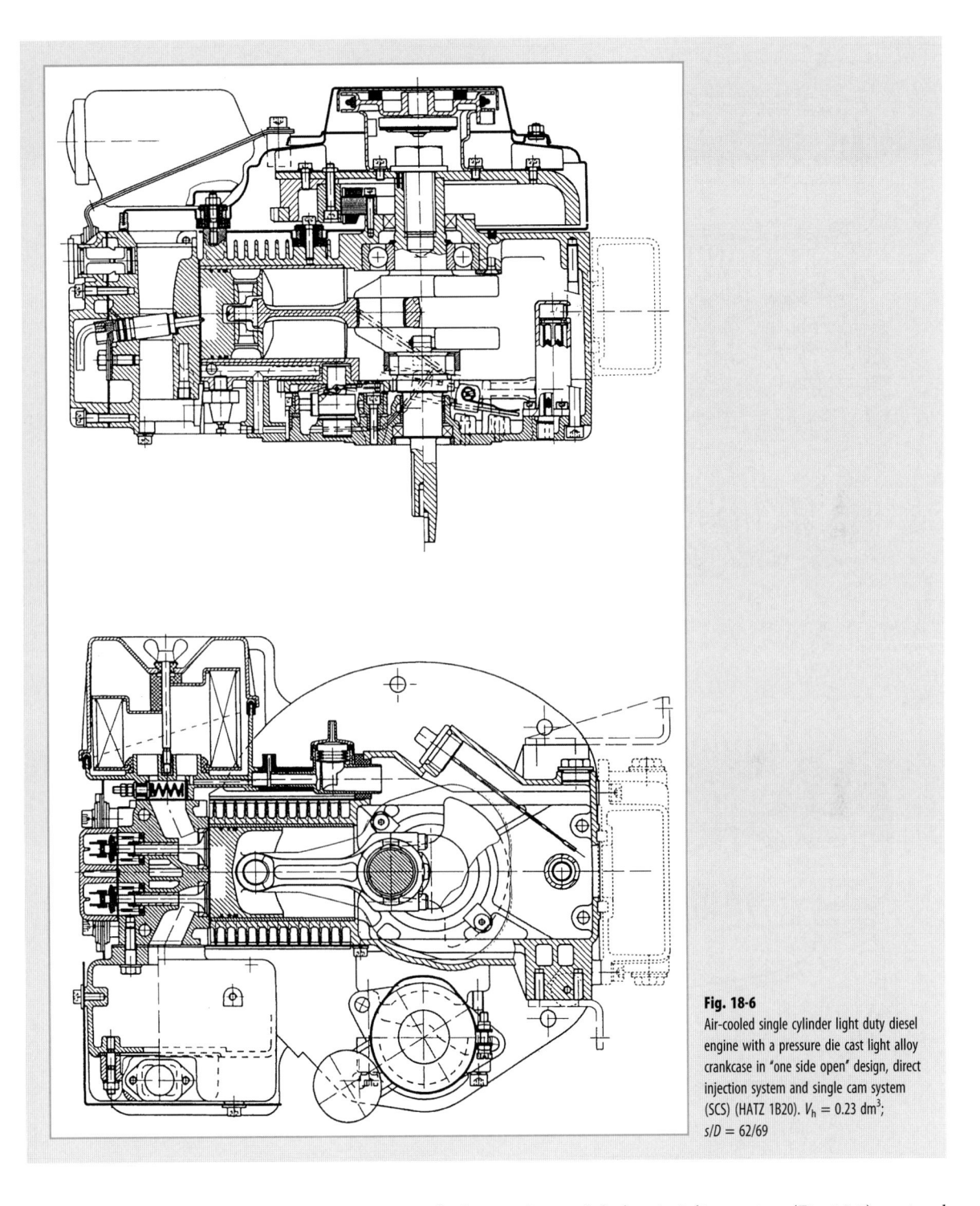

Fig. 18-6
Air-cooled single cylinder light duty diesel engine with a pressure die cast light alloy crankcase in "one side open" design, direct injection system and single cam system (SCS) (HATZ 1B20). $V_h = 0.23$ dm^3; $s/D = 62/69$

which a bolted-on oil pan seals the case as in multiple cylinder engines, is suitable for larger single cylinder diesel engines.

Camshaft and governor. The camshaft may be positioned in a variety of ways. It may be arranged laterally parallel to the crankshaft as in inline engines (Fig. 18-7), centered above the crankshaft or, optionally, on the control or flywheel side (Fig. 18-6). The transmission element of the P-governors utilized to control injection pumps are usually driven by the camshaft or oil pump gear. Stepped

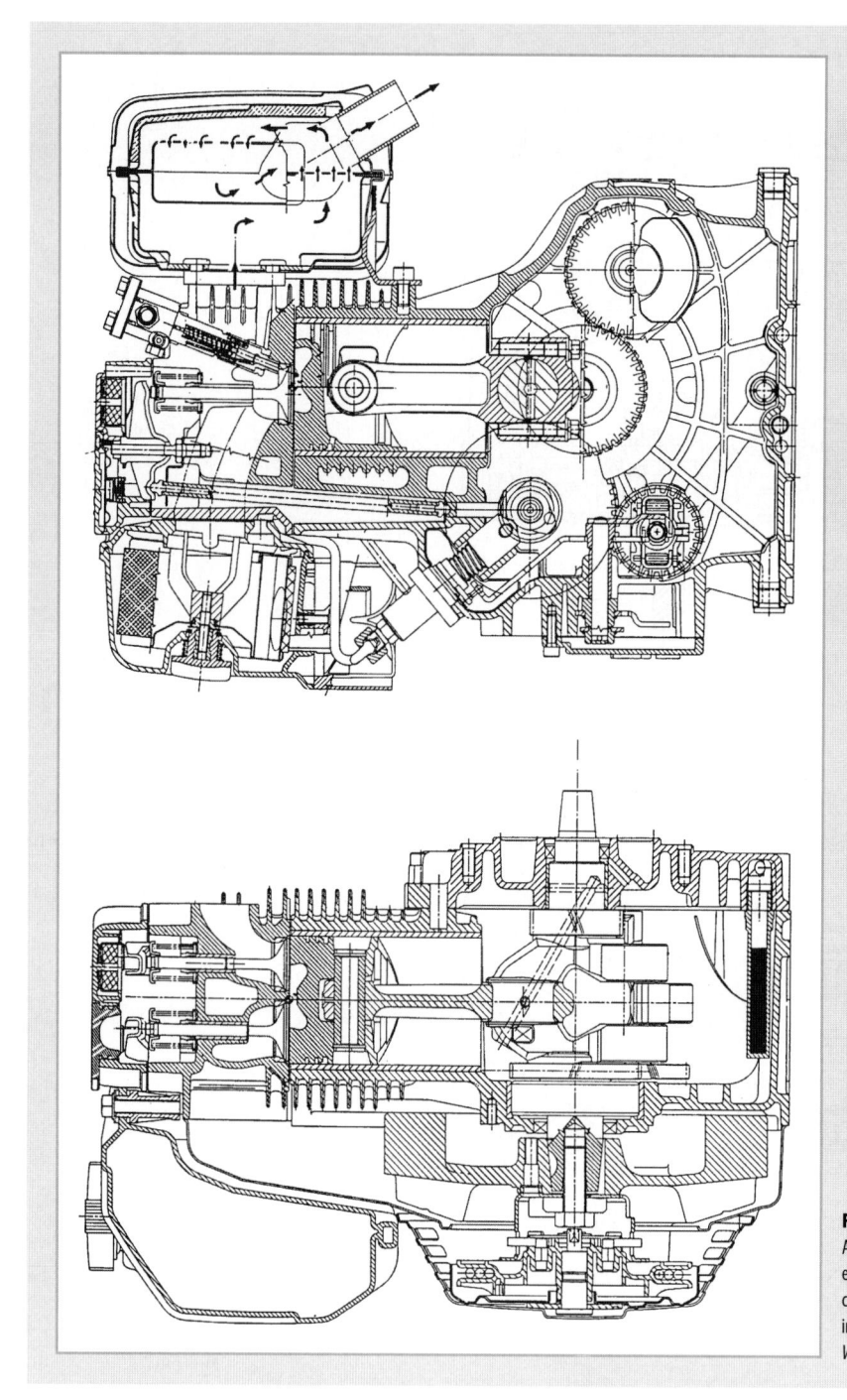

Fig. 18-7
Air-cooled single cylinder light duty diesel engine with pressure die cast light alloy crankcase in "one side open" design and direct injection system (LOMBARDINI 15LD315).
$V_h = 0.315$ dm³; s/D = 66/78

up transmission ratios are advantageous for control and reduce the space required for a governor. Every engine manufacturer has its own systems, which are more or less flexible depending on the use. A variable-speed governor, ± torque control and a freely selectable proportional degree of 3–10% (3–5% for electric generators) are standard. Camshafts with injection cams and valve gear assemblies are usually housed in the open space between the control cover and crankshaft web together with the injection pump, speed governor and oil pump drive (see

Fig. 18-7). Accepting a free moment of inertia, only the balance shaft for 100% first order mass balancing is located on the flywheel side.

Crankshaft Assembly

Crankshaft. A forged crankshaft is quenched and tempered and hardened in the region of the main bearing journal and the crank pin. Plain bearings still stand for long service life, roller bearings for low friction. Recently, PTFE coated plain bearings have also been being used to reduce friction, above all cold friction. Deep-groove ball bearings now reach calculated expected service lives of 4,000–5,000 hours of operation, which is sufficient for standard single cylinder diesel engine applications. Roller and plain bearings are often combined when the roller bearing is mounted on the flywheel side. Nodular cast iron crankshafts are conceivable for single cylinder engines but still remain to be proven in use in rough construction equipment. Composite crankshafts are too unreliable even in micro diesel engines and remain reserved for gasoline engines.

Connecting rod. A standard connecting rod is forged of steel, split in the connecting rod big end and given a bronze bushing in the piston pin boss. When they are appropriately designed, forged aluminum connecting rods may be implemented in short-stroke engines with displacements of less than 0.3 l. Connecting rods made of sintered material may be expected in the future. GGG 60 may also be used as a connecting rod material with the advantage of operation without a bushing in the small piston pin boss.

Pistons. Only aluminum alloy full skirt pistons are employed because of their minimum of oscillating masses (see Sect. 8.6). Control pistons are the exception and only expedient in encapsulated single cylinder engines.

A standard assembly consists of three rings:
- a usually chromium plated compression ring that is a rectangular or keystone ring, convexly finish machined and tapered,
- another tapered compression ring that is a Napier ring or a ring with an inner bevel and
- an oil scraper ring that is a top beveled or double-beveled ring with or without spring support (spiral expander ring).

Flywheel. Flywheels predominantly consist of gray cast iron with cast-in blading. Plastic fan rings may also be bolted onto flywheels. Consisting of several layers of plate and thus anechoic, deep drawn steel flywheels are the trend in small engines (see Fig. 18-6).

Mass balancing. Comprising 35–70% of the oscillating masses in addition to the rotating mass fraction, bolted-on counterweights are common for mass balancing (see Sects. 8.1 and 8.3). Fifty percent of first order mass balancing (normal balancing) usually suffices below a displacement of 0.5 l. One hundred percent mass balancing of first order

Fig. 18-8 Arrangement of balancing masses and balance shafts to completely balance first order inertial forces (HATZ-SUPRA concept line; see Fig. 18-3)

inertial forces is frequently connected with an additional free moment (see Sect. 8.3). The engine pictured in Fig. 18-3 (HATZ SUPRA series 1D30/40/60/80) achieves comparatively ideal 100% first order mass balancing. Figure 18-8 illustrates the principle: A single small counterweight at the flywheel side crankshaft web makes small overall engine height possible and simultaneously frees space on the control side to operate a balance shaft system that rotates counter to the direction of rotation in an almost ideal position (see Sect. 8.3.6). All other counterweights are arranged in the flywheel without interfering inside the housing.

Cylinder Liner, Cylinder Head and Valve Gear Assembly

Cylinders. Finned, compact gray cast iron cylinders attached to the crankcase with a continuous tensioning bolt design are employed because of their dimensional stability. Fitted aluminum finned cylinders with cast-in gray cast iron liners are also widespread.

Crankcases with integrated cast-in gray cast iron cylinder liners raised up to the top face of the cylinder head are also state-of-the-art for engines with $V_h < 0.4$ l and preferred for reasons of cost (see Fig. 18-7). In its B engine series, Hatz employs a centrifugally cast liner floating on an oil cushion. Its advantage is the option of replacement during maintenance (see Fig. 18-6). Single cylinder air-cooled diesel engines predominantly have cast aluminum alloy cylinder heads with spiral swirl ports inserted in the Croning shell molding to generate air swirl for mixture formation during direct injection (DI). The methods of core splitting and extraction suited for pressure casting employed by Hatz for the cylinder heads of its B series provide a very interesting solution (Fig. 18-6). The valves are solely actuated by rocker arms and push rods, either the cam followers or the interconnected drag levers having direct contact with the cam. Forged or deep drawn sheet metal arms are employed as rocker arms. Hatz has effectively implemented its patented purely mechanical valve clearance compensation (see Fig. 18-9) together with rocker arms fabricated from sheet metal in its B series engines. The intake and exhaust valve and the injection pump drive in these engines are actuated by only one cam profile path (the patented single cam system SCS in Fig. 18-10), thus saving space.

Injection system. The PLN (pump-line-nozzle) system is dominant. Direct injection (DI) requires short injection lines to minimize dead space volumes in the high pressure system and maximize the drive's dynamic rigidity. Driven by roller tappets or overhead camshafts, the UPS (unit pump system) particularly meets these requirements and, hence, is considered an alternative for future small single cylinder diesel engines in terms of installation space too.

Shaft nozzle holders and P type nozzles with needle diameters of 4 mm and thus less moving mass are standard for

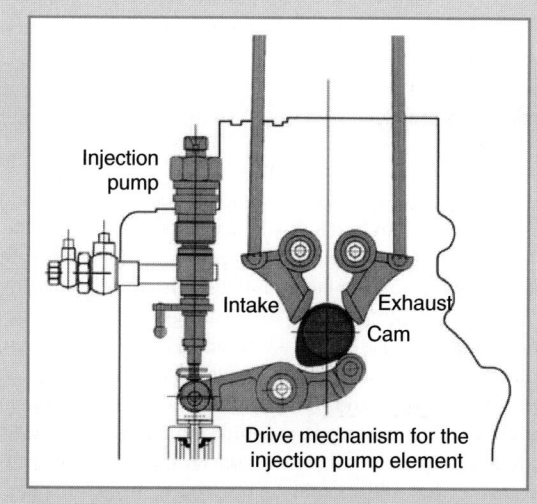

Fig. 18-10 Schematic of the patented single cam system SCS (HATZ)

PLN systems. Dual spring nozzle holders allow pilot injection that reduces combustion noise at light load but with the drawback of high smoke values in the full load range. Multijet injection would also be desirable for small diesel engines but must be ruled out because of the high costs. RSN nozzles have proven to be very suitable. They attain variable choke control at a larger needle stroke and thus reduce both combustion noise and NOx and CO emissions. Even though they are suboptimal, simple hydraulic measures have to be relied on in the pump element and in the injection valve since electronic components that control the start of injection either have to be ruled out for reasons of cost or are unavailable for small single cylinder diesel engines.

18.2 Stationary and Industrial Engines

18.2.1 Definition and Classification

The term stationary and industrial engines refers to nearly every combustion engine that has been modified and certified for use outside of road traffic, so-called nonroad applications. While vehicles with industrial engines may be operated partly on public roads, e.g. tractors, street cleaners or front loaders, the proportion of road traffic is only of secondary importance for these applications. They are certified in a cycle of levels based on the application in compliance with ISO 8178 (see Sect. 15.2). Typically, stationary and industrial engines are often implemented in a multitude of applications in very low quantities. This compels manufacturers to produce engines in a modular design in order to satisfy the conditions of a particular installation. A modular system that covers a maximum number of potential applications without allowing the

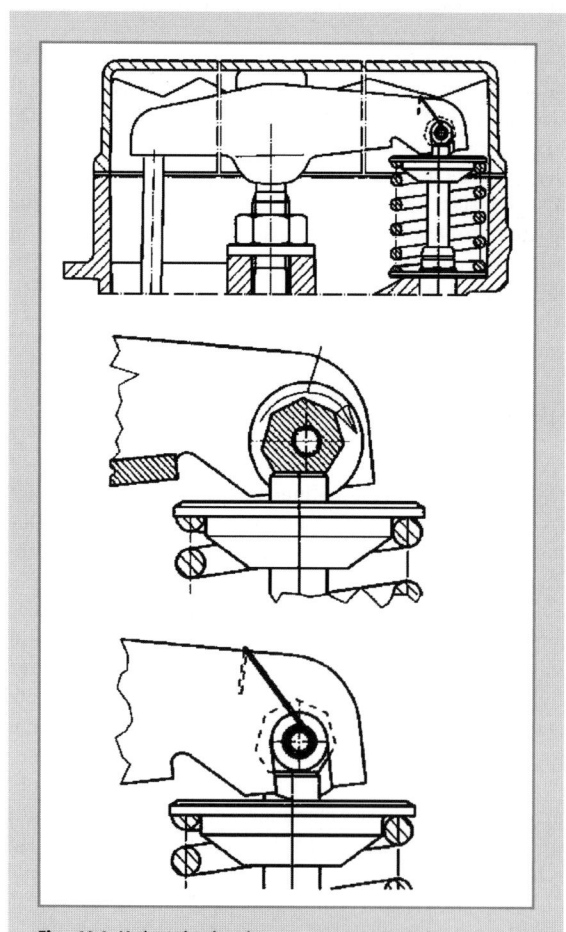

Fig. 18-9 Mechanical valve clearance compensation with an automatic eccentric multistep catch (torsion spring driven)

diversity of variants to grow to uneconomical scales is essential for add-on parts. The applications can be roughly classified in three groups:
– stationary engines,
– mobile machinery and
– agricultural equipment.

Stationary engines basically serve to generate power (generating sets) but are also employed for other units, e.g. cooling systems, pumps and compressors. Depending on the application, they are in continuous operation that may largely consist of light load operation or intensely intermittent operation followed by high load, e.g. emergency power units. Stationary engines are operated with variable loads but at constant speed. This particularly pertains to power units, which are operated in Europe at 1,500 rpm to guarantee a constant alternating current frequency of 60 Hz and in the USA at 1,800 rpm to guarantee 50 Hz. They are certified in the D2 cycle in compliance with ISO 8178 (see Sect. 15.2).

The domain of mobile machinery applications covers the large sector of construction equipment such as excavators, front loaders and bulldozers as well as forklifts, rail vehicles and airfield tractors. Depending on the application, engines in mobile machinery are operated in the entire map as well as at a fixed operating speed when the performance requirement is controlled by a hydraulic unit. The operating speed usually corresponds to the engine's rated speed. Train engines are either certified like industrial engines or in accordance with emission legislation for on-road commercial vehicle engines.

Agricultural equipment actually corresponds to mobile machinery in its applications but constitutes a distinctive category in terms of performance requirements and installed components. Therefore, agricultural equipment engines also constitute their own group technologically. Frequently, a distinctive design feature of agricultural equipment engines is their assumption of a stiffening function of the vehicle.

Based on their design, industrial engines may be classified as modified vehicle engines and as engines systematically developed for industrial application. Modified vehicle engines include both modified car engines for ratings of up to approximately 100 kW and commercial vehicle engines for ratings of up to approximately 500 kW. Industrial engines generally cover a power range from 2 to approximately 500 kW. Applications above 1,000 kW are operated by medium speed and large low speed engines (see Sects. 18.3 and 18.4). Only very few engines are in the power range between 500 and 1,000 kW.

Just as for car and commercial vehicle engines, electronically controlled injection systems are also becoming established for industrial engines, above all for engine powers above 75 kW. In effect since 2006, level 3 emission control legislation for industrial engines has made it impossible to meet the requirements in this power range with mechanically controlled injection systems at a justifiable cost. Unit pumps, unit injectors and, increasingly, common rails vie with one another as injection systems. Moreover, the electronics also allows the integration of customer functions and intelligent linkage of engine electronics with vehicle or equipment electronics (see Chaps. 5 and 6).

Many engines below 75 kW are equipped with mechanically controlled injection systems because the emission requirements are less stringent. Since an engine's purchase price plays a more dominant role than operating costs in this market segment, the advantages of engine electronics are foregone.

Industrial engines were originally only constructed as air-cooled engines (see Sect. 9.1.4). Dispensing with an additional cooling medium translates into an indisputable advantage for handling and maintenance in terms of reliability, especially in sometimes extremely harsh operating conditions, e.g. under extreme climatic conditions. The gradual tightening of emission limits is causing air cooling to be edged out partly by water cooling since lower component temperatures give the latter an advantage in terms of nitrogen oxide emissions as well as power density. Contrary to original expectations, developers have always successfully complied with every limit level in the past years, even with air-cooled engines. Thus, a market for air-cooled engines may be expected to continue to exist even in the near future. The manufacturer Deutz is very successfully marketing a concept for oil-cooled engines (see Sect. 9.1.3). The basic concept very closely approximates the design of water-cooled engines but avoids the additional cooling medium of water. Since oil temperatures are generally higher than the coolant temperature of water-cooled engines, oil-cooled engines are also operated with somewhat higher component temperatures (see Sect. 9.1.3).

18.2.2 Range and Selection

Despite the concentration process witnessed in the automotive and engine industry in recent years, the range of engines worldwide remains extremely large. Thus, this can neither be treated in detail nor in full here. Table 18-1 conveys an impression of the fields of application for diesel engines worldwide. The significant applications come from the car, commercial vehicle and agricultural equipment sectors, thus documenting the small quantities the multitude of applications for stationary and industrial engines normally entail. Nevertheless, numerous well known vehicle manufacturers also offer industrial engines. Figure 18-11 provides an overview – without any claim of completeness – of the global range of industrial engines.

The use of modified vehicle engines in the industrial engine range particularly provides advantages generated by mass production. However, this is frequently accompanied by little flexibility in terms of variations in installation and flexibility is precisely what manufacturers specialized in the production of industrial engines are strong in.

A distinction is made between so-called captive and non-captive manufacturers of industrial engines. The core

Table 18-1 Use of manufactured diesel engines (100 units)

Region	Japan	East Asia	North America	Western Europe	Eastern Europe	Worldwide total
Passenger cars	323	167	0	4,383	1,013	6,209
Commercial vehicles	774	1,047	693	2,328	277	5,853
Agricultural machinery	590	7,156	42	340	67	8,792
Construction equipment	299	61	112	271	22	812
Other industrial engines	140	47	31	186	9	482
Power units	204	179	17	247	28	711
Marine engines and auxiliary marine engines	38	203	13	35	3	297
Total	2,368	8,860	908	7,790	1,1419	23,156

business of captive manufacturers is the production of industrial machinery or vehicles. They cover the range of engines necessary for this machinery with their own engine production. Some typical manufacturers in this market are Caterpillar, John Deere and Yanmar. This market segment is called captive because it is unattainable for other engine manufacturers. However, the aforementioned manufacturers also sell their engines on the non-captive market and thus compete with pure engine manufacturers such as Cummins or Deutz.

In light of the increasing complexity also driven by emission control legislation, a concentration process also observable in the automotive industry has been emerging among the manufacturers of industrial engines in recent years. In particular, small captive manufacturers – whose annual engine production is 20,000–30,000 units – are increasingly switching to purchasing their engines on the non-captive market.

18.2.3 Applications

The market for stationary and industrial engines is characterized by a multitude of applications that are very frequently connected with low quantities down to job production. The art of engine manufacturing consists of covering this multiplicity of applications without sinking into an unmanageable and commercially unsustainable diversity of variants in one's own production in the process. The solution is a platform strategy based on a basic engine and a modular concept for the add-on components as pictured in Fig. 18-12 [18-1]. Installation conditions have given rise to a particular diversity of variants for oil pans, intake manifolds and exhaust manifolds. The latter must accommodate the various mounting positions for exhaust gas turbochargers. Alternators may be mounted just as variably when their installation is required in the first place.

The multiplicity of applications not only ensues from the installation restrictions but also the load profiles, climatic operating conditions, fuel grades, different emission standards, fuel consumption requirements and the sales price obtainable for an engine.

While only engines that meet the requirements of emission level 3 (see Sect. 15.2) are still sold in the European Union and North America, large parts of Africa and the Middle East have no emission standards of any sort in force. The sale of engines compliant with emission level 3 is impossible in these regions for commercial reasons and because of the engines' technological complexity. Air or oil-cooled engines with mechanically controlled injection systems are also preferred here for climatic and logistical reasons.

The widely varying grades of fuel throughout the world are another aspect that already has to be taken into account during development. While a fuel standard compliant with EN590 with a cetane number of at least 51 applies in Western Europe, diesel fuels in the USA have an average cetane number of 40–42. Experience has shown that this increases nitrogen oxide emissions by approximately 0.2 g/kWh. Since the same limits for industrial engines apply in the USA and Europe, this must be factored in when an engine is modified. Furthermore, the fuel's sulfur content – above 5,000 ppm in some regions – represents a challenge both in terms of the effects on particulate emissions and damage by sulfuric acid corrosion. Some EU countries allow the use of fuel oil for industrial engines. This is not compliant with the EN590 standard either. Biofuels are increasingly being used. Subsumed under the term FAME (fatty acid methyl ester), such esterified vegetable oil-based fuels are normally not approved by manufacturers of injection systems. Thus, engine manufacturers bear the risk of approval. Rape oil methyl ester (RME) is the biodiesel common in Germany (see Sect. 4.2).

Since an engine's useful life also depends on its operational demands, manufacturers calibrate them in performance classes (see Table 18-2). An engine's power is reduced to prevent thermal overloading by steady full load operation.

The cooler package for a tractor illustrates the complexity of installation in Fig. 18-13. The installation accommodates the tractor's down swept front end, the design of which improves the driver's field of view. Since cooler efficiency quite heavily depends on configuration and flow, installation must be simulated before technical approval to ensure that the cooler package conforms to the engine's design data. This is necessary to comply with emission requirements as well as to prevent the engine from thermally overloading.

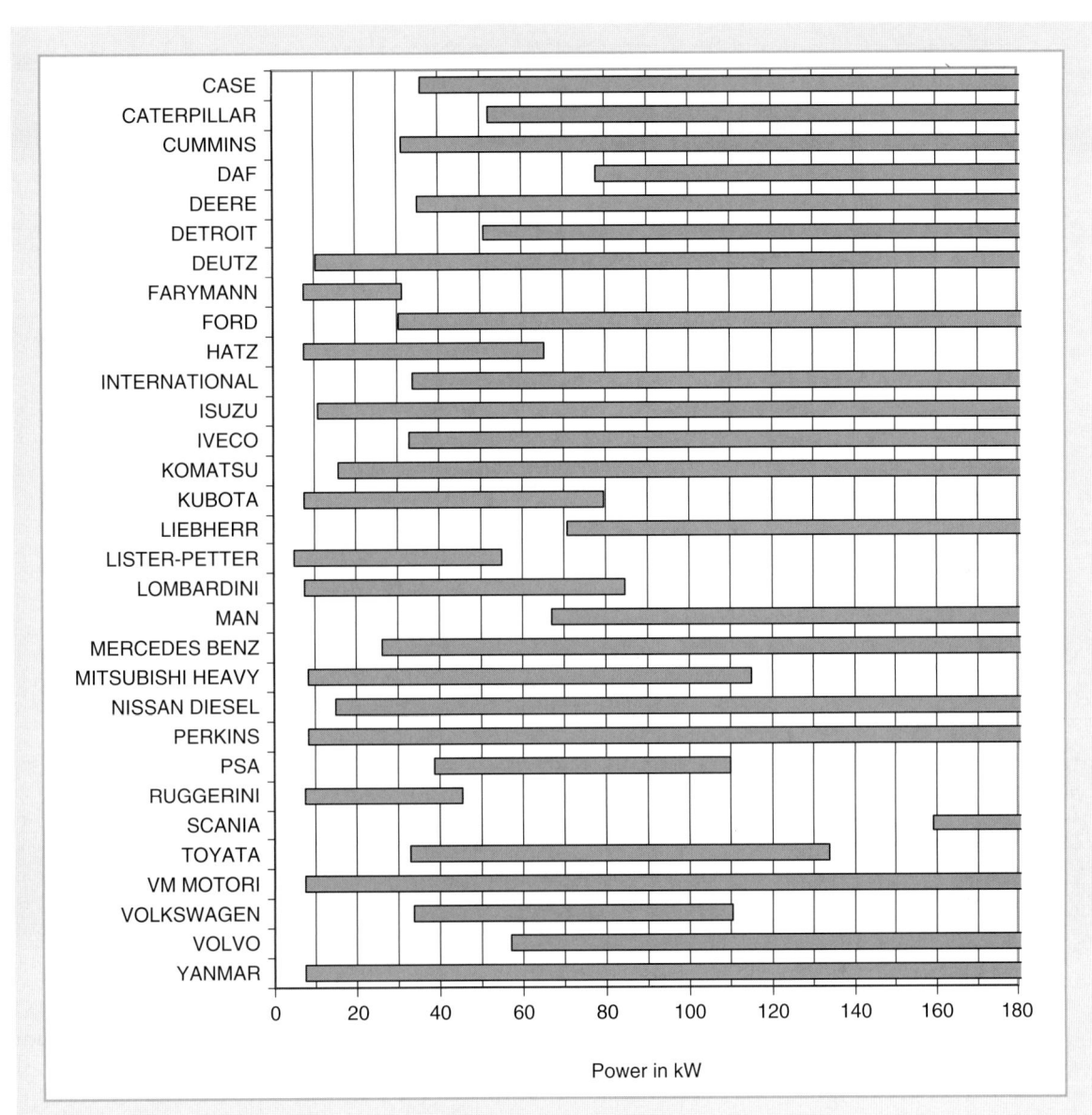

Fig. 18-11 The range of industrial engine power from the most important (German and foreign) manufacturers (source: DEUTZ AG market analysis)

18.2.4 Modified Vehicle Engines

18.2.4.1 General Remarks

Along with the special diesel engines developed solely as industrial engines, other modified car or commercial vehicle engines also end up in general industrial use. The indisputable advantages of modified vehicle engines are the cost advantage generated by synergies with mass production and lightweight construction with a good weight-to-power ratio.

Limits on their application arise whenever cost effectiveness or technical features suffer, e.g. overloading of the crankshaft assembly, which is distinguished by lightweight construction in vehicle engines. Therefore, the suspension of a weight-optimized engine ought to be treated just as if it were installed in a vehicle. For reasons of strength, an engine crankcase should not to be drawn on for system-supporting functions as is frequently common in the design of agricultural and construction equipment. Since a weight-optimized vehicle engine has less mass, its noise and vibration damping measures already require somewhat more complex treatment

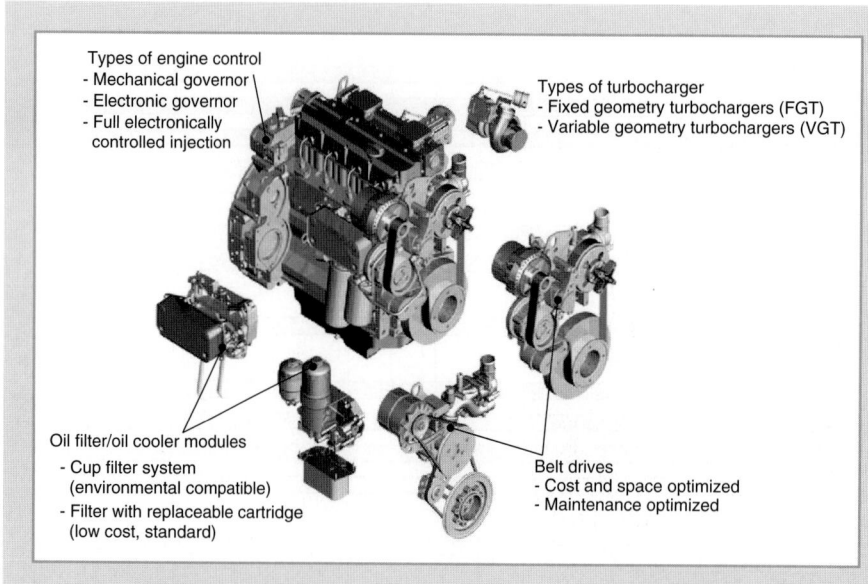

Fig. 18-12 Modular concept of an industrial engine consisting of the basic engine and various add-on part options

Engine labels:
- Types of engine control
  - Mechanical governor
  - Electronic governor
  - Full electronically controlled injection
- Types of turbocharger
  - Fixed geometry turbochargers (FGT)
  - Variable geometry turbochargers (VGT)
- Oil filter/oil cooler modules
  - Cup filter system (environmental compatible)
  - Filter with replaceable cartridge (low cost, standard)
- Belt drives
  - Cost and space optimized
  - Maintenance optimized

Table 18-2 Definition of performance classes and examples of respective applications

Power class	Power reduction %	Vehicle engines	Stationary engines		
			Construction equipment	Agricultural and forestry equipment	Pumps and compressors
I	0	Construction site vehicles Fire vehicles Dump trucks Crane vehicles Street cleaners	Front loaders Backhoe loaders Graders Earth moving equipment Road rollers Concrete and mortar mixers	Combine harvesters	Fire pumps Emergency pumps
II	5	Snow blowers Snowplows	Hydraulic excavators Blacktop paver Concrete and road milling machines	Four-wheel tractors Skidders Pruning platforms Chopper forage harvesters Harvesters	Sprinkling and irrigation systems High pressure compressors up to 10 bar
III	10		Trench diggers Drilling equipment		High pressure compressors over 10 bar

than a heavy industrial engine's. A vehicle engine's power is adjusted to requirements specific to the vehicle and, in accordance with DIN/ISO 3046, should be selected significantly lower for industrial engines in continuous operation to facilitate engine service life with low wear (see Sect. 18.2.3). When high starting torque is required, a better commercial vehicle engine rather than a car diesel engine ought to be the stated basis for engine selection from the start: Car engines normally do not reach their maximum torque until the upper speed range. While special adjustment of injection pumps enables shifting the maximum torque to the lower speed range within certain limits, any deviation from standard vehicle equipment becomes expensive and unprofitable when it requires too much special equipment, which may well be desirable but is not absolutely necessary.

In order to be able to assess them, the options for modifying a standard vehicle engine for needs when used as an industrial engine have to be harmonized with the requirements.

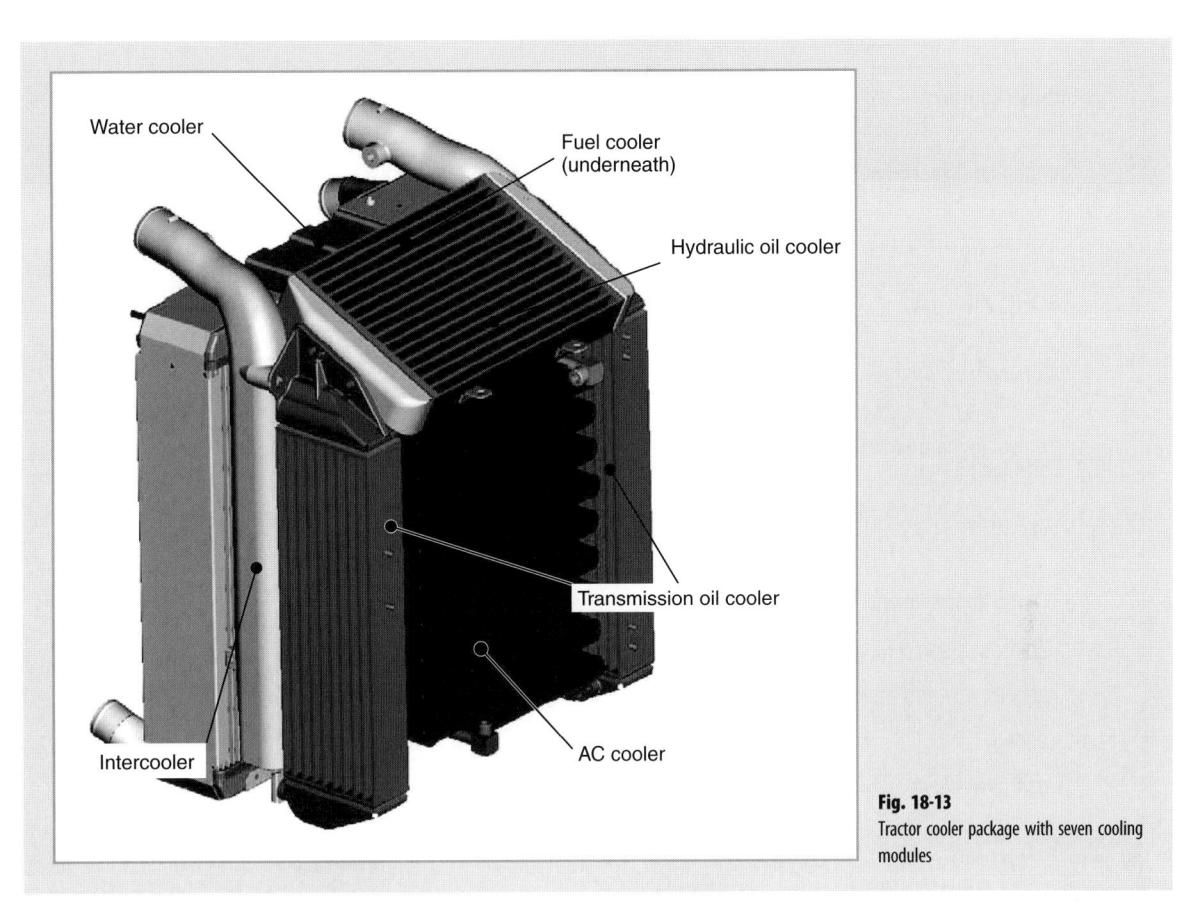

Fig. 18-13 Tractor cooler package with seven cooling modules

Service life and maintenance intervals play a crucial role for industrial engines. Such installation often requires larger engine oil volumes to facilitate longer oil maintenance intervals. Matched to the conditions of installation, special oil pans with an oil volume of up to 20 l are employed. The car version is approximately 4 l. Not only the extra costs, but also existing manufacturing facilities and manufacturing options have to be weighed to determine whether such a modification still leads to the desired aims or is even affordable.

18.2.4.2 Implemented Engines

Figure 18-14 pictures a modified vehicle engine manufactured by Volkswagen [18-2]. The engine is based on a car engine, the only difference being is its dataset to obtain maximum synergy with mass car production. Swirl chamber engines are still manufactured as naturally aspirated engines only in the power range up to 37 kW. However, stage 3A emission legislation requires that they be replaced by direct injection engines, which are manufactured both as naturally aspirated engines and as turbocharged engines with and without intercooling up to approximately 80 kW with displacements of 1.9 and 2.5 l.

A dataset concept consisting of a block of the particular basic engine functions independent of the application and a block of industrial engine functions was developed to modify these basic engines developed for vehicle use for the widest variety of requirements for industrial applications. The engine control unit stores the industrial engine functions in seven specific datasets. These differ in their type of engine control:

– torque control specified by the accelerator pedal,
– power control specified by the accelerator pedal,
– operating speed control specified by the accelerator pedal taking the form of a proportional governor,
– operating speed control by a 0–5 V interface taking the form of a proportional integral governor with or without a safety concept,
– automotive driving by converting driver demand by the accelerator pedal into an injected fuel quantity and
– stationary operation by an externally connected fixed speed governor.

Diverging from vehicle engines, the electronics has a few actuators and sensors specifically for industrial applications. The variance described here has only been producible in engines with a mechanical-hydraulic injection pump by

- Direct injection
- Distributor injection pump VP 37
- ECU Bosch EDC 15V
- SOHC valve train
- Aluminum cylinder head
- Vertical rechargeable cartridge oil filter
- VGT
- TDI 2.5 specifications:
  - 80kW at 3.500 rpm
  - 280Nm at 1,400 - 2,400 rpm
  - 200kg
  - 211 g/kWh(bsfc $_{best}$)

Fig. 18-14 VW supercharged 2.5 l TDI® engine with electronically controlled distributor injection pump, two-valve cylinder head, variable turbine geometry turbocharger and a maximum power of 80 kW at 3,500 rpm

setting the injection pump as a function of the application in conjunction with special equipment such as a variable-speed governor and frequently only together with an additional electronic governor.

18.2.5 Industrial Engines

18.2.5.1 Product Concept

Diesel engines exclusively developed as stationary and industrial engines can be found in the performance class up to approximately 75 kW. All numbers of cylinders between one and four are common. Fuel consumption and power density play a subordinate role; procurement costs, ruggedness and versatility are more important. Naturally aspirated engines are increasingly being replaced by supercharged engines because of emission requirements but frequently without an intercooler for reasons of installation and cost. However, naturally aspirated engines still dominate the power range up to 37 kW. Several power take-offs and the option of 100% power take-off on the damper side crankshaft end are common. The low maintenance requirements have led to a very large proportion of engines in this performance class being air or oil/air-cooled.

Also implementable as vehicle engines, industrial engines in the power range above 75 kW – designed as four, six or eight cylinder engines – are usually either derived from commercial vehicle engines or industrial engines. The requirements for exhaust and noise emission, maintenance, service life, weight-to-power ratio, power density, etc. for industrial engines in this performance class are similar to those for commercial vehicle engines. The only difference from commercial vehicle engines are special design features such as power take-offs including take-offs on the damper side crankshaft end too, particularly rigid engine blocks, stiffened oil pans and engine balancers when installed in tractors, additional cooling systems for hydraulic systems and generators.

The variety of potential uses for industrial engines and thus potential customer demands necessitate just as many equipment options, which often may only be produced in relatively small quantities. Hence, industrial engines are normally more expensive than corresponding car or commercial vehicle engines.

18.2.5.2 Implemented Engines

Given the broad range of industrial engines, this section presents the basic features of a few examples implemented in the segment for displacements between 200 and 400 cm^3 and the segment in the range of one liter per cylinder. Space limitations preclude coverage of the vast variety.

Virtually only designed as inline engines, a multitude of multiple cylinder water-cooled engines occupy the segment for small displacements up to 400 cm^3. Important manufacturers are Kubota, Yanmar, Daihatsu, Isuzu, Lister Petter and Deutz. Their basic design principles are all very similar. However, the injection equipment varies widely from pump-line-nozzle systems to distributor pumps and inline pumps up through unit pump systems. The direct injection swirl chamber principle is frequently employed. Along with its noise advantage and lower manufacturing costs, it also has a lower level of nitrogen oxide emission than the direct injection principle. The disadvantageous fuel consumption connected with it is accepted in this market segment.

A very interesting example of design is Hatz's W35 series (see Fig. 18-15) with a displacement of 350 cm^3 per cylinder and available as an inline engine with two, three and four cylinders. This inline engine has a vertical parting plane running longitudinally. Both halves of the pressure die cast aluminum crankcase already contain an oil pan, timing case, gear case, flywheel housing, cylinder head and cylinder liner holder. The cylinder head and the thin-walled centrifugally cast cylinder liner are modules. The water pump is integrated in the timing case's gear drive. The engine is already completely enclosed with merely three components, i.e. the left and right crankcase half and the valve cover. The injection system and its timing are modularly designed. A mechanically controlled, very compact unit injector is driven by the overhead camshaft and the driving lever. Figure 18-16 pictures the design of a completely equipped 4W35 naturally aspirated engine.

The implemented examples in Figs. 18-17 and 18-18 illustrate the ways different market and statutory requirements influence the design of an engine series. An emission limit of 4.7 g/kWh for NO_X+HC and 0.3 g/kWh for particulate matter in the C1 cycle compliant with ISO 8178 applies to the 64 kW version (see Sect. 15.2). For reasons of space and cost, this engine does not have an intercooler and is equipped with a mechanically controlled unit pump injection system. The emissions goal was achieved by appropriately designing the combustion chamber and injection nozzle geometry, injection timing and gas exchange. The disadvantages of fuel consumption connected with this concept are more than compensated by the low manufacturing costs. The 113 kW version in Fig. 18-18 was modified for high capacity utilization in a tractor. An emission limit of 4.0 g/kWh for NO_X+HC and 0.2 g/kWh for particulate matter in the C1 cycle applies to this power. Fuel consumption plays a significantly larger role because of the load profile and the high number of operating hours per year. More technological complexity and thus a higher engine price are accepted. Both intercooling and cooled exhaust gas recirculation are suitable measures to obtain low nitrogen oxide emissions in conjunction with good fuel consumption. In addition, the common rail injection system allows optimizing fuel consumption, combustion noise and emissions in the entire engine map [18-3, 18-4].

These typical stationary engines furnish more options for power take-off. The timing gear drive can drive not only the camshaft but also one or more hydraulic pumps and/or compressors. Alternatively, up to 100% of the engine power can also be taken off the front side. These water-cooled engines are also provided with integrated cooling systems, the oil and water cooler being attached on the sides of the engine. An additional cooler for hydraulic oil can also be provided as desired. This not only facilitates extremely compact design but also simplifies the assembly of a complete engine delivered with integrated cooling.

Agricultural machinery is a typical case of industrial engine application. The high rigidity of the engine block not only diminishes noise emission but also enables utilizing the engine as a supporting tractor component. In the four cylinder version, a balancing differential gear eliminates the free inertial forces and thus guarantees the smooth operation necessary for rigid installation in a tractor.

18.2.6 Outlook

Like the development of vehicle engines, the further development of industrial engines will be characterized in coming years by a further drastic tightening of emission control legislation (see Sect. 15.2). The related technological requirements in stage 3A emission legislation have already led to the replacement of mechanically controlled injection systems by electronically controlled systems in many engines. Mechanically controlled injection systems only continue to be dominant in the power range below 75 kW.

Compliance with limit level 4 requires the introduction of exhaust aftertreatment technologies such as particulate filters and nitrogen oxide aftertreatment by SCR. Since these technologies require monitoring and control, integrated engine management that also takes over the control of the exhaust temperature for exhaust aftertreatment is indispensible. In view of the higher emission limits, low cost mechanically controlled injection systems will probably only be applied in the power range below 56 kW in the future too.

The common rail injection system is considered to have the greatest future potential because of its flexibility, particularly the option of multiple injections. This also simplifies the modification of dynamic operating performance, which

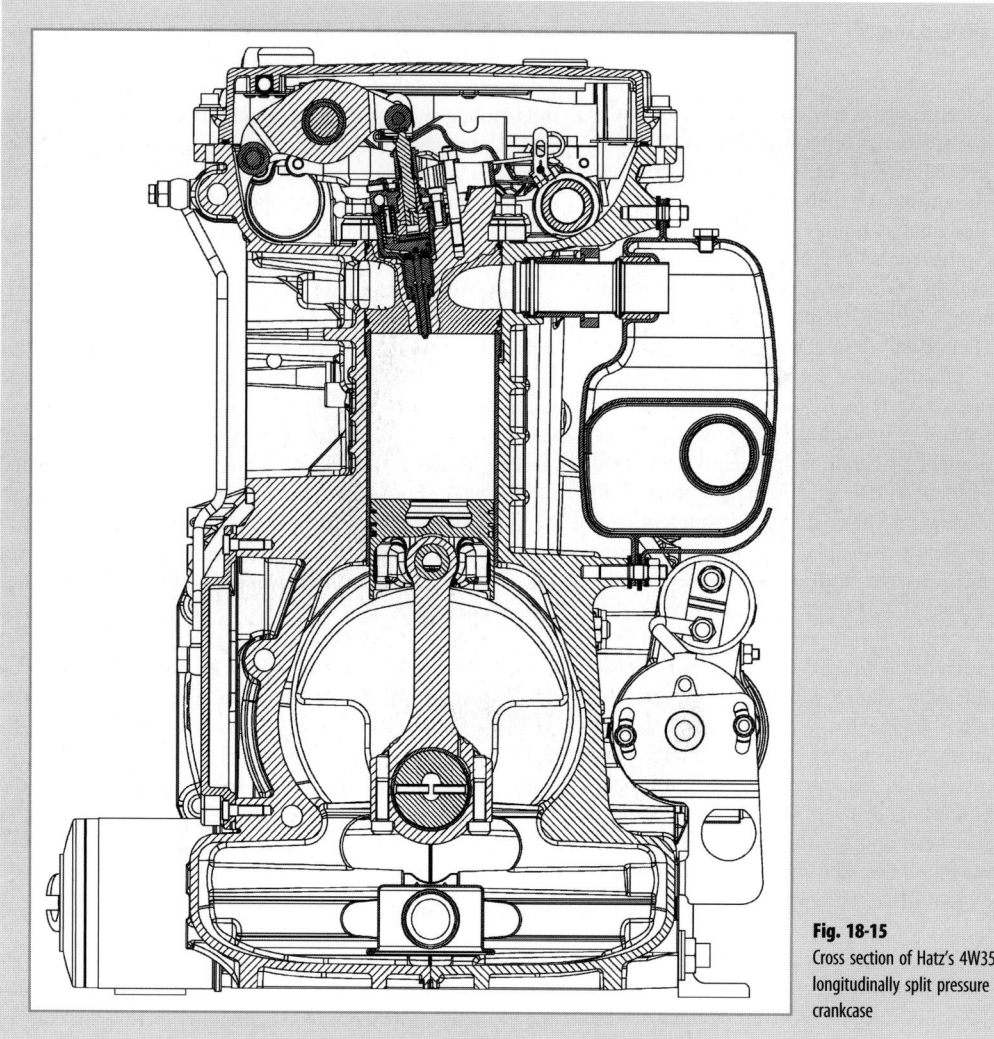

Fig. 18-15 Cross section of Hatz's 4W35NA engine. Vertical longitudinally split pressure die cast aluminum crankcase

must not only satisfy customer demands but also emission control legislation by introducing a transient cycle. In addition, electronic engine management furnishes the option of integrating additional customer functions and allows data exchange with other vehicle or machine systems.

This increasing complexity of technologies and their integration in the overall engine system not only demands engine manufacturers have expertise in the field of injection and combustion but also additional expertise in electronics and exhaust aftertreatment. Machinery manufacturers that produce engines in relatively small quantities for internal need are hardly able to do this. They are increasingly switching to buying engines from manufacturers that, by virtue of their size, have core competencies in the aforementioned fields. Thus, a continuation of the concentration process among engine manufacturers may also be expected in the coming years.

18.3 Medium Speed Four-Stroke Diesel Engines

18.3.1 Definition and Description

18.3.1.1 Classification of Medium Speed Four-Stroke Diesel Engines

The term medium speed engines refers to trunk-piston engines that now almost exclusively operate with the four-stroke process. Still sporadically available on the market, two-stroke trunk-piston engines are of secondary importance and are not examined here any more closely.

Medium speed four-stroke engines have speeds of between approximately 300 and 1,200 rpm. Their cylinder dimensions range from under 200 to over 600 mm. In recent years, mean piston velocities have settled at approximately 9–11 m/s and even more in individual cases.

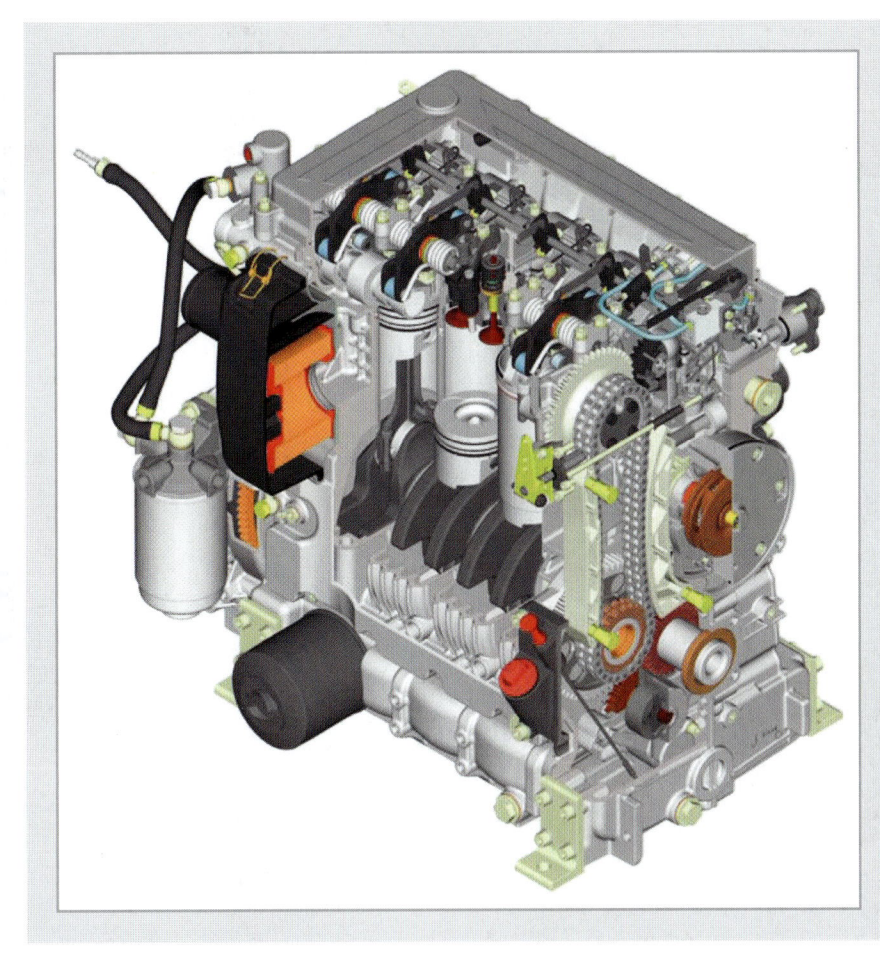

Fig. 18-16
Hatz W35 series water-cooled four cylinder OHC inline engine with mechanically controlled unit pump

In addition, brake mean effective pressures p_e of up to 29 bar are being attained. This corresponds to effective brake work of $w_e \leq 2.9$ kJ/dm^3.

These values yield a power range for modern medium speed four-stroke engines of approximately 100 to over 2,000 kW/cylinder.

Inline engines are now built with six to ten cylinders (sometimes even fewer) and V engines with twelve to twenty cylinders.

18.3.1.2 Medium Speed Diesel Engine Use

Medium speed four-stroke engines are employed as main marine engines, auxiliary marine engines and stationary engines to drive generators. Smaller medium speed engines are also implemented to drive pumps and compressors, in combined heat and power stations or as traction motors.

In the last forty years, diesel engines have established themselves in civil shipping over other propulsion alternatives (steam or gas turbines). Medium speed four-stroke engines have gained great importance and become established for types of ships that were previously reserved for two-stroke low speed engines (Fig. 18-19). Medium speed engines are the only option for many cases of application from the outset for reasons of space utilization, e.g. ferries, RoRo ships and other special ships. Diesel-electric drives with medium speed engines are frequently implemented in passenger and cruise ships because of their high flexibility. Smaller medium speed engines are predominantly employed as driving engines in inland shipping too. At least in larger ships, medium speed engines are used almost exclusively as auxiliary marine engines (Fig. 18-20).

Another field of application is diesel power stations that generate electricity, which are extremely widespread, particularly in industrial threshold countries without interconnected power systems covering entire regions. An appropriate number of machine sets with unit outputs of 10–20 MW can be used to cost effectively erect power stations of up to 100 MW and higher in short time, especially since they can be progressively expanded based on demand.

Fig. 18-17 DEUTZ TD2012L04 2 V turbocharged four cylinder engine without intercooler and with mechanically controlled pump-line-nozzle injection system, two-valve cylinder head, side-mounted exhaust gas turbocharger. Maximum power of 67 kW at 2,200 rpm

18.3.1.3 Fuels

Heavy Fuel Oil Operation

Advanced larger medium speed engines are able to process even poor grades of heavy fuel oils as defined in CIMAC H/K55 (see Sect. 4.3). This has been achieved by consistently developing components, above all the combustion chamber, valves, pistons and cylinder liners, etc. Heavy fuel oil is used almost exclusively in large medium speed engines that are utilized both as main marine engines and in stationary plants, provided environmental requirements do not impose any restrictions.

The time available for combustion is a significant variable that influences the combustion of heavy fuel oils. Understandably, larger medium speed engines with speeds of up to 750 rpm are more easily built to be suitable for heavy fuel oil than smaller engines with speeds in the range of 1,000 rpm and above. The permissible grade of heavy fuel oil may generate certain restrictions for such higher speed engines. Nevertheless, auxiliary marine engines are also increasingly being designed for heavy fuel oil operation to be able to supply main and auxiliary machines with the same fuel (unifueled ships).

Appropriate precautions must be taken for the combustion of heavy fuel oil, both to process the fuel and to select the lubricating oil (see Sect. 4.3). Engine design must ensure that the components influenced by the heavy fuel oil have the "right" temperatures to prevent harmful effects of corrosion and deposits (see also Sects. 18.3.3 and 7.1).

Hence, smaller higher medium speed engines are often run with diesel oil. The time and effort required to process heavy fuel oil is not worthwhile in most cases.

Gas Operation

Along with heavy fuel oils and the widest variety of grades of diesel fuels, medium speed four-stroke engines are also run with different fuel gases. Both spark ignited and dual fuel systems are employed (see Sect. 4.4).

Unlike diesel operation, conventional dual fuel design in which a homogeneous air/fuel mixture is compressed and ignition is triggered by injecting a small quantity of ignition

Fig. 18-18
DEUTZ TD2012L04 2 V supercharged four cylinder engine with intercooling and electronically controlled common rail injection system, four-valve cylinder head, centered overhead exhaust gas turbocharger and external cooled exhaust gas recirculation. Maximum power of 113 kW at 2,200 rpm

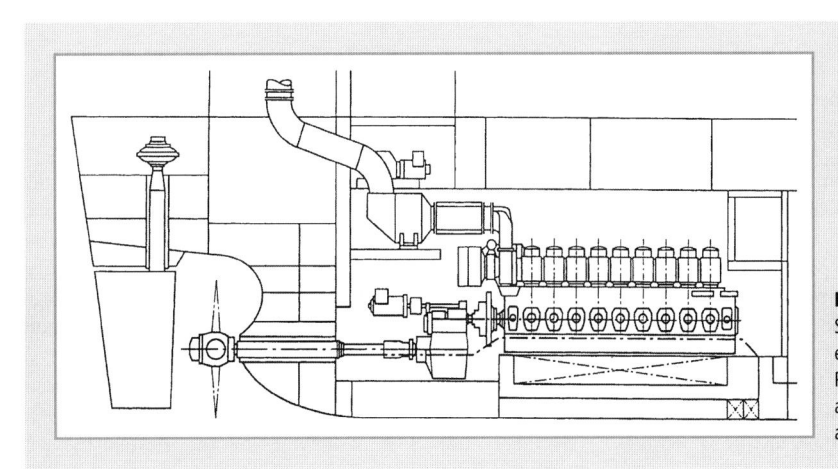

Fig. 18-19
Single engine system with a medium speed engine, reduction gear and generator driven by PTO (power take-off). Engine power of 12,500 kW at 428 rpm to drive a refrigerated cargo ship with a rated capacity of 400,000 cft

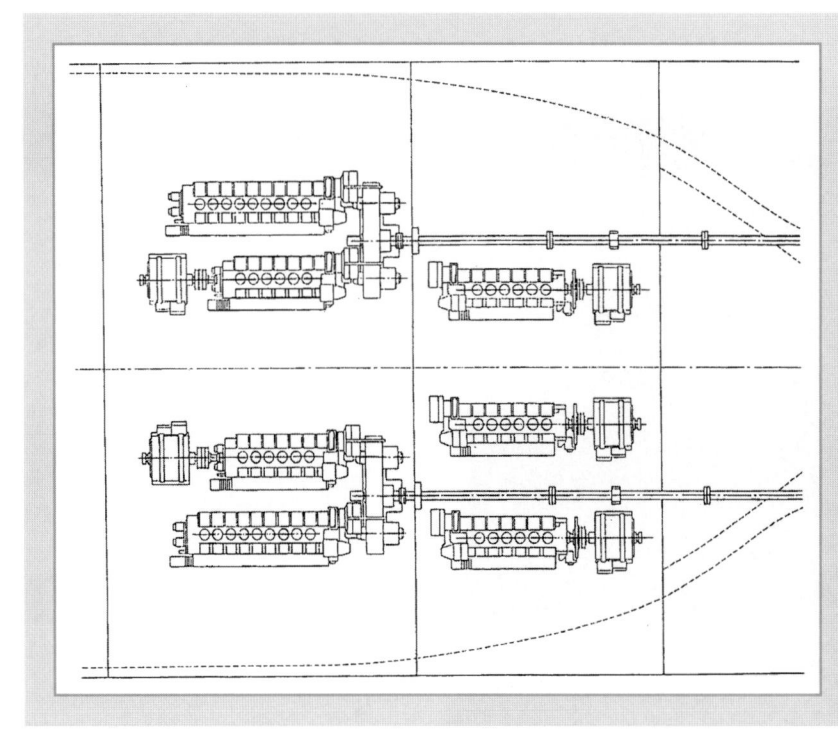

Fig. 18-20 Cruise ship machine room configuration: each of the main engines is in a double "father-and-son" configuration using a reduction gear to operate two propellers. There are three auxiliary engines of the same type to generate on-board power and additionally generators driven by the 'son' engines

oil must recover the brake mean effective pressure and the power. As recent developments have demonstrated [18-5], appropriately designed dual fuel engines have come close to the power of diesel operation without having to apply the dual fuel process with high pressure gas injection (see Sect. 4.4.3.1).

The lean burn process was developed to keep emission values low. Ignition oil ignites a lean air/fuel mixture in a secondary combustion chamber. Upon being discharged from the prechamber, it serves as a high energy ignition aid for the lean mixture in the main combustion chamber [18-6].

Both the lean burn system and spark ignited engines have become increasingly widespread among medium speed four-stroke engines in recent years.

18.3.1.4 Advantages of Medium Speed Engines

Medium speed four-stroke diesel engines are situated between high speed high performance engines and two-stroke low speed engines. In terms of use, the transitions are fluid.

The basic advantages of medium speed engines over two-stroke engines are their lower space requirement and comparatively low weight-to-power ratio in conjunction with better specific costs [18-7].

This holds true even though a medium speed propulsion engine is always equipped with a reduction gear (Fig. 18-21).

Apart from the space advantages, there are other points in favor of medium speed engines:

- free selection of the optimal propeller speed,
- good suitability for elastic installation to insulate structure-borne noise,
- very simple option to recover waste heat,
- speeds common to generators,
- simple shaft generator attachment to generate power in heavy fuel oil operation,
- good prerequisites for measures that reduce pollutants,
- easy attachment of power and compound turbines to increase cost effectiveness and
- good suitability for engine management systems and remote monitoring.

18.3.2 Design Criteria

18.3.2.1 Specific Power

Triggered by competitive pressure and facilitated by the further development of supercharging equipment, specific powers have continuously been increased over the course of time. Brake mean effective pressures and specific work have partly reached the limit of what is achievable with one-stage supercharging. Mean piston velocity has also been increased continuously.

Proportional to the product of specific work and mean piston velocity, the specific power per unit piston area P_A, (see Sect. 1.2) is a parameter for the state-of-the-art. Today's medium speed four-stroke diesel engines attain

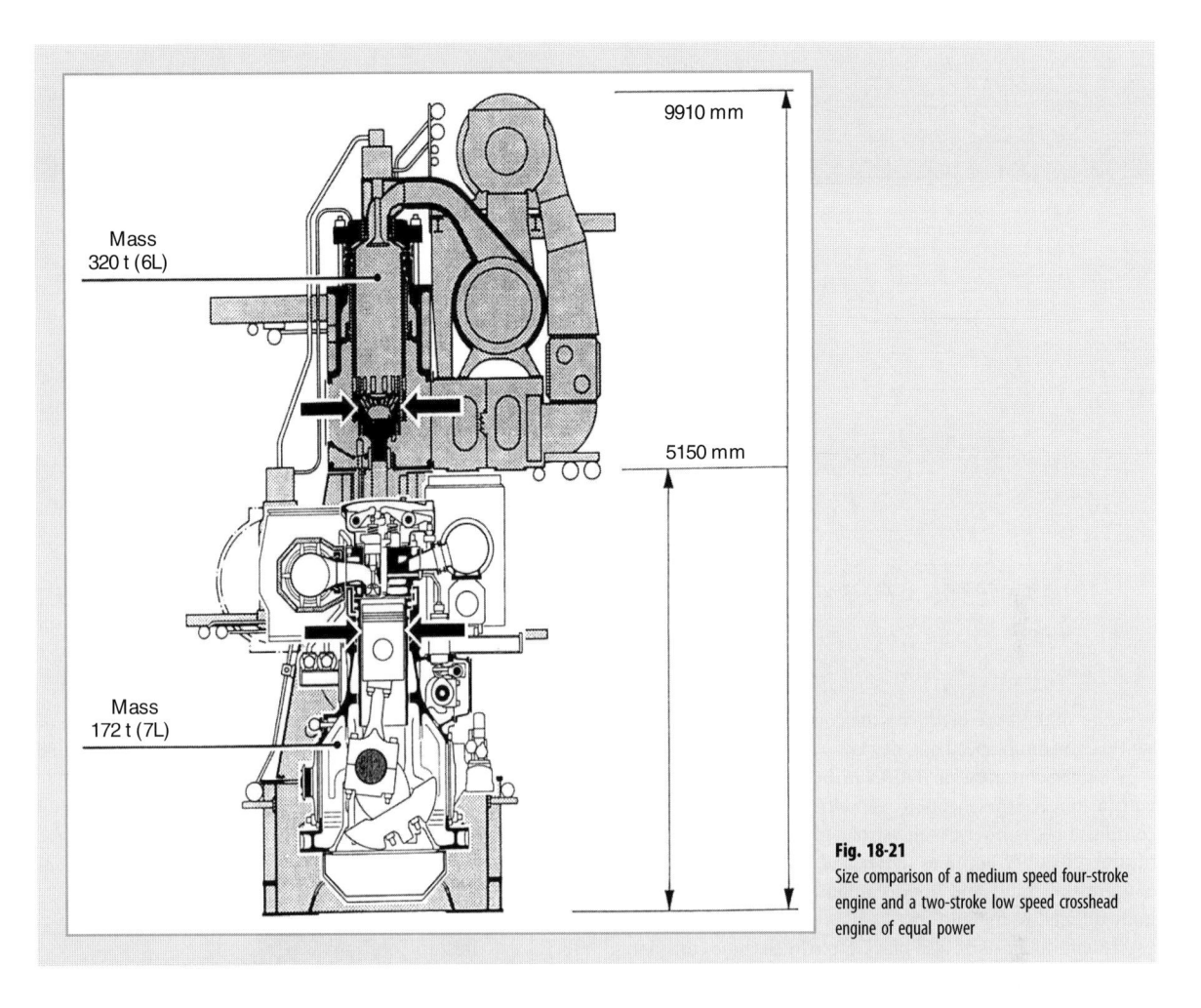

Fig. 18-21 Size comparison of a medium speed four-stroke engine and a two-stroke low speed crosshead engine of equal power

powers per unit piston area of 5 W/mm² at peak values of approximately 7 W/mm².

18.3.2.2 Maximum Cylinder Pressure

Parallel to increasing specific power per unit piston area, great efforts have been made to lower fuel consumption, not least motivated by the oil crises in the 1970s and 1980s and the steady rise of fuel prices. The ratio of maximum cylinder pressure to effective brake work, also expressed by the ratio p_{Zmax}/p_e, has proven to be an important value to characterize efficiency. When possible, peak pressure also had to be elevated disproportionately as mean pressure increased to assure a sufficiently high p_{Zmax}/p_e ratio of approximately 7–8 and, thus, minimum fuel consumption. Hence, maximum cylinder pressures have reached a remarkable level in recent years. Engines with peak pressure of 200 bar are already in operation and the trend is continuing. Figure 18-22 presents the development of maximum cylinder pressure and specific fuel consumption over the past decades.

18.3.2.3 Stroke to Bore Ratio

It is important that the rate of pressure increase $dp_Z/d\varphi$, i.e. the interval between final compression pressure and maximum cylinder pressure or so-called ignition jump, does not grow too large, particularly in heavy fuel oil operation. It follows from this that high maximum pressures also require significantly higher final compression pressures. The final compression pressure is influenced by the boost pressure and compression ratio. The boost pressure level is limited for reasons of thermodynamics (see Sect. 2.2). The permissible boost pressure can be expediently described with a parameter that specifies the ratio of boost pressure to brake mean effective pressure. A p_L/p_e ratio of 0.15–0.17 has proven optimal for medium speed engines to ensure consumption is low on the one hand and to keep the temperature level of the combustion chamber components in heavy fuel oil operation in a safe operating range on the other. Finally, it follows that the compression ratio must be elevated commensurately to obtain the desired final compression pressure. Thus, as a

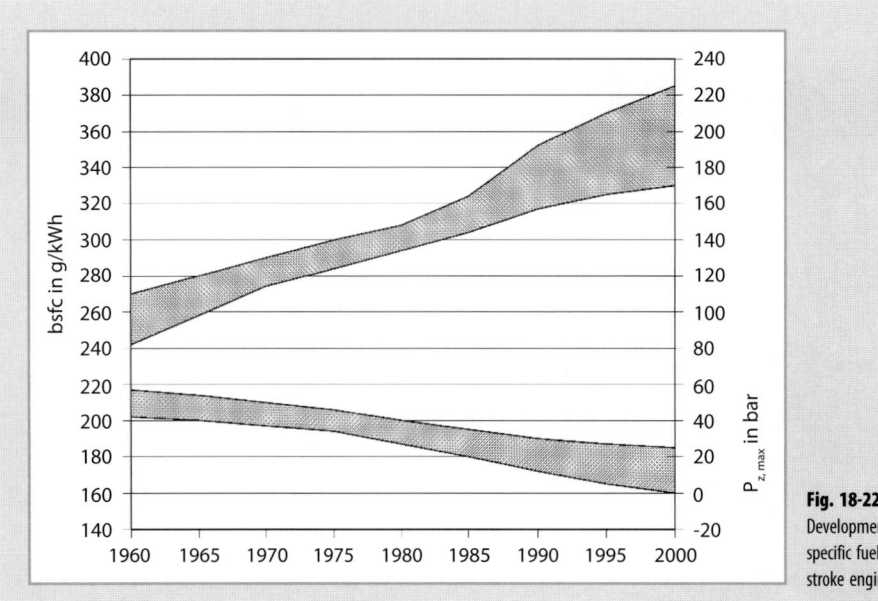

Fig. 18-22 Development of maximum cylinder pressure and specific fuel consumption in medium speed four-stroke engines over the last 40 years

function of the bore diameter, the compression ratio of present day medium speed engines is $\varepsilon = 13\text{--}16$.

A higher compression ratio can be obtained more easily with a longer stroke engine as well as with a well shaped combustion chamber. As the stroke grows shorter, the combustion chamber becomes flatter and flatter at the specified compression ratio and good combustion becomes increasingly difficult to obtain.

Absolute engine size also plays a role in all these considerations. The smaller the cylinder dimensions become, the more adversely detrimental spaces around the valves make themselves noticeable. They increase disproportionately as dimensions grow smaller. Logically, a smaller stroke to bore ratio suffices to obtain a specific compression ratio when the cylinder diameter is larger but not when it is smaller.

18.3.2.4 Speed

Piston stroke and mean piston velocity produce the appropriate engine speed. Speeds in the range of approximately 300–1,200 rpm can be produced depending on the cylinder diameter, stroke to bore ratio and maximum permissible mean piston velocity. Thus, appropriate generator speeds can power engines that generate three-phase current with 50 Hz or 60 Hz (see Sect.1.2).

18.3.2.5 Other Criteria

In addition to medium speed four-stroke engines' low fuel and lubricating oil consumption, suitability for heavy fuel oil, good manufacturing costs, etc., operators also place great value on simplicity of assembly and ease of maintenance.

This is one reason complex technical solutions, e.g. one or multiple stage supercharging, have not become established among medium speed engines.

Emphases of development are suitability for heavy fuel oil even under high specific loads as well as cost effectiveness, reliability and improved exhaust emission. Examples of design solutions illustrate this in the following section.

18.3.3 Design Solutions

18.3.3.1 Basic Engine Design

Space limitation only permit touching on a few basic components and describing their principle features here.

The formerly frequently common crankcase design with a bedplate, a crankshaft inserted from above and an externally mounted cylinder block bolted to the bedplate with tensioning bolts has been replaced in most cases with newer designs with a one-piece frame design with an overhead crankshaft. This design assures a very good load transfer, eliminates additional loaded interfaces and is inexpensive.

MAN Diesel chose an interesting solution for medium speed engines. Elongated main bearing bolts running to the top edge of the one-piece frame and cylinder cover bolts extending deep into the frame significantly relieve the load on the cast structure (Fig. 18-23).

18.3.3.2 Crankshaft Assembly

Crankshaft and crankshaft bearing. In addition to appropriate oil care, the sizing of the crankshaft bearing is extremely important for the prevention of bearing problems in

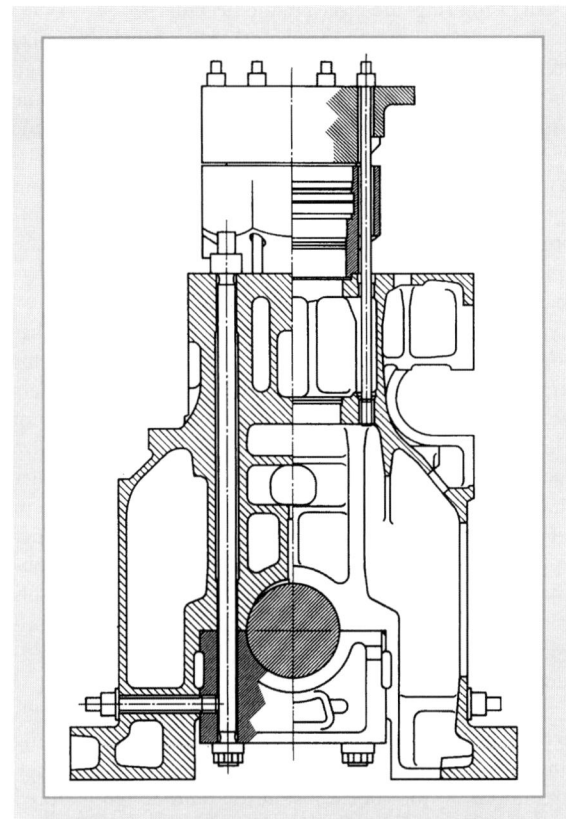

Fig. 18-23 Engine frame with single cylinder jackets, elongated main bearing bolts and elongated cylinder cover bolts (source: MAN Diesel)

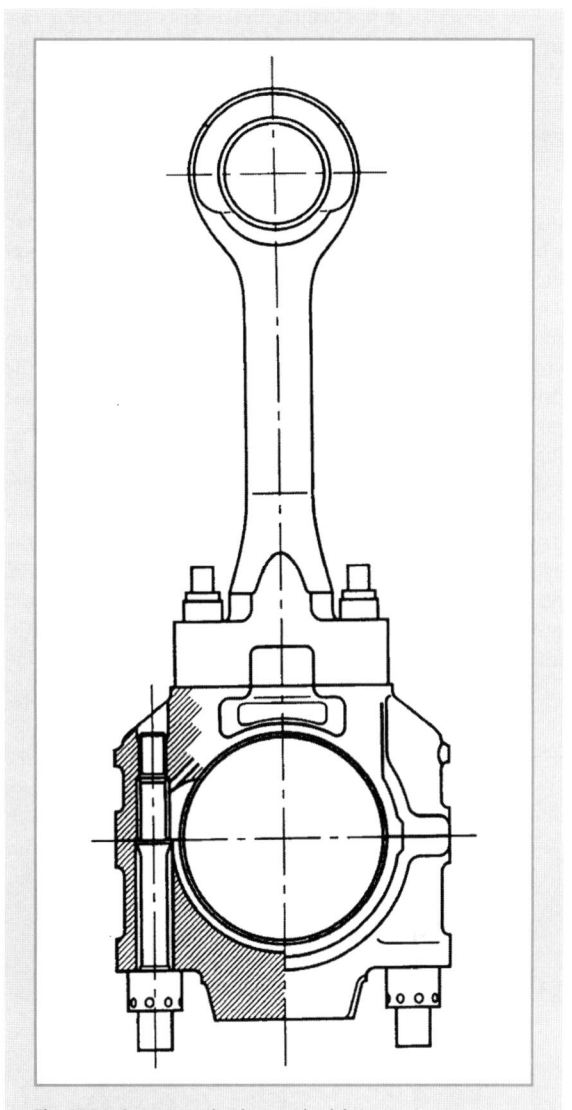

Fig. 18-24 Connecting rod with marine head design

heavy fuel oil operation, which are caused by corrosive or abrasive wear. Practice has demonstrated that a certain minimum residual gap in the lubricating film may not be undershot if satisfactory service lives of the bearing shells are to be obtained. Significantly higher than earlier, the cylinder pressures frequently necessitate stronger basic bearings and crankshaft journals that enlarge the bearing area for reasons of strength. Moreover, the introduction of new bearing technologies, e.g. grooved bearings or sputter bearings, has increased stability considerably. Despite the higher gas forces, this has even enabled significantly increasing the operational reliability of the suspension and the service lives of the bearings in many cases (see Sect. 8.5).

Connecting rod. The crankshaft journal bearing may be split simply in a straight line only in the fewest cases when engines have comparatively low loads. As a rule, stronger crankshaft journals matched to higher loads require that the connecting rod be split obliquely in order to at least be able to guide the connecting rod and the connecting rod shank through the cylinder liner when the piston is pulled. A marine head design in which the shank is bolted with its own two-piece bearing body is employed in many cases (Fig. 18-24). This additional parting line has the advantage of fewer spatial limitations when sizing the bearing body and, hence, facilitates a rigid low-deformation design. Moreover, the bearing does not have to be opened when a piston has to be disassembled.

18.3.3.3 Combustion Chamber Components

Piston. In addition to monoblock nodular cast iron pistons for smaller cylinder dimensions, medium speed four-stroke engines have composite pistons in most cases. The piston top part is steel and the ring grooves are often hardened or chrome plated to reduce wear (see Sect. 8.6).

The piston skirt predominantly consists of nodular cast iron because of the increased load or less frequently of a light alloy. The piston skirt and piston crown are sporadically made of steel [18-8]. Such composite steel/nodular cast iron pistons make it possible to control firing pressures of up to over 200 bar.

The high thermal load requires optimal cooling of the piston top part (see Sect. 7.1). Oil from the circulation system, which is usually supplied to the piston through the connecting rod, is employed as the cooling medium (see Sect. 8.6). The shaker effect during the piston's upward and downward motion flings the cooling oil onto the inner walls of the piston crown where it absorbs the heat and then returns to the driving chamber through appropriate return bores in the piston skirt. The piston top part is often furnished with cooling bores to enlarge the area of heat transfer (Fig. 18-25).

In conjunction with a piston top land ring on the cylinder liner, the piston is designed as a stepped piston to prevent deposits of combustion residues and thus bare spots on the cylinder liner and to additionally reduce oil consumption. Narrow piston clearance that traps abrasive particulates and protects the lubricating film reduces the piston rings' mechanical load.

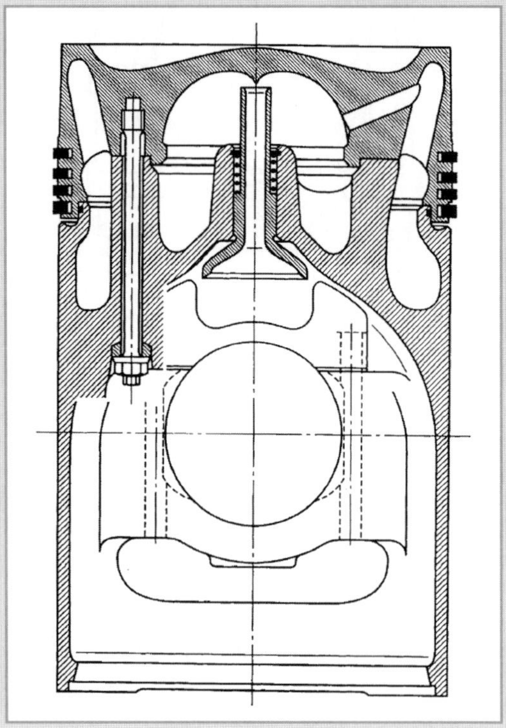

Fig. 18-25 Composite stepped piston with steel top part and modular cast iron skirt for very high cylinder pressures (source: MAN Diesel)

All piston rings are placed in the steel top.

The development of chrome-ceramic coated rings, i.e. chrome rings with ceramic inclusions, combined the high stability of plasma rings with the low wear of chrome rings. Thus, low rates of wear in an order of magnitude of 0.01–0.02 mm/1,000 h were attainable even for the poorest fuels (Fig. 18-26). Hence, modern medium speed diesel engines have chrome-ceramic coated compression rings and chrome coated second and, where applicable, third rings, thus making the ring package highly stable.

Cylinder liner. Separate, single vertical cylinder jackets that hold the cylinder liners provide advantages primarily for larger engines since they reduce actions of adjacent cylinders or ship deformations and thus optimize the roundness of cylinder liners during operation. The water flow and intensive cooling are limited to the upper region of the cylinder liner since they are only required there. The goal is a uniform temperature distribution over the entire surface of the liner to prevent cold corrosion and ensure good lubrication conditions. Together with the stable cylinder geometry, this establishes the prerequisites for low lubricating oil consumption, which should not exceed 0.5–1 g/kWh in modern medium speed engines.

The introduction of piston top land rings in the 1990s heralded a significant advance. Also called anti-polishing rings among other names, they have become widespread. In addition to the cooled piston top land design pictured in Fig. 18-27, uncooled and indirectly cooled designs are employed in which a relative thin-walled ring is inserted directly in the cylinder liner. Ring diameters somewhat smaller than the cylinder liner's actual cylinder bore surface are common to all these designs. In combination with a stepped piston, this effectively prevents "bore polishing" (bare spots caused by hard coke deposits on the piston crown or pitting in the center of the cylinder bore surface). Thus, the cylinder liner, piston top land ring and stepped piston attain a service life of up to 80,000 hours while oil consumption is low. This also includes the low wear values presented in Fig. 18-28 of approximately 0.01 mm/1,000 h attained in conjunction with chrome-ceramic coated rings and measured at the uppermost piston ring's reversal point. The greatest wear, known as bore wear usually occurs at this spot on the cylinder liner.

Cylinder head with valves. As loads increase, nodular cast iron is increasingly being employed for cylinder heads. By virtue of its significantly higher mechanical strength than laminar gray cast iron and in conjunction with a design that facilitates loading, it significantly contributes to the operational reliability of this highly mechanically and thermally loaded component.

Four valves are employed in more highly loaded medium speed engines (Fig. 18-29). Valve cages are increasingly being dispensed with even in larger engines [18-9]. Operational

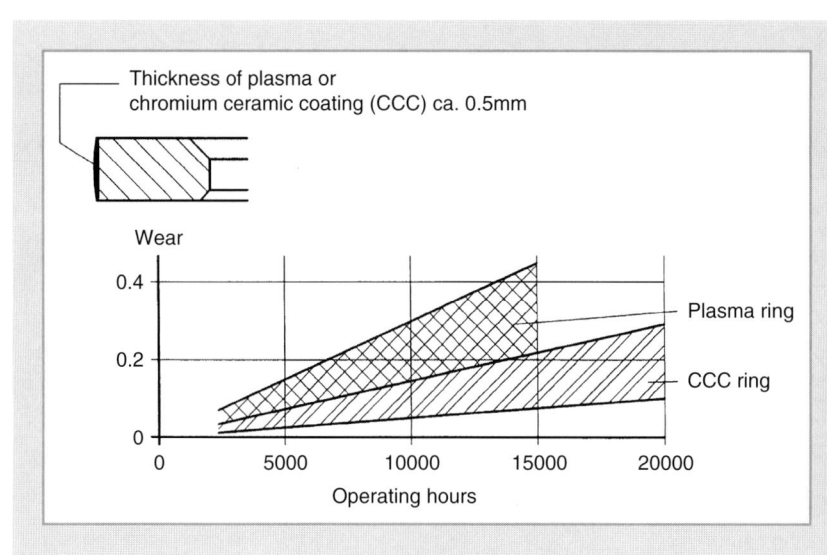

Fig. 18-26
Average wear of the first piston ring in medium speed four-stroke engines operated with heavy fuel oil

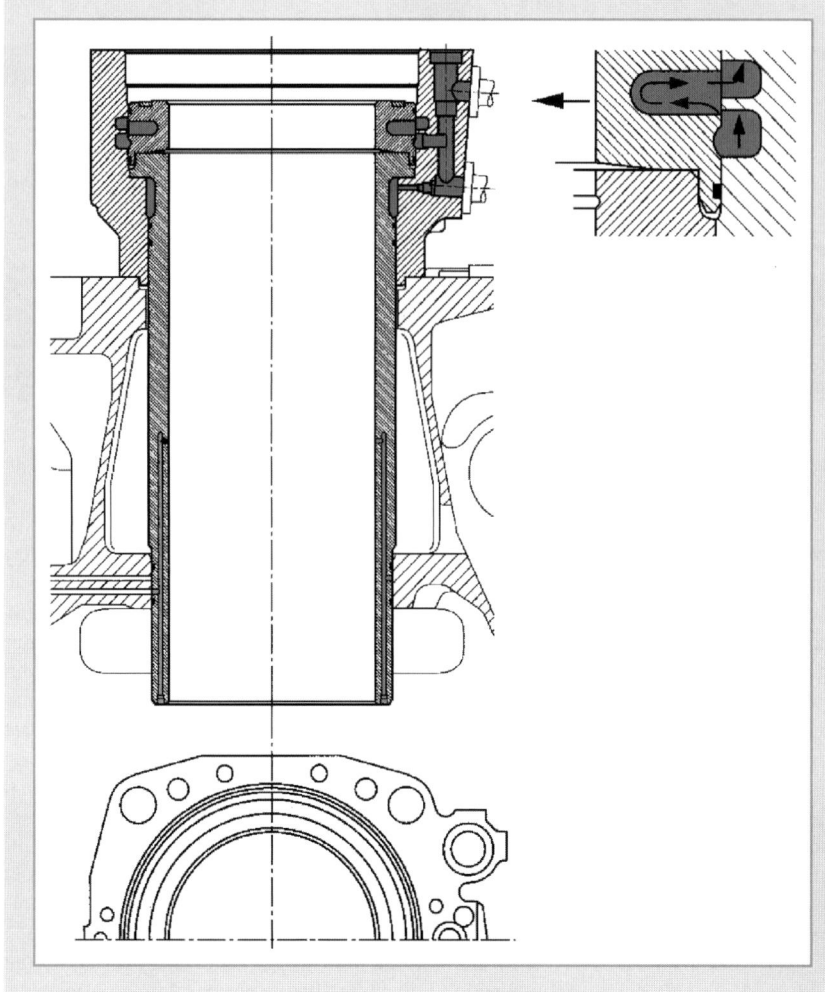

Fig. 18-27
Cylinder liner with water jacket and piston top land ring with bore cooling

Fig. 18-28 Average wear of the cylinder liner in medium speed four-stroke engines operated with heavy fuel oil

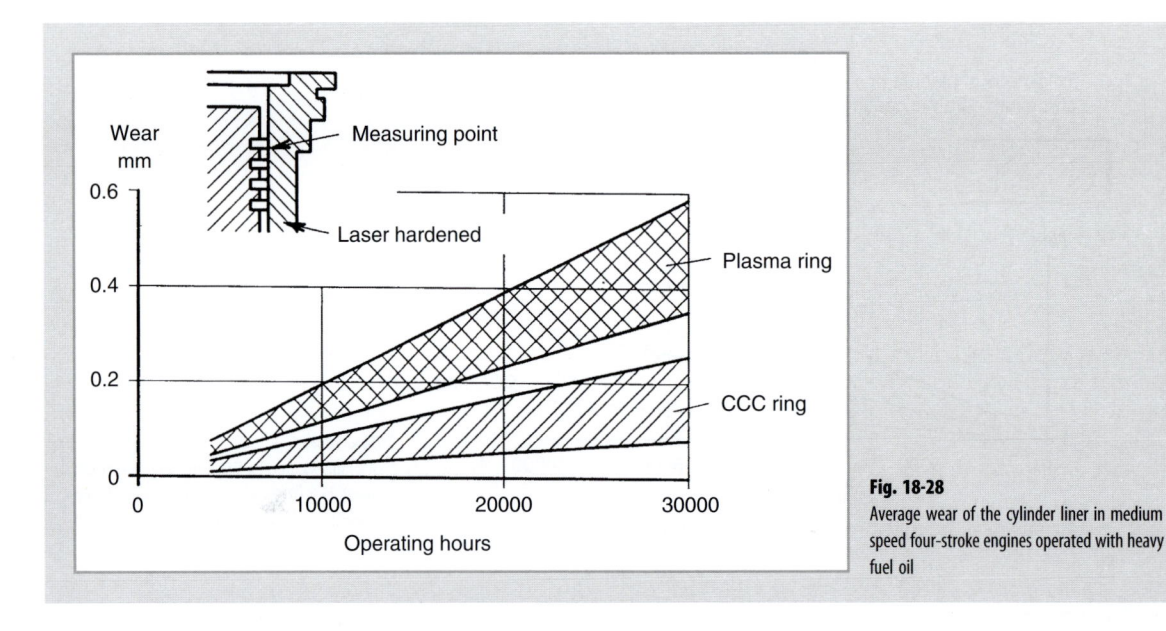

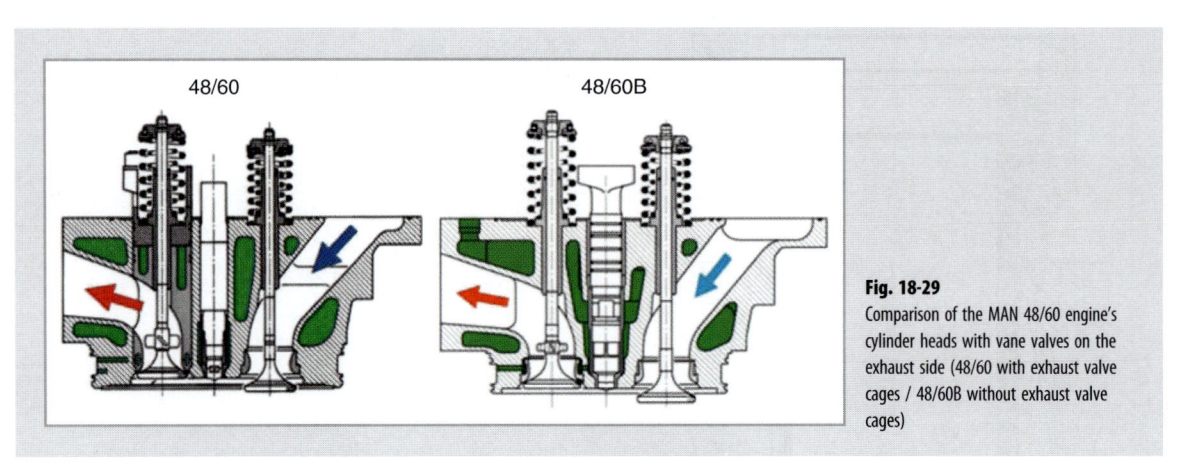

Fig. 18-29 Comparison of the MAN 48/60 engine's cylinder heads with vane valves on the exhaust side (48/60 with exhaust valve cages / 48/60B without exhaust valve cages)

reliability has been increased and valve service lives have been extended so long that a cylinder head must be disassembled for other maintenance work too (e.g. piston rings). This eliminates the maintenance advantage of the valve cage and the disadvantages predominate, e.g. structural complexity, reduced cylinder head stiffness and additional potential leak points. The valve seat inserts are often cooled, at least on the exhaust side.

Both the valve cone and the seat inserts usually have a hard seat facing, e.g. stellites (carbide metals), which assures high wear resistance and prevents pocketing by combustion particulates in heavy fuel oil operation and thus burnouts caused by inadequate valve seat sealing and the resultant discharge of heated combustion gases. Nimonic valves with and without hard seat facing are also used on various occasions.

Practice has demonstrated that valve rotation is absolutely essential during operation when heavy fuel oil is burned. Valves may be rotated by mechanical rotators, e.g. rotocaps. Individual manufacturers employ rotating vanes in the valve stem on the exhaust side. Significantly more intense rotation is obtained from discharging exhaust than mechanical rotators. In the process, the mass inertia of the valve causes rubbing seating on the seat.

18.3.3.4 Injection System

Readers are referred to Sect. 5 for the basics. Limitations of space only permit briefly touching on common configurations and designs for medium speed engines.

Since not only the ratio p_{Zmax}/p_e but also the duration of combustion influences fuel consumption considerably, attempts have also been made to optimize the duration of combustion in medium speed engines and thus keep it as brief as possible. A relatively close relationship between the duration of combustion and the duration of injection exists in injection systems used solely for direct injection: A brief duration of combustion also requires a correspondingly brief duration of injection. Thus, many manufacturers are now employing high intensity injection systems. Ultimately, this results in comparatively high pressures in the injection system, which in turn has to be appropriately factored into the design of components. In virtually every design, the injection valve is centered in the cylinder head and a multi-hole nozzle conveys the fuel to the combustion chamber.

The ability to adjust injection to the sometimes widely differing combustion characteristics is advantageous when heavy fuel oils are burned. Thus, for example, there are designs that aim to positively influence the combustion cycle by a low pilot injected fuel quantity. Other manufacturers have created the option of influencing the ignition point or injection timing to be able to appropriately respond to different ignition delays.

In conjunction with the existing statutory limits on emissions, such measures will increasingly gain importance since this method, among others, can influence NO_x emission by shifting the ignition point.

Given their variability, common rail systems furnish engine developers a broad range and higher flexibility of injection parameters. In addition to variable start of injection and injection pressure, multiple injections may be necessary for optimal combustion with low pollutant content.

Common rail systems are increasingly also being implemented in medium speed engines despite the problems that arise in heavy fuel oil operation from utilizing heavy fuel oils with a viscosity of up to 700 cSt (at 50°C) since these fuels must be preheated to a temperature of up to 150°C to obtain the requisite injection viscosity. These problems are compounded by the high content of abrasive particulates and aggressive constituents present in heavy fuel oils. Injection components must function reliably at high temperatures under these operating conditions.

A pressure accumulator (rail) extending along the entire length of the engine is problematic for large diesel engines because of thermal expansion and the options to manufacture such a component for 1,600 bar with radial bores. Hence, the pressure accumulators are divided into several segments in the systems introduced so far by Wärtsilä and MAN Diesel [18-10, 18-11]. The fuel supply can also be spread to several high pressure pumps. Supplying high pressure fuel to the accumulator system through two or more high pressure pumps has the additional advantage of enabling engine operation even when one of the pumps fails.

Based on concepts with a segmented rail, MAN Diesel developed a modular system for several engine types (Fig. 18-30).

Along with the advantage of greater flexibility to adapt to different numbers of cylinders and better utilization of

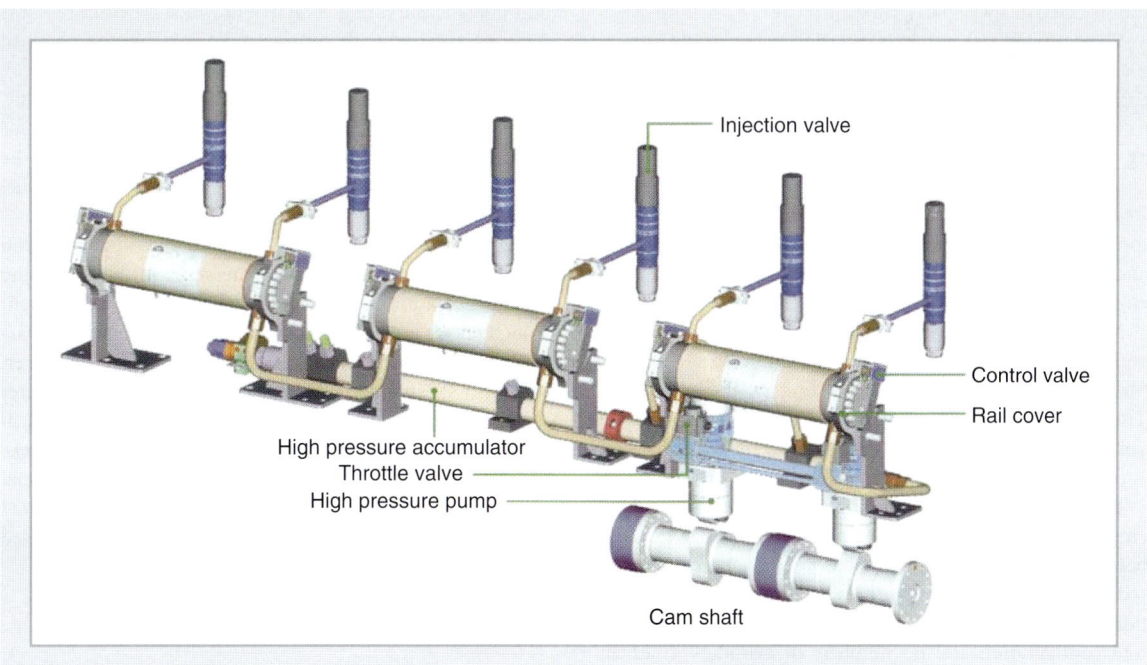

Fig. 18-30 Configuration of MAN Diesel's common rail system

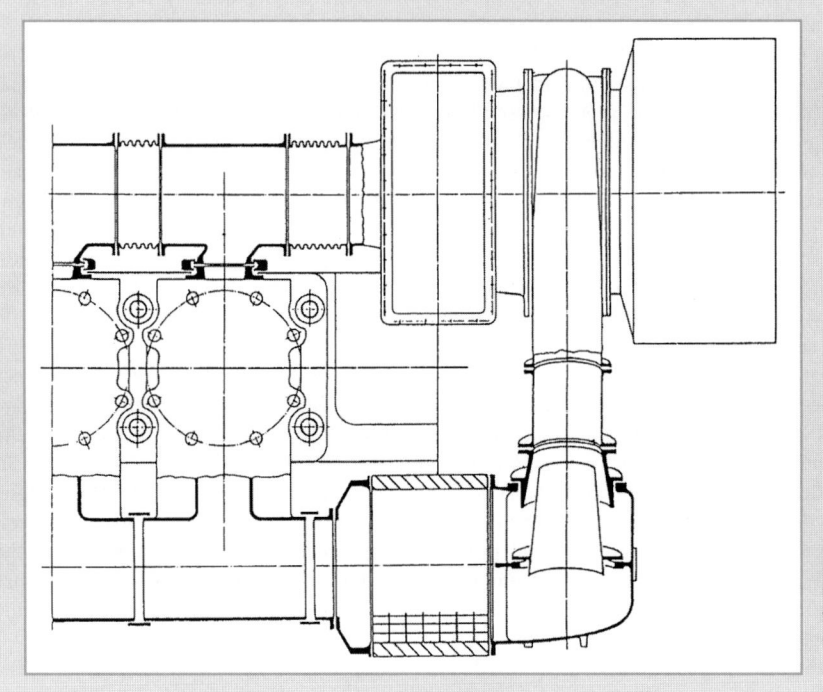

Fig. 18-31 Constant pressure supercharging system with flow-optimized ports and diffusers between the cylinder head and exhaust manifold and after the compressor for pressure recovery

existing space by compact units, segmentation into individual rail modules provides further advantages for assembly and the stocking of spare parts [18-12].

The common rail system is expected to supersede classical mechanical injection with a single plunger pump driven by a camshaft in the future.

18.3.3.5 Supercharging System

Readers interested in the theory of supercharging are referred to Sect. 2.2. This section touches on different aspects of design and implementation, which are typical for medium speed engines.

Modern medium speed engines are almost exclusively equipped with exhaust gas turbocharging. Axial or radial turbochargers are employed depending on the engine size. The advances achieved in turbocharger engineering in recent years now allow a pressure ratio of five and above in one stage. A further increase is in development.

Both pulse and constant pressure supercharging are employed in engines. Constant pressure supercharging has become increasingly established for medium speed engines in recent years since the advantages, e.g. lower fuel consumption, uniform turbine pressurization, simpler exhaust manifold configuration and no numbers of cylinders disadvantaged by supercharging, outweigh the disadvantage of poorer accelerating performance. This can be mitigated by narrow exhaust manifolds and – when necessary – harmonized with the conditions in pulse operation by appropriate additional measures (e.g. Jet Assist in which the compressor is briefly pressurized with compressed air during the phase of revving up).

Designing the ports that conduct air and exhaust to facilitate the flow and correctly configuring the diffusers for pressure recovery beneficially influences the efficiency of the entire supercharging system. This has a positive effect on fuel consumption (Fig. 18-31).

One problem with one-stage supercharging in conjunction with high brake mean effective pressures is the increasing difficultly to optimally cover an engine's air requirement throughout the entire load range because of the different characteristic curves of the engine and turbocharger. However, thoroughly satisfying results can be obtained by such measures as recirculating charge air in the lower load range and, if necessary, discharging exhaust gas at excess load (wastegate).

Infinitely variable turbine geometry would be the optimal solution. However, tests with adjustable guide vanes resulted in considerable problems in heavy fuel oil operation caused by contamination.

18.3.4 Operational Monitoring and Maintenance

In addition to electronically assisted control, timing and monitoring systems, which serve to optimize operation, diagnostic and trend systems are also employed. Thus, an operator receives information on the system's current status, which is intended to facilitate decisions on measures to be taken.

The latest state of development is the use of expert systems that not only display a system's instantaneous status but also quite specifically inform the operator which component has to be serviced or replaced because of changed engine operating characteristics. This is allowing a switch from scheduled to condition-based maintenance [18-13].

Development in recent years has made it now possible to plan long maintenance intervals for components of modern medium speed engines despite increased loads. Good accessibility and the use of appropriately designed special tools make the work performed on wear parts, e.g. piston rings, injection nozzles, intake and exhaust valves, main and connecting rod bearings, etc., substantially easier. Potential errors during the performance of maintenance work have largely been minimized, a fact that contributes considerably to an engine system's operational reliability.

18.3.5 Exhaust Emission

Measures that improve exhaust emission in medium speed four-stroke diesel engines are primarily aimed at reducing nitrogen oxides NO_x and soot production during combustion. The latter not only causes exhaust blackening but is also responsible for the emission of particulate matter (see Sect. 15.3). This is exacerbated by the operation of large diesel engines predominantly with sulfurous heavy fuel oils (see Sect. 4.3.4.2). In addition, increased soot formation, made noticeable by heavy plumes of smoke, particularly occurs at light load. This is a problem for seagoing ships when they are maneuvering in ports.

In-engine measures are initially an expedient remedy, e.g. improved injection, modified valve timing, etc. [18-8, 18-14, 18-15]. External measures, e.g. the use of water/fuel emulsions or particulate filters (see Sect. 15.5), generally complicate the engine system's handling and increase its susceptibility to faults.

Following this principle, MaK's first step during its development of low-emission large diesel engines was to reduce NO_x emission in compliance with IMO specifications (see Sect. 15.2.3.3). Then, smoke emission was lowered below the visibility limit. The long-term goal was a low-emission engine (LEE) with an eye toward future requirements [18-16].

It turned out that several measures have to be combined as a function of load. The Miller cycle (see Sect. 2.2.4) reduces the maximum combustion temperature in the upper load range and thus the formation of NO_x. The charge loss connected with this can be compensated by increased boost pressure, provided the limits of one-stage supercharging are not reached (see Sect. 2.2.3). A larger compression ratio in conjunction with a longer stroke ($s/D = 1.5$) can reduce NO_x emissions significantly but often with heavier smoke emission at light load. However, the use of flexible camshaft technology (FCT) at lower power can keep exhaust blackening below the visibility limit (smoke number SN $\leq 0.4 \ldots 0.5$; see Sect. 15.6 and Fig. 18-32). To do so, the injection cams' rated power is advanced at light load upwards of approximately 25% power so that, in conjunction with a modified injection pump, atomization is improved and combustion is low in soot. At the same time, the intake valve opens and closes later, thus dispensing with the Miller effect, while the exhaust valve opens earlier to increase the boost pressure by a larger exhaust gradient.

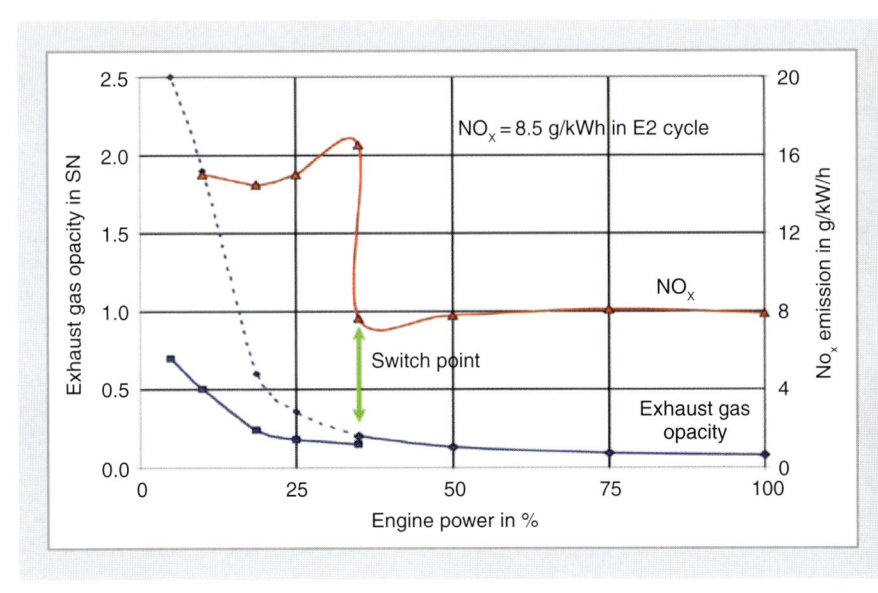

Fig. 18-32 NO_x emission and exhaust gas opacity (SN) of an exhaust-optimized Caterpillar M 43 C marine diesel engine (current IMO limit: $NO_x = 12.9$ g/kWh, SN: smoke number based on the Bosch filter method)

18.3.6 Implemented Engines

Given the abundance of different medium speed engines from the widest variety of manufacturers (the diversity of types is especially great in the lower power range), only a few examples can be singled out here.

Figure 18-33 pictures MAN Diesel's family of large medium speed engines, consisting of the L58/64, L48/60, L40/54 and L32/40 models with cylinder outputs of 1,400, 1,200, 720 and 500 kW at speeds of 428–750 rpm. The standardized engineering of these four engines, each of which is

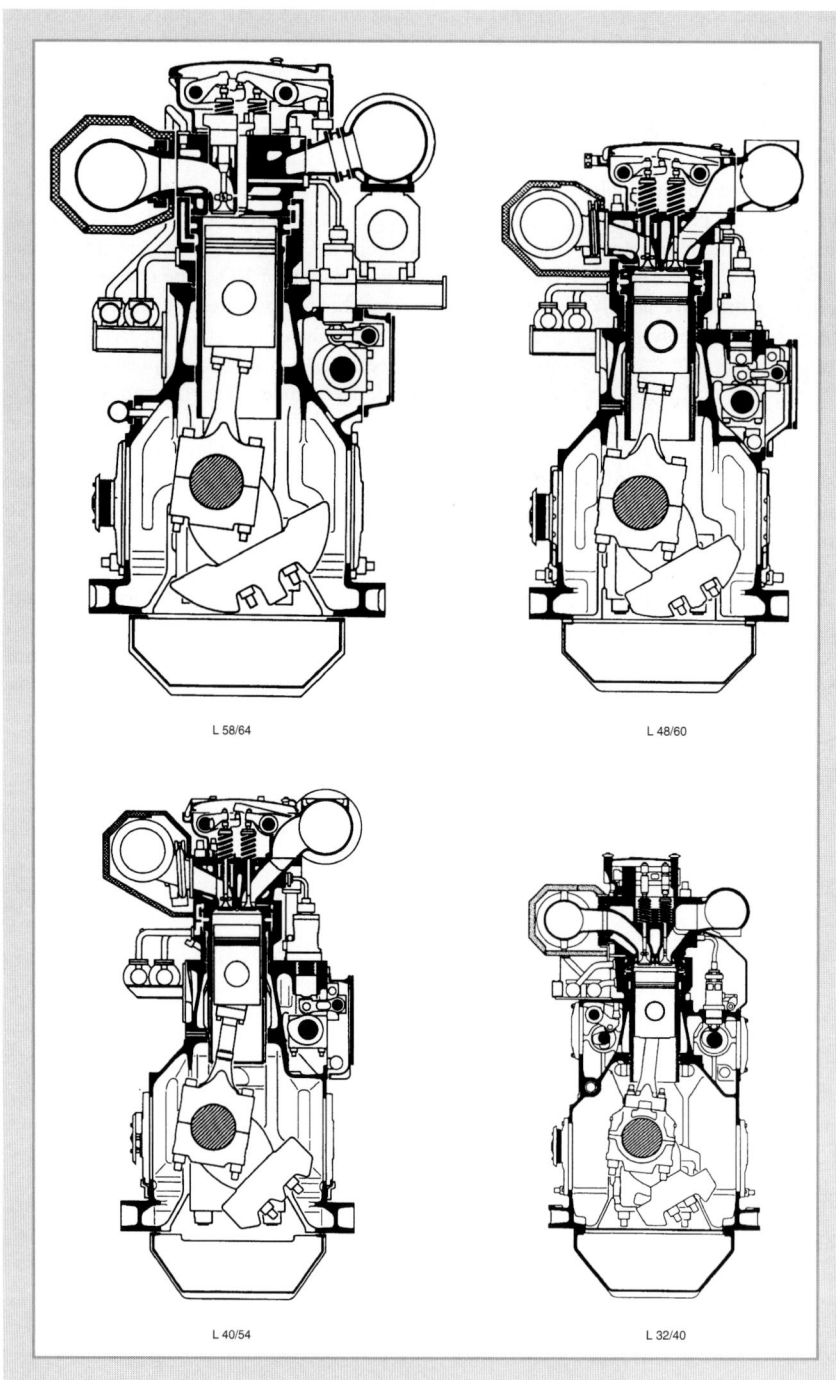

Fig. 18-33
Upper power range of MAN Diesel's medium speed engine family, consisting of four inline engines of largely identical design

Fig. 18-34 Wärtsilä 46

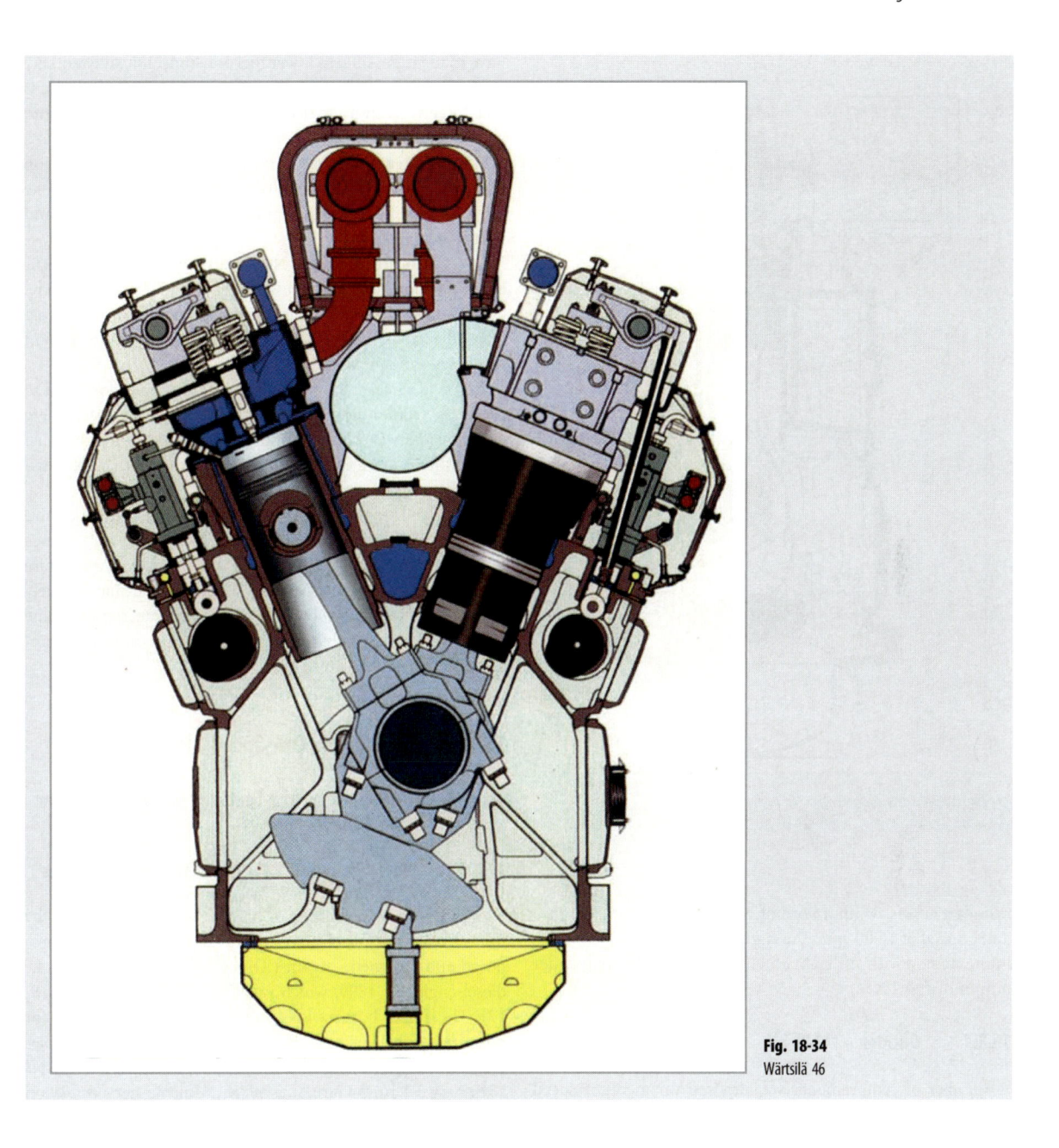

constructed in an inline configuration, is remarkable. In addition to the type with 320 and 480 mm, a V version is also available.

A broad range of medium speed engines in the bore range of 250–350 mm is commercially available. Among others, the Wärtsilä 32 has been very successful in recent years. With a cylinder diameter of 320 mm and a stroke of 400 mm, a cylinder output of 500 kW is reached at a speed of 750 rpm. This engine is also built in an inline and V design with six, seven, eight and nine or twelve, sixteen and eighteen cylinders.

Pictured in cross section in Fig. 18-34, the larger Wärtsilä 46 has a cylinder diameter of 460 mm and a stroke of 580 mm. It is available with cylinder outputs of 975, 1,050 and 1,155 kW at 500 rpm. For the W46F stage of development, the speed was increased to 600 rpm and thus a cylinder output of 1,250 kW was attained.

MaK's M20 engine is representative of smaller dimensions in the range of a cylinder bore of 200 mm (Fig. 18-35). The engine has a number of design features that had been reserved for larger medium speed engines, e.g. individually attached

Fig. 18-35 Cross section of MaK's M20 engine

have made it possible to lower fuel consumption substantially in recent years. Today, a medium speed four-stroke engine is able to convert more than 50% of the fuel-based energy into mechanical work.

Along with a further concentration of power, the reduction of emissions will be the priority as medium speed four-stroke engines are refined in the coming years (see Sect. 18.3.5 and Part IV of this book).

In addition to reducing NO_x emission, intensive work is also being done on further lessening particulate emission. The goal for medium speed engines is to produce invisible exhaust from idle to full load by suppressing soot production, even in heavy fuel oil operation.

In the ideal case, this would prevent sulfurous combustion products from binding to soot and increasing particulate emission. However, as mentioned in Sect. 4.3, low exhaust blackening is not a criterion for equally low particulate emission when sulfurous fuel is burned.

Elastic and semi-elastic installations in ships, which reduce structure-borne noise, will continue to grow in importance for noise emission. In addition to noise reducing and noise absorbing measures at the engine itself, which have largely been exhausted, appropriately sound engineered machine rooms or, where feasible at these engines' dimensions, encapsulation will reduce noise even further.

18.4 Two-Stroke Low Speed Diesel Engines

18.4.1 Development and Features of Two-Stroke Low Speed Diesel Engines

18.4.1.1 Development of Two-Stroke Low Speed Engines

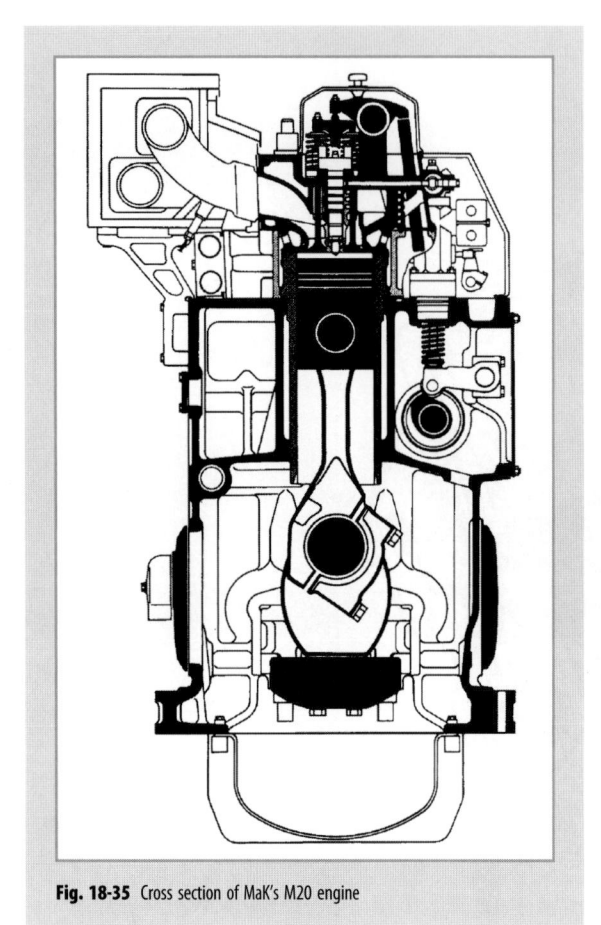

cylinder jackets. With a bore of 200 mm, stroke of 300 mm and a speed of 1,000 rpm, the engine presently available in an inline version with six, eight and nine cylinders has a cylinder output of 190 kW.

18.3.7 Outlook

A number of requirements are imposed on medium speed four-stroke diesel engines and will be in the future too. Naturally, demands for higher cost effectiveness, reliability and simplicity of maintenance are foremost from the perspective of operators. The technical concepts of modern medium speed four-stroke diesel engines largely meet these demands. High service life primarily means low wear values for the most important components. Simplicity of maintenance, i.e. the use of easily handled hydraulic tools and good accessibility of serviced components are part of cost effectiveness.

Of course, cost effectiveness also means low specific consumption of fuel and lubricating oil. Elevation of the maximum cylinder pressure, optimization of the combustion process and developments in supercharging technology

Soon after the introduction of the first diesel engine with the four-stroke principle envisioned by Diesel, Hugo Güldner proposed and designed – against Diesel's counsel – a two-stroke diesel engine in 1899, which was denied success though [18-17]. The Sulzer bothers in Winterthur deserve the credit for introducing the first operable two-stroke diesel engine implemented as a marine engine in 1906. Other firms such as MAN-Nürnberg, Krupp-Germania-Werft, Burmeister & Wain (Copenhagen) soon followed. Together with larger cylinder dimensions, the two-stroke principle was viewed as an opportunity for the marine engine to compete against reciprocating piston steam engines with their large energy units.

While a two-stroke engine with its two working strokes was theoretically able to produce twice the power of an equally large four-stroke engine, the real increase was only approximately 60% because of losses due to the lower purity of the charge and the required compression of the scavenging air.

This has led to a great diversity of engine concepts over decades, which, on the one hand, are characterized by generally similar basic features and, on the other hand, also bear features specific to manufacturers, e.g. Doxford, Grandi

Motori Trieste (formerly Fiat), Götaverken, Stork, Werkspoor, etc., and are characterized by the following criteria:

Working principle:
- single-acting piston,
- double-acting piston and
- opposed piston.

Methods of scavenging:
- uniflow scavenging,
- loop scavenging and
- cross flow scavenging.

Methods of supercharging:
- mechanical supercharging,
- exhaust gas turbocharging and
- combined mechanical supercharging and exhaust gas turbocharging.

Combinations of the individual principles yielded the widest variety of designs, some of which were able to hold their own into the 1970s. All manufacturers worked with direct fuel injection instead of air injection, which had been common at first.

On the one hand, the two "oil crises" in the second half of the 1970s and the early 1980s once again triggered great strides in development in terms of diesel engines' fuel economy. On the other hand, they set off a concentration of the only three firms now remaining worldwide, which develop and engineer two-stroke low speed diesel engines and, in addition to licensing them, also partly manufacture them themselves. Since the mid 1990s, these (Fig. 18-36) have been:
- MAN Diesel SE (formerly MAN B&W Diesel AG),
- Wärtsilä (formerly Gebrüder Sulzer/New Sulzer Diesel) and
- Mitsubishi Heavy Industries (MHI).

All three suppliers are pursuing the same concept: a low speed single-action, exhaust gas turbocharged, uniflow scavenged two-stroke diesel engine.

Thus, the once great diversity of concepts has essentially given way to one concept that appears quite logical today.

18.4.1.2 The Transition to Uniflow Scavenging

Until the mid 1970s, stroke to bore ratios among all the still commercially active manufacturers fluctuated between 1.7 and 2.1. These allowed loop or cross flow scavenging without losses of scavenging efficiency. The distinctive feature of these two scavenging systems was their particular simplicity of design, which functioned without an exhaust valve in the cylinder cover. This made maintenance exceptionally simple and user-friendly, which helped this engine type to particular commercial success in the 1960s and 1970s.

However, the first of the oil crises (1973) subsequently triggered a clear turning point in development. The astronomical increase of the share of fuel costs in total operating costs after 1973 was crucial to this. Thus, not only was the maximum cylinder pressure increased in quick succession to save fuel but developments in the direction of fuel economy also commenced on the shipbuilding side. Propeller speeds and propeller diameters grew smaller and thus propeller efficiencies larger.

This inevitably resulted in larger stroke to bore ratios for two-stroke engines in order to be able to maintain the mean piston velocity and thus the power output.

MAN-Augsburg and Sulzer were able to keep pace with this development with their simple valveless engines with a stroke to bore ratio of approximately 2.1 until the end of the 1970s. Then, however, market demand for even lower speeds forced a transition to uniflow scavenging. MAN solved this problem in early 1980 by taking over the diesel operations of the Danish firm Burmeister & Wain (B&W) since large B&W two-stroke engines with a stroke to bore ratio of approximately 2.4 had already been operating with uniflow scavenging for a long time.

Combining the proverbial reliability of valveless, loop scavenged engines with the longer strokes being demanded from uniflow scavenged engines subsequently became essential to development and design. Introducing its (super-long stroke) RTA series [18-18] in the early 1980s, Sulzer mastered this challenge first by adopting uniflow scavenging with a central exhaust valve and increasing the stroke to bore ratio to approximately 3.0 for the first time. This made substantially lower rated speeds possible (67 rpm in the largest engines).

Such engine models were available from MAN B&W (L...MC/MCE) and Mitsubishi (UEC...L) soon afterward. Determinative for operating costs, these extremely long-stroke engines with concentrated power reduce the fuel consumption of marine systems not only by their better combustion process but also by their better propulsion efficiency.

Today, none of the three manufacturers' advanced two-stroke low speed engines differ from one another in terms of their basic principle.

Observable in other industrial products at the close of the twentieth century, "natural" selection had occurred here too, not least because of the advanced design and computational tools that had become available and permit selecting the logically correct solution rapidly and efficiently. Despite the strongly increased specific powers, lower engine weights and related higher load in recent years, the reliability of these machines has been improved to the point that overhaul intervals of three years are now feasible. Today, the basic difference between the three manufacturers is their different concepts for electronically controlled injection.

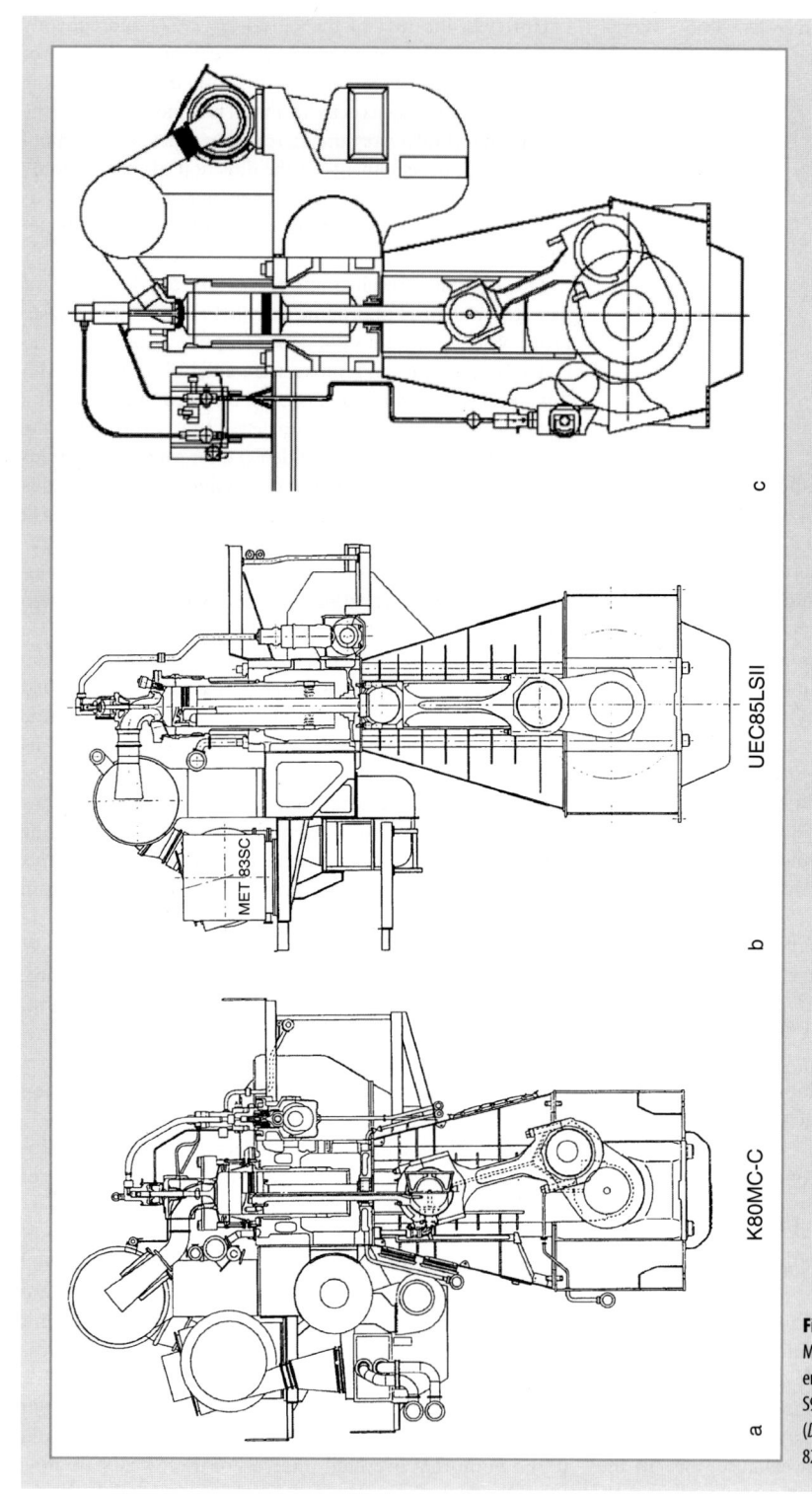

Fig. 18-36
Modern designs of two-stroke low speed diesel engines with uniflow scavenging: (**a**) MAN B&W: S90MC-C ($D = 900$ mm); (**b**) Mitsubishi: UEC85LsII ($D = 850$ mm); (**c**) Wärtsilä RT-flex82C ($D = 820$ mm)

18.4.1.3 Features of Modern Two-Stroke Low Speed Engines

In its more than ninety year history, the low speed diesel engine has experienced tremendous technical development made possible by:
- ongoing conceptual development,
- the transition to uniflow scavenging,
- advances in supercharging equipment and turbocharger design,
- new knowledge in materials technology,
- the utilization of advanced theoretical and experimental methods of development and
- the transition to electronically controlled injection and valve timing.

The *brake mean effective pressure* p_e of approximately 5 bar in the naturally aspirated engines of the early 1950s have now risen to 20 bar in supercharged engines. This corresponds to an increase of *effective brake work* w_e from 0.5 to approximately 2.0 kJ/dm³. *Mean piston velocity* has increased from approximately 5 to over 9 m/s.

In the process, the *maximum cylinder pressure* p_{Zmax} has increased from approximately 50 bar to more than 160 bar. Pronounced *long strokes* with stroke to bore ratios of between approximately 3.0–4.2 may be regarded as the typical design feature of today's large two-stroke diesel engines. Their cylinder diameter D can be between 260 and 980 mm.

Uniflow scavenging with an actively controlled exhaust valve is now implemented almost exclusively as the scavenging process.

Characterized by the *specific power per unit piston area* P_A, the power density reaches values of more than 790 W/cm². Thus even large two-stroke low speed engines constitute a demonstrably "high-tech" product (see Sect. 1.2).

At the same time, in conjunction with turbocompounding and exhaust gas heat recovery, *specific fuel consumption* has dropped from approximately 220 to 156 g/kWh. This corresponds to an effective efficiency of approximately 55%. Thanks to the ruggedness of the basic concept, a two-stroke low speed diesel engine is now also able to burn the poorest grades of fuels.

During this period, the cylinder outputs of the largest engines rose from a few 100 kW to over 5700 kW. This has made power outputs of over 80,000 kW possible with one engine. The trend toward ever larger container ships still only equipped with one propeller and expected to operate at identical or even higher speed while allowing for increased time requirements for loading and unloading as well as liner traffic has already given rise to demands for power outputs of 100,000 kW and more in recent years.

According to Eq. (1-13), a linear correlation of specific work w_e (or brake mean effective pressure p_e), mean piston velocity c_m and the number of cylinders z exists for engine power. However, engine power increases with the square of piston diameter D. Logically, manufacturers have so far attempted to meet demands for greater power by a steady, incremental increase of specific power data as well as with engines with larger piston diameters. The upper limit is now at $D \leq 1$ m but a further increase of D at an appropriately designed thermal load (see Sect. 7.1), component mass, mixture formation and combustion is definitely conceivable. Increasing the numbers of cylinders above the formerly common maximum number of twelve remains another option.

Wärtsilä already offers large two-stroke engines with a piston diameter of 960 mm (RTA96C and RT-flex96C) and up to $z = 14$ cylinders inline, thus, obtaining a cylinder output of 5,720 kW and a total engine power of 80,080 kW at values of $p_e = 18.6$ bar or $w_e = 1.86$ kJ/dm³ and $c_m = 8.5$ m/s. This engine type is the largest and most powerful diesel engine ever built. The first engines of this type were commissioned in 2006.

Higher numbers of cylinders than $z = 14$ can hardly be considered realistic at present since engine mass and overall length inevitably increase as z increases. This particularly raises problems with the engine foundation and the load of a ship's hull.

For this reason, MAN has been carrying an even larger engine, the K108MC-C, in its program since 2002. With a nominal cylinder output of 6,950 kW, this engine would achieve a total power of 83,400 kW with a maximum number of cylinders of $z = 12$.

Further considerations to reduce overall length and engine mass have led to the option of a V configuration of the cylinders. Relevant designs and simulations by MAN revealed that the engine mass of a twelve cylinder V engine with a piston diameter of 900 mm would be reduced by 15% and the length by 6.8 m, i.e. by approximately 30%. It also turned out that no major problems in terms of maintenance and accessibility are to be expected. However, none of these two engine variants with a 108 cm bore or V configuration has been produced so far [18-19].

Figure 18-37 charts the chronological development of the important engine parameters of Wärtsilä's large two-stroke diesel engines.

18.4.2 Modern Two-Stroke Low Speed Engine Design

18.4.2.1 Engine Families, Power Map

All commercially available two-stroke low speed engines have the same concept and thus similar design features. Thus, a detailed description of one manufacturer's engine is sufficient.

Since engine power and speed are firmly interrelated for a direct drive propeller, a closely graduated engine family of differing bores and stroke to bore ratios is a necessity for every manufacturer. Figure 18-38 presents partly overlapping maps of one manufacturer's offerings as an example. This makes it possible to select the optimal engine while considering criteria such as installation dimensions, number of cylinders, fuel consumption, etc.

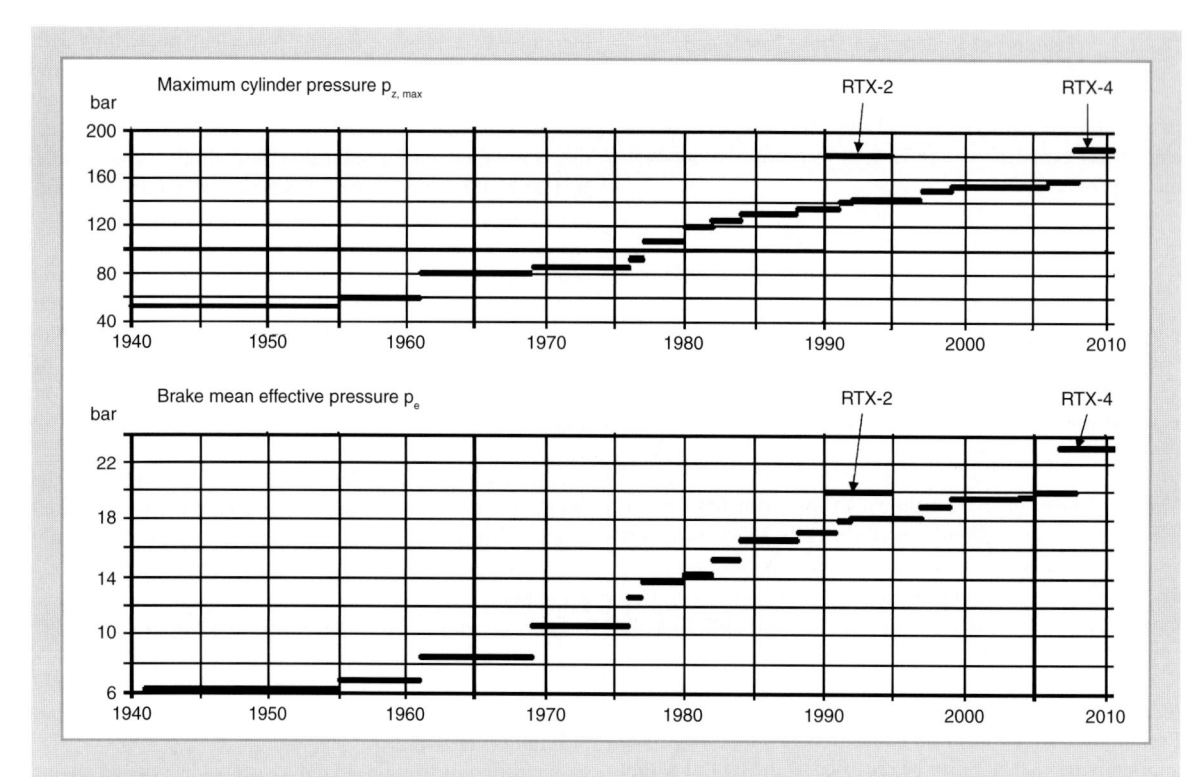

Fig. 18-37 Development of the engine parameters of maximum cylinder pressure $p_{Z,max}$ and brake mean effective pressure p_e over the last 65 years (Wärtsilä RTX-2 and RTX-4 research engines)

Different engines in the Wärtsilä family described here include:
– engines with stroke to bore ratios of up to approximately 3.5 (RTA52U, RT-flex60C, RTA62U, RTA72U, RTA82C/RT-flex82C, RTA96C/RT-flex96C), which, produce high powers at relatively high speed, above all for faster ships, e.g. container ships or car carriers, and
– engines with stroke to bore ratios higher than 4.0 (RTA48T, RTA58T/RT-flex58T, RTA68/RT-flex68, RT-flex82T), which, at correspondingly lower speeds, are intended for slower ships with larger propellers, e.g. tankers and cargo ships.

Every engine's power map is specified by the vertices R1/R2 to R3/R4 (Fig. 18-38), the engine's rated power being freely selectable within this map for a particular application. Depending on demand, this allows full utilization of the maximum power (point R1) or a variant with reduced power with the advantage of lower consumption and/or lower propeller speed.

The R1+ map concept represents another design variant, which delivers the same propulsion power in conjunction with fuel consumption reduced by 2 g/kWh at increased speed and reduced mean pressure.

18.4.2.2 Engine Design

Engine frame. The 14RT-flex96C (80,080 kW, 102 rpm) is currently the world's most powerful diesel engine in operation. The typical features of a modern two-stroke low speed are explained using the example of the Sulzer RTA96C diesel engine illustrated in Fig. 18-36. The following engine parameters apply to this engine:

stroke/bore (mm/mm)	2,500/960 (= 2.6)
cylinder output (kW)	5,720
rated speed (rpm)	102
brake mean effective pressure (bar)	18.6
specific work (kJ/dm^3)	1.86
mean piston velocity (m/s)	8.5
specific consumption (g/kWh)	171
maximum cylinder pressure (bar)	145
power per unit piston area (W/cm^2)	790

The engine frame consists of a rigid, welded structure with a bedplate and an A-frame in which white metal crosshead guide rails are integrated. In modern engines, the overlying cast cylinder jacket, which holds the cylinder liners, is "dry", i.e. does not contain a cooling water

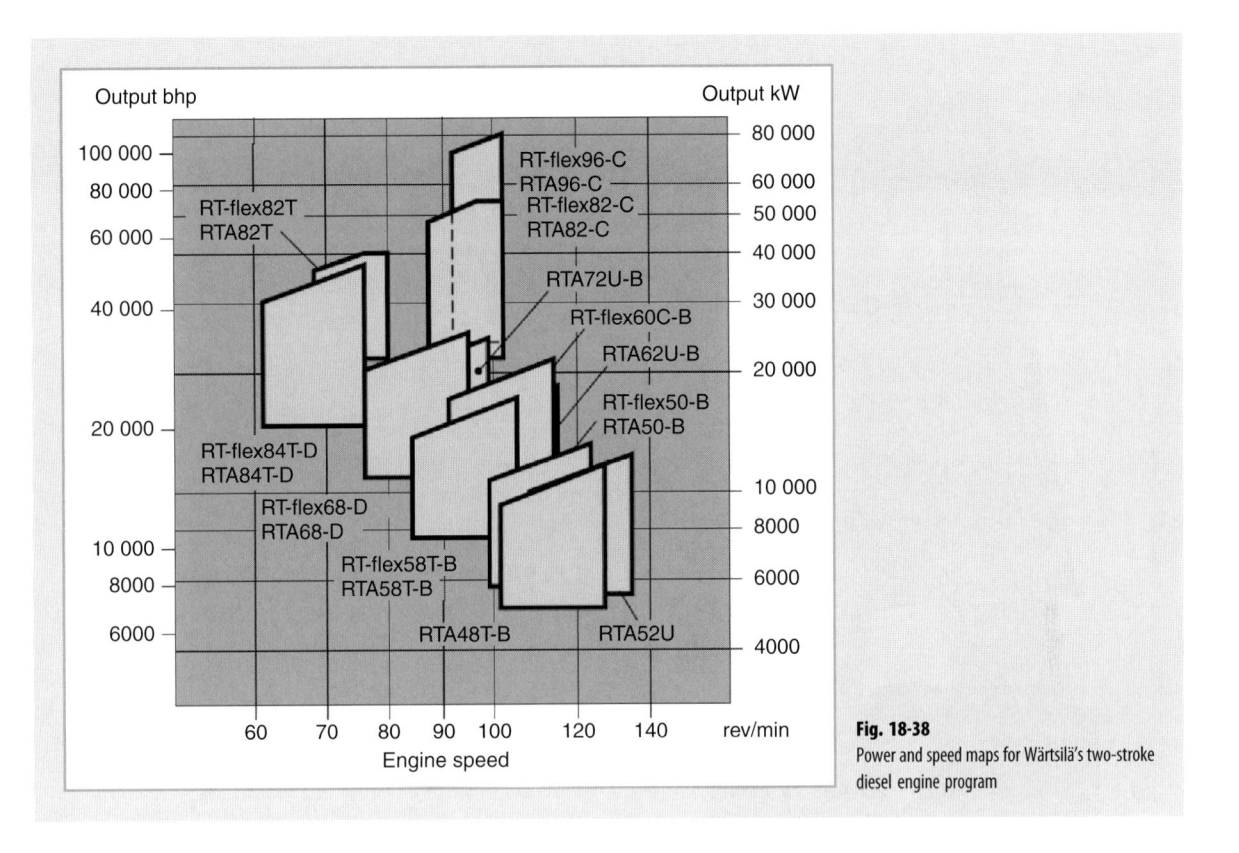

Fig. 18-38 Power and speed maps for Wärtsilä's two-stroke diesel engine program

space. All three components are bolted together and thus provide the stable structure demanded. All the bolted joints are easily accessible from the outside. The stresses and deformations in these important engine components are very low and assure high reliability (Fig. 18-39). This basic structure is similar in all Wärtsila two-stroke engines and its principle is also found among the other manufacturers.

Crankshaft assembly. Crankshaft assembly is the central domain of engine design. The design must ensure that every crankshaft assembly element, e.g. crankshaft, push rod (connecting rod), crosshead, piston rod, bearing etc., functions properly during the entire service life of sometimes more than 25 years.

A crankshaft consists of forged single throws connected with the main bearing journals by a transverse press fit (shrink fit). Apart from shrinking, throws and main bearing journals may also be connected by narrow gap welding. The stiffness and loads of the very slender throws are meticulously optimized by finite element calculations and dynamic measurements (Fig. 18-40). The calculations utilize dynamic analyses of the crankshaft and bearing loads, taking into account the stiffness and damping of the radial and axial bearing structures including the effects of cylinder damping. This makes small distances between cylinder bore center lines (approximately 1.75 x bore) possible with high reliability.

A very short connecting rod is employed to limit engine height despite the large stroke/bore ratio. The push rod ratio r/l (crank radius/connecting rod length) is 0.5–0.45, i.e., approximately twice as large as usual. However, generously sized crosshead shoe surfaces can readily absorb high side forces.

The actual crosshead bearing constitutes a distinctive feature of this type of engine [18-20]: Only it oscillates and its load vector always points downward. This interferes with reliable hydrodynamic lubrication and the supply of oil. One corrective is hydrostatic lubrication with approximately 12 bar oil pressure in oil chambers specially provided for this. They briefly elevate the crosshead journal during every revolution (Fig. 18-41) to assure oil is supplied.

The particularly high reliability of the main bearing, connecting rod bearing and crosshead bearing is a characteristic of this engine design. The following factors contribute to this:
- The specific loads are kept low by appropriate sizing.
- The white metal wear layer has excellent emergency running properties and is very flexible.
- Unique to two-stroke engines, the clear division between the combustion chamber and crank chamber protects these bearings against the effect of combustion products.

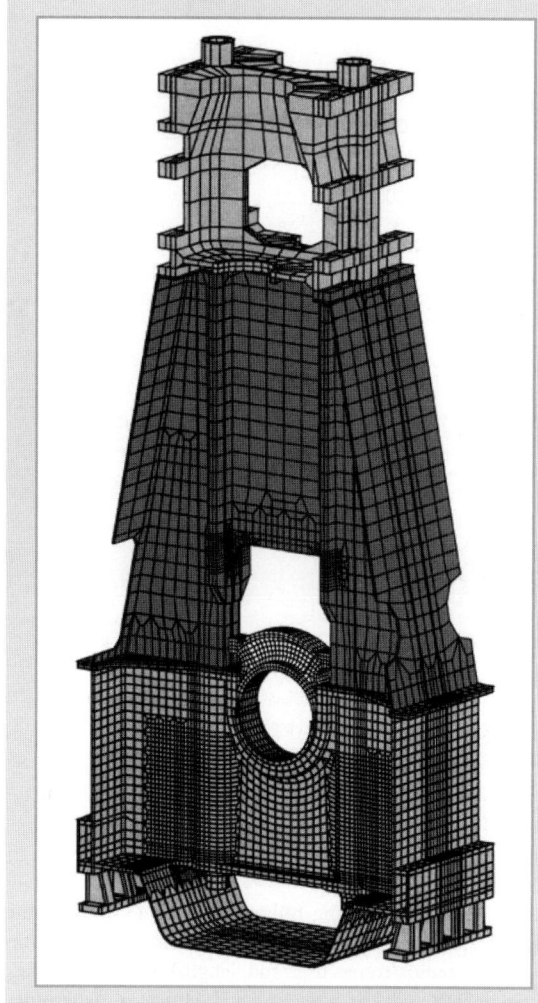

Fig. 18-39 Structural model of an engine frame (RTA96C) consisting of welded bedplate and A-frame with cast single cylinder blocks for FE calculation

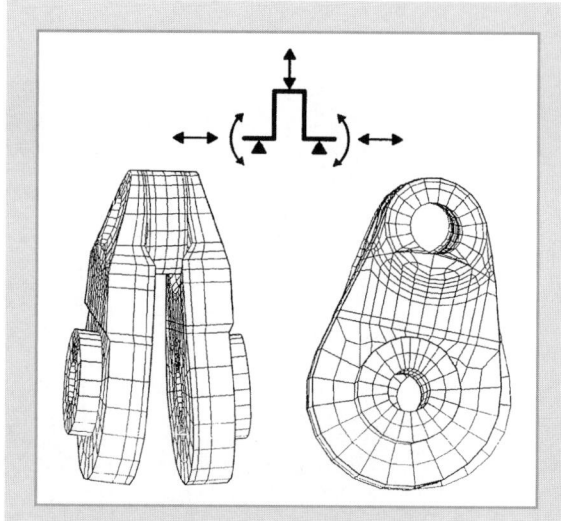

Fig. 18-40 FE calculations and measurements of deformations for a single crankshaft throw

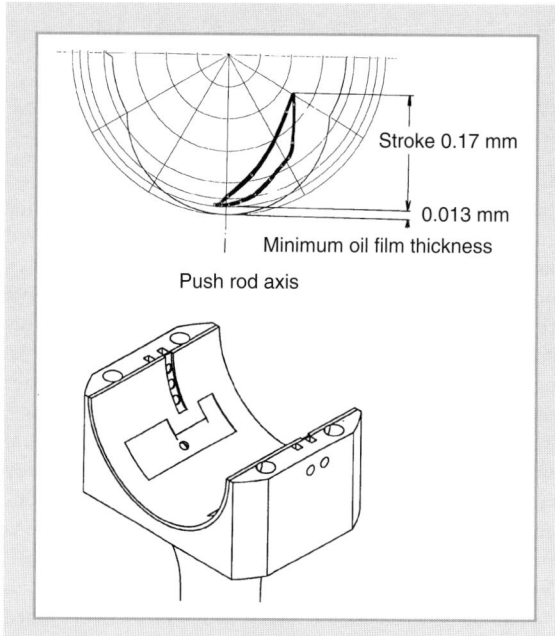

Fig. 18-41 Simulation and measurement of oil film thickness for a crosshead bearing

Combustion Chamber. Since energy (fuel mass) conversion is high when cylinder bores are large, meticulous design of the combustion chamber is the prerequisite for operating reliability. The simultaneous absorption of high thermal and mechanical loads in modern supercharged diesel engines led to the introduction of bore cooling at the end of the 1970s (Fig. 18-42), which Sulzer had already patented at the end of the 1930s. It provides effective, precisely metered cooling of the combustion chamber components (piston, cylinder liner, cylinder cover, valve and valve seat) at simultaneously higher rigidity without the necessity of thermal coatings, so-called cladding. This made it possible to keep the temperatures of the combustion chamber walls within permissible limits despite the steadily increased specific power (Fig. 18-43). Moreover, the water cooling initially employed as piston cooling was replaced in all large two-stroke engines by operationally simpler oil cooling. In particular, spray nozzles (jet shaker) increased the heat transfer in the cooling bores by 50% over the previous "shaker effect". Other engine manufacturers have adopted this principle.

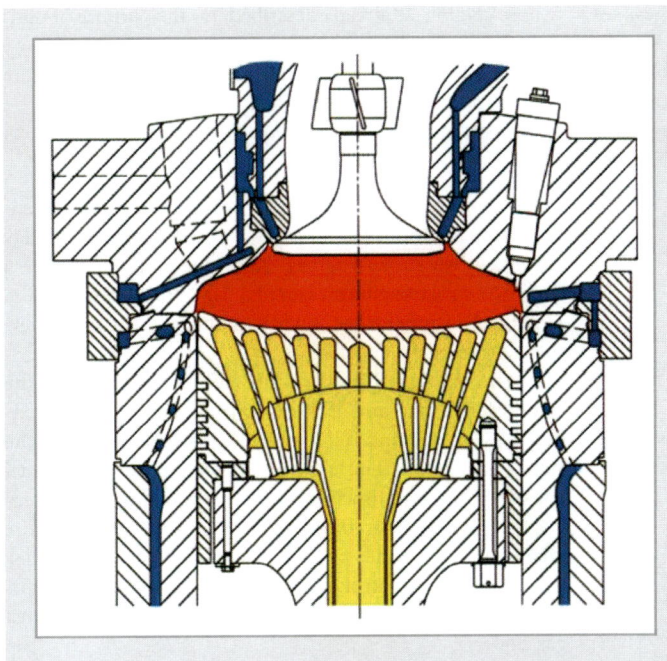

Fig. 18-42 Bore cooling

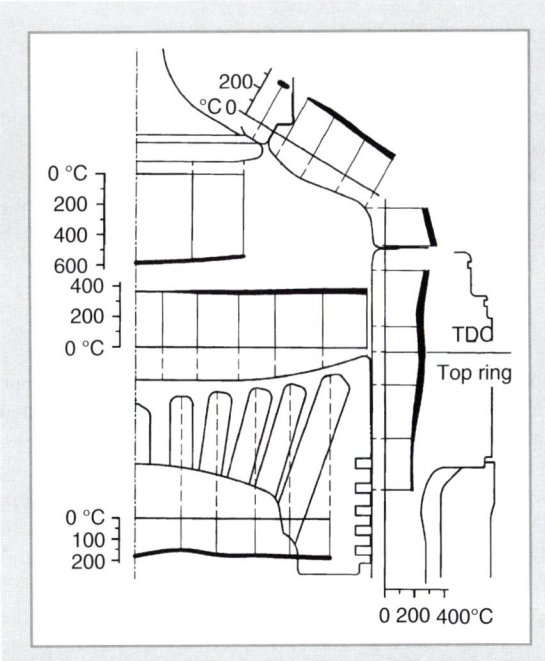

Fig. 18-43 Surface temperatures measured inside the combustion chamber of an 11RTA96C engine with a power of $P_e = 54,340$ kW, a brake mean effective pressure of $p_e = 18.2$ bar ($w_e = 1.82$ kJ/dm^3) and a speed of $n = 90$ rpm (R3 power level; see Sect. 18.4.4.1)

Exhaust Valve. Formerly a frequent source of disturbances during heavy fuel oil operation, the exhaust valve has attained remarkable reliability in the engine described here. Eighty thousand hours of operation are reached without maintenance. This can be attributed to the corrosion resistant valve material Nimonic 80A in particular. In addition to the optimal surface temperatures obtained by bore cooling (Fig. 18-43), the valve seat's self-cleaning of combustion residues, effected by the rubbing seating of the rotating valve, also contributes to this. The discharging exhaust gas periodically generates valve rotation by means of an impeller mounted on the valve stem. Hydraulic valve actuation eliminates mechanical vibrations in the valve drive. In addition, the closing force for the valve is generated pneumatically by compressing a pneumatic spring.

Camshaft. The camshaft is driven very precisely by gears, which, unlike a chain drive, also guarantee consistent timing after years of operation. Injection pumps and actuators for the hydraulic valve drive for two cylinders apiece are placed atop the engine A-frame. The engine can be reversed by turning the injection cams with individual servo pumps. The injection pumps in Wärtsilä engines are valve controlled; they are helix controlled in MAN B&W engines. Variable injection timing (VIT) can be applied to optimize part load operation. Injection is performed by three injection valves located symmetrically on the circumference of the cylinder cover. This assures surface temperatures at the piston are optimal (see Fig. 18-43). First implemented by Wärtsilä, the application of common rail technology is new in large two-stroke diesel engines (see Sect. 18.4.5.2).

Scavenging and Supercharging. A modern two-stroke low speed engine requires a turbocharger with high overall efficiency of up to 72% and a high pressure ratio of up to 4.2. A more compact configuration locates the intercooler near the cylinder jacket. Condensate, which is unavoidable during intercooling, must still be separated before the cylinders to prevent cold corrosion and any disruption of the lubricating oil film.

The symmetrical configuration of the scavenging ports and the exhaust valve facilitates efficient scavenging. The result is a volumetric scavenging efficiency of over 95% compared with values of approximately 85% in loop scavenged engines.

18.4.3 Operating Performance of Two-Stroke Low Speed Engines

The two-stroke low speed diesel engine has earned the reputation of being the most reliable internal combustion engine for good reason. During the last three decades, various factors have induced diesel engine manufacturers to additionally increase the engine's inherent reliability:
- The quality of heavy fuel oil has noticeably diminished since the "oil crisis" in the 1970s because oil refineries have intensified their production of lighter distillate from crude oil by new processes (see Sect. 4.3).
- Despite the demanding operating conditions it entails, customers expect longer overhaul intervals of three years, which corresponds to a ship's usual cycle for periodic maintenance in dry-dock.

Large two-stroke engines have undergone further improvements in recent years to achieve this. Only two of the most important component assemblies, which are decisive for the overhaul intervals, are treated here as examples:

Piston ring and cylinder liner wear. Advances in development in this area are based on:
- fully honed cylinder liners with clearly defined hard phase fractions that optimally distribute the operating load,
- chrome ceramic piston rings that improve running-in performance and operational reliability,
- anti-polishing rings on the upper edge of cylinder liners, which scrape off potential coke deposits on the piston skirt,
- improved lubricating film and reduced mixed friction zones [18-20] by electronically controlled cylinder lubrication systems that precisely meter the quantity of lubricating oil, which is particularly effective for hydrodynamic lubrication between rings and the cylinder liner,
- bore cooling, as already described, for the purpose of optimal component temperatures,
- three injection nozzles for the purpose of optimal mixture formation and combustion temperatures and
- prevention of material attrition on the piston.

This has made overhaul intervals of three years or approximately 18,000 hours of operation possible in present day advanced two-stroke engines. With operating values for a large two-stroke engine measured during propeller operation, Fig. 18-44 documents the conditions under which these intervals have to be produced.

The maximum firing pressure at full load reaches the value of 142 bar and is kept constant between approximately 80% load and full load by the variable timing of the start of injection. Thus, the characteristic of specific fuel consumption as a function of load remains flat. The very low exhaust temperatures (approximately 450°C before and 300°C after the turbine) are also noteworthy. They indicate the two-stroke engine's particularly high efficiency, sufficient energy still being available for the exhaust gas boiler.

18.4.4 Two-Stroke Low Speed Engines as Marine Engines

18.4.4.1 Propulsion System Tuning

Since the diesel engine is part of a system, optimal design of the propulsion system is particularly important. Important parameters are:
- optimal adjustment of the ship–propeller–engine,
- optimization of the auxiliary systems necessary for the engine (lubricating oil, fuel, cooling system, etc.),
- prevention of disturbing or harmful vibrations,
- optimal generation of auxiliary power on board and
- optimal recovery of waste heat.

Optimally adjusting the engine, propeller and ship to one another requires first selecting the diesel engine connected with a fixed propeller from the available models of a series, allowing for the propeller and ship characteristics. Taking the ship's shape and the propeller data as the starting point, the propulsion power can be determined at the selected ship speed on the basis of model tests, simulations and previous examples. The ship's various operating modes, e.g. loaded, ballasted, clean or dirty outer hull, must be considered. Once power and speed are established, the engine is selected. Since the power maps of the individual engines from the series overlap, several engines may often deliver the desired power-speed combination for a particular case. Then, other criteria, e.g. number of cylinders, dimensions, specific consumption, etc., may be referenced for the final decision.

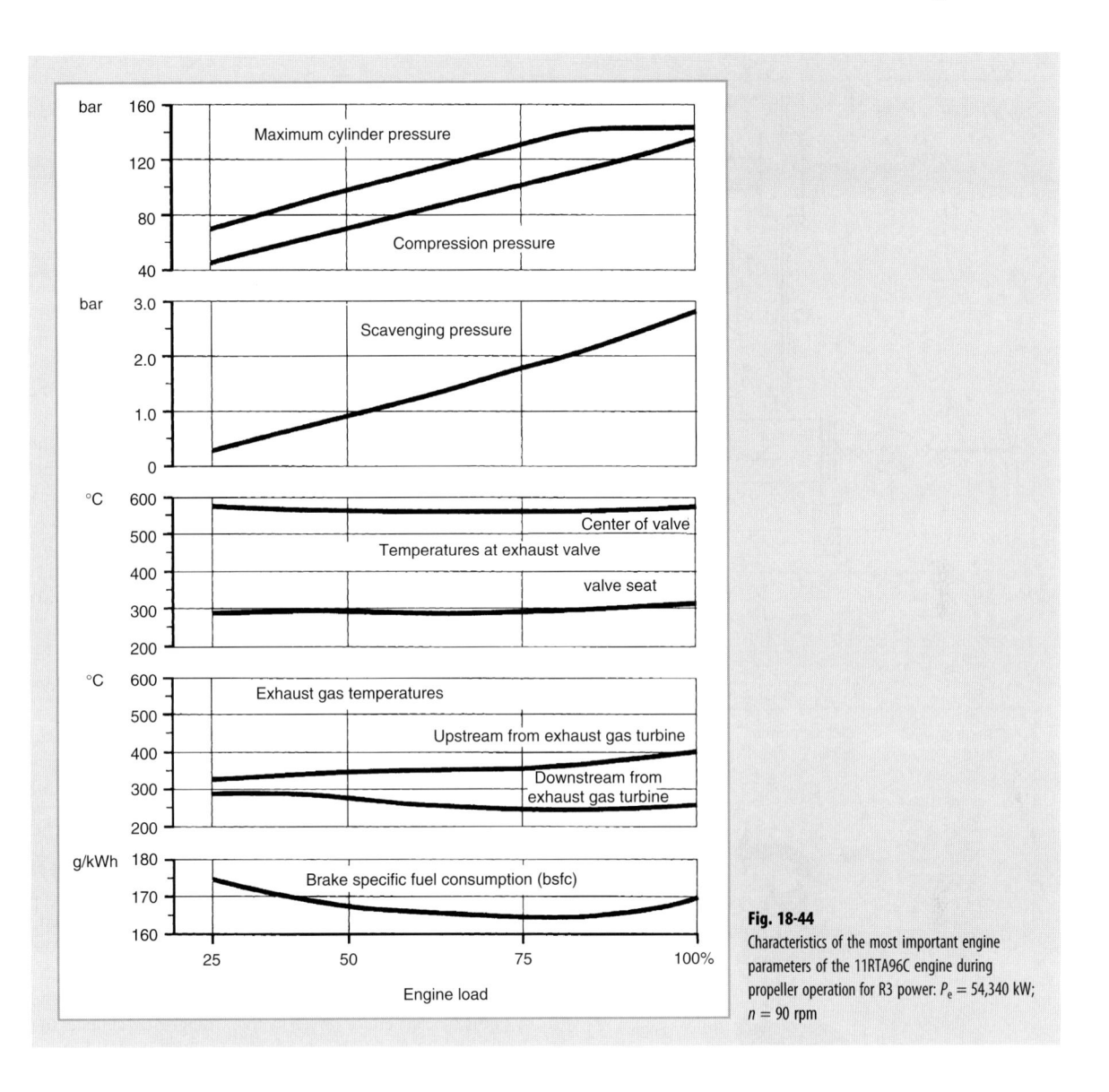

Fig. 18-44
Characteristics of the most important engine parameters of the 11RTA96C engine during propeller operation for R3 power: $P_e = 54{,}340$ kW; $n = 90$ rpm

When the engine is intended to drive other units, e.g. shaft generators, they must be additionally factored in when determining the required rated power. Just like on-board diesel generators, shaft generators can be drawn on to generate the electrical power required on board a ship. The advantage of auxiliary diesel engines is their great operating flexibility. However, the operating costs are somewhat higher when they are not also operated with cheaper heavy fuel oil instead of diesel oil (unifuel concept). By comparison, shaft generators connected with the main engine by gears have lower fuel costs because of the PTO concept (power take-off) but require higher investments in the system. A significant cost factor is the necessary tuning of the power frequency by a gear with a variable transmission ratio or by a thyristor inverter when propeller speed changes.

The low exhaust temperatures of 270 to 300°C at the outlet of an exhaust gas turbine impose narrow limits on the energetically advantageous use of turbo generators.

The high thermal efficiency of a two-stroke low speed diesel engine causes over 50% of the thermal energy to be converted into mechanical work. Thus, less exhaust gas heat is available for recovery than in a four-stroke diesel engine. The exhaust gas heat is primarily utilized to generate steam. The heat extracted from the intercooler's first stage as well as part of the cooling water heat is primarily utilized to

Fig. 18-45 Schematic of a marine propulsion system with a low speed large two-stroke diesel engine and exhaust heat recovery (total heat recovery)

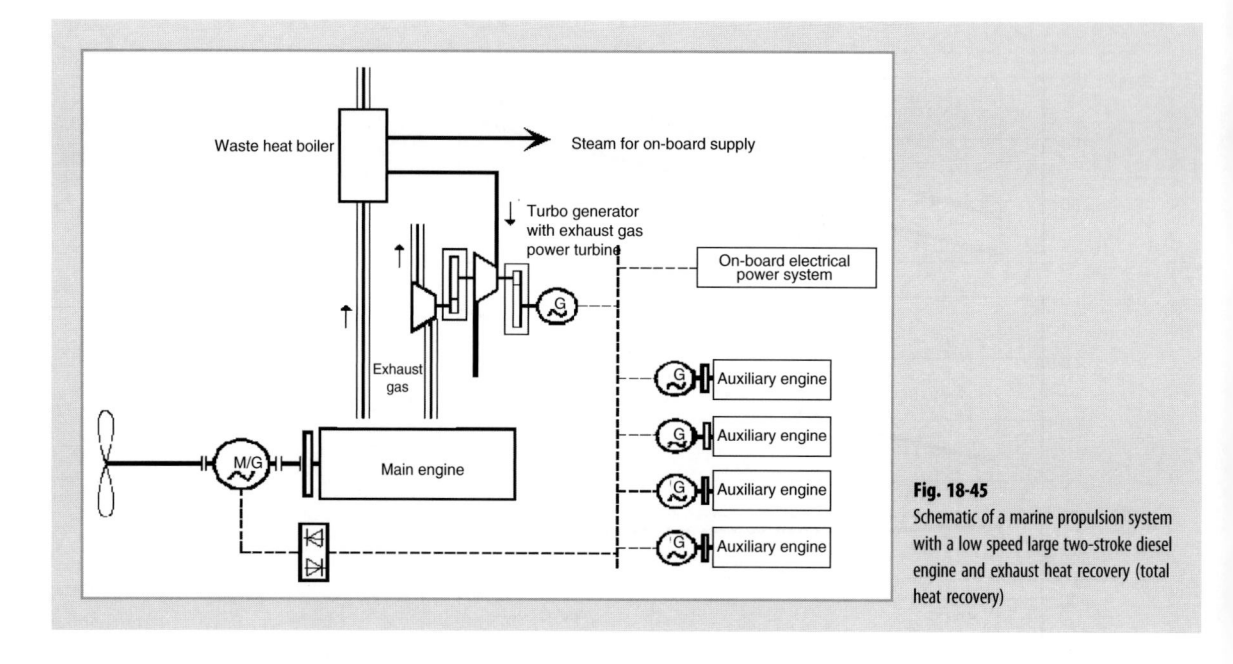

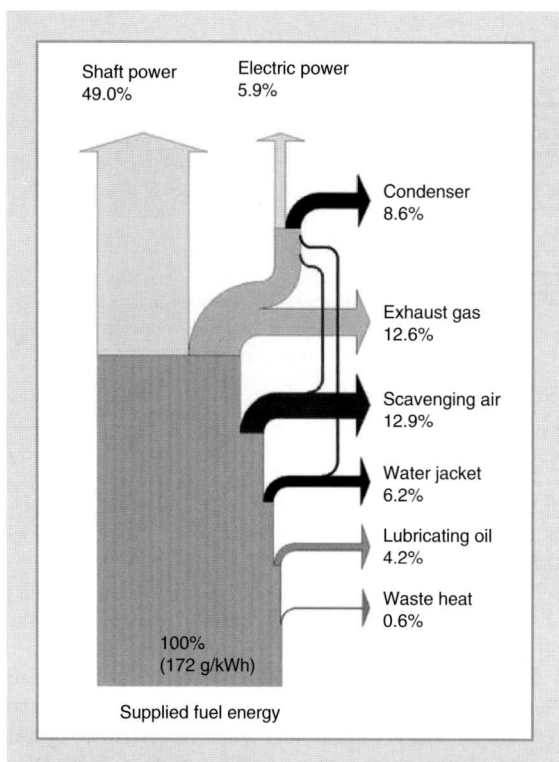

Fig. 18-46 Heat balance of a Wärtsilä 12RT-flex96C two-stroke low speed diesel engine with an overall efficiency of 54.9% by recovering waste heat by means of a turbo generator to generate electrical power

generate warm water or fresh water in freshwater generators (see Chap. 14).

Wärtsilä's expanded turbocompound concept, the waste heat recovery (WHR) system (Fig. 18-45) furnishes an attractive option to recover exhaust heat. The development of highly efficient turbochargers that now have efficiencies of up to 72% makes this concept possible.

After exiting the turbocharger, a majority of the quantity of exhaust gas is fed to a two-stage steam generator that supplies a steam turbine. The turbochargers' high efficiency allows diverting approximately 10% of the quantity of exhaust gas before the turbocharger and supplying it to an exhaust gas turbine, the shaft of which is connected with the shaft of the steam turbine by a gear. The generator driven by this arrangement supplies the ship's onboard electrical system with additional electrical power that a shaft motor can even convert into increased propulsion power for the ship. Thus, in conjunction with the waste heat recovery system, a large two-stroke engine's overall efficiency can be increased to approximately 55% (Fig. 18-46). This reduces CO_2 emissions over a standard engine by approximately 11% and the remaining exhaust emission also decreases in the same ratio relative to the engine power delivered [18-21].

MAN Diesel SE offers a similar concept called the Thermo Efficiency System (TES).

18.4.4.2 Vibration Damping in the Drive Train

The pursuit of higher propulsion efficiency has resulted in low engine speeds for a certain engine power. This was achieved by

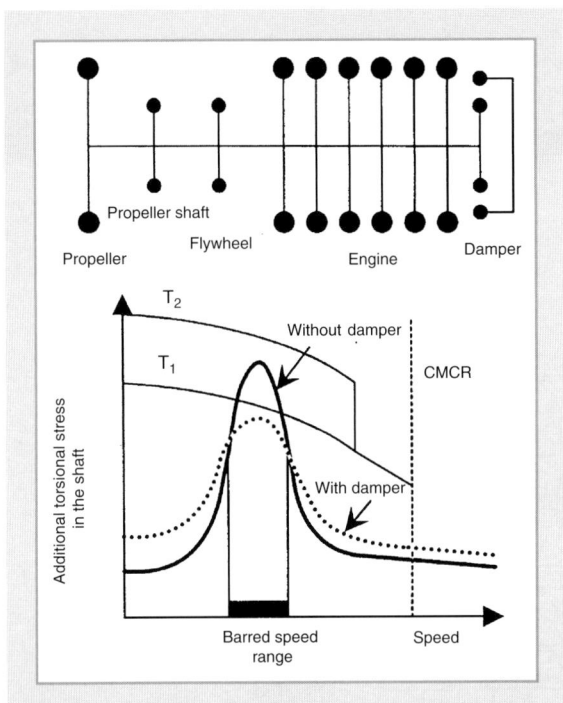

Fig. 18-47 Calculated vibration system of engine–propeller shaft–propeller. Reduction of the torsional vibration amplitude by means of a damper. T_1 is the permissible limit for continuous mode T_2 the permissible limit for transit mode

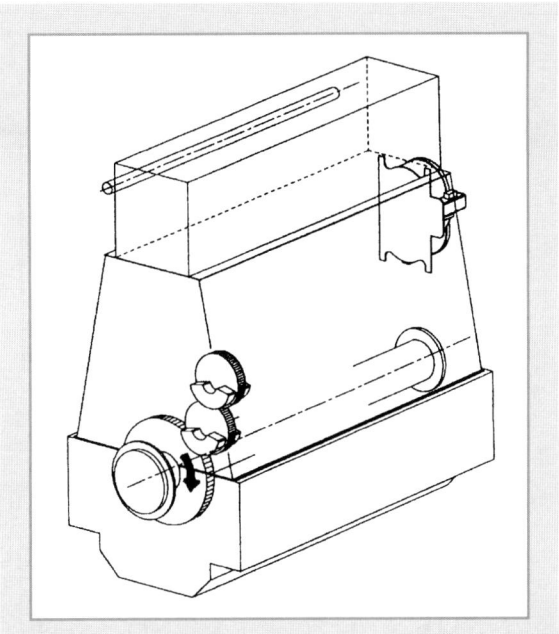

Fig. 18-48 Balancing of the second order moment by electrically driven unbalancers arranged on the free engine end and gear driven unbalancers on the driving end

- increasing the stroke to bore ratio from approximately 2.0 to over 4.0 and
- increasing the use of low speed engines with large bores and consequently lower numbers of cylinders, e.g. four and five cylinder engines.

The trend toward higher s/D ratios has led to a reduction of the natural frequencies of the crankshaft, which is becoming slenderer. Thus, critical engine speeds approach the propulsion engine's operating speed more often, unless this has been prevented by meticulously tuning the vibration system of the engine–propeller shaft–propeller (Fig. 18-47). If fully eliminating the resonances from the operating range is impossible, then the vibration amplitude is reduced by means of a vibration damper. Similar methods are also applied when damping or "detuning" a shaft system's axial vibrations.

The transition to smaller numbers of cylinders requires additional provisions to cancel the unbalanced first and second order free moments in four and five cylinder engines. Optimal placement of the engine in a ship can serve as a corrective: to prevent the excitation of hull vibrations, an engine should not to be placed in a node of the oscillating hull. If the solution is still unsatisfactory, then counterweights on the crankshaft "detune" the first order free moments' phase relation and amplitude. Second order free moments' can usually be balanced by "Lanchester" balancers running at twice the engine speed (see Sects. 8.1 and 8.2). Placing an electrically driven unbalancer on the free end of the engine in combination with the balancing mass integrated in the engine on the drive side and driven by a gear has proven itself for the vertical second order moment $M_{2\,V}$ that frequently occurs in two-stroke engines with four to six cylinders. Thus, the effect of balancing the moment can be inexpensively tested and adjusted in marine operation independent of load and speed (Fig. 18-48). Mounting an electrically driven second order balancer, which is located as far astern as possible and also covers potentially disturbing effects of the propeller moment, furnishes another option to damp vibrations, especially in ships with bridges located far aft (Fig. 18-49).

18.4.5 Outlook

18.4.5.1 Trends in Future Development

General remarks. The advances in the efficiency and reliability of large two-stroke engines in just the last few decades do not mean that development has already reached its limits. Theoretical analyses by Eberle [18-22] demonstrate the definite possibility of thermal efficiencies above 60% and even higher specific powers. However, since particularly high reliability continues to be required of marine diesel engines,

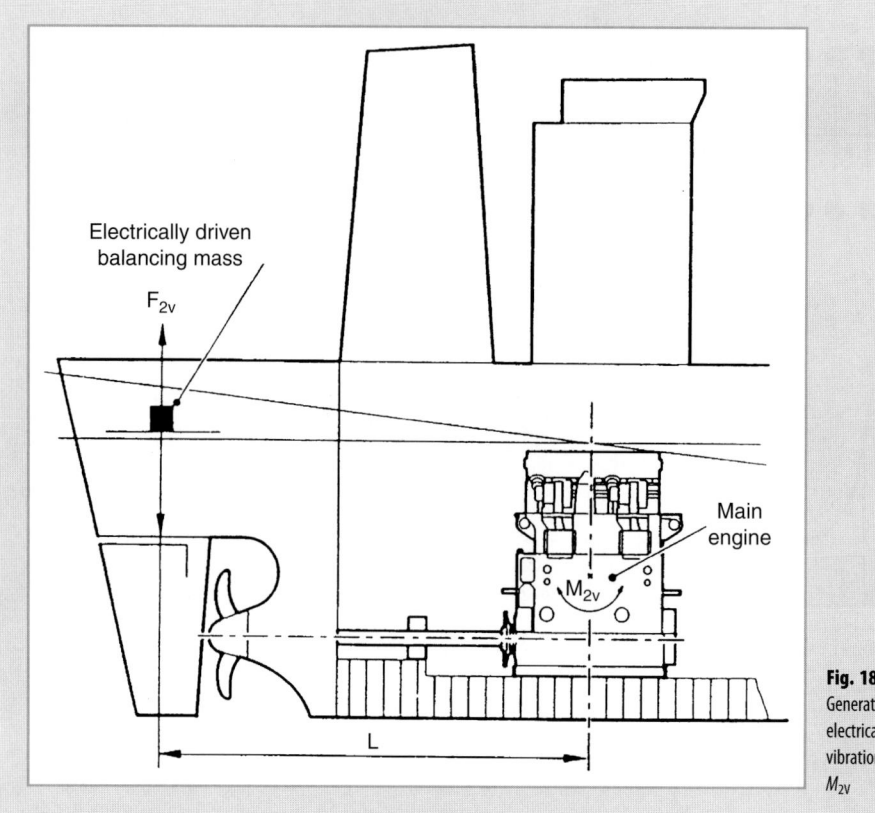

Fig. 18-49 Generation of a free second order force by an electrically driven balancing mass to damp vibrations caused by the moment of inertia M_{2v}

compromises that benefit the power concentration or detract from operational reliability are unfeasible.

The following discusses potential development scenarios for engine characteristics. However, the conclusions contain a level of uncertainty since they may be influenced by unquantifiable constraints, e.g. economic growth, environmental requirements, crude oil availability or technical trends.

Brake mean effective pressure and maximum cylinder pressure. Thermodynamically, a constant ratio of maximum cylinder pressure and brake mean effective pressure (or specific work) corresponds to an approximately constant level of thermal efficiency. Thus, at a proven optimal ratio of $p_{Zmax}/p_e = 8.0$, the maximum cylinder pressure in current large two-stroke engines increases linearly with the brake mean effective pressure when there is no intention to accept any efficiency loss (Fig. 18-50):

– Since high brake mean effective pressures implicitly require high maximum pressures in the cylinder, at a $p_e = 21$ bar, the pressure p_{Zmax} increases to 168 bar.
– Similarly, higher boost pressures of approximately 4.2 bar and correspondingly high turbocharger efficiencies of approximately 72% are essential (see Sect. 2.2).

Such values are now thoroughly realistic. The principle of bore cooling (see Sect. 18.4.2.2) still holds substantial potential for the component temperature level. Tribological development of the mating of the piston ring-cylinder liner will likewise have to follow this trend to maintain the two-stroke diesel engine's traditional reliability.

Stroke to bore ratio and mean piston velocity. When the general definition of engine power P_e (see Sect. 1.2) as a function of the number of cylinders z, mean piston velocity c_m, engine speed n, stroke to bore ratio $s/D = \zeta$ and brake mean effective pressure p_e is examined in a less common formulation

$$P_e \sim z \cdot p_e \cdot c_m^3 / (n^2 \cdot \zeta^2),$$

where the power P absorbed by the propeller is a function of design speed $n_p = n$, then the following ensues

$$P_e \sim n^a,$$

where $a = 0.3$ is a function of the stroke to bore ratio of the aforementioned parameters

$$\zeta_2 = \zeta_1 (n_1/n_2)^{1.15} \cdot (p_{e2}/p_{e1})^{0.5} \cdot (c_{m2}/c_{m1})^{1.5} \cdot (z_2/z_1)^{0.5}$$

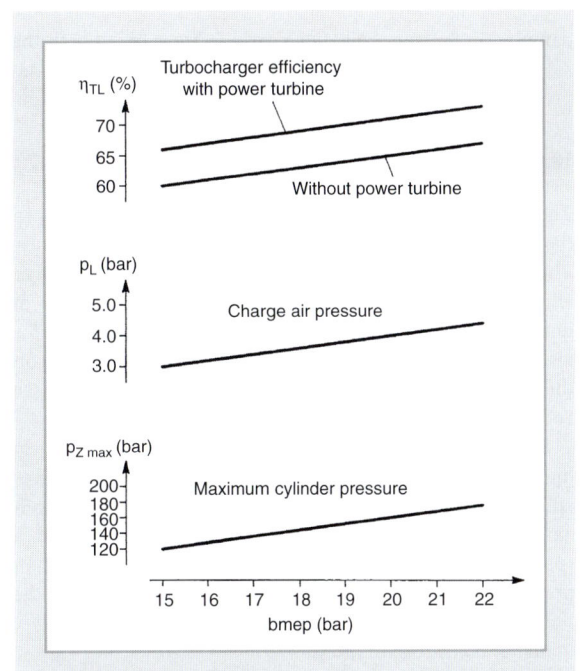

Fig. 18-50 Maximum cylinder pressure p_{Zmax}, boost pressure p_L and turbocharger efficiency η_{TL} as a function of brake mean effective pressure p_e at constant thermal efficiency

This will allow:
1. optimally adjusting engine performance for consumption, load and emissions at any load and
2. increasing the reliability of components by monitoring and controlling the most important engine functions.

Developed in recent years, common rail injection satisfies the requisite conditions.

18.4.5.2 Common Rail Technology in Large Two-Stroke Diesel Engines

The introduction of the common rail fuel injection system suitable for heavy fuel oil for large two-stroke engines constituted a technological leap in the direction of the aforementioned development. Wärtsilä's RT-flex concept had been tested on a laboratory engine since 1998 and was implemented in a standard engine of the 6RT-flex58T type in 2001. The camshaft, which previously controlled injection and exhaust valve actuation and greatly restricted the potential for optimization, was replaced with electronic timing that furnishes great flexibility to optimize an engine under the widest variety of operating conditions. A central pump unit (supply unit) supplies both heavy fuel oil and hydraulic oil that actuates exhaust valves to the two accumulators in the rail unit located at the level of the cylinder covers (Fig. 18-51). From there, the

The index "1" corresponds to the current "state-of-the-art" reference engine, the index "2" to the potential next stage of development.

The stroke to bore ratio may be expected to increase further
– when the power per unit piston area $P_A \approx p_e \cdot c_m$ is higher and
– even larger, lower speed propellers with higher efficiency are employed.

However, an increase in the stroke to bore ratio is at odds with the higher specific costs and the engine's height and width.

In conjunction with anticipated emission regulations, further increases in the power of highly supercharged two-stroke low speed engines by increasing the brake mean effective pressure to 21 bar and above will make flexible adjustment of the engine with electronically controlled injection and exhaust valve timing imperative. Operating parameters that correspond to an optimal setting at full load will no longer be optimal in the part load range without intervention. In addition, this will necessitate adjustment of the
– injection parameters,
– valve timing and
– cooling and lubrication

as a function of load as well as a high performance supercharging system capable of delivering the necessary pressure ratio in conjunction with the higher efficiencies.

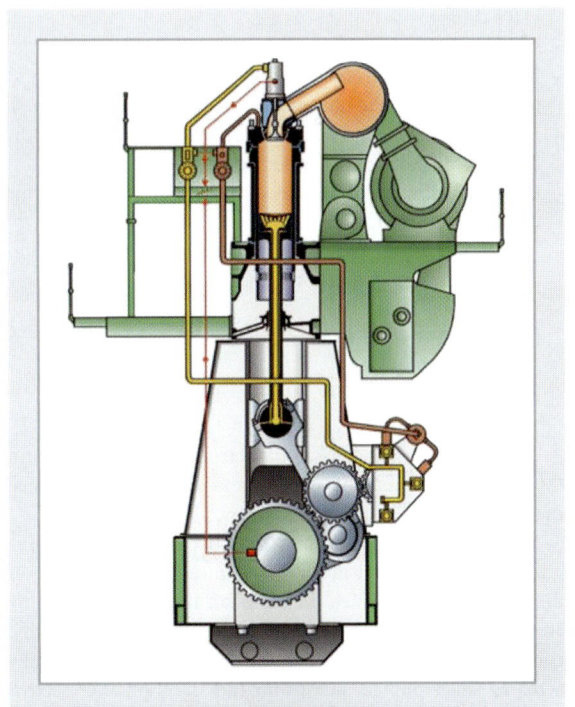

Fig. 18-51 Configuration of a Wärtsilä RT-flex96C two-stroke diesel engine with a common rail injection system suitable for heavy fuel oil and an electronically controlled exhaust valve

fuel and hydraulic oil each travel through a control unit to the three peripheral injection valves or to the central exhaust valve on every cylinder. Fully electronic timing of both functions makes the operating parameters of injection timing and duration as well as the movement of the exhaust valve freely adjustable. While the servo system that actuates the valves operates with a pressure of 200 bar, the maximum injection pressure in the fuel rail is 1,000 bar. A metering piston integrated in the injection control unit controls the volume of the injection for each cylinder. Its position at any time is measured electronically, thus facilitating exact metering of the injected fuel quantity for each cylinder. The three injection nozzles per cylinder are individually controllable and thus also allow cutting off individual nozzles at low loads to improve injection with the smallest quantities of fuel and prevent the formation of smoke at part load [18-23].

The fundamental advantages of the RT-flex system are its reduced fuel consumption and lower exhaust emission. In addition, the technology enables automatically adjusting engine operating parameters to an engine's current state. This constitutes an elementarily important step toward an "intelligent" engine.

Specifically, common rail technology provides the following advantages for the operating performance of large two-stroke engines:
- lower fuel consumption in the middle and upper part load range by variable injection pressure and freely selectable valve timing,
- smokeless operation under all operating conditions,
- precise speed control and stable engine operation even at lowest speeds in the range of 10–15 rpm,
- lower mechanical and thermal load by more uniform combustion and balanced cylinder pressure level,
- easier adjustment of and less maintenance required for common rail components,
- higher operational reliability and availability through integrated monitoring functions and the redundancy of key components,
- lower engine weight (approximately 2 t per cylinder at average bore sizes) and
- lower vibrations.

The starting air system is also electronically controlled and thus allows better engine starting and braking performance than mechanically controlled systems.

MAN Diesel SE also expanded its two-stroke engine program in recent years with an electronically controlled series designated "ME" in which the camshaft has been replaced by an electro-hydraulic servo oil system. Not to be equated with a common rail system, this system utilizes electro-hydraulically actuated single plunger pumps for fuel injection and electronically controlled valve actuators for the exhaust valve function [18-24]. The flexibility of injection pressure, timing and valve timing is comparable to that of common rail systems. Basic differences are the increased requirements for hydraulic damping in the servo oil system, which are satisfied with the aid of pneumatic accumulators, and the elimination of independent activation and cutoff of individual injection nozzles.

18.4.5.3 Exhaust Emissions

The reduction of exhaust emissions is now the top priority in the further development of large two-stroke marine engines. The inherent advantage of a low speed diesel engine is its high efficiency. Carbon dioxide emissions and the fraction of unburned hydrocarbons are very low. This holds true for visible smoke too but not for particulate emission based on the ISO standard, particularly in heavy fuel oil operation (see Sect. 4.3.4.2).

By comparison, the fraction of nitrogen oxides is relatively high compared with other combustion engines with lower efficiency. Regulated by IMO Marpol Annex VI, the nitrogen oxide limit in effect is 17.0 g/kWh for marine engines with a speed below 130 rpm. The following measures are applied in present day large two-stroke engines to comply with the limits:
- higher compression ratios,
- optimized geometry of injection nozzles (number of spray holes, orifice diameter, angle of spray) and
- retarded injection timing.

The potential to further reduce NO_X emission by exhausting all the options listed here remains low (see Fig. 18-52 for optimized NO_X reduction).

A further reduction of the limit for nitrogen oxides in the IMO regulation or local emission control legislation is foreseeable. The following additional options are available to reduce NO_X further (see Fig. 18-52):
- optimizing the injection parameters by common rail injection (CR injection).
- reducing NO_X by between 20 and 50% by engine internal measures employing water (wet technologies), scavenge air humidification, water/fuel emulsions and direct water injection also in combination with common rail injection (Wärtsilä's RT-flex system).
- combining water injection with internal exhaust gas recirculation by shortening the scavenging process (WaCoReG) and thus making it possible to reduce NO_X by up to 70% and
- aftertreating exhaust with an SCR catalyst into which ammonia or a urea solution is injected (see Sect. 15.6) and thus making it possible to convert up to 95% of the nitrogen oxides.

The SCR catalyst in large two-stroke engines has to be placed before the exhaust gas turbocharger since the exhaust temperature level there is high enough to obtain optimal conversion rates and simultaneously prevent corrosion problems. Particularly compact designs allow installation of the catalyst in the exhaust pipe at the engine.

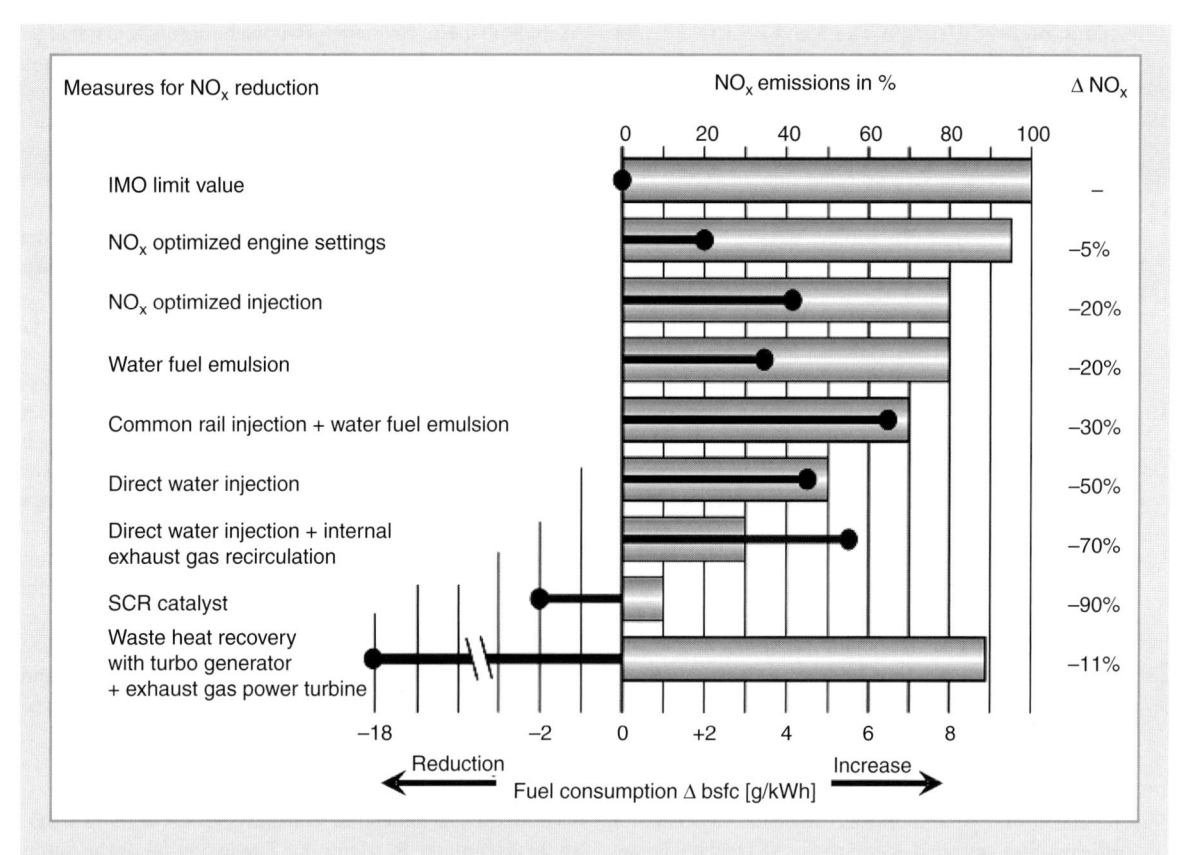

Fig. 18-52 Measures for a two-stroke low speed engine to reduce nitrogen oxides and its influence on fuel consumption b_e, based on the currently effective NO_x limit of 17 g/kWh according to the IMO regulation

Whereas the reduction measures discussed thus far always result in higher fuel consumption (Fig. 18-52), the combustion cycle can be optimized to reduce fuel consumption when an SCR catalyst is used.

Waste heat recovery by means of a steam generator in conjunction with a turbo generator and an exhaust gas power turbine (total heat recovery; see Fig. 18-45) delivers overall drive system power boosted by a total of 11%. This translates into a reduction of nitrogen oxide emission relative to power of the same magnitude and, at the same time, a corresponding increase in efficiency, thus reducing fuel consumption by 18 g/kWh (see Fig. 18-52).

18.4.5.4 Concluding Thoughts

Vying with other combustion engines as the dominant drive source for oceangoing vessels, the two-stroke low speed diesel engine has so far been able to maintain and even consolidate its edge in terms of thermal efficiency, suitability for heavy fuel oil and reliability. Its future prospects are also encouraging since, thanks to technological advances in materials technology, turbocharging and electronically controlled common rail injection, considerable potential still exists in terms of power density, cost effectiveness, reliability and emission performance. The use of state-of-the-art technologies to further reduce exhaust emissions, particularly carbon dioxide, nitrogen oxides, black smoke and particulate matter, will be crucially important [18-25].

Literature

18-1 Lingens, A.; Feuser, W.; Bülte, H.; Münch, K.-U.: Evolution der Deutz – Medium Duty Plattform für zukünftige weltweite Emissionsanforderungen. 12th Aachen Colloquium Automobile and Engine Technology (2003)

18-2 Wegener, U.: Elektronisch geregelte Dieselmotoren – Status und Ausblick. 13. Heidelberger Flurförderzeug-Tagung 2005. VDI-Berichte (1879)

18-3 Bülte, H.; Beberdick, W.; Pütz, M.; Kipke, P.: DEVERT® – The DEUTZ Concept to Fulfill the

Emission Level U.S. EPA Tier 3 and EU COM 3A for Non-Road Engines. ICES2006-1441, Aachen: ASME Spring Conference (2006)

18-4 Lingens, A.; Bülte, H.; Münch, K-U.; Hülsmann, B.: Fuel Injection Strategies for Medium Duty Engines to Meet Future Emission Standards. Fisita Congress (2002)

18-5 Schiffgens, H.J. et al.: Die Entwicklung des neuen MAN B&W Diesel Gas Motors 32/40 DG. MTZ 58 (1997) 10

18-6 Koch, F.; Hanenkamp, A.: Moderne Gasmotoren auf Basis der erfolgreichen MAN B&W Schweröl-Dieselmotoren. 13th Aachen Colloquium Automobile and Engine Technology (2004)

18-7 Syassen, O.: Der konsequente Viertakt Dieselmotor. Hansa 126 (1989) 112

18-8 Lausch, W.; Fleischer, F.; Maier, L.: Möglichkeiten und Grenzen von NO_X Minderungsmassnahmen bei MAN B&W Viertakt-Grossdieselmotoren. MTZ 54 (1993) 2

18-9 Koch, F.; Hollstein, R.; Imkamp, H.: Weiterentwicklung des mittelschnelllaufenden 4-Takt-Schweröl-Dieselmotors MAN B&W 58/64. 14th Aachen Colloquium Automobile and Engine Technology (2005)

18-10 Vogel, C.; Wachtmeister, G.; Maier, L.: New concept of HFO common rail injection system for MAN B&W MS-Diesel engines. CIMAC Congress Kyoto (2004) 136

18-11 Ollus, R.; Paro, D.: Experience and development of world's first Common-Rail Injection System for Heavy-Fuel operated Medium-Speed Diesel Engines. CIMAC Congress Kyoto (2004) 114

18-12 Vogel, C.; Haas, S.; Tinschmann, G.; Hloussek, J.: Die Motorenfamilie mittelschnelllaufender Dieselmotoren mit schweröltauglichem Common Rail Einspritzsystem von MAN B&W. 14th Aachen Colloquium Automobile and Engine Technology (2005)

18-13 Lausch, W.; Perger, W.V.; Schmidt, H.: Systeme müssen selbst drohende Störungen schnell lokalisieren. Schiff & Hafen (1994) 11

18-14 Rulfs, H.: Grossmotorenforschung am Einzylindermotor. 50 Jahre FVV. MTZ Sonderheft (2006) pp. 66–68

18-15 Marquardt, L.; Berndt, B.: Untersuchung zur innermotorischen Stickoxidminderung in mittelschnelllaufenden Viertakt-Schwerölmotoren. FVV-Heft R531(2005) S. 185–203

18-16 Schlemmer-Kelling, U.: Entwicklungstendenzen bei mittelschnelllaufenden Grossdieselmotoren. 6. Dresdner Motorenkolloquium (2005)

18-17 Sass, F.: Bau und Betrieb von Dieselmaschinen. Berlin/Göttingen/Heidelberg: Springer (1957)

18-18 Briner, M.; Lustgarten, G.: Design Aspects of the new Sulzer RTA Superlongstroke. Sulzer-interne Schrift (1981) 12

18-19 Pedersen, S.; Groene, O.: Design Development of Low Speed Engines. 21. Marine Propulsion Conference Athen (1999) 3

18-20 Lustgarten, G.: Zweitakt-Kreuzkopf-Dieselmotor, Reife Technologie oder High-Tech? Vortrag an der TH Hannover (1989)

18-21 Heim, K.: New Technologies in Sulzer Low-Speed Engines for Improving Operational Economy and Environmental Friendliness. 7th International Symposium on Marine Engineering Tokyo (2005) 10

18-22 Eberle, M.K.; Paul, A.: Possible ways and means to further develop the diesel engine in view of economy. CIMAC Conference Warsaw (1987)

18-23 Demmerle, R.; Heim, K.: The Evolution of the Sulzer RT-flex Common Rail System. CIMAC Conference Kyoto (2004) 6

18-24 Egeberg, C.; Knudsen, T.; Sorensen, P.: The Electronically Controlled ME/ME-C Series Will Lead the 2-Stroke Diesel Engine Concept into the Future. Kyoto: CIMAC Conference (May 2004)

18-25 Holtbecker, R.: Taking the Next Steps in Emissions Reduction for Large Two-Stroke Engines. CIMAC Conference Vienna (2007)

Standards and Guidelines for Internal Combustion Engines

German Standards		
Standard	Title	Date of publication
DIN 1940	Verbrennungsmotoren; Hubkolbenmotoren, Begriffe, Formelzeichen, Einheiten (Reciprocating internal combustion engines; terms, formulae, units)	1976-12
DIN 6261	Verbrennungsmotoren; Teile für Kreiskolbenmotoren, Äußerer Aufbau, Triebwerk, Begriffe (Internal combustion engines; components of rotating piston engines, setup of engine, driving mechanism, definitions)	1976-02
DIN 6262	Verbrennungsmotoren; Arten der Aufladung, Begriffe (Internal combustion engines; methods of pressure-charging, definitions)	1976-06
DIN 6267	Verbrennungsmotoren; Arten der Ölschmierung, Begriffe (Methods of lubrication for internal combustion engines)	1971-01
DIN 6271-3	Hubkolben-Verbrennungsmotoren; Anforderungen; Leistungstoleranzen; Ergänzende Festlegungen zu DIN ISO 3046 Teil 1 (Reciprocating internal combustion engines; performance; power tolerances; supplementary stipulations to DIN ISO 3046 Part 1)	1991-04
DIN 6274	Verbrennungsmotoren für allgemeine Verwendung; Druckluftbehälter mit Ventilkopf; 38 mm Durchgang; Zusammenstellung (Internal combustion engines for general purposes; Compressed air containers with valve block; 38 mm bore; Assembly)	1982-04
DIN 6275	Verbrennungsmotoren für allgemeine Verwendung; Druckluftbehälter für zulässigen Betriebsüberdruck bis 30 bar (Internal combustion engines for general purposes; Compressed air containers for permissible working overpressures up to 30 bar)	1982-04
DIN 6276	Verbrennungsmotoren für allgemeine Verwendung; Ventilköpfe für Druckluftbehälter; 38 mm Durchgang (Internal combustion engines for general purposes; Valve blocks for compressed air containers; 38 mm bore)	1982-04

Standard	Title	Date of publication
DIN 6280-10	Hubkolben-Verbrennungsmotoren; Stromerzeugungsaggregate mit Hubkolben-Verbrennungsmotoren; Stromerzeugungsaggregate kleiner Leistung; Anforderungen und Prüfung (Reciprocating internal combustion engines; generating sets with reciprocating internal combustion engines; small power generating sets; requirements and tests)	1986-10
DIN 6280-12	Stromerzeugungsaggregate – Unterbrechungsfreie Stromversorgung – Teil 12: Dynamische USV-Anlagen mit und ohne Hubkolben-Verbrennungsmotor (Generating sets – Uninterruptible power supply – Part 12: Dynamic UPS systems with and without reciprocating internal combustion engines)	1996-06
DIN 6280-13	Stromerzeugungsaggregate – Stromerzeugungsaggregate mit Hubkolben-Verbrennungsmotoren – Teil 13: Für Sicherheitsstromversorgung in Krankenhäusern und in baulichen Anlagen für Menschenansammlungen (Generating sets – Reciprocating internal combustion engines driven generating sets – Part 13: For emergency power supply in hospitals and public buildings)	1994-12
DIN 6280-14	Stromerzeugungsaggregate – Stromerzeugungsaggregate mit Hubkolben-Verbrennungsmotoren – Teil 14: Blockheizkraftwerke (BHKW) mit Hubkolben-Verbrennungsmotoren; Grundlagen, Anforderungen, Komponenten, Ausführung und Wartung (Generating sets – Reciprocating internal combustion engines driven generating sets – Part 14: Combined heat and power system (CHPS) with reciprocating internal combustion engines; basics, requirements, components and application)	1997-08
DIN 6280-15	Stromerzeugungsaggregate – Stromerzeugungsaggregate mit Hubkolben-Verbrennungsmotor – Teil 15: Blockheizkraftwerke (BHKW) mit Hubkolben-Verbrennungsmotoren; Prüfungen (Generating sets – Reciprocating internal combustion engines driven generating sets – Part 15: Combined heat and power system (CHPS) with reciprocating internal combustion engines; tests)	1997-08
DIN 6281	Stromerzeugungsaggregate mit Kolbenkraftmaschinen; Anschlussmaße für Generatoren und Kolbenkraftmaschinen (Generator sets with reciprocating internal combustion engines; Connection dimensions for generators and reciprocating internal combustion engines)	1978-04
DIN 6288	Hubkolben-Verbrennungsmotoren – Anschlussmaße und Anforderungen für Schwungräder und elastische Kupplungen (Reciprocating internal combustion engines – Dimensions and requirements for flywheels and flexible couplings)	2000-07
DIN 24189	Prüfung von Luftfiltern für Verbrennungskraftmaschinen und Kompressoren; Prüfverfahren (Testing of air cleaners for internal combustion engines and compressors; test methods)	1986-01

ISO Standards		
ISO TC 22 Road Vehicles		
Standard	Title	Date of publication
ISO 7648	Flywheel housings for reciprocating internal combustion engines; Nominal dimensions and tolerances	1987-07
ISO TC 70 Internal Combustion Engines		
Standard	Title	Date of publication
ISO 1204	Reciprocating internal combustion engines; designation of the direction of rotation and of cylinders and valves in cylinder heads, and definition of right-hand and left-hand in-line engines and locations on an engine	1990-12
ISO 2261	Reciprocating internal combustion engines – Hand-operated control devices – Standard direction of motion	1994-12
ISO 2710-1	Reciprocating internal combustion engines – Vocabulary – Part 1: Terms for engine design and operation	2000-09
ISO 2710-2	Reciprocating internal combustion engines – Vocabulary – Part 2: Terms for engine maintenance	1999-12
ISO 3046-1	Reciprocating internal combustion engines – Performance – Part 1: Declarations of power, fuel and lubricating oil consumptions, and test methods; Additional requirements for engines for general use	2002-05
ISO 3046-3	Reciprocating internal combustion engines; performance; Part 3: Test measurements	1989-11
ISO 3046-4	Reciprocating internal combustion engines – Part 4: Speed governing	1997-03
ISO 3046-5	Reciprocating internal combustion engines – Performance – Part 5: Torsional vibrations	2001-12
ISO 3046-6	Reciprocating internal combustion engines; performance; Part 6: Overspeed protection	1990-10
ISO 6798	Reciprocating internal combustion engines – Measurement of emitted airborne noise – Engineering method and survey method	1995-12
ISO 6826	Reciprocating internal combustion engines – Fire protection	1997-02
ISO 7967-1	Reciprocating internal combustion engines – Vocabulary of components and systems – Part 1: Structure and external covers	2005-06

Standard	Title	Date of publication
ISO 7967-2	Reciprocating internal combustion engines; Vocabulary of components and systems; Part 2: Main running gear Trilingual edition	1987-11
ISO 7967-2/AMD	Reciprocating internal combustion engines – Vocabulary of components and systems – Part 2: Main running gear; Amendment 1	1999-12
ISO 7967-3	Reciprocating internal combustion engines; Vocabulary of components and systems; Part 3: Valves, camshaft drive and actuating mechanisms Trilingual edition	1987-11
ISO 7967-4	Reciprocating internal combustion engines – Vocabulary of components and systems – Part 4: Pressure charging and air/exhaust gas ducting systems	2005-06
ISO 7967-5	Reciprocating internal combustion engines – Vocabulary of components and systems – Part 5: Cooling systems	2003-02
ISO 7967-6	Reciprocating internal combustion engines – Vocabulary of components and systems – Part 6: Lubricating systems	2005-06
ISO 7967-7	Reciprocating internal combustion engines – Vocabulary of components and systems – Part 7: Governing systems	2005-06
ISO 7967-8	Reciprocating internal combustion engines – Vocabulary of components and systems – Part 8: Starting systems	2005-06
ISO 7967-9	Reciprocating internal combustion engines – Vocabulary of components and systems – Part 9: Control and monitoring systems	1996-11
ISO 8528-1	Reciprocating internal combustion engine driven alternating current generating sets – Part 1: Application, ratings and performance	2005-06
ISO 8528-2	Reciprocating internal combustion engine driven alternating current generating sets – Part 2: Engines	2005-06
ISO 8528-3	Reciprocating internal combustion engine driven alternating current generating sets – Part 3: alternating current generators for generating sets	2005-06
ISO 8528-4	Reciprocating internal combustion engine driven alternating current generating sets – Part 4: Controlgear and switchgear	2005-06
ISO 8528-5	Reciprocating internal combustion engine driven alternating current generating sets – Part 5: Generating sets	2005-06
ISO 8528-6	Reciprocating internal combustion engine driven alternating current generating sets – Part 6: Test methods	2005-06

Standard	Title	Date of publication
ISO 8528-7	Reciprocating internal combustion engine driven alternating current generating sets – Part 7: Technical declarations for specification and design	1994-11
ISO 8528-8	Reciprocating internal combustion engine driven alternating current generating sets – Part 8: Requirements and tests for low-power generating sets	1995-12
ISO 8528-9	Reciprocating internal combustion engine driven alternating current generating sets – Part 9: Measurement and evaluation of mechanical vibration	1995-12
ISO 8528-10	Reciprocating internal combustion engine driven alternating current generating sets – Part 10: Measurement of airborne noise by the enveloping surface method	1998-10
ISO 8528-12	Reciprocating internal combustion engine driven alternating current generating sets – Part 12: Emergency power supply to safety services	1997-09
ISO 8861	Shipbuilding – Engine-room ventilation in diesel-engined ships – Design requirements and basis of calculations	1998-05
ISO 8665	Small craft – Marine propulsion engines and systems – Power measurements and declarations	1994-08
ISO 8999	Reciprocating internal combustion engines – Graphical symbols	2001-03
ISO 10054	Internal combustion compression-ignition engines – Measurement apparatus for smoke from engines operating under steady-state conditions – Filter-type smokemeter	1998-09
ISO 11102-1	Reciprocating internal combustion engines – Handle starting equipment – Part 1: Safety requirements and tests	1997-10
ISO 11102-2	Reciprocating internal combustion engines – Handle starting equipment – Part 2: Method of testing the angle of disengagement	1997-10
ISO 13332	Reciprocating internal combustion engines – Test code for the measurement of structure-borne noise emitted from high-speed and medium-speed reciprocating internal combustion engines measured at the engine feet	2000-11
ISO 14314	Reciprocal internal combustion engines – Recoil starting equipment – General safety requirements	2004-03
ISO 14396	Reciprocating internal combustion engines – Determination and method for the measurement of engine power – Additional requirements for exhaust emission tests in accordance with ISO 8178	2002-06

Standard	Title	Date of publication
ISO 15550	Internal combustion engines – Determination and method for the measurement of engine power – General requirements	2002-05
ISO 21006	Internal combustion engines – Engine weight (mass) declaration	2006-05

ISO TC 70/SC 7 Internal combustion engines – Tests for lubricating oil filters

Standard	Title	Date of publication
ISO 4548-1	Methods of test for full-flow lubricating oil filters for internal combustion engines – Part 1: Differential pressure/flow characteristics	1997-09
ISO 4548-2	Methods of test for full-flow lubricating oil filters for internal combustion engines – Part 2: Element by-pass valve characteristics	1997-09
ISO 4548-3	Methods of test for full-flow lubricating oil filters for internal combustion engines – Part 3: Resistance to high differential pressure and to elevated temperature	1997-09
ISO 4548-4	Methods of test for full-flow lubricating oil filters for internal combustion engines – Part 4: Initial particle retention efficiency, life and cumulative efficiency (gravimetric method)	1997-12
ISO 4548-5	Methods of test for full-flow lubricating oil filters for internal combustion engines; Part 5: Cold start simulation and hydraulic pulse durability test	1990-12
ISO 4548-6	Methods of test for full-flow lubricating oil filters for internal combustion engines; Part 6 : Static burst pressure test	1985-12
ISO 4548-7	Methods of test for full-flow lubricating oil filters for internal combustion engines; Part 7: Vibration fatigue test	1990-11
ISO 4548-9	Methods of test for full-flow lubricating oil filters for internal combustion engines – Part 9: Inlet and outlet anti-drain valve tests	1995-07
ISO 4548-11	Methods of test for full-flow lubricating oil filters for internal combustion engines – Part 11: Self-cleaning filters	1997-09
ISO 4548-12	Methods of test for full-flow lubricating oil filters for internal combustion engines – Part 12: Filtration efficiency using particle counting, and contaminant retention capacity	2000-02

ISO TC 70/SC 8 Internal combustion engines – Exhaust emission measurement

Standard	Title	Date of publication
ISO 8178-1	Reciprocating internal combustion engines – Exhaust emission measurement – Part 1: Test-bed measurement of gaseous and particulate exhaust emissions	2006-09

Standard	Title	Date of publication
ISO 8178-2	Reciprocating internal combustion engines – Exhaust emission measurement – Part 2: Measurement of gaseous and particulate exhaust emissions under field conditions	2008-04
ISO 8178-3	Reciprocating internal combustion engines – Exhaust emission measurement – Part 3: Definitions and methods of measurement of exhaust gas smoke under steady-state conditions	1994-09
ISO 8178-4	Reciprocating internal combustion engines – Exhaust emission measurement – Part 4: Steady-state test cycles for different engine applications	2007-12
ISO 8178-5	Reciprocating internal combustion engines – Exhaust emission measurement – Part 5: Test fuels	2008-10
ISO 8178-6	Reciprocating internal combustion engines – Exhaust emission measurement – Part 6: Report of measuring results and test	2000-11
ISO 8178-7	Reciprocating internal combustion engines – Exhaust emission measurement – Part 7: Engine family determination	1996-11
ISO 8178-8	Reciprocating internal combustion engines – Exhaust emission measurement – Part 8: Engine group determination	1996-11
ISO 8178-9 Amd 1	Reciprocating internal combustion engines – Exhaust emission measurement – Part 9: Test cycles and test procedures for test bed measurement of exhaust gas smoke emissions from compression ignition engines operating under transient conditions	2004-10
ISO 8178-10	Reciprocating internal combustion engines – Exhaust emission measurement – Part 10: Test cycles and test procedures for field measurement of exhaust gas smoke emissions from compression ignition engines operating under transient conditions	2002-11
ISO 8178-11	Reciprocating internal combustion engines – Exhaust emission measurement – Part 11: Test-bed measurement of gaseous and particulate exhaust emissions from engines used in nonroad mobile machinery under transient test conditions	2006-04

European Standards (CEN/TC 270)		
Standard	Title	Date of publication
EN 1679-1	Reciprocating internal combustion engines. Safety. Part 1: Compression ignition engines	1997
EN 1834-1	Reciprocating internal combustion engines. Safety requirements for design and construction of engines for use in potentially explosive atmospheres. Part 1: Group II engines for use in flammable gas and vapor atmospheres	2000-01
EN 1834-2	Reciprocating internal combustion engines. Safety requirements for design and construction of engines for use in potentially explosive atmospheres. Part 2: Group I engines for use in underground workings susceptible for firedamp and/or combustible dust	2000-01
EN 1834-3	Reciprocating internal combustion engines. Safety requirements for design and construction of engines for use in potentially explosive atmospheres. Part 3: Group II engines for use in flammable dust atmospheres	2000-01
EN 10090	Valve steels and alloys for internal combustion engines	1998-02
EN 12601	Reciprocating internal combustion engine driven generating sets. Safety	2001-02

CIMAC Recommendations		
Number	Title	Year
No 2	Recommendations for gas turbine acceptance test	1968
No 3	Recommendations of measurement for the overall noise of reciprocating engines	1970
No 4	Recommendations for SI units for diesel engines and gas turbines	1975
No 5	Recommendations for supercharged diesel engines. Part I: Engine de-rating on account of ambient conditions; Part II: Engine acceptance tests	1971
No 6	Lexicon on combustion engines – Technical terms of the IC engine and gas turbine industries	1977
No 7	Recommendations regarding liability – Assured properties, publications, fuels for diesel engines	1985
No 10	Recommendations regarding liability – Assured properties, publications, fuels for gas turbines	1988

Number	Title	Year
No 11	Recommendations regarding fuel requirements for diesel engines. Updated version of No 8	1990
No 12	Exhaust emissions measurement recommendations for reciprocating engines and gas turbines	1991
No 14	Standard method for the determination of structureborne noise from engines	1994
No 15	Guidelines for the lubrication of two-stroke crosshead diesel engines	1997
No 16	Guidelines for operation and/or maintenance contracts	1999
No 17	Guidelines for diesel engines lubrication – Oil consumption of medium speed diesel engines	1999
No 18	Guidelines for diesel engines lubrication – Impact of fuel on lubrication	2000
No 19	Recommendations for the lubrication of gas engines	2000
No 20	Guidelines for diesel engines lubrication – Lubrication of large high speed diesel engines	2002
No 21	Recommendations regarding fuel quality for diesel engines	2003
No 22	Guidelines for diesel engines lubrication – Oil degradation	2004
No 23	Standards and methods for sampling and analyzing emission components in non-automotive diesel and gas engine exhaust gases – Marine and land based power plant sources	2005
No 24	Treatment of the system oil in medium speed and crosshead diesel engine installations	2005
No 25	Recommendations concerning the design of heavy fuel treatment plants for diesel engines. Updated version of No 9	2006
No 26	Guidelines for diesel engine lubrication – Impact of low sculpture fuel on lubrication of marine engines	2007
No 27	Turbo charging efficiencies – Definitions and guidelines for measurement and calculation	2007
No 28	Guide to diesel exhaust emissions control of NO_x, SO_x, particulates, smoke and CO_2 – Seagoing ships and large stationary diesel power plants	2008
No 29	Guidelines for the lubrication of medium speed diesel engines. Updated version of No 13	2008

SAE Standards

Motor Vehicle Council – Powertrain Systems Group

Number	Title	Publication date
J604	Engine terminology and nomenclature – General	Jun 1995
J824	Engine rotation and cylinder numbering	Jun 1995
J922	Turbocharger nomenclature and terminology	Jun 1995
J1004	Glossary of engine cooling system terms	Aug 2004
J1515	Impact of alternative fuels on engine test and reporting procedures	Jun 1995

Diesel fuel injection equipment

Number	Title	Publication date
HS3458	SAE Fuel injection systems and testing methods standards manual (34 documents)	Jan 1997
J830	Fuel injection equipment nomenclature	Apr 1999
J968/1	Diesel injection pump testing. Part 1: Calibrating nozzle and holder assemblies	Dec 2002
J968/2	Diesel injection pump testing. Part 2: Orifice plate flow measurement	Dec 2002
J1668	Diesel engines – Fuel injection pump testing	Apr 1999

Engine power test code

Number	Title	Publication date
J228	Airflow reference standards	Apr 1995
J1349	Engine power test code – Spark ignition and compression ignition – Net power rating	Mar 2008
J1995	Engine power test code – Spark ignition and compression ignition – Gross power rating	Jun 1995
J2723	Engine power test code – Engine power and torque certification	Aug 2007

Ignition systems		
J139	Ignition system nomenclature and terminology	Nov 1999

Piston ring		
J2612	Internal combustion engines – Piston vocabulary	Jan 2002

Fuels & lubricants. TC 1 – Engine lubrication		
J300	Engine oil viscosity classification	Jan 2009

TC 3 – Driveline and chassis lubrication		
J306	Automotive gear lubricant viscosity classification	Jun 2005

TC 7 – Fuels		
J313	Diesel fuels	Jul 2004
J1297	Alternative automotive fuels	Jul 2007
J1498	Heating value of fuels	Dec 2005
J1616	Recommended practice for compressed natural gas vehicle fuel	Feb 1994
J1829	Stoichiometric air-fuel ratios of automotive fuels	Oct 2002
J2343	Recommended practice for LNG medium and heavy-duty powered vehicles	Jul 2008

Index

A

Absorption mass, 257
Absorption mufflers, 394–395
Acceptance test, 16
ACEA (Association des Constructeurs Européen d'Automobiles), 364
Acidification, 370
Acoustic comfort, 519–520
Additives, 77, 79, 82–85, 90, 93, 335
Additive systems, 460
Aerodynamic engine noises, 498–499
– intake noise, 499
– muffler (systems), 499
AFP steel, 339, 354
Afterglow, 379
Agricultural and forestry tractors, 440
Air cleaners, 387–393
– cyclone, 390–391
– cyclone cells, 390–391
– design, 389–391
– dry air cleaners, 387
– filtration efficiency, 388–389
– initial filtration efficiency, 390–391
– oil wetted air bath cleaners, 389
– overall filtration efficiency, 388
– restriction, 389
– service life, 387
Air content, 211
Air efficiency, 31, 37
Air flow rate, specific, 16
Air to fuel equivalence ratio, 11
Air mass flow characteristic, 38, 41, 72
Alfin process, 273, 305
Alkalinity, 360
Alternative fuels, 94–103
– biodiesel, 96
– biofuels, 95
– BTL process, 99
– dimethyl ether, 98
– ethanol, 97
– fossil sources (GTL), 98–99
– fuel paths, 94
– hydrogen, 94–95, 98
– Iogen process, 97
– life cycle assessment, 101–102
– liquefied gas, 98
– liquid fuels, 95–97
– methanol, 97
– natural gas, 97–98
– NExBTL, 101–102
– pure vegetable oils, 96
– renewable sources (BTL), 99–101
– sustainability, 94
– synthetic fuels, 94–95, 98–101
Aluminum die cast, 510–511
Ammonia slip, 462
Amplitude frequency responses, 253, 255
Analysis, 248
– harmonic, 248
ANC, 396–397
– system, 397
Angle of connecting rod offset, 239
Angular momentum, 243, 252
Antifoaming agents, 83
Antioxidants, 84, 360
Anti-thrust side, 223
Anti-wear additives, 83
API, 364
Application engineering, 189–190
– basic engine tuning, 190
– control functions, 190
– engine and transmission protection functions, 190
– scope of diagnostics, 190
Aromatic content, 79
Aromatics, 78, 81, 87, 91–92, 111
Autofrettage, 150, 350
Auto-ignition, 80, 377, 378
– temperature, 377
Auxiliary connecting rod, 224, 240

B

Balance shaft, 225–227, 236, 243, 246, 511
Balancing, 242
- dynamic, 242
- of free inertial forces, 237–240
- of longitudinal yawing moment, 238
- of moments of inertia, 241
- of transverse forces, 238
Base oil, 359
Basic engine, 510–512
Bearing forces, 222
Bending moment, 232, 247
Bending moment of the housing, 247
Bending stress, 232–233
Blow-by gases, 370, 373
Boiling characteristics, 88
Boost efficiency, 31
Boost pressure control, 48–52
- engine operating line, 48–49
- sequential turbocharging, 50–51
- variable turbine geometry (VTG), 50
- wastegate, 49–50, 52
Bore cooling, 296
Bore ratio, 7
Bottoming cycle, 404
Brake mean effective pressure (bmep), 15
Bypass filters, 375

C

Calcium, 333
Calorific value, 11, 92–93, 115
- of the air/fuel mixture, 12, 115
Camshaft, 599
- drive, 511
Carbon monoxide (CO), 449
- production, 417
Car engine, 525
Carnot cycle, 13
Catalyst
- deterioration, 458
- loading, 457
- poisons, 457
Catalytic burner, 457
Cavitation, 202, 334
CDPF, 461
Ceramic heater, 381
Ceramic materials, 291
Ceramic sheathed-element glow plug, 381–382, 384
- deterioration, 382
- heater resistance, 381
- heating time, 381
- ignition switch, 382–383
- overheating, 383
- sheathed-element glow plugs, 380
- single glow plug monitoring, 382
Changes of states of gases, 12
- adiabatic, 12
- ideal, 12
- isobaric, 13
- isochoric, 13
- isothermal, 12
Charge air cooling, 407
Climatic comfort, 507, 519
Coating
- lubricating film, 281
- methods, 352
- protective, 280–281
- skirt, 281–282
- sliding properties, 281
- surface, 280
- surface roughness, 281
CO_2 emission, 518–519
- energy efficiency, 518
- fuel consumption, 518
Cogeneration, 405–410
Cold idle emissions, 384
Cold start, 383–386
Combined heat and power stations (CHPS), 405–407
- generating sets, 405–406
- modules, 405–407
- parameters, 407–409
Combustion, 61, 67
- air utilization, 63
- auto-ignition, 61
- compression ratio, 62
- constant pressure, 68
- diffusion flame, 68
- direct injection engine, 62
- efficiency, 66
- evaporation rate, 64
- injection rate, 62
- injection time, 64
- internal, 61
- prechamber, 62
- spray penetration velocity, 64
- swirl chamber, 62
Combustion chamber, 512–513
- recess, 513–514
- surface, 203, 205
Combustion cycle, real, 12
Combustion noise, 489, 491, 523
- alternative fuels, 491
- common rail injection systems, 490
- cylinder pressure excitation spectrum, 489
Combustion simulation, 11
Combustion systems, 509

Combustion systems, alternative, 73, 95
- CCS (combined combustion system), 103
- dilution controlled combustion system (DCCS), 73
- dimethyl ether (DME), 73
- gas-to-liquid, 73
- GTL fuels, 73
- HCCI system (homogeneous charge compression ignition), 73
- HCLI (homogeneous charge late injection), 73
- homogenization, 74
- HPLI (highly premixed late injection), 74
- methanol, 73
- multifuel engines, 74
- post-injection, 73
- rape oil methyl ester (RME), 73
Comfort, 519–521
- dual-mass flywheel, 520
- encapsulation, 519
- mass balancing, 520
- source of vibration, 520
- vibration excitation, 520
- vibration exciter, 520
- vibrations, 519–520
Commercial vehicle engines, 531
- braking, 534
- development, 534
- drive away performance, 532–534
- exhaust emissions, 536
- starting performance, 532–534
Commercial vehicles, 428–437, 478, 521–527, 528–544
- classification, 528
- cost analysis, 529
- delivery traffic, 528–529
- driving comfort, 534
- heavy duty, 528–529
- light duty, 528–529
- power requirements, 532
Common rail high pressure inline pump, 153
Common rail one and two-piston high pressure radial pump, 152
Common rail system, 145–170, 176
- fuel filters, 147–148
- for heavy fuel, 112
- high pressure pump, 150–153
- high pressure system, 149–150
- injectors, 156–159
- low pressure system, 147–148
- metering unit, 147–148
- pressure control, 147–149
- presupply pumps, 147–148
- rail, 149
- system pressures, 146–147
Common rail three-piston high pressure radial pump, 152
Compound engine, 3–5
Compression ratio, 7, 386, 403

Comprex, 44, 56
Connecting rod, 236, 512, 597
- bearings, 387
- eye, 236, 512
- force, 230
- offset, 239
- shank, 228
- small end, 238
Constant pressure turbocharging, 45
Continuous regenerating trap (CRT®), 460–461
Control components, 354, 371
Coolant, 309, 333–336
- physical properties, 333
Coolant circuit, 309–310
Cooling airflow routing, 302–303
Cooling fan, 307–308
- sound power, 307
Cooling fin, 305
- fin efficiency factor, 301
- fin height, 301–302
- number of fins, 301–302
Cooling system, 309–336
- commercial vehicle, 317
- dual circuit system, 313
- fan, 317
- heat exchanger, 320–333
- high temperature circuit, 313
- intercooling, 326–328
- locomotives, 317–318
- low temperature circuit, 313
- sea water operation, 314
- ships, 312
- single circuit system, 312
- tanks, 317
- thermostat, 317
Cooling water, 333–336
- cooling water care, 334–336
- glycol, 333
- OAT (organic acid technology), 335
- water hardness, 333–334
Coordinates, 256
- modal, 256
Corrosion, 202
- fretting, 202
- high temperature, 202
- inhibitors, 370
- low temperature, 202
- protection, 84, 359
- surface, 202
- vibration corrosion cracking, 202
- wet, 202
Counterweight mass, 236–238, 240
Crank angle, 9, 222
Crankcase, 344–345, 370, 373, 510–511, 541–542
- cast aluminum, 344

- cast steel, 344
- gray cast iron, 344
- nodular graphite cast iron, 344
- vermicular graphite cast iron, 344

Crank pin offset, 225
Crank pin transition, 231
Crankshaft, 221–287, 304–305, 339–342, 541–542, 567, 582–583, 597–598
- AFP steel (precipitation-hardened ferritic pearlitic steel), 339
- assembly, 250, 567, 582–583
- bearing, 582–583
- composite, 230
- fiber flow forging, 340–341
- induction hardening, 341
- nodular graphite cast iron types, 341
- offset, 237
- pearlitic iron, 341–342
- support, 222
- throws, 228

Crank spacing, 228–229
Crank throw diagram, 241
Crosshead, 596–598
- bearings, 597

Crosshead bearings, 597
Crosshead crankshaft assembly, 223
Cross shaft, 227
Crude oil, 77, 104
Cycle, 12–14
- beating, 430
- ideal, 12–14
- rate, 15

Cylinder bank offset, 225
Cylinder charge, 11
Cylinder heads, 345, 567, 611
- cast aluminum, 344
- cast steel, 344
- gray cast iron, 344
- nodular graphite cast iron, 344
- vermicular graphite cast iron, 344

Cylinder liner, 567, 583
Cylinder pressure indication, 22

D

Damper, 257–258
Damping, 252
- external, 252
- internal, 252
- matrix, 252–253, 256

Dead zones, 293, 302
Decoupling elements, 397
Degree of cyclic irregularity, 249
Density, 90
Detergent additives, 83–84

Detergents, 360
Diagnostics, 211
- diagnostic system management (DSM), 186–187
- EOBD, 187
- garage, 187–189
- OBD (on-board diagnostics), 186–187

Diesel-electric drives, 577
Diesel engine heat pump, 408–410
- coefficient of performance, 408–410
- heat factor, 409

Diesel fuel, 77
- availability, 77–78
- distillation, 78

Diesel oxidation catalysts (DOC), 457
- catalytic burner, 458
- differential pressure sensor, 461
- lubricating oil dilution, 458
- temperature sensor, 461

Diesel particulate filters, 182, 393
Direct injection, 447
Dispersants, 360
Displacement specific power output, 16
Distributor injection pump, 143
Downsizing, 52
Drive train configuration, 508
Driving comfort, 507, 534
Dry air cleaner, 387, 389
3D torsional model, 231
1D torsional vibration model, 231
Dual-fuel engine, 118
Dual-mass flywheel, 259
Dust capacity, 389
Dust concentration, 391
Dynamic viscosity, 203

E

Ease of operation, 507
eBooster, 54
Eccentric masses, 246
Eccentric shafts, 246
Effective brake work, 15
Efficiency, 14
- carnot, 13
- conversion factor, 15
- factor, 15
- internal/indicated, 15
- mechanical, 15
- net, 14
- thermal, 13

EGR cooling, 455
Eigenvalue problem, 253
Electrically assisted turbocharger (EAT), 53
Electric fuel pump, 147
Electric start, 562–563

Electronic control, 176–183
- ambient conditions, 177
- assembly, 177
- AUTOSAR, 180
- central processing unit, 177, 179
- common rail system, 176
- computing power, 176
- digital controller, 180
- engine control, 176
- input circuit, 177–178
- interconnection techniques, 177
- main injections, 176
- memory, 176–177
- output circuit, 177–178
- pilot injections, 176
- post-injections, 176
- real-time operation, 176, 180
- software architecture, 178–180
- start of injection, 176
- torque requirements, 176
- unit injector system, 176

Electronic engine management, 515–516
- air mass, 516
- air mass sensor, 516
- engine control unit, 516

Electroslag remelting, 286, 350
Elemental analysis, 11
Emission control legislation, 426–443
- for car engines, 426
- diesel particulate filters, 428
- diesel smoke measurement, 426
- driving cycles, 426
- emission control standards, 426
- on-board diagnostics, 428
- particulate filters, 427

Emission limits, 428
Emission reduction, in-engine, 466
Emissions, 417
- performance, 112–113
- reduction, 455
- testing, 469–483

Encapsulation, 499, 502
- maintenance points, 502

Enclosure ventilation, 502
Energy balance, 202, 204, 209
Engine, 507–508
- auxiliary units, 511

Engine brake, 540
- constant throttle, 540
- exhaust valve, 540
- intebrake, 540
- jake (Jacobs), 540
- turbobrake, 540

Engine characteristic map, 17–18
Engine cooling, 291–336, 402, 501–502
- air, 300–309
- air bleeding, 315
- coolant, 333–336
- direct, 292
- exhaust gas heat exchanger, 328–333
- fan power, 311–312
- indirect, 292
- intercooler, 326–328
- liquid, 291–292
- load, 309
- module, 318–320
- oil cooler, 324–326
- radiator, 321–324
- vehicles, 315–316

Engine cycle simulation, 26
Engine functions, 181–183
- AdBlue, 183
- air management, 182
- air mass, 182
- diesel particulate filters, 182
- exhaust gas management, 182
- lambda closed-loop control, 182
- NO_X storage catalyst (NSC), 183, 185
- regeneration, 183
- regenerative operation, 183
- SCR, 183
- torque, requirements, 181, 183

Engine map, 430
Engine mount, 563
Engine noise, 487–489
- power take-offs, 493

Engine noise emission, 487–502
- cooling system, 489

Engine oil, 359, 361
Engine operating line, 2, 40, 49
Engine parameters, 15–16
Engine process simulation, 19
Engine test, 384
Equations of state, 20
- calorific, 20
- thermal, 20

Equivalent 1D rotational vibration model, 252
Equivalent rotating mass, 250–256
Equivalent torsional vibration model, 250–251
Evaporation cooling, 298
- closed system, 298
- open system, 299

Evaporative losses, 360
Excess work, 249
Excitation torque, 252–253, 255–256
- harmonic, 255–256

Exhaust back pressure, 456
Exhaust emissions, 417–483, 518–519, 536, 589, 606–607
- carbon monoxide (CO), 519, 536
- emission, 417

- hydrocarbons (HC), 519, 536
- impact, 417
- nitrogen oxides NO_X, 425, 536
- pollutant input, 417
- transmission, 417

Exhaust gas aftertreatment, 466–469, 474, 516–518
- active regeneration, 517
- CRT effect, 467–468
- exhaust back pressure, 468
- exhaust gas mass flow, 466
- exhaust gas temperature, 517
- oxidation catalytic converters, 516–519
- particulate filters, 516–519
- post-injection, 517
- soot ignition temperature, 517
- soot particles, 517
- space velocity, 466

Exhaust gas dilution systems, 474–475
- CFV (critical flow venturi), 476
- CVS full flow dilution, 476
- PDP (positive displacement pump), 476
- sampling, 476–477

Exhaust gas heat exchanger, 328–333
- fouling, 328–329

Exhaust gas opacity, 384, 431

Exhaust gas recirculation (EGR), 71, 452–453
- combustion temperature, 71–73
- discharge valve, 72
- EGR rate, 72
- exhaust gas aftertreatment, 71
- NO_X emission, 72
- recirculation rate, 72
- sequential turbocharging, 73
- supercharging, 72–73
- turbine geometry, variable, 72
- Zeldovich mechanism, 71

Exhaust gas turbocharging, 554

Exhaust gas valve, 395–396

Exhaust measurement systems for gaseous emissions, 469–478
- calibration, 470
- chemiluminescence detector (CLD), 471–472
- diagnostic test, 470
- flame ionization detector (FID), 470–471
- linearization, 470
- nondispersive infrared analyzers (NDIR), 472–473
- paramagnetic detector (PMD), 473–474

Exhaust muffler, 501

Exhaust noises, 501

Exhaust system splitting, 398

Expert systems, 589

F

Fatigue fracture, 202

Fatigue limit, 221

Fatigue notch factor, 233

Fatigue strength, 201, 234–236
- diagram, 201, 233, 235–236

Fillets, 229

Filling (and emptying) method, 25

Filterability, 79

Final compression temperature, 377–378

Fine particulate matter, 420–424

Finite element method (FEM), 196

Firing sequence, 245, 249

First law of thermodynamics, 21

First order secondary exciter, 249

Flame starter plug, 383

Flame velocity, laminar, 116

Flap, 395–396

Flash point, 91

Flowability, 89–90

Flow
- characteristics, 81
- improvers, 81–83
- noises, 395
- simulation, 371

Foot balancing, 238

Forked connecting rod, 224–225, 240

Four-stroke cycle, 14

Fracture-split connecting rod, 228

Fracture splitting technology, 342

Frictional work, 24

Friction modifier, 361

Fuel, 309

Fuel consumption, 389, 518, 529–530
- CO_2 emission, 518
- distance-based, 529–530
- energy efficiency, 518
- specific, 16

Fuel delivery control, 151

Fuel economy engine oil, 359, 361, 369

Fuel gases, 114–124

Fuel ignition, 378–379
- auto-ignition temperature, 378
- cold start assist, 378
- flame propagation, 378
- glow control units, 378
- glow, 378
  - plugs, 378
  - systems, 378
- sheathed-element glow plugs, 378

Fuel mean value adaptation (FMA), 160, 163

Fuel standards, 85–87

Fuel, sulfur free, 83

Full flow filters, 374

Fundamental diesel engine equation, 15

G

Gas engine, 114, 117, 431
- dual fuel, 117–119
- gas-diesel, 117–120
- spark ignited, 117–119

Gas exchange, 21–22, 31–59, 215–216, 514–515, 537–540
- downsizing, 537
- exhaust gas return rate, 514
- exhaust gas turbochargers, 516
- four-stroke cycle, 31–36
- simulation, 56
  - filling and emptying method, 57
  - method of characteristics, 57–58
  - one-dimensional unsteady pipe flow, 25
  - zero-dimensional method, 57
- intercooling, 515
- lean operation, 514
- two-stroke cycle, 36–38
- variable turbine geometry, 514

Gas force, 221, 248–249
- deformation, 271–272
- stress, 271–272

Gas mixer, 120–121
Gas torque, 248
Gear pump, 147
Generator operation, 18
Glow plugs, 383, 513–514
Glow systems, 379, 513
Gray cast iron, 345

H

Hand crank start, 561–562
HCCI mode, 212–213
Heat balance, 204, 310, 402
- cooling capacity Φ_K, 402
- exhaust heat output, 402
- external, 314
- fuel power, 402
- intercooler, 402
- internal, 204

Heat engine, rational, 4
Heat exchange, 203
- convection, 203
- radiation, 203
- thermal conduction, 203

Heat exchangers, 308, 317, 330
- brazed, 324
- heat capacity flows, 321
- heat transfer capacity, 321
- mechanically joined, 324
- pack construction, 326
- plate design, 326
- prandtl number, 333
- radiator, 324
- tube bundle, 326

Heat flux density, 204, 206–207
Heat flux measurement, 204
Heat flux sensor, 204–205, 207
Heating heat, 408
Heat penetration coefficient, 206
Heat transfer, 202, 204, 211, 215
Heat transfer coefficient, 203, 207, 209–212, 214, 216–217
Heat transfer equation, 203, 208, 212, 214, 216–217
- Bargende equation, 211, 214–216
- Hohenberg equation, 214–216
- Woschni equation, 210–211, 214, 216

Heat transmission, 291
Heavy fuel oil, 582
Heavy fuel oil operation, 599
Helical port, 70
- tangential port, 70

Helmholtz resonator, 391
High cycle fatigue, 201
High pressure injection, 513–514
- common rail system, 513
- pressure control valve, 513–514
- pressure sensor, 513
- rail, 513
- unit injector system, 513

High pressure pumps, 150–153
High pressure system, 149–150
High temperature cooling, 296
Hole geometry, 132
Hole-type nozzles, 130
Hot film air-mass flowmeter, 452
Hot isostatic pressing, 351
HT/HS viscosity, 361
Hydrocarbon formation, 479
Hydrocarbons (HC), 449
Hydrogen power, 120

I

Ignition, 66
- cetane number, 66
- diffusion zone, 67
- ignition quality, 66
- start of ignition, 66

Ignition assist system, 379, 383
- afterglow phase, 379
- car, 379
- intermediate glow, 379
- low-voltage glow systems, 379
- preglow, 379
- sheathed-element glow plugs, 379
- software module, 379
- standby glow, 379
- start glow, 379

Ignition delay, 66
- cetane number, 67
- diffusion zone, 67

– ignition quality, 66
– start of ignition, 66
Ignition improvers, 83
Ignition interval, 224
Ignition limits, 117
Ignition quality, 66, 74, 79–82
– cetane index, 87–88
– cetane numbers, 81–83, 87–88
Ignition system, 121
– laser ignition, 122
– precombustion chamber spark plug, 122
– spark plug, 121
Indicator diagrams, 4
Inertial force, 221, 228
– deformation, 228
– oscillating, 196
– rotating, 196, 222
– stress, 228
Inertial torque, 226
Influence coefficients of the free inertial forces and moments of inertia, 244
Influence coefficients of the inertial forces, 239
Injection, 137
– lift controlled, 138
– pressure controlled, 138
Injection characteristic, 74–75
– CFD modeling, 74
– 3D modeling, 74
– 3D simulation, 74
– model calculation, 74
– system configuration, 74–75
Injection hydraulics, 127–129
– cavitation, 129
– chamber, 128
– gap flow, 129
– lines, 128
– pressure forces, 129
– short pipe, 128–129
– throttles, valves, 129
Injection management, 183
– main injections, 183
– microcontroller, 183
– pilot injections, 183
– post-injections, 183
– start of injection, 183
Injection nozzle, 70–71, 129–137
– nozzle design, 71
– nozzle projection dimension, 71
– spray propagation, 71
– squish flow, 70–71
Injection pressure, 451–452
Injection system, 137–170, 175–190, 449, 568, 586–588

– common rail, 140–141, 569, 587–588
– distributor pump
  – axial, 139
  – radial, 140
– flow measurement, 170
– high pressure accumulator (rail), 587
– high pressure pump, 587
– injector, 569
– injector testing, 171–172
– inline pump, 137
– large diesel engines, 163–170
– measurement of the injected fuel quantity, 170–171
– measurement of the injection characteristic, 171
– metrology, 170–172
– pump line nozzle, 140
– pump nozzle, 138
– unit injector, 140
– unit pump, 140
Injector, 140
Injector quality adaptation (IQA), 160–161, 183
Inline pump, 141
Inner connecting rod, 240
Intake air, 387
– dust concentration, 387
– dust content, 387
– particle size distribution, 387
Intake air heaters (IAH), 383
Intake noise damping, 391–393
Intake noises, 391
Intake swirl, 35
Intercooler, 48, 326–328
– density recovery, 326
Intercooling, 48
– efficiency, 48
– intercooler, *see* Intercooler
Intermetallic phases, 354
Internal heat balance, 204
Intrinsic damping, 258
Isentropic exponents, 12

J

Jet Assist, 588
Journal overlap, 231

K

Kinetic energy, 64
– cavitation, 64–65
– fuel spray, 64
– injection pressure, 64
– internal nozzle flow, 64
– pressure gradient, 64
– spray penetration depth, 64
K-ε model, 211

L

Lanchester system (twin-balance shaft), 246
Law of the wall, 217
– logarithmic, 217
Length reduction, 251
Light duty vehicles, 521, 528, 532
– chassis dynamometer, 522–523
– classification, 521–522
– common rail, 523
– cylinder crankcase, 523
– displacement, 523
– engines, 521–522
– exhaust aftertreatment, 521
– exhaust emissions, 521–522
– four-cylinder engine, 525–527
– gray cast iron, 526
– gross vehicle weight rating, 522–523
– intercooling, 523
– noise emission, 523
– piezo actuators, 526
– rated power, 523–524
– requirements, 522–523
– torque, 522
– variable turbine geometry, 523
Light-off temperature, 468
Load, 195–196
– dynamic, 195–196
– mechanical, 195
– static, 195
– thermal, 195
Longitudinal force component, 238
Longitudinal forces, 237–238
Longitudinal and partial tilting moment compensation, 241
Longitudinal yawing moment, 236
Low cycle fatigue, 200
Low speed two-stroke diesel engines, 595–600
– brake mean effective pressure (bmep), 595
– common rail, 605–606
– crankshaft assembly, 597–598
– cylinder output, 595
– engine frame, 596–597
– exhaust emissions, 606–607
– exhaust valve, 595, 599
– features, 595
– firms, 595
– fuel consumption, 595
– intercooling, 600
– maximum cylinder pressure, 595
– mean piston velocity, 595, 604–605
– number of cylinders, 600
– power per unit piston area, 595–596
– reversing, 599
– scavenging, 595
– scavenging ports, 600
– starting air system, 606
– stroke to bore ratios, 595–596, 604–605
– supercharging, 600
– uniflow scavenging, 595, 600
– valve rotation, 599
– water-fuel emulsion, 606
– water injection, 606
Low temperature performance, 79, 81–82
Low temperature resistances, 90
Low-voltage glow systems, 379
– pulse width modulation, 380
– rated voltage, 379
– voltage, 379
Lubricant, 359
Lubricating oil system, 530
– change interval, 530
– oil regeneration systems, 530
Lubrication system, 359
Lubricity, 92

M

Main exciter order, 249
Main oil gallery, 370
Marine engine, 665
– exhaust heat recovery, 602
– heavy fuel oil, 601
– hull vibrations, 603
– propeller characteristic, 601
– propulsion engine, 603
– shaft generator, 601
– turbocompound, 602
– vibrations, 602
Marine head design, 583
Mass average temperature, 216
Mass balancing, 246
– internal, 246
Mass moment of inertia, 249, 251
Mass reduction, 251
Mass torque, 247
Master connecting rod, 224, 240
Materials, 265
– fatigue strength, 265
– high temperature strength, 266
– lightweight alloys, 264
– monoblock, 276
– monotherm, 276
– nodular cast iron, 276
– piston materials, 265
– thermal conductivity, 265
– wear resistance, 265
Maximum stress, 233
Mean piston velocity, 596
Mean reference stress, 234
Measurement of particulate and dust emissions, 478
– alternative methods of soot measurement, 480

- condensation particle counter (CPC), 479
- dilution tunnel, 478
- dust measurement, 479
- particle measurement programme (PMP), 479
- particle number counting, 479

Mechanical governors, 175
- RQ, 175
- RQV(K), 175

Mechanical noise, 520
- oil pump, 493
- valve gear assembly, 520
- valve noise, 493
- water pump, 493

Medium speed diesel engine, 576–592
- common rail system, 587–588
- cylinder dimensions, 576
- effective brake work, 577
- exhaust emissions, 589
- gas power, 578–580
- heavy fuel oil operation, 578
- maximum cylinder pressure, 581
- mean piston velocity, 580–581
- power per unit piston area, 580–581
- speeds, 582
- stroke to bore ratio, 581–582

Metallurgy, 359

Metal sheathed-element glow plugs, 380, 386
- afterglow time, 381
- coil, 380
- control coil, 380
- glow tube, 380
- heating coil, 380
- tubular heating element, 380

Metering functions, 160
Methane number, 115
Method of calculation, 196
Micropilot engine, 119
Miller cycle, 72
Mineral oils hydrocrack oils, 359
Minimum stress, 233
Mixed friction, 361

Mixture, 11
- heterogeneous, 11
- homogeneous, 11
- stoichiometric, 11

Mixture formation, 11–12, 62–66, 120–122, 509, 512–513
- air movement, 62
- air swirl, 62–64
- air utilization, 62
- auto-ignition, 61
- bowl geometry, 64
- central, 120
- combustion chamber, 61–62
- compression ratio, 61–62
- direct injection engine, 62
- external, 11
- helical ports, 63
- internal, 11, 61
- prechamber, 62
- squish flow, 64
- swirl chamber, 62
- umbrella valves, 63

Modal analysis, 253
Model, quasidimensional, 26
Modified vehicle engines, 573
Moments of inertia, 236
Moments of rotational inertia, 243
Mufflers, 397, 398
Multigrade engine oils, 362

N

Natural frequency, 170, 257
Natural vibration mode, 252–254, 256
Needle guide, 130
Needle lift, 131
Net power, 15
Nitrogen oxide, 370, 425, 445–446

Noise, 391
- Helmholtz resonator, 391
- reflection mufflers, 391
- generation, 383, 386
- quality, 397
- radiation, 498
  - encapsulation measures, 498
  - V-belt pulley, 498
- reduction, 489
  - potential, 489

Nonroad engines, 437–443, 439
- agricultural equipment, 439
- air cooling, 446
- applications, 439
- cogeneration plants, 438
- combined heat and power stations, 438
- emergency power units, 438
- further development, 440
- marine engines, 440–443
- performance classes, 513
- power range, 439
- product concept, 574
- range, 439
- uses, 439
- water cooling, 455

Nonroad mobile machinery (NRMM), 440
- excavators, 440
- forklifts, 440
- front loaders, 440

Normal balancing, 238, 240, 243, 245, 246
Normal paraffins, 80
Normal stress, 235
NO_X production, 445

NO$_X$ reduction catalysts, 461
NO$_X$ storage catalysts (NSC), 463
– desulfation, 466
– LNT, 463
– regeneration, 466
– sensor, 463
Nozzle design, 71
Nozzle holders, 129–137
– dual spring holder, 134
– single spring holder, 134
Nozzle needle, 129–130
Nozzle projection, 71
Nucleate boiling, 294–298, 333
Nusselt number, 203, 216

O
OBD, 437
Offset (piston), 237
Oil aging, 374
Oil change, 370, 375
Oil circuit, 370
Oil cooler, 326, 370, 373
Oil filters, 373
Oil flow, 370
Oil pans, 354
Oil pressure, 370, 376
Oil pump, 370, 372–373
Oil reservoir, 370
Oil scraper rings, 285
– contact surface, 286
– flanks, 286
– run-in, 286
– spiral expander, 385
– surface treatments, 286
– wear reduction, 287
Oil separators, 373
Oil spray, 371
Oil volume, 370
Oil wetted air bath cleaner, 387, 389–390
Oil wetted air cleaners, 390
Operating error, 201
Operating point change, 509
Optical spray pattern analysis, 133
Orders of excitation, 244–245
Organic Rankine cycle, 405
Oxidation catalytic converters, 329
Oxidation products, 360
Oxidation stability, 93
Ozone O$_3$, 425

P
Paper dry type air cleaners, 387, 390
Parameter studies, 26
Partial sound sources, 499
– combustion air intake, 499
– cooling system noise, 500
– exhaust outlet, 499
– surface noise, 499
Particulate composition, 446
Particulate filters, 428, 458
– DPF, 458
– regeneration, 459
Particulate matter, 393, 401, 438, 461
– formation, 462
Particulate measurement, 476
– diffusion charging sensor, 481
– laser induced incandescence, 481
– MASMO, 480
– opacimeter, 481
– photoacoustic soot sensor, 481
– photoelectric aerosol sensor, 481
– smokemeter, 481
– TEOM, 483
Particulate NO$_X$ hyperbola, 452
PEMS, 437
Performance, 519
– responsiveness, 519
Performance comparison, 16
Petroleum (crude oil), 77–78, 103
– diesel fuel DF, 102, 119
– distillates, 98
– fuel oil, 103
– gas oil, 104, 108, 112
– gasoline, 91
– heavy fuel oil, 103–114
– residual oil, 103
Phase relationships, 241
Phase shift, 256
Phosphorus, 361, 464
Piezo injector, 156
Pilot fuel, 118, 119
Pilot injection, 453
Pintle nozzles, 130
Piston, 270, 541–542, 567
– cooling, 270
– forced oil cooling, 270
– function, 270
– mechanical load, 270
– oil consumption, 270
– piston crown, 270
– seizure resistance, 270
– spray cooling, 273
– stepped piston, 584
– structural strength, 270
– surface temperature, 270
– temperatures, 270–271
– three-dimensional temperature field, 270
Piston cleanliness, 369
Piston cooling, 275, 401, 598

- cooled ring carrier, 275
- cooling gallery, 275
- cooling oil, 276
- heat dissipation, 277

Piston crown, 270
- loading, 272–273
  - bowl bottom, 272
  - bowl rim, 272–273
  - piston alloy, 273
  - strength, 272–273
  - thermal cycling, 272

Piston designs, 271–282
- AlSi alloys, 273
- aluminum, 275–276
- articulated, 275
- cast, 275
- composite, 276
- compression height, 271
- cooling gallery, 271, 273
- eutectic, 275
- forced oil cooling, 275
- main dimensions, 271
- monoblock, 276
- oil spray cooling, 275
- piston cold clearance, 277
- piston crown, 275–277
- piston mass, 271
- piston pin, 271
- piston pin bore, 271
- piston top land, 271
- ring carrier, 275
- ring zone, 271
- skirt, 271
- steel, 275–277
- thermal conductivity, 275

Piston force, 221

Piston load, 302
- compression height, 271
- cooling gallery, 271
- main dimensions, 271
- piston mass, 271
- piston pin, 271
- piston pin bore, 271
- piston top land, 271
- ring zone, 271
- skirt, 271

Piston movement, 237

Piston pin, 270, 286–287
- connecting rod, 286
- cooling, 273–274
- expanding snap ring, 286
- forced oil cooling, 275
- force transfer, 286
- function, 270
- mechanical load, 273

- oil consumption, 270
- profiled piston pins, 286
- seizure resistance, 270
- spray cooling, 273
- structural strength, 270
- surface temperature, 270
- temperatures, 270–271
- three-dimensional temperature field, 287
- tubular, 286

Piston ring, 282–286, 309, 583–584
- blow-by, 283
- bore wear, 584
- chrome rings, 584
- chromium-ceramic coated rings, 584
- compression ring, 283
- cooling, 270
- forced oil cooling, 275
- function, 270
- heat dissipation, 273
- mechanical load, 270
- mixed lubrication, 285
- oil consumption, 270
- oil control ring, 285
- piston crown, 270
- plasma rings, 584
- rate of wear, 584
- sealing function, 283
- seizure resistance, 270
- spray cooling, 270
- structural strength, 270
- surface temperature, 270
- tapered compression ring, 284
- temperatures, 270
- three-dimensional temperature field, 270
- zone, 270

Piston side thrust, 221–223

Piston stroke, 9

Piston top land rings, 584

Piston velocity, 9
- instantaneous, 9
- mean, 10

Pivoting angle of the connecting rod, 222

Plain bearings, 259
- clearance, 264–265
- connecting rod, 264–265
- crankshaft, 264
- fatigue, 270
- hydrodynamic, 260
- maximum operating value, 263
- one-layer, 266–267
- plain bearing material, 266
- properties, 266
- simulation, 261
- stresses, 261

- three-layer, 267–269
- thrust, 255
- two-layer bearings, 267
- wear, 269

Pollutant production, 68
- edge of the spray, 69
- fuel droplets, 69
- hydrocarbons, 69
- spray core, 69
- temperature, 69

Pour point improvers, 361
Power per unit piston area, 16, 544–547, 580–581
Power take-offs PTO, 561
Power to weight ratio, 349
Prandtl number, 203
Precession, 230
Prechamber engine, 449
Precombustion chamber, 119, 122
Pressure control valve, 154–155
Pressure curve analysis, 22
Pressure limiting valve, 156
Pressure wave correction (PWC), 160–162
Pressure wave supercharging, 56
Primary breakup, 65
- droplet formation, 65
- fuel density, 65
- spray breakup, 65
- spray hole diameter, 65
- spray propagation, 65
- spray velocity, 65
- surface tension, 65
- weber number, 65

Primary drive noise, 493
- toothed belts, 493
- torsional vibration dampers, 493

Primary particulates, 448
Process heat, 405
Process simulation, 74
- 3D modeling, 74
- system configuration, 74

Product defect, 201
Production, 78–80
- atmospheric distillation, 78
- atmospheric vacuum distillation, 78–79
- cracking, 78–79
- desulfurization, 78, 89
- hydrocracking, 78, 88
- vacuum distillation, 78

Propeller operation, 18
Pulsation, 390
Pulse turbocharging, 45
Purity, 101
Push rod ratio, 597

Q

Quality control, 12
Quantity control, 12

R

Radial force, 231
Radiator, 321–324
Radii of gyration, 243
Rail, 165
Rail pressure, 153
- high pressure lines/pipes, 153
- injectors, 153
- sensor, 154

Rated voltage, 385
Rate of heat release, 22–23, 67–68, 74–75
- CFD modeling, 74
- constant pressure combustion, 68
- diffusion flame, 68
- double Vibe function, 24
- 3D modeling, 74
- 3D simulation, 74
- efficiency, 67
- evaporation rate, 67
- injection rate, 67
- injection time, 68
- model calculation, 74
- spray penetration velocity, 67
- system configuration, 74–75
- Vibe function, 23

Reciprocating piston engines, 252
Recoil start, 562
Reduction formula, 251
Reference alternating stress, 234
Reference stress, 233
Relative motions, 197
Residual force vector, 238, 246
Residual imbalance, 236
Residual torque vector, 246
Residue formation, 370
Resonance tuning, 257
Resonant frequency, 395
Retention rate, 31
Reynolds number, 203
Rigid body motion, 254
Ring sticking, 369
Ring zone, 273
- aluminum pistons, 273
- bore geometries, 273
- boss support, 273
- coking, 273
- pin bore, 273
- piston pin bore, 273
- ring groove wear, 273

Roller bearing drag levers, 511

Rolling resistance, 18
Rotational irregularities, 258–259
Rotational speed, 9
Rudolf Diesel, 3, 6
- patent DRP 67207, 3
- patent DRP 82168, 4

S

Sac hole nozzle, 131, 449
SAE classification, 361
Safety factor, 199, 235
Scavenging, 36
- after exhaust, 36
- bypass/short-circuit flow, 37
- displacement, 37
- efficiency, 31, 37
- loop, 36
- total mixture, 37
- uniflow, 36
SCR catalyst, 607
Seat geometry, 130
Secondary breakup, 65
- air density, 64
- air to fuel equivalence ratio, 66
- droplet diameter, 64
- evaporation, 64
- fuel, 65
- fuel evaporation, 66
- injection pressure curve, 64
- reaction zone, 66
- spray breakup, 62
- spray cone angle, 64
- vapor state, 65
Secondary piston motion, 198–199, 277, 492
Seiliger (dual combustion) cycle, 13
Selective catalytic reduction (SCR), 461
- AdBlue, 462
- ammonia, 461
- NO_X sensor, 463
- temperature sensor, 463
Sensors, 184
- accelerator pedal module, 184
- boost pressure, 185
- butterfly valve/Regulating throttle, 184
- differential pressure, 186
- exhaust temperature, 186
- hot film air mass, 185
- lambda oxygen, 186
- NO_X, 186
- phase, 185
- rail pressure, 185
- speed, 185
- temperature, 184
Service life, 395

Servicing, 387
Sheathed-element glow plug temperatures, 384
Shot peening, 342
Similarity theory, 208, 210–211, 214–216, 295, 301
Single cam system SCS, 568
Single cylinder diesel engine, 564–568
- cooling systems, 560
- electric start, 562–563
- engine compartment ventilation, 563
- engine power, 559
- hand crank start, 561–562
- mass balancing, 567
- output potentials, 561
- performance specifications, 559–564
Single degree of freedom system, 256
- modal, 256
Single plunger pump systems, 144–145
Single zone model, 19–20
Sintered metal filters, 459
Skirt side force
- deformation, 271–272
- stress, 271–272
Sludge formation, 360, 370
Smoke limit, 17
Solenoid valve controlled pumps, 149
Solenoid valve injector, 156
Soot, 62, 445, 447, 517
- afterburning, 448
- agglomerates, 448
- production, 447
- size distribution, 448
Sound
- character, 393
- design, 397
- power level, 520
- pressure level, 520
Soundproofing, 502
- engine mount, 502
Spark ignition, 11
Special measurement systems, 474
- diode laser spectroscopy (DIOLA), 474
- Fourier transform infrared spectroscopy (FTIR), 474
- mass spectrometer (MS), 474
- nondispersive ultraviolet analyzer (NDUV), 474
Specifications, 362
Specific work, 550
Speeds
- critical, 250
Spray configuration, 133
Spray force analysis, 133
Spray hole length, 132
Spray propagation, 74
Squish flow, 70

Standard cycle, 12
- constant pressure cycle, 13
- constant volume cycle, 13
- seiliger (dual combustion) cycle, 13
Star diagram, 224
Start assist system, 379–383, 534
- afterglow phase, 379
- intermediate glow, 379
- low-voltage glow systems, 379
- preglow, 379
- sheathed-element glow plugs, 379
- software module, 379
- standby glow, 381
- start glow, 379
Starter systems, 561
Starting speed, 378
Start of injection, 451
Steels, 380
- pearlitic malleable iron, 342
Stiffness matrix, 253
Stoichiometric air requirement, 11
Stress analysis, 195–197, 199–201
Stress concentration factors, 232–233
Stresses, 199
- actual, 199
- effective, 199
Stress gradient, 235
- relative, 235
Stroke/bore ratio, 7
Stroke/connecting rod ratio, 7, 237
Structural ceramics, 354
Structure-borne noise
- excitation, 489–494
- insulation, 498
- radiation, 489
- transmission, 494–498
  - air cowling, 498
  - cylinder head, 495–496
  - cylinder head covers, 496
  - engine block, 495–496
  - intake manifold, 496–498
  - oil pans, 495–498, 502
  - spur gear covers, 498
  - valve covers, 498
Suction side control, 149
- control unit, 150
Sulfate ash, 361, 364
Sulfur, 346, 364
- content, 77, 85–86, 89, 92, 102
Supercharger characteristic curve, 39–40
Supercharger types, 39–40
- positive-displacement supercharger, 39–40
Supercharging, 38, 514–515, 600
- downsizing, 515
- efficiency, 44

- exhaust gas recirculation, 514
- exhaust gas turbocharging, 38–39, 41–42, 515
- intercooling, 515
- lean burn operation, 514
- mechanical supercharging, 38–41, 56
- variable turbine geometry, 514
Surface filtration layer, 459
Surface noise, 489
Surface temperature methods, 204, 206–207, 351
Surge line, 39, 42–43, 49
Synthesized oils, 359

T
Tailpipe noise, 395, 537
TA Luft (Technical Instructions on Air Quality Control), 119
Tangential force, 222
Tangential pressure, 249
Temperature distribution, 199
Temperature gradients, 207
Test cycle, 431
- C1, 440
- D 13, 431
- ELR, 431
- ESC, 431
- ETC, 431
- JE 05, 431
- US-FTP, 430
- WHSC, 437
- WHTC, 437
Thermal compression cracks, 202
Thermal conductivity, 203
Thermal dethrottling, 455
Thermal load, 202
Thermophoresis, 332
Thread rolling, 541
Three-way catalytic converter, 361, 364, 418–420, 461, 465
Throw angle, 224, 226–228, 239
Throw diagrams, 241
Thrust side (piston, cylinder), 222
Tilting moment, 236, 241
Torque, 15, 230, 247
- characteristic, 230
- internal, 247–249
Torsional angle, 256
Torsional forces, 222, 248–249
- amplitude of harmonic excitation, 248
Torsional moment, 233, 256
Torsional stress, 232
Torsional vibration, 256–259
- absorber, 257
- dampers, 256–259
- resonances, 256–257
Total base number, 360

Transfer function, 256
- matrix of the, 256
Transition radii, 232
Transmission design, 508
Transverse bending, 197
Transverse force component, 237
Transverse forces, 239
Trigeneration, 410–411
Trunk-piston engines, 223
Turbine map, 44
- effective turbine cross section, 44
- reduce mass flow, 44
- turbine efficiency, 44, 46
Turbobrake, 56
Turbocharger, 393
Turbocharger fundamental equations, 42
- first, 42
- second, 44
Turbocompounding, 55–56, 404
Turbo lag, 56
Two-stage turbocharging, 52, 57
Two-stroke cycle, 14
Two-zone model, 20

U

Uniflow scavenging, 593–595
Unit injector system, 144
Unit pump system, 145

V

Valve bridge cracks, 306
Valve covered orifice (vco) nozzle, 130
Valves, 32, 346, 567
- bimetallic, 346
- cages, 584
- cross section, 33
- exhaust, 606
- flow coefficient, 33
- lift curves, 26
- monometal, 346
- overlap, 32
- rotation, 586
- self-cleaning, 599
- timing, 36
Vegetable oils, 4
Vehicle drive, 18, 531
Vehicle engines, 571–574
- modified, 571–574
- use, 572
Vehicle system voltage glow systems, 379
- ignition switch, 379
- self-regulating function, 379
V engines, 246
Vibration absorber, 257–258
Vibration absorber natural frequency, 258
Vibration cavitation, 202
Vibration nodes, 252
VI improvers, 369
Viscosity, 90, 361, 369
Volumetric efficiency, 31

W

Wall heat flux, 204–205, 209
- density, 204
Wall heat loss, 15, 202, 204, 208–213, 216–217
Wall insulation, 501
Washcoat, 458
Waste heat, 401–413
- convection, 401–403
- cooling energy, 401–403
- exhaust gas, 401–409
- intercooling, 401–403
- radiation, 401–403
Wear, 340, 359, 387
- protection, 387
Winglets, 332
Work, specific, 15
Woschni equation, 216

Z

Zeldovich, 446
Zero fuel quantity calibration (ZFC), 160, 162
Zinc dithiophosphate, 361

Printing and Binding: Stürtz GmbH, Würzburg